The Tautomerism of Heterocycles

Advances in Heterocyclic Chemistry

Edited by

A. R. Katritzky

and

A. J. Boulton

SUPPLEMENT 1

The Tautomerism of Heterocycles

THE TAUTOMERISM OF HETEROCYCLES

Advances in Heterocyclic Chemistry
Supplement 1

JOSÉ ELGUERO and CLAUDE MARZIN

Département de Chimie Organique
Université du Languedoc
Montpellier, France

ALAN R. KATRITZKY

School of Chemical Sciences
University of East Anglia
Norwich, England

PAOLO LINDA

Istituto di Chimica Organica
Università di Perugia
Perugia, Italy

Academic Press · New York · San Francisco · London · 1976

A Subsidiary of Harcourt Brace Jovanovich, Publishers

ACADEMIC PRESS, INC.
111 Fifth Avenue, New York, New York 10003

United Kingdom Edition published by
ACADEMIC PRESS, INC. (LONDON) LTD.
24/28 Oval Road, London NW1

LIBRARY OF CONGRESS CATALOG CARD NUMBER: 75-21965

ISBN 0-12-020651-X

PRINTED IN THE UNITED STATES OF AMERICA

Contents

Preface

Volumes 1 and 2 of *Advances in Heterocyclic Chemistry*, which appeared in 1963, included a total of four chapters that covered the prototropic tautomerism of heteroaromatic compounds, and which were written by one of us (A.R.K.) together with Dr. Jeanne Lagowski. This was the first, and has remained the only, detailed survey of the subject. It can justifiably be claimed to have exerted a considerable influence on a field that had expanded in a haphazard and random manner. Before the appearance of these chapters numerous contributions included inadmissible pieces of evidence, *non sequiturs*, and logical inconsistencies. The quality of contributions to the field has since improved significantly.

The success of these chapters, and the work that they have helped to stimulate, has hastened their obsolescence. In the last ten years an enormous quantity of work has appeared, and the up-dating of this work is now imperative. The present volume represents such a revision. The volume of new work is such that we have chosen to up-date rather than to rewrite completely, in order to keep the project within reasonable bounds.

Scope

As before, we are concerned with the tautomerism of heterocyclic compounds in which at least one of the possible tautomeric structures is aromatic. We take our definition of aromaticity in a wide sense to be any structure that possesses uninterrupted cyclic conjugation; we have also included a number of cases of the tautomerism of nonaromatic heterocycles where it is felt that they are informative to the general discussion. Again, as before, we have restricted our discussion to the prototropic tautomerism of heterocycles.

In three respects the present work exceeds the scope of the previous one: (a) we have not restricted ourselves to five- and six-membered rings; (b) we have included cases of ring–chain tautomerism that involve the movement of a proton; and (c) instead of limiting ourselves to N, O, and S as heteroatoms, we have also considered others, e.g., Se, P. Except for these three areas, the coverage is generally limited to papers that have appeared since 1962 and were thus not covered in the previous review. We have attempted to cover all the published literature available through December 1973; however, since it is difficult to locate papers which deal only incidentally with tautomerism, there are undoubtedly pertinent references we have missed. We would appreciate having such omissions brought to our attention. A considerable volume of work published in 1974 is also included.

Arrangement of Material

Chapters 1–4 correspond to the four chapters of the previous review and deal, respectively, with methods of study, six-membered rings, and five-membered rings with one heteroatom, or with several. Chapter 5 contains material on purines and other condensed five-six and five-five membered ring systems: previously this was included with the five-membered rings with two heteroatoms, but this field has now grown to such an extent as to justify a separate chapter. Chapter 6 covers compounds with three-, four-, seven-, and more-membered rings.

With great progress in physical methods of organic chemistry it has been necessary to reorganize the presentation of Chapter 1 compared with that previously adopted. The new arrangement is discussed at the beginning of the chapter.

Within Chapters 2, 3 and 4, the arrangement of the material follows rather closely that previously adopted: tautomerism not involving functional groups is first discussed, and then in turn compounds are dealt with that contain potential OH, SH, NH_2 and CH_3 groups. In Chapter 4 (five-membered rings with several heteroatoms) compounds with two potential tautomeric groups are dealt with separately; with the increased volume of material, this is now necessary for clarity. Within each major subdivision, compounds are subdivided according to the number and type (O before S before N) of heteroatoms in the ring. Selenium and tellurium compounds are treated along with their sulphur analogues and phosphorus compounds with their nitrogen analogues. In view of the logical arrangement of the material, and the detailed Subject Contents (p. xiii), it has not been felt necessary to provide a subject index.

References

We have cross-referenced each section with the review published in 1963. These cross-references are given to actual page numbers in the following format: (I-331) and (II-27) refer, respectively, to *Advances in Heterocyclic Chemistry* Vol. 1, p. 331 and Vol. 2, p. 27. Cross-references within the present book are given to sections and are in the format Section 4-2Bf (i.e., Chapter 4, Section 2Bf): chapters are assigned Arabic numerals and subsections Arabic numerals, capital and small letters in decreasing order of importance.

All other references are dealt with by a new system which was devised for use in the monograph *Chemistry of the Heterocyclic N-Oxides* by A. R. Katritzky and J. M. Lagowski, Academic Press, London, 1971. This system, which simplifies the preparation of the manuscript and, we hope, decreases the number of errors in the bibliography, has proved popular with the users of the monograph just mentioned. The system is fully explained in Chapter 7.

Conventions

Tables, equations, schemes and figures are given in Arabic numbers, commencing anew in each chapter, and the chapter number is also given thus 1-21, 2-34. Formulae are given in boldface Arabic numbers, starting afresh in each chapter, and are enclosed in square brackets, thus: [**24**].

Acknowledgements

The preparation of this book would not have been possible without the help of a great many people. In particular we would like to thank Mrs Janet Dennis and Mrs Christine Jenvey for invaluable help in the preparation and checking of the manuscript, Mrs Stephanie Craggs, Mrs Margaret Livock, Mrs Betty Ross, Miss Giovanna Rossi and Mrs Susan Lemon for their careful and accurate typing, and Mrs Devi Muthyala for corrections.

Dr A. J. Boulton has read the whole of the manuscript, and we thank him for giving us the benefit of his sharp criticism. We are grateful to Professor M. Begtrup (Technical University of Denmark), Dr A. J. H. Summers (Oxford University), Professor Maquestiau (Mons) and Dr J. Arriau (Pau), for communicating unpublished work and to Professor Jacquier (Montpellier) and Professor Marino (Perugia) for their interest in this venture.

J. Elguero and C. Marzin (Montpellier)
A. R. Katritzky (Norwich)
P. Linda (Perugia)

Aims and Purpose

The study of the tautomeric structure of heterocycles is fascinating in its own right, but it is also of immense practical and theoretical importance. To rationalize chemical and physical properties we must be able to write down 4-pyridone as [**1**] and 4-aminopyridine as [**3**], rather than to refer incorrectly to these compounds as "4-hydroxypyridine" or "pyridin-4-onimine" or to write them incorrectly as [**2**] or [**4**]. For a proper understanding this is essential not only for the heterocyclic chemist interested in structure, but also for every chemist who deals with the reactions of heterocycles and for every biochemist who wishes to be able to explain the workings of his biochemicals.

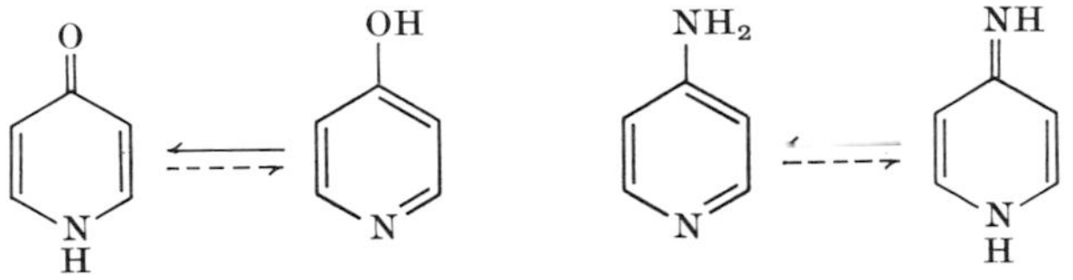

O NH OH N NH$_2$ N NH N H

[**1**] [**2**] [**3**] [**4**]

A good example of the potential dangers involved is the case of the nucleic acids. These compounds consist of sugar phosphate chains in which various heterocyclic bases are linked to the sugar molecules. Hydrogen bonding between base pairs links together two such chains in a double helix. It is clearly important that, e.g., uracil exists in the dioxo form and that adenine exists in the amino form, in order for the double helix to have the correct hydrogen bonding, as shown in Scheme 0-1. Considerable confusion was initially engendered because uracil was frequently written in the hydroxy form by chemists: a graphic account has been given (68MI1) of the difficulties encountered in sorting this out.

Difficulties of this type have been increased because of the unfortunate tendency of some otherwise highly intelligent and reasonable chemists to represent tautomeric heterocyclic compounds by other than the predominantly existing tautomer. As it is hoped to make clear during the course of this book, the information presently available enables us to predict with a good chance of success the predominant tautomeric structure for nearly all the heterocyclic compounds that we are likely to encounter. Surely *this* is the structure by which they should be named and, more important, depicted in reaction schemes. It is essential that trained heterocyclic chemists should use the correct structure, and it is even more important that good habits be engendered in nonchemists. It seems very difficult to expect biochemists and

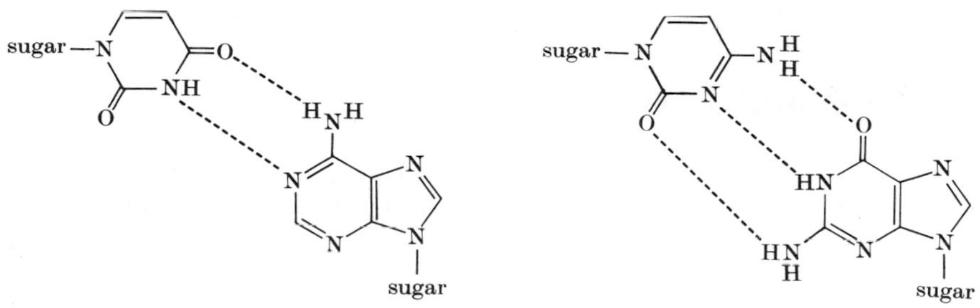

Uracil/adenine

Cystosine/guanine

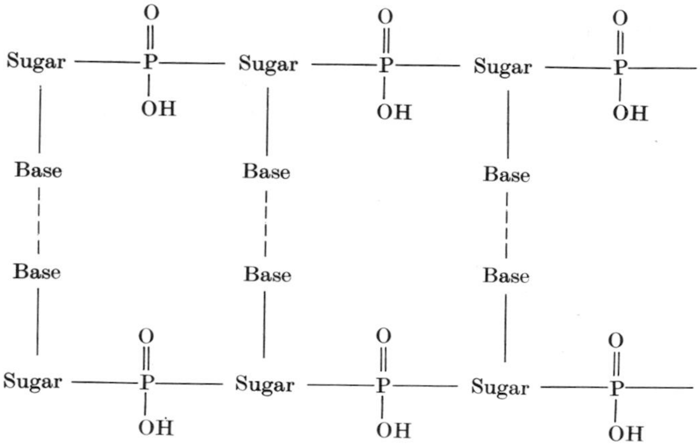

SCHEME 0-1

biologists to have any understanding of the subject at all if we do not write down what we mean. It is hoped that this book will help increase our understanding of chemical and biochemical processes.

Detailed Subject Contents

The Tautomerism of Heterocycles

CHAPTER 1

The Nature of Heteroaromatic Tautomerism

This chapter is concerned with the nature of heteroaromatic tautomerism and with the methods by which the subject can be studied. We present here a general survey of the types of tautomerism falling within the scope of this book and the methods by which the phenomenon may be defined and discussed. The influence of external factors is also covered. Then we consider in more detail the individual methods of investigation; first the chemical methods and then the physical methods, grouped according to the type of technique.

The arrangement of the first part of this chapter follows fairly closely that of the previous review (I-311), but this does not apply to the sections dealing with physical methods. The number and scope of the physical methods applied to tautomerism has so increased over the past decade that it is now found preferable to group them rather differently.

1. General Discussion

A. THE PRINCIPAL TYPES OF TAUTOMERISM (I-313)

The prototropic tautomerism of heteroaromatic compounds comprises all those cases where a mobile hydrogen atom can move (as a proton) from one site to another in a heteroaromatic molecule. The most common type involves the movement of a proton between a cyclic nitrogen atom and a substituent atom directly connected to the ring. A comprehensive classification of the possible types is shown in Scheme 1-1. The top row of formulae in Scheme 1-1 shows the various sites available: (A) a cyclic sp^2-hybridized nitrogen, (B) an sp^2-hybridized ring carbon; (C) an sp^2-hybridized atom directly attached to a ring carbon atom; (D) an atom directly attached to a ring nitrogen; and (E) a side-chain atom not directly attached to the ring. The lower row of formulae in Scheme 1-1 shows the corresponding potential sites from which a proton can be removed. Any heteroaromatic molecule which contains at least one site of the type shown in the upper row together with one of those shown in the lower row of formulae in Scheme 1-1 is capable of prototropy. We must add to these the possibility of ring-opening occurring simultaneously with proton transfer.

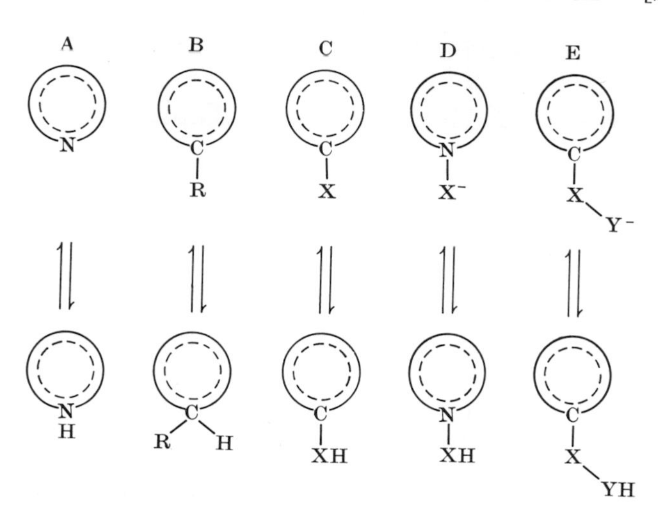

SCHEME 1-1

The classification of the previous survey (I-313) has been extended by two further sections: (vii) which covers certain rearrangements, and (viii) which deals with ring–chain tautomerism. For each class the relationship to the sites outlined in Scheme 1-1 is also given.

i. Annular nitrogen and an atom adjacent to the ring (cf. [**1a**] ⇌ [**1b**] or [**2a**] ⇌ [**2b**]) (types AC and AD of Scheme 1-1):

[**1a**] [**1b**] [**2a**] [**2b**]

ii. Annular carbon and an atom adjacent to the ring (cf. [**3a**] ⇌ [**3b**] or [**4a**] ⇌ [**4b**]) (types BC and BD of Scheme 1-1):

[**3a**] [**3b**] [**4a**] [**4b**]

iii. Two atoms adjacent to the ring (cf. [**5a**] ⇌ [**5b**], [**6a**] ⇌ [**6b**] or [**7a**] ⇌ [**7b**]) (types CC, CD and DD of Scheme 1-1):

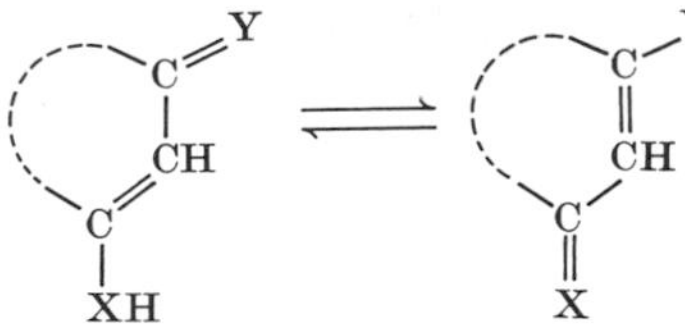

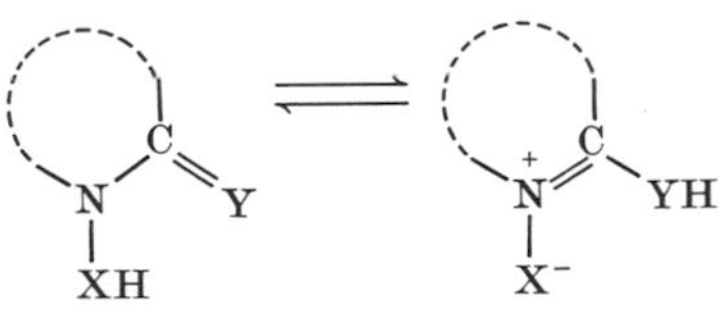

[**5a**] [**5b**] [**6a**] [**6b**]

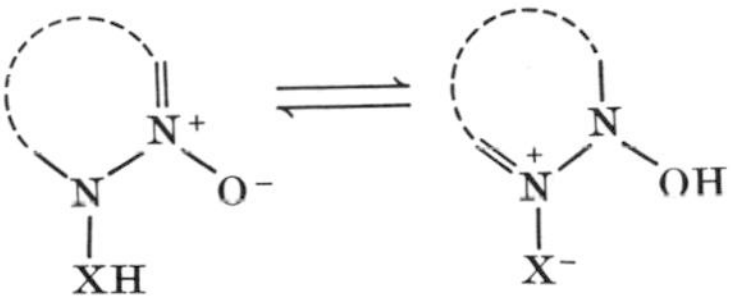

[**7a**] [**7b**]

iv. Two annular nitrogen atoms: neutral molecules (cf. [**8a**] ⇌ [**8b**]) and conjugated acids (cf. [**9a**] ⇌ [**9b**]) (type AA of Scheme 1-1):

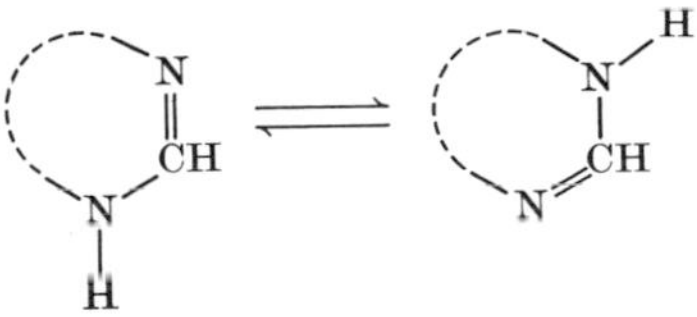

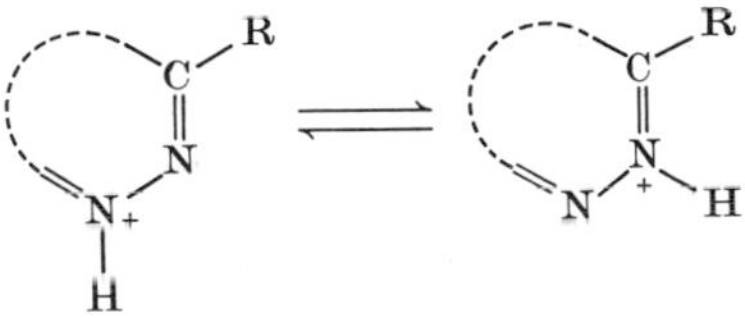

[**8a**] [**8b**] [**9a**] [**9b**]

v. Two annular carbon atoms (cf. [**11a**] ⇌ [**11b**]) (type BB of Scheme 1-1):

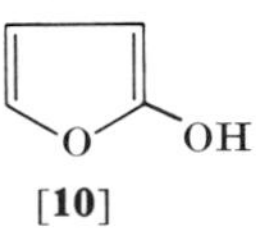

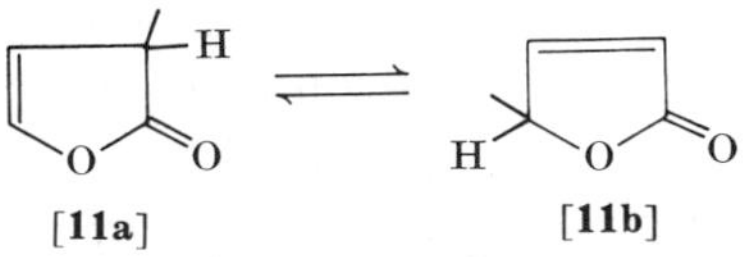

[**10**] [**11a**] [**11b**]

vi. Annular carbon and nitrogen atoms: neutral molecules (Cf. [**12a**] ⇌ [**12b**]), and conjugated acids (cf. [**13a**] ⇌ [**13b**]) (type AB of Scheme 1-1):

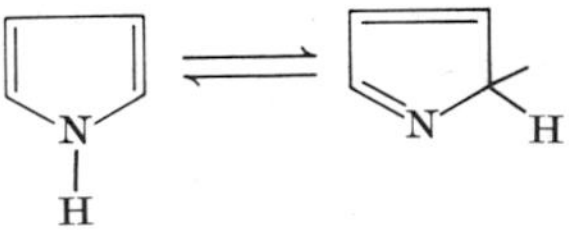

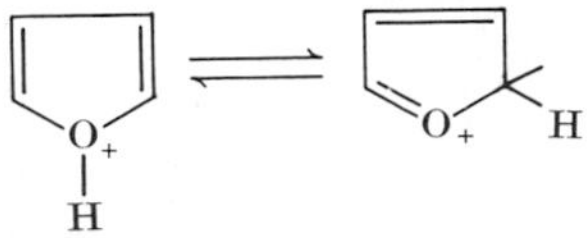

[**12a**] [**12b**] [**13a**] [**13b**]

vii. Dimroth (cf. **[14a]** ⇌ **[14b]**), and other rearrangements (cf. **[15a]** ⇌ **[15b]**):

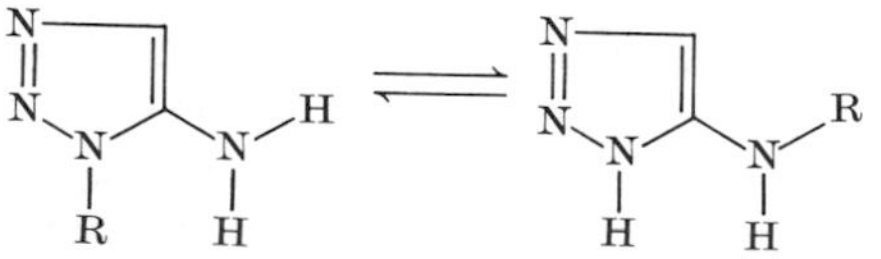

[14a] [14b] 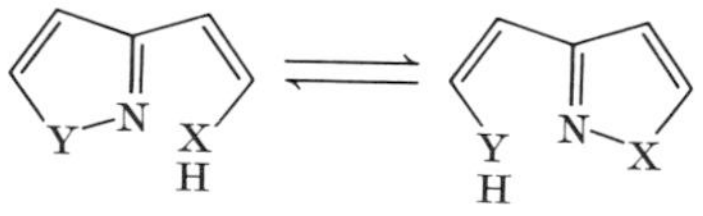 [15a] [15b]

viii. Ring–chain tautomerism (cf. **[16b]** ⇌ **[16c]**):

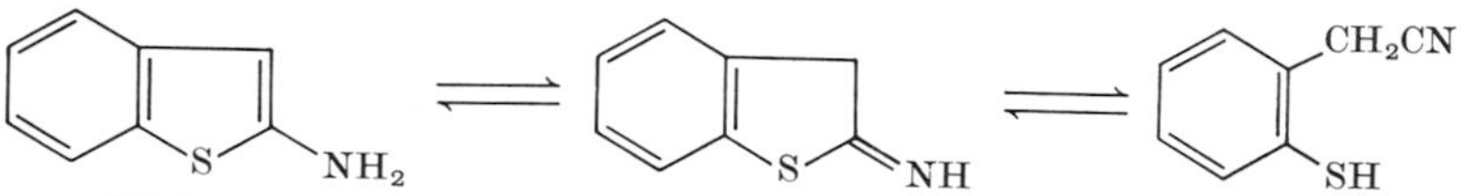

[16a] [16b] [16c]

B. Nomenclature (I-315)

In general, we use the nomenclature of the previous survey (I-315). The only important modification occurs in the case of the annular tautomerism (Section 4-1). Instead of naming the two tautomers as, for example, 4-methyl-1*H*-imidazole **[17a]** and 4-methyl-3*H*-imidazole **[17b]** as done previously (I-316), we now prefer to name them as 4-methylimidazole and 5-methylimidazole, i.e., commencing the numbering at the pyrrole-like nitrogen atom. The notation "4(5)-methylimidazole" refers to any mixture of **[17a]** and **[17b]**. Using this same nomenclature, the fixed compounds **[18]** and **[19]** are named 1-methyl-1,2,4-triazole and 1-methyl-1,3,4-triazole, rather than being treated both as 1,2,4-triazole derivatives. This latter system has the blessing of the rules of the International Union of Pure and Applied Chemistry (IUPAC); we feel, however, that the presently adopted scheme has the merits of greater clarity

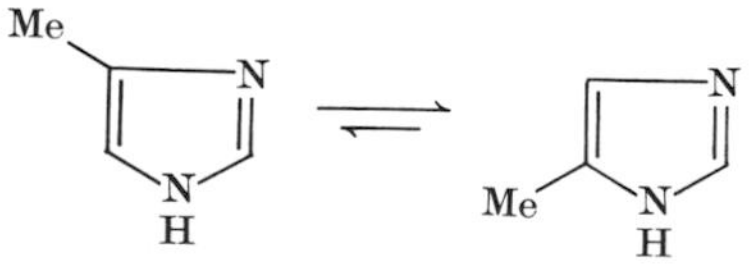

[17a] [17b] 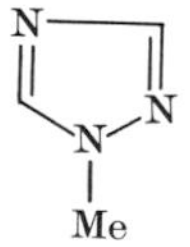 [18] 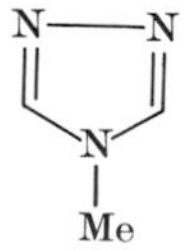 [19]

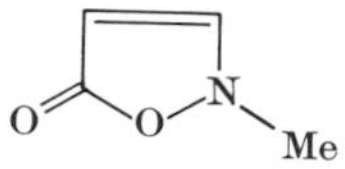

[19/1]

and convenience. For oxo derivatives problems of nomenclature are serious; for a thorough discussion see Liebscher (74ZC49). We have utilized the previous method (based on Chemical Abstracts) in which the suffix "in" is retained for most six-membered rings but not introduced for five-membered rings. We refer to individual tautomers as CH-, OH-, NH-forms for convenience. When referring to a tautomeric substance we generally use that tautomeric structure which predominates in aqueous solution (this is usually the form in the crystalline compound).

C. The Existence of Individual Tautomers (I-316)

The theories due to Hunter (I-316) and to other authors, generally proposed in connection with annular tautomerism (Section 4-1), have been abandoned, and the existence of individual tautomers is now universally accepted. However, it is necessary to stress the fundamental and classical difference between mesomerism, involving a canonical form with separated charges, and tautomerism, involving a zwitterion (sometimes called a mesoionic compound), because we continue to notice examples of apparent confusion of these fundamentally different situations.

Frequently, one of the tautomers can only be written as a zwitterion; this occurs, for instance, if a potential hydroxy group is *meta* to a pyridine-like nitrogen, as with β-hydroxypyridines (Section 2-3D) [**20**] and with 4-hydroxypyrazoles (Section 4-3Ad) [**21**]. Furthermore, heteroaromatic tautomers structures are always mesomeric, and usually at

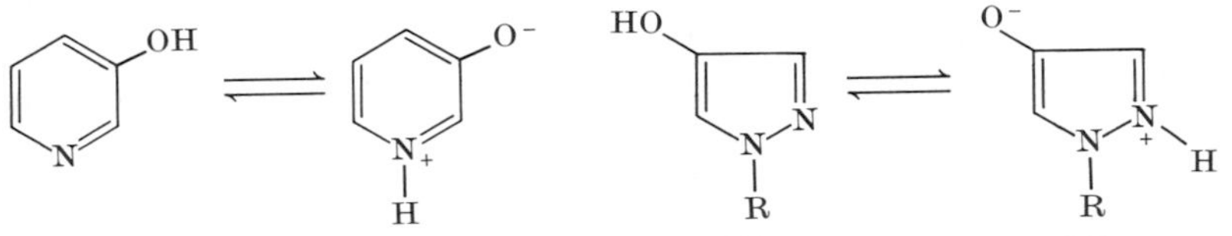

[**20a**] [**20b**] [**21a**] [**21b**]

least one of the contributing canonical forms is zwitterionic. Mistakenly, and misleadingly, in some such cases the authors write a resonance form as if it were a different tautomer: for example, for the pyrazol-3-ones, certain authors (64T531) have considered three "forms" [**22**], [**23**], and [**24**], and it was not made clear that [**22**] and [**24**] are merely two canonical forms of a same molecular species (Section 4-2F). Many further such examples could be quoted (see, for example, 73J(PII)2036).

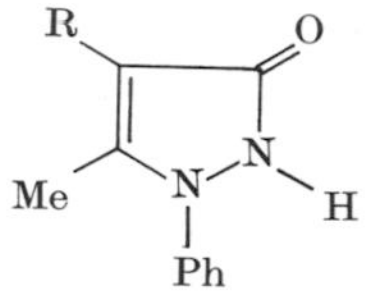

[22] 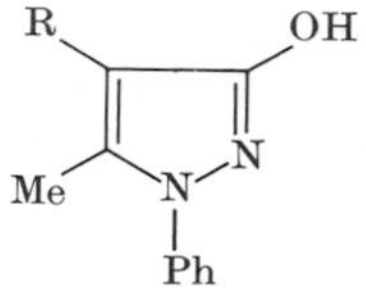 [23] 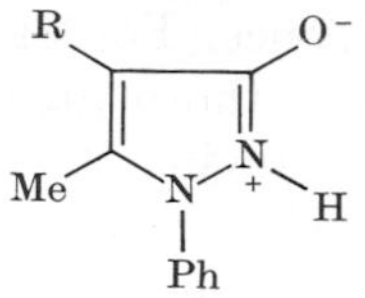 [24]

Interatomic distances obtained from X-ray or other physical measurements enable calculations of the contributions of the various canonical forms to the resonance hybrid (Section 1-4Ib). However, these remain canonical forms, and it is, for example, highly misleading (70CH706) to write (69CH811): "the zwitterionic structure [**25a**] of the five-membered ring in 2-imino-5-phenyl-4-thiazolidinone is clearly demonstrated for the first time." Structure [**25a**] is neither a zwitterion nor an imino compound; it is merely a contributing canonical form of the amino–oxo tautomer [**25b**]. The situation is comparable to that for acetamide, where canonical form [**26a**] contributes to [**26b**]. Another similarly misleading example has also got into print (69CH1038).

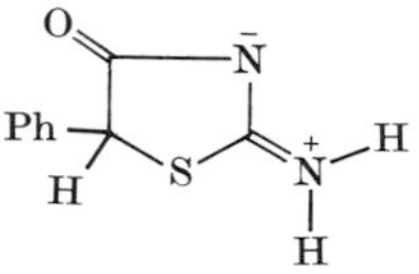

[25a] 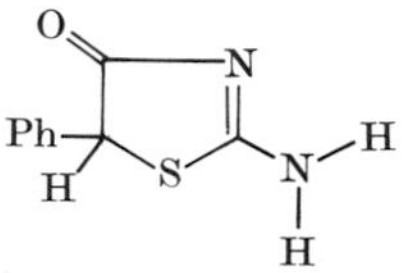 [25b] 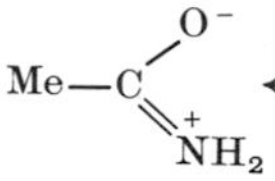 [26a] 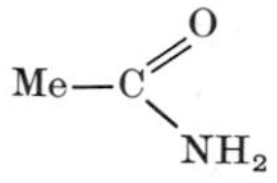 [26b]

D. The Mechanism and Rate of Tautomeric Change (I-317)

As regards the mechanism of tautomeric change, we have nothing to add to the previous discussion (I-317).

The possibility of isolating individual tautomers (the separate existence of such tautomers is called desmotropy by German authors) depends on the energy barrier which separates them (see the discussion under nuclear magnetic resonance (NMR): Table 1-4, Section 1-5Aa). It is a useful if oversimplified generalization that tautomers will be separable (the rate of interconversion being sufficiently slow) if a CH bond has to be cleaved and a new CH bond formed in the prototropy (I-317, 70C134). For example, for α-hydroxy-furans (II-5), -thiophenes (II-9) and -pyrroles (Section 3-2Ec): the two oxo tautomers [**27a**] and [**27b**] not only are separable but undergo interconversion sufficiently slowly in solution to enable measurement individually of their physical properties. Another of many similar examples is that of the 1,2-diazepinones (Section 6-2Bb).

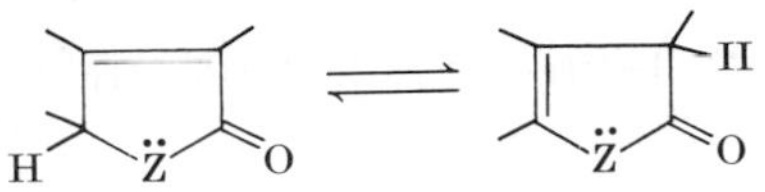

[**27a**] [**27b**]

If the labile proton is attached to carbon in only one tautomer of a pair, it is still sometimes possible to separate them or at least to obtain mixtures enriched in one tautomer. However in such cases equilibration rates in solution are generally considerably faster, so that it is often difficult to obtain the physical characteristics of one pure tautomer. Examples of this type are the indole–indolenine pair (Section 3-1Ac), the 3-hydroxybenzothiophene–3-thianaphthenone pair (Section 3-2Cc) and the benzodiazepinones (Section 6-2Bd). If the interconversion of two tautomers involves cleavage and formation of hydrogen to heteroatom bonds (i.e., OH, SH, NH) only, the rate of tautomeric change in solution is so great that a mixture of tautomers in equilibrium will always be observed.

The situation is different for the solid state. In general the crystal lattice prevents interconversion between tautomers, and the crystal is composed of a single tautomeric species. However, this picture is often complicated by the existence of strong inter- or intra-molecular hydrogen bonding. If such hydrogen bonding exists, three possibilities may be distinguished. Most commonly, the hydrogen atom is covalently bonded to one heteroatom, and hydrogen-bonded to another heteroatom—i.e., a single defined tautomeric form occurs in the crystal (this is the case, e.g., for 2-pyridone, I-351). Secondly, proton transfer may occur in the crystal between two sites—i.e., of the type $X{-}H\cdots Y \rightleftharpoons X\cdots H{-}Y$ (no well authenticated case is known for heteroaromatic compounds, but the ice crystal is a good illustration of the general principle). Thirdly, a symmetrical hydrogen bond may be formed in which the proton lies in a single potential well at a point nearly equidistant from the two heteroatoms—this structure corresponds to a true intermediate between the two tautomeric structures [a possible example is 4-hydroxypyridine 1-oxide (I-359)]. The only rigorous method to distinguish between the above possibilities is accurate X-ray or neutron diffraction with the precise location of all the hydrogen atoms; this is rarely achieved.

Infrared (IR) spectroscopy, the technique most commonly used in the solid state, does not easily distinguish between the possibility of two crystalline forms being (a) two tautomers [**28a**] and [**28b**] (B and X are heteroatoms) or (b) two polymorphs with different crystal packing.

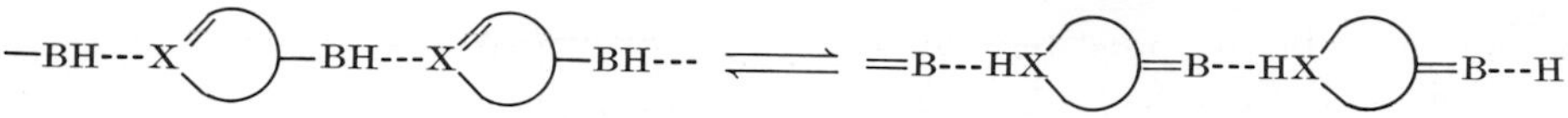

—BH---X(ring)—BH---X(ring)—BH--- ⇌ =B---HX(ring)=B---HX(ring)=B---H

[**28a**] [**28b**]

For an example of such a difficulty in the benzimidazole 3-oxide series, see (71J(B)2350) and (Section 4-7Ad).

Strictly speaking, the interconversion of two CH tautomers, for instance, [**27a**] and [**27b**], should frequently be classified as isomerization rather than as an example of prototropy (cf. Table 1-4, Section 1-5Aa), because the activation energy of the process is too high for the two tautomers to be in mobile equilibrium in solution. However, it is obviously necessary to discuss such cases in the present work to achieve reasonable overall completeness and consistency. We have limited our consideration of ring–chain tautomerism (Section 4-9C) by excluding cases in which the ring and chain tautomeric (or isomeric) forms are interconverted only by an acidic or basic catalyst, such as occurs for the compounds [**29a**] and [**29b**].

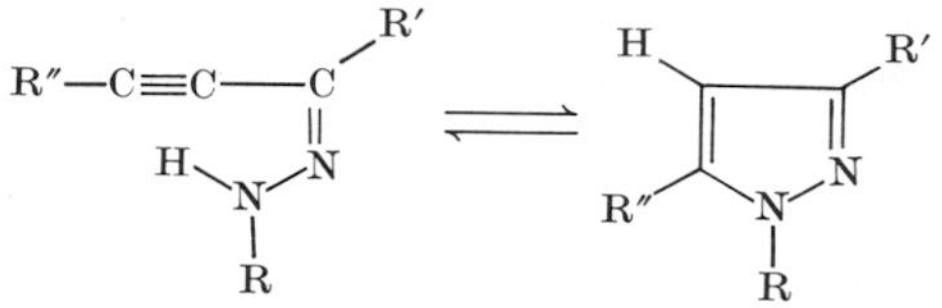

R″—C≡C—C(R′)=N—NH—R ⇌ (pyrazole ring with H, R′, R″, N—R)

[**29a**] [**29b**]

Primary isotopic effects on equilibria are usually small, and it was reported in the previous survey (I-319) that the replacement of hydrogen by deuterium had a small effect on the tautomeric equilibrium. Primary isotope effects on kinetic rates are known to be larger, and tautomeric interconversion rates appear to be no exception: see the porphyrins (Section 6-3).

E. Influence of External Factors on the Equilibrium Position of a Tautomeric Mixture (I-318)

Throughout this survey particular attention will be given to the effect of external factors for systems with small differences of energy between tautomers because the equilibrium constant for these is very sensitive to such effects. When the energy difference ΔG° between two tautomers is large, as is the case for most methyl derivatives (Sections 2-8, 4-8), we cite merely the most significant and well-studied results, for example α-picoline (Section 2-8Ba) [**30a**] ⇌ [**30b**].

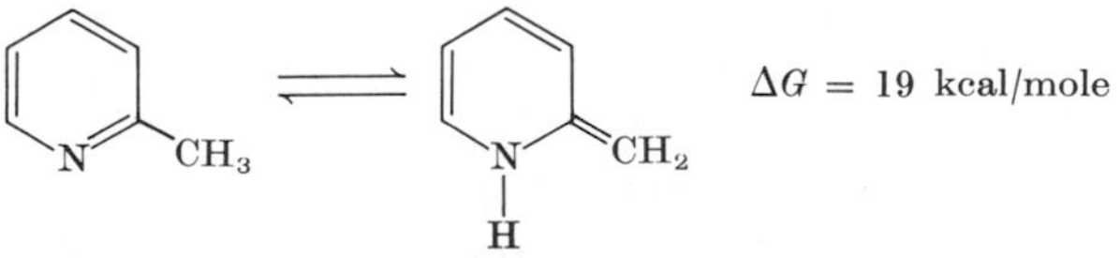

$$\Delta G = 19 \text{ kcal/mole}$$

[**30a**] [**30b**]

Tautomeric equilibrium constants depend on two major external factors: the phase [solid, solution (again depending on solvent and concentration) or vapour] and temperature. Unfortunately many results in the literature are incomplete in this respect. In particular IR and NMR spectroscopic data are still published without any indication of concentration and even without mentioning the physical state of the sample (solid or solution) or the solvent utilized.

Work on the temperature variation of K_T (I-319) is still rare (see however, Section 2-3Bh), and this usually precludes a theoretical approach to the rationalization of tautomerism (Section 1-7B), in terms of ΔH° rather than ΔG°.

F. Importance of the Tautomerism of Heteroaromatic Compounds (I-319)

The importance of heteroaromatic tautomerism to the practicing chemist has already been emphasized by one of us in review articles (65CI331, 70C134), and the following account is based on these.

A tremendous increase in factual knowledge has been made recently in all branches of chemistry; in this general advance heterocyclic chemistry has certainly not been left behind. It is no longer possible to try to commit to memory all the methods of synthesis and all the reactions of even a fairly small class of heterocyclic compounds. It is now essential to place the old facts and the new ones together in a proper systematic scheme and to use the logic behind the scheme to remember and extrapolate from them. Several recent textbooks have in fact taken this approach (e.g., 67MI1). It is quite fundamental to a rational organization of the reactions of heterocyclic compounds that they should be depicted in the reaction mechanism scheme in the correct tautomeric form. For this, it is necessary to know the position of equilibrium between the two or more tautomers that may be present or, more precisely, to know the energy difference between the alternative forms. In addition to tautomeric forms, it is important to consider, in any study of reactivity, the concentrations of the various ionic species which exist, and thus pK_a values are also highly significant. One example of the importance of tautomeric structure in the interpretation of reaction mechanisms will be given (Scheme 1-2). 2-Aminopyridine can exist as such [**31a**] or in the imino form [**32a**]; both of these tautomeric forms are mesomeric, thus [**31a**] has contributions from the charge-separated structure [**31b**] and the imino form [**32a**] from the charge-separated structure [**32b**]. A cursory glance at the structures indicates

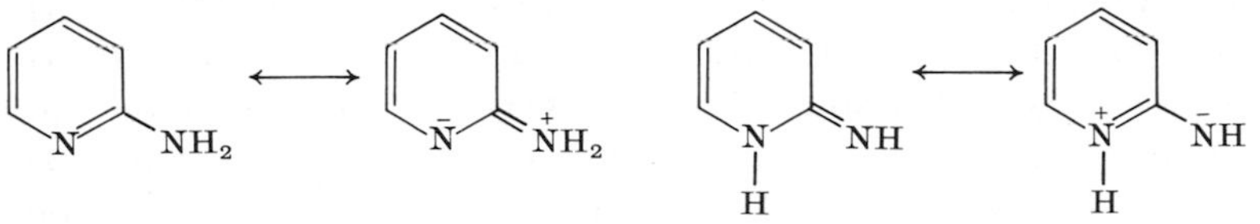

NH_2 ⟷ $\overset{-}{N}$ $\overset{+}{N}H_2$ N–H NH ⟷ $\overset{+}{N}$–H $\overset{-}{N}H$

[**31a**] [**31b**] [**32a**] [**32b**]

that the amino compound should be attacked by electrophiles at the ring nitrogen atom, whereas the imino compound should be attacked by electrophiles at the exocyclic nitrogen atom. A correct interpretation of the chemistry of 2-aminopyridine clearly requires knowledge of which of the two forms is entering into the reaction.

2-Aminopyridine in fact occurs very predominantly in the amino form [**31a**], which is preferred over the 2-imino form [**32a**] by a large

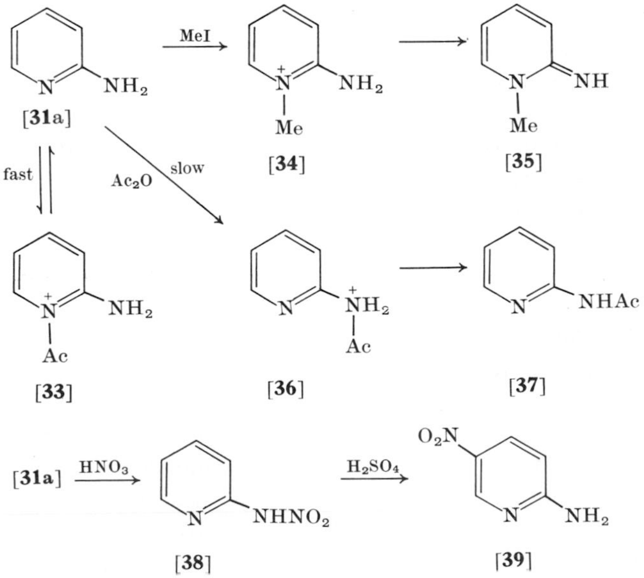

[**31a**] NH_2 —MeI→ [**34**] $\overset{+}{N}$–Me NH_2 → [**35**] N–Me NH

fast ⇌ Ac$_2$O slow

[**33**] $\overset{+}{N}$–Ac NH_2 [**36**] $\overset{+}{N}H_2$–Ac → [**37**] NHAc

[**31a**] —HNO_3→ [**38**] $NHNO_2$ —H_2SO_4→ [**39**] O_2N NH_2

SCHEME 1-2. Reactions of 2-aminopyridine

factor (Section 2-6Ba). This is reflected in its reactions (Scheme 1-2). 2-Aminopyridine [**31a**] does react with electrophiles at the ring nitrogen; thus, with methyl iodide cation, [**34**] is formed, which with base yields [**35**]. Note that the fact that [**35**] is a derivative of the imino form [**32a**] is emphatically not evidence for the interaction of [**32**] in the formation of [**35**]. Reaction of 2-aminopyridine with acetic anhydride yields 2-acetamidopyridine [**37**], which must have been formed via [**36**], itself produced by electrophilic attack on the exocyclic nitrogen. However, [**37**] is produced only because the initial product [**33**] of attack at the cyclic nitrogen atom is formed reversibly and because [**33**] is unstable under the conditions of the reaction. The reaction of 2-aminopyridine with acetic anhydride illustrates the point that knowledge of tautomeric composition, although necessary for a rational interpretation of reactivity, is not of itself sufficient to predict the course of all reactions. The nitration of 2-aminopyridine yields first 2-nitroamidopyridine [**38**] which subsequently rearranges to 2-amino-5-nitropyridine [**39**]: however, the initial reaction probably takes place not on either neutral species {[**31a**] or [**32a**]}, but on the monocation of 2-aminopyridine.

There are numerous examples in the older literature where wrong conclusions have been drawn by correct reasoning on incorrect structures. Sometimes correct conclusions have even been drawn by incorrect reasoning on incorrect structures. The classical chemists were frequently confused by the lack of accurate data on tautomeric equilibria, and their progress was made in spite of the lack of modern theory; we can only admire the advances made at that time. However, it is unfortunately true that incorrect inferences are still being drawn today, and we believe that such publications deserve criticism.

Reasoned predictions of the predominant tautomeric forms of alternative reaction products can sometimes be used to distinguish between two isomeric compounds. For example, the action of substituted hydrazines on ethyl acetoacetate can lead either to a pyrazol-3-one or to a pyrazol-5-one; from the known tautomerism of these compounds (Sections 4-2G and 4-2H), the presence of a carbonyl band in solution is a good indication of the pyrazol-5-one structure [**40**], and the lack of such a band points to a pyrazol-3-one [**41**].

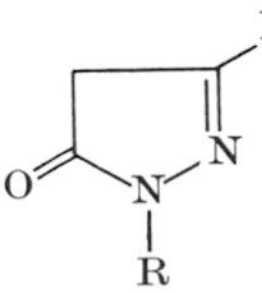

Me O N N R

[**40**]

OH Me N N R

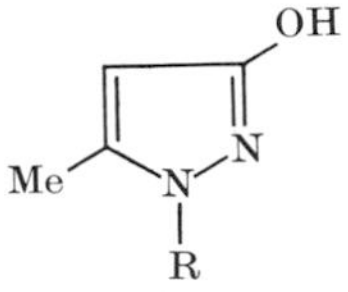

[**41**]

The importance of heteroaromatic tautomerism to the correct interpretation of biochemical processes has been emphasized in the previous review (I-319) and in the introduction to this book.

2. Chemical Methods Used to Study Tautomerism

Sharp criticisms of incorrect applications of chemical methods and emphasis of the difficulty, or even impossibility, to obtain K_T using them was made in the previous survey (I-320) and in shorter reviews (70C134). It is encouraging to report that there has been a big reduction in the amount of dubious work of this type published in the last decade.

As is easy to demonstrate [see, e.g., Charton (69J(B)1240)], product composition alone cannot be used to determine tautomeric composition. If a compound, which exists as a mixture of two tautomers XH and HX, reacts with a reagent R according to Scheme 1-3, then the product

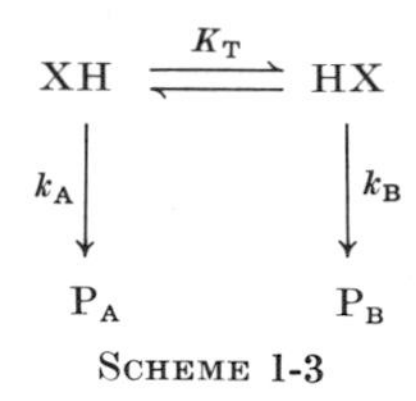

SCHEME 1-3

composition will be determined by Eqs. (1-1) and (1-2), where P_A and

$$\mathrm{XH} + \mathrm{R} \xrightarrow{k_A} P_{\mathrm{A}} \tag{1-1}$$

$$\mathrm{HX} + \mathrm{R} \xrightarrow{k_B} P_{\mathrm{B}} \tag{1-2}$$

P_{B} are the products and k_{A} and k_{B} the corresponding rate constants. Let us assume that P_{A} is formed only from XH,* P_{B} only from HX, that P_{A} and P_{B} are not interconvertible under the reaction conditions, and that both reactions are first order in each component. (If these assumptions do not hold, a valid conclusion with regard to the composition of the tautomeric compound requires a more elaborate treatment.) The rate laws are given by Eqs. (1-3) and (1-4). By defining C_{A} and C_{B} as in Eqs. (1-5) and (1-6), it is possible to derive Eq. (1-7).

$$V_{\mathrm{A}} = k_{\mathrm{A}}[\mathrm{XH}][\mathrm{R}] \tag{1-3}$$

$$V_{\mathrm{B}} = k_{\mathrm{B}}[\mathrm{HX}][\mathrm{R}] \tag{1-4}$$

* Skulski (72RC2139) published an important paper in which he shows that each tautomer may give both derivatives P_{A} and P_{D}, either directly or through an anion X^- or a cation HXH^+.

$$C_A = [P_A]/([P_A] + [P_B]) \quad (1\text{-}5)$$

$$C_B = [P_B]/([P_A] + [P_B]) \quad (1\text{-}6)$$

$$\frac{C_A}{C_B} = \frac{V_A}{V_B} = \frac{k_A[XH]}{k_B[HX]} = \frac{k_A}{k_B} \cdot K_T \quad (1\text{-}7)$$

Thus the product composition is a function of the rate constants k_A and k_B as well as the composition of the tautomeric mixture. The less abundant tautomeric species is also frequently the more reactive.

However, there are examples where sufficient information is available for chemical methods to be used to study tautomeric equilibria legitimately. Thus the rate of hydrogen exchange at the 5-position of 6-hydroxyquinoline [**42**] is much greater than that of 6-methoxyquinoline [**43**], which indicates that the former compound undergoes exchange as the zwitterionic tautomer [**42b**] (71J(B)11). Comparison of the measured rate [**42**] with that for the fixed model [**44**] gives a method for determination of the equilibrium constant [**42a**] $\rightleftharpoons$ [**42b**].

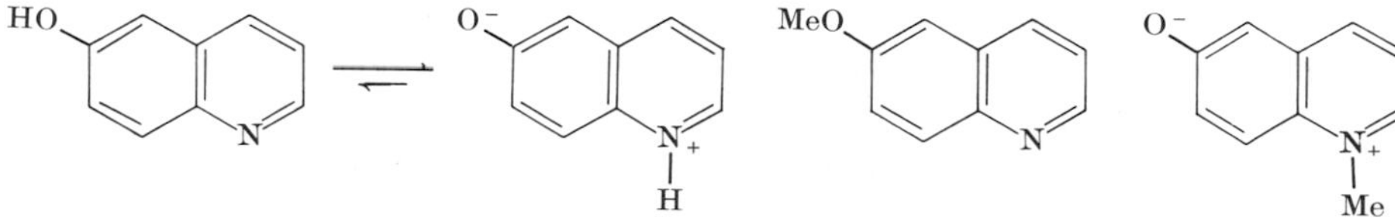

[**42a**] [**42b**] [**43**] [**44**]

Another example where a chemical method can validly be used is in cation formation. 4(5) Bromoimidazole [**44/2**] on dissolution in D_2SO_4 yields cation [**44/3**] resulting from protonation of [**44/2b**] (63AG300). This is good evidence for the predominance of the 4-bromoimidazole structure [**44/2b**] because interconversion of [**44/2a**] $\rightleftharpoons$ [**44/2b**] must take place either through the anion, which is unlikely in the acidic medium, or through the cation, which would lead to a scrambling of [**44/1**] and [**44/3**].

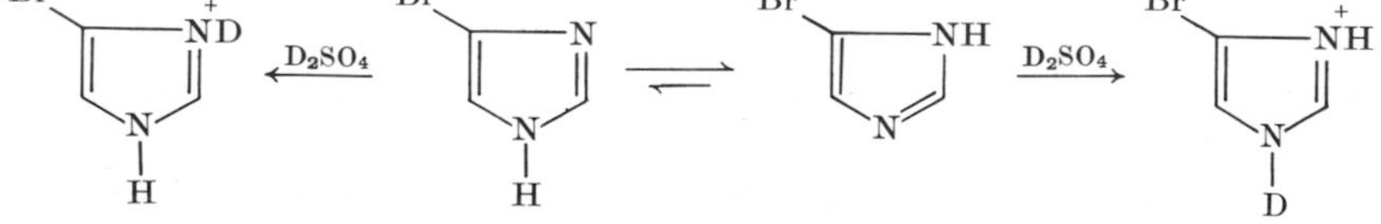

[**44/1**] [**44/2a**] [**44/2b**] [**44/3**]

Another possibility is to calculate K_T from the variation of the ratio C_A/C_B as a function of the substituent. Thus, for the annular tautomerism of a 4(5)-substituted imidazole, Charton (69J(B)1240) showed change of the substituent affects both K_T and k_A/k_B.

Despite the preceding three paragraphs, we believe that in general a total inversion of the problem is needed: rather than use chemical methods for the study of tautomerism, tautomeric equilibrium constants (determined by physical methods) should be used to help our understanding of the chemical reactivity (see Section 1-1F).

3. Physical Methods Used to Study Tautomerism (I-325)

A. Classification

a. Classification of Methods Chosen

Since the previous review there has been a large expansion of the number and scope of physical methods and a considerable variation in their relative importance. We have found it advantageous to use a classification different from that given previously (I-325).

The new classification is shown in Table 1-1. The methods are divided into four important groups (i) non-spectroscopic methods; (ii) NMR spectroscopy; (iii) other spectroscopic methods; (iv) theoretical methods.

Each group is then further subdivided in a logical manner. In the first group subdivision is according to the foundation of the methods: thermodynamic, kinetic, electrical, optical or diffraction. The spectroscopic methods are classified by increasing frequency or time scale: this allows correlation with time-averaging, thus two rapidly equilibrating tautomers appear as a single time-averaged species with a "slow" method, such as NMR spectroscopy, whereas they will produce individual signals in UV spectroscopy.

The theoretical methods are classified into the four most frequently used procedures (Section 1-7B): the direct calculation of K_T. In the three other cases a property is calculated for each tautomer which is then compared with the experimental value: (Section 1-7C), heat of combustion; (Section 1-7D), dipole moment; and (Section 1-7E), electronic spectrum. The following pairs of methods are therefore related (Section 1-4D with 1-7C; Section 1-4G (1-6B) with 1-7D; Section 1-6D with 1-7E) (see Table 1-1).

We believe that readers will find useful the following data indicated in different columns of Table 1-1:

i. The physical state of the sample,
ii. Whether or not fixed model compounds are used (cf. 70C134),
iii. Whether qualitative (L) or quantitative (T) results are generally obtained by the method; this column is left blank for the X-ray

and neutron diffraction methods because they are applied to crystals normally composed of a single tautomer,

iv. The importance of the method: the most important methods are given three asterisks and the least important no asterisks. This classification reflects the present state, and some of the methods may well increase in importance,

v. A cross reference to the volumes of "Physical Methods in Heterocyclic Chemistry" in which the method in question is described.

b. Classification of Methods by Compounds Required

This classification we have chosen is only one of several possibilities. Of great interest is a classification based, as follows, on the compounds required for a study:

i. No compound at all: Section 1-7B, theoretical calculations of K_T.

ii. Only model compounds, but not the tautomeric compound. Section 1-4A, basicity measurements; Section 1-4B, Hammett equation.

iii. The tautomeric compound, but no model compounds: Section 1-4E, relaxation time, Sections 1-4Ib, 1-4Ic, 1-4Id, diffraction methods; Sections 1-5, 1-6A, 1-6B, 1-6C, 1-6E, 1-6F, 1-6G, spectroscopic methods; Sections 1-7C, 1-7D, 1-7E theoretical methods.

iv. The tautomeric compound and model compounds: Section 1-4A, basicity measurements; Section 1-4D, heat of combustion; Section 1-4F, polarography; Section 1-4G, dipole moment; Section 1-4Ha, 1-4Hb, 1-4He, optical methods; Section 1-5, nuclear magnetic resonance; Section 1-6C, infrared spectroscopy; Section 1-6D, UV and visible spectroscopy.

B. Precautions Required in the Comparison of a Tautomer with a Fixed Derivative

In the alternative classification just given, methods of types (ii) and (iv) need the use of model compounds. In such work, it is generally assumed that a certain property of the tautomers P_{H} is identical to that of the fixed derivative P_{Me}. In some cases it is possible to verify the validity of such an assumption (e.g., for basicity measurements, Section 1-4A), and even to introduce a corrective term (Section 1-4A, 1-4G). Nevertheless, in most examples in the literature, the authors did not or could not verify the validity of the hypothesis $P_{\mathrm{H}} = P_{\mathrm{Me}}$.

Moreover, in two not uncommon sets of circumstances, we should

TABLE 1-1

PHYSICAL METHODS

Method	Section	Rate	Sample state	Without fixed derivatives	With fixed derivatives	L/T[a]	Importance of the method[a]	Physical methods in heterocyclic chemistry
Nonspectroscopic methods								
Thermodynamic								
Basicity measurements	1-4A	Averaged	Solution	–	+	*T*	***	63PM(1)1, 71PM(3)1
Hammet equation	1-4B	Averaged	Solution	–	+	*T*	**	—
Heats of combustion	1-4D	Averaged	Liquid, solid	–	+	*L*		74PM(6)199
Kinetic								
Relaxation times	1-4E	(63MI895)	Solution	+	–	*T*	*	—
Electric								
Polarography	1-4F	Averaged	Solution	–	+	*L*	—	63PM(1)217
Dipole moment (dielectric constant method)	1-4G	Averaged	Solution	–	+	*T*	*	63PM(1)189, 71PM(4)237
Optic								
Refractive index	1-4Ha	Averaged	Liquid, solution	–	+	*L*	—	—
Molar refractivity	1-4Hb	Averaged	Liquid, solution	–	+	*T*	—	—
Optical rotation	1-4Hc	Averaged	Liquid, solution	+	–	*L*	—	71PM(3)397
Diffraction								
X-ray diffraction	1-4Ib	10^{18} Hz	Solid (crystal)	+	–		**	63PM(1)161, 72PM(5)1

Electron diffraction	1-4Ic	10^{19} Hz	Vapour	+	−	*T*	—	71PM(3)27
Neutron diffraction	1-4Id	10^{13} Hz	Solid (crystal)	+	−		*	—
Spectroscopic methods								
Nuclear magnetic resonance	1-5	10^{6} to 10^{7} Hz	Solution	+	+	*T*	***	63PM(2)103 71PM(4)121
Nuclear quadrupole resonance	1-6A	—	Solid	+	−	*L*	—	63PM(2)89, 71PM(4)21
Microwave (dipole moment)	1-6B	10^{10} to 10^{11} Hz	Vapour	+	−	*T*	**	74PM(6)53
Infrared and Raman	1-6C	10^{13} to 10^{14} Hz	Solid, liquid solution, vapour	+	+	*L*(*T*)	***	63PM(2)161, 71PM(3)53, 71PM(4)265
UV and visible	1-6D	10^{14} to 10^{15} Hz	Solution, vapour	−	+	*T*	***	63PM(2)1, 71PM(3)67
Ultraviolet, photoelectron spectroscopy	1-6F	10^{17} to 10^{18} Hz	Vapour	+	−	—	—	74PM(6)1
ESCA	1-6F	10^{19} to 10^{20} Hz	Vapour, solid	+	−	—	—	74PM(6)1
Mass spectrometry	1-6G	—	Vapour	+	−	*L*	*	71PM(3)223
Theoretical methods								
Equilibrium constants	1-7B	—	—	+	−	*T*	***	—
Heats of combustion	1-7C	—	—	+	−	*T*	*	—
Dipole moments	1-7D	—	—	+	−	*T*	**	—
Electronic spectrum	1-7E	—	—	+	−	*T*	*	—

[a] For explanation see text Section 1-3Aa.

expect this hypothesis to be seriously in error. This situation occurs if the replacement of a proton by a methyl group either introduces or removes an interaction which significantly modifies the properties of the molecule. Interactions introduced by the methyl group are generally steric in nature, whereas those removed are generally hydrogen bonds.

An example of steric hindrance in a methyl derivative occurs in molecules bearing a phenyl group *ortho* to the methyl. One of the present authors (71J(B)2350) warned explicitly of the errors implicit in using *N*-methyl compounds of this type as models in the imidazole *N*-oxide series (Section 4-7Ab) [**47**]. The magnitude of the deviation of the phenyl ring from coplanarity induced by the methyl is illustrated experimentally by a comparison of X-ray determined torsion angles of approximately 0° and 50° given for structures [**48**] (69CX(B)1050) and [**49**] (69CX(B)192) (cf. Section 4-2Bd).

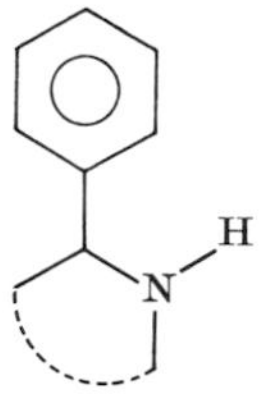

Tautomer NH [**45**]

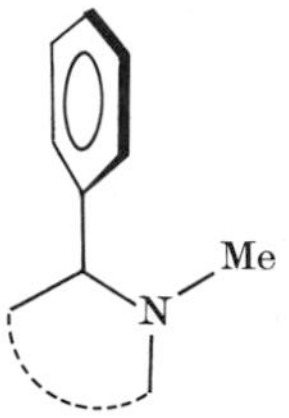

Fixed compound NMe [**46**]

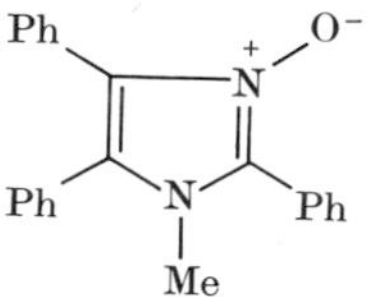

[**47**]

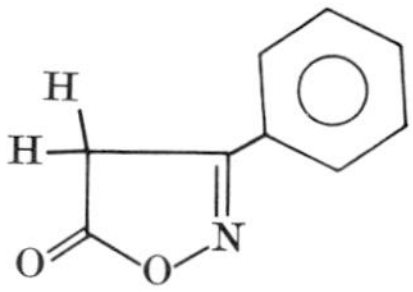

$\theta = 2.45°$ [**48**]

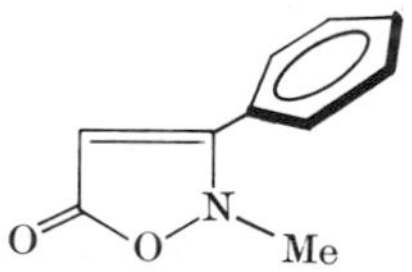

$\theta = 50.4°$ [**49**]

In a series of papers Sélim and Sélim (68BF3268, 68BF3270, 68BF-3272, 69BF823) have attempted to use the nature of the signal of the phenyl protons in the NMR spectrum to determine the structure of the predominant tautomer. For example, for the 2-amino-4-phenyl-thiazoles (68BF3268), they reasoned as follows. The phenyl protons of the tautomeric compound [**50**] and those of the model compound [**52**] appear as a multiplet whereas in the other model [**51**] the phenyl protons appear as a sharp signal. The authors conclude from this that the amino tautomer [**50a**] predominates. Although this conclusion happens to be correct as shown by other evidence (Section 4-5Dc), the reasoning

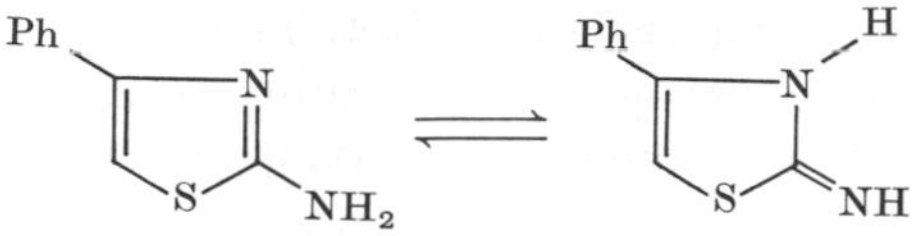

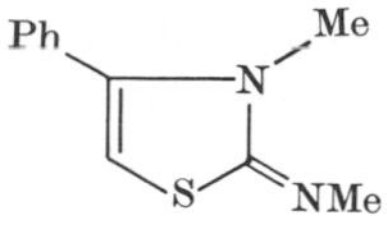

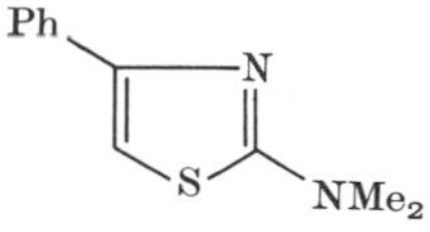

[50a] [50b] [51] [52]

is wrong: the modification in the appearance of the phenyl proton signal probably arises from an interaction with the *N*-methyl group rather than from any change of gross structure.

Other authors (68BF2868) had already reported this phenomenon in the closely related series of the thiazole-2-thiones (Section 4-4Cc). 3-Methyl-4-phenylthiazoline-2-thione [**53**] exhibits a singlet for the phenyl protons, and as the authors pointed out, this is due to significant twisting between the two rings. As both the *S*-methyl derivative [**54**] and the parent compound [**55**], which is known to exist in the thione form (Section 4-4Cc), exhibit complex multiplets for the phenyl protons (68BF2868, 72PSI: solvent DMSO), this proves the steric origin of the phenyl proton singlet for [**53**]. The UV absorption of such compounds is also affected by steric hindrance as illustrated by the hypsochromic effect from [**56**] to [**57**] (69AS2888).

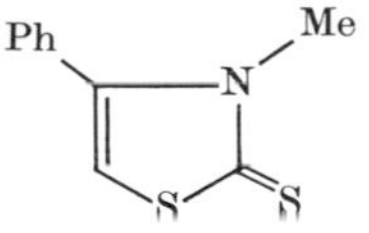

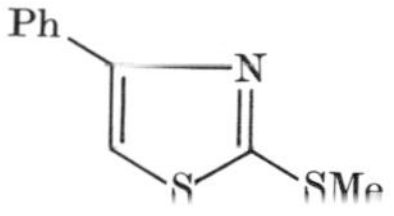

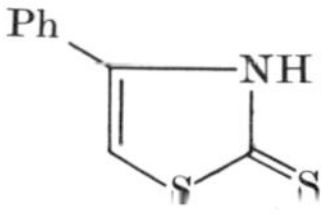

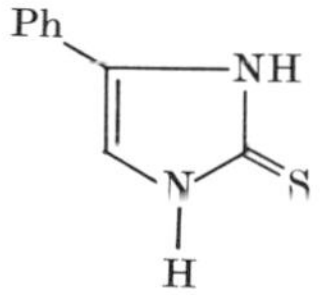

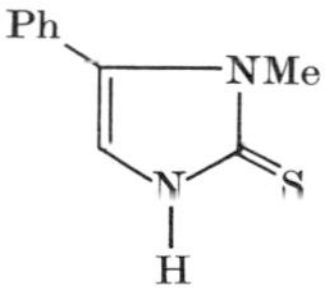

[53] [54] [55] [56] [57]

Clearly, model compounds should be chosen with low steric hindrance, as far as possible, so that the assumption $P_{H} = P_{Me}$ is met as nearly as possible.

An illustration of the removal of a hydrogen bonding interaction in the model compounds is provided by 2-hydroxypyridine 1-oxide. This compound contains a strong intramolecular hydrogen bond in both the tautomeric forms [**58a**] and [**58b**], and also in the cation [**59**]. Simila hydrogen-bonded structures are not possible in either the 1-methoxy

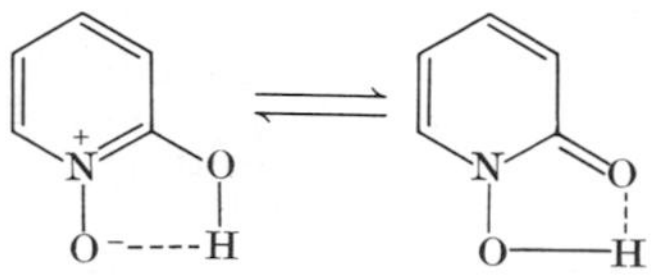

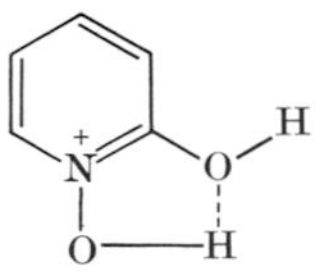

[58a] [58b] [59]

the 2-methoxy models, nor in the cation derived from the 1-methoxy model. In such circumstances, great care must be taken in interpreting the results of methods employing model compounds. (cf. I-359).

4. Nonspectroscopic Physical Methods

A. Basicity Measurements (I-325)

a. Examination of Inherent Approximations

The basis of the method was fully explained in the previous review (I-326). We follow the usual approach (Scheme 1-4), as previously described (I-326), and limit discussion to proton addition.

$$\begin{array}{ccccc} H^+ + XMe & & H^+ + XH \underset{}{\overset{K_T}{\rightleftharpoons}} HX + H^+ & & MeX + H^+ \\ \Big\updownarrow K_{MeA} & & K_A \searrow \quad \swarrow K_B & & \Big\updownarrow K_{MeB} \\ HXMe^+ & & HXH^+ & & MeXH^+ \end{array}$$

Scheme 1-4

We now wish to examine closely some of the assumptions and approximations of the method. If K_1 is the experimentally determined value, then Eqs. (1-8) and (1-9) follow.

$$K_1 = K_A + K_B \tag{1-8}$$

$$K_T = [XH]/[HX] = \frac{K_A}{K_B} = \frac{K_1}{K_B} - 1 = \frac{K_A}{K_1 - K_A} \tag{1-9}$$

As K_A and K_B cannot be measured experimentally, recourse is needed to "fixed" model compounds which are derived by the replacement of a tautomerizable hydrogen by an alkyl group, usually methyl. We define a proportionality constant f between the equilibrium constants for the individual tautomers and those of the corresponding methyl derivative, by Eqs. (1-10) and (1-11) from which Eq. (1-12) follows.

$$K_A = f_A \cdot K_{MeA} \tag{1-10}$$

$$K_B = f_B \cdot K_{MeB} \tag{1-11}$$

$$K_T = (f_A/f_B)(K_{MeA}/K_{MeB}) \tag{1-12}$$

As f_A and f_B are unknown, Eq. (1-12) cannot be used as such, and it is necessary to make one of two alternative simplifying assumptions:

i. First simplification: $f_A = f_B = 1$. Relations in Eq. (1-13) follow. If both f_A and $f_B = 1$, we can then write: $K_T = K_{MeA}/K_{MeB}$, but since experiments using a single model compound are frequently used, Eqs. (1-14)* and (1-15) are often utilized in practice. Equations (1-14) and (1-15) assume that $f_A = 1$ and $f_B = 1$, respectively, but the use of one of these equations does not require that *both* f_A and f_B be equal to unity.

$$K_A = K_{MeA}, \qquad K_B = K_{MeB}, \qquad K_1 = K_{MeA} + K_{MeB} \tag{1-13}$$

$$K_T = \frac{K_1}{K_{MeB}} - 1 \tag{1-14}$$

$$K_T = \frac{K_{MeA}}{K_1 - K_{MeA}} \tag{1-15}$$

ii. Second simplification: $f_A = f_B = f$. In this case K_T is given by Eq. (1-16). However it is illuminating to calculate f, as given by Eq. (1-17) and compare its value with unity, i.e., to test the validity of the first simplification.

$$K_T = \frac{K_{MeA}}{K_{MeB}} \tag{1-16}$$

$$f = \frac{K_1}{K_{MeA} + K_{MeB}} \tag{1-17}$$

It is useful in the consideration of this second simplification to introduce the concept of a statistical factor. For example, certain azoles show degenerate equilibria. For pyrazole [**60a**] $\rightleftharpoons$ [**60b**], $K_T = 1$, and K_1 (the experimentally determined pK_a) is K_{NH}. From the N-methyl derivative [**61**], we have $K_{MeA} = K_{MeB} = K_{NMe}$. Hence for

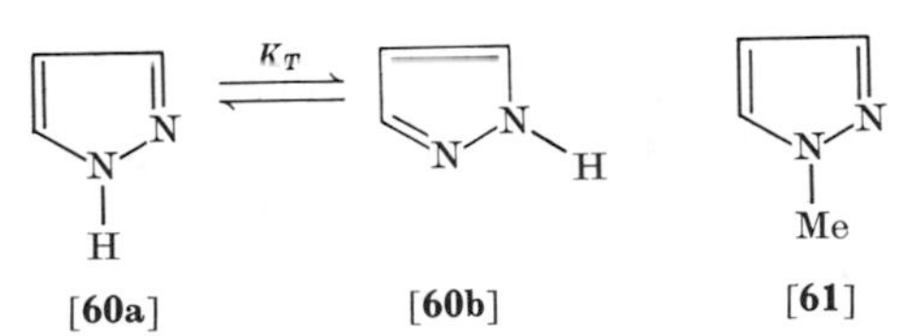

[**60a**] [**60b**] [**61**]

pyrazole and similar compounds, $f = K_{NH}/2K_{NMe}$, the statistical factor S is 2, and Eq. (1-18) follows.

$$\log f = pK_{NMe} - pK_{NH} - 0.30 \tag{1-18}$$

* In papers dealing with a tautomeric equilibrium of the type neutral species $\rightleftharpoons$ zwitterionic species, this equation is written as K_T = antilog $(pK_{MeB} - pK_1) - 1$, and is known as Ebert's equation (63PM(1)1 page 31). Concerning this equation, and tautomeric equilibria in general, others (73J(PII)557) warn against the approximation which assumes that the concentrations of the ionized species (cation and anion) are negligible at the isoelectric point; it is the sum of the molar fractions of the four species XH, HX, X^- and HXH^+, which must be equated to the stoichiometric concentration.

For a compound whose equilibrium is displaced largely towards one of the two tautomeric forms, $S = 1$. Hence Eqs. (1-19) and (1-20) follow for all compounds

$$\log f = pK_{\text{XMe}} - pK_{\text{XH}} - \log S \tag{1-19}$$

$$S = K_T + 1/K_T \tag{1-20}$$

An examination of the literature reveals that the first simplification has generally been used where K_1, and either K_{MeA} or K_{MeB}, but not both, have been determined. Under such conditions the validity of simplification (i) cannot be tested, and a method based on simplification (ii) is not applicable.

However, in a fair number of examples K_1, K_{MeA} and K_{MeB} have all been determined. It then usually becomes apparent that $K_1 \neq K_{\text{MeA}} + K_{\text{MeB}}$ and hence that one of Eqs. (1-14) or (1-15) will give $K_T < 0$, a result that has no physical meaning. For example, in the case of 4(5)-nitroimidazole [**62a**] ⇌ [**62b**] (60J1363), the K_a values for the parent and the methyl derivatives [**63, 64**] are as shown under the formulae.

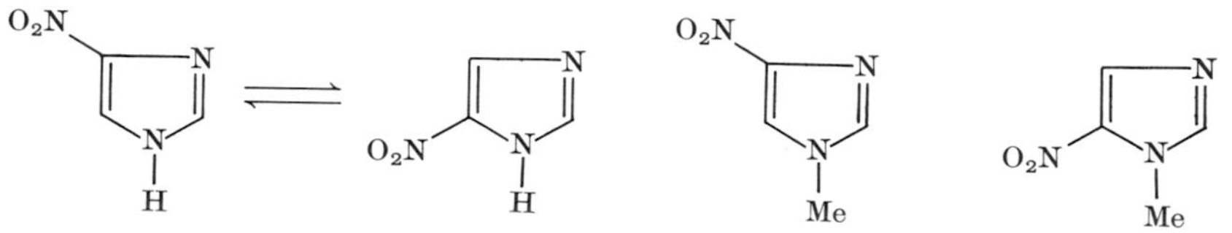

$K_1 = 1.12$

[**62a**] [**62b**]

$K_{\text{MeA}} = 3.39$

[**63**]

$K_{\text{MeB}} = 0.0074$

[**64**]

Since $K_1 < K_{\text{MeA}}$, the first simplification, $K_1 = K_{\text{MeA}} + K_{\text{MeB}}$, is untrue; the use of Eqs. (1-14) and (1-15) gives quite inconsistent values for K_T of 150 and below zero. The use of Eq. (1-16) gives $K_T = K_{\text{MeA}}/K_{\text{MeB}} = 460$ and $\log f = -0.48$.

In Table 1-2 are shown the values of $\log f$ for a series of symmetrically substituted azoles ($K_T = 1$, $\log S = 0.30$).

b. Three Tautomeric Forms with a Common Cation

1,2,4-Triazole and its symmetrically 3,5-disubstituted homologues exist as three tautomers, two of them degenerate (Scheme 1-5). If we assume that the three tautomers lead to a common cation of unique structure (see Section 4-1Eb), then the relations shown in Scheme 1-5 follow.

$$f = K_1/(2K_{\text{NMe(A)}} + K_{\text{NMe(C)}}) \tag{1-21}$$

TABLE 1-2

LOG f VALUES FOR SYMMETRICALLY SUBSTITUTED AZOLES

Compound	Reference	pK_{NH}	pK_{NMe}	$\log f = pK_{NMe} - pK_{NH} - 0.30$
Pyrazole	68BF5009	2.52	2.09	−0.73
3,5-Dimethyl pyrazole	68BF5009	4.12	3.80	−0.62
Imidazole	57JA1656	7.04	7.00	−0.34
	72PS2	7.02	7.06	−0.26
Benzimidazole	51PP420	5.48	5.57	−0.21

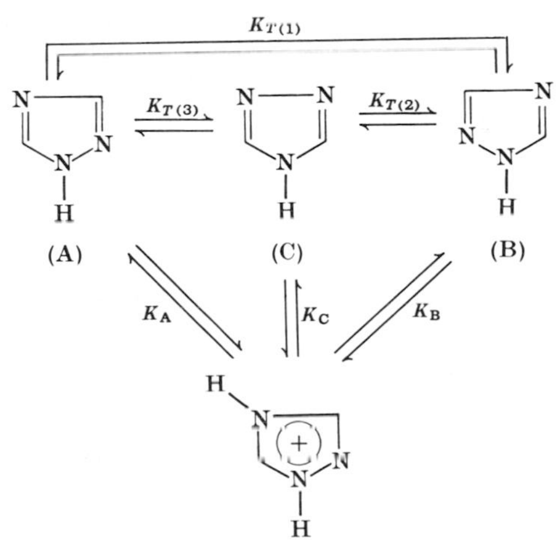

SCHEME 1-5

For 1,2,4-triazole, two sets of results are available. (i) (65TH01), $pK_{NH} = 2.21$, $pK_{NMeA} = 2.05$, $pK_{NMe(C)} = 2.64$, and hence $K_T = 3.9$, $f = -0.51$. (ii) (67C161), $pK_{NH} = 2.27$, $pK_{NMe(A)} = 1.7$, $pK_{NMe(C)} = 2.7$, and hence $K_T = 10$, $f = -0.89$.

For 3,5-dimethyl-1,2,4-triazole, the following has been found (65TH01), $pK_{NH} = 3.97$, $pK_{NMe(A)} = 3.69$, $pK_{NMe(C)} = 4.34$, and hence $K_T = 4.5$, $f = -0.63$.

c. *Application of Basicity Method to the Case of Three Different Tautomeric Forms*

An unsymmetrically substituted 1,2,4-triazole possesses three tautomeric forms [**65a,b,c**]. The corresponding monocation also possesses three tautomeric forms [**66a,b,c**] and the monocations derived for each

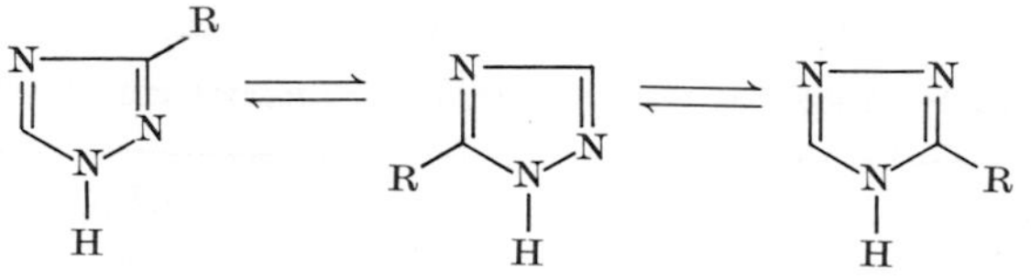

[65a] ⇌ [65b] ⇌ [65c]

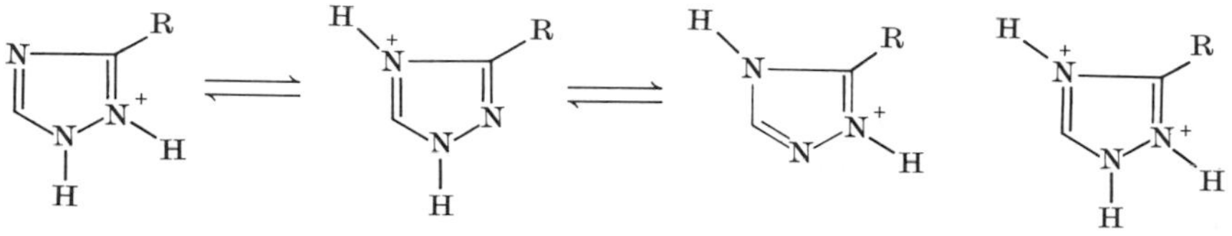

[66a] ⇌ [66b] ⇌ [66c] [67]

of the three monomethyl derivatives possess two tautomeric forms each. The complexity of the problem usually discourages the use of basicity measurements in the tautomeric examination (but see 67BF3780).

However, the dication [**67**] possesses a unique structure, and the measurement of pK_a values for second proton addition of the parent, together with second pK_a values for mono- and dimethyl derivatives offers a possible solution. There are, however, considerable difficulties in the measurement and interpretation of pK_a values at very high acidities. No example of the use of this method in the exact form described is available, but pK_a values for second protonation have been utilized in the determination of the tautomeric structures of 10-hydroxy-1,7-phenanthrolines (71J(B)2339) (cf. Section 2-3Bd), and to find the proportions of the amino and zwitterionic forms which coexist with 4-aminocinnoline (71J(B)2344) (cf. Section 2-6Eb).

d. General Conclusions

In concluding this survey of the approximations involved in the basicity method, the values of log f show sufficient deviation from unity to render suspect the first simplification mentioned above. It is difficult to test the validity of the second simplification: use of a differently substituted heterocycle having appreciably different K_T values and comparison of the agreement of the log f values would imply that the effect of N-methylation on the basicity of a pyridine-like nitrogen atom is independent of the nature of the C-substituents, which is itself not true, especially for substituents capable of delocalizing the excess charge.

Experimentally, the values found by the two methods are often at least in reasonable agreement; one of the least favourable cases is found for two substances already mentioned: imidazole, $\log f = -0.3$, and 4(5)-nitroimidazole, $\log f = -0.5$.

Spinner and Yeoh (71J(B)279) consider that both Eq. (1-9) and Eq. (1-16) give K_T values which are distorted because the base-weakening effect of *O*-methylation is considerable, and greater than that of *N*-methylation. They suggest Eq. (1-21a), where a value of 1 for the correction term c is proposed.

$$\log(1 + 1/K_T) = pK_{OMe} - pK_1 + c \tag{1-21a}$$

Finally, it should be emphasized that despite all its imperfections the method of calculating K_T from basicity measurements is one of the most accurate: other methods tend to be either purely qualitative, or can be used only when equilibria are relatively equally balanced ($0.1 < K_T < 10$), or involve making assumptions about the "fixed" compounds that are at least as questionable as those we have mentioned here.

B. The Hammett Equation (I-335)

In the previous review (I-335) some applications of the Hammett equation were criticized. The methods then outlined do not seem to have been followed up [see, however, Section 2-6Ca (66JO1199) for an example of the application of the Hammett ρ coefficient to a study of the tautomerism of aminopyrimidines]. Recent work on the application of the Hammett equation to heteroaromatic tautomerism has been due largely to Charton, and we wish to comment on two important publications (65JO3346, 69J(B)1240) [for some general criticism of this work see (70QR433)].

a. Imidazoles (65JO3346)

Consider a 4(5)-substituted imidazole XH, a 4-substituted 1-methylimidazole MeA and a 5-substituted 1-methylimidazole MeB. Experimentally, Charton has found that Eqs. (1-22) and (1-23) connect the pK values indicated. For each tautomer of an NH imidazole, relations (1-24) and (1-25) follow. However, $pK_{XH}^{\circ} = pK_{HX}^{\circ} = pK$ of imidazole itself, therefore Eq. (1-26) follows. We are now obliged to make one of the two simplifications, Eqs. (1-27) or (1-28), as discussed in Section 1-4Aa and then arrive at Eq. (1-29).* Qualitatively this suggests that

* For 6-substituted quinoxaline cations (Section 2-2Gd), Vetešnik *et al.* (68CZ566) proposed the relation: $\log K_T = 0.713\sigma_p - 0.196\sigma_m$.

electron-donating groups ($\sigma_m < 0$) will favour the 5-substituted tautomer HX, and electron-withdrawing groups ($\sigma_m > 0$) the 4-substituted tautomer XH.

$$pK_{MeA} = -10.5\sigma_m + 6.88 \qquad \rho_{MeA} = -10.5 \tag{1-22}$$

$$pK_{MeB} = -7.27\sigma_m + 7.39 \qquad \rho_{MeB} = -7.27 \tag{1-23}$$

$$pK_A = \rho_A\sigma_m + pK^\circ_{XH} \tag{1-24}$$

$$pK_B = \rho_B\sigma_m + pK^\circ_{HX} \tag{1-25}$$

$$\log K_T = pK_B - pK_A = (\rho_B - \rho_A)\sigma_m \tag{1-26}$$

$$\rho_A = \rho_{MeA}, \qquad \rho_B = \rho_{MeB} \tag{1-27}$$

$$\rho_B - \rho_A = \rho_{MeB} - \rho_{MeA} \tag{1-28}$$

$$\log K_T = 3.2\sigma_m \tag{1-29}$$

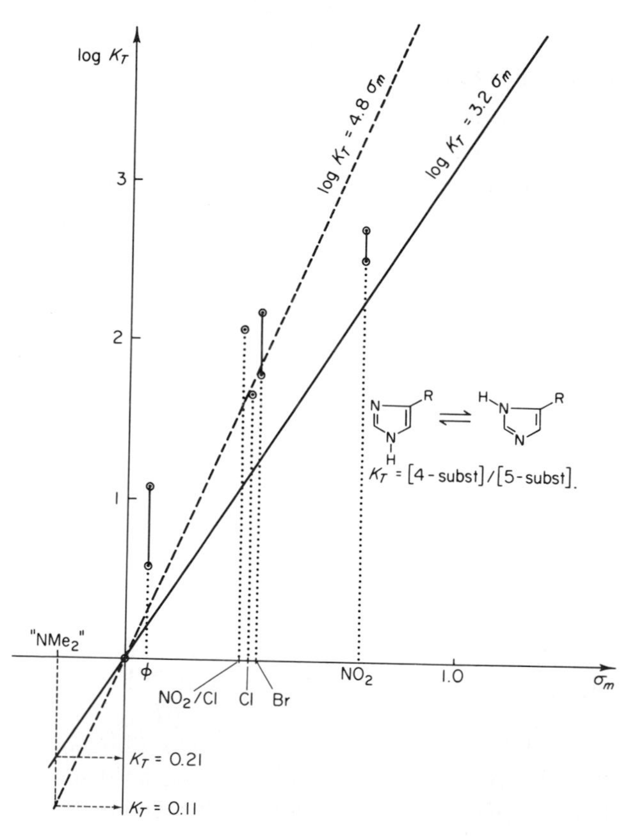

FIG. 1-1. Tautomeric equilibria and Charton's equation for 4(5)-substituted imidazoles.

Charton's equation (Eq. 1-26) is of great importance in calculating azole equilibria, because once $\Delta\rho$ is known the large number of σ_m values available in the literature allows prediction of the equilibrium position for any asymmetrically substituted azole. If we assume that the effect of substituents is additive, then the equation is also applicable to 4,5-disubstituted imidazoles.

The experimental validity of Eq. (1-26) is tested in Fig. 1-1, where the line of Eq. (1-29) is plotted together with individual points obtained from basicity measurements of K_T. The agreement is not excellent, and a somewhat higher value of $\Delta\rho$, about 4.5, would be preferable, but the general feature is correct. Moreover, the points determined experimentally probably contain large errors, as they are very sensitive to errors in pK_a's.

For further discussion of the annular tautomerism of imidazoles, see Section 4-1C.

b. Tetrazoles (69J(B)1240)

The method of the preceding section cannot be conveniently applied to tetrazoles, which are very weak bases. However, tetrazoles are relatively strong acids, and it is possible to determine their acidity constants, pK_a (proton loss) (cf. Scheme 1-6). The experimental acid-

H_2O + [X-tetrazole, 1-H] $\underset{}{\overset{K_T}{\rightleftharpoons}}$ [X-tetrazole, 2-H] + H_2O

K_C ⇅ K_D

[X-tetrazolate anion]

SCHEME 1-6

base equilibrium constant K_1 and the tautomeric equilibrium constant K_T can be expressed as functions of the individual acid dissociation constants K_C and K_D by Eq. (1-30). For tetrazoles, Charton has used "the extended Hammett equation" (1-31), but in this case the constant h does not cancel since, in contrast to the imidazoles, $K_T \neq 1$, even in the unsubstituted parent compound. Hence Eq. (1-32) follows.

$$K_1 = K_C K_D/(K_C + K_D); \qquad K_T = K_C/K_D \tag{1-30}$$

$$\text{p}K = \alpha\sigma_1 + \beta\sigma_R + h \tag{1-31}$$

$$\log K_T = (\alpha_C - \alpha_D)\sigma_I + (\beta_C - \beta_D)\sigma_R + h \tag{1-32}$$

Charton proposes that values for α and β may be obtained from three models, which consist of pK values for protons loss from imidazoles, imidazolium cations and 1-methylimidazolium cations.

In particular, the pK_a values for proton loss from 4- and 5-substituted 1-methylimidazolium cations are proposed as models for K_D and K_C, respectively. Data are sparse, and it is difficult to assess the method objectively. Charton concludes that tetrazole itself exists 80% in the 1(H)-form and that this form is always favoured even with electron-donating groups (e.g., alkyl groups) since log K_T remains positive. These conclusions, which apply only to aqueous solutions, will be examined in a later section (4-1C), and it may be anticipated that their reliability is not beyond doubt.

c. Comparison of Basicity and Hammett Equation Methods

Following this examination of the methods of basicity measurements (Section 1-4A) and the Hammett equation (Section 1-4B), the application of Eqs. (1-14) and (1-15) to the basicity measurements of tautomeric compounds is reconsidered. Correlation by the Hammett equation of experimental values of pK_1 and various σ coefficients have been published for several series of tautomeric compounds, such as pyrazoles (Section 4-1Bb), imidazoles (Section 4-1Cb), tetrazoles (Section 4-1Fa), benzimidazoles (Section 4-1Hb). However, log K_T and log K_1 cannot both simultaneously be proportional to σ, except in special cases, since $K_T = K_A/K_B$ and $K_1 = K_A + K_B$. Writing Eqs. (1-14) and (1-15) as (1-33) and (1-34), it is evident that if $K_T \gg 1$, $1 + K_T \simeq K_T$, and

$$K_1 = K_B(1 + K_T) \tag{1-33}$$

$$K_1 = \frac{K_A(1 + K_T)}{K_T} \tag{1-34}$$

therefore $K_1 = K_B \cdot K_T$. Similarly if $K_T \ll 1$, $1 + K_T \simeq 1$, and hence $K_1 = K_A/K_T$. This means that if the equilibrium is greatly shifted one way or the other, K_1 and K_T are directly proportional, and in this case Hammett correlation between K_1 and σ of type Eq. (1-35) should be observed.

$$pK_1 = \rho\sigma + b \tag{1-35}$$

Such a correlation can be expected in another special case. If the two individual unsubstituted tautomers are not equivalent, then $pK_{XH}{}^\circ \neq pK_{HX}{}^\circ$ and Eq. (1-26) must be written as (1-36). If now K_T

$$\log K_T = (\rho_B - \rho_A)\sigma + c \tag{1-36}$$

$$pK_1 = \rho\sigma + b \tag{1-37}$$

is constant and does not vary with substitution, i.e., if $\rho_A = \rho_B$, then $\log K_T = c$, and $K_T = C$. In those conditions, $K_A = C \cdot K_B$ and therefore $pK_1 = -\log (K_A + K_B) = -\log [(1 + C)K_B] = pK_B + b$. Hence Eq. (1-37) follows.

Such problems are discussed in a paper by Jaffé and Jones (64HC(3) 209); unfortunately, in place of Eq. (1-36) they give Eq. (1-38) [numbered (21) in their paper], which is incorrect.

$$\log K_T = \frac{\rho_A}{\rho_B} \sigma + c \tag{1-38}$$

C. General Introduction to Other Methods Utilizing Fixed Derivatives

Based on the treatment of the basicity method above (Section 1-4Aa), we now outline a general treatment which can be applied to methods which use "fixed" derivatives and in which an "averaged" phenomenon is observed. A similar method of calculation to that given above is used, and the same precautions must be taken.

Compound:	XMe	XH $\underset{K_T}{\rightleftharpoons}$ HX		MeX
Designation of property:	P_{MeA}	P_A	P_B	P_{MeB}

Scheme 1-6a

The general system is defined as in Scheme 1-6a; here P is the particular molecular property under discussion and N_A and N_B are the mole fractions of XH and HX, respectively, defined such that Eq. (1-39) holds.

$$K_T = \frac{[XH]}{[HX]} = \frac{N_A}{N_B} \geq 1 \tag{1-39}$$

If P is an *additive property*, the experimental value P_X can be expressed by $P_X = N_A P_A + N_B P_B$, and since $N_A + N_B = 1$, Eq. (1-40) follows. As P_A and P_B are not directly measurable, we assume that $P_A = P_{MeA}$ and $P_B = P_{MeB}$ to get relation (1-41), which, however, will be valid only if Eq. (1-42) holds.

$$K_T = \frac{P_X - P_B}{P_A - P_X} \tag{1-40}$$

$$K_T = (P_X - P_{MeB})/(P_{MeA} - P_X) \tag{1-41}$$

$$|P_{MeA} - P_A| \text{ and } |P_{MeB} - P_B| \ll |P_{MeA} - P_{MeB}| \simeq P_A - P_B \tag{1-42}$$

In certain cases it is possible to determine the effect introduced by replacing H with Me (or any alkyl group in the general case) assuming that this effect is, to a first approximation, equal for the two tautomers. This effect is called P^*. Then using relations (1-43) we find Eq. (1.44).

$$P_{\mathrm{MeA}} = P_{\mathrm{A}} + P^* \quad \text{and} \quad P_{\mathrm{MeB}} = P_{\mathrm{B}} + P^* \tag{1-43}$$

$$K_T = (P_{\mathrm{X}} - P_{\mathrm{MeB}} + P^*)/(P_{\mathrm{MeA}} - P^* - P_{\mathrm{X}}) \tag{1-44}$$

As we have stated in Section 1-4Aa, in certain examples of annular tautomerism XH = HX; under these conditions $K_T = 1$, $P_{\mathrm{A}} = P_{\mathrm{B}}$, $N_{\mathrm{A}} = N_{\mathrm{B}} = 1/2$, $P_{\mathrm{X}} = P_{\mathrm{A}} = P_{\mathrm{B}}$, and $P_{\mathrm{MeA}} = P_{\mathrm{MeB}}$. This allows the determination of P^*, using Eq. (1-45).

$$P^* = \frac{P_{\mathrm{MeA}} + P_{\mathrm{MeB}} - 2P_{\mathrm{X}}}{2} = P_{\mathrm{MeA}} - P_{\mathrm{X}} = P_{\mathrm{MeB}} - P_{\mathrm{X}} \tag{1-45}$$

We find that this type of calculation is applicable to the following physical methods: Section 1-4D, heats of combustion; Section 1-4G, dipole moments; Section 1-4Hb, molar refractivity; Section 1-5, nuclear magnetic resonance; Section 1-6D, UV and visible spectra,* and finally methods (Section 1-7C and 1-7D) which involve making comparisons between certain calculated and measured quantities.

For an equilibrium involving three or more tautomers, it is not possible to calculate the tautomeric equilibrium constants using only one method since there are three or more unknowns and only two equations. For example, with three tautomers A ⇌ B ⇌ C, Eqs. (1-46) and (1-47) only are available (cf., e.g., 68T151). It is therefore necessary to measure another physical quantity and obtain another equation, such as Eq. (1-48).

$$P_{\mathrm{X}} = N_{\mathrm{A}}P_{\mathrm{A}} + N_{\mathrm{B}}P_{\mathrm{B}} + N_{\mathrm{C}}P_{\mathrm{C}} \tag{1-46}$$

$$1 = N_{\mathrm{A}} + N_{\mathrm{B}} + N_{\mathrm{C}} \tag{1-47}$$

$$P_{\mathrm{X}}' = N_{\mathrm{A}}P_{\mathrm{A}}' + N_{\mathrm{B}}P_{\mathrm{B}}' + N_{\mathrm{C}}P_{\mathrm{C}}' \tag{1-48}$$

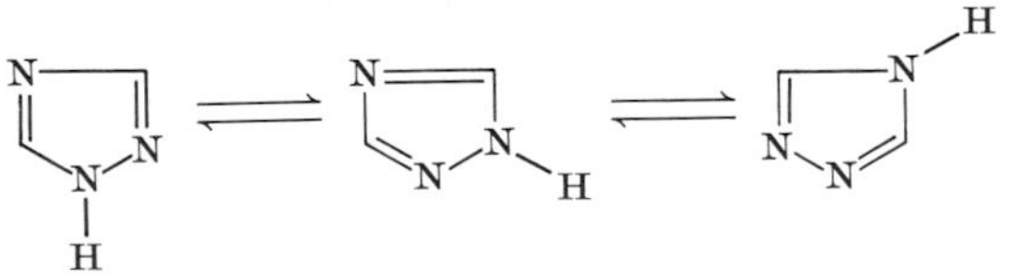

[68a] [68b] [68c]

* In this case the experimental property P_{X} is not averaged, but it is the weighted sum of the properties of the individual tautomers.

Statistical factors must be considered. Thus if A ≡ B ≠ C, as for example for the annular tautomerism of *s*-triazole [**68**], we have $K_T = N_A/N_C = N_B/N_C$ and not $K_T = (N_A + N_B)/N_C$. If it is found experimentally that [**68**] exists 67% as 1,2,4-triazole and 33% as 1,3,4-triazole, then $K_T = 1$ rather than 2. The two tautomers have the same stability, $\Delta G = 0$, but the statistical factor (the probability or entropy of mixing) makes the asymmetrical tautomer predominant. In such a case, Eqs. (1-41) and (1-44) are modified to Eqs. (1-49) and (1-50).

$$K_T = (P_X - P_{MeC})/2(P_{MeA} - P_X) \tag{1-49}$$

$$K_T = (P_X - P_{MeC} + P^*)/2(P_{MeA} - P^* - P_X) \tag{1-50}$$

D. Heats of Combustion (I-338)

There are no reports of tautomerism studies using heats of combustion or formation, although Dewar has studied tautomeric equilibria by comparing calculated heats of formation for each tautomer with that found experimentally for the tautomeric compound itself (see Section 1-6V).

A priori a purely experimental method based on heats of combustion would be very imprecise since the conditions imposed by Eq. (1-42) are not fulfilled. For example, the actual heats of combustion of different azole tautomers are not known, but calculations due to Dewar and Gleicher (66JT759) indicate that the difference $(P_A - P_B)$ for two tautomers can be in the region of 15 kcal/mole. However, replacing NH by NMe $(P_{MeA} - P_A)$ raises the heat of combustion by an amount of the order of 160 kcal/mole (55MI88, 99). The quantitative effect produced by *N*-methylation is expected to be only slightly different for the two tautomers, and results would thus be quite useless.

E. Relaxation Times

Measurements of relaxation times were introduced by Eigen *et al.* (65CB1623) to calculate the interconversion rates k_1 and k_{-1} of a pair of tautomers possessing a common anion as in Scheme 1-7. This method

$$\begin{array}{ccc} XH & \underset{k_{-1}}{\overset{k_1}{\rightleftharpoons}} & HX \\ & \nwarrow\!\searrow \quad \swarrow\!\nearrow & \\ & X^- + H^+ & \end{array}$$

Scheme 1-7

allows direct calculation of the value of the equilibrium constant using the equation $K_T = k_1/k_{-1}$ and without the need for model compounds. In this way a value of $K_T = 1.3 \times 10^{-2}$ (K_T = [enol]/[keto]) (65CB1623) is found for barbituric acid (Section 2-2N). The method has been but little utilized up to now, but shows considerable promise: it avoids the approximation of model compounds and yields a *true* value for K_T.

F. Polarography (I-334)

This method, one example of which was given in the previous review (I-334), has been but little developed since Laviron (63BF2840) compared at different pH values the polarograms of 4(5)-nitroimidazole with those of the two *N*-methylated derivatives. The shapes of the polarograms, and the identity of the reduction potential of 4(5)-nitroimidazole in acid solution to that of 1-methyl-4-nitroimidazole, indicate that 4(5)-nitroimidazole exists mostly in the 4-nitro form (Section 4-1Cb). The absence of a kinetic current when the wave due to the 5-nitro tautomer appeared was interpreted by Laviron as due to a "relatively slow" rate of transformation of 4-nitro → 5-nitroimidazole.

The polarographic technique has been used to the study of the tautomerism of 2-aminothiazol-4-ones (67AS1437). The polarograms of fixed amino and imino compounds exhibit important differences which enable, by analogy, the determination of the tautomerism of the NH_2, NHMe and NHPh derivatives (4-6Ffii).

G. Dipole Moments (Dielectric Constant Method) (I-333)

Considerable progress has been made with this method since the previous review (I-333). An exhaustive review of the dipole moments of azoles, including various studies on tautomerism, is available (71KG867). We consider here dipole moments obtained from dielectric constant measurements (I-333); for moments obtained from microwave spectra see Section 1-6B. Although the procedure for studying the equilibrium is the same in both cases, dipole moments calculated from dielectric constants are usually determined in solution whereas those from microwave spectra generally apply to the gas phase, and in a few cases to solutions in nonpolar solvents (71PM(4)237).

The interpretation of dipole moments measured in solution is made difficult by molecular association. Tautomeric forms possess acidic hydrogens together with basic sites, and are in consequence considerably associated in solution compared to the corresponding N-CH_3 model compounds. Occasionally, dipole moment values can be extrapolated to infinite dilution: e.g., imidazole and 1-methylimidazole have been measured in benzene as a function of concentration (61ZE821).

Two procedures have been used for calculating K_T from dipole moment measurements; that dependent only on experimental values is described here whilst the second approach, which uses both experimental and calculated values, will be considered in Section 1-7D. The experimental method consists in using the square of the dipole moment (71PM(4)237) in Eqs. (1-41) and (1-44), i.e., using relation (1-51). Normally it is assumed that the *increase* in dipole moment

$$P = \bar{\mu}^2 \tag{1-51}$$

introduced by *N*-methylation does not exceed 0.4D (69IQ103). Thus good results will be obtained from Eq. (1-41) provided that condition (1-42) is satisfied. This holds for tetrazole (61T237) for which the data are given in Scheme 1-8. Hence, K_T is found to be 4.4 which corresponds

[**69**] C_2H_5

μ_{EtA} = 5,50 D (benzene) (56JA4197)
P_{EtA} = 30,25

[**70a**] ⇌ [**70b**]

μ_X = 5,10 D (dioxan) (43MI1)
P_X = 26

[**71**] C_2H_5

μ_{EtB} = 2,7 D (benzene) (56JA4197)
P_{EtB} = 7,3

SCHEME 1-8

to 81% of the 1,2,3,4-tetrazole tautomer [**70a**], not 70% as given by Owen (61T237). If correction factors are introduced to allow for the effects of the ethyl groups (cf. Scheme 1-8) a value of 95% is found for the percentage of 1,2,3,4-tetrazole [**70a**].

A serious criticism of many literature applications of the dipole moment method, including that just discussed, is that results which were determined in two different solvents are directly compared. Dipole moments can vary considerably with solvent, especially those of compounds containing NH groups.

If two tautomers have similar dipole moments, the calculation becomes subject to large errors because condition (1-42) is no longer fulfilled. Thus for certain pyrazoles, Hiller *et al.* (65KG107) found the values in benzene given in Scheme 1-9. If no correction is introduced for the effects of *N*-methyl groups on the dipole moments the observed moment for 3(5)-phenylpyrazole [**74**] indicates 100% of the 3-phenyl tautomer [**74a**]. If a correction of 0.22 D is made for the effect of methyl, as indicated by the moments of [**72**] and [**73**], 100% of the 5-phenyl tautomer [**74b**] appears to be found.

[**72**] 2,06 D; [**73**] 2,28 D; $\Delta\mu$ = 0,22 D

[**74a**] 2,28 D ⇌ [**74b**]

[**75**] 2,30 D (2,08 D); [**76**] 2,45 D (2,23 D)

SCHEME 1-9

A less rigorous variation of this experimental approach consists in utilizing another heterocycle or other model in a manner reminiscent of Charton's Hammett equation approach (69J(B)1240) to tetrazole discussed in Section 1-4B. Jensen and Friediger (43MI1) have deduced that the 1,2,5-triazole isomer predominates for *v*-triazole since the dipole moment (1.77 D, in benzene) is close to that of pyrazole (1.57 D, benzene). Benzotriazole was assigned the 1-N(H) form since its dipole moment (4.07 D, dioxan) differs considerably from that of *v*-triazole. Dal Monte *et al.* (58G977) reached the same conclusion by comparing the dipole moments of benzotriazole (4.02 D) and benzimidazole (3.84 D); both are considered to exist as the 1-N(H) tautomer. Although this method is clearly nonquantitative and highly approximate, it can give qualitative results in favourable circumstances.

A variant on this purely experimental method involves comparison of the experimentally determined dipole moment for a tautomeric compound with values calculated for each tautomer. This method will be discussed in the Section 1-7D of "Theoretical Methods." For recent papers on azoles (experimental and calculated dipole moments) see references 75BF(B)ip2 and 73TE238.

H. OPTICAL METHODS

These methods are included for completeness, but have little importance.

a. Refractive Index (I-337)

The lack of interest in this method was remarked in the previous review (I-337). There is still no application to the study of heterocyclic tautomerism, and it seems unlikely to be developed as it is not competitive with other techniques.

b. Molar Refractivity (I-338)

Although this method (I-338) has also been abandoned, some classical results obtained by Auwers (34LA(508)51, 37LA(527)291, 38CB604) in the field of annular tautomerism merit attention. The method is based on the hypothesis that the molar refractivity Rx of a mixture is related to those of the components by expression (1-52).

$$R_{\mathrm{x}} = N_{\mathrm{A}}R_{\mathrm{A}} + N_{\mathrm{B}}R_{\mathrm{B}} \tag{1-52}$$

This approximation breaks down if there are strong interactions between A and B. Unfortunately, this probably applies to tautomeric systems since the molecules HX and XH are frequently intermolecularly hydrogen-bonded.

Auwers used the exaltation of molar refractivity E. Values of E for certain pyrazoles (34LA(508)51) are shown in Scheme 1-10 together

				K_T	
				Eq. (1–41)	Eq. (1–44)
R = CO_2Et	47	39	14	3.1	2.3
R = CH_3	35	30	23	1.4	0.7
R = H	37	29	19	1.2	0.8

molecular refractivity, E

SCHEME 1-10

with alternative figures for K_T calculated from Eq. (1-41), (with $P = E$) and Eq. (1-44) in which E is corrected by E^*, the effect of the *N*-methyl group on E. E^* is calculated [Eq. (1-45)] by a comparison of pyrazole and 1-methylpyrazole $E_{\mathrm{MeA}} - E_{\mathrm{A}} = 2$. Auwers drew only qualitative conclusions in his paper, and in the first case, it is seen to be correct, i.e., the 3-phenyl-5-carbethoxypyrazole tautomer predominates. However, no reliable quantitative conclusions can be drawn regarding the position of equilibrium in the other two compounds, in contrast to reports found in the literature (cf. Fusco, 67MI6).

Auwers has also used the refractivity method qualitatively to show the predominance of the 1-N(H) isomers in indazole, in 3-substituted indazoles (37LA(527)291) and in benzotriazole (38CB604). For benzotriazole, recalculation using Eq. (1-44) shows 100% of the 1-N(H) tautomer with $E^* = -2$. The results are more difficult to interpret for indazole and 3-chloroindazole which possess exaltation values such that $E_x < E_{MeA} < E_{MeB}$. Since $K_T > 0$, Eq. (1-41) implies that the value of E_x must lie between E_{MeA} and E_{MeB}. Eq. (1-44) can be used, but to obtain 100% of the 1-N(H) isomer, E^* then has to be between $+3$ and $+11$.

This discussion of the results of molar refractivity measurements illustrates the unreliability of the conclusions and justifies the abandonment of the method.

c. *Optical Rotation* (I-338)

Optical activity in a heterocyclic natural product usually requires the presence of an asymmetric carbon atom. In this way some oxazol-5-ones were proved to exist in the CH-form (II-50). The method is of little importance because it applies only to CH tautomers which are more easily identified by NMR spectroscopy (Section 1-5). The previous review also discussed the use of optical activity to study the rate of tautomeric interconversion (I-338).

I. Diffraction Methods

a. *Introduction*

Diffraction methods are already of great importance in the study of heteroaromatic tautomerism and an increase in this importance can be predicted. Energy characteristics of the three methods are listed in Table 1-3.

TABLE 1-3

Characteristics of Diffraction Methods

Parameter	X-ray	Electron	Neutron
Energy (eV)	8×10^3	1.2×10^5	2.5×10^{-2}
Wavelength (Å)	1.5	0.1	1.8
Frequency (Hz)	2×10^{18}	3×10^{19}	10^{13} Å/sec

b. X-Ray Diffraction (I-332)

X-ray diffraction (I-332) is the most widely used diffraction method (72PM(5)1). It is limited, as is neutron diffraction, to crystalline solids. Relatively numerous cases are known in which a crystal unit cell contains two different molecules, but only two examples of two tautomeric forms occurring in a single crystal have been reported (65CX (19)797; 73MI469); some examples are known of the isolation of two different crystalline forms corresponding to different tautomeric forms (desmotropy, see Section (1-1D).

The main limitation of the application of X-ray diffraction to a study of prototropic tautomerism is that the location of the hydrogen atoms is difficult. However, the precise location of the tautomeric proton is now normally achieved, either directly (by working at low temperatures to reduce thermal agitation) or indirectly (by measurements of intra and/or intermolecular bond angles and distances).

These diffraction methods are, together with infrared spectra (Section 1-6C), the only physical methods to give information about the solid state.

The numerous tautomeric structures established by X-ray diffraction (72PM(5)1) cannot all be cited here, but two aspects of the results obtained by this method are worthy of detailed consideration. Firstly, it is possible to calculate from the experimentally determined bond distances, using the valence-bond approach, the contribution of each canonical structure; see, for instance, cytosine (64CX1581) and *s*-triazole (69CX(B)135). This information is useful in the explanation of other properties, such as infrared spectra; however, we emphasize again the care needed not to confuse resonance and tautomerism in such discussions (cf. Section 1-1C).

Structure determination using X-ray measurements has the great advantage, over that utilizing IR studies (Section 1-6C), in that clear differentiation is possible (always in principle and often in practice), for example, between an O-H---N bond and an O---H-N bond. For example, in the 4-pyridone series (Section 2-2Bb) it has been possible to show that the 3,5-dichloro-2,6-dimethyl-4-pyridone [**77**] exists as such, while the tetrachloro derivative exists in the pyridinol form [**78**] (72CH573). Other structural parameters (C–O distance, C–N–C angle) support the result deduced from the OH and NH interatomic distances shown (in Å) in structures [**77**] and [**78**]. Many as yet unsolved problems could be settled by accurate X-ray diffraction, for example this has recently been achieved for the structure of pyrazol-5-ones in the solid state (Section 4-2Ae).

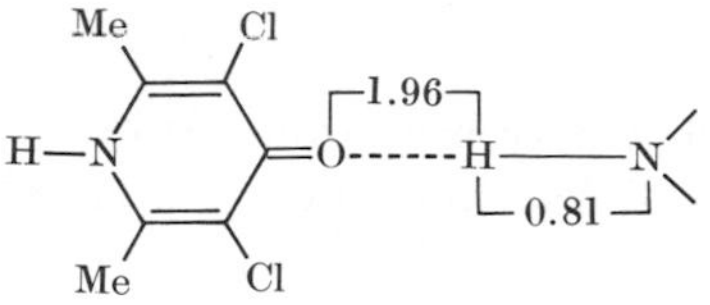

[77] 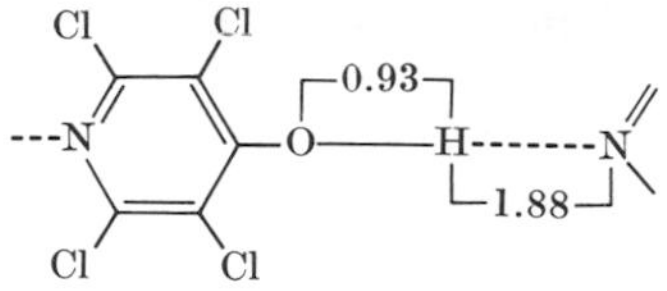 [78]

In the field of annular tautomerism, crystallographic measurements have demonstrated the asymmetric nature (C_S symmetry and not C_{2v}) of certain ring systems as pyrazole (70CX(B)1880) (Section 4-1Ba), imidazole (66CX(20)783, 69ZX(129)211) (Section 4-1Ca), 1,2,4-triazole (69CX(B)135) (Section 4-1Ba), and the hydrate of 5-amino-1,2,3,4-tetrazole (67CX(22)308) (Section 4-5Ff) [see, however, (73J(PII)2036) for a paper claiming the nonlocalization of the tautomeric proton in 5-bromotetrazole]. X-ray crystal structures of the three benzazoles (indazole, Section 4-1G; benzimidazole, Section 4-1Ha, and benzotriazole, Section 4-1I) have been recently investigated,* and the location of the proton on the nitrogen atom N(7) of purine (Section 5-1Aa) was established in the now classical work of Watson *et al.* (65CX(19)573).

c. *Electron Diffraction*

Electronic diffraction (71PM(3)27), can be used only in the vapour phase because of the low penetration power of the electron beam in solids. This method has the advantages that virtually free molecules are observed and that hydrogen atoms are located with a higher accuracy than by X-ray diffraction. Unfortunately the vapour pressure required (about 10 mm Hg) generally greatly exceeds that of tautomeric heteroaromatic compounds, which usually possess strong intermolecular hydrogen bonds.

No example of the application of this technique to the study of the prototropic tautomerism of a heteroaromatic compound is yet available.

d. *Neutron Diffraction*

Neutron diffraction (69AG307) is still a technique little used, mainly owing to the problem of obtaining the source (nuclear reactor). The intensity of a neutron beam is normally very low compared to the intensities produced by X-ray tubes, and the times required for neutron diffraction measurements are therefore rather long (about 1 hour per degree). Nevertheless, neutron diffraction is of interest as it permits

* These three benzazoles have been studied at Montpellier (74T2903); in each case the proton is bound to the nitrogen atom N(1).

the location of light atoms (especially hydrogen atoms) in a molecule containing heavy atoms.

The "asymmetrical" pyrazole structure (Section 4-1Ba) has been demonstrated by this method (70AS3248) and the molecular structure of the 2-hydroxypyridinium hydrochloride (Section 2-2C) has been established by a simultaneous use of data obtained from neutron and X-ray diffraction (69TL5219).

5. Nuclear Magnetic Resonance Spectroscopy (I-336)

A. Proton Magnetic Resonance (I-336)

a. Introduction

In recent years proton magnetic resonance (PMR) has been the physical method that has overall produced the greatest progress in organic chemistry, and this statement clearly applies in the field of heterocyclic chemistry (71PM(4)121), as will be readily apparent if the present section is compared to the corresponding section in the previous review (I-336).* Nevertheless, PMR is by no means an ideal method for the study of tautomeric equilibria as it is a "slow" method of observation, which in many cases yields merely an "averaged" spectrum. Thus the important advantage of directly obtaining quantitative results is lost, and the method becomes one of interpolation with all the disadvantages mentioned above, particularly the need for model compounds and the possibility of errors inherent in the calculations. Table 1-4 gives kinetic information which enables precise definition of the problems posed by tautomerism in various systems. To simplify matters, we note that most examples of prototropic tautomerism in heterocyclic compounds belong to the kind of process designated as cases 3 or 5.

The rates of tautomeric change have been discussed in Section 1-1D. As pointed out there, the interconversion of two different CH-tautomeric forms, which can be separated, should strictly be considered isomerization. Such processes constitute Case 2 of Table 1-4. Case 3 covers many prototropic equilibria which involve a single C–H bond breakage. For such compounds the energy barrier is often too low to allow the separation of the two tautomeric forms and too high to allow coalescence to be reached at easily attainable laboratory temperatures.

* Recently (72RU452) Kol'tsov and Kheifets published a review on "study of tautomerism by NMR spectroscopy" which includes numerous examples of tautomerism in the heteroaromatic series.

TABLE 1-4

GENERAL KINETIC DATA

Isomer separation	Equilibrium in solution	Kind of process	$\Delta G^{\ddagger}$ (kcal/mole)	k (sec^{-1})	τ (sec)	Example
The isomers can be separated	No equilibrium[a]	1. Isomerism	> 40	$< 10^7$	$> 10^7$	$\Delta G^{\ddagger} \sim 45$
	Slow equilibrium	2. Equilibration	20–40	10 to 10^{-7}	10^{-1} to 10^7	 $\Delta G^{\ddagger} \sim 30$
The isomers cannot be separated	Fast equilibrium, $\Delta G^{\ddagger} < 30$ kcal/mole	3. Separate signals in NMR spectroscopy ($T = 500°K$)	> 20	< 10	$> 10^{-1}$	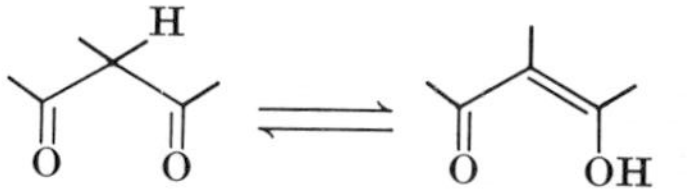 $\Delta G^{\ddagger} \sim 20$
		4. Coalescence	10–25	10^4 to 10	10^{-1} to 10^{-4}	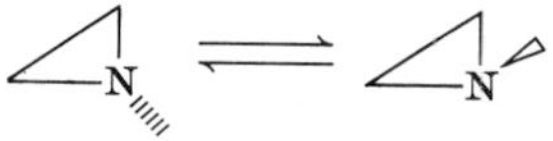 $\Delta G^{\ddagger} \sim 10$
		5. Averaged signals in NMR spectroscopy ($T = 150°K$)	< 10	$> 10^4$	$< 10^{-4}$	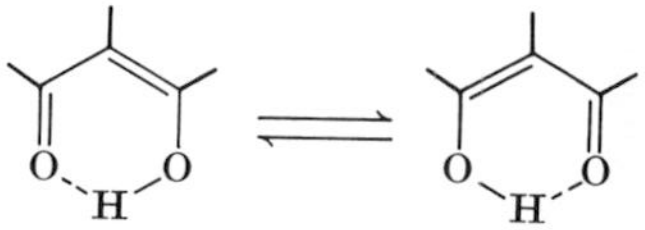 $\Delta G^{\ddagger} \sim 5$

[a] Strictly speaking, very slow equilibrium, arbitrarily, $\tau > 10^7$ seconds.

Case 5 covers many examples of prototropic equilibria which involve only the breakage of heteroatom-hydrogen bonds, including annular tautomerism. Frequently the intermediate situation is disclosed by separate signals at lower temperatures, which coalesce on heating to give time-averaged signals at higher temperatures: this constitutes Case 4.

Cases 2 and 3 pose no problems, as K_T can be found by simple integration of the NMR spectrum, but Cases 4 and 5 need detailed consideration.

b. Determination of K_T by Interpolation (Case 5)

In NMR spectroscopy the chemical shift δ is the molecular property P usually used in the interpolation method. Theoretically, coupling constants could equally well be used, for example $P = J(^1H - {}^{15}N)$ (66JA2407), but this method has been little utilized in the quantitative study of heterocyclic tautomerism [see, however, (70CZ2936, Section 4-8B) and (74OR224, Section 4-2Be) for two examples and (74CH702, Section 4-1Da) for a very interesting utilization of $J(^1H - {}^{13}C)$] and our further discussion applies to chemical shifts.

For the degenerate rearrangements which frequently occur in annular tautomerism, the chemical shifts of two types of atom must be considered separately, depending on whether the environment varies with the tautomerism. Thus, for pyrazole [**79a** ⇌ **79b**] the environment

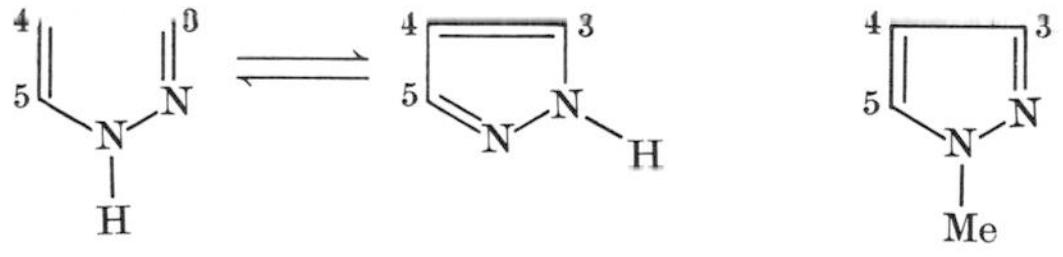

[**79a**] ⇌ [**79b**] [**80**]

of the proton in position 4 remains constant, $P = \delta$, and Eq. (1-53) holds. However, although the signals of protons in positions 3 and 5 appear together in pyrazole because of a fast exchange [**79a** ⇌ **79b**], here $P = \Sigma\delta$ and Eq. (1-54) applies.

$$\delta^4_{NMe} = \delta^4_{NH} + \delta^* \qquad (1\text{-}53)$$

$$(\delta^3_{NMe} - \delta^5_{NMe}) = (\delta^3_{NH} + \delta^5_{NH}) + \Sigma\delta^* = 2\delta_{NH} + \Sigma\delta^* \qquad (1\text{-}54)$$

This requirement for an equation of type (1-54) also applies to the ^{13}C resonance carbons in position 3 and 5, and to the ^{14}N resonance nitrogens in positions 1 and 2. Several examples show that the method is unreliable as P^* is comparable in magnitude to $|P_{MeA} - P_{MeB}|$ and thus condition (1-42) is not satisfied.

In Scheme 1-11, $\Delta\delta$ values of $\delta H_{3(5)} - \delta H_4$ are given for solutions in

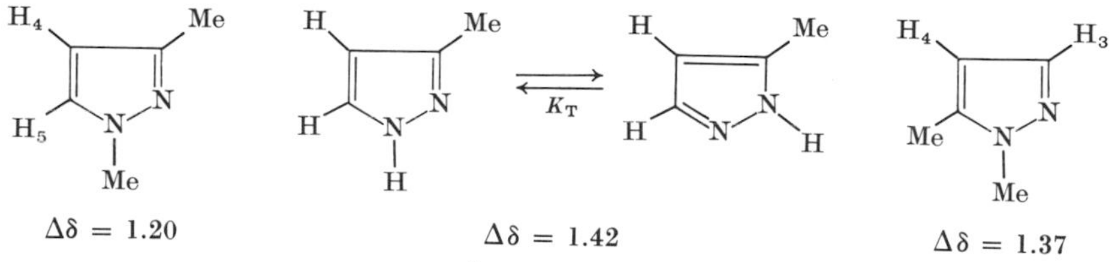

$\Delta\delta = 1.20$ $\Delta\delta = 1.42$ $\Delta\delta = 1.37$

SCHEME 1-11

$CDCl_3$ (65JO1892). The authors concluded that "the 5-methyl structure is the predominant component of the tautomeric equilibrium" [Eq. (1-26) cannot be applied here since $\Delta\delta_X$ does not lie between $\Delta\delta_A$ and $\Delta\delta_B$]. If P is defined by Eq. (1-55), then literature values of chemical

$$P = \Sigma\Delta\delta = (\delta H_3 - \delta H_4) + (\delta H_5 - \delta H_4) \quad (1\text{-}55)$$

shifts in $CDCl_3$ (66BF3727) give $P_X = 2.60$ for pyrazole and $P_{MeA} = P_{MeB} = 2.40$ for 1-methylpyrazole. A simple application of Eq. (1-45) thus gives $P^* = -0.20$ ppm; it is impossible to split the effect of -0.20 ppm between positions 3 and 5 (there is no reason to assume that the *N*-methylation effect is the same on the chemical shift of proton 3 as on that of proton 5), and this precludes the calculation of K_T. However, the important conclusion can be drawn that the condition of Eq. (1-42) is not fulfilled since P^* and $P_{MeA} - P_{MeB}$ are of the same order of magnitude. Hence the conclusion of (65JO1892) is not justified on this evidence.

Tensmeyer and Ainsworth (66JO1878) attempted to determine the equilibrium position in NH pyrazoles using empirical constants calculated from the *N*-methyl derivatives, with the implicit assumption that $P^* = 0$. Their calculations resemble those of Auwers discussed in Section 1-4Hb, and the increments for positions 3 and 5 differ significantly only for substituents such as phenyl. They conclude that equilibrium [**81a**] ⇌ [**81b**] is displaced towards [**81a**] when R = H,

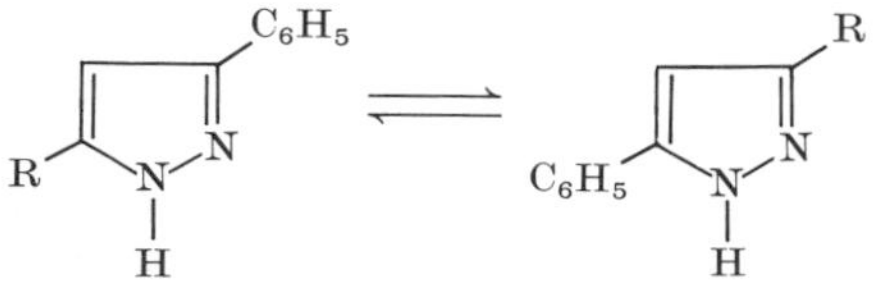

[**81a**] [**81b**]

CH_3, or CO_2CH_3, in accordance with Auwers' findings (34LA(508)51). The authors do not comment on cases where R = $CO_2C_2H_5$ or R = $CO_2C_3H_7$, but the values merit discussion. Scheme 1-12 gives values of

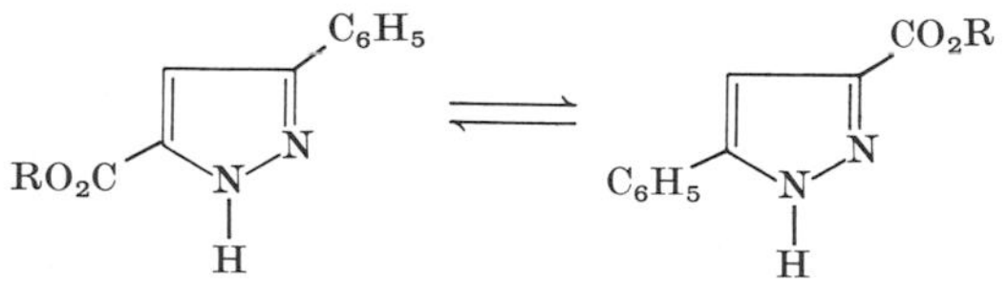

R	CH_3	C_2H_5	n-C_3H_7
K_T	∞	0.28	0.48

SCHEME 1-12

K_T calculated using Eq. (1-44) with $P = \delta_4$ and $P^* = -0.03$ ppm as given by the authors for the effect of *N*-methylation. The minor modifications in the alkyl group R clearly cannot account, either sterically or electronically, for such a large variation in tautomer composition. It must be concluded that, for proton resonance, the differences in chemical shift of the tautomers are too small compared to other influences on chemical shift to permit even qualitatively reliable conclusions.

c. *Temperature Coalescence Studies* (Case 4)

We will examine in detail the four known examples of coalescence phenomena since this type of study is capable of considerable future development. All the examples involve annular tautomerism.

i. *s-Triazole* (68JO2956). The tautomerism of *s*-triazole involves three species [**82a, 82b, 82c**], the first two of which are identical. The

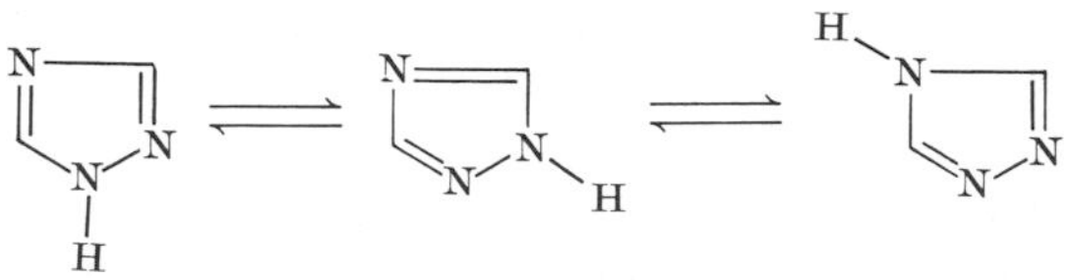

[**82a**] [**82b**] [**82c**]

spectrum of this compound as a 4.3% solution in hexamethylphosphoramide (HMPT) at 37° shows two signals, one broad at 15.0 ppm due to the NH, and one narrow at 8.17 ppm due to the two CH protons. On cooling, the NH peak sharpens considerably whilst that of the CH proton broadens, to a coalescence about 10°, and then separates on further cooling into two peaks, which are separated by 56 Hz. The spectra of *s*-triazole and two "fixed" *N*-methyl derivatives in the CH region at various low temperatures are shown in Scheme 1-13. The assignments of the H_3 and H_5 protons have been reversed from those of (68JO2956) to be consistent with assignments in DMSO solution (67BF2630).

H_3 H_5 H_3

N N

H_5 N 8.85 7.92 −30°

H

H_3 H_5 H_3

N N

H_5 N 8.93 8.02 −40°

CH_3

N—N

H N H 8.34 40°

CH_3

SCHEME 1-13

Clearly, the 1,3,4-tautomer [**82c**] is not significantly involved in the equilibrium, which is essentially an autotropic rearrangement of the 1,2,4-tautomer [**82a**] ⇌ [**82b**]. At the coalescence temperature, the half-life is given by the expression $\tau = \sqrt{2}/2\pi\Delta\nu$; in the present case, $\Delta\nu = 56$ Hz, thus $\tau = 4 \times 10^{-3}\ \text{sec}^{-1}$. Using the Eq. (1-56) a coales-

$$\Delta G_c{\ddagger} = 4.57\,T_c(10.32 + \log T_c/k) \qquad (1\text{-}56)$$

cence temperature lying between 10° and 20° indicates $\Delta G_c^{\ddagger}$ to be ca. 14 kcal/mole, which is considerable.

Two critical remarks are necessary with regard to this work; firstly, the published chemical shifts at 0° and 10°C are incorrect, although the true values are to be found in the original thesis (67TH01).* Secondly, great care is needed when a solvent as basic as HMPT is used because it is extremely difficult to obtain it completely dry; since proton transfer is catalysed by water, it is possible that the coalescence temperature (and hence $\Delta G_c^{\ddagger}$) is dependent on the presence of water.

* A more complete picture (72PS14) than that published (68JO2956) for the variation of the NMR spectrum of *s*-triazole in HMPT as a function of temperature shows at −30° the same separation of 56 Hz which gives a value of $\Delta G_c^{\ddagger} =$ 14.3 kcal/mole for a coalescence temperature of 20°.

It is more important to consider exactly the significance of the measured $\Delta G_c^\ddagger$. Whereas it could correspond to the simple autotropic rearrangement, other possibilities include an equilibrium between free molecules and molecules associated with the solvent or reaction with the solvent [see, e.g., Eq. (1-57)]. The experiment quoted does not permit us to distinguish between these possibilities.

(1-57)

ii. *4-Nitropyrazole* (69TL495). In acetone solution, the NMR spectra of most azoles undergo changes because of reversible addition of the azole to the acetone carbonyl group. This is a "slow" process on the NMR time-scale, as is shown by the presence of narrow peaks with intensities that are temperature dependent. In addition to this phenomenon, 4-nitropyrazole and 3,5-dimethyl-4-nitropyrazole show coalescence, and at lower temperature ($-95°$ to $-100°$) resolution of the signals due to protons or methyl groups at positions 3 and 5.

For 4-nitropyrazole, coalescence is observed at $-70°$ and the peak separation is 33 Hz; this gives $k_c = 75 \text{ sec}^{-1}$ and thus $\Delta G_c^\ddagger = 10$ kcal/mole. For 3,5-dimethyl-4-nitropyrazole, $T_c = -90°$, $\Delta\nu = 8$ Hz, $k_c = 18 \text{ sec}^{-1}$, and thus $\Delta G_c^\ddagger = 9.5$ kcal/mole. Although these figures are very approximate, they give an order of magnitude for the activation energies. In the same solvent, 1,2,4-triazole and 3,5-dimethyl-1,2,4-triazole show a clear broadening at $-70°$, but since $\Delta\nu$ cannot be measured, the rate of interconversion cannot be calculated.

Some of the criticisms made in the previous section also apply here, particularly with regard to water in the acetone and association with the solvent. At least part of the 10 kcal/mole necessary to transfer a proton from one nitrogen to another is needed to break up azole–acetone association, and the energy diagram will be as shown in Fig. 1-2. Thus in a less "basic" solvent, such as $CHCl_3$, CS_2 or CCl_4, the activation energy will be much less. Experimentally no line broadening has been observed in these solvents until recently (71DA110), when, in a mixture CH_2Cl_2–$CHCl_3$, 3,5-dimethylpyrazole was found to broaden at $-80°$; at $-110°$ the broadening reaches 4.0 to 4.4 Hz.

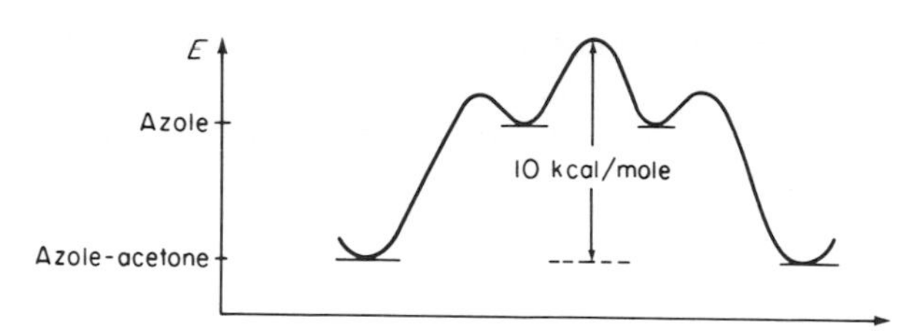

FIG. 1-2. Schematic energy diagram for annular tautomerism of an azole dissolved in acetone.

iii. *Benzotriazole* (69T4667). Study of the NMR of benzotriazole in dry tetrahydrofuran between 0° and −100° shows that the AA′BB′ system of the benzene ring protons broadens on cooling and ultimately becomes an ABCD system. Hence the AA′BB′ system at room temperature is due not to the 2-N(H) isomer [**83a**], but to an autotropic rearrangement of the two 1-N(H) structures [**83b**] ⇌ [**83c**].

[**83a**] [**83b**] [**83c**]

The NH signal becomes narrower on cooling, values for the line width being 56 Hz at 56° and 2 Hz at −100°. The authors have used the modified Arrhenius equation (1-58) where ν is the width of the signal

$$\log \nu = \log \nu_0 - \frac{E_a}{4.57T} \qquad (1\text{-}58)$$

in hertz. This gives $\nu_0 = 2.2\ 10^3$ Hz and $E_a = 2.4$ kcal/mole (and not $\nu_0 = 1.5 \times 10^3$ and $E_a = 2$ kcal/mole as indicated in the publication). Since frequency factors for the Arrhenius equation are generally between 10^{12} and 10^{14} $\sec^{-1}$, the authors consider that the extremely low value of 10^3 Hz "shows a low rigidity of the NH bond."

iv. *2-Chlorobenzimidazole* (71TL3299). The tautomeric equilibrium [**84a**] ⇌ [**84b**] has been studied in tetrahydrofuran dried on a sodium mirror. When R = H, the authors find that the signals due to the benzene ring protons start to broaden at −100°, whereas if R = OCH_3, the coalescence temperature is −35°. Application of the "total line shape method" and the Arrhenius equation gives $E_a = 4.25$ kcal/mole and $A = 1.95 \times 10^5$ $\sec^{-1}$, and from these values the following

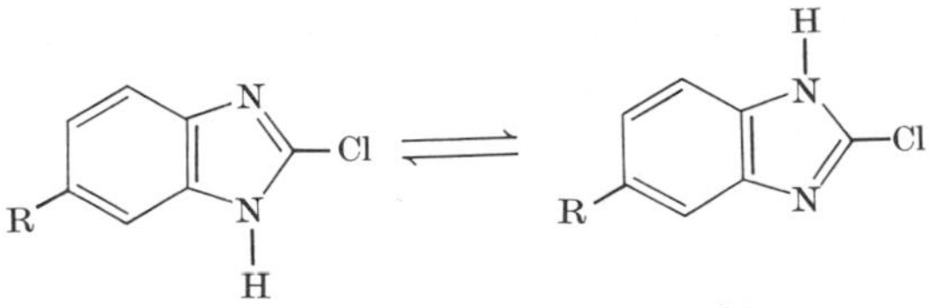

[84a] [84b]

quantities are found for a temperature of 198°K: $k = 4.12 \text{ sec}^{-1}$; $\Delta G^{\ddagger} = 10.9$ kcal/mole; $\Delta H^{\ddagger} = 3.9$ kcal/mole; $\Delta S^{\ddagger} = -35$ e.u.

The authors state that "in the compound examined here the frequency factor has a small value and consequently the entropy of activation is highly negative. This is presumably connected with a highly ordered transition state, which may involve a polymeric structure of the compound with nitrogen atoms linked by hydrogen bridges." (cf. Section 4-1b). They also indicate that "concentration effects probably play an important role in determining thermodynamic parameters." In this article there is a criticism of the previous method based on the line width of the NH signal: "The line shape of the NH proton is also temperature dependent, but it was not employed since it is too sensitive to solvent effects; solvent–solute and solute–solute associations in fact modify the line shape of this signal."

If the equilibrium constant is defined such that $K_T = [\text{XH}]/[\text{HX}]$, from the value of $\Delta G = 0.22$ kcal/mole determined by the authors, we find that at 27°, $K_T = 1.44$, which corresponds to about 60% of 2-chloro-6-methoxybenzimidazole in the equilibrium mixture.

This last example illustrates the usefulness of the coalescence method, which enables calculation of both ΔG and $\Delta G^{\ddagger}$ and thus to obtain complete information about the compound being studied. It is to be hoped that later work will clarify quantitatively the influence of solvent and structure on the spectral changes observed.

B. Carbon-13 Magnetic Resonance

The magnetic resonance of nuclei other than hydrogen is already an important technique and its use is likely to increase as a result of improved sensitivity in multichannel spectrometers and the use of Fourier transforms; these factors will allow routine work at natural abundance levels. Carbon-13, the most widely studied of stable isotopes occurring in small amounts, has the enormous advantage over the proton in that its resonance zone covers 250 ppm (compare 15 ppm for the proton), but the disadvantage that its sensitivity is about 10^4 times less than that of the proton.

Several applications of ^{13}C NMR to heteroaromatic tautomerism have already been made. The chemical shift of the carbon bonded to oxygen of 5-pyrazolones permits an easy distinction between the OH and NH structures (70J(C)1842). Values of δC_5 in DMSO, expressed in ppm relative to CS_2, are given in Scheme 1-14. Application of Eq.

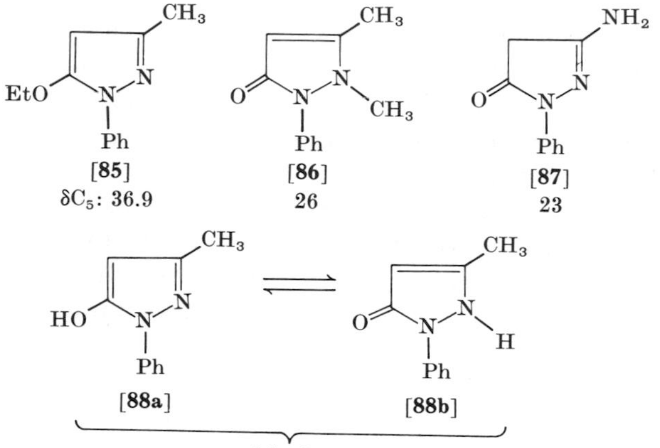

[**85**] δC_5: 36.9 — [**86**] 26 — [**87**] 23

[**88a**] ⇌ [**88b**]

δC_5: 35

SCHEME 1–14

(1-41) yields $K_T = 4.7$ for [**88a**] ⇌ [**88b**], i.e., 83% is in the OH form [**88a**] and 17% in the NH form [**88b**]. The CH-structure [**87**] is easily distinguished from the others by the signal due to the sp^3 carbon at position 4.

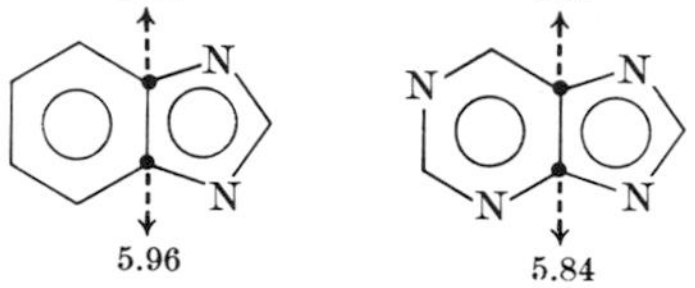

[**89**] [**90**]

$0.5(5.84 + 6.07) = 5.95$

^{13}C NMR has been used to study the annular prototropy of the N(7) and N(9) atoms in purine [**90**] (71JA1880); chemical shifts of the neutral molecule and the anion in aqueous solution were compared with those for benzimidazole [**89**]. For the ring junction carbons, the values of $\Delta\delta = \delta_{XH} - \delta_{X^-}$ found were as given in the formulae [**89, 90**]. The mean value is the same for purine and benzimidazole, and the effect produced by protonation of the anion is thus independent of the heterocycle (or at most only slightly dependent). In fact, the shifts of the two bridgehead carbons on the protonation of the purine anion are nearly equal, and the authors conclude that: "in aqueous solution the labile proton probably spends almost equal time at N-7 and N-9". It would be interesting to confirm this result by a comparative study of 1-methylbenzimidazole and 7-methyl- and 9-methyl-purine.

The ^{13}C chemical shift-pH profiles of the imidazole ring carbons of L-histidine have been compared with those of both N-methyl derivatives to show the predominance of the 1H-tautomer (Section 4-1Cbv) of histidine in basic solutions (73JA328).

A comprehensive study on annular prototropy of azoles (74JO357) in dioxan and methylene chloride indicates that, contrary to what was expected, the spread of the scale from proton to carbon-13 spectroscopy does not for these compounds improve appreciably the accuracy of the interpolation method. Calculations must be carried out on carbon atoms in positions the chemical environment of which varies with the tautomeric structure, and the use of $P = \Sigma\delta$ is required for application of Eq. (1-45), as already pointed out for proton NMR (Section 1-5Ab). Various examples show that the condition (1-42) is not fulfilled since $P_{\mathrm{MeA}} - P_{\mathrm{MeB}}$ is of the same order of magnitude as P^*.

Values of ^{13}C chemical shifts and P^* calculated using Eq. (1-45) for the pyrazole series are given in Scheme 1-15. Evidently P^* depends

59.0 59.0 H | 53.0 54.0 CH$_3$ | 48.3 48.3 H$_3$C CH$_3$ H | 53.8 46.1 H$_3$C CH$_3$ CH$_3$

$P_X = 118$ $\quad P_{\mathrm{MeA}} = P_{\mathrm{MeB}} = 117$ $\quad P_X = 96.6$ $\quad P_{\mathrm{MeA}} = P_{\mathrm{MeB}} = 99.9$

$P^* = -1.0$ ppm $\quad P^* = 3.3$ ppm

SCHEME 1-15

upon the example selected. Some unsymmetrically substituted pyrazoles are treated in Scheme 1-16 by applying Eq. (1-44). Obviously K_T is so sensitive to the P^* value chosen that not even a qualitative evaluation of the equilibrium position can be made.

Results of the application of Eqs. (1-45), (1-44), and (1-50) to other azoles are given in Scheme 1-17; they disclose a similar sensitivity to

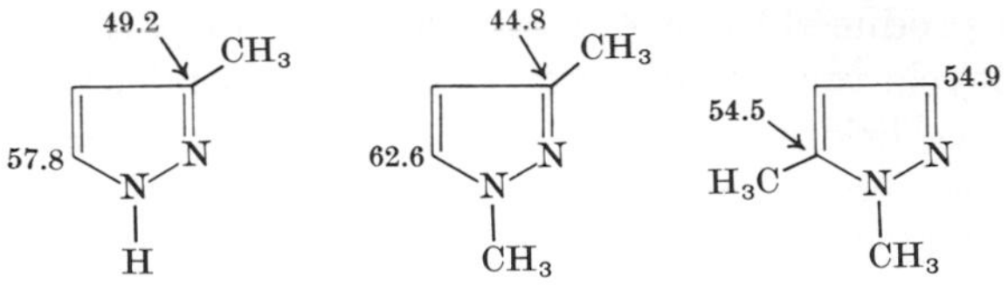

$P_X = 107.0$ $P_{MeA} = 107.4$ $P_{MeB} = 109.4$

$$K_T = \frac{-2.4 + P^*}{0.4 - P^*}$$

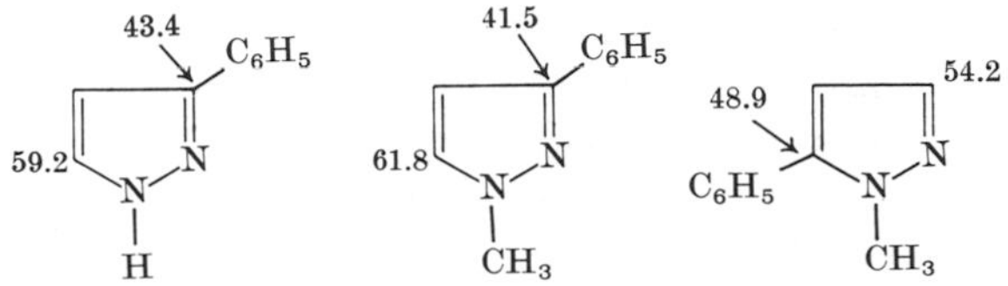

$P_X = 102.6$ $P_{MeA} = 103.3$ $P_{MeB} = 103.1$

$$K_T = \frac{-0.5 + P^*}{0.7 - P^*}$$

SCHEME 1-16

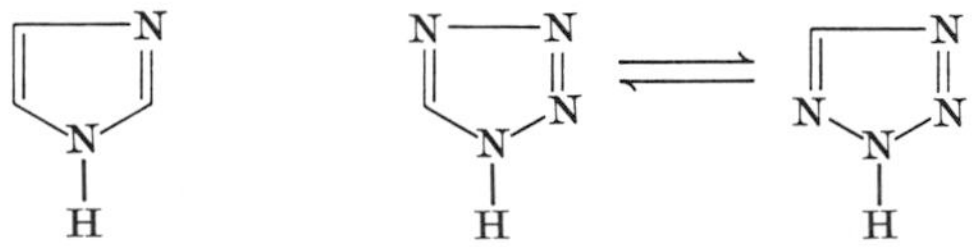

$P^* = -5.5$

$$K_T = \frac{10.1 + P^*}{-0.9 - P^*}$$

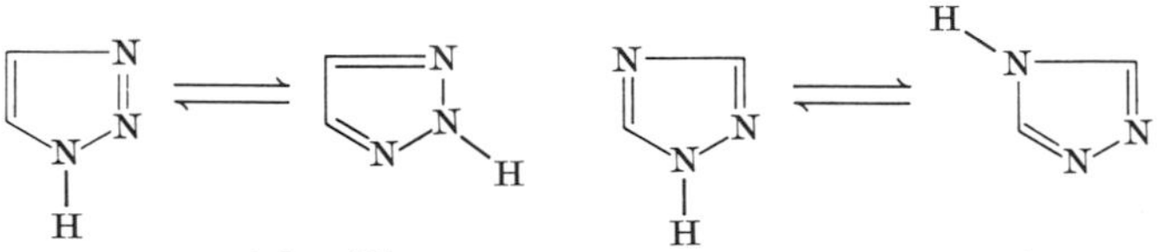

$$K_T = \frac{6.2 + P^*}{6.8 - 2\,P^*} \qquad K_T = \frac{-7.6 + P^*}{-3.0 - 2\,P^*}$$

SCHEME 1-17

the value of P^* chosen and give no reliable evidence on the position of tautomeric equilibrium.

Tetrazole ($P = \delta$; there is only one carbon atom) is the only case for which the numerator and denominator are different enough so that one can conclude qualitatively that $K_T > 1$. This indicates that the 1,2,3,4-tautomer predominates. This is true not only for dioxan, but also for aqueous solution, which is inconsistent with Charton's conclusions (69J(B)1240) (Section 1-4Bb).

It is worth mentioning that although very few dynamic studies with ^{13}C NMR have so far been reported (71JA4297), it is hoped that the method can be employed to determine the low potential barriers for certain prototropic equilibria, now that superconducting magnets are becoming available.

C. Nitrogen-14 Magnetic Resonance

^{14}N chemical shifts can be obtained either by heteronuclear double resonance, or by direct measurement. Heteronuclear double resonance is the more accurate of the two methods, but it is limited to compounds possessing a proton signal broadened by nuclear quadrupole relaxation of the ^{14}N (71J(B)397). The direct measurement of ^{14}N chemical shifts has disadvantages connected with the width of signals from nuclei of spin $I = 1$, and the fact that these peaks, proportional to the number of atoms present, are 10^3 times less intense than those of protons. Both methods have been applied to the study of heterocyclic tautomerism.

Heteronuclear double resonance has been used to study 2-quinolone [**92a** ⇌ **92b**] (67CH371). Although the choice of model compounds [**91**], [**93**] and solvents (acetone and $CDCl_3$) leaves something to be desired, the enormous separation of about 150 ppm between the pyridine and amidine nitrogen atoms shown in formulae [**91–93**] leaves no doubt that 2-quinolone possesses the NH structure [**92b**].

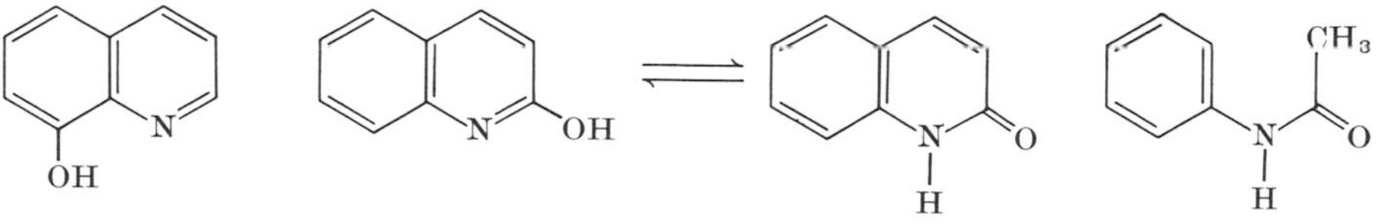

[**91**] $\delta^{14}N$: 93–95 ppm; [**92a**] ⇌ [**92b**] 230–240 ppm; [**93**] 243 ppm

Direct measurements by Witanowski *et al.* (71T3129) of ^{14}N chemical shifts of hydroxy- and amino-pyridines were compared with those calculated by molecular orbital methods for the alternative tautomeric forms (cf. Section 2-2Ba).

Witanowski *et al.* (72T637) have also studied annular tautomerism in azoles and benzazoles using ^{14}N NMR by comparing the chemical shifts of NH and NMe compounds. If P is the sum of the ^{14}N chemical shifts of an azole, P^* can be calculated using Eq. (1-45) and data from the "symmetrical" azoles: pyrazole, imidazole and benzimidazole.

From this, $P^* = -10 + 2$ ppm. However, investigation of the error inherent in ^{14}N chemical shift measurements gives a more realistic range for $P^* = -10 \pm 10$ ppm.

The application of Eqs. (1-44) and (1-50) to the published values (72T637) for "unsymmetrical" azoles leads to values of K_T as shown in Scheme 1-18. The two last entries in Scheme 1-18 are not significant,

System	Equation for K_T	Value of K_T for P* = −10
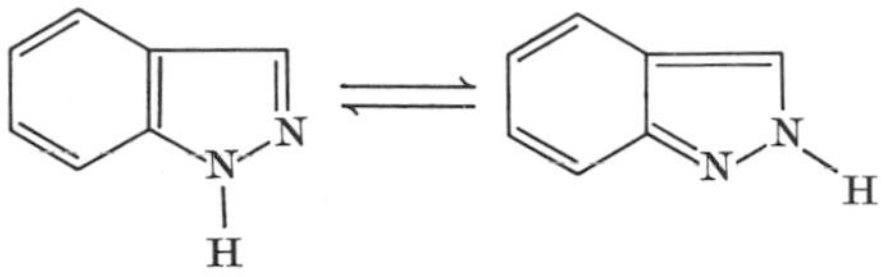	$K_T = \dfrac{25 + P^*}{-4 - P^*}$ (†)	2.5
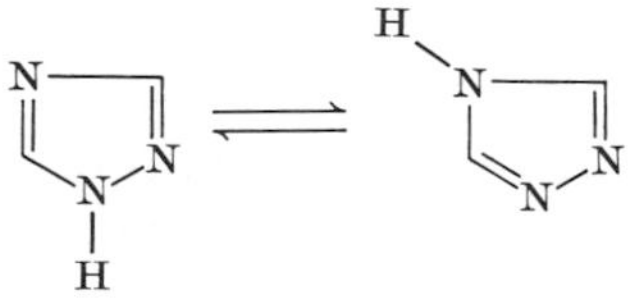	$K_T = \dfrac{28 + P^*}{44 - 2P^*}$	0.75
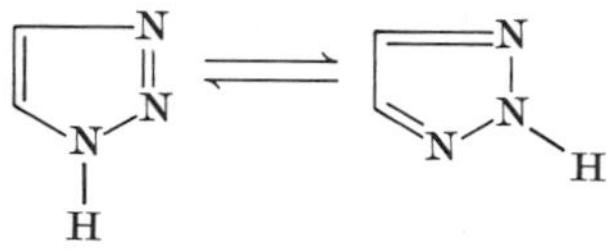	$K_T = \dfrac{10 + P^*}{-96 - 2P^*}$	0.0
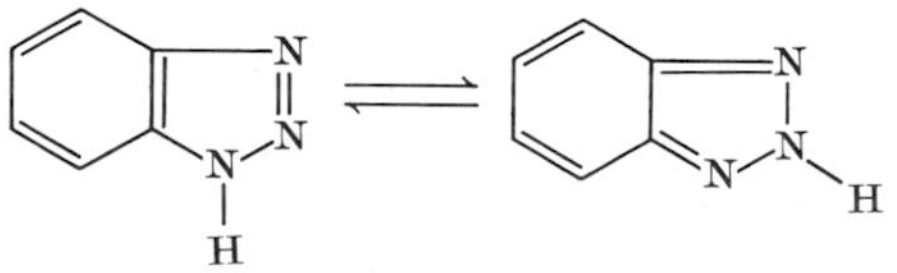	$K_T = \dfrac{49 + P^*}{-108 - 2P^*}$	—
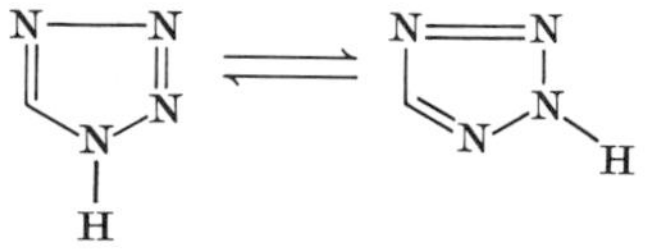	$K_T = \dfrac{84 + P^*}{-61 - P^*}$	—

SCHEME 1-18

as large errors occur in $\delta^{14}N$ if two or more signals overlap, as happens for benzotriazole and tetrazole. For the others, the value of K_T corresponding to $P^* = -10$ ppm must be considered as only a semiquantitative indication.

D. EVALUATION OF THE INTERPOLATION METHOD

The results of the application of the interpolation method to the chemical shifts of proton as well as carbon-13 or nitrogen-14 are very

† In all these equilibria, the left-drawn tautomer corresponds to the numerator in the definition of K_T.

disappointing, at the very least, as far as annular tautomerism is concerned.

If the interconversion frequency between the two tautomers is such that an averaged signal is observed (Case 5, Table 1-4) it is necessary to use the sums of the chemical shifts, $P = \Sigma\delta$; these sums are of similar magnitudes for the compounds XMe and MeX, whatever the type of resonance under consideration. Thus $P_{MeA} - P_{MeB}$ is small and has the same order of magnitude as the perturbation P^* arising from the replacement of NH by N–CH_3. Hence the condition of Eq. (1-42) is not fulfilled. The determination of K_T from an interpolation of chemical shifts cannot be considered a reliable or even a viable method.

E. Magnetic Resonance of Other Nuclei

We examine successively tautomerism studies using the magnetic properties of ^{15}N, ^{17}O and ^{19}F nuclei.

Nitrogen-15 resonance is due for important development as the improvement in the spectrometer sensitivity (Fourier transform) enables the determination of NMR spectra of ^{15}N nuclei in natural abundance. This technique will in time replace ^{14}N magnetic resonance, which has an unsatisfactory precision (Section 1-5C).

Results already reported concern PMR studies on ^{15}N-enriched molecules: valuable information can be obtained from ^{1}H–^{15}N coupling. Unlike an ^{14}N nucleus ($I = 1$) the ^{15}N nucleus with a ($I = \frac{1}{2}$) does not possess a quadrupole moment, and, unless exchange processes occur, it is easy to observe the direct ^{15}N–^{1}H coupling. A line width analysis (69OR481) shows that the lack of ^{1}H–^{14}N coupling in barbituric acid in DMSO solution is due to the quadrupole relaxation of the ^{14}N nucleus in the lactam tautomer [greatly predominant in DMSO (Section 2-4Dg)] and not to proton exchange with the lactim tautomer. The lack of ^{1}H–^{15}N coupling can be used to rule out certain tautomers (see Sections 2-6Ba, 5-54b).

Two important papers (65JA5439, 65JA5575) on the structure of cytosine and its conjugated acid (Sections 2-6Cb, 2-6Cd) are considered in more detail. The PMR spectrum (DMSO solution) of cytosine ^{15}N labelled at the exocyclic nitrogen shows a doublet corresponding to two protons with a coupling constant $J^{1}H$–$^{15}N = 90$ Hz. This result proves the amino structure [**94**] for this compound (65JA5575). The spectrum of cytosine hydrochloride labelled at the exocyclic nitrogen, in SO_2 at $-60°$, exhibits two nonequivalent protons coupled with ^{15}N (65JA5575). Cytosine hydrochloride ^{15}N labelled at each of the three nitrogen atoms shows in DMSO in addition a single proton coupled by 94.0 Hz (65JA5439), which proves definitively structure [**95**] [this last coupling has also been observed in ^{15}N resonance (65MI269)]. The predominance of the dilactam tautomer [**96**] for uracil (Section 2-2M) has also been demonstrated using the ^{1}H–^{15}N couplings present in the compound labelled at the two nitrogen atoms (65JA5439).

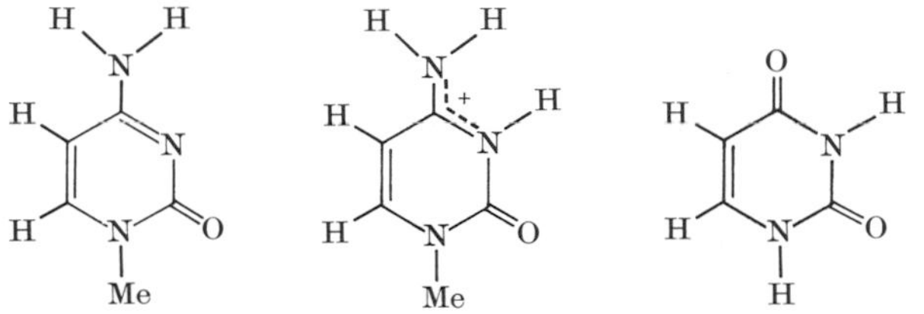

[**94**] [**95**] [**96**]

Oxygen-17 nuclear magnetic resonance has not yet been used in any study of heteroaromatic prototropy, but studies of the tautomerism of azonaphthols (65BG155) and of β-diketones (67JA1183) suggest that this technique should be very useful in spite of the line width (the oxygen-17 nucleus has a spin $I = \frac{5}{2}$ and hence a quadrupole moment) and of the low sensitivity of this isotope.

Fluorine-19 resonance has so far been used only qualitatively; from the similarity in the chemical shifts and in the ^{19}F–^{19}F coupling constants of the 2,3,5,6-tetrafluoro-4-hydroxypyridine [**97a** ⇌ **b**] and of the fixed *O*-methyl derivative [**98**], the authors (65J575) deduced the predominance of the pyridinol tautomer [**97a**] (Section 2-2Bb). The

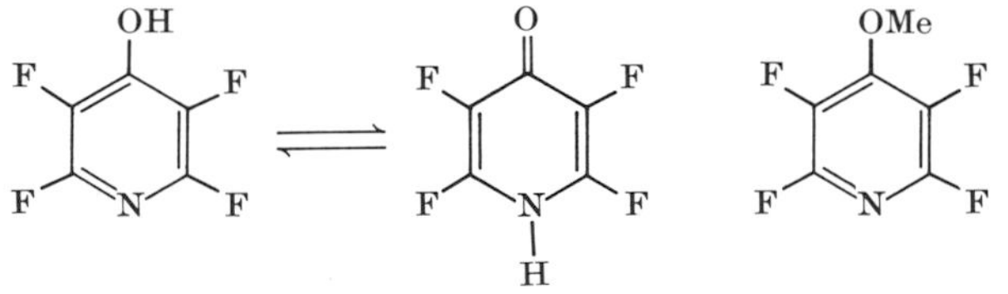

[**97a**] [**97b**] [**98**]

^{19}F-chemical shifts of 1-(*p*-fluorophenyl)tetrazolethiones (Section 4-4Fe) (67JO3580) were used to give information regarding the electronic character of the tetrazole ring, and thus indirectly on its tautomeric structure.

6. Other Spectroscopic Methods

A. Nuclear Quadrupole Resonance

^{14}N Nuclear quadrupole resonance offers a method to study nitrogen heterocycles in their crystalline state; the experiments are usually done at 77*K*. The low quadrupole moment of the ^{14}N nucleus makes it one of the most difficult to study, but it is at the same time one of the most rewarding (71PM(4)21).

The only examples in which this technique has been applied to tautomeric heterocycles are in the annular tautomerism field: the "asymmetric" nature of pyrazole (Section 4-1Ba), of imidazole (Section 4-1Ca) and of *s*-triazole (Section 4-1Eb) has been demonstrated (68MP73). For *s*-triazole, the three nitrogen atoms appear at different resonance frequencies, which is consistent only with the 1,2,4-triazole structure [**99a**], not with 1,3,4-triazole [**99b**].

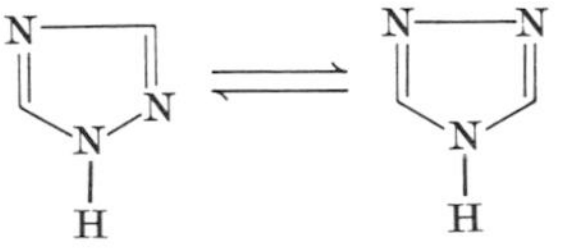

[**99a**] ⇌ [**99b**]

B. Microwave Spectroscopy

Provided that a sufficient number of isotopically substituted species are available, microwave spectroscopy allows calculation of molecular dipole moment and geometry in the vapour phase. In addition, microwave spectra of unsubstituted molecules can give information about molecular symmetry, which is sometimes able to allow distinction between tautomeric forms. Bond lengths obtained from microwave measurements are shorter than those determined by diffraction methods: the deviations for the bond lengths between atoms other than hydrogen are on average 0.006 Å (71JA1637).

Microwave spectroscopy studies of several azoles have been reported or are in progress (Table 1-5). The position of the tautomerisable hydrogen of *s*- and *v*-triazoles, and consequently their annular tautomerism in the vapour state, is thus determined (Table 1-5).

C. Infrared and Raman Spectroscopy (I-330)

a. Introduction

These two complementary methods will be discussed together though they have been unequally used. Whereas infrared spectroscopy,

TABLE 1-5

MICROWAVE SPECTRA OF AZOLES

Azole	Author	Reference	μ	Results
Pyrazole (Section 4-1B)	W. H. Kirchhoff	67JA1312[a]	2.21	Planar molecule with no in-plane symmetry axis
	R. D. Brown *et al.*	70JM528	—	Nuclear quadrupole coupling constants are given
	L. Nygaard *et al.*	72PS3, 74JL(22)401	—	Complete molecular geometry
Imidazole (Section 4-1C)	J. H. Griffiths *et al.*	67NA1301	3.8	—
	J. H. Griffiths	69TH02	—	—
v-Triazole (Section 4-1D)	O. L. Stiefvater *et al.*	70SA(A)825	—	The assigned spectrum arises from 1,2,3-triazole, rather than the tautomeric 1,2,5-triazole form
s-Triazole (Section 4-1E)	R. D. Brown *et al.*	71CH873	2.72	The molecule is planar and exists predominantly in the unsymmetrical tautomeric form

[a] Correction: 67JA2242. See also (74CH605)(1-deuterio-1,2,3-triazole) and (74JM(49)423)(tetrazole).

which was considered in detail in the previous review (I-330), is one of the most important methods used in tautomerism studies, applications of Raman spectroscopy (I-338) are rare. The introduction of Raman lasers should allow some expansion in the use of this method: however, the high price of these instruments compared with infrared spectrometers will hinder this.

It is impracticable to summarize all the examples of applications of infrared spectroscopy to tautomerism studies described in Chapters 2–6, but some aspects are worth being pointed out here.

b. Experimental Conditions

Of all the physical methods, infrared spectroscopy is the only one which enables study of a sample in all its physical states: vapour phase, solution, liquid, solid. Cells constructed of calcium fluoride or thallium bromide-iodide permit studies in aqueous solutions, which is useful for comparison with results obtained from UV spectroscopy (Section 1-6D) or from basicity measurements (Section 1-4A). Spectra of hydrochloric acid solutions can be obtained with sodium chloride cells (69AJ2595).

As regards solid state infrared spectra, we have pointed out (Sections 1-1D and 1-4Ib) that it is sometimes difficult or even impossible to determine by IR spectroscopy the tautomeric structure of certain strongly hydrogen-bonded compounds: pyrazol-5-ones (Section 4-2Fc are a typical case. Also, if the compound crystallizes in two different forms, it is difficult to determine by IR spectroscopy whether two tautomers are concerned or a single one exhibiting polymorphism (Section 1-1D): see the debated case of the imidazole *N*-oxides or *N*-hydroxyimidazoles (Section 4-7Ab). An ingenious simultaneous use of IR spectroscopy and differential-scanning calorimetry for 2-alkylthio-4-oxoquinoline 3-carboxylate ester (68J(C)2656) (Section 2-2Bb) demonstrated that the two crystalline modifications of this compound were indeed individual and distinct tautomers.

IR spectroscopy does not deal with single crystals, as is the case in X-ray diffraction (Section 1-4Ib) so that IR can show the solid under investigation to be composed of a mixture of tautomers. 1,3-Diphenylpyrazol-5-one presents a peculiar case (Section 4-2Ge) (72TH01): its IR spectrum as a suspension in Nujol (no absorption between 1800 and 1600 cm^{-1}) is different from that in a KBr disk (carbonyl band at 1710 cm^{-1}). Hydroxyacridines show similar behaviour for which an explanation has been proposed (I-381).

The technique using an argon matrix (20°K) has the great advantage of allowing a study of the monomeric species at low temperature: thus pyrazole (Section 4-1B), which is strongly associated in the solid state as well as in solution, exhibits in argon matrix a spectrum close to that in the vapour phase [(70PH2133); see also (73JM319) which corrects errors in the preceding paper].

c. Symmetry Considerations and Group Frequencies

In the present survey we will discuss the bands associated with C=X and C—XH groups (X = O, S, NR, CR_2), bearing in mind that the group frequency approach is an approximation which varies in its validity in different parts of the vibrational spectrum. Indeed considerable progress has been made since the previous review (I-330) in complete assignments for the IR spectra of heterocycles (see 71PM (4)265).

Two characteristics of the IR absorptions of amino groups are worth citing, because they are very useful in the identification of such a group and thus in a study of amino ⇌ imino tautomerism: (i) frequently the two NH_2 stretching vibrations follow relation (1-59) established by

$$\nu_{sym}\text{NH} = 345.5 + 0.876\nu_{asym}\text{NH} \tag{1-59}$$

Bellamy and Williams (63PM(2)161, p. 326, 71PM(4)265, p. 428). (ii) An NHD group (obtained by partial exchange of NH_2) exhibits a new νNH band located between the ν_{sym}NH and ν_{asym}NH bands of the original group and it shows also a νND band located between the ν_{sym}ND and ν_{asym}ND of the ND_2 group arising from a total exchange of NH_2 (63PM(2)161, p. 323; 71PM(4)265, p. 428). In some cases two central νNH bands appear, which have been attributed to the existence of a cis-trans isomerism in the NHD group (Section 4-5Fai) (68CO (266C)1587).

Symmetry considerations have demonstrated the asymmetry of certain azoles [pyrazole (Section 4-1A), imidazole (Section 4-1B)] and confirmed the existence of the tautomer of symmetry C_S in the case of triazoles [**100**] (Section 4-1D) and [**101**] (Section 4-1E).

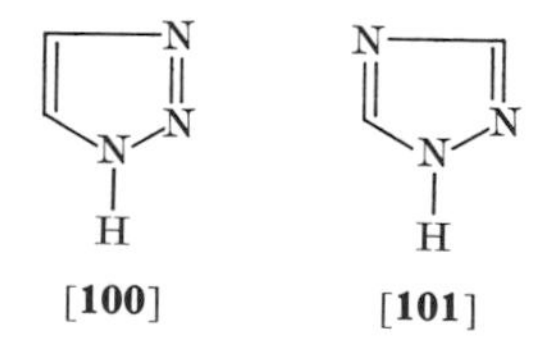

[**100**] [**101**]

d. Equilibrium Constant Measurement

Most tautomerism studies by IR spectroscopy are qualitative or at most semiquantitative. However, the measurement of band intensities, either as extinction coefficients ε_A or as integrated areas A, (71PM(4)265,

p. 269) allows determination of the tautomeric equilibrium constant, by applying Eq. (1-41) with $P = \varepsilon_A$ (61T41, 62T777) (I-331) or $P = A$ (63ZO2597). However, it is difficult to obtain P^* (the effect due to the substitution of H by Me) and to be sure that the measured intensity comes from a single vibration.

It is also possible to determine the intensity ratio of two bands and to obtain in this way a K_{rel} proportional to K_T, which allows the comparison of a series of differently substituted compounds (72JH25) (Section 2-8Dbi) or the study of K_{rel} as a function of the temperature, which enables calculation of $\Delta H°$ (see 68J(B)1470 for a similar case in conformational equilibrium).

Spectral variations as a function of the concentration, particularly intensity changes, can correspond to different types of association, and not to a modification of the tautomeric equilibrium.

e. *Raman Spectroscopy*

Vibrations inactive in IR are active in Raman spectroscopy, and this phenomenon has recently been used in a tautomerism study of the 2-amino-2-oxazolin-4-ones (72JH285). It is easy to get Raman spectra in water because this solvent gives rise to only very weak Raman scattering. This is why this technique is particularly suitable for studies of biopolymers, which are stable only in aqueous medium (71PM(3)53, p.65), and for studies of heterocycle protonation in aqueous or acidic solution: see, e.g., the 4-aminopyridine cation (69AJ2595) (Section 2-6Bb).

D. Ultraviolet and Visible Spectroscopy (I-328)

Electronic spectroscopy is one of the most widely used methods to study tautomeric equilibria (I-328) both qualitatively and quantitatively; this is a consequence of the simplicity of the method. The method is fully explained, with examples, in the previous review (I-328). The K_T determination can be done using Eqs. (1-41) and (1-44) and taking as the property P the extinction coefficient ε or the absorbance A.

To apply those equations, some conditions must be fulfilled: Beer's law must be valid for the tautomers as well as for the model compounds*; the spectra arising from different tautomers must be sufficiently different to avoid a large error in the intensity measurements from

* As Beer's law is usually followed, many authors neglect to verify it; even if the law is not obeyed, UV spectroscopy can be still used quantitatively, providing that a calibration curve is established.

overlapping (see 73CB956 for an example of a failure of this method when applied to a study of the tautomerism of cyanaminoquinazolines, Section 2-7Dc). Most UV studies are carried out in solution, but it is possible to measure UV in the vapour phase [see Section 2-3Bg (73JA 1700)] or in the solid state, as KBr disks for example (67AC877). The great sensitivity of the method allows work in dilute solution, which is useful for sparingly soluble compounds [e.g., some *N*-unsubstituted pyrazolones (Section 4-2I)].

E. Fluorescence Spectroscopy and Excited State Equilibrium (I-333)

In comparison with UV spectra, fluorescence spectra are still hardly used (see 74PM(6)147). This is perhaps to be expected, as many heterocycles do not fluoresce and further the technique is particularly delicate and can yield misleading results. However, application of the methods usually used for the absorption state, especially Eqs. (1-41) and (1-44), to the fluorescence spectra enables measurement of the interesting but still relatively unexplored equilibrium constants in the excited state. This treatment assumes that the excited state lifetime is long enough to permit the equilibrium to be reached. There has been some progress since the small amount of work reported in the previous review (I-333).

Najer *et al.* (64CO(258)4579, 68BF4568) find evidence for the predominance of the amino tautomer in the case of the 2-amino-1,3,4-oxadiazoles (Section 4-5Faii), by both UV and fluorescence spectra comparisons of the tautomeric species with the fixed derivatives. This indicates either that the equilibrium favors the amino form in the excited state as well as in the ground state or that the equilibrium is not reached in the excited state.

A qualitative tautomerism study of 4-hydroxy-1,5-naphthyridine (Section 2-3Ba) using fluorescence spectra, takes advantage of the fact that only the keto form fluoresces (67AC877). The equilibrium constant between the neutral [**102a**] and zwitterionic [**102b**] forms of 8-hydroxyquinoline (Section 2-3Ha), in the ground state, K_T = [**102a**]/[**102b**] = 30 differs greatly from that in the excited state, $K_T^* = 10^{-19}$! (64J4868); there is an enormous increase in the acidity of the hydroxyl group and

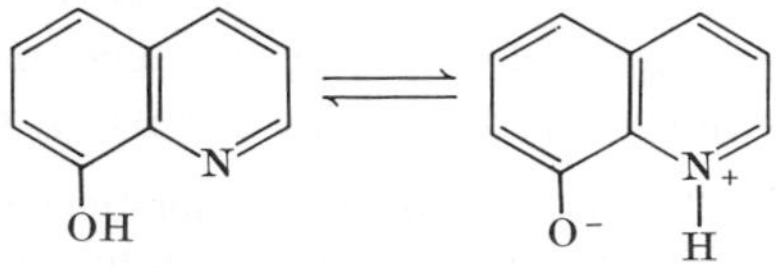

OH N ⇌ O⁻ N+ H

[**102a**] [**102b**]

in the basicity of the nitrogen atom in the excited state. Comprehensive studies are available on 7-hydroxylumazine (Section 2-4Hc, 65BG458) and the hydroxyacridines (Section 2-3Ha, 69T1001) for which the polarization of electronic transitions has been used.

7-Azaindole is a particularly interesting case in which a modification of the tautomeric equilibrium occurs from the ground state to the excited state (Section 5-2Aa). Study of the fluorescence spectra of 7-azaindole and of the 1-methyl and 7-methyl fixed derivatives shows that the stable form in the ground state is the tautomer [**103a**] whilst in the excited state it is the dimer [**103c**]; the phototautomerism takes place according to the biprotonic process [**103a**] ⇌ [**103b**] ⇌ [**103c**] (71JA5023).

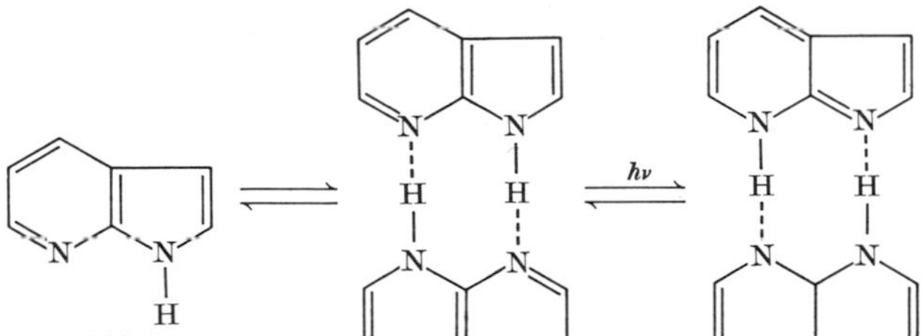

[**103a**] [**103b**] [**103c**]

Fluorescence spectroscopy is not the only means for reaching the equilibria in the excited state. It is possible from pK_a measurements and UV spectra of the bases and of their conjugated acids to calculate pK_a^* with the aid of Förster's cycle (63MI301) and hence obtain K_T^* using the methods described in Section 1-4A. This method has been applied by Sandström (69AS2888) to the case of the oxazole-2-thiones (Section 4-4Ca); this very good paper gives a comparative discussion of the two methods of measuring K_T^*.

F. Photoelectron Spectroscopy and ESCA

Recently two new physical techniques for the study of organic structures have appeared: ultraviolet photoelectron spectroscopy (PES or UPS) and X-ray photoelectron spectroscopy (ESCA or XPS). They are too recent to have yet been much used in tautomerism studies, but in view of the great interest shown in them, it can be predicted that they will be applied in the near future and that they will be very useful. Photoelectron spectroscopy involves precise measurements of the kinetic energy of photoelectrons emitted from a sample which is irradiated with X-rays or vacuum ultraviolet photons. Numerous reviews are available (70AO17, 71CR295, 71JE712, 72AG144, 72CS355).

Ultraviolet photoelectron spectroscopy allows study of the electrons of the valence shell, which affords information about inductive and mesomeric substituent effects. The experiment is carried out in the vapour phase at low pressure ($<10^{-3}$ torr), and the source is usually a helium resonance lamp (584 Å). Spectra with sharp lines are obtained. Unfortunately, for moderately sized molecules, UPS gives complicated spectra which can be interpreted only by molecular orbital methods (71CR295). ESCA deals with the core electrons and thus with the atomic structure of the sample. Soft X-rays are used as a source, and the sample is studied either in the vapour phase or as a crushed powder in the solid state. The main disadvantage of ESCA is the width of the bands obtained so that in certain cases the resolution may be insufficient to detect easily a shift between two lines, in spite of the use of deconvolution methods. Clark studied by ESCA the structures of adenine (70CH23), cytosine (70CH24), thymine (70CH24), and pyrazole (71MI234). In the first three cases, the experimental values are consistent with those calculated by Pullman (69TC(13)278) for the predominant tautomer, e.g., the amino-oxo form for cytosine (Section 2-6Cb).

The results for pyrazole and imidazole are of great interest: the differentiation between "isomeric" positions (carbons as well as nitrogens) proves the asymmetric structure of these compounds in the solid state (Sections 4-1A and 4-1B). The ESCA method (71MI234) provides important evidence that strong intermolecular hydrogen bonding in the solid state renders the molecule more symmetrical or, in other words, tends to blur the difference between the tautomers. For example, in pyrazole the shift between the ^{15}N levels is 1.3 eV compared to a calculated value of 2.7 eV: but in *N*-methylpyrazole, where hydrogen bonding no longer occurs, the corresponding shift is 2.3 eV.

G. Mass Spectrometry

a. Introduction

The idea of studying tautomerism in the vapor phase by mass spectrometry is relatively recent. A major problem is to distinguish between prototropic equilibria occurring before and after ionization. If prototropy occurs after ionization, there will be no simple quantitative relation between the tautomer existing in the vapour phase and the molecular ion involved in the fragmentation process. Two principal methods are available* to investigate the tautomerism in mass spectrometry.

i. A study of the ratio of the intensities of two daughter ions, one associated exclusively with the XH tautomer and the second with the HX tautomer, as a function of the temperature (source and system) (69OM49).

This method was used to investigate the ring–chain tautomerism of the 4,4-dimethyl-2-phenyloxazolidine [**104a**] ⇌ [**104b**] (71T-4407), to determine the enthalpy difference and hence obtain an estimate of the equilibrium constant.

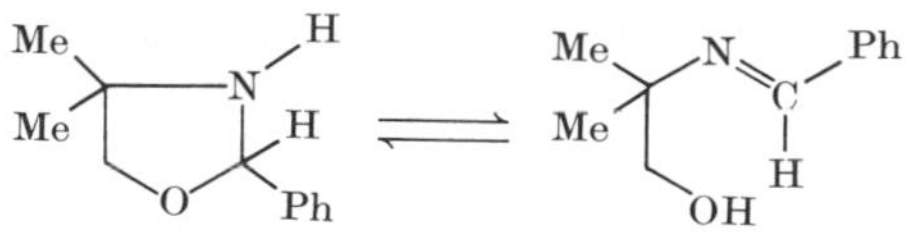

[**104a**] ⇌ [**104b**]

ii. Generation *in situ* (e.g., through a McLafferty rearrangement) of an ion of unambiguous structure and to compare its fragmentation with that of the compound whose tautomerism is under study. Two examples are worth citing: the first one deals with ketones, studied by Nakata and Tatematsu (69BJ1678; see also 67T2095): no common peak of appreciable intensity was observed in the spectra of compounds [**105**] and [**106**], and it was therefore concluded that keto–enol interconversion of these two m/e 216 ions [**105**] and [**107**] did not occur in the mass spectrometer.

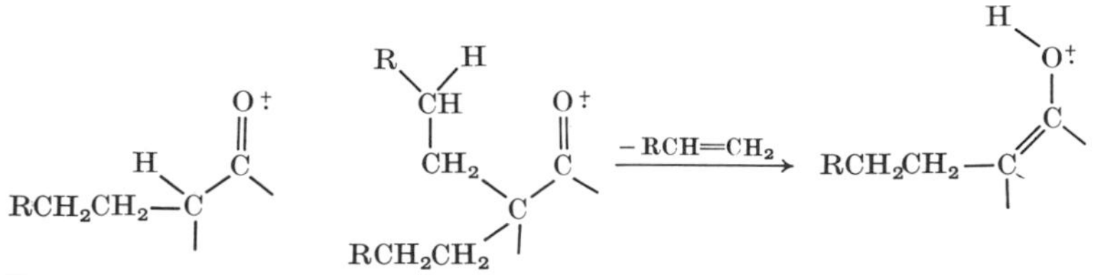

XH(keto) $m/e = 216$ [**105**]

[**106**] $\xrightarrow{-RCH=CH_2}$

HX(enol) $m/e = 216$ [**107**]

A second example concerns phenol, the fragmentation of which initially appeared to indicate that mass spectrometry could not provide evidence on the tautomerism equilibrium because the molecular ion loses carbon monoxide through a cyclohexadienone intermediate

* It is also possible to use ionization potentials; see examples in Sections 2-5Ba (72OM823) and 2-6Ba (72TL3193).

[**109b**]. However Djerassi *et al.* (67T2095) have shown that even here it is possible to determine the tautomer from which the fragmentation occurs and thus the structure of the predominant tautomer in the vapour phase. As the loss of carbon monoxide from [**109a**] occurs only above 15 eV, phenol probably occurs as the hydroxy form in the vapour phase as expected and considerable energy is necessary to tautomerize the molecular ion.

The phenol fragment ion ($C_6H_5OH^+$) derived from electron impact induced cleavage of phenyl *n*-butylether [**108**], shows much less of the *m/e* 66 species [**110**] (5% of the mass 94 phenol ion compared with 30% in the mass spectrum of phenol itself), probably because the internal energy of the phenol ion is less following the initial rearrangement process of the ether.

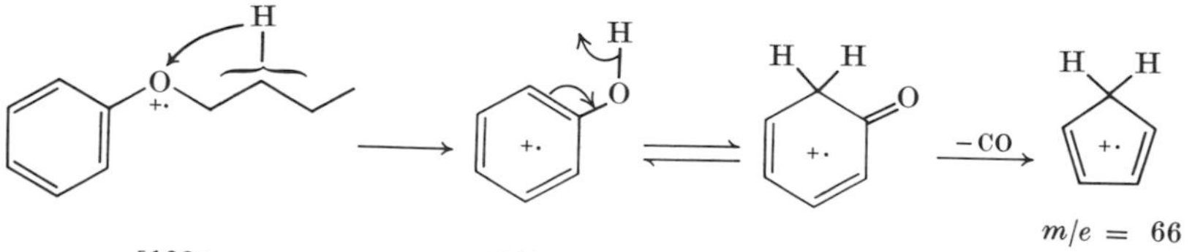

−CO

m/e = 66

[**108**] [**109a**] [**109b**] [**110**]

b. Applications to Heteroaromatic Tautomerism

Those methods have been applied successfully to the heteroaromatic tautomeric equilibria. Nakata *et al.* (72J(PI)1924) showed by mass spectroscopy that the molecular ion of 4-hydroxycoumarin [**111a**] exists as such and not in the dioxo form [**111b**] (Section 4-2A) or as a mixture of [**111a**] and [**111b**]. If the two species were in equilibrium, variation of inlet temperature would have an effect upon the spectral patterns (method i above) which is not found. The choice between the two tautomers [**111a**] or [**111b**] was made by generating *in situ* the deuterium-labelled hydroxy tautomer through loss of ketene from an *O*-acylated derivative (method (ii)).

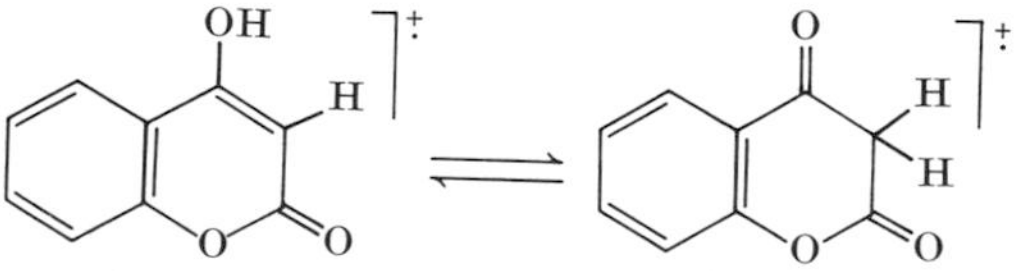

[**111a**] [**111b**]

Maquestiau *et al.* investigated the annular tautomerism of triazoles. In the case of the *s*-triazole (72OM1139) the results obtained after fragmentation can be explained easily on the basis of a tautomeric equilibrium between the two nonsymmetrical forms [**112a**] and [**112b**],

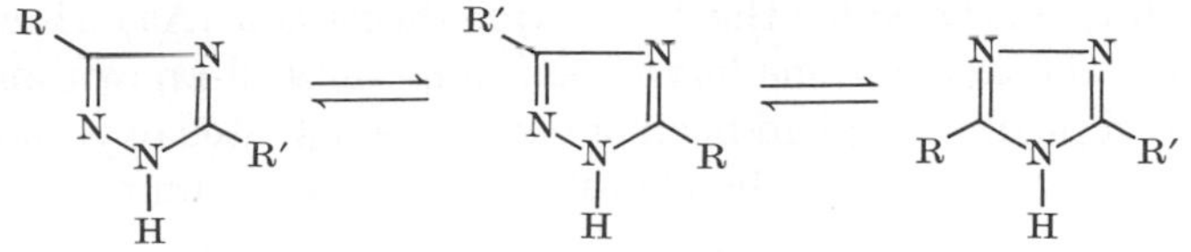

[112a] [112b] [112c]

the contribution of the 1,3,4-triazole form [**112c**] being zero or negligible (4-1E). The authors used the *N*-methyl derivatives as model compounds for the fragmentation. The mass spectrum of *v*-triazole, especially the metastable transitions, are explained on the basis of a mixture of tautomers [**113a**] and [**113b**] in the vapour phase (73OM271) (4-1D).* The application of method (ii) to alkylated and acetylated derivatives of these triazoles is foreseen (72PS4).

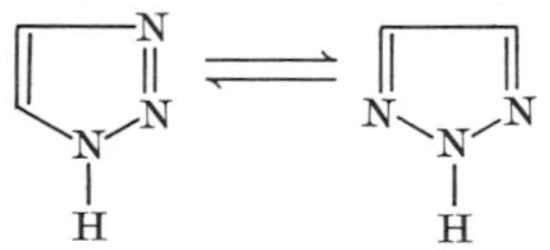

[113a] [113b] 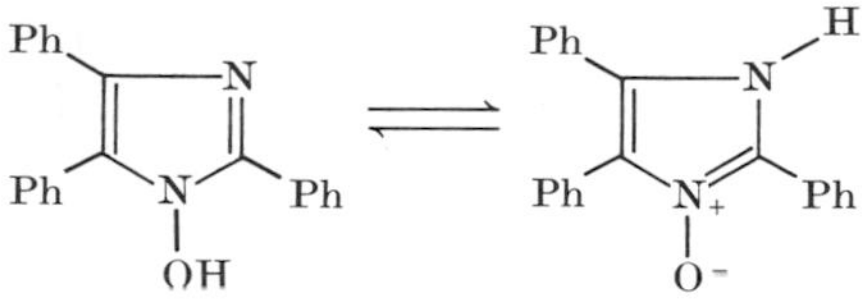 [114a] [114b]

The imidazole *N*-oxides may exist in two tautomeric forms, e.g., [**114a**] and [**114b**] 2,4,5-triphenylimidazole *N*-oxide (Section 4-7Ab). Volkamer and Zimmermann (69CB4177) claim to have isolated these two forms in the solid state, and that the mass spectra of the two tautomers differ considerably: the *N*-hydroxy form [**114a**] exhibits a metastable peak (*m/e* 279) corresponding to the loss of a fragment of mass 17 but does not show any M-16 signal, whereas the *N*-oxide form [**114b**] shows a M-16 and no M-17 peak. However, a second group have questioned these results (71J(B)2350, 74BB105).

7. Theoretical Methods

A. Introduction to the Quantum Mechanical Approach (I-334)

Since the previous survey (I-334), the application of quantum mechanical methods to tautomerism studies has greatly progressed; nowadays these methods may be considered to be a tool as useful as the physical methods described earlier in this chapter. As these methods do not require any experimental data, they are of great scope: it is possible to study the tautomerism of unknown hypothetical molecules

* Evidence of the exclusive presence of 1*H*-benzotriazole tautomer in the gaseous phase is now available (73OM1267).

as Clark did (70MI238) for the 1*H*-azirine (Section 6-1Aa). However, in most cases, the calculations have been done on well-known molecules, the equilibrium of which had been or could be evaluated experimentally.

Calculation methods can be classified into three main groups:

i. *Ab initio* methods: the large amount of computer time necessary for such calculations on a complex molecule prohibits this method. The tautomerism of small molecules only can be studied, as is the case for the azirines just cited (70MI238).
ii. Methods involving all the valence electrons, of which complete neglect of differential overlap (CNDO), extended Hückel theory (EHT), intermediate neglect of differential overlap (INDO), and modified INDO (MINDO) are among those currently used.
iii. Methods involving only π-electrons: the simplest methods of this type, Hückel molecular orbital (HMO) and its derivatives, do not need any geometrical data. Among the more elaborate π-electron methods can be cited the Pariser-Parr-Pople (PPP) and improved linear combination of atomic orbitals (LCAO) methods.

Generally the most serious errors arise from the geometry chosen, as even in the most favourable case only that of the tautomer present in solid state is accurately known from X-ray determination. For the geometry of the other tautomeric forms hypotheses are necessary.* Reference (71HC(13)77) gives a detailed discussion of the geometries of the various tautomers of purines (Section 5-1Aaiii) and (69BF1097) for calculations of ΔE taking into account modifications in the geometry from one tautomer to the other one (Sections 4-1D, 4-1E).

Studies of the prototropic equilibria of heteroaromatic compounds by quantum mechanical methods can be done in two ways. The first is to calculate directly the constant of the tautomeric equilibrium from the total energies of each tautomer (Section 1-7B). In the second approach, a chosen experimental physicochemical property of a molecule capable of existing in several tautomeric forms is compared with those calculated theoretically for each tautomer. The following properties have been used: heat of combustion (Section 1-7C); dipole moment (Section 1-7D), electronic spectrum (Section 1-7E).

For methods of the second type, some papers simultaneously discuss evidence from available fixed derivatives and from theoretical calculations (68T151) and others use calculations of the properties of unavailable fixed forms.

* It can be hoped that those difficulties will be overcome by using methods minimizing the energy as a function of the geometry.

B. Equilibrium Constants (I-334)

Among the numerous theoretical studies, we mention here, because of their systematic character, only those of Pullman and Pullman (71HC(13)77) on purines (Section 5-1Aaiii), Sandström (66AS57, 69AS2888) on tautomeric cyclic thiones (Sections 4-4C, 4-4E), Arriau and Deschamps (71T5779, 71T5795, 71T5807, 71TH01) on pyrazolones (Sections 4-2G, 4-2I) and of Roche and Pujol (69BF1097, 70TH02, 71CP465) on the annular tautomerism of azoles (Sections 4-1D, 4-1E, 4-1F).

The energies of the different tautomers are calculated, and it must be emphasized that it is not possible to compare energies obtained from different methods. A disadvantage common to all such methods is that they deal with very high energies, several hundreds of electron volts (1 eV = 23.063 kcal/mole), whilst the energy difference between two tautomers of $K_T = 100$ at 25°, i.e., one of which exists to the extent of 99%, is only 2.7 kcal/mole, representing less than 1% of the total energy of the molecule. Because of this situation, energy differences between two tautomers, ΔE_π or ΔE_{total} according to the methods, are usually too large; however, the direction (the sign of ΔE) and the calculated relative order of abundance if more than two tautomers are concerned are generally correct.

Another limitation is that most calculations deal with the enthalpy differences of the isolated molecule [$\Delta H^\circ_{\text{eq}}$ (vapour)], whereas most experimental results give energy differences in solution, and although it is possible to evaluate $\Delta H^\circ_{\text{eq}}$ (soln) by study of the variation of the equilibrium constant as a function of the temperature (Section 1-1E), it is difficult to relate results obtained in solution to the equilibrium between isolated, unperturbed molecules (72TL775). Nevertheless some authors have taken into consideration effects of the solvents (71T5779) and of crystalline structure (71HC(13)77) on the position of the tautomeric equilibrium.

A theoretical study of the tautomerism of all the fundamental five-membered ring heterocycles substituted by a BH group (Me, NH_2, SH, OH) is in progress (72UP1); some preliminary results are given in Section 4-5H.

C. Heats of Combustion (I-338)

This method has been developed largely by Dewar, who deduces the most stable tautomer by comparison of experimental heats of atomization with those calculated for some of the possible tautomers. Among the most significant of numerous publications, we mention (66JT759) and (69JA796) on the annular tautomerism of azoles and benzazoles (Section 4-1B to 4-1I) and (70JA2929) on the tautomerism of heteroaromatic hydroxy and amino derivatives and nucleotide bases.

D. Dipole Moments (I-333)

Dipole moments are among the physical properties that can be calculated by semiempirical methods with good precision. With knowledge of the experimental dipole moment of the molecule under consideration and the moments calculated for the various tautomeric forms, it is possible to deduce the equilibrium constant, proceeding as previously indicated (Section 1-4G). An example of this method is work by Cencioni, Franchini and Orienti (68T151) on the tautomerism of isoxazol-5-ones (Section 4-2B) (71PM(4)237, p. 253).

E. Electronic Spectra

This method is analogous to the two previous ones, but frequently it cannot be used for a tautomerism study because of similarity of the electronic spectra of two tautomeric species (Section 1-6D). This occurs, e.g., for the lactam–lactim tautomerism of some purine derivatives (Section 5-1Bb) (71HC(13)77).

Other applications of this method include the work of Kwiatkowski (71JL(10)245, 71MI587, 72MI217) on the functional derivatives of azines and that of Nishimoto (67BJ2493) on lumazine: the electronic spectra of the nine possible tautomers were calculated and compared with the experimental spectrum: the spectrum of the dioxo tautomer (I-391) gives the best fit with the experiment.

8. Criteria for Choice of Physical Methods

A. Introduction

Tautomeric equilibria are very dependent on state and solvent, and to a lesser extent on concentration and temperature. Hence if the results of different methods are to be directly compared, they must be carried out under similar conditions. Conversely, if the tautomeric behaviour of a compound is to be fully described it must have been examined under a variety of different conditions.

The modern methods of characterization of a compound generally include infrared, ultraviolet and NMR spectra, and such routine data generally afford valuable data on the tautomeric structure. If further study is required then it is generally useful to make model compounds for as many of the possible tautomeric forms as is practical, bearing in mind the reservations mentioned in Section 1-3B. Further study now depends on the state in which information is desired and whether quantitative information is required.

B. Solid State

If the compound is crystalline, full information can be obtained by X-ray analysis (Section 1-4Ib). The usual method, however, is IR (Section 1-6C), and the value of this can be increased by use of model compounds, and by Raman spectroscopy (Section 1-6Ce). Photoelectron spectroscopy (Section 1-6F) and also nuclear quadrupole resonance (Section 1-6A) hold future hopes and the possibility of using UV in the solid phase (e.g., KBr disk) should not be forgotten.

C. Vapour Phase

Theoretical methods (Section 1-7) deal mainly with isolated molecules and are most applicable to the vapour phase.

Mass spectroscopy (Section 1-6G) shows much promise for vapour phase work, and for the simpler molecules microwave spectroscopy (Section 1-6B) and electron diffraction (Section 1-4Ic) are available. It is possible to apply both IR (Section 1-6C) and UV (Section 1-6D) to the vapour phase, but the experimental requirements are demanding.

D. Liquid Phase

We must distinguish between the pure liquid (melt) and solution in various solvents. For the melt, IR (Section 1-6C) is usually the most convenient method.

The basicity method (Section 1-4A) is limited in practice to aqueous solutions, but is of great importance for such solutions. The other most important methods for solutions are UV (Section 1-6D), IR (Section 1-6C) and NMR (Section 1-5) spectroscopy. Also of potential value in special cases are methods involving the Hammett equation (Section 1-4B) and dipole moments (Section 1-4G).

E. Quantitative Determination of K_T Values

It is relatively simple to determine the predominant form of a tautomeric equilibrium; and if K_T lies between 0.1 and 10, then it too can usually be found without difficulty. However, quantitative determination of K_T values where they are either large (>100) or small (<0.01) is difficult. The only general methods available are basicities (Section 1-4A), limited to aqueous solution, the Hammett equation (Section 1-4B), and, in principle if not often in practice, theoretical methods (Section 1-7).

Chapter 2

Six-Membered Rings

1. Tautomerism in Pyridines and Azines: General Discussion

A. A Qualitative Approach to the Rationalization of Functional Group Tautomerism

As already outlined, the majority of the important tautomeric processes occurring in heteroaromatic compounds involve substituent groups XH directly attached to the aromatic ring where X is oxygen, nitrogen, or sulphur. Such tautomeric processes generally involve the transfer of the proton of the XH group to a ring nitrogen or to a ring carbon atom.

In the five-membered rings with one heteroatom (Chapter 3), the process almost always involves transfer to a ring carbon atom, cf. [**1a**] $\rightleftharpoons$ [**1b**], as transfer to the ring heteroatom [**1a**] $\rightleftharpoons$ [**1c**] would involve an unfavourable high-energy zwitterionic form [**1c**].

By contrast, substituted pyridines and azines nearly always undergo tautomerism involving transfer of the XH group proton to a ring nitrogen atom. Thus processes of type [**2b**] $\rightleftharpoons$ [**2a**] are important, and not [**2b**] $\rightleftharpoons$ [**2c**]. The reason for this difference in behaviour can be

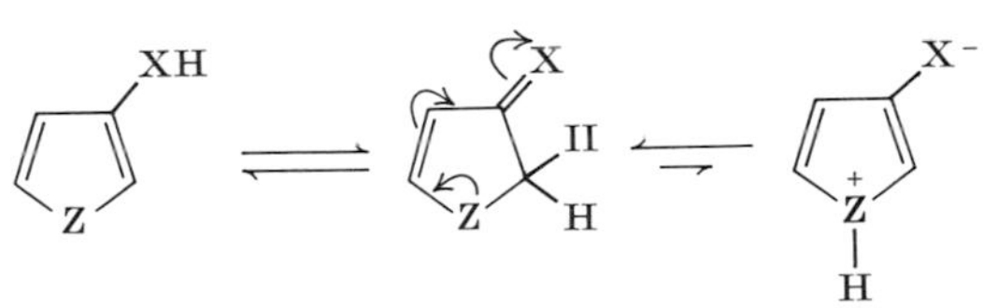

[**1a**] [**1b**] [**1c**]

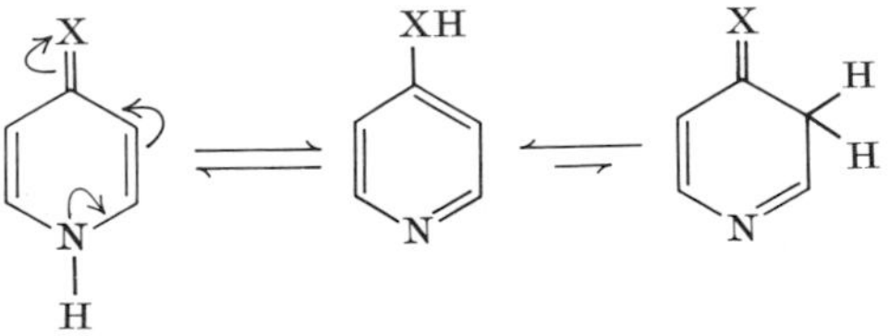

[**2a**] [**2b**] [**2c**]

understood from a consideration of the stabilization of the various structures. Let us consider the case where X = oxygen in both [**1**] and [**2**] and Z = NH in [**1**]. An enol [**3a**] is unstable with respect to the corresponding ketone [**3b**], in simple cases pK_T is ca. 8, corresponding

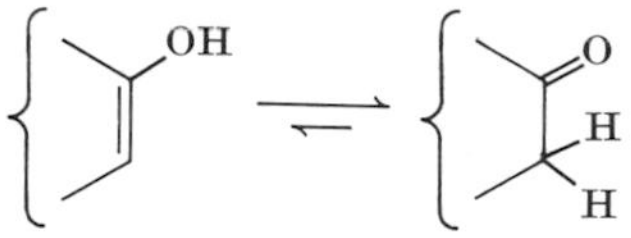

[**3a**] [**3b**]

to ΔG° of ca. 11 kcal mole^{-1} (72TL5019). The reason for [**2c**] being of lower stability than [**2b**] lies in the large loss of resonance energy involved: the loss of the pyridine ring resonance is compensated but little by the poorish conjugation, C=N—C=C—C=X, of [**2c**]. The picture differs when considering [**1a**] ⇌ [**1b**] in two important ways: the pyrrole ring resonance energy is considerably lower than that of the pyridine ring (74HC255) and the mesomeric stabilization of [**1b**] in the sense of the curly arrows is significant. Turning back to pyridine, the equilibrium [**2b**] ⇌ [**2a**] is much more evenly balanced, the loss of the pyridine resonance energy in [**2b**] is offset by the gain of delocalisation energy in the sense of the curly arrows in [**2a**], and more importantly the pyridonoid ring of [**2a**] is an aromatic ring, of considerable resonance energy (72J(PII)1295).

Thus the tendency of the XH proton to move to a ring nitrogen rather than to a ring carbon atom in pyridines and azines is explained. However, under some circumstances, nonaromatic structures corresponding to [**2c**] do become stable: such cases nearly always involve *polysubstitution*, usually potential di- or tri-hydroxy compounds, and the tendency is further helped by benzene annelation. For example homophthalimide is stable in the dioxo form [**4**] (cf. Section 2-3Ec).

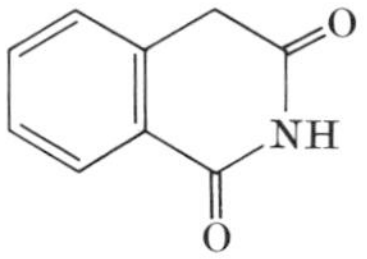

[**4**]

B. A Quantitative Approach to Functional Group Tautomerism

The qualitative approach just outlined has been developed quantitatively (71MI395, 72J(PII)1295, 72KG1011, 73J(PII)1080) by comparing the tautomeric equilibria of, e.g., [**5a**] ⇌ [**5b**] with [**6a**] ⇌ [**6b**]. The basic equation of this method is Eq. (2-1), which is an expression

$$RT \ln (K_u/K_s) + T(\Delta S_s - \Delta S_u) = A_{\text{pyridine}} - A_{\text{pyridone}} \qquad (2\text{-}1)$$

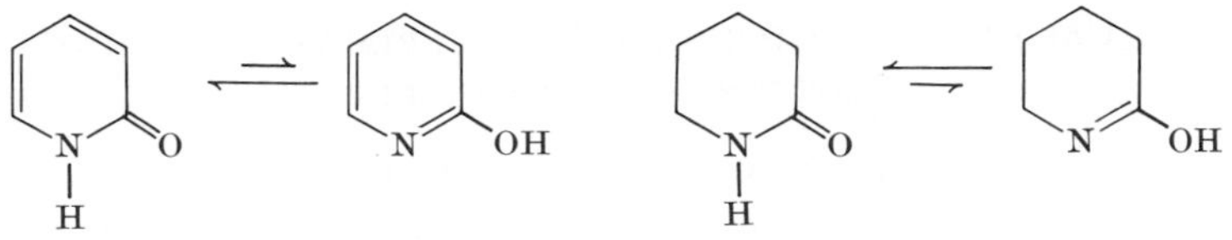

[5a] ⇌ [5b] [6a] ⇌ [6b]

of the basic assumption that the $\Delta\Delta H^\circ$ for equilibria [**5**] and [**6**] depends solely on the difference in aromaticities of rings [**5a**] and [**5b**]. This assumption is set out diagrammatically in Scheme 2-1 (72J(PII) 1295), and evidence was adduced to show that the assumption is justified and that the ΔS term can be approximated satisfactorily.

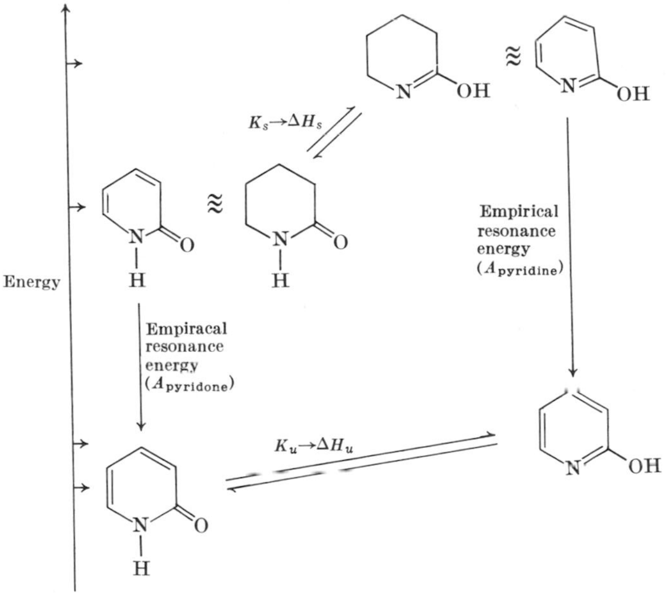

Energy

$K_s \rightarrow \Delta H_s$

Empirical resonance energy (A_{pyridine})

Empiracal resonance energy (A_{pyridone})

$K_u \rightarrow \Delta H_u$

SCHEME 2-1

So far this approach has been used to calculate aromatic resonance energies from equilibrium data rather than vice versa. However, as further methods for the calculation of aromaticities in other ways are developed (cf. 72TL5019, 74HC(17)255), it will become feasible to estimate tautomeric equilibrium constants from such data. Results already available do allow extension of probable trends in aromatic resonance energies which may be used to correlate and rationalize experimental results: this is done in Section (2-10), where the available results are summarized.

C. Use of Nonprototropic Equilibria as a Guide to Prototropic Tautomerism

Groups other than a proton can migrate from one site to another in a molecule. The activation energy for migration of, for example, a methyl group is generally quite high, and such processes we usually refer to as isomerization, or rearrangement, rather than tautomerization. However, comparison of isomerization equilibria of methyl compounds with the corresponding tautomeric equilibria is illuminating.

Beak and his co-workers have investigated the ΔH differences for a number of equilibria of type [**7**] by measuring the equilibrium constants at a variety of temperatures and/or by calorimetry. They have obtained quantitative information regarding alkylthiopyridine–pyridothione (69JO2125), imidate–amide, and alkoxypyridine–pyridone equilibria (66CH631, 68JA1569) and have also studied the effect of substituents on alkoxypyridine–pyridone equilibria (72T5507). This work allows the calculation of aromaticity differences rather more rigorously than that discussed in (Section 2-1B) because extrapolations to the gas phase are simplified. However, there is evidence that in the comparison of similar prototropic tautomeric equilibria (e.g., for [**5**] and [**6**]) there is a cancellation of enthalpy differences arising from solvation (72J(PII)1295, 72TL775). Since equilibria [**7**] and [**8**] are each related to the same

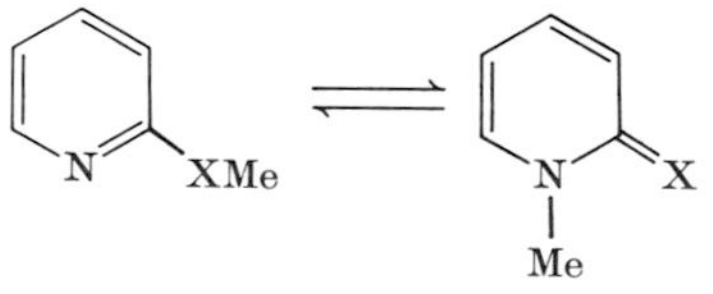

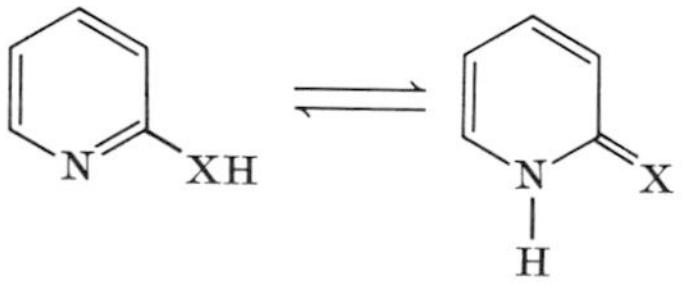

[**7a**] [**7b**] [**8a**] [**8b**]

aromaticity differences, there must be a relation between them. More exactly, consideration of the relevant equation indicates that there is a relation between the difference of K_T for [**7**] and [**9**] to the difference in K_T for [**8**] and [**10**], and that direct comparison of K_T for [**7**] and [**8**] is unjustified without consideration of [**9**] and [**10**]. This is because X–H and X–Me bond energies differ considerably. Insufficient quantitative relationships are yet available for this approach to be of great utility, but this situation will change. Meanwhile the interesting studies of

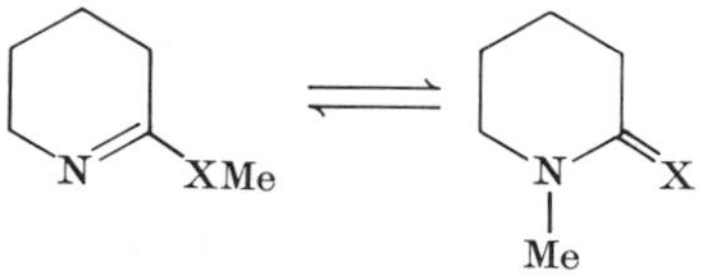

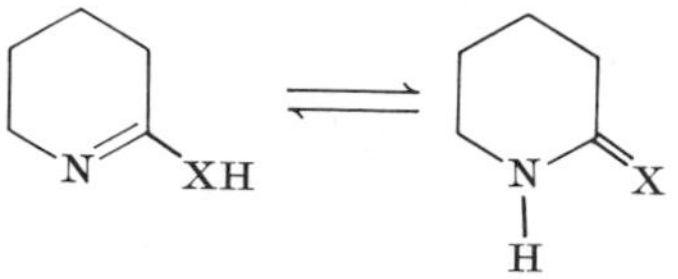

[**9a**] [**9b**] [**10a**] [**10b**]

Paoloni's group (68JH533) on the successive isomerization of tri-methoxy-*s*-triazine may be noted. For the equilibrium of 4-acetoxy-pyridine with 1-acetyl-4-pyridone see (70J(C)2426).

D. Annular Tautomerism in Pyridines and Azines

In contrast to its importance in azoles (Chapter 4), annular tautomerism is of relatively low importance in six-membered rings. An important exception to this statement is in azinones, e.g., [**11a**] ⇌ [**11b**]; however, this is dependent on the presence of a potential OH functional group and is considered under the tautomerism of potential hydroxypyrimidines [**11c**].

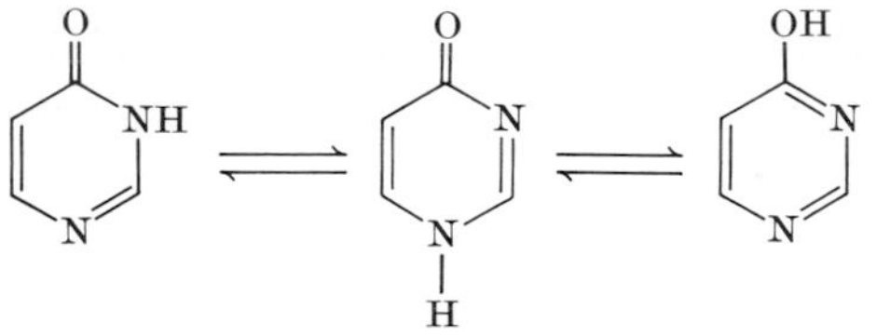

[**11a**] [**11b**] [**11c**]

Annular tautomerism does occur in dihydro derivatives of aromatic six-membered rings [**12a**] ⇌ [**12b**], and such compounds (which are "homoaromatic", cf. [**12c**]) are considered briefly. Dihydroazines can indeed exist in fully conjugated forms, as [**13**]; however, such tautomers are usually *destabilised* with respect to the nonaromatic (or homoaromatic forms), which is readily understandable, as [**13**] is antiaromatic.

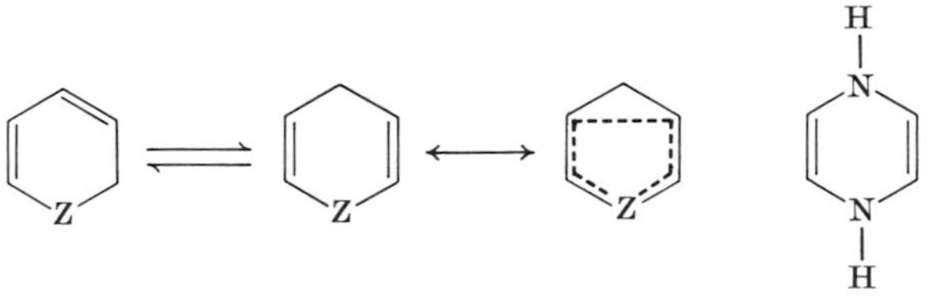

[**12a**] [**12b**] [**12c**] [**13**]

2. Tautomerism Not Involving Functional Groups

Almost the only class of compound included in this section which strictly comes within the scope of a review of hetero*aromatic* tautomerism is the diazine and triazine protonated cations. However, for completeness, and because of their relation to other classes of tautomeric heterocycles, we also discuss briefly the homoaromatic compounds comprising the pyrans and thiapyrans, the dihydropyridines and the dihydroazines.

A. Pyrans

4*H*-Pyran [**14b**] is known (62AG465, 62JA2452, 65BS131, 69T2023), as are various substituted derivatives (see, e.g., 67BS51, 68DA(180)473, 69JO3169, 71BF4059). The possibility of tautomerism of these compounds with the unknown 2*H*-pyran [**14a**] has apparently not been investigated. In the corresponding benzo series, both 2*H*-1-benzopyran [**15a**] (62JA813, 64JA2744) and 4*H*-1-benzopyran [**15b**] (54MI431, 62JA813, 69BF1715, 69T2023) have been isolated, but apparently they do not undergo interconversion. Tautomeric equilibria of 2*H*-1-benzopyrans of type [**16a**] with [**16b**] have been demonstrated by IR (68BF-4203); [**16b**, R = Me] is stabilized by hyperconjugation of the methyl group.

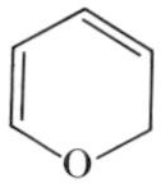

[**14a**] 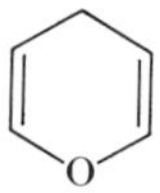 [**14b**] 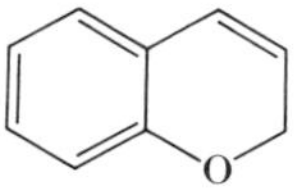 [**15a**] 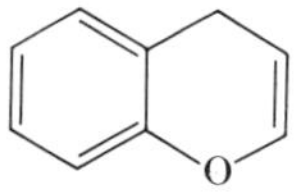 [**15b**]

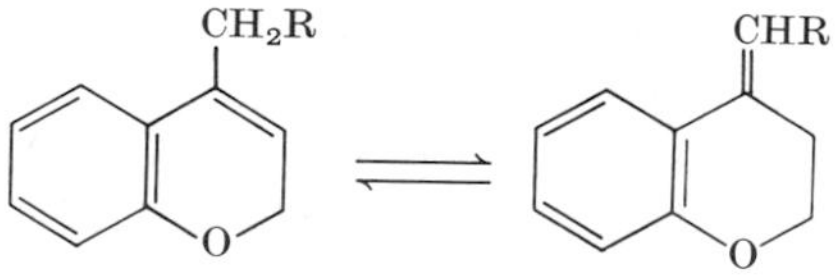

CH_2R CHR

[**16a**] R = H or Me [**16b**]

B. Thiopyrans

In the sulfur series both 2*H*-[**17a**] (69R30) and 4*H*-thiopyrans [**17b**] (62AG465, 65BS131) are known; here a 1*H*-tautomer [**17c**] could also exist, but is still unknown. Although tautomerism of the parent compound has not been discussed the 2,4-diphenyl derivative [**18a**] was

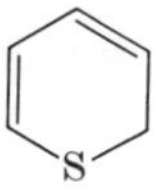

[**17a**] 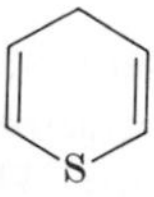 [**17b**] 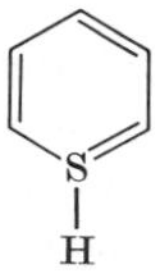 [**17c**] 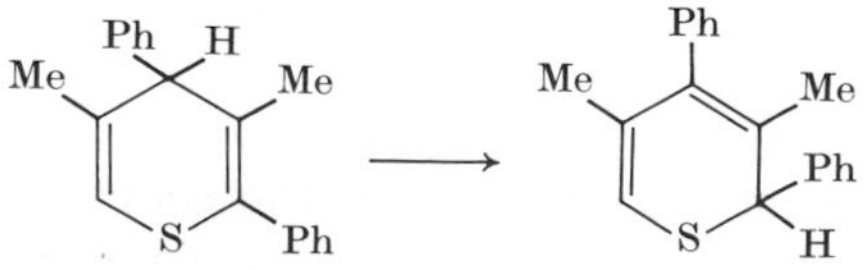 [**18a**] [**18b**]

found to be tautomerized to [**18b**] by acid (69RO1711). As previously mentioned for the 2*H*-1-benzopyran analogues (Section 2-2A) the 2*H*-thiopyran [**19a**] coexists with its tautomer [**19b**], but corresponding equilibria were not found for [**20**] and [**21**] (72R785).

In the benzo series, both 2*H*-1-benzothiopyran [**22a**] and 4*H*-1-benzothiopyran [**22b**] have been isolated [61JA4034). However, the phenyl compounds [**23a**] and [**23/1a**] were each tautomerized, to yield [**23b**] and [**23/1b**], respectively, by alumina or acids (71AN793).

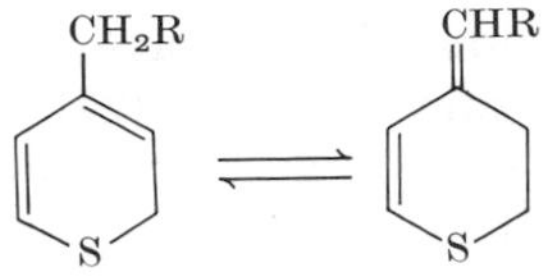

CH_2R CHR RCH_2 RCH_2 R H

[**19a**] [**19b**] [**20**] 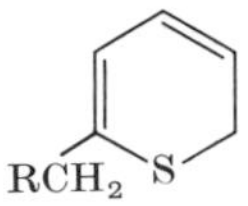 [**21**] 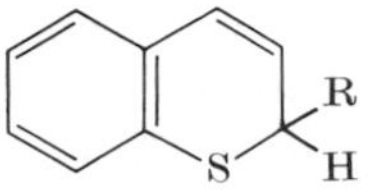 [**22a**] R = H

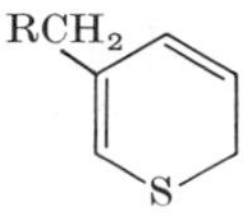

R = H or Me [**23a**] R = Ph

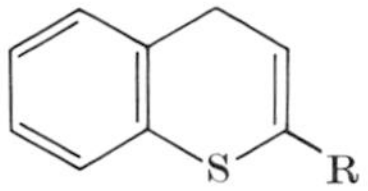

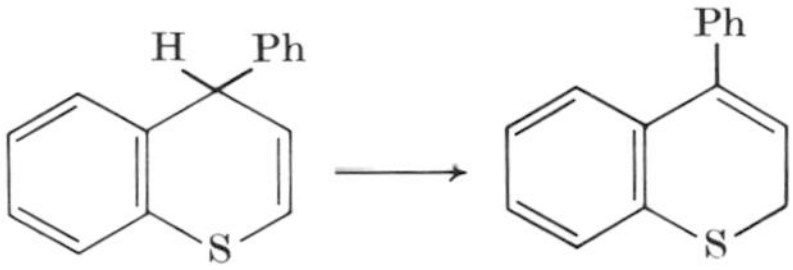

H Ph Ph

[**22b**] [**23/1a**] [**23/1b**]

[**23b**]

C. Dihydropyridines

These compounds have recently been reviewed (72CR1), and the rather scattered data on their interconversion are summarized; this will not be repeated. Five tautomeric forms exist for a dihydropyridine [**24a–e**]. Most of the dihydropyridines so far examined have either structure [**24a**] or [**24b**]. 2,5-Dihydropyridines [**24d**] are unstable (64JH13), and 3,4-dihydropyridines [**24e**] exist only if stabilized by

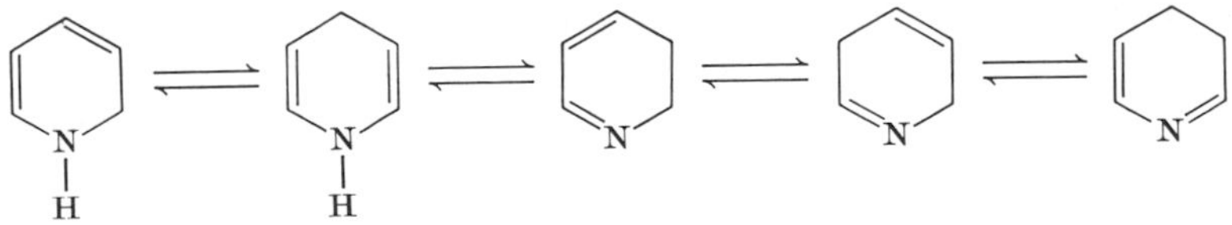

[**24a**] [**24b**] [**24c**] [**24d**] [**24e**]

resonance with a substituent, as e.g., in [**25**] (67T4517, 69JO3672). 1-Methyl-1,4- [**26a**] and -1,2-dihydropyridine have been separately isolated: the equilibrium between these favours [**26a**] by 2.3 kcal mole^{-1} (72JA5926).

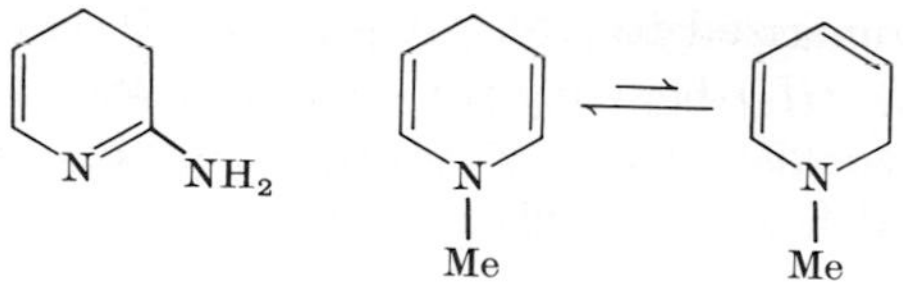

[25] [26a] [26b]

D. Dihydroazines

The known 1,2-, 1,4- and 4,5-dihydropyridazines [**27a–c**] have been reviewed (68HC(9)211). The 1,4-dihydropyridazine structure [**27b**] is probably the most stable tautomer (68AS1669, 68AS2700, 70RO1349). Simple dihydrocinnolines [**28**, R = H, Me, Ph] exist in CCl_4 and in dioxan in the 1,4-dihydrocinnoline form [**28b**] on NMR evidence: e.g., [**28**, R = Me] has NH singlet at 8.25 δ, 4-*H* octet at 3.21 δ (63J2867).

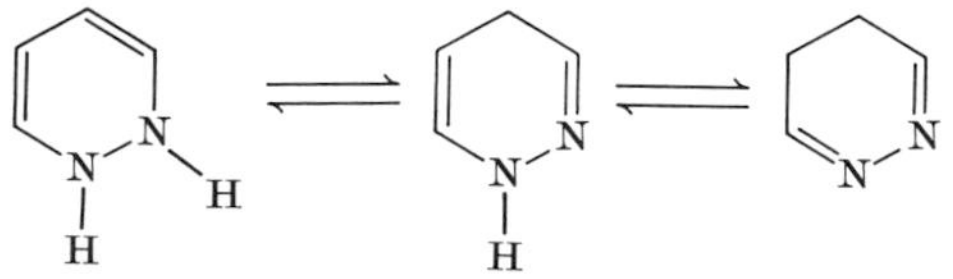

[27a] [27b] [27c]

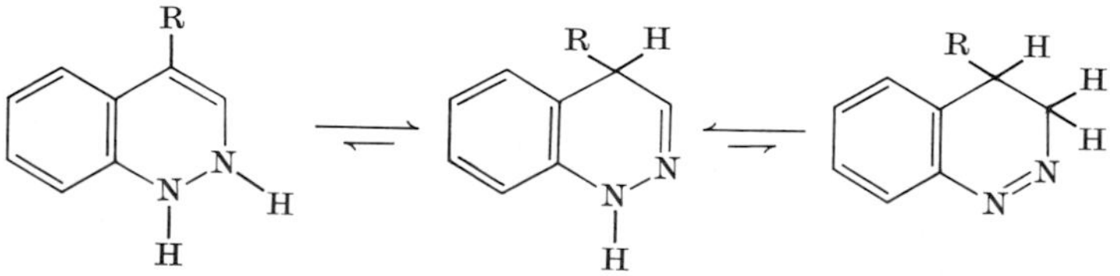

[28a] [28b] [28c]

Recently, structure [**29a**] was assigned by NMR (neat liquid) to the dihydrodimethylpyrazine [**29**] (70CH25), previously assumed (49J263) to have structure [**29b**]. Small peaks at 1.63 and 5.6 δ (3:1) were tentatively ascribed to 10% of tautomers [**29b**] and/or [**29c**] (70CH25). An H-bonded νNH band (IR) was previously interpreted as due either to a mixture of [**29a**] and [**29b**] or to [**29c**] alone (58J1174);

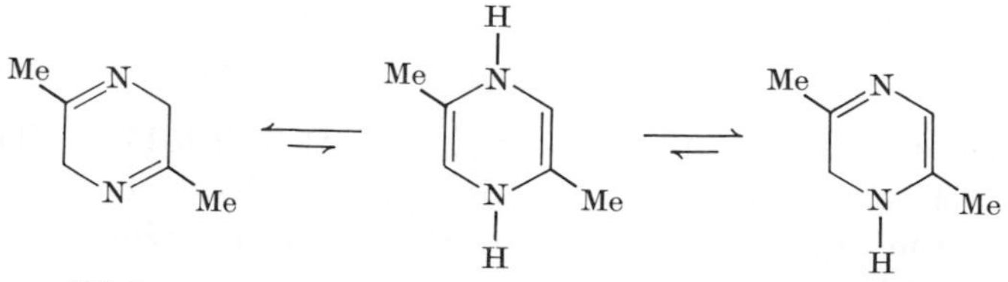

[29a] [29b] [29c]

this effect is now considered to be caused by water in the sample (70CH25). 1,4-Dihydro-2-phenylquinoxaline [**30a**] has been isolated but rearranges easily to the thermodynamically more stable 1,2-dihydro tautomer [**30b**] (70H1151).

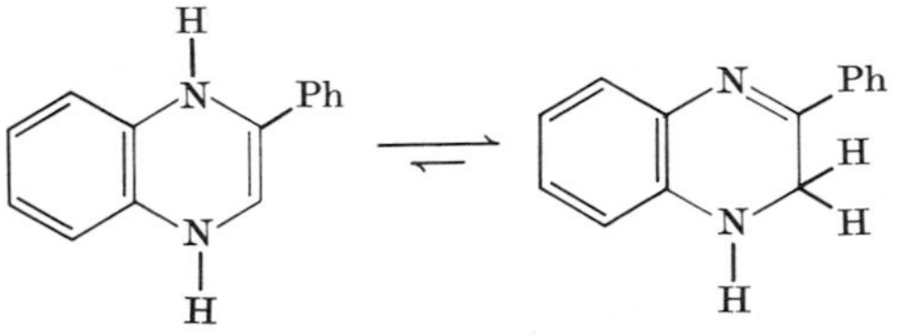

[**30a**] [**30b**]

The 1,4-diphosphoniacyclohexadiene system [**31a**] is readily converted into the isomer [**31b**] (69JO4024).

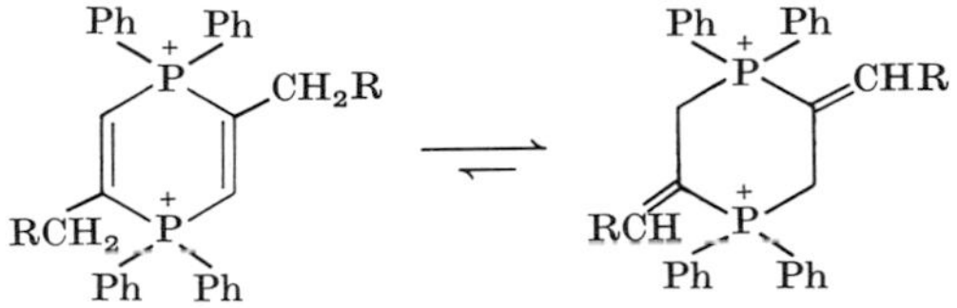

[**31a**] [**31b**]

Equilibrium [**32**] is shifted towards the nonaromatic dihydrotetrazine [**32b**], and from the antiaromatic tautomer [**32a**] (70H251).

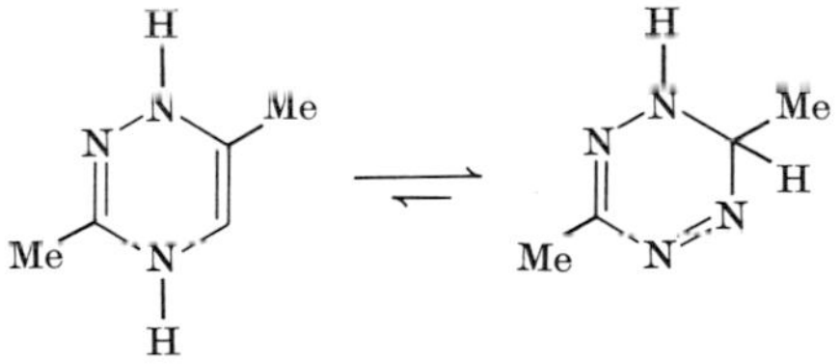

[**32a**] [**32b**]

E. Oxazines and Thiazines (I-341)

NMR spectra in DMSO or in $CDCl_3$ demonstrate that 4,1,2-benzoxadiazines can be isolated in the 3*H* form [**33a**] but are easily transformed into the tautomeric 1*H*-4,1,2-benzoxadiazines [**33b**] on treatment with

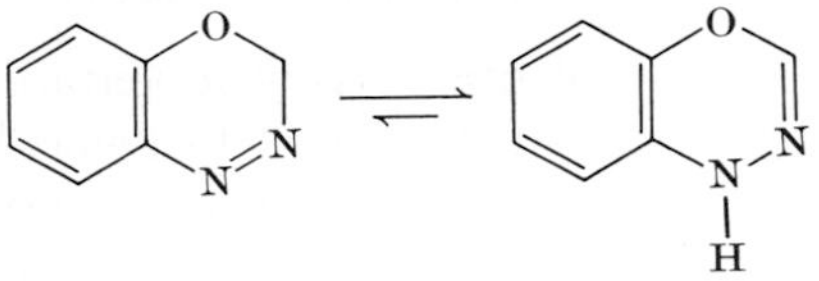

[**33a**] [**33b**]

alkali alkoxides (70CB331): CH (6.75 δ) and NH (8.35 δ) absorptions replace the CH_2 signal at 5.6 δ.

The νNH at 3400 cm^{-1} (Nujol) or 3300 cm^{-1} (CCl_4) and the absence of a C(2)-proton signal in the NMR spectrum in $CDCl_3$ show that the benzo-1,4-thiazine [**34**] exists as the 4*H* form [**34a**], not as the 2*H* tautomer [**34b**] (66CT770). This preference for [**34a**] is probably due to the ethoxycarbonyl group: various other derivatives of the pyrimidinothiazine system [**35**] exist in the 2*H* form shown (64JO2121, 72DA1366). 1,4-Thiazine [**36a**] itself is formulated on the basis of chemical reactivity in the 2*H* form [**36a**] (I-341).

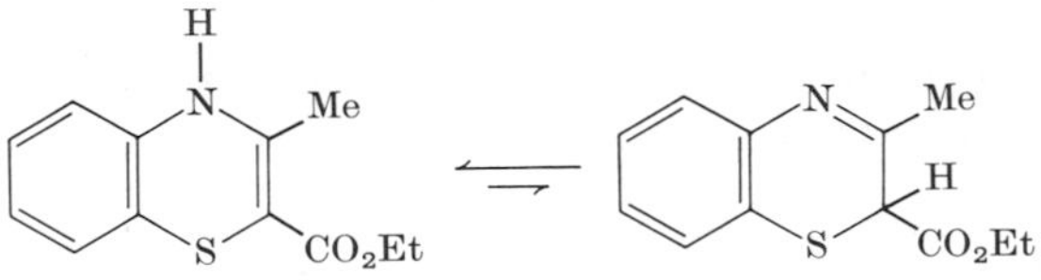

[**34a**] [**34b**] 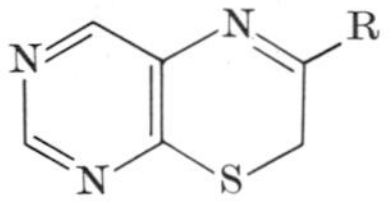 [**35**]

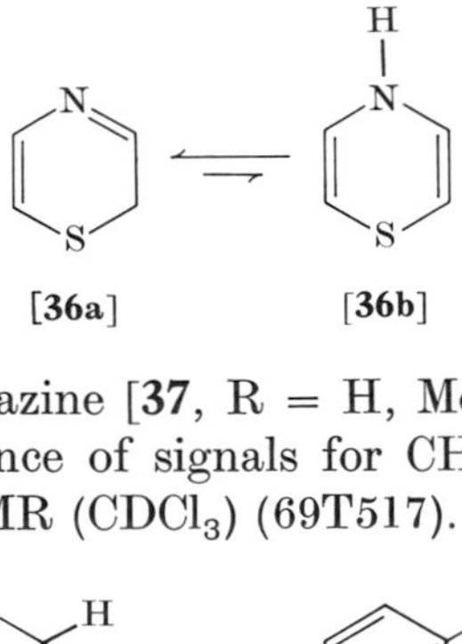

[**36a**] [**36b**]

The analogous benzoxazine [**37**, R = H, Me] also exists in the 2*H* form [**37a**] on the evidence of signals for CH_2 or CHMe and of the absence of NH in the NMR ($CDCl_3$) (69T517).

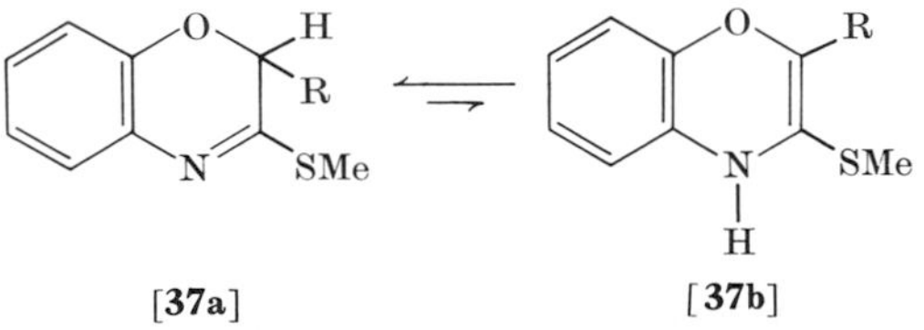

[**37a**] [**37b**]

F. Other Neutral Species

The 5-, 6- and 7-protons of dimethylcyclopenta[*d*]pyrazine form an AB_2 system which suggests either complete delocalization [**38a**] or a time-average of rapidly interchanging tautomers [**38b**] ⇄ [**38c**] (70J(C)-610): in the present authors' opinion, the second alternative is much the more likely.

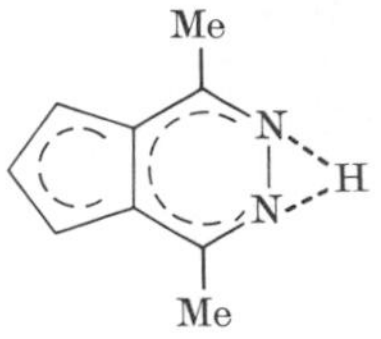

[38a] 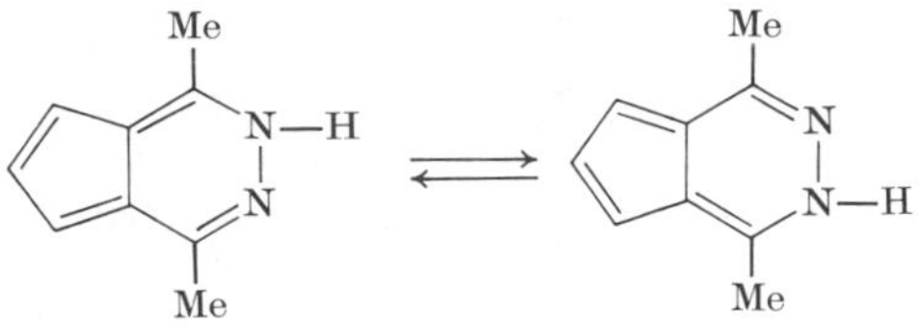 [38b] [38c]

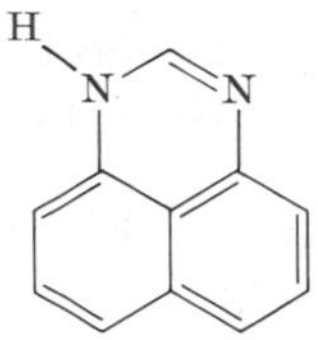

[39] 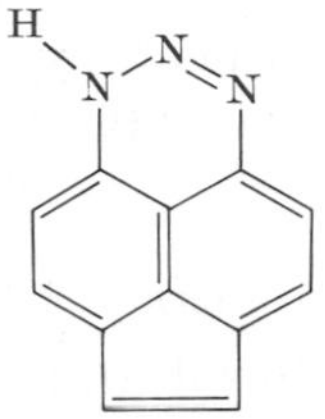 [40] 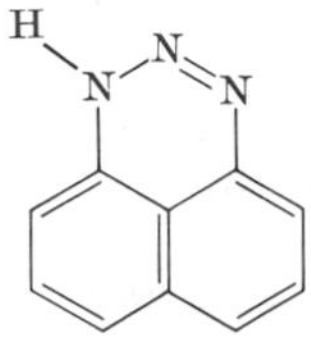 [41]

Similar cases of autotropic tautomerism occur for compounds [**39**] and [**40**] (70J(C)290) and [**41**] (64J3005).

G. Azine Cations (I-341)

a. Introduction

Annular tautomerism in azine cations resembles that in the corresponding *neutral* species of azoles [see (Section 4-1) for a comprehensive discussion]. Thus the annular tautomerism of each of the diazine monocations and that of *s*-triazine is "autropic" involving two identical species, e.g., [**42**], unless the diazine is unsymmetrically substituted, e.g., [**43**]. The 1,2,4-triazine cation occurs in three nonidentical forms [**44a,b,c**].

Relatively little systematic work has been done on this subject, although in principle many of the methods discussed in (Section 4-1) would be applicable. A complication is that, particularly for compounds containing three or more heterocyclic nitrogen atoms, covalent hydration competes with simple protonation. This aspect has been

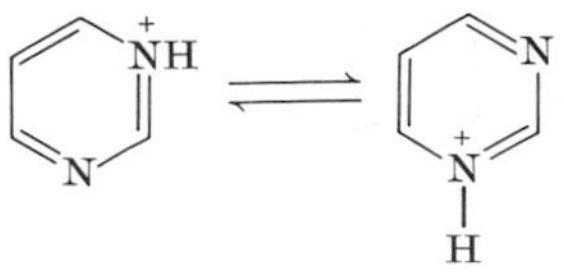

[42a] [42b] 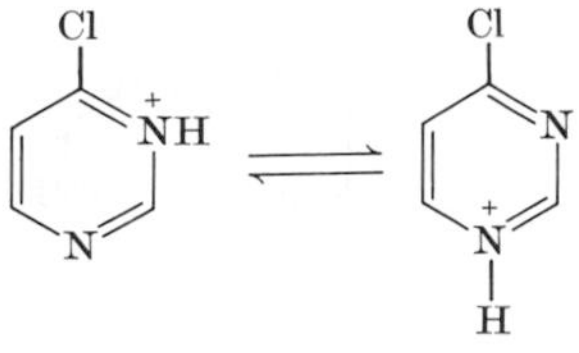 [43a] [43b]

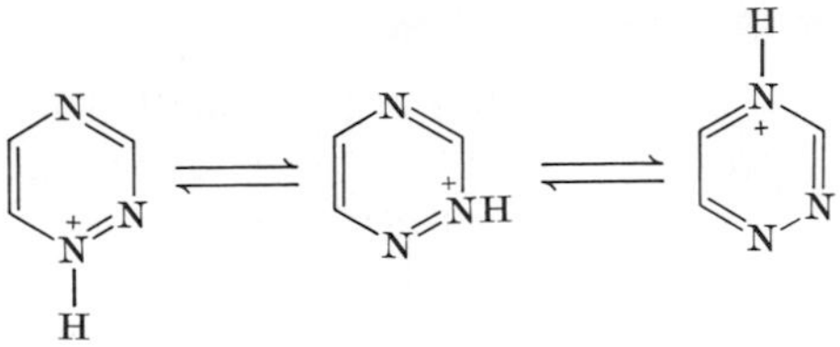

[44a] [44b] [44c]

extensively studied and covalent hydration in diaza and polyaza derivatives has been well reviewed (65HC(4)1, 65HC(4)43, 67AG(E)919, 70HC(11)123).

b. Pyridazines and Benzopyridazines (I-341)

The basicities of a series of 6-substituted 3-dimethylaminopyridazines were found to be correlated by σ_m constants. However, the basicities of 6-substituted 3-chloropyridazines were correlated by σ_p substituent constants. This suggested that cations of types [**45**] and [**46**], respectively, were formed in the two series (72J(PII)392).

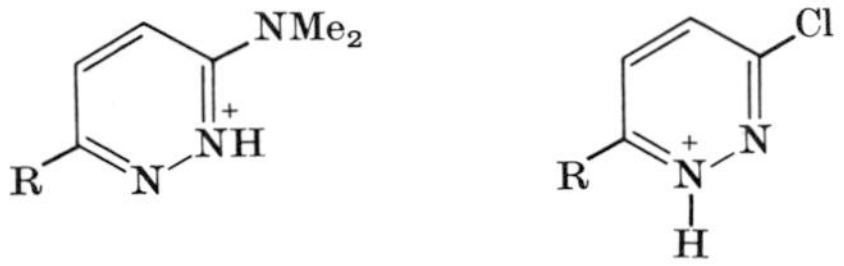

[**45**] [R = various] [**46**] [R = various]

The protonation site of cinnolines has been the subject of considerable study: the previous review (I-341) reported evidence for N(1)-protonation [**48b**], but cautioned that this was not compelling. Correlation of very few pK_a values of 4-substituted cinnolines also led to the conclusion that N(1) was the protonation site [**48b**] (64J5884), but a definitive UV comparison with fixed models showed that protonation of cinnoline and its 3-amino and 4-methyl derivatives occurs predominantly at N(2) [**48a**] (66CI458, 67J(B)748), contrast protonation of 4-aminocinnoline at N(1) (Section 2-6Db). Previous evidence for N(2)-protonation from NMR (65CI1766) is inconclusive (67J(B)748), but MO calculations support N(2)-protonation (71T2921).

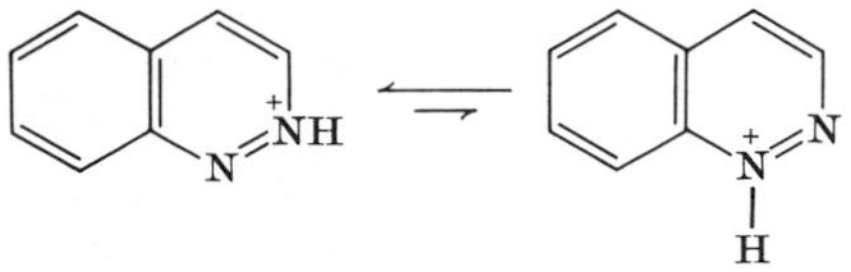

[**48a**] [**48b**]

c. Pyrimidines and Benzopyrimidines

Evidence derived from basicity correlations has been presented (69JO821) that pyrimidines substituted in the 4-position with an amino, methoxy or methylthio group are protonated at N(1) [**49a**] rather than at N(3) [**49b**] (69JO821) (cf. Section 2-6Ca) for 4-amino derivatives. Although covalent hydration is involved in protonation of quinazoline (I-341), the anhydrous quinazoline also participates, and it was assigned the H(3) (50**b**] structure, rather than H(1) [**50a**] on substituent effect analysis (66J(B)436). Further physical data are needed to confirm and quantify these findings.

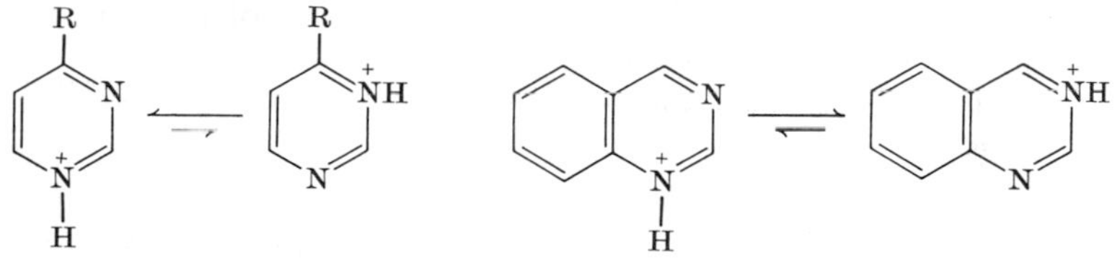

[**49a**] [**49b**] [**50a**] [**50b**]

The different biological behaviour of thiamine and cocarboxylase has been attributed to differences in the site of protonation of the two compounds (73OR573, 74MI1269).

d. Pyrazines and Benzopyrazines

It has been suggested that NMR chemical shift differences between DMSO and CF_3CO_2H solution indicate that 2-chloro-3-methyl- and 2-amino-3-methyl pyrazine give mixtures of the two possible cations in comparable amounts but that 2-methoxy-3-methyl pyrazine is preferentially protonated at N(1), to give the cation [**52b**] stabilized by resonance (68PH1642). These unlikely conclusions require reinvestigation.

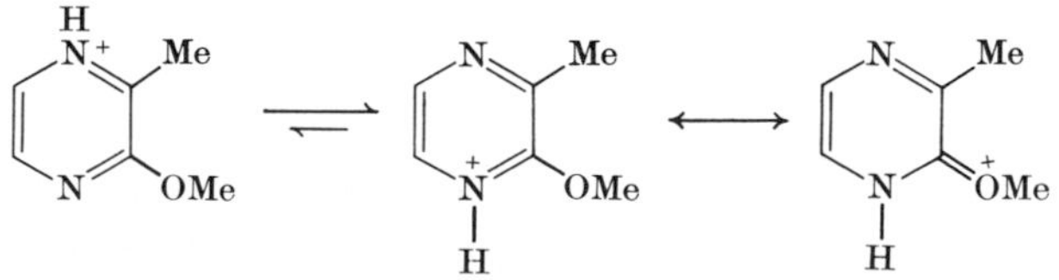

[**52a**] [**52b**]

The effect of substituents (R = from NH_2 to NO_2) on the tautomerism of quinoxaline cations [**53a**] ⇌ [**53b**] has been investigated using the Hammett equation and statistical analysis (68CZ566). The tautomeric equilibrium [**53a** ⇌ **b**] is strongly dependent on the nature of the 6-substituent: log K_T range from -0.44 for NH_2 to $+0.42$ for

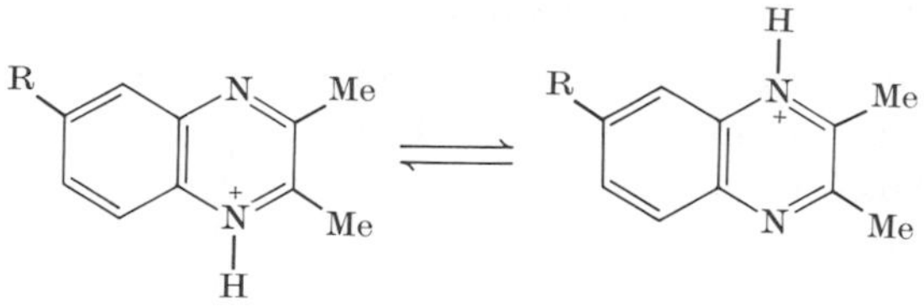

[**53a**] [**53b**]

NO_2. More recently, Russian authors have presented further experimental and theoretical data to show that 2-amino and 2-substituted amino substituents favour 1-protonation, but most other 2-substituents favour 4-protonation (73KG398). For a discussion of the structure of protonated pteridines see (74JH7).

e. Triazines

The cations of unsymmetrically substituted 1,3,5-triazines, and of all 1,2,4-triazines, should show tautomerism. However, there appear to be no available data on the structures of these cations.

3. Compounds with Potential Hydroxy Groups and One Heteroatom (I-341)

A. Introduction

a. Arrangement

We discuss first compounds in which the potential hydroxyl function is α or γ to a ring nitrogen: the 2- and 4-pyridones and their benzologues. The tautomerism of these compounds is distinct from the β-hydroxy derivatives next considered. This is followed by the potential polyhydroxypyridines, and then by hydroxypyridine *N*-oxides. Compounds in which the ring heteroatom is oxygen or sulphur (the hydroxypyrones and hydroxythiapyranones) are then treated. Previously the hydroxypyrones and hydroxythiapyrones were dealt with before the potential hydroxypyridines: however, the former are more closely analogous to hydroxypyridones, i.e., to potential dihydroxypyridines.

b. Tautomeric Behaviour of Hydroxy Groups

We have discussed in general terms the functional group tautomerism in six-membered rings (Section 2-1A). The tendency for a proton to move from a functional group to a cyclic nitrogen is strong for pyridines, because of the marked acidity of the OH group, and the basicity of N. If the potential oxygen function is α or γ to the ring nitrogen, as in [57], then the zwitterion [**57b**] produced by proton transfer is stabilized by the uncharged canonical form [**57c**]. This is not the case with the β-hydroxy compounds [**58**], and for this reason the two classes show different behaviour.

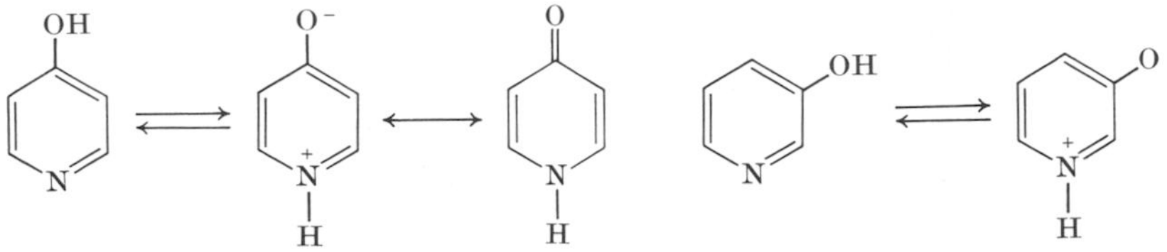

[**57a**] [**57b**] [**57c**] [**58a**] [**58b**]

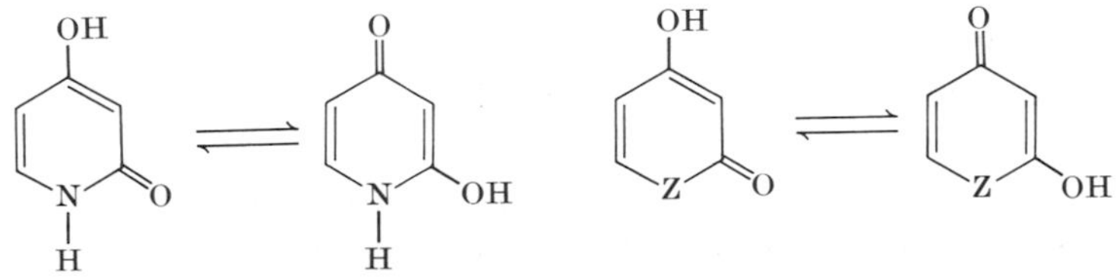

[**59a**] [**59b**] [**60a**] [**60b**]

If two potential hydroxy groups are present, tautomerism involving only the oxygen functions becomes possible for potential dihydroxypyridines [**59a**] $\rightleftharpoons$ [**59b**], and also for hydroxypyrones and thiapyrones [**60a**] $\rightleftharpoons$ [**60b**] (Z = O or S).

c. *A Comparison of Hydroxypyridine and Aminopyridine Tautomerism*

It is of fundamental importance to the understanding of heteroaromatic tautomerism that 4-pyridone [**61**] exists mainly in the oxo form [**61b**], and not as 4-hydroxypyridine [**61a**], whereas 4-aminopyridine [**62**] exists mainly in the amino form [**62a**], and not as the imino form [**62b**]. Similarly, the enaminone system NH—C═C—C═O is considerably more stable than the alternative N═C—C═C—OH system (by ca. 10^8) (69J(B)299). This difference in behaviour can be understood on simple

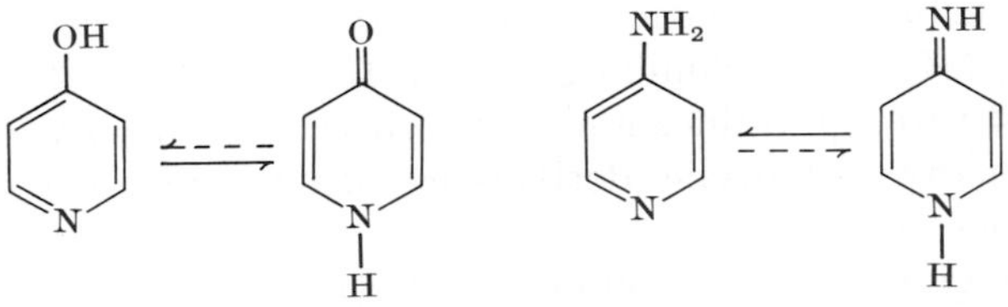

[**61a**] [**61b**] [**62a**] [**62b**]

valence bond considerations. The individual tautomeric forms all possess mesomerism and are stabilized by contributions from charge-separated forms as shown in structures [**63a–d**] for 4-pyridone and [**64a–d**] for 4-aminopyridine.

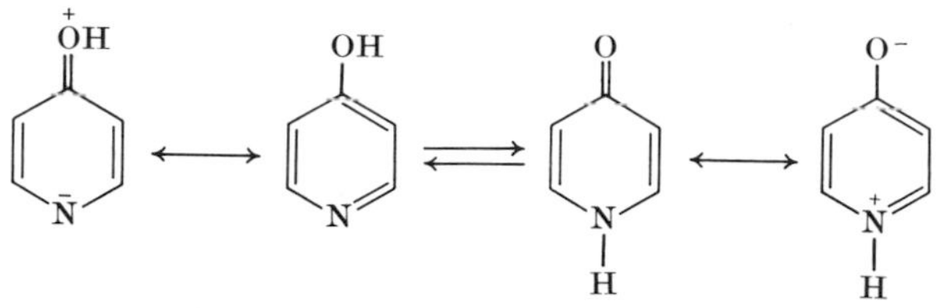

[**63c**] [**63a**] [**63b**] [**63d**]

In the case of 4-aminopyridine the contributions from the charge-separated forms [**64c, 64d**] are of moderate importance; although the imino form [**64b**] has considerable aromatic stabilization energy, that of the amino form is ca. 12 kcal greater (73J(PII)1080) and is responsible for the predominance of the amino form. However, in the tautomerism between 4-hydroxypyridine and 4-pyridone, the charge-separated form [**63c**] in which positive charge resides on oxygen and negative on nitrogen is of much less importance than the other charge-separated form [**63d**]. For this reason the equilibrium [**63a**] $\rightleftarrows$ [**63b**]

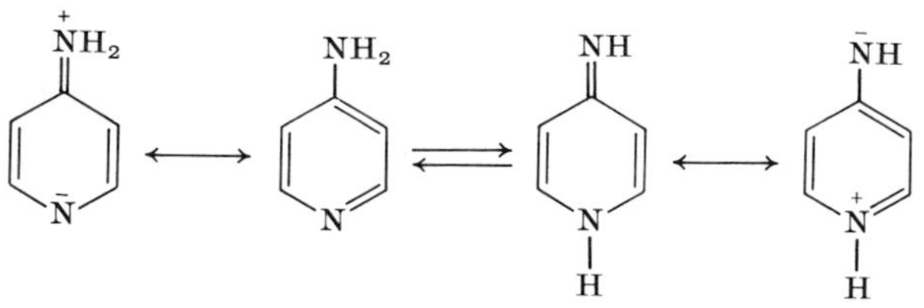

[**64c**] [**64a**] [**64b**] [**64d**]

swings over in favour of 4-pyridone, the aromatic stabilization energy of which is only about 8 kcal less than that of pyridine (73J(PII)1080). Molecular orbital calculations lead to similar conclusions (68ZC305, 70JA2929, 71CZ1413). Similar generalizations apply to the corresponding 2-substituted pyridines and to potential α- and γ-hydroxy derivatives of many polycyclic derivatives of pyridine, e.g., quinolines, isoquinolines, acridines. Many experimental data supporting this conclusion are given in the previous review (I-339 ff) and the subsequent experimental data will now be summarized.

B. 2- and 4-Pyridones: α- and γ-Hydroxypyridines

a. 2- and 4-Pyridone and Benzologues (I-347)

The previous review (I-347) summarizes the compelling evidence for the predominant oxo structure of these compounds. Subsequent work, of lesser significance on the whole, fully confirms the earlier conclusions.

The hydrogen bonding properties of 2-pyridone and related compounds have been widely studied and this work incidentally confirms the oxo structures (66JA1621, 69JA6090, 70JA7578, 71JA6387, 71PH1129), as does further proton NMR (72KG95). There has been a special proton NMR study of the monomer–dimer equilibrium of 2-pyridone in various solvents: both monomer and dimer exist in the oxo form (72AS2255). Pyrido[2,3-*c*]-2,6-naphthyridin-5-one [**64/1**] has been claimed to exist in the hydroxy form [**64/1b**] from a supposed νOH band at 2900 cm^{-1} (solid) (72JH1021); this conclusion is probably incorrect.

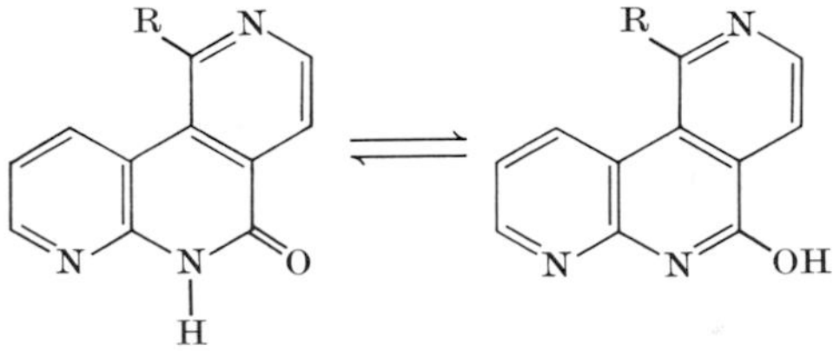

[**64/1a**] [**64/1b**]

Nitrogen-14 NMR demonstrates the oxo structures for 2- and 4-pyridone in acetone (71T3129), and for various quinolones (67CH371), as did proton NMR for 2-pyridone in $CDCl_3$ (69PH2465).

The earlier assignment of the 4-quinolone structure based on IR data (I-350) is confirmed by UV comparisons with fixed forms of its 2-methyl derivative (69T255).

MO calculations are available on pyridones (70JA2929, 69TE160, 71JL(10)245) and quinolones (70TE250).

It is noteworthy that the anion radicals both of 2- and 4-pyridone, generated in argon matrices, possess structures [**64/2a**] or [**64/2b**] and [**64/3**] as shown by ESR (74JA2342).

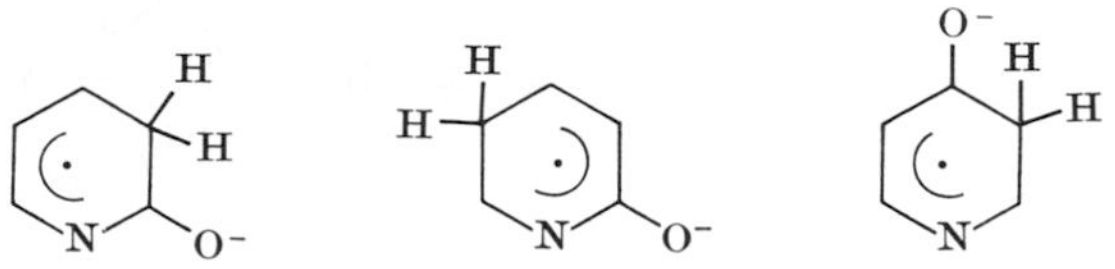

[**64/2a**] [**64/2b**] [**64/3**]

b. Substituent Effects: General Survey

In many cases, further substitution of the pyridine ring does not alter greatly the position of a tautomeric equilibrium, and many substituted 2- and 4-pyridones exist largely in the oxo forms. However, there are four important structural influences which can counter this tendency and swing the equilibrium towards the hydroxypyridine form:

i. Strongly electron-withdrawing substituents *alpha* to the nitrogen atom, which tend to shift the equilibrium to favour the pyridinoid form by altering the basicity of the nitrogen atom.
ii. Substituents which stabilize one or other of the tautomeric forms by hydrogen bonding.
iii. Ring strain effects caused by the annelation of saturated rings.
iv. Partial bond fixation (usually caused by benzene ring fusion) strongly influences the 3-hydroxyisoquinoline to 3-isoquinolone equilibrium position.
v. The effects of solvent, concentration, phase, and temperature, which can sometimes be very significant.

Relevant data on the effect of substituents on pyridone–hydroxypyridine tautomerism are summarized in Table 2-1: a widely varying situation is revealed. We now consider successively the effects mentioned above. Most quantitative data are available for aqueous solutions, and our initial discussion concentrates on these results: the effect of phase change is dealt with later.

c. Electronic Effects of Substituents

Gordon and Katritzky (68J(B)556) have shown how such effects can be predicted. Each of the tautomeric forms oxo [**65**] and hydroxy [**67**] of a substituted 4-pyridone is in equilibrium with the same mesomeric cation [**66**]. The position of the tautomeric equilibrium [**65**] ⇌ [**67**] is thus determined by the relative acidity of the OH and the NH protons in the conjugate acid: if the OH proton is more acidic than the NH proton, then [65] will predominate, and vice versa. Quantitatively, Eq. (2.2) applies.

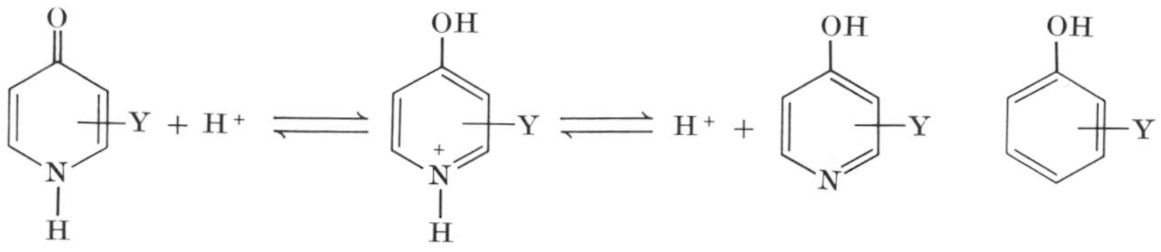

O, N, H, Y + H^+ ⇌ OH, $\overset{+}{N}$, H, Y ⇌ H^+ + OH, N, Y   OH, Y

[**65**]   [**66**]   [**67**]   [**67/1**]

TABLE 2-1

RATIO OF THE OXO[a] FORM TO THE HYDROXY[a] FORM (pK_T) FOR SUBSTITUTED α- AND γ-PYRIDONES

Substituent	Solvent:	H_2O		EtOH	CCl_4	Solid state	
	Method:	pK_a	UV	UV	IR	IR	X-ray
Pyrid-2-one		3.0[e]	oxo[c]	oxo[c]	oxo[c]	oxo[d]	oxo[d]
5-Chloro		1.3[e]	—	—	—	—	—
6-Chloro		0.2[illegible]	1.3[f]	OH[f,g]	oxo + OH[g]	oxo + OH[g] oxo[h]	oxo[i]
5-Methyl		2.6[e]	—	—	—	—	—
6-Methoxy		0.7[e,j]	> 1.3[k]	ca 0[k]	—	oxo[l]	—
6-Amino		2.5[m]	—	oxo[g]	—	oxo[g]	—
6-Chloro-4-methyl		0.2[f]	1.2[h]	OH[h]	OH[h]	—	—
3,4,5,6-Tetrachloro		−1.3[f]	OH[f]	—	OH[f]	—	—
5,6-Dimethoxycarbonyl		oxo[n]	oxo[n]	oxo[n]	oxo[n]	oxo[n]	—
Pyrid-4-one		3.3[c]	oxo[b]	oxo[b]	oxo[b]	—	—
2-Chloro		−0.3[f]	0.2[f]	ca −2[f]	—	—	—
3-Amino		4.6[o]	—	—	—	—	—
3-Nitro		3.4[o]	—	—	—	—	—
2,6-Dichloro		OH[f]	OH[f]	—	OH[f]	—	—
3,5-Dichloro		2.4[f]	0.9[f]	oxo[f]	—	—	—
2,3,5-Trichloro		0.0[e]	0.6[e]	(?)OH[e]	OH[e]	OH[e]	—
2,3,5,6-Tetrachloro		OH[f]	OH[f]	—	OH[f]	—	OH[p]
2,6-Dimethyl		3.7[e]	—	—	—	—	—
3,5-Dinitro		—	oxo[e]	oxo[e]	oxo[e]	oxo[e]	—
2,6-Dimethoxycarbonyl		−1.1[e]	−0.4[e]	OH[e]	oxo + OH[e]	oxo + OH[e]	—

[a] Oxo = pyridone form; OH = pyridinol form.
[b] I-350.
[c] I-350.
[d] I-351.
[e] 68J(B)556.
[f] 67J(B)758.
[g] 67DA(177)592.
[h] 70CB398.
[i] 69AK71.
[j] 66J(B)562; cf. also 71J(B)279.
[k] 71J(B)289.
[l] 71J(B)279.
[m] Calculated from 71J(B)1425.
[n] 71J(B)296.
[o] 67J(B)84.
[p] 72CH573.

The effect of the substituent Y on the basicity of the 4-pyridone [**65**], and hence on the acidity of the OH proton in the cation [**66**], can be estimated by a Hammett-type equation (Eq. 2.3) where σ_s refers to the substituent constants listed by Clark and Perrin (64QR295). Similarly, the basicity of [**67**] is given by Eq. (2.4) (for detailed discussion, see (68J(B)556); hence Eq. (2.5) follows.

$$\mathrm{p}K_T = \mathrm{p}K[\mathbf{65}] - \mathrm{p}K[\mathbf{67}] \quad (2\text{-}2)$$

$$\mathrm{p}K[\mathbf{65}] = 5.25 - 5.9\sigma_s \quad (2\text{-}3)$$

$$\mathrm{p}K[\mathbf{67}] = \mathrm{p}K[\mathbf{67/1}] - 6.7 \quad (2\text{-}4)$$

$$\therefore \mathrm{p}K_T = 1.5 - 5.9\sigma_s - \mathrm{p}K[\mathbf{67/1}] \quad (2\text{-}5)$$

Table 2-2 compares experimental and calculated pK_a and pK_T values for 4-pyridones: agreement is encouraging though by no means complete. It is clear that substituents α to the nitrogen atom affect the

TABLE 2-2

CALCULATED AND FOUND VALUES OF pK_a FOR THE *N*- AND *O*-METHYL DERIVATIVES AND TAUTOMERIC RATIOS OF 4-PYRIDONES (68J(B)556)

C-Substituent(s)	pK_a of *N*-methyl derivative		pK_a of *O*-methyl derivative		$\log_{10} K_T$[a]	
	Calc.	Found	Calc.	Found	Calc.	Found
H	—	3.33	—	6.62	—	3.3
3-Methyl	3.33	—	6.90	—	3.6	—
3-Amino	2.74	3.91	7.43	7.52	5.0	3.6
3-Chloro	2.63	—	4.31	—	1.7	—
3-Methoxycarbonyl	3.52	—	4.65	—	1.1	—
3-Nitro	0.53	−0.80	2.30	2.63	1.8	3.4
3,5-Dichloro	−0.05	−1.22	2.12	1.14	2.2	2.4
3,5-Dinitro	−2.97	−5.25	−1.89	—	1.1	—
3,5-Dimethyl	3.93	—	7.32	—	3.4	—
2-Methyl	3.43	—	7.25	—	3.8	—
2-Amino	3.32	—	8.08	—	4.8	—
2-Chloro	2.43	2.27	2.33	1.93	−0.1	−0.3
2,6-Dimethyl	3.48	4.12	6.49	7.86	2.1	3.7
2,6-Dichloro	1.76	—	−2.81	−1.36	−4.6	—
2,6-Dimethoxycarbonyl	0.94	1.29	0.47	0.18	−1.4	−1.1
2,3,5-Trichloro	−0.54	−2.06	−2.54	−2.06	−1.7	0.0
2,3,5,6-Tetrachloro	−2.15	—	−7.21	−4.50	−5.6	—

[a] K_T = [oxo form]/[hydroxy form].

pK_a of [**67**] far more than that of [**65**] whereas β-substituents have roughly comparable effects in both systems. Hence it is expected that 2-substituents in 4-pyridone should affect the tautomeric equilibrium far more than 3-substituents. Similar considerations should apply to 2-pyridones.

The experimental results (Table 2-1) are now discussed in the light of these findings. The tautomeric equilibrium of pyridones is indeed displaced significantly in favour of the hydroxypyridine form by chlorine and fluorine atoms, and by methoxy and amino groups when substituents are alpha to the nitrogen atom, but is little affected by β-chlorine atoms and β-amino and β-nitro groups. Many examples are given in Table 2-1; some of these and some further ones will be discussed. Additional examples are given in (66J(B)991) and in (71J(B)1425).

That substitution at the β-position to nitrogen has much less effect is shown, e.g., by the pK_T value for 3-amino-4-pyridone ($pK_T = 4.6$) and 3-nitro-4-aminopyridine ($pK_T = 3.4$) in comparison with $pK_T = 3.3$ for the unsubstituted compound (67J(B)84). Again, 3,5-dichloro-4-pyridone [**68**], in which the chlorine substituents are both β to the ring nitrogen atom, exists largely in the pyridone form in water as shown by pK and spectral methods (67J(B)758). As expected, X-ray crystallography shows that 5-chloro-2-pyridone [**68/1**] exists as such (72CX3405).

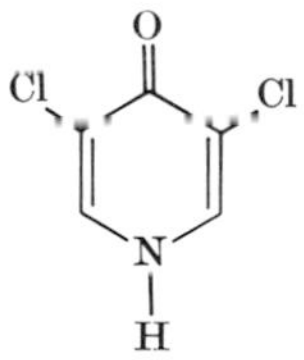

[**68**] 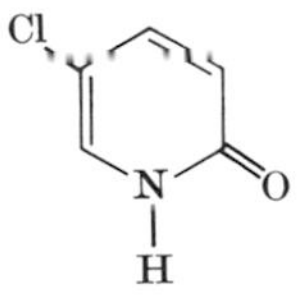 [**68/1**]

In marked contrast, for 2-chloro-4-pyridone comparable amounts of tautomers [**69a**] and [**69b**] coexist in aqueous solution (67J(B)758) and 2,6-dichloro-4-pyridone exists in H_2O very largely as 2,6-dichloro-4-hydroxypyridine [**70**] (67J(B)758). Single crystal X-ray diffraction

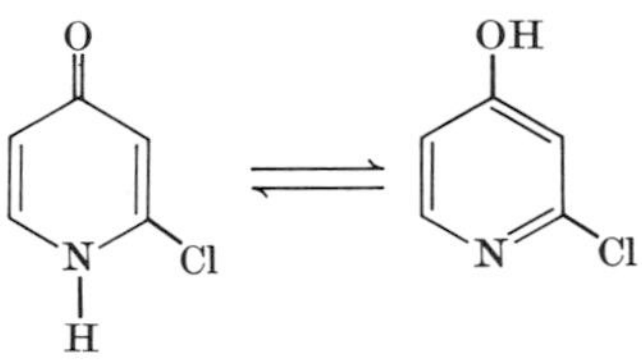

[**69a**] [**69b**] 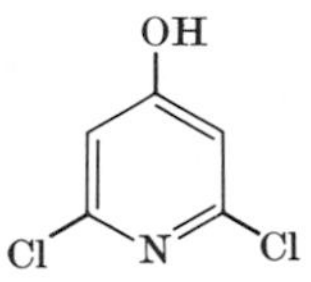 [**70**]

confirms the oxo form for 3,5-dichloro-2,6-dimethyl-4-pyridone but the OH form for tetrachloro-4-hydroxypyridine in the solid (72CH573). X-ray crystallography shows that an intermolecular complex is formed with the oxo form of 2-pyridone and the hydroxy form of 6-chloro-2-hydroxypyridine (71CX(B)1201) as previously demonstrated for the individual compounds (I-351, 69AK71).

Polyhalogeno derivatives exist as 2- and 4-hydroxypyridines in the solid state and in H_2O, EtOH and hexane (64J(S1)5634, 65J575, 74J(PI)2307) as shown by UV and the characteristic νOH in the IR. Although 2-hydroxyhexafluoroquinoline and 1-hydroxyhexafluoroisoquinoline give mixtures of *O*- and *N*-methyl derivatives with CH_2N_2 (67J(C)53), this is *not* good evidence for their tautomeric composition (cf. Section 1-2).

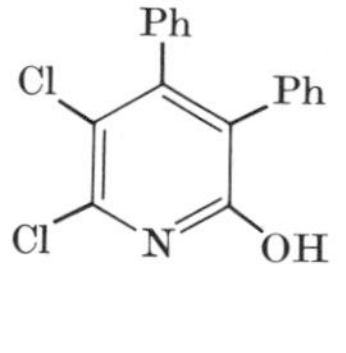

[**71**] 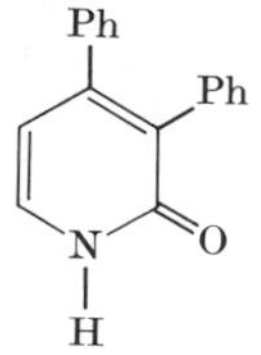 [**72**]

The νOH at 3100 cm^{-1} and λ_{max} at 303 nm compared with λ_{max} for the OMe derivative at 298 nm and λ_{max} for the NMe derivative at 337 nm demonstrate the phenolic structure of 3,4-diphenyl-5,6-dichloropyridin-2-ol [**71**]. As expected, the corresponding compound without chlorine takes the pyridone form [**72**] (νC = O 1645 cm^{-1} and νNH at 3450 cm^{-1}) (68AJ467).

5,6-Bis(methoxycarbonyl)-2-pyridone [**73**] exists predominantly in the oxo form [**73b**] in water whereas 70% of the OH form [**73a**] exists in dioxan (71J(B)296). There may well be additional effects of N-pair/O-pair electron repulsions in the OH form [**73a**] which are relieved in [**73b**].

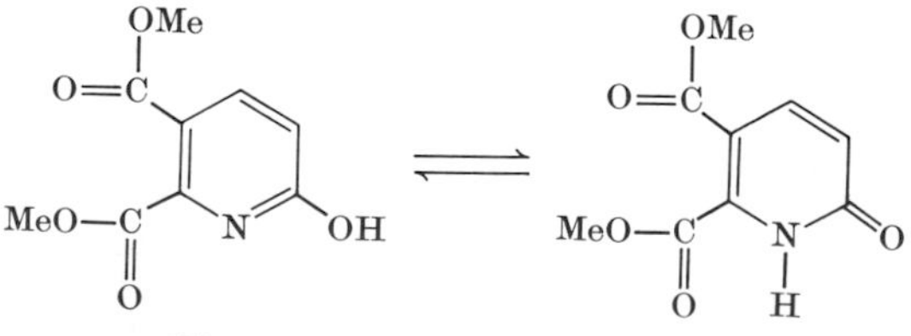

[**73a**] [**73b**]

The available evidence indicates that similar considerations govern the electronic effects of substituents in benzopyridone tautomerism.

UV and NMR spectral comparisons with fixed forms showed that 3-alkyl and 3-aryl-1-isoquinolones [**74**] exist in the oxo form [**74a**] in EtOH and $CDCl_3$ (66JO2090). The pK_a method was used to determine the pK_T for equilibrium [**74a**] $\rightleftarrows$ [**74b**] of several substituted isoquinolin-1-ones in 10% MeOH/90% H_2O mixtures (69CB3666): thus the pK_T is 4.9 in favour of the oxo form for 1-isoquinolone itself, but this is reduced to 2.0 by a 3-chloro substituent.

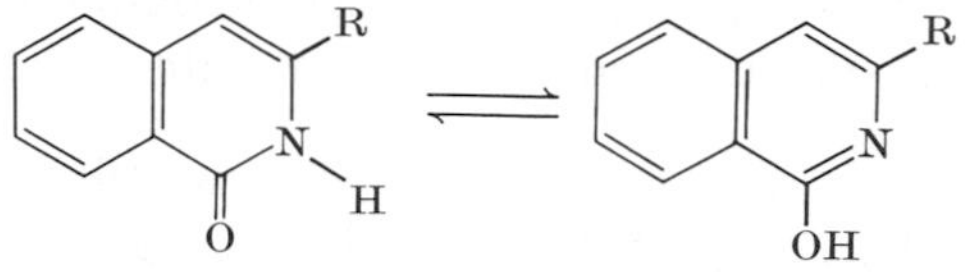

[**74a**] [**74b**]

3-Cyano-4-quinolone exists predominantly in the oxo form in the solid, but in the hydroxy form in acetonitrile (68J(C)2656), whereas the corresponding 3-ethoxycarbonyl compound is predominantly in the oxo form even in nonpolar solvents. Kay and Taylor (68J(C)2656) point out that these, and other examples discussed, illustrate the complexity of substituent effects on tautomerism, and correctly indicate that the quantitative treatment outlined at the beginning of this section needs refinement.

Spinner and Yeoh (68TL5691, 71J(B)296) have criticised in stronger terms the quantitative treatment reported at the beginning of this section and state: "There is no unique correlation between pyridol content and electron withdrawing effect of a substituent either for 5- or 6-substituted...." Given that one accepts the laws of thermodynamics, the relation Eq. (2-2) is absolutely true; however, the estimation of the pK values by the Hammett treatment depends on the validity of the Hammett equation, which is of course an approximation. However, it seems clear to us that a better understanding of the rationale of tautomeric equilibria in the pyridone–hydroxypyridine series (and also for heterocycles in general) will become possible by refining relations of the Hammett type which relate pK_a values of individual tautomers to substituent parameters.

d. Hydrogen-Bonding Effects of Substituents

An alternative way in which hydroxy–oxo equilibria can be displaced in favour of the hydroxy form is by selective stabilization of this form by intramolecular hydrogen-bonding. There are considerable scattered examples in the literature, but there have been few systematic investigations.

Hydrogen bonding of OH to an *oxygen* atom is not usually sufficiently strong to tip the equilibrium: thus the diethoxycarbonyl derivative [**75**] exists in the solid (and probably therefore also in polar solvents) mainly in the oxo form [**75a**], although mainly as the hydroxy form [**75b**] in the nonpolar solvents CCl_4, $CHCl_3$ (66CB445). However, if the phenyl groups in [**75**] are replaced by 2-pyridyl (as in [**76**]) or by 2-quinolyl groups (as in [**77**]), then the greater electron-withdrawing effect assists the tautomeric swing, and these compounds now exist in the hydroxy forms shown, even in the solid state (65TL3175).

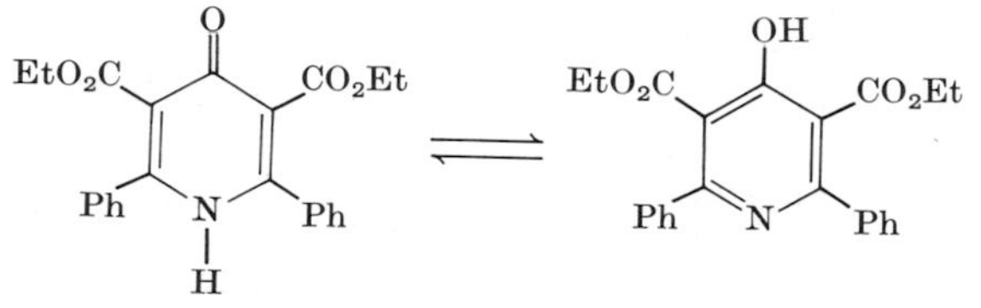

[**75a**] [**75b**]

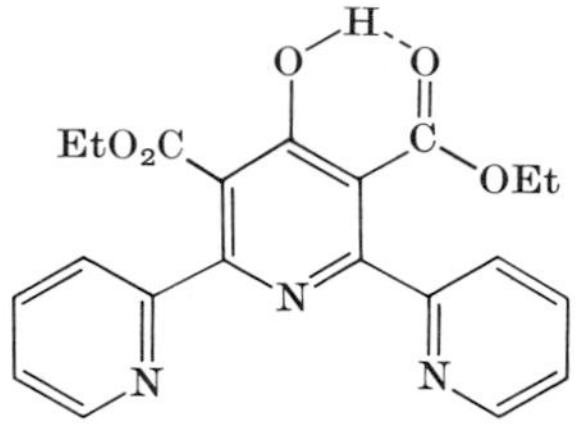

[**76**] 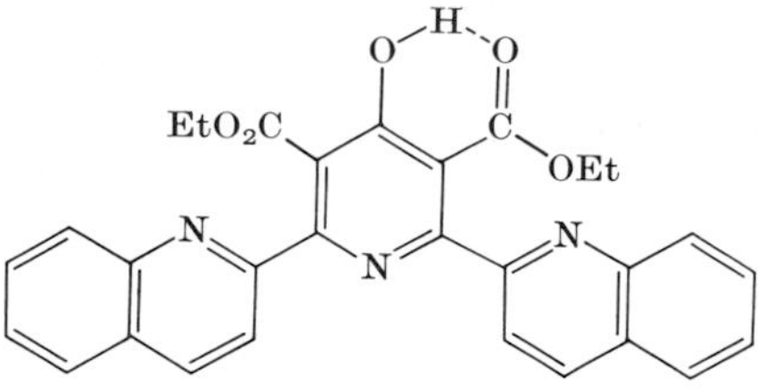 [**77**]

A similar reinforcement of hydrogen bonding and electronic effects is found in the 4-quinolone series. 2-Alkylthio-4-quinolone-3-carboxylate esters [**78**] exist as tautomeric mixtures with predominance of the enol forms [**78a**] in CCl_4 and of the oxo forms [**78b**] in MeOH (68J(C)2656).

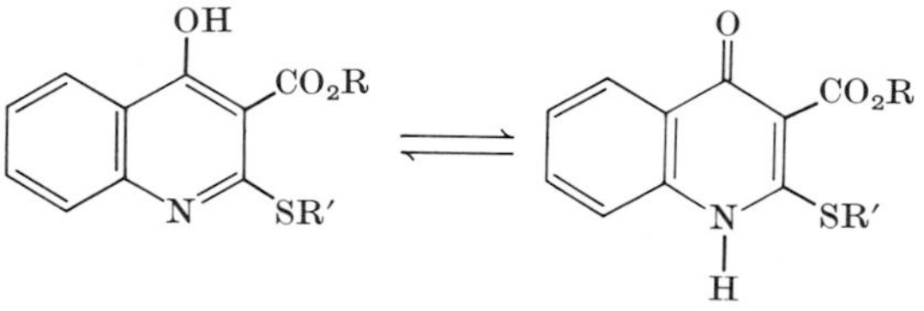

[**78a**] [**78b**]

Tautomerism in this series was also studied in the solid state by infrared spectroscopy and differential-scanning calorimetry (Section 1-6Cb) (68J(C)2656). The results showing the tautomers present under various conditions are collected in Table 2-3. In one case, the two tautomers [**78**, R = R′ = Me] may be obtained individually by crystallization

TABLE 2-3

TAUTOMERISM IN 2-ALKYLTHIO-4-QUINOLONE-3-CARBOXYLIC ESTERS (68J(C)2656)

Compound [**78**]	Stable form in solid	Solid crystallized from MeOH	Stable form in MeOH	Stable form in CCl_4
R = R′ = Me	OH	C=O	C=O	OH
R = Et, R′ = Me	OH	C=O	C=O	OH
R = Et, R′ = *n*-Pr	OH	C=O	C=O	OH
R = Et, R′ = *i*-Pr	C=O	C=O	C=O	OH
R = Et, R′ = CH_2Ph	OH	OH	C=O	OH

from appropriate solvents (light petroleum for the OH form, MeOH for the oxo form) (69T4649).

The influence of the 2-alkylthio group of [**78**] in swinging the tautomerism towards [**78a**] is demonstrated by 3-ethoxycarbonyl-4-quinolone itself [**79**] (68J(C)2656) and by 3-acyl-2-aryl-4-quinolones [**80**] (61JO2791, 65JO3033), which all show characteristic νNH and νC=O bands and thus exist predominantly as the oxo forms under all conditions studied.

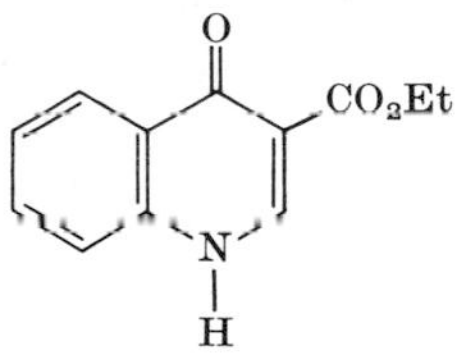

[**79**] 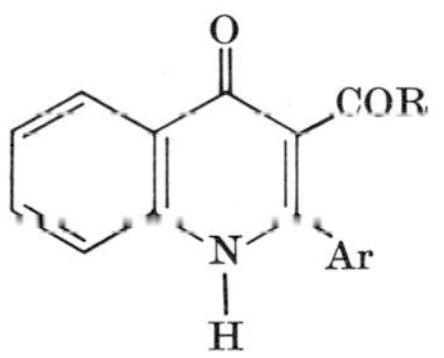 [**80**]

Hydrogen bonding of the OH group with a nitrogen atom is considerably more effective. Already in a five-membered chelate ring, UV and fluorescence spectral comparisons with 8-hydroxyquinoline and 4-quinolone suggest that 4-hydroxy-1,5-naphthyridine [**81**] changes from the predominantly oxo form [**81b**] in polar solvents to the enol form [**81a**] in nonpolar solvents (67AC877).

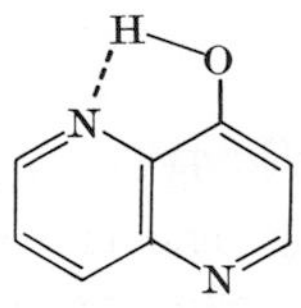

[**81a**] 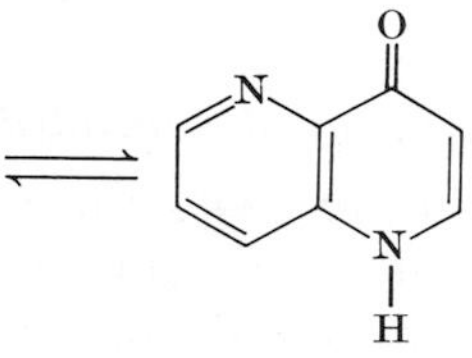 [**81b**]

Perhaps the most extensive investigation of the quantitative effects of hydrogen bonding on tautomerism concerns phenanthrolines of type [**82**]. It was shown (71J(B)2339) that, together with stabilization by intramolecular hydrogen bonding, destabilization of the pyridone form by electron repulsion between the π-electrons of the O atom and adjacent N plays an important role in stabilizing the hydroxy form [**82a**] over the oxo form [**82b**] in 4-substituted 10-hydroxy-1,7-phenanthrolines [**82**, R = H, Cl, OEt, NHPh]. Structures of type [**82c**] are

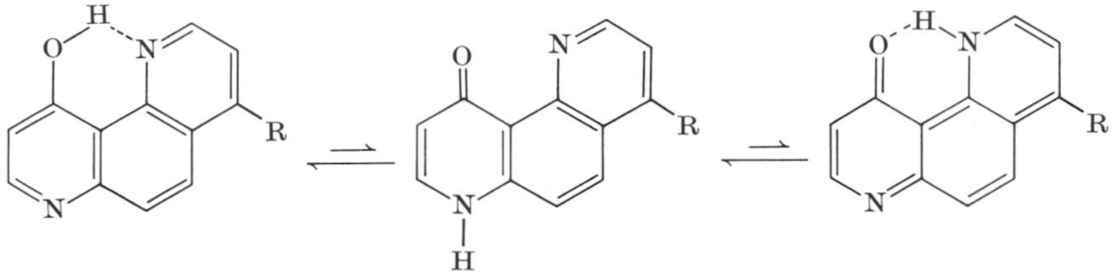

[**82a**] [**82b**] [**82c**]

expected to be of considerably higher energy. All these compounds exist predominantly in the OH form [**82a**] in the solid state and in both polar and nonpolar solvents as shown by UV and IR comparisons and by pK_a values (71J(B)2339). The potential 4-hydroxy analogue [**82**, R = OH] exists neither as [**82a**] nor as [**82b**], but in the dioxo form [**83**]. The equilibrium constant in favour of the hydroxy form [**82a**, R = Cl] was estimated as ca. 25 for the 4-chloro derivative by a comparison of the pK values and those of fixed forms with analogous data for simple 4-quinolone. Similar reasoning gives a pK_T = 6.4 for the dioxo form [**83**] to dihydroxy form [**82a**, R = OH] equilibrium (71J (B)2339).

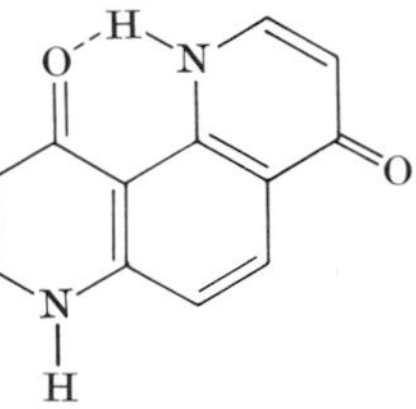

[**83**]

e. Strain Effects of Annelated Saturated Rings

In compounds such as [**84**], the ring size can exert a significant effect on the tautomeric equilibrium [**84a** $\rightleftharpoons$ **b**]: replacement of a five-membered ring by a six-membered one reduces the hydroxy to oxo ratio by a factor of ca. 10–30 in all solvents examined (from H_2O

to cyclohexane). Thus, for [**84**, $n = 1$], K_T = [**84a**]/[**84b**] = 3.0 and for [**84**, $n = 2$], K_T = [**84a**]/[**84b**] = 0.2 in EtOH. This behaviour has been attributed (68TL5691, 71J(B)279) to two distinct ring strain effects (conformational differences between five- and six-membered rings being considered to be of minor importance).

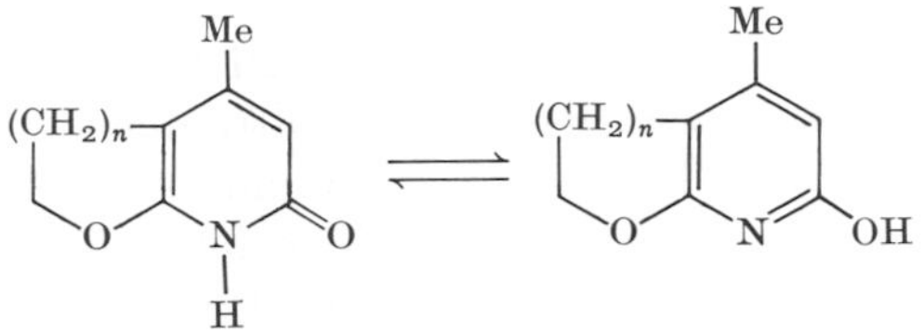

[**84a**] ⇌ [**84b**]

The first effect is "direct" Mills–Nixon–Brown ring strain, present in the five-membered ring. This is considered to be partially relieved by conversion to the hydroxy form [**84b**, $n = 1$] in which the order of the endocyclic bond is lower than for the oxo form [**84a**]. In the six-membered ring [**84**, $n = 2$], the effect works in the opposite direction.

The second is "*strain-induced electron withdrawal*" and is ascribed to the tendency of smaller rings to be more electron-withdrawing; 3-membered > 4 > 5, etc.

A detailed discussion (71J(B)279) leads the authors to conclude that it is the five-membered ring compound [**84**, $n = 1$] which is anomalous, whereas the six-membered ring shows more expected behaviour.

To separate the two main factors of direct strain and strain-induced withdrawal, the tautomeric behaviour of [**85, 86, 87**] were studied (71J(B)289). Direct ring strain was assumed to be equal in [**86**] and [**87**], but greater in [**85**]. Comparison of K_T and the individual

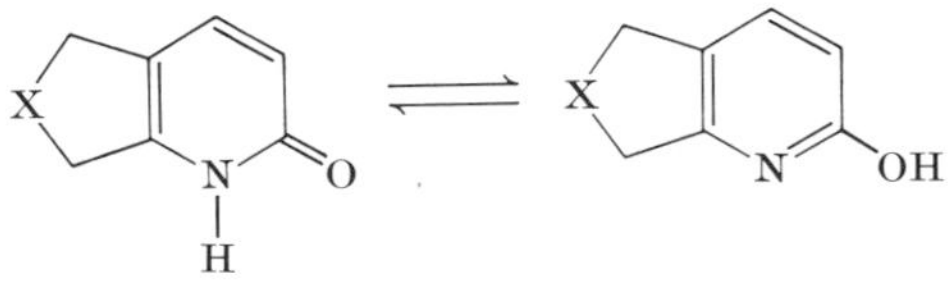

[**85a**]	X = O	[**85b**]
[**86a**]	X = S	[**86b**]
[**87a**]	X = SO_2	[**87b**]

pK values with those of the open-chain analogue [**88**] (and suitable fixed models), suggested that a third of the total effect is due to strain-induced electron withdrawal and two thirds to direct ring strain.

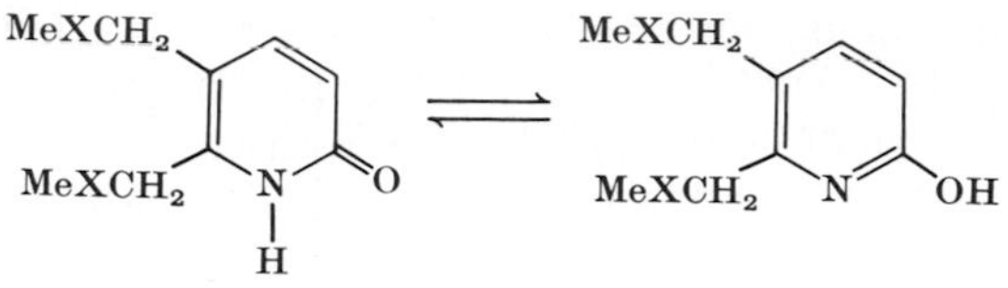

MeXCH$_2$ MeXCH$_2$ MeXCH$_2$ MeXCH$_2$ O OH N N H

[**88a**] [**88b**]

Ring strain effects in nitrogen-annelated systems have also been studied. The K_T was determined by the ultraviolet method in various solvents for several 6-hydroxy-7-azaindolines. The NH compound [**89**] exists predominantly in the pyridone form [**89b**] in ethanol, but K_T values of [**90b/90a**] = 2.97 and [**91b**]/[**91a**] = 0.26 were found for the *N*-substituted derivatives (66T3233, 67DA(177)592). Similar studies of the corresponding 6-hydroxy-7-azaindoles [**92**] gave [**92b**]/[**92a**] = 5.94 and [**93b**]/[**93a**] = 1.51 (67DA(172)118).

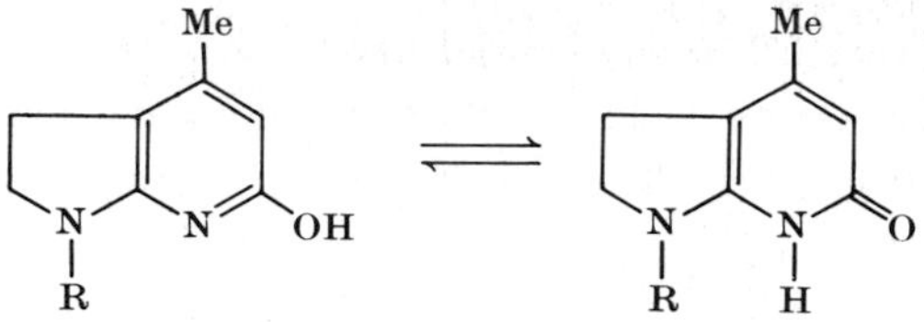

[**89a**]	R = H	[**89b**]
[**90a**]	R = *n*-Bu	[**90b**]
[**91a**]	R = Ph	[**91b**]

[**92a**]	R = H	[**92b**]
[**93a**]	R = *n*-Bu	[**93b**]

Compound [**93/1**] has been shown to exist as such by NMR (74J-(PI)1531).

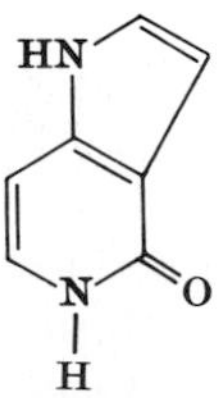

[**93/1**]

f. Effect of Partial Bond Fixation

In isoquinoline, partial bond fixation occurs in the sense of [**94**], just as in naphthalene. This factor is clearly expected to favour the 3-hydroxyisoquinoline tautomer [**95a**] over the 3-isoquinolone form [**95b**]. At the time of the last review (I-352), the experimental situation was confused, both the oxo and hydroxy forms being favoured: a little later, further fragmentary work appeared concerning the 6,7-dimethoxy derivative (64CB667).

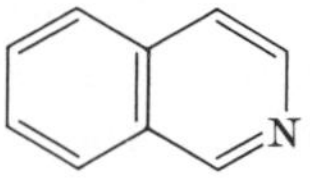

[**94**] 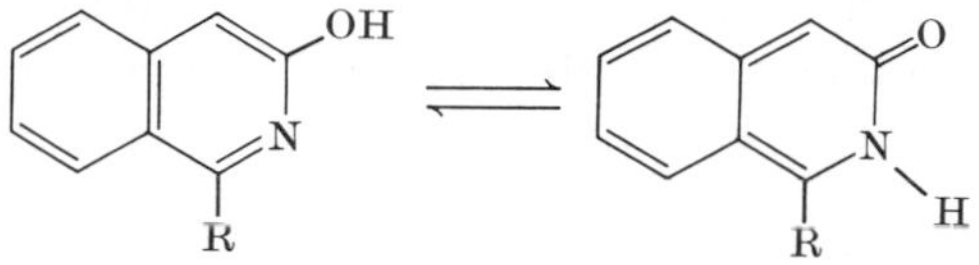 [**95a**] R = H [**95b**]
[**96a**] R = Me [**96b**]

Careful UV comparisons with fixed forms (67J(B)590) now show definitively that 3-isoquinolone [**95**] and the 1-methyl [**96**] and 6,7-dimethoxy-1-methyl derivative possess finely balanced tautomeric equilibria with the isoquinolin-3-ol form [**95a**] predominating in most nonhydroxylic solvents and the 3-isoquinolinone form [**95b**] in H_2O, whereas in EtOH and $CHCl_3$ the amounts of the two forms are comparable. IR and NMR work supports the above results (71T4653), as do theoretical calculations (70JA2929). Similar general conclusions based on UV evidence were reported later (69J(C)1729) without reference to (67J(B)590).

1-Phenyl- [**97**] and 1,4-diphenyl-3-hydroxyisoquinoline exist predominantly in the hydroxy form in ethanol, whereas the isoquinolone form predominates in the 1-methyl analogues (69J(C)1729). 1-Chloro-3-hydroxyisoquinoline [**98**] exists in the hydroxy form in all solvents, as

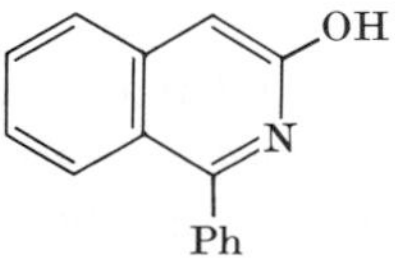

[**97**] 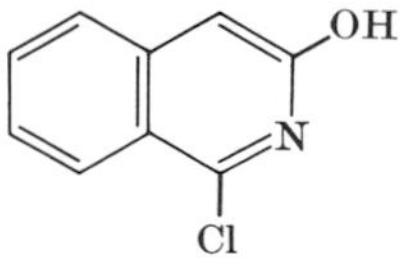 [**98**]

shown independently by two groups by UV and IR comparisons (67J(B)590, 68AG(E)464) and in the solid state as shown by X-ray crystallography (74CX(B)1146).

g. Effects of Phase

In solvents of varying dielectric constant the tautomeric equilibrium can change considerably as shown in Table 2-1 (Section 2-3Bb), and this is especially noticeable when the equilibrium is already finely balanced in aqueous solution.

The general preference of the oxo form in pyridone–hydroxypyridine equilibria is due to the contribution to the resonance hybrid of charge-separated canonical forms of type [**63d**] (see discussion in Section 2-3Ac). The stabilization afforded by such structures should be less in media of low dielectric constant. Furthermore, the hydrogen bonding ability of the solvent plays an obvious role as hydrogen bond donors stabilize the oxo form, and hydrogen bond acceptors stabilize the hydroxy form.

Some examples of the strong dependence on solvent in substituted 2- and 4-pyridones will now be given. While the oxo form of 6-chloro-4-methyl-2-pyridone predominates in water (Table 2-1, Section 2-3Bb), mainly the hydroxy form exists in various other media: 67% in MeOH, 56% in $CHCl_3$, 96% in DMSO and 95% in C_6H_{12} (70CB398; see also 70DA(192)1295). Similar strong solvent effects, in which K_T changes by a factor of ca. 300–500 from H_2O to C_6H_{12}, were also found for 6-methoxy-2-pyridone (71J(B)279), 6-amino-2-pyridone (67DA(177)592), 2,3-dihydro-6-hydroxy-4-methylfuro[2,3-*b*]pyridine (67DA(176)613,71J (B)279), 3,4-dihydro-7-hydroxy-5-methylpyrano[2,3-*b*]pyridine (71J(B) 279), 2,3-dihydro-6-hydroxy-4-methylpyrrolo[2,3-*b*]pyridine (66T3233), 6-hydroxy-7-azaindoles (67DA(172)118) and 3-hydroxyisoquinoline (67J(B)590).

Recently (68TL2767) a quantitative relationship was established between heteroaromatic tautomeric equilibrium constants and solvent polarity Z values (Fig. 2-1) for solutions in EtOH:H_2O, MeOH:H_2O, ethylene glycol, acetonitrile, $CHCl_3$ and isooctane. There is a definite trend to smaller slope for compounds which tend to exist more in the oxo form. This reasoning indicates that 2- and 4-pyridone could themselves change over in isooctane.

The weakness of any use of a single solvent parameter is the implicit assumption that all solute–solvent interactions are proportional to those between the compound used to set up the scale: we cannot expect to understand more fully the effects of solvent on tautomeric equilibria until solvent–solute interactions are better understood. A possible method lies in the development of multiparameter treatments, (cf. 71J(B)460). The tautomeric equilibrium between 6-chloro-2-pyridone and 6-chloro-2-hydroxypyridine is little affected by anionic and cationic surfactants (75G431).

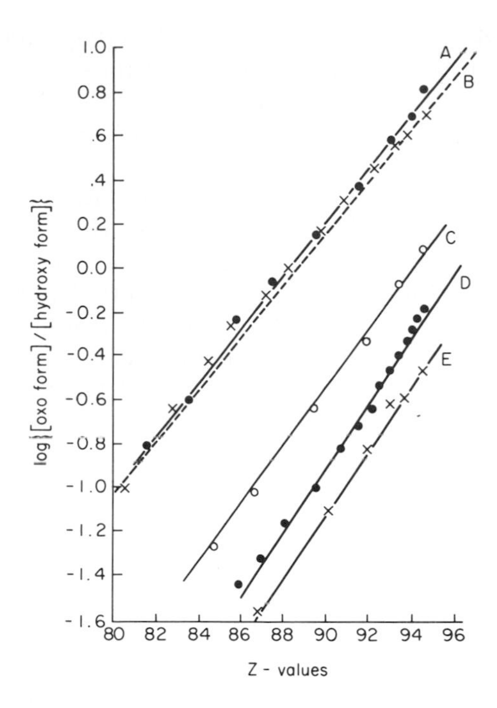

FIG. 2-1. Plots of log [oxo form]/[hydroxy form] measured in ethanol water mixtures against Z values for (A) 6-chloro-2- (B) 2,3,5-trichloro-4- (C) -2-chloro-4-pyridone; (D) 3-hydroxypyridine; and (E) 2,6-dimethoxycarbonyl-4-pyridone. The lines shown for each compound were fitted by the least squares method; the slopes are as follows: Curve A, 0.125; B, 0.116; C, 0.133; D, 0.151; E, 0.147.

There has been considerable confusion regarding the vapour phase equilibrium of 2-pyridone–2-hydroxypyridine. Initial work by Russian authors (65DA(164)584) suggested that 2-pyridone existed in the gas phase in comparable proportions of the hydroxy and oxo forms because both the νNH and νOH could be detected. Mass spectroscopic data appeared to agree with these assessments (72OM823). However, subsequently (69DA(189)326) the same Russian group reported that the apparent K_T values which they had previously determined were not those applying to systems at equilibrium. For the tautomerism of the 2-pyridone derivative [**85**, $n = 2$] at 370° ca. 20% of the hydroxy form was found at equilibrium.

Very recently the equilibrium ratio at 120–140° of the simple system 2-hydroxypyridine-2-pyridone was determined as 2.5 ± 1.5 by UV (comparison with the fixed forms) (73JA1700) and about 10 for equilibrium [**84**, $n = 2$] in favour of the OH form. The authors suggest that the discrepancy with earlier work (69DA(189)326) is probably due to extensive decomposition at higher temperature.

An interesting IR study (72T5859) of the tautomerism of chlorinated 2-pyridones isolated in an argon matrix at low temperature (20°K) is unfortunately only semiquantitative, because of lack of knowledge of the absorption coefficients of the tautomeric compounds examined. 3,5,6-Trichloro-2-pyridinol exists predominantly in the hydroxy form and that the hydroxy/oxo ratio decreases in the order 3,5,6-trichloro ≥ 6- > 5- > 4- > 3- monochloro-2-pyridinols. This variation is correlated with increase in the NH acidity (cf. 67J(B)758) as the hydroxy to oxo ratio increases.

h. Temperature Effects

The temperature variation of tautomeric equilibrium for several 2-pyridones in various solvents was obtained by Russian workers using UV spectroscopy, and the thermodynamic quantities were calculated by Arrhenius plots (70DA(192)1295). Increase in temperature decreases the proportion of the oxo form in the more polar solvents while the reverse is true in the less polar solvents.

Conversion of the hydroxy into the oxo form is exothermic ($\Delta H < 0$) in water, the reaction becomes less exothermic going to less polar solvents and is endothermic ($\Delta H > 0$) in dioxan. However, peculiar behaviour is found for mixed solvents. A recent publication discusses further the thermodynamic quantities involved (72J(PII)1295).

i. The Tautomerism of Pyridones: Summary and Conclusions

Qualitatively, our picture of pyridone–hydroxypyridine tautomerism is clear: the influences set out in Table 2-4 can all be rationalized in an empirical manner.

Quantitatively, we are still groping for more precise treatments, although promising progress has been made in dealing with the electronic effects of substituents and solvent polarity.

C. Pyridone Cations (I-352)

The controversy concerning the structure of the protonated form of pyridones (I-352) is now settled in favour of *O*-protonation. The IR spectra of hexachloroantimonates of 2-pyridone (63CI1353) and of 2- and 4-pyridone (70JH479) and various substituted derivatives

TABLE 2-4

INFLUENCES FAVOURING OH FORMS IN PYRIDONE–HYDROXYPYRIDINE TAUTOMERISM

Section	Effect
2-3Bc	Electron attracting substituents near nitrogen atom
d	Substituents hydrogen-bonding to OH group
f	Benzene ring annelation causing low N—CO bond order
e	Fusion with a five-membered or smaller saturated ring
g	Nonpolar solvents or gas phase
h	High T in polar or low T in nonpolar solvents

(66J(B)996), and of *N*-methyl-2-pyridone salts (65CC749) all indicate oxygen protonation (see also 63JO2883, 69AJ2595, 71PM(4)359).

NMR spectra for 4-pyridone cations in sulphuric acid (64R186; see also I-353) also indicate oxygen protonation, and 2-pyridone cations give similar results (63J753). Analysis of the NMR parameters of 2-pyridone and related compounds is in full agreement (69PH2465) as expected for *O*-protonation; the coupling constants J(3H4H) and J(5H6H) of the 2-pyridones are decreased on conversion into cations, whereas the vicinal coupling constants of the normal 2-substituted pyridines are increased. UV spectra of 2- and 4-pyridone cations also agree (63J3855, 66J(B)996). 8-Ethoxycarbonyl-2-methoxycarbonyl-4-quinolone is protonated at the 4-oxo group (73CI182).

X-ray structures of 2-hydroxypyridinium and 2,6-dihydroxypyridinium chloride and a neutron diffraction study of the former are also available (69TL5219).

3- and 5-Amino-2-pyridone are protonated first at the amino group, but 4- and 6-amino-2-pyridone and 2- and 3-amino-4-pyridone are protonated first at the pyridone oxygen atom (71J(B)1425).

2,3-Dihydro-4-pyridones are probably protonated at least in part on nitrogen (69BJ2690).

2-Pyridone forms complexes with metal ions by coordination at the oxygen atom (69R1139).

D. β-HYDROXYPYRIDINES (I-353)

It was already clear at the time of the previous review (I-353) that these compounds exist as zwitterion/hydroxy mixtures in H_2O, and mainly in the hydroxy form in nonpolar solvents. No further substantial advances have been made, but many details have been filled in, and MO calculations are available (68TE379, 69TC(14)221, 70JA2929, 70TC(16)243, 70TC(16)316, 71JL(10)245, 72KG197).

Paoloni and his group have suggested that canonical forms of type [**101c**] should be considered as contributing to [**101b**] (69TC(14)221).

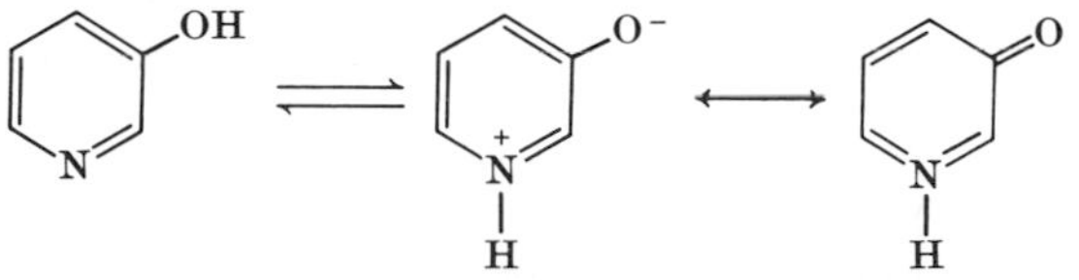

[**101a**] [**101b**] [**101c**]

This conclusion has led to some controversy in the literature (73OR551, 74OR469).

NMR (65TE28, 70IZ25) indicates an equilibrium between [**101a**] and [**101b**] for several β-hydroxypyridines in water and the predominance of form [**101a**] in dioxan solutions (cf. I-356). Many substituted 3-hydroxypyridines have also been investigated (69TE247, 71ZO2520, 72TE224, 73KG810). The tautomerism of 3-hydroxyquinoline has been studied in the excited (68J(A)3051) as well as the ground state (68J(A)3051, 71KG1540, 72KG191).

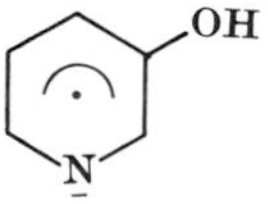

[**101/1**]

Application of the Hammett equation gives some indication that the hydroxy rather than the zwitterionic form predominates in aqueous solution for 4-hydroxyisoquinoline (65JO3341), but further study is still needed here (cf. I-345).

Absorption and emission spectra of aqueous solutions of 3-hydroxyquinoline (63J4897) show that in the first excited state the zwitterionic form predominates even more than in the ground state.

Although 2- and 4-pyridone anion radicals exist in the keto structures [**64/2a** or **b**] and [**64/3**] (cf. Section 2-3Ba), the corresponding 3-hydroxypyridine radical exists in the hydroxy structure [**101/1**] (74JA2342). Protonation and deprotonation equilibria for pyridoxamine were studied by NMR (68T4477), and evidence was found for [**102a**] $\rightleftarrows$ [**102b**]. Equilibrium [**103a**] $\rightleftarrows$ [**103b**], similar to that previously described (I-355) for pyridoxal, has now been demonstrated for the pyridoxal monoanion.

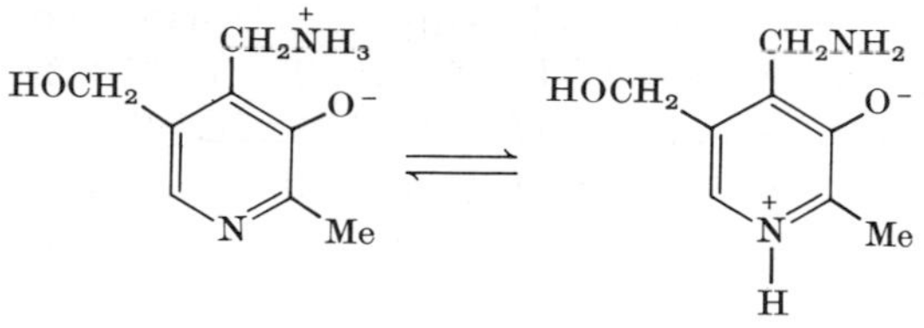

[**102a**] [**102b**]

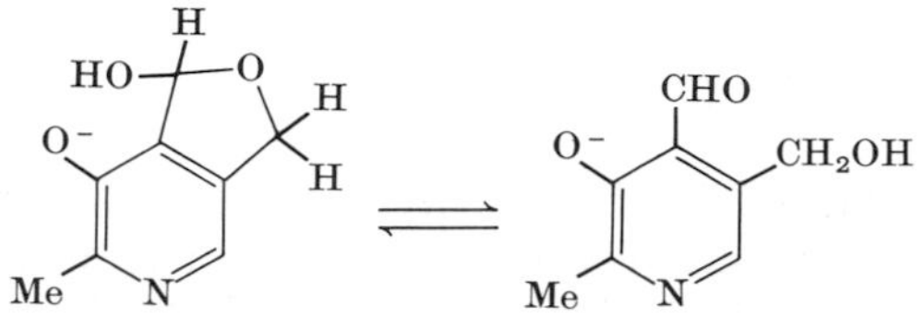

[**103a**] [**103b**]

A considerable contribution of a quinonoid tautomer [**104b**] to the structure of 2,2′-bis(6-methyl-3-pyridinol) [**104a**] has been deduced from an X-ray diffraction study (71JA5402). The high fluorescence of [**104**] is ascribed to tautomeric shifts analogous to those proposed for salicylanils (68JP2092).

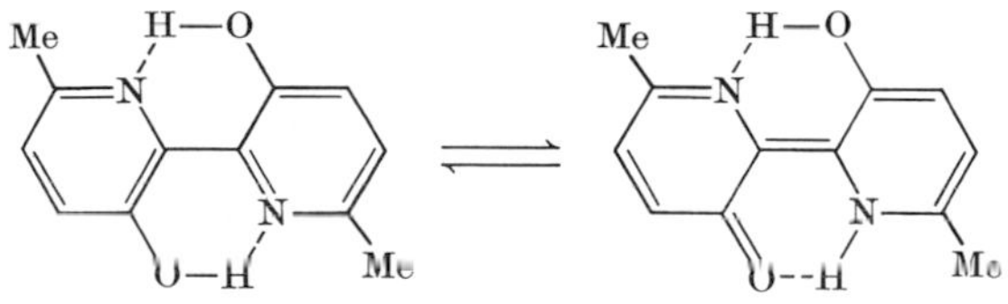

[**104a**] [**104b**]

E. Hydroxypyridones (I-356)

Considerable work has appeared since the last review: we now classify the material according to the orientation of the oxygen functions.

a. Oxygen Functions in the 2,4-Positions (I-356)

The previous review reported (I-356) that 4-hydroxy-2-pyridone [**105b**] predominated in the equilibrium [**105**]. Likewise, the structure [**106b**] is now demonstrated for 4-hydroxy-6-methyl-2-pyridone [**106**] by UV comparisons in EtOH with model compounds (70JH389) as previously suggested from other spectral evidence (64M1247) for [**106**] and many derivatives. IR and NMR (DMSO) spectra (70JH389) and calculations (70JA2929) are in agreement. X-ray (73CX(B)61) shows that the 4-hydroxy-2-oxo tautomer is found in the crystalline state for

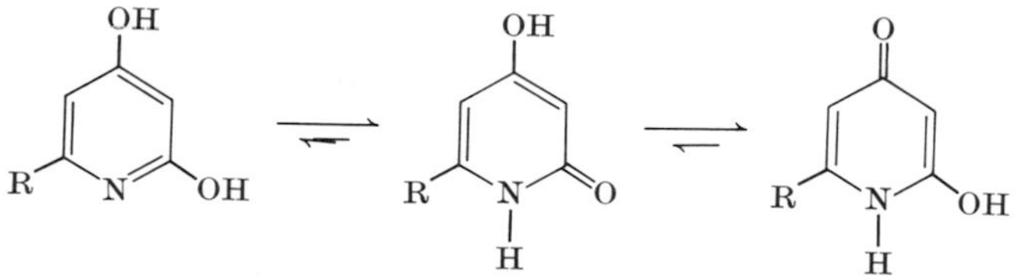

[**105a**] R = H [**105b**] [**105c**]
[**106a**] R = Me [**106b**] [**106c**]

3-deazauridine; it also predominates in 50% EtOH on NMR evidence (70JH323).

Compound [**107**] has been assigned the 4-hydroxy-2-pyridone structure shown on deshielding effects on the C(3) quinoline ring hydrogen in the NMR (71JO354), but the alternative 2-hydroxy-4-pyridone structure is not excluded on this evidence.

Dubious chemical evidence was previously reported (see I-357) for the coexistence of 4-hydroxy-2-quinolone and 2-hydroxy-4-quinolone, but the IR indicates only the latter (67M100; for a different point of view see 67BF3367). The 4-hydroxy-2-oxo structure [**108**] in DMSO

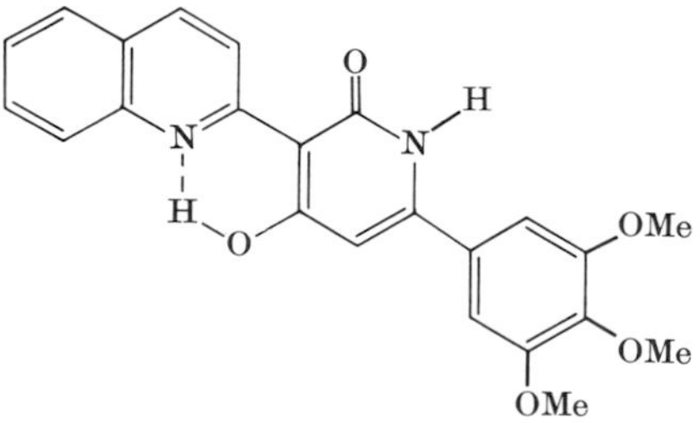

[**107**] 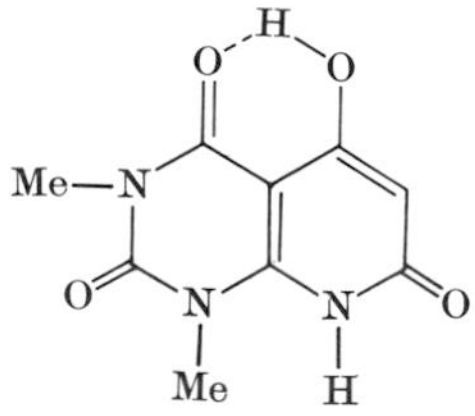 [**108**]

was assigned on the basis of NMR comparisons (64JO219). 8-Aza derivatives of 2-hydroxy-4-quinolone retain the structure of the parent (73CB3533).

b. Oxygen Functions in the 2,6-Positions: Glutaconimide (I-357)

Five tautomeric forms [**109a–e**] exist for glutaconimide. Only preliminary investigations on the tautomerism of [**109**] were reported at the time of the previous review (I-357), but considerably more work has now been done. No evidence for the existence of forms [**109d**] and [**109e**] has been found, but the other three forms are all implicated in the tautomeric equilibrium. Of special interest is the appearance of the nonaromatic form [**109c**] as a significant contributor. UV and pK_a comparisons with fixed derivatives suggest that in H_2O the energy

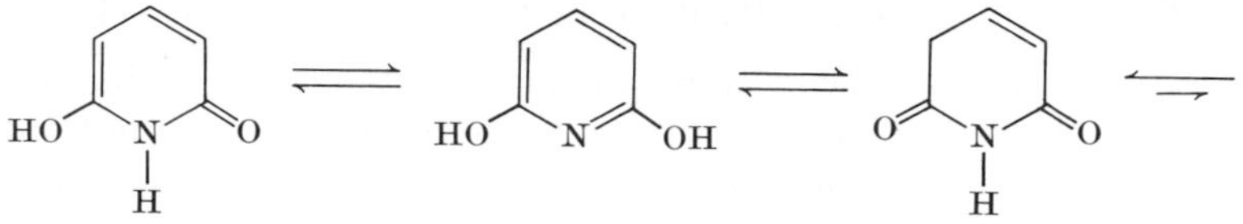

HO N O H [109a] ⇌ HO N OH [109b] ⇌ O N O H [109c] ⇌

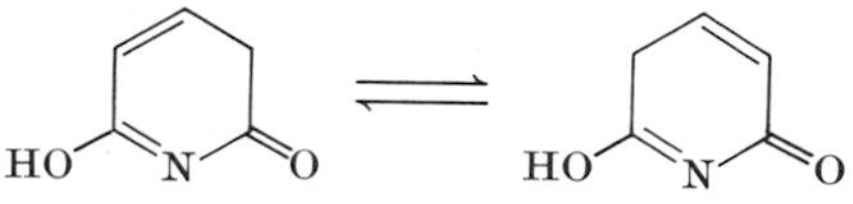

HO N O [109d] ⇌ HO N O [109e]

differences between forms [**109a**], [**109b**] and [**109c**] of glutaconimide are small. A mixture of [**109a**]:[**109b**]:[**109c**] in the approximate ratio of 60:25:15 was proposed (66J(B)562). However, further studies indicate that the greater base weakening effect of methoxy as compared to hydroxyl may have caused the diol component [**109b**] to have been overestimated: this contention was supported spectroscopically for H_2O and EtOH solutions (71AJ2557, 71J(B)279). In dioxan solution the diol form predominates, and the ratio 2% [**109a**]: **85%** [**109b**]: 13% [**109c**] was deduced (71AJ2557). These results contrast with those for 2-hydroxy-6-methoxypyridine for which K_T is 1.1 in EtOH and 19 in dioxan favouring the OH form (Table 2-1; Section 2-3Bb) (71J(B)279). The replacement of the 6-methoxy group by a 6-hydroxy group in [**109**] reduces significantly the pyridinol content (71AJ2557); differential hydrogen-bonding and steric effects evidently cause OMe to disfavour the attachment of a proton to the adjacent nitrogen atom more than does an OH group.

For *N*-methylglutaconimide [**110**], pK_a comparisons indicate 40% of [**110a**] and 60% of [**110b**] (66J(B)562); other work gives 67% of [**110a**] in H_2O from UV data and the proportion of [**110a**] falls to 15% in EtOH and 35% in dimethyl sulphoxide, whereas [**110a**] largely predominates in dioxan (98.8%) and chloroform (99.8%) (71AJ2557).

IR spectra of glutaconimide and of its *N*-methylated derivative in the solid state are interpreted in terms of the hydroxy–oxo forms [**109a**], [**110b**] linked by short unsymmetrical hydrogen bonds (71AJ 2557).

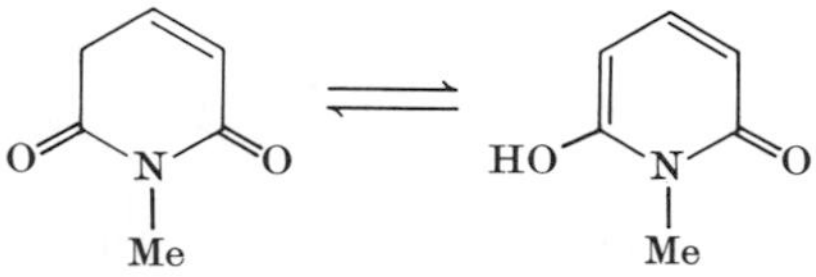

O N O Me [110a] ⇌ HO N O Me [110b]

c. Oxygen Functions in the 2,6-Positions: Other Neutral Species (I-357)

Considerable work has been done on substituted glutaconimides. However, the general pattern of tautomerism in this series is not yet clear as much of the work has not considered the full range of possible forms, and/or has not taken into account the strong solvent dependence of the tautomeric equilibria in this series.

Spectral studies suggest that citrazinic acid esters [**111**] exist in the monohydroxy form [**111a**] in MeOH but in the dioxo form [**111b**] in nonpolar solvents (63CZ1408). The IR spectra (solid) of citrazinic acid [**112**] were also studied, but no choice was made between the oxo–hydroxy [**112a**] and the diol form [**112c**] (63CZ1408).

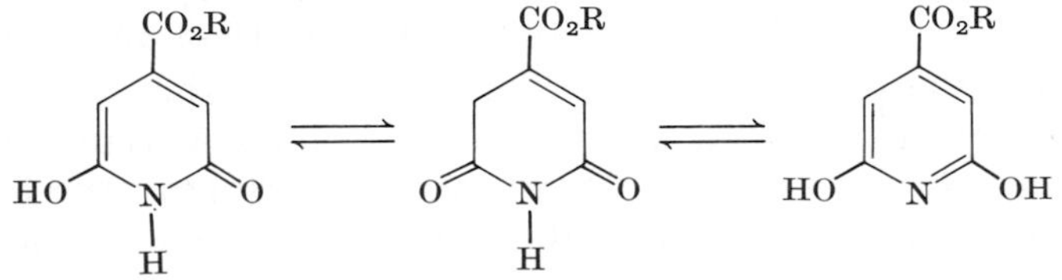

[**111a**] R = alkyl [**111b**] [**111c**]
[**112a**] R = H [**112b**] [**112c**]

The dioxo form of the 3,5-dicarboxylic esters [**113**] was suggested for the solid state and in MeOH and DMSO from IR and UV comparisons (63CZ1625).

A dione structure was tentatively assigned to the spiro-cyclopropane derivative [**114**] on spectral data (69J(C)1678).

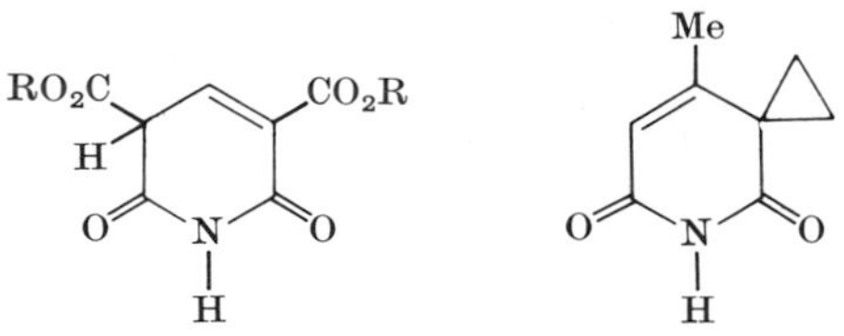

[**113**] R = Me or Et [**114**]

For the 4-ethoxy derivative [**115**], the dioxo form [**115b**] appears to be preferred to the oxo–hydroxy form [**115a**] on IR evidence (phase not

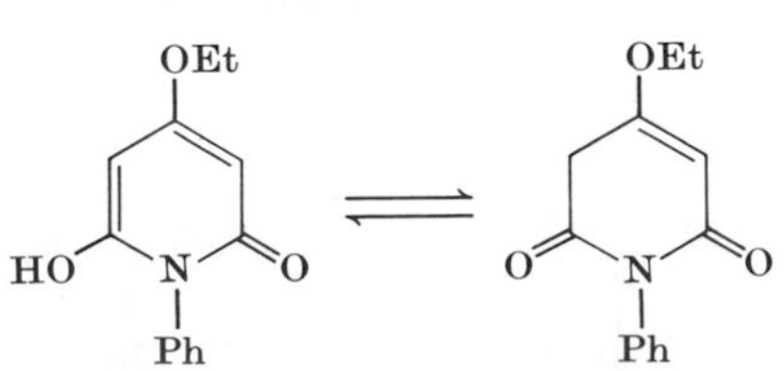

[**115a**] [**115b**]

given) (63J3069) although in similar compounds the existence of an equilibrium mixture in $CHCl_3$ was previously postulated (cf. I-357).

IR (KBr, also some in $CHCl_3$ and MeOH) and UV (H_2O and EtOH) studies of glutazine [**116**, R = H] and its derivatives [**116**, R = CN, $CONH_2$, CO_2Et] are best explained by the coexistence of the oxo–hydroxy [**116a**] and dioxo forms [**116b**] (67M1763). The bicyclic analogue [**117**] has been formulated as written (65M2046).

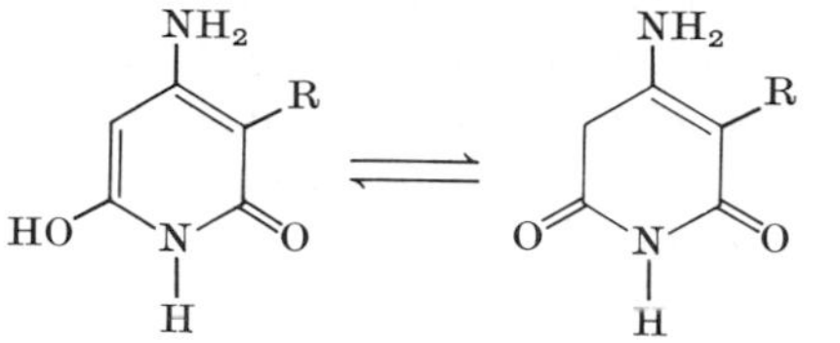

[**116a**] [**116b**] 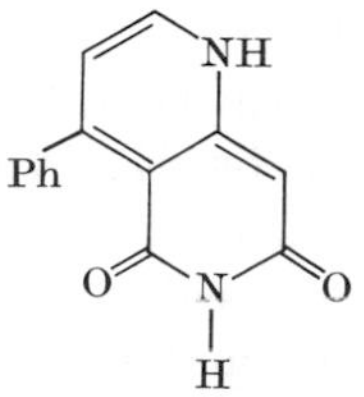 [**117**]

Unsymmetrically substituted 2,6-dihydroxypyridines are more complicated systems, because two tautomers are now possible for each of the forms corresponding to [**109a**] and [**109c**] in the parent and symmetrically substituted derivatives. UV comparisons with fixed forms suggest that [**118**] exists predominantly in the dioxo form [**118a**] in dioxane, but in water (pH 3) as 80% [**118a**] together with the oxo-hydroxy forms [**118b**] and/or [**118c**], while in EtOH [**118b**] and/or [**118c**] predominate. No diol form [**118d**] was detected in H_2O, EtOH or dioxan (71AJ2557).

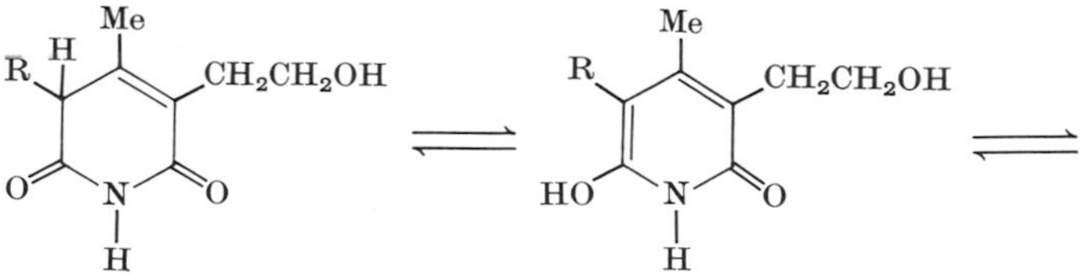

[**118a**] R = H
[**119a**] R = CN

[**118b**]
[**119b**]

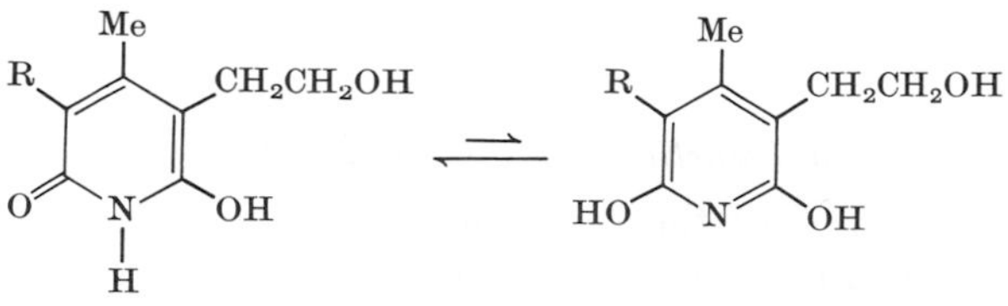

[**118c**]
[**119c**]

[**118d**]
[**119d**]

By contrast, the cyano derivative [**119**] shows no evidence for the existence of the dioxo form [**119a**] in H_2O, EtOH or dioxan. In H_2O and EtOH the oxo–hydroxy forms [**119b**] and/or [**119c**] predominate, while in dioxan 75% of the diol form [**119d**] is present (71AJ2557). Electron withdrawal by the CN group probably stabilizes the completely conjugated forms relative to the classical imide system.

The trifluoromethyl derivative [**120**] is an interesting example of the solid phase interconversion of two tautomeric forms. The transformation on IR irradiation in the solid state first reported as from one oxo–hydroxy form [**120a**] into the other [**120b**] (65JO3377) was later reinterpreted on IR evidence as a diol [**120c**] to oxo-hydroxy [**120b**] or [**120a**] conversion (71AJ2557); the spectra are typical of a 6-hydroxy-2-pyridone spectrum for the product and of a dihydroxypyridine for the starting material.

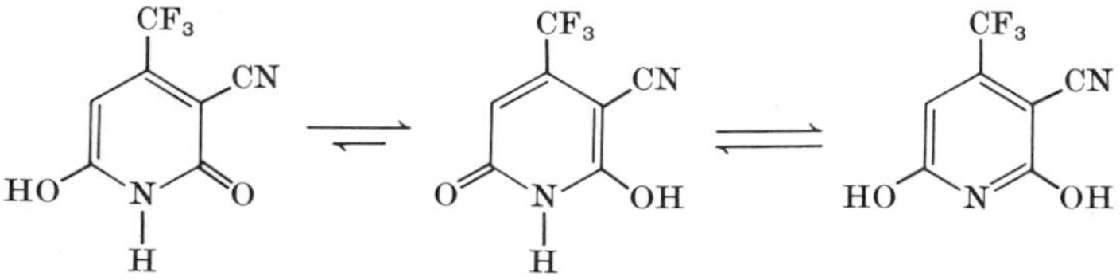

CF_3 CN HO N H O ⇌ CF_3 CN O N H OH ⇌ CF_3 CN HO N OH

[**120a**] [**120b**] [**120c**]

A review on azaquinones, including many potential tautomeric compounds derived from 2,6-dihydroxypyridine, has appeared (73-AG(E)139).

The earlier view based on IR and UV evidence (I-358) that homophthalimide [**121**] exists in the dioxo form has been reinforced by NMR evidence from the CH_2 singlet at 4.04 δ in DMSO (70BF1991), by further IR data (67BF3367) and by X-ray crystallography (74CX-(B)1146).

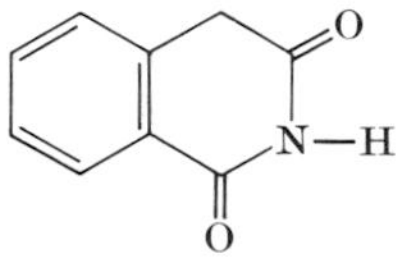

O N—H O

[**121**]

d. Oxygen Functions in the 2,6-Positions: Anions and Cations

The UV of the glutaconimide anion together with pK_a values for proton loss demonstrated structure [**122a**], which is of the same type as the anion of the *N*-methyl derivative [**123**]. For [**122a**]/[**122b**], a pK_T of ca. 5 was estimated (66J(B)562).

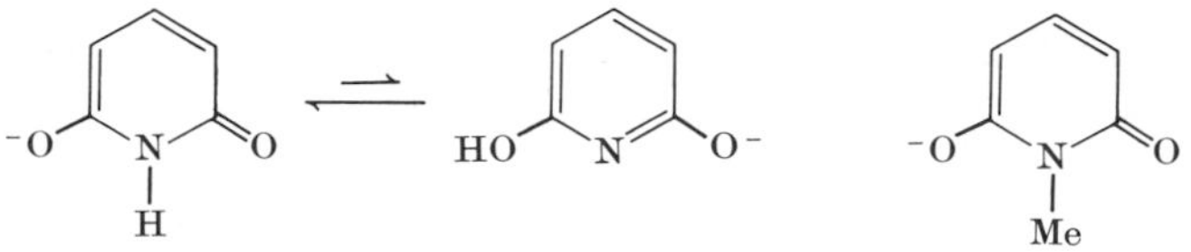

[**122a**] [**122b**] [**123**]

The UV spectra of the glutaconimide, *N*-methylglutaconimide and 6-methoxy-2-pyridone cations [**124**, R, R′ = H or Me] indicate that they all possess structures [**124a**]. 3,3-Dimethylglutaconimide gives a cation of type [**124b**] (66J(B)562).

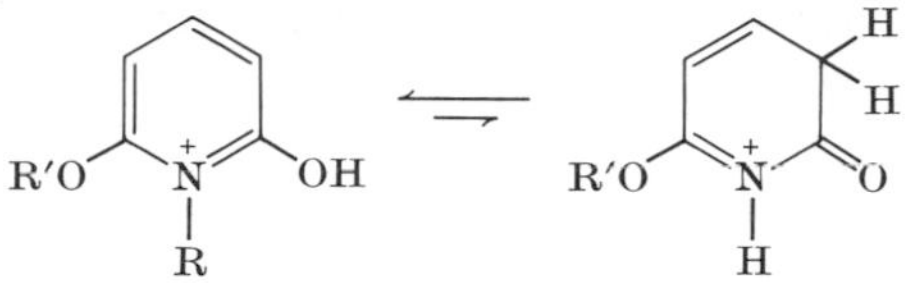

[**124a**] [**124b**]

e. Oxygen Functions in the 2,3- and 2,5-Positions and Ring–Chain Tautomerism

3-Hydroxy-2-pyridone exists as the oxo–hydroxy form [**125a**] in preference to the diol form [**125b**] in the solid and in EtOH, as shown by νOH at 3270 cm^{-1}, νNH at 3150 cm^{-1} and νC=O at 1660 cm^{-1} and by the UV similarity to 3-methoxy-2-pyridone (68CT1466). This has been confirmed by UV comparisons, and the cation shown to be [**126**]: the work was extended to show that 2,5-dihydroxypyridine exists as [**127**] (73KG60; cf. also 73KG56).

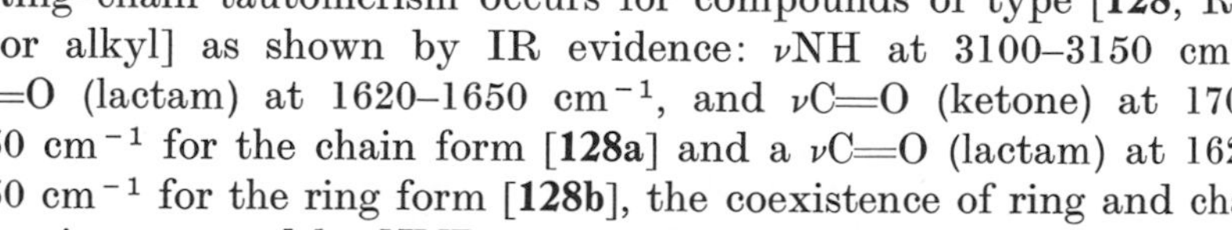

[**125a**] [**125b**] [**126**] [**127**]

Ring–chain tautomerism occurs for compounds of type [**128**, R = H or alkyl] as shown by IR evidence: νNH at 3100–3150 cm^{-1}, νC=O (lactam) at 1620–1650 cm^{-1}, and νC=O (ketone) at 1700–1750 cm^{-1} for the chain form [**128a**] and a νC=O (lactam) at 1620–1650 cm^{-1} for the ring form [**128b**], the coexistence of ring and chain forms is supported by NMR spectra: NH at 8.55 δ in the chain form [**128a**] and aliphatic OH at 6–7 δ in the ring form [**128b**]. For [**128**, R^3 = H] the ring form [**128b**] is preferred but for ketones [**128**, R^3 = alkyl] the chain form [**128a**] is the preferred one. However for R^2

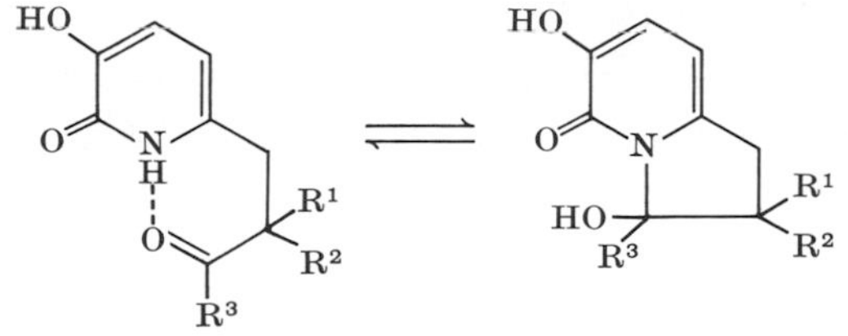

HO O N H O R¹ R² R³ ⇌ HO O N HO R³ R¹ R²

[**128a**] [**128b**]

larger than n-C_5, steric hindrance appears to favour the chain form [**128a**]. When R^2, R^3 = $-(CH_2)_n-$ the chain form [**128a**] is preferred except for $n = 4$ (68CT1466, 69CT425).

f. Oxygen Functions in the 2,3,6-Positions

NMR shows that 2,3,6-trihydroxy-5-methylpyridine [**129**] exists in DMSO mainly in the dihydroxy form [**129a**] with K_T [**129b**]/[**129a**] = 0.4. However dioxo forms of type [**129b**] predominate for the corresponding unsubstituted, 4-methyl and 4,5-dimethyl derivatives for

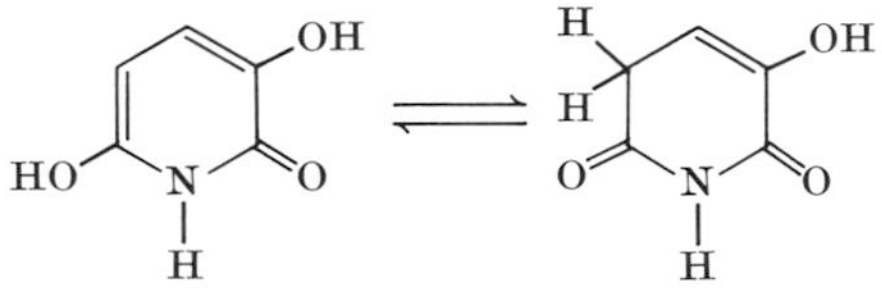

OH HO N O H ⇌ H H OH O N O H

[**129a**] [**129b**]

which K_T = 1.7, 4.2 and 2.0, respectively; UV and IR spectra agree. Equilibration between these tautomers in MeOH requires 80–100 minutes (68CB2679, 73AG(E)139).

Analogous compounds in the quinonoid oxidation state are known. Thus 4-chloro-5-methylamino-2,3,6-pyridinetrione [**129/1**] was shown by X-ray analysis to exist as pyridinetrione [**129/1a**] rather than as a hydroxypyridine tautomer [**129/1b**] (73CX(B)1971).

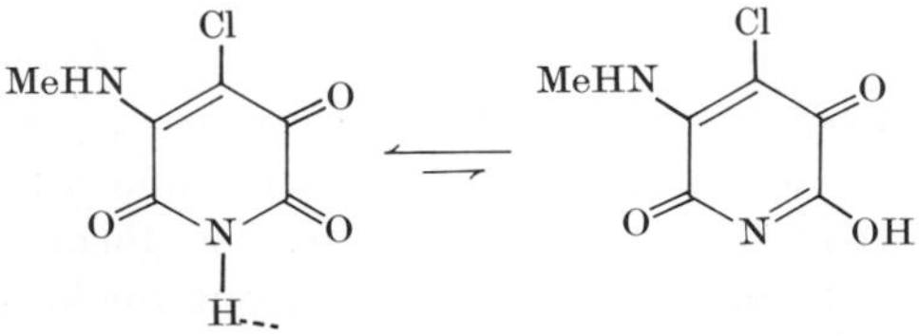

Cl MeHN O O N O H... ⇌ Cl MeHN O O N OH

[**129/1a**] [**129/1b**]

g. Oxygen Functions in the 2,4,6-Positions

Little is known about compounds in this class. Derivatives of the type [**130**] (R = aryls) are written as oxo–dihydroxy compounds (66T455) from the similarity of the UV and IR to "true" 2-pyridones.

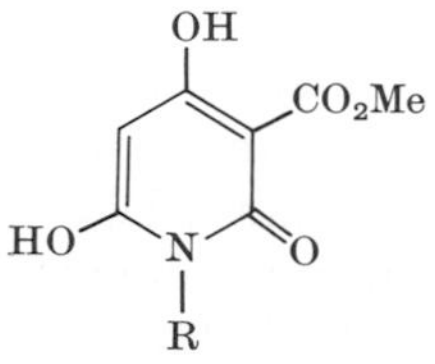

OH, CO_2Me, HO, N, O, R

[**130**]

F. HYDROXYPYRIDINE 1-OXIDES (1-HYDROXYPYRIDONES) (I-359) AND 1-IMIDES

The previous review (I-359) summarized the evidence which clearly demonstrates that for 2- and 4-hydroxypyridine 1-oxide the tautomeric equilibria are considerably more displaced towards the 2- and 4-hydroxy forms than for the corresponding "non-oxides." This behaviour was rationalized. Subsequent work has been concerned with 3-hydroxypyridines 1-oxides, and the corresponding benzopyridine *N*-oxides.

a. Pyridine, Quinoline and Isoquinoline Derivatives

IR investigations of 3-hydroxypyridines 1-oxides [**131**] show that these compounds exist in the *C*-hydroxy form [**131a**] although strong association occurs, especially in the solid state (72KG962). For azo derivatives of 3-hydroxypyridine 1-oxides, see (73KG810).

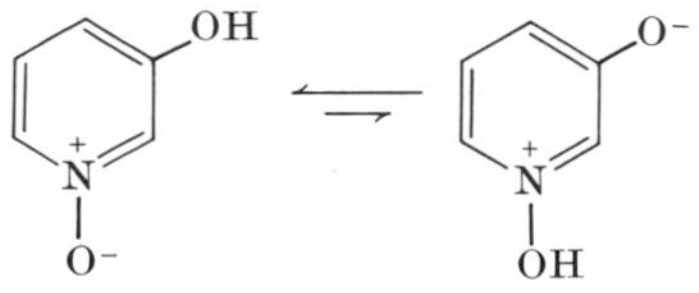

OH, $\overset{+}{N}$, O^- ⇌ O^-, $\overset{+}{N}$, OH

[**131a**] [**131b**]

Absence of the *N*-oxide absorption in the IR ($CHCl_3$) of 1-hydroxyisoquinoline 2-oxides [**132**, R = alkyl or Ph] and the similarity of the UV spectra (EtOH) with those of 3-methyl-1-isoquinolones rather than 3-methylisoquinoline 2-oxides suggests that the oxo forms [**132a**] predominate over the *N*-oxide tautomers [**132b**] (63JO2215); NMR in $CDCl_3$ confirms this (66JO2090).

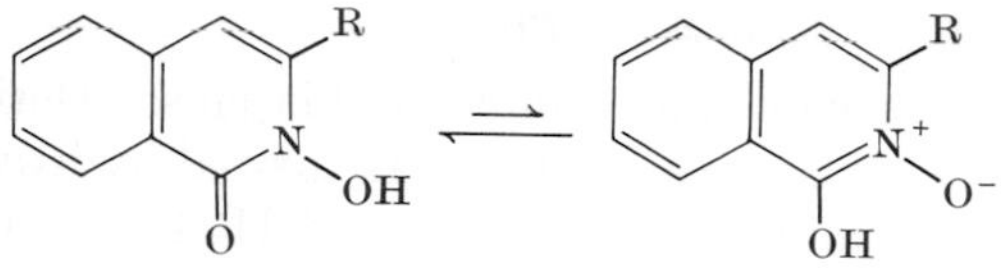

[132a] [132b]

UV comparisons of 3-isoquinolinones with potential 3-hydroxy-isoquinoline 2-oxides [**133**] were tentatively interpreted in favour of the *N*-hydroxy form [**133b**] for dioxan, DMSO and EtOH (71T4653), but spectral properties of similar compounds were assigned in terms of the *N*-oxide structures [**133a**] (61JO3761). In view of the tautomerism of 3-hydroxyisoquinoline itself (Section 2-3Bf), it is likely that the *N*-oxides exist in form [**133a**], but further work is needed here.

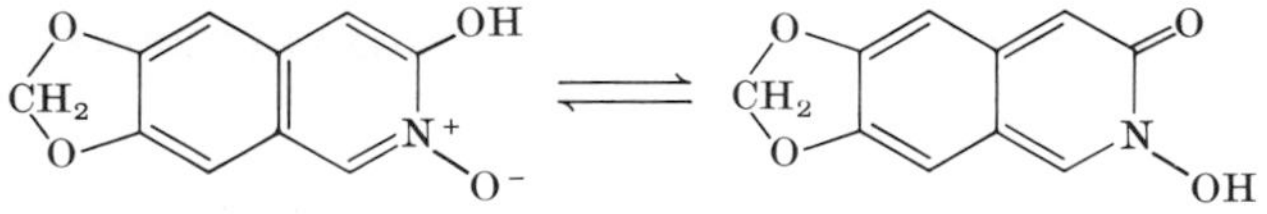

[133a] [133b]

b. Acridine and Phenanthrene Derivatives (I-360)

The tautomerism of 9-hydroxyacridine 10-oxide [**134**] has been the subject of considerable controversy (see I-360 and discussion in 66T-3227). IR evidence suggested that in the crystalline state it exists as the *N*-oxide [**134a**], strongly hydrogen-bonded (63CB1726). A definitive UV and pK_a study has now demonstrated that in H_2O the two forms [**134a,b**] exist in approximately equal amounts (66T3227), behaviour

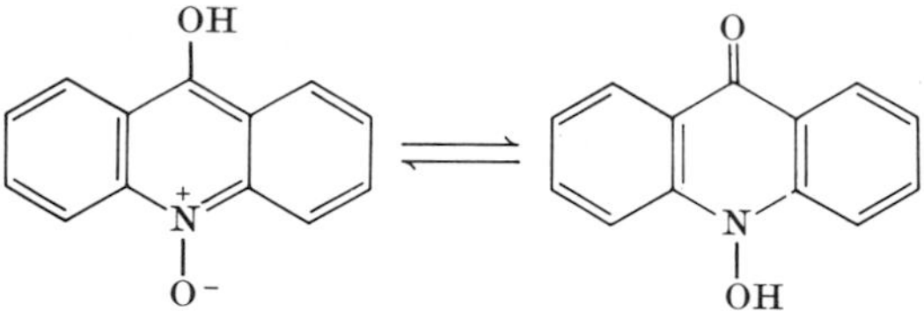

[134a] [134b]

resembling that of 4-hydroxypyridine 1-oxide (I-359). Comparing 4-pyridone and acridone, the annelation of two benzene rings considerably increases the relative stability of the oxo form; explicable by the stabilization of the uncharged canonical form [**135a**] relative to [**135b**] (although the effect on the charge-separated forms [**135c**] and [**135d**] is in the opposite direction). Annelation of the two benzene rings to 4-hydroxypyridine 1-oxide does *not* significantly alter the equilibrium

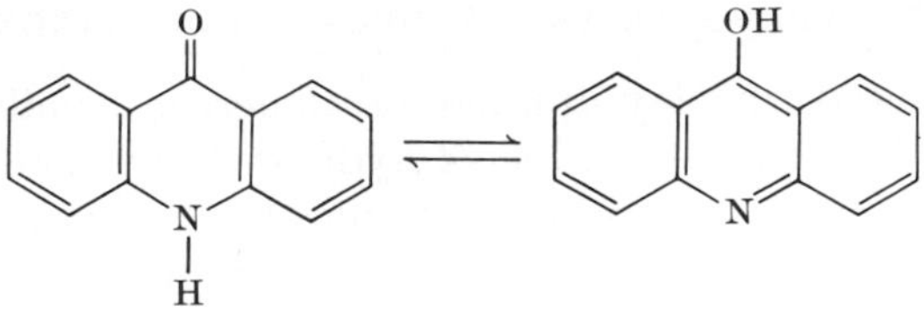

[**135a**] [**135b**]

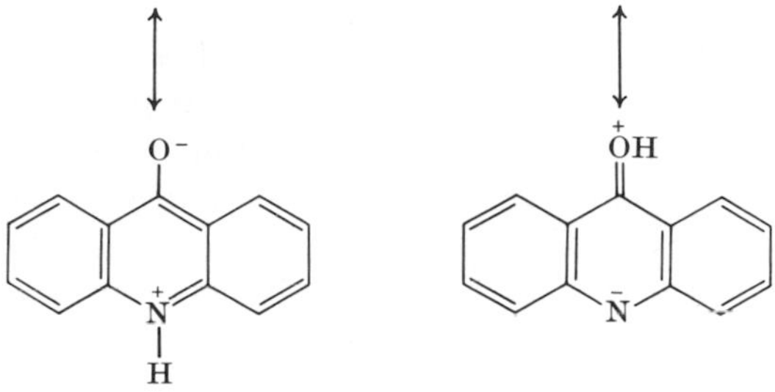

[**135c**] [**135d**]

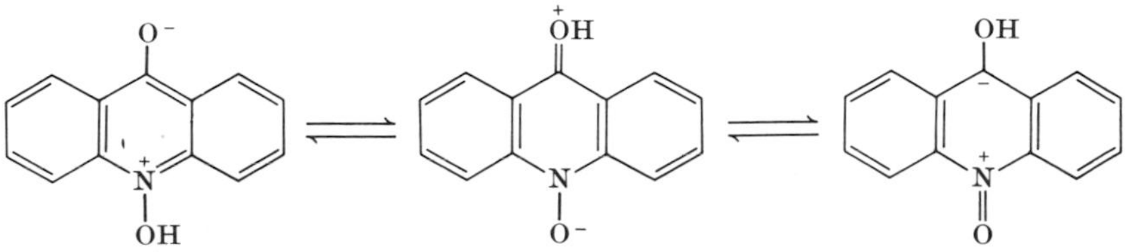

[**136a**] [**136b**] [**136c**]

position: canonical forms of type [**136a**] are less important than those of [**135c**], but those of type [**136b**] are more important than those of type [**135d**] and there is a further type of charge-separated form [**136c**] which will also be stabilized by the annelation of benzene rings. For similar conclusions reached later, see (68J(C)1045).

Structure [**137**] has been assigned to *N*-hydroxyphenanthridone from UV similarity (in EtOH) with phenanthridone and difference from phenanthridine 5-oxide. IR data support this (67JO1106). The similar structure [**138**] was tentatively assigned (67JO1106).

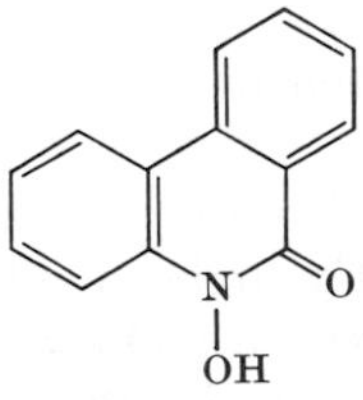

[**137**] 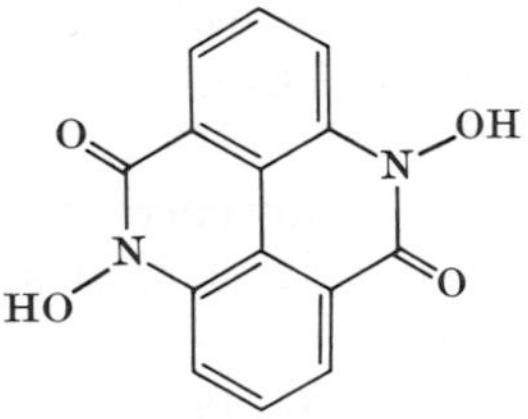 [**138**]

G. HYDROXYPYRANONES AND HYDROXYTHIOPYRANONES (I-342)

These compounds display annular tautomerism, similar to that of, for example, 2-hydroxy-4-pyridone/4-hydroxy-2-pyridone type, but no form corresponding to 2,4-dihydroxypyridine is now possible. In general 2-oxo rather than 4-oxo forms are found to be the most stable.

a. Monocyclic Compounds (I-342)

The view that 4-hydroxypyran-2-ones as [**139**] exist predominantly in the 4-hydroxy form as [**139a**] in MeOH (I-343) has been reinforced by spectral investigation of the 3,5-dimethyl derivative (68AN664); the presence of a small amount of the 2-hydroxy tautomer [**139b**] cannot be excluded by the spectral evidence. The dioxo tautomer [**139c**] is unimportant as shown by the absence of CH–CH_3 signals in

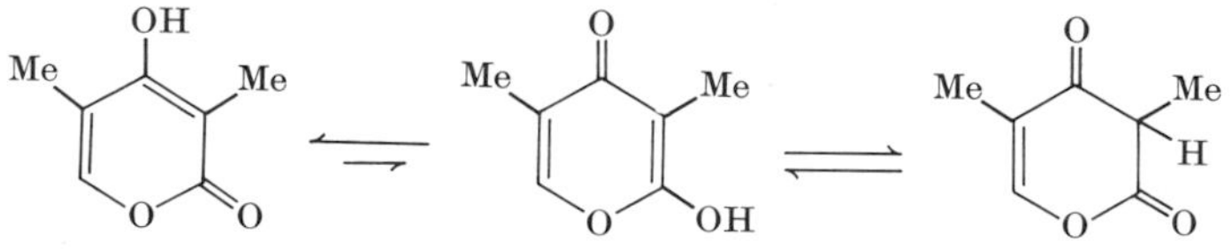

[**139a**] [**139b**] [**139c**]

the NMR spectrum (68AN664). Equilibration studies showing that 4-methoxy-2-pyrone [**140**] is more stable than 2-methoxy-4-pyrone [**141**] by at least 3 kcal/mole at 140° (63TL863) are consistent with these results.

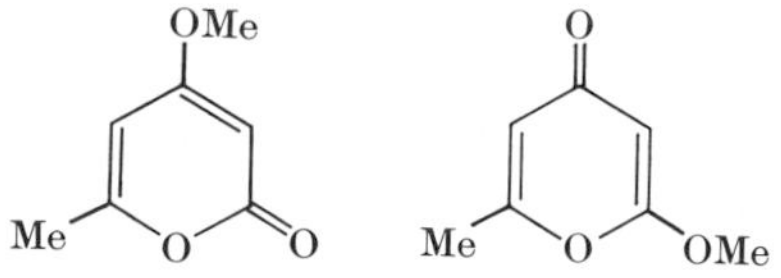

[**140**] [**141**]

The previous assignment of structure [**142**, R = Me] to dehydro-acetic acid based on IR and NMR (61AK(17)523) has been confirmed (65JO1255), and the analogous structure [**142**, R = Ph] verified by IR and NMR evidence (64J5200). Similarly, NMR and IR show that compound [**143**] exists in the form written in $CDCl_3$ and in the solid state (62J2606).

Surprisingly, the absence of νC=O > 1700 cm^{-1} and the presence of an OH signal at 12.21 δ, at significantly higher field than for [**142**], suggests that [**144**] exists in the 2-hydroxy form [**144a**], and not in the chelated form [**144b**], probably because of greater steric hindrance between the 6-methyl and the 5-acetyl group in [**144b**] (63J4483).

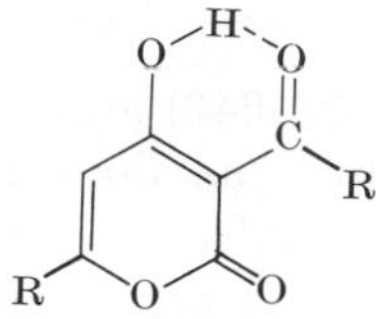

[142] 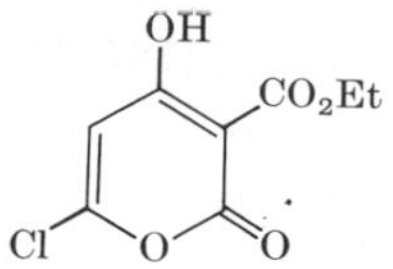 [143]

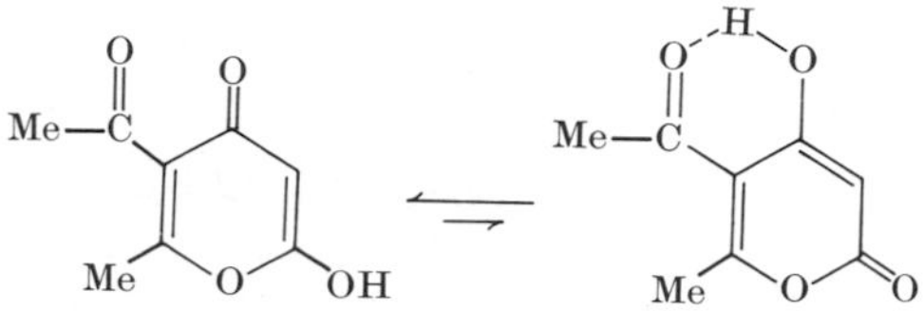

[144a] ⇌ [144b]

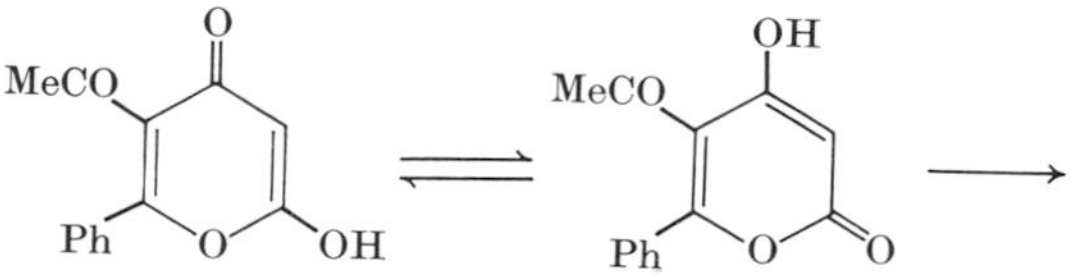

[145a] ⇌ [145b] ⟶

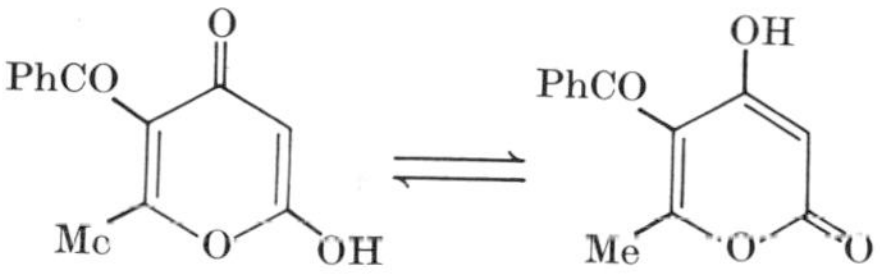

[146a] ⇌ [146b]

Heating the acetyl compound was originally described as giving the tautomer ([**145a**] → [**145b**]) (63J4483), but the thermal reaction product has been shown to result from rearrangement and to have structure [**146**] (66M710). The Austrian authors assign the 4-hydroxy structures [**145b**] and [**146b**], but in view of the other work mentioned above this could be incorrect. Potential aminomethylene compounds [**146/1**] exist in the dioxo form [**146/1a**] rather than in a hydroxy–oxo form, e.g., [**146/1b**], as shown by IR and NMR evidence (63JO1886, 64J5200).

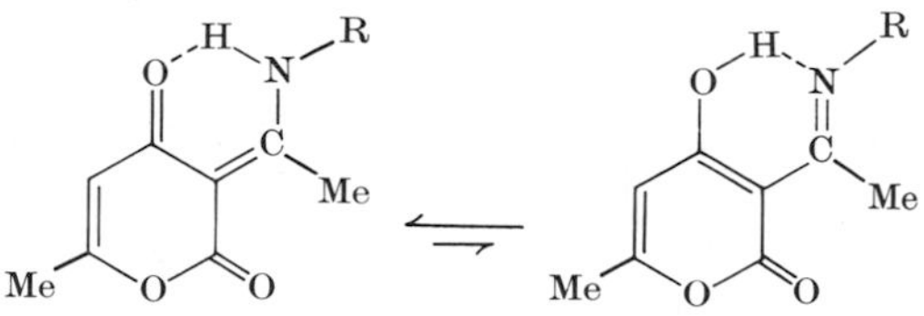

[146/1a] ⇌ [146/1b]

b. Polycyclic Compounds (I-342)

4-Hydroxycoumarin [**146/2**] (66CX(20)646) appears to exist as such in the solid state on X-ray evidence, as does dicoumarol [**146/3**] (68CO(267C)1790); however, strong hydrogen-bonding renders the exact location of all hydrogen atoms, and hence the determination of the tautomeric forms, difficult. MO calculations have been reported for [**146/2**] and its anion (68TE184).

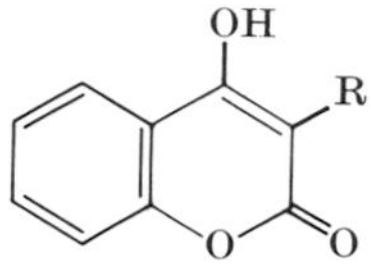

[**146/2**] 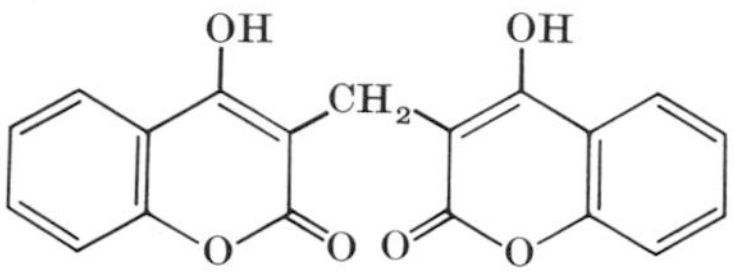 [**146/3**]

Tautomerism of the molecular ion of 4-hydroxycoumarin has been discussed in Section 1-6Gb.

2-Benzyl-3-benzoylchromones [**147**, R = H, Ph] undergo photoenolization [**147a**] ⇄ [**147b**] (65JA5424).

The phototautomerism of 7-hydroxy-4-methylcoumarin has been studied (72AC1044, 72NA(235) 53, 74JA4699).

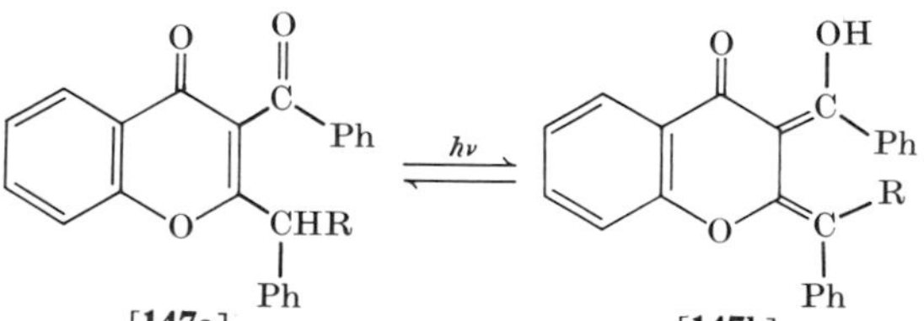

[**147a**] [**147b**]

Spectroscopic evidence has been interpreted as indicating that the isocoumarins [**148**] exist as the 3-hydroxy form [**148a**] in the solid, but as the 1,3-dioxo form [**148b**] in $CHCl_3$ and MeOH (65AP4, 65AP411).

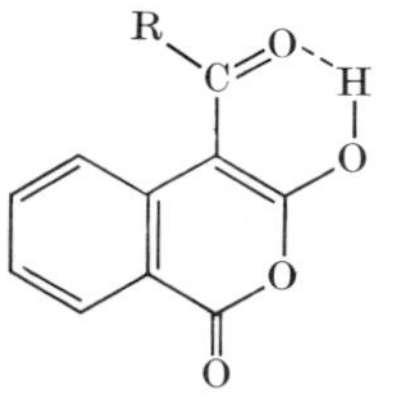

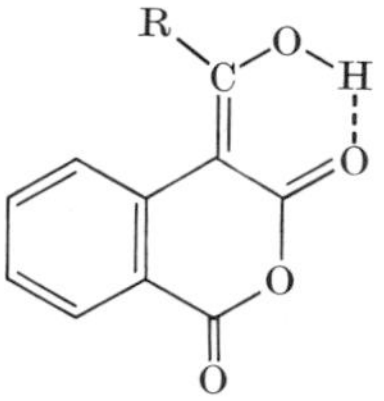

[**148a**] R = Me, Ph, etc. [**148b**]

H. Pyridines in Which a Fused Benzene Ring or a Phenyl Group Carries a Hydroxy Group (I-381)

The conclusion of the previous review that such compounds exist largely in the hydroxy form in the ground state has been amply confirmed. However, tendencies to proton transfer in excited states exist, and this phenomenon has attracted considerable attention.

a. Quinolines and Isoquinolines (I-384)

K_T for the 8-hydroxyquinoline equilibrium [**149a**] ⇌ [**149b**] in the excited state has been evaluated and the fluorescence phenomena were

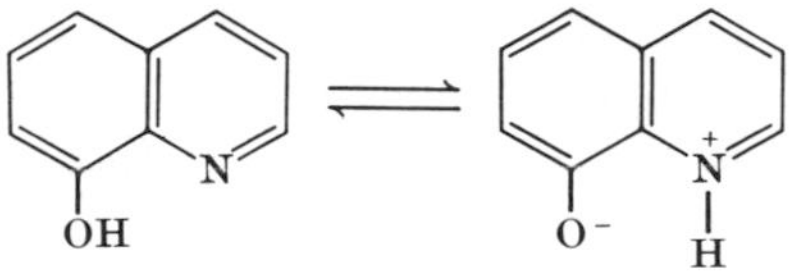

[**149a**] [**149b**]

interpreted in terms of tautomerism (64J4868). Various halogenated (67PH2668), 5-substituted (68PH3692), and many other substituted 8-hydroxyquinolines (72AC1240) have been investigated similarly. Recently, substituent effects on excited state pK_a values, and hence on tautomeric equilibria, have been treated by the Taft equation for 8-hydroxyquinolines (73PH1595). The effect of solvent is considerable: neither phosphorescence nor intersystem crossing can be detected for 5- or 8-hydroxyquinoline in hydrocarbon solvents or in EtOH and the relevance of this to tautomerism is discussed (70AC1178).

6- and 7-Hydroxyquinoline have also been studied (68J(A)3051, 68T1777), together with 5- and 8-hydroxyquinoline and 5-hydroxyisoquinoline (68T1777).

5-Hydroxy-6-methoxy-8-nitroquinoline exists in the oxo form [**149/1**] as demonstrated by X-ray crystallography (69CX(B)362).

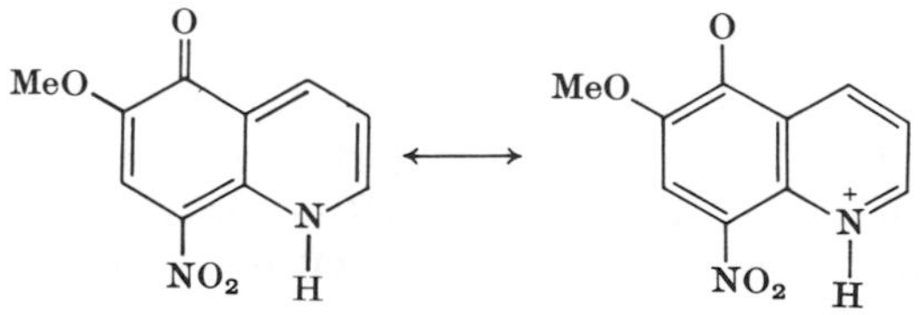

[**149/1**]

b. Acridines and Phenazines (I-381)

UV and fluorescence-polarization spectra and theoretical calculations confirm previous ideas (I-381) on the constitution of 1-, 2-, 3-, and

4-hydroxyacridines, i.e., that the 2- and 4-derivatives are in tautomeric equilibrium with the oxo forms [**150**] and [**151**], respectively (69T1001).

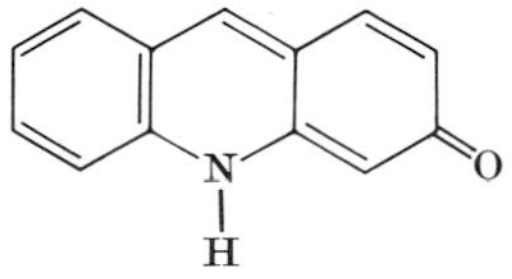

[**150**] 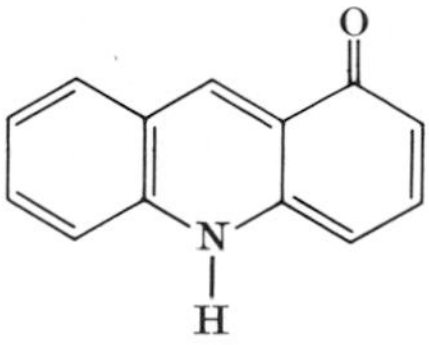 [**151**]

Some of the previous views (see I-382) on hydroxyphenazine tautomerism have been changed by IR and UV and comparisons with fixed forms. Both 1- and 2-hydroxyphenazine are now considered to exist predominantly in the hydroxy forms in MeOH, CCl_4 and in the solid state. However, some more complex derivatives such as [**152**] (violet form) and [**153**] exist in the oxo forms (CCl_4 and in the solid state) (64SA1665).

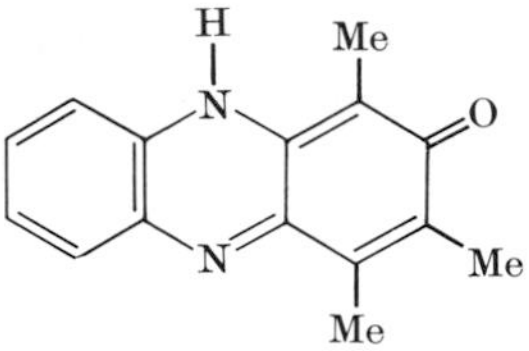

[**152**] 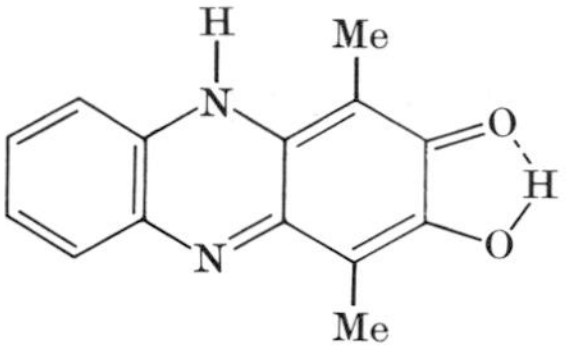 [**153**]

4. Compounds with Potential Hydroxyl Groups and Two or More Heteroatoms

For MO calculations on all the monohydroxydiazine tautomeric forms see (72G325).

A. Oxazinones (I-363) and Thiazinones

Little is known about the thiazine derivative [**154**]; it was referred to as [**154a**] (64M147), and previous IR results appear to eliminate the dioxo structure [**154b**] (absence of $\nu C{=}O$) but there is another possible oxo–hydroxy form [**154c**] (60CB671).

1,3-Thiazine-2,4-dione (1-thiauracil) [**154/1**] and the corresponding 5-methyl derivative exist in the dioxo forms on IR evidence (70AJ51). Further IR and UV, and especially the NMR CH_2 singlet at 5.0 δ, support the oxo structure [**155a**] (I-363) for 1,4-benzoxazin-3-one [**155**]

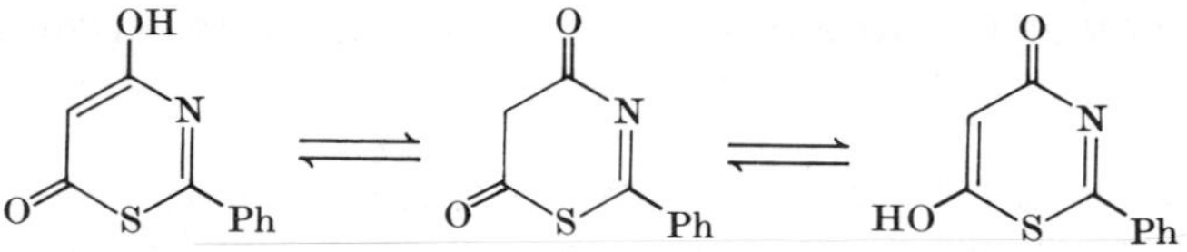

[154a] [154b] [154c] 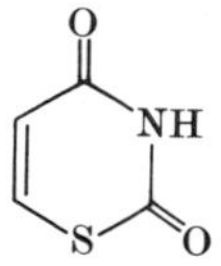 [154/1]

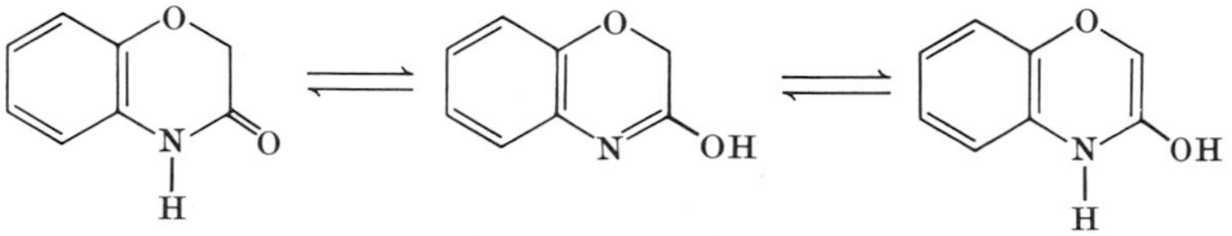

[155a] [155b] [155c]

(69T517) in EtOH, $CHCl_3$ and solid state. More IR evidence is available for the dioxo structure [**156**] of isatoic anhydride (**64AH(40)317**) (cf. I-363).

The 3-oxobenzo-1,2,4-thiadiazine-1,1-dioxide shows νC=O at 1700 cm^{-1} and therefore exists in the oxo form [**157**] in the solid state (63JO2313); the previous review considered only the structure of the corresponding 3*H* derivative (I-387).

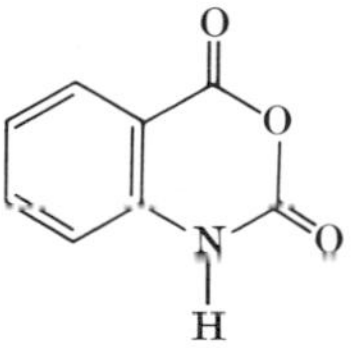

[156] 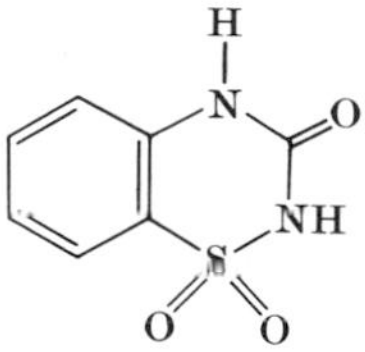 [157]

B. Azinones: Hydroxy Azines: General Discussion (I-363)

Tautomeric equilibria are here more complex, with at least 3 tautomeric forms. In, for example, a monohydroxy diazine, the hydroxy form is usually the least stable and hence the pK_a method does not give K_T for the first proton addition of the model compounds. Although rarely applied, the pK_a value for the second proton addition can, however, be used since the doubly protonated dication of a compound which can exist in three tautomeric forms is unambiguous.

Most work has been concerned with demonstrating the predominance of an oxo form and distinguishing between the various possible oxo azines. However, true hydroxy derivatives exist, particularly in polyoxygenated derivatives and much work on such complex tautomerism is now available.

C. Pyridazinones, Cinnolinones and Phthalazinones (I-364)

a. Potential Monohydroxypyridazines (I-364)

The predominance of the oxo forms [**158**] and [**159**] in H_2O for 3- and 4-pyridazinone shown previously (I-364) by pK measurements has been confirmed, and for 3-pyridazinone [**158**] reinforced by study of the spectral and pK properties of the model anhydro base [**160**] (71J(B)1261).

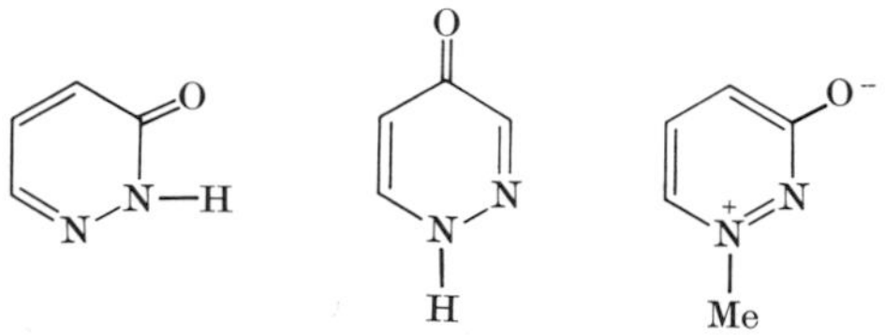

[**158**] [**159**] [**160**]

UV comparisons with fixed forms and IR and NMR spectra demonstrate the predominance of oxo structures for compounds [**161**] and [**162**, R = H, Et] in $CHCl_3$ and EtOH (70BF4011).

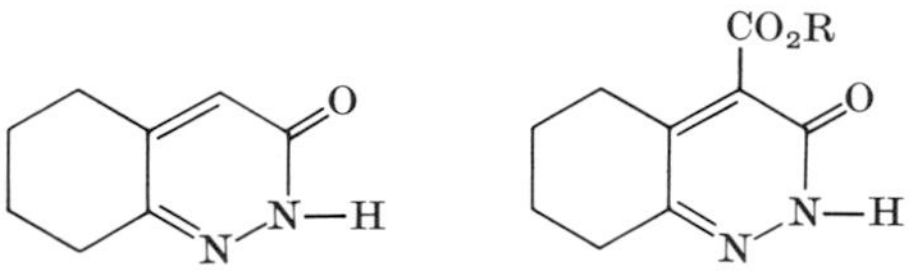

[**161**] [**162**]

Polyfluoropyridazin-3-ones exist as such in acetone, EtOH and the solid state as shown by UV, IR and ^{19}F NMR comparisons with fixed models (68J(C)2989) (cf. Section 2-3Bc). Structure [**162/1**] has been assigned by X-ray crystallography (63CX318). It has been suggested on UV evidence (71J(B)1261, 72J(PII)392) that 3-pyridazones probably undergo *O*-protonation to give cations of type [**163a**], but this statement needs reconfirmation in view of the peculiar behaviour recently found in determination of the pK_a values of pyridazinones (74J(PII)1199).

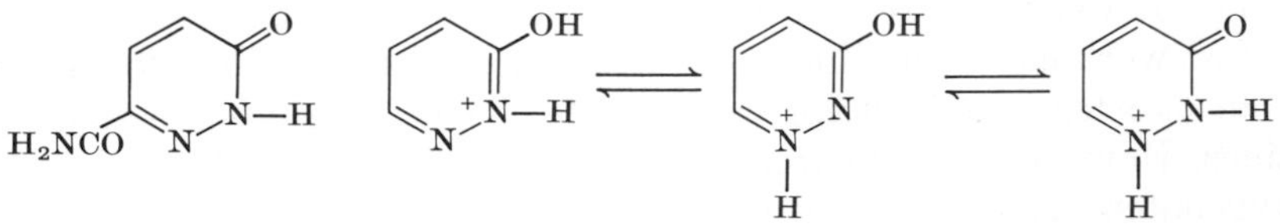

[**162/1**] [**163a**] [**163b**] [**163c**]

b. Potential Monohydroxycinnolines (I-365)

The discrepancy between pK and UV studies of 4-hydroxycinnoline [**164**] tautomerism (I-365) is now explained: the "reference" compound originally formulated as 1-methylcinnolin-4-one [**165**] is the isomeric anhydro base [**166**] (63J4924). The UV of authentic [**165**] does indeed resemble that of cinnolin-4-one [**164**] (65J5391), thus confirming the pK conclusion (cf. I-365) that the 4-oxo structure [**164b**] predominates over the 4-hydroxy form [**164a**] with pK_T = 3.6 in aqueous solution: for a further discussion, see (65J2260, 71J(B)2344); it is not possible to estimate the contribution of [**164c**]. The IR spectra of a series of 4-cinnolinones support the 4-oxo formulation (64JH221).

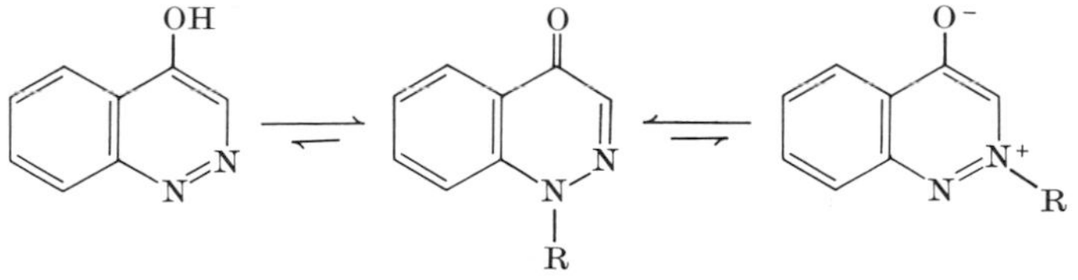

[**164a**] [**164b**] R = H [**164c**] R = H
[**165**] R = Me [**166**] R = Me

The tautomerism of 3-hydroxycinnoline [**167**] has been studied in H_2O by the UV and pK methods. The 3-cinnolinone form [**167b**] predominates by pK_T = 2.58 over [**167a**]: the zwitterionic form [**167c**] is also a minor component. The comparatively low value for K_T is

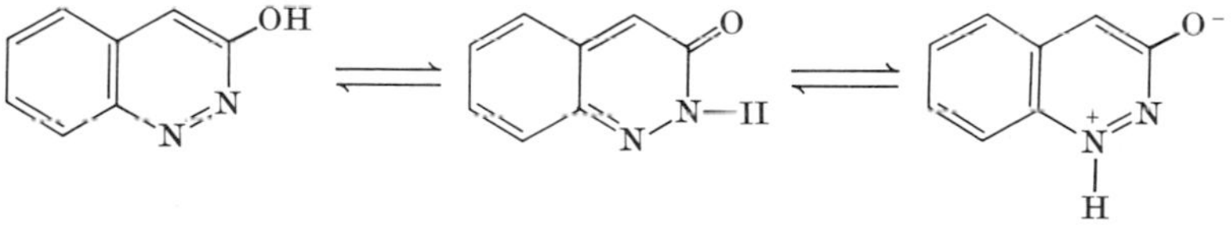

[**167a**] [**167b**] [**167c**]

attributed to destabilization of the oxo form [**167b**] by the *ortho*-quinonoid structure in which the benzenoid character of the homocyclic ring is partially disrupted (71J(B)2344). (cf. Section 2-3Bb and discussion on 3-isoquinolinol tautomerism) (Section 2-3Bf). As in the 3-pyridazinone series (Section 2-4Ca), the cation of 3-cinnolinone possesses the 2H-3-hydroxy form [**167/1**] (71J(B)2344).

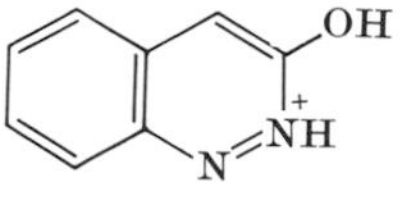

[**167/1**]

c. Other Fused Hydroxy Pyridazines

Structures [**168**]–[**171**] have been demonstrated by IR comparisons and confirmed by NMR and UV spectra (72G169).

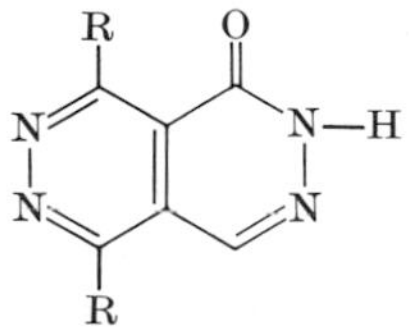

[**168**] R = H
[**169**] R = Me

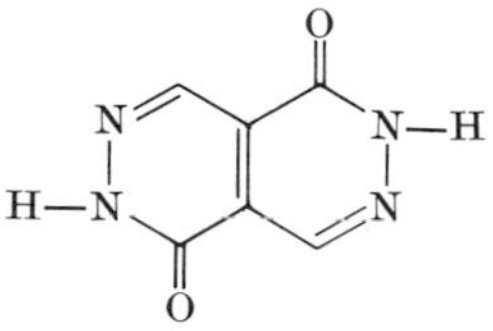

[**170**]

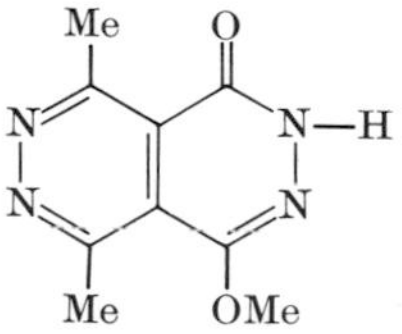

[**171**]

d. Maleic Hydrazide (Pyridazine-3,6-dione) and Monocyclic Derivatives (I-367)

Extensive investigations of the tautomerism of maleic hydrazide were reported in the previous review (I-367): they showed convincingly the predominance of the oxo–hydroxy form [**172b**]. One discordant NMR report (59ZA(170)205) in the previous review was suggested (I-367) to be the result of neglect of fast proton exchange (cf. also 62CI695). Further NMR work has indeed confirmed this, and has presented chemical shift evidence in favour of the oxo–hydroxy form [**172b**] (64J1523), as did UV comparisons with fixed models in water (74J(PII)1199, see also 64MI81). An independent NMR study (64CC-970) and Hückel MO calculations are in agreement (72ZC230), as are ^{13}C NMR studies (73CB2918). No quantitative estimate for the contribution for the minor tautomers [**172a**] and [**172c**] is available: the present authors believe that Hammett treatments (72J(PII)392) are insufficiently developed to provide this.

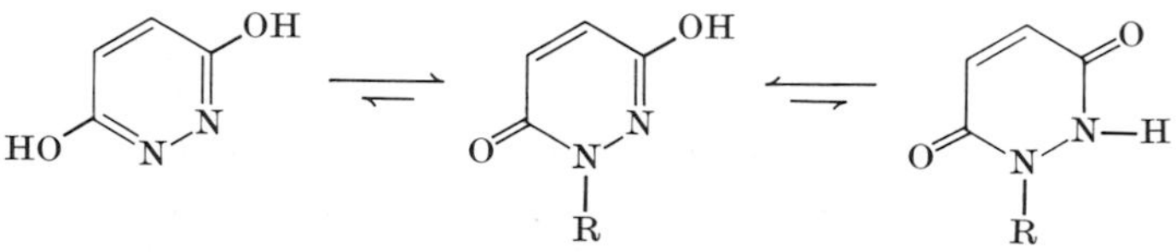

[**172a**]

[**172b**] R = H [**172c**]
[**173a**] R = Me [**173b**]

The mono-*N*-methyl derivative [**173**] also occurs in the hydroxy–oxo form [**173a**] (64J1523; see also 70BF4376, 71JO3372, 73CB2918) as do *O*-substituted derivatives [**174**] (70BF4376) and [**175**] (63CT669).

Maleic hydrazide derivatives are probably *O*-protonated to give cations of type [**176**] (70BF4376).

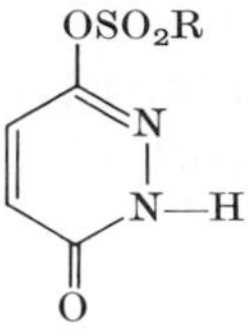

[174] 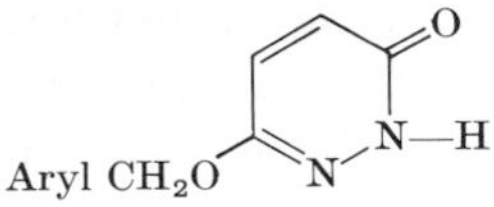 [175] 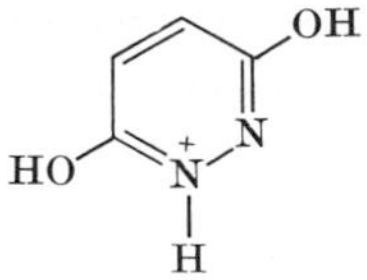 [176]

4,5-Dichloro-3-hydroxypyridazin-6-one [**177**] also has the oxo–hydroxy structure as shown by X-ray analysis (73AS797); the crystal structure of the corresponding *N*,*N*′-dimethyl derivative is quite different (73AS835).

3-Hydroxy-1-(4-nitrophenyl)pyridazin-6-one [**178**] (68RC1867) and the 1-(3-nitrophenyl) analogue (69RC315) also exist in the oxo–hydroxy forms depicted, on IR (solid state) evidence. However, the corresponding 1-(2-nitrophenyl)- [**179**] and 1-(2,4-dinitrophenyl)- compounds exist in the dioxo form probably owing to NH-hydrogen bonding to the *ortho* NO_2 group (69RC1187).

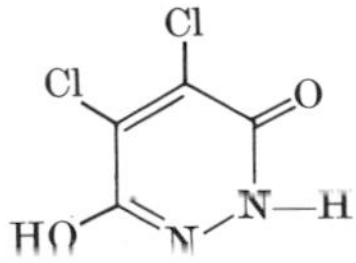

[177] 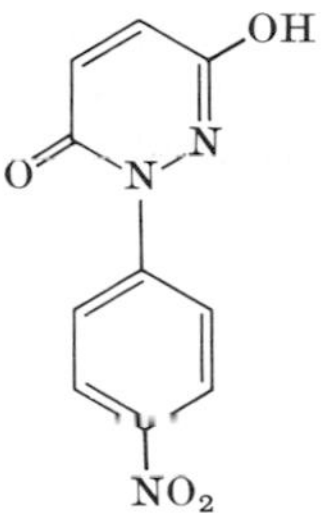 [178] 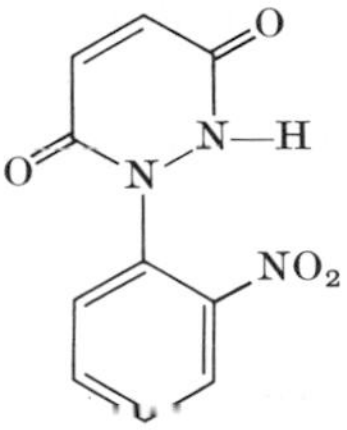 [179]

e. Polycyclic Derivatives of Maleic Hydrazide (I-368)

The 1-hydroxyphthalazin-4-one structure is confirmed for phthalic hydrazide by IR (64AH317). IR and UV comparisons with fixed models suggested that compounds [**180**] and [**181**] exist in the oxo–hydroxy forms shown in the solid and in MeOH (68J(C)2857). The same applies

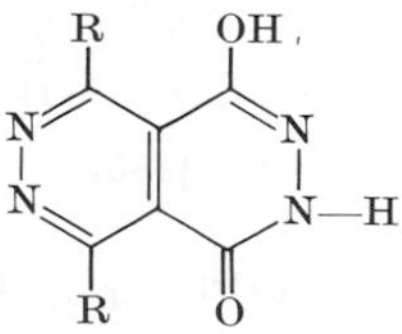

[**180**] R = H
[**182**] R = Me

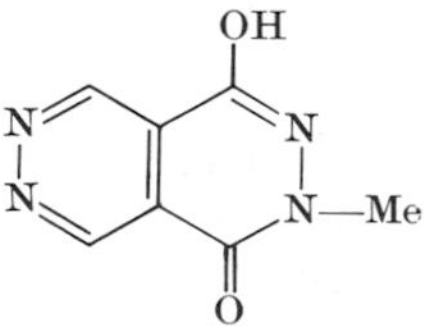

[**181**]

to the corresponding dimethyl derivative [**182**] (72G169), while the dihydroxy–dione structure [**183**] is suggested by IR (72J(PI)953). The chelating ability of [**184**] has been offered in support of the oxo–hydroxy structure (67RC1241), but chelation may well be a function of the *anion*.

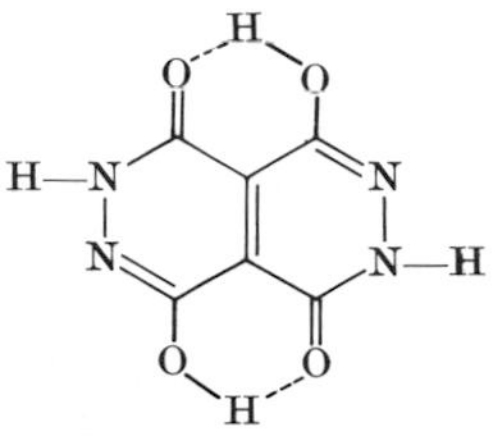

[**183**] 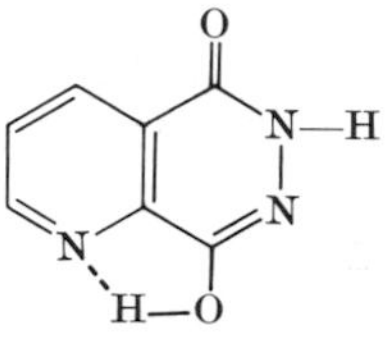 [**184**]

The νC=O at 1635–1675 cm^{-1} excluded the dihydroxy tautomer [**185a**] of the thiazolo [4,5-*d*]pyridazines [**185**] in the solid state without distinguishing between the two oxo–hydroxy [**185b**], [**185c**] and the dioxo form [**185d**] (70BF4317). IR and UV comparisons with methylated models demonstrate the oxo–hydroxy form for pyridazoquinoxaline derivatives of type [**186**] (73KG556).

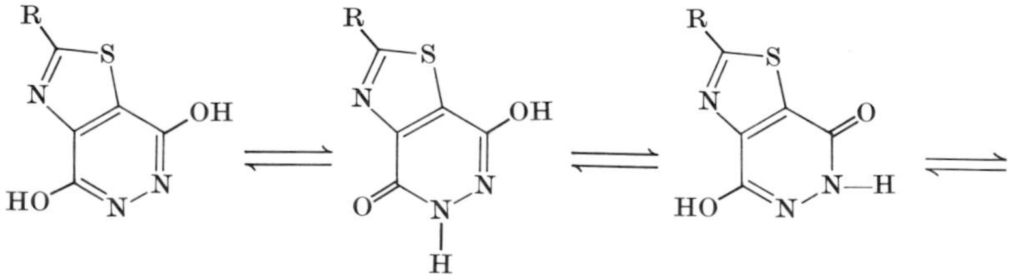

[**185a**] [**185b**] [**185c**]

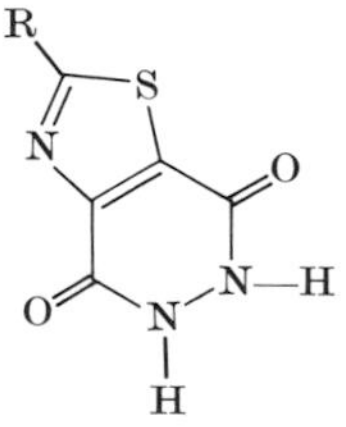

[**185d**] 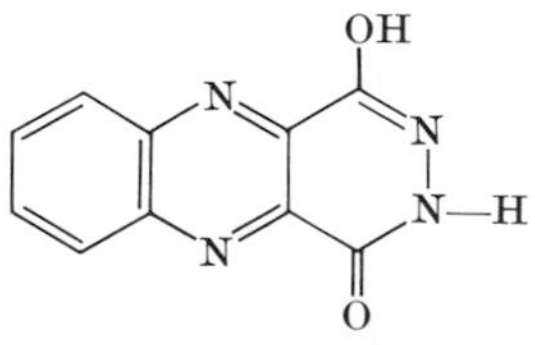 [**186**]

Compounds [**187, 188**, X = OH, OPh, morpholino, $\bar{N}Et_2$] also appear from IR results to exist in the oxo–hydroxy form as shown (70BF4376).

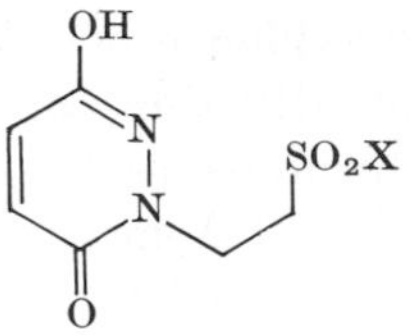

[187] 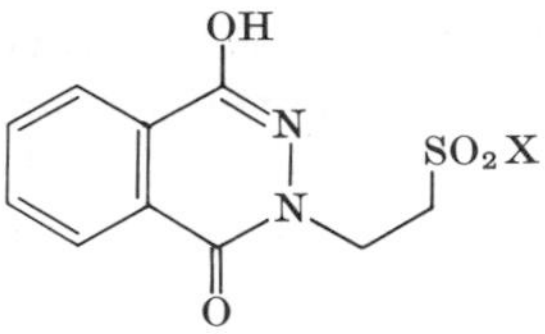 [188]

D. Pyrimidinones and Quinazolinones (I-368)

a. Potential Monohydroxypyrimidines and Monohydroxyquinazolines (I-368)

NMR comparisons with fixed models confirm the earlier conclusion from UV, IR, and pK_a methods (I-370) that the *o*-quinonoid form [**189b**] predominates for pyrimidin-4-one [**189**] in D_2O (66JO175) and DMSO (65JH447, 71BF1858). A recent quantitative study of the tautomerism of 5-substituted pyrimidin-4-ones shows that the two NH forms [**189a**] and [**189b**] coexist in approximately equal amounts in water: the 5-substituent effect is small for the equilibrium [**189a** ⇌ **b**] but tautomeric form [**189c**] is favoured (from 0.5 to 3.4%) by electron withdrawing substituents (74CT1239). Analysis of the shape of the UV spectrum in H_2O–EtOH indicates that the percentage of *o*-quinonoid form increases with the EtOH content (66JO175). French authors suggest that 2-phenyl-4-pyrimidinone [**190**] exists as a mixture of hydroxy [**190c**] and *o*-quinonoid form [**190b**] with [**190c**] predominating (71BF1858). The same group later reported (73CO(276C)1341) that dipole moment measurements in benzene indicated that for 2-phenyl-4-pyrimidinone the *o*-quinonoid form [**190b**] predominates (93%) over

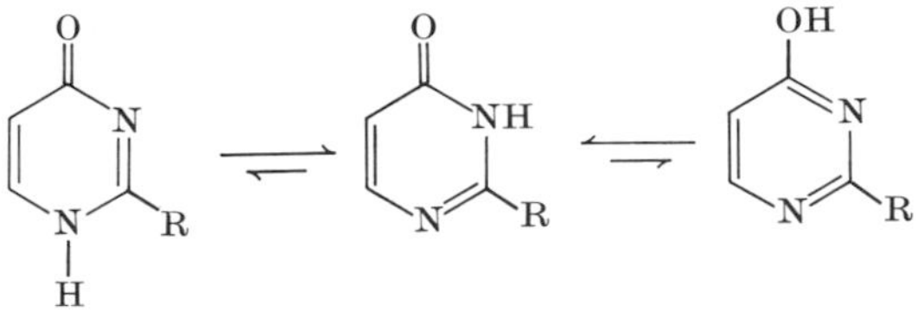

R = H	[**189a**]	[**189b**]	[**189c**]
R = Ph	[**190a**]	[**190b**]	[**190c**]

the *p*-quinonoid form [**190a**] and that [**189b**] predominates (82%) over [**189a**] for the parent compound. The apparent discrepancy may be due to a solvent effect on *o*-/*p*-quinonoid tautomerism which could possibly be explained by preferential stronger intermolecular hydrogen bonding of the *o*-quinonoid form in less polar solvents (cf. similar effect in 4-pyrimidthiones). However, the evidence of NMR J values is not conclusive and the matter needs reinvestigation, especially as the NMR of other 4-pyrimidinones has been interpreted on the oxo formulation (67T2657) and in view of the likely effect of N-methylation on the shape of the phenyl NMR absorption (see Section 1-3B).

X-ray crystallographic studies confirm the oxo structure [**191**] (I-368) of pyrimidin-2-one in the solid state (70AS3230), and have demonstrated the oxo structure [**192**] for 5-fluoro-2-pyrimidinone (72AS760). NMR chemical shift comparisons with methyl derivatives (64AK(22)65) also demonstrate the predominance of the oxo form in solution. Calculated UV spectra of 2- and 4-pyrimidinones agree with experimentally measured values (73CT1474). Pyrimidin-2- and 4-one cations are formed by N-protonation and possess structures of type [**193**] and [**193/1**] (70H299, 70JH487). (See also 74CT1239.)

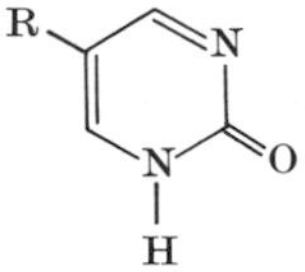

[**191**] R = H
[**192**] R = F

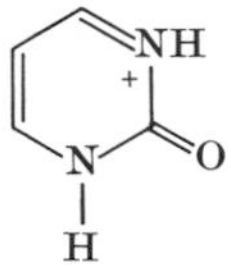

[**193**]

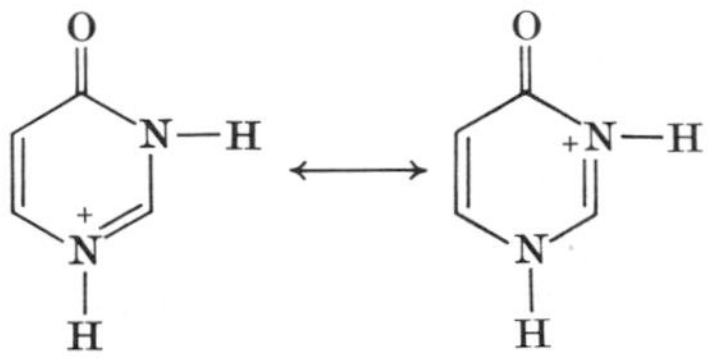

[**193/1a**] ⟷ [**193/1b**]

Little work has appeared dealing with substituent effects on pyrimidinone tautomerism. Marked differences between the UV of 5-cyano-4-pyrimidinone [**194**] and the 4-methoxy compound [**195**] in MeOH suggest that the former exists as an oxo form, probably [**194**] (67CB3664). Trifluoro-4-hydroxypyrimidine appears to exist in the enol form [**196**] in $CHCl_3$ from a νOH band (67J(C)1822): this is not unexpected in view of the effect of α-halogen substituents in stabilizing the hydroxy forms of pyridones (cf. Section 2-3Bc). For work on alkoxypyrimidinones see Section 2-4Db.

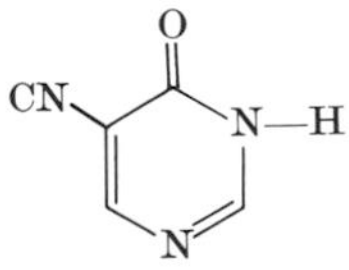

[**194**] 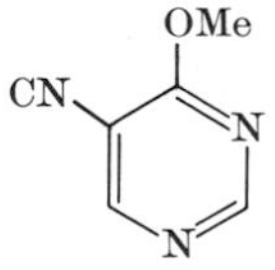 [**195**] 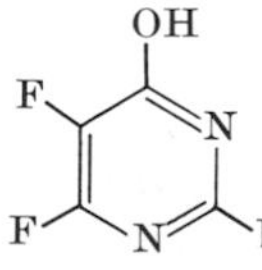 [**196**]

The predominance of the phenolic [**197a**] over the zwitterionic form [**197b**] of 5-hydroxypyrimidine [**197**] (I-371) is confirmed for polar solvents by NMR (64AK(22)65), and some MO calculations are available (70JA2929, 70TC(16)243, 71JL(10)245).

[**197a**] [**197b**]

The prevalence of the *o*-quinonoid forms [**198, 199**] in tautomeric quinazolin-4-ones (I-371) is further supported by IR, UV and NMR studies on the 2-phenyl [**198**] and 2-benzyl compounds [**199**] (69T783).

Although it has been suggested that NMR indicates structure [**200a**] for the 1-methylquinazolin-4-one cation (63T1011), the present authors consider the oxo structure [**200b**] to be not excluded by the evidence given and also more likely.

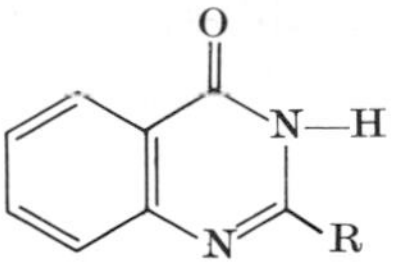

[**198**] R = Ph
[**199**] R = $PhCH_2$

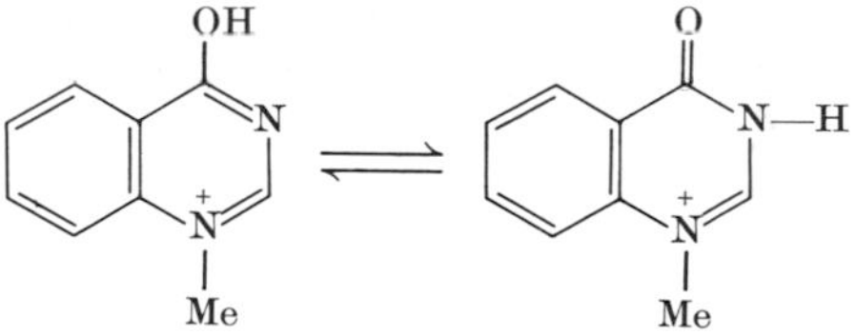

[**200a**] [**200b**]

b. Uracils: Neutral Species (I-371)

The previous review presented (I-371) conclusive evidence for uracil as pyrimidine-2,4-dione [**201a**]; this is now further supported by Raman spectroscopy (67SA2551), by NMR on ^{15}N-labelled uracil in DMSO (65JA5439) and by X-ray analysis (67CX(23)1102). X-ray results on 5-methyluracil (thymine) (61CX333, 69CX(B)1038) 5-ethyl-6-methyluracil (66CX(20)703), *N*-methyluracil (63CX28) and *N*-methylthymine (62JB3573) all indicate oxo structures. (See also **74MI790**.)

Anomalous excitation spectra for triplet state formation and fluorescence emission from uracil and thymine in neutral aqueous solution when compared to their absorption spectra have been interpreted to suggest that the fluorescing tautomer has the hydroxy–oxo structure [**201b**], while the triplet-forming tautomer possesses the usual dioxo structure [**201a**] (72PN2488).

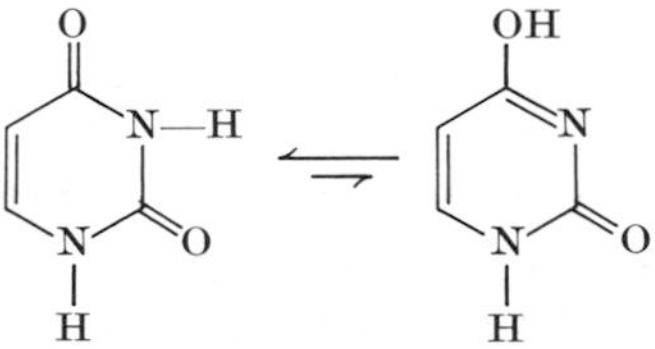

[**201a**] [**201b**]

For ^{13}C NMR studies on uracil nucleosides see (70JA4079) and for ^{14}N NQR see (72JT(57)5087, 72MI265). Papers do continue to appear attempting to relate tautomeric structure to position of methylation of mesomeric anions: (71JO848) is an example of this unsound practice.

X-ray analyses have demonstrated the dioxo structure for 5-fluoro- (73CX(B)2549) and 5-nitrouracil (67CX(23)376), and UV comparisons demonstrate similar 2,4-dioxo structures for several further uracil derivatives: thymine and its *N*-methyl derivatives (66CB2391), 6-fluorouracil and 6-chlorouracil (64JA2474) and 5,6-dihalouracil (68IS-603). The pyrimidine-2,4-dione structure has also been confirmed for various nucleosides by X-ray methods: 5-fluoro- (64MI203), 5-chloro- (73CX(B)1259), 5-iodo- (65CX(18)203) and 5-bromo-2′-deoxyuridine (66MI320), 2′-deoxyuridine itself (72CX(B)2260), 5-chloro- (69PN(63)-1359), 5-bromo- (66MI320) and 5-methyl-uridine (69CX(B)2144), and uracil-β-D-arabinofuranoside (73CX(B)1641). Thymidine was previously (I-375) considered to exist "at least in part," in the monohydroxy form on UV and IR evidence, but X-ray methods disclose the dioxo form (69CX(B)1423).

The dioxo structures have been also determined by X-ray analysis for dihydrouracil (68CH746), dihydrothymine (68JA470) and dihydrouridine (72CX(B)596).

That 2-ethoxy-4-pyrimidinone is indeed an oxo compound, as asserted in the previous review (I-372), has been confirmed by comparative UV studies (70JO903). These have shown that in H_2O the *o*- [**202a**] and *p*-quinonoid forms [**202b**] are present in comparable amounts: cf. the fine structure of pyrimidin-4-one (Section 2-4Da and I-370) and of isocytosine (Section 2-6Cc). Increase of temperature favours the proportion of [**202a**] in H_2O. UV and IR show that the *o*-quinonoid form [**202a**] greatly predominates in $CHCl_3$ at all temperatures studied. The solvent effect is expected as the more polar *p*-quinonoid form [**202b**] is favoured by the more polar solvent (70JO903).

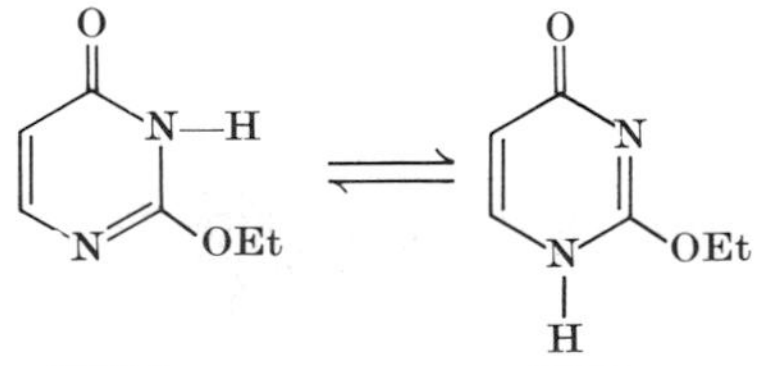

O N—H N OEt ⇌ O N N OEt H

[**202a**] [**202b**]

The presence of some 5-oxo tautomer [**203a**] in 5-hydroxyuridine [**203b**] is suggested by Japanese authors (71CT564) on chemical evidence, including hydrogen exchange at the 6-position. In any event the equilibrium appears to favour [**203b**] heavily.

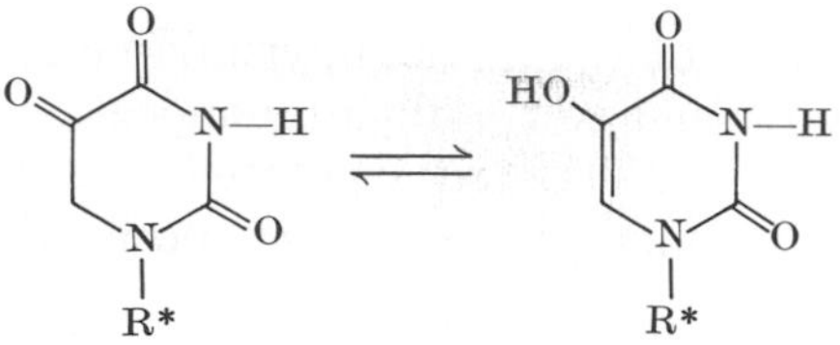

[**203a**] ⇌ [**203b**]

R* = β-D-ribofuranosyl

The energy difference between the "usual" dioxo forms [as **201a**] and the "rare" oxo–hydroxy forms [as **201b**] of uracil, fluorouracil and thymine have been calculated by the CNDO method (69BJ1467), and the data were related to mutagenetic theories (see, e.g., 64MI135, 70C134, and references therein).

Quinazoline-2,4-dione was shown in the previous review (I-373) to possess the dioxo structure. Some derivatives have now been studied. Glycosmicine derivatives all exist as the dioxo form [**205**]: νC=O at 1701 cm^{-1} and 1661 cm^{-1}; NH at 8.59 δ ($CDCl_3$) (63T1011).

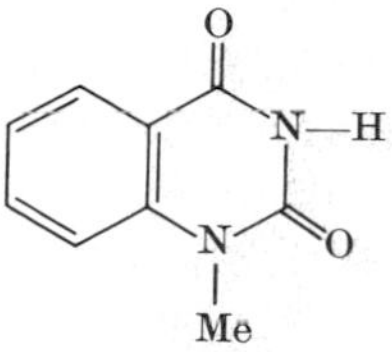

[**205**]

c. Uracil Cations

The NMR of uracil in strongly acidic media reveals the sites of protonation (70H299). The monocation [**206**] has structure [**206b**]. In FSO_3H, the dication [**207**] is formed and is deduced to exist in the structure shown (70H299; see also 72TL3823). X-ray crystallography of 1-methyluracil hydrobromide confirms the site of protonation (64CX122).

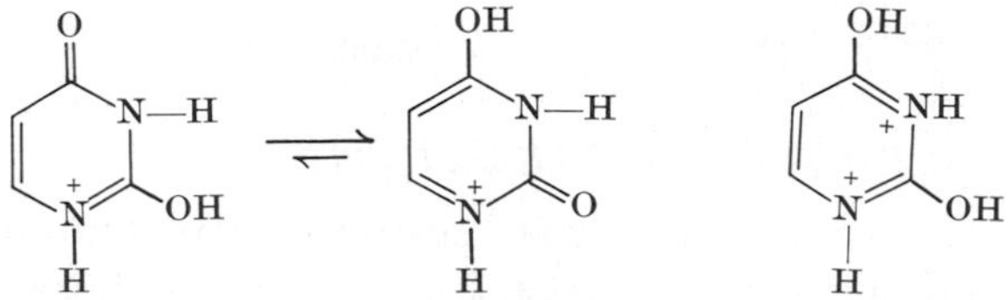

[**206a**] ⇌ [**206b**] [**207**]

The UV similarity of the 5-aminopyrimidine-2,4-dione monocation and 4,5-diaminopyrimidine-2,6-dione monocation with the neutral species uracil and 6-aminouracil suggested that protonation in each case involves the 5-amino group to give species [**208**] and [**209**] (71J-(B)1425).

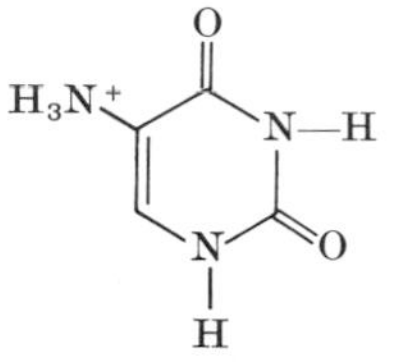

[**208**] 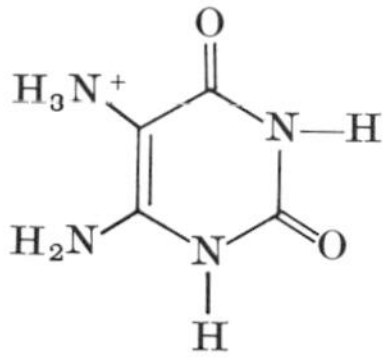 [**209**]

d. Uracil Anions (I-372)

The uracil monoanion [**210**] and its derivatives can exist in two tautomeric forms, each of which is mesomeric. Although earlier UV studies (I-372) suggested that the preferred form was [**210b**], the spectra at pH 11–13 of uracil (61BJ53), halouracils (64JA2474, 65JA4621) and thymine (65JA4621, 66CB2391) are usually interpreted as resulting from the presence of tautomeric mixtures. Pfleiderer and Deiss (68IS603) concluded from the UV absorption that halogenouracils formed the single anion [**210b**]. Very recently a UV study of the monoanion equilibrium in buffers showed that tautomeric form [**210a**] is favoured in high dielectric constant media but form [**210b**] predominates in solvents of low dielectric constant (71BY(232)1). The degree of substitution is also important: UV comparisons show that forms of type [**210b**] predominate for 5-*t*-butyl uracil monoanion, but those of type [**210a**] for 5,6-dialkylated anions (71RC211).

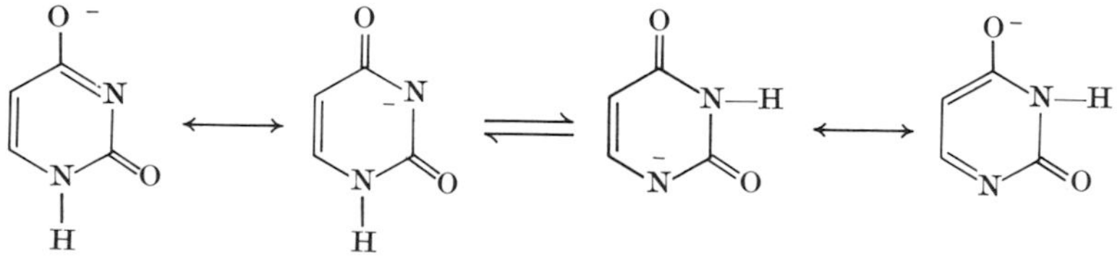

[**210d**] [**210a**] [**210b**] [**210c**]

e. Potential 4,6-Dihydroxypyrimidines (I-377)

4,6-Dihydroxypyrimidine [**211**] possesses six tautomeric forms [**211a–f**]. The previous review recorded (I-377) a tentative assignment of the predominant form as [**211b**]. Much further work has appeared subsequently. Australian workers (64AJ567) compared the UV of [**211**] with *O*- and *N*-alkyl models for the forms [**211a, b, c** and **e**]: they concluded that [**211**] in H_2O exists as a mixture of two tautomers, with the dioxo form [**211e**] predominating over the oxo–hydroxy form [**211c**]. Russian investigators initially (64ZO3134) decided from UV comparisons and from NMR measurements that the hydroxy–oxo form [**211b**] predominated with some contribution from the dioxo form [**211e**]; a similar conclusion was reached from NMR by Japanese workers (66JO175).

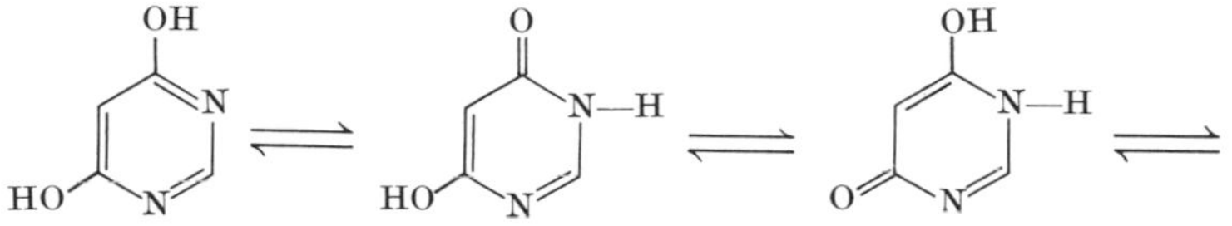

[**211a**] [**211b**] [**211c**]

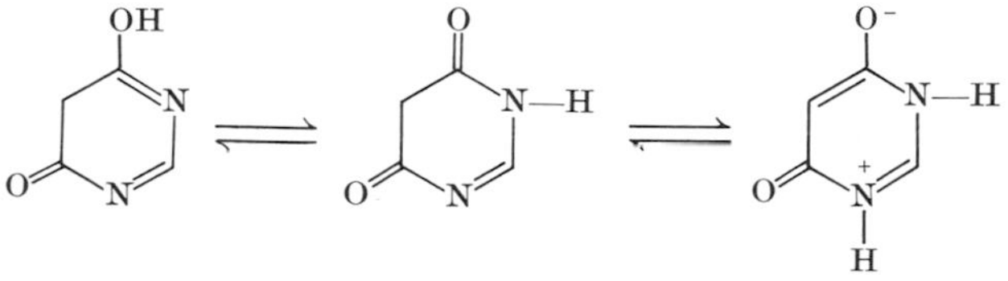

[**211d**] [**211e**] [**211f**]

Meanwhile, workers at Norwich (66J(B)565) demonstrated by UV, IR, NMR and pK_a comparisons that [**211**] exists in H_2O predominantly as the betaine form [**211f**] together with a substantial amount of the oxo–hydroxy form [**211b**]. Later the same Russian group (66DA(166) 635, 67T1197) also showed that the predominant form in H_2O is the betaine [**211f**]. They also investigated a series of 2- and 5-substituted derivatives of [**211**] and reached similar conclusions.

UV (66J(B)565) and NMR (67T1197) comparisons of 4-methoxypyrimidin-6-one [**212**] demonstrate that form [**212b**] predominates for H_2O.

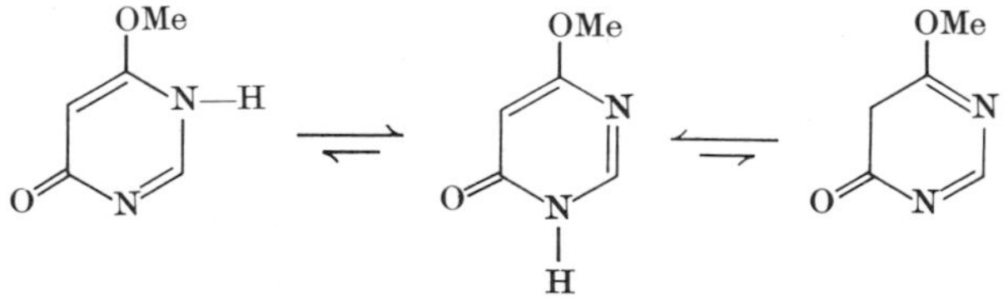

[**212a**] [**212b**] [**212c**]

UV spectra show that [**211**] and its mono-*O*-methyl and mono-*N*-methyl derivatives and also the 1,3-dimethyl derivative all form cations of the same type [**213**, R, R′ and R″ = H or Me]. C-protonation is eliminated by two NMR peaks of equal intensity in the cation of [**211**] in H_2SO_4. The UV of the 4,6-dimethoxy-pyrimidine cation is of different type [**214**], as expected (66J(B)565, 67T1197). The NMR spectra in D_2SO_4 are in essential agreement although fast D-exchange of the C_5–H of [**211**] and of 1-methyl-4-hydroxy-6-pyrimidinone indicates some C-protonation to give [**215**] (67T1197).

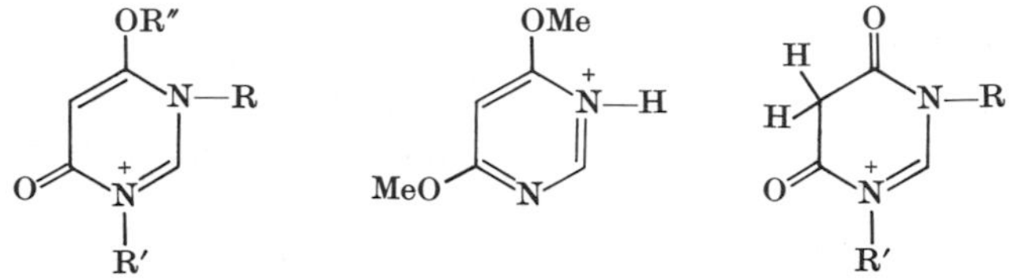

[**213**] [**214**] [**215**]

The monoanion [**216**] of 4,6-dihydroxypyrimidine [**211**] on the evidence of UV comparisons probably exists as [**216b**], possibly with a contribution from [**216c**] (66J(B)565).

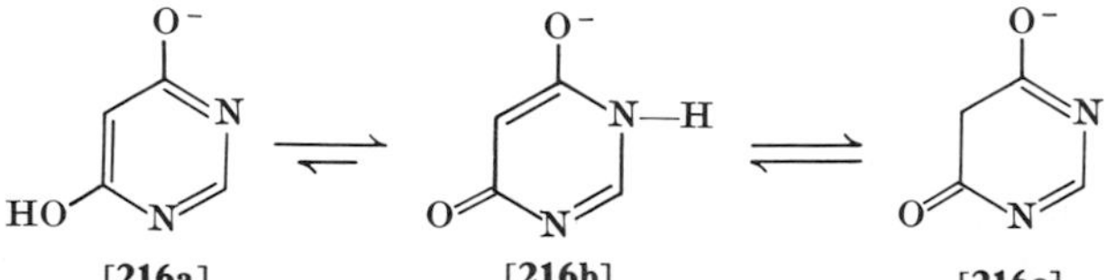

[**216a**] [**216b**] [**216c**]

The betaine structure [**217**] for "malonylaminopyridine," reported in the previous review (I-378) has been confirmed (67JH523, 70H905, 71M412).

Similar spectral and other physical properties of the pyrimidopyrazine [**218**] with those of malonylaminopyridine (I-378) is evidence for the mesomeric betaine structure [**218**] (68MC1045). Spectral evidence has

[**217**] [**218**]

been given for the betaine structures [**219**, R = H, Alk, Ph] in various solvents (69BF3133) (see also 66J(B)565 for [**219**, R = H]) and [**220**] (72M426).

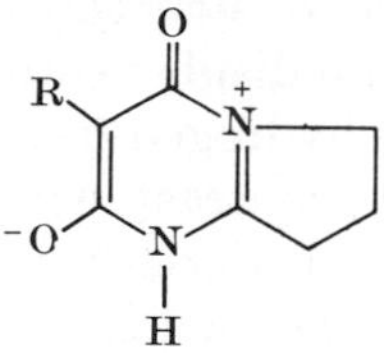

[**219**] 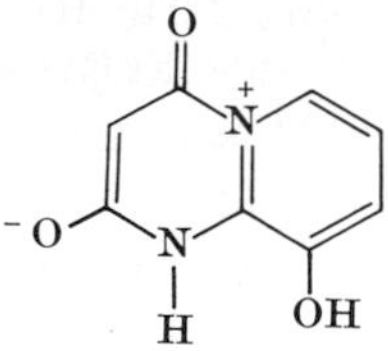 [**220**]

f. Other Potential Dihydroxypyrimidines

2,5-Dihydroxypyrimidine [**221**] was suggested tentatively to exist as a mixture of the oxo–hydroxy [**221a**] and betaine forms [**221b**] (65J7116); later IR and UV work supports the oxo–hydroxy form [**221a**] (67CZ1637). 4,5-Dihydroxypyrimidine, described as "possessing phenolic properties" (63J5590), appears to exist as [**222**] from IR spectra (67CB2280).

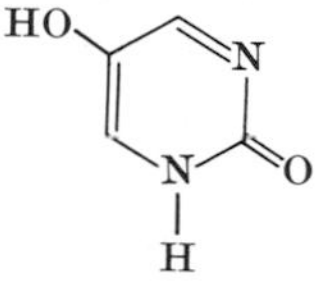

[**221a**] 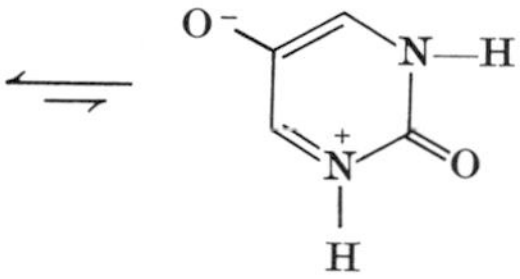 [**221b**] 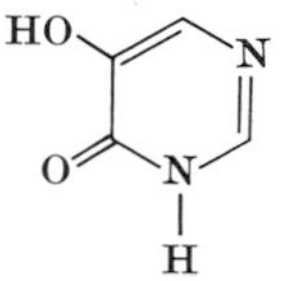 [**222**]

g. Barbituric Acid (I-375)

The well-established (I-375) trioxo structure [**223a**] for 5-alkyl- and 5-arylbarbituric acids [**223**] is further confirmed by NMR data (69CC-743, 69OR481) and by X-ray results for the unsubstituted compound in the solid state (63CX166) as did ^{13}C NMR in DMSO (70PH7684). Relaxation methods (cf. Section 1-4E) (65CB1623) give $pK_T = 1.9$ for [**223a**]/[**223b**] (R = H) in water. Further studies by the pK_a method are in general agreement and demonstrate that substitution in the 5-position decreases the predominance of the oxo-form [**223a**] (pK_T = 1.30, 0.88, 0.76 and 0.21, respectively, for R = H, Me, Cl, Br); the effect of solvent on the equilibrium [**223a**]/[**223b**] is less easy to rationalize (74RO113).

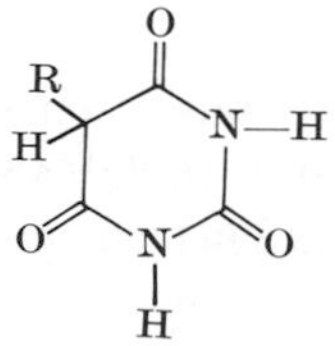

[**223a**] 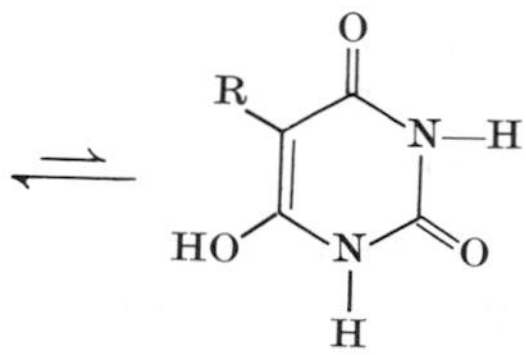 [**223b**]

5,5-Diethylbarbituric acid exists in the trioxo structure as shown by X-ray methods (69CX(B)1978), but some 5,5-disubstituted barbituric acids of type [**224**] are claimed to exist in the hydrogen-bonded hydroxy form on IR evidence (solid state) and by contrast with the IR of *N*-methyl derivatives of type [**225**] (65AP885). Further studies on the tautomerism of 5,5-disubstituted barbituric acids are available (69ZO2568, 70ZO669).

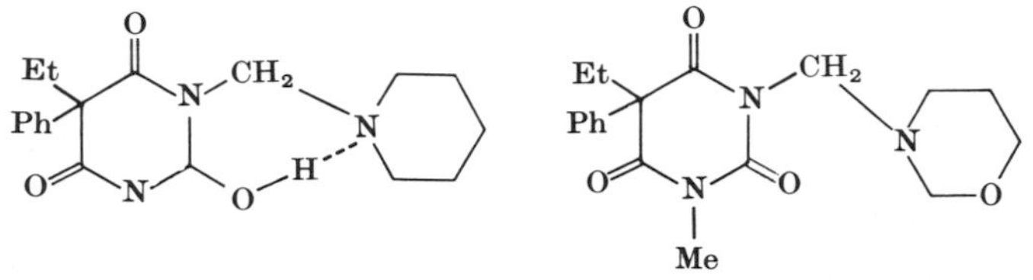

[**224**] [**225**]

X-ray analysis shows that 5-nitrobarbituric acid [**226**] exists when anhydrous in the 2,6-dioxo-4-hydroxy form [**226a**] (63CX950), but the trihydrate is in the aci-nitro form [**226b**] (64CX891): IR and UV spectral studies agree with these conclusions (68RR39), as well as dipole moment results (69RR1435). MO calculations indicate that the nitro-hydroxy structure [**226a**] is the most stable, and agree well with the UV spectra for [**226a**] in H_2O (68RR147). These calculations have been extended to postulate structures for the monoanion [**227**] and dianion: it is suggested that the trihydrate of [**226**] may exist in the crystal as an oxonium H_3O^+ salt containing the monoanion [**227**] rather than structure [**226b**]—an interesting suggestion (69RR311). Previously structure (**227/1**] had been demonstrated by X-rays for the monoanion in ammonium barbiturate (64CX282).

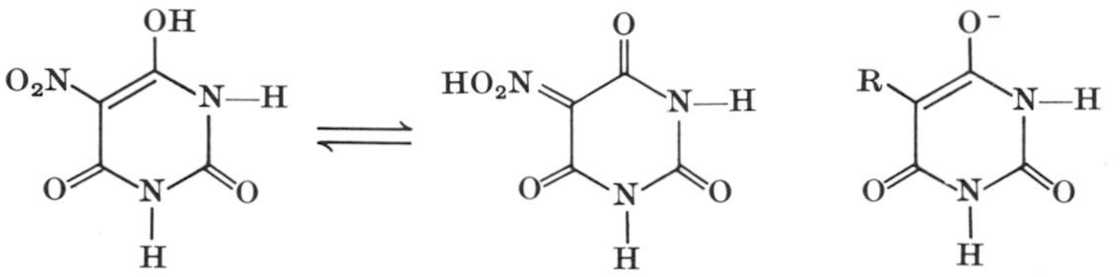

[**226a**] [**226b**] [**227**] R = NO_2
[**227/1**] R = H

The hydroxy–trioxo formulation [**228a**] has been substantiated for dialuric acid in DMSO solution by NMR (69OR481); the previous review (I-377) had indeed suggested that [**228a**] was more probable than the previously proposed tetrahydroxy structure [**228b**]. However, crystal data (65CX(19)1051, 69CX(B)1970) indicate that it exists in

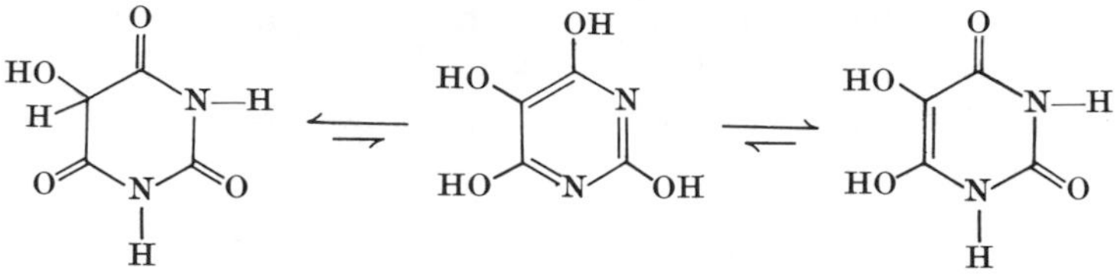

[228a] ⇌ [228b] ⇌ [228c]

the dihydroxy–dioxo form [**228c**] in the solid state. The ^{15}N NMR spectrum of 4-phenylazo-1,3-diphenylbarbituric acid has been investigated (72JO4121).

h. Other Potential Trihydroxypyrimidines

Although no study on the tautomerism in solution of 2,4,5-trihydroxypyrimidine (5-hydroxyuracil or isobarbituric acid) is available, νOH at 3398 cm^{-1} suggested for the solid state the monohydroxy–dione

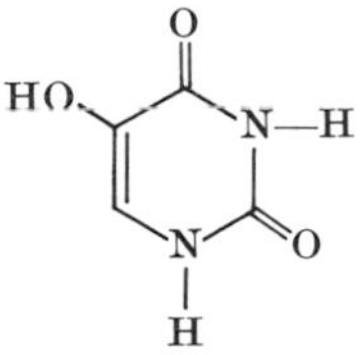

[229]

form [**229**] (59JA3786), as does the crystal structure determination of 5-hydroxyuridine (73CX(B)1393).

E. Pyrazinones and Quinoxalinones (I-378)

Quantitative data are difficult to obtain in this series since monocations of a common type are rarely formed by *O*- and *N*-methyl model compounds (I-378).

Further confirmation of the 2(1*H*)-oxo structure [**230a**] (I-378) for pyrazin-2-one [**230**] is available from NMR comparisons (68PH1642, 68PH1646), and from the comparison of the MO-calculated and experimental electronic spectra (73BA405). Earlier conclusions (I-378) on protonation of [**230**] are also confirmed (73BA405). However, 2,3,5-trifluoro-6-hydroxypyrazine [**231**] displays no νC=O (solid state) and UV comparisons (EtOH) with fixed forms suggest the hydroxystructure [**231b**] (70J(C)1023). This swing to the hydroxy form caused by perfluoro substitution also occurs in the pyridine (Section 2-3Bc), and pyrimidine series (Section 2-4Da); in the pyridazine series (Section

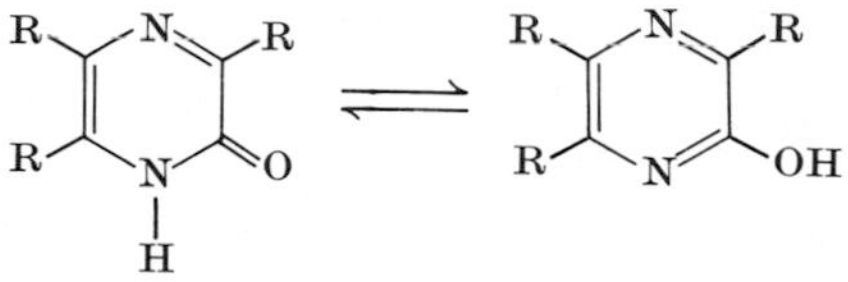

[**230a**] R = H [**230b**]
[**231a**] R = F [**231b**]

2-4Ca), the oxo form persists because no fluorine atom is adjacent to the nitrogen atom in question.

Although broad absorption at 2500–3500 cm^{-1} was interpreted as indicating that the 2,5-dihydroxypyrazine [**232**] exists in the solid state in the dihydroxy form [**232b**], this seems unlikely and needs confirmation. In EtOH the dihydroxy [**232b**] and oxo–hydroxy forms [**232a**] are considered to coexist (70J(C)980), but again this conclusion must be considered very tentative and in need of further investigation.

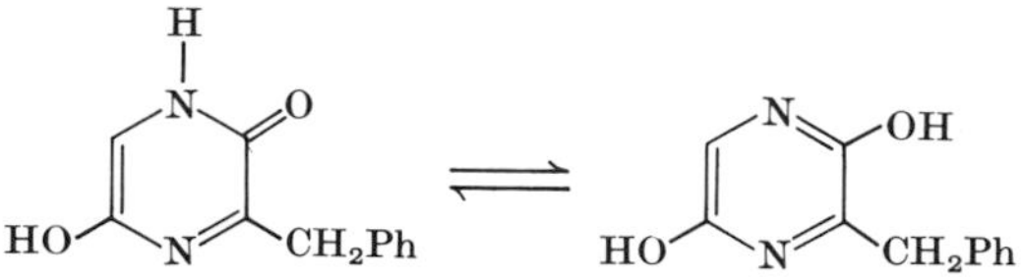

[**232a**] [**232b**]

2,3-Dihydroxypyrazine and some 5- and 6-alkyl and -aryl derivatives [**233**] were assigned the dihydroxy form [**233a**] on the basis of a proton NMR signal at 11–12 δ (72JO221). The assignment is most unlikely, and the compounds probably exist as pyrazine-2,3-diones [**233c**] cf. the analogous quinoxaline-2,3-diones (I-379).

The νC=O at 1665–1695 cm^{-1} (solid state) of 3-hydroxypyrazine 1-oxide suggests the oxo form [**234**] (64JO2623).

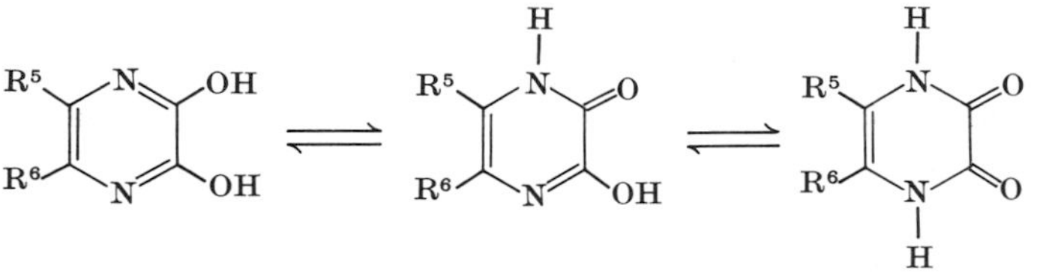

[**233a**] [**233b**] [**233c**] 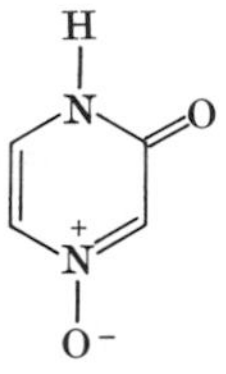 [**234**]

F. Triazinones (I-387)

a. Derivatives of 1,3,5-Triazine (I-387)

Further confirmation of the trioxo [**235**] structure (I-387) for cyanuric acid is deduced from UV spectra (64SA211) and from X-ray data (64NA(202)1206, 71CX(B)134).

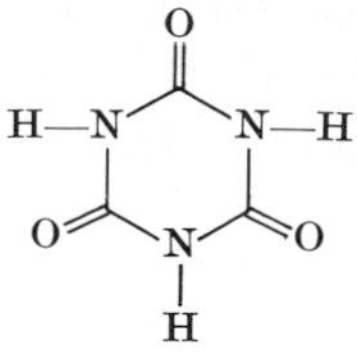

[235] 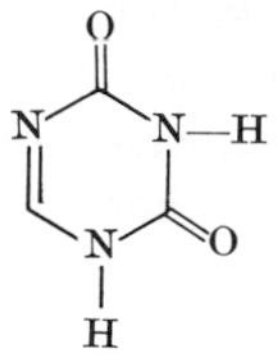 [236]

IR spectra of 5-azauracil in dioxan and EtOH indicate the dioxo form [**236**] (62CZ2754) (cf. I-388). Interpretation of the UV of [**236**] in H_2O is difficult: the spectrum is similar to both the *N*-alkyl and *O*-alkyl models (62CZ2754), and the situation is complicated by covalent hydration (65CZ90). Thus, 5-azauridine exists entirely in the crystal and partly in solution in form [**237**] (64CZ2060).

In the 5-azauracil monoanion [**238**], the symmetrical form [**238b**] is preferred by $pK_T \sim 1.5$ (62CZ716); this agrees with MO calculations (63CZ1499).

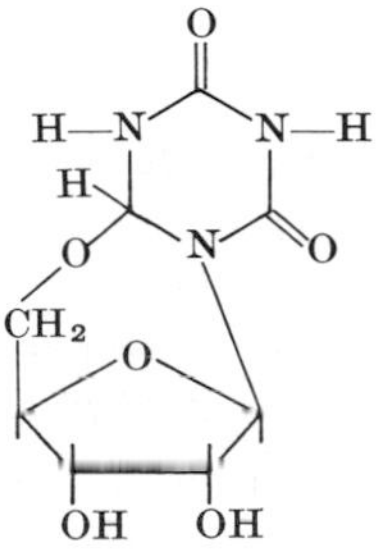

[237] 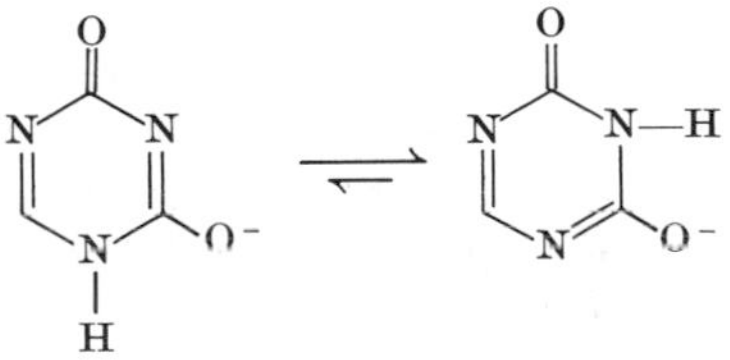 [238a] [238b]

b. Derivatives of 1,2,4-Triazine (I-388)

5,6-Diphenyl-1,2,4-triazin-3-one [**239**] occurs in the 2*H*-form [**239a**], by UV comparisons (in EtOH) with the corresponding 2- and 4-methyl derivatives, (71JO3921). A similar conclusion was reached for the 3-methylthio-1,2,4-triazin-5-one [**240**], which also exists mainly in the 2*H*-form [**240a**] (62CZ1886, 63CZ3392, 66CZ1864). However, for 1,2,4-triazin-3-one itself, the tautomeric equilibria could be complicated by the occurrence of covalent hydration (2-4Fa).

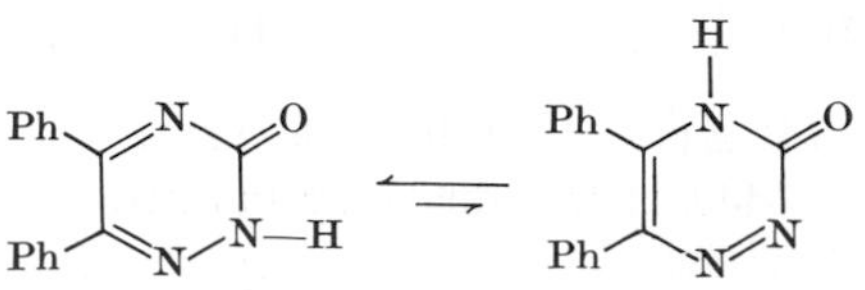

[239a] [239b] 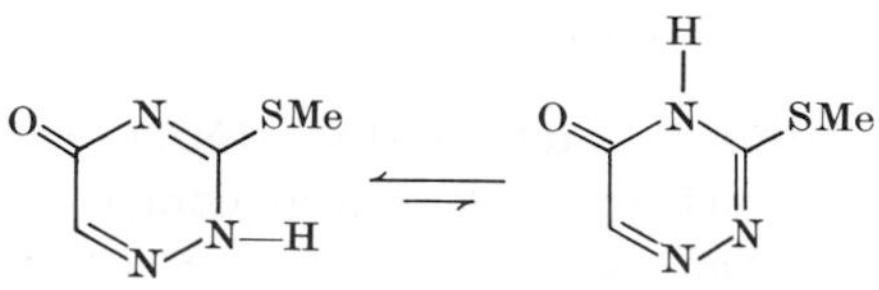 [240a] [240b]

In the 1,2,4-triazin-5-one series [**241**], the OH form [**241a**] is eliminated by IR and the detailed IR favours [**241b**], the *p*-quinonoid form in solution. This is confirmed by the UV spectra. In $CHCl_3$ ca. 20% of the minor form [**241c**, R and R′ = Me or Ph] was indicated by IR. The structure of the monocation was shown by UV to be [**242**] (71CZ1955).

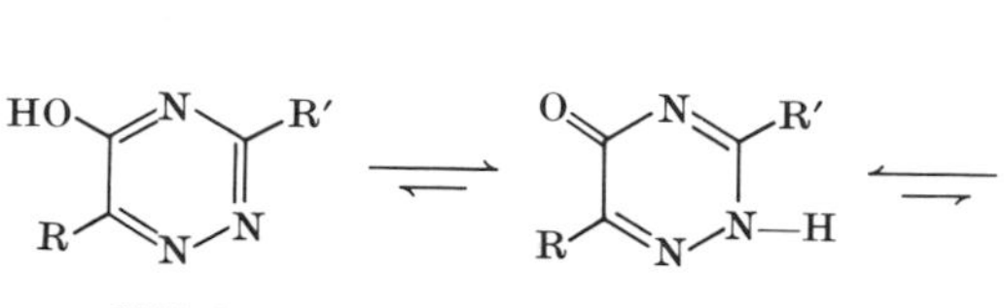

[**241a**] [**241b**] [**241c**] 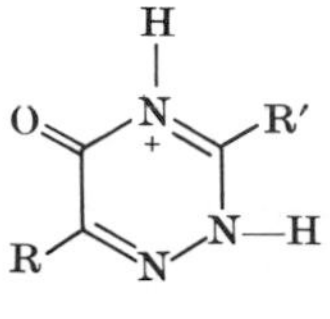 [**242**]

These conclusions are in full agreement with similar work by American (72JH995), French (73BF(2)2126) and Australian groups (72AJ2711). Although German workers (72LA(758)111) concluded from UV comparisons that the 4*H*-oxo form [**241c**] was predominant for 1,2,4-triazin-5-ones, reconsideration of this UV evidence quoted demonstrates that it is not unambiguous, and the other work quoted clearly indicates that the 2*H*-oxo form [**241b**] predominates for 1,2,4-triazin-5-ones. However, recently the French group concluded from IR evidence that certain of these compounds exist as 5-hydroxytriazines in the solid state and that the tautomeric equilibrium in solution favours oxo forms but is influenced by substitution (74T3171). Compound [**242/1**] exists in the form shown (69AH(61)181) by analogy with 6-azaisocytosine (Section 2-6Ec). However, the same paper claims protonation of an analogous compound on ring nitrogen to yield [**242/2**]; we believe this to be wrong, and to be derived from an erroneous interpretation of the IR spectra.

The previous review (I-388) reported evidence for the dioxo formulation of 6-azauracil [**243**, R = H]; this work was later extended to 6-azauridine derivatives [**243**, R = sugar] with similar conclusions for a

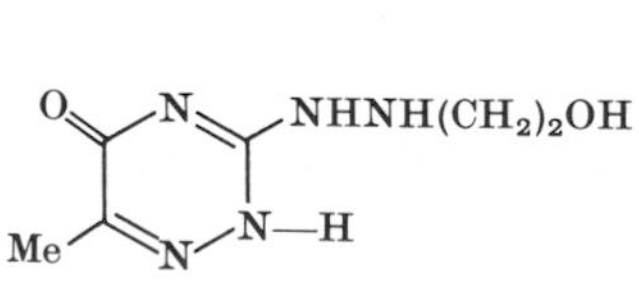

[**242/1**] 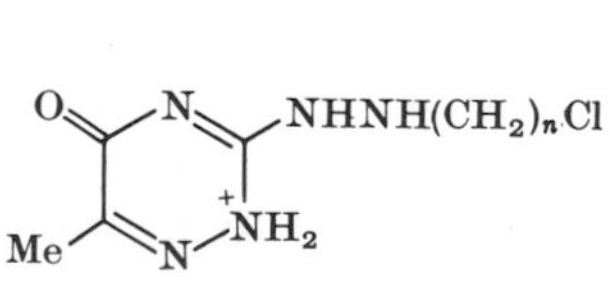 [**242/2**] 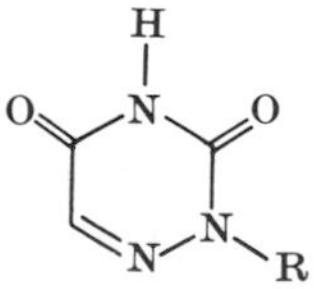 [**243**]

dioxo structure (63CZ1507). See also 74CX(B)1430) for the crystal structure of the title compound and (74JA1239) for 6-azacytidine structure.

The 6-azauracil monoanion prefers the structure [**244a**] with pK_T 2.5 against [**244b**] (62CZ716; cf. also 63CZ1499).

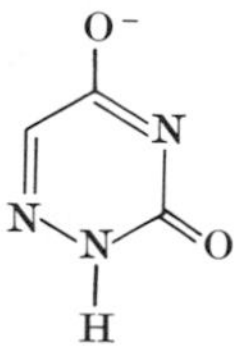

[**244a**] 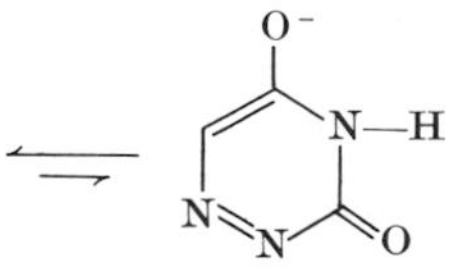 [**244b**] 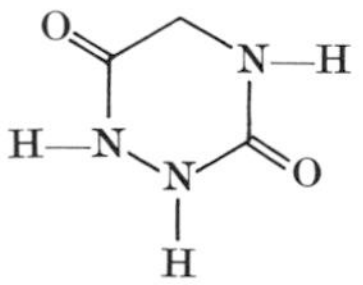 [**245**]

A fragmentary IR investigation of [**245**] was interpreted as showing the presence of some OH form (solid state) (64CB994), but a definitive investigation is required.

c. Derivatives of 1,2,3-Triazine

The oxo form [**246a** or **b**] is demonstrated for 1,2,3-benzotriazin-4-one (solid state) by νC=O at 1690 cm^{-1} (69JH779), and recently X-ray evidence has confirmed the 3*H* form [**246a**] (73CX(B)1916).

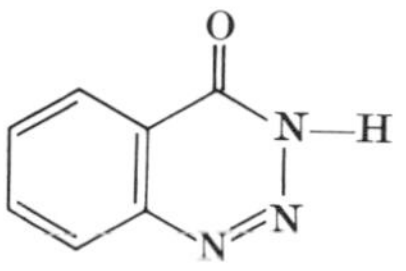

[**246a**] 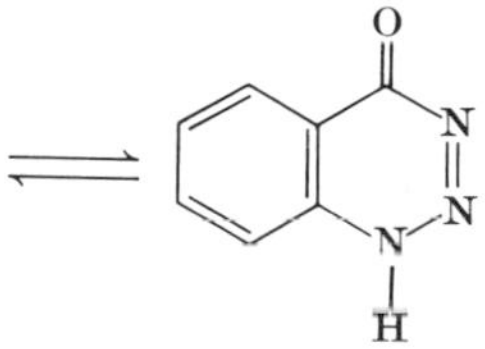 [**246b**]

G. Oxopolyazanaphthalenes (I-388)

a. Oxo- and Dioxotriazanaphthalenes (I-388)

Spectral evidence is available for the carbonyl structures shown of many oxo- and dioxoazanaphthalenes: [**246/1**] (69HC149), [**246/2**] (69AJ1759, 69CT2266), [**246/3**] (69AJ1759), [**246/4**] (69HC149) and [**246/5**] (67BJ153).

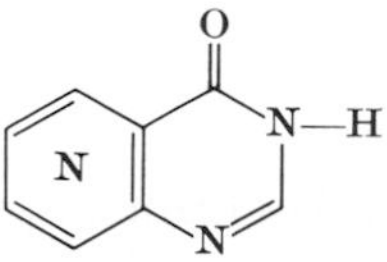

[**246/1**] 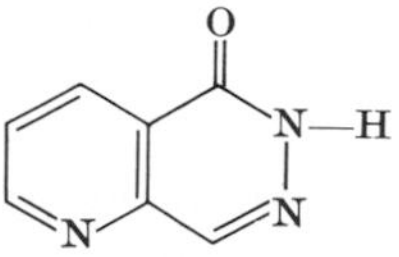 [**246/2**] 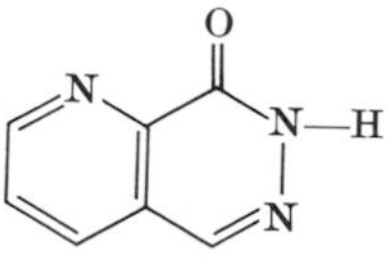 [**246/3**]

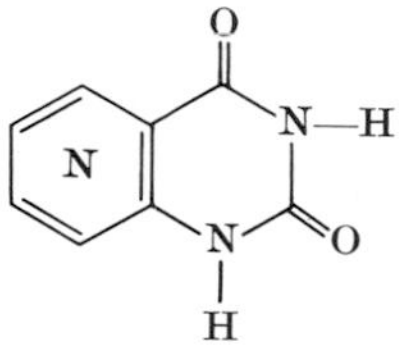

[**246/4**] 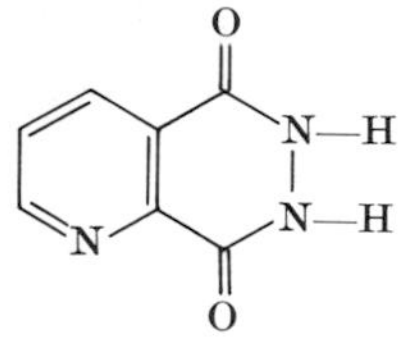 [**246/5**] 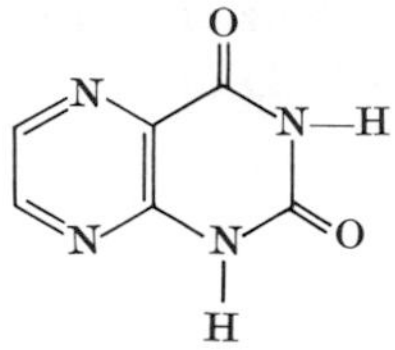 [**246/6**]

b. Monooxo Derivatives of Pteridines (I-389)

The covalent hydration of pteridinone has been reviewed thoroughly from both the qualitative (65HC(4)1) and the quantitative point of view (65HC(4)43). Since then some MO calculations have appeared (69NK769). More recently the oxo-amino structure of pterin (I-391) and of some derivatives has been confirmed (73H2680, 74CB3275) by ^{13}C NMR and the structure of their cations and anions studied by the same technique (73H2680, 74CB876).

c. Polyoxo Derivatives of Pteridines (I-391)

The dioxo structure [**246/6**] of 2,4-dihydroxypteridine (lumazine) has been confirmed by X-ray crystallography (72CX(B)659). For MO calculations see (67BJ2493) and for ^{13}C NMR of the anionic (73H2680), neutral and cationic species (73H2680, 74CB3275).

Previous work on 2,4,7-trihydroxypteridine (7-hydroxylumazine) (I-394) is confirmed (65BG458); K_T [**247a**] ⇌ [**247b**] for several 7-hydroxylumazines was measured in H_2O by the UV method. K_T depends on the position and number of methyl substituents: methyl at N(1) sterically hinders the amide form [**247b**], while a methyl group at C(6) hinders the hydroxy form [**247a**]. The tautomerism of the monocations, monoanions and excited electronic states was also studied by UV and fluorescence spectra (65BG458). 6,7,8-Trialkylpteridine-2,4-diones have also been studied (73H1908).

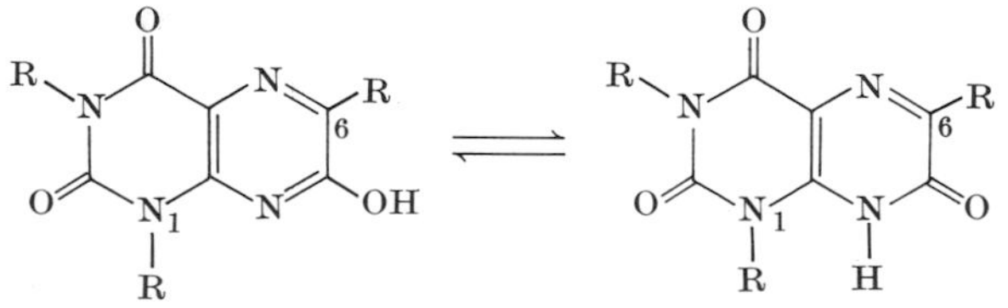

[**247a**] [**247b**]

The tetraoxo structure (I-395) of pteridine -2,4,6,7-tetraone appears to be confirmed by ^{13}C NMR and its dianion structure determined (74CB876). Analogously pteridine-6,7-dione exists as such (73CB3951, 74CB876).

d. Alloxazines and Isoalloxazines (I-426)

The previous conclusions (I-462) that alloxazines exist in the alloxazine form [**247/1a**] rather than as isoalloxazine [**247/1b**] have been reinforced by UV and fluorescence spectroscopy (66DA(171)1101) and by IR (68ZO2449). The same authors demonstrated that the *N*-hydroxy form [**247/2a**] predominates in solution, but the *N*-oxide [**247/2b**] is predominant in the solid state. For MO calculations on alloxazine, see (67BJ2493). Recently phototautomerism of alloxazines has been studied (74JA4319).

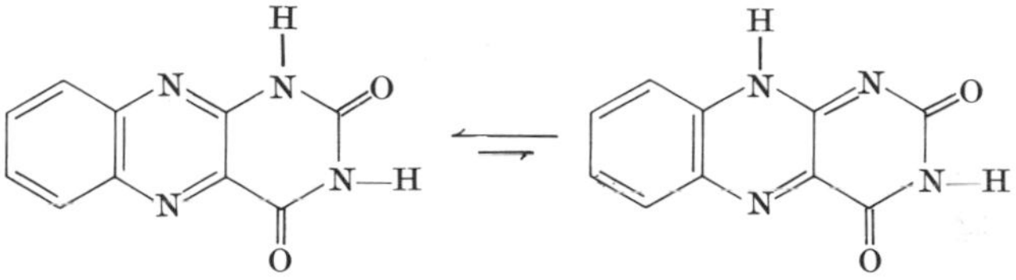

[**247/1a**] [**247/1b**]

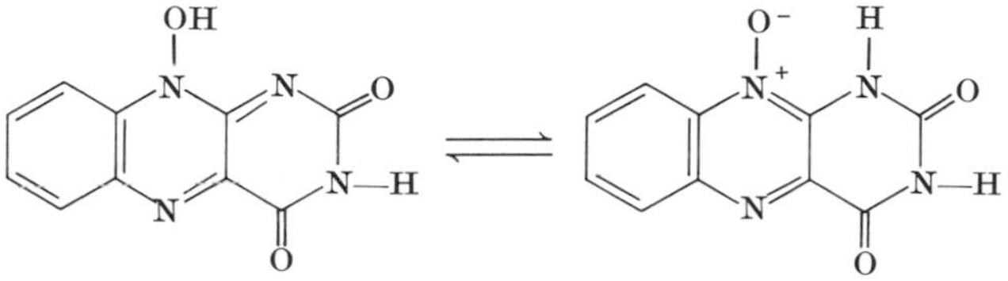

[**247/2a**] [**247/2b**]

H. Azines Carrying a Hydroxy Group on a Phenyl Substituent or on a Fused Benzene Ring

Differently coloured crystalline modifications have been claimed to be the tautomeric forms [**248a**], [**248b**] and [**248c**] on IR evidence (62LA(657)131).

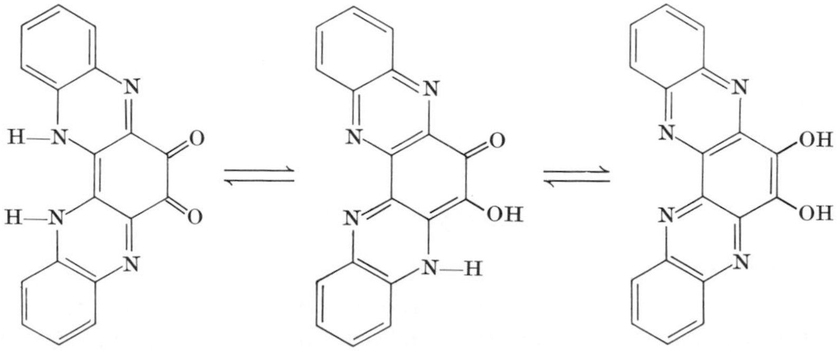

[**248a**] [**248b**] [**248c**]

UV comparisons of 2-hydroxybenzo[*c*]cinnoline with fixed forms demonstrate the hydroxy form [**249**] in EtOH (70AJ619).

UV comparisons with fixed forms show that tautomer [**250a**] is favoured over [**250b**] in aqueous solution (63JO2394).

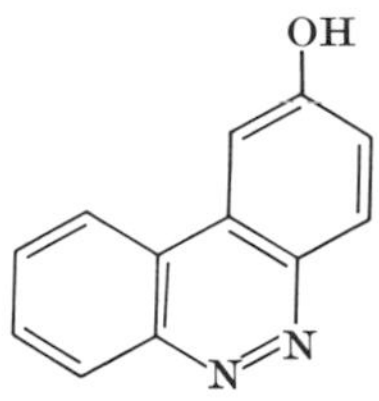

[**249**]

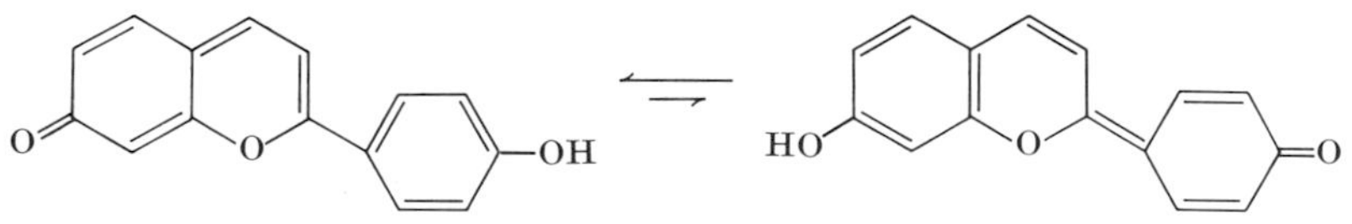

[**250a**] [**250b**]

Further work on hydroxyphenoxazones has appeared (cf. discussion and references I-383) (71CZ143, 72CZ1905), but the previous conclusions stand.

5. Compounds with Potential Mercapto Groups (I-396)

A. General Introduction

It was pointed out in the previous review (I-396), that the similarity generally found for the tautomerism of potential mercapto-heterocycles to the analogous potential hydroxy-heterocycles derives from a cancellation of the opposing influence of the greater acidity of SH over OH, but the greater bond strength of C=O over C=S. This assessment still stands, and much further evidence is now available for the quantitative similarity in tautomeric behaviour of the two groups of compounds: the α- and γ-thiones are usually somewhat more favoured than the corresponding α- and γ-oxo compounds.

B. Pyridinethiones and Benzologues

a. Pyridine-2- and -4-thiones (I-396)

Further confirmation for the thione character of 2-pyridinethione comes from the NMR spectrum, which shows that the C=S form applies both to monomers and hydrogen-bonded dimers (72AS2255). However, ionization potentials have been interpreted as suggesting that mercaptopyridines exist at least in part in the SH form in the gas phase (72OM823) (cf. discussion on corresponding pyridone/hydroxy-pyridines, Section 2-3Bg).

Polyhalogenated 2- and 4-mercaptopyridines are true thiols as shown by IR and/or NMR (69J(C)1660, 69TL1507) and more recently by UV (74J(PI)2307); this resembles the behaviour in the pyridone series (Section 2-3Bc).

Pyridine-2- and 4-thiones are shown by NMR to be protonated on sulphur (65J3825).

b. Quinoline-2- and 4-thiones and Isoquinoline-1- and 3-thiones (I-398)

Very little new work has appeared. The potential tautomerism of homothiophthalimide has been mentioned (66JH282), but apparently not investigated.

c. Acridinethiones (I-398)

The thione form for acridinethione [**256**] (cf. I-398) has been confirmed by UV and IR comparisons (70KG191).

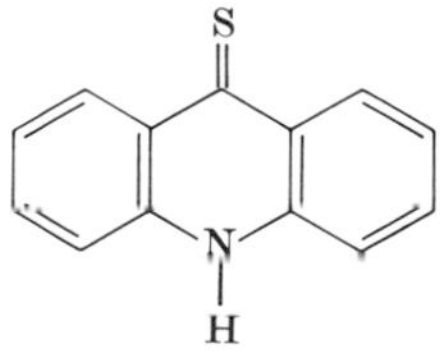

S

N

H

[**256**]

d. Quinolines Carrying a Mercapto Group on the Benzo Ring (I-400)

The concentration of the 8-mercaptoquinoline zwitterion decreases as the dielectric constant of the solvent decreases: H_2O > MeOH > EtOH > Bu^tOH (I-400). Further UV spectral studies showed that the zwitterionic form, which possesses a high intensity band at long wavelengths, is even less favoured in various aprotic solvents (0.5% in CH_3CN, 0.14% in acetone, 0.09% in $CHCl_3$, > 10^{-5}% in Et_2O and isooctane) (66AC1702).

C. 1-Hydroxypyridinethiones (Mercaptopyridine 1-Oxides) (I-399)

pK_a values of 9-mercaptoacridine 10-oxide [**257**] and of the 9-methylthio analogue indicate the thiol structure [**257a**] in H_2O in equilibrium with up to 50% of the thione form [**257b**] (66T3227) (cf. the oxygen analogues: Section 2-3Fb).

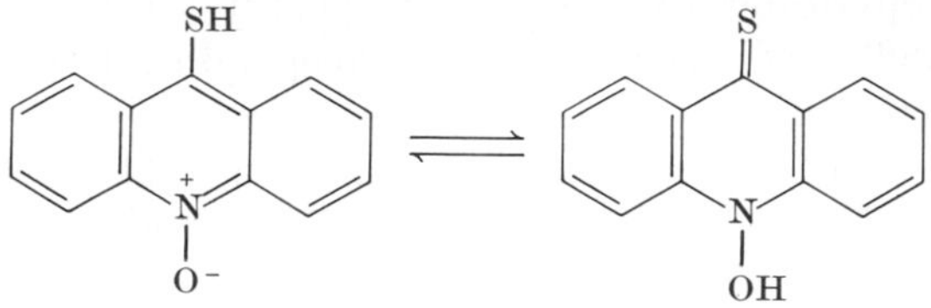

[**257a**] [**257b**]

D. Oxazinethiones and Thiazinethiones (I-400)

Spectral comparison with the analogous oxygen derivatives (Section 2-4A) establish the thione structure [**258**] in EtOH and $CHCl_3$ and in the solid state (69T517). Structure [**258/1**] has been assigned on IR evidence (70AJ51).

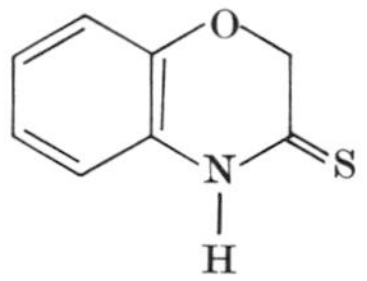

[**258**] 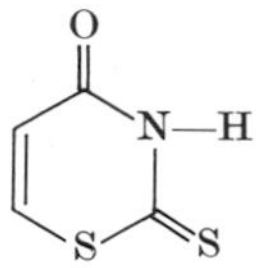 [**258/1**]

E. Diazinethiones (I-400)

Thione structures have been assigned to (62J3129; cf. also 71J(B)1261), or confirmed for (cf. I-400), pyrimidine-2- and -4-thione, pyridazine-3- and 4-thione, cinnoline-4-thione, phthalazine-1-thione, pyrazine-2-thione (see, for X-ray results, 66CX(21)249), quinoxaline-2-thione, and quinazoline-4-thione and for 8-methoxycinnoline-4-thione (cf. 71J(C)3088) by UV comparisons with models. For pyrazine-2-thione and its cation, MO-calculated and experimental UV spectra agree on the thione formulation (73BA405).

IR spectra are also consistent with the thione structure for cinnoline-4-thione (64JH221), and 2-thiouracil is a thione on various evidence (61AK(16)459) as is 4-thiouridine as shown by X-ray methods (69AG-(E)139).

As Albert and Barlin point out (62J3129), pK_T values for thiol:thione equilibria can only be calculated from pK_a values if cations of similar structures are formed. These workers originally assigned structure [**259**] to the pyridazine-4-thione monocation and structure [**260**] to the cinnoline-4-thione monocation. Although considerable further work has been done (65J2260, 65J5391, 71J(B)1261), these cation structures are still not completely settled. A complete quantitative elucidation of the tautomeric structure of the monocations, and hence of the tautomeric neutral species, requires the measurement of pK_a values for second proton addition.

In the pyrimidine-4-thione and quinazoline-4-thione series, the monocations are unambiguously of structure type [**261**]; pK_a measurements thus allow the measurement of the equilibrium constant:

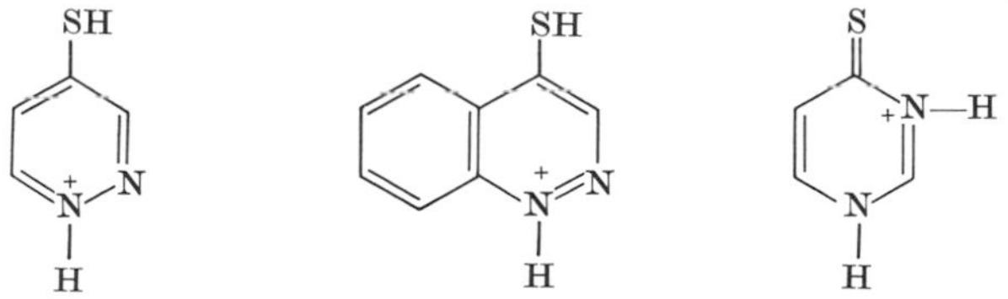

[**259**] [**260**] [**261**]

[**262a**]/[**262b**] = 2 (and 30 for the analogous quinazoline-4-thione case), but give no quantitative information on the minor mercapto contributor [**262c**] (62J3129). By contrast, dipole moment comparisons with the fixed forms indicate that in benzene the *o*-quinonoid form [**263a**] predominates by a much greater factor (97%) over the *p*-quinonoid form [**262b**] in pyrimidine-4-thione and that the 2-phenyl analogue exists essentially completely in the 3*H* form (100%) (73CO(276C)1341). For a possible explanation in terms of solvent effects, see Section 2-4Da.

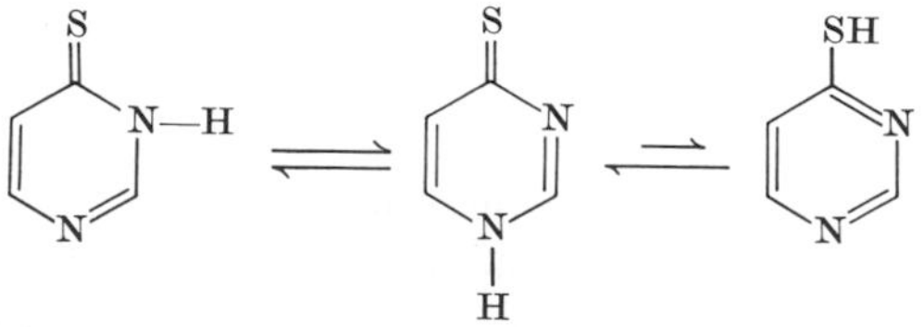

[**262a**] [**262b**] [**262c**]

A thione structure is demonstrated for tetrahydrocinnoline-3-thione [**263**] and some derivatives, on IR, UV and NMR spectral evidence (70BF4011). The suggestion (71BF1858) that 2-phenylpyrimidine-4-thione [**264**] exists mainly as the thiol [**264c**] in DMSO on NMR evidence is unlikely and needs reinvestigation (cf. Section 2-4Da).

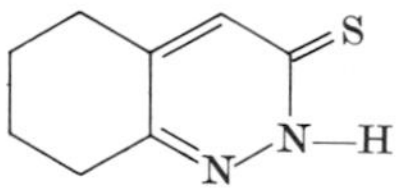

[263] 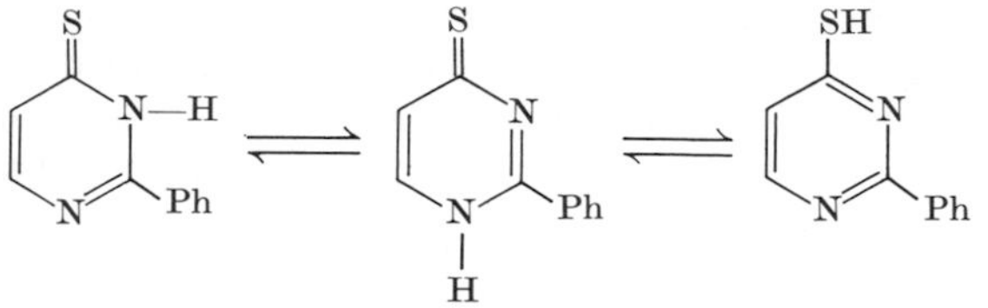 [264a] [264b] [264c]

The formulation [**265**] (63J4333) is probably correct in view of the methoxy-substituent effect on the basicity of N(1). UV comparison with the corresponding *S*-methyl derivative indicates a thione structure [**266a**] or [**266b**] in MeOH (67CB3664). Polysubstituted 4-thiopyrimidines as [**267**, R = Ph, CH_2Ph or *p*-BrC_6H_4] show νNH bands and thus exist in a thione form: UV comparisons with N(1) substituted derivatives suggested the presence of both [**267a**] and [**267b**] (64JO1115).

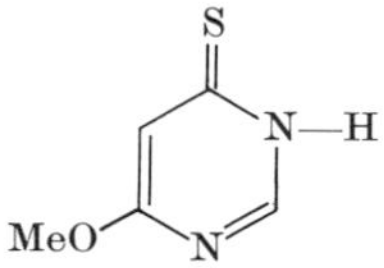

[265] 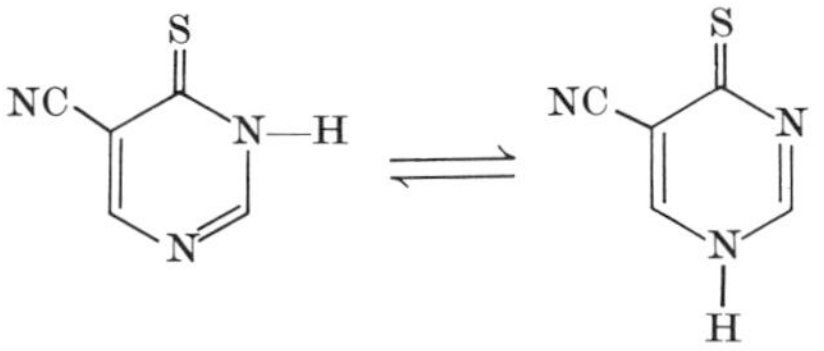 [266a] [266b]

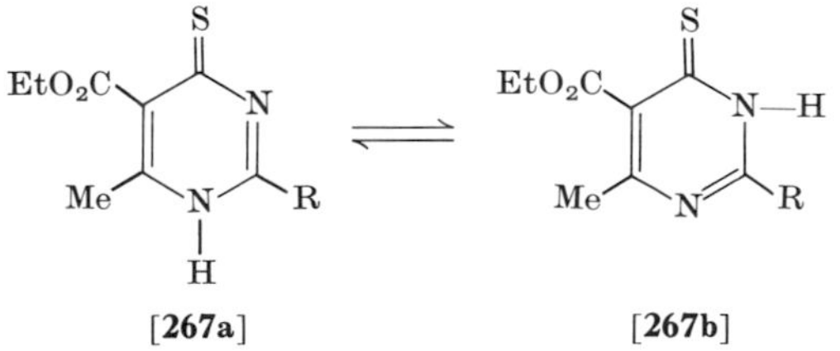

[267a] [267b]

Very recently UV and IR demonstrated that 4-thiouracil, 4-thiouridine and 5,6-dihydro-4-thiouracil exist in the 2-oxo-4-thione form both in aqueous and nonaqueous media and that the mono anion of 4-thiouracil behaves similarly to the uracil anion (cf. Section 2-4Dd) (74JA6832). Previous X-ray diffraction shown that 4-thiouracil nucleosides possess the 2-oxo-4-thione structure in the solid state (70MB(50)153, 72JA621, 73MI473b).

The 2-thione (I-401) and 2-selenone structures of 2-thio- and 2-seleno-uracil and the dithiouracil dithione structure (I-401) are confirmed by X-ray (67JA1249). Previous X-ray studies demonstrated the diselenone character of diselenouracil (66JS643). The 2,4-dithione structure was confirmed by X-ray evidence for 2,4-dithiouridine by two independent groups (71CX(B)961, 71CX(B)1178). Although the X-ray structure of 2-thiobarbituric acid has been investigated (67CO-(265C)631), the data do not allow a definitive choice between the dioxo-thione [cf. barbituric acid (Section 2-4Dg) and the hydroxy-oxo-thione structures of which the latter was previously chosen (I-401). An analogous study on 1,3-diethyl-2-thiobarbituric acid also does not appear to be conclusive (68CO(266C)1281). The 2-thione-4-hydroxy-6-oxo formulation has been demonstrated by IR for anhydrous 2-thio-5-nitrobarbituric acid, while, rather surprisingly, the trihydrate is concluded to exist in the 2-mercapto aci-nitroform (74RR679) (cf. 2-4Dh).

UV comparisons with fixed forms suggested the 4-hydroxy-6-mercaptopyrimidine [**268**] exists in the oxo–thione form [**268a**] in H_2O and EtOH possibly with contribution from forms [**268b**] and [**268c**] (63J4333, 64J3204); however, the betaine form [**268d**] was not considered, and the present authors consider that it is the most likely structure.

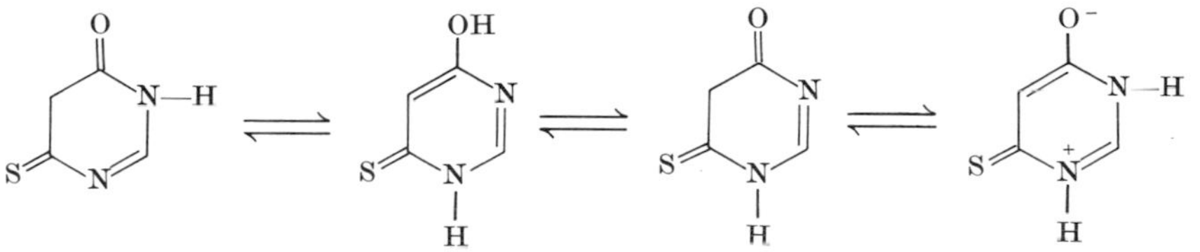

[268a] [268b] [268c] [268d]

UV comparisons with fixed models (H_2O) and IR (solid state) suggested structures [**269**], and [**271**] for the pyridazines with two potential tautomeric functions (64BJ1107). The monothione-mono-

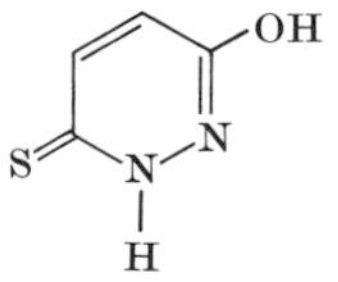

[**269**] 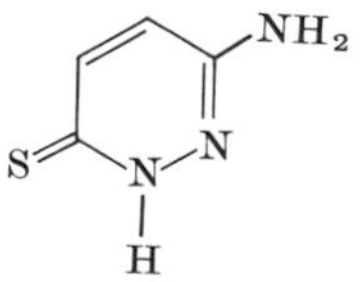 [**271**]

mercapto structure of 3,6-dimercapto-pyridazine, previously (I-402) demonstrated by IR, has been recently confirmed by comparison of the UV spectra and pK_a data with those of the fixed forms in water (74J(PII)1199, see also 64MI81).

F. Triazinethiones (I-402)

UV comparisons with fixed models demonstrate structures of type [**272**, X, Y = S or O] for 2-thio-, 4-thio-, and 2,4-dithio-6-azauracil in dioxan, EtOH and H_2O (62CZ1886); IR show that similar structural conclusions also apply in $CHCl_3$ and the solid state (63CZ3392). This work has been extended to 4-thio-6-azauridine (63CZ1507). In all these compounds the monoanion [**273**] is formed by loss of the 3-position*

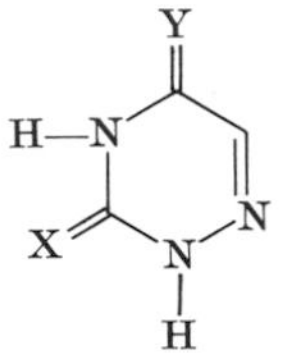

[**272**] 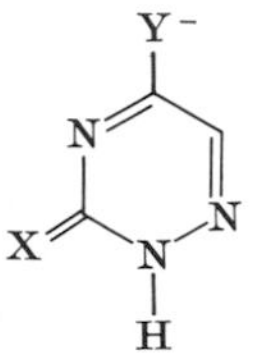 [**273**]

hydrogen (62CZ1886). Later detailed spectral comparisons have confirmed these conclusions for the compounds [**274**] (71BF3658), [**275**] (72BF1511), [**276**] (72BF1975): however, evidence was also given for some significant proportion of a mercapto structure for compound [**277**] (72BF1975).

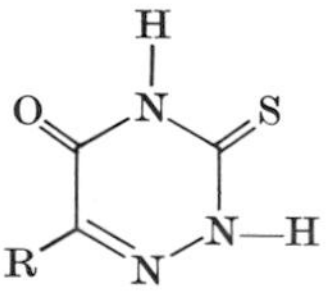

[**274**] R = Me
[**275**] R = Ph
[**276**] 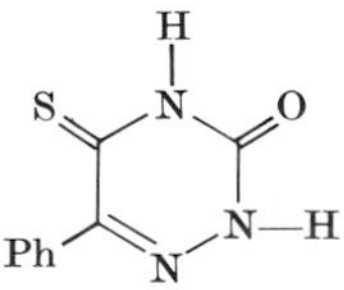
[**277**] 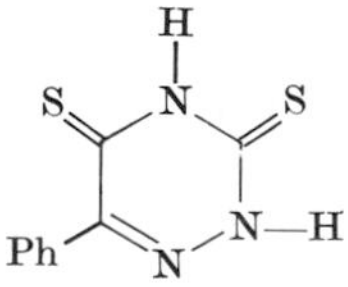

Fragmentary IR studies on [**278**] and [**279**] (solid state) are interpreted as showing tautomerism (64CB994): the evidence is thin.

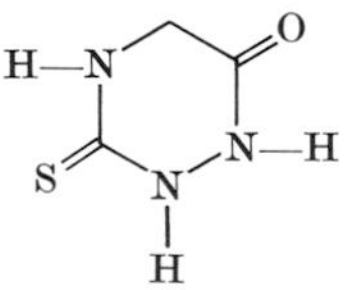

[**278**] 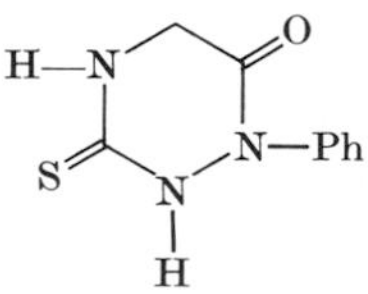 [**279**]

4-Methyl-1,2,4-triazine-5,6-dione-3-thione [**280**] and partially methylated derivatives have been studied by IR and by UV comparisons (70BF1599): the predominant form (solid, EtOH, dioxan) is [**280c**].

* 4-Position on systematic numbering.

The IR complexity [**280**] was taken as indicating the presence of various tautomeric forms in the solid state (70BF1590). Structures [**281**] and [**282**] are shown to predominate for these dimethyl derivatives, while [**283**] exists mainly in the dioxothione form shown, at least in EtOH and the solid state; changes in the spectra in dioxan may be due to the thiol tautomer.

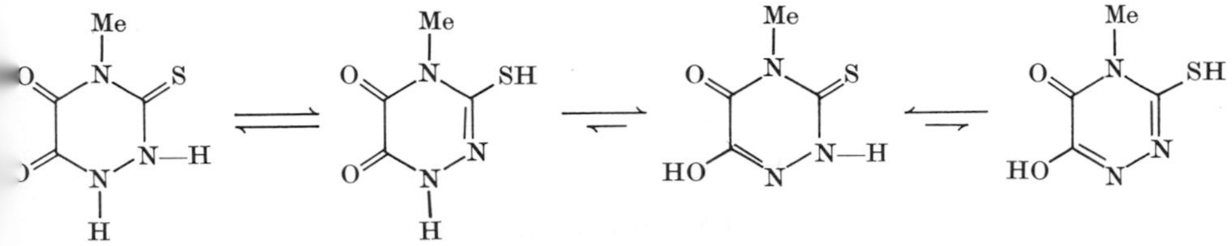

[**280a**] [**280b**] [**280c**] [**280d**]

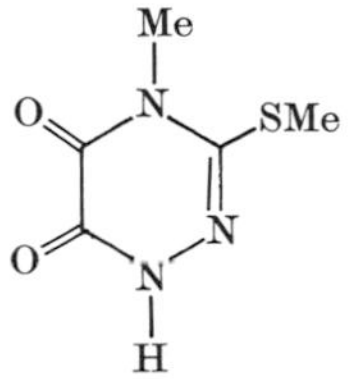

[**281**] 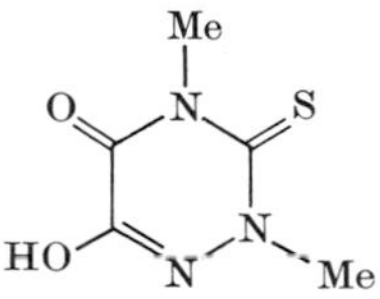 [**282**] 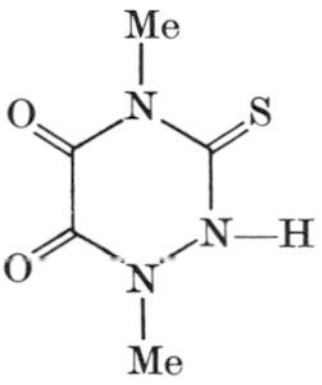 [**283**]

The absence of νOH and the presence of νC=O points to the existence of structure [**285**] in the solid state (68JO888).

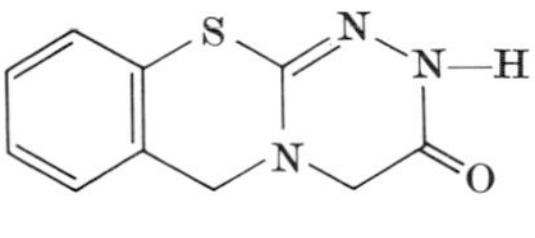

[**285**]

A thione structure [**286**] for 1,2,3-benzotriazine-4-thione is shown by νNH at 3100 cm^{-1} (solid) (69JH779): no distinction is presently possible between [**286a**] and [**286b**].

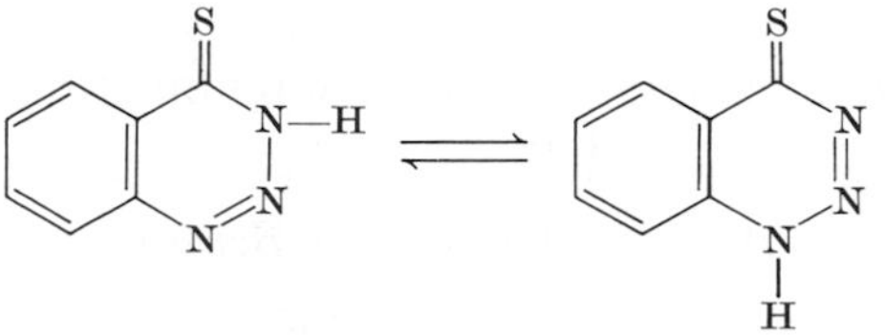

[**286a**] [**286b**]

Bands attributed to N—C═S at 1115, 1330, 1540 cm^{-1} and to NH at 2900, 3200 cm^{-1} support the trithione structure [**287**] for trithiocyanuric acid (66J(C)909).

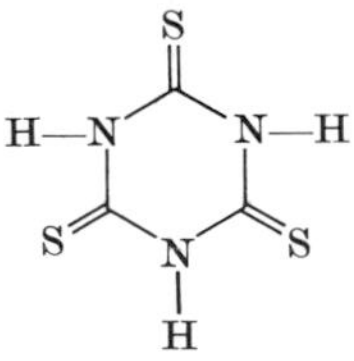

[**287**]

G. Pteridinethiones

Pteridine-6-thione undergoes a large bathochromic shift on anion formation characteristic of loss of H_2O from a covalently hydrated thione [**288**] (65J27) (cf. oxygen derivatives I-391).

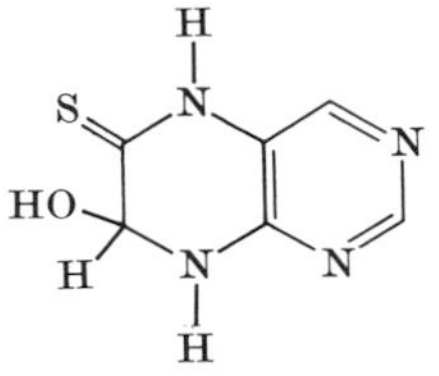

[**288**]

6. Compounds with Potential Amino Groups (I-403)

A. General Discussion

It was concluded in the previous review that amino forms should be, and indeed were, more favoured than corresponding hydroxy forms. Subsequent work has fully borne this out. Reasons have already been discussed in Section 2-1A.

A great deal more work is now available on the more complex tautomerism of potential aminohydroxy compounds, and rationalization of these data is, for the most part, quite straightforward.

B. Amino Derivatives of Six-Membered Rings with One Heteroatom

a. Neutral Species of Aminopyridines and Benzopyridines (I-404)

It was already clear at the time of the previous review that for most aminopyridines and their benzologues the amino structure greatly predominates (I-404). Qualitatively this conclusion has been amply confirmed. However, the most significant advance in the tautomerism of aminopyridines has been the realization that quantitatively the preference for the amino form is considerably greater than was previously thought (Table 2-5). This arises mainly from the great basicity of the 1-methylpyrid-2- and -4-onimines, which necessitates the use of special H acidity functions for their measurement (71CH510, 72J(PII)-1295, 73J(PII)1080). CNDO calculations are in agreement (73ZO2730).

TABLE 2-5

TAUTOMERIC RATIOS FOR AMINOPYRIDINES

Compound	pK_T	pK_T[b]	pK_a of amino	pK_a of 1-methyl imino
2-Aminopyridine	6.2[a]	5:3	6.86[a]	13.02[a]
4-Aminopyridine	8.7[c]	3.3	9.12[c]	17.87[c]
2-Aminoquinoline	4.3[c]	—	7.34[c]	11.68[c]
1-Aminoisoquinoline	3.8[c]	—	7.62[c]	11.38[c]

[a] 72J(PII)1295.
[b] pK_T previously reported in 52J1461.
[c] 73J(PII)1080.

There has been considerable further qualitative work. NMR in DMSO confirms the amino structure for 4-monoalkylamino-pyridines and -quinolines by disclosing coupling between the NH and the α-H of the alkyl group (68JH631). Similar conclusions were reached from ^{14}N NMR (71T3129). Previously some authors (66CO(262C)1161) had interpreted the spectra as showing 4-aminoquinoline in the imino form in neutral media and in the amino form in the presence of alkali amide, but this is incorrect. The ionization potentials of 2-, 3- and 4-aminopyridines and of their methylamino derivatives indicate that all these aminopyridines exist in the gas phase predominantly in the amino form (72TL3193), as did X-ray studies of 2-amino-5-chloropyridine (74CX(B)474) for the solid state.

The amino structure for 9-aminoacridine was established earlier (I-407); renewed doubts since expressed (63CZ1651) were shown groundless by the IR partial deuteration method (1-6Cc) which confirmed the amino structure for various solvents (65J5230).

The effect of other ring substituents on the tautomeric equilibria of aminopyridines has not been extensively investigated: qualitatively they do not appear to have much influence. NMR and IR spectra demonstrate that 6-alkoxy-2-aminopyridines [**289**] exist as such in $CHCl_3$ (65BF52). NMR spectral data in DMSO agree with the diamino structures [**290**] and [**291**] (64J1423).

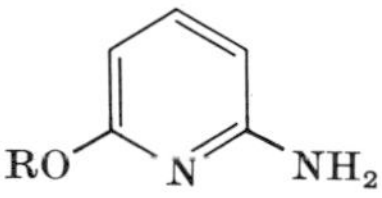

RO N NH_2 H_2N N NH_2 NH_2 N NH_2

[**289**] 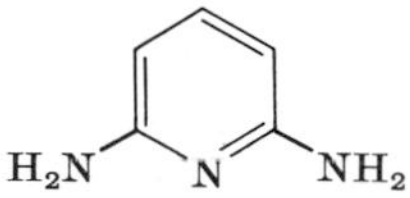 [**290**] 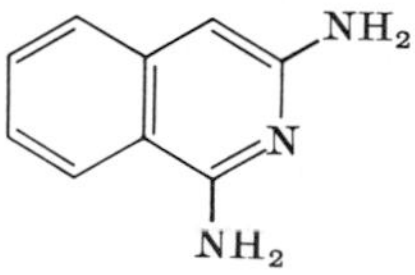 [**291**]

UV spectral differences between 6-amino-4-methyl-7-azaindoline [**292**] and the corresponding 6-oxo-7-azaindoline in solvents of various polarity and IR bands for NH_2 (solid and $CHCl_3$) demonstrate the amino structure [**292**] (66T3233). Mass spectrometry supports the amino structure for 2-, 3-, and 4-aminopyridine in the gas phase (72TL3193).

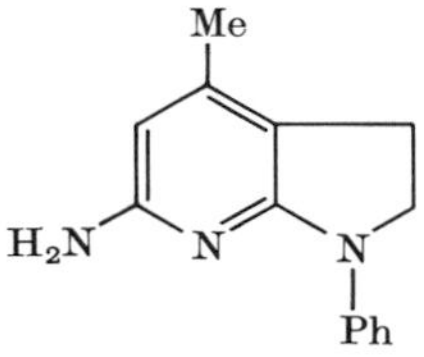

Me H_2N N N Ph

[**292**]

Examples of muddled thinking regarding the effect of tautomeric structure on chemical reactivity are still found. While the suggestion (72JH1039) that the reaction of 2-aminopyridine [**293**] with picryl fluoride gives [**295**] by initial formation and subsequent rearrangement of [**294**] is probably correct, it is *not* correct for these authors to conclude that in the initial production of [**294**] the 2-aminopyridine [**293b**] reacts via the imino form [**293a**]. Reaction of 2-aminopyridine with an electrophile is expected to occur at the ring nitrogen atom (see also 73JH167.)

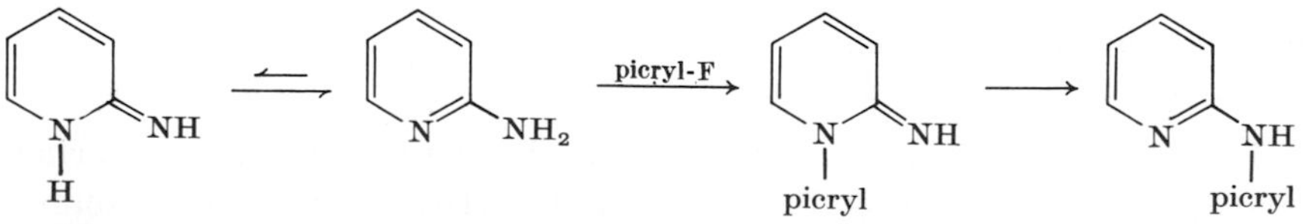

N NH H ⇌ N NH_2 —picryl-F→ N NH picryl → N NH picryl

[**293a**] [**293b**] [**294**] [**295**]

b. Cations of Aminopyridines and Aminobenzopyridines (I-409)

NMR confirms (see I-409) protonation on *ring* nitrogen in the cations of 2- and 4-aminopyridine (cf. [**296**]) since coupling is found between α- CH and NH^+ in the hydrochlorides in liquid SO_2 (65J3825; see also 69AJ2595). UV spectra show that the 2-aminoquinoline and 4-amino-2-methylquinoline cations are also formed by preferential protonation at the ring nitrogen in both the ground and excited states (72AC1611).

While 3-aminopyridine also undergoes predominant protonation at the ring nitrogen atom (I-409), 3- and 5-amino-2-pyridone form the cations [**297**] and [**298**]. However, if the amino and carbonyl groups of

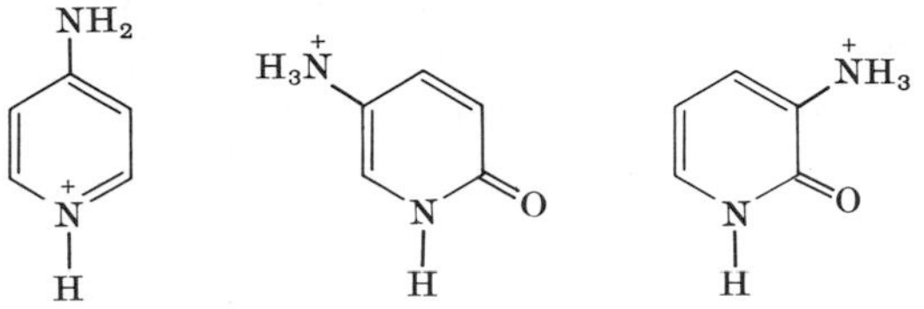

[**296**] [**297**] [**298**]

an aminopyridone are directly conjugated, protonation occurs (as demonstrated by UV) at the oxygen atom to give [**299**]–[**301**]. However, 3-amino-4-pyridone is protonated at NH_2 (71J(B)1425). 2-Amino-4-quinolone undergoes *O*-protonation, as expected (73J(PI)1314).

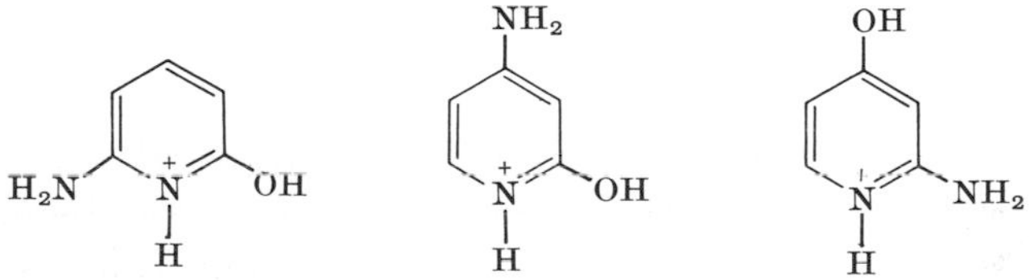

[**299**] [**300**] [**301**]

The corresponding aminopyridinethiones behave similarly on protonation: formation of the cations of 3- and 5-aminopyridine-2-thione and 3-aminopyridine-4-thione involves proton addition to the NH_2 group, whereas 6-aminopyridine-2-thione and probably 2-aminopyridine-4-thione add the proton at the C=S group (72J(PII)1459). Most aminoacridines are protonated at the ring nitrogen atom. However 4,5-diaminoacridine undergoes successive protonations at the amino groups to give [**302**] as shown by UV, probably because of steric

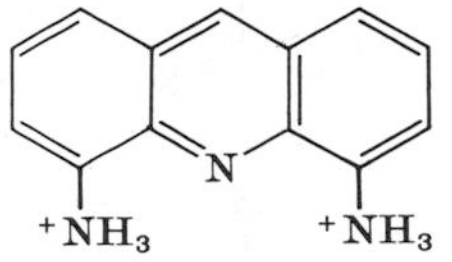

[**302**]

hindrance to protonation at ring nitrogen; 4-aminoacridine is also protonated at the amino group (65J4653) (cf. I-409).

c. *Aminopyridine 1-Oxides* (I-410)

These compounds exist very predominantly in the amino form (I-410). No further work of significance has appeared: despite some authors' (67JO1106) contrary opinion, formation of coloured chelates with Fe^{3+} and Cu^{2+} is *not* evidence for any significant proportion of [**303b**] in 6-aminophenanthridine 5-oxide [**303**], and this compound certainly exists very predominantly as [**303a**].

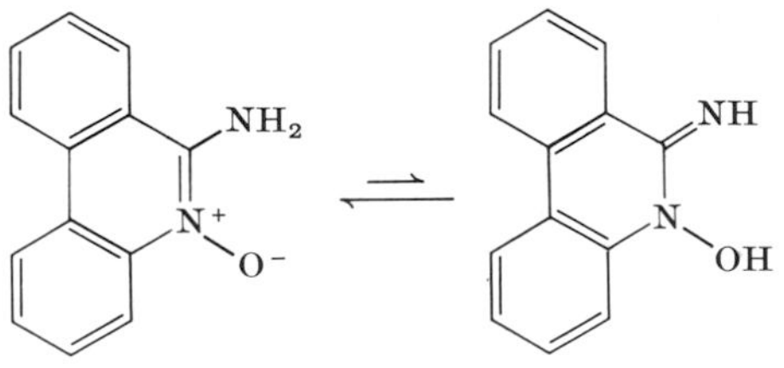

NH_2 N^+ O^- ⇌ NH N OH

[**303a**] [**303b**]

C. Aminopyrimidines (I-412)

a. *Aminopyrimidines Not Cytosine Derivatives* (I-412)

The previous review (I-413) reported applications of the pK_a method which showed that for 2- and 4-aminopyrimidine the amino forms predominated over the imino in H_2O by factors of ca. 10^6. Further pK_a measurements and the closeness of the ρ factors in the Hammett equation for 5-substitution in 4-amino- and in 4-dimethylamino-pyrimidines have since shown that the amino tautomer is predominant (I-413) for various 5-substituted 4-aminopyrimidines [**304**] in H_2O (66JO1199; see also 71BF1858). Correlation with σ_m and σ_p constants has been considered evidence for N(1) protonation to give cations of type [**305**] (66JO1199).

UV similarities of the cyano [**306**] and aldehyde [**307**] derivatives

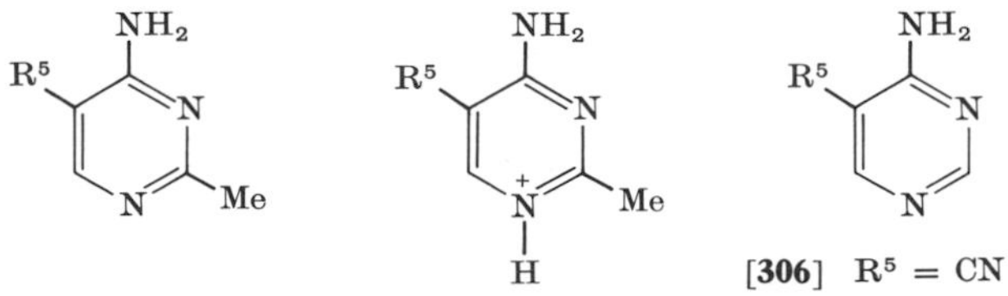

NH_2 R^5 N N Me NH_2 R^5 N N^+ Me H NH_2 R^5 N N

[**304**] [**305**] [**306**] R^5 = CN
[**307**] R^5 = CHO

in MeOH, and the bathochromic shift in the NMe_2 derivative of [**306**], support the amino structures shown (67CB3664).

pK_a and UV comparisons (65J755) and NMR (64AK(22)65) of 2,4-diaminopyrimidine with fixed models demonstrate the diamino tautomer [**308**]. It is uncertain (65J755) whether the 1-methyl derivative has structure [**309a**] or [**309b**].

The imino structure [**310a**] was suggested on the IR evidence of $\nu C{=}O$ at 1700 cm^{-1} (KBr) (71J(C)3040), but the alternative [**310b**] seems just as probable; although the NMR in $CHCl_3$ shows a signal for CH_2, the IR is unfortunately not reported in this solvent.

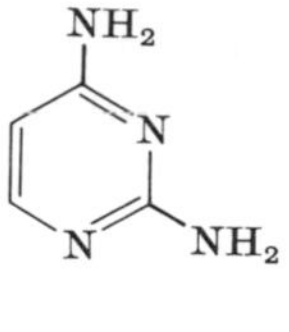

[**308**] 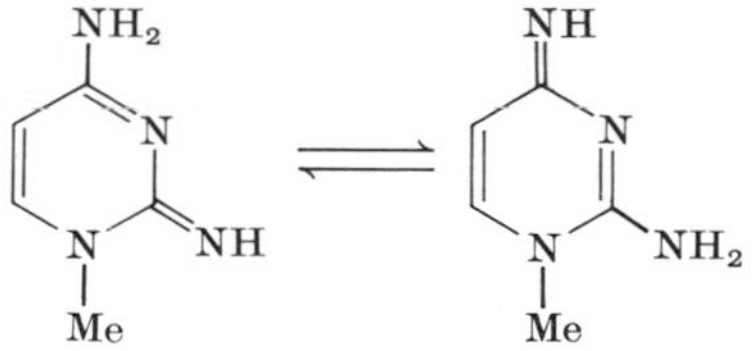 [**309a**] [**309b**]

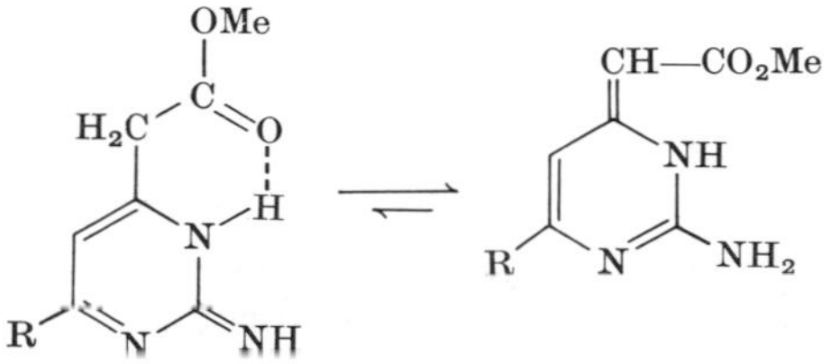

[**310a**] [**310b**]

1-Methyl-2-aminoperimidine [**310/1**] and the analogous compound [**310/2**] exist predominantly in the amino form [**310/1a, 310/2a**] as shown by IR (solid state and $CHCl_3$). UV spectra of the alternative models are too similar to allow easy differentiation, but pK_a measurements give K_T values of ca. 50 (in H_2O), considerably smaller than for most amino:imino equilibria (71KG807).

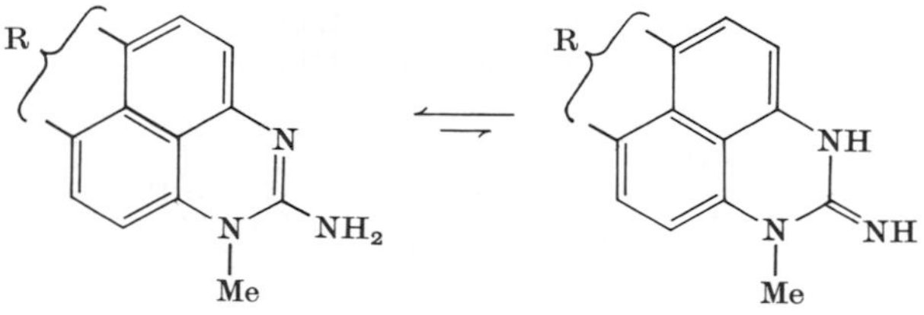

[**310/1a**] R = H,H [**310/1b**]
[**310/2a**] R = CH_2—CH_2 [**310/2b**]

b. Cytosine Derivatives: Neutral Species (I-374, I-414)

The previous review (I-374, I-414) summarized the compelling evidence for the predominant oxo–amino structure [**311a**] for cytosine [**311**]: further supporting work has been reported for both H_2O and DMSO solutions (62AJ851, 63J3046); the pK_T favours [**311a**] over [**311c**] by 3 units. Further Raman spectra (67SA2551) and ^{14}N nuclear quadrupole resonance spectra (72JP(57)5087, 72MI265) support the structure [**311a**] of cytosine and cytidine.

The X-ray structure of cytosine (63CX20, 64CX1581, 73CX(B)1234), confirms structure [**311a**] for the solid state. For MO calculations see (71MI5).

As previously reported (I-414), the pK_a method had demonstrated the predominance of oxo–amino form [**312a**] over [**312b**] for 1-methyl cytosine with a pK_T ca. 5; this is confirmed by NMR in DMSO on ^{15}N-labelled compounds (63SC(142)1569, 65JA5575) and by X-ray studies (64NA(201)179).

6-Chloro- and 6-fluoro-cytosine also exist in 4-amino-2-oxo forms from UV comparisons (64JA2474). The oxo–amino structures shown for 3-methylcytosine [**315**] and of some cyclic derivatives [**316**] were established similarly (63JA4024).

The oxo–amino form [**313a**] of cytidine, both as the monomer (cf. X-ray (I-414, 65CX(18)313) and bound in helical polynucleotide structures, is supported by IR (in D_2O) (59BY274, 61PN791). For an X-ray investigation of the monoclinic form of cytidylic acid *b* see (67MB(25)67). At one time, the imino form [**314b**] was claimed for

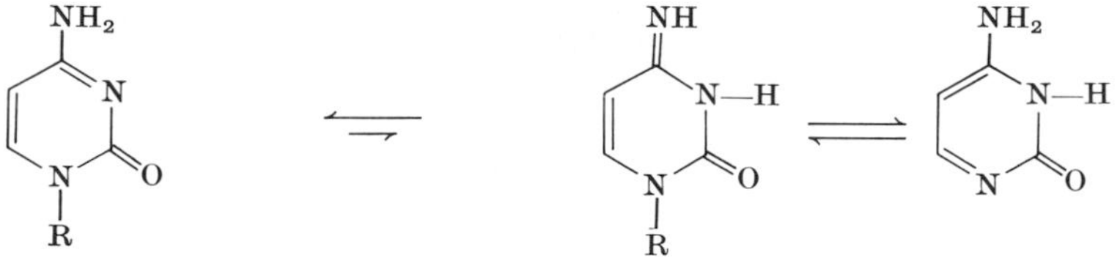

[**311a**] R = H [**311b**] [**311c**]
[**312a**] R = Me [**312b**]
[**313a**] R = β-D-ribofuranosyl [**313b**]
[**314a**] R = 2′-deoxy-β-D-ribofuranosyl [**314b**]

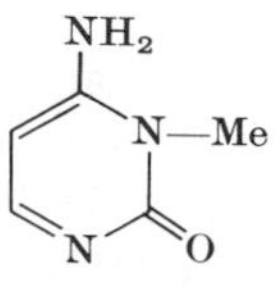

[**315**] 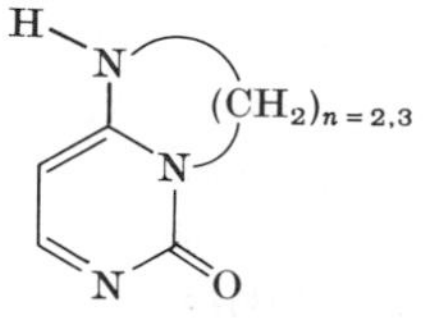 [**316**]

2-deoxycytidine on NMR evidence (62JA4464). Subsequent work (63JA1007, 63JA1657, 63TL1027, 65JA5575) showed that the NMR spectrum of authentic deoxycytidine in DMSO or D_2O is consistent with the amino structure [**314a**]: the discordant conclusion was probably concerned with deoxycytidine hydrochloride.

Recent claims (71BO435, 72JA951) to demonstrate 15 ± 3% of the imino form in cytidine 5′-monophosphate have incorrectly criticized the previous work (I-414, 61PN791) on cytosine tautomerism. A bogus enthalpy difference was deduced suggesting that the imino form [**311b**] was only 1.1 kcal/mole less stable than the amino form [**311a**]. This difference was compared with theoretical calculations (70JA2929) which suggested a difference of 2.2 kcal/mole, and was discussed in terms of possible pairing of the imino tautomer [**311b**] with adenine (instead of the amino form [**311a**] pairing with guanine) to lead to genetic mutation (72JA951). This bubble was finally pricked by the senior author (73JA3408) and others (73JA3511). Helene *et al.* (64CO(259)3385, 64CO(259)4853) believe from spectral studies that whereas tautomer [**311a**] dominates in aqueous solution for cytosine, its 1-methyl derivative and cytidine, tautomers [**311c**] and [**311d**] also exist in small amounts, which increase with temperature and in EtOH. Different authors (68TC(11)279) report from similar studies that cytosine exists as a mixture of [**311a**] and [**311b**], the first predominating in trimethylphosphate and water at room temperature, but the second at high temperature and in acetonitrile. From the temperature dependence of the absorption spectrum of cytosine in aqueous solution they give tautomeric ratio K [**311a**]/[**311b**] of 33, 14 and 8 at 30°, 50° and 70°C, respectively. So-called energy and entropy differences between the imine form [**311b**] and the amine form [**311a**] were claimed as ~5.5 kcal/mole and ~12 cal/mole deg, respectively. However such changes in the UV absorption spectra of cytosine or cytidine from an aqueous to aprotic medium are probably due to hydrogen bonded solvent–solute complexes (71MI923). Observed spectral changes of cytidines showed that the imine tautomer of cytidine does not exist in an appreciable amount in organic solvents; the changes upon heating the aqueous solutions of cytosine or cytidine reflect the stability of the hydrogen bonded complex (71MI923).

Displacements of the near ultraviolet absorption bands of cytosines from the neutral to the ionic form provide the ionization constants in the excited state: there was no considerable shift in the tautomeric equilibrium of the forms [**311a**] and [**311b**] (K [**311a**]/[**311b**] = 1.55×10^4). The concentration of form [**311b**] in both the ground and the excited state is small. The temperature dependence of the pK values suggests that rise in temperature increases the relative amount of forms [**311b**] and [**311c**]. The free energy change and the heat of the tautomeric conversion from [**311a**] to form [**311b**] were determined as 6.73 and 5.72 kcal/mole, respectively (72MI5). Certainly, cytosine and its 1-substituted derivatives [**311**] exist very predominantly in the amino–oxo form [**311a**]. For a sounder discussion of energy differences between the amino and imino forms of cytosine and their relation to mutagenesis see e.g., (64MI135, 70C134, 71MI923). A comprehensive review is now available on the tautomeric structures of pyrimidines of biological importance (see 75HC(18)199).

c. *Dihydrocytosine, 6-Oxocytosines and Isocytosine* (I-374)

5,6-Dihydrocytosine [**317**] shows $K_T \simeq 25$ in favour of the amino form [**317a**] in H_2O, but in $CHCl_3$ the imino form [**317b**] predominates

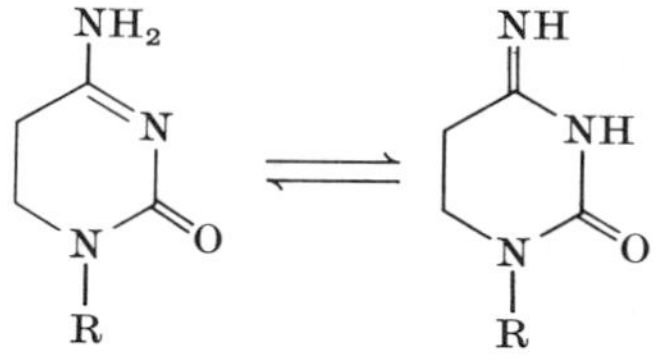

[**317a**] ⇌ [**317b**]

by a factor of 10 (68J(C)2050). Attempted extrapolation (72JA951) from the nonaromatic dihydrocytosine case to cytosine tautomerism is misguided (see Section 2-6Cb). ^{13}C NMR shows that 6-oxocytosine nucleosides exist as such in DMSO.

Fragmentary IR evidence has been advanced for the imino structure [**318b**] (68CB512), but this unlikely conclusion needs further investigation.

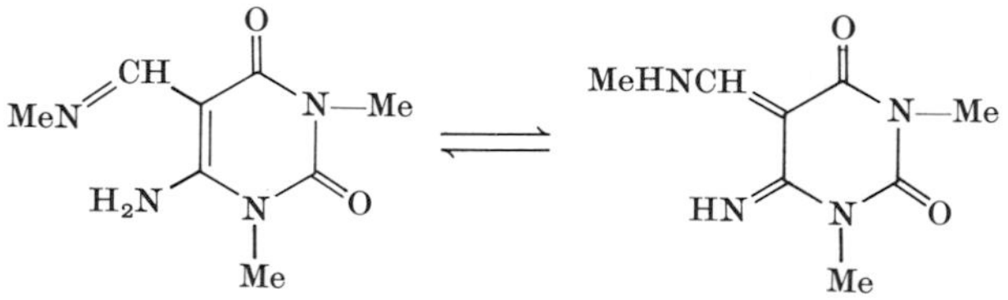

[**318a**] ⇌ [**318b**]

Isocytosine [**319**], previously assigned a single oxo–amino structure [**319a**] (I-374) is now known to exist in H_2O as a mixture of the two oxo–amino forms [**319a**] and [**319b**] (64CO(259)4387, 64CO(259)4853, 65AJ559) from UV comparisons with the 1- and 3-methyl derivatives, and interestingly enough this is true also in the solid state, where a

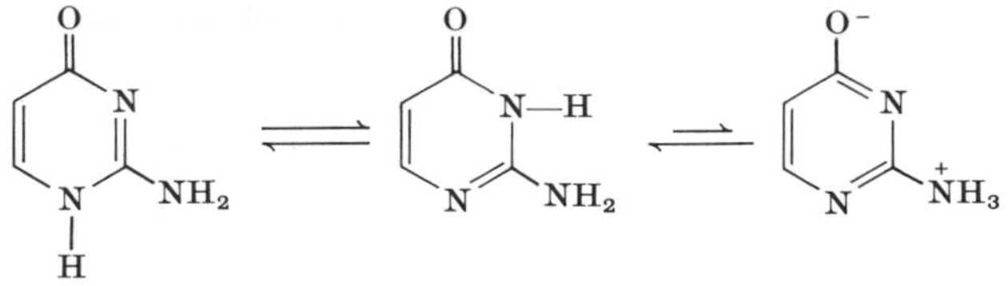

[319a] [319b] [319c]

1:1 mixture of the two tautomers has been found by X-ray structure determination (65CX(19)797). The (very unlikely) zwitterionic form [**319c**] was eliminated by spectral comparison with the 4-pyrimidone anion.

UV and NMR demonstrate structure [**320**] (72MC727).

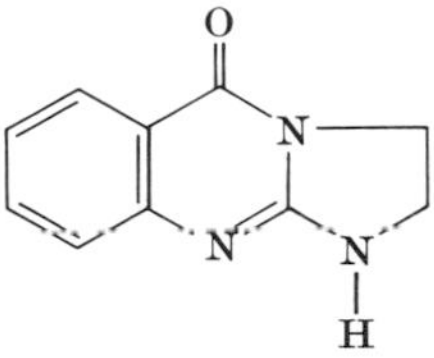

[320]

d. Cytosine Cations (I-374)

The proposal that the cytosine cation existed as [**321a**] was criticised in the previous review (I-374), and it has since been shown by NMR (distinct NH and NH_2 signals) and UV comparisons that the cation is indeed [**321b**] (63J3046). Despite a curious counter argument (66J(B)-210), this conclusion of *N*(3)-protonation is amply confirmed by NMR

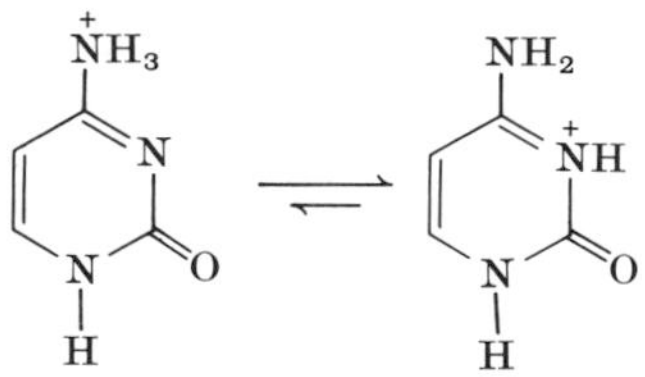

[321a] [321b]

spectral features of monoprotonated 1-methyl-^{15}N(3)-cytosine (65JA-5439) and of 1-methyl-($^{15}NH_2$)-cytosine (65JA5575) and by X-ray structure of 1-methylcytosine hydrobromide (62CX1174); see further (70H299), which discusses also the structure of the cytosine dication

[**322**] and the structure of the isocytosine cation [**323**] and dication [**323/1**]. See also (73H2680).

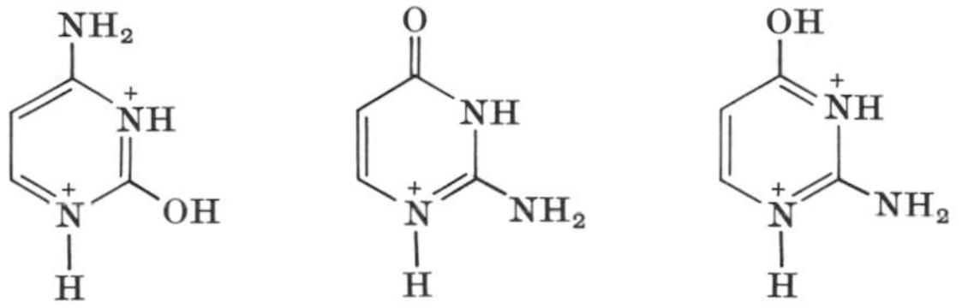

[**322**] [**323**] [**323/1**]

D. Other Aminodiazines and Aminobenzodiazines (I-415)

a. Aminopyridazines (I-415)

IR spectral assignments and comparison with aminopyridines confirm the amino structures for 3- and 4-aminopyridazine (I-415) in the solid and $CHCl_3$ (63CT744).

b. Aminocinnolines and Aminophthalazines (I-415)

The previous review reported earlier work suggesting that 4-aminocinnoline [**324**] existed in the imino form [**324b**] and pointed out that this conclusion was probably incorrect [I-415). These doubts have now been substantiated, and a definitive UV and pK_a investigation (71J(B)2344) demonstrates that 4-aminocinnoline exists as the amino form [**324a**] with pK_T = 4.11 for [**324a**] $\rightleftharpoons$ [**324b**], whilst the zwitterionic form [**324c**] is considerably less stable even than the imino form [**324b**]; pK_T = 7.61 for [**324a**] $\rightleftharpoons$ [**324c**].

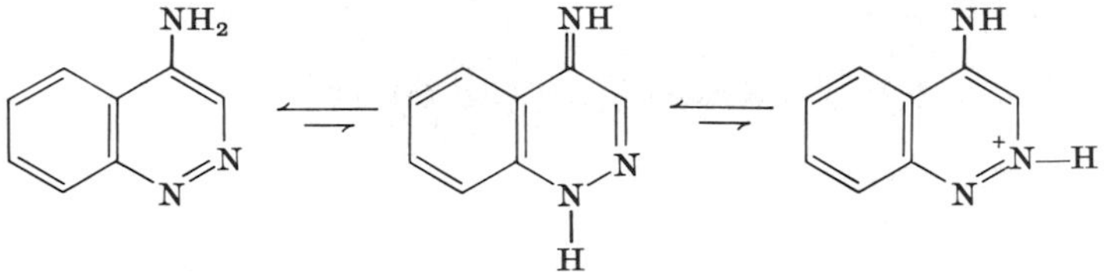

[**324a**] [**324b**] [**324c**]

The monocation of 4-aminocinnoline is predominantly [**325a**] (67J(B)748) with pK_T = 2.0 for the equilibrium [**325a**] $\rightleftharpoons$ [**325b**] (71J(B)2344).

3-Aminocinnoline [**326**] prefers the amino structure [**326a**] by a still larger factor: a pK_T difference of 8.3 with respect to the imino form [**326b**] and a pK_T difference of ca. 9 from the zwitterionic form [**326c**]

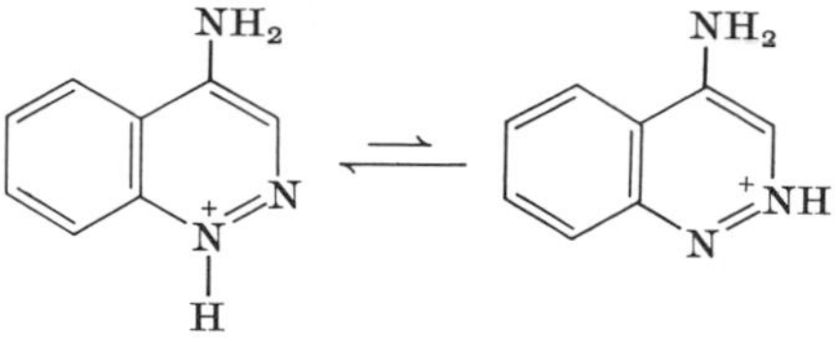

[325a] [325b]

have been found by UV and pK_a measurements. The higher preference for the amino form [**326a**] as compared to that for [**324a**], reflects the *ortho*-quinonoid system which destabilizes the imino structure [**326b**] (cf. Sections 2-4Cb and 2-5E) (71J(B)2344).

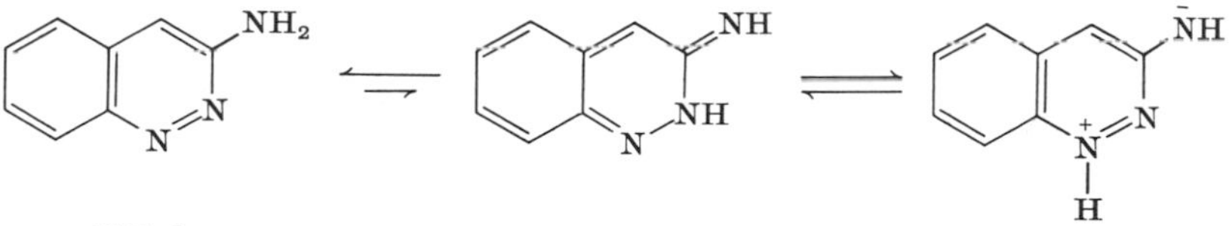

[326a] [326b] [326c]

The 3-aminocinnolinine monocation [**327**] exists predominantly as [**327a**] but the 1-protonated tautomer [**327b**] may contribute up to 20% in H_2O (71J(B)2344).

UV shows the diaminophthalazine structure [**328**] (72J(PI)2820).

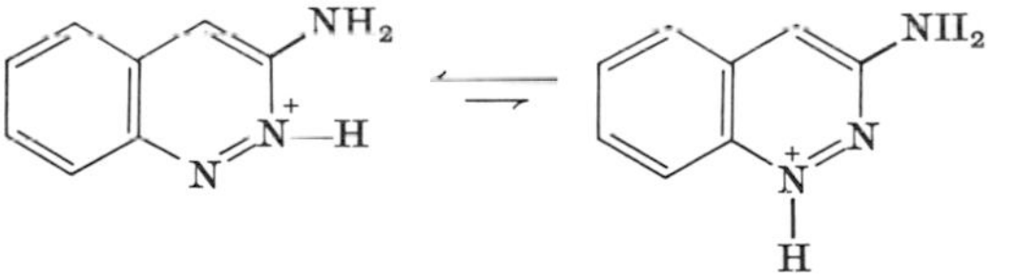

[327a] [327b] 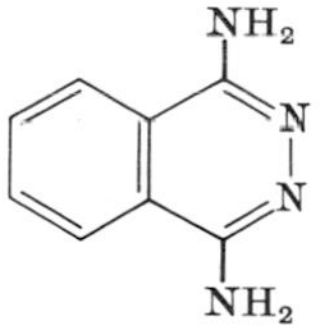 [328]

c. *Aminopyrazines* (I-415)

NMR chemical shift comparisons of the NH_2 signal (6.32 δ) in 2-aminopyrazine with the corresponding 2-aminopyridine NH_2 signal (6.21 δ) in DMSO show the amino form to predominate (68PH1642, 68PH1646), in agreement with previous IR studies (I-415), and with very recent comparison of theoretical and experimental electronic spectra (73BA405).

d. *Aminoquinoxalines*

UV spectra show that the tetracyclic diaminoquinoxaline [**329**] exists as such (63AJ445). Early UV evidence has been interpreted in terms of structure [**330**] for the pentacyclic compound (51J3211).

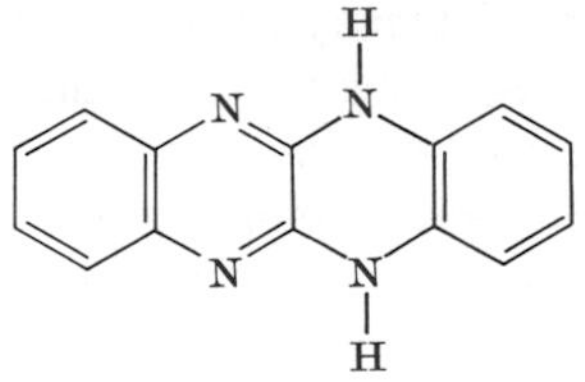

[**329**] 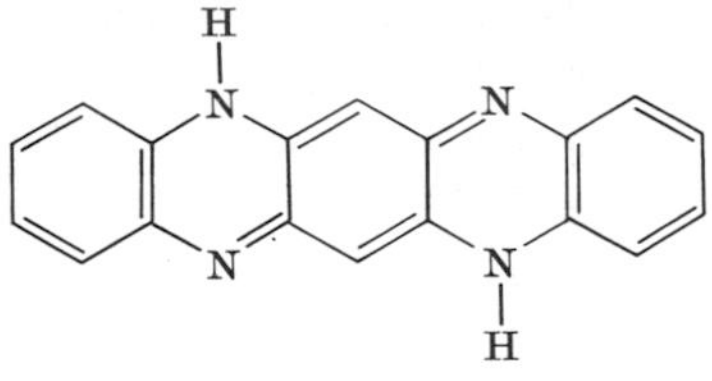 [**330**]

UV spectral comparisons with fixed models demonstrate that indolo-[2,3-*b*]quinoxaline, capable of the three tautomeric forms [**331a–c**] possesses the aminoquinoxaline structure [**331a**] in EtOH (62J3926).

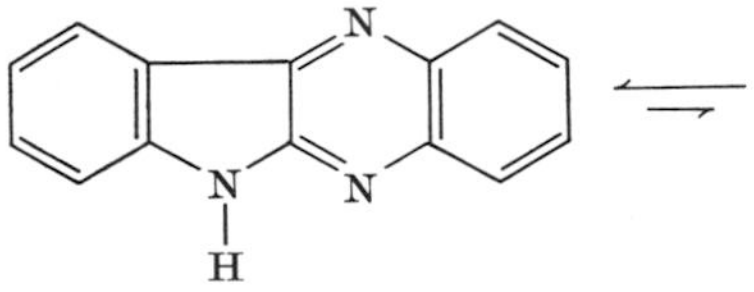

[**331a**]

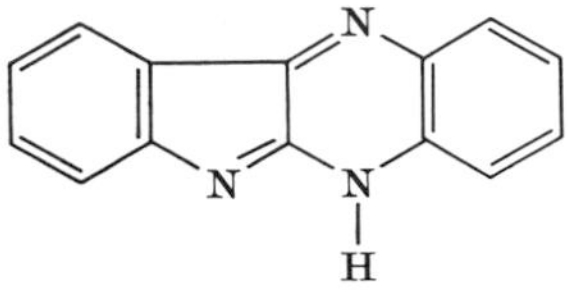

[**331b**] 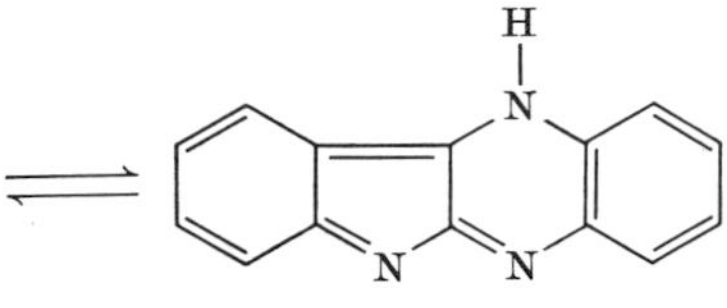 [**331c**]

The monocations of 2-aminopyrazine and its mono- and di-*N*-oxides have been investigated by NMR (73KG1115).

e. Aminopteridines (I-417)

pK_a and UV comparisons with fixed forms demonstrate that 2,4-diaminopteridine [**332**] (70CB722) and its 6-oxo [**333**] and 7-oxo derivatives [**334**] (70CB735) exist as shown in water as does ^{13}C NMR for the parent compound (73CB3951). The predominant monocations are found by protonation at the position marked by arrows in structures [**332**] and [**333**]. Molecular orbital calculations on 2- and 4-amino-pteridines and 2,4-diaminopteridines indicate that *N*(1) protonation is to be expected on cation formation (73IQ27).

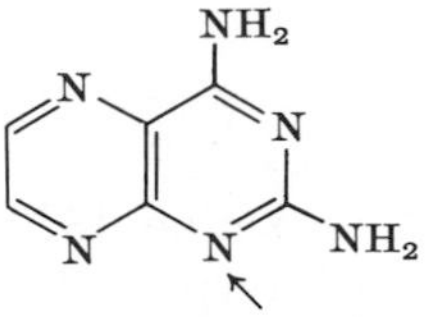

[**332**] 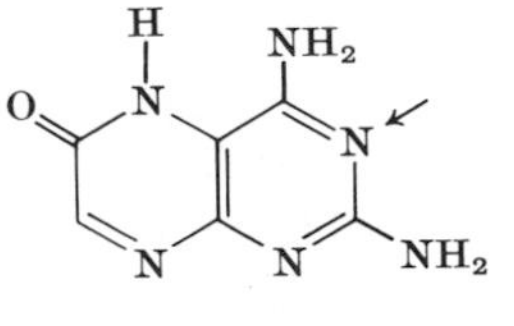 [**333**] 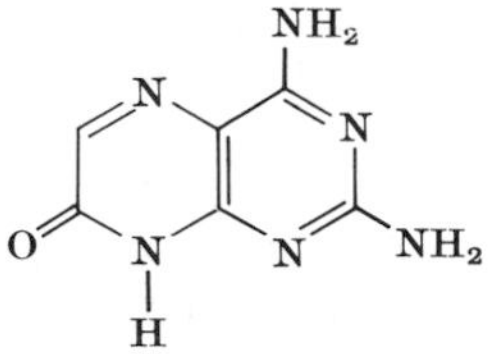 [**334**]

pK_a data and UV comparisons with fixed forms indicated structure [**334/1**] for 4-amino-2-oxodihydropteridine (isopterine) and similar structures for various 6- and/or 7-methyl derivatives (63CB2950). Structures [**334/2**] and [**334/3**] have been assigned similarly (63CB2977). The sites of protonation of these compounds are discussed in the original papers (see also 71CB2273); in some cases covalent hydration is disclosed.

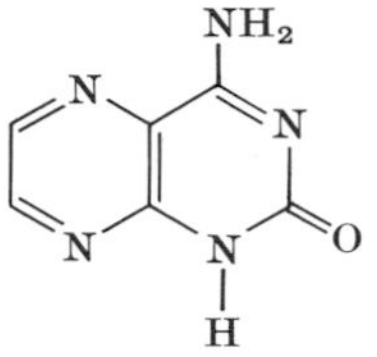

[**334/1**] 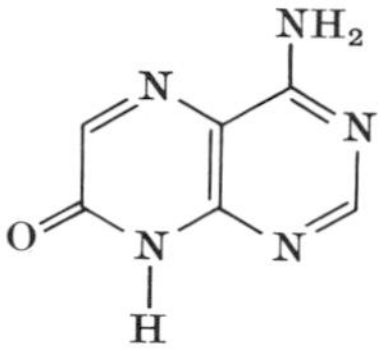 [**334/2**]

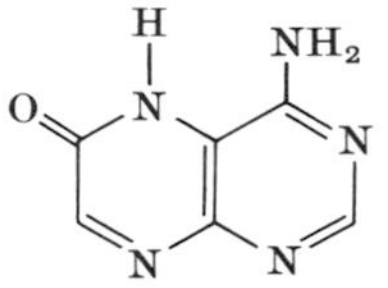

[**334/3**] 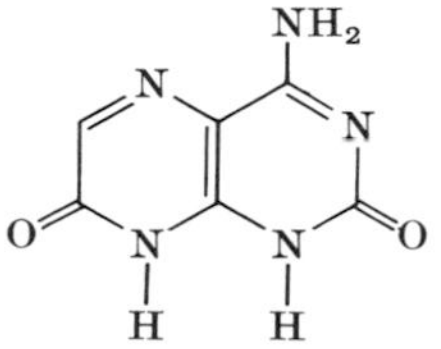 [**334/4**]

pK_a data and UV comparisons demonstrate structure [**334/4**] for 7-hydroxyisopterin (63CB2964). The sites of protonation and covalent hydration are also discussed (see also 71CB2273).

For further work on aminopteridines see (64J4769).

E. Aminotriazines (I-415)

a. 3-Amino-1,2,4-triazines (I-417)

The earlier review (I-417) classified as "probably erroneous" the claimed isolation of the two aminotriazine tautomers [**335a**] and [**335b**]. This criticism has now been justified and the isomeric amino-1,2,4-triazine structures [**335b**] and [**336**] are established for these two substances by NMR (64J4157).

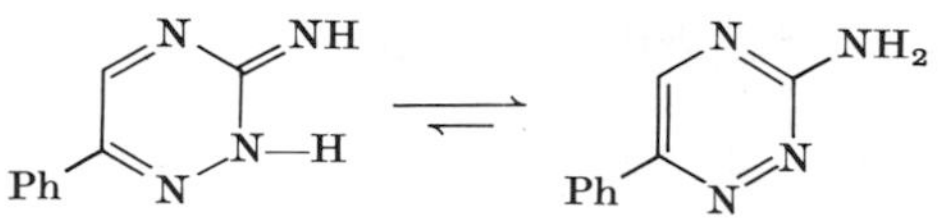

[**335a**] [**335b**] 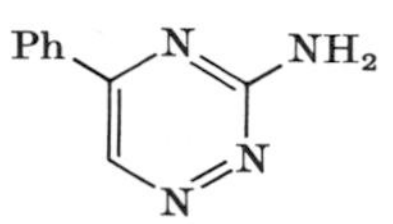 [**336**]

3-Amino-5-ethyl-6-methyltriazine exists as [**337**] in neutral DMSO. However, ca. 30% of the corresponding monocation [**338**] exists in CF_3CO_2H in the ethylidine iminium form [**338a**] by NMR and for the rest probably as [**338b**] (68TL2747) (cf. Sections 2-7Da and 2-8Bd).

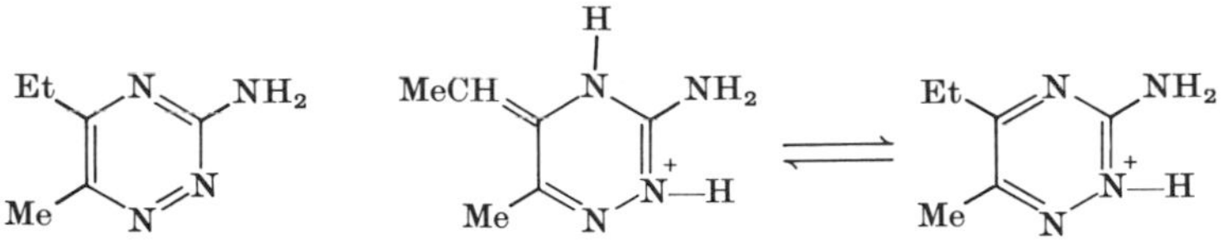

[**337**] [**338a**] [**338b**]

b. 4- and 5-Amino-1,2,4-triazin-2-ones (except 6-Azacytosines)

IR and UV spectral comparisons with dimethylated models show that the triazacarbazole [**339**] exists in the oxo form in the solid and in solution (69ZO2339). Work on the thione analogue [**340**] leads to similar conclusions (69ZO640), and this work has also been extended to the isomeric series [**341, 342**] (71ZG179).

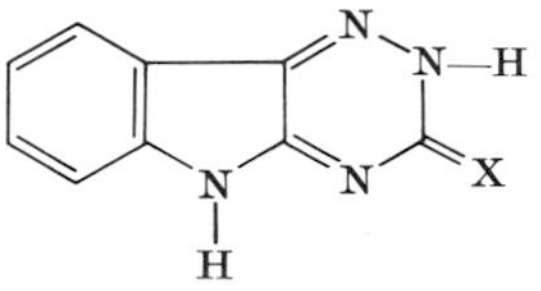

[**339**] X = O
[**340**] X = S

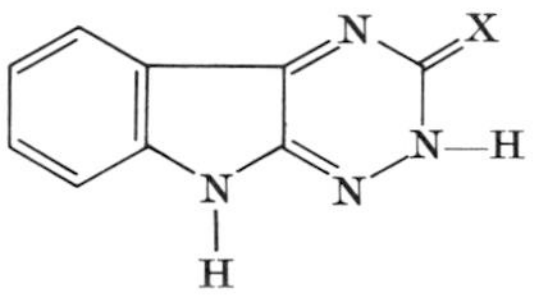

[**341**] X = O
[**342**] X = S

c. 5-Aza- and 6-Aza-cytosines and -isocytosines

The tautomerism of 6-azaisocytosine and its 5-alkyl derivatives has been controversial. Definitive comparisons of UV and IR spectra and pK_a with fixed models show the 2*H*-oxo form [**343a**] to predominate with K_T for equilibrium [**343a**] $\rightleftharpoons$ [**343b**] ca. 10^2 in H_2O (66CZ1864). Earlier an erroneous IR interpretation had indicated the imino–oxo structure [**343c**] (64CT100); while a little later UV and IR spectra were correctly interpreted to indicate an amino-oxo form, structure [**343b**]

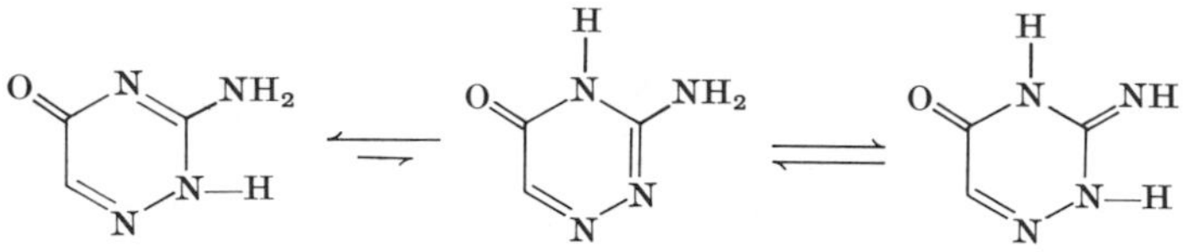

[**343a**] [**343b**] [**343c**]

was chosen (64CT1329), but was subsequently shown to be wrong, as mentioned above.

For a discussion of MO calculation on this tautomerism, see (66CZ-1864). The analogous derivative [**344**] appears from UV and IR comparisons to exist as depicted (69AH(61)181).

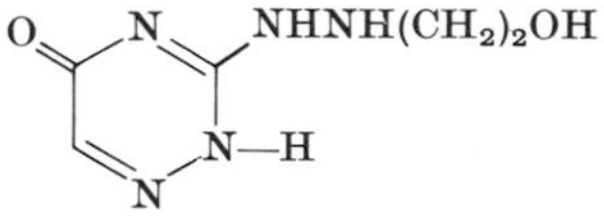

[**344**]

Definitive UV spectral comparisons and IR evidence demonstrate that 6-azacytosine [**345**] and its 1-methyl derivative [**346**] occur in the amino-oxo form shown but 3-methyl-6-azacytosine [**347**] exists in the unusual imino form [**347a**] (64CZ1394) and spectra of partially deuterated specimens confirm this conclusion (65J5230). The corresponding monocations all have structure of type [**348**] (64CZ1394). Nucleoside analogues such as 6-azacytidine also exist in oxo–amine structures cf. [**346**] (63CZ1507).

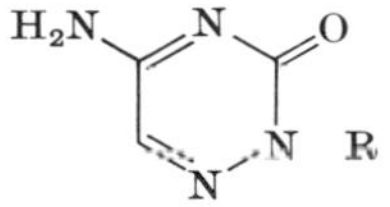

[**345**] R = H
[**346**] R = Me

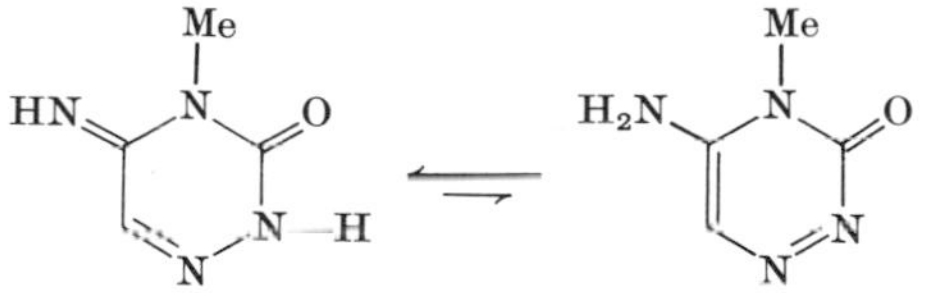

[**347a**]

[**347b**]

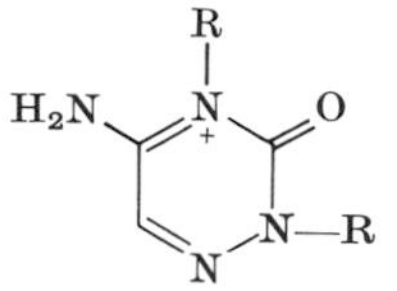

[**348**]

1-Methyl-5-azacytosine [**349**, R = Me] and derivatives such as 5-azacytidine [**349**, R = sugar] exist in the amino–oxo form as they show νNH_2 bands ($CHCl_3$, DMSO and solid). Quantum chemical calculations agree, and the results are discussed in terms of pairing with purine derivatives (65CZ1626).

NMR measurements establish unequivocally structures [**350**] and [**351**] (71OR689).

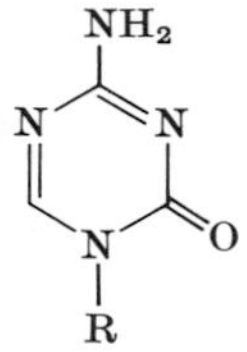

[**349**] 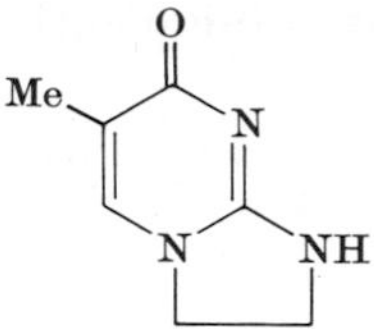 [**350**] 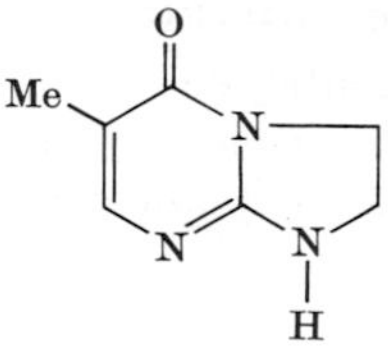 [**351**]

d. Amino-1,3,5-triazines (I-416)

Amino-*s*-triazine structures of type [**352**] are demonstrated by IR (71RR1447). However, IR studies, including the partial deuteration technique, indicated an imino–oxo form, probably [**352/1a**], for various 3-aryl-*s*-triazine derivatives, rather than the amino form [**352/1b**] (73ZC298).

The protonation of 2,4-diamino-*s*-triazines of type [**352/2**] occurs on a cyclic nitrogen atom (73ZO1556).

Russian authors have calculated K_T values for various amino-*s*-triazines by MO methods (71ZF197).

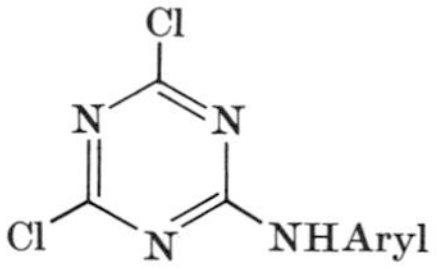

[**352**]

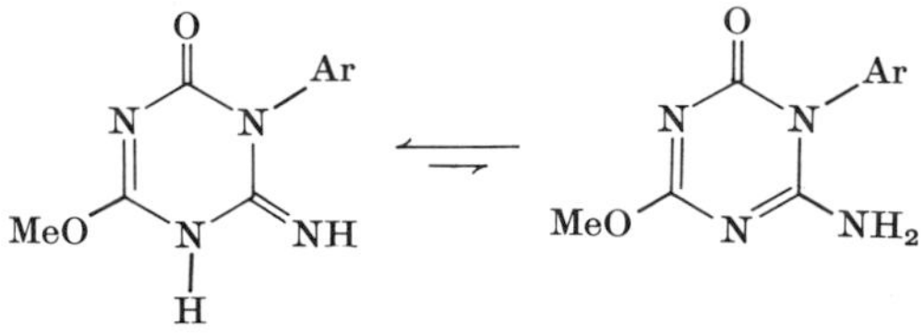

[**352/1a**] [**352/1b**] 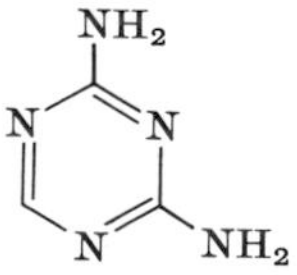 [**352/2**]

e. 4-Amino-1,2,3-benzotriazines

4-Amino-1,2,3-benzotriazines [**353**] have been studied by IR (69JH779). For the parent compound [**353**, R = H], the only bands in the νNH region at 3320 and 3100 cm^{-1} (solid) are attributed to two distinct NH groups of one [**353b**] or other [**353c**] of the imino forms, and the imino structure is assigned on similar grounds to some, but not all, of the substituted derivatives [**353**] R $\neq$ H. Differences in crystal packing can affect IR spectra, and the above evidence is not considered by the present authors to be definitive: further investigation is required.

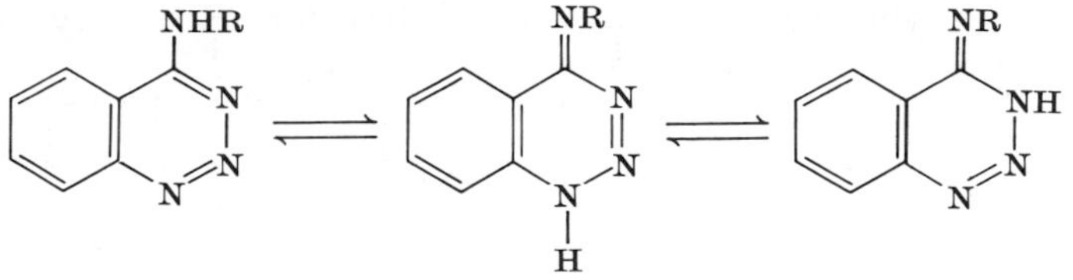

[353a] [353b] [353c]

F. Aminooxazines and Aminothiazines

The 2-amino-1,3-thiazin-4-one [**354**] exists in the amino form shown on UV spectroscopic evidence (67CB3671); however, structure [**355**], listed in the same paper, is probably incorrect—an oxo–amine form is more probable. For the equilibrium [**356a**] ⇌ [**356b**] see (64M950).

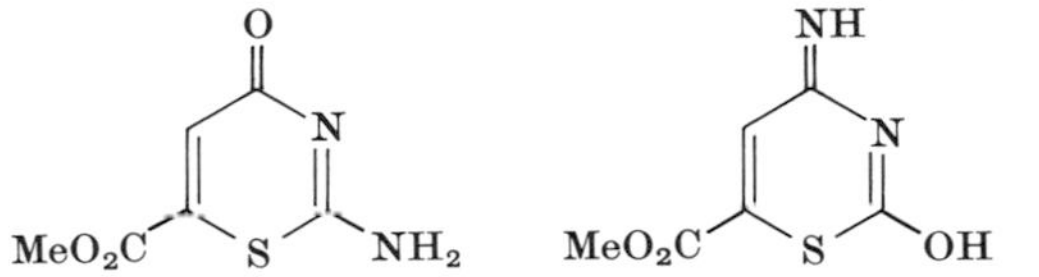

[354] [355]

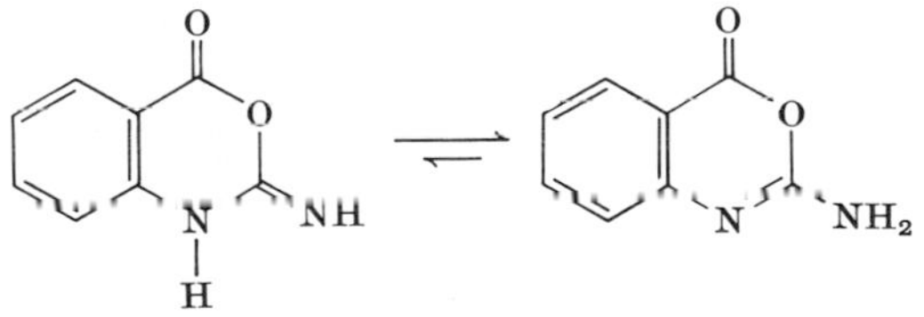

[356a] [356b]

In the nonaromatic aminobenzoazine series, tentative NMR evidence (range of chemical shifts for *H*5–*H*8 protons) is given for the predominance of the form [**357a**] in the cyclohexyl, but [**358b**] in the phenyl series (69T517).

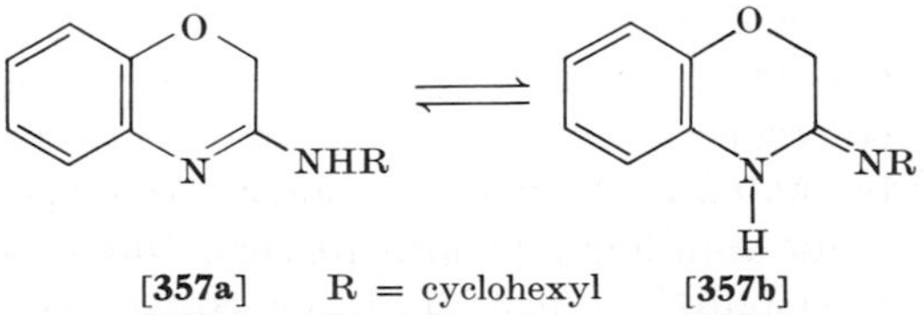

[**357a**] R = cyclohexyl [**357b**]
[**358a**] R = phenyl [**358b**]

UV (EtOH) and IR (solid) comparisons show that [**359a**] predominates for the dihydrothiazine [**359**] (65BF2120). Similar work, including NMR, demonstrates the predominance of [**360a**] for R = alkyl or aryl (71KG946).

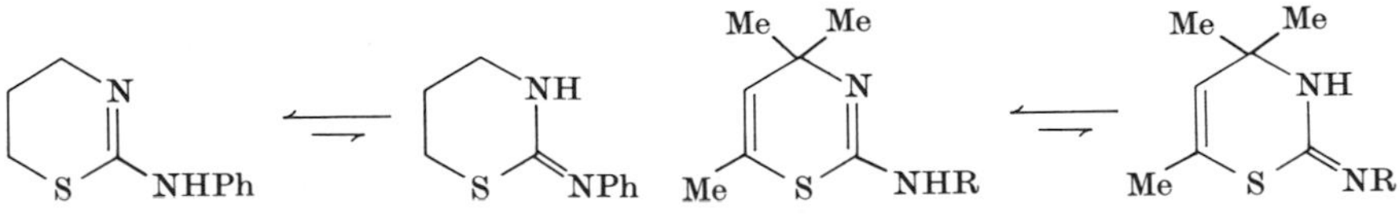

[**359a**] [**359b**] [**360a**] [**360b**]

Although [**361a**] is favoured on IR evidence (64JO245), this does not appear conclusive, and [**361b**] is perhaps more probable. The dithiazine [**362**] has been formulated as the imino form [**362a**] on IR evidence, νC=N at 1667 cm^{-1} (63JO2313); however, this is most unlikely, and the probable structure is [**362b**].

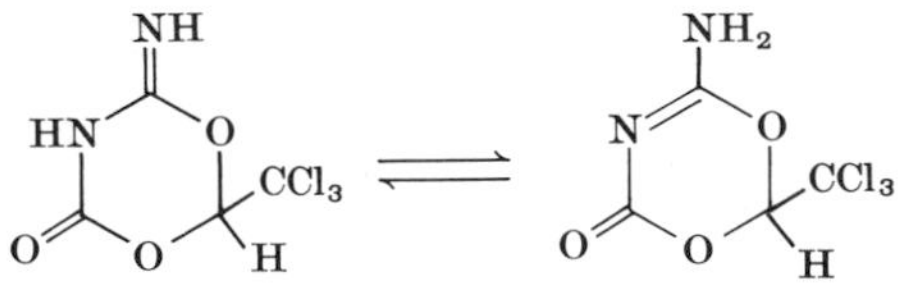

[**361a**] [**361b**]

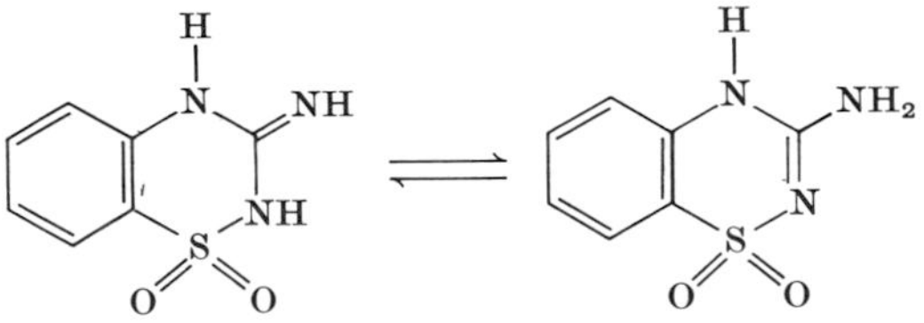

[**362a**] [**362b**]

7. Compounds with Potential Substituted Amino Groups (I-417)

A. General Introduction

Substitution of one of the hydrogen atoms in an NH_2 group can have a dramatic effect on the tautomeric behaviour of the remaining hydrogen atom. It is convenient to discuss the subject in terms of the substituent, for different substituents can modify the behaviour of the corresponding parent amino compounds in very different ways.

Alkyl groups have little effect, and for convenience alkylamino compounds (NHR) have already been discussed in the preceding section together with the parent amino compounds: this was also done in the previous review.

The effects of other groups are expressed by the sequence:

aryl < COR, COAr < SO_2R, NO_2, CN.

The effect obviously depends to a considerable extent on NH-acidity; if X increases the acidity of the NHX-proton there is a larger tendency for the imino form to occur. Within each type of group, the effect can be varied by further substitution: thus, for example, the influence of $COCCl_3$ is far greater than that of $COCH_3$.

B. Arylamino Compounds

Nothing was reported on this type of tautomerism in the previous review, and relatively little work has been performed since, except for 2-anilinopyridine and di-(2-pyridyl)amine.

There has been controversy regarding the tautomeric structure of anilinopyridine. It was first suggested (68MI1275) that the imino form [**363b**] was responsible for νCN at ca. 1650 and νNH (bonded) at 3220–3255 cm^{-1}. Later authors attributed the complex νNH absorption to two rotational isomers of the pyridinoid form [**363a**] and suggested from IR comparisons and the invariance of the UV spectra in various solvents that anilinopyridines exist essentially as such [**363a**] under all conditions examined (71T6011). Recent quantitative data (73J(PII)-2111) confirm this: $pK_T = 4.3$ for equilibrium [**363a**] $\rightleftharpoons$ [**363b**]. Compared with the corresponding aminopyridine, delocalization of the exocyclic nitrogen lone pair onto the ring nitrogen will be less favoured in the anilinopyridine, whereas negative charge at a =NPh group will be stabilized: these factors do reduce the predominance of the amino form (pK_T for 2-aminopyridine = 6.2) by a factor of ca. 10^2 in 2-anilinopyridine, but the amino form [**363a**] does still predominate (73J(PII)2111).

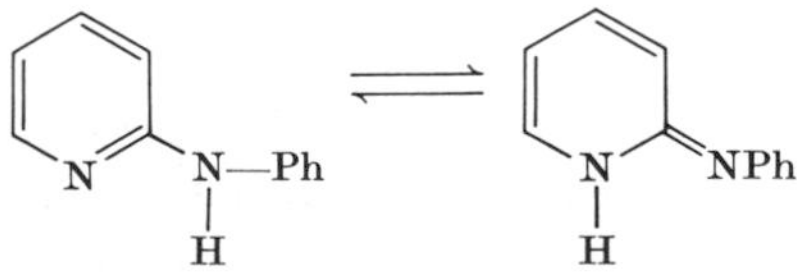

N N—Ph H ⇌ N NPh H

[**363a**] [**363b**]

X-ray results indicated the amino structure for di-(2-pyridyl)amine (73CX(B)1669). Interestingly, IR spectra show that hexa-aza-macrocycle [**363/1**] exists in the solid state in the fully conjugated form [**363/1a**] rather than [**363/1b**] (72CH577, 74J(PI)976).

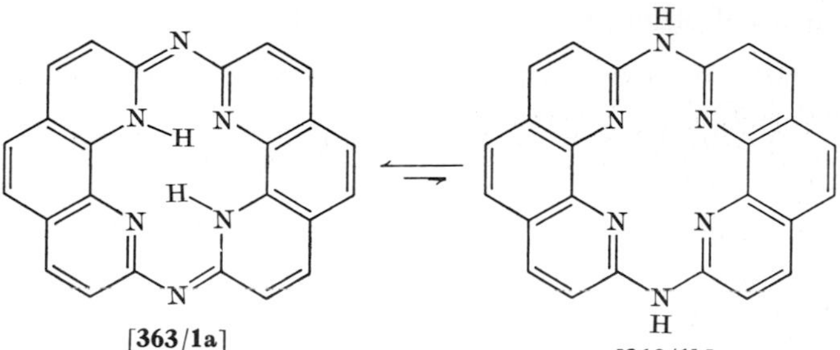

[363/1a] ⇌ [363/1b]

C. Acylamino Compounds (I-418)

a. Acylaminopyridines (I-418)

The tautomerism of acylamino-pyridines and -azines had already been widely studied at the time of the previous review. The main conclusions (I-420) were (a) the "acylamino" form largely predominates over the "acylimino" form much as the "amino" form predominates over the "imino" in the corresponding parent compounds for reasons which were discussed in detail (I-418 ff); (b) both the introduction of halogens in the acyl group, and (c) increasing polarity of the solvent increases the proportion of acylimino form.

Evidence that the tautomeric equilibria of acylamino compounds are sensitive to the same electronic influences of other ring substituents as is hydroxy–oxo tautomerism (see Section 2-3Be) is provided by Yakhontov, Sheinker, and their group (66T3233): the acylamino derivatives [**364**] exist essentially completely as such even with R = $COCCl_3$ or SO_2Me. The reasons are exactly similar to those discussed in

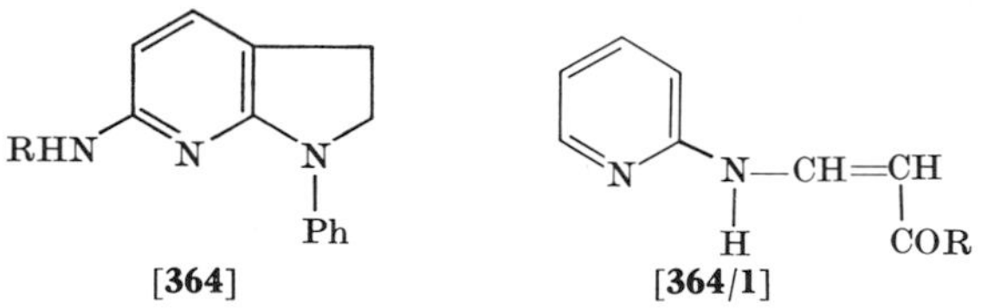

[**364**] [**364/1**]

detail in Section 2-3Be. Recent IR and NMR results have reinforced all of these conclusions except that, in contrast to earlier evidence (I-419), even 2-trifluoroacetamidopyridine exists predominantly in the acylamino form (73TH01).

Compound [**364/1**] has been formulated as such on NMR evidence [74PC469). Some authors continue to confuse tautomeric equilibria with ionization processes; e.g., acylamino derivatives are incorrectly stated to change towards the imido form in basic media (74RR671).

b. Acylaminoazines

The isolation of two substances claimed to be individual tautomers [**365a**] and [**365b**] of 1,3-dimethyl-6-benzimido-5-methyliminomethyl-dihydrouracil has been claimed (68CB512). Although the authors provide IR spectra of the two modifications and of deuteriated derivatives, the present authors believe form [**365a**] to be unlikely: at least, X-ray analysis is needed to prove that the substances are not crystalline modifications with different H-bonding, or structural isomers, or *possibly* an alternative tautomeric pair [**365b**] and [**365c**].

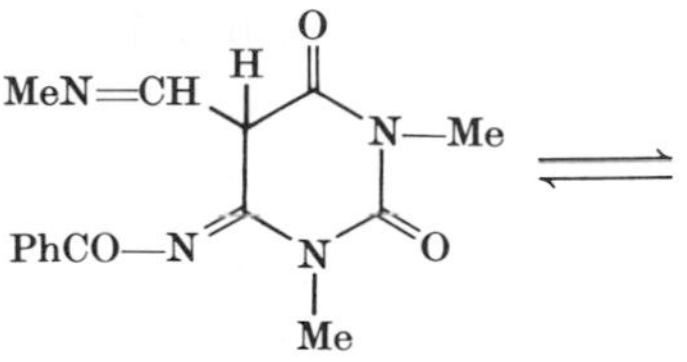

MeN=CH, H, O, N—Me, PhCO—N, N, O, Me

[**365a**]

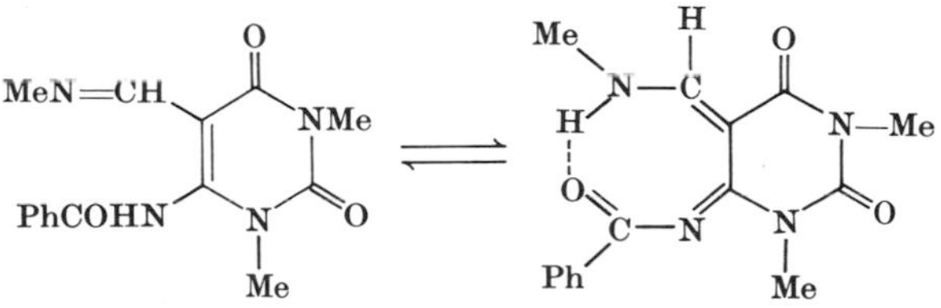

O, MeN=CH, NMe, PhCOHN, N, O, Me

[**365b**]

Me, H, O, N—C, H, N—Me, O, C—N, N, O, Ph, Me

[**365c**]

The tautomerism of various thiocarbonylaminoquinazolines [**366**] has recently been investigated by IR and NMR (73CB471). The amino [**366a**] and chelated imino forms [**366b**] are of comparable stability: for R = OMe, both can be isolated crystalline, and the NMR spectra show a mixture to be present in $CDCl_3$. The thioamides [**366**, R = H or alkyl] and thioureas derived from primary amines [**366**, R = NHR′] tend to favour the amine form [**366a**], but those from secondary amines [**366**, R = NR′R″] prefer the imino structure [**366b**].

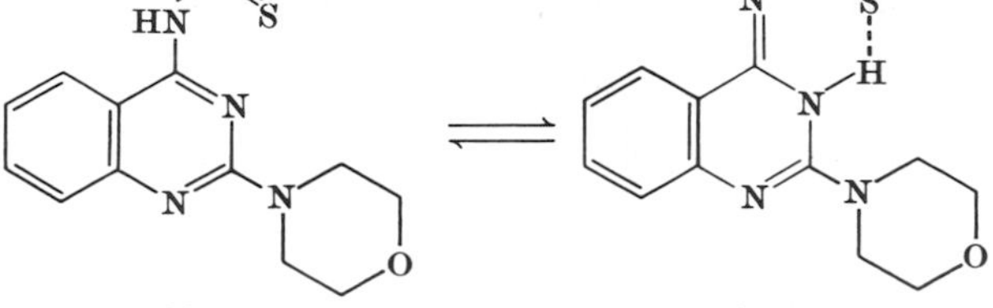

R, C, HN, S, N, N, N, O

[**366a**]

R, C, N, S, H, N, N, N, O

[**366b**]

c. Acylaminopyridine 1-Oxides (I-422)

The conclusion that acylaminopyridine 1-oxides exist predominantly as such and not in the alternative imino form applies also to 2-acylaminoquinoxaline 1-oxides on IR and UV evidence (66KG101).

D. Sulfonamido, Nitramido and Cyanamido Compounds

a. Sulfonamido Compounds (I-422)

The earlier statement (I-422) that hydroxylic solvents strongly favour the sulphonylimino forms has been reinforced and a linear correlation between log K_T and E_T (the Dimroth–Reichardt solvent parameter, cf. 2-3Bg) found for 2-methyl and 2-phenyl sulphonamidopyridines and for 2-methylsulphonamidopyrimidine (73TH01). Further interpretation and measurement of IR spectra, in contrast to evidence given earlier (I-423), indicates that 2-methylsulphonamidopyrimidine exists as the sulphonylamido tautomer in the solid state (73TH01).

The 5-ethyl-3-sulfonamido-1,2,4-triazine [**367**] exists in DMSO as a mixture of [**367a**] and [**367b**] forms with the latter predominating, as shown by NMR (cf. Sections 2-6Ea and 2-8Bd) (68TL2747).

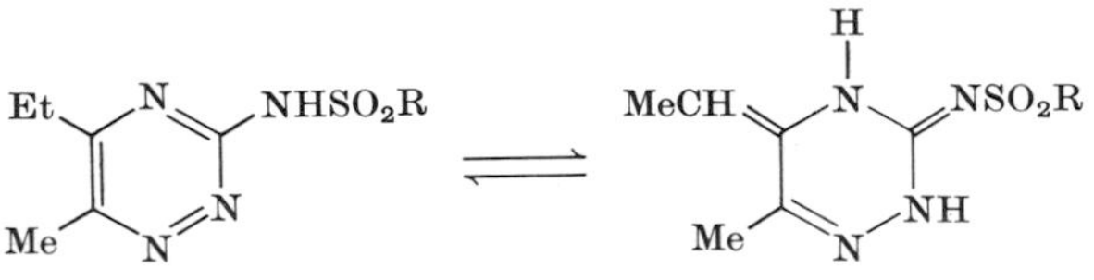

Et, Me, N, N, N, $NHSO_2R$ ⇌ H, MeCH, N, NSO_2R, NH, Me, N

[**367a**] [**367b**]

b. Nitramido Compounds (I-424)

The NO_2 group swings the equilibrium towards the imino form even more than SO_2R, and the compounds mostly exist in the nitrimino form: previous indications to this effect are now confirmed.

Previous conclusions that 3-nitramidopyridine exists as the zwitterion [**368**] in the solid state (I-425) have been confirmed by IR (69AJ2611, 70RC1447). For 2-nitramidopyridine [**369**] the imino form [**369a**] was known to predominate in ethanol (pK_T = 0.2), with the proportion of

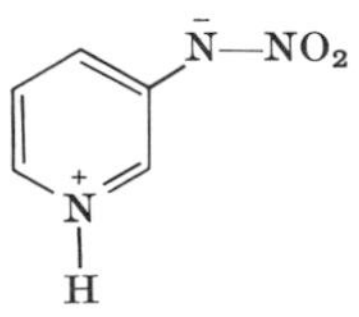

$\bar{N}$—NO_2, $\overset{+}{N}$, H

[**368**]

the amino form [**369b**] increasing in dioxan ($pK_T = -0.8$) (I-425): similar behaviour has since been disclosed for 4-nitramidopyridine (pK_T in water = 2, in dioxan $\simeq -1.2$) by spectral and pK methods (69AJ2611). Nitronic acid forms such as [**369c**] have been ruled out on pK_a reasoning: the —N=NO_2H group should be much more highly acidic than the *aci* form of nitroethane, which has a pK_a of 4.4, whereas the acidic pK_a of 4-nitroaminopyridine is 6.87 in water (69AJ2611). IR and NMR data show that the cation of 4-nitramidopyridine is mainly [**370**] (69AJ2611).

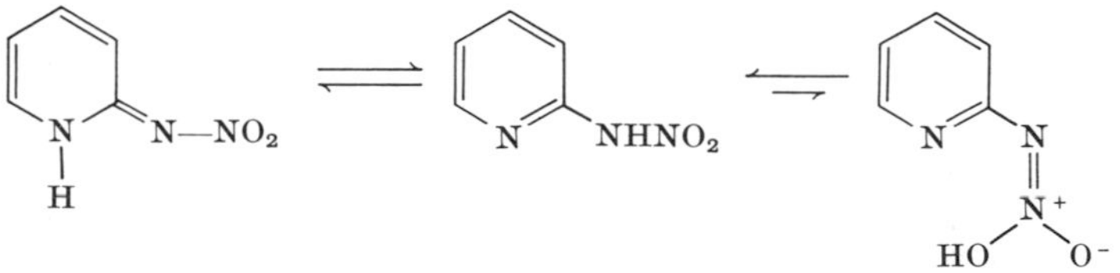

N—NO_2 / N / H $NHNO_2$ / N   N / N / N^+ / HO / O^-

[**369a**] [**369b**] [**369c**]

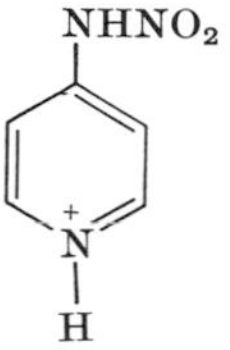

$NHNO_2$ / $\overset{+}{N}$ / H

[**370**]

c. Cyanamido Compounds (I-425)

IR comparisons show that the cyanamidoquinazoline [**371**] occurs (solid) in the cyanimino form [**371b**] (73CB956).

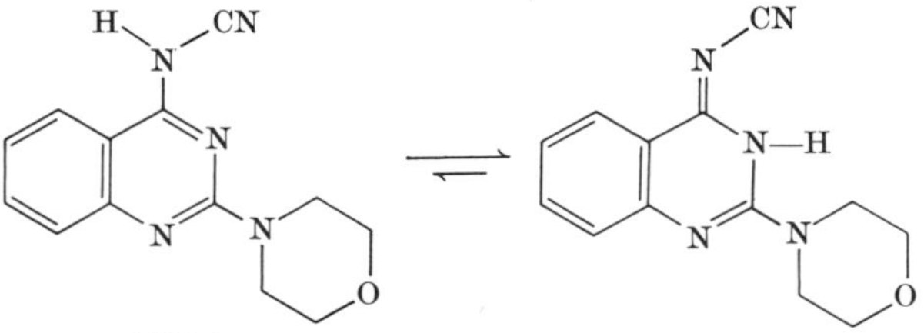

H / CN / N / N / N / N / O   CN / N / N—H / N / N / O

[**371a**] [**371b**]

E. Hydrazino and Hydroxylamino Compounds (I-423)

a. Hydrazino-pyridines and -benzopyridines

The previously suggested (I-423) predominance in 4-hydrazinopyridine of the hydrazine tautomer [**372**] is confirmed by spectra; the monocation was ascertained to be [**373**] and the dication [**374**] (69AJ-2611). The isomeric hydrazinoquinolines exist as such and form monocations of type [**375**] by protonation at the ring nitrogen atom. The UV spectra of the dications have been interpreted in terms of mixtures of species of type [**376a**] and [**376b**] (67J(C)1533). The monocation is also formed by proton addition at the ring nitrogens of 9-hydrazinoacridine (65J4653).

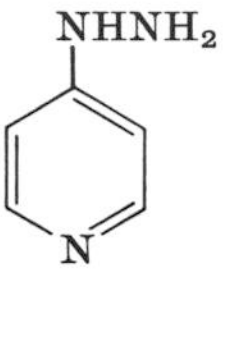

[**372**] 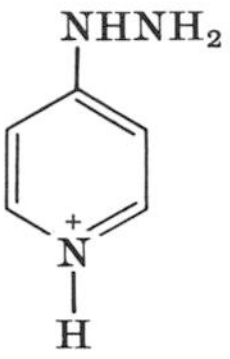 [**373**] 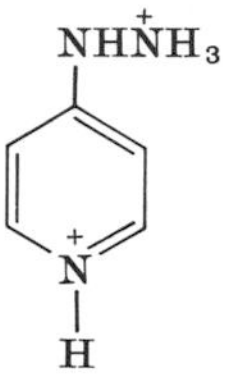 [**374**]

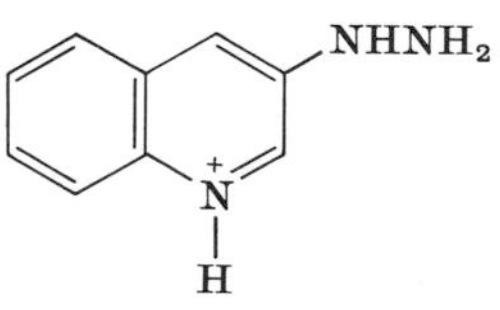

[**375**] 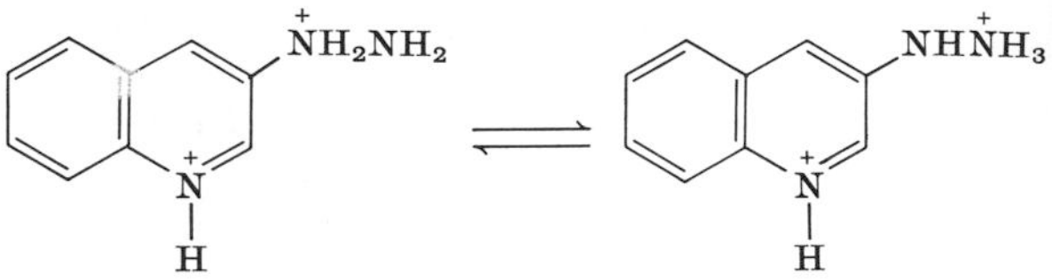 [**376a**] [**376b**]

However, UV and IR comparisons demonstrate [**377**] for the thiosemicarbazide (69ZO1156) and the photochromism of quinolylhydrazones has been explained in terms of the induced tautomerism [**378a**] ⇌ [**378b**] (68TL4593). More recently the Z,Z-structure [**378c**] has been postulated by the same authors (73CH684), by analogy with the structure [**378/1**] suggested for ethyl pyruvate 2-quinolylhydrazone (71CB2793).

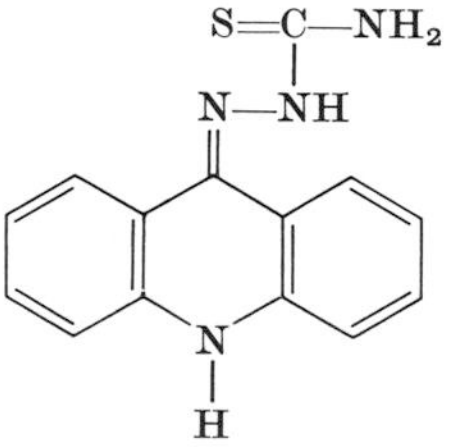

[**377**]

b. Hydrazinoazines

Scattered work suggests that the hydrazine structures predominate for a wide range of structural types. Thus UV comparisons support the

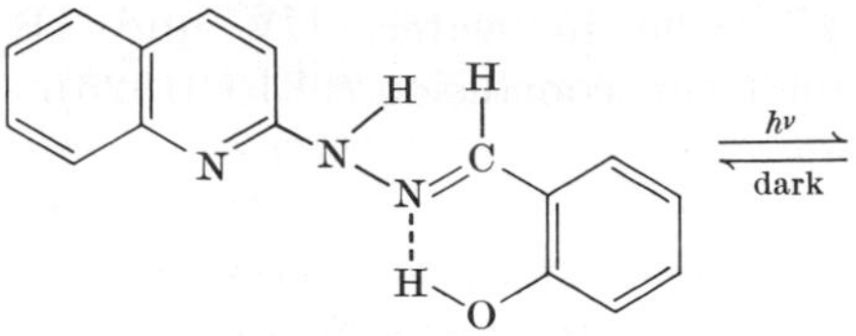

[378a]

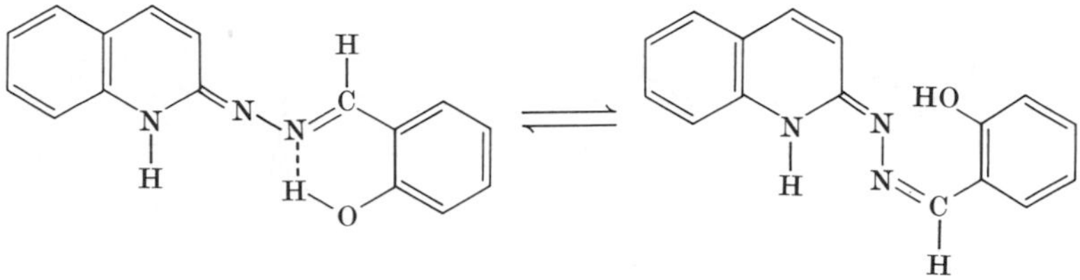

[378b] [378c]

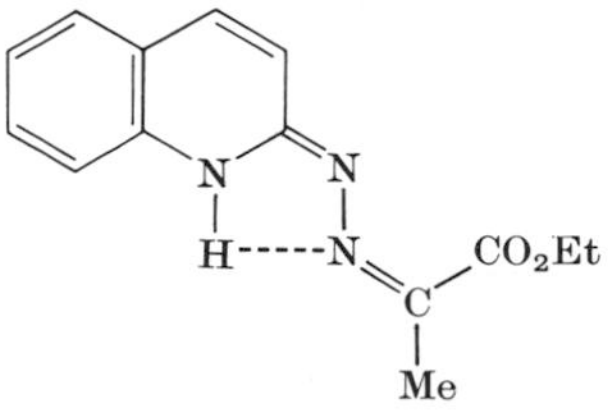

[378/1]

hydrazino formulations [**379**] (72J(PI)2820), and [**380**] (71RO173). Structure [**381**] is assigned by comparison of IR and UV spectra with related 1,2,4-triazin-5-ones (68CZ2513).

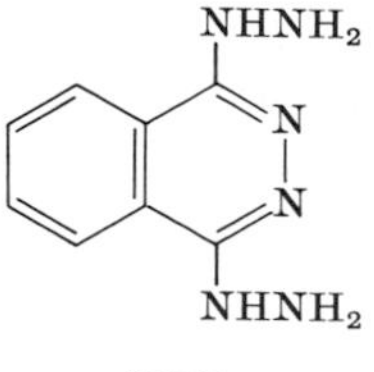

[379] 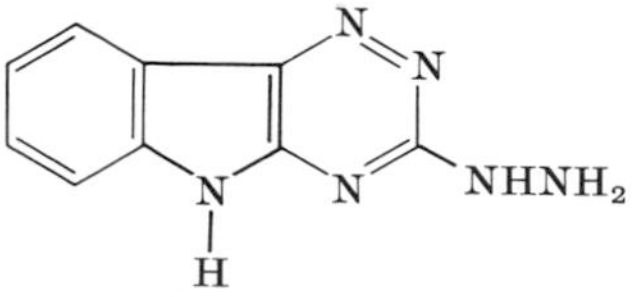 [380] 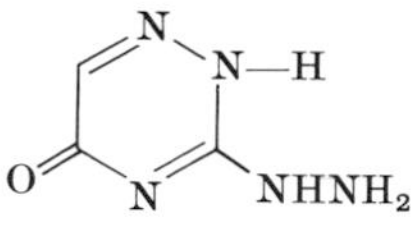 [381]

The pK_a method discloses that 4-hydrazino-1-methylpyridimidin-2-one [**382**] exists preferentially in the hydrazine form [**382a**], the

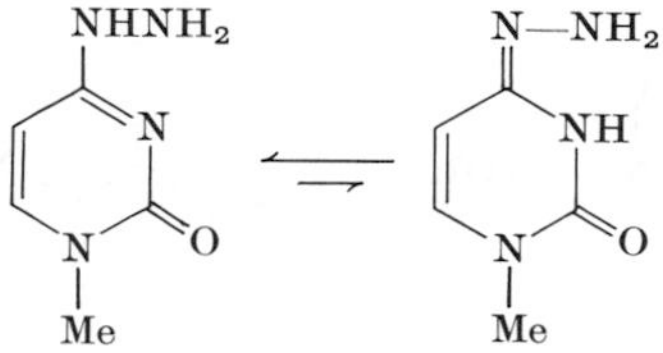

[382a] [382b]

equilibrium [**382a**] ⇌ [**382b**] has $K_T \simeq 30$ in water. UV and IR (H_2O, EtOH, $CHCl_3$) evidence supports this conclusion (68J(C)1925).

c. Hydrazinooxazines

Little work is available, but in the 1,4-benzoxazine series structures [**383**] and [**384**] were assigned from NMR results (69T517).

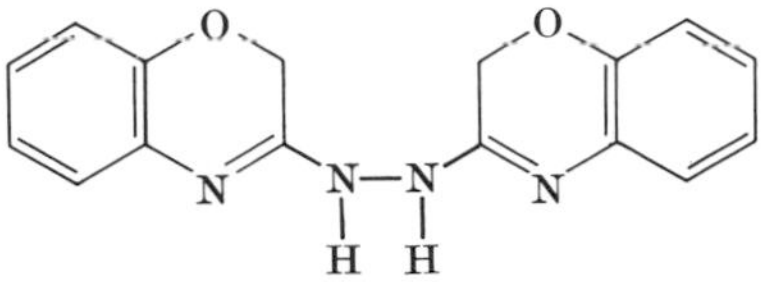

[**383**] 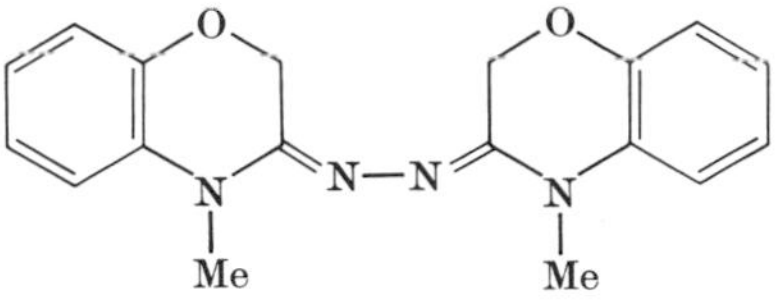 [**384**]

d. Hydroxylamino Compounds

Far less is known about hydroxyamino/hydroxyimino tautomerism, and the subject was not mentioned in the previous review. However, it is now clear that the tendency to exist in the hydroxyimino or oxime form is considerable.

In this connection its is significant that (in contrast to the corresponding hydrazino compound [**382a** ⇌ **b**]) for the hydroxylamino derivative [**385**] the oximino form [**385a**] is predominant with $K_T \simeq 10$. UV and IR spectral comparisons confirmed this for H_2O, EtOH and $CHCl_3$ solutions (68J(C)1925). For similar work see (72MI261) and for a discussion on its relevance to mutagenesis see (69RV187).

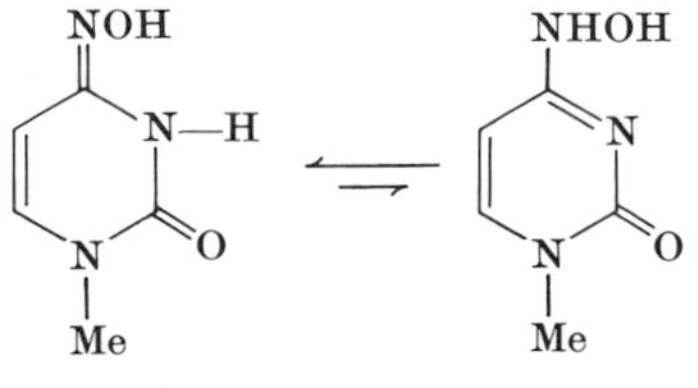

[**385a**] [**385b**]

Unspecified spectral evidence is stated in a preliminary report on potential hydroxylaminoquinoxalines to favour the hydroxyimino form [**386a**] (71MI379).

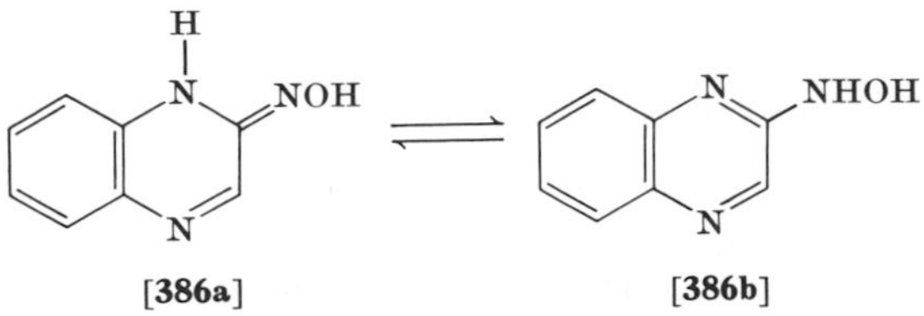

[**386a**] [**386b**]

While the type of tautomerism is really somewhat different, it is significant that potential α-hydroxy-β-nitroso compounds tend to exist in oxime tautomeric forms. Thus, "2,6-dihydroxy-3-nitrosopyridine" [**387**] is now known to exist as [**387a**], and "nitrosohomophthalimide" is assigned structure [**388**] by spectral comparison with homophthalimide. However, the corresponding diamino-nitroso compounds have structures of type [**389**] (64J1423).

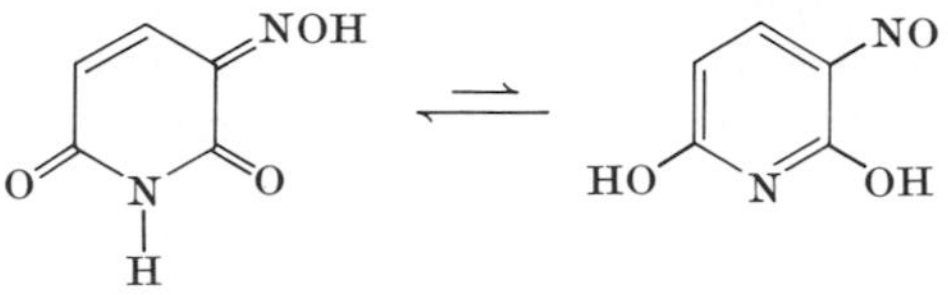

[**387a**] [**387b**]

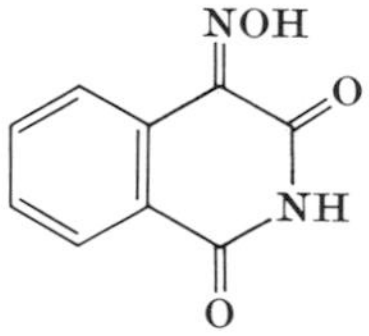

[**388**] 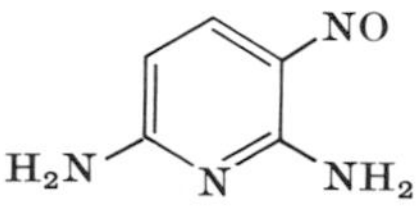 [**389**]

8. Potential Methyl or Substituted Methyl Compounds (I-426)

A. General Survey

Clearly, the general tendency is for potential methyl compounds to exist overwhelmingly in the methyl or CH form and not in the alternative methylene, methide or methine form (which we shall denote NH form), for reasons already discussed in Section 2-1A. This behaviour is indeed found for the vast majority of cases; however, certain modifications to the heterocyclic ring structure, or (more usually) substitution of one or two of the CH_3-hydrogen atoms by electron-withdrawing substituents, can swing the equilibrium towards, or even in favour of, the corresponding NH form. Indeed, MO calculations indicate that any type of substituent in an acetyl group increases the stability of the enol tautomer (72CH771), and similar considerations should apply to substitution in the heteroaryl methyl groups.

We discuss the compounds according to the nature of substituents in the methyl group, adopting a classification similar to that used for substituted amino compounds. Thus, we deal first with unsubstituted (including alkyl substituted) methyl groups, then consider arylmethyl (e.g., benzyl) groups, followed by acylmethyl, sulphonylmethyl, etc.

B. Methyl and Alkyl Compounds

a. Pyridines and Benzopyridines (I-427)

It was already clear at the time of the previous review from pK_a considerations that equilibrium [**390a**] ⇌ [**390b**] greatly favoured the methyl form [**390a**] (I-427). Later, the pK_a of a model indicated pK_T > 12, and also for the 4-isomer (67MI325). More recently precise measurements of the pK_a of pyridine methides allows the determination of pK_T values for these equilibria (71CH510, 72J(PII)1295, 73J(PII)1080) (see Table 2-6).

N Me ⇌ N CH$_2$ H

[**390a**] [**390b**]

TABLE 2-6

Methide Tautomerism Equilibria[a]

System	pK_a (pyridine)	pK_a[b] (methide)	pK_T
2-Methylpyridine	5.97	19.25	13.3
4-Methylpyridine	6.02	19.46	13.4
2-Methylquinoline	5.83	15.23	9.4
1-Methylisoquinoline	6.42	15.89	9.5

[a] 72J(PII)1295, 73J(PII)1080.
[b] See also (72BF1903) for alternative values.

b. Pyrindines

It is convenient to deal with pyrindine tautomerism at this point: these compounds can be considered as cyclized allylpyridines in which the NH forms are stabilized by cyclic conjugation.

The spectra of benzo[*b*][1]pyrindine show a mixture of three tautomers [**391a**], [**391b**] and [**391c**] in proportions which vary with the physical state and temperature, but with comparable amounts of

[**391a**] and [**391b**] (the former predominating) whereas the proportion of [**391c**] was low, at ca. 0.1%. The cation in CF_3CO_2H is predominantly [**392a**] with ca. 20% [**392b**] (71JO2065; see also 72TL3273).

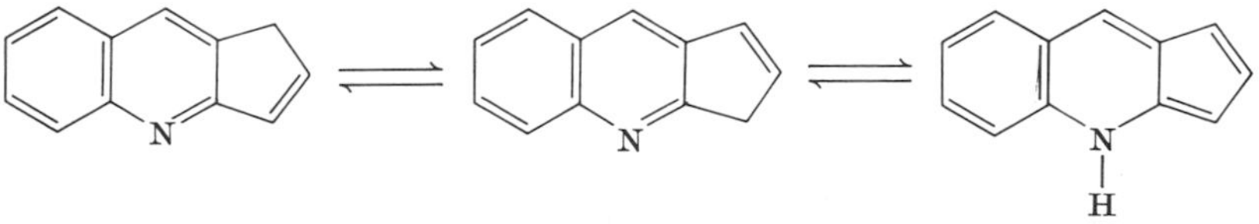

[**391a**] [**391b**] [**391c**]

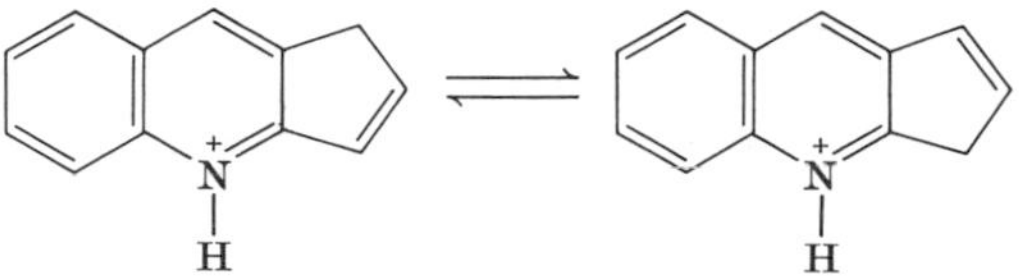

[**392a**] [**392b**]

Trace amounts of the NH tautomer [**393c**] in simple 1-pyrindines (as [**393**]) are also detected by UV and pK_a methods (62JA3979). IR and NMR spectral results agree with the existence of a mixture of [**393a**] and [**393b**] and with little [**393c**] (66TL2579, 67T3601); later the NMR spectrum was interpreted to show a slight predominance of [**393a**] over [**393b**] (71JO2065). The subject has been reviewed (73AH187).

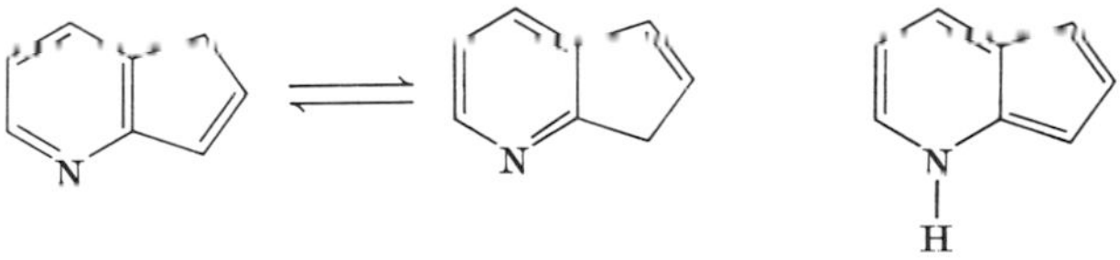

[**393a**] [**393b**] [**393c**]

c. Methyl Azines

Countless papers include qualitative spectral evidence for the predominant existence of methyl azines in the methyl form. In all this work there is only one area where NH forms are at all important: the 1,2,4-triazines.

Methylene forms [**393/1a**] also predominate in certain pyrimidines of type [**393/1**] (74CI659), but here the tautomerism in question is not simple methylazine:methylenedihydroazine, because one of the pyrimidine nitrogen atoms is substituted and the tautomeric proton moves between exocyclic C and N (i.e., type [**5a**] ⇌ [**5b**], see Section 1-1a). It is difficult to rationalize the predominance of [**393/1a**] although

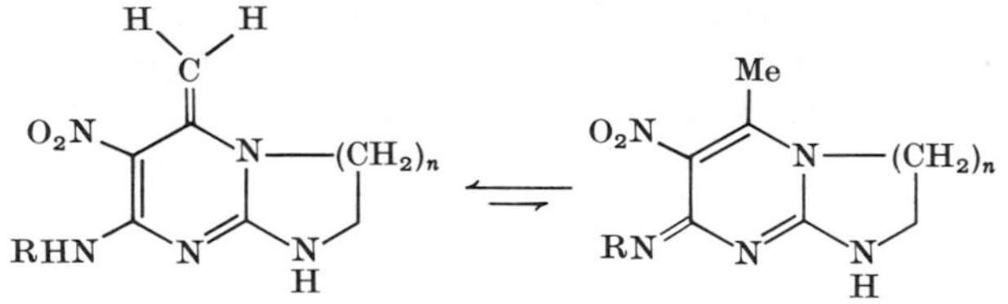

[393/1a] [393/1b]

the evidence is convincing, especially in view of the normal behaviour of [**393/2**] and [**393/3**] (74CI659).

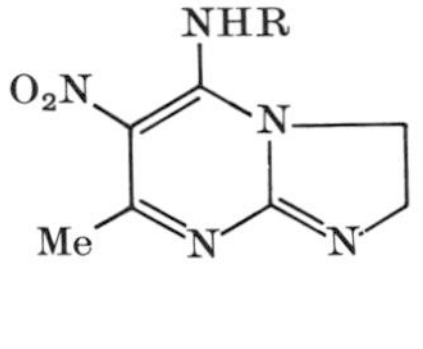

[393/2] 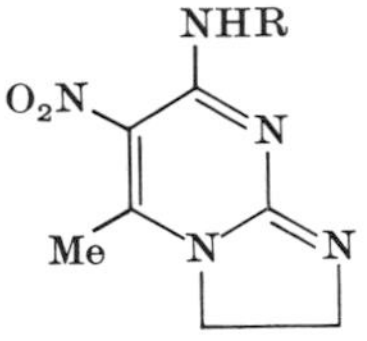 [393/3]

d. 1,2,4-Triazines

In the 1,2,4-triazinone series, for equilibrium [**394a**] $\rightleftharpoons$ [**394b**] UV comparisons with fixed forms give $K_T = 0.71$ (95% ethanol), NMR gives $K_T = 0.83$ (DMSO): the methylene form [**394b**] is in each case the minor component. In DMSO the analogue [**395**] has $K_T = 0.20$ also in favour of the methyl form [**395a**] (cf. Sections 2-6Ea and 2-7Da). NMR shows that whereas the 4-*N*-methyl derivatives exist essentially completely in the NH or *methylene* form [**396**], the 2-*N*-methyl derivatives largely favour the CH or methyl form [**397**]. This tautomerism is subject to additional complications because of covalent hydration and dimer formation (71JO3921).

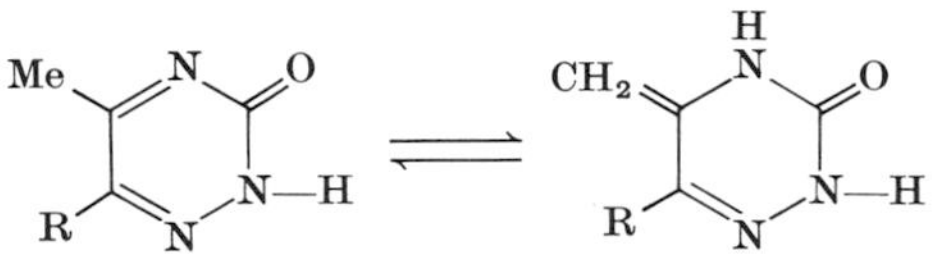

[394a] R = Me [394b]
[395a] R = H [395b]

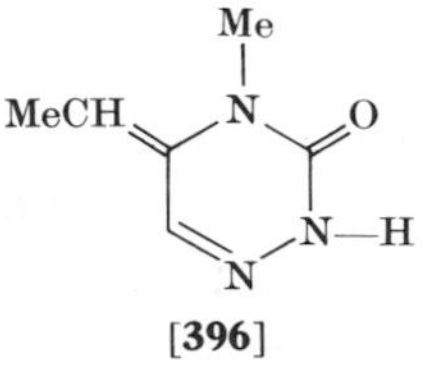

[396] 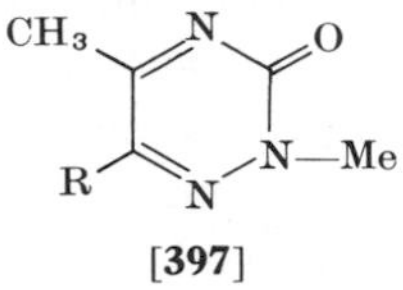 [397]

Methylene structures [**398**, R and R′ = alkyl] were also assigned to the corresponding sulphur compounds (68TL2747) by unambiguous NMR evidence (DMSO) and X-ray crystallography. The 3-methylthio analogue exists completely in the expected form [**399**].

H
R′CH N S
N—H
R N

[**398**]

R′CH₂ N SMe
N
R N

[**399**]

A similar picture is disclosed in the nitrogen series. Alkylidene structures predominate for certain potential aminotriazines as already discussed for [**367**] (cf. Section 2-7Da) and for cation [**338**] (cf. Section 2-6Ea), but not for [**337**].

In summary, compounds of type [**403a**], where the XH-proton is strongly acidic, tautomerize not to [**403b**] (disfavoured because of the high N=N bond order) nor to [**403c**], but all the way to [**403d**]. The instability of [**403c**] compared with [**403d**] is presumably related to the tendency of acylimides [**404a**] to exist in the acylvinylamine form [**404b**] (71TL4897).

RCH_2 N XH
N
N

[**403a**]

⇌

H
RCH_2 N X
N
N

[**403b**]

⇌

RCH_2 N X
N—H
N

[**403c**]

⇌

H
RCH N X
N—H
N

[**403d**]

$RCON{=}CH{-}CH_2R'$ ⇌ $RCONH{-}CH{=}CHR'$

[**404a**] [**404b**]

C. Arylmethyl Compounds

a. Benzyl and Diphenylmethyl Derivatives (I-431)

Tautomeric ratios were originally estimated as $pK_T = 11.5$ for compound [**405**] and 11.7 for [**406**] in favour of pyridinoid forms [**405a, 406a**] by the pK_a method (67MI325). A recent reevaluation

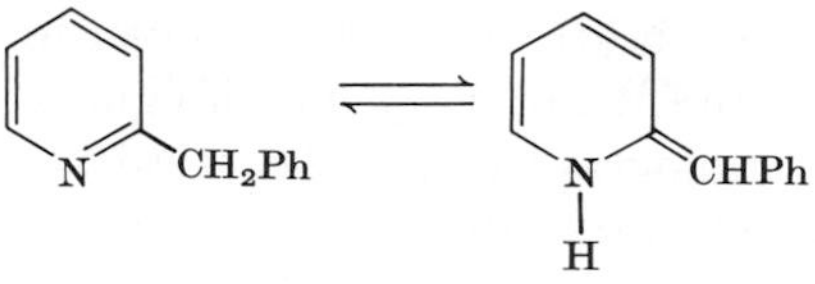

[**405a**] [**405b**] 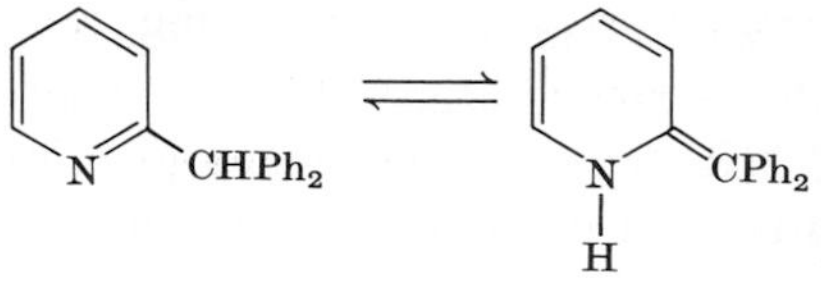 [**406a**] [**406b**]

(H_2O) gives 11.9 and 12.8, respectively (73J(PII)2111). The preference for the NH form in the order $CH_2Ph > CHPh_2 > CH_3$ demonstrates that whereas one phenyl substituent stabilizes the formal negative charge in [**407b**, Z = CHPh or CPh_2], with two phenyl groups steric congestion occurs and causes noncoplanarity of the three rings and consequent loss of stability (73J(PII)2111).

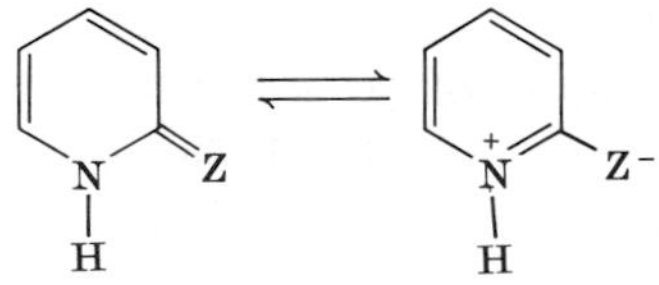

[**407a**] [**407b**]

b. Photochemical Tautomerization of o-Nitrobenzyl Compounds (I-431)

The photochromism of 2-(2,4-dinitrobenzyl)pyridine [**408**], thought to depend on tautomerism of [**408a**] to [**408b** or **c**], was reported in the last review (I-431). Since then much further work has appeared. A detailed study of the structural requirements for this property (62PH-2434) shows that a 2-pyridyl ring is not a requirement for photochromism. Indeed, no heterocyclic ring is required at all: the photo

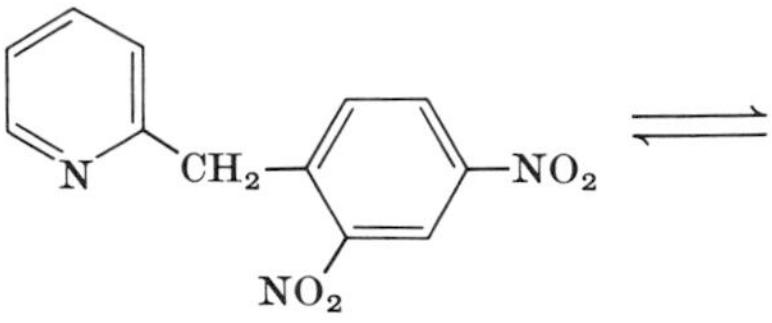

[**408a**]

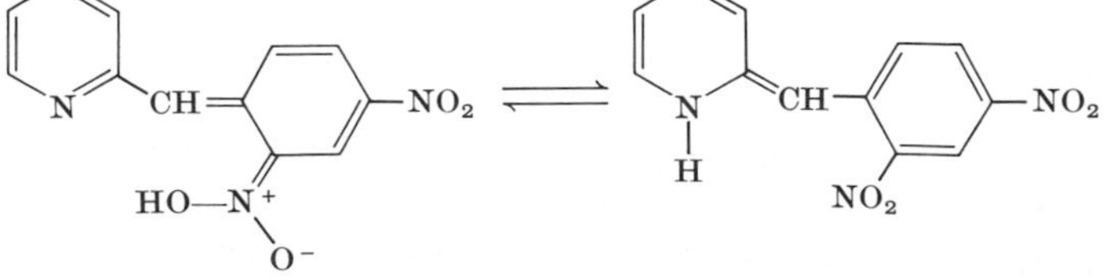

[**408b**] [**408c**]

tautomer is therefore likely to be the OH form [**408b**] rather than the NH form [**408c**], and this is supported by calculations (71TL2467).

The following structural features were needed for such photochromic behaviour: (i) a hydrogen or a benzyl carbon sufficiently activated by *ortho* or *para* substituents and (ii) a nitro group *ortho* to the benzyl carbon: a survey of the properties of a large range of compounds is available to support this (62PH2434; see also 63G1530).

Detailed kinetics have been reported for 2-(2,4-dinitrobenzyl)- and 2-(2-nitro-4-cyanobenzyl)-pyridine (62JO3155, 63PH874), and the effect of 4′-carbonyl substituents has been studied (63JO1989).

4-(2,4-Dinitrobenzyl)pyridine [**409**] is also photochromic; presumably the OH tautomeric form [**409c**] is involved. However, the molten compound is blue, and has an absorption spectrum similar to that of the *N*-methyl derivative [**410**], suggesting that the blue compound is the NH form [**409b**], not the OH form [**409c**]. From extinction coefficients, the (thermal) pK_T for [**409b**]/[**409a**] was estimated as ca. 3, and ca. 7 for the corresponding α-pyridyl analogues (64JT(41)2568). Nitro derivatives of 4-benzhydrylpyridine have also been investigated for photochromism (73ZC375).

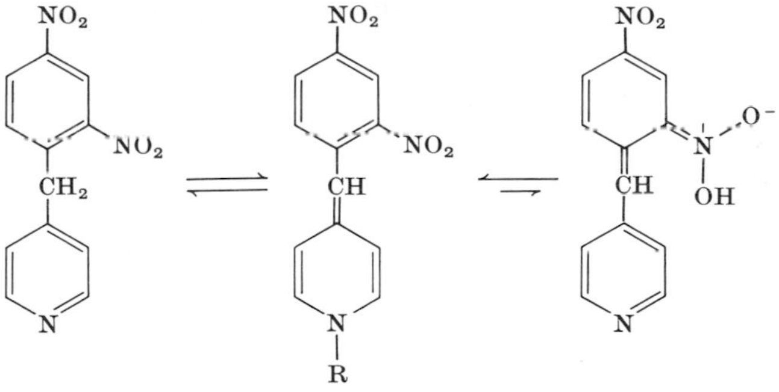

NO_2 NO_2 CH_2 N ⇌ NO_2 NO_2 CH N R ⇌ NO_2 CH N O⁻ N OH N

[**409a**] [**409b**] R = H [**410**] R = Me [**409c**]

c. Bisheteroarylmethanes (I-433)

X-ray analysis of 8-bromo-2,2′-methylenediquinoline discloses [**411**] (63R1026) and confirms the suggestion (I-433) that red diquinolylmethane crystals comprise the NH tautomers.

The tautomerism of a series of bis-2-quinolylmethanes [cf. **412**] has been thoroughly studied by Scheibe *et al.* (65BG190, 65C325, 66HC(7)-153). For most solvents and most systems studied, the colourless diquinolylmethane form [**412a**] predominates over the NH form [**412b**].

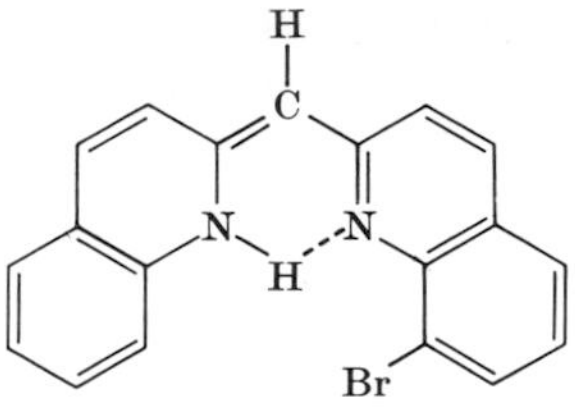

[**411**]

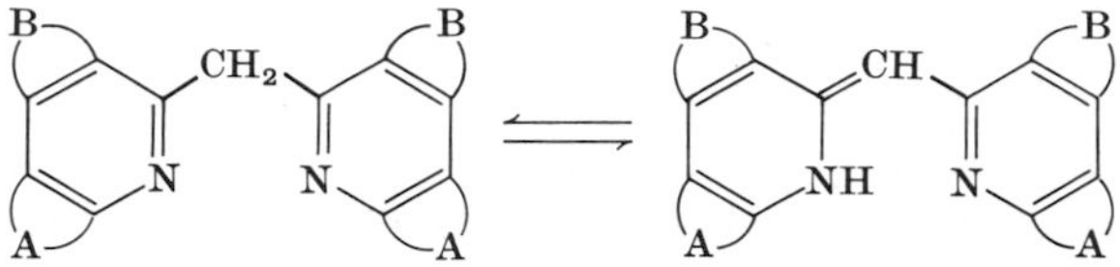

[**412a**] [**412b**]

Solvent influence is not great for the solvents benzene, dioxan, CCl_4, DMF and heptane, but EtOH tends to favour [**412a**] especially, while CS_2 favours it least. Structural influences can also be discussed: for the polybenzo derivative [**412**, A = B = C_6H_4] the methane form is most favoured, for the bisnaphtho derivatives [**412**, A = C_6H_4: C_2H_2. B = no substitution], the methane form is least favoured. The authors discuss the explanation of the results in terms of MO calculations and steric interactions.

If a cyano group is substituted at the bridging carbon atom, as in [**412/1**], the methine form predominates by a considerable factor [65JO243).

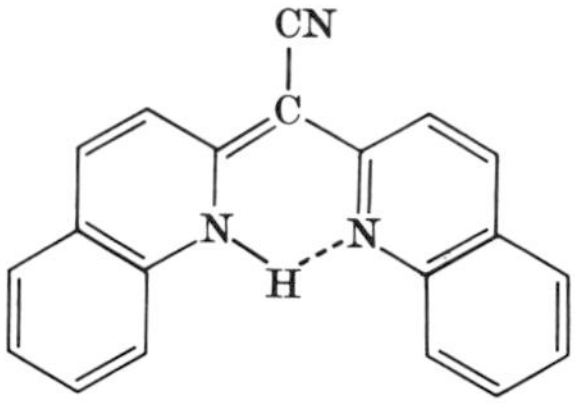

[**412/1**]

D. Methyl Substituted with C=O, C=N or C≡N

a. Introduction

Structural variation possibilities are numerous, and the tautomerism is complex in this series. In addition to the three basic tautomeric forms [**413a,b,c**] which we shall designate as CH, NH and XH (usually OH), respectively, both [**413b**] and [**413c**] can exist for α-substituted

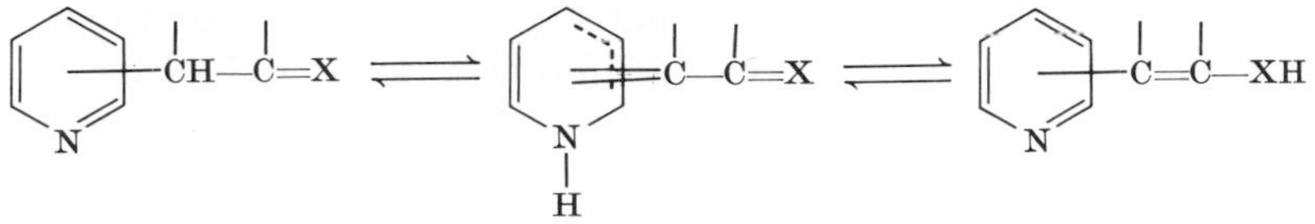

[413a] [413b] [413c]

derivatives in cyclic hydrogen-bonded modifications (see later [**414**]), *except* for cyano compounds.

However, not all the forms are equally likely in each subgroup of compounds. Thus, OH forms [**413c**] are disfavoured for esters where C═X is part of CO_2R, but the NH forms are particularly favoured for imides where C═X is C═NR. The relative importance of NH forms [**413b**] is much less for monocyclic than for polycyclic compounds; [**413b**] has lost part of the aromaticity of the ring and such a loss is more important for the monocyclic derivatives.

α- or γ-(*o*-Hydroxyphenyl) compounds can be considered as derivatives of the OH forms [**413c**] where the C═C is part of a benzene ring. Tautomerism can occur here with NH forms [**413b**] but effectively not with [**413a**].

b. Acylmethyl Compounds (I-428)

i. *General.* Much early work was carried out on these compounds by Japanese investigators under Iwanami and published in Japanese: this has been abstracted in the sequel, but was apparently not available to some of the subsequent workers (for a summary of this Japanese work, see 71BJ1311). An important paper by Mondelli and Merlini (66T3253) emphasized the three possible tautomers for α-substituted derivatives [**414a, b, c**], and pointed out that for esters (Y═O alkyl) the OH tautomeric form [**414c**] was unlikely. However, [**414b**] is less aromatic than [**414c**], and hence for Y ≠ OR [**414b**] is expected to be favoured for the monocyclic pyridines where this large aromaticity loss would be incurred (cf. Section 2-10). Tautomers [**414b, c**] may be distinguished by allylic coupling (see later).

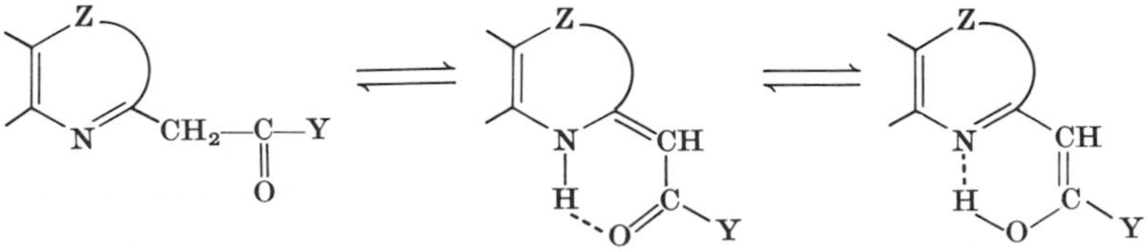

[414a] [414b] [414c]

ii. *Substituted Pyridines*. 2-Acylmethylpyridines could exist in CH or keto [**415a**], NH methide-enamine [**415b**] or OH enol [**415c**] forms. As previously pointed out (I-429) the strong H-bonds in [**415b**] and [**415c**] render the distinction between them difficult. Similar possibilities exist for e.g., [**416**] and [**417**], but without intramolecular hydrogen bonding.

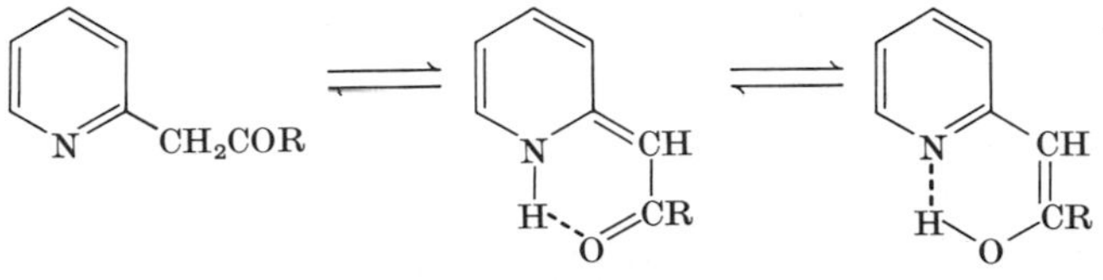

CH_2COR CH CR CH CR

[**415a**] [**415b**] [**415c**]

Earlier (I-428) 2-phenacylpyridine was concluded to exist in non-polar solvents and in the solid as the hydrogen-bonded enol [**415c**, R = Ph]; this has been confirmed, and quantitative data have been obtained (65J3093) from spectral data and pK_a measurements for H_2O for each of the 2-, 3- and 4-isomers (see Table 2–7). The resonance stabilized NH or methide form [**415b**] in the 2-isomer is claimed on NMR evidence to become increasingly favoured at high temperatures (in CS_2) (66MP1, 66T1373). However, von Philipsborn (63H2592) and Mondelli and Merlini (66T3253) suggest the use of allylic coupling between CH/CH_3 in 2-pyridylacetone [**418**] to distinguish between [**418a**], where it should be present, and [**418b**], where it should be absent. On this criterion, [**418a**] is favoured, in agreement with the earlier conclusions.

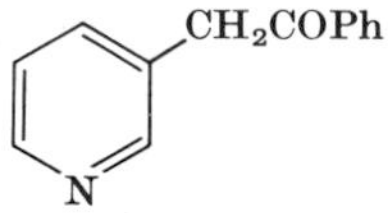

CH_2COPh CH_2COPh

[**416**] 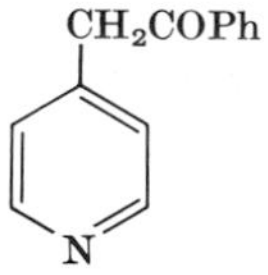 [**417**]

CH C CH_3 CH C CH_3

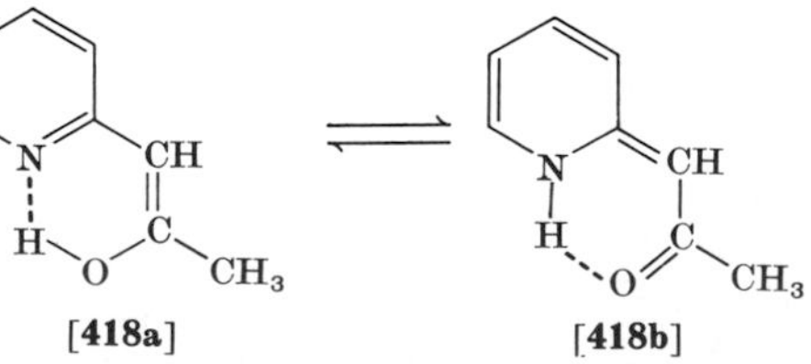

[**418a**] [**418b**]

The above discussion amply demonstrates the considerable tendency toward enolization when a 2-pyridyl group is substituted for an α-hydrogen in a ketone. However, in a suitable steric environment a

TABLE 2-7

VALUES OF pK_T OR PREDOMINANT FORM FOR PHENACYLPYRIDINE TAUTOMERISM [**415a**] ⇌ [**415b–c**]

Compound	Ref.	pK_T[c] (H_2O)	UV method (nonpolar solvents)	IR method ($CHCl_3$)
2-Phenacylpyridine	[a,b]	2.34	[**415a**]	[**415b**]
3-Phenacylpyridine	[b]	6.30	[**416**]	[**416**]
4-Phenacylpyridine	[b]	2.65	[**417**]	[**417**]

[a] 63T401, 63T413.
[b] 65J3093.
[c] For keto:methide equilibrium.

2-pyridyl group can promote *ketonization*: thus the crowded molecule [**419**] exists in the keto form [**419b**] (71JO2986).

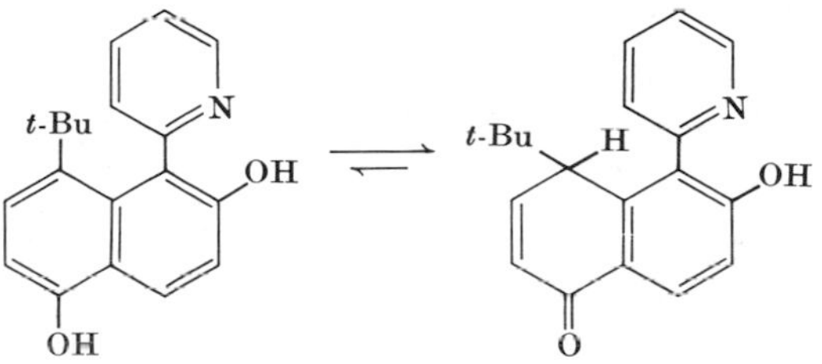

[**419a**] [**419b**]

The keto–enol ratios [**415c**]/[**415a**] for a series of alkyl 2-picolyl ketones have been measured in various solvents (CCl_4, $CHCl_3$, DMSO) by IR and NMR. Under all conditions studied, the keto form predominates; however, there is a significant dependence on the solvent polarity. While results are similar for CCl_4 and the liquid state, the equilibrium is shifted by the polar solvents $CDCl_3$ and especially DMSO towards the keto form. The K_T values were correlated with Taft steric substituent constants (72JH25). Further studies of substituent effects on the tautomerism [**415c**] ⇌ [**415a**] (R = Alk, Ar etc.) by NMR (65MI237) in non-protic solvents demonstrate that the enol form [**415c**] is favoured by R = aryl or CF_3 compared to R = alkyl.

In alkyl 4-picolyl ketones [**420**] the keto form is more favoured (unpublished data quoted in 72JH25) by an order of magnitude over the 2-analogue, demonstrating the influence of the position of the "aza" group.

2-Picolyl thioketones [**421**] occur predominantly in the enol form [**421b**] as shown by NMR and UV (although νSH bands are not detected in the IR) (64ZN952, 64ZN962, 65MI237, 65PC(30)163). Although N---H—S bridges are probably weaker than N---H—O bridges, the decreased bond energy of the C═S bond evidently outweighs this.

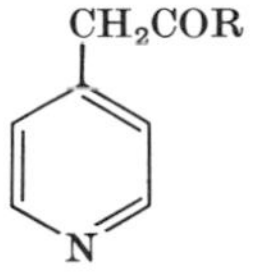

[**420**] 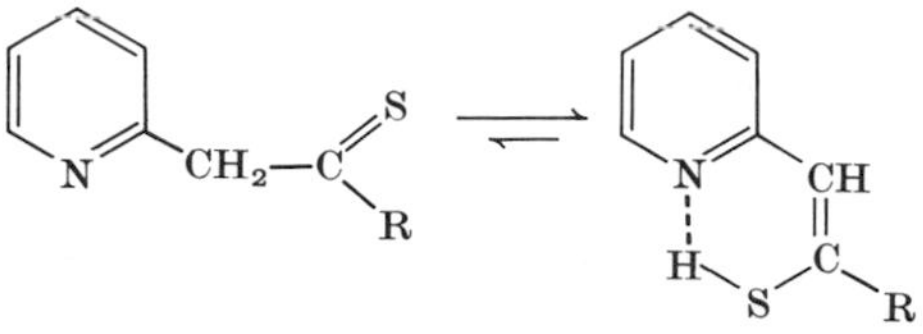 [**421a**] [**421b**]

iii. *Substituted Quinolines.* 2-Acetonylquinoline [**425**, R = Me] exists 85% in the NH methide form [**425b**] with 15% of the CH acylmethyl form [**425a**] in CCl_4 on NMR evidence (66T3253). The NH form [**425b**] was preferred to the OH form [**425c**], suggested earlier on UV evidence (63CT514), because IR bands typical for 2-substituted quinoline were missing (66T3253). A more recent detailed IR, UV and NMR investigation supports the view that 2-quinolyl ketones exist predominantly in the NH methide form [**425b**] (70CT908).

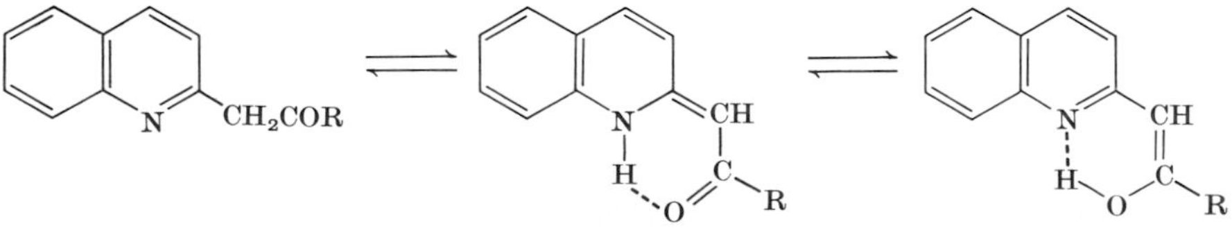

[**425a**] [**425b**] [**425c**]

The 3-acyl-β-quinindanes [**426**, R = Me, Ph, pyridyl] are likewise shown by spectra to exist in the chelated NH or enamine form [**426**], completely in the solid and predominantly in EtOH (70DA(195)868). In this case, steric factors are favourable. The structure of the chelated NH form [**427**] (R = Me or aryl) is also shown by IR (70CT901).

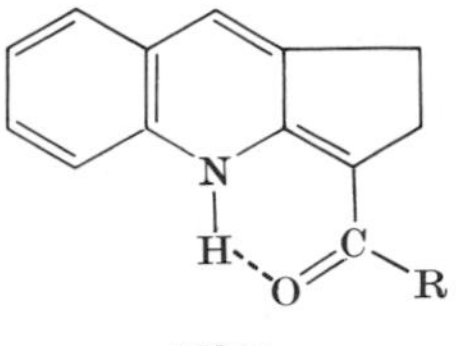

[**426**] 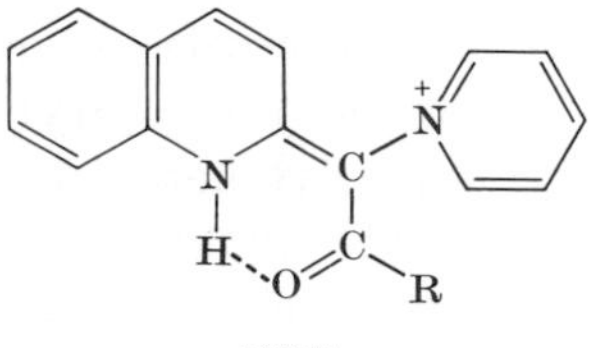 [**427**]

The two 4-quinolyl ketones [**428**] and [**429**] have been isolated crystalline (66JH272) in the CH keto forms [**428a**] and the OH enol forms [**428b**] as shown by IR. In solution the CH forms appear to be the more stable.

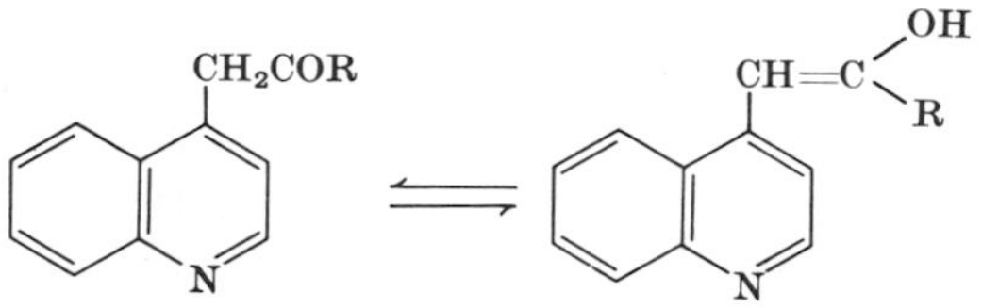

[**428a**] R = Et [**428b**]
[**429a**] R = *i*-Pr [**429b**]

iv. *Other Six-Membered Rings.* Nothing is available on the simplest acylmethylpyrazines, but the more complex (diacylmethyl)pyrazine [**430**] has been studied. Although the authors allege the existence of the equilibrium [**430a**] ⇌ [**430b**] (67JH109), the spectral data quoted (67JH109) show clearly that if any form is important in addition to the NH [**430b**], it must be the OH form [**430c**], as would be expected.

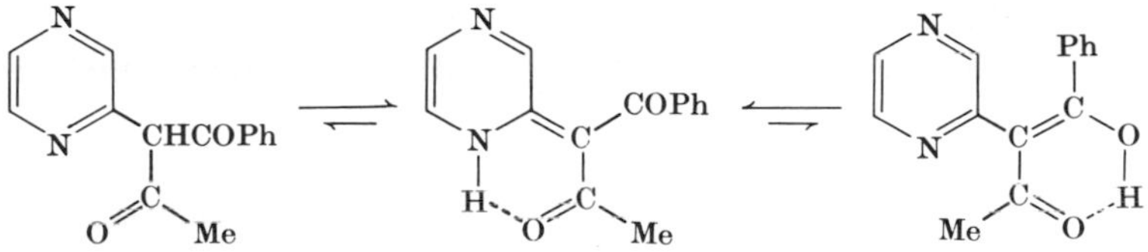

[**430a**] [**430b**] [**430c**]

2-Acetonylquinoxaline exists largely in the CH keto form [**431**] on NMR evidence (66T3253). However, the equilibrium swings to the NH methide form for the quinoxalinone derivatives [**432**] (71BJ1316), and this persists for a range of more complex derivatives including the pteridines [**433**], [**434**] (71BJ1314) and [**435**] (63H2592) and the benzoxazines [**436**] (71BJ1316).

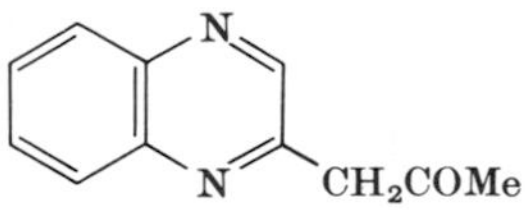

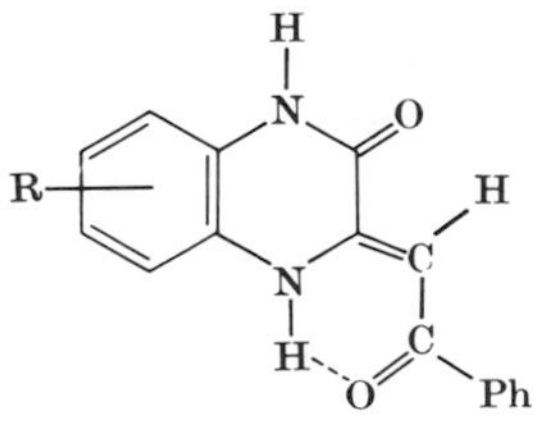

R = H, Me, Cl

[**431**] [**432**]

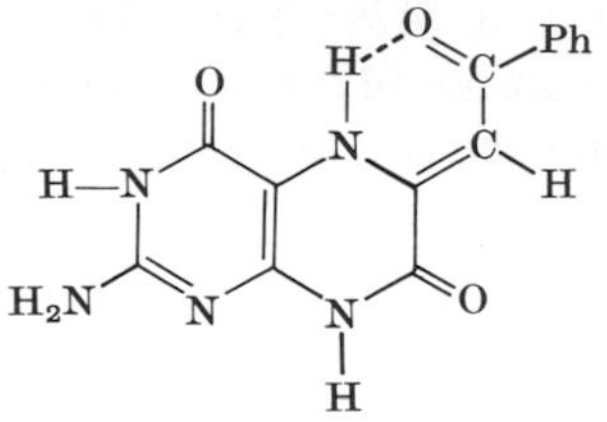

[**433**] 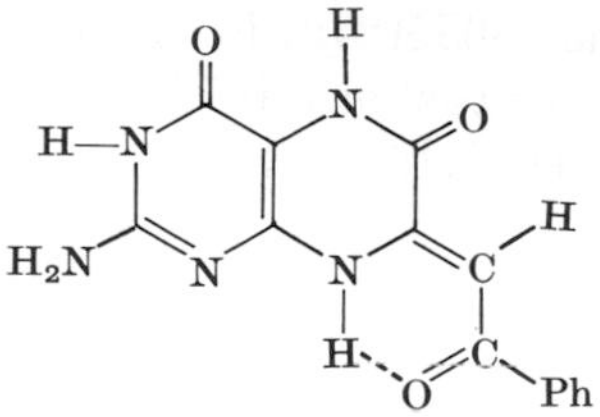 [**434**]

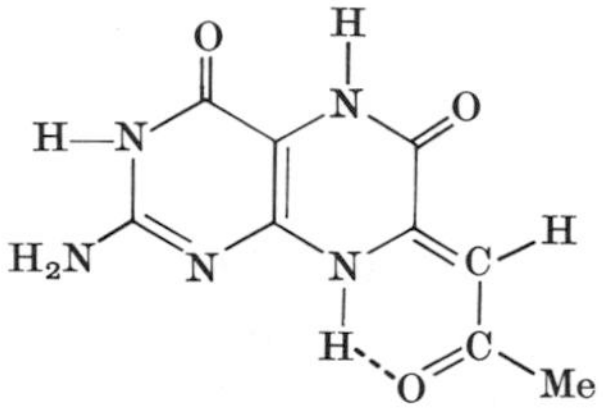

[**435**] 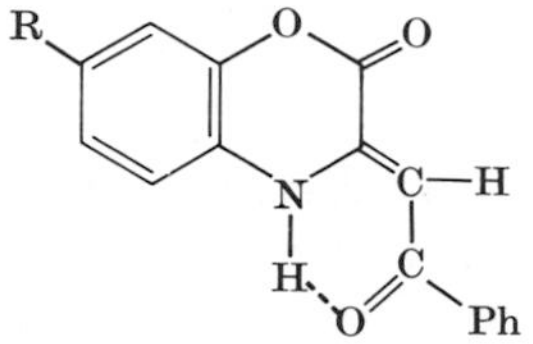 [**436**] R = H, CH_3, Cl

The hydrogen-bonded OH or enol forms written predominate for the quinazoline ketones [**437**, R = alkyl or Ph] (70CT1262), the pyrido-[2,3-*d*]pyrimidine ketones [**438**, R = alkyl or Ph] (70CT1457) and triazine ketones of type [**439**] (73BF(2)2039), as shown by spectral evidence.

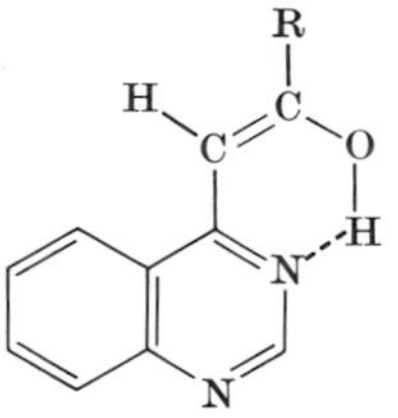

[**437**] 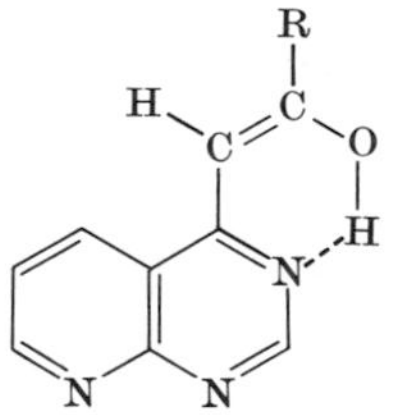 [**438**] 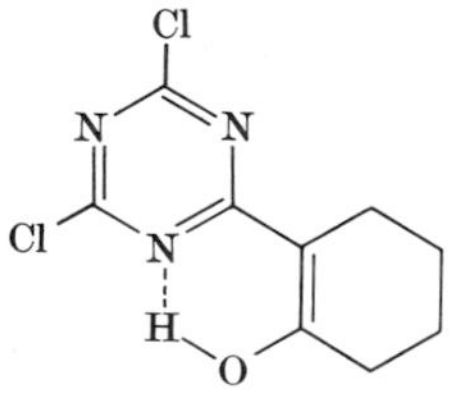 [**439**]

c. Hydroxyphenyl and Aminophenyl Derivatives

Although 2-(*o*-hydroxyphenyl)pyridines, and corresponding pyrimidines, etc., exist very predominantly in the phenolic hydrogen-bonded OH forms [as **440a**], light absorption causes phototautomerism. Thus

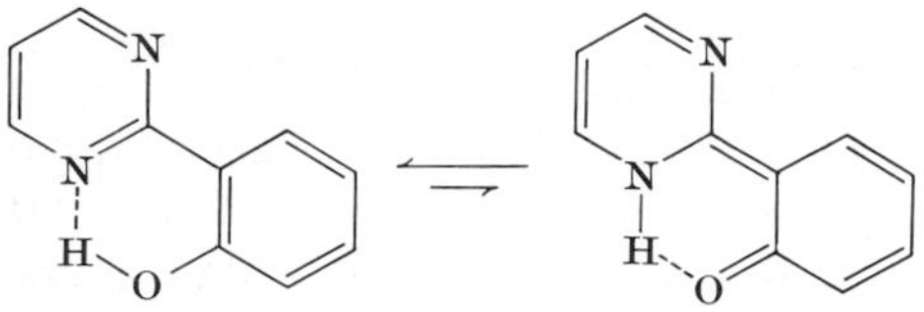

[**440a**] [**440b**]

[**440a**] is converted into the photoexcited keto or NH tautomer [**440b**] (70JH1113). This work has been extended to 2- and 4-(*o*-hydroxyphenyl)quinazolines (70JH1113), to bis-derivatives of types [**441**], [**442**] and [**443**] (72JH225), and to 2-(*o*-hydroxyphenyl)quinoline (74MI291).

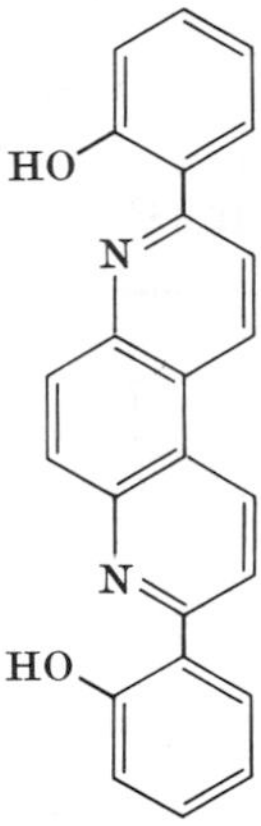

[**441**]

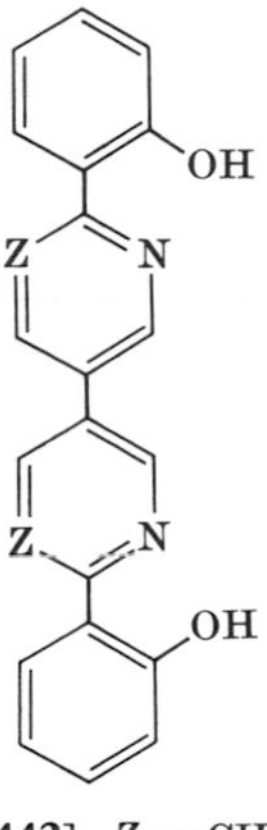

[**442**] Z = CH
[**443**] Z = N

There is no evidence for any significant proportion of alternative tautomers for other hydroxyphenyl or for aminophenyl derivatives. The OH form [**444a**] is strongly favoured for [**444**] (I-386), and it is not surprising that the amine form [**445a**] is preferred for [**445**] (73KG535).

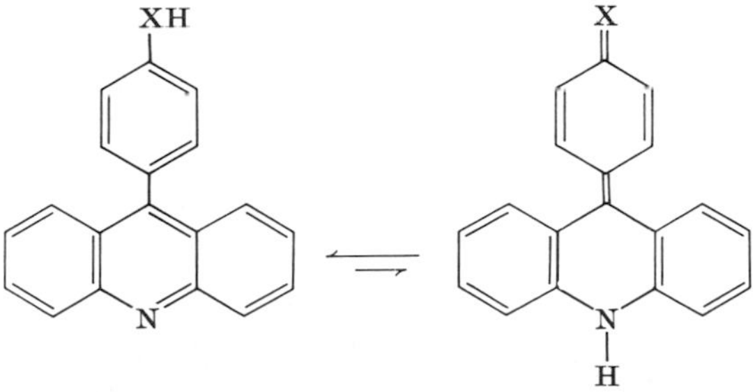

[**444a**] X = O [**444b**]
[**445a**] X = *N*-Acyl [**445b**]

d. Alkoxycarbonylmethyl Derivatives (I-430)

The ethyl pyridylacetates [cf. **446**] all exist predominantly in the acylmethyl form [as **446a**] from UV and IR spectral measurements (H_2O, $CHCl_3$). The pK_T values were calculated from basicity comparisons with [**447**] to be as given in Table 2-8 (64AJ455).

TABLE 2-8

TAUTOMERIC EQUILIBRIUM FOR ETHYL PICOLYL ACETATES[a]

Picolyl acetate	pK_a	pK_a methide (as [**447**])	pK_T
2-	4.02	10.34	6.3
3-	4.67	>11	>6
4-	4.86	10.18	5.3

[a] 64AJ455.

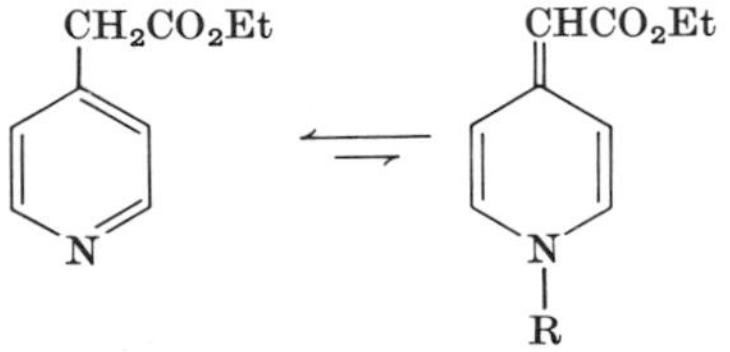

[**446a**] [**446b**] R = H [**447**] R = Me 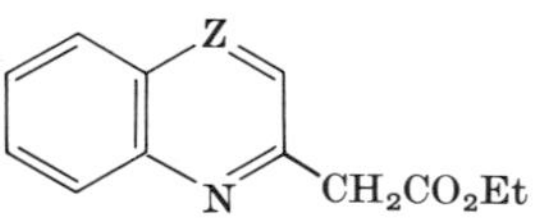 [**447/1**] Z = CH [**447/2**] Z = N

The NH form [**448a**] is less unfavoured in 3-α-(methoxycarbonylmethylene)pyridazine derivatives (pK_T = 2.98) than in 2-pyridine (pK_T = 5.61) or in 2-pyrimidine (pK_T = 6.20) analogues (73J(PII)557). This reflects the destabilization of [**448b**] by two adjacent lone-pair orbitals.

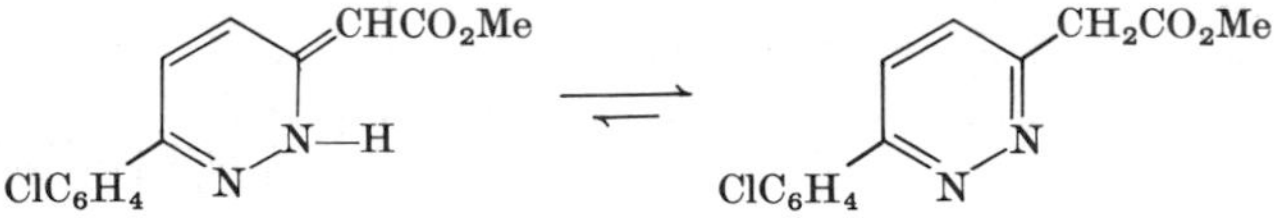

[**448a**] [**448b**]

An NMR investigation has shown that ethyl-2- [**447/1**] and 4-quinolylacetate and ethyl-2-quinoxalylacetate [**447/2**] also exist very largely in the CH or methylene forms, as written), for all media studied (66T3253). However α-(alkoxycarbonylmethylene) derivatives of the less aromatic bicyclic rings show more tendency to occur in the corresponding NH form. This tendency increases as the solvent is changed from $CDCl_3$ to DMSO to CF_3CO_2H (the possibility of protonation in this solvent is discussed in the original paper): it is already quite large for quinoxalinones [**449, 450**] and still greater for 1,4-benzoxazinones [**451**] and benzoquinoxalinones [**452**] (66T3253).

Other work supports these conclusions: in the *N*-methylquinoxalinone series, the NH or methide form **[450b]** occurs as shown by IR (KBr) (61NK778, 66J(C)806, 71BJ1311), for the benzoxazinone the equilibrium 30% **[451a]** ⇌ 70% **[451b]** was demonstrated by NMR ($CDCl_3$) (66J(C)806). Further substituted benzoxazinones occur mainly in the NH or methide form **[451b]** by IR and UV (64BJ1745) and NMR evidence (71BJ1311).

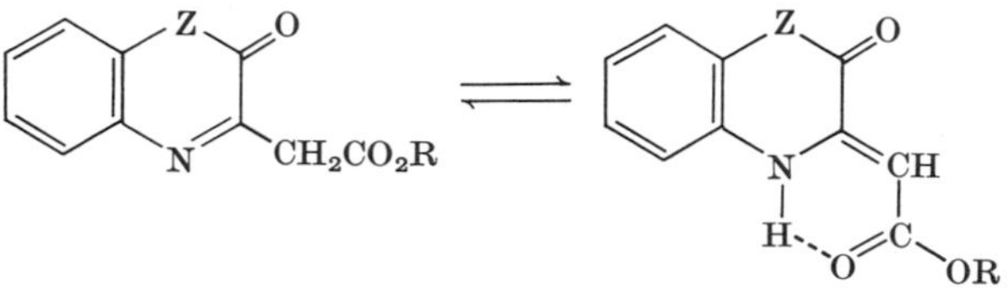

[449a]	Z = NH	**[449b]**
[450a]	Z = NMe	**[450b]**
[451a]	Z = O	**[451b]**

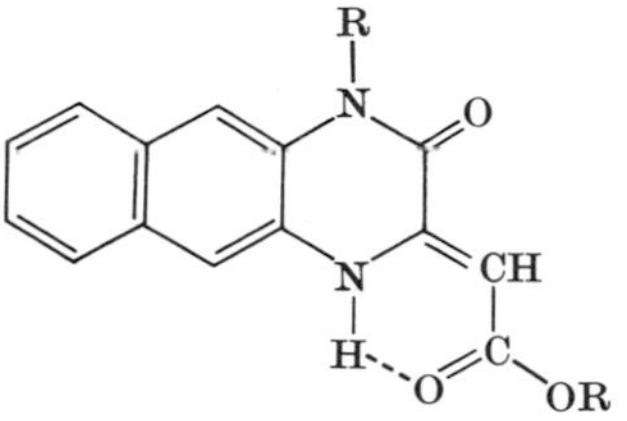

[452]

Not surprisingly, the nonaromatic monocyclic ring derivatives **[453, 454]** are also enamines (64BJ1740, 71BJ1311), and an α-amino group probably still further favours the NH or enamine forms for **[455]** (72JO2498) and **[456]** (63LA(662)83). The tricyclic compound **[457]** can be considered as a cyclic ester (72JO2498). However, the α-methyl group apparently favours the CH form in **[458]** (72JO2498), and in the pteridinone series the tautomeric equilibrium is critically dependent on the orientation of the fused ring: the methide or NH form predominates for **[459]** (63H2592, 63H2597), but CH or methylene forms for **[460]** and **[461]** (71BJ1314).

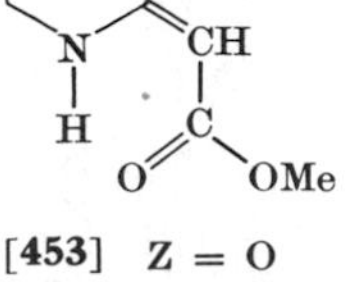

[453] Z = O
[454] Z = NH

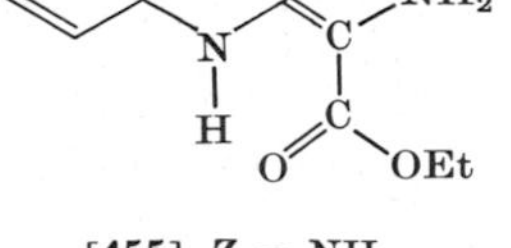

[455] Z = NH
[456] Z = O

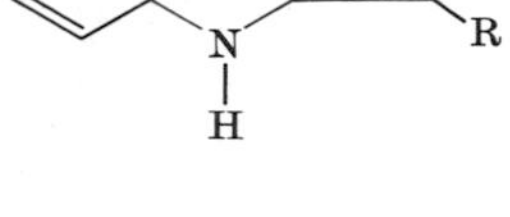

[457]

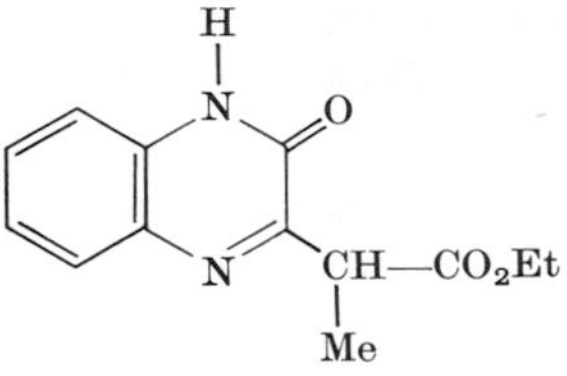

[458] 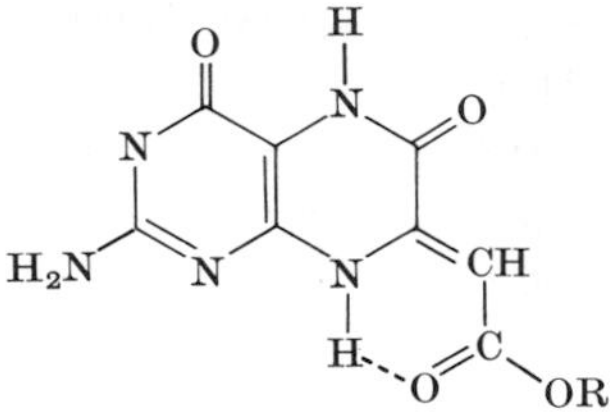 [459]

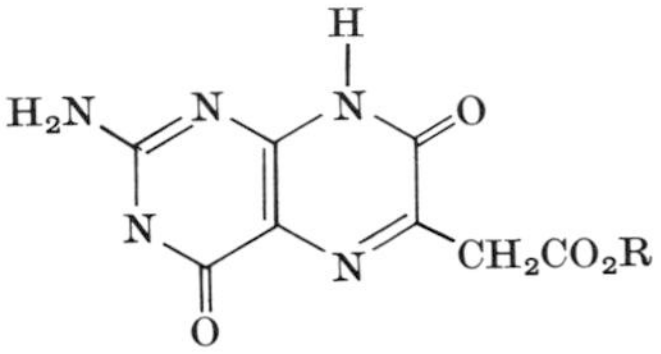

[460] 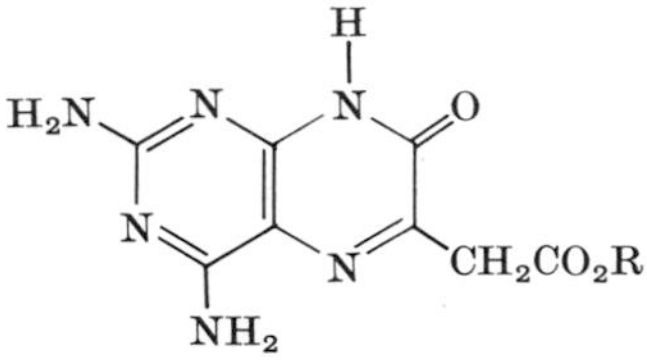 [461]

Triazinones of type [**462**] exist in the enamine form [**462b**] in the solid but mainly as the CH or methylene form [**462a**] in $CHCl_3$ (72LA(757)-100). In the condensed triazinone series [**463**], both the tautomers [**463a**] and [**463b**] have been isolated crystalline (73TL2905).

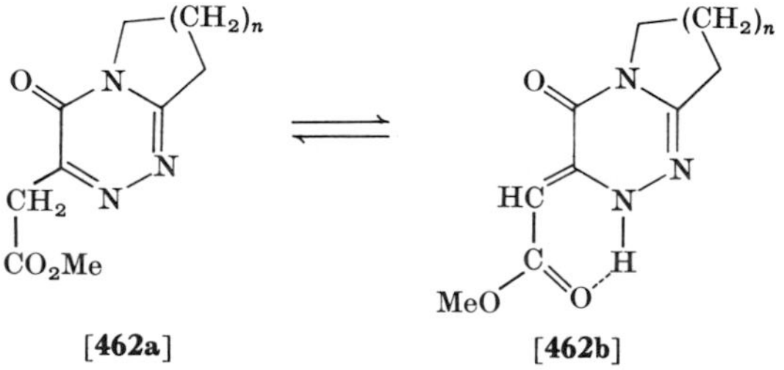

[462a] [462b]

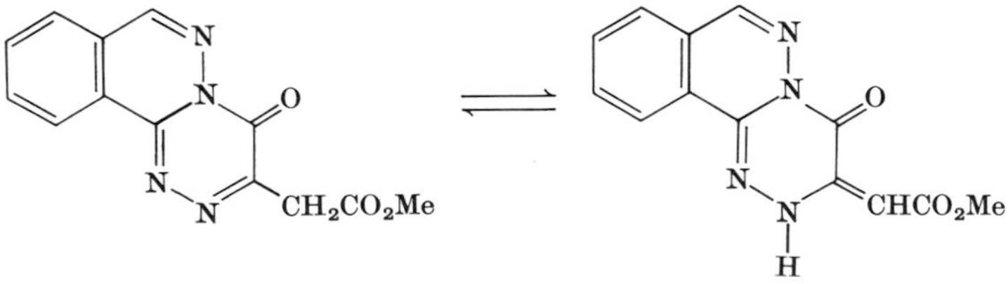

[463a] [463b]

e. Heteroaryl Derivatives of Pyruvic Acid and Esters

A very considerable tendency for tautomeric shift exists for the heteroaryl derivatives of pyruvic acid and pyruvic esters, cf. [**464**]

(62CB2195, 63JA770, 66T3253). For the 2-pyridyl compound, the OH form [**465**] is predominant, but for all the other ring systems studied enamine tautomers of type [**464a**] appear to be favoured: this includes the 2- and 4-quinoline, quinoxalinone, and 1,4-benzoxazin-3-one systems (66T3253). The pteridinyl pyruvate ester [**466**] exists in the NH enamine form as shown (62CB2195, 63H2597).

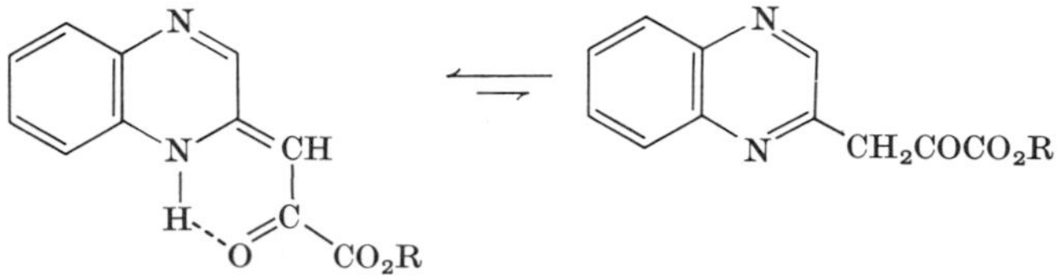

CH_2COCO_2R

[**464a**] [**464b**]

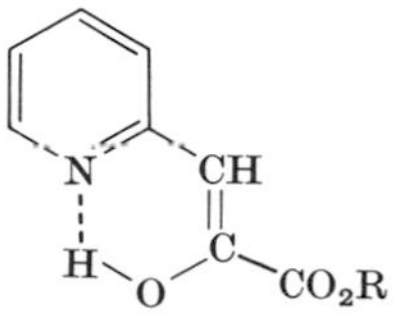

[**465**] 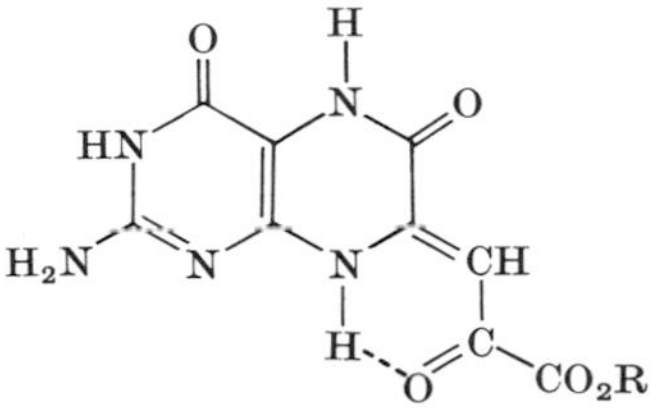 [**466**]

f. Iminoylmethyl Compounds

Picolylimine tautomerism [as **467a**] to pyridylenamine [as **467b**] was studied quantitatively by NMR. The 2-picolyl isomer exists completely as the N′H or enamine form [**468**] (C_6D_6 and DMSO) whereas for the 3-isomer [**467**] both forms coexist with ca. 30% of [**467b**] in C_6D_6 and ca. 60% of [**467b**] in DMSO. For the 4-isomer, the N′H or enamine form [**469**] dominates in DMSO, but 40% of the imino tautomer is present in C_6D_6 (70T4777).

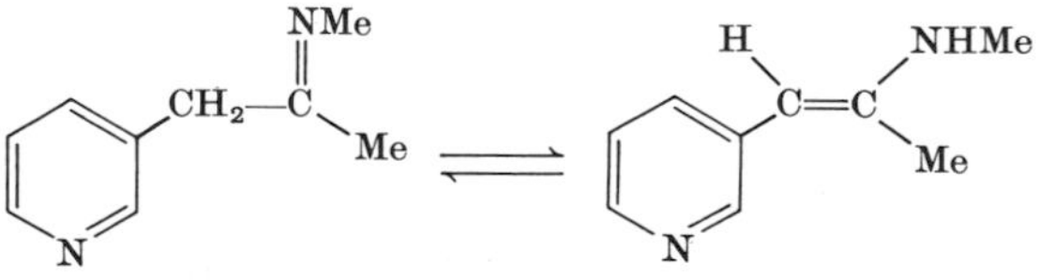

[**467a**] [**467b**]

Two crystalline modifications of [**470**] were isolated; these give different spectra in solution which suggest their formulation as the *cis* and *trans* forms [**470a**] and [**470b**] (64AP10).

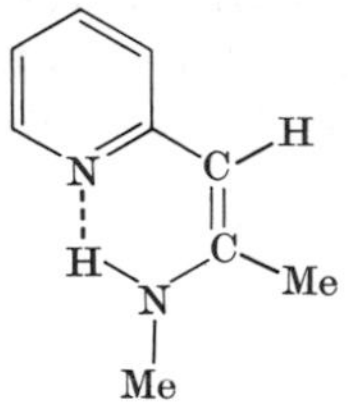

[468] 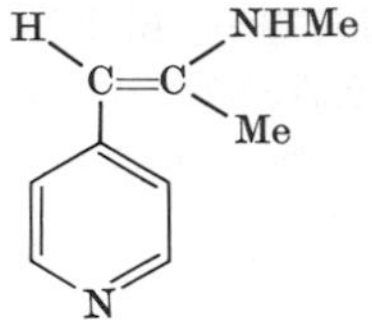 [469]

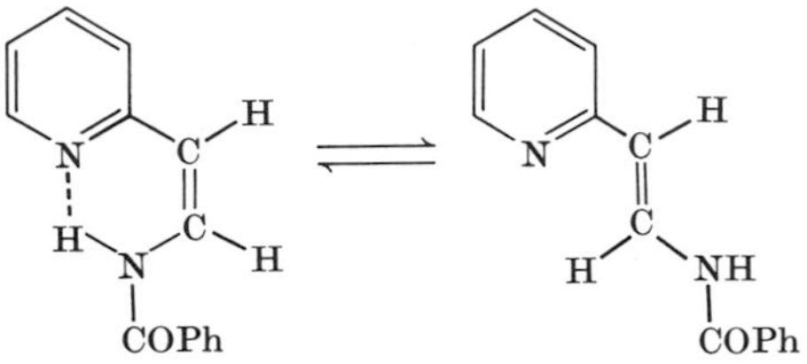

[470a] [470b]

g. Cyanomethyl Derivatives

Simple 2- [**471**] and 4-cyanomethylpyridine [**472**] exist in the CH forms shown (liquid and in CCl_4 or $CDCl_3$) from IR and NMR spectra (67JO2685), and this also applies to the 2- and 4-quinoline and 2-quinoxaline analogues, although for the quinoxalinone system both the CH [**473a**] and NH tautomers [**473b**] are found by NMR in DMSO (66T3253).

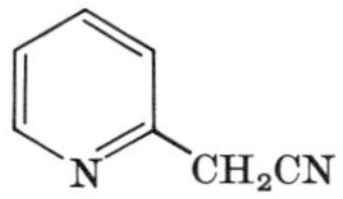

[471] 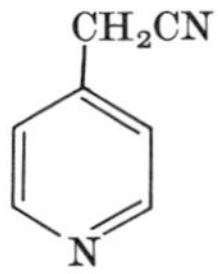 [472]

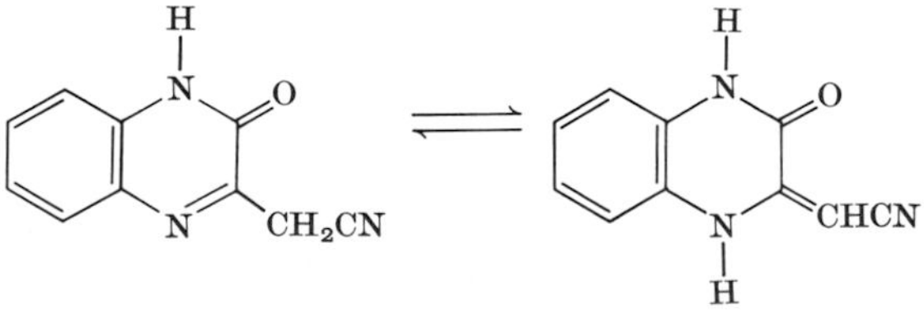

[473a] [473b]

However, the situation is changed for dicyanomethyl compounds. Thus νNH at 3100–3150 cm^{-1} and an intense, split $\nu C{\equiv}N$ show the predominance of the NH methide form [**474a**] over the dicyanomethyl tautomer [**474b**] and zwitterionic form [**475**] for the 3-isomer (solid state) (69AG(E)986), see also (72JO1047). The 2-quinolyl analogue [**476**] also takes up the NH or methide form (65JO243).

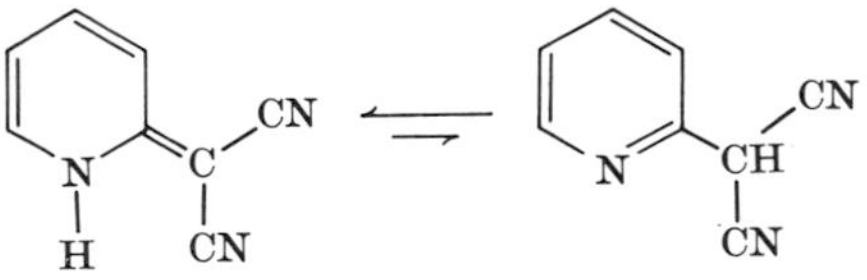

[**474a**] [**474b**]

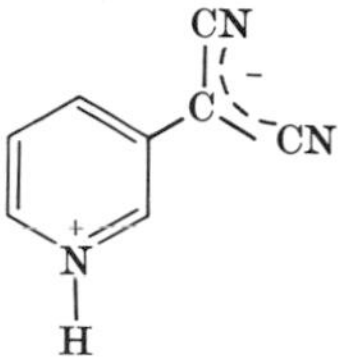

[**475**] 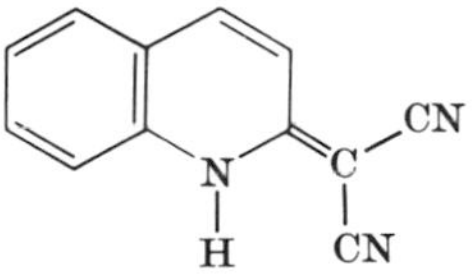 [**476**]

Series of compounds of type [**477**], [**478**] also exist essentially completely in the NH form shown, on NMR and IR evidence (64M1201; cf. also 64M1473, 65M2046). Further substituted derivatives of type [**477**] have been reported (72JO1047). Mixed cyano–carbamoyl derivatives [**479**] also take up the NH form (65M2046). In general, compounds

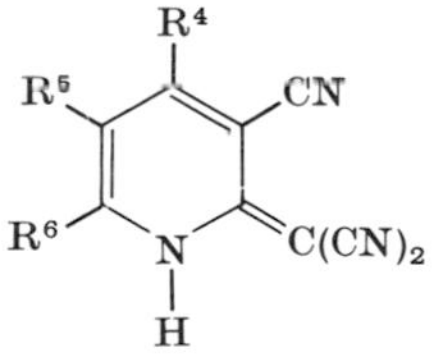

[**477**] 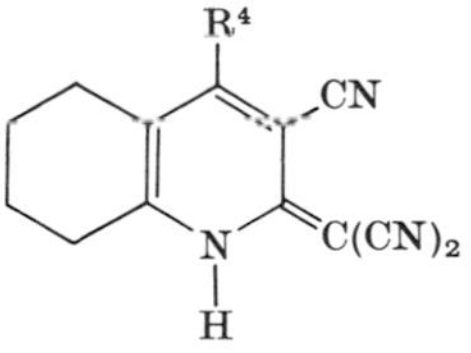 [**478**] 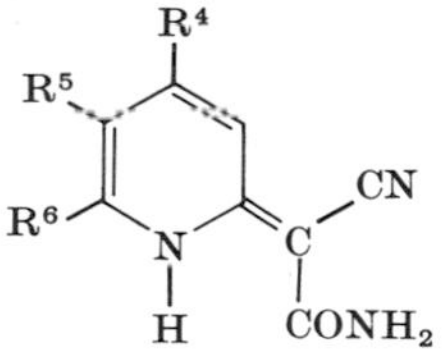 [**479**]

containing the CH(CN)X group, when X is COR, CO_2R, or C(:NR)R′ also exist essentially completely in a conjugated tautomeric form, either the NH or OH. This has been shown by spectral evidence for [**481**], [**482**] and [**483**] (67JO2685), [**484**], [**485**] and [**486**] (65JO243), [**487**]

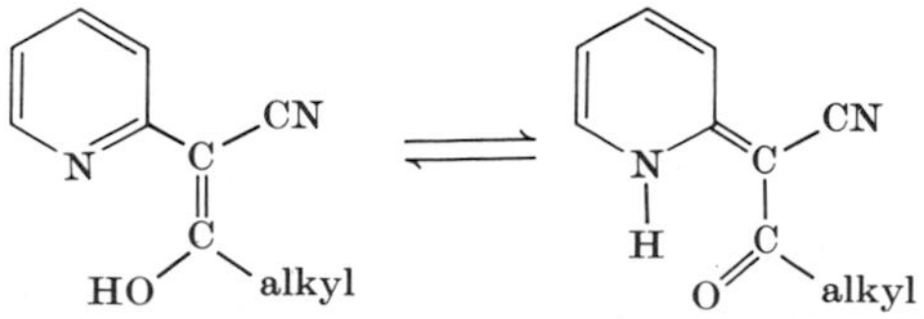

[**481a**] [**481b**] 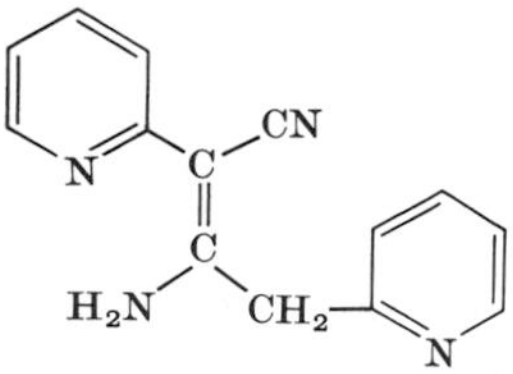 [**482**]

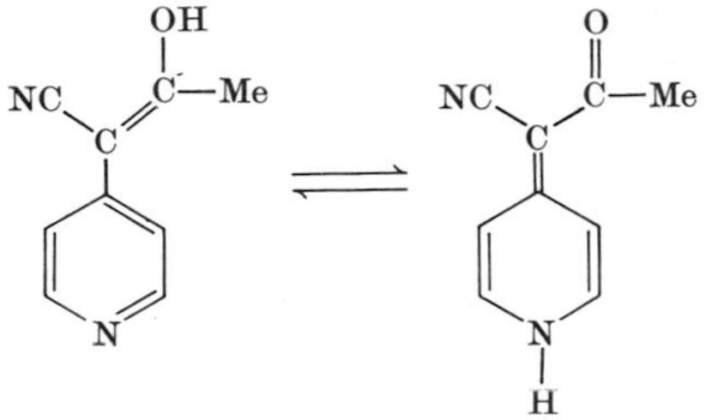

[483a] [483b]

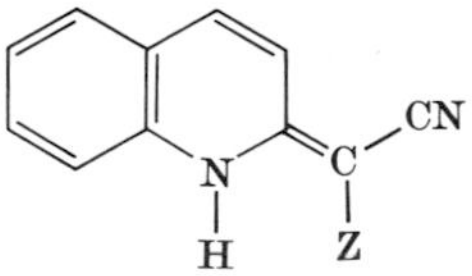

[484] Z = CO_2Bu^t
[485] Z = CO_2Et

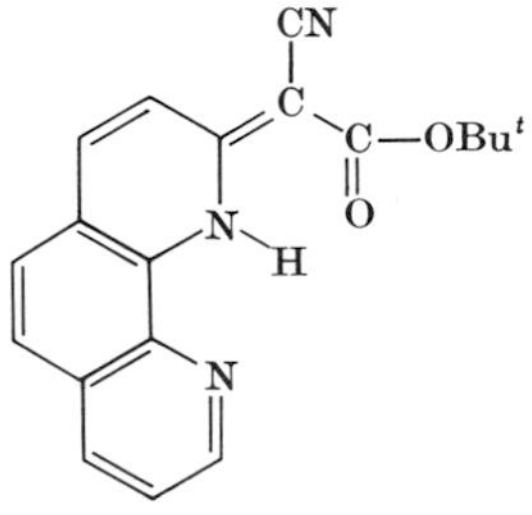

[486] 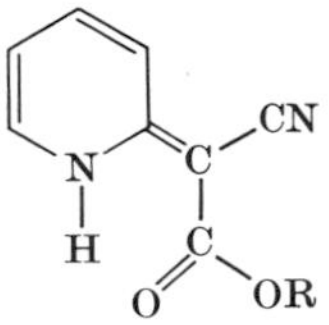 [487] 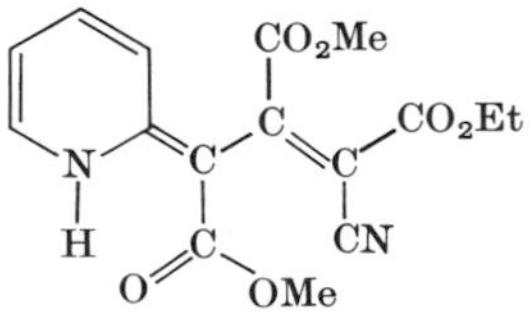 [488]

and [**488**] (71JO1165). UV spectral changes with different solvents for [**481**] have been interpreted in terms of the stabilization of [**481b**] in some solvents (67JO2685).

In the 4-quinolyl system, whereas [**489b**] exists in the solid, IR data indicate an equilibrium between [**489a**] and [**489b**] in $CHCl_3$. The fully conjugated form [**490**] predominates for the 1,8-naphthyridine analogue (68MC731).

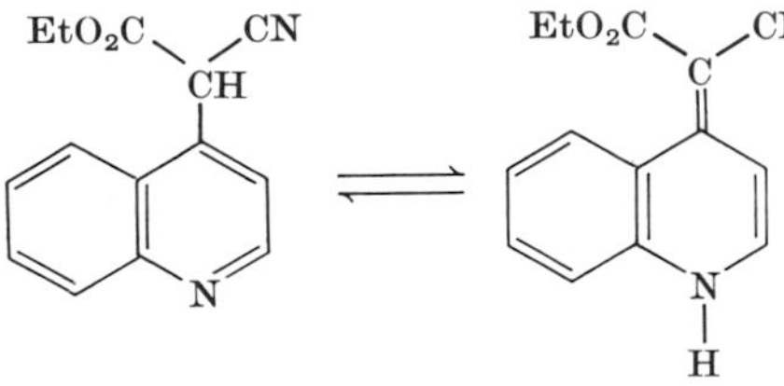

[489a] [489b] 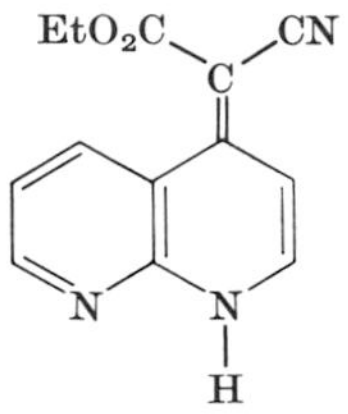 [490]

h. Azomethyl Derivatives

Potential azomethyl derivatives [cf. **490/1**] exist in the hydrazone form [cf. **490/1a**] rather than either of the two possible azo forms [**490/1b, 490/1c**]. Such systems have been little investigated, but see [73KG852) for work in the quinoxaline field.

N CH H N N R ⇌ N CH H N N R ⇌ N CH_2 N N R

[**490/1a**] [**490/1b**] [**490/1c**]

E. Methyl Substituted with SO_2R, NO_2 or POR_2 Groups

Tautomerism in this series is usually considerably simpler than for the acylmethyl compounds just discussed, because two forms only need be considered. The equilibria are of type [**491a** ⇌ **b**]: the contribution of the methine form is greatest for nitro compounds (X=N), smaller in the sulphones (X=SPh) and least in the phosphinyl derivative.

N CH_2—XO_2 ⇌ N H =CH—XO_2

[**491a**] [**491b**]

a. Sulphonylmethyl Derivatives

Basicity, UV, NMR and IR comparisons demonstrate the predominance of the CH or sulphonylmethyl form [as **492a**] for the phenyl 2-, 3- and 4-picolyl sulphones [**492**] in H_2O, DMSO, $CHCl_3$ and the solid state. pK_T data are given in Table 2-9 (65J3090).

CH_2SO_2Ph N ⇌ $CHSO_2Ph$ N H $CHSO_2Ph$ N Me

[**492a**] [**492b**] [**493**]

b. Nitromethyl Derivatives

UV, IR (NH at 3640–2200 cm^{-1} and $\nu C{=}N$ at 1502 cm^{-1}) and NMR (CH_2 at 5.9 δ and CH at 7.0 δ) show that 4-nitromethylpyridine [**494**] exists as a mixture of CH or methylene [**494a**] and NH or methide

TABLE 2-9

TAUTOMERIC EQUILIBRIA FOR PICOLYL PHENYL SULPHONES[a]

Phenyl sulphone	pK_a	pK_a methide (as [**493**])	pK_T in H_2O
2-Picolyl	2.54	10.45	7.9
3-Picolyl	3.67	13.27	9.6
4-Picolyl	3.75	11.37	7.6

[a] 65J3090.

forms [**494b**] ↔ [**494c**] (69JA1856). Further extensive work (72JO3662) has assessed the contributions of methide [**495**] and dipolar [**496**] or CH tautomers for the 2- and 3-pyridyl series.

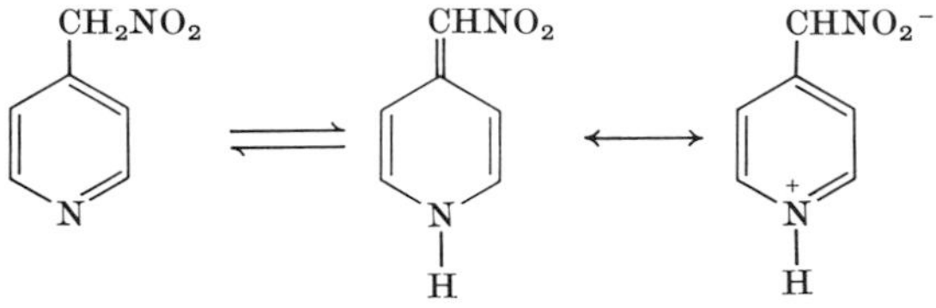

CH_2NO_2 N [**494a**] ⇌ $CHNO_2$ N H [**494b**] ⟷ $CHNO_2^-$ $\overset{+}{N}$ H [**494c**]

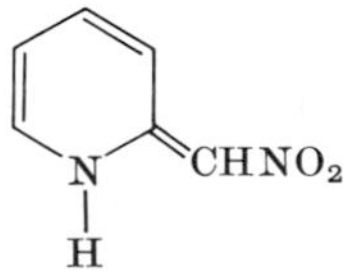

N H $CHNO_2$ [**495**] $CHNO_2^-$ $\overset{+}{N}$ H [**496**] $MeCHNO_2$ N [**497**]

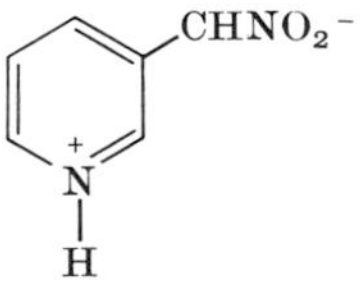

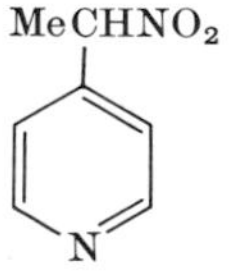

Tautomeric equilibrium constants (DMSO) are given in Table 2-10.

The greater NH or methide contribution for 2-nitromethylquinoline than for the 4-isomer is ascribed to steric interference in the 4-isomer between the nitro group and the 5-position hydrogen in the methide form.

The contribution of methide tautomers to secondary α-nitroalkyl heteroaromatics of type [**497**] is evidently considerably smaller, and they could not be detected by spectral methods.

c. Phosphinylmethyl Compounds

Basicity, UV, IR and NMR comparisons show that for each of the 2-, 3- and 4-picolyldiphenylphosphine oxides [as **498**] the CH or

TABLE 2-10

TAUTOMERIC EQUILIBRIUM CONSTANTS FOR NITROMETHYL HETEROCYCLES IN DMSO[a,b]

Nitromethyl derivative			
Position	Ring system	Other substituents	K([NH form]/[CH form])
2	Pyridine	—	0.09
3	Pyridine	—	0.02
4	Pyridine	—	0.50
4	Pyridine	2-Me	0.46
4	Pyridine	2,6-diMe	0.67
2	Quinoline	—	1.63
4	Quinoline	—	0.50
1	Isoquinoline	—	3.56
4	Pyrimidine	—	0.50
4	Cinnoline	—	10.70

[a] Calculated from NMR peak intensities.
[b] 72J03662.

methylene form [**498a**] predominates under all conditions studied. The pK_a measurements indicated p$K_T \geq 8$ for these equilibria (66J(B)631).

[**498a**] ⇌ [**498b**]

9. Miscellaneous Cases of Substituent Tautomerism

A. ALDEHYDES AND KETONES (I-435)

2-Acetylpyridine forms neutral complexes with Cu^{2+}, Ni^{2+} and Co^{2+}, which are assigned structure [**499**]. A new band at 1625 cm^{-1}, attributed to $C{=}CH_2$, appears in place of the $C{=}O$ band at 1700 cm^{-1} (68IS1).

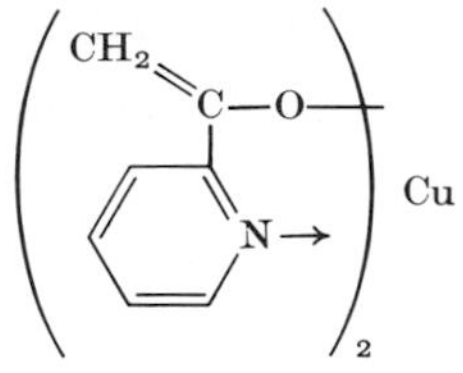

[**499**]

The enediol structure of α-pyridoin (I-434) is confirmed by X-ray crystallography (65CX(18)22).

B. Carboxylic Acids (I-435)

Thermodynamic data, especially the ionization entropies of pyridine monocarboxylic acids in aqueous solution, confirm the high percentage (I-435) of the zwitterionic form in pyridine acids (64PH3435) compared with those of amino acid zwitterions. The low entropy factor for the pyridine-2-carboxylic acid in comparison with those of the 3- and 4-isomers reflects the proximity of the charges.

10. Pyridine and Azine Substituent Tautomerism: Conclusions

A. Compounds with One Tautomeric Functional Group

The effects of the relative position of the functional group and ring nitrogen atom(s) and of fused benzene rings are summarized in Tables 2-11 to 2-13.

a. Compounds with One Cyclic Nitrogen Atom

Compounds with one nitrogen atom are dealt with in Table 2-11. Here two tautomeric forms only are involved, and the pK_T values show clearly that (in aqueous solution) the oxo, thione, but amino and methyl forms predominate. Quantitatively there is a relation between the values for the C=S to SH and the C=O to OH tautomeric ratios, which is shown graphically in Fig. 2-2 and which gives Eq. (2-6):

$$pK_T(\mathrm{S}) = 0.8pK_T(\mathrm{O}) + 1.7 \tag{2.6}$$

Available data for C=NH to NH_2 and for C=CH_2 to CH_3 tautomerism are also plotted in Fig. 2-2, but the significance of the apparently higher slopes is doubtful: these data are probably quantitatively unreliable. However, the trend is probably real.

The effect of fused benzo rings is thus similar on the various types

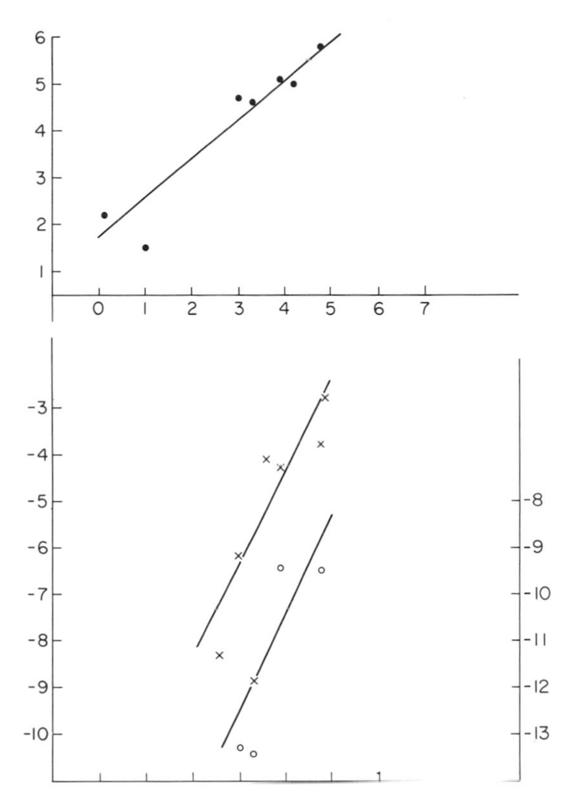

FIG. 2-2. Plot of pK_T values for S(●), NH(×) (left-hand Scale) and CH_2 compounds (○) (right-hand scale) against corresponding O compounds (see text). (The same horizontal scale applies to all graphs).

of substituent group tautomerism. For the 2-substituted pyridines, benzo fusion at [**a**] increases somewhat, and at [**c**] considerably, the proportion of the pyridonoid form. The effect of dibenzo-fusion at [**a, c**] is less clear, but it is not cumulative. Benzo fusion at [**b**] decreases the proportion of pyridonoid form, as expected on bond localization grounds (cf. discussion in Section 2-3Bf).

For the 4-substituted compounds, benzo-fusion at [**a**] and especially dibenzo-fusion at [**a, b**] increases the amount of pyridonoid form.

b. Compounds with Two or Three Cyclic Nitrogen Atoms

For compounds with two or three nitrogen atoms (Table 2-12), the position is considerably more complicated, because at least three tautomeric forms must be considered, and discussion is further hampered by considerably less quantitative data.

TABLE 2-11

TAUTOMERIC EQUILIBRIA OF MONOFUNCTIONAL PYRIDINES AND BENZOPYRIDINES[a,b]

	X	O	S	NH	CH_2
2-XH-pyridine ⇌ 1H-pyridin-2-X (ring positions 1–6; fusion bonds a, b, c)	Parent	1-H(3.0), I-350	1-H(4.7), I-397	XH(5.3–6.2), I-407, T2-5	XH(13.3), T2-6
	Benzo [*a*]	1-H(3.9), I-350	1-H(5.1), I-397	XH(4.3), T2-5	XH(9.4), T2-6
	Benzo [*b*]	1-H(low), 2-3Bf	1-H(3.0), I-397	—	—
	Benzo [*c*]	1-H(4.8), I-350	1-H(5.8), I-397	XH(3.8), T2-5	XH(9.5), T2-6
	Dibenzo [*a*, *c*]	1-H(3.9), I-350	—	XH(2.8), I-409	XH, 2-8Ba
3-XH-pyridine ⇌ 3-X^--pyridinium (N$^+$–H) (ring positions 1–6; fusion bonds a, b)	Parent	1-H(0.1), I-354	1-H(2.2), I-397	XH, I-408	XH } 2-8Bb
	Benzo [*a*]	XH(≃1), I-354	1-H(1.5), I-397	XH, I-406	XH } 2-8Bb
	Benzo [*b*]	1-H(~0.5), I-354 (cf. 2-3D	—	XH, I-406	XH } 2-8Bb

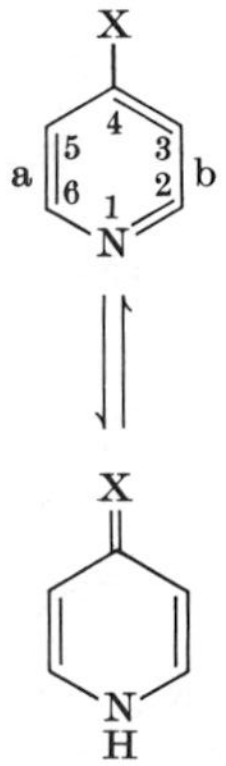

Structure	Compound					
4-X-pyridine (positions 1–6; rings a, b) ⇌ 4-X=-1H-pyridine (N–H)	Parent	1-H(3.3), 1-350		1-H(4.6), I-397	XH(3.3, 8.7), I-407, T2-5	XH(13.4), T2-6
	Benzo [*a*]	1-H(4.2), I-350		1-H(5.0), I-397	XH(3.2), 1-407	XH, 2-8Bb
	Dibenzo [*a*, *b*]	1-H(7.0), I-350		1-H, I-399	XH, 2-6Ba	XH, 2-8Bb

[a] Results are expressed as log {[major]/[minor]}.
[b] Reference given to results in previous review or present book.

TABLE 2-12

TAUTOMERIC EQUILIBRIA OF MONOFUNCTIONAL AZINES AND BENZOAZINES[a,b]

	X	O	S	NH
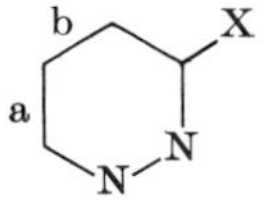	Parent	2-H(XH, 4.3), I-365	2-H, 2-5E	XH, 2-6Da
	Benzo [*a*]	2-H(XH, 2.6), 2-4Cc	—	XH(2H, 8.3), 2-6Db
	Benzo [*b*]	2-H, I-366	2-H, 2-5E	—
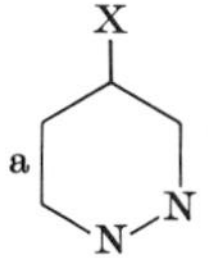	Parent	4-H(XH, 2.6), I-364	4-H, 2-5E	XH, 2-6Da
	Benzo [*a*]	4-H(XH, 3.6), I-364	4-H, 2-5E	XH(1H, 4.1), 2-6Db
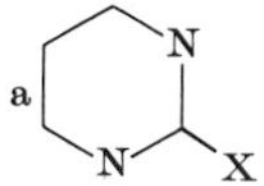	Parent	1-H, I-369	1-H, 2-5E	XH(1H $\simeq$ 6), I-413
	Benzo [*a*]	1-H, I-371	—	—
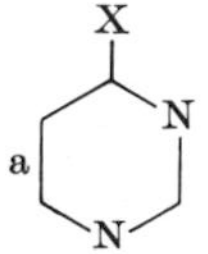	Parent	3-H(1-H, 0.40), I-370	3H(1-H, 0.30), 2-5E	XH(1 or 3H $\simeq$ 6), I-413
	Benzo [*a*]	3H(1-H, 0.85), I-370	3H(1H, 1.4), 2-5E	XH, I-415
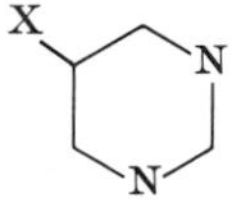	Parent	XH, 2-4Da	—	—

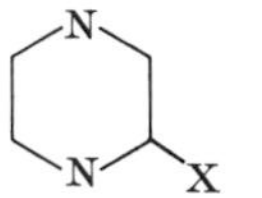	Parent	1-H, 2-4E	1-H, 2-5E	XH, 2-6Dc
	Benzo [*a*]	1-H(XH, > L6), I-378	1-H, 2-5E	XH, I-415
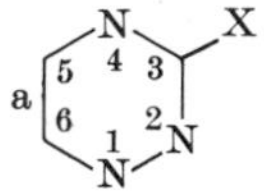	Parent	2-H(substituted), 2-4Fb	—	XH, 2-6Ea
	Benzo [*a*]	(?), I-388	—	—
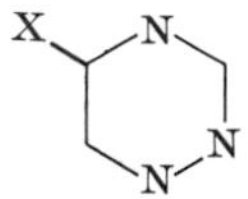	Parent	2-H, 2-4Fb	—	—
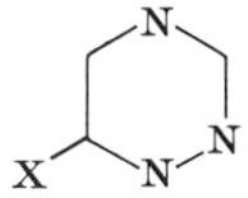	Parent	—	—	—
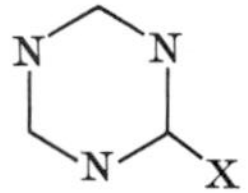	Parent	—	—	XH, 2-6Ed
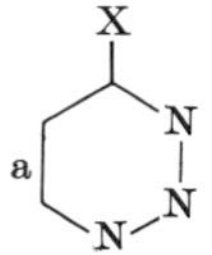	Parent	—	—	—
	Benzo [*a*]	NH, 2-4FcX	—	NH(?), 2-6Ee
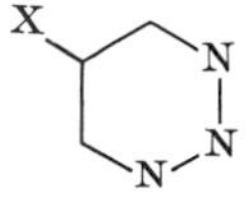	Parent	—	—	—

[a] Results are expressed as log {[major]/[minor]}. [b] Reference given to results in previous review or present book.

Qualitatively the position is again clear: for the oxygen and sulphur compounds with an α- or γ-cyclic nitrogen, *this* nitrogen exists as NH. There are two such nitrogens in 4-substituted pyrimidines, and here the α-NH form predominates by a rather small factor. More important in the 1,2,4-triazines is that the $2H$ form predominates, whether this is α-NH in the 3-substituted or γ-NH in the 5-substituted derivatives.

Quantitatively, far fewer data are available in this series. The effect of benzo-fusion appears to be of the same order of magnitude as in the mono–aza series. Where comparisons can be made, the S-compounds tend to resemble the O-analogues, and the pK_T of the NH_2-compounds show trends which resemble those found in Fig. 2-2.

B. Compounds with Two or More Functional Groups

Results are so fragmented that no generalizations can be attempted. Available results are summarized in Table 2-13: because of lack of data, only potential OH and NH_2 are considered as substituents, and even here very many gaps are to be found.

C. Pyridine and Azine Tautomerism in Relation to Aromaticity

The tautomerism of pyridines and azines is quantitatively related to the tautomerism of aliphatic functional groups by aromatic resonance energy differences. This approach has already been outlined in Section 2-1B. As already explained, the basic assumption is that the difference between the tautomerism of a pyridine derivative [**5**] and that of the corresponding aliphatic derivative [**6**] depends solely on the difference in the aromatic stabilization of the rings in [**5a**] and [**5b**]. If a and i denote, respectively, the intrinsic energies of the —HN—C(=O)— and the —N=C(OH)— bond systems, Eqs. (2-7) and (2-8) follow for

$$\Delta H_u = [A_{\text{pyridone}} + a] - [A_{\text{pyridine}} + i] \tag{2-7}$$

$$\Delta H_s = a - i \tag{2.8}$$

ΔH_u and ΔH_s, the differences in heat content for the tautomeric equilibria in the unsaturated [**5**] and saturated [**6**] series, respectively.

If the difference $(a - i)$ is the same in the two series, and there is evidence that variations in $(a - i)$ are less than 1 kcal mole^{-1} (72J-(PII)1295), then Eq. (2-9) follows. Using Eq. (2-10), we obtain Eq. (2-11).

$$\Delta H_s - \Delta H_u = A_{\text{pyridine}} - A_{\text{pyridone}} \tag{2-9}$$

$$\Delta H = \Delta G + T\Delta S = -RT \ln K + T\Delta S \tag{2-10}$$

$$RT \ln (K_u/K_s) + T(\Delta S_s - \Delta S_u) = A_{\text{pyridine}} - A_{\text{pyridone}} \tag{2-11}$$

TABLE 2-13

TAUTOMERIC EQUILIBRIA OF DIFUNCTIONAL PYRIDINES AND AZINES

N at	X = Y = O	X = O, Y = NH	X = NH, Y = O	X = Y = NH
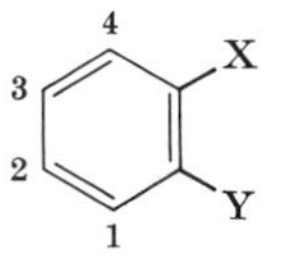 1	XH, NH, 2-3Ed	[a]	YH, NH, T2-1	—
2	—	YH, NH, T2-1	—	—
1,2	—	—	—	—
1,3	XH, N¹H, 2-4Df	—	—	XH, YH, I-414
1,4	XH, YH, 2-4E	—		—
2,3	—	—		—
1,2,4	—	—	—	—
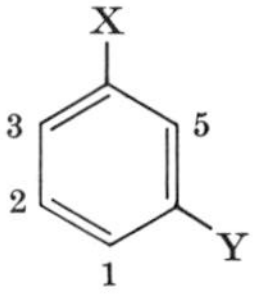 1	XH, NH, 2-3Ea	—	[a]	—
2	—	[a]		—
5	mixture, 2-3Eb	XH, NH, T2-1		XH, YH, 2-6Ba
1,2	—	—	—	—
1,3	NH, NH, 2-4De	N¹H, XH, I-414		XH, YH, I-413
1,5	NH, NH, 2-4Db	N¹/N⁵H, YH, 2-6Cc	N¹H, XH, 2-6Cb	XH, YH, 2-6Ca
2,5	—	—		—
1,2,5	1H, 5H, I-388	N¹H/N⁵H, YH, 2-6Ea	NH, XH, 2-6Eb	—
1,3,5	1H, 3H, 2-4Fa	—		XY, YH, I-416
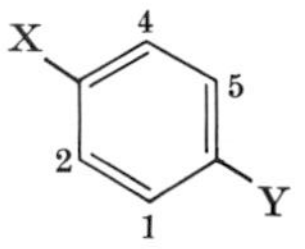 1	XH, NH, 2-3Ee	[a]	[a]	—
1,2	XH, N¹H, 2-4Cd	—		—
1,4	?, 2-4E	—		—
1,5	XH, N¹H, 2-4Df	—	—	—
1,2,4	—	—	—	—

[a] No spectroscopic data available (see 74MI01).

TABLE 2-14

AROMATIC ENERGY DIFFERENCES[a]

Series	X	ΔG_u	ΔG_s	$\Delta G_u - \Delta G_s$	$\Delta H_u - \Delta H_s$
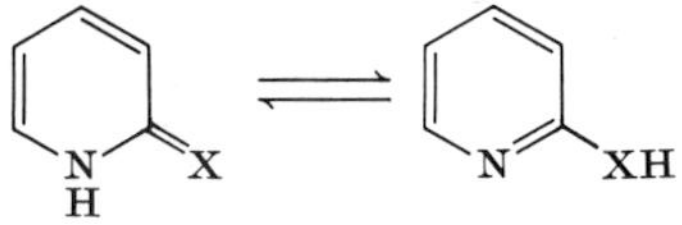 N H, X ⇌ N, XH	O	−4.1	−10.6	6.5	7.5
	S	−6.7	−12.0	5.3	6.0
	NH	8.5	0	8.5	10.0
	CH_2	18.2	2.7	15.5	17.5
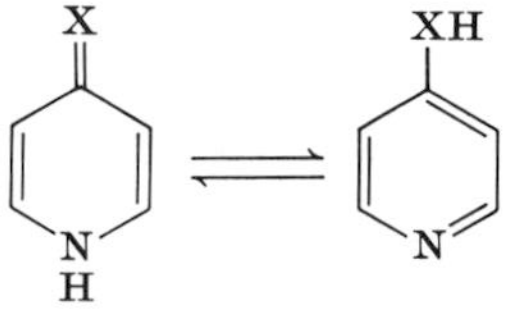 X, N H ⇌ XH, N	O	−4.5	−11.1	6.6	7.7
	S	—	—	—	—
	NH	12.0	0	12.0	14.5
	CH_2	18.4	1.1	17.3	19.5
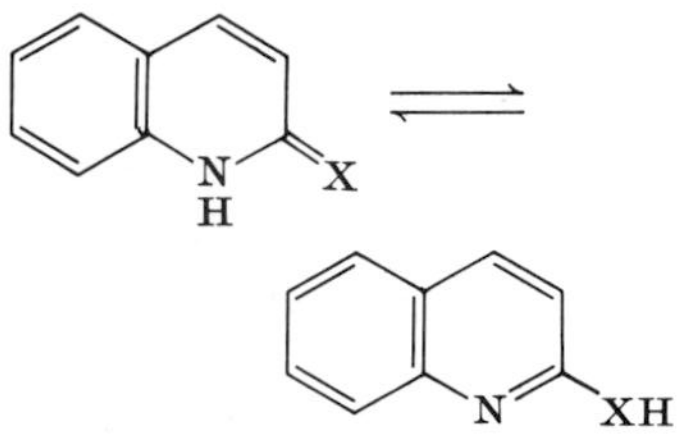 N H, X ⇌ N, XH	O	−5.3	−7.0	1.7	2.0
	S	−7.3	−9.9	2.6	2.9
	NH	5.9	1.5	4.4	5.0
	CH_2	12.9	5.8	7.1	7.9
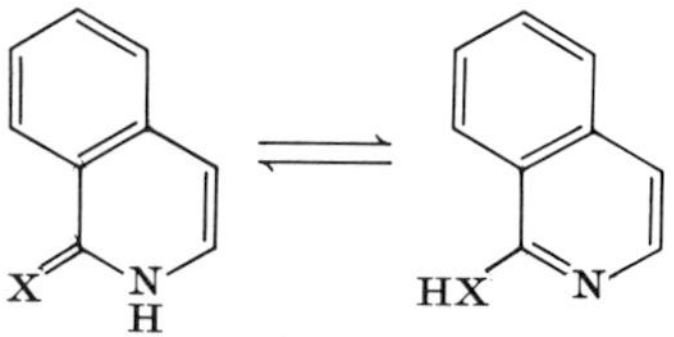 X, N H ⇌ HX, N	O	−6.6	−10.2	3.6	4.4
	S	−8.3	−11.9	3.6	4.0
	NH	5.1	0	5.1	6.2
	CH_2	13.0	6.5	6.5	7.4

[a] In kcal mole^{-1}.

Methods for correcting for ΔS have been discussed in detail (72J-(PII)1295), and the equation has been applied to measure tautomeric equilibria to obtain the aromatic energy differences of Table 2-14. Energy differences of this type will be significant in the rationalization and prediction of tautomeric equilibria in future work.

Chapter 3

Five-Membered Rings and One Heteroatom

The arrangement of this chapter follows that of the original review (II-1): we first discuss tautomerism which does not involve any functional group and then consider in turn, by substituent OH, SH, NH_2, CH_3, etc., the various classes of substituted furans, thiophenes and pyrroles.

The tautomerism of furans, thiophenes and pyrroles differs fundamentally from that of the pyridines and azines. Whereas for the compounds of Chapter 2, typical tautomeric equilibria involve two or more *aromatic* structures [cf. **1a** $\rightleftharpoons$ **1b**], the tautomeric equilibria of furans, thiophenes and pyrroles almost always involve at least one nonaromatic tautomer [cf. **2a** $\rightleftharpoons$ **2b**]. A true understanding of the tautomeric

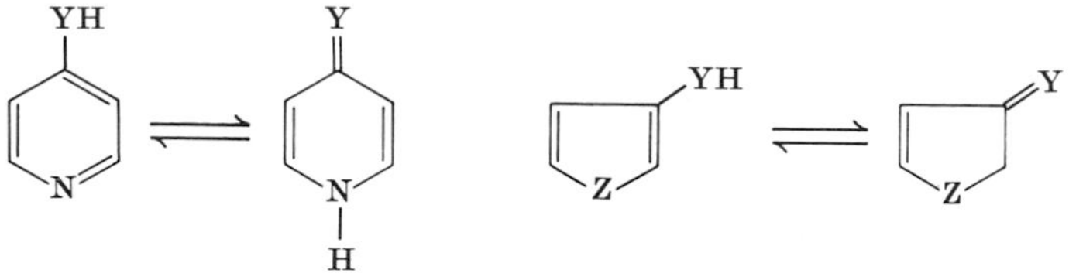

[**1a**] [**1b**] [**2a**] [**2b**]

equilibria in this chapter therefore requires quantitative knowledge of the aromaticity of these compounds. Although great progress has been made in the last ten years covered by this review, it has mainly been of a qualitative nature, and we are only now commencing to unravel the quantitative energetics involved.

1. Tautomerism of Five-Membered Rings Not Involving the Functional Group (II-3)

For structural reasons tautomerism not involving a functional group is allowed in the neutral species only of pyrroles, but not of furans or thiophenes. However, the cations of all types are capable of displaying such tautomerism.

A. Neutral Molecules (Pyrroles, Indoles and Isoindoles) (II-3)

a. Introduction

For pyrrole itself, and for simple derivatives, pyrrolenine forms, such as [**3b**] and [**3c**], are highly destabilized and hence the tautomeric equilibrium is strongly displaced in favour of [**3a**]. The available experimental evidence supports this view, and we reported in the last review

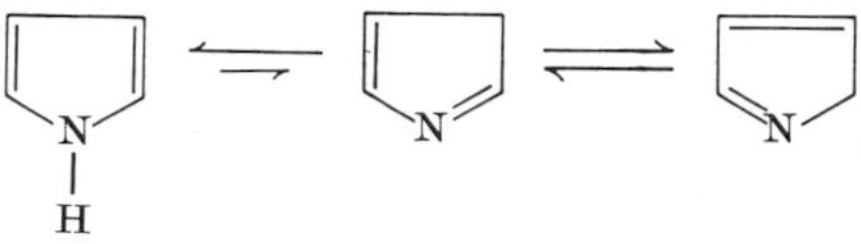

[**3a**] [**3b**] [**3c**]

that no well authenticated case of a pyrrolenine tautomeric form predominating was known. More recently, it has become clear that if additional stabilization is available in the pyrrolenine form [**3b**] or [**3c**] by the participation of substituents, then the tautomeric equilibrium can indeed be displaced towards the pyrrolenine.

Benzene ring annelation has particularly large effects: this is understandable as the loss of aromaticity of the heterocyclic ring is much less in indole than in pyrrole and less still in isoindole.

The view (II-4) that dipyrromethenes rapidly exchange the hydrogen atom attached to nitrogen has been substantiated by NMR spectral evidence (74M853).

b. Pyrrole–Pyrrolenine (II-3)

Authentic cases of stable pyrrolenines are still unknown. As expected, pyrrolylmethyleneimines [**4a**] (R = aryl) exist as such and not in the pyrrolenine form [**4b**] as shown by comparison of the UV of the potential tautomeric derivatives with those of the corresponding 1-methyl compounds (64AJ894); infrared spectra support this conclusion.

However, pyrrole anion radicals, $C_4H_5N^-$, isolated in an argon matrix, do apparently exist in the isomeric pyrrolenine form [**5**] (73JA27).

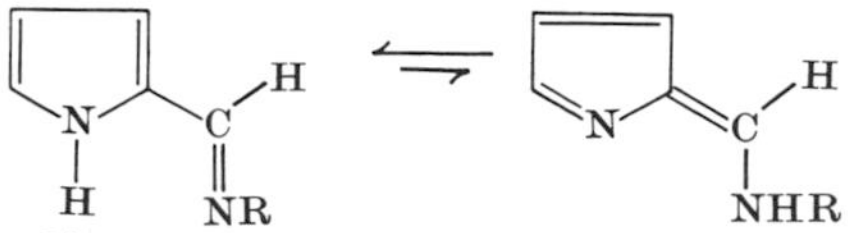

[**4a**] [**4b**] 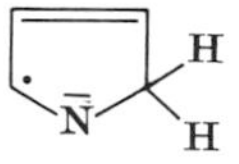 [**5**]

c. Indole–Indolenine

Most indoles exist overwhelmingly in the indole form. The same is true for various metal salts of indoles, where NMR demonstrates that the metal is bonded to the indole nitrogen atom (71JO3091, 72JO3066).

However, the tendency for indolenines to exist is considerably greater than that for pyrrolenines. The loss of aromatic resonance energy indole to indolenine is much less than that for pyrrole to pyrrolenine. The first stable indolenine tautomer [**6a**] was reported by Harley-Mason and Leeney (64P368) who found $K_T = 0.67$ for the equilibrium [**6a**] $\rightleftharpoons$ [**6b**] from the area ratio of the CH_2 peak at 3.46 δ and the CH peak at 5.40 δ in CCl_4 at 35°. Infrared spectra in the same solvent confirmed the predominance of [**6a**].

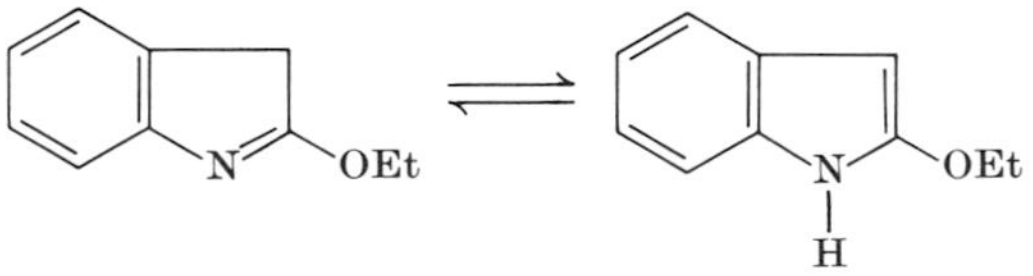

OEt N OEt N H

[**6a**] [**6b**]

This work has been extended by Japanese authors, who demonstrated that whereas 2-ethylthioindole exists largely in the indole form [**7**] (νNH = 3400 cm^{-1} in Nujol, and NH at 2.01 δ in $CDCl_3$) (70T4491), 2-piperidinoindole exists substantially in the indolenine form [**8**] as shown by the absence of NMR signals corresponding to indole β–H and NH in $CDCl_3$ (71T775), Ultraviolet spectral data agree with these conclusions in both cases.

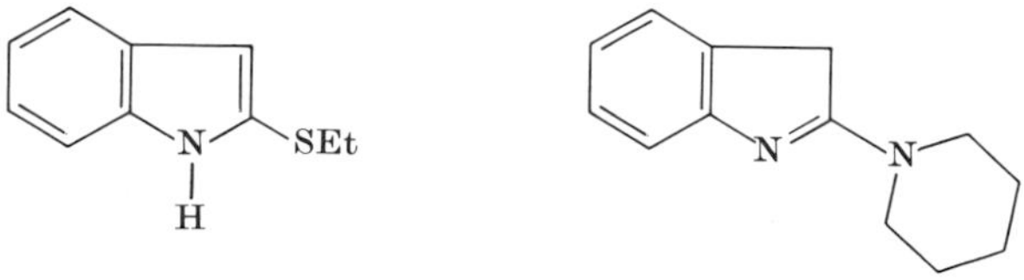

N SEt H N N

[**7**] [**8**]

This work indicates (71T775) that the stability of the indolenine form in the equilibrium [**9a**] $\rightleftharpoons$ [**9b**] (Z = SEt < OEt < NR_2) depends on the increased importance of the zwitterionic form [**10**] due to the

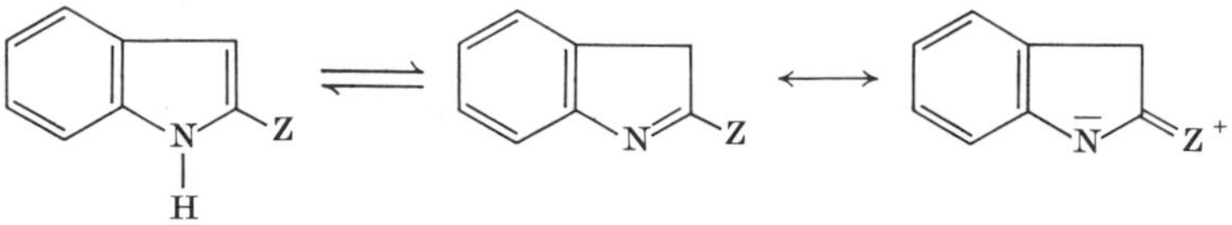

N Z H N Z N Z $^+$

[**9a**] [**9b**] [**10**]

increasing electron-donor effect of the substituent Z. This is further supported by the fact that compound [**10/1**] on NMR [$CDCl_3$ and $(CD_3)_2SO$] evidence exists entirely in the indole form shown (73J(PI)-1602).

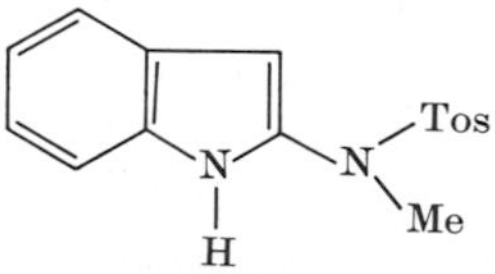

[**10/1**]

A claim (66CO(262C)1204) that NMR spectra indicate structures [**11a**] (Z = O, S) in trifluoroacetic acid is probably incorrect. The indole structure [**11b**] was found for $CDCl_3$ solution, whereas in CF_3CO_2H the compounds almost certainly exist as the cations [**12**]. For further examples of indole–indolenine tautomerism see the 1-hydroxyindoles (Section 3-2Eb).

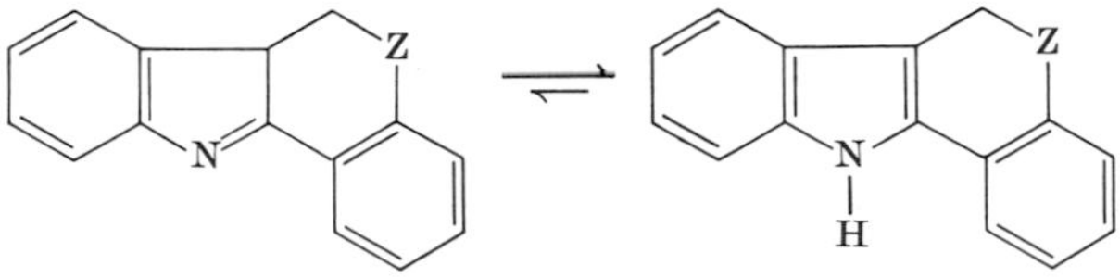

[**11a**] [**11b**]

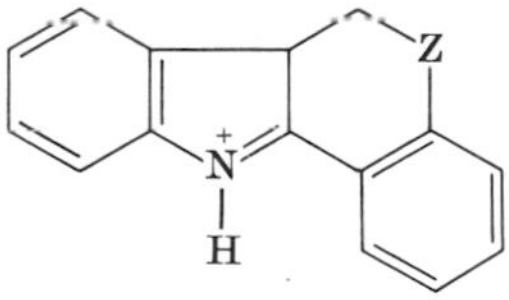

[**12**]

d. Isoindole–Isoindolenine

Tautomeric equilibria for isoindole–isoindolenine are frequently balanced or displaced in favour of the isoindolenine form. This is because of the low resonance energy of isoindoles. Isoindole–isoindolenine tautomerism has been recently reviewed by White and Mann (69HC132).

Isoindole [**13a**] itself is predicted by two MO calculations (64JA4152, 67TL3669) to exist as such rather than in the isoindolenine form [**13b**],

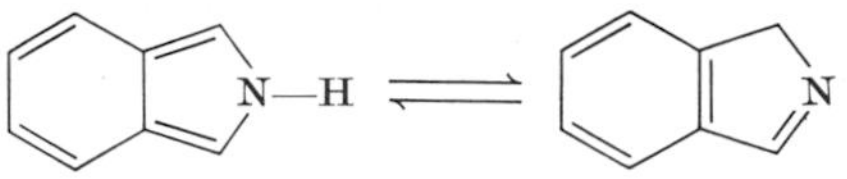

[**13a**] [**13b**]

and this has been shown experimentally (72CH393, 72TL4295, 73J(PI)1432). In practice it is found that substitution can alter the equilibrium to favour either the isoindole or the isoindolenine tautomer. For 1-arylisoindoles [**14a**] (63JA646) the proportion of the isoindolenine form [**14b**] as shown by NMR and UV data depends on the ability of the aryl group to conjugate with the C=N bond and (for $CDCl_3$ solution) increases with variation of the *para*-substituent from 9% for X = H to 31% for X = OMe and to 50% for X = NMe_2 (64JA4152): no evidence was found for the existence of the alternative isoindolenine form [**14c**].

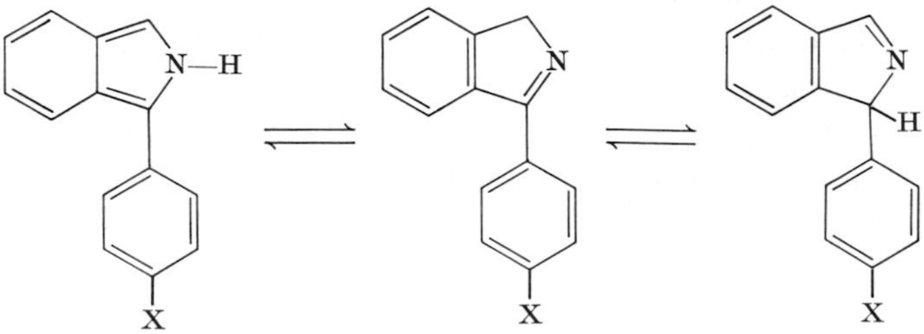

[**14a**] [**14b**] [**14c**]

1-Ethoxyisoindolenine has been described as existing in the form shown [**15**], but no physical evidence was given (59LA(623)166). Alkyl substitution also favours the isoindolenine tautomer. The equilibrium [**16a**] ⇌ [**16b**] in various solvents was studied by comparing the areas of the NMR signals at ca. 6.4 and ca. 7.0 δ (66CH198, 70J(C)1251). The proportion of the isoindole form varies with the solvent type: it varies from ca. 15% in hydrogen-bonding polar solvents like $CDCl_3$, methanol or ethanol, through about 25% in nonpolar solvents such as CCl_4 or CS_2, to 40–95% in basic polar solvents like acetone, pyridine and $(CD_3)_2SO$. This behaviour correlates with the solvent–solute hydrogen bonding. For the solid state the absence of strong IR bands attributable to νNH in the 3300 cm^{-1} region (68J(C)3036) shows that the tetramethyl derivative [**16**] exists predominantly as the isoindolenine form [**16b**].

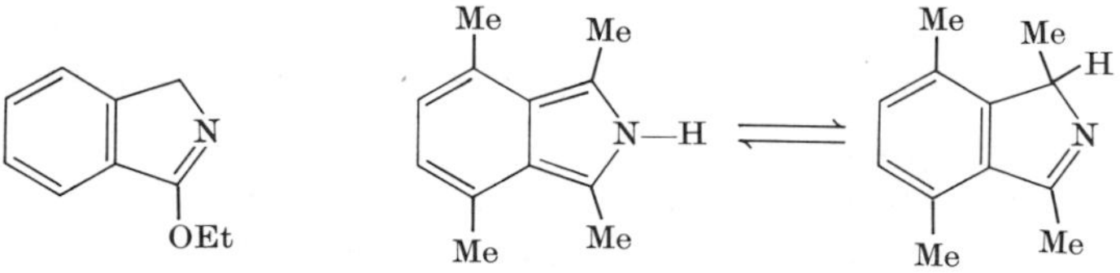

[**15**] [**16a**] [**16b**]

The effect of substituents in increasing the proportion of the 1-H-isoindolenine form has been attributed to both the stabilization of the C=N linkage by electron-releasing groups (69HC132, 70J(C)1251) and to the lessening of peri-interactions which occur in the isoindole form (70J(C)1251).

B. Conjugated Acids of Pyrroles, Indoles, Indolizines, Furans and Thiophenes (II-3)

Furans, thiophenes and pyrroles are weaker bases than vinyl ethers, vinyl sulphides and enamines because protonation interrupts the cyclic conjugation and hence destroys the aromaticity. Indeed the aromaticity of pyrrole, indole, carbazole and indolizine can be estimated quantitatively from pK_a comparisons (72CI335, 72TL5019). Their cations possess at least three possible structures, corresponding to protonation at the heteroatom, and at the α- or β-carbon. Generally, carbon-protonation is favoured.

The previous review reported only on pyrrole and indole cations. Since then much work has appeared, encompassing substituted derivatives and also furans and thiophenes.

a. Pyrrole Cations (II-3)

These cations have been reviewed recently (70HC(11)406). The delocalization of the electron pair on the N atom into the aromatic sextet suggests that protonation will not occur on the N atom [**17a**]. In fact there is no evidence for N-protonation in solution, although the IR spectra of the hydrochlorides of alkylpyrroles in the solid state have been interpreted in such terms (II-4). In aqueous solution the *C*-protonated form [**17b**] predominates rather than [**17c**]. Molecular orbital calculations are in agreement that [**17b**] is the most stable form (68IQ165).

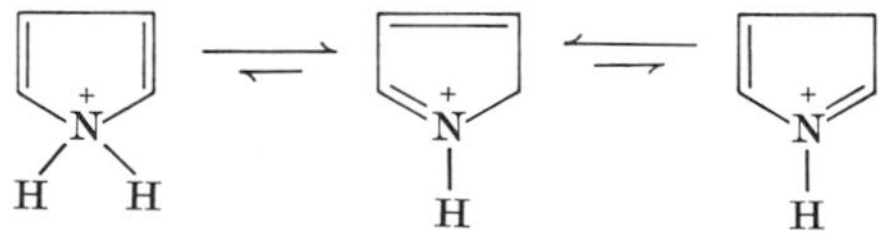

[**17a**] [**17b**] [**17c**]

The sites of protonation of a variety of substituted pyrroles have been determined both from the spin–spin coupling of the ring protons and from a comparison of the chemical shifts of ring protons in the free base with those of the conjugate acid (II-4, 63JA26, 63JA2763). An alkyl group in the 2-position directs protonation to the 5-position. For 3-alkyl pyrroles, protonation occurs in the adjacent 2-position. Both 2- and 3-protonated forms have been found for 2,5-dialkylpyrroles in a ratio of approximately 2:1. For 2,5-dimethyl-1-phenylpyrrole the 2-protonated form is favoured by a factor of 5:1 over the 3-protonated form (63JA26), the greater amount of 2-protonation is probably because of steric interactions (67T745). The NMR results are supported by ultraviolet data (63JA26, 67T745): a band characteristic of the 3-protonated form is found in the 270 nm region in addition to one at 235–250 nm, which has been assigned to the 2-protonated conjugate acid.

Protonation of pyrroles containing electron-withdrawing substituents has been studied less thoroughly. Salts of acyl and nitroso pyrroles appear to be protonated in the substituent group on the acyl or nitroso oxygen atom; a strong band between 1620 and 1630 cm^{-1} in the IR spectra has been ascribed to the exocyclic C=C double bond in [**18**]

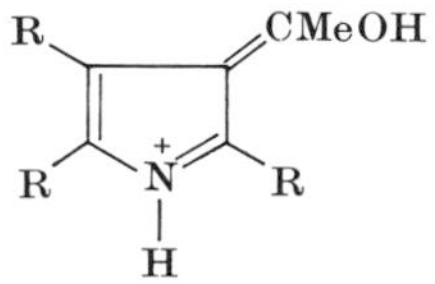

R, CMeOH, R, $\overset{+}{N}$, R, H

[**18**]

(64DA(157)367, 66KG216). However, an NMR study (71RO179) indicates that 3-substituted 2,4-dimethylpyrroles are protonated in the 5-position, for both electron-donor and electron-acceptor 3-substituents.

b. Indole Cations (II-3)

The conclusion (II-3) that the 3-protonated structure [**19**] is strongly preferred for both 2- and 3-alkylindole cations has been further supported (70MI3).

UV and NMR spectra of several indoles demonstrate that the indole ring is protonated at C-3. Thus the spin–spin splitting of the Me group (doublet) in the 3-methylindole cation is sure evidence for 3-protonation (62JA2534, 64JA3796). Compound [**20**] provides a model for *N*-protonation and the examination of its UV spectrum excludes any substantial proportion of *N*-protonation for simple indoles (64JA3796, 64JO1449).

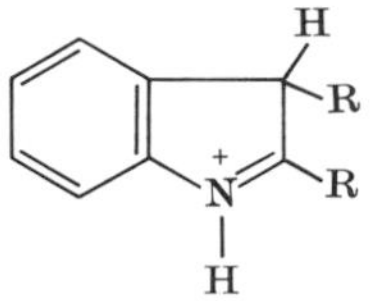

H, R, $\overset{+}{N}$, R, H

[**19**]

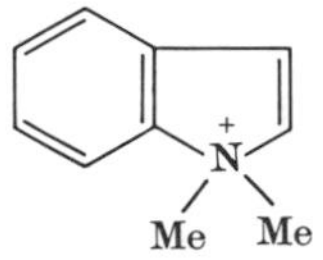

$\overset{+}{N}$, Me, Me

[**20**]

c. *Indolizine Cations*

NMR studies show that in CF_3CO_2H the indolizinium cation and methyl-substituted indolizinium cations carry the proton at C(3) [**21a**] (62J3288). However, if such protonation at C(3) is sterically hindered, then C(1) protonation [**21b**] can occur (cf. Table 3-1). In

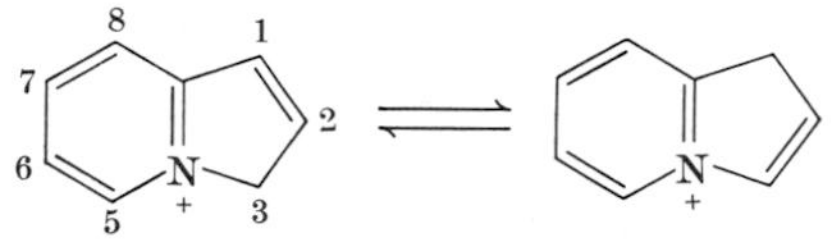

[**21a**] [**21b**]

3,5-disubstituted indolizines, intramolecular overcrowding is relieved by C(3) protonation, and exclusive protonation at C(3) occurs for these compounds and also for 1,3-disubstituted analogues (66J(B)44). Similar conclusions were reached by another author (66J(B)191), who previously had suggested this behaviour on pK_a and ultraviolet spectral arguments (64J4226).

d. *Furan Cations*

Furan and many of its simple alkyl derivatives are unstable under strongly acidic conditions, and therefore the structure of such furan cations are difficult to determine. However, *t*-butyl groups stabilize the molecule, and the furanonium ions [**22**], [**23**] and [**24**] are stable for several days in H_2SO_4 solution. A peak attributable to one proton at the 2-position for [**22**] and [**24**] and for two at the 5-position for [**24**]

TABLE 3-1

PERCENTAGE COMPOSITION OF SOLUTION OF 3-SUBSTITUTED INDOLIZINES IN CF_3CO_2H (66J(B)44)

Indolizine	3*H*-Cation	1*H*-Cation
3-Me	21	79
2,3-Me_2	41	59
3-Me-2-Ph	72	28
3-Et-2-Me	78	22
3,5-Me_2	100	—
1,2,3-Me_3	100	—
1,3-Me_2-2-Ph	100	—

together with the general feature of their NMR spectra allowed assignment of the α-protonated structures to all these cations as well as to [**25**], [**26**] and [**27**] (67TL2951).

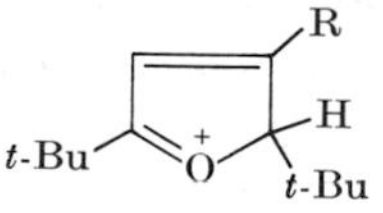

[**22**] R = H
[**23**] R = *t*-Bu

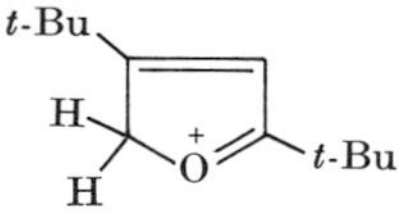

[**24**]

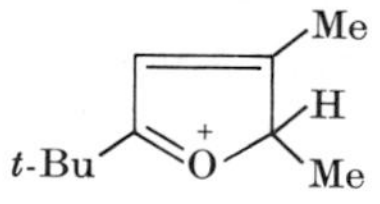

[**25**]

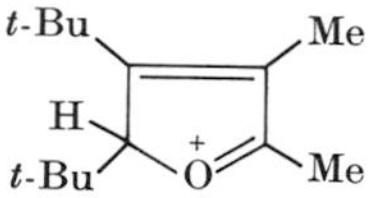

[**26**]

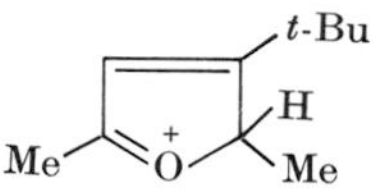

[**27**]

e. Thiophene Cations

Analysis of the proton NMR spectra demonstrated 2-protonation for the thiophenes [**28**], [**29**], [**30**] and [**31**] in HF and HF-BF_3 solution at $-90°$ to $-20°$ (66R1072). 2,5-Di-*t*-butylthiophene [**32**] is completely converted in sulfuric acid into the 5-protonated form of the isomeric 2,4-di-*t*-butylthiophene [**33**] as shown by the similarity of its NMR spectrum with that of the cation [**24**] (67TL2951).

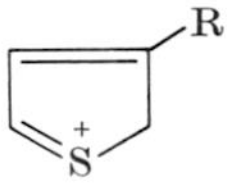

[**28**] R = H
[**29**] R = Me

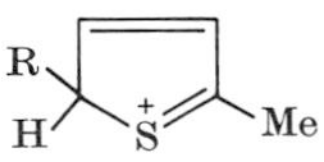

[**30**] R = H
[**31**] R = Me

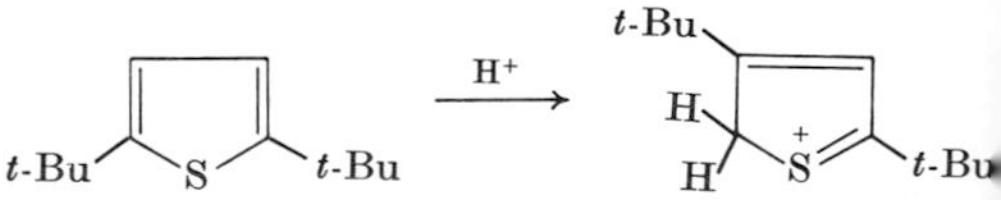

[**32**]

H^+

[**33**]

2. Compounds with Potential Hydroxy Groups (II-5)

Considerable further work on these compounds has, on the whole, confirmed what was becoming clear at the time of the previous review (II-5): potential hydroxy compounds usually exist in oxo forms, unless considerable stabilization of the hydroxy form occurs by further electron-withdrawing, or chelating, substituents.

A. Monohydroxyfurans (II-5)

a. α-Hydroxyfurans (II-5)

As already reported in the previous review (II-5), most potential 2-hydroxyfurans [**34a**] exist as 2-oxo derivatives and usually the

Δ^3-oxo form [**34b**] is more stable than the Δ^4-oxo form [**34c**]. However, in this series there is a considerable energy of activation for conversion [**34b**] $\rightleftharpoons$ [**34c**] and the less stable tautomers [**34c**] are easily obtained by ring closure (II-5, 64CR360). Conjugation of the double bond with the oxygen atom is evidently less stabilizing than the conjugation of the double bond and carbonyl group. Thus the less stable tautomers [**34c**] are usually easily transformed into [**34b**] by heating with bases or acids (64CZ1663); rates of the base-catalyzed tautomerization [**34c**] → [**34b**] for (R = 5-Me) and (R = 5-*t*-Bu) have been measured (68AK229).

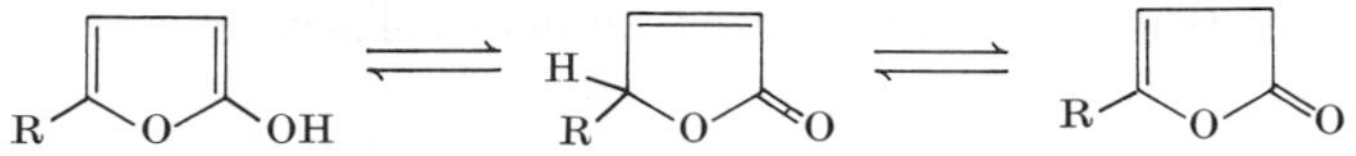

[**34a**] [**34b**] [**34c**]

However, the Δ^4-oxo structure can evidently be stabilized by suitable substituents; thus [**35**] has been assigned (01LA(319)196, 49J118), and UV data are given. By contrast, the triphenylfuran-2-one [**36**] has

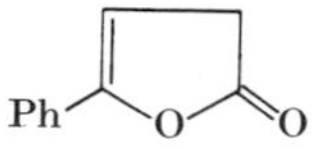

[**35**] 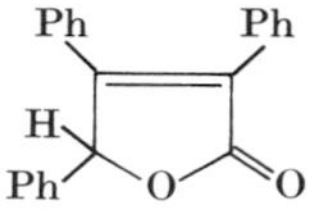 [**36**]

νC=O 1755 cm^{-1}, νC=C 1645 cm^{-1} and λ_{max} at 284 nm (in EtOH) which indicates that the compound exists as [**36**] (70BF3572). These conclusions must be considered provisional: we consider that the available data do not yet allow conclusive generalizations from compounds [**35**] and [**36**].

b. α-Hydroxybenzofurans

2-Benzofuranone [**38**] exists in the carbonyl form shown with νC=O 1818 cm^{-1} (69JO4164). Dipole moments in benzene (70JA4447) and electronic spectra in EtOH (70ZF16) indicate the oxo structure written for benzo[*c*]furanone [**39**].

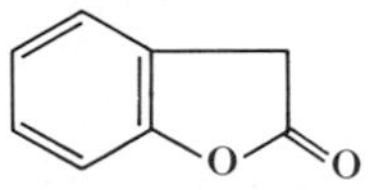

[**38**] 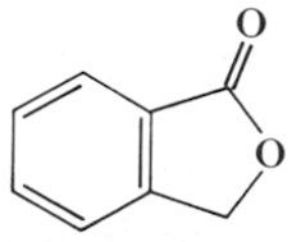 [**39**]

The ^{15}N NMR spectrum of 3-phenylazo-2-benzofuranone has been interpreted in terms of syn–anti isomerism, rather than azo-hydrazone tautomerism (72JO4121) (cf. [**98/1a**] ⇌ [**98/1b**], Section 3-2 cf).

c. β-Hydroxyfurans (II-6)

At the time of the previous review, very little was known about β-hydroxyfurans (II-6). More recently much work has helped to clarify their tautomeric behaviour. Eugster and co-workers (65H1322) concluded from a detailed spectroscopic analysis that β-hydroxyfuran [**40**] and many of its simple derivatives exist in ether, EtOH and CCl_4 as furan-4-ones [**40b**] (α,β-unsaturated ketones). These conclusions have been confirmed (73R731).

The oxo structure [**41b**] has also been assigned to the 4-dialkylamino analogues [**41**] mainly on the basis of solid-state IR bands at 1620 and 1690 cm^{-1} characteristic of C=O groups conjugated with double bonds. The NMR spectra in $CDCl_3$ are consistent with the structure [**41b**] (68R1011). Similar features occur in the IR spectra in CH_2Cl_2 of

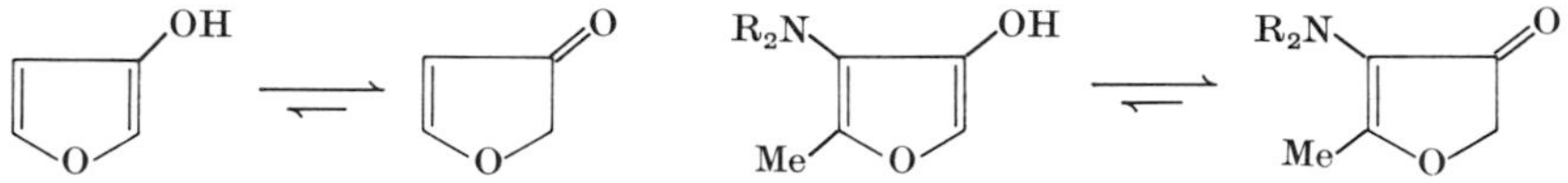

[**40a**] [**40b**] [**41a**] [**41b**]

compounds [**42**, **43a**]. The absorbance at 235 and 316 nm (in CH_2Cl_2) suggests extensive conjugation, and indicates oxo structures as shown with contributions from the mesomeric form [**43b**] (63JA2943). Calculations agree that 3-hydroxyfurans should exist largely in the 3-oxo form, cf. [**40b**] (70JA2929).

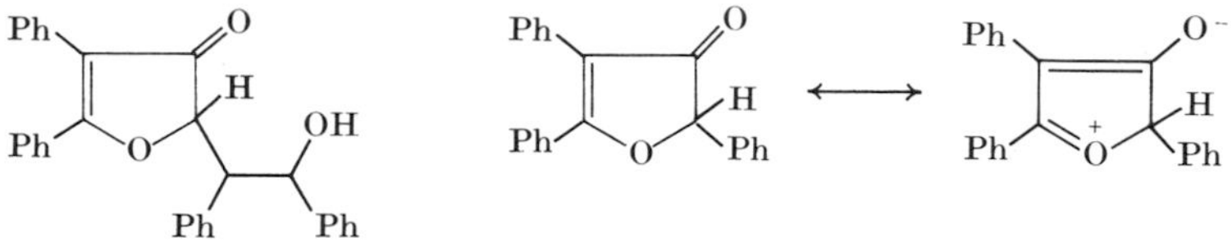

[**42**] [**43a**] [**43b**]

On the other hand, the infrared spectrum of isomaltol (2-acetyl-3-hydroxyfuran) shows exclusively the hydroxy form [**44**] probably existing as a strongly *H*-bonded dimer in the solid state (νOH at 3100–2650 cm^{-1}, νC=O at 1610–1550 cm^{-1}) and as either a dimer or an intramolecular *H*-bonded stabilized form in $CHBr_3$, CCl_4 and C_2Cl_4 (νOH ca. 3295 cm^{-1}, little change on dilution). The chemical shift at 8.95 δ in CCl_4 supports structure [**44**] (63H1259, 64JO776, 65H1322).

Study of some 4-halo derivatives of [**44**] led to similar conclusions (65J2543).

Rather surprisingly, the NMR spectra of 3-hydroxy-4-acylfurans [**45**, R = Me and OEt] disclose CH_2 at 4.50 δ in CCl_4 demonstrating their existence in the oxo forms [**45b**] (65H1322). Other workers have

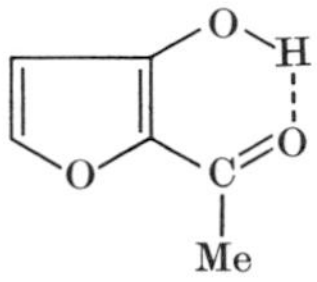

[**44**] 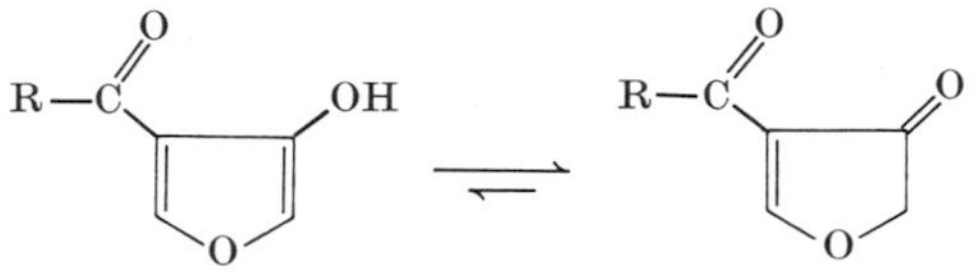 [**45a**] [**45b**]

also shown that a 4-ethoxycarbonyl group does not alter the tautomeric equilibria away from the 3-oxo form (73R731). For a similar situation in pyrrole analogues see (Section 3-2Ee) and for a contrasting one in thiophenes see (Section 3-2Cd).

H-Bonding stabilization is evidently important in the true hydroxyfuran [**46**]: analysis of its NMR spectrum in $CDCl_3$ (64AJ1438) confirms earlier infrared and ultraviolet evidence for the hydroxy form [**46**] (55MI1955).

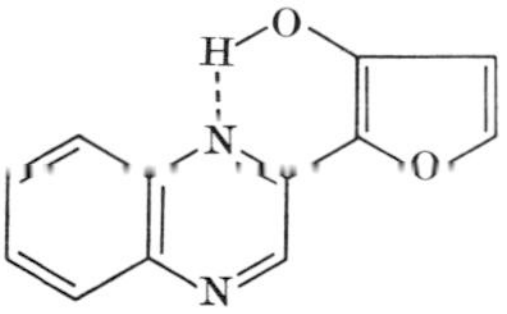

[**46**]

Enolization can also be quite extensive in 3-oxotetrahydrofurans; for the considerable investigations in this field see (71KG1313, 71SA2119).

d. β-Hydroxybenzofurans (II-6)

UV evidence for the oxo structure of 3-coumaranone [**47a**] (II-6) has been confirmed by NMR in $CDCl_3$ and DMSO (CH_2 at 4.79 δ) (66CB3076, 69AP423). Analogous 3-oxo structures (cf. [**47a**]) have been proved for various 3-coumaranone derivatives including: 2-alkyl (66CB3076),

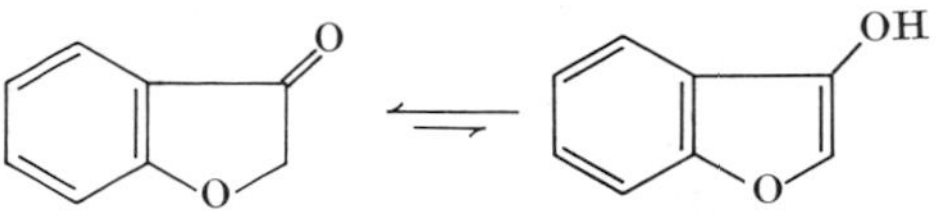

[**47a**] [**47b**]

2-aryl (70BF2309), hydroxy (69AP423) and acetoxy derivatives (63CB1680).

However, an OH signal at 0.65 δ in the NMR spectrum of 2-acetyl-3-hydroxybenzofuran in CCl_4 confirms (65T3331) an earlier conclusion, based on the interpretation of the ultraviolet spectra, that this compound exists in the enol form [**48**] (55JA1623). Nevertheless, a 2-iminomethyl group does not stabilize the hydroxy form: instead, as in the benzothiophene series, a 3-oxo-enamine structure [**49**] is favoured, on evidence from UV comparisons (73KG154).

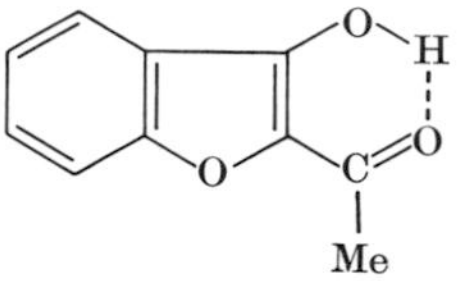

[**48**] 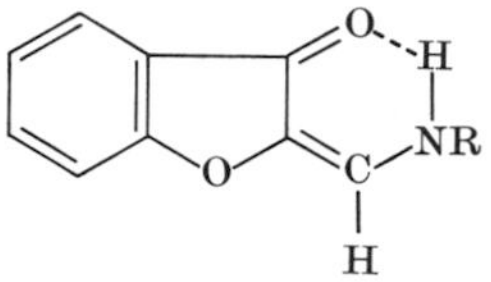 [**49**]

A νOH at 3448 cm^{-1} and νC=O at 1712 cm^{-1} were cited as evidence for the existence of a tautomeric mixture of [**50a**] and [**50b**] in the solid state (62T853); if true this would be unusual, and the subject deserves reinvestigation.

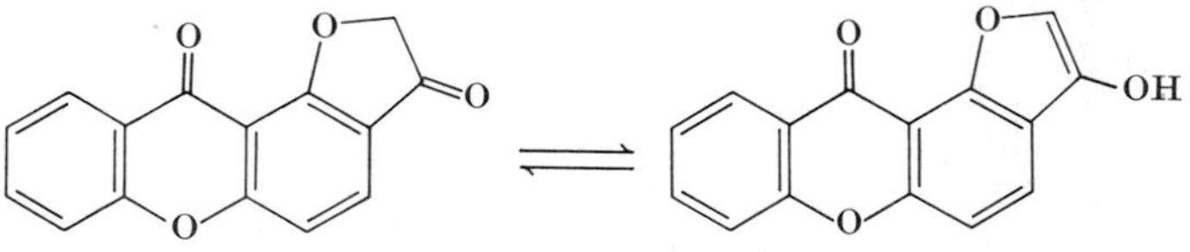

[**50a**] [**50b**]

B. Potential Dihydroxyfurans (II-6)

Much recent work has extended the scattered data reported previously and clarified the main features of the tautomeric behaviour of several types of potential dihydroxyfurans.

a. Oxygen Functions in 2,3-Positions (II-6)

Contrary to a previous suggestion (43H687, cf. II-6) that unsubstituted α-keto-γ-lactones [**51**] exist in the dioxo form [**51a**], the parent and some methyl and bromo derivatives prefer the monoenol form [**51b**] in DMSO, $CDCl_3$ and the solid state as demonstrated by characteristic

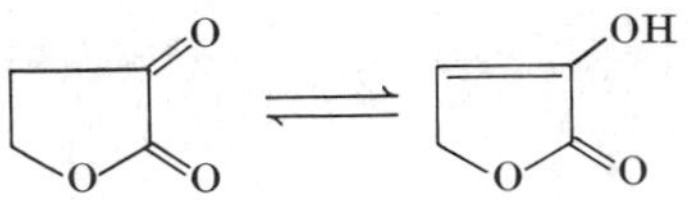

[**51a**] [**51b**]

νC=O and νOH bands. Three nonequivalent hydrogens in the NMR and absorbance at 230–240 nm confirm [**51b**] although the parent compound surprisingly does not absorb above 220 nm (in EtOH) (49H998, 64J783, 71T3839).

Annelation effects appear to influence the situation strongly; IR and NMR measurements showed (68M2223) that the equilibrium in the corresponding benzofuran [**52a**] ⇌ [**52b**] is strongly temperature dependent (100% [**52a**] at 20°; 75% at 60°; 25% at 90° and 5% at 100°). This surprisingly large variation deserves further investigation.

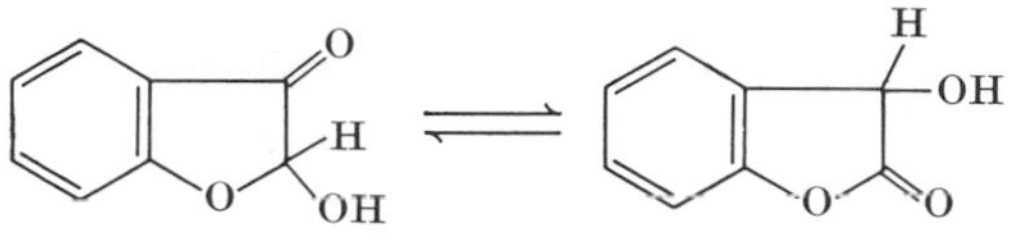

[**52a**] [**52b**]

b. Oxygen Functions in 2,4-Positions (II-7)

The previous conclusion (II-7), that the tetronic acids, which are potentially 2,4-dihydroxyfurans, generally exist predominantly in the dioxo form [**53a**] in solvents of low polarity and in the monoenol form [**53b**] in polar solvents has been confirmed by spectroscopic data (71T3839). As well as the parent [**53**], the substituted compounds [**54**], [**55**] and [**56**] exist in the monohydroxy-forms shown in the solid state and polar solvents, and this is true also for the solid 3-methyl (63CX520) and for 3,5 dimethyltetronic acid (69CX(B)1247) as shown by X-ray crystallography.

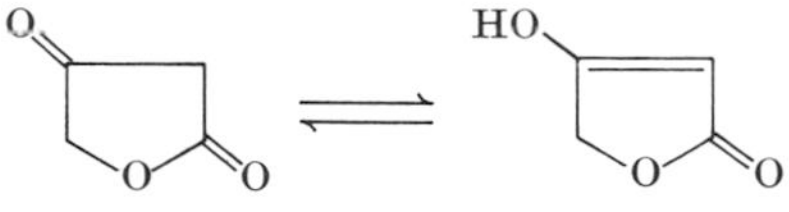

[**53a**] [**53b**] 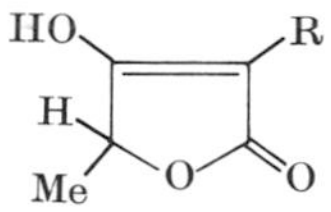 [**54**] R = Br [**55**] R = H [**56**]

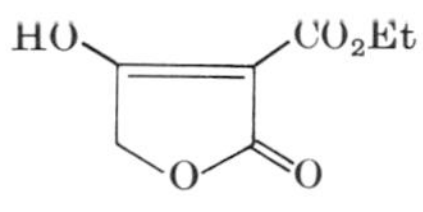

5-Alkylidene-4-hydroxy-3-furan-2-one structures were assigned to compounds [**57, 58, 59**] by IR comparisons with simple potential 2,4-dihydroxyfurans (cf. 53J1207, II-7), and confirmed by NMR (63J4778).

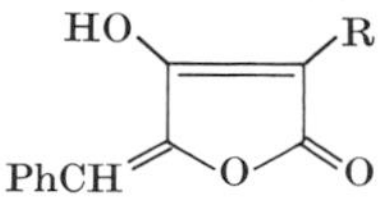

[**57**] R = H
[**58**] R = CO_2Et
[**59**]

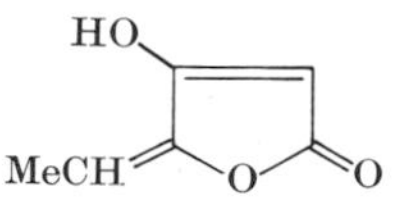

c. Oxygen Functions in 2,5-Positions (II-7)

Recent infrared assignments (70BF4505; see also 71PM(4)284) confirm the dioxo structure [**60**] of succinic anhydride (cf. II-7) as did X-ray data (65CX(19)698).

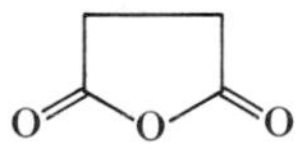

[**60**]

d. Oxygen Functions in 3,4-Positions (II-8)

Available spectroscopic data for compounds [**61**] (67MI230, 68R1011) and [**62**] (66H53, 66JO2391) have been collected and discussed (71T3839). Absorbance at 1645, 1670 and 3300 cm^{-1} and at 289 nm and three nonequivalent protons in the NMR demonstrated the monohydroxy structures shown [**61, 62**] in $CDCl_3$ and in the solid state. Compound [**62/1**] exists as a mixture (ca. 2:1) of the two oxo–hydroxy tautomers [**62/1a, 62/1b**] as shown by NMR in CCl_4 (73H1882).

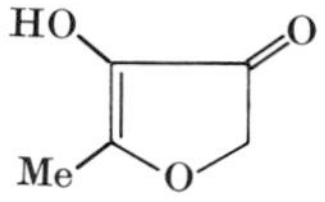

[**61**] 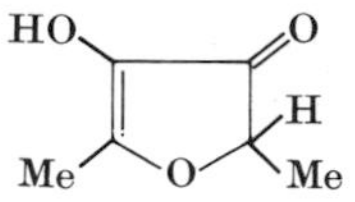 [**62**]

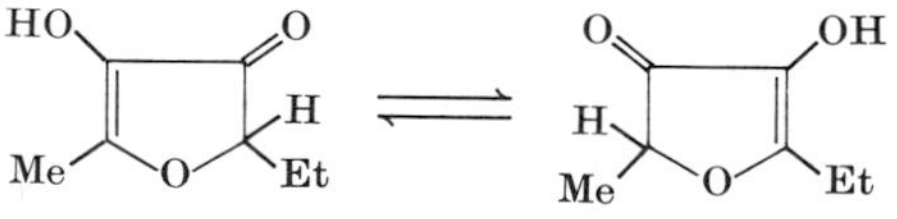

[**62/1a**] [**62/1b**] 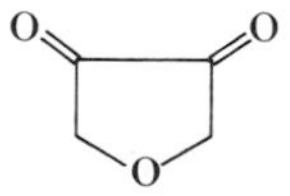 [**63**]

Rather surprisingly the dioxo structure [**63**] of the parent compound (II-8) is also confirmed by spectral comparisons (νC=O at 1780 cm^{-1}; λ_{max} at 520 nm; single *H*-resonance in the NMR) with [**61**] and [**62**] (71T3839).

e. Ascorbic Acid (*Vitamin C*) (II-7)

Ascorbic acid (vitamin C) has been extensively studied: X-ray work (68CX(B)23, 1431) confirms (II-7) the dienol–oxo form [**63/1**].

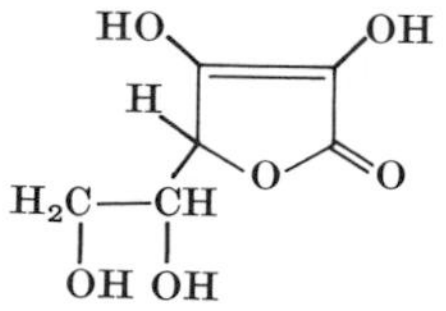

[**63/1**]

C. MONOHYDROXYTHIOPHENES (II-8) AND SELENOPHENES

Our knowledge of the thiophenones, the potential hydroxythiophenes, has been transformed since the last review (II-8), largely as a result of the work of Gronowitz and his school, the first results of which were reported in the previous review.

a. α-Hydroxythiophenes (II-8)

i. *Introduction.* Extensive investigations have now disclosed the complex pattern of tautomerism of 2-hydroxythiophenes with Δ^3- and Δ^4-thiolen-2-ones: with different types of substitution, each of the three possible tautomers may be favoured at equilibrium. The tautomeric behaviour of 2-hydroxythiophenes has been reviewed by Hörnfeldt (68MI343). The results of Gronowitz discussed in our previous survey (II-9) on the constitution of 2-hydroxythiophene (very predominantly 3-thiolen-2-one [**64c**, R = H]) and of 5-methyl-2-hydroxythiophene (85% 5-methyl-3-thiolen-2-one [**64c**, R = Me], 15% 5-methyl-4-thiolen-2-one [**64b**, R = Me]), have been verified by a complete assignment of their NMR and IR spectra (63T1867).

R S OH [**64a**] ⇌ R S O [**64b**] ⇌ H R S O [**64c**]

The tautomeric ratio for further 5-substituted 2-hydroxythiophenes was determined in CCl_4 by NMR (Table 3-2) (63T1867). These and other results will now be discussed systematically according to the type and orientation of substituent.

TABLE 3-2

PERCENTAGE OF TAUTOMERIC FORMS FOR POTENTIAL 2-HYDROXYTHIOPHENES (63T1867)

5-Substituent	OH Form [**64a**]	Δ^4 Form [**64b**]	Δ^3 Form [**64c**]
Me	—	20	80
$CH_2{=}CH{-}CH_2$	—	15	85
$C_6H_5CH_2$	—	30	70
SMe	—	15	85
CO_2Et	85	15	—

ii. *Substitution at the 5-Position.* The two thiolenone forms [**64b**], [**64c**] in the 5-alkyl series (Me, Et, *i*-Pr, *n*-Pr, *n*-Bu, *t*-Bu) can be individually isolated but are interconverted by acids or bases to give mixtures of both forms. The composition at equilibrium depends partly on hyperconjugation with the alkyl group: the proportion of the Δ^3 form [**64c**] decreases in the series Me, Et, *i*-Pr, *t*-Bu (64AK(22)211).

Polar solvents appear to favour slightly the Δ^4-unsaturated form [**64b**] compared with nonpolar solvents (64AK(22)211).

NMR spectra of 5-bromo- [**65**] and 5-chloro-3-thiolen-2-one [**66**] in CCl_4 show the Δ^3 forms [**65a**, **66a**] almost exclusively (67T3737). Typical IR absorption of 3-thiolen-2-ones (64AK(22)211) at 790–805 cm^{-1} confirms these conclusions (67T3737), and further studies on the 5-chloro derivative [**66**] agree (68AK427). The tautomeric 5-chloro-4-thiolen-2-one [**66b**] has been detected by NMR in low concentration (2%) only in fresh prepared solutions (67T3737).

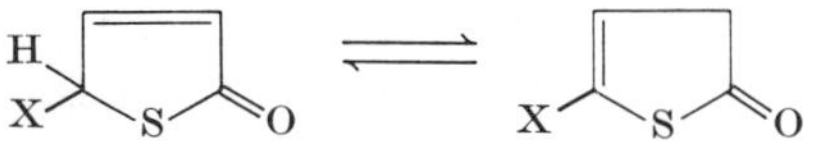

[**65a**] X = Br
[**66a**] X = Cl

[**65b**] X = Br
[**66b**] X = Cl

A 5-aryl substituent also shifts the equilibria essentially completely to the 4-thiolen-2-one structure [**67a**, Ar = Ph or 2-thienyl] in the solid state (characteristic νC=O at 1715–1725 cm^{-1}) and in CCl_4 as shown by the NMR spectrum; evidently aryl conjugation stabilizes the Δ^4 form (64AK(22)211). However, in MeOH solution the oxo form [**67a**] is in equilibrium with significant amounts of the enol form [**67b**] (25–30%) partly because the contribution of resonance structures such as [**67c**] and partly because of solvent hydrogen-bonding (64AK(22)211).

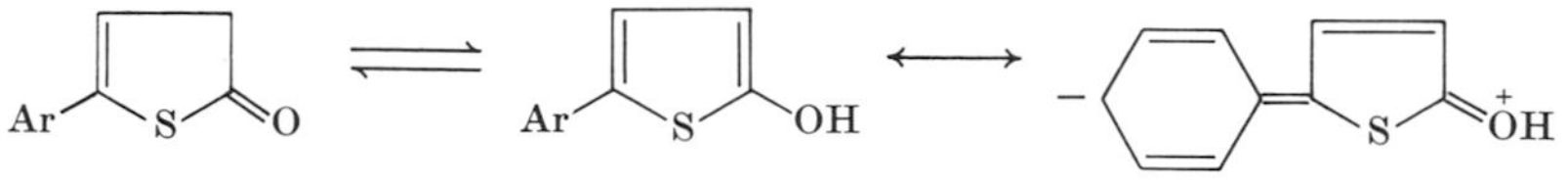

[**67a**] [**67b**] [**67c**]

The hydroxy form is favoured even more by electron-withdrawing substituents at the 5-position, such as carbethoxy (see Table 3-2). This is due to stabilization by extended conjugation and also possibly to intermolecular hydrogen bonding.

iii. *Substitution at the 3- and/or 4-Position.* The tautomeric equilibria in 3- and 4-substituted thiolen-2-ones usually favour the Δ^3-tautomeric form. Thus NMR spectral comparisons with the corresponding 5-isomers (reference in II-9) show that 3-methyl-2-hydroxythiophene exists as 3-methyl-3-thiolen-2-one [**68**] and 4-methyl-2-hydroxythiophene as 4-methyl-3-thiolen-2-one [**71**] in $CDCl_3$ and in cyclohexane (62AS789). Even when alkyl groups are present in both the 3- and the 5-positions, as in 3,5-dimethylthiolen-2-one, the equilibrium greatly favours the Δ^3 form, showing the strong influence of the 3-methyl group (71T3861). The 3- and 4-bromo [**69**], [**72**] and 3-methoxy [**70**] substituted thiolen-2-ones also exist entirely in the 3-thiolen-2-one form (64AK(21)239).

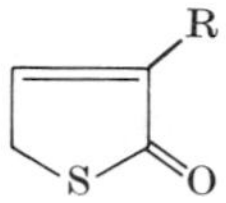

[**68**] R = Me
[**69**] R = Br
[**70**] R = OMe

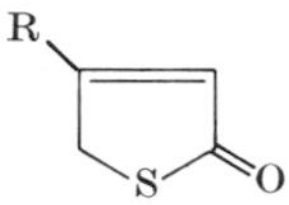

[**71**] R = Me
[**72**] R = Br

Just as for the corresponding 5-substituted derivatives, NMR and IR show that 3-acetyl-2-hydroxythiophene and 3-carbethoxy-2-hydroxythiophene [**73**] exist in the hydroxy form [**73a**] in the liquid state and in solutions in $CDCl_3$ and CCl_4 (67T871). Here, intramolecular hydrogen bonding may be an important factor stabilizing the enol form [**73a**].

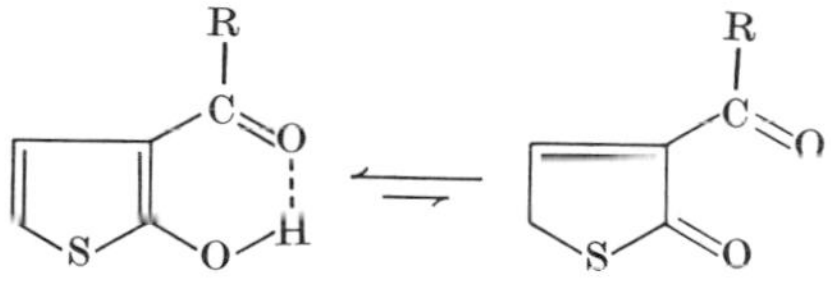

[**73a**] [**73b**]

iv. *Miscellaneous Substitution.* We also have to consider the potential tautomeric character of alkylidene derivative related to [**74**]. The 5-isopropylidene-3-thiolen-2-one structure [**74a**] was assigned to the exclusion of other possible structures [**74b–d**] by the NMR spectrum

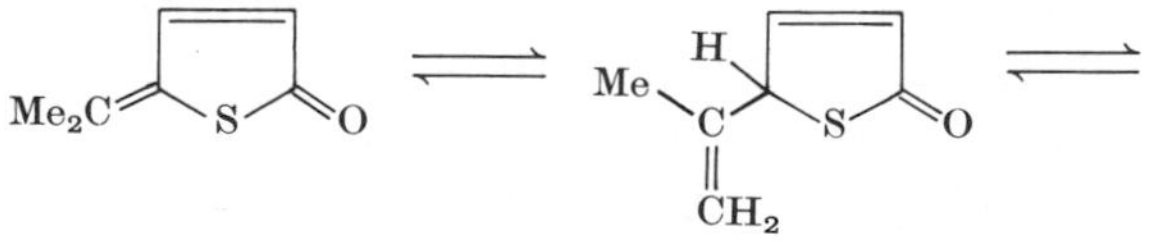

[**74a**] [**74b**]

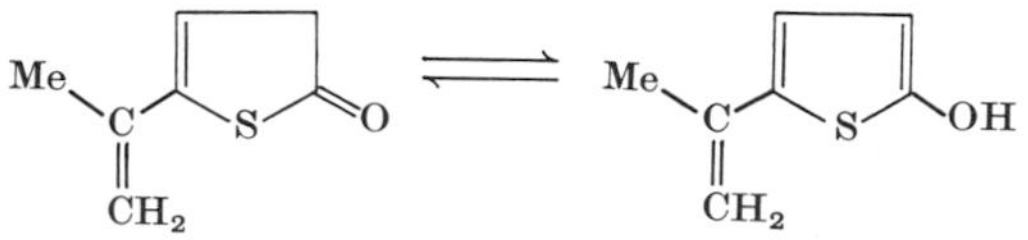

[**74c**] [**74d**]

(63T1867). Similar structures were assigned to the 5-ethylidene, 5-cyclopentylidene (63T1867) and 5-benzylidene analogues (50JA5543, 63JO733, 63T1867). The amino–oxo tautomer [**75**] predominates in solution and in the solid state (70IZ2413, 71ZO1953).

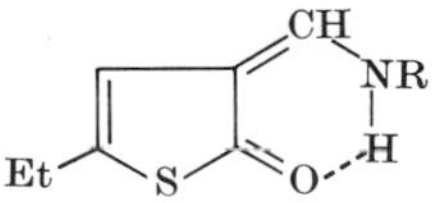

[**75**]

v. *Rates of Tautomer Interconversion.* The tautomerization rates of several substituted thiolen-2-ones has been measured and found to be 10^3 to 10^4 times higher than that of the corresponding butenolides (3-2Aa) (67AS673, 68AK427, 68AK455, 68AK461), possibly because of *d*-orbital participation by the sulphur atom (68AK229).

b. α-Hydroxyselenophenes

The tautomerism of the selenophenes [**76a**] ⇌ [**76b**] (R = H, Me, SMe), studied by IR and NMR, gave results close to those obtained

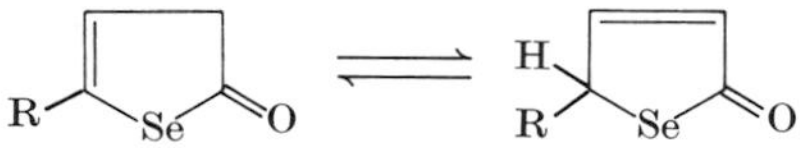

[**76a**] [**76b**]

for thiophene analogues (70CO(270C)825, 71BF3547) in line with the generally similar behaviour of thiophenes and selenophenes. For dipole moments see (70CO(271C)1481).

c. α-Hydroxybenzothiophenes (II-9)

A strong peak at 1700 cm^{-1} in the solid state (characteristic of thiolactone carbonyl) and the absence of any hydroxyl bond suggest predominantly the oxo structure [**77a**] for 2-hydroxybenzothiophene [**77**]. The UV spectrum (67JO3028) and the NMR signal at $\delta = 3.79$ in CCl_4 confirm this result (70J(C)1926). The earlier report of the isolation of both tautomeric forms [**77a**] and [**77b**], which was reported in the earlier review, but considered as requiring further investigation (II-9), must now be considered as erroneous.

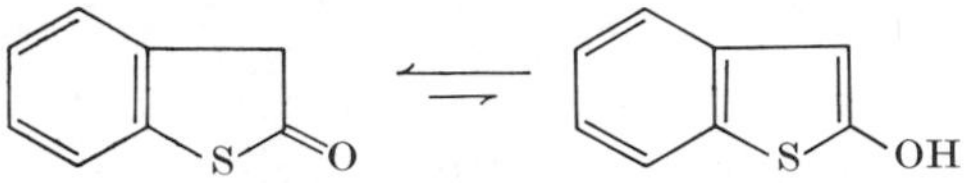

[**77a**] [**77b**]

As in the thiophene series (Section 3-2Ca(iv)) the amino–oxo tautomer [**77/1a**] of 2-hydroxybenzo[*b*]thiophene-3-aldimines [**77/1b**] predominates in solution (74PC970).

Dipole moment measurements in benzene indicate the oxo structures [**78**] and [**79**] (70JA4447) and X-ray crystallography confirmed [**78**] (68JS175).

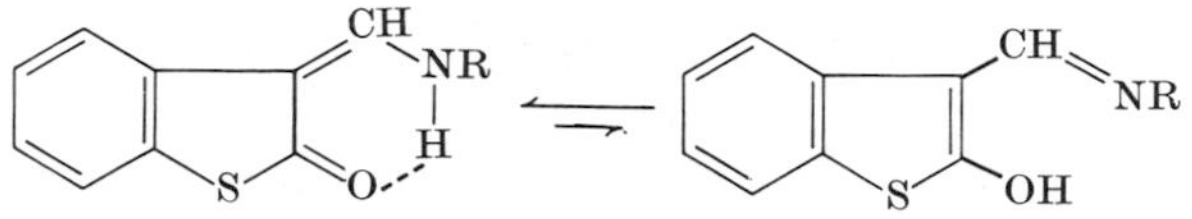

[**77/1a**] [**77/1b**]

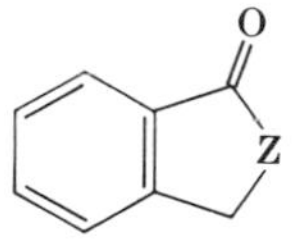

[**78**] Z = S
[**79**] Z = Se

d. β-Hydroxythiophenes (II-9)

When the oxygen function is placed at *β*- rather than *α*- to the thiophene sulphur, then the hydroxy form becomes considerably more favoured. For simple alkyl and aryl derivatives both oxo and hydroxy forms generally coexist, but electron-withdrawing groups tend to cause the hydroxy form to predominate greatly.

The NMR of 3-hydroxy-2-methylthiophene [**80**] shows a tautomeric mixture of 80% of the hydroxy form [**80a**] with 20% of 2-methyl-4-thiolen-3-one [**80b**] in CS_2 (65AS1249) in agreement with earlier infrared studies on the less stable parent [**81**] (II-9). For the 2-*t*-butyl analogue [**82**] the amount of the enol form [**82a**] is reduced to 55%. The oxo form is still more favoured in 3-hydroxy-2,5-dimethylthiophene (70%), probably because of hyperconjugation effects (65AS1249).

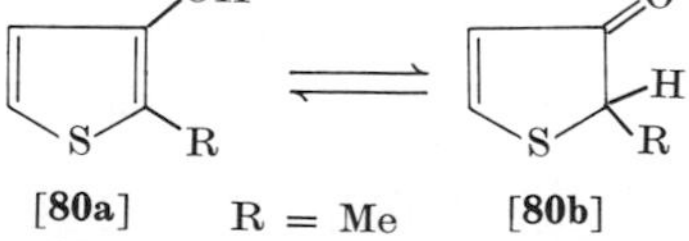

[**80a**]	R = Me	[**80b**]
[**81a**]	R = H	[**81b**]
[**82a**]	R = *t*-Bu	[**82b**]

More recently, the effects of solvent variation and of alkyl substitution in the 2- and 5-positions on the tautomerism of 3-hydroxythiophene has been studied systematically (72CQ9). The following are found to increase the proportion of OH form: (i) *t*-butyl rather than methyl at the 5-position, (ii) methyl rather than *t*-butyl at the 2-position, (iii) *H*-bond acceptor power and no *H*-bond donor power of solvent, i.e., $Me_2CO > C_6H_{12} \sim CS_2 > MeNO_2 > CHCl_3$. Tautomeric equilibration between the two forms is rapid (72CQ9). However, the 2,5-di-*t*-butyl compound exists completely in the oxo form, apparently because of steric hindrance (72AS31).

NMR studies in $CDCl_3$ were interpreted as indicating that compound [**83**] exists in the hydroxy form [**83a**] when freshly prepared, but after a month signals at 5.90 and 2.30 δ were assigned to 16% of the oxo form [**83b**] (71T3853). This behaviour seems curious and would be worth reinvestigation.

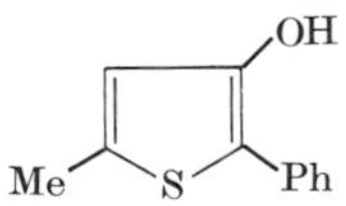

[**83a**] 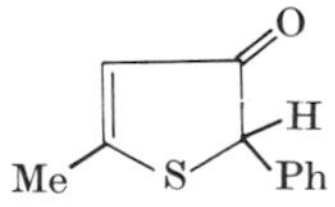 [**83b**]

Several studies have been devoted to the tautomerism of carbonyl derivatives of 3-hydroxythiophene. IR spectra confirmed previous conclusions (II-10) that many 5-alkyl- and 5-aryl-3-hydroxythiophene-2- [**84**] and -4 [**85**] -carboxylic esters exist as enols in CCl_4 (64H1748).

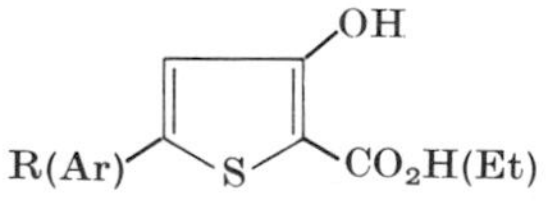

[**84**] 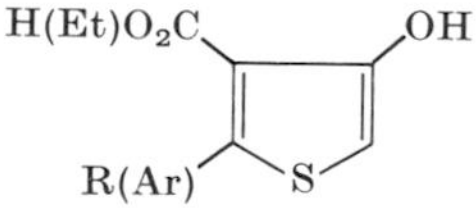 [**85**] 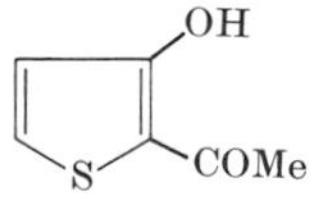 [**85/1**]

The NMR spectra in acetone of the aldehydes [**86**], [**87**] and of the carboxylic acids [**88**] and [**89**] show solely signals due to the enol forms

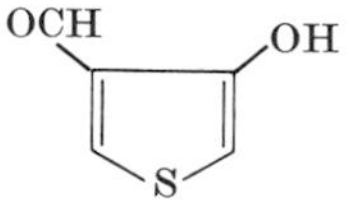

[**86**] 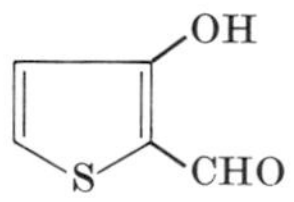 [**87**] 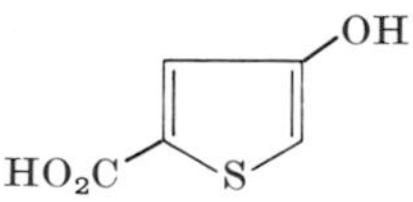 [**88**] 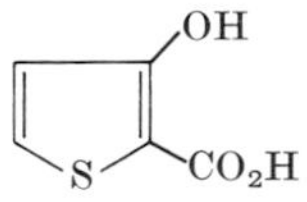 [**89**]

(66AS261). This confirms earlier IR indications that all these compounds exist as hydroxythiophenes (54CB841). NMR and IR studies show that many further 3-hydroxy-2-, -4-, and -5-carbonyl-substituted thiophenes exist exclusively as enols in CCl_4 and in $CDCl_3$ (65T3331) and X-ray confirmed the structure (65T3331) of [**85/1**] in the solid state (69AS2031).

The stabilization of the aromatic enol form by carbonyl substituents has been explained on the basis of the electron-withdrawing inductive and mesomeric effect of the groups, together with intramolecular and intermolecular hydrogen bonding (65H617, 65T3331). That the tendency to enolize is greatest for thiophenes is shown by the existence of [**90**] in the enolic form (65T3331) in contrast with the analogous furan (Section 3-2Ac) and pyrrole compounds (Section 3-2Ee).

Signals at $\delta = 8.9$ to 8.1 for OH show that some 5-alkyl-2-cyano-3-hydroxythiophenes [**91**] (71T3853] exist in the hydroxy form in

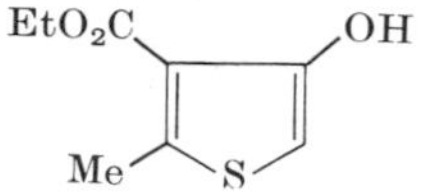

[**90**] 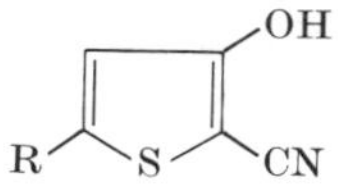 [**91**]

$CDCl_3$ (R = Me, *i*-Pr, *t*-Bu). However, signals for OH at 8.36 δ and for CH_2 at 3.86 δ in the spectrum of 4-cyano-3-hydroxy-5-isopropylthiophene (70J(C)2409) in $CDCl_3$ indicate a hydroxy and oxo tautomeric mixture (40 : 60) [**92a**] ⇌ [**92b**]. Spectral evidence demonstrates

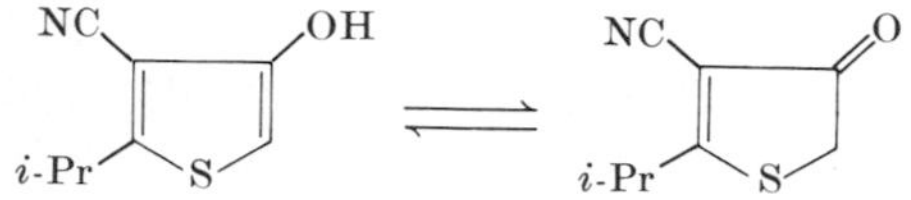

[**92a**] [**92b**]

the hydroxy form for 2,4-dicyano-3-hydroxy and 4-cyano-2-ethoxycarbonyl-3-hydroxythiophene (72T875).

Keto–enol equilibria have been studied for various 3-oxotetrahydrothiophenecarboxylic esters (71KG1473).

e. β-Hydroxyselenophenes

3-Oxoselenophene [**93**, R = H] from the NMR spectrum in $CDCl_3$ (CH_2 δ at 3.7) is an oxo compound [**93a**] (71BF3547), but 2-acylated analogues [**93**, R = CHO, COMe] exist as expected in the 3-hydroxyselenophene structure on spectral evidence (72JH355).

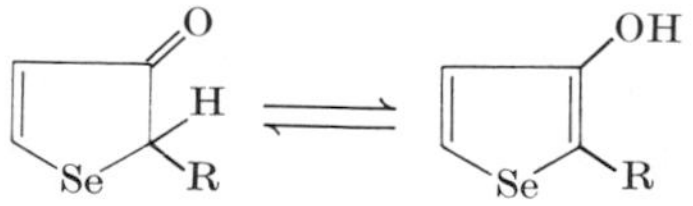

[**93a**] [**93b**]

f. β-Hydroxybenzothiophenes (II-10)

The tautomerism of 3-hydroxybenzothiophene [**94**], uncertain at the time of the previous review (II-10), has been clarified by the work of

Rubaszewska and Grabowski (69T2807). These authors were able to isolate and record spectra individually for both the hydroxy [**94a**] and oxo forms [**94b**]. Indeed they found that the tautomeric equilibration occurs quite slowly in the solid and in nonpolar solvents, and explain previous inconsistent results on this basis. In water, the oxo form [**94b**] predominates (85%, $K_T = 0.18$), but in less polar solvents the enol is favoured.

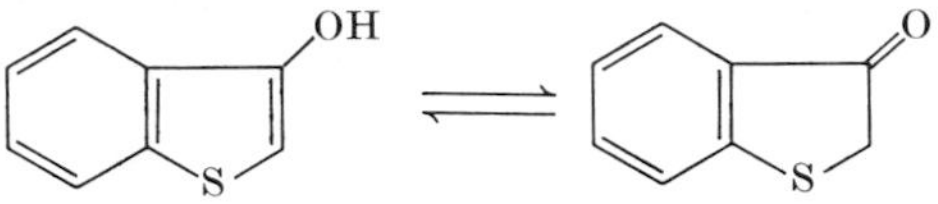

[**94a**] [**94b**]

Recently this picture has been confirmed and the tautomeric constants determined in various solvents for the parent and some chloro substituted derivatives; the enol form is favoured by chlorine substituents (74CQ(6)184) to variable extents depending on the orientation.

Infrared evidence (56CB1897, 58J1217), quoted in the previous review (II-10), that the oxo form [**94b**] predominates in $CHCl_3$ and CS_2 accordingly refers to *nonequilibrated* solutions prepared from the crystalline oxo form, whereas the opposite conclusion reached (65BF2658) for $CDCl_3$ and CF_3CO_2H applies to solutions at or near equilibrium. These conclusions have been confirmed (73PS1).

Other recent work (69AP423) which has reported a strong solvent effect on the 3-hydroxybenzothiophene equilibrium [**94a**] $\rightleftharpoons$ [**94b**] (found by NMR) from CCl_4 (100% 3-oxo form) to dimethyl sulphoxide (71% 3-oxo form), probably also refers to solutions which had not reached equilibrium.

NMR spectra 12.1–12.4 δ show that 2-acetyl-3-hydroxy-benzothiophene exists as the enol [**95**] in CCl_4 and $CDCl_3$ as do [**96**] and [**97**] (65T3331) (cf. II-10).

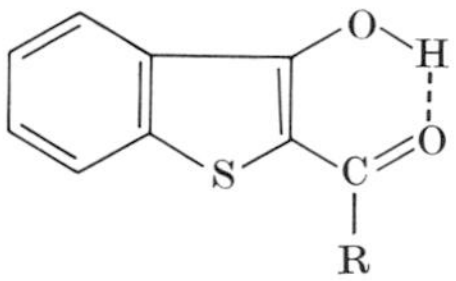

[**95**] R = Me
[**96**] R = OEt

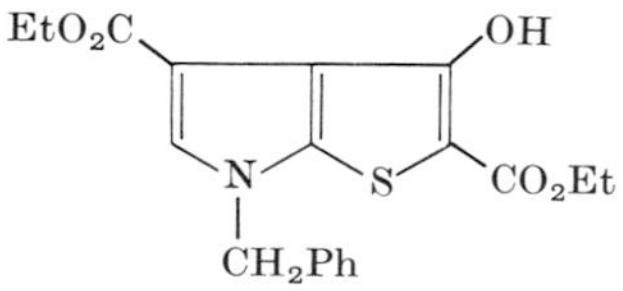

[**97**]

UV comparison (MeOH) with fixed forms demonstrates that the potential iminomethylene-hydroxybenzothiophenes [**98**] exist in the oxo–amino form [**98b**]; however, the aldehyde is confirmed to exist in a structure analogous to [**95**] (72KG920).

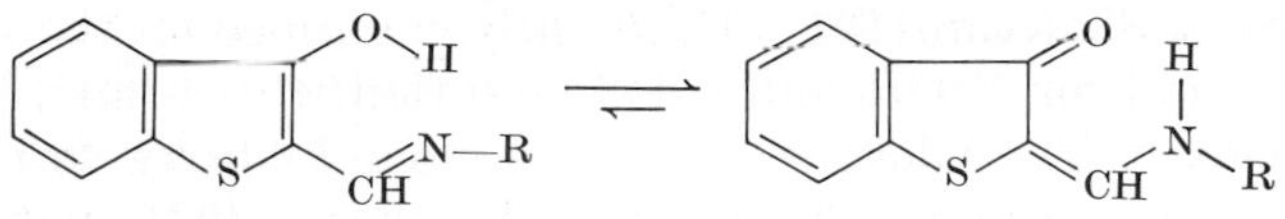

[**98a**] [**98b**]

The existence of a tautomeric equilibrium [**98/1a**] ⇌ [**98/1b**] is supported by spectroscopic evidence (72JO4121). 3-hydroxy-4*H*-thieno[3,2-*b*]pyrroles exist in the oxo form [**99**] on NMR evidence (64JO2725).

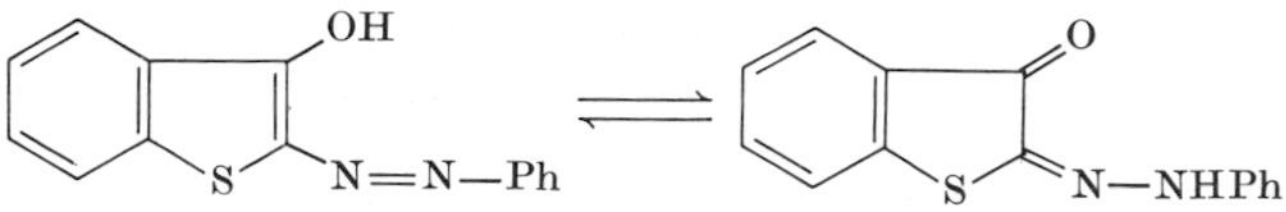

[**98/1a**] [**98/1b**]

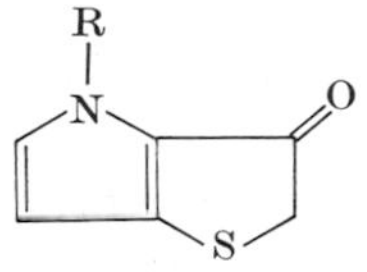

[**99**]

g. β-Hydroxybenzoselenophenes

3-Hydroxybenzoselenophene appears to exist in the 3-oxo form [**100**] in both $CDCl_3$ and CF_3CO_2H as shown by the CH_2 singlet at 4.04–3.92 δ

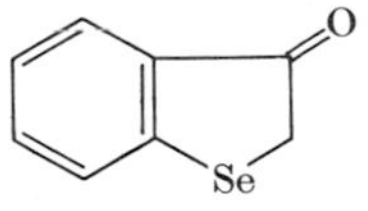

[**100**]

(68JH133; see also 71KG1640): however, in view of the work reported for the benzothiophene analogues, these conclusions must be treated with caution.

D. Potential Dihydroxythiophenes (II-11)

Little was known about the tautomerism of these compounds at the time of the previous review; now the position is considerably more clear.

a. Oxygen Functions in 2,3-Positions

The NMR of 3-hydroxy-3-thiolen-2-one [**101**] in CD_3COCD_3 (66AS261) is similar to that of 3-methoxy-3-thiolen-2-one described earlier

(64AK(21)239). Structure [**101**] was recently confirmed for the solid and liquid state, and for EtOH and $CDCl_3$ solution; the 4- and 5-methyl derivatives exist in analogous mono-oxo-mono-hydroxy forms.

Unexpectedly, 4-bromo-2,3-dihydroxythiophene [**102**] exists in the dioxo form [**102b**] in CCl_4 as shown by the ABX system of its NMR spectrum; the IR spectrum in CCl_4 confirmed that the 4-bromo-3-hydroxy-3-thiolen-2-one form [**102a**] is present only in traces (71T3839).

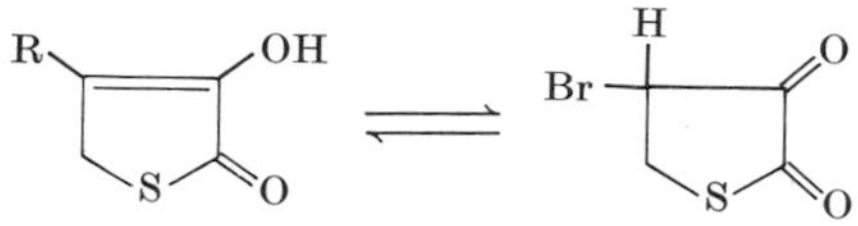

[**101**] R = H
[**102a**] R = Br

[**102b**]

The 4-thiolen-2-one structure [**103a**] was originally assigned to potential 5-ethoxycarbonyl-2,3-dihydroxythiophene [**103**] (65T3331). Tautomers [**103c**] and [**103d**] were excluded by the AB system (J_{AB} = 3.0–3.2 c/s) in the NMR (65T3331). However, the choice between [**103a**] and [**103b**] was based only on the expected stability of substituted γ-thiolactones (63T1867, 64AK(21)239, 64AK(22)211), and later structure [**103b**] was preferred after spectral comparisons (71T3839).

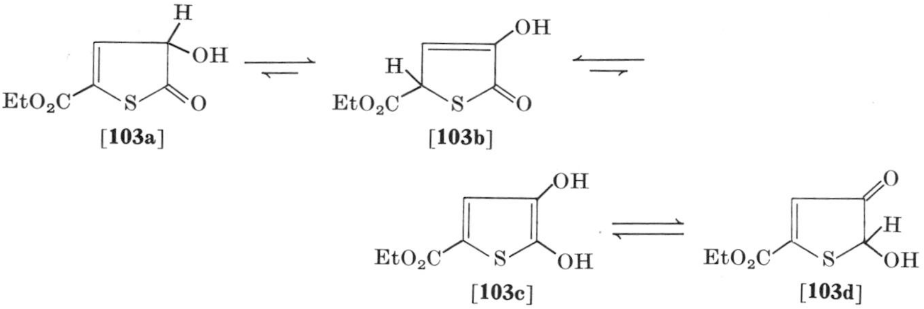

[**103a**] [**103b**]

[**103c**] [**103d**]

b. *Oxygen Functions in 2,4-Positions*

Many substituted 2,4-dihydroxythiophenes were demonstrated spectroscopically (UV, NMR and IR) to exist in the solid state and in MeOH, DMSO and $CHCl_3$ as "thiotetronic acids," 4-hydroxy-3-thiolen-2-ones [**104**] (71T3839), as suggested earlier (13CB2107); see also 73RC1735.

The chemical behaviour of compound [**105**] was interpreted in terms of a tautomeric equilibrium [**105a**] $\rightleftharpoons$ [**105b**] or of complete predominance of [**105a**] (68J(C)1501).

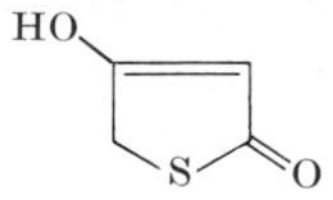

[104] 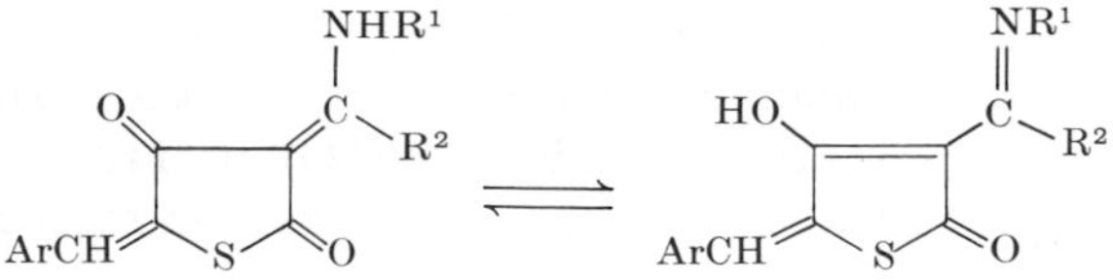 [105a] ⇌ [105b]

c. Oxygen Functions in 2,5-Positions

"2,5-Dihydroxythiophene" exists as the thioanhydride, 2,5-dioxotetrahydrothiophene [**106**]; it shows only one NMR peak (63T1867, 71T3839). The analogous seleno derivative correspondingly exists as 2,5-dioxotetrahydroselenophene [**107**] with a singlet at 2.94 δ in the NMR in $CDCl_3$ [**107**] (70CO(270C)825, 71BF3547).

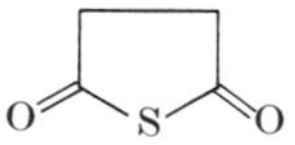

[106] 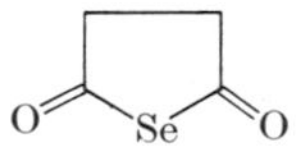 [107]

d. Oxygen Functions in 3,4-Positions (II-11)

Although early investigators postulated that 3,4-dihydroxythiophenes exist as diols (cf. II-11), the 4-hydroxy-3-oxo-2-thiolene structure [**108**, $R_2 = R_5 =$ H or Me, $R_2 =$ Me and $R_5 =$ H, $R_2 = CO_2Et$ and $R_5 =$ H] has now been shown to predominate by NMR. However, the 2,5 diethoxycarbonyl compound [**109**] does indeed exist in the dienol form (71T3839).

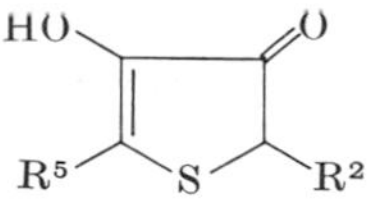

[108] 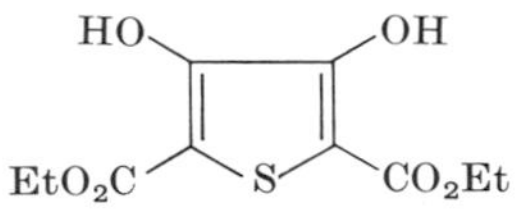 [109]

E. Monohydroxy-pyrroles and -indoles (II-11)

Owing partly to their importance in natural products, interest in hydroxypyrroles has been intense, and much further work has been added to the substantial results already reported in the previous review (II-11 to II-18). In addition to the α- and β-hydroxypyrroles, corresponding to the furan and thiophene analogues, *N*-hydroxypyrroles exist, and their tautomerism, not investigated at the time of the previous review, has since received attention.

a. N-Hydroxypyrroles

These compounds possess three tautomeric forms [cf. **110a–c**]; however, both the *N*-oxide forms [**110b**], [**110c**] have lost the resonance energy of the pyrrole ring, and the gain in resonance for the formation

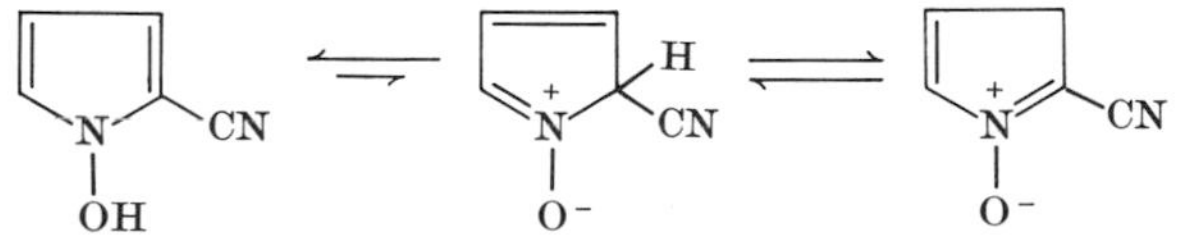

[**110a**] [**110b**] [**110c**]

of a nitrone group is far less than that for an amide link. Hence, just as for the pyrrolenines (Section 3-1Ab), we expect the *N*-hydroxypyrrole form to be the stable tautomer. This is so for the single compound so far investigated. Only the *N*-hydroxy form [**110a**] was observed for 1-hydroxy-2-cyanopyrrole in the infrared (νN–OH at 2800–3400 cm^{-1} in the liquid) and NMR spectra (N–OH at 7.56 δ, and three aromatic-type hydrogens) (73JO173).

b. N-Hydroxyindoles

The equilibrium is much more finely balanced in the indole series, when the aromatic resonance energy loss is less on formation of the *N*-oxide tautomer [**111a**], as already discussed for the indole-indolenine tautomerism itself (Section 3-1Ac). Thus 1-hydroxyindole exists as a tautomeric mixture of the *N*-oxide form [**111a**] and the hydroxy form [**111b**] in $CDCl_3$ and CCl_4 as shown by NMR and IR spectra (67BF1296). The equilibrium [**112a**] $\rightleftharpoons$ [**112b**] was studied quantitatively by comparison of intensities of the methyl NMR signals (67BF1296). It is strongly dependent on the solvent; the indolenine *N*-oxide [**112a**] is favoured in hydrogen bond donor solvents such as phenol while 1-hydroxy-2-methyl indole [**112b**] is the major species in pyridine and acetonitrile (67BF1296, 67SA717); 20–40% of *N*-oxide exists in nitromethane, anisole, CCl_4 and $CDCl_3$.

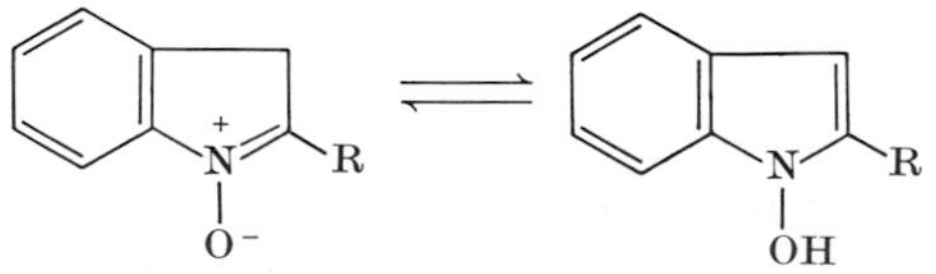

[**111a**] R = H [**111b**]
[**112a**] R = Me [**112b**]

The similarity of the NMR spectra of 1-hydroxy-2-methoxycarbonyl-indole [**113**] and of its 1-methoxy analogue suggests the hydroxy form for [**113**] in $CDCl_3$ (68J(C)504). A CH_2 peak in the NMR spectrum shows that 5-bromo-1-hydroxyindole-2-carboxylic acid exists as 5-bromo-3*H*-indole-1-oxide [**114**] in $CDCl_3$, probably because an intramolecular hydrogen bond stabilizes the *N*-oxide form.

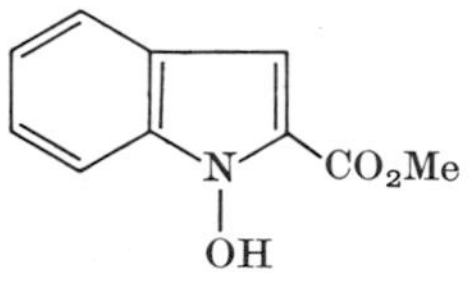

[**113**] 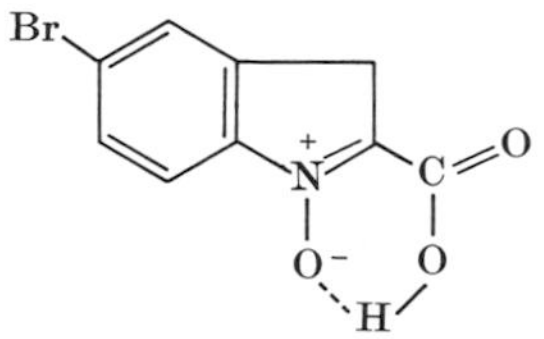 [**114**]

c. α-Hydroxypyrroles (II-11)

The conclusions drawn for substituted 2-hydroxypyrroles and described in the previous review (II-11) have been for the most part confirmed and also extended to the parent compound. Δ^4-Pyrrolin-2-one [**115a**] and Δ^3-pyrrolin-2-one [**115b**] have been prepared separately (65JO3824), and their structures have been confirmed by a complete NMR analysis (71OR7, 72G91). Deuterium exchange experiments indicate that [**115a**] and [**115b**] are in equilibrium at room temperature in polar solvents but that the uncatalyzed transformation of [**115a**] into the more stable tautomer [**115b**] is slow (71OR7). The existence of small amounts of the enol tautomer [**115c**] was postulated but not detected directly (71OR7). These conclusions agree with theoretical predictions (70JA2929).

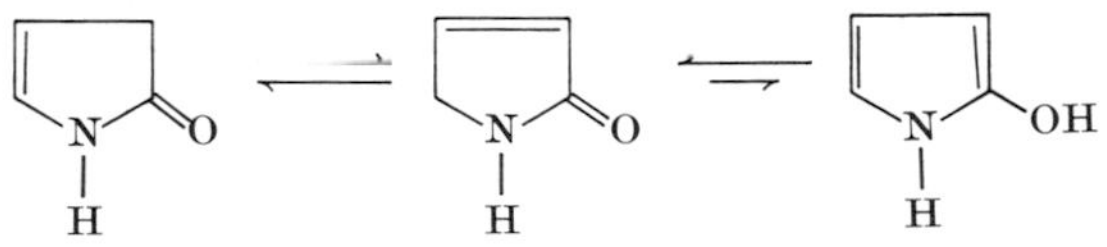

[**115a**] [**115b**] [**115c**]

The 1-ethoxycarbonyl derivative [**116a**] is easily transformed into [**116b**] by heating (65JO3824).

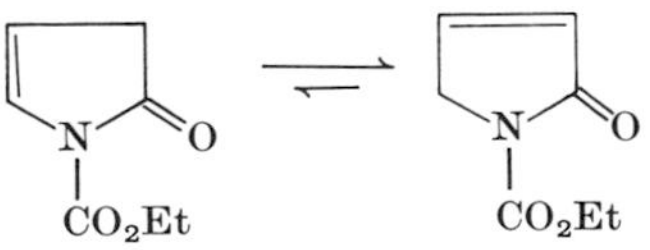

[**116a**] [**116b**]

Recently structures [**117a**] and [**118a**] were assigned from spectral behaviour (in $CDCl_3$ and in the solid state) (73JO173): the preference of the Δ^4 forms to the Δ^3 alternative is curious, but the spectral evidence appears convincing. Perhaps equilibrium was not attained. The spectra indicate that the triphenylpyrrolin-2-one [**119**] exists in the Δ^3 form shown (70BF3572) as do 3,5-dimethyl analogues (73CB3753) and [**119/1**] (71CH1099).

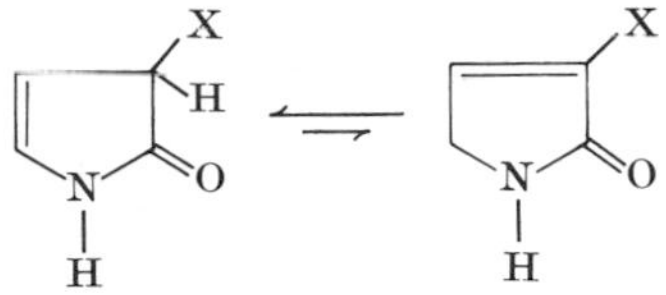

[**117a**] X = OMe [**117b**]
[**118a**] X = NHPh [**118b**]

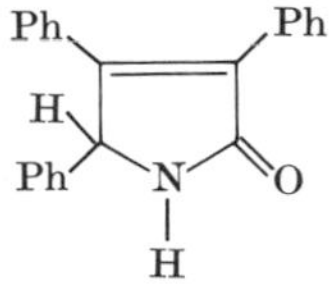

[**119**]

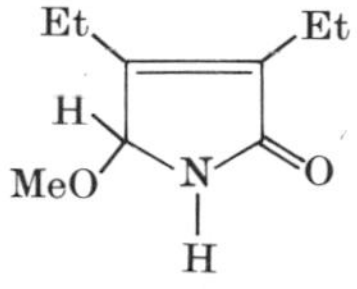

[**119/1**]

At the time of the previous review, it was thought that hydroxypyrrole structures could be stabilized by electron-withdrawing substituents in the 5-position. However, a supposed example has been proved incorrect as re-examination of the NMR pattern of 5-formyl-2-hydroxypyrrole [**120**] shows that the compound exists as 5-hydroxymethylene-3-pyrrolin-2-one [**120b**] (66J(C)40), not as [**120a**] (II-13) or as [**120c**] (65MI200). The assignment is based on: (i) NH at 0.33 δ, broad (in DMSO) exchanged by D_2O; (ii) :C*H*OH at 6.59, sharp, not exchanged by D_2O; (iii) the product of methylation of [**120/1**] retains the signal at 6.45 δ associated with the hydroxymethylene methine proton.

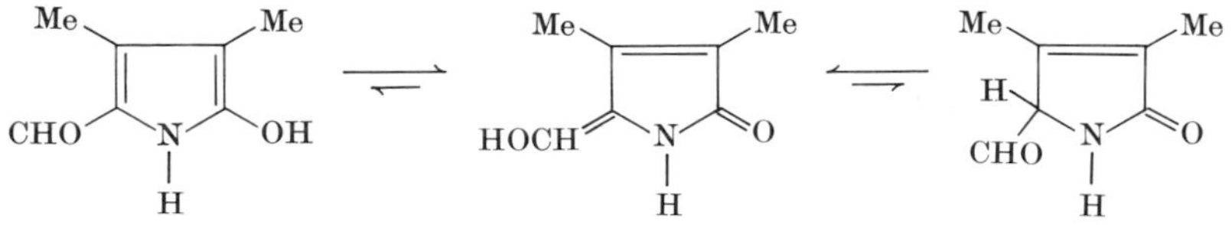

[**120a**] [**120b**] [**120c**]

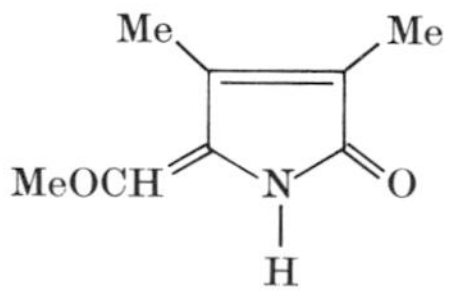

[**120/1**]

The Δ^3-structures [**121**] and [**122**, R = cyclo-C_6H_{11}, CH_2Ph or Ph] are supported by νNH at 3509 cm^{-1}, νC═O at 1721 cm^{-1} and νNO_2 at 1515 cm^{-1} together with consistent NMR spectra (69JO3279).

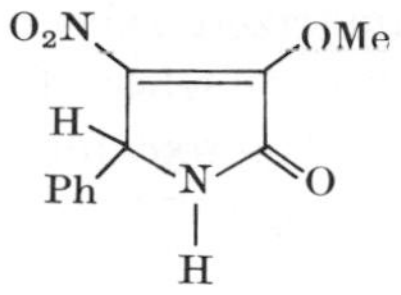

[121] 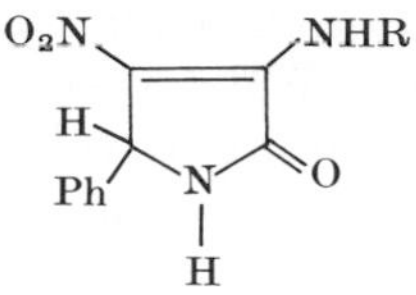 [122]

d. α-Hydroxybenzopyrroles (II-18)

Previous evidence for the predominance of the oxindole "amide" form [**123a**] over the alternative hydroxy form [**123b, 123c**] has been summarized (II-18). Since then, NMR, IR (intense absorption in the 1690–1752 cm^{-1} region) and UV spectra of hydroxy-, methoxy-, halogeno-, amino- and nitro-oxindoles confirms that all these 2-hydroxyindoles

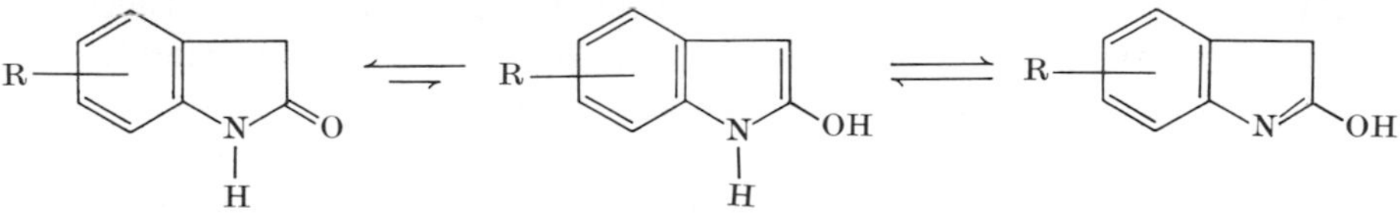

[**123a**] [**123b**] [**123c**]

exist entirely in the oxo forms [**123a**] in $CHCl_3$ and DMSO and in the solid state (68T6093). Similar results were previously obtained for 3-phenyloxindole (57J4789) and later for 3-(*p*-nitrophenyl)oxindole (74CT1053). A weak band at 3608 cm^{-1} in 5-nitrooxindole was considered as possibly due to the presence of small amount of the enol tautomer [**123b**] (68T6093), but this needs confirmation. See also Section 5-2Ba.

The $\nu C{=}O$ of phthalimidine in dioxan and in the solid state is consistent with the oxo structure [**124**] expected (63PM(2)196, 71PM(4)289).

The tautomeric equilibrium of the cycloheptapyrrole [**125a**] $\rightleftharpoons$ [**125b**] was studied by ultraviolet comparisons with fixed forms: the aza-azulenone form [**125a**] predominates in EtOH (54CI1356).

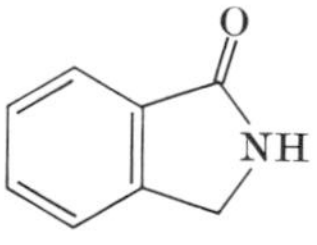

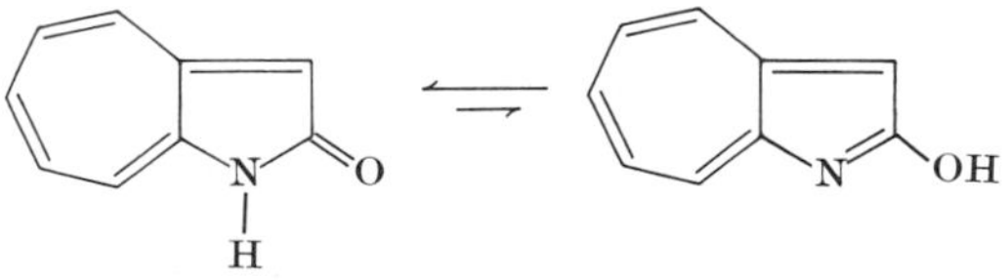

[**124**] [**125a**] [**125b**]

e. β-Hydroxypyrroles (II-14)

At the time of the previous review, little was known of the tautomeric structure of β-hydroxypyrroles, but this position has now changed:

work on this tautomerism was summarized recently (70LA(736)1). According to the NMR spectrum (68AG(E)734) (CH_2 at $\delta = 3.58$), 3-hydroxy-4,5-dimethylpyrrole exists in the oxo form [**126**] in DMSO. Structures [**127**] were assigned similarly (69T5721; cf. also 69CI1077).

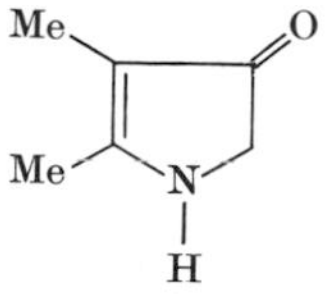

[**126**] 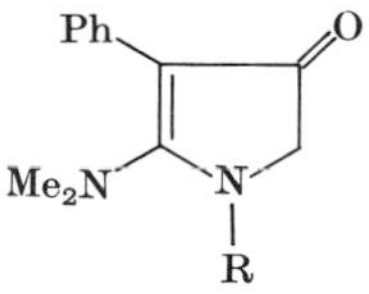 [**127**] R = Me or Ph

Previous conclusions (53J3802) that potential 4-alkoxycarbonyl-3-hydroxypyrroles [**128**] existed as [**128b**] have been confirmed by NMR, UV and IR (63CC625); some authors (cf. II-15 and references therein) have emphasized the zwitterionic form [**128c**], but this is merely an alternative canonical form of structure [**128b**]. Similarly IR (νC=O at 1660 and νC≡N at 2190 cm^{-1}) and NMR (CH_2 at $\delta = 3.40$) spectra indicate the cyano compound [**129**] exists entirely in the oxo form in the solid state and in $CDCl_3$ (70J(C)2409).

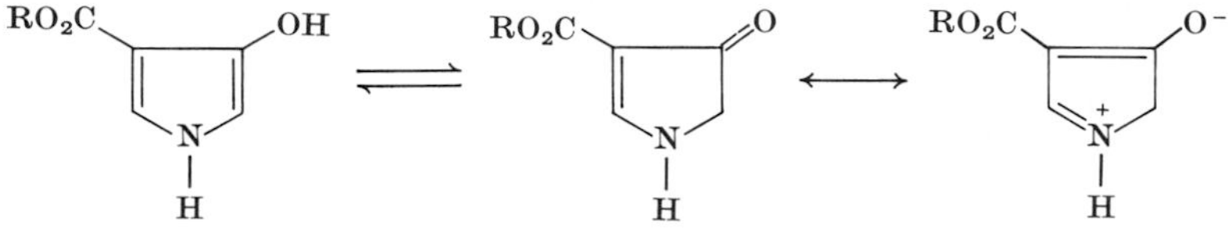

[**128a**] [**128b**] [**128c**]

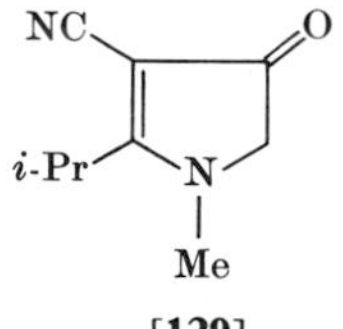

[**129**]

However, UV and IR data do suggest an enol structure for 2-alkoxycarbonyl-3-hydroxypyrroles [**130**] (67AJ935). Similarly the enol [**131a**] coexists with only small amounts of the 4-oxo-Δ^2-pyrroline form

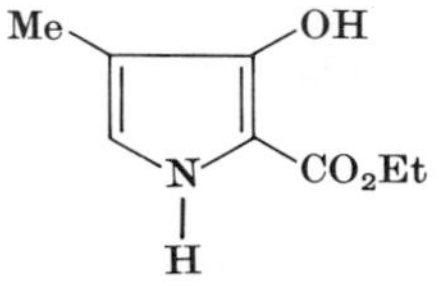

[**130**] 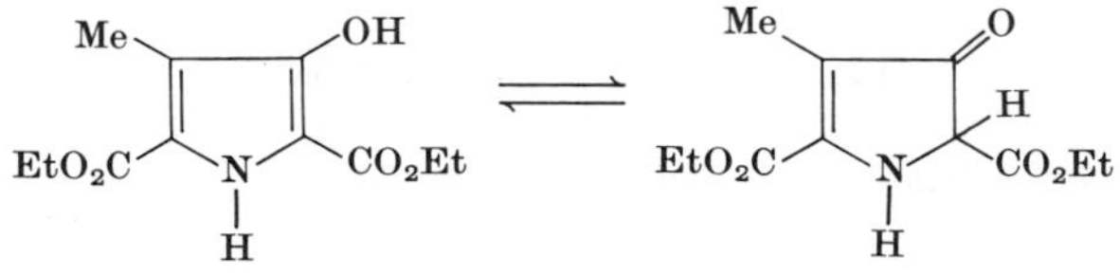 [**131a**] [**131b**]

[**131b**] (67AJ935) and the enol structure [**131/1**] has been recently suggested on IR evidence (73CT2571).

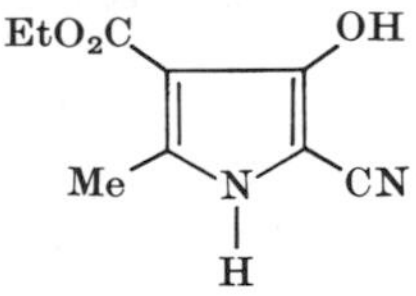

[**131/1**]

Hydroxypyrrolenine structures [**132a**] have been assigned to many derivatives of 5-dialkylamino-3-hydroxypyrrole [**132**] (69T5721). The alternative oxo form [**132b**] was excluded mainly because of the apparent absence of νC=O in the IR spectrum (present in the fixed form of [**132b**]) in the solid state. The presence of a CHR peak in the 4.0–3.6 δ region of the NMR spectrum in $CDCl_3$ eliminated the pyrrole structure [**132c**]. We believe this matter needs reinvestigation as the reported preponderance of [**132a**] over [**132b**] is unexpected and νC=O may be much affected by *H*-bonding.

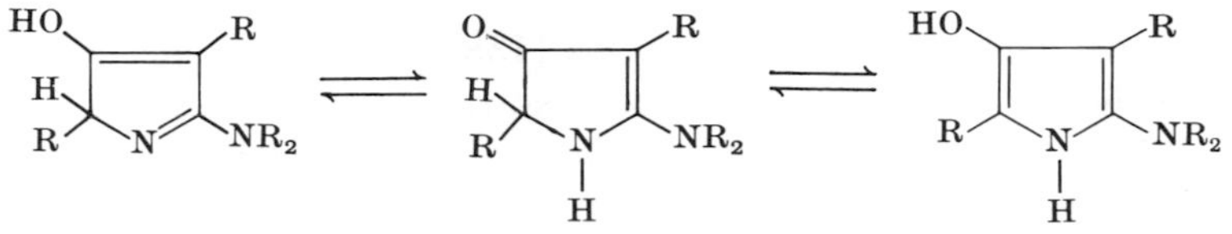

[**132a**] [**132b**] [**132c**]

R = alkyl or aryl

The keto–enol tautomerism of various pyrrolidin-3-one carboxylic esters has been studied (69KG978).

f. β-Hydroxyindoles (II-19) *and Hydroxyindolizines*

Infrared evidence had previously showed that 1-acetyl- and 1-methylindoxyl exist in the oxo forms [**133**] (R = Ac, Me), but that substituents capable of strong intramolecular hydrogen bonding stabilize the hydroxy form as in [**134**] (II-18). Similarly 2-(2-pyridyl)-3-hydroxyindole [**135**] is a true hydroxy compound in the solid state and in $CHCl_3$ from the absence of νC=O; however, a weak νC=O was found in

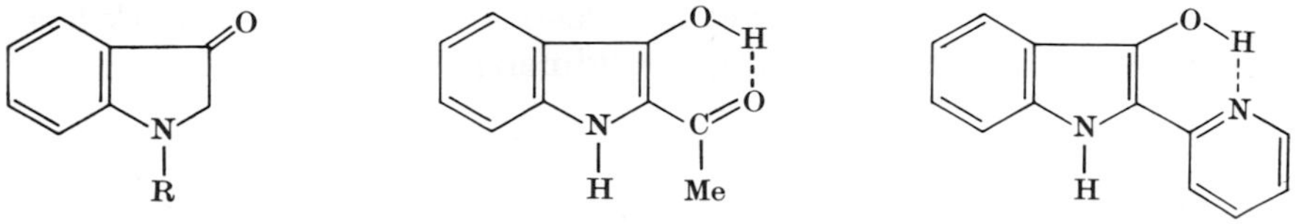

[**133**] [**134**] [**135**]

dioxan (65J1706). While adrenolutin [**135/1**] exists largely in the oxo form [**135/1a**] in pyridine and DMF, 10–15% of the hydroxy form [**135/1b**] was detected in DMSO by NMR (72MI133).

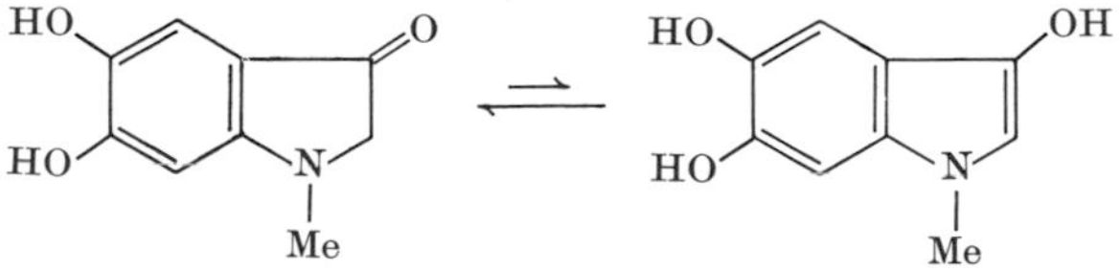

[**135/1a**] [**135/1b**]

No ring carbonyl absorption occurs in the IR of hydroxyindolizines [**136**] (63J3277) and [**137**] (65J2948).

[**136**] [**137**]

The hydroxy [**138a**] and oxo forms [**138b**] have both been isolated (63G383, 63H1030). The IR of [**138a**] shows bands at 1739, 1718 and 1678 cm^{-1} whereas the enol form [**138b**] shows νOH and is favoured when R is electronegative. Recrystallization from EtOH gives the hydroxy and from C_6H_6 the oxo form.

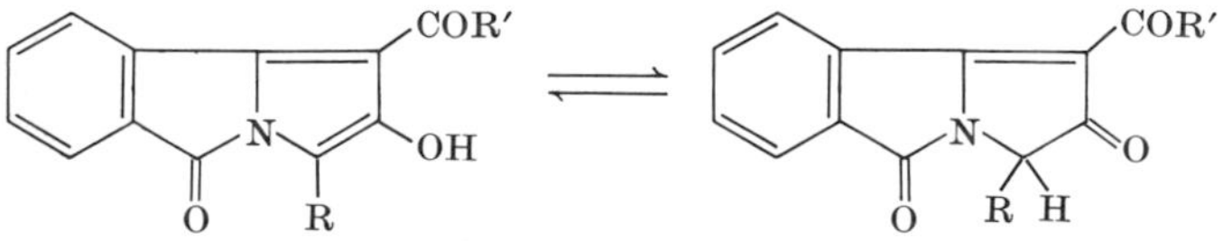

[**138a**] R = aryl [**138b**]
R′ = alkyl or aryl

F. Potential Dihydroxy-pyrroles and -indoles (II-15)

Surprisingly little has appeared on these compounds since the last review, and several types are yet uninvestigated. No further work has appeared on isatin (II-16) or on pyrrolidinetriones (II-17).

a. Oxygen Functions in the 2,3-Positions (II-15)

Potential 2,3-dihydroxypyrroles were shown in the last review (II-15) to exist in the monoenol form, and this is confirmed by work on one more compound. The similarity of the UV (in EtOH) and IR (solid) of [**139**] with those of the 3-methoxy derivative suggested the enol structure [**139a**] rather than the acinitro structure [**139b**]. The NMR spectrum in DMSO was inconsistent with the dioxo structure [**139c**] (69JO3279). Similarly the monoenol structure of 4,5,6,7-tetrahydroisatin, previously demonstrated by IR spectral studies (II-15), was confirmed by NMR (73J(PI)2814).

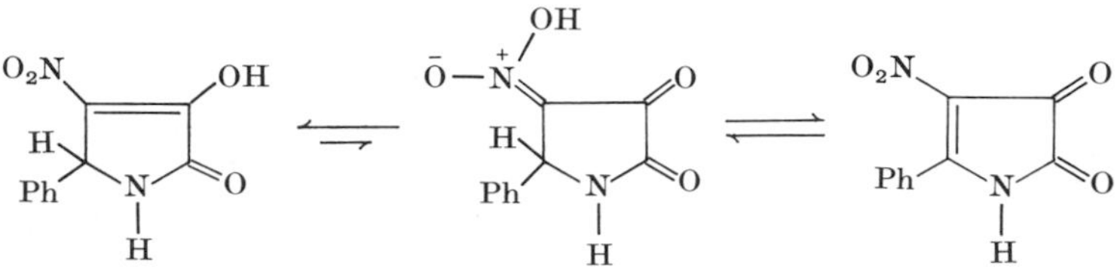

[**139a**] [**139b**] [**139c**]

Alterations in the NMR and IR spectra with temperature have been interpreted (68M2223) as evidence for the mobile equilibrium [**140a**, **140b**]. The IR band at 1700 cm^{-1} (for [**140a**]) in $CHBr_3$ is said to weaken on heating to give a band at 1610 cm^{-1} (for [**140b**]), and this is said to be confirmed by the NMR spectra. This surprising conclusion needs confirmation in view of the proven oxo structure [**141**] for dioxindole (II-16).

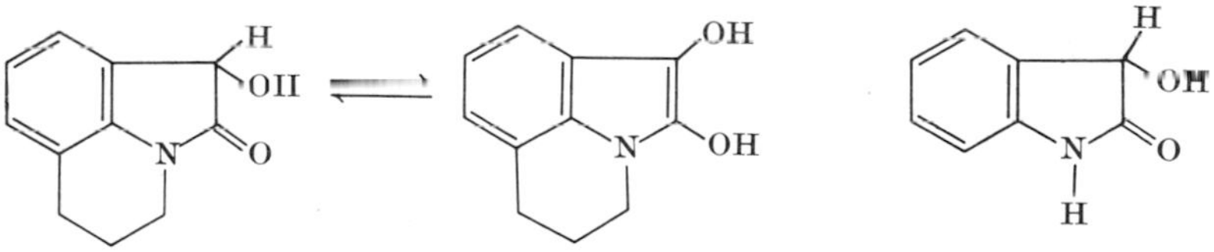

[**140a**] [**140b**] [**141**]

b. Oxygen Functions in 2,5-Positions

Infrared spectra and assignments confirm the dioxo structure—[**142a**] for succinimide and many derivatives (71PM(4)291)—and X-ray supports the structure of the parent compound (61CX720). The potential tautomer [**142b**] of succinimide has been isolated (62CI1576) and is moderately stable.

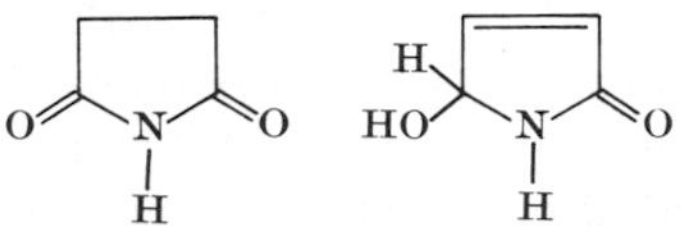

[**142a**] [**142b**]

NMR and IR investigations of the ring–chain tautomerism [**143 a**⇌ **b**] of 1-hydroxy-3-isoindolinones show only the cyclic form [**143a**] (70CB3205, 73CB1423).

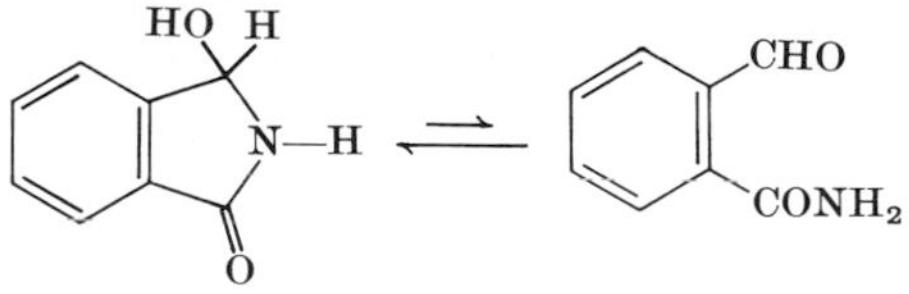

HO H N—H ⇌ CHO CONH$_2$ O

[**143a**] [**143b**]

G. Other Compounds with Potential Hydroxy Groups (II-19)

The tautomeric equilibria [**144, 145, 146**] have been measured and compared with the corresponding anthrol–anthrone equilibrium. For [**144** and **145**], the keto [**144b**, **145b**] forms are of comparable stability with the enol [**144a**, **145a**] but for [**146**] the keto form [**146b**] greatly predominates (71JO3999).

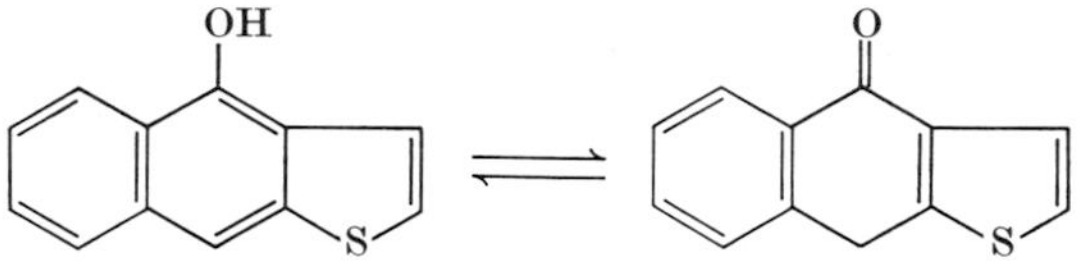

OH S ⇌ O S

[**144a**] [**144b**]

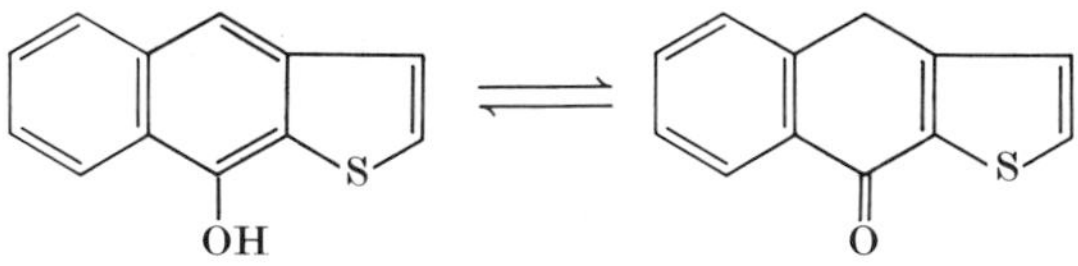

S ⇌ S OH O

[**145a**] [**145b**]

OH S ⇌ O S

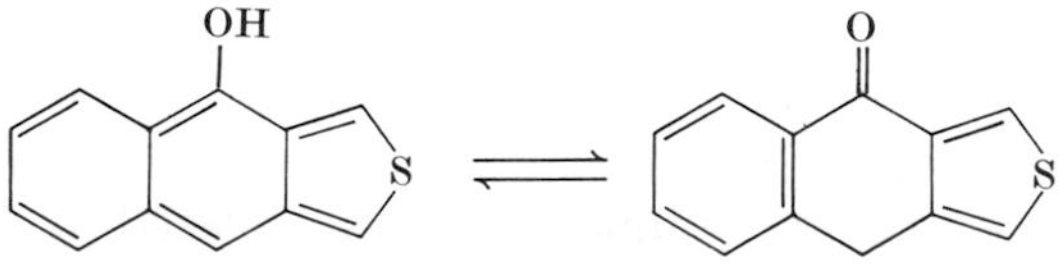

[**146a**] [**146b**]

NMR (4H at 2.97 δ) and IR (νC=O at 1655 cm^{-1}) show that 4,7-dihydroxyindole [**147a**] exists as 4,7-dioxo-5,6-dihydroindole [**147b**] in $CDCl_3$ and in the solid state (69TL4173). However, the 2-phenyl analogue has been isolated both as [**148a**] and [**148b**]; but the former is easily converted into the latter (70F972).

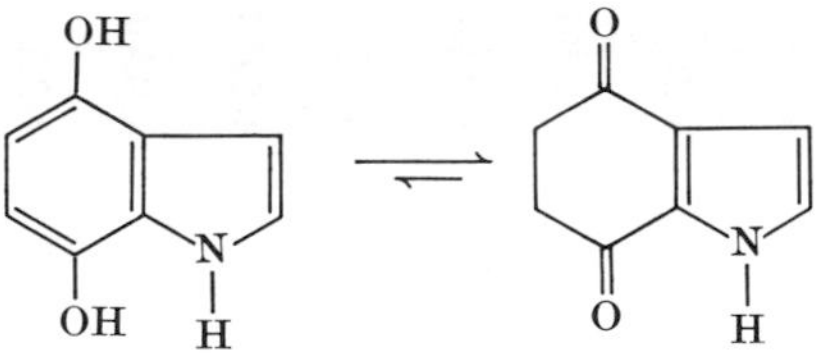

[147a] ⇌ [147b]

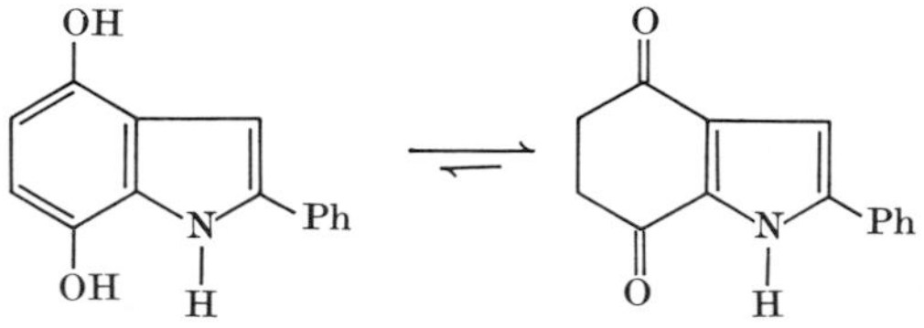

[148a] ⇌ [148b]

H. Compounds with Potential Hydroxy Groups—Conclusions

a. Introduction

Table 3-3 summarizes schematically the results available on monohydroxy furans, thiophenes, and pyrroles. The sign indicates the possibility of existence of the various tautomers in detectable amounts: + + +, very high; + +, medium; +, low; −, very low. In the section "substituted parent" the effects of various types of substituents are lumped together, and the probabilities given refer to finding a particular type of tautomer within the whole range of substituted compounds.

The tendency for enolic hydroxy compounds to revert to the oxo form is easily understood with reference to simple aliphatic compounds: the keto–enol tautomeric equilibrium constants for simple ketones are ca. 10^8. To understand the effects of oxygen function orientation, the various heteroatoms and benzo substitution on the equilibrium position, we must consider the further features of each of the structures [**149**] and [**150**]; such *extra* stabilizations of the various structures are set out in Table 3-4.

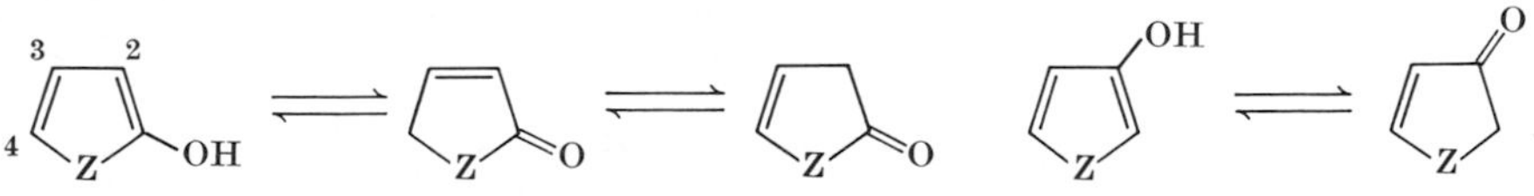

[149a] ⇌ [149b] ⇌ [149c]   [150a] ⇌ [150b]

b. Influence of Heteroatom

Aromatic resonance energies are in the order pyrrole ≥ thiophene ≫ furan (74HC(17)255) whereas the resonance energies of the open-chain

TABLE 3-3
HYDROXY-OXO TAUTOMERISM IN FIVE-MEMBERED RINGS WITH ONE HETEROATOM

Type	Z	α-Hydroxy compounds					β-Hydroxy compounds			
		Compound	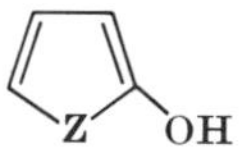 Z, OH	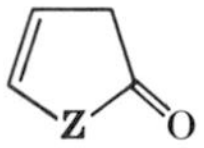 Z, O	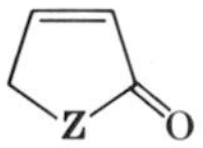 Z, O	Section	Compound	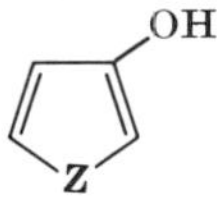 Z, OH	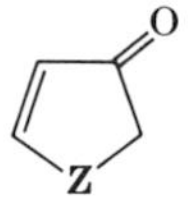 Z, O	Section
Parent	O	2-Hydroxyfuran	–	+	+ +	3-2Aa	3-Hydroxyfuran	–	+ + +	3-2Ac
	S	2-Hydroxy thiophene	–	+	+ +	3-2Ca	3-Hydroxy-thiophene	+ +	+ +	3-2Cd
	NR	2-Hydroxy-pyrrole	–	+	+ +	3-2Ec	3-Hydroxy-pyrrole	–	+ + +	3-2Ed
Substituted parent	O	Subst. 2-hydroxy-furans	–	+ +	+ +	3-2Aa	Subst. 3-hydroxy-furans	+	+ +	3-2Ac
	S	Subst. 2-hydroxy-thiophenes	+ +	+ +	+ +	3-2Ca	Subst. 3-hydroxy-thiophenes	+ +	+ +	3-2Cd
	NR	Subst. 2-hydroxy-pyrroles	–	+ +	+ +	3-2Ec	Subst. 3-hydroxy-pyrroles	+	+ +	3-2Ed
Benzo [*b*]	O	2-Hydroxy-benzo[*b*]furans	–	+ + +		3-2Ab	3-Hydroxy-benzo[*b*]furans	+	+ +	3-2Ad
	S	2-Hydroxybenzo-[*b*]thiophenes	–	+ + +		3-2Cc	3-Hydroxybenzo-[*b*]thiophenes	+ +	+ +	3-2Ae
	NR	2-Hydroxyindoles	–	+ + +		3-2Ed	3-Hydroxyindoles	+ +	+ +	3-2Ef
Benzo [*c*]	O	1-Hydroxybenzo-[*c*]furans	–		+ + +	3-2Ab				
	S	1-Hydroxybenzo-[*c*]thiophenes	–		+ + +	3-2Cc				
	NR	1-Hydroxyiso-indoles	–		+ + +	3-2Ed				

TABLE 3-4

RESONANCE STABILIZATION OF INDIVIDUAL TAUTOMERIC FORMS OF POTENTIAL HYDROXY COMPOUNDS

Structure	Furan series	Thiophene series	Pyrrole series
[**149a**] + [**150a**]	Furan aromaticity + OH-furan conjugation	Thiophene aromaticity + OH-thiophene conjugation	Pyrrole aromaticity + OH-pyrrole conjugation
[**149b**]	Ester resonance + $\alpha\beta$ unsaturated ester conjugation	Thiolester resonance + $\alpha\beta$ unsaturated thiol ester conjugation	Amide resonance + $\alpha\beta$ unsaturated amide conjugation
[**149c**]	Ester resonance–cross conjugation term	Thiolester resonance–cross conjugation term	Amide resonance–cross conjugation term
[**150b**]	Vinylogous ester $\equiv$ $\alpha\beta$ unsaturated ketone conjugation + enol ether conjugation + through conjugation	Vinylogous thiolester $\equiv$ $\alpha\beta$ unsaturated ketone conjugation + vinylthioether conjugation + through conjugation	Vinylogous amide $\equiv$ $\alpha\beta$ unsaturated ketone conjugation + enamine conjugation + through conjugation

groups are amide $\gg$ ester $\geq$ thiolester (60MI25), and the same order is expected for the vinylogous groups. We should expect the tendency to exist in an oxo form to be significantly less for thiophene (which has much to lose and relatively less to gain) than for pyrrole (a lot to lose but also a lot to gain) and furan (less to lose but also less to gain). Examination of Table 3-3 shows the great parallelism between furans and pyrroles and the greater tendency to enolize for the thiophenes.

c. *Influence of α- or β-Orientation of the Oxygen Function*

Examination of Table 3-3 shows that hydroxy forms are much more likely to occur for β- than for α-orientation.

As [**149b**] is the more stable of the two oxo structures [**149b**] and [**149c**] and as the structures [**149a**] and [**150a**] probably have similar energies, the reason must be sought in the stability difference between [**149b**] and [**150b**]. Evidently the grouping Z—CO—C═C is more stable than Z—C═C—C═O; presumably the direct conjugation Z–CO is greater than the extended conjugation of the latter group.

d. *Influence of Substitution*

As might be anticipated, substitution at the 5-position of [**149a**] stabilizes [**149c**] relative to [**149b**] whereas substitution at the 3- or 4-position has the reverse effect. This has been particularly well examined in the thiophene series (Section 3-2Ca). Similarly, substitution at the 2-position in [**150**] stabilizes the hydroxy form [**150a**]. Also, as the aromatic ring systems in question are all electron-excessive, any electron-withdrawing groups especially tend to stabilize the aromatic hydroxy forms [**149a**] and [**150a**].

e. *Influence of Benzo Substitution*

Fusion of a benzene ring has a drastic effect in swinging the tautomerism of α-hydroxy compounds even more in favour of the oxo form. This is because of the smaller loss of resonance energy in forming an oxo form in which the aromaticity of the benzene ring is retained.

For β-hydroxy compounds the effect is certainly much less, possibly even in the other direction.

f. *Potential Dihydroxy Compounds*

Information relating to these compounds is summarized in Table 3-5. The aromatic dihydroxy form is very rarely found; most common is the monoenol form except for the 2,5-orientation when the dioxo form predominates.

TABLE 3-5

SUMMARY OF TAUTOMERIC FORMS FOR POTENTIAL DIHYDROXY-FURANS, -THIOPHENES, AND -PYRROLES

Orientation	Structure	Series		
		Furan	Thiophene	Pyrrole
2,3		(3-2Ba)[a]	(3-2Da)	(3-2Fa)
2,4		Polar solvent (3-2Bb)	(3-2Db)	?
		Nonpolar solvent (3-2Bb)	—	?
2,5		(3-2Bc)	(3-2Dc)	(3-2Fb)
3,4		—	2,5-diacyl only (3-2Dd)	?
		Subst. compounds (3-2Bd)	Most compounds	
		Parent (3-2Bd)	—	?

[a] Section number is given in parentheses.

g. General Conclusions

This discussion has had to be conducted in a very qualitative manner. It is hoped that the satisfactory qualitative agreement will encourage quantitative work firmly to establish the basis for oxo–hydroxy tautomerism in these compounds.

3. Compounds with Potential Mercapto Groups (II-20)

These compounds are still underinvestigated, although the position has certainly improved since the time of the last review (II-20). It is now clear that there is a far greater tendency for mercapto compounds to exist as such in the thiol form than for the corresponding hydroxy compounds to exist in the OH form. The reasons for this are discussed later (Section 3-3D).

A. Mercaptofurans

The νSH at 2570 cm^{-1} in the infrared spectrum of 2-mercaptofuran suggested that it exists as such [**151**, R = H] in the liquid state (66CB3215), and the thiol structure [**151**, R = MeCO] has been proved for 5-acetyl-2-mercaptofuran by NMR, UV and IR comparisons (66KG149, 70KG723). However, the thione form [**152b**] is favoured by hydrogen bonding in the 2-mercaptofuran-3-aldimines [**152**, R = H,

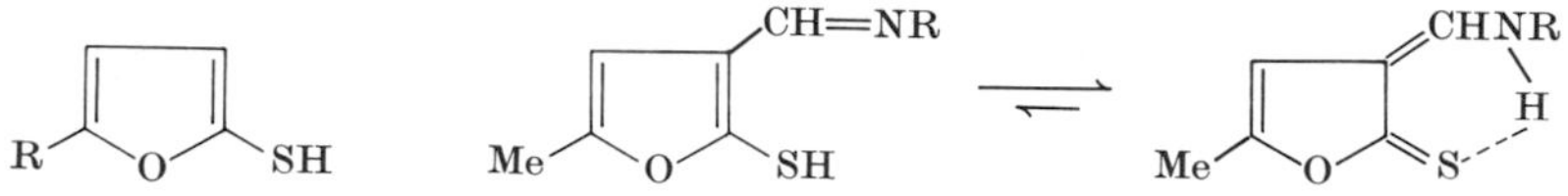

R O SH [**151**] Me O SH CH=NR [**152a**] ⇌ Me O S CHNR H [**152b**]

cyclo-C_6H_{11}, Ph, $PhCH_2$, β-naphthyl as shown by NMR in various solvents (67IZ2783, 70IZ675). The IR and NMR of 3-mercapto-2-methylfuran are recorded in a patent (70GP1).

B. Mercapto-thiophenes (II-20) and -selenophenes

Dipole moment calculations and measurements in benzene (70BF1720) confirm that 2- [**153**] and 3-mercaptothiophene [**154**] exist as such as previously demonstrated by spectroscopy (II-20; cf. also 65AK(23)483, 65AK(23)501) and suggest the analogous 3-mercaptobenzo[*b*]thiophene structure [**155**]. νSH bands at 3500–3550 cm^{-1} for 3-mercapto- [**155**] and 2-mercapto-benzo[*b*]-thiophene [**155/1**] demonstrate the thiol structure for these and for some simple derivatives (substituted in the benzo ring) (70J(C)2431). 2-Chloro-5-mercaptothiophene is also a thiol by NMR and IR (65T1333).

Just as was shown in the preceding section for [**152**], the thione form is also favoured in the analogous 2-mercaptothiophene-3-aldimines (67IZ2783, 70IZ675) and 2-mercaptobenzo[*b*]thiophene-3-aldimines (74PC970).

The complex NMR and IR (characteristic νSH at 2500 cm^{-1} and νC=S at 1200–1300 cm^{-1}) show that 2-mercaptoselenophene exists in solution as the tautomeric mixture [**156a**] ⇌ [**156b**]. However, 3-mercaptoselenophene [**157**] is a true thiol (71BF3547).

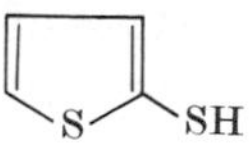

[**153**] 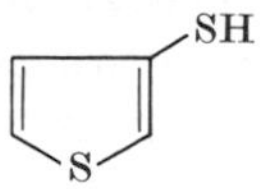 [**154**] 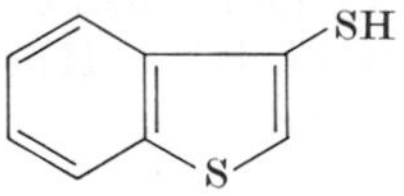 [**155**] 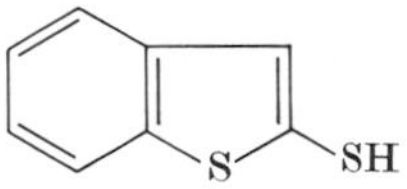 [**155/1**]

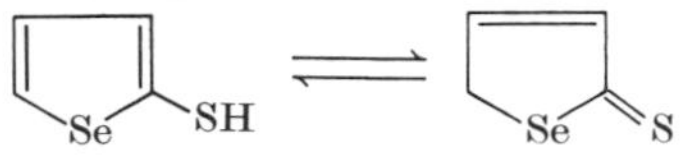

[**156a**] ⇌ [**156b**] 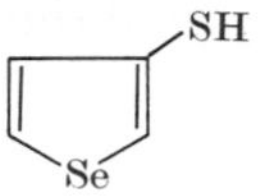 [**157**]

Dipole moment measurements in benzene indicate thione and selenone structures for benzo[*c*] derivatives of type [**158**, Z = O, S, Se] (70JA4447). Here the gain of benzene ring conjugation offsets the small loss of the overall delocalization energy.

Little is known regarding more complex potential mercaptothiophenes. "2-Hydroxy-5-mercaptothiophene" exists, as expected, in the dithioanhydride structure [**159**] (63T1867).

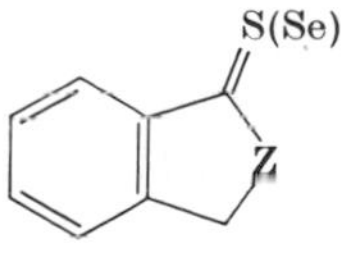

[**158**] 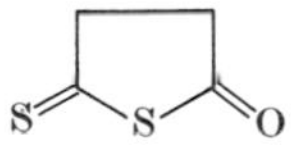 [**159**]

C. Mercapto-pyrroles and -indoles (II-20)

Several 2-pyrrolethiols have been obtained recently. Although the simplest derivatives are too unstable for isolation and have not been investigated by physical methods, their chemistry suggests that they exist in the thiol structure. This is supported by IR(νSH) and NMR spectra of [**161**] and [**162**] which are clearly thiols (72AJ985); in confirmation, spectral data also show the thiol structure [**163**] (73LA207).

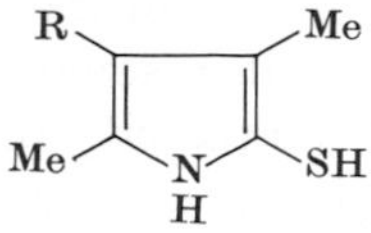

[**161**] R = CO_2Et
[**162**] R = COMe

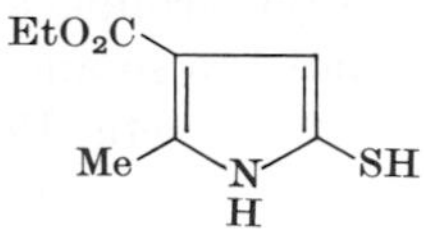

[**163**]

NMR spectra and UV comparisons show that methyl-substituted indoline-2-thiones [**164**, R^1 = H, Me; R^3 = H, Me, $PhCH_2$, *t*-Bu] exist predominantly in the thione form [**164a**] in solution (69CT550). However, a 3-aryl substituent considerably stabilizes the thiol form [**164b**] and *N*-methylation the thione form [**164a**] (71CH836). The 3-phenyl compound [**164**, R^3 = aryl, R^1 = H] exists as a tautomeric mixture of [**164a**] and [**164b**] in EtOH, but the thiol form [**164b**] is favoured in MeOH.

The tautomeric interconversion of [**164a**,**b**] is slow in nonpolar solvents; at equilibrium the thiol form [**164b**] is favoured for 3-aryl-1-

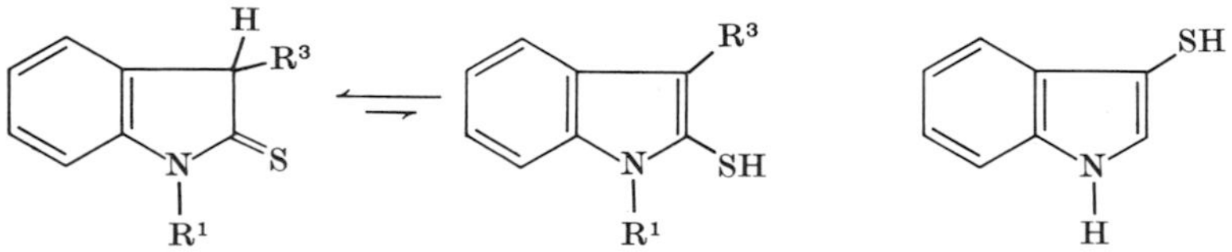

[**164a**] [**164b**] [**164/1**]

unsubstituted compounds, but the thione form for all 1-methyl derivatives. A *p*-methoxy substituent has little effect on tautomerism (69CT550, 71CH836, 74CT1053). The large effect of *N*-methylation is probably due to breakdown of *H*-bonding rather than simple electronic effects. For a review of indoline-2-thiones see (72MI603). 3-Indolylthiol [**164/1**] shows the NMR signals for an α-CH of an indole and here exists in the SH form (69TL4465; cf. also 69IM41).

D. Compounds with Potential Mercapto Groups: Conclusions

Table 3-6 summarizes the results for the mercapto compounds discussed in the preceding sections. Comparison with Table 3-3 for the oxo–hydroxy tautomerism reveals that the mercapto form is very much more strongly favoured than the hydroxy form in the corresponding oxygen compounds. This can be rationalized by an extension of the qualitative explanation given in Section 3-2H. A major difference is that enethiols are much less prone to tautomerism into thiocarbonyl compounds than vinyl alcohols are to ketonize. This factor is probably more important than the smaller differences in resonance stabilization of thioamides *vs*. amides, etc. Because of the paucity of data it is not yet possible to discuss the effects of the heteroatom on substitution. However, the effect of benzo annelation can be clearly discerned, and its influence parallels that in the oxygen series.

TABLE 3-6

THIOL-THIONE TAUTOMERISM IN FIVE-MEMBERED RINGS WITH ONE HETEROATOM

Type			α-Mercapto series	β-Mercapto series
Parent	O	Mercaptofuran	SH (3-3A)[a]	SH(?) (3-3A)
	S	Mercaptothiophene	SH (3-3B)	SH (3-3B)
	NR	Mercaptopyrrole	SH (3-3C)	?
Substituted parent	O	Substituted mercaptofurans	SH, thione (3-3A)	?
	S	Substituted mercaptothiophenes	SH (3-3B)	?
	NR	Substituted mercaptopyrroles	SH (3-3C)	?
Benzo [b]	O	Mercaptobenzo[b]furans	?	?
	S	Mercaptobenzo[b]thiophenes	SH (3-3B)	SH (3-3B)
	NR	Mercaptoindoles	SH, thione (3-3C)	SH (3-3C)
Benzo [c]	O	Mercaptobenzo[c]furans	thione (3-3A)	—
	S	Mercaptobenzo[c]thiophenes	thione (3-3B)	—
	NR	Mercaptoisoindoles	?	—

[a] Section number is given in parentheses.

4. Compounds with Potential Amino Groups (II-20)

Our knowledge of amino-thiophenes and -pyrroles has improved very considerably, while that of aminofurans remains poor. It is clear that most amino derivatives do prefer the NH_2 form, at least for the monocyclic ring systems, although in benzo derivatives the imino forms become more stable.

A. Aminofurans (II-21)

Aminofurans remain a very underinvestigated class. Thus, no physical study is yet available on the unsubstituted 2- and 3-aminofurans. 2-Amino-3-cyanofurans [**165**, R, R′ = Me or Ph] on IR (νNH 3350–3500 cm^{-1}, δNH_2 1645–1665 cm^{-1}) and UV evidence exist in the amino form in the solid state and in MeOH (66CB1002).

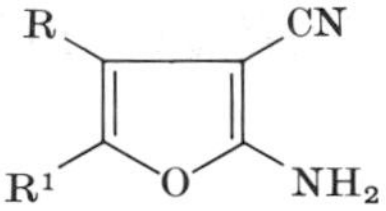

R CN R[1] O NH₂

[**165**]

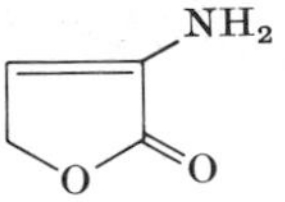

NH₂ O O

[**165/1**]

The chemical behaviour of 3-aminofurans suggests that they exist in the imino form (66HC(7)470) whereas theoretical calculation (70JA2929) predict 2- and 3-aminofurans might exist as tautomeric mixtures. 2-Oxo-3-amino derivatives possess the Δ^3 structure [**165/1**] as shown by IR and NMR (71CB2458). The 2-aminobenzofuran structure [**166**] is supported by spectral data (71T5873).

The presence of νNH_2 at 3295–3450 cm^{-1} indicates that the 2-acyl 3-aminobenzo[*b*]furan derivatives [**166/1, 2, 3**] exist in the amino form

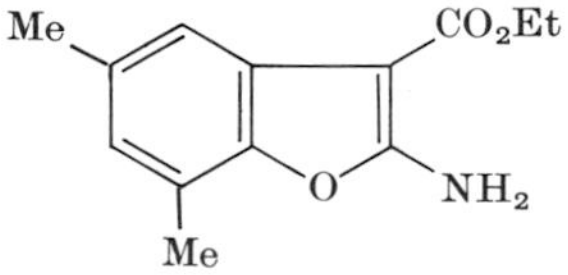

Me CO_2Et O NH_2 Me

[**166**]

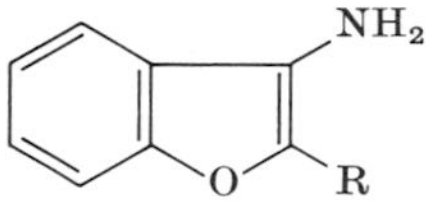

NH_2 O R

[**166/1**] R = COMe
[**166/2**] R = COPh
[**166/3**] R = CO_2Et

in the solid state. UV in EtOH and NMR spectra in $CDCl_3$ are in agreement with that formulation also in solution (73PC(315)779). There is little to add to the previous conclusions (II-21) regarding acetamidofurans and aminobenzofurans.

B. Aminothiophenes and Aminobenzothiophenes (II-22)

a. Aminothiophenes (II-22)

The amino structure of 2-aminothiophene [**167**] reported in the previous review (II-22) is reaffirmed by more NMR in $CDCl_3$, an amino (3.78 δ) to ring proton (6.07–6.66 δ) ratio of 2:3, and IR νNH_2 bands at 3360 and 3420 cm^{-1} (67TL5201, 69JH147). NMR, IR and UV confirm the amino structures [**168**] and [**169**] (71T5873).

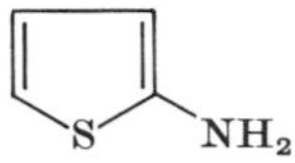

[**167**]

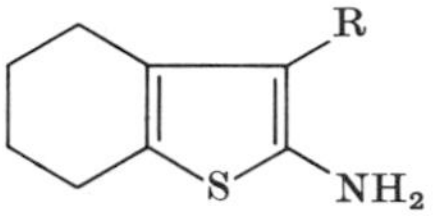

[**168**] R = CN
[**169**] R = CO_2Et

3-Aminothiophene was shown previously (II-22) to exist as such and further IR and NMR spectra confirm its structure (73JH1067) and β-aminothiophene structures have since been proved by IR for some ethoxycarbonyl derivatives (72T875). The imino structure [**170**] (R = Ph or Me) was assigned from the infrared (imine NH at 3480 cm^{-1} in $CHCl_3$) and NMR spectra (71CC1372). The trifluoroacetyl derivative [**171**] (71CC1372) as well as some 5-acyl analogues of [**170**] (70CC2709) exist in the di-amido forms by NMR comparisons.

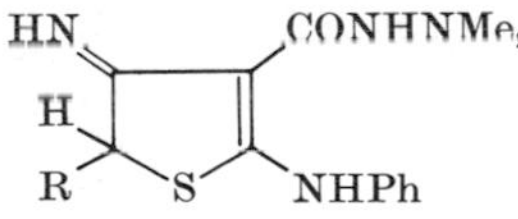

[**170**]

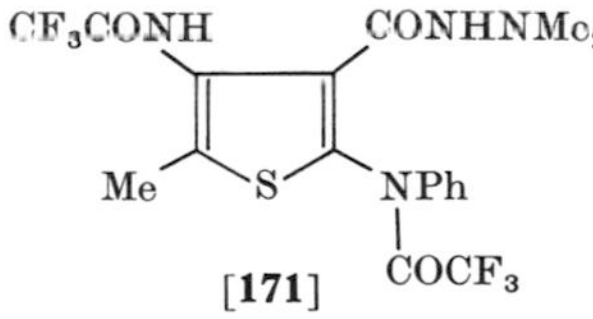

[**171**]

b. Aminobenzo[b]thiophenes

The amino structure [**172**] is established for 2-aminobenzothiophene by νNH bands at 3300–3400 cm^{-1} in the infrared spectrum in CCl_4, together with deuteration experiments (cf. Section 1-6Cc, 65JO4074). A broad NMR peak at 3.83 δ for aromatic NH_2 and the ultraviolet similarity in EtOH (maximum at 281 nm) with benzothiophene itself (maximum at 288 nm) support the amino structure [**172**]. 2-Acetamidobenzothiophene [**173**] also exists as such (65JO4074).

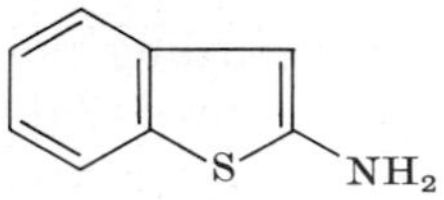

[**172**]

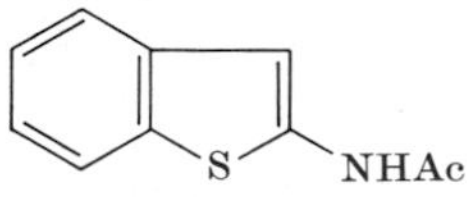

[**173**]

NMR spectra of 3-phenylaminobenzothiophene [**174**] show the 2-position CH peak in $CDCl_3$ expected for the amino form [**174a**], but the imino form [**174a**] was considered to predominate in trifluoroacetic acid on the basis of a CH_2 peak (65BF2658); however, the possibility of a *cation* of structure [**175**] being formed in CF_3CO_2H was not apparently considered, and this provides a more plausible explanation of the spectrum.

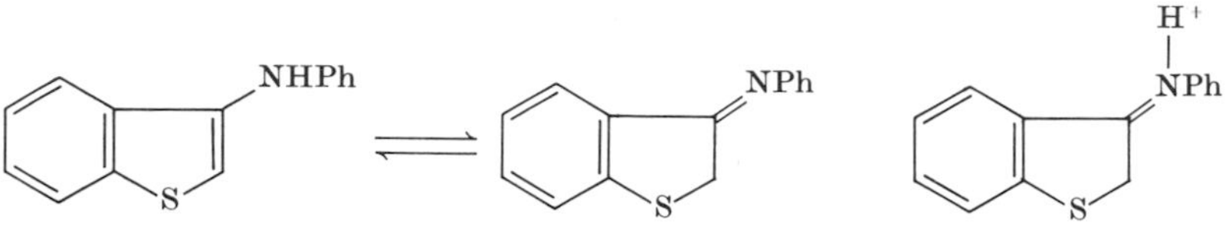

[**174a**] [**174b**] [**175**]

c. Aminobenzo[c]thiophenes

The isobenzothiophene derivative [**176**] exists in the imino structure shown (64JO607). Although early reports (90CB2478, 98CB2646) and also a later one (64BB491) suggested that compound [**177a**, R = H] may exist as tautomeric mixture [**177a**] ⇌ [**177b**], other authors (70JO3495) could not detect the open-chain tautomers [**177a**, R = H or Ph] by NMR or infrared (cf. 64BB491), nor did they find any evidence for the existence of the amino form [**177c**, R = H or Ph].

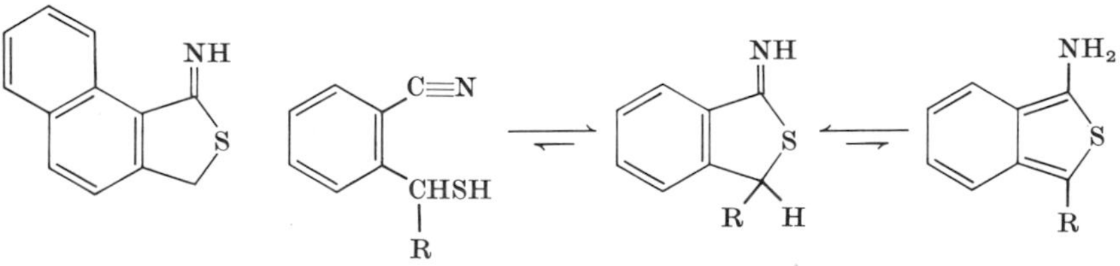

[**176**] [**177a**] [**177b**] [**177c**]

C. Amino-pyrroles (II-22) and -indoles (II-23)

a. Aminopyrroles (II-22)

Little more is available to add to the conclusions of the previous review that these compounds exist in the amino forms. Theoretical calculations predict the predominance of these forms in simple aminopyrroles (70JA2929, cf. II-22).

However, some work is now available regarding potential diaminopyrroles. Recently (73JO173) the imino structure [**178**] was proposed to exist, on unquoted spectral data: possibly phenyl conjugation stabilizes the imine.

NMR indicates that succinimidines [**179**, R = H, OH, Ph] exist in the diimino form shown (71BF671). Analogous structures [**180**] and [**181**] (R, R^1 = Ph, CH_2Ph) are suggested by IR (64BF123), as previously shown by UV methods for the parent compound (54J442).

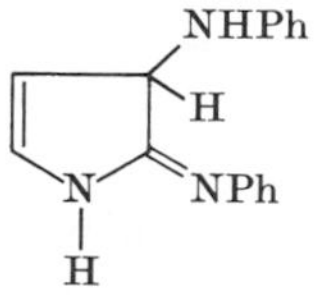

[**178**] 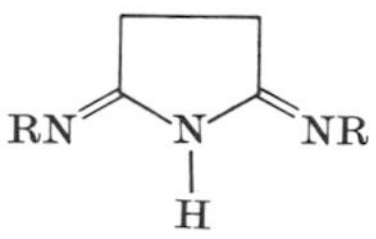 [**179**] 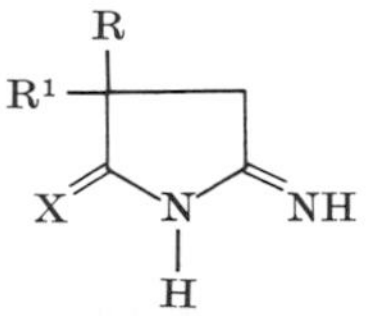 [**180**] X = NH
[**181**] X = O

b. α-Aminoindoles (II-23)

The tautomeric composition of 2-aminoindoles [cf. **182**, **183**] is sensitive to substitution and solvent (see II-23); the general pattern of these effects is now discernible.

NMR and ultraviolet spectral comparisons with fixed forms indicate that tautomeric mixtures [**182a**] ⇌ [**182b**] exist in $CDCl_3$ and EtOH. For R = *n*-Pr, *i*-Pr and *n*-Bu the 2-aminoindole [**182a**] predominates, but for R = $-CH_2-CH_2-OH$ the indoline form [**182b**] is preferred (71T775).

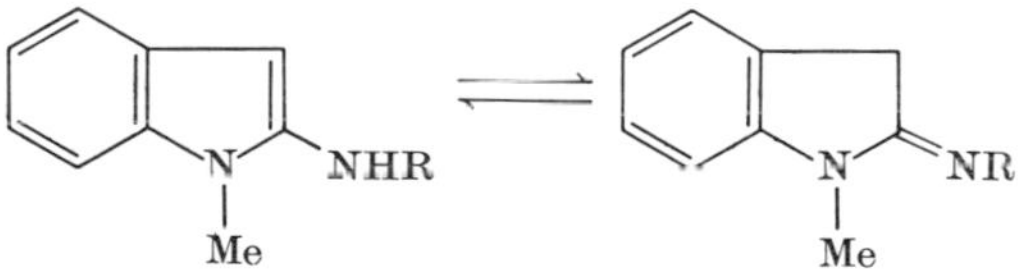

[**182a**] [**182b**]

2-Picramidoindoles [**183**, R^1 = H or Me] exist in the aminoindole form [**183a**] in DMSO as shown by the absence of a CH_2 NMR signal, but as the iminoindoline form [**183b**] in the solid state from X-ray crystallography (70J(C)956).

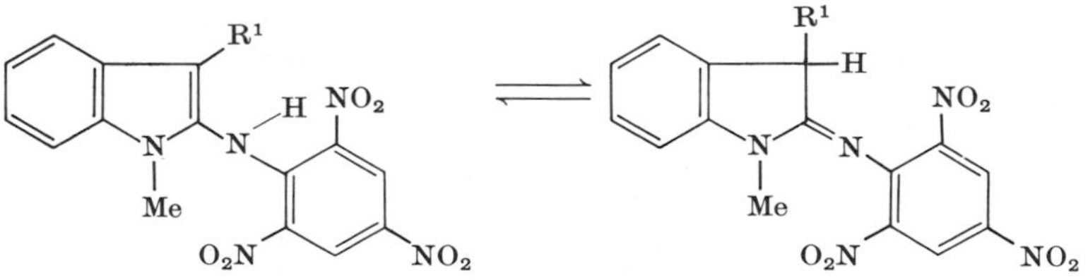

[**183a**] [**183b**]

2-*p*-Toluenesulphonamidoindole [**184**, R = H] exists at 35° as a mixture of [**184b**, R = H] and [**184a**, R = H] in ratio 1.08 in DMSO by

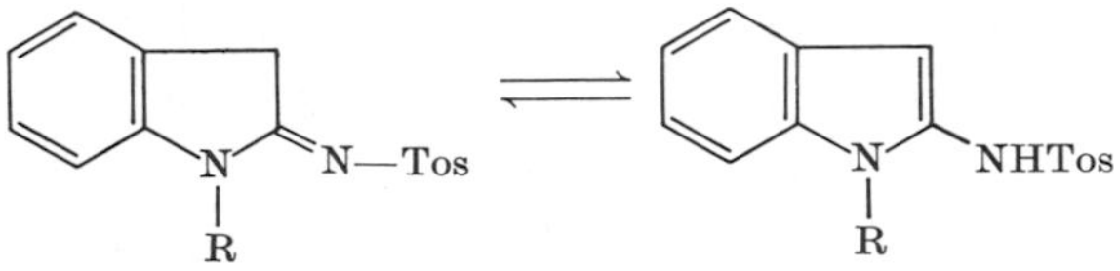

[184a] [184b]

comparing the NMR signal at δ 5.77 for 3-CH in [**184b**] with that at 4.13 for 3-CH_2 in [**184b**] (72J(PI)2411). The proportion of the tautomers varies a little with temperature and the imino form [**184a**] becomes favoured at higher temperatures. K = [**184b**, R = H]/[**184a**, R = H] 1.2 (20°), 1.09 (40°), 0.96 (60°), and 0.91 (80°) (73J(PI)1602). The same authors demonstrated (73J(PI)1602) that apparent discrepancies in the NMR spectra of [**184**, R = H] obtained by other authors (73JO11) were due to traces of solvent in their sample. The corresponding imino-indoline forms are more favoured in the N–Me derivatives [**184**, R = Me]; 85% of [**184a**, R = Me] and 90% of the *p*-chlorobenzenesulphonyl analogue exist in DMSO, while in $CHCl_3$ no amino form [**184b**] was found by NMR (72J(PI)2411). Curiously, a 3-methyl group also appears to favour the imino form: [**185**] is reported to exist in $CHCl_3$ as [**185a**] although in the amino form [**185b**] in DMSO (70TL2979).

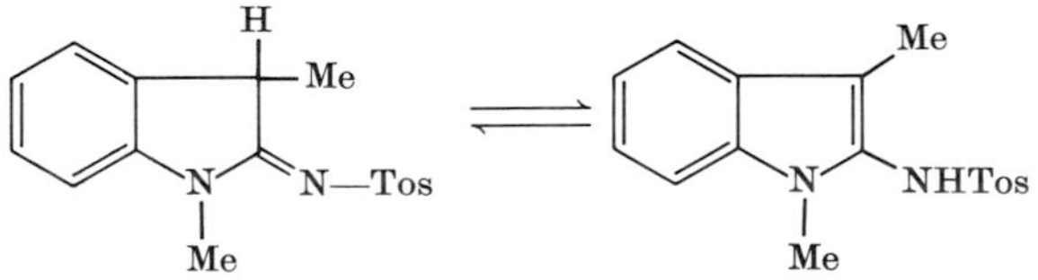

[185a] [185b]

Recently the effect of substituents in the arylsulphonyl group (Me, OMe, NHAc, halogens, NO_2), on the tautomeric equilibrium [**186a**] $\rightleftharpoons$ [**186b**] has been studied. Electron-withdrawing substituents favour the imino form [**186a**]; the amino-tautomer content is stated to range from 20% for [**186**, R = *p*-OMe] to 3% for [**186**, R = *o*-NO_2] (73JO11).

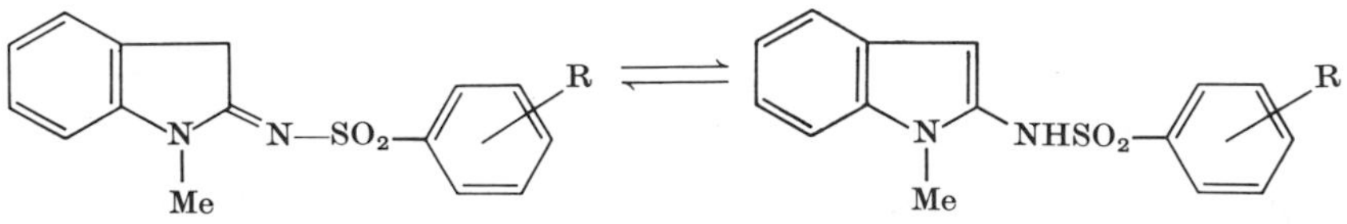

[186a] [186b]

c. β-Aminoindoles

Much less work is available on 3-aminoindoles, but there seems to be little doubt that they are true amino compounds (cf. 69BF2004). The NMR of [**187**] in $CDCl_3$ and DMSO, and X-ray crystallography in the solid state all indicate the 3-aminoindole structure shown (72J(PI)2411). Recent independent work on derivatives of [**187**] substituted in the arylsulphonyl group has confirmed and extended these conclusions (73JO11).

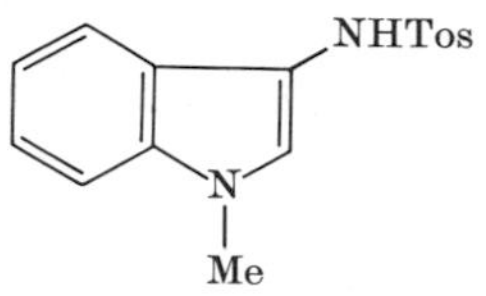

NHTos

Me

[**187**]

d. Potential Amino Derivatives of Isoindole (II-24)

Earlier ultraviolet spectral comparisons indicated that although structure [**187/1a**] predominates over [**187/1b**] when R = H or OH, the amino form [**187/1b**] is predominant when R = aryl (II-24). However, recent infrared studies appear to show that even when R = aryl form [**187/1a**] predominates (68THO1). This apparent discrepancy could be due to the use of different solvents.

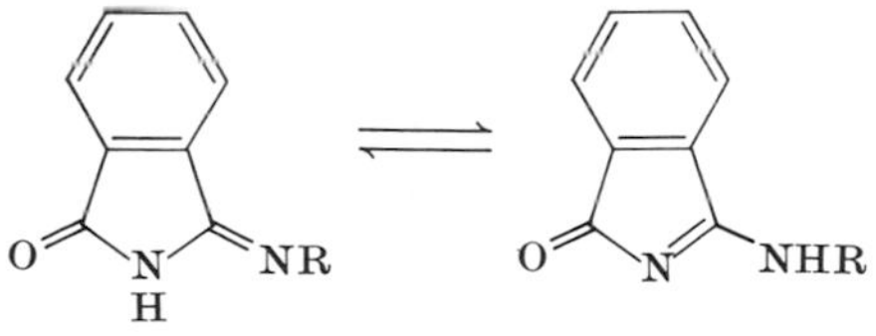

[**187/1a**] [**187/1b**]

The AA′BB′ pattern for the benzene ring protons in phthalimidines [**188**, R = H, OH, alkyl, aryl] suggests the diimino form [**188a**] for MeOH and DMSO (71BF671). However, previous ultraviolet evidence had indicated a tautomeric equilibrium [**188a**]/[**188b**] depending on the

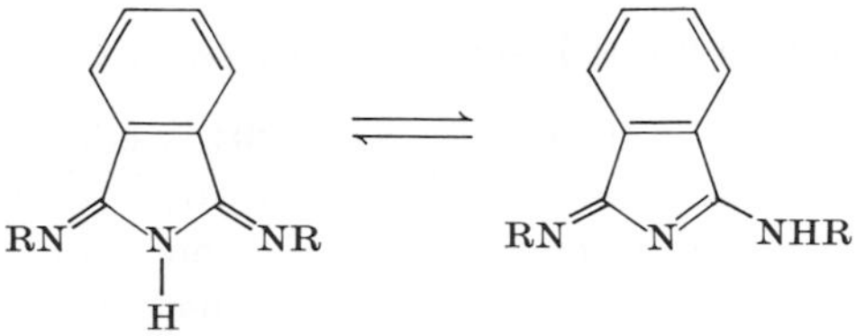

[**188a**] [**188b**]

substituent (II-24) and in the NMR study apparently no allowance was made for possible rapid exchange, so this matter requires further investigation.

e. Aminoindolizines

νNH_2 at 3320 and 3400 cm^{-1} in $CHCl_3$ for [**189**, R = H] and νNH at 3240 cm^{-1} and νC=O at 1665 cm^{-1} in the solid state for [**189**, R = COMe] show that these 1-amino-2-phenylindolizines exist in the amino forms written (65J2948).

NHR
N
Ph

[**189**]

D. Compounds with Amino Groups: Conclusions

Monoamino compounds generally do exist in the amino form. Although data are scanty for many classes of compound, the only real exceptions to this generalization are benzo[*c*]-annelated compounds and those in which the NH_2 group carries an electron-withdrawing substituent, as in NHTs. The equilibrium for 2-aminoindole is also finely balanced. The tendency for imino forms to occur for potential diamino derivatives is much greater.

Unfortunately it is not possible on the data available to discuss the tautomeric influence of the heteroatom, orientation of the amino group, and substituents as has been done for the corresponding hydroxy compounds (Section 3-2H), but many of the generalizations made there should apply bearing in mind that there is a large swing towards the amino form which is also exemplified by the smaller tendency of enamines to tautomerize to imines as compared to the strong tendency for enols to revert to ketones.

5. Other Substituted Furans, Thiophenes and Pyrroles (II-24)

A. Compounds with Potential Methyl Groups (II-24)

While 2-methylene-2,5-dihydrofuran [**190**] is converted by acid into 2-methylfuran (II-24), the sterically crowded derivative [**191a**] is recovered as such from a strong acid solution (67TL2951). Unquoted NMR data are stated to agree. Steric repulsions probably cause the furan tautomer [**191b**] to be less stable than [**191a**].

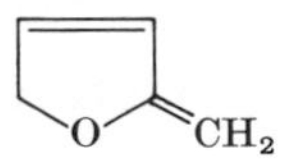

CH_2 t-Bu Me t-Bu H O CH_2 ⇌ t-Bu Me t-Bu O Me

[**190**] 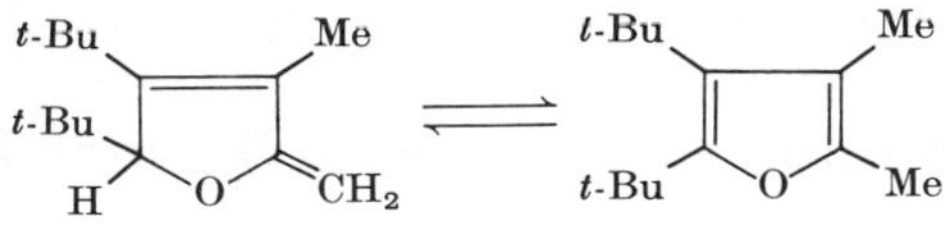 [**191a**] [**191b**]

B. Alkenylthiophenes

The triethylamine-catalyzed tautomerization of [**192a**, R = Me] with [**192b**, R = Me] was studied in pyridine solution by NMR. At equilibrium [**192a**, R = Me] is favoured (76%) indicating easier conjugation of

S R ⇌ S R

[**192a**] [**192b**]

a substituent at the 2- than at the 3-position of thiophene (66AS1733). This work has been extended to the corresponding nitro compound [**192**, R = NO_2]: the tautomeric equilibrium is far more rapid here, but the position is not very different; 70% of [**192a**, R = NO_2] is present (72AS556).

C. Dipyrrylketones

Comparison of the UV spectrum of dipyrrylketone [**193**] with its di-*N*-methylated derivative show the ketone form [**193a**] in ethanol. The NMR in DMSO (two NH protons at low field) agrees with this assessment (65AJ1977).

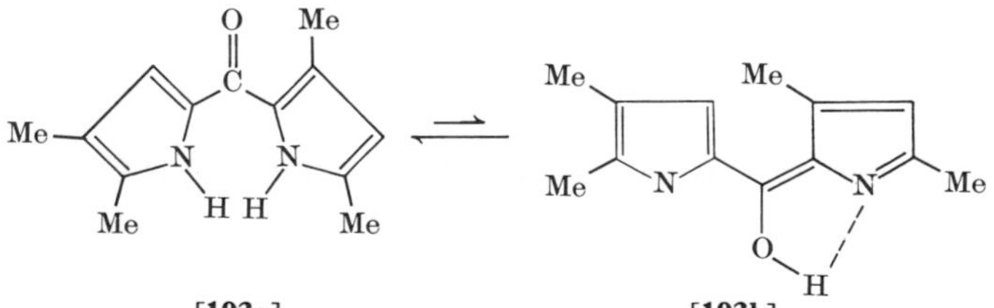

O Me C Me N N H H Me Me ⇌ Me Me Me N N Me O H

[**193a**] [**193b**]

CHAPTER 4

Five-Membered Rings with Two or More Heteroatoms

We have attempted in this chapter to follow as closely as possible the classification of the previous review (II-27). Our consideration is hence based on a distinction between tautomerism involving only annular nitrogen (or carbon) atoms, which is considered first, and that involving substituent groups. We consider functional group tautomerism first by sequence of functional group, O–, S–, N–, and C– linked, and within each group the ring systems are classified successively by increasing numbers of heteroatoms, by increasing separation of ring heteroatoms, and then by the nature of the heteroatoms in the sequence O, S, N.

However, some modifications have been needed to the scheme previously adopted. Much recent work on the more complex compounds has led us to transfer to a separate Chapter 5 [5,5]- and [5,6]-bicyclic compounds with heteroatoms in both rings, such as the purines and their analogues. For the same reason, five-membered ring compounds with two potential tautomeric substituents (OH, SH or NH_2) have been placed in a separate section (4-6) of the present chapter, and alkyl derivatives of difunctional compounds are considered with the difunctional compounds; e.g., alkoxyisoxazolones should be considered under potential dihydroxyisoxazoles. Heterocycles with two ring sulphur atoms have also been given their own section (4-9A), and new sections have been added which deal with certain rearrangements (4-9B) and ring–chain tautomerism (4-9C). On the other hand, substituted amino groups are now dealt with together with unsubstituted amino groups (Section 4-5).

1. Tautomerism Involving Annular Nitrogen Atoms Only (II-28)

A. INTRODUCTION (II-28)

a. The Nature of Annular Tautomerism

Theoretical interest in, and experimental knowledge of, annular tautomerism has increased enormously since the last review. This applies, however, only to annular tautomerism between nitrogen atoms, where both tautomers are "aromatic." Just as the transfer for pyrroles of the NH proton to a ring carbon atom to yield a pyrrolenine [**1a**] →

[**1b**] is highly unfavoured (Section 3-1Ba), so the participation of pyrazolenine form, e.g., [**2b**] in the pyrazole [**2**] tautomerism is very small indeed (67CB3097, 69JA706). A rare exception occurs in radical anions: e.g., that of pyrazole has structure [**2/1**] (73JA27).

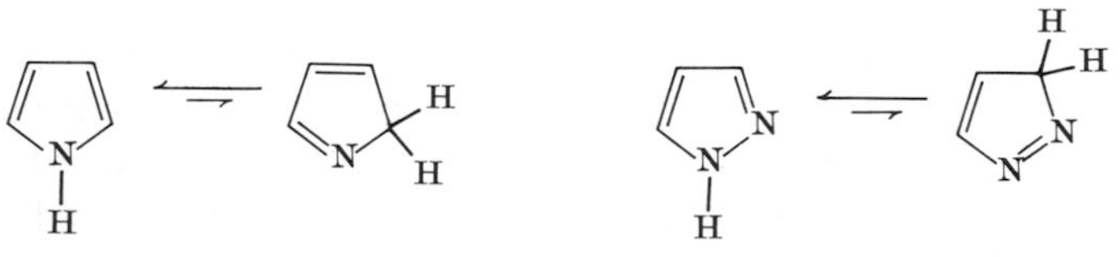

[**1a**] [**1b**] [**2a**] [**2b**]

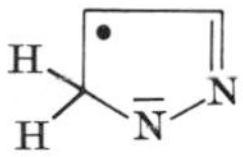

[**2/1**]

In past years, the existence of annular tautomerism has been denied, or considered to require some special explanation distinct from common prototropy. The chief reason for this is the very fast rate of proton transfer in cases of annular tautomerism relative to many other types of prototropy. This is due to several different factors: firstly, hetero-atom-hydrogen bonds are easier to break and form than the carbon-hydrogen bonds which are involved in some of the other types of tautomerism; secondly, the partial or total charges resulting from proton gain or loss are delocalized in an aromatic ring; and thirdly, there are intermolecular associations [**3**] which facilitate proton transfer in these systems.

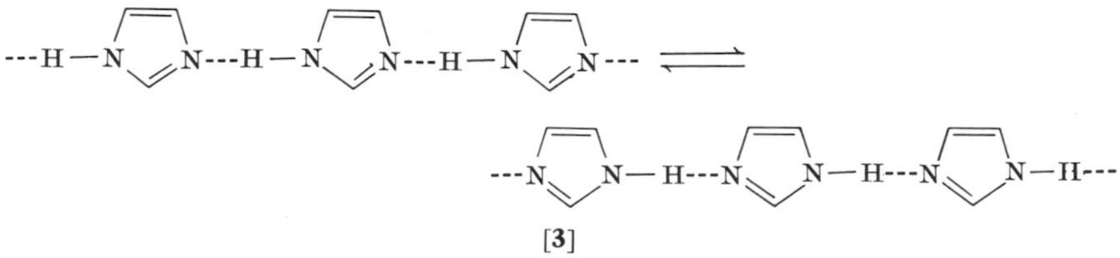

[**3**]

The high rate of proton transfer in azoles gives them pseudo C_{2v} symmetry, and this has led successive authors to introduce new ideas or terminology to explain this fact, some false, such as "mesohydric tautomerism" (I-316), a π-complex between the imine hydrogen and the π-electrons of the heterocyclic ring (II-34), the complete transfer of an imine proton to a second molecule (II-30), and others nearer the truth: the concept of "protomerism" introduced by Zimmerman (64AG1), and "a physical mixture" ("miscela fisica," 56G797). Such hypotheses will be examined several times later in this chapter, but for clarity we can anticipate the final conclusion: annular tautomerism is a true example of classical prototropy with a particularly low activation energy. This was the opinion Auwers gave forty years ago (34LA(508)51), and the firm conclusion reached in the previous review.

Nevertheless, we need to consider in detail whether or not the well defined asymmetry of azoles in the gaseous state is altered in solution or in the crystal by hydrogen bonding of type [**3**], as this can lower the activation energy for prototropy and affect many of the methods by which it can be investigated.

b. Annular Tautomerism and Autotropic Rearrangements

In contrast to other types of tautomerism, annular tautomerism often involves two identical entities as in pyrazole [**4a**] ⇌ [**4b**], imidazole and their symmetrically substituted derivatives. This is included by Dewar (69MI336) in the definition of "autotropic rearrangements."

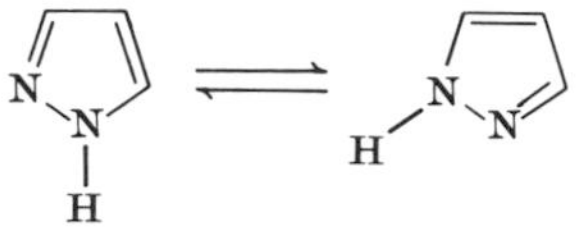

[**4a**] [**4b**]

This fact has two important consequences. In the first place, an analysis of the point group to which such a molecule belongs, usually by IR (Section 1-6Cc) but in certain cases by microwaves (Section 1-6B), allows the rejection of theories other than prototropy for annular tautomerism by showing that a molecule has C_s symmetry, and therefore that the NH–hydrogen is localized. In the case of 1,2,3- and 1,2,4-triazoles it is possible to differentiate between the two tautomers by simple symmetry considerations [**5, 6**].

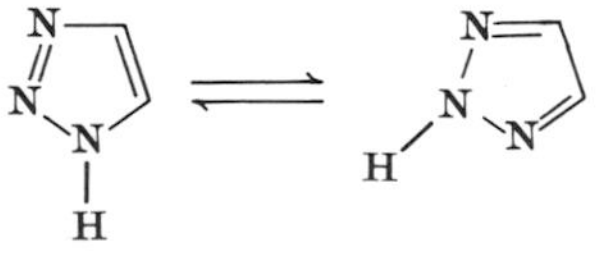

C_s [**5a**] C_{2v} [**5b**] 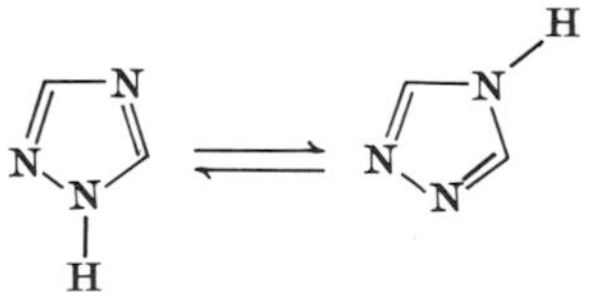 C_s [**6a**] C_{2v} [**6b**]

The second consequence is that since the equilibrium constant for tautomerism of such two identical entities is by inspection equal to 1, such compounds can be used to test those physicochemical methods which necessitate the use of "fixed" derivatives. Such methods assume that the NH compound and the N–CH_3 compound have properties which differ only slightly; see detailed discussion in Section 1-4C.

c. Annular Tautomerism in Cations

All the tautomers of a given azole generally share a common anion. However, this is not the case for cations, and annular tautomerism involving two or more nonidentical monocations always occurs if the azole possesses more than two nitrogen atoms as in triazoles, benzotriazoles and tetrazoles (see Scheme 4-1). Pyrazoles and imidazoles, whether unsymmetrically substituted or not, form but a single cation.

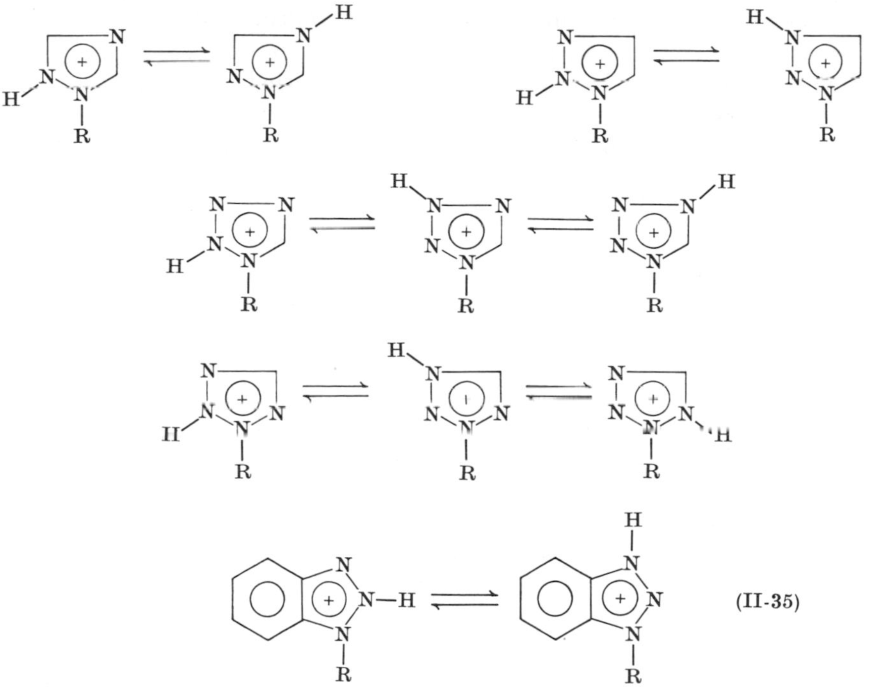

(II-35)

SCHEME 4-1

B. PYRAZOLES (II-31)

a. Symmetry of Pyrazole and Symmetrically Substituted Pyrazoles

The two nitrogen atoms of the pyrazole molecule possess different environments, i.e., pyrazole has symmetry C_s and not C_{2v}. This has been amply proved for the gaseous state, the solid state, and solutions by the following methods: IR spectroscopy (67J(B)1363, 69TH01,

70PH2133, 74MI677), ^{14}N quadrupolar resonance (68MP73), microwaves (67JA1312, 72PS3, 74JL(22)401), X-ray diffraction (60CX946, 67PH2375, 70CX(B)1880, 74AS1845) and neutron diffraction (70AS3248).

Comparison of experimental dipole moments measured in benzene with those calculated for a symmetric structure and for a structure in which the proton is localized on one nitrogen, shows that a series of 3,5-dimethylpyrazoles, carrying various substituents in the 4-position, possess asymmetric structures (67KG130).

Most theoretical calculations on pyrazole have been done using the C_s structure: it has been demonstrated that such an assumption gives better results for electronic spectra than an approximately C_{2v} structure of type [**3**] (68J(B)725). However, the bond lengths originally reported for pyrazole in the first X-ray analysis by Ehrlich (60CX946) are now known to be incorrect. This structure included proton positions which were not determined experimentally. Many authors (69BF1097, 69JA796, 70CB3289) used for their calculations, or to compare with their calculations, a geometry similar to that of Ehrlich but with the NH attached to the alternative nitrogen, according to a suggestion first made by Mighell and Reimann (67PH2375; see also 70AS3248). However, see (69BB407) for a calculation using the original (60CX946) geometry.

Table 4-1 gives the measured and calculated bond lengths of pyrazole; the values obtained from X-ray spectroscopy and from neutron diffraction are averages of those obtained for two crystallographically independent molecules.

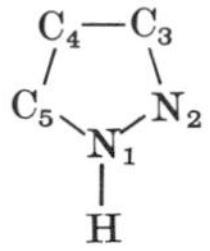

C_4—C_3, C_5, N_2, N_1, H

[**7**]

As previously indicated, time-averaged behaviour is usually observed in the nuclear magnetic resonance spectra of pyrazoles (Section 1-5Ac). See also (73MI130) for a study of ^{15}N mono-labelled pyrazole. The (^{1}H–^{15}N) coupling constants observed correspond to a mixture of the two tautomers. Two identical substituents in positions 3 and 5 thus give rise to one signal only (or to one set of signals). Variable temperature experiments provide a way of proving the asymmetry of the molecule by slowing down sufficiently the rate of proton exchange (69TL495).

TABLE 4-1

BOND LENGTHS OF PYRAZOLE [7] (Å)

Method	Reference	N(1)–N(2)	N(2)–C(3)	C(3)–C(4)	C(4)–C(5)	C(5)–N(1)
X-ray	60CX946	1.36_1	1.34_6	1.33_5	1.41_4	1.31_4
	70CX(B)1880	1.35	1.33_5	1.37_5	1.36	1.33
	73AS1845[a]	1.352	1.328	1.389	1.371	1.337
Neutron diffraction	70AS3248[b]	1.365	1.350	1.407	1.398	1.356
Microwaves	72PS3, 74JL(22)401	1.349	1.331	1.416	1.373	1.359
LCAO[c]	69BF1097	1.37	1.30	1.42	1.37	1.37_5
SCF–MO[d]	69JA796	1.356	1.309	1.423	1.369	1.367
CNDO[e]	73TC145	1.318	1.339	1.407	1.370	1.372

[a] At 108K.

[b] Corrected for rigid body motion.

[c] LCAO, linear combination of atomic orbitals.

[d] SCF–MO, self-consistent field-molecular orbitals.

[e] CNDO, completely neglecting differential-overlap. The equilibrium geometries (bond distances and bond angles) are also calculated for imidazole (Section 4-1Ca) and *s*-triazole (Section 4-1Ea).

b. Unsymmetrically Substituted Pyrazoles (II-31)

For pyrazoles carrying different substituents in positions 3 and 5, the K_T for the equilibrium [**8a**] ⇌ [**8b**] is not in general equal to unity.

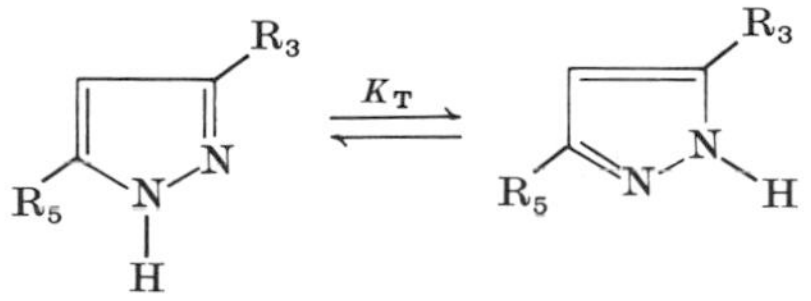

[**8a**] [**8b**]

In the previous review (II-31) it was reported that this problem had received little attention: Auwers's studies on molecular refractivity (Section 1-4H) were the only significant results. Even now, little systematic work is available but many individual results are scattered in the literature. These will be reviewed according to the method employed, with an attempt to draw a general conclusion at the end.

i. *Basicity Measurements* (*Section 1-4A*). The use of Eq. (1-16), $K_T = K_{\mathrm{MeA}}/K_{\mathrm{MeB}}$, enabled determination of $K_T = 30$ (water) (65T-1693) for equilibrium [**9**].

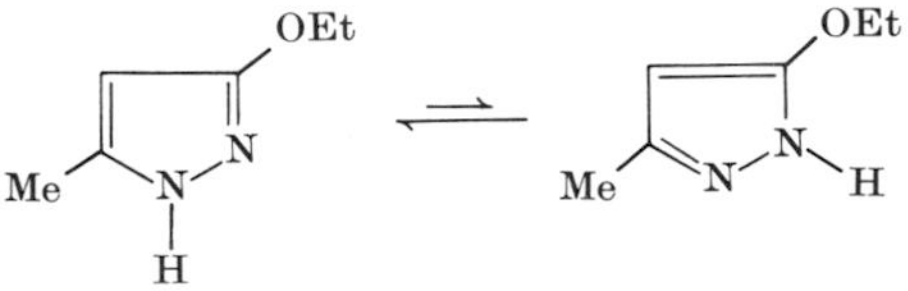

[**9a**] [**9b**]

$K_{\mathrm{MeA}} = 8.9.10^{-3}$ $K_{\mathrm{MeB}} = 3.1.10^{-4}$

From the pK_a values of a number of NH and NMe pyrazoles (68BF5009) it is possible to calculate using Eqs. (1-16) and (1-19), the values of K_T shown in Table 4-2. Values of log f, calculated by Eq. (1-19,) are also given in Table 4-2. The factor f, defined in Section 1-4A, provides a measure of the amount by which the basicity of a particular form of a tautomeric compound is altered by methylation in the fixed model. This methylation effect is relatively constant, and the average magnitude of log f is -0.66 pK_a units.

ii. *The Hammett Equation* (*Section 1-4B*). No treatment has been attempted in the literature. On the basis of values of K_T derived from various methods, Section 4-1Bc discusses a possible correlation founded on the Hammett equation. We note here that any attempt at a correlation of the form p$K_a = \rho\sigma_m$ for the pyrazole NH (68BF5009) is inconsistent with Charton's hypothesis (Section 1-4B) because it is not possible for log $K_{\mathrm{NH}} = \rho'\sigma_m$ to be true simultaneously with log $K_T = \rho\sigma_m$.

TABLE 4-2

LOG f VALUE FOR SUBSTITUTED PYRAZOLES

Pyrazole	Reference	K_T	$\log f = pK_{NMe}\text{-}pK_{NH}\text{-}\log S$
Unsubstituted	68BF5009	1	−0.73
3(5)-Methyl	68BF5009	1.17 (3-methyl)	−0.77
3,5-Dimethyl	68BF5009	1	−0.62
5(3)-Bromo-3(5)-methyl	68BF5009	23 to 31 (3-bromo)	−0.61 to −0.75
4,5(3)-Dibromo-3(5)-methyl	68BF5009	8.3 (3-bromo)	−0.61
5(3)-Ethoxy-3(5)-methyl	65T1693 67BF3772	30 (3-ethoxy [**9a**])	−0.58

iii. *Dipole Moments* (*Section 1-4G*). As already discussed in detail (Section 1-4G) the two model *N*-methyl derivatives of 3(5)-phenylpyrazole have dipole moments too close to allow a calculation of K_T by interpolation (65KG107). In the case of the 3(5),4-dibromopyrazole [**10**], comparison of dipole moments calculated for each tautomer using a vector addition method, with the experimental moment (in benzene) led the authors (67KG130) to assert that the tautomeric equilibrium is completely displaced towards the 3,4-dibromo form [**10a**].

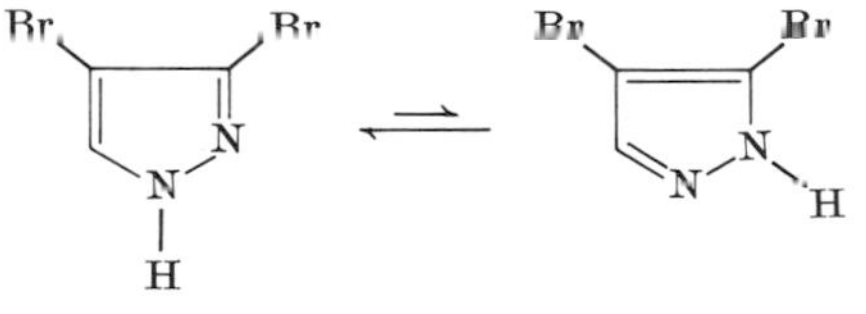

[**10a**] [**10b**]

iv. *Molar Refractivity* (*Section 1-4Hb*). Auwers's results in the pyrazole series (34LA(508)51) have already been discussed in detail (Section 1-4Hb). The only significant conclusion regarding the tautomeric equilibrium concerns 5(3)-ethoxycarbonyl-3(5)-phenylpyrazole [**11**] where $K_T = 2.3–3.1$ in favour of [**11a**].

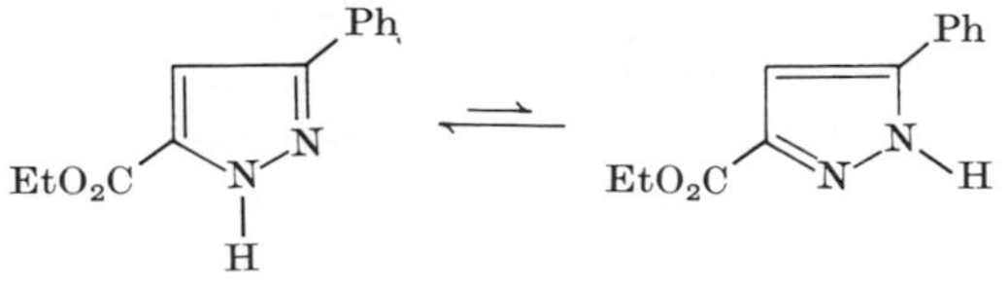

[**11a**] [**11b**]

v. *Nuclear Magnetic Resonance* (*Section 1-5A*). We have criticized (Section 1-5Ab) the use of interpolation methods in proton NMR studies to calculate K_T. Although we do not, for reasons already given, believe that the results are of great significance, we summarize them: the equilibrium of 3-methyl- $\rightleftharpoons$ 5-methyl-pyrazole (H or C_6H_5 at position 4) is shifted in favour of the 5-methylpyrazole (64JA1456, 65JO1892); the equilibrium 3-phenyl- $\rightleftharpoons$ 5-phenyl-pyrazole (H, CH_3 or CO_2CH_3 at position 5) is shifted in favour of the 3-phenylpyrazole (66JO1878). Two methods of applying ^{14}N NMR to 5-ethoxycarbonylpyrazole failed because of poor resolution (73CP697). Line broadening of the ^{13}C resonances of pyrazole and 3-methylpyrazole decreases with increasing temperature and with addition of H_2O or D_2O (73JA4761).

vi. *Infrared and Raman Spectra* (*Section 1-6C*). Comparison (65T1693) of the infrared spectrum of the ethoxymethylpyrazole [**9**] with those of the methylated compounds [**12, 13**] (CCl_4, $CHCl_3$), indicates a close similarity between [**9**] and 3-ethoxy-1,5-dimethylpyrazole [**12**]; this indicates predominance of the form [**9a**] and is consistent with the results deduced from basicity measurements.

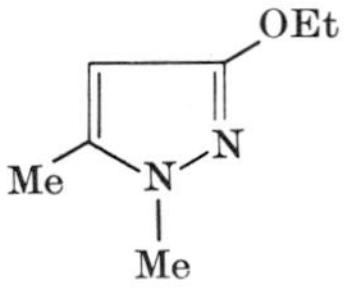

[**12**] 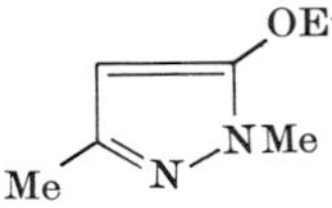 [**13**]

vii. *Ultraviolet and Visible Spectra* (*Section 1-6D*). The predominance previously noticed of [**9a**] is confirmed for the equilibrium [**9a** $\rightleftharpoons$ **9b**] by comparing the ultraviolet spectra of [**9**] with [**12**] and [**13**] in cyclohexane (65T1693); in water all the spectra are so similar that no conclusion can be drawn.

Using, instead of absorption band position, the bathochromic or hypsochromic shifts $\Delta\lambda$ observed when the spectra are taken in cyclohexane (or alcohol) and then in dilute hydrochloric acid, some investigators (56G797) claimed that equilibria [**14**] and [**15**], are shifted towards [**14a**] and [**15a**]. A survey of the experimental results (Scheme 4-2) shows that these conclusions must be considered tentative at best.

viii. *Mass Spectrometry* (*Section 1-6G*). Pyrazoles behave as though a mixture of two isomers exists in the gaseous state (67J(B)885); thus if $R^3 = D$ and $R^5 = H$ (69TH01) the fragments $m/e = 41$ and $m/e = 42$, which arise from the loss of DCN and HCN from the molecular ion, possess the same intensity.

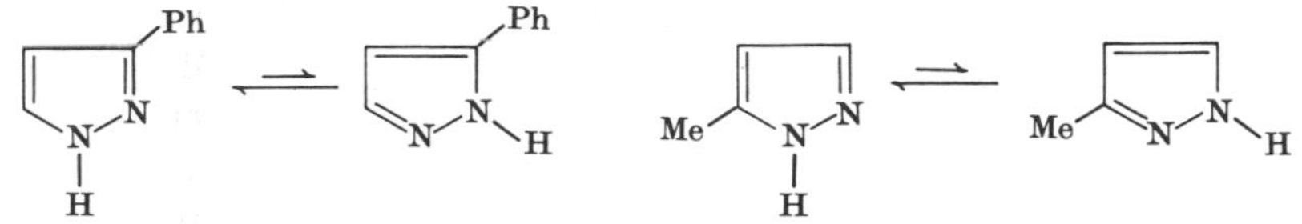

[14a] ⇌ [14b]   [15a] ⇌ [15b]

+5[b]   +2[a]   −3[b]   +4[b]   +5[a]   +2[b]

[a] $\Delta\lambda$ for tautomeric cpd(nm)
[b] $\Delta\lambda$ for N—Me derivative (nm)

Scheme 4-2

ix. *Calculated Equilibrium Constants* (*Section 1-7B*). Finar (68J(B)-725) points out that there is no connection between the π energy calculated by the HMO method and the relative stability of the pyrazole tautomers. As pyrazoles are associated in solution (as dimers for instance), he suggests that the negative charge density at N(2) influences the position of the equilibrium, because of the importance of stabilization due to the hydrogen bonds in [**16**]. Table 4-3 gives the results from both methods.

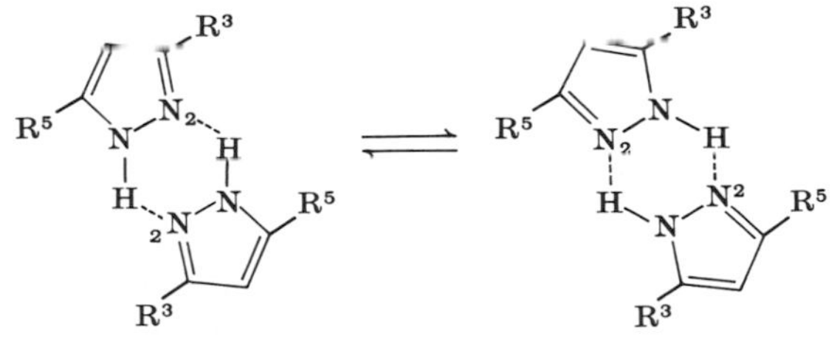

[16a] ⇌ [16b]

It is most unfortunate that the results from the calculations are compared with experimental data which are all doubtful, being obtained from ozonolysis experiments [(61H1171), criticized as unacceptable in the previous review, II-31), molecular refraction (34LA(508)51) (an ambiguous case), proton NMR (65JO1892, 66JO1878) and UV spectra (56G797). The last two methods are inconclusive, as has just been discussed.

The equilibrium [**17a**] ⇌ [**17b**] has been studied by PPP, EHT and CNDO methods (71T5779): in each case, the investigators found that

TABLE 4-3

MOLECULAR ORBITAL CALCULATIONS OF PYRAZOLE TAUTOMERISM

Pyrazole	Predominant tautomer according to the	
	Total E_π	Net charge on N(2)
3(5)-Methyl	5-Methyl	3-Methyl
3(5),4-Dimethyl	4,5-Dimethyl	3,4-Dimethyl
3(5)-Phenyl	5-Phenyl	3-Phenyl
3(5)-Methyl-4-phenyl	3-Methyl-4-phenyl	3-Methyl-4-phenyl
3(5)-Methyl-5(3)-phenyl	Equal	3-Methyl-5-phenyl
4-Methyl-3(5)-phenyl	4-Methyl-3-phenyl	4-Methyl-3-phenyl
3(5),4-Dimethyl-5(3)phenyl	3,4-Dimethyl-5-phenyl	3,4-Dimethyl-5-phenyl

the 3-hydroxy form [**17a**] is more stable than the 5-hydroxy form [**17b**], in accordance with the experimental results (65T1693, 4-2Ib).

In considering the equilibrium [**18a**] ⇌ [**18b**] other authors, using a semi-empirical SCF–MO method, claimed greater stability for the 5-amino tautomer [**18b**] (70JA2929) (see Section 4-5Bi). Calculations of heats of atomization were performed, but they were not compared with experimental data.

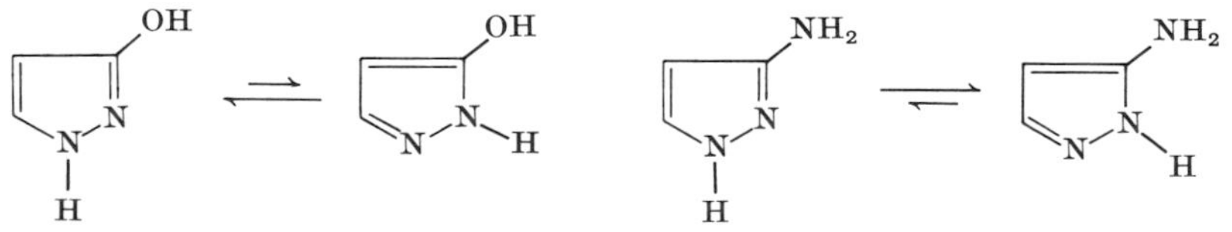

OH OH NH_2 NH_2
N N N N N N N N
H H H H

[**17a**] [**17b**] [**18a**] [**18b**]

x. *Theoretical Methods: Dipole Moments* (*Section 1-7D*). For 3(5)-methylpyrazole, the dipole moments calculated by the CNDO/2 method (70J(B)1692) for the two tautomeric forms are close to each other and differ considerably from the experimental value, precluding any conclusions being drawn.

xi. *Theoretical Methods: Electronic Spectra* (*Section 1-7E*). HMO calculations of the π–π^* transitions of a number of pairs of isomeric pyrazoles (68J(B)725) usually give calculated values between which the experimentally determined value lies. However, it would be hazardous to draw conclusions about the positions of equilibrium from these data.

xii. *X-ray Crystallography* (*Section 1-4Ib*). In the 3-(3-pyrazolyl)-L-alanine (72JA1717) and in the 1,1,1-trimethylhydrazinium 3-carbomethoxy-5-pyrazole-carboxylate (74CX(B)2505) the proton has been located.

c. Summary and Conclusions

The mainly qualitative nature of the results just enumerated precludes any attempt at a qualitative correlation of Hammett type (Section 1-4B). However, it is possible to compare the tendency of a substituent to occur in position 3 or 5 with the sign of the σ_m for this substituent (Table 4-4). Considering the common cation [**19**], an

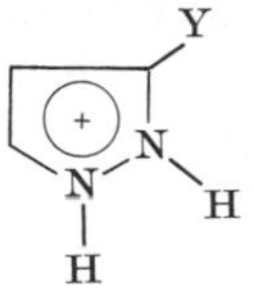

[**19**]

inductive electronic-withdrawing effect would be expected to increase the acidity of the 2-proton over that at the 1-position, whereas an electron donor inductive effect should act in the reverse sense.

It can be observed that there is a correlation with two exceptions: the hydroxy group, for which case σ_m is very small, and the marked discrepancy for the carbethoxy group. A possible explanation of the latter discrepancy is chelation [**20**] which provides enough energy to

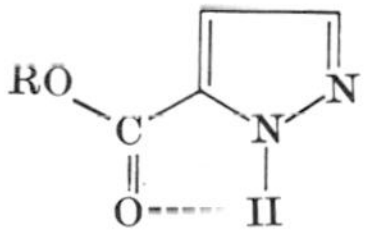

[**20**]

compensate for the electronic effects of the carbethoxy group. However, studies of the degree of association in benzene (41J1) indicate that carbethoxypyrazoles generally are more intermolecularly associated than the other pyrazoles, which seems to be inconsistent with a strong

TABLE 4-4

Pyrazole Tautomerism and Substituent σ_m

	NH_2	CH_3	OH	OR	C_6H_5	Br	CO_2Et
Position	5	5	3	3	3	3	5
σ_m	−0.161	−0.069	−0.002	0.115	0.060	0.391	0.398

intramolecular hydrogen bond. However, chelated dimer formation could occur.

Further discussion of the dependence of the annular tautomerism of pyrazoles on substitution must await a more detailed experimental study.

C. Imidazoles (II-32)

a. Symmetry of Imidazole and Symmetrically Substituted Imidazoles

As for pyrazole, studies by many methods prove that, irrespective of the physical state of the sample, the imino proton of imidazole is localized on one of the nitrogens; thus imidazole possesses symmetry C_s. Methods used include IR spectroscopy (65CP1334, 65CP1344, 70PH2133); ^{14}N quadrupolar resonance (68MP73); microwaves (67-NA1301); and X-ray diffraction (63NA575, 63ZX1, 66CX(20)783, 69ZX211, 71KX115). However, one group (68SA237) has attempted an analysis of the IR spectrum of imidazole in the solid state taking group symmetry C_{2v}; these authors do not mention the two papers published in 1965 and cited above.

Many theoretical calculations have been carried out on imidazole assuming C_s symmetry. The results obtained are very satisfactory, especially the agreement with the dipole moment; experimentally it has been reported as 3.8 ± 0.4 *D* for the gas phase by microwave measurement (67NA1301), 3.4 *D* in benzene (extrapolated to infinite dilution) (61ZE821) and 3.99 *D* in dioxan solution at 25° (61DA1374); a complete bibliography is given in (71KG867). Values calculated for the dipole moment include the following: 4.09 (67JA6835); 3.55 (67TC182); 3.75 and 3.82 (67TC259); 3.91 (69BF1097); 3.63 (70BF273); 3.17 and 4.01 (70MI40); 4.19 (71CP465); 3.99 (73TC145). The EHT calculations (67TC342) produce, as expected, a calculated moment in poor agreement, 7.17 *D*.

b. Unsymmetrically Substituted Imidazoles

Much work has been carried out to determine K_T for equilibria of type [**21a**] ⇌ [**21b**] for imidazoles unsymmetrically substituted in

R4 R5 N N H R2 ⇌ R4 R5 N N H R2

[**21a**] [**21b**]

positions 4 and/or 5. Without doubt, imidazoles now provide the best-studied case of annular tautomerism. Most of this work has appeared during the past decade although the previous review (II-32) included the classic work by Ridd and his associates on this problem using basicity measurements. A more recent publication (70KG1683) has surveyed the whole problem, but we have considered it useful to summarize all the data now available.

i. *Basicity Measurements* (*Section 1-4A*). Table 4-5 summarizes the data for the calculation of the K_T values used to draw the lines of Fig. 1-1 (Section 1-4Ba).

TABLE 4-5

K_T FOR ANNULAR TAUTOMERISM OF IMIDAZOLES

Imidazole	Reference	K_T	$\log f = pK_{NMe} - pK_{NH} - \log S$
Unsubstituted	57JA1656	1	−0.34
	72PS2	1	−0.26
2-Bromo	67J(B)641	1	−0.27
4(5)-Nitro	60J1363	457	−0.48
	63BF2840	320	−0.64
	64JO862	500	−0.42
4(5)-Phenyl	60J1363	10 to 37	[−0.26 to −0.48][a]
4(5)-Bromo	63BF2840	63 to 152	[−0.26 to −0.64][a]
4(5)-Chloro	64JO862	44.5	
5(4)-Chloro-4(5)-nitro	64JO862	118	+0.12
2,4(5)-Dinitro	64JO862	1.1 to 3.1	[−0.26 to −0.42][a]
2-Bromo-4(5)-nitro	67J(B)641	210	—

[a] Assumed range, see text.

The following comments refer to the values reported in Table 4-5. For 4(5)-chloroimidazole and 2-bromo-4(5)-nitroimidazole, log f cannot be calculated because the authors did not report the pK_a of the NH compound. For 4(5)-phenyl-, 4(5)-bromo- and 2,4(5)-dinitro-imidazole, the authors give the pK_a values of the imidazole N–H and of only one of the methylated derivatives: we used Eqs. (1-14) and (1-15), assuming a range of values for the correction term log f between the minimum value found for imidazole itself, −0.26, and the value the same authors found for 4(5)-nitroimidazole. The result for 2,4(5)-dinitroimidazole is indicative of the large error range. We re-emphasize that Eq. (1-15) (Section 1-4Aa) must be used cautiously, because of its great sensitivity to the difference $K_1 - K_{MeA}$, especially if this value is very low; thus if log $f = 0.14$, then $K_T = 200$. The result obtained for the 4(5)-chloro-5(4)nitro-imidazole, log $f = +0.12$, if confirmed, would be an exception to the rule that *N*-methylation decreases the basicity (see Table 1-2, Section 1-4A): but the pK_a values of 4(5)-chloro-5(4)-nitro-imidazole (p$K_a = -3.62$) and of 5-chloro-1-methyl-4-nitro-imidazole (p$K_a = -3.49$) (64JO862) need verification.

ii. *Hammett Equation* (*Section 1-4B*). In Section 1-4Ba, we have discussed in detail Charton's method (65JO3346) as applied to substituted imidazoles (see Fig. 1-1 in Section 1-4Ba). The conclusion was that electron donor groups favour the 5-substituted, and electron acceptor the 4-substituted imidazoles, according to an equation of type log $K_T = \rho\sigma_m$; ρ was found as 3.2 theoretically and ca. 4 experimentally. Recently (70JH227) the authors attempted to correlate the pK_a values of imidazoles with the sum of the Hammett coefficients $\sum\sigma^*$; the σ values of the substituents are expressed as 0.5 ($\sigma_0 + \sigma_m$). This procedure is not compatible with that of Charton (1–4B) because it implicitly supposes $K_T = 1$, an inadmissible assumption.

iii. *Polarography* (*Section 1-4F*). The only available example (63BF-2840) qualitatively supports the predominance of 4-nitroimidazole in the equilibrium [**22a**] $\rightleftharpoons$ [**22b**], as already discussed (Section 1-4F).

iv. *Ultraviolet and Visible Spectra* (*Section 1-6D*). In acid media the spectra of 1-methyl-4-nitroimidazole and of 4(5)-nitroimidazole are almost superimposable.

1-Methyl-5-nitroimidazole shows a different pattern; in particular a third band appears at 345 nm (63BF2840), although curiously this band was not noted in two later publications (64JO862, 67J(B)641). This result appears to again confirm that form [**22a**] predominates in solutions of 4(5)-nitroimidazole.

v. *Nuclear Magnetic Resonance*. Comparison of ^{13}C chemical shift–pH profiles for histidine and its 1- and 3-methyl derivatives shows that in basic solution the preferred structure is [**22/1**]. This tautomeric form is maintained in a number of histidine derivatives including the polypeptide bacitracin (73JA328).

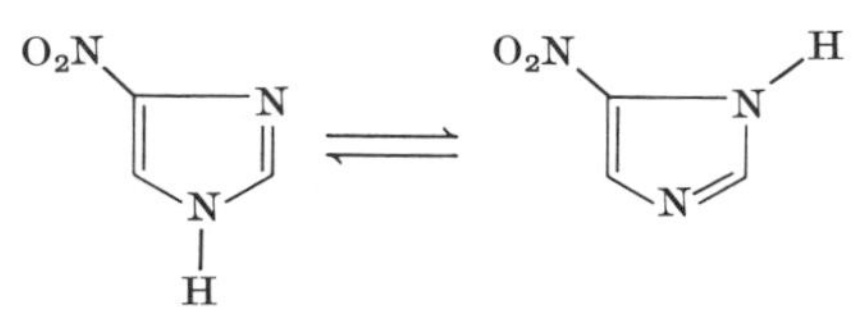

[**22a**] [**22b**] 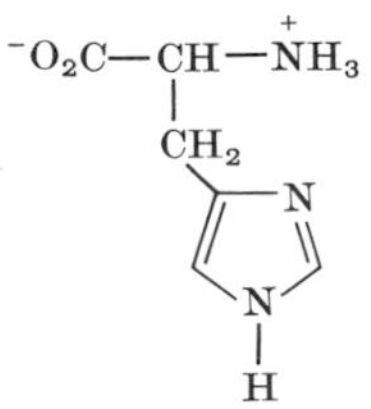 [**22/1**]

vi. *Theoretical Methods: Heats of Combustion* (*Section 1-7C*). Dewar's calculations (70JA2929) indicate a slight predominance of 4-amino-imidazole in the equilibrium [**23a**] ⇌ [**23b**]; this is inconsistent with Charton's hypothesis (Section 1-4B), since σ_m (NH_2) = −0.161.

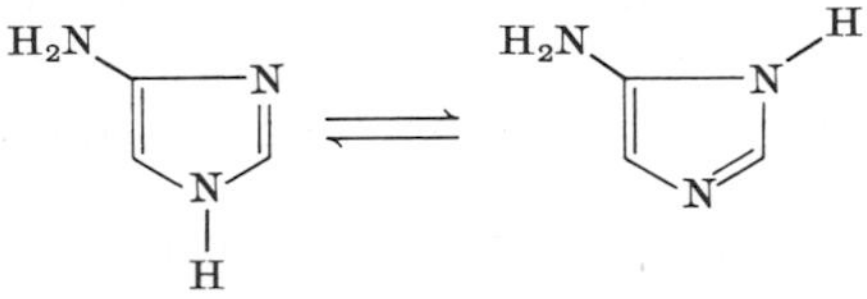

[**23a**] ΔH = 47.775 eV ⇌ [**23b**] ΔH = 47.753 eV

vii. *X-ray Crystallography* has been used to determine the structure of histamine as 5-(2-aminoethyl)imidazole (73JA4829).

viii. *Site of Protonation*. The site of D^+ uptake by 4(5)-bromo-imidazole and the 2-methyl analogue by D_2SO_4 demonstrates the predominant 4-bromo structure for these compounds (63AG300); see discussion in Section 1-2.

c. *Summary and Conclusion*

In general, all the work is in good agreement with the theoretical treatment in terms of the Hammett equation as discussed in Section 1-4B, and summarized in the preceding section.

D. *v*-Triazoles (II-34)

a. *v-Triazole*

Although much work has appeared in the past decade, the overall position regarding the annular tautomerism of *v*-triazole itself remains unsatisfactory, and nothing is known regarding substituted derivatives.

The previous survey (II-34) summarized (see also 74HC(16)33) the conclusion of Jensen and Friediger (43MI1) that the symmetrical tautomer [**24b**] predominated. However, contrary to the report in the previous survey Jensen and Friediger did not use the comparison of the dipole moment of *v*-triazole with those of its methylated derivatives, but a much less reliable method, the comparison of the *v*-triazole

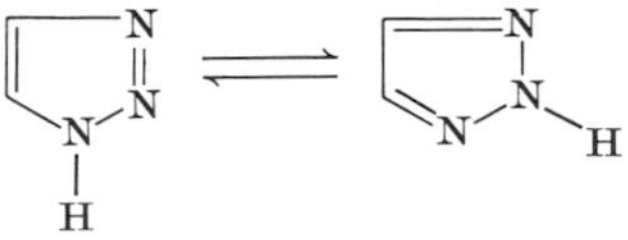

[**24a**] ⇌ [**24b**]

moment with imidazole and pyrazole (cf. Section 1-4G). In more recent work (73CP1483), the use of methylated derivatives demonstrated that the percentage of tautomer [**24b**] in benzene at 25° is 80–85%; this percentage decreases as the temperature increases.

In the vapour phase the presence of the asymmetric structure [**24a**] has been established by IR spectroscopy (69J(B)307) as well as by microwaves (70SA(A)825) (for a more complete discussion see 74CH605). Mass spectrometry (Section 1-6G) requires that both tautomeric forms exist in the vapour phase to explain the behaviour of *v*-triazoles upon electron impact (73OM271).

The IR spectrum in dilute solution (CCl_4, $CHBr_3$) is similar to the vapour phase spectrum (69J(B)307); therefore the C_s symmetry is apparently retained. Neither UV spectra (58G977) nor proton NMR (67BF2998) provide useful evidence. From (^1H–^{13}C) coupling constants Begtrup (74CH702) calculated that the asymmetric tautomer [**24a**] predominates (60%) in deuteriochloroform solution. However, ^{14}N resonance (72T637) suggests that in a concentrated methanolic solution the symmetrical tautomer [**24b**] predominates, $>70\%$ or $>90\%$, according to the method of calculation (Section 1-5C).

A theoretical approach to the problem consists in comparing the experimental dipole moment 1.77 *D* in benzene at 25° (43MI1) with the theoretically calculated moments for both tautomers [**24a**] and [**24b**]. Table 4-6 contains all the results, except the unreliable moments calculated by the EHT method (67TC342), which are obviously too large. In all cases, the symmetrical tautomer [**24b**] is predicted to predominate; the values quoted are obtained using Eq. (1-49) (Section 1-4B) and Eq. (1-51) (Section 1-4G).

Several direct theoretical approaches (Section 1–7B) to the equilibrium constant (69BF1097, 70TH02, 71CP465) again predict the

TABLE 4-6

CALCULATED DIPOLE MOMENTS FOR *v*-TRIAZOLE TAUTOMERS[a]

	Method			
	CNDO/2	LCAO–A	+Electrons σ	LCAO–SCF
Reference	67JA6835	69BF1097	70BF273	71CP465
1,2,3-Triazole [**24a**]	4.30	4.47	4.82	4.42
1,2,5-Triazole [**24b**]	0.20	0.36	1.03	1.74
K_T	10	11	20	275

[a] See text.

predominance of tautomer [**24b**] but a CNDO/2 calculation predicts the unsymmetrical tautomer [**24a**] (68TL3727, 73T3285).

The indications from the evidence concerning the predominance of the symmetrical [**24b**] or unsymmetrical tautomer [**24a**] seem to depend upon the method used to approach the problem. Studies at varying temperatures [in order to separate ΔH° and ΔS° terms (73CP1483), see discussion (Section 1-7B)] and with different concentrations and solvents are needed to solve this problem, still the most confused of all the simple cases of annular tautomerism.

b. Substituted v-Triazoles

No studies have been reported of the annular tautomerism of C-substituted *v*-triazoles [**25**].* However, some deductions can be made from literature data. For the case of [**25**, Y = Br] the pK_a values of the *N*-methylated derivatives corresponding to [**25a**] and [**25c**] are −1.67 [**26**] and −0.47 [**28**] (67J(B)641). The pK_a of 1-methyl-3-bromo-1,2,5-triazole [**27**] was not determined because of its too low value. In

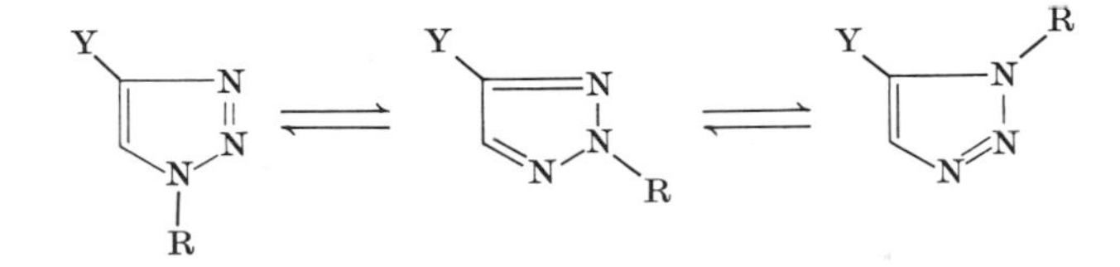

[**25a**] R = H   [**25b**] R = H   [**25c**] R = H
[**26**] R = Me   [**27**] R = Me   [**28**] R = Me

this particular case we cannot assume that the least basic tautomer is predominant as cations of similar structure are not formed. For *v*-triazole itself [**31**], although the dication is of unique structure [**29**], two distinct monocations exist [**30a, 30b**]. Of these [**30a**] probably predominates by a considerable factor in the tautomeric equilibrium between them, as the charge is usually spread more effectively between equivalent structures, and inductive effects should favour [**30a**]. The basicities of the methyl derivatives just discussed are evidence that [**31a**] is more basic than [**31b**]; however the pK_T between them is given by (pK_a [**31a**] − pK_a [**31b**] − x). In the absence of any indication of the value of x, no conclusion can be drawn regarding the value of pK_T for equilibrium [**31**].

With these results in mind, the fact that the *v*-triazole itself is such a weak base that it is not protonated in dilute aqueous acid (70JH991)

* Recently (74J(PII)1849) the crystal structure of 5-amino-1H-1,2,3-triazole-4-carboxamide has been reported.

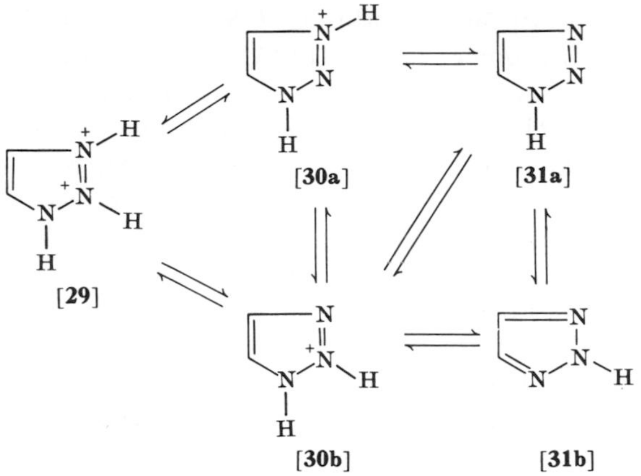

[29] [30a] [31a] [30b] [31b]

SCHEME 4-3; $pK'_T = \frac{[\mathbf{30a}]}{[\mathbf{30b}]} = x$

does not necessarily show that the structure [**24b**] predominates in solution. Still less can any conclusion be reached regarding the tautomerism of the bromo-*v*-triazole [**25**, Y = Br].

A recent paper (74JO940) described the obtention of two separate tautomers of a substituted *v*-triazole, but there are some doubts about the structure of one of the alleged tautomers.

E. *s*-TRIAZOLES (II-34)

At the time of the previous review (II-34) no specific studies dealt with the annular tautomerism of *s*-triazoles, and all that could be done was to confirm that for *s*-triazoles a classical equilibrium exists (see Section 4-1Aa), for which the equilibrium constant or at least the predominant tautomer is capable of determination. It can be indicated immediately that all experiments and all calculations show for *s*-triazoles without ambiguity a higher stability for the unsymmetrical tautomers [**32a**, **32c**].

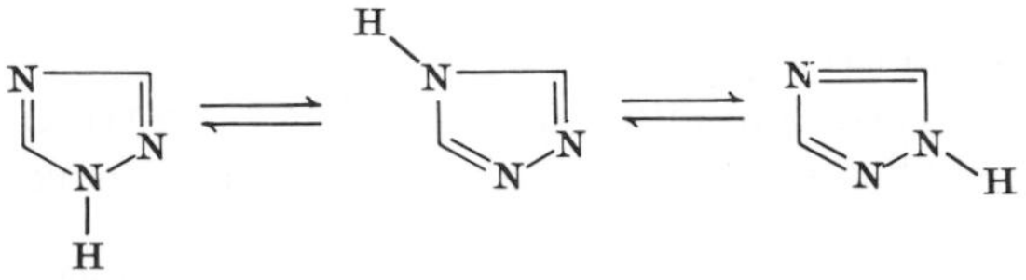

[32a] [32b] [32c]

a. s-Triazole in the Vapour Phase and Solid State

In vapour phase, IR (72PS8) and microwave spectra (71CH873) are consistent with the symmetry C_s. In the solid state a preliminary X-ray study suggested structure [**32a**] (65BG550) which was confirmed by a later work at low temperature (−155°) (69CX(B)135). The same conclusion is reached in a study of the ^{14}N quadrupolar resonance spectrum at 77 K (68MP73, 1-6A). For an IR and Raman study of *s*-triazole in the vapour and solid phase see (72TH02).

Mass spectral data (68CH727) based on a wrong interpretation of NMR spectra, originally gave the opposite result. As already indicated in the discussion on the use of mass spectroscopy (Section 1-6G), recent work (72OM1139) has cleared up this question, showing that the fragmentation is well explained assuming the autotropic rearrangement [**32a**] $\rightleftharpoons$ [**32c**], but authors (73OM57) consider that the observed fragmentation does not allow a choice between [**32a**] and [**32b**].

b. s-Triazole in Solution

i. *Basicity Measurements* were discussed in Section (1-4Ab). A value of K_T between 4 and 10 in favour of tautomer [**32a**] is obtained.

ii. *Dipole Moment* (*Section 1-4G*). Comparison of the experimental dipole moment, 3.29 *D* in dioxan, with those obtained from a vectorial calculation for both tautomers ([**32a**]: 3.24 *D* and [**32b**]: 5.36 *D*) indicates a predominance of the unsymmetrical tautomer [**32a**] (66KG-776; (see also 75BF(2)ip2)). The authors verified their calculation by measuring the dipole moment of the 1-methyl-1,2,4-triazole: 3.50 *D*.

iii. *Nuclear Magnetic Resonance Spectra* (*Section 1-5A*). Interpolation methods cannot be applied reliably to proton NMR, because displacements of the tautomeric equilibrium do not modify the chemical shifts sufficiently (67TL2109). However, the dynamic experiments discussed in detail in Chapter 1 (Section 1-5Ac) demonstrate the presence of the form [**32a**] in HMPT (68JO2956). Experimental results obtained by ^{14}N resonance spectroscopy (72T637) were rationalized on the basis of the presence of a mixture 60% of [**32a**] and 40% of [**32b**] in a concentrated solution of methanol (Section 1-5C, Scheme 1-18; see however, 73JA324).

iv. *Theoretical Methods.* Theoretical calculations of the equilibrium constant by a direct method (Section 1-7B) (66JT759, 69BF1097, 69JA796, 70TH02, 71CP465) as well as by comparison of calculated and experimental dipole moments (Section 1-7D) (Table 4-7) indicate the higher stability of tautomer [**32a**]; see also (73KG707). A CNDO/2 calculation predicts approximately equal stability for the two forms [**32a**] and [**32b**] (68TL3727).

TABLE 4-7

CALCULATED DIPOLE MOMENTS FOR *s*-TRIAZOLE TAUTOMERIC FORMS

Method	Dielectric constant	Micro-waves	VESCF	CNDO/2	LCAO–A	+ Electrons σ	LCAO–SCF	All electrons
Reference	66KG776	71CH873	71CH873	71CH873	69BF1097	70BF273	71CP465	70MI40
1,2,4-Triazole [**32a**]	3.29[b] (dioxan)	2.72[b]	2.33	3.07[a]	3.07	3.09	3.69	—
1,3,4-Triazole [**32b**]			—	—	5.73	5.68	5.37	5.63

[a] 3.02 utilizing a fully optimized geometry (73TC145).
[b] Experimental results.

c. Substituted s-Triazoles

Three tautomeric forms can be expected for C-substituted *s*-triazoles [**33a–c**].

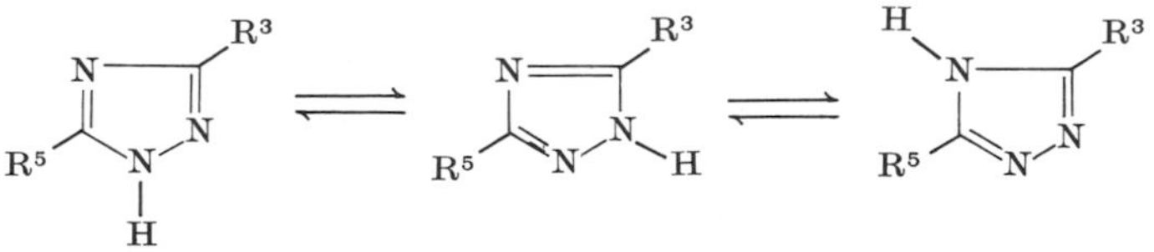

[**33a**] [**33b**] [**33c**]

If $R^3 = NH_2$ (see also Section 4-5Fc), basicity measurements (67J(B)641) as well as dipole moment ones (66KG776), give the order of decreasing stability for the three tautomers as [**33b**] > [**33a**] > [**33c**]. The methylation of nitro-1,2,4-triazoles has been discussed with reference to tautomerism, but definite conclusions were not reached (70KG265). In a later paper, the same authors (70KG558) showed, from dipole moment and basicity measurements, that tautomer [**33a**, $R^3 = NO_2$, $R^5 = H$] is predominant.

For 3-methylthio-5-(2-pyridyl)-*s*-triazole [**33**], UV comparison with all the *N*-methyl derivatives shows that structure [**33a**, R^3 = SMe, R^5 = 2-pyridyl] predominates in EtOH and cyclohexane (72CT2096); the equilibrium here between [**33a**] and [**33b**] is perhaps influenced by the formation of an intramolecular hydrogen bond in [**33a**]. For more complete results see (75CTip1).

F. Tetrazoles

For this heterocycle, the two tautomers [**34a, b**] are both asymmetric (symmetry C_s). Therefore we do not discuss separately the case of tetrazole itself [**34**, R = H] as it does not form a special case. A consequence is that methods which rely on determination of the symmetry cannot be used; this hinders the study of the tautomerism of tetrazole, especially in the gaseous state (IR, microwaves).

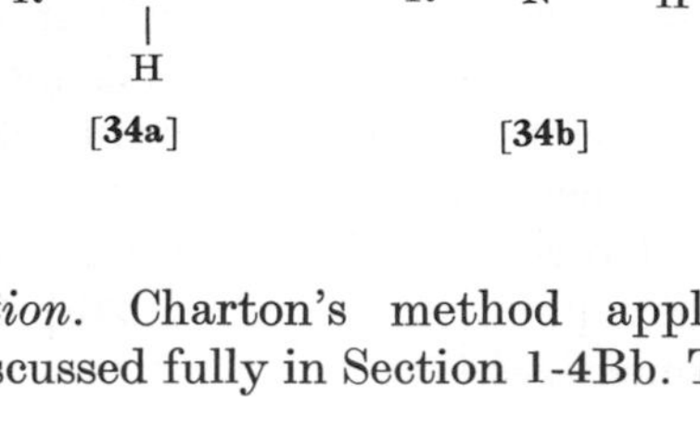

[**34a**] [**34b**]

a. Results

i. *Hammett Equation.* Charton's method applied to tetrazoles (69J(B)1240), was discussed fully in Section 1-4Bb. The conclusion was that in water, if K_T = [**34b**]/[**34a**], then log $K_T = 5.3\,\sigma_I + 0.6\,\sigma_R + 0.6$. For R = H this gives a mixture of 80% of form [**34b**] and 20% of form [**34a**]. However, this result is probably incorrect; see later. As already mentioned for pyrazoles (Section 4-1Bb) and imidazoles (4-1Cb, an equation of the type $pK_a = \rho\sigma^* + b$ (67PH1756) is inconsistent with Charton's hypothesis.

ii. *Dipole Moments* (*Section 1-4G*) (*II-35*). As already pointed out in the previous survey (II-35) and as discussed in detail in Chapter 1 (see Scheme 1-8, Section 1-4G), dipole moment measurements (56JA-4197, 61T327) indicate a large preference (80–95%) for tautomer [**34a**, R = H] in dioxan, see also (75BF(2)ip2).

The method of vectorial summation leads in the case of R = H to the value 5.22 *D* for [**34a**] and 1.63 *D* for [**34b**]. If the experimental value of 5.11 *D* (benzene) (43MI1) is interpolated, tautomer [**34a**] is found to be predominant to the extent of 99%, not 97% as claimed by the author (63PH721), who erroneously took $P = \mu$ instead of μ^2 in Eq. 1-51.

iii. *X-Ray Diffraction* (*Section 1-4Ib*). The only two tetrazoles to have been determined are aminotetrazole monohydrate (67CX(22)308) and 5-bromotetrazole (73J(PII)2036). The first has the structure [**34a**, R = NH_2] 5-amino-1,2,3,4-tetrazole (Section 4-5Ff). This conclusion is consistent with that drawn from the IR spectrum in the solid state (65RO2236). In the case of 5-bromotetrazole [**34**, R = Br] due to the presence of the bromine atom it was impossible to locate the hydrogen atoms. Ansell (73J(PII)2036) claims "Ring distances clearly indicate the tetrazole ring to be a resonance hybrid [**34a**] $\leftrightarrow$ [**34b**]" (Section 1-1C). However in the case of tetrazole itself the structure [**34a**] has recently been established (74MI321).

iv. *Nuclear Magnetic Resonance* (*Section 1-5Ab*). As previously noted (II-35), the comparison of the CH chemical shifts of tetrazole itself with those of N-substituted derivatives, indicates a predominance of the form [**34a**, R = H] (60JA5007). In addition to the inherent inaccuracy of the method (Section 1-5Ab), the authors determined the chemical shifts of tetrazole as a solution in dimethyl formamide and those of *N*-substituted derivatives as neat liquids. This method cannot be applied to 5-methyltetrazole; the difference in chemical shifts is too small (65JO3472). Comparison of the ^{13}C chemical shifts of tetrazole with its two *N*-methylated derivatives (74JO357) establishes the predominance of tautomer [**34a**, R = H].

^{14}N resonance spectroscopy (72T637) does not lead to any definitive conclusion about the equilibrium [**34a**] $\rightleftharpoons$ [**34b**]; however the original authors prefer tautomer [**34b**].

v. *Nuclear Quadrupole Resonance* (*Section 1-6A*) applied to tetrazole (68MP73) gives no conclusion.

vi. *Mass Spectrometry* (*Section 1-6G*). Tetrazole [**34**, R = H] and 5-methyltetrazole [**34**, R = CH_3] tautomerism has been studied by comparing their fragmentation with those of the 1- and 2-methyl derivatives (69OM433). The NH compounds show similarities to *both* of the two *N*-methylated model compounds, but no firm conclusion regarding the gaseous phase tautomerism of N–H tetrazoles was reached.

vii. *Theoretical Methods: Equilibrium Constants* (*Section 1-7B*). The relative stability of tautomers [**34a**] and [**34b**] depends on the method used: the LCAO-A method (69BF1097) shows a predominance of the tautomer [**34b**] (there is an error in this paper, see 70TH02), the LCAO-SCF method (71CP465) leads to a comparable stability of both tautomers and the CNDO method (67JA6835) is also in favour of tautomer [**34b**].

viii. *Theoretical Methods: Heats of Combustion* (*Section 1-7C*). The experimental heat of combustion, 28.11 eV, is closer to the one calculated for tautomer [**34b**], 27.99 eV, than to the one calculated for tautomer [**34a**], 27.72 eV (66JT759). Calculations by the same authors (70JA2929) indicate a predominance of the 5-amino-1,2,3,4-tetrazole [**34a**, R = NH_2] (Section 4-5Ff).

ix. *Theoretical Methods: Dipole Moments* (*Section 1-7D*). Table 4-8 compares calculated dipole moments for each tautomer with the experimental moment. Tautomer [**34a**] has a calculated moment close to the experimental one, and thus presumably predominates in dioxan solution. A microwave study (74JM(49)423) concludes that both tautomeric forms are probably present in significant amounts in the vapor of tetrazole.

b. Conclusions

The conclusion of most of the results just summarized, is that tautomer [**34a**] seems to be predominant if R = H. The major contradictory result is that derived from the Hammett equation treatment by Charton (69J(B)1240). It is unlikely that this is simply explained by a solvent effect, because the ^{13}C resonance spectrum (74JO357) of tetrazole in water is similar to those in dioxan (used for dipole moment measurements) and dimethyl sulfoxide. Taking into account the approximations that Charton used (Section 1-4Bb) for tetrazoles, it seems likely that his treatment is in error in this particular case and that tetrazole predominates under all conditions in the 1,2,3,4-tetrazole form [**34a**].

TABLE 4-8

DIPOLE MOMENTS CALCULATED FOR INDIVIDUAL TAUTOMERS OF TETRAZOLE

Method	Dielectric constant	HMO	CNDO/2	LCAO–A	+Electrons σ	LCAO–SCF
Reference	43MI1	61T237	67JA6835	69BF1097	70BF273	71CP465
1,2,3,4-Tetrazole [**34a**]	5.11[a] (dioxan)	4.8	5.23	5.26	5.53	5.28
1,2,3,5-Tetrazole [**34b**]		2.15	2.35	2.04	1.96	3.22

[a] Experimental results.

G. Indazoles (II-31)

As in the case of tetrazole, both tautomers [**35a**] and [**35b**] have symmetry C_s, whether or not position 3 is substituted.

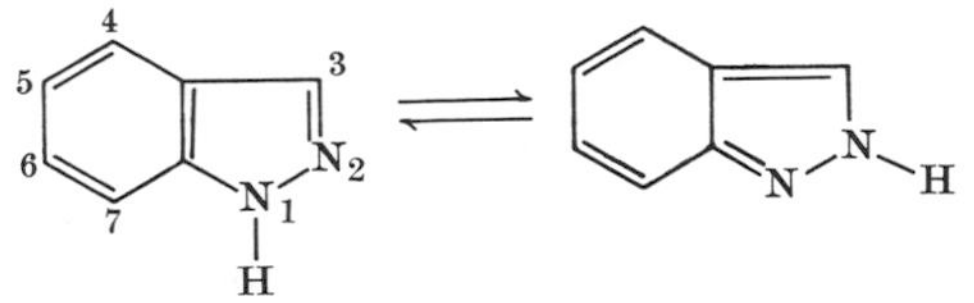

[**35a**] [**35b**]

In the previous survey (II-31), a predominance of tautomer [**35a**] was established on the basis of Auwers's studies on molecular refractivity (37LA(527)291) and of an ultraviolet study of indazole and its 1-methyl and 2-methyl derivatives (42LA(550)31). Many further results now support this qualitative conclusion, which is true also for the solid state from a recent X-ray structure determination (74T2903, 74CX(B)-2009).

i. *Basicity Measurements* (*Section 1-4A*). Literature (67BF2619) pK_a values allow calculation of the K_T and $\log f$ values shown in Table 4-9. The benzenoid tautomer [**35a**] clearly predominates. The values of $\log f$ are typical of a 1,2-diazole (see Table 1-2, Section 1-4Aa), but higher than the $\log f$ for pyrazole [-0.73 (Section 4-1Bbi)] and *s*-triazole [-0.64 (Section 1-4Ab)].

ii. *Dipole Moment* (*Section 1-1G*). Comparison of the experimental dipole moment of indazole in benzene at 25°, 1.60 *D* [an earlier determination gave 1.85 *D* (43MI1)] with the experimental values for the *N*-methyl derivatives [**36**], 1.5 *D* and [**37**], 3.4 *D*, Mauret *et al.* (75BF-(2)ip2) indicate the predominance ($> 95\%$) of tautomer [**35a**]. Utilizing the experimental geometry determined by X-ray, a CNDO calculation

TABLE 4-9

pK_T for Indazoles from Basicities

Compound	K_T = [**35a**]/[**35b**]	$\log f = pK_{NMe} - pK_{NH} - \log S$
Indazole	40	-0.90
5-Nitroindazole	36	-0.86
6-Nitroindazole	36	$[-0.82]$[a]

[a] This value has been established assuming $K_T = 36$, because the pK_a of only one of the methylated derivatives, 1-methyl 6-nitroindazole, has been determined (67BF2619).

of the dipole moment of tautomer [**35a**] gives 2.09 *D* (CNDO/S) or 1.67 *D* (CNDO/2) (74T2903).

iii. *Nuclear Magnetic Resonance* (*Sections 1-5A*, *1-5C*). Although a proton resonance study was claimed (64BF2019) to show that the benzenoid structure [**35a**] was predominant, the results are also consistent with structure [**35b**]. However, a rigorous analysis of the spectra observed for indazole and its methyl derivatives [**36**] and [**37**] in acetone (66BF2075) shows that the coupling constants of indazole more closely resemble those of 1-methylindazole [**36**].

The ^{14}N spectra of indazole in comparison with its two methyl derivatives (72T637) suggest that [**35a**] predominates largely in solution, since the shifts of both nitrogen atoms are within the experimental error of those of [**36**] and quite different from those of the other isomer [**37**].

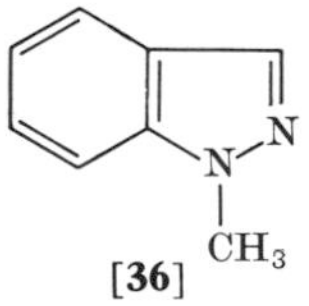

N N CH_3 [**36**]   N N CH_3 [**37**]

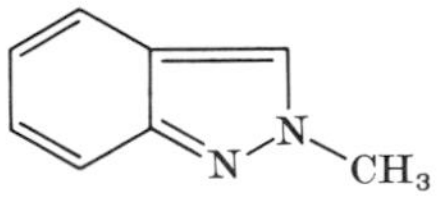

iv. *Infrared and Raman* (*Section 1-6C*). It is not possible to use symmetry arguments. Qualitative work demonstrates the greater similarity of the Raman spectrum of indazole to that of the 1-methyl compound [**36**], which again suggests the benzenoid structure [**35a**] for the tautomeric compound (40CB162).

v. *Ultraviolet and Visible* (*Section 1-6D*). The UV spectrum of indazole in water or ethanol resembles the 1-methyl derivative [**36**] but differs from that of the 2-methyl derivative [**37**]; this shows the predominance of tautomer [**35a**] (41J113, 50JA3047, 61M1131).

vi. *Theoretical Methods: Electronic Spectrum* (*Section 1-7E*). The electronic spectra of tautomers [**35a**] and [**35b**] have been calculated theoretically using the LCAO method (61M1114) and the PPP method with configuration interaction (70BJ3344). In both cases the experimental spectrum is in better accord with that calculated for [**35a**] than for [**35b**]. Provided that π energies reflect tautomer stabilities (Section 1-7B), calculations from the two theoretical studies quoted also indicate that tautomer [**35a**] is the preferred one.

H. Benzimidazoles

a. Unsubstituted Benzimidazoles

Benzimidazole is a case of autotropic rearrangement between two forms [**38a**] and [**38b**] each of symmetry C_s (74T2903, 73MI23, 74CX(B)-

1647). Indeed Cordes and Walter (68SA1421) have (erroneously) analysed the IR spectrum of benzimidazole in the solid state assuming a symmetry C_{2v} [**39**]; their results show the risks inherent in this method (Section 1-6C).

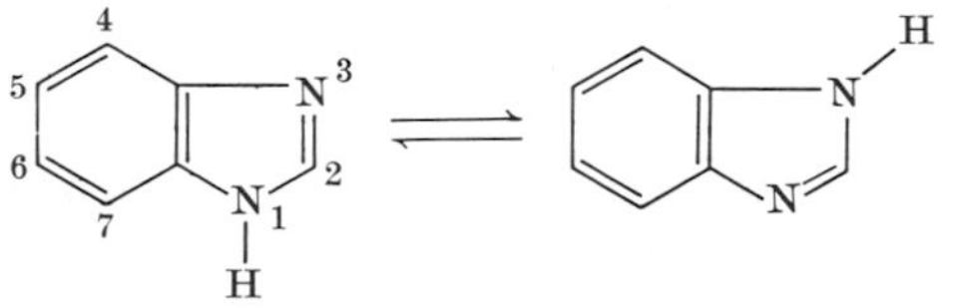

[**38a**] [**38b**] 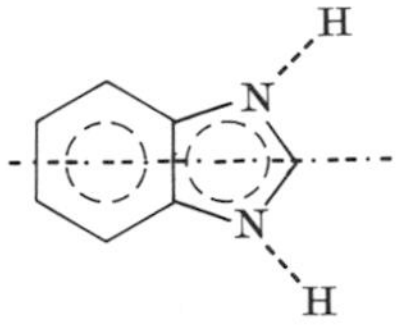 [**39**]

b. Substituted Benzimidazoles

A substituent in position 2 has no influence on K_T. An X-ray study shows that in that case also the symmetry is C_s (73CX(B)2298). Substituents in positions 4(7) and 5(6) exercise a relatively remote perturbation, therefore K_T does not differ much from 1, unless an unusual (attractive or repulsive) interaction occurs between, say, a substituent in the 7-position and the proton bound to the nitrogen atom N(1).

i. *Basicity Measurements* (*Section 1-4A*) (*II-33*). Available studies of benzimidazole equilibria using the pK_a method are listed in Table 4-10; some (60J1363) were already considered in the first survey (II-33). By definition K_T is taken as [**40a**]/[**40b**] and [**41a**]/[**41b**].

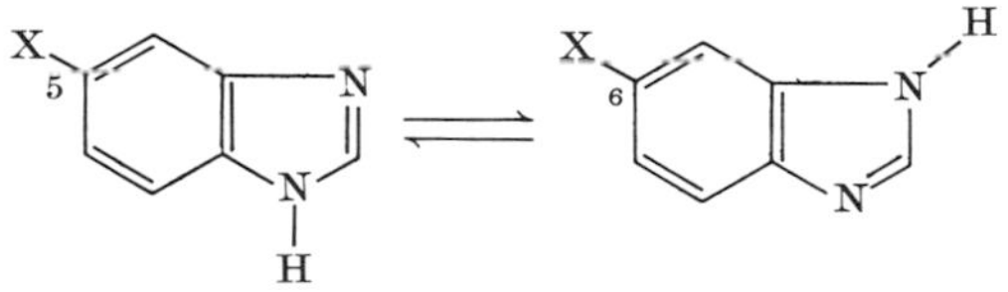

[**40a**] [**40b**]

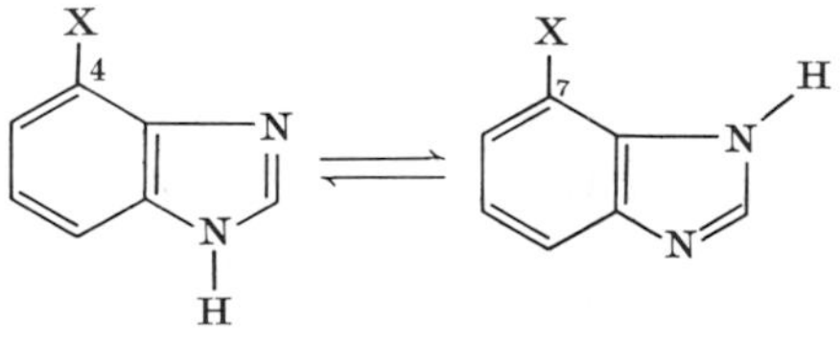

[**41a**] [**41b**] 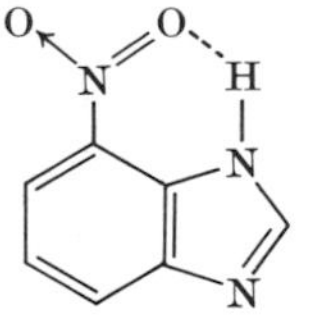 [**42**]

As the difference between the K_T values of 5(6)-nitrobenzimidazole and 2-methyl-5(6)-nitrobenzimidazole must be due to a second-order

TABLE 4-10

K_T AND LOG f VALUES FOR BENZIMIDAZOLES

Benzimidazole	Solvent	Reference	K_T	$\log f = pK_{NMe} - pK_{NH} - \log S$
Unsubstituted	OH_2	51PP420	1	−0.21
	CH_3OH, 50%	51PP420	1	−0.34
5(6)-Nitro	OH_2	60J1363	1.85	−0.28
2-Methyl-5(6)-nitro	OH_2	60J1363	0.90	−0.38
5(6)-Chloro	EtOH, 50%	60J1363	1.00	−0.34
5(6)-Chloro-2-methyl	EtOH, 50%	60J1363	1.00	−0.26
4(7)-Nitro	OH_2	60J1363	0.24	−0.18

that the effects of 5(6)-substitution on equilibrium [**40a**] $\rightleftharpoons$ [**40b**] cannot easily be rationalized and that the constant K_T is close to 1. For 4(7)-nitrobenzimidazole, tautomer [**41b**] is favoured (~80%); this is attributed to an intramolecular hydrogen bond [**42**]: many measurements (volatility, polarography, cryoscopy, etc.) provide evidence for such a hydrogen bond (51JA3030).

A direct correlation between the pK_a of a tautomeric compound and a Hammett sigma value (σ_m, for example) follows only if K_T does not depend upon the substituent, i.e., $K_T \sim 0$, 1 or ∞ (see discussion in Section 1-4C). Such a correlation exists (64J915) for 4(7)-halobenzimidazoles: $\Delta pK_a = 2.13\ \sigma_m$; the authors suggest that $K_T \sim \infty$; i.e., these compounds exist virtually only in the form [**41b**, X = halogen]. They found that no such simple correlation of pK values exists for substituted 5(6)-benzimidazoles and presumed this to be due to variation of K_T from compound to compound.

However, later work (67JO1954) on 5(6)-substituted benzimidazoles and the corresponding N-methyl models of the individual tautomers [**40a**] and [**40b**] established equations such as $\Delta pK_a = \alpha\sigma_1 + \beta\sigma_R^\circ + C$, for the pK_a values in water, and for pK_a of 1-methyl-5-substituted benzimidazoles (60J1363) determined in 50% EtOH. The authors use the correction term $\log f = -0.35$ for their results in H_2O (see Table 4-10). These good correlations and the apparent insensitivity to tautomerism is correctly attributed by the authors (67JO1954) to near equivalence in thermodynamic stability of the tautomers.

Another reported (66BS249) correlation $\Delta pK_a = -1.03\ (\sigma_m + \sigma_p)$ for 5(6)- and 4(7)-substituted benzimidazoles assumes implicitly that in all cases $K_T \sim 1$ (even for the 4(7)-nitrobenzimidazole).

ii. *Nuclear Magnetic Resonance* (*Section 1-5A*). The variable temperature proton resonance of 2-chloro-5(6)-methoxybenzimidazole in tetrahydrofuran cited in Chapter 1 (1-5Ac), gives $K_T = 0.69$, corresponding to a mixture 40% [**40a**] and 60% [**40b**] (71TL3299). This unambiguous determination demonstrates the advantage of direct methods compared with indirect methods, such as basicity measurements, for which a good approximation to this equilibrium would be $K_T = 1$.

I. Benzotriazoles (II-34)

At the time of the previous survey (II-34), considerable evidence was already available regarding benzotriazole tautomerism [**43**]; it was then concluded that the 1*H* form [**43a**] predominates by a considerable factor, and the later work is in complete agreement.

Further UV work has been reported (61M1131, 63J5556); this and comparison of the proton NMR of the NH compound with its *N*-methyl derivatives (63J5556, 68LA(716)11) supports the predominance of [**43a**]

More decisively, a variable temperature study in anhydrous tetrahydrofuran (69T4667) (see discussion in Section 1-5Ac) establishes structure [**43a**] in solution. However ^{14}N resonance (72T637) gave no answer because the three nitrogen atoms exhibit only one signal. In recent work (74CP115) the use of methylated derivatives allowed determination, by dipole moments, of the percentage of tautomer [**43a**] in benzene at 25° as near 100%; this does not vary when the temperature increases. Mass spectrometry (73OM1267) also showed only the presence of tautomer [**43a**] in the vapour phase.

Although the symmetry of the two tautomers of benzotriazole is different: [**43a**] C_s, [**43b**] C_{2v}, no microwave or infrared vapour phase

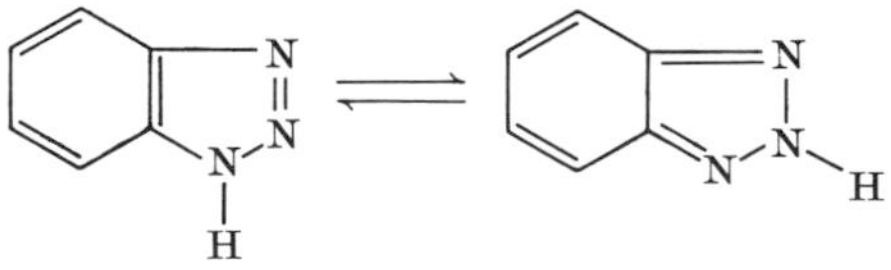

[**43a**] [**43b**]

study has yet been reported. In the solid state, structure [**43a**] has been established by an X-ray study (74T2903, 74CX(B)1490).

In so far as the π energy can be used as a measure of stability, LCAO (61M1114) and PPP (70BJ3344) calculations indicate a predominance of the benzenoid tautomer [**43a**], and the calculated spectrum of [**43a**] is in good agreement with the experimental spectrum.

J. General Conclusions on Annular Tautomerism

The conclusions of Sections 4-1B to 4-1I are summarized with simplifications in Table 4-11. The tautomer with the highest dipole moment (and hence the most likely to be associated) and that with the lowest basicity are listed. Provided two different tautomers form a common cation, the less basic tautomer is necessarily the more abundant.

The results for the substituted pyrazoles and imidazoles have been rationalized (Sections 4-1Bc and 1-4B) in terms of the effects of substituents on the acidity of the alternative protons in the common cation. For indazole and benzotriazole, the aromatic tautomer predominates, and K_T is near unity for most benzimidazoles.

TABLE 4-11

Summary of Annular Tautomerism of Azoles

Series	Azole	Definition of K_T	Conclusion	Tautomer with highest μ	Tautomer less basic
B	Pyrazole	3-Subst[a] / 5-Subst	$\log K_T = \rho\sigma_m$ ($\rho > 1$)	—	—
C	Imidazole	4-Subst / 5-Subst	$\log K_T = 4.5\ \sigma_m$	—	—
D	*v*-Triazole	1,2,3 / 1,2,5	1,2,5 in solution (uncertain in vapour phase)	1,2,3	1,2,5[b]
E	*s*-Triazole	1,2,4 / 1,3,4 i.e. **[32a]** / **[32b]**	1,2,4 ($K_T \gg 1$)	1,3,4	1,2,4
F	Tetrazole	1,2,3,4 / 1,2,3,5 i.e. **[34a]** / **[34b]**	1,2,3,4 ($K_T > 1$)	1,2,3,4	—
G	Indazole	1*H* / 2*H* i.e. **[35a]** / **[35b]**	1*H*, $K_T \ggg 1$	—	1*H*
H*a*	Benzimidazole	5-Subst / 6-Subst i.e. **[40a]** / **[40b]**	$K_T \sim 1$	—	—
H*b*	Benzimidazole	4-Subst / 7-Subst i.e. **[41a]** / **[41b]**	See text	—	—
I	Benzotriazole	1*H* **[43a]** / 2*H* **[43b]**	1*H*, $K_T \ggg 1$	1*H*	2*H*[c]

[a] Subst = substituted.
[b] No common cation, see discussion in Sections 4-1D and 4-1K.
[c] No common cation (II-35), see discussion in Section 4-1K.

It remains to rationalize the results for the triazoles and for tetrazole. These compounds may be considered as azapyrazoles or azaimidazoles, as follows. *v*-Triazole is 3-aza- [**44a**] or 5-aza-pyrazole [**44b**]: as the aza substituent has a positive σ_m value, [**44a**] should be preferred; however, this is not the case in solution, where [**44b**] seems to predominate; for the vapour phase the situation is uncertain. Tetrazole [**45**] is 2,4-diaza- [**45a**] or 2,5-diaza-imidazole [**45b**]: the 2-aza substituent is constant, and the 4(5)-aza substituent should favour the 4-aza structure, as is indeed found. On the above rationalization, using σ-values *s*-triazole

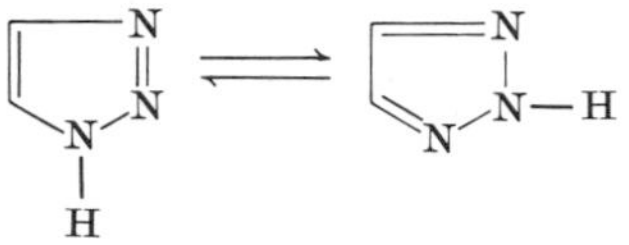

[**44a**] [**44b**] 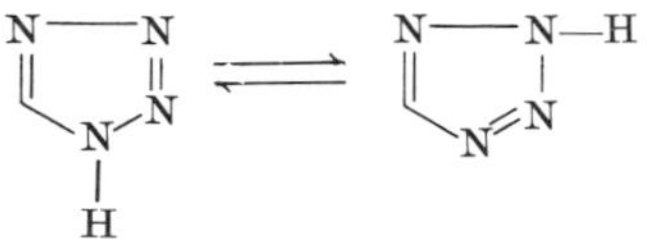 [**45a**] [**45b**]

should exist as 4-aza- [**46a**] rather than as 5-aza-imidazole [**46b**] whereas the reverse is found. Clearly other effects play their part; at least in six-membered rings, structures with lone electron pairs on each of two adjacent pyridine-like nitrogen atoms appear to be disfavoured (Section 2-4Cd), and this may influence [**44**] and [**46**]. It is curious that tetrazole favours the structures which maximize the number of adjacent pyridine-like nitrogens while *s*-triazole appears to minimize them. (The question of *v*-triazole remains confused, see Section 4-1D). The field awaits further investigation.

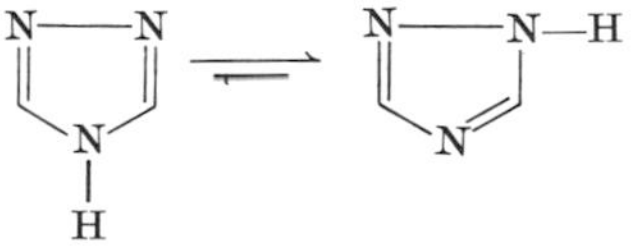

[**46a**] [**46b**]

K. Conjugated Acids: Protonation Sites

a. Diazoles

Pyrazole, imidazole, indazole and benzimidazole all undergo protonation at the pyridine-like nitrogen to give aromatic cations of type [**47b**], [**48b**]. Other nonaromatic cations exist, with protonation having occurred at the pyrrole-like nitrogen [**47c**], [**48c**] (see Section 3-1Ba) or at carbon [e.g., **47a, 48a**]. However, general considerations of basicity indicate (68BF5009) that for the pyrazole cation [**47c**] represents less than 0.1% in the equilibrium mixture: the actual contribution for all these nonaromatic species is probably far less than this.

Some completely bogus interpretations of tautomeric equilibria continue to appear: a horrifying example discusses equilibrating cations of type [**47c**] (3,4- or 4,5-disubstituted) (74JH189).

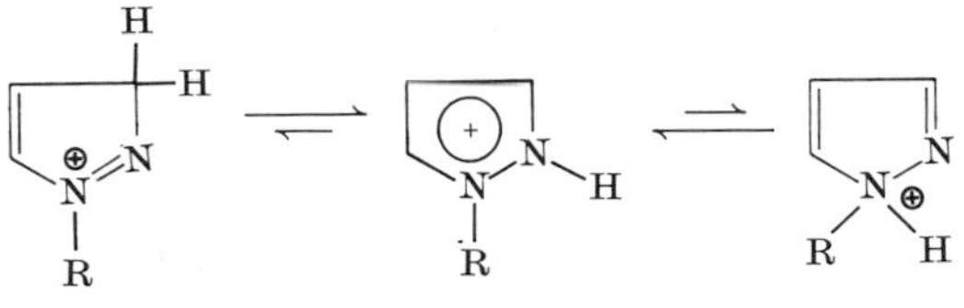

[**47a**] [**47b**] [**47c**]

Experimentally, for the parent compounds and for 1-alkyl-pyrazoles, -indazoles, -imidazoles and -benzimidazoles, only the forms [**47b**], [**52**], [**48b**] and [**49**] have been detected in the NMR from the NH/CH coupling in sulfuric or fluorosulfonic acid (62TL913, 63AG300, 63BG470, 67BF2998, 67J(B)1251, 68BF5009, 69JL(4)108, 74OR272).

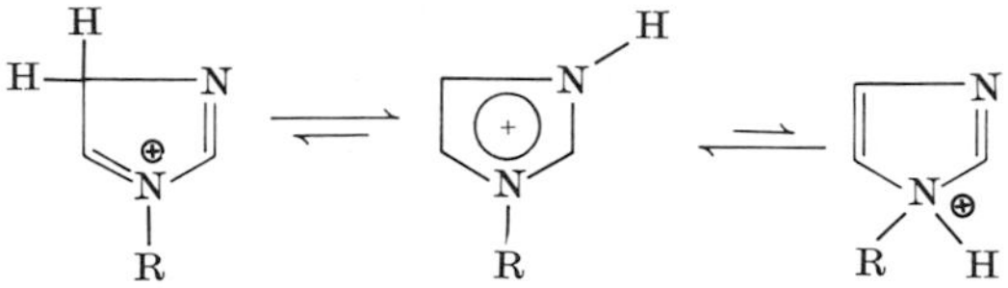

[**48a**] [**48b**] [**48c**]

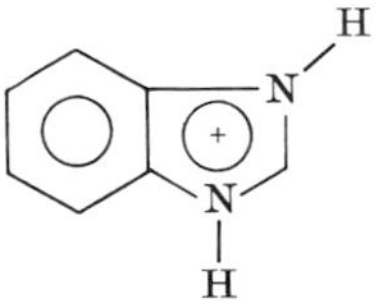

[**49**]

An interesting result is obtained by dissolving *N*-deuteriated 4(5)-bromoimidazole and 4(5)-bromo-2-methylimidazole in sulfuric acid (or alternatively by dissolving the nondeuteriated imidazoles in deuteriosulfuric acid). NMR shows that the cation [**51**] corresponding to protonation of the 4-bromo tautomer of the original imidazole [**50**] is obtained (63AG300). See discussion in Sections (1-2) and (4-1Cb).

A recent X-ray study (72JA4034) shows the imidazole cation to have C_{2v} symmetry which is consistent only with structure [**48b**].

Infrared studies of cations of type [**48b**] (64CC2292, 70CP951) and a complete UV study (absorption, fluorescence, phosphorescence, polarized UV) of benzimidazolium [**49**], indazolium [**52**], and benzotriazolium

[**53**] perchlorates (63BF54) afford further evidence for the structures given.

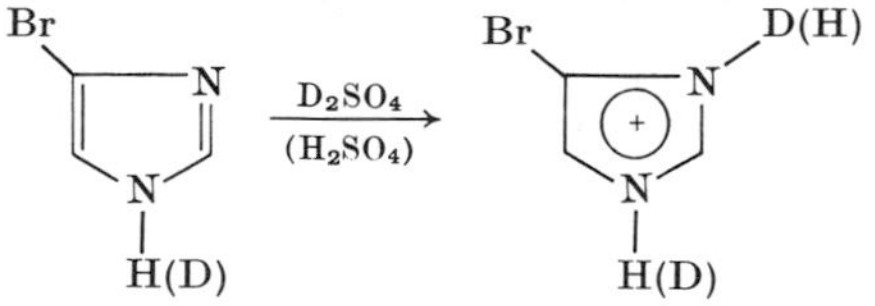

$$\xrightarrow[(H_2SO_4)]{D_2SO_4}$$

[**50**] [**51**]

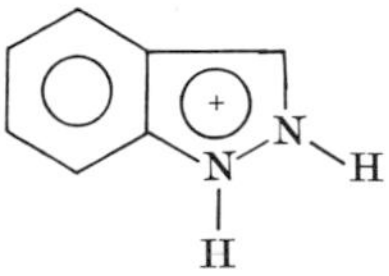

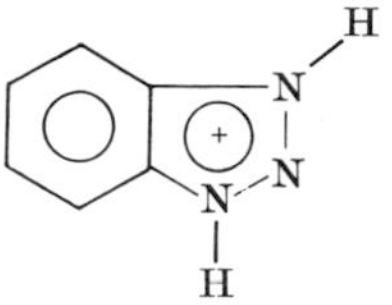

[**52**] [**53**]

b. Triazoles

For triazoles, even if the nonaromatic cations are left out, two possible protonation sites still remain. The few experimental results are consistent with protonation on a β-nitrogen atom with respect to the pyrrole-like nitrogen being more favoured than protonation at an α-nitrogen: CNDO/2 calculations (68TL3727) show that the tautomer with the protons maximally separated is more stable.

i. *v-Triazole and Benzotriazole.* An attempt to resolve NH/CH coupling, for *v*-triazole in fluorosulfonic acid at $-17°$, failed (67BF2998). The connection between the K_T for neutral species and monocations, and the basicities of the neutral species, is discussed in Section 4-1Db for *v*-triazole and in (II-35) for benzotriazoles.

ii. *s-Triazoles.* Comparison of the protonation shifts from deuteriochloroform to trifluoroacetic acid (67J(B)516) of 1- and 4-methyl-*s*-triazole with those for 1-methylimidazole indicates that protonation of the triazoles gives the cations stabilized by amidinium resonance [**54a, 55**] and not the alternative cations of type [**54b**]. As 1- and 4-methyl-*s*-triazole yield cations of similar structure, K_T can be calculated from the pK_a measurements (Section 4-1Ebi).

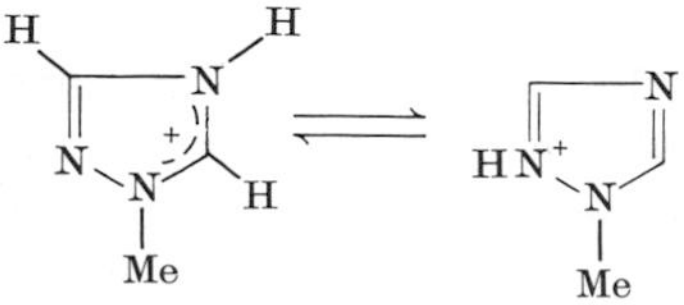

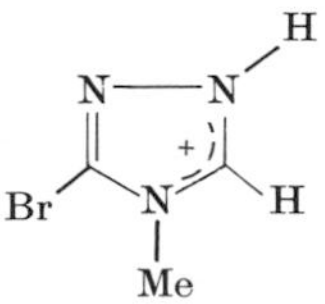

[**54a**] [**54b**] [**55**]

c. Thiadiazoles and Tetrazoles

UV and basicity measurements of the 1,2,3-benzothiadiazole cation (66J(B)469) suggest protonation at the 3-position [**56a**], but the presence of a small amount of the 2-protonated species [**56b**] cannot be ruled out. From pK_a measurements Russian authors (72RO2414) conclude that 4-position protonation occurs for 1-phenyl-1,2,3,4-tetrazoles.

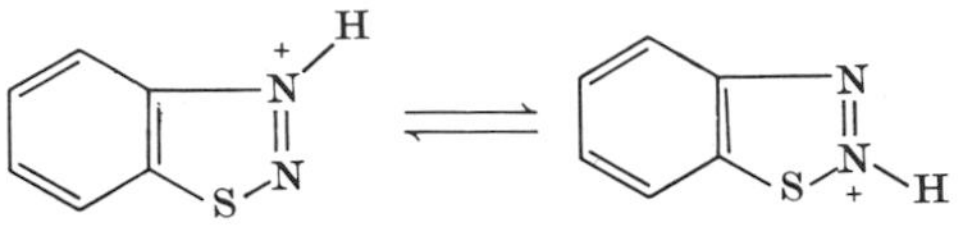

[**56a**] [**56b**]

2. Compounds with Potential Hydroxyl Groups: Heteroatoms-1, 2 and a Potential 3-(or 5-) Hydroxyl Group

A. Introduction

Probably more work has appeared on potential hydroxyazoles during the last decade than on any other class of heteroaromatic tautomerism. Azoles with two potential tautomeric groups* are now considered in a separate section (4-6A to G); it is also convenient to consider 1,2-heteroazoles with a potential 3- or 5-hydroxy group separately, owing mainly to intense activity in the investigation of isoxazolones and pyrazolones. Sufficient work is now available for the section dealing with potential hydroxyazoles with three ring heteroatoms to be arranged rationally (see Section 4-3Da).

In other respects the organization of the previous review has been largely retained: we consider successively compounds with heteroatoms-1,2, -1,3, -1,2,4 and -1,2,3. Within each classification by the arrangement of heteroatoms, we subdivide first by the orientation of the oxygen function, and finally by whether an oxygen, sulphur, or wholly nitrogen heterocycle is under consideration.

Results are summarized in Section 4-3F.

B. Isoxazol-5-ones (II-37)

a. Introduction

Isoxazol-5-ones can exist in three tautomeric structures which are denoted the CH [**57a**], NH [**57b**] and OH [**57c**] forms. At the time of

* This is taken to include derivatives alkylated at a site of potential tautomerism: thus, e.g., alkoxyisoxazolones are considered as derivatives of potential dihydroxyisoxazoles.

the previous survey (II-37) the general outlines of the tautomeric behaviour of isoxazol-5-ones had already become recognized: the rare occurrence of the OH form [**57c**] which was detected only when chelation took place between the hydroxyl group and the 4-substituent; the CH-form [**57a**] predominates in nonpolar solvents, but its percentage decreases in favour of the NH form [**57b**] as the medium becomes more polar. These results were based on dipole moments, and IR and UV spectra.

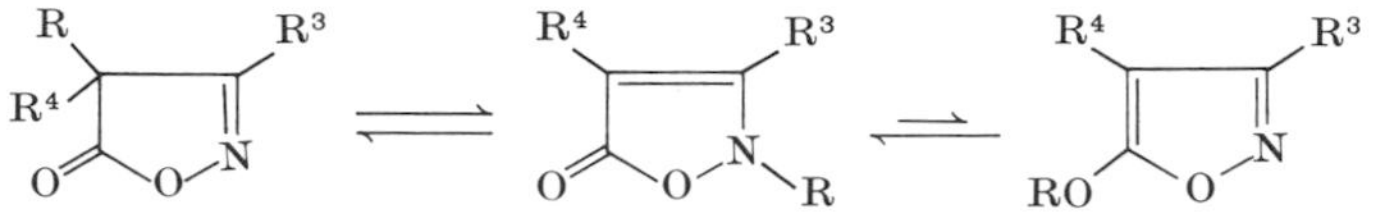

[**57a**] R = H [**57b**] R = H [**57c**] R = H

[**58**] R = Me [**59**] R = Me [**60**] R = Me

Since then intense activity in this field has utilized nearly all the available methods. We first indicate the advantages and limitations of the more usual methods as applied to the study of isoxazolones.

i. UV spectroscopy (Section 1-6D) of an isoxazol-5-one [**57**] and its three methylated derivatives [**58–60**] enable the evaluation of the percentages of the three tautomeric forms. If one form can be eliminated, the dipole moment measurements on the parent and model compounds (Sections 1-4G, 1-7D) can give the proportions of the two other tautomers.

ii. IR spectroscopy (Section 1-6C) cannot easily detect the OH form [**57c**] in a mixture with [**57a**] and/or [**57b**] because the other forms give rise to absorptions in the same region. However, the CH form [**57a**] and NH form [**57b**] are easily differentiated by their C—O bands.

iii. NMR spectroscopy (Section 1-5A) does not differentiate between the NH [**57b**] and OH forms [**57c**] because of too fast interconversion, but these are clearly distinguished from the CH form [**57a**].

The results summarized in Table 4-12 show that the solvent influences the equilibrium markedly for all the isoxazol-5-ones. Until a recent paper (70BB343), the dielectric constant and the aprotic nature of a solvent were considered the main solvent factors. However, although hydroxylic solvents favour the NH form [**57b**], so do some aprotic solvents, and it was shown that increasing the basicity and hydrogen bonding power of the solvent shifts the equilibrium from the CH form [**57a**] to the NH form [**57b**] (70BB343, 71J(C)86).

b. Occurrence of the OH form [**57c**]

This form does not normally contribute significantly to the equilibrium, but it does become the predominant form when stabilized

TABLE 4-12

TAUTOMERIC COMPOSITION OF ISOXAZOL-5-ONES

5-Isoxazolone							
R^3	R^4	Solvent	Method	Reference	%CH [**57a**]	%NH [**57b**][a]	%OH [**57c**][a]
H	H	$CHCl_3$	IR, UV, NMR	71J(C)86	100	—	—
		Dioxan	IR, UV, NMR	71J(C)86	90	10[b]	
Me	H	CCl_4	NMR	62T777	100	—	—
		DMSO	NMR	70BF2690	62	38	
		H_2O	UV	62T777	70	30	—
Ph	H	$CDCl_3$	NMR	70BF2690	100	—	—
		Dioxan	Dipole moment	68T151	90–100	0–3	7–0
		DMSO	NMR	70BF2690	40	60	
		H_2O	UV	62T777	70	30	—
H	Me	$CHCl_3$	pK_a, UV	67JH533	100		—
		aq. H_2SO_4	pK_a, UV	67JH533	—	100	—
H	Ph	$CHCl_3$	pK_a, UV	67JH533	—	100	—
		Dioxan	Dipole moment	68T151	0–5	50–56	49–51
		aq. H_2SO_4	pK_a, UV	67JH533	—	100	—

Me	Me	$CDCl_3$	IR, UV, NMR	70BF2690, 70BB343	70–80			30–20	
		Dioxan	Dipole moment	68T151	28–30		48–58		23–13
		DMSO	IR, UV, NMR	70BF2690, 70BB343	—			100	
		H_2O	IR, UV	67T831	—		100		—
Me	Ph	$CDCl_3$	pK_a, NMR	67JH533, 70BF2690	—		100		—
		Dioxan	Dipole moment	68T151	0–5		74–78		24–19
		DMSO	NMR	70BF2690	—			100	
		aq. H_2SO_4	pK_a, UV	67JH533	—		100		—
Ph	Me	C_6H_{12}	UV	62T777	100		—		—
		$CHCl_3$	NMR	62T777, 70BF2690	40		60		—
		Dioxan	Dipole moment	68T151	37–39		53–59		9–4
		DMSO	NMR	70BF2690	50			50	
		H_2O	UV	62T777	—		100		—
Ph	Ph	$CHCl_3$	UV	67JH533		100			—

[a] A percentage given in the middle between two columns means that the authors could not determine whether they had one of the two forms or a mixture of them.

[b] Both OH and NH forms participate.

by chelation with a suitable 4-substituent, e.g., a ketone or ester carbonyl as in [**61a**] (II-37, 62T777, 63G964). This has been confirmed for R = alkoxy, but when R = alkyl, two recent papers have shown that the enolization takes place in the 4 acyl group; according to one report (73T4291) the predominant tautomer has a chelated structure [**61b**], but the other (75JH85) presents convincing spectroscopic data in favour of tautomer [**61c**] (comparable to 4-acyl-pyrazolones, Section 4-2Gk). Recent IR and UV evidence (72J(PI)90) shows that isoxazolone phosphonate esters [**62**] exist in $CHCl_3$ mostly in the OH form [**62a**] if R^3 = Ph or 2-thienyl, but in a mixture of NH [**62b**] and OH forms [**62a**] if R^3 = Me, showing the influence of the 3-substituent.

Careful combination of NMR and dipole moment evidence suggests that nonchelated isoxazol-5-ones exhibit considerable proportions of the OH form [**57c**] in dioxan solution (68T151); see Table 4-12.

The IR and NMR arguments, utilized by Wamhoff and Korte (66CB2962, 66TL3919) to assign compound [**63**] the OH form shown, are inconclusive. Basic solvents favour the OH form (74BB263, 74OR224).

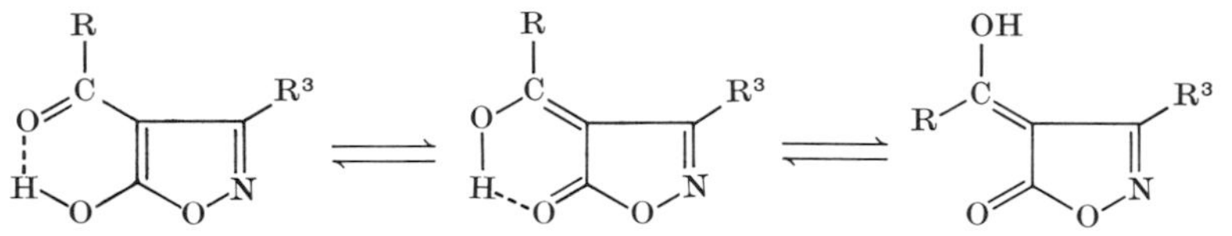

[**61a**] [**61b**] [**61c**]

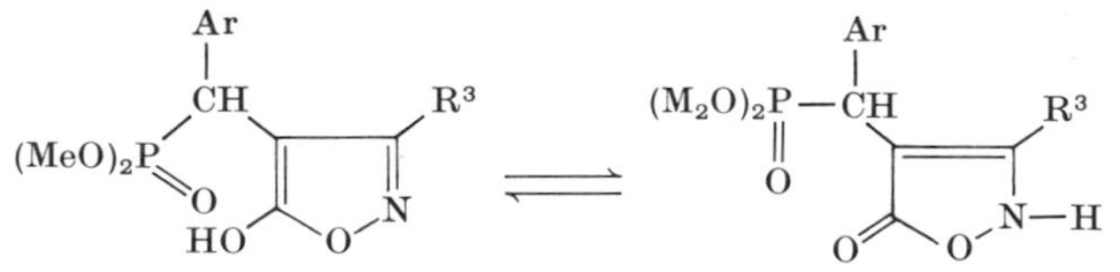

[**62a**] [**62b**]

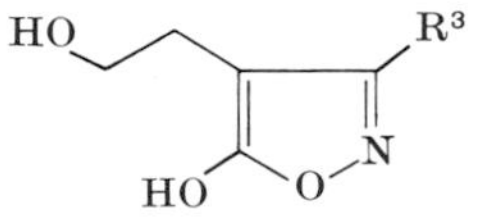

[**63**]

c. *Unsubstituted Isoxazol-5-one* ($R^3 = R^4 =$ H)

De Sarlo *et al.* (71J(C)86) used pK_a, UV and NMR spectral evidence to show that in nonpolar solvents ($CHCl_3$, cyclohexane) the CH form [**57a**] exists almost exclusively whilst in aqueous solution an equilibrium between the CH form [**57a**] and the NH form [**57b**] occurs (see Table

4-12). In dioxan all three tautomers are probably present. The authors do not use their coupling constant magnitudes: J(3H4H) is larger in the *N*-methyl compound [**59**; J = 3 Hz] than in the *O*-methyl derivative [**60**; $J \sim$ 2 Hz] and the higher value for J(3H4H) occurs again for the NH compound, which is evidence for the NH form [**57b**; J = 3 Hz]. (See discussion on this use of coupling constants in Section 4-2Fc.)

French authors have investigated the tautomeric equilibrium in isoxazol-5-one by several semi-empirical methods (72JL(12)191). In the absence of specific interaction with the solvent, the CH form [**57a**] is energetically slightly favoured over the NH form [**57b**]: calculated dipole moments and electronic spectra agree with experimental data.

d. 3-Substituted Isoxazol-5-ones (R^4 = H, $R^3 \neq$ H)

An X-ray study (69CX(B)1050) shows that 3-phenylisoxazol-5-one exists in the CH form [**57a**] in the solid state. In spite of the coplanarity of the two rings,* there is no evidence from bond lengths of conjugation between them. Mass spectroscopic fragmentation of 3-phenylisoxazol-5-one has been described in terms of the OH form [**57c**, R^3 = Ph] (69AJ563) and also in terms of the CH structure [**57a**, R^3 = Ph] (69T747).

In solution, all the methods used agree: dipole moments (68T151) UV, IR and NMR spectroscopy (63G964, 70BF2690) and pK_a measurements (67JH533). The CH form [**57a**] always predominates in solution, but its percentage decreases as the polarity of the solvent increases (EtOH to H_2O to DMSO) or if R^3 is a cycloalkyl group (67BF3003, 70BF2690).

e. 4-Substituted Isoxazol-5-ones (R^3 = R = H, $R^4 \neq$ H).

The tautomerism of two 4-substituted isoxazol-5-ones, R^4 = Me or Ph has been investigated. All the methods, pK_a, UV (67JH533, 71J(C)86) and dipole moments (68T151), indicate that these 4-substituted isoxazol-5-ones show behaviour distinct from the 3-substituted compounds: a decrease in the stability of the CH form [**57a**] is observed in all solvents. For 4-phenylisoxazol-5-one, the CH form [**57a**] has not been detected, probably because in the CH form the phenyl ring cannot conjugate with the isoxazolone ring (68T151).† For 4-substituted

* For *N*-methyl-4-bromo-1-phenylisoxazol-5-one (67AL538, 69CX(B)192), the dihedral angle between the two rings is $\sim 50°$ because of steric hindrance between the methyl and phenyl groups (for a discussion of steric hindrance in model compounds, see Section 1-3B).

† An X-ray study (69CX(B)182) indicates possible conjugation in the *N*-methyl derivative as the two rings are almost coplanar.

isoxazol-5-ones, the solvent appears only to influence the equilibrium between the OH and the NH forms; in dioxan the OH and NH forms exist in equal proportions (68T151). In basic solvents the OH form predominates (74OR224).

f. 3,4-Disubstituted Isoxazol-5-ones (R^3, $R^4 \neq$ H)

In general 3,4-disubstituted isoxazol-5-ones resemble 4-substituted isoxazol-5-ones more closely than the 3-substituted analogues: the NH form [**57b**] is significantly favoured (67JH533, 67T831, 68T151, 70BB-343, 70BF2690, 71J(C)86).

Two papers deal with solvent and concentration effects upon the equilibrium of 3,4-dimethylisoxazol-5-one (67T831, 70BB343): the results follow the same pattern, but the second gives more accurate data using UV, IR and NMR spectroscopy. In concentrated solutions in nonpolar solvents the NH tautomers associate, so that the total equilibrium occurs between associated NH form, free NH form and CH form; at a high enough dilution the associated form can be neglected and the CH form [**57a**] becomes predominant. The direction of this tautomeric equilibrium is reversed in solvents in which specific association can take place between the isoxazol-5-one and the solvent. In dioxan, dipole moments indicate the presence of the OH form [**57c**] alongside the two other forms (68T151); Maquestiau *et al.* (70BB343) did not observe the OH form, but this is inherent in the methods used.

3-Methyl-4-phenylisoxazol-5-one favours the NH form (67JH533, 68T151, 70BF2690, 71J(C)86) in all solvents studied. The behaviour of 4-methyl-3-phenylisoxazol-5-one is less easily rationalized: 50% of CH form is found in DMSO (70BF2690), a solvent which usually favours the NH form. For other 3,4-disubstituted isoxazolones, see (73AS2802, 74BB263). The same authors have demonstrated the zwitterionic structure of 5,6,7,8-tetrahydro-4*H*-isoxazolo [3,4-*d*]azepin-3-ol (73AS3251) (see also Section 4-2Ca).

g. 4-Arylazoisoxazol-5-ones

4-Arylazoisoxazol-5-ones [**64**] are of interest because of their use as agricultural fungicides. They can exist in four tautomeric structures, the hydrazone form [**64a**] in addition to NH [**64b**], OH [**64c**] and CH forms [**64d**].

Their tautomerism has been studied by two groups who reach different conclusions; Summers *et al.* (64CI1264, 65J3312, 66E499) suggest that these 4-arylazoisoxazol-5-ones exist in the hydrazone form [**64a**] whereas Cum, Lo Vecchio *et al.* propose the NH form [**64b**]

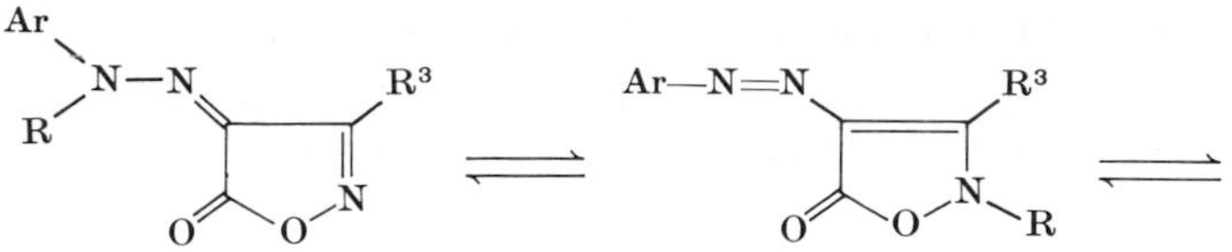

[**64a**] R = H
[**65**] R = Me

[**64b**] R = H
[**66**] R = Me

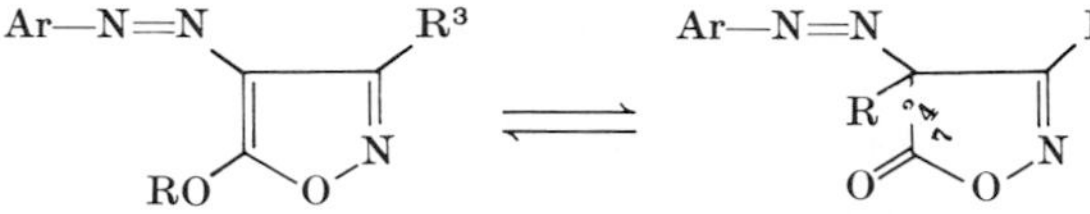

[**64c**] R = H
[**67**] R = Me

[**64d**] R = H
[**68**] R = Me

(65G583, 67G346). The first group draws its conclusion from UV, IR and NMR studies and comparison with the methyl derivatives [**65**], [**66**] and [**68**]. The νC=O at 1710–1730 cm^{-1} and an NH signal in the NMR spectrum near 12.7 ppm, independent of the concentration, eliminate the OH [**64c**] and CH [**64d**] forms. UV comparisons with fixed methyl derivatives makes them prefer the hydrazone form [**64a**] to the NH form [**64b**]; this seems a reasonable conclusion.

The Italian authors favour the NH form [**64b**]. Their first paper (65G583) dealt with a comparative IR and NMR study of methyl compounds, but they confused one of the methylated derivatives [**66**] with an isomeric triazole [**69**] formed during the reaction (66E499). Later on (67G346), they remained in favour of the NH form [**64b**] using IR evidence from a series of 4-arylazo-3-substituted isoxazol-5-ones. We do not consider their arguments convincing. The very complicated problem of 4,4′-azoisoxazol-5-ones (eleven tautomers) has been resolved (73G1045).

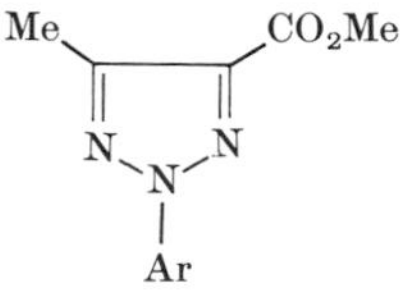

[**69**]

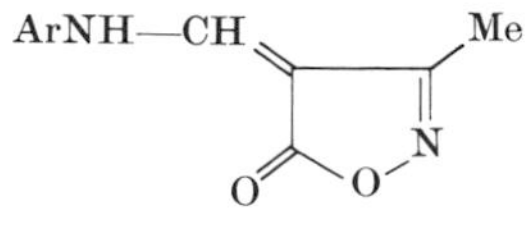

[**69/1**]

h. 4-Arylaminomethylene-3-methylisoxazol-5-ones

These compounds also possess four tautomeric forms, analogous to [**64a,b,c,d**] for the 4-arylhydrazonoisoxazolones just discussed. IR, NMR and UV data indicate that they exist as 4-methylene compounds [**69/1**], again similar to [**64a**] (73AJ889, 75JH27) (see for a related compound, 73T4291).

i. 5-(3-Methyl-5-oxoisoxazoliden-4-yl)-3-methylisoxazol-5-one

Compound [**70**] from hydroxylamine and ethylacetoacetate can exist in five tautomeric forms [**70a–e**]. French authors (67BF3003) concluded from IR and NMR data that the OH form [**70e**] was favoured, and this was supported by Italian (67IM1335; cf. 71T379) spectral evidence. However, Nishiwaki (69J(C)245; cf. also 69T747) in a further study including mass spectrometry suggested that the NH form [**70a**] and the OH form [**70e**] could exist either alone or as a mixture. The French author (70TH01) found convincing evidence for the OH structure [**70e**]: there is no measurable allylic coupling between the ring CH and the isoxazole ring methyl, whereas 2,3-dimethylisoxazol-5-one shows such coupling with $J = 0.8$ Hz. It appears to us that the OH form [**70e**] is predominant, but that some NH form [**70a**] could exist in rapid equilibrium without greatly modifying the proton spectrum.

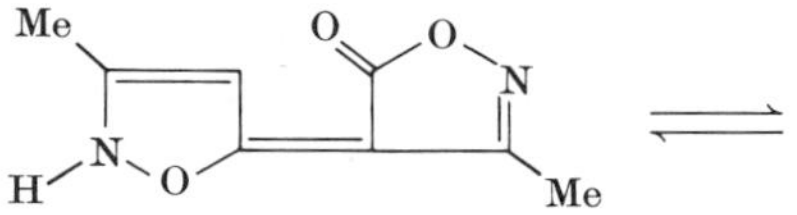

[**70a**] 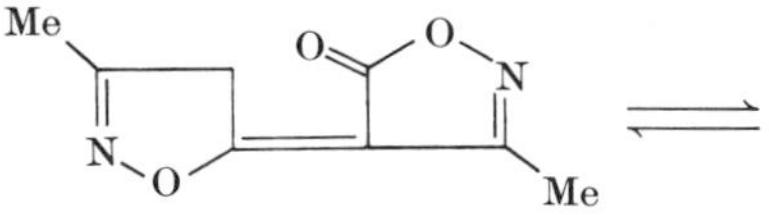 ⇌ [**70b**] 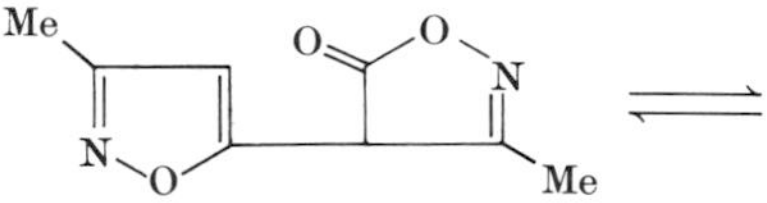 ⇌ [**70c**] 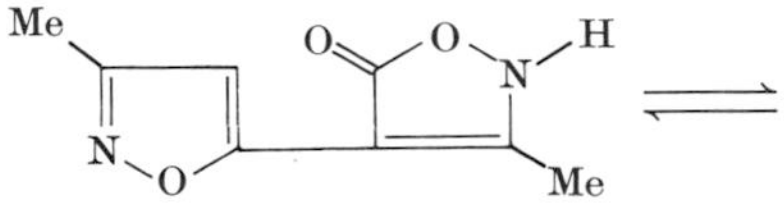 ⇌ [**70d**] 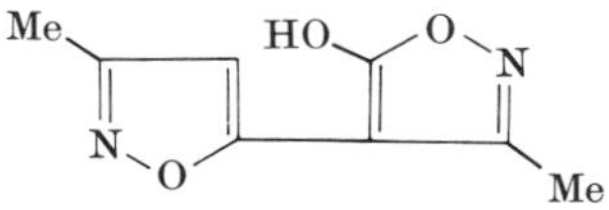 ⇌ [**70e**]

C. 3-Hydroxyisoxazoles (II-38) and 3-Hydroxy-1,2-benzisoxazoles

a. Introduction

The isoxazol-3-ones can exist in tautomeric OH [**71a**] and NH forms [**71b**]; for the benzo derivatives the corresponding forms are [**72a**], [**72b**]. At the time of the previous review (II-38) the few available results indicated the presence only of the OH form [**71a**]. This conclusion is confirmed by all the papers published since then, and it has also been shown that the benzoisoxazolones exist in form [**72a**]. In a extended study Krogsgaard-Larsen has shown, using a broad range of

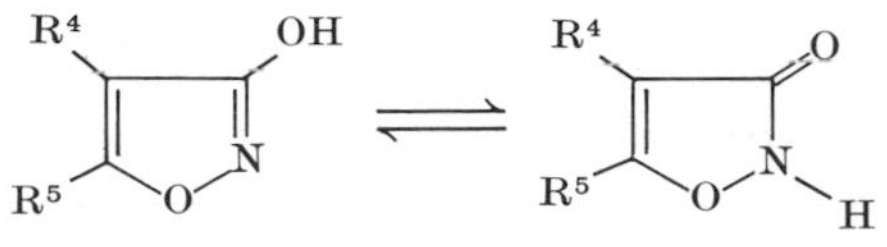

[**71a**] [**71b**]

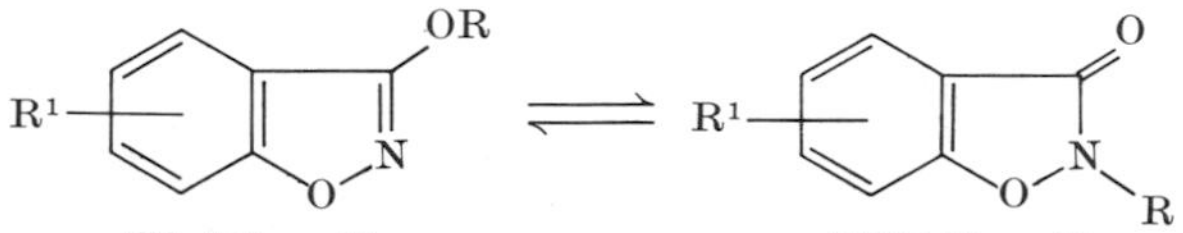

[**72a**] R = H [**72b**] R = H
[**73**] R = Me [**74**] R = Me

physical methods (including X-ray diffraction) that 3-hydroxyisozazoles carrying in the 4- or 5-position a polymethyleneamino chain exist as zwitterions with proton transfer from the hydroxy to the amino group (72AS1298, 74AS308, 74AS533).

b. 3-Hydroxyisoxazoles

i. *Basicity Measurements* (*Section 1-4A*). pK_a values (64T2835) for 4,5-dimethylisoxazol-3-one and its methyl derivatives indicate K_T = 5.2 in favour of the OH form [**71a**, R^4 = R^5 = Me]. As the authors point out, this value appears to be low, as the NH form has not been observed.

ii. *X-Ray Diffraction* (*Section 1-4Ib*). X-ray diffraction (69CX(B)-2108, 69JH901) shows that 3-hydroxy-5-phenylisoxazole exists in the OH form [**71a**, R^5 = Ph]. The two rings are slightly twisted; two different crystalline forms α and β obtained by crystallization in *n*-hexane or sublimation arise from crystal packing differences.

iii. *NMR Spectroscopy* (*Section 1-5A*). The OH [**71a**] and NH forms [**71b**] interconvert rapidly.

In CCl_4 and $CDCl_3$ (Table 4-13) a peak between 10 and 12 ppm is universally assigned to the hydroxy proton of the OH form [**71a**]. In DMSO or H_2O this peak disappears or becomes very broad because of a fast exchange in these solvents (64T2835, 66CT1277, 70BF1978).

iv. *IR Spectroscopy* (*Section 1-6C*). This method, which can often easily differentiate between lactim and lactam forms, is more difficult to apply here. Assignment based on νNH or νOH bands is often doubtful. No absorption (in the case of R^4 = R^5 = Me) above 3000 cm^{-1} in $CHCl_3$ was taken as proof of the OH form (64T165); but associated OH groups usually absorb above 3000 cm^{-1} in solution (64T2835, 67H137, 70BF1978) as well as in the solid state. More conclusively, 4,5-dimethyl-3-hydroxyisoxazole exhibits a band at 3550 cm^{-1} (64T2835, 70BF1978),

TABLE 4-13

3-HYDROXYISOXAZOLES: NMR INVESTIGATIONS

Solvent	Substituents		References
	R_4	R_5	
CCl_4	H	Ph	64T2835
CCl_4	Me	Me	
$CDCl_3$	H	H, Me	66CT1277, 67H137
$CDCl_3$	H	*i*-Pr, Ph	67BF3003, 70BF1978
$CDCl_3$	H	cyclo C_nH_{2n-1}	
$CDCl_3$	Me	Me, Et	
$CDCl_3$	—$(CH_2)_4$—		

corresponding to a free OH absorption, in highly dilute solution. Further literature confusion arose because the νC=O absorption of the *N*-methyl derivatives appears near 1660 cm^{-1} where the ring νC=C and νC=N of 4-substituted 3-hydroxyisoxazoles and the *O*-methylated derivatives also absorb (70BF1978). 4-Unsubstituted 3-hydroxyisoxazoles show this vibration at 1625 cm^{-1} (66CT1277, 70BF1978). A ring mode at 1530 cm^{-1} seems typical of 3-hydroxyisoxazoles and is absent in the *N*-methyl derivatives (70BF1978).

v. *UV Spectroscopy* (*Section 1-6D*) is not applicable because the fixed *O*-methylated and *N*-methylated derivatives absorb in the same region (63CB1088, 64T2835, 70BF1978).

c. *3-Hydroxy-1,2-benzisoxazoles*

In the solid state, IR spectra of the parent [**72**] and methyl derivatives [**73**] and [**74**] show that the OH tautomer [**72a**] is predominant both for the 5-chloro- (67CB954, 69CB3775) and the parent 3-hydroxybenzisoxazole (69JH123, 71JO1088).

In methanol, the UV spectra of the tautomeric compounds are similar to those of their fixed *O*-methyl derivatives [**73**] (67CB954, 71JO1088), which is consistent with the previous result. However, in $CHCl_3$, the IR spectrum of the tautomeric compound [**72**] appears to exhibit a νC=O absorption similar to that of the *N*-methyl derivative [**74**] in KBr (71JO1088) alongside the bands arising from [**72a**].

The NMR spectrum in $CDCl_3$ shows the peaks of the aromatic protons and one further signal only derived from the OH group of [**72a**] either alone or exchanging fast with the NH form [**72b**] (71JO1088).

D. Isothiazol-5-ones

In contrast to isoxazol-5-ones, investigation of isothiazol-5-ones has been undertaken only recently (71J(C)1314, 72TL327). Three tautomers exist: the CH [**75a**], NH [**75b**], and OH forms [**75c**]. In EtOH the UV spectra of the tautomeric compounds (R = Br or NO_2) absorb at longer wavelengths than the *O*-methyl derivatives. This suggests that the NH form [**75b**] is favoured and rules out the predominance of both the OH form [**75c**] and the less conjugated CH tautomer [**75a**] (71J(C)-1314). IR measurements in solid state are interpreted (71J(C)1314) to indicate that isothiazol-5-ones exist predominantly in the NH form [**75b**] with an important contribution of [**76**], a canonical form of [**75b**], to the resonance hybrid. The presence of νNH and the lack of νC=O near 1700 cm^{-1} led to this suggestion, but we believe that there is a problem in differentiating the OH and NH forms in solid state, as is discussed in detail for the 1-substituted pyrazol-5-ones (Section 4-2Fc). NMR spectra in DMSO ruled out the CH form [**75a**]. The authors

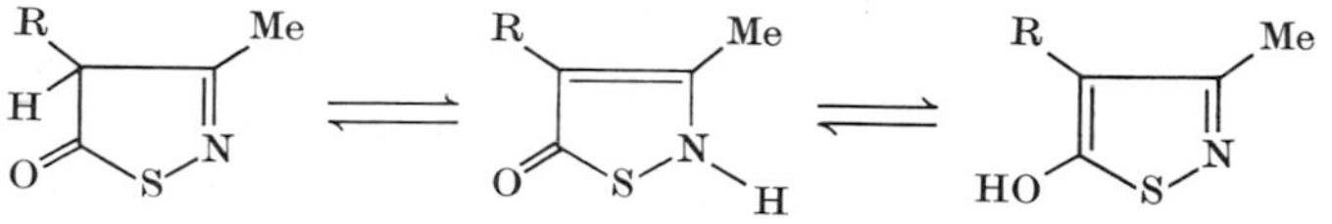

[**75a**] [**75b**] [**75c**]

conclusion that "these observations are in general agreement with those reported for the analogous 5-hydroxyisoxazoles" seems premature because at present no detailed comparison can be made: the results obtained for the isothiazol-5-ones are qualitative only and cover too narrow a range of solvents. Furthermore few isoxazol-5-ones substituted at the 4-position by a bromine or a nitro group have been studied (Section 4-2Be), and we have noted the importance of the nature of the 4-substituent on the equilibrium position of isoxazol-5-ones. Comparison of the available results with those for 4-bromo- and 4-nitro-1-substituted pyrazol-5-ones (Section 4-2Ge) is more justified: in both series the CH form is not favoured in DMSO.

The same authors (72TL327) have synthesized 4-carbethoxy-3-methyl-isothiazol-5-one but did not give any information about its tautomerism. Benzisothiazol-5-ones of type [**76/1**] exist, as expected, in the NH form [**76/1a**] on IR evidence (73CB376).

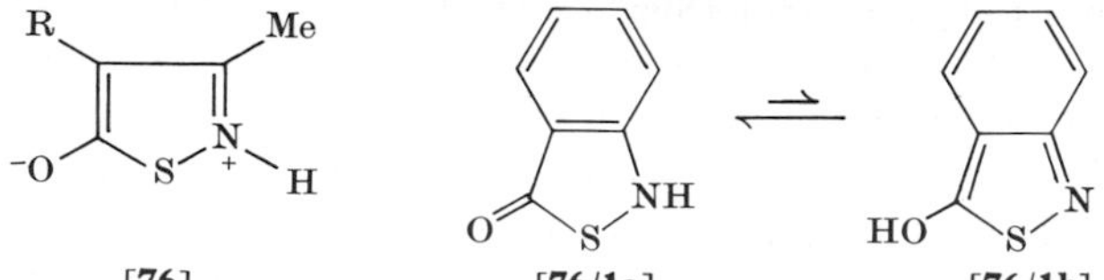

[**76**] [**76/1a**] [**76/1b**]

E. 3-Hydroxyisothiazoles and Benzisothiazol-3-one

In the previous survey no compounds of these series were reported, but the pattern of their tautomerism is now clear. Comparison of the IR and UV spectra of 5-phenyl-3-hydroxyisothiazole (63CB944) and its *O*-methyl [**78**, R^4 = H, R^5 = Ph] and *N*-methyl [**79**, R^4 = H, R^5 = Ph] derivatives indicates that the OH form [**77a**] predominates in the solid state and in methanol. After fragmentary work (64TL1477) on the parent 3-hydroxyisothiazole and its *N*-methyl derivative, a more complete study (70T2497) of UV and NMR results and comparison with both the fixed derivatives gave K_T in several solvents. In nonpolar solvents the OH form [**77a**] is greatly predominant (cyclohexane, $CDCl_3$) but its percentage decreases as the dielectric constant of the solvent increases (EtOH, $K_T = 4$; MeOH; DMSO; H_2O, $K_T = 0.3$). The investigators attempted to determine the equilibrium position by chemical methods, but there is no direct correlation between the ratios *O*-acyl/*N*-acyl or *O*-alkyl/*N*-alkyl and the tautomeric ratio [**77a**]/[**77b**].

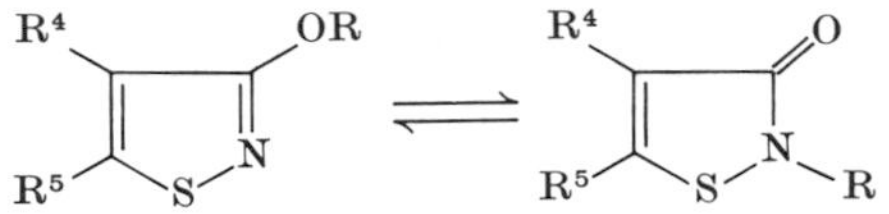

[**77a**] R = H
[**78**] R = Me

[**77b**] R = H
[**79**] R = Me

More recently (71JH571, 71JH581) American authors have prepared several 3-hydroxyisothiazoles; they draw the tautomeric compounds in the oxo form [**77b**] without any discussion of the tautomerism; however, their UV and NMR results show clearly that the tautomeric compounds and their *N*-methylated derivatives have different structures, which is evidence for the hydroxy form [**77a**]. Similarly, 3,4-diphenyl-3-hydroxyisothiazole is a true OH compound on spectral data (72J(PI)1432).

A theoretical approach to the tautomerism of 3-hydroxyisothiazoles (68AJ1113) by LCAO–SCF calculations gives total π-energies favouring the OH form [**77b**, $R^4 = R^5$ = H]; however, this is reversed when atomic cores are included. Introduction of sulfur 3d orbitals also affects the results.

Up to now X-ray diffraction has provided the only evidence about the predominant tautomer [**80a**] or [**80b**] of benzoisothiazol-3-ones. In

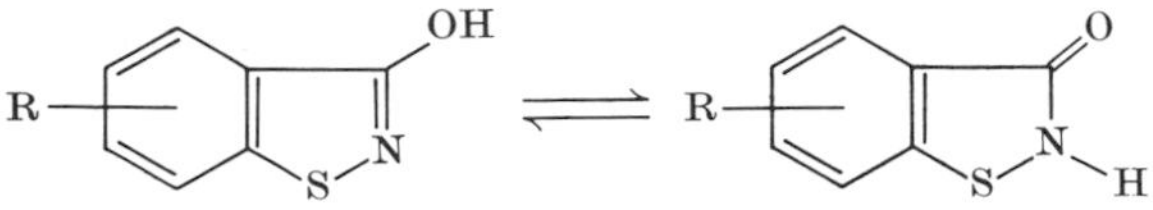

[**80a**]

[**80b**]

solid state the parent compound (70G629) and its 7-chloro derivative (69CX(B)2349) both exist in the NH form [**80b**]: this result is consistent with the isothiazol-3-ones previously described as the solid state tautomer is usually that which is predominant in aqueous solution.

F. 1-Substituted Pyrazol-5-ones (II-38): Individual Tautomeric Forms

a. Introduction

The tautomerism of 1-substituted pyrazol-5-ones is probably the most studied of any heteroaromatic system. The previous review (II-38) reported the fragmentary results of some twenty-five papers at a time when order was just being introduced into the field. In the ten years since then, some fifty further papers have completely transformed the picture and made this field well understood. The approaches used for the pyrazol-5-ones provide a good model for a tautomeric investigation in heterocyclic chemistry.

From the available results, we first indicate those authors and papers reporting a systematic study of the tautomerism of 1-substituted pyrazol-5-ones: Grandberg (63ZO2597, 67M174); Katritzky (64T299) (the results of this paper were mentioned in the previous survey: II-40, cf. reference 52); Jacquier and Elguero (66BF775, 67BF3772, 67BF3780, 68BF5019); Newman and Pauwels (69T4605, 70T1571, 70J(C)1842, 70T3120); Maquestiau (71DD17, 71TL2929, 73BB215, 73BB233, 73BB747, 73BB757, 74T1225, 75JH85) and Deschamps (71T5779, 71T5795, 71T5807, 71TL2929, 74T1225). The main results for the 1-substituted pyrazol-5-ones, together with the procedure followed and the precautions required in tackling problems of tautomerism of heteroaromatic compounds in general, can be found in these papers. Very recently, the subject has been reviewed, with additional experimental results, by Dorn (73PC382).

Of all the physical methods available (Section 1-3), the four classical ones were used most frequently: basicity measurements (Section 1-4A), proton NMR (Section 1-5A), IR (Section 1-6C) and UV spectroscopy (Section 1-6D). The only papers dealing with other methods are the following: dipole moments (Section 1-4G) (67MI74), X-ray diffraction (Section 1-4I) (71CB2694, 73MI469, 73MI473), ^{13}C NMR (Section 1-5B) (70J(C)1842, 74JH135, 75UP1), ^{15}N NMR (75UP1). Very recently various quantum chemical approaches to the problem have been used successfully (papers of Deschamps previously cited).

The nomenclature of the previous review (II-40) for the CH [**81a**],

OH [**81b**] and NH [**81c**] forms has been adopted by most authors to name the three tautomeric structures of 1-substituted pyrazol-5-ones. By analogy we name the three corresponding fixed forms CR [**82**], OR [**83**] and NR [**84**]. We now describe for each of the tautomeric forms the physical properties used to identify it and at the same time discuss those of the corresponding model derivative to point out possible discrepancies between the NH and NR compounds.

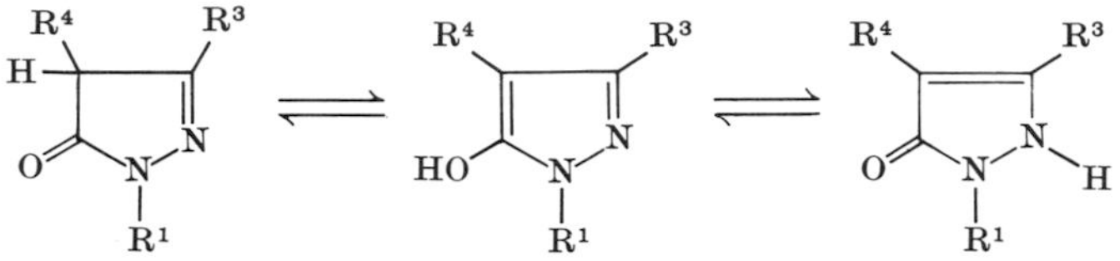

[**81a**] [**81b**] [**81c**]

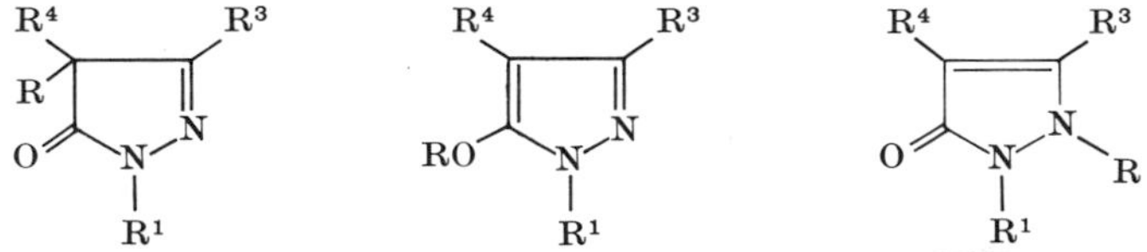

[**82**] [**83**] [**84**]

b. The CH Form [**81a**]

The CH form [**81a**] is the easiest to observe using NMR spectroscopy. The equilibrium between the CH form and the others falls into case 3 of Table 1-4 (Section 1-5Aa): equilibrium is reached immediately* but the CH species gives NMR signals distinct from those of the other forms: thus it is possible to estimate it quantitatively by integration. As previously discussed (Section 1-5Aa), this behaviour is typical for equilibria involving the cleavage of a CH bond.

In PMR the CH form [**81a**] is distinguished by the signal arising from the 4-proton (or CH_2 if $R^4 = H$) as well as by the spin–spin coupling of this proton with substituents at the 3- and/or 4-position. In ^{13}C NMR (70J(C)1842) the CH form is distinguished by the upfield signal arising from the sp^3 carbon at the 4-position (~ 150 ppm; value in DMSO upfield from CS_2), whereas the 5-carbon appears at about the same

* The only result which appeared not to agree with the generalization that equilibrium is reached immediately in solution, concerns 1-phenyl-3,4-dimethyl-pyrazol-5-one [**85**], the tautomeric proportions of which were thought to change with time (64T299). It was shown later that this compound underwent oxidation in these conditions (67BF3780). Because of the lack of spectroscopic data, the claimed isolation of diketo and dienol tautomers of a dipyrazolonylmethane (63JI833) cannot be accepted.

field as that of the NR form = ca. 25 ppm (both are part of a C=O group).

The IR spectra of the CH form [**81a**] exhibit a strong νC=O band at 1700–1720 cm^{-1}: this band is at the same frequency for the CH [**81a**] and CR [**82**] compounds (see Table 4-14 for examples) which demonstrates the lack of hydrogen bonding in the CH tautomer. The νC=O depends slightly on the nature of the 3- and 4-substituents and on the solvent: for 1-phenyl-3-methylpyrazolone [**86**], the absorption

TABLE 4-14

νC=O FOR PYRAZOLONES IN CH AND CR FORMS IN $CHCl_3$ (64T299)

Substituent	1,3-diMe	1,3,4,4-tetraMe	1-Ph-3-Me	1-Ph-3,4,4-triMe
νC=O (cm^{-1})	1694	1695	1706	1705
Structure	[**87**]	[**88**]	[**86**]	[**89**]

moves from 1715 cm^{-1} in acetonitrile to 1697 cm^{-1} in MeOH (73BB215). The νC=N absorption can also be observed at 1600–1620 cm^{-1}, provided the pyrazolone does not bear any aromatic substituent.

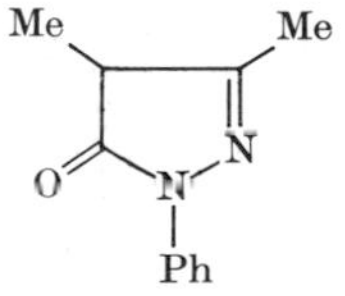

[**85**] 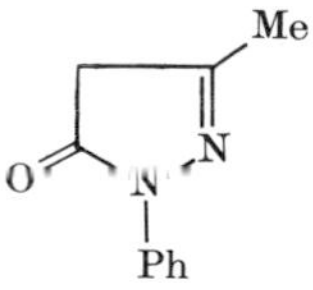 [**86**] 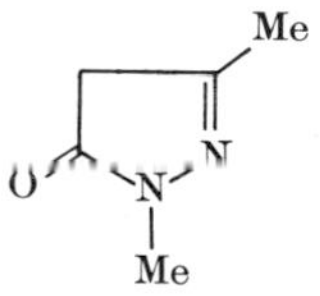 [**87**]

Cogrossi (66SA1385) using CCl_4 solutions in the near infrared assigned to the CH tautomer (R^4 = H) the bands at 5270 cm^{-1} (second overtone of νC=O), 5730 cm^{-1} (first overtone νCH of aliphatic CH_2) and 4290 cm^{-1} (CH_2 combination band).

In the UV spectrum and in a solvent such as cyclohexane the CH form [**81a**], and the CR model [**82**], are distinguished by a maximum around 240–250 nm with an extinction coefficient of about 5000 in the *N*-methyl series [**81a**, R^1 = Me] and 15,000–20,000 in the *N*-phenyl series [**81a**, R^1 = Ph] (64T299). These absorption maxima are not sufficiently far from those of the other tautomeric forms (taking into account the band-width in solution) to resolve the maxima characteristic of each tautomer; nevertheless it is possible to evaluate from extinction coefficients the proportions in a mixture of two tautomers (67BF3780). It is not possible to differentiate the CH and OH forms of

N-aryl derivatives [**81a**, **81b**, R^1 = Ph] by UV spectroscopy because their absorption is similar, CH $\lambda_{max} \sim 250$ nm, $\varepsilon_{max} \sim 19{,}000$ (in cyclohexane); OH, $\lambda_{max} \sim 235$ nm, $\varepsilon_{max} \sim 12{,}500$ (in H_2O) (64T299).

The CR model compound [**82**] is clearly the least basic among the three fixed compounds as shown by data of Scheme 4-4. It would be erroneous to deduce from this that the CH form must be predominant.

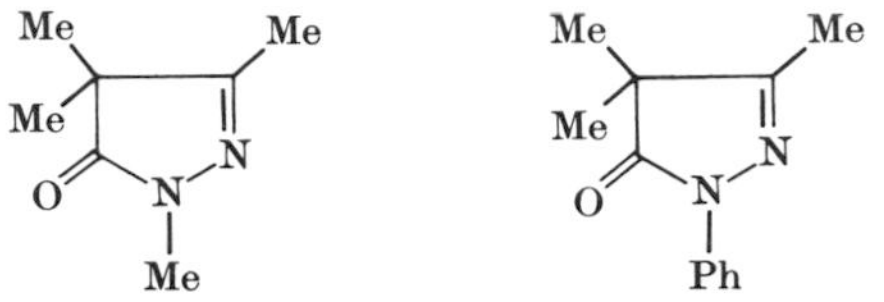

	[**88**]	[**89**]
pK_a:	−3.79 (64T299)	−4.02 (64T299)
μ(D):	—	2.83 (dioxan) (67MI74)

SCHEME 4-4

Such a deduction is valid only if a common cation is formed, and protonation of the CH-tautomer gives a cation different from that arising from the other two tautomeric forms (67BF3780).

Dipole moments in the *N*-aryl series (the only one to have been described) are close to each other for the CR [**82**, R^1 = Ph] and OR forms [**83**, R^1 = Ph] but that of the NR form [**84**, R^1 = Ph] is different.

c. The OH Form [**81b**]

In NMR spectroscopy, it is necessary to consider simultaneously the OH [**81b**] and NH forms [**81c**], because they undergo the fast exchange (Section 1-5Aa, case 5, Table 1-4) expected as the cleavage and formation of hydrogen-heteroatom bonds only is concerned. Up to the present nobody has claimed to have slowed down the interconversion rate sufficiently to observe signals arising from both OH and NH forms; in the literature all the examples deal with averaged signals and an interpolation method must be applied. In proton resonance the chemical shifts of the fixed derivatives are usually too close to allow an interpolation,* although differences in chemical shift between pyridine and $CDCl_3$ solution were used to prove the presence of the OH form for 1,4-dimethylpyrazol-5-one [**90**] in pyridine (67BF3780).

If the 3- and 4-positions are unsubstituted, the coupling constant J(3H4H) can be used (71LA(750)39); see Scheme 4-5. The coupling

* See (70ZC475) for an attempted correlation between the chemical shifts ($CHCl_3$) of various tautomeric forms and the π-density calculated by the HMO method.

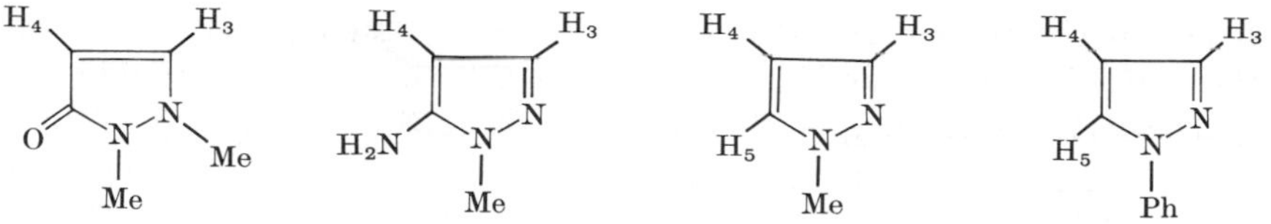

[91]

$^{3}J_{(3H4H)} = 3.50$ Hz (71LA(750)39) $^{3}J_{(3H4H)} = 1.9$ Hz (71LA(750)39) $^{3}J_{(3H4H)} = 2.0$ Hz $^{3}J_{(3H4H)} = 1.9$ Hz (66BF3727)

$^{3}J_{(3H4H)} = 3.60$ Hz (70AS1819) (*)

* The NMR spectra of compounds OR (R^1 = Me or Ph) unsubstituted in positions 3 and 4 are not known.

Scheme 4-5

constant between the 3–Me and the 4–H is 0.4 Hz for the OH and 0.8 Hz for the NH tautomer (74JH135). From the data cited, the coupling constant of 2.0 Hz observed for compound [**92**] (69T4605, 73BB215) is in favour of the OH structure [**92a**] rather than the NH form [**92b**]. This coupling constant, which does not vary much with the solvent and is, for instance, 2.2 Hz in nitrobenzene (72TH01, 73BB215), increases to 3.2 Hz in $CDCl_3$–TFA (72TH01, p. 59); in the cation [**93**] the C(3)–C(4) bond order is increased in agreement with theoretical prediction (71T5807).

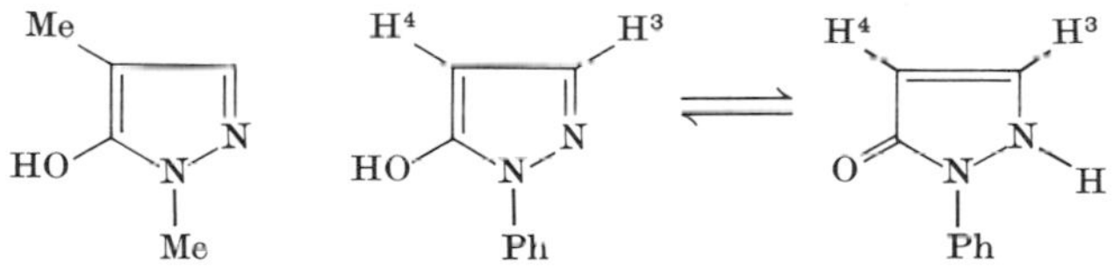

[**90**] [**92a**] [**92b**]

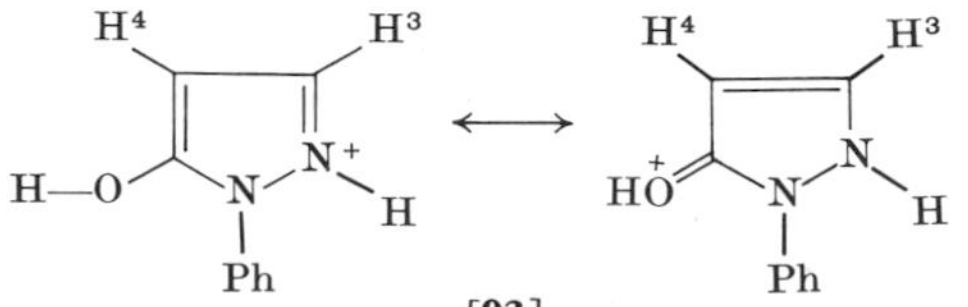

[**93**]

In ^{13}C resonance (Scheme 4-6) (70J(C)1842, 74JH135) the OR [**83**] and NR [**84**] model compounds exhibit very different signals for the carbon which bears the oxygen atom, and as 1-phenyl-3-methyl-pyrazol-5-one (no CH form in DMSO) gives a signal closer to that of the OR derivative, the authors conclude that the OH form predominates. The correct procedure is to apply Eq. (1-44) (Section 1-C) for P = δ ^{13}C; this gives $K_T = (9 + P^*)/(2 - P^*)$. It is difficult to estimate P^*, but

$\delta = 37$ $\delta = 26$ $\delta = 35$

(solvent DMSO, chemical shift in ppm from CS_2)

SCHEME 4-6

it is probably positive (the substitution of H by Me generally gives upfield shifts) and therefore K_T is large, in agreement with the authors cited. An ^{1}H, ^{13}C and ^{15}N study of the 1-phenyl-3-methylpyrazol-5-one labelled with ^{15}N in both nitrogens has been done (75UP1). ^{13}C NMR signals corresponding to the minor tautomer, the CH form [**86**] have been observed in DMSO.

In IR spectroscopy, it is necessary to consider separately the results obtained in solid state and in solution. In solution, the OH form is distinguished by the lack of a carbonyl band and by the presence of νCH near 3150 cm^{-1} (70T3429 [**81b, 83,** R^3 or R^4 = H] and a ring stretching vibration near 1560 cm^{-1} (63ZO2597, 64T299, 73BB215); this ring stretching is weaker for the 4-substituted compounds (64T299). The intensity of the 1560 cm^{-1} band has been used (63ZO2597) to estimate quantitatively the amount of OH form, assuming that $\varepsilon_{OH} = \varepsilon_{OR}$, but this hypothesis has been criticized (67BF3772). Normally, the OH and NH forms cannot be differentiated on the basis of OH and NH group vibrations (67BF3772).

In the solid state the differentiation of the OH and NH forms by IR spectroscopy becomes very difficult. As Katritzky and Maine (64T299) pointed out, strong hydrogen bonding blurs the distinction between the two forms: the NH form exists as [**94a**] and the OH form as [**94b**]. Proton transfer may occur in the crystal, which, in the limit, could render the two designations equivalent. This point of view has been assumed by most subsequent authors who frequently give the results in solid state in the form "OH/NH" which could indicate either an intermediate structure or the impossibility of differentiating the two structures by IR spectroscopy (67BF3772, 67BF3780, 69CT1485, 69T4605). However, replacing [**81b**] $\rightleftharpoons$ [**81c**] by [**81b**] $\leftrightarrow$ [**81c**] (67BF74) or even [**94a**] $\rightleftharpoons$ [**94b**] by [**94a**] $\leftrightarrow$ [**94b**] (69T4605), is not justified. Even if some proton tunnelling occurs, another technique (X-ray or better neutron diffraction) should differentiate between [**94a**] and [**94b**] by

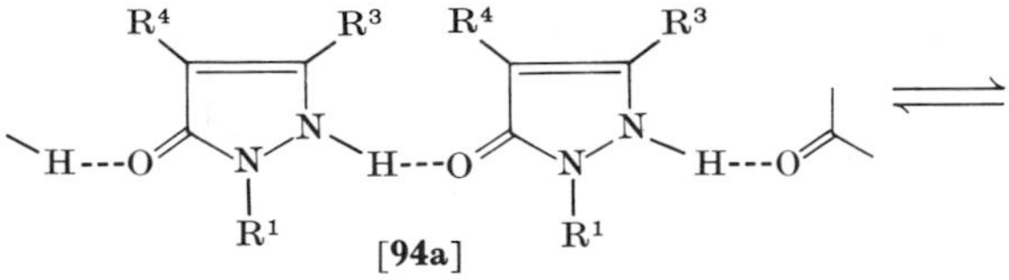

[**94a**]

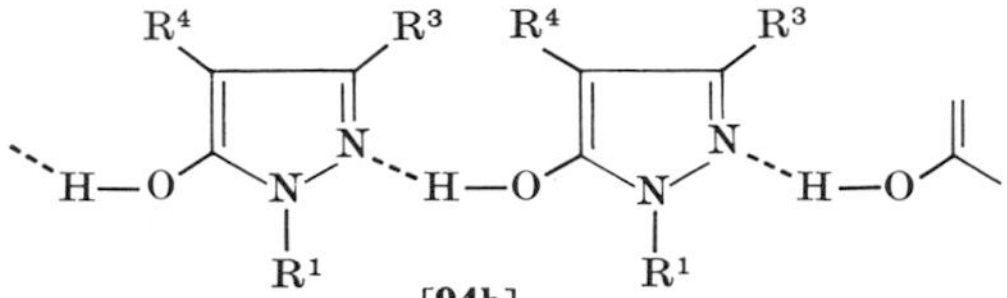

[**94b**]

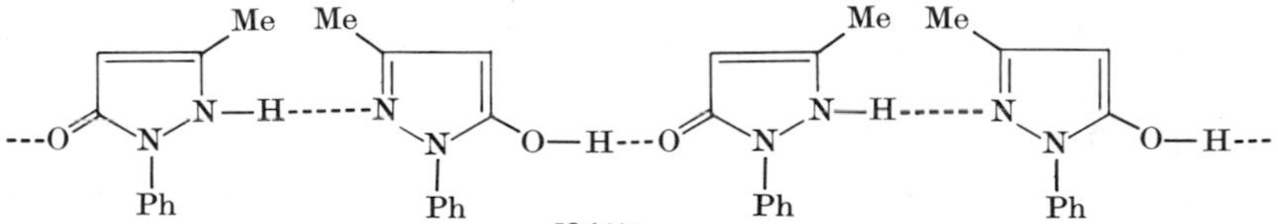

[**94/1**]

locating the tautomeric proton. Recently, this has been achieved, see later (Section 4-2Gb).

Newman and Pauwels (70T3429) thought they could solve the problem by IR spectroscopy, using a νCH band at $\sim 3150\ \mathrm{cm}^{-1}$ in the spectra of 1-substituted pyrazol-5-ones in the solid state; the authors proved it was a νCH band by partial deuteriation. However, as the model compounds OR [**83**] and NR [**84**] were not examined in that region, we do not consider their conclusion as definitive and we shall continue to discuss the tautomerism in terms of forms CH and OH/NH in the solid state.

In UV spectroscopy the OH form is easily distinguished for pyrazolones bearing only alkyl substituents, because it absorbs at short wavelength (maximum is hidden in some solvents) ~ 220 nm, $\varepsilon_{\max} \sim 5000$ (67BF3772). However, as we already mentioned, the CH and OH forms of the *N*-arylpyrazolones absorb similarly (67BF3772).

Basicity measurements (64T299) show that while the OR compounds [**83**] (Scheme 4-7) are the most basic of the fixed derivatives, the difference from the NR derivatives [**84**] is not large. We will discuss later the calculation of K_T from pK_a values. The dipole moment (67MI74) of the OR form is, as already mentioned, close to that of the CR form (Schemes 4-7 and 4-4).

d. The NH Form [**81c**]

The NMR behaviour and the IR spectra in the solid state of the NH tautomer [**81c**] has already been discussed in connection with that

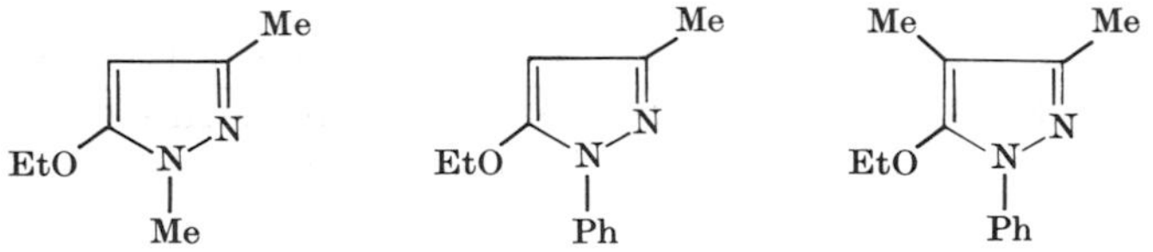

	[95]	[96]	[97]
pK_a:	+3.51	+2.34	+2.55
μ(D):	—	2.65 (dioxan)	—

SCHEME 4-7

of the OH form [**81b**]. The molecular structure of antipyrine [**84**, R = Me, R[1] = Ph, R[3] = R[4] = H] a classical model for the NH form has been determined (73CX(B)714). IR spectroscopy in solution can distinguish the NH tautomer because of the presence of νC=O at 1630–1680 cm^{-1} (63ZO2597, 64T299, 67BF3772, 71BB17, 73BB215). Some authors (64T299, 73BB215) report also a νC=C band characteristic of the NH or NR structure near 1590 cm^{-1}. The main disadvantage of the νC=O band of the NH form is its great sensitivity to solvent variation (67BF3772, 71BB17, 73BB215): νC=O for 1,2,3-trimethylpyrazol-5-one [**98**] moves from 1668 cm^{-1} in CCl_4 to 1578 cm^{-1} in D_2O. Thus care is required not to confuse the νC=O band with the νC=N of the CH form and/or the bands due to aromatic substituents: nonidentification of the NH tautomer of 3,4-dimethyl-1-phenylpyrazol-5-one [**85**] (63T1497) has been explained in this way (67BF3780) (see also 73BB215, 73BB233).

In UV spectroscopy the absorption of the NH tautomer [**81c**] and of the NR model [**84**] has been studied in detail (67BF3780); the bathochromic (+) and hypsochromic (−) effects of methyl substituents at different positions are shown in structure [**99**], where, for example, +9 nm thus corresponds to a comparison between N–H [**81c**] and N–Me [**84**]. Similar differences are found by calculation (71T5795).

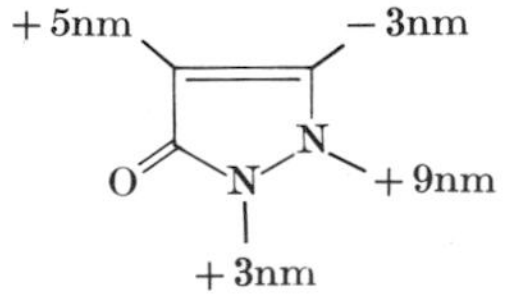

+5nm −3nm +9nm +3nm

[**99**]

For the *N*-alkylpyrazol-5-ones the NH tautomer absorbs at 290–260 nm with $\varepsilon_{max} \sim 10{,}000$ (67BF3772, 67BF3780, 71BB17), which is the same region as the CH tautomer but with double the extinction coefficient. For *N*-arylpyrazol-5-ones, the NH tautomer exhibits two maxima near 245 and 270 nm with comparable intensities ($\varepsilon_{max} \sim$ 8000) (67BF3780), the second one varies with the nature of the solvent, moving from 280 nm in $CHCl_3$ to 255 nm in water (67BF3772). The results from basicity and dipole moment measurements for the N–R derivatives [**84**] are given in Scheme 4-8. The basicity of the NH form is close to that of the OH form (Schemes 4-7 and 4-8). The dipole moment of the NH form is much higher than the two other tautomeric forms as predicted (64T299).

	[**98**]	[**100**]	[**101**]	[**102**]	
pK_a	+2.22	—	+1.40	+1.24	(64T299)
	+2.14	—	—	—	(67BF3772)
μ(D)	5.62 (dioxan)	5.03 (benz)	5.47 (benz)	—	(67MI74)

SCHEME 4-8

e. Properties of Individual Tautomeric Forms: Summary

In Table 4-15 we gather, with simplifications, the characteristics of the three tautomers of pyrazol-5-ones. Semi-empirical calculations justify the experimental results obtained from electronic spectra (71T5795), pK_a measurements (71T5807) and dipole moments (71T-5779).

G. 1-SUBSTITUTED PYRAZOL-5-ONES: TAUTOMERIC EQUILIBRIA

a. Calculation of K_T from Basicity Measurements

The calculation of K_T for pyrazolones from pK_a measurements is complicated by the existence of three bases and two conjugate acids.

i. *The Structure of the Pyrazol-5-one Conjugated Acids and Bases.* UV spectroscopy shows that the OR and NR forms yield a common cation with a structure of type [**104a**]* (64T299; also see references cited in 67BF3772) whilst the CR form gives a different cation (64T299). The structure of the cation formed by the CR model is not known with certainty; [**104b**] was proposed (67BF3780), and a later theoretical

* This is no longer true if the molecule carries a basic substituent; 4-dimethylaminoantipyrine is protonated at the dimethylamino group, not on the oxygen atom (65CC3322).

TABLE 4-15

CHARACTERISTICS OF THE THREE TAUTOMERIC FORMS OF 1-SUBSTITUTED PYRAZOL-5-ONES[a]

Tautomer	NMR (δppm)		IR (cm^{-1})	UV (nm)	pK_a	μ (D)
	^1H	^{13}C				
CH [**81a**]	δCH_2	$\delta C_4 \sim 150$	$\nu C{=}O$, 1700–1720	240–250 nm		
	J(3H4H)	$\delta C_5 \sim 25$	$\nu C{=}N$, 1600–1620	N-Me, $\varepsilon \sim 5000$	−4	N-Ph, 2.8(dioxan)
				N-Ph, $\varepsilon \sim 15{,}000$		
OH [**81b**]	J(3H4H) = 2.0	$\delta C_5 \sim 37$	Solution: absence of $\nu C{=}O$; pyrazole band 1560	N-Me, 220 ($\varepsilon \sim 5000$)	N-Me, +3.5	
			Solid: OH/NH	N-Ph 250 ($\varepsilon \sim 20{,}000$)	N-Ph, +2.5	N-Ph, 2.6(dioxan)
NH [**81c**]	J(3H4H) = 3.5	$\delta C_5 \sim 26$	Solution: $\nu C{=}O$, 1630–1680, $\nu C{=}C$, 1590	N-Me, 250–260 ($\varepsilon \sim 10{,}000$)	N-Me, +2.2	N-Me, 5.6(dioxan)
			Solid: OH/NH	N-Ph, ~245 and ~270	N-Ph, +1.3	N-Ph, 5.3(benzene)

[a] For the behaviour under electron impact (mass spectrometry) of the model compounds CR, OH and NR see (73BB747, 73BB757).

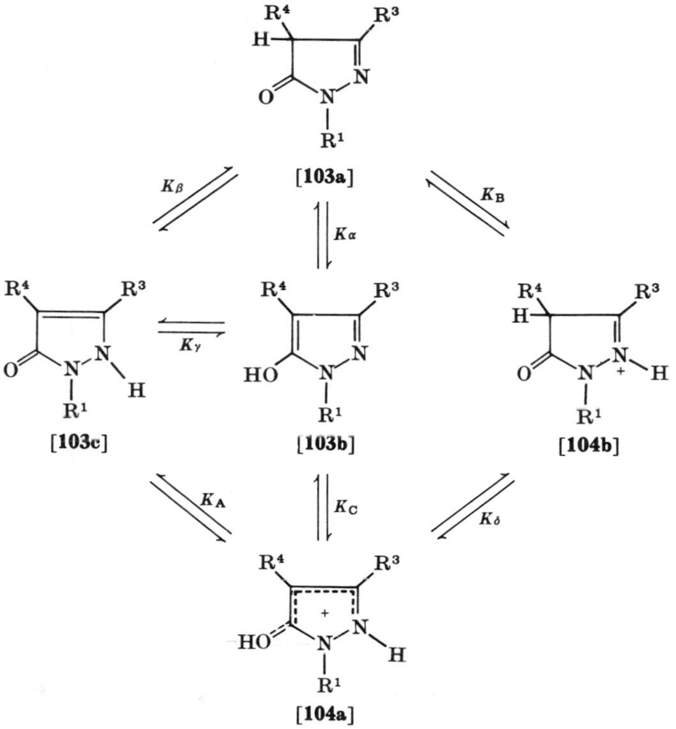

[**103a**]

[**103c**] [**103b**] [**104b**]

[**104a**]

SCHEME 4-9

calculation (71T5807) supported this. Recently (72CZ656), without any experimental support, structure [**105**] was proposed for the conjugate acid of the NH form; this is almost certainly incorrect. The structure of

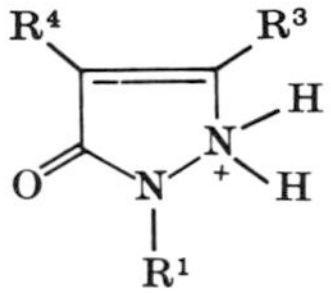

[**105**]

the 2-methyl derivative corresponding to [**104a**] has been established by IR, UV and NMR (70KG202). All three tautomeric forms possess a common anion for which calculations (71T5807) show an electronic structure closest to that of the OH tautomer (see 65T3351, 69CZ3895, 72CZ656).

ii. *Acid–Base Equilibria for a Tautomeric Pyrazol-5-one.* Scheme 4-9 shows the five species (three neutral tautomers, two cations) in equilibrium in acid solution. The complete equations for these equilibria have been given (67BF3780), but because the CH form is about 10^6 times less basic than the OH and NH forms, the equations can be simplified if only $K\gamma$ = [**103b**]/[**103c**] is needed. Eqs. (1-14), (1-15) and (1-16) can then be used (Section 1-4Aa). Applying these, Eqs. (1-14) and (1-15) often lead to negative values of $K\gamma$, whereas Eq. (1-16) usually gives reasonable values of $K\gamma$: for 1,3-dimethylpyrazol-5-one [**106**] $K\gamma$ is near zero from UV spectroscopy and $K\gamma = 0.043$ from Eq. (1-16) (67BF-3780), other authors found $K\gamma = 0.11$ (64T299).

A correlation was reported (72CZ656) between pK_a values in 20% ethanol, including proton loss from the neutral molecule as well as proton addition, and Hammett σ constants for a series of 1-aryl-3-methylpyrazol-5-ones substituted in the phenyl ring. We have already twice indicated (Sections 4-1Bb and 4-1Cb) that equations of the type $pK_a = \rho\sigma + b$ are incompatible with a Hammett dependence of the tautomeric equilibrium such as is suggested by Charton, $pK_T = \rho'\sigma' + b'$ (Section 1-B). Hence, the first of the two hypotheses which suggests (72CZ656) that the substituents affect the position of the tautomeric equilibrium in the same way as they affect the basicity must be rejected. The second hypothesis, that only one of the two possible tautomeric forms (OH and NH) is significantly populated, is difficult to verify, because the authors used 20% ethanol, but literature results for methanol and water solutions for the tautomeric equilibrium of 1-phenyl-3-methylpyrazol-5-one [**112**] are given in Table 4-16.

TABLE 4-16

TAUTOMERIC COMPOSITION OF 3-METHYL-1-PHENYLPYRAZOL-5-ONE [**112**]

Reference	Methanol			H_2O		
	%CH	%OH	%NH	%CH	%OH	%NH
64T299	—	—	—	—	10	90
67BF3780	—	—	100	—	—	100
72BB215, 72TH01	10	30	60	—	—	>60

b. The K_T Values of 1-Substituted Pyrazol-5-ones: Medium Effects

It is impossible to gather all the results into one table of reasonable length so Table 4-17 is restricted to the results needed later for discussion of the substituent effects. The solvent effects will be discussed later, relying essentially on Maquestiau's work (71BB17, 71TL2929, 73BB215). No study is available of the variation of K_T with temperature.

Studies in the vapour phase are rare. The presence of a νC=O band at 1720 cm^{-1} in the IR spectrum of the 1-phenyl-3-methylpyrazol-5-one [**112**] indicates that it exists in the vapour as the CH form (71T5779, 73BB757). A series of pyrazolones have been studied by mass spectrometry (69OM697), but the authors do not attempt to draw any conclusion regarding the tautomerism; it is not clear why they chose the OH structure (existing as such in solid state according to them) in representing the fragmentation processes. Maquestiau *et al.* (73BB747, 73BB757) in a very careful study of this problem concludes that the CH form predominates in the vapour phase.

We have already commented upon the structure of pyrazolones in the solid state in connection with the IR spectra of the OH [**103b**] and NH [**103c**] forms. Except for the three compounds of Scheme 4-10 which exhibit a C=O absorption characteristic of a CH structure

[**115**] KBr: 1715 cm^{-1} (*) (67ZO2487)

[**128**] nujol: 1730 cm^{-1} (68BF5019)

[**130**] nujol: 1730 cm^{-1} (68BF5019)

* The case of compound [**115**] is more complex, since it shows no carbonyl band in nujol (64T299); this difference of behaviour between nujol and KBr has been confirmed (72TH01).

SCHEME 4-10

[**103a**], all the other crystalline pyrazolones studied in the literature exist in the so-called NH/OH structure (see above). Recently, a detailed X-ray analysis of 1-phenyl-3-methylpyrazol-5-one has demonstrated equal numbers of the OH and NH forms, hydrogen-bonded with each other [**94/1**] (71MI469). The CH structure of the 1-(2′,4′-dinitrophenyl)-pyrazol-5-ones [**128**] and [**130**], could be due to the low basicity of the "pyridine-like" nitrogen atom which is therefore less able to participate in hydrogen bondings of the type [**94b**].

Most studies of the tautomerism of pyrazolones have been carried out in solution; we now recount the variation of K_T with the nature of the solvent and with the concentration.

TABLE 4-17

TAUTOMERIC COMPOSITION OF 1-SUBSTITUTED PYRAZOL-5-ONES

Compound	R¹	R³	R⁴	Solvent	Reference	CH [**103a**]	OH [**103b**]	NH [**103c**]
[**106**]	Me	Me	H	$CDCl_3$	71BB17	100	—	—
				DMSO	71BB17	8	80	12
				CH_3OH	71BB17	5–10	15–20	70–75
[**107**]	Me	H	Me	$CDCl_3$ (10^{-1})[a]	73BB233	7	93	
				DMSO	73BB215	0	85	15
				MeOH	73BB215	0	20	80
[**108**]	Me	Me	Me	$CDCl_3$ (10^{-1})[a]	73BB233	50	—	50
				DMSO	73BB215	0	50	50
				MeOH	73BB215	0	0–5	95–100
[**109**]	Me	Ph	H	$CDCl_3$	64T299	100	—	—
[**110**]	Me	Me	Ph	$CDCl_3$	73BB233	—	Low	High
				DMSO	73BB233	—	Low	High
[**111**]	Ph	H	H	$CDCl_3$	69T4605	90	10	—
				THF	71TL2929	12	88	
				DMSO	69T4605	—	100	—
[**112**]	Ph	Me	H	$CDCl_3$	64T299	100	—	—
				THF	71TL2929	61	39	
				DMSO	68JO3336 73BB215	20	High	Low
				MeOH	73BB215	10	30	60
[**113**]	Ph	H	Me	$CDCl_3$ (5×10^{-2})[a]	73BB233	50	50	
				THF	73BB233	—	100	
				DMSO	73BB233	—	High	Low
				MeOH	73BB233	—	High	Low
[**114**]	Ph	Me	Me	$CDCl_3$ (10^{-1})[a]	73BB233	66	—	34
				THF	73BB215	17	83	
				MeOH	67BF3780	—	Low	High

[115]	Ph	Ph	H	$CDCl_3$	64T299	100	—		—
				THF	71TL2929	38		62	
				DMSO	72TH01	—		100	
[116]	Ph	Me	Ph	$CDCl_3$	73BB215	—	—		100
				DMSO	73BB215	—	—		100
[117]	Ph	Me	Br	$CDCl_3$ (5×10^{-2})[a]	73BB233	82		18	
				THF	73BB215	13		87	
				DMSO	73BB215	—		100	
[118]	Ph	Me	CN	$CDCl_3$	73BB233	—	100		—
				DMSO	73BB233	—		100	
[119]	Ph	CO_2Et	H	$CDCl_3$	70T1571	30	70		—
				DMSO	70T1571	—	100		—
[120]	Ph	NH_2	H	$CDCl_3$	70T1571	100	—		—
				DMSO	73BB233	100	—		—
				CF_3CH_2OH	73BB233	100	—		—
[121]	2,4,6-Cl_3-C_6H_2	EtO	H	$CDCl_3$	70T1571	100	—		—
				DMSO	70T1571	High		Low	
[122]	Ph	H	CO_2Et	$CDCl_3$	73BB233	—	100		—
				DMSO	73BB233	—	100		—
[123]	*o*-Cl-C_6H_4	Me	H	DMSO	72TH01	Minor	Major		Minor
[124]	*m*-Cl-C_6H_4	Me	H	$CDCl_3$	73BB215	100	—		—
				DMSO	73BB215	13	Major		Minor
[125]	*p*-Cl-C_6H_4	Me	H	DMSO	72TH01	27	Major		Minor
[126]	*o*-NO_2-C_6H_4	H	H	$CDCl_3$	69T4605	High		Low	
				DMSO	69T4605	—	100		—
[127]	*p*-NO_2-C_6H_4	Me	H	$CDCl_3$	73BB215	100	—		—
				DMSO	73BB233	Very low	High		Low
[128]	2,4-$(NO_2)_2$-C_6H_3	Me	H	$CDCl_3$	66BF775	100	—		—
				DMSO	68BF5019	—		100	
[129]	*p*-NO_2-C_6H_4	Me	NO_2	$CDCl_3$	73BB233	—	100		—
				DMSO	73BB233	—		100	

[a] The proportions depend on the molar concentration (in parentheses).

c. Solvent Effects on K_T

Almost all authors dealing with the tautomerism of pyrazolones have noticed the important variation of K_T with the nature of the solvent and several (63ZO2597, 64T299, 67BF3780, 69T4605) have suggested partial explanations based on the dielectric constants of the solvents or on their basicity. However, it is only recently that the fundamental studies of Maquestiau, Van Haverbeke and Jacquerye (71BB17, 73BB215), have provided a complete rationalization. Their conclusions, valid especially for 1-monosubstituted and 1,3-disubstituted pyrazolones [1,3,4-trisubstituted pyrazolones are more complicated (73BB233)], are summarized.

i. *Nonpolar Solvents Favour the CH Form.* The CH form reaches 100% for 1,3-dimethylpyrazol-5-one [**106**] in CH_2Cl_2, $CHCl_3$ and CCl_4.

ii. *Dipolar Aprotic Solvents.* The percentage of NH form varies little, but as the basicity of the solvent increases, the CH form changes into the OH form.* For 1,3-dimethylpyrazolone [**106**], the CH form is

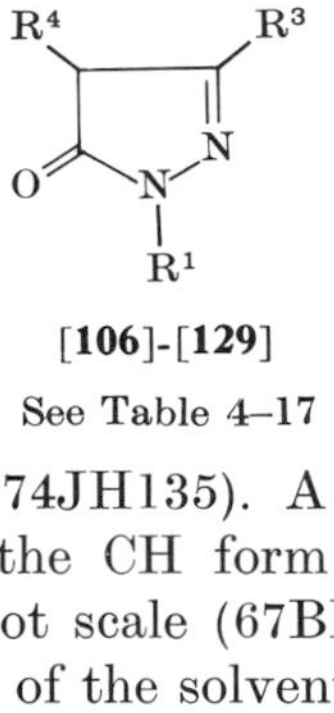

R⁴ R³ N O N R¹

[**106**]-[**129**]

See Table 4–17

absent in HMPT (see also 74JH135). A decreasing linear relation connects the percentage of the CH form with the solvent basicity according to the Agami-Caillot scale (67BF3780). The dielectric constant and the dipole moment of the solvent do not play an important role.

iii. *Protic Solvents.* The proportion of CH form is low and constant, but as the acidity of the solvent increases, the percentage of the NH form increases to the detriment of the OH form. The solvent acidity is the determining factor and not the dielectric constant as shown for 1,4-dimethylpyrazolone [**107**] where the percentage of NH form increases from 73% in EtOH to 100% in 2,2,2-trifluoroethanol, more acidic than ethanol but with the same dielectric constant.

These correlations have been justified theoretically for 4-unsubstituted compounds (71T5807, 71TL2929). For an isolated molecule in the vapour phase, and to a first approximation in solutions in a nonpolar

* That a basic solvent, such as pyridine, favours the OH form had already been pointed out (67BF3780) and explained by a specific acid–base solvation, without ionization.

solvent, the CH tautomer is favoured. Basic hydrogen bond acceptor, aprotic solvents should favour the OH form which has an electronic structure close to the anion and acidic protic solvents should favour the NH form whose structure is close to the cation.

The authors also point out (71T5779) that the NH tautomer has the highest dipole moment (see also Table 4-15): an increase in the polarity of the solvent should favour this tautomer; it is difficult experimentally to separate this effect from that of acidity which seems to dominate.

d. Concentration Effects on K_T

The important variation of K_T with concentration and the errors likely to occur if this factor is neglected were first pointed out (67BF-3780) for 4-substituted pyrazol-5-ones. Recently undertaken studies go further (73BB233). As the concentration increases, the percentage of CH form decreases in favour of the NH form; this is seen in nonpolar solvents (e.g., CH_2Cl_2, $CHCl_3$) for those pyrazolones where the CH form is not too highly favoured in these solvents. This is just the case for 4-substituted pyrazolones (see below for the effect of 4-substituents on K_T). The high concentration favours the intermolecular association of the NH form and effectively increases the polarity of the medium in use.

e. Electronic Substituent Effects on K_T: Systematic Work

We discuss successively the effect of substituents in the 1-, 3- and 4-positions.

i. *Substituents in the 1-Position.* This has received little attention. Data in Table 4-17 shows that the change from the N–Me to the N–Ph series increases the percentage of the CH form to the detriment of the NH form: cf. the pairs [**106**]–[**112**]; [**107**]–[**113**]; [**108**]–[**114**].

In the *N*-phenyl series (compounds [**112**], [**123**], [**124**], [**125**], [**127**] and [**128**]) the quantitative substituent effects are difficult to rationalize; for example, the percentage of CH form in DMSO tends to decrease if the molecule bears electron-withdrawing substituents (cf. compounds [**127**] and [**128**]); however, there is no linear variation with the Hammett σ constants.

ii. *Substituents in the 3-Position.* That the substitution of a 3–H by a 3–Me increases the proportion of the CH form is well recognized (63ZO2597, 67BF3780, 70T1571, 73BB215); this is confirmed by comparison of the pairs of pyrazolones [**107**]–[**108**], [**113**]–[**112**] and [**113**]–[**114**] (Table 4-17); the first pair shows also that the percentage of NH form increases to the detriment of the OH form.

This observation can be extended (70T1571) to a whole series of electron-donor and electron-acceptor substituents, among them NH_2 (β-amino-1-phenylpyrazol-5-one Section 4-6Fd, [**120**]) and OEt (3-ethoxy-1-trichlorophenylpyrazol-5-one (Section 4-6Bb, [**121**]). An examination of compounds [**111**], [**112**], [**115**], [**119**], [**120**] and [**121**] (Table 4-17) shows that the percentage of CH form in the medium decreases in the order $NH_2 > OEt > Me > Ph > H > CO_2Et$. Note that the Hammett σ constants vary in the same order (the phenyl group can be classified as a weak electron-acceptor or -donor).

iii. *Substituents in the 4-Position.* The effect of 4-substituents is more complex; it was known (67BF3780) that the introduction of a 4-methyl group decreased the proportion of the CH form, but only recently (73BB233) has the effects of other substituents been discussed. Comparison of the results in Table 4-17 for the following sets of pyrazolones [**106**]–[**108**]–[**110**] in the N–Me series and [**112**]–[**114**]–[**116**]–[**117**]–[**118**]; [**111**]–[**122**]; and [**127**]–[**129**] in the N–Ph series, leads to the following conclusions:

I. 4-Substituents decrease the percentage of CH form in the following order: $H > Me \sim Br \gg Ph, CO_2Et,^* CN, NO_2$.

II. A 4-methyl or 4-phenyl group increases the proportion of NH form, but the groups CO_2Et, CN or NO_2 favour the OH form.

Grandberg's conclusion (63ZO2597) that 3-methyl-1-phenyl-4-*n*-butyl-pyrazol-5-one [**131**] always exists in the CH form even in the solid state, seems doubtful, especially in view of the great solvent sensitivity of 3,4-disubstituted pyrazolones (67BF3780, 73BB233).

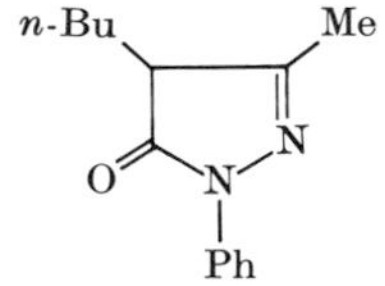

n-Bu Me
N
O N
Ph
[**131**]

Presently available theoretical calculations (71T5779) do not reproduce very satisfactorily the effects of 3- and 4-substituents. Recently the same authors (74T1225) described another approach to this problem as outlined in Fig. 4-1, Section 4-5H.

f. Miscellaneous Electronic Substituent Effects on K_T

Less complete literature results regarding the tautomeric equilibrium of 1-substituted pyrazol-5-ones can be correlated with the pattern just discussed.

* Although CO_2Et can participate in hydrogen bonding it is included here because the authors (73BB233) believe its effect here is electronic in character.

Compounds [**132**, X = S or O] (66CB2962, 66TL3919) are claimed to exist only in the OH form in the solid state and in DMSO by IR spectroscopy. Reappraisal of their evidence shows that only the absence of CH form was proved as is indeed found in the solid state for almost all pyrazolones and in DMSO for the 4-substituted derivatives.

Comparison of [**115**] and [**133**] shows (67ZO2487) that the substitution of phenyl by perfluorophenyl decreases somewhat the proportion of the CH form. Results for *N*-phenyl- and *N*-naphthyl-pyrazol-5-ones (69BF4159) are consistent with previous work. Compounds [**134**] and [**135**] (69JH723) have an OH/NH structure in KBr; in $CHCl_3$ [**134**]

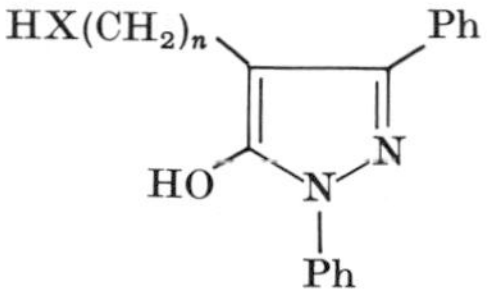

$HX(CH_2)_n$ Ph HO N N Ph

[**132**]
$n = 2,3$

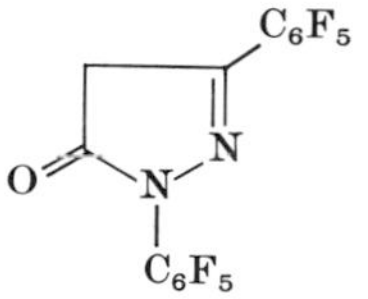

C_6F_5 O N N C_6F_5

[**133**]

$(CH_2)_n$—CO_2R O N N Ph

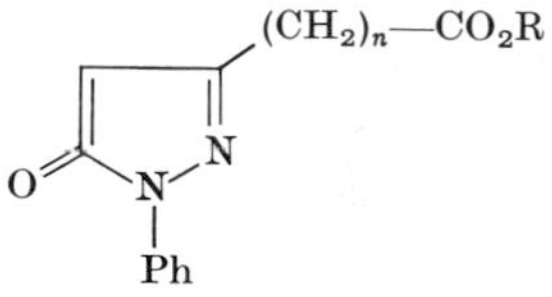

[**134**]

exists totally in the CH form (cf. [**112**] whereas [**135**] is a mixture of CH and OH/NH forms (cf. [**114**]). Compound [**136**] (70J(C)881) has a NH structure (cf. [**116**] in DMSO). Compound [**137**] possesses, not unexpectedly, the OH/NH structure in the solid state (70ZO1050), as does the 1-methyl-3,4-diphenyl derivative (73CC338).

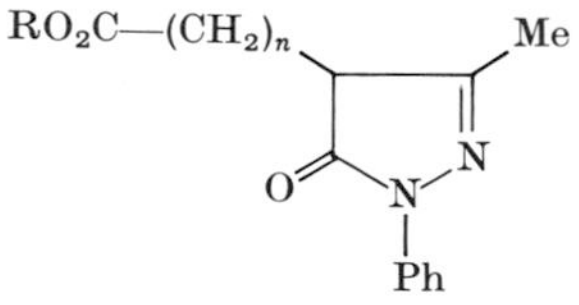

RO_2C—$(CH_2)_n$ Me O N N Ph

[**135**]

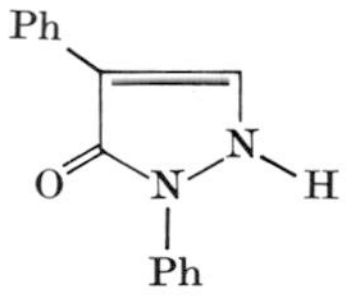

Ph O N N H Ph

[**136**]

$OP(OMe)_2$ Ar—CH Me HO N N Ar′

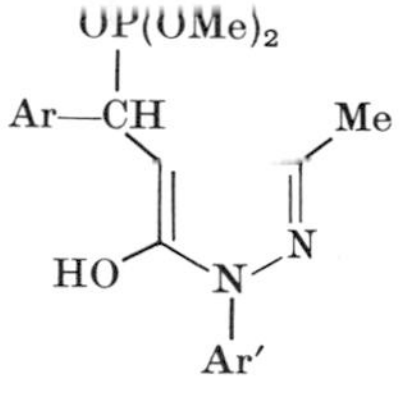

[**137**]

Compounds with a polymethylene chain connecting the 3- and 4-carbon atoms were studied (64T299, 69BF4159); 3,4-tetramethylene-1-phenylpyrazol-5-one [**138**] is closely similar to the 1-phenyl-3,4-dimethyl analogue [**114**]. This study has been repeated (70MC773), the authors, obviously unaware of reference (64T299), report that compounds [**138**] and [**139**] exist in the form OH/NH in the solid state and in the CH form in $CHCl_3$, in agreement with the former conclusions. Analogous compounds in the *N*-methyl series [**140**] (70TL3155) exist exclusively in the NH form in $CHCl_3$; it is difficult to make quantitative comparisons with [**108**] since the proportions depend on concentration.

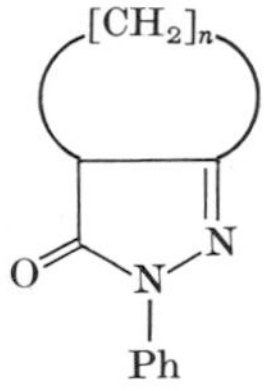

[**138**] $n = 4$

[**139**] $n = 3$

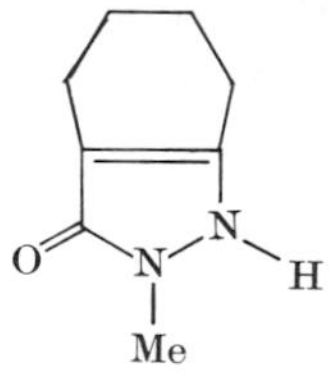

[**140**]

g. Substituent Effects Involving Hydrogen Bonding

The interesting possibility of stabilizing a tautomeric form by hydrogen bonding has not received much attention and the results are often inconclusive because the substituent electronic effects, and the medium, could rationalize the result without recourse to hydrogen bonding.

N-Acetylpyrazol-5-ones [**141**, R = Me or Ph] exhibit no signals for the CH form, either in the solid state or, more significantly, in $CHCl_3$ (65T3351). Since an electron-attractor group as 1-substituent favours the CH form as summarized above, the lack of CH form in solution is explained by chelation [**142**] stabilizing the OH form.*

The phosphorus analogue [**143**] has been examined only in the solid state and the only argument in favour of a chelated structure is the associated νP=O band in the IR spectrum (Nujol mull) (67CB919).

Structure [**144**], with a hydrogen bond in a seven-membered ring, was suggested (69LA(724)159) on the basis of a broad band between

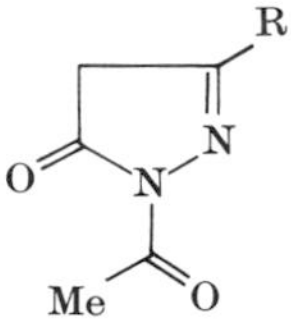

[**141**] 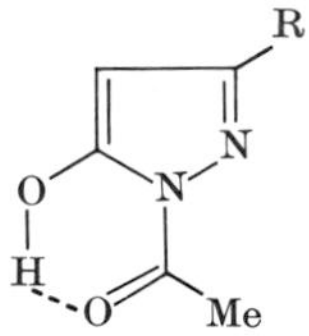 [**142**] 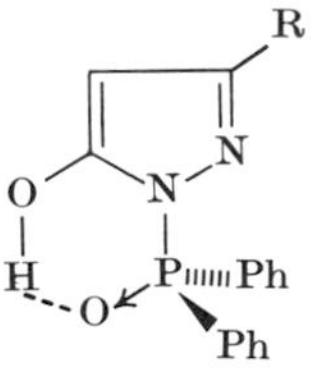 [**143**] 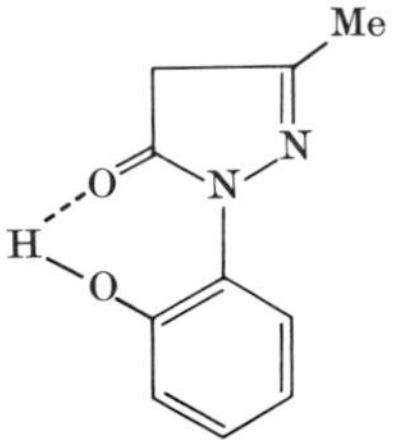 [**144**]

2300 and 3200 cm^{-1}. However, as the IR spectrum was in KBr, and as practically all 5-pyrazolones possess the OH/NH structure in the solid state with just such a broad band (see especially 63ZO2597), the

* This explanation is not sufficient in the case of the 2-acetylindazolone (Section 4-2Jc) which exists in the oxo form in the solid state (69CX(B)2355). A recent paper (74CT207) claims that 1-acetyl-3-methylpyrazol-5-one [**141**, R = Me] in solution is a mixture of NH and OH tautomers in equal proportion (no proof is given).

CH structure [**144**] is no longer accepted (72PS12). In the carbonyl region, the highest frequency band is near 1610 cm^{-1} and for 3-methyl-1-phenyl-5-pyrazolone in Nujol the corresponding band is found at 1606 cm^{-1} (s) (64T299).

Interesting hydrogen bonding can occur between the acidic proton of the OH form and a pyridine nitrogen of a substituent in the 1-position [**145**] or 4-position [**146**]. The existence of such hydrogen bonding was demonstrated experimentally by X-rays for compound [**147**] (71CB2694); however, the fact that the NMR spectrum (in DMSO) of [**147**] exhibits no signal of the CH form, does not allow any conclusion regarding the influence of H-bonding on the tautomerism since 1,3-diphenylpyrazolone [**115**] in the same solvent (Table 4-17) shows the same behaviour. Compounds [**148**] and [**149**] probably occur

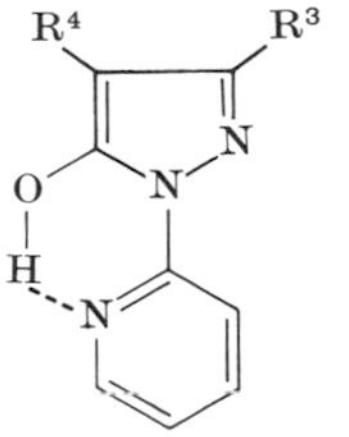

[**145**] 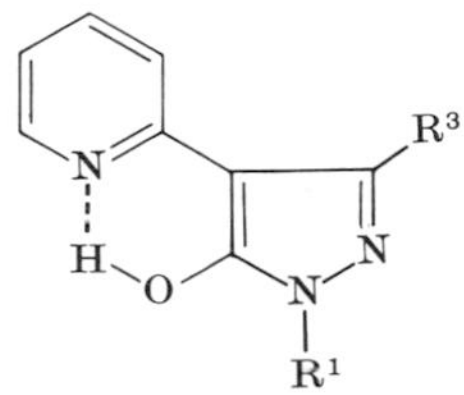 [**146**] 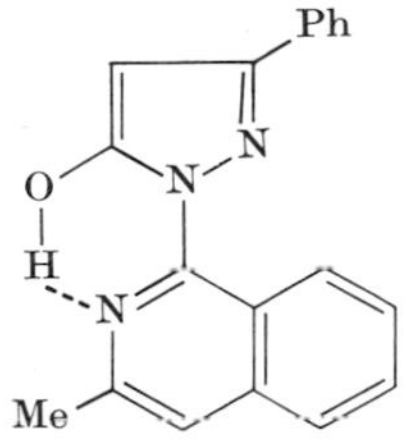 [**147**]

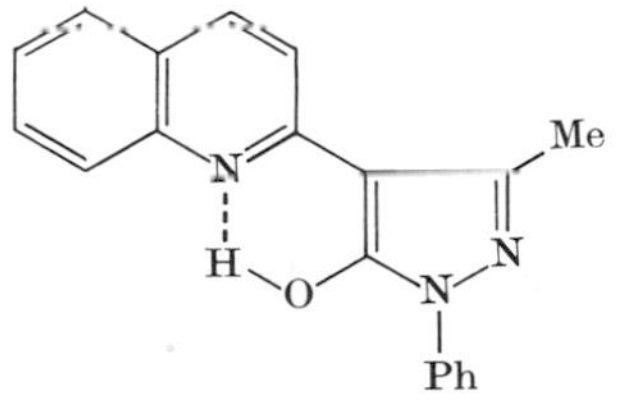

[**148**] 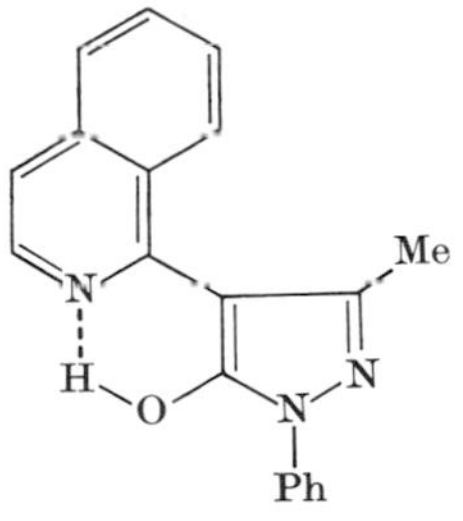 [**149**]

in the OH form with hydrogen bonding but this has not been proved (66G1410). However, recently Venkataratnam and co-workers (72TL-3937) have provided good evidence for the existence of H-bonded OH forms of type [**145**] in solution, thus correcting the NH structures proposed previously by Nair (71IJ104). 4-Methyl-1-(3-quinolyl)-pyrazol-5-one resembles the 1-phenyl analogue in existing in the OH/NH form in the solid, but in the CH form in $CHCl_3$. However, [**145**, R^3 = Me or CO_2Me; R^4 = H or CMe:$CHCO_2Et$], and analogues with a 1-(2-quinolyl) or 1-(1-isoquinolyl) group occur in the OH form in solution as well as in the solid state.

A structure similar to [**145**] occurs if the substituent at position 1 is a 2-benzothiazolyl or a 2-benzimidazolyl group (70KG660).

An important study (69CT1485) of series of 1-pyrimidinylpyrazolones [**150**] and [**151**] using CR, OR and NR model compounds and IR, NMR and UV spectroscopy, led the authors to conclude that these compounds exist in mixtures of the OH and NH forms in cyclohexane, $CHCl_3$, MeOH and H_2O (pH = 7.0) and in the OH/NH form in the solid state. The authors (69CT1485) do not consider intramolecular hydrogen bonding of type [**145**] which would stabilize the OH form relative to the two other forms. They postulate strong intermolecular hydrogen bonding to lower νC=O from near 1670 cm^{-1} for the N–Me derivatives to ca. 1630 cm^{-1} for the tautomeric compounds.

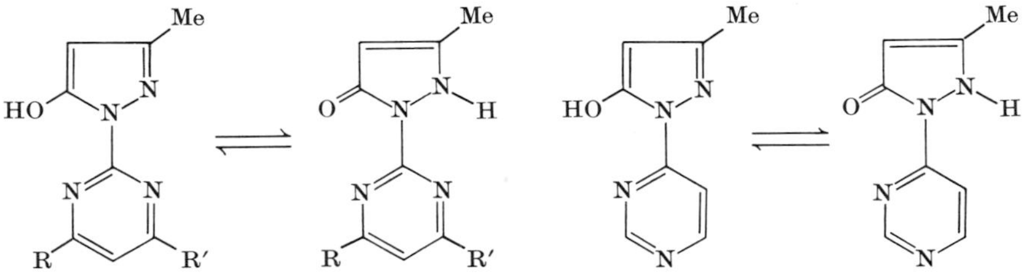

[**150a**] [**150b**] [**151a**] [**151b**]

We believe, however, that these compounds may exist entirely in the OH forms [**150a**] and [**151a**]. The Japanese authors themselves point out that NMR and UV results support the OH structure. The band appearing near 1630 cm^{-1} for the tautomeric derivatives but absent in the OR compounds remains to be explained: we believe it is a pyrimidine ring mode shifted by a strong intramolecular hydrogen bonding and not a carbonyl frequency. It is significant that its position does not shift significantly from $CHCl_3$ to KBr disk. The H-bonded structure [**145**] better explains, because of the energy gained by hydrogen bonding, the fact that in $CHCl_3$ the percentage of CH form decreases from 100% for 3-methyl-1-phenylpyrazol-5-ones [**112**] down to 0% for the compounds [**150**] and [**151**]. Furthermore this explanation is consistent with the solvent effect of pyridine which favours the OH tautomer by intermolecular hydrogen bonding (67BF3780).

h. 4-Acyl and 4-Alkoxycarbonyl 1-Substituted Pyrazol-5-ones

A 4-carbethoxy group in 1-phenylpyrazol-5-one increases the number of tautomers to four (69T4605) excluding the CH form which has never been observed. We consider the NH [**152a**], chelated OH [**152b**], free OH [**152c**] and exocyclic double bond forms [**152d**]. The first conclusion

(69T4605) was that compound [**152**] possessed the NH structure [**152a**] in $CHCl_3$ and the nonchelated OH form [**152c**] in DMSO. This study has been repeated in more detail (73BB233) and from better IR spectra and using the fixed NMe derivative, the authors demonstrated an equilibrium to occur between the chelated [**152b**] and free [**152c**] OH forms; basic solvents such as DMSO break the intramolecular hydrogen bond so that the free OH form [**152c**] is stabilized by intermolecular hydrogen bonds.

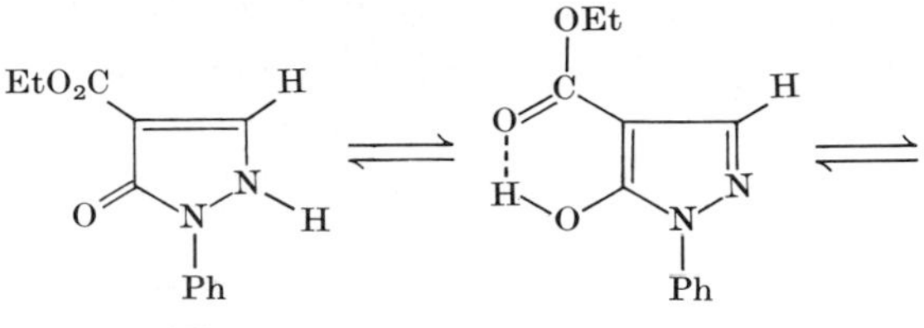

[**152a**] [**152b**]

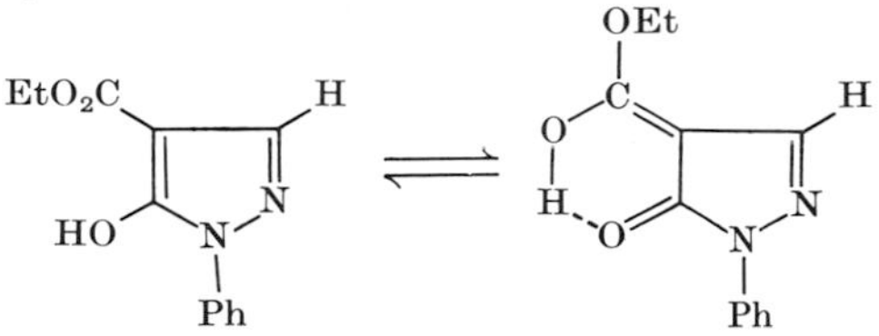

[**152c**] [**152d**]

A similar situation arises with the 4-acylpyrazolones [**153**]. UV (H_2O–EtOH) and IR ($CHCl_3$) studies on the 4-formyl-3-methyl-1-phenylpyrazol-5-one [**153**; R = H; R^3 = Me] and comparison with the OR derivative [**154**] led the authors (64ZO3005) to suggest the predominance of the hydroxymethylene form [**153b**]. Study of various 4-acetyl- and 4-benzoyl-1-pyrazolones [**153**; R^1 = Me, Ph; R^3 = Me, Ph], by NMR in DMSO and pyridine, only enables elimination of the CH form [**153c**] (71MI216). These authors do not mention the existence of two forms in the solid state previously assigned (59AS1668) as the CH form [**153c**] and the enol form [**153b**], as was mentioned in the previous review (II-43). The older work (59AS1668) must now be

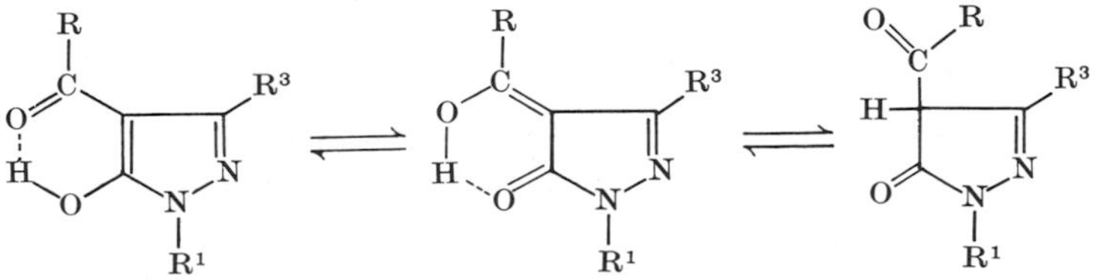

[**153a**] [**153b**] [**153c**]

considered superseded (see reference 75JH85). Differences in behaviour of 4-carbethoxy- [**152b**] and 4-acyl-pyrazolones [**153b**] parallels the nonenolisation of the ester carbonyl in β-ketoesters (66JO171). In a recent paper Maquestiau *et al.* (75JH85) gave conclusive evidence in favour of a nonchelated hydroxymethylene tautomer (as in the case of 4-acylisoxazol-ones, Section 4-2Bb).

A comparative IR study (CCl_4) of the compounds [**155**] and [**156**] demonstrated the chelated 5-hydroxypyrazole structure [**155**] (absence of a C=O band, presence of an intramolecularly bonded OH band) (68G245): the effects of the CO_2Et [**152b**] and CS_2Et [**155**] groups are thus similar.

i. 4-Iminomethylpyrazol-5-ones

These compounds, nitrogen analogues of the 4-acylpyrazol-5-ones, can also exist in four forms [**156/1a**–**156/1d**] (excluding the unlikely CH form). Russian investigators (73RO821) report spectral evidence for the occurrence of both [**156/1a**] and [**156/1d**] under different conditions of temperature and solvent.

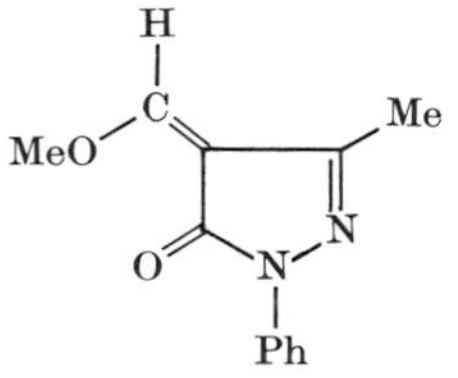

[**154**] 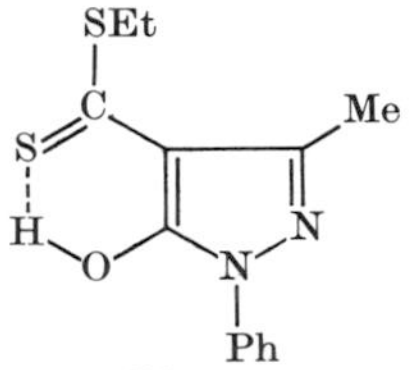 [**155**] 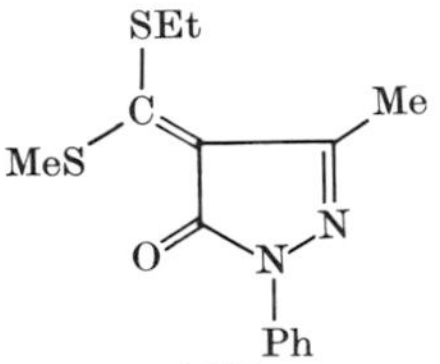 [**156**]

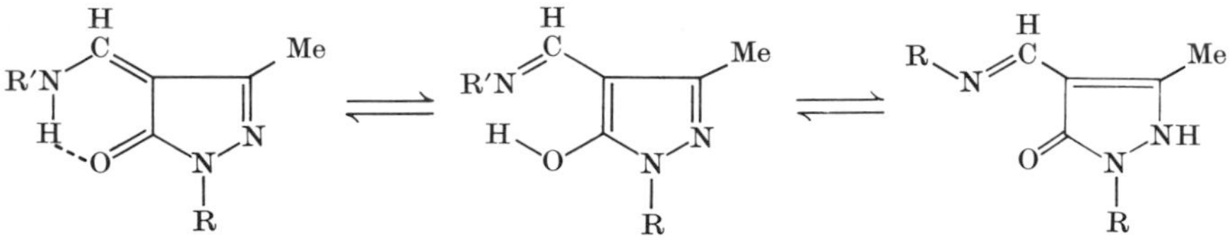

[**156/1a**] with H-bond
[**156/1b**] without H-bond

[**156/1c**]

[**156/1d**]

j. 4-Arylazopyrazol-5-ones (II-43)

Throughout this section results refer to the solid state and/or $CHCl_3$ except where otherwise stated. At the time of the previous survey (II-43) the structure of the phenylazopyrazolones in $CHCl_3$ and in the solid state was far from being solved since the three most important papers each suggested a different predominant structure: [**157a**] (59JO2039; see also 62JO994), [**157b**] (59NK402) and [**157c**] (60AG967).

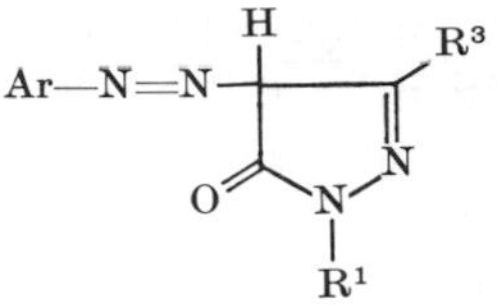

[157a] 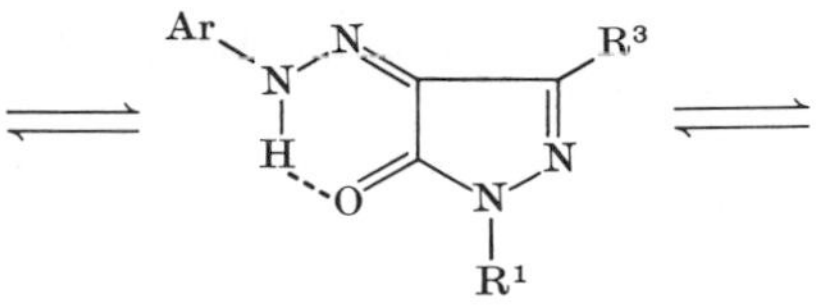 [157b]

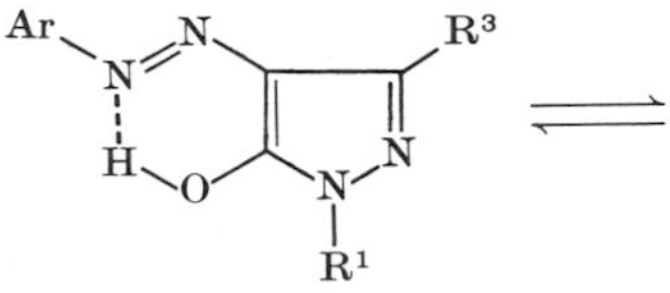

[157c] 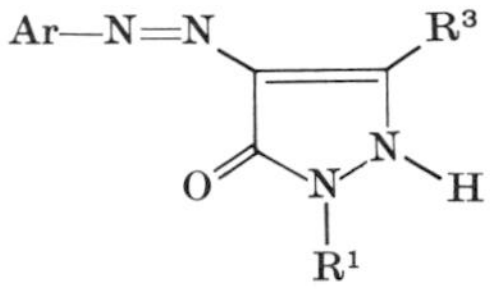 [157d]

Shortly afterwards, a further paper (63T1497) supported the chelated phenylhydrazone structure [**157b**], and three definitive papers later appeared simultaneously (66BF2990, 66JO1722, 66TL3897) which establish definitely the structure [**157b**] in its *Z* configuration for various substituents in the 1- and 3-positions and variously substituted phenylazo groups (Table 4-18). Structure [**157b**] was proved by comparison of the spectroscopic properties (PMR, IR, UV) of the tautomeric compound with those of the four fixed derivatives [**158–161**].

TABLE 4-18

4-Arylazopyrazol-5-ones, References with Proof of Structure [**157b**]

R^1	R^{3a}	References
Ph	Me	63T1497 66BF2990 66TL3897 68JO513 69MI22
Ph	Ph	66BF2990 66JO1722 69MI22
Ph	CO_2Et	63T1497 69MI22
Ph	t-C_4H_9	66BF2990
Me	Me	66BF2990

[a] For R^3 = H see reference 74T1345.

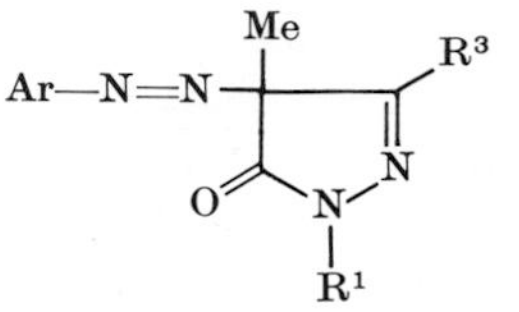

[**158**] 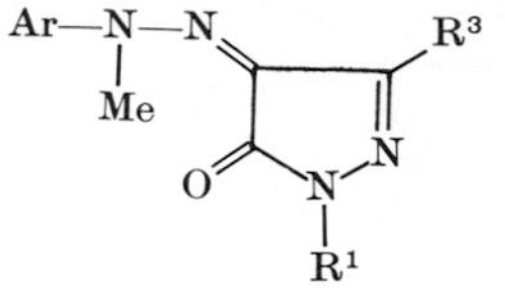 [**159**]

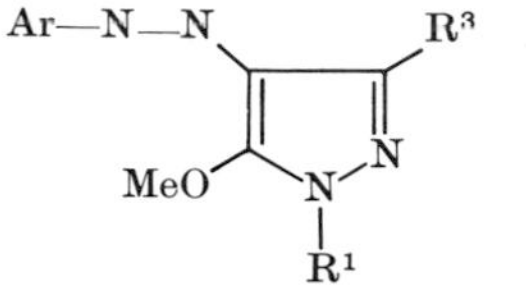

[**160**] 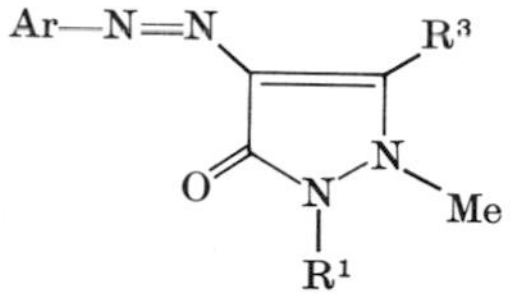 [**161**]

UV shows that structure [**157b**] applies also to EtOH and *n*-hexane solutions (66JO1722). Later publications support structure [**157b**] and extend the result to other substituents (see Table 4-18). In $CHCl_3$, the PMR signal at 13.4 ± 0.2 ppm arising from NH is significant (68JO513). Other work (69MI22) concerned an IR study in KBr of 76 arylazopyrazolones. UV (HCl 0.1*N*) and IR (Nujol and KBr) studies, by analogy with the results of the 3-methyl-1-phenyl-4-phenylazopyrazolone [**157b**, R^1 = Ar = Ph, R^3 = Me] (63T1497), indicate structure [**162**] for the compound obtained from the action of 2-azido-3-ethylbenzothiazolium fluoroborate on the 3-methyl-1-phenylpyrazol-5-one (66LA(699)133).

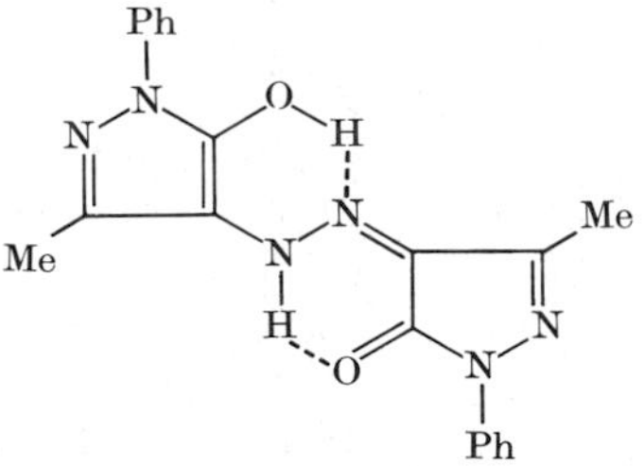

[**162**]

So far, we have implied that the stability of the chelated hydrazone form [**157b**] is so great compared with the other forms that, in contrast to the pyrazolones, solvent and substituent effects on the equilibrium need not be considered. Recently, the whole question has been treated from a theoretical viewpoint (74T1345); earlier, this view had been questioned, so far as solvent effects are concerned, but the arguments given (69JO1685) were not conclusive. The specifically labelled compound [**163**] showed ^{15}N–H coupling in $CDCl_3$ (J = 96 Hz), which (as the authors point out) is further strong support for structure [**163a**].* In DMSO and pyridine, the NH signals are very broad and difficult to integrate, although the chemical shifts are about the same as in $CDCl_3$ ($\sim$13 ppm). The authors write: "This suggests that the collapse of the ^{15}NH doublet and the peak broadening are both the result of proton exchange between the ^{15}N and the carbonyl oxygen. This means that the product probably consists of a mixture of forms [**163a**] and [**163b**] in these solvents." Dudek's work (66JA2407) suggests that an equilibrium of [**163a**] $\rightleftharpoons$ [**163b**] should merely decrease the coupling

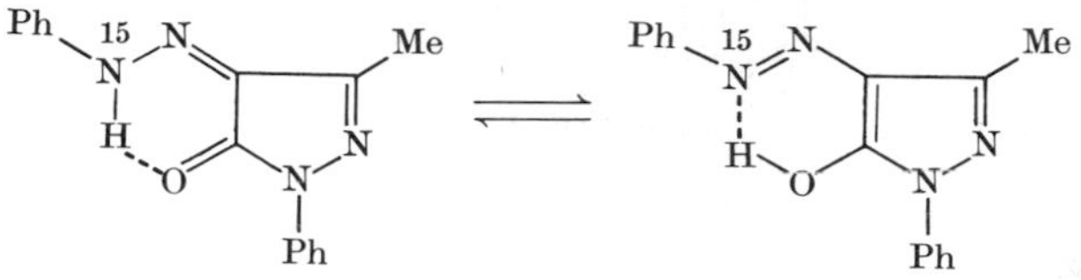

[**163a**] [**163b**]

constant ^{15}N–^{1}H. Such a coupling can be expressed by Eq. (4-1) where P_N and P_0 are the molar fractions of the NH [**163a**] and OH [**163b**] compounds, and J_N and J_0 are the coupling constants for ^{15}N–^{1}H, which are 96 Hz and assumed to be zero for [**163a**] and [**163b**] respectively.

$$J_{obs} = J_N P_N + J_0 P_0 \qquad (4\text{-}1)$$

The behaviour observed in DMSO and pyridine is better explained as an exchange with the solvent arising from cleavage of the intramolecular hydrogen bond; cf. the behaviour of 4-carbethoxy-1-phenylpyrazolone in DMSO (Section 4-2Gh). A definite conclusion cannot yet be reached, but recent IR spectroscopy (74T1345) in DMSO supports this contention.

H. 1-Substituted 3-Hydroxypyrazoles (II-44)

a. Introduction

The problem of the tautomerism of pyrazol-3-ones is simpler than that of the pyrazol-5-ones because only two tautomers can exist: the OH [**164a**] and NH forms [**164b**]. Nevertheless the situation was quite confused at the time of the previous survey (II-44): basicity measurements (in acetic acid) were in favour of the OH structure [**164a**] (61SA-40), but IR studies were interpreted alternatively to show the OH

* This result has been confirmed for the same compounds and in CH_2Cl_2 and CCl_4 as well as $CDCl_3$: a coupling of 96.5 Hz is observed which is not temperature dependent (70CZ1406).

tautomer [**164a**] (61SA40) or the NH form [**164b**] (59JA6292). However, in 1962 two papers on the IR spectra both confirmed the structure [**164a**] (62BF1707, 62BJ747) for Nujol and in $CHCl_3$ (R^5 = H, R^4 = H, NO_2, NH_2, N_2Ph). Much subsequent work shows that the OH form invariably predominates except for hydroxylic solvents where the NH form [**164b**] competes strongly. Evidence in this field has recently been summarized by Dorn (73PC(315)382). (See also 74CB1318).

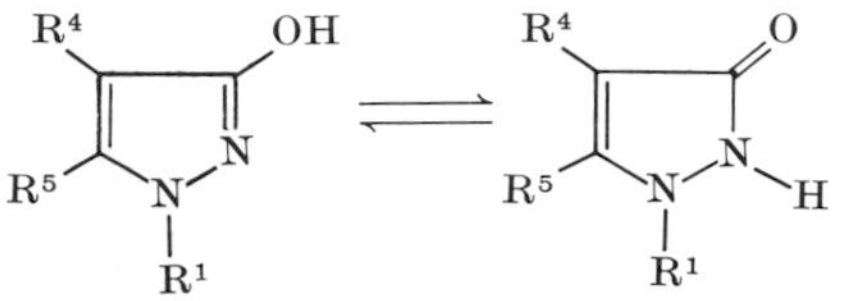

[**164a**] [**164b**]

b. Physical Characteristics of the Two Tautomeric Forms

Table 4-19 gathers the physical characteristics of the tautomers and of the model derivatives [**165**] and [**166**]. We discuss successively the results of five physical methods.

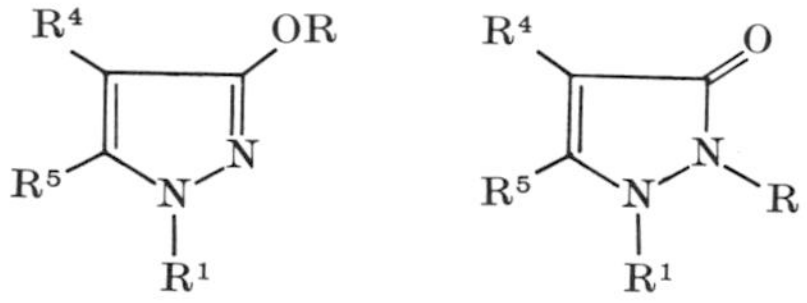

[**165**] [**166**]

i. *Nuclear Magnetic Resonance* (*Section 1-5A*). Just as with the OH and NH forms of pyrazol-5-ones (Section 4-2Ae), only averaged signals are observed. Several authors used interpolation methods based on slight differences in the chemical shifts of the 4-proton in 4-unsubstituted compounds and thus deduced qualitatively the predominance of the OH tautomer [**164a**] in pyridine (67BF3780) and $CHCl_3$ (69CT1485). Dorn (73PC(315)382; 74PC705) recommended the use of J(4H5H) in the case of 4,5-unsubstituted compounds.

ii. *IR Spectroscopy* (*Section 1-6C*) is the best method to differentiate the two structures because of the presence [**164b**] or the absence [**164a**] of a carbonyl band. The νC=O is sensitive to solvent effects; in protic solvents, it can occur as low as 1580 cm^{-1} (67BF3780). Although the pyrazole ring mode near 1560 cm^{-1} (H_2O, MeOH) is said to be distinctive (63ZO2597) the band is not always reported (70TL3155) (here a 4,5-disubstituted compound is concerned).

TABLE 4-19

CHARACTERISTICS OF THE TWO TAUTOMERS OF 1-SUBSTITUTED PYRAZOL-3-ONES

Tautomer	PMR	IR (cm^{-1})	UV (nm)			pK_a	μ(D)
				C_6H_{12}	H_2O		
OH [**164a**]	Averaged signals	Absence of νC=O	1-Me:	225	220	1-Me, 2.0	
		Pyrazole band, 1560–1570	1-Ph:	260	250	1-Ph, 1.2	1-Ph, 1.9($CHCl_3$)
NH [**164b**]		νC=O, 1650–1670	1-Me:	260	245	1-Me, 2.1	
			1-Ph:	270	260	1-Ph, 1.7	1-Ph, 5.0(dioxan)

3-Hydroxypyrazoles are strongly associated. In 0.004 *M* solutions in $CHCl_3$, a small proportion of free OH is observed at 3575 cm^{-1}, but most of the compound is still associated (64T315). The solid state association of 3-hydroxypyrazoles with no absorption in the 1800–2500 cm^{-1} region (63ZO2597) differs from polymeric pyrazol-5-ones (Section 4-2Fc, [**94a**], [**94b**]). Cyclic dimers [**167**] have been suggested

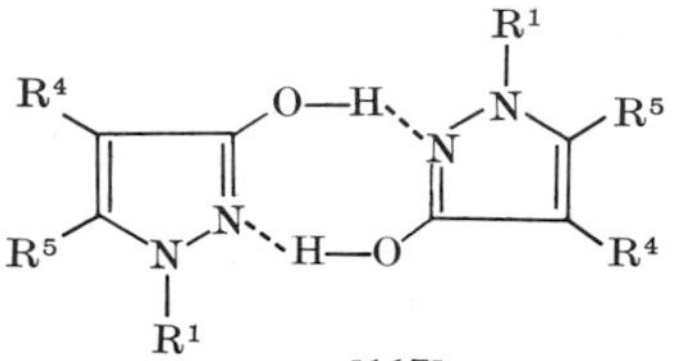

[**167**]

(64T315) and proved by dimeric molecular weights by osmometry (70TL3155). The OH structure thus applies to the solid state as well as to $CHCl_3$ solution. Such a structure has been established very recently by X-ray measurements for 5-methyl-1-phenylpyrazol-3-one (73MI473).

iii. *UV Spectroscopy* (*Section 1-6D*). The differentiation by UV spectroscopy of the OH and NH tautomers (Table 4-19) is very easy for 1-methyl compounds, but it becomes uncertain, or even impossible, for 1-phenyl compounds, especially in aqueous solution (64T315, 67BF3780). Other aromatic groups at the 1-position make the absorptions of the OR [**165**] and NR [**166**] derivatives more distinct, allowing a determination of the tautomerism; e.g., the 1-dinitrophenyl derivatives absorb differently (in EtOH or MeOH): [**169**] has λ_{max} 342 nm (ε = 7000), and [**170**] has λ_{max} 250 nm (ε = 17,500) and 340 nm (ε =

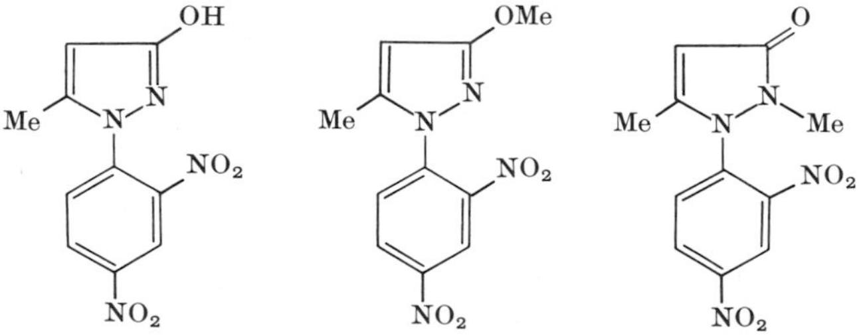

[**168**] [**169**] [**170**]

3000). This allows proof of the predominance of the OH structure in MeOH for the tautomeric compound [**168**] with λ_{max} 342 nm (ε = 7000) (68BF5019). The same applies to the pyrimidylpyrazoles [**183**] and [**184**] (69CT1485) in cyclohexane, $CHCl_3$, MeOH and even H_2O though this is less clear.

iv. *Basicity Measurements* (*Section 1-4A*). The use of Eqs. (1-14) and (1-15) (Section 1-4Aa) is impossible because K_1 is smaller than $K_{MeA} + K_{MeB}$ (67BF3780), so that it is necessary to use Eq. (1-16), but the results are not very satisfactory, especially in the N–Me series (67BF-3780). This arises from the fact that $f_A/f_B \neq 1$. Unlike the pyrazol-5-ones (Section 4-2Ga), the methyl or ethyl groups of the model derivatives are close to the site of protonation, cf. structures of the common cation [**171**] (64T315, 71T5807).

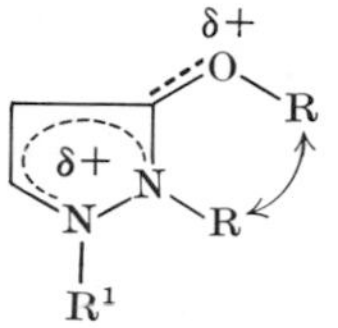

[**171**]

In the *N*-phenyl series, the difference between the pK_a of the two fixed derivatives (ca. 0.5 pK_a units, Table 4-19) gives K_T ca. 3, corresponding to the qualitative conclusion of the authors (64T315).

v. *Dipole Moments* (*Section 1-4G*). The determination of K_T from dipole moment measurements (Section 1-4b) gives a clear result for the 3-hydroxypyrazoles (Scheme 4-11) despite a regrettable change of

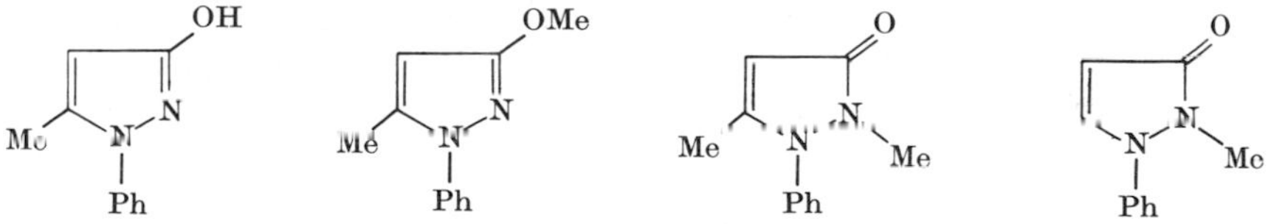

[**172**]	[**173**]	[**174**]	[**175**]
μ(D): 1.87 ($CHCl_3$) 2.11 (dioxan)	1.80 ($CHCl_3$)	5.23 (benzene)	4.99 (dioxan)

SCHEME 4–11

solvent (67MI74). Equation 1-44 (Section 1-4C), using $P = \mu^2$ (Eq. 1-51 of Section 1-4G) gives Eq. (4-2). As P^* is positive though small (ca. 0.1, Section 1-4G), it follows that the OH form predominates greatly in these solvents.

$$K_T = [\text{OH}]/[\text{NH}] = (23.8 + P^*)/(0.3 - P^*) \qquad (4\text{-}2)$$

c. K_T *Values of 1-Substituted Pyrazol-3-ones*

In the solid state and in solvents such as cyclohexane, dioxan, C_2Cl_4, $CHCl_3$, acetonitrile and pyridine, all the pyrazol-3-ones exist in the OH form [**164a**] whatever the N substituent (62BF1707, 62BJ747,

63ZO2597, 64T315, 66JO1538, 67BF3780, 67MI74, 68BF5019, 68T6809, 69J(C)836, 70TL3155, 71JH999, 71F1017, 72J(PI)777, 74CT207). The following apparent exceptions in the literature need reinterpretation:

i. The dipolar structure [**176**], claimed to exist in the solid state from IR spectra (64T531) is, as already pointed out (Section 1-1C), merely the NH form [**164b**]. However, the reported results are also easily compatible with an associated OH form [**167**].

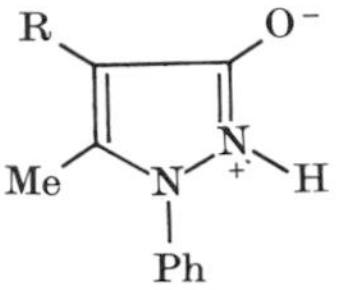

[**176**]

ii. A compound which exhibits an IR band at 1675 cm^{-1} (w) and a J(4H5H) of 3.5 Hz is claimed (68JA5273) to exist in the NH form [**177**]. Although the solvents are not given, this result is so surprising that the gross structure is doubtful, especially as the UV spectrum in EtOH [λ_{max} 230 nm (sh)] is not consistent with such an NH structure (cf. Table 4-19).

iii. It has been claimed (71JH999) that 1,4-diphenylpyrazol-3-one [**178**] exhibits a weak νC=O at 1710 cm^{-1} in $CHCl_3$. This band probably arises not from the NH tautomer but from an impurity.

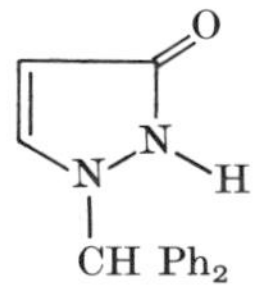

[**177**] 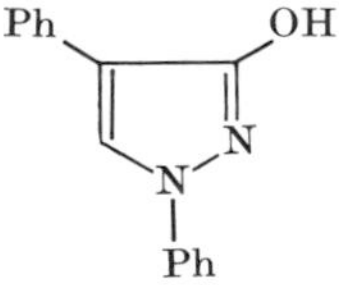 [**178**]

For MeOH and H_2O solutions, the equilibrium position is much more evenly balanced between the OH and NH forms. Table 4-20 records the literature data; those from references (67BF3780) and (69CT1485) are deduced from UV spectroscopy. The presence of the NH form has

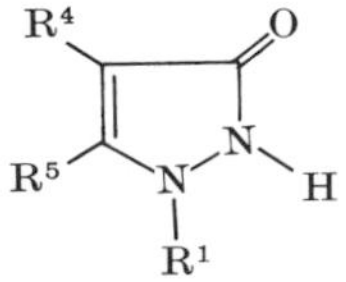

[**179**]–[**184**]
See Table 4–20

TABLE 4-20

TAUTOMERIC COMPOSITION OF 1-SUBSTITUTED PYRAZOL-3-ONES IN AQUEOUS AND METHANOLIC SOLUTIONS[a]

Compound	R^1	R^4	R^5	Solvent	Reference	OH [**164a**]	NH [**164b**]
[**179**]	Me	H	Me	MeOH	67BF3780	85	15
				H_2O	67BF3780	30[b]	70[b]
[**180**]	Me	Me	H	MeOH	67BF3780	90	10
				H_2O	67BF3780	40	60
[**181**]	Me	Me	Me	MeOH	67BF3780	70	30
				H_2O	67BF3780	5	95
[**182**]	Ph	H	Me	MeOH	67BF3780	Present	Present
				H_2O	64T315	Major	Minor
[**183**]	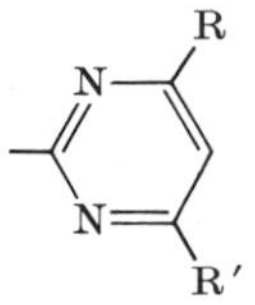	H	Me	H_2O	69CT1485	Minor	Major
[**184**]	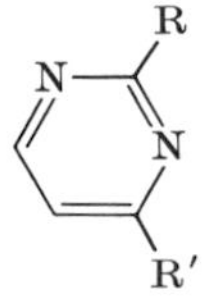	H	Me	H_2O	68CT1485	Minor	Major

[a] For further results, see reference 73PC(315)382.
[b] NH predominates by a small factor (64T315).

been demonstrated qualitatively by IR spectroscopy (67BF3780): 1,4-dimethylpyrazol-3-one [**180**] shows νC=O at 1585 cm^{-1} in 1:1 CD_3OD–D_2O and at 1575 cm^{-1} (stronger) in D_2O; for 1-phenyl-5-methylpyrazol-3-one [**182**] νC=O appears at 1595 cm^{-1} (weak) in MeOD. Table 4-20 indicates that the change from methanol to water favours the NH tautomer, which is similar to the pyrazol-5-one series (Section 4-2Gc).

d. 4-Arylazopyrazol-3-ones

The compounds are less studied than the 4-arylazopyrazol-5-ones; the two available literature studies both favour the 4-arylazo-3-hydroxy-1-phenylpyrazole type structure [**185**]. IR spectra in Nujol

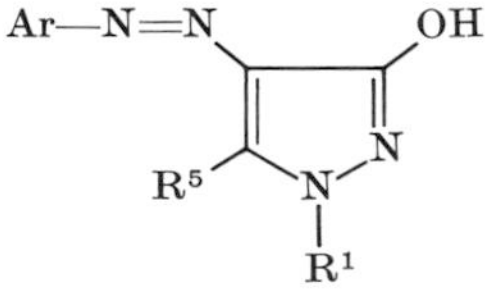

Ar—N=N OH

R^5 N N

R^1

[**185**]

(62BF1707) indicated an OH-structure for the 4-phenylazo derivative [**185**, R^5 = H] by analogy with the spectrum of the 3-hydroxy-5-phenylpyrazole. A study in $CHCl_3$ (68JO513) by IR and NMR spectroscopy leads to the same conclusion for a series of 4-arylazo-3-hydroxy-5-methyl-1-phenylpyrazoles; in $CHCl_3$ the OH-proton appears at 9.2 $\pm$ 0.3 ppm to be compared with 13.4 $\pm$ 0.2 ppm for the arylazopyrazol-5-ones (Section 4-2Gi). Hydrogen bonding between the hydroxyl proton and the 4-arylazo group has been suggested. Proton NMR spectra of ^{15}N specifically labelled compounds are in agreement with the above conclusions (72JO4121).

I. N-Unsubstituted Pyrazol-3(5)-ones (II-44)

a. Introduction

The N-unsubstituted pyrazol-3(5)-ones [**186**] constitute a complex case of tautomerism because of the large number of tautomeric forms [**186a**] to [**186h**], and also because of the insolubility of these compounds in aprotic solvents of low polarity. However, structures [**186f**], [**186g**] and [**186h**] can be eliminated rapidly as significant contributors. They correspond to pyrazolenine and isopyrazole structures, which do not contribute to the annular tautomerism of pyrazole (Section 4-1B). Such structures exist only if they are fixed by substituents as in [**187**], and even then they are unstable and easily oxidized (65T1693).

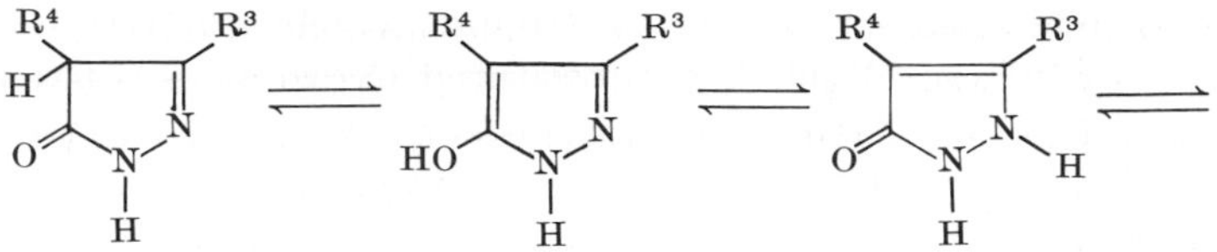

CH [186a] ⇌ 5—OH [186b] ⇌ NH [186c] ⇌

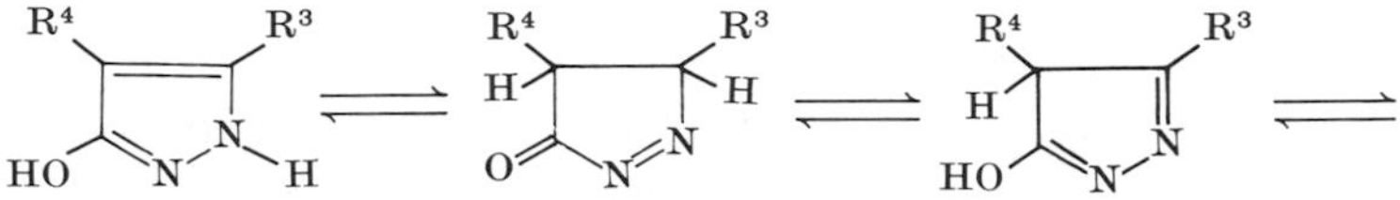

3—OH [186d] ⇌ [186e] ⇌ [186f] ⇌

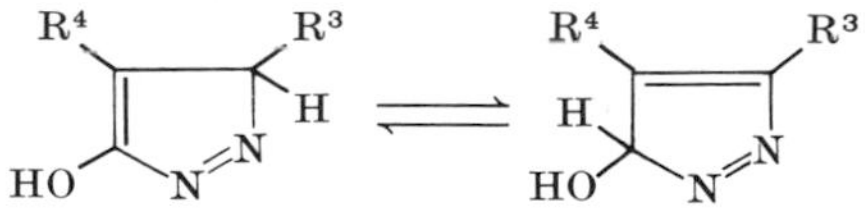

[186g] ⇌ [186h]

Structure [**186e**] is a 1-pyrazoline; such compounds can be isolated, but are easily isomerized into 2-pyrazolines (67MI211). Compound [**188**] was recently synthesized (70J(C)540): it isomerises into the pyrazolone [**189**] in the presence of triethylamine in CH_2Cl_2, the activation energy is relatively high (Table 1-4, case 2, Section 1-5Aa), probably because two CH bonds must be broken to obtain an aromatic NH or OH tautomer. Compound [**188**] shows $\nu C{=}O$ at 1800 cm^{-1} in $CHCl_3$, and an AA′BB′ system in $CDCl_3$ (which proves that it does not exist as [**186g**] in these solvents), and a weak absorption in the visible (in CH_2Cl_2 λ_{max} 436 nm, $\varepsilon = 250$).

No study of the tautomerism of N-unsubstituted pyrazolones has revealed absorption corresponding to the forms [**186e**] to [**186h**]. The

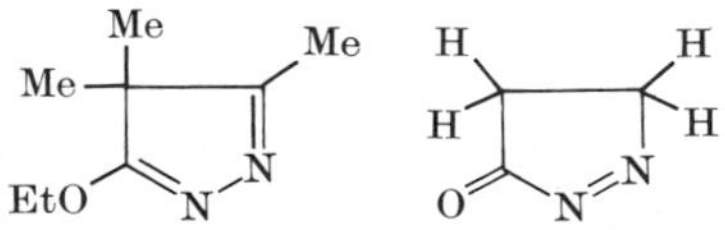

[187] [188]

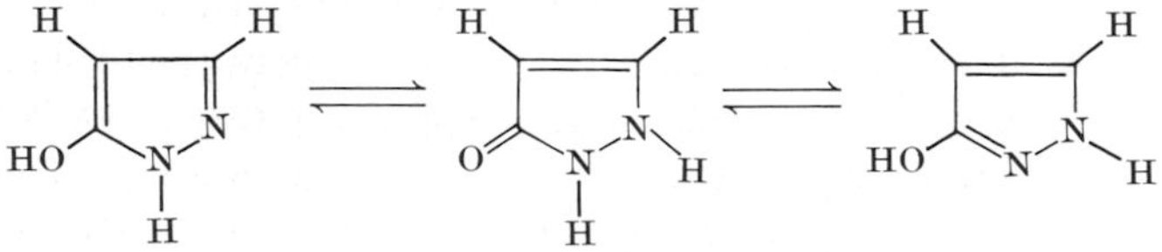

[189a] ⇌ [189b] ⇌ [189c]

same applies to semi-fixed forms, i.e., compounds in which one of the two tautomeric hydrogens of the N-unsubstituted pyrazolones is fixed by methylation. Thus 3,4,4-trimethylpyrazol-5-one can exist in three forms [**190a**], [**190b**] and [**190c**], the last two corresponding to the tautomers [**186e**] and [**186f**]. It has been shown (65T1693, 67BF3772) for a series of solvents that compound [**190**] has the structure [**190a**]

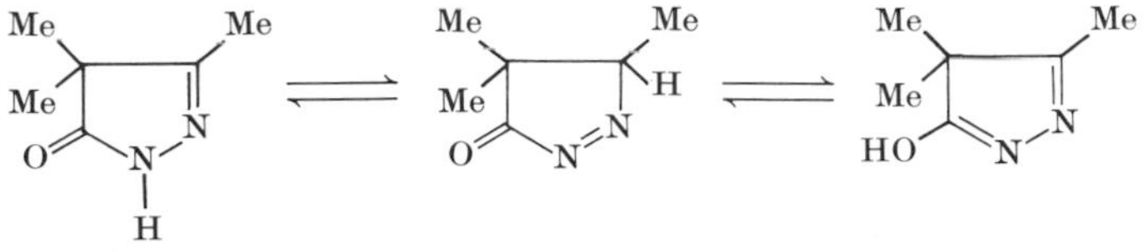

[**190a**] [**190b**] [**190c**]

and that it is a good model for the CH tautomer [**186a**]. Henceforth we consider only the first four tautomers [**186a–d**].

The equilibrium between the two OH tautomers [**186b**] and [**186d**] and also that of the corresponding semi-fixed model [**191**] are cases of annular tautomerism that have already been discussed (Section 4-1B): all the spectroscopic results (65T1693, 72TH01, 73PC(315)382) and the calculations (71T5779) are in favour of the 3-OH or 3-OR forms [**186d**] and [**191b**] (at least with R^3 = Me in [**186**]).

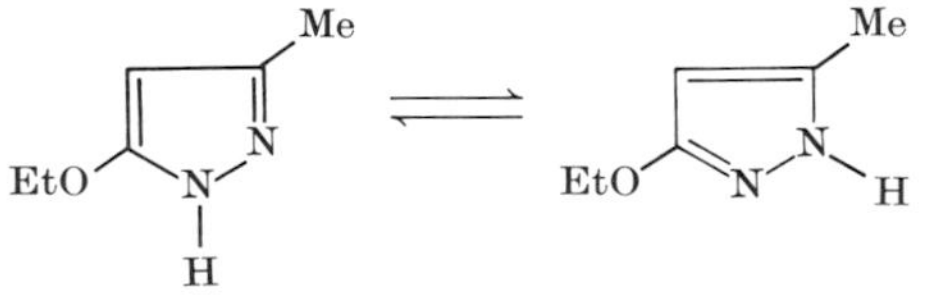

[**191a**] [**191b**]

The characteristics of the various tautomeric forms can be obtained from Tables 4-15 and 4-19 assuming that replacement of NH for NMe does not change these properties. Thus the CH form [**186a**] corresponds to [**81a**], the 5-OH form [**186b**] to [**81b**], the NH form [**186c**] to [**81c**] or [**164b**], and the 3-OH form [**186d**] to [**164a**]. For example the pK_a of the 3,4,4-trimethylpyrazolone [**190**] and that of the 1,3,4,4-tetramethylpyrazolone [**88**] are identical $pK_a = -3.8$ (64T299, 65T1693).

b. K_T Values of N-Unsubstituted Pyrazol-3(5)-ones

We base the discussion on three definitive papers (65T1693, 67BF-3780, 72TH01) in which two other publications are discussed (63T1497, 65T3351). The compounds studied are summarized in Table 4-21. Dorn has very recently reviewed this field, with additional experimental data (73PC(315)382).

TABLE 4-21

N-UNSUBSTITUTED PYRAZOL-3(5)-ONES

R^3	R^4	References	R^3	R^4	References
H	H	71F1017	$—(CH_2)_n—$		65T1693
Me	H	63T1497	Me	R	66CB2962
		65T1693			66TL3919
		65T3351	Me	Br	65T3351
		67BF3780	Ph	H	63ZO1092
		69OM697			65T3351
		70BF247			69BF4159
		71CX(B)1227			69J(C)836
		72TH01			69OM697
Cycloalkyl	H	70BF247			70BF247
H	Me	67BF3780	Ph	Me	69BF4159
		72TH01	$PhCH_2$	H, Me	69BF4159
Me	Me	63T1497			
		65T1693			
		65T3351			
		67BF3730			
		69OM697			
		72TH01			

In the solid state, from IR studies, the OH/NH structure was proposed for pyrazol-3(5)-ones, as for the pyrazol-5-ones (Section 4-2Fc, [**94a**] and [**94b**]). Some authors interpreted their IR spectra in terms of the OH structure solely because of the absence of a carbonyl band (63ZO1092, 65T3351, 66CB2962, 66TL3919, 69J(C)836*), and that conclusion has been used to justify fragmentation derived from the OH form in mass spectrometry. Fortunately, X-ray diffraction of 3-methylpyrazolone [**194**] (71CX(B)1227) proves the dimeric NH structure [**192**] with a type of hydrogen bonding different from that of

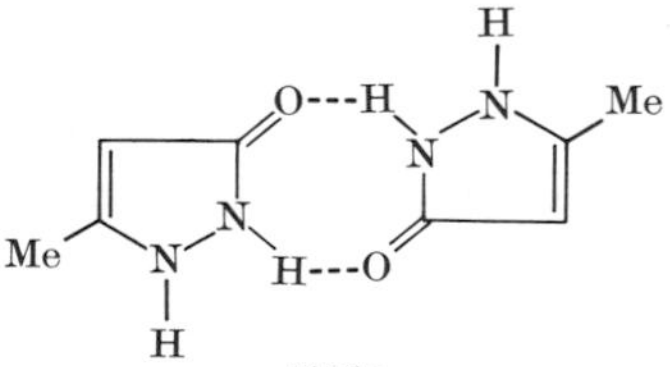

[**192**]

* The authors consider the participation of a form [**186f**], besides the 5-OH [**186b**] and 3-OH [**186d**] forms, but without any convincing evidence.

pyrazol-5-ones [**94a**] and [**94b**] (Section 4-2Fc) and pyrazol-3-ones [**167**] (Section 4-2Hb).

In nonpolar solvents (C_6H_{12}, CH_2Cl_2, $CHCl_3$), only the CH [**186a**] and 3–OH [**186d**] forms have been found. The CH form [**186a**] is best identified by IR spectroscopy with νC═O (occasionally split) bands near 1720 cm^{-1} (67BF3780, 72TH01) and the 3–OH form [**186d**] by UV spectroscopy (65T1693). The ratio 3–OH/NH appears to be 5:1 in cyclohexane (65T1693) whereas in $CHCl_3$ the OH form is present in a very low proportion only (no band at 1560–1680 cm^{-1}) (67BF3780, 72TH01). This result should be verified by another method; unfortunately the low solubility of unsubstituted pyrazolones in these solvents has so far prevented NMR study.

In aprotic dipolar solvents (pyridine, DMSO) an OH tautomer is predominant (65T3351, 66CB2962, 66TL3919, 67BF3780), more precisely the 3–OH tautomer [**186d**] (72TH01). The presence of the NH tautomer [**186c**] in DMSO (cf. 70BF247) can be eliminated, as such an assignment located the νC═O absorption at 1580 cm^{-1} whereas the model compound 1,2,3-trimethylpyrazol-5-one νC═O is at 1641 cm^{-1} in DMSO (71BB17). Furthermore the NMR spectrum of pyrazolone [**189**] is in favour of an OH structure: as already pointed out (Section 4-2Ae), the fixed NR form, 1,2-dimethylpyrazol-5-one [**91**] has J = 3.5 Hz, whereas for a hydroxypyrazole, coupling constants of about 2.0 Hz are expected. For pyrazolone [**189**] the J(4H5H) of 2.3 Hz (72UP2),* favours either of the two hydroxy structures [**189a**] or [**189c**]. In less basic solvents (dioxan, THF) a small amount of CH form [**186a**] is observed by IR spectroscopy (72TH01).

Finally in protic solvents (H_2O, various alcohols) only the 3–OH [**186d**] and NH forms [**186c**] are present (67BF3780, 72TH01; see also 69BF4159, 70BF247); the acidity of the solvent favours the NH tautomer. In agreement with the result from basicity measurements (65T1693), the OH form seems to be the 3–OH [**186d**] (65T1693, 69J(C)836, 72TH01).

A series of theoretical calculations has been done on the pyrazolone itself [**189**] (71T5779); in all the cases the CH form [**186a**] is the most stable and the 5–OH form [**186b**] the least stable with the 3–OH form [**186d**] and the NH form [**186c**] occupying an intermediate position; this is in agreement with the experimental results for nonpolar solvents. A Hückel ω calculation (71T5779) shows that the forms [**186e**] to [**186h**] are greatly unfavoured.

* The NMR spectrum of the pyrazolone [**189**] in DMSO has been described (71F1017), but without the coupling constants.

In conclusion, the N-unsubstituted pyrazolones occupy a position intermediate between the pyrazol-5-ones and the pyrazol-3-ones as indicated schematically in Table 4-22.

TABLE 4-22

SUMMARY OF PYRAZOLONE TAUTOMERISM

Form	1-Substituted pyrazol-5-ones	N-unsubstituted pyrazol-3(5)-ones	1-Substituted pyrazol-3-ones
CH	Frequently	Seldom	Never
NH	Frequently	Frequently	Seldom
OH	Seldom	Frequently	Usually

This intermediate position, at least as far as OH and NH forms are concerned, can be clearly observed in the Fig. 1 of (67BF3780).

Few conclusions can be drawn about substituent effects. In MeOH changes in the percentage of NH form are given in Scheme 4-12. The same order is found for pyrazol-3-ones (Section 4-2Hc, Table 4-20) but with higher percentages. If the effect of methyl groups is additive, pyrazolone itself [**189**] should exist in MeOH as 15% NH and 85% 3–OH form.

	[**193**]	[**194**]	[**195**]
% of NH form in MeOH:	35%	45%	65% (67BF3780)
	30%	40%	65% (72TH01)

SCHEME 4–12

c. 4-Acyl- and 4-Phenylazo-N-Unsubstituted Pyrazol-3(5)-ones

For the 4-acyl derivatives, an equilibrium between structures [**196a**] and [**196b**] in DMSO and pyridine has been proposed (71MI216) (see Section 4-2Gh). The 4-ethoxycarbonyl derivative probably has a structure of type [**196a**] (72J(PI)1022). The structure [**197**] of the

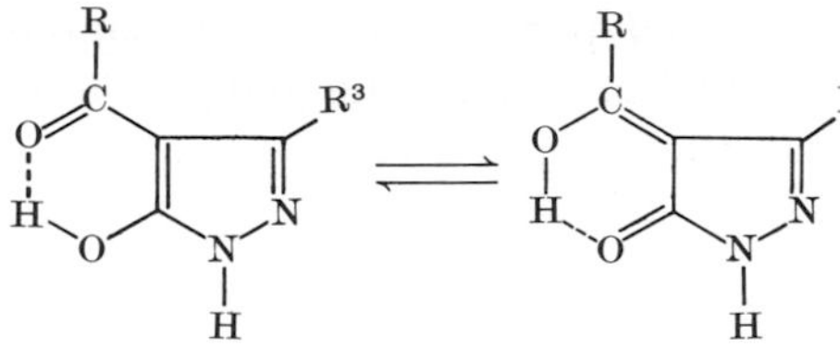

[**196a**] [**196b**] 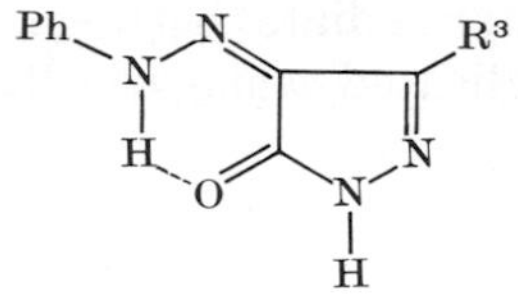 [**197**]

4-phenylazo compound has been demonstrated (66BF2990, 74T1345) in the same way as for the 4-phenylazo derivatives of 1-substituted pyrazol-5-ones (4-2Gj) [**157b**]).

J. Indazolones (II-44)

a. Introduction

The tautomerism of indazolones is not yet clear: the many studies are incomplete and often in disagreement. However, the following tentative conclusions can be drawn: 1-substituted indazolones exist in the OH form [**198**], the 2-substituted indazolones in the NH form [**199**] whereas the structure of unsubstituted indazolone varies with the physical state: in the solid state in the oxo form [**200a**] and in EtOH in the OH form [**200b**].

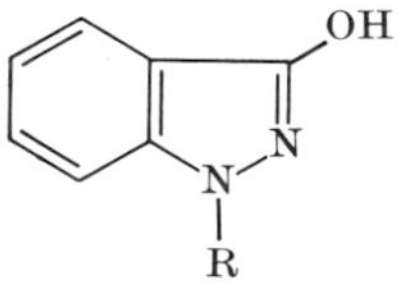

[**198**] 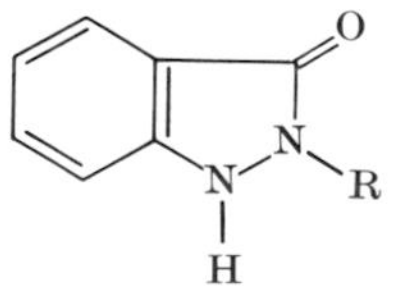 [**199**]

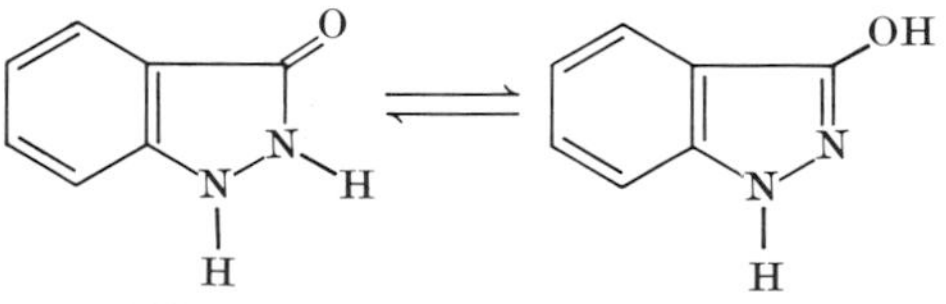

[**200a**] [**200b**]

b. 1-Substituted 3-Hydroxyindazoles

The IR spectra of 1-substituted 3-hydroxyindazoles [**198**], in the solid state as well as in solution ($CHCl_3$, CCl_4), show no νC=O but possess νC=N near 1550 cm^{-1} and νOH near 3450 cm^{-1} (in solution) or 3200–2500 cm^{-1} (solid state). The νOH disappears in the corresponding *O*-alkyl derivatives [**201**] (64H1986; see also 64PC329, 70JH807). This pattern applies to 1-alkyl-, 1-aryl- and 1-carbethoxy-3-hydroxyindazoles.

Compounds were assigned the 1-methylsulfonyl- and 1-phenylsulfonyl-indazol-3-one structures [**202**] on the basis of νC═O absorption at 1690 cm^{-1} and νNH at 3250 cm^{-1} (71J(C)3313); however, the possibility arises that these compounds are 2-substituted indazol-3-ones [**199**, R = SO_2CH_3, $SO_2C_6H_5$].

Compounds [**198**], [**201**] and [**203**], with R = alkyl, show similar UV spectra in EtOH: two peaks near 220 and 300 nm; this is consistent with the IR results (64H1986).

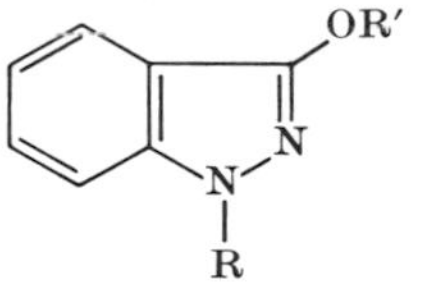

[**201**] 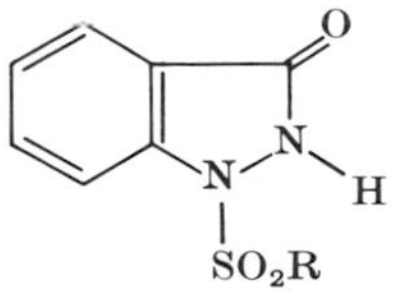 [**202**] 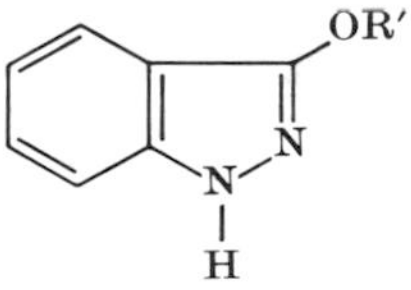 [**203**]

c. 2-Substituted Indazolones

2-Substituted indazolones [**199**] possess the NH structure [**199**] as demonstrated by associated νNH bands at 3000–3200 cm^{-1} in solution; and 3200–2500 cm^{-1} in the solid state together with a strong νC═O band at 1665–1670 cm^{-1} in solution and 1020–1650 cm^{-1} in the solid state. The νC═O is about 40 cm^{-1} lower than for model compounds [**204**], probably because of intermolecular hydrogen bonds (64H1986; see also 67J(C)1792). For 1-carbethoxy-2-methylindazolone [**204**,

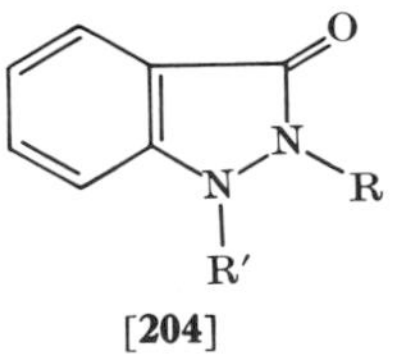

[**204**]

R′ = CO_2Et, R = CH_3], νC═O is found at 1685 cm^{-1} (Nujol) (64H-1986) or at 1710–1720 cm^{-1} (KBr) (64PC329). An X-ray study of 2-acetylindazolone (69CX(B)2355) demonstrates the oxo structure [**199**, R = $COCH_3$].

The UV for [**199**] and [**204**] show a third peak near 240 nm between the peaks at 220 and 310 nm present in these structures and also in those of type [**198**] and [**201**] (62MI125, 64H1986).

d. Unsubstituted Indazolones (II-46)

The IR of indazolone itself in the solid state (there is no study in solution) shows a very strong peak at 1626 cm^{-1} and a broad complex absorption at 2700–3100 cm^{-1}: according to O'Sullivan (60J3278, II-46) this proves the oxo structure [**200a**] for this compound; this is in agreement with the νC═O absorption of 2-alkyl- [**199**] and 1,2-dialkyl [**204**] indazolones (64H1986). Russian authors (66KG96) give νC═O 1627 cm^{-1} for indazolone (KBr).

Evans, Whelan and Johns (65T3351) dispute this interpretation, believing that indazolone in the solid state (KCl) exists in the OH form [**200b**]; they assign the band at 3413 cm^{-1} to νNH and the broad band at 2725 cm^{-1} to νOH. They do not specifically assign the peak at 1616 cm^{-1} (shoulder at 1642 cm^{-1}) but claim that 3-methoxyindazole [**203**, R′ = Me] shows such a band. However, there is a contradiction with the previous results (64H1986) concerning this last compound (see Scheme 4-13). The conclusion drawn by Evans *et al.* (65T3351) is probably erroneous, because it relies on incorrect experimental results.

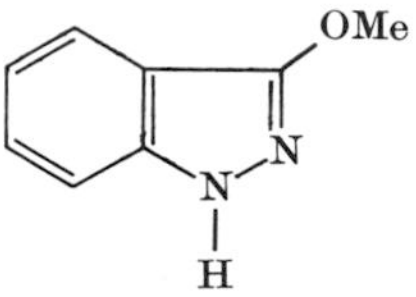

OMe N N H

KCl: 1621 cm^{-1}
$CHCl_3$: 1621 cm^{-1}
(65T3351)

KBr: 1530 cm^{-1}
$CHCl_3$: 1535 cm^{-1}
(64H1986)

SCHEME 4-13

The UV results (62MI125, 64H1986, 65T3351), indicate that indazolone exists in the 3-hydroxy form [**200b**] in ethanolic solution: compounds [**198**], [**200b**], [**201**] and [**203**] absorb similarly and differently from compounds [**199**] and [**204**] (vide supra).

In a mass spectral study (69OM37), the different fragmentation of compounds [**198**] and [**199**] reflects different ground state structures (lactim and lactam). On the basis of the previously cited work (65T-3351), a fragmentation pattern of indazolone is described from tautomer [**200b**].

Comparing indazolones with pyrazolones, the fused benzene ring makes the quinoid form [**205**] unfavourable but does not influence very much the other forms.

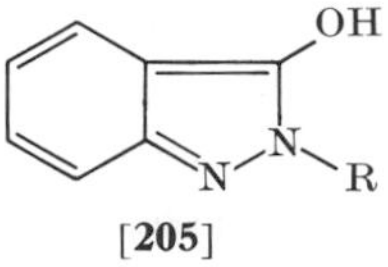

OH N N R

[**205**]

3. Compounds with a Potential Hydroxy Group: Others

For a general introduction to hydroxyazoles see Section 4–2A.

A. Heteroatoms-1,2 and a Potential 4-Hydroxy Group

a. Introduction

4-Hydroxy-isoxazole [**206**, Z = O] and -isothiazole [**206**, Z = S] and 1-substituted 4-hydroxypyrazoles [**206**, Z = NR] can exist in three tautomeric forms, which we designate as the OH [**206a**], CH [**206b**] and NH [**206c**] form. The NH form is a betaine, or zwitterion, but this does not affect fundamentally the procedure for their investigation.

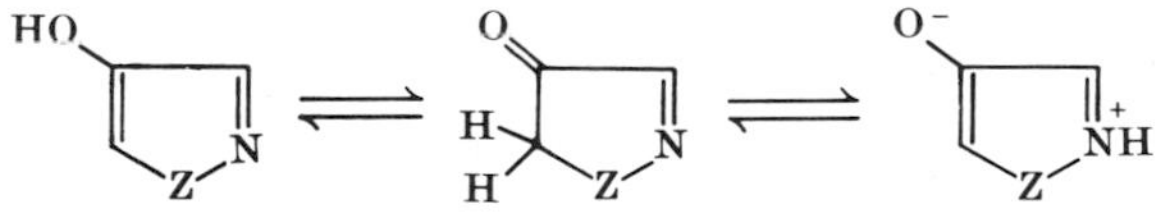

OH [**206a**] ⇌ CH [**206b**] ⇌ NH [**206c**]

Practically nothing was known of the tautomerism of these compounds at the time of the previous review (II-47), but now with the appearance of much fragmentary, and some definitive, work, the picture is quite clear: all the compounds exist predominantly in the OH form [**206a**].

b. 4-Hydroxyisoxazoles (II-47)

Until recently only scattered results were available in the literature; these indicated the predominance of the OH structure [**207a**] for 4-hydroxyisoxazoles from IR evidence in the solid state (56JA2532, 66AL457, 68AN1363) and polarographic methods (48JA3385). In 1971 Bianchi, Cook and Katritzky (71T6133) and in 1972 Nye and Tang (72T455) published definitive studies of a series of tautomeric 4-hydroxyisoxazoles and their fixed derivatives. The results summarized below show clearly the predominance of the OH form [**207a**] and did not detect any CH form [**207b**] which some earlier authors thought to have found.

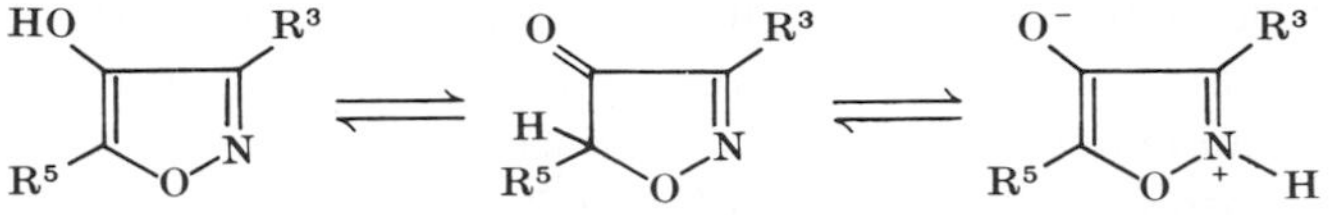

[**207a**] ⇌ [**207b**] ⇌ [**207c**]

i. Basicity Measurements and UV Spectra in Aqueous Solution. No long wavelength band at about 360 nm expected for the NH form [**207c**] is observed. The pK_a values of the tautomeric compounds together with those of the *O*-methyl derivatives indicate a great preponderance of the OH form [**207a**] (71T6133). The UV spectra in 20 *N* H_2SO_4 (71T6133) show similar cation structures [**208**] for the tautomeric compound and its *O*-methyl derivative. An increase in the conjugation of the cations with a 5-phenyl group supports structure [**208**].

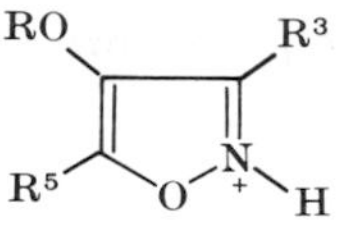

[**208**]

Nye and Tang (72T463) evaluate K_T for the equilibrium between the OH [**207a**] and NH form [**207c**] (for a discussion about the method of calculating K_T, see Section 4-3Ad) and thus confirm that the zwitterionic NH form [**207c**] is quite unstable relative to the OH form [**207a**].

ii. *UV Spectra in Nonaqueous Media.* The UV spectra of all the tautomeric compounds studied ($R^3 = R^5 =$ Me; $R^3 =$ Me, $R^5 =$ Ph; $R^3 =$ Ph, $R^5 =$ Me; $R^3 = R^5 =$ Ph) (71T6133) are similar to those of corresponding *O*-methyl derivatives whatever the polarity of the solvent.

iii. *IR Spectra.* No νC=O is observed either in solution or in the solid state (71T6133); in $CHCl_3$ both free νOH at 3580 cm^{-1} and associated νOH at 3100–3200 cm^{-1} are found, establishing the OH structure [**207a**] of the 4-hydroxyisoxazoles considered.

iv. *NMR Spectra.* The NMR spectra in $CDCl_3$ (71T6133) or in acetone (72T455) show no CH ring proton signal, eliminating the CH structure [**207b**]. The chemical shift for a 5-methyl group is the same for a tautomeric compound as for the corresponding methoxy derivative, again good evidence for the OH form [**207a**].

c. *4-Hydroxyisothiazoles*

The single paper dealing with these compounds (68BJ959) records the synthesis of a series of 4-hydroxyisothiazoles together with their *O*-methyl and *O*-acetyl derivatives. Tautomerism is not discussed; the compounds are written in the OH form [**209**], and this is consistent with the quoted NMR in DMSO and UV in EtOH.

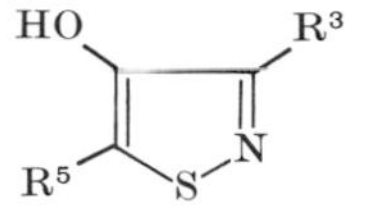

HO R³ R⁵ S N

[**209**]

d. 4-Hydroxypyrazoles (II-47)

Up to the time of the previous review (II-47) the tautomerism of 4-hydroxypyrazoles had been investigated only by chemical methods. A series of subsequent papers using fragmentary IR and UV data (63ZO2597, 70CB3885, 70CC3563, 71BF1038) concluded in favour of the OH form [**210a**] in solution as well as in the solid state. Recently, Nye and Tang (72T455, 72T463) solved the problem definitively.

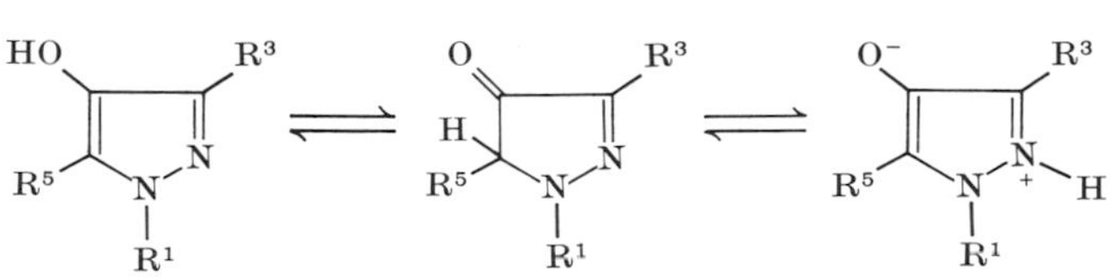

[**210a**] [**210b**] [**210c**]

Nyc and Tang's studies deal mainly with the 3,5-diphenyl 1-unsubstituted [**210**, R^1 = H] and 1-methyl [**210**, R^1 = Me] series. We take the last case as an example; the conclusions are valid for other 4-hydroxypyrazoles. The three tautomeric forms have the structures [**211a**], [**211b**] and [**211c**]. They utilized, as the corresponding fixed models, [**212**], [**213**] and [**214**]. The following experimental techniques demonstrate the predominance of the OH form [**211a**].

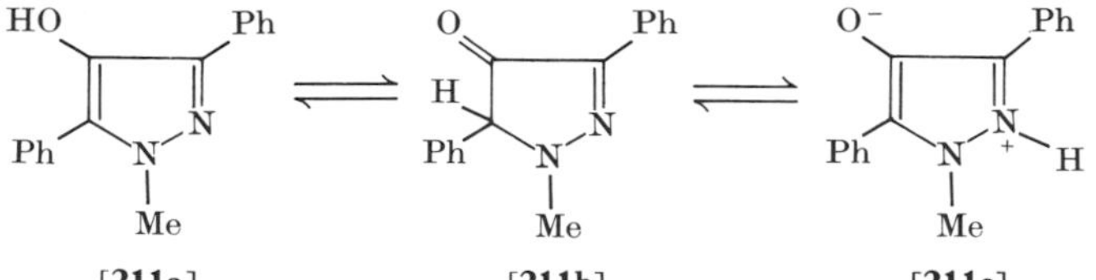

[**211a**] [**211b**] [**211c**]

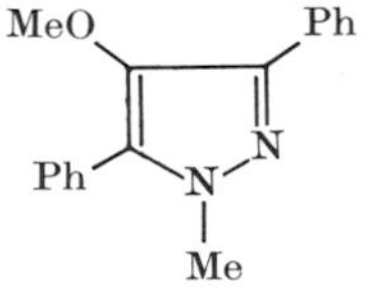

[**212**] 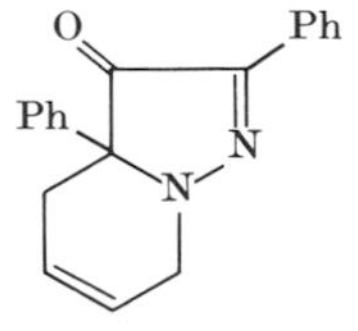 [**213**] 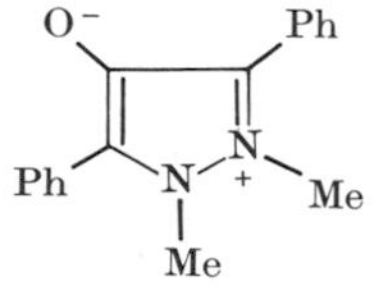 [**214**]

i. *IR Spectroscopy* ($CHCl_3$ and dioxan solutions or solid state: the spectrum of [**210**] exhibits no carbonyl band, whereas the CR model [**213**] shows νC=O at 1680 cm^{-1} (Nujol) and the NR model [**214**] a band at 1446 cm^{-1}: thus the OH structure exists alone or is greatly predominant.

ii. *UV Spectroscopy* easily distinguishes the three tautomers: the absorptions of the tautomeric compound are compared in Table 4-23

TABLE 4-23

UV ABSORPTION MAXIMA (nm) OF 4-HYDROXYPYRAZOLE DERIVATIVES (IN 95% EtOH)

Compound	Type	$\lambda_{max}(\varepsilon)$
[**211a**]	OH	245(21,500), 276(sh)(15,200), 290(sh)(10,600)
[**213**]	CR	270(17,200), 392(8,200)
[**212**]	OMe	251(24,100)
[**214**]	NMe	226(14,200), 248(sh)(6,740), 336(11,400)

with those of the three fixed derivatives. The OH and OR structures show no absorption at wavelengths longer than 300 nm, unlike the NR and CR model compounds.

iii. *NMR Spectroscopy* ($CDCl_3$, DMSO) again confirms the OH structure [**211a**]. No signal arises for the CH form [**211b**], and for 3,5-diphenyl-4-hydroxypyrazole [**215a**] two signals are exchangeable with D_2O, at 8.40 ppm (NH) and at ~7.50 ppm (OH) (DMSO): here the zwitterionic structure [**215b**] is symmetrical (C_{2v}), and only one signal would be expected from the two NH groups.

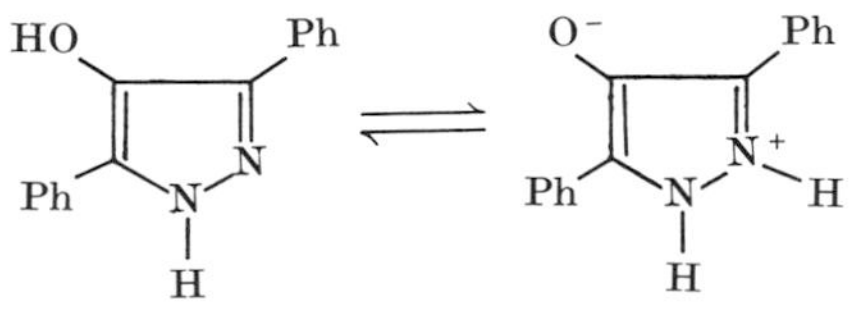

[**215a**] [**215b**]

iv. *Basicity Calculations.* In their second paper (72T463) the authors calculated K_T, for the tautomeric equilibrium between the OH form [**211a**] and the zwitterionic NH form [**211c**]. They did not measure the pK of the OR model [**212**], but used Eq. (1-15) (Section 1-4A) assuming that $K_A = K_{MeA}$.* They obtain K_T = [**211c**]/[**211a**] = 3.2×10^{-5} (ΔG = 6.15 kcal/mole at 25°), an equilibrium strongly in favour of the hydroxy form [**211a**].

They calculated also the constant K_i (named K_{ACNZ} by them) of the equilibrium of Eq. (4-3) and find the value given in Eq. (4-4).

$$XH + HX \rightleftharpoons X^- + HXH^+ \tag{4-3}$$

$$K_i = \frac{[X^-][HXH^+]}{[XH][HX]} = 1.7 \times 10^{-3} \tag{4-4}$$

The corresponding ΔG_i = 3.8 kcal/mole at 25° can be considered as an approximate indication of the degree of interaction between the positive and negative charges within the zwitterion.

e. Acyl- and Phenylazo-4-hydroxypyrazoles

The tautomerism of 3,5-dibenzoyl-4-hydroxypyrazole has been investigated (68CB1473) by an IR study in the solid state, but not solved because of its complexity. Insolubility prevented effective use of NMR; nevertheless a 4-hydroxy chelated structure [**216**] was suggested.

The 4-hydroxy-5-phenylazopyrazoles [**217**] (68JO513) exist as such as shown by IR in $CHCl_3$ where the only νC=O observed is that of the

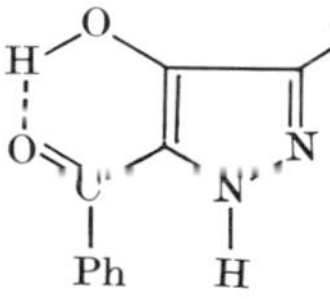

[**216**] 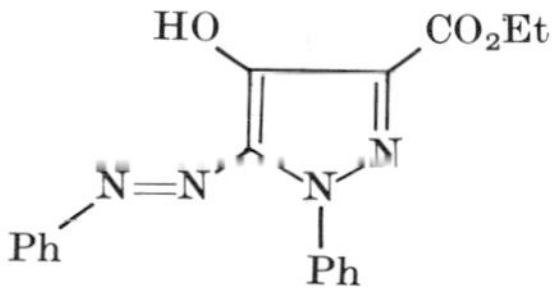 [**217**]

CO_2Et group, and by NMR in $CHCl_3$: the OH group exhibits a peak at 9.3 ppm comparable with those of 3-hydroxy-4-phenylazopyrazoles (Section 4-2Hd, OH at 9.2 ppm) and different from those of 4-phenylazopyrazol-5-ones (Section 4-2Gj, NH at 13.4 ppm). This conclusion is confirmed by proton NMR of ^{15}N-labelled derivatives (72JO4121).

B. Heteroatoms-1,3 and a Potential 2-Hydroxy Group

a. Introduction

At the time of the previous review (II-48), it could be stated firmly that oxazol-2-ones [**218**, Z = O], thiazol-2-ones [**218**, Z = S] and

* In reference (72T463) a different nomenclature is used; the relation to that previously used here is $K_C \equiv K_1$, $K_{ZN} \equiv K_T$, $Z \equiv XH$, $N \equiv HX$, $A \equiv X^-$, $C \equiv HXH^+$ and $K_{ZC} \equiv K_{MeA}$.

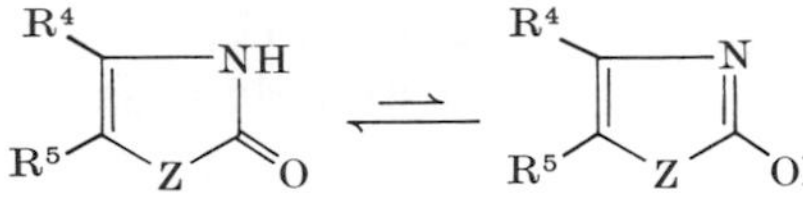

[218a] [218b] 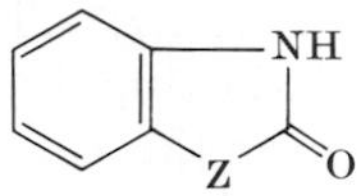 [219]

imidazol-2-ones [**218**, Z = NR] all existed in the oxo form [**218a**]. The corresponding benzothiazolone and benzimidazolone were also shown to exist in the corresponding oxo forms [**219**].

The work of the last decade has included the benzoxazolones in this pattern, and has provided considerable confirmatory evidence, but the statement of the previous review still adequately summarizes the position: "all the available evidence suggests that five-membered heterocyclic compounds containing a potential hydroxyl group between the two heteroatoms in the 1- and 3-position predominantly exist in the oxo form."

b. Oxazol-2-ones (II-48) and Benzoxazolones

Little further work has appeared on oxazol-2-ones (see 74HC(17)99) but NMR data support the NH form [**220**] for [**220**, R^4 = H, R^5 = CH_2NH_2] and [**220**, R^4 = H, R^5 = CH_3] (67H137). The methylene derivative [**221**] is, as expected, unstable with respect to [**223**]; the conversion is acid catalysed, doubtless via [**222**] (64JO978).

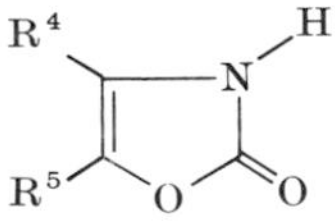

[220]

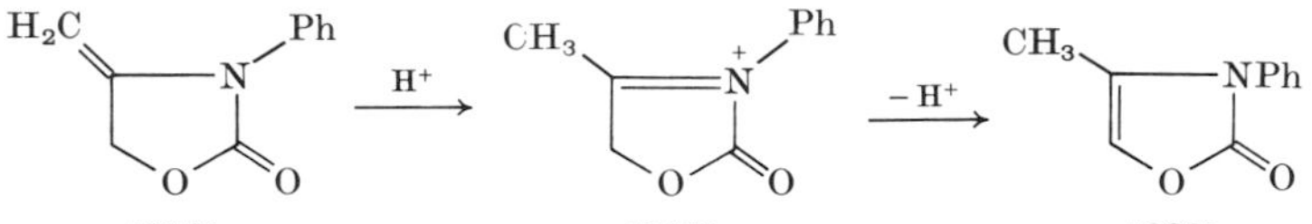

[221] [222] [223]

In 1968, Russian authors (68OG665) using Jaffé and Hammett equations (see Section 1-4B) found a good correlation between the ionization constants (determined in aqueous acetone) of benzoxazol-2-ones variously substituted on the phenyl ring and the substituent σ_I values: this indicates (cf. discussion, Section 1-4Bc) that the tautomeric equilibrium is strongly shifted towards one of the two forms [**224a**] or [**224b**]. Later they confirmed (70ZO1872) the predominance of the NH structure [**224a**] by IR spectroscopy: in all cases $\nu C{=}O$ absorbs at 1800–1750 cm^{-1} in the solid state and in solution; in dilute solution the

[224a] [224b]

spectra exhibit both free and associated νNH bands. IR comparison with fixed methyl derivatives supports the predominance of tautomer [**224a**] but UV spectra cannot be used because of the similarity of the absorptions.

Photochemical rearrangement of benzisoxazol-3-one (69JH123, 71JO1088) produces benzoxazol-2-one. The tautomerism of benzoxazolone is not discussed, but the NH structure [**224a**, R = H] follows from quoted NMR and IR data of the tautomeric compound and its methylated derivatives. The NH structure [**224a**, R = H] exists also in the solid state as shown by an X-ray determination (73AS945).

c. Thiazol-2-ones (II-49) *and Condensed Analogues*

Structures [**225**] and [**226**] were established previously (II-49). UV spectra of the benzobisthiazol-2-one [**227**] and its methyl derivatives in EtOH show the predominance of the di-NH form [**227**] (64AN80). Lactam structures for 2-oxothiazolo[4,5-*h*]- [**228**] and 2-oxothiazolo-[5,4-*f*]-isoquinoline [**229**] are suggested by the νC=O (71CC4054).

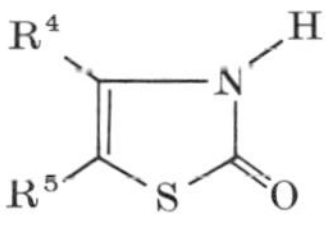

[225] 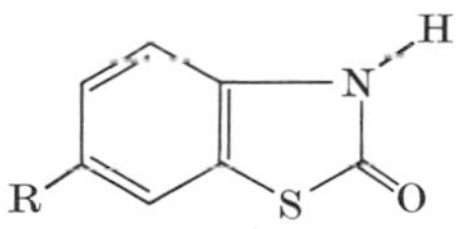 [226] 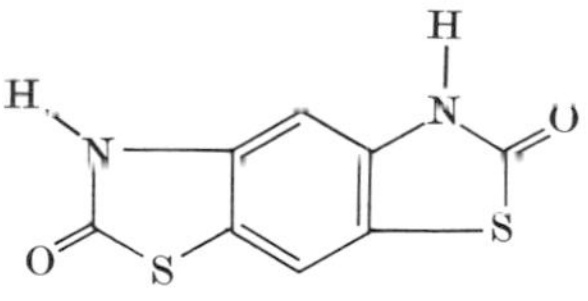 [227]

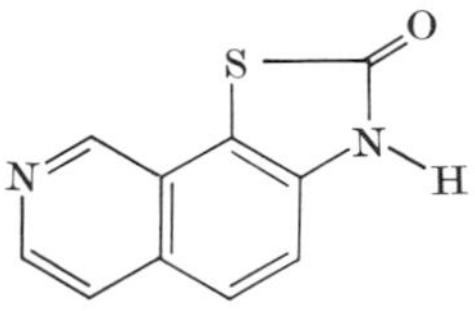

[228] 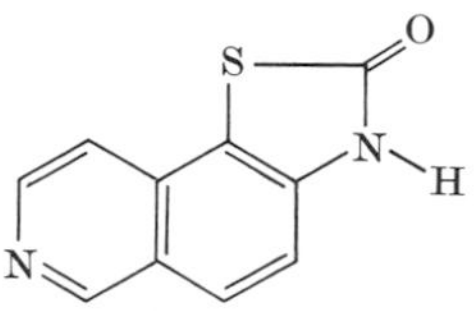 [229]

d. Imidazol-2-ones (II-50)

Structures [**230**], [**231**] and [**232**] were established previously (II-50). IR and UV spectroscopy have since been used to prove the oxo structure of further imidazol-2-ones [**230**, $R^4 = R^5$ = COAr] (68KG698), and benzimidazol-2-ones [**231**, R = H, R^3 = H, alkyl/or acyl] (61J4827) [**231**, R = COAr] (68KG698) [**231**, R = H, 5–Cl, 5–Me, 5–NO_2, 5-alkoxy

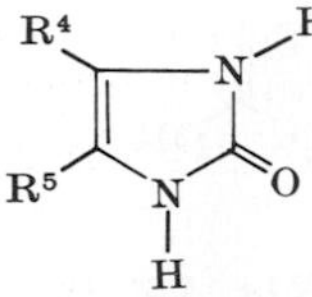

[230] 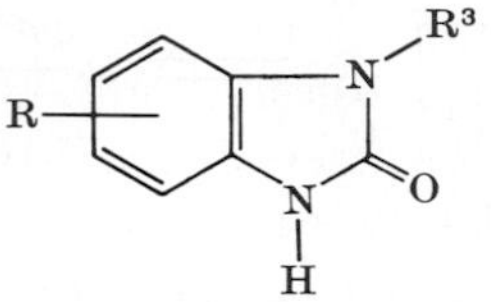 [231] 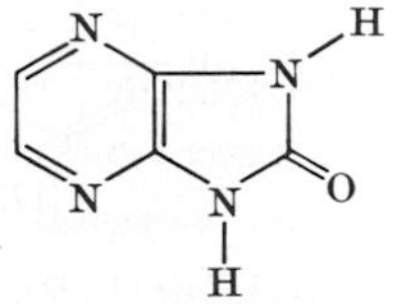 [232]

7–Cl, R^3 = CH_2Ph] (65AN116, 65MI117, 70JH807). The authors rely on the presence of νC=O and νNH absorptions and the lack of νOH in the IR spectra in the solid state. This has been confirmed by an independent X-ray study (73CX(B)2311). IR results in the solid state on 2-oxo-5,6-dihydro-1*H*,4*H*-imidazo[4,5,1-*ij*]quinoline (63JO2581) confirm the oxo structure [**233**] already indicated in the previous survey [II-50].

Dipole moment comparisons with mono- and di-methyl derivatives have been carried out for imidazol-2-one and benzimidazolone (72J-(PII)2045). Unfortunately interpretation of the results is hampered by strong association, and although the authors believe that some OH form is probably present, especially for monocyclic compounds, this conclusion must be treated with caution.

The imidazol-2-ones derivatives [**234**] and [**235**] with R = H or Me seem to exist as such (64ZO197). If R = H two carbonyl absorptions arise from the two different imidazolone rings besides the quinone νC=O band.

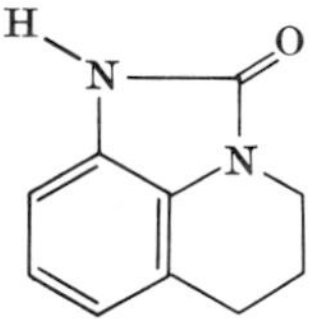

[233] 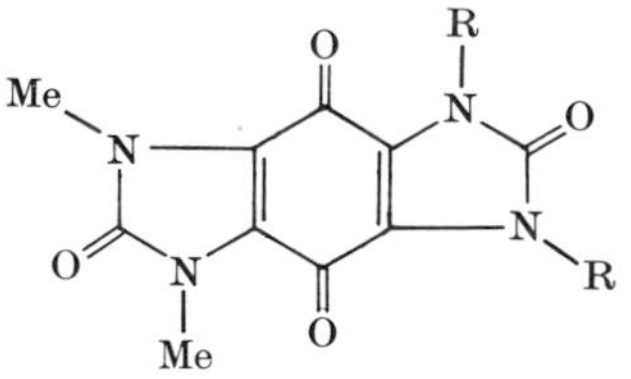 [234] 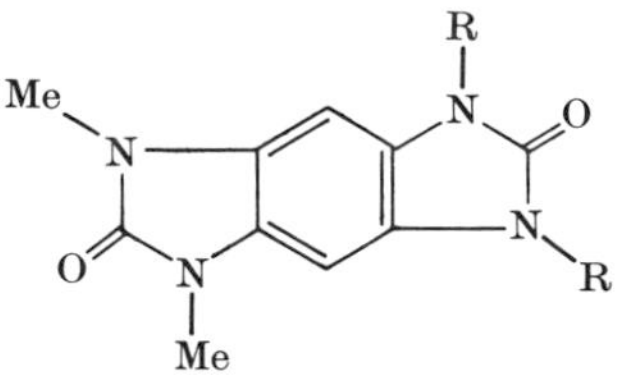 [235]

The 2-oxo-1,3-diazaazulene [**236a**] was originally considered from UV spectra in MeOH to exist as a mixture of the two tautomers [**236a**] and [**236b**], the first being predominant (63MI271). However, a careful

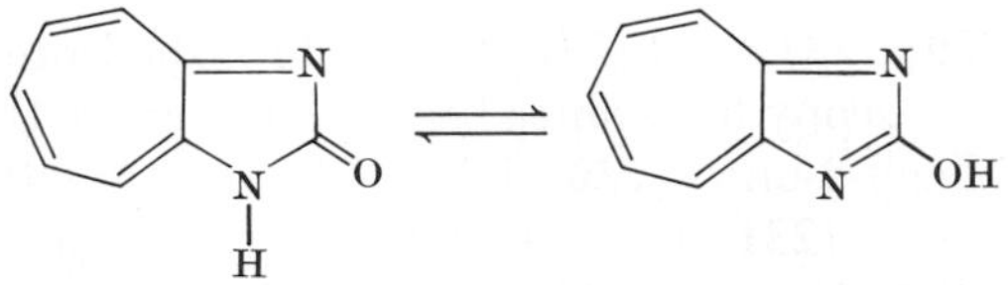

[236a] [236b]

UV study of this equilibrium in aqueous solutions over a wide pH range (70BJ2283), with comparisons with *O*-ethyl and *N*-ethyl derivatives, indicates only the presence of the NH form [**236a**] which undergoes a first protonation at the aza nitrogen.

For the 2,6-dihydroxy-1,3-diazaazulene several forms can be written; of these [**237a**] and [**237b**] are the most probable: structure [**237a**] was suggested on inconclusive chemical evidence (65CT473).

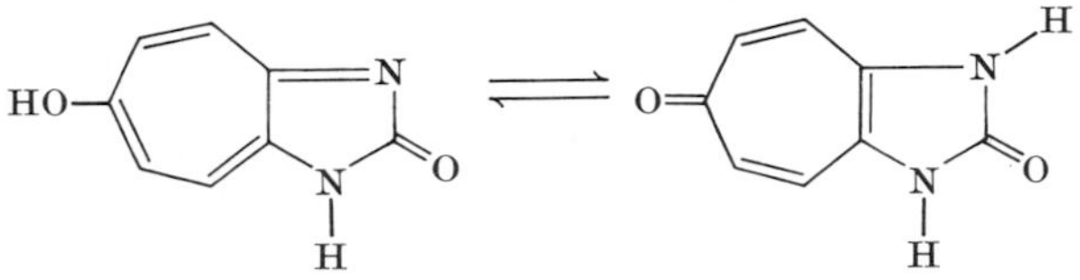

[**237a**] ⇌ [**237b**]

C. Heteroatoms-1,3 and a Potential 4-(or 5-)Hydroxy Group

a. Introduction

Tautomeric equilibria in this series are complex. Oxazol-5-ones [**238**, Z = O], thiazol-5-ones [**238**, Z = S] and 1-substituted imidazol-5-ones [**238**, Z = NR] can exist in four tautomeric forms: C4H, OH, NH and C2H [**238a–d**]. Oxazol-4-ones [**239**, Z = O], thiazol-4-ones [**239**, Z = S], and 1-substituted imidazol-4-ones [**239**, Z = NR] can exist in three tautomeric forms: CH, OH, NH [**239a–c**]. Unsubstituted imidazol-4(5)-ones can exist in eight tautomeric forms (Section 4-3C1), and

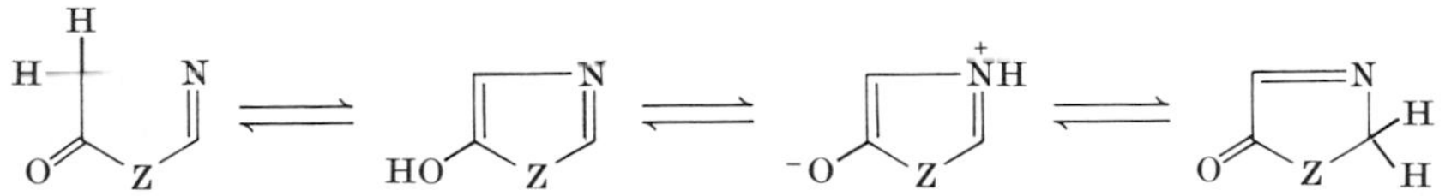

C4H [**238a**] ⇌ OH [**238b**] ⇌ NH [**238c**] ⇌ C2H [**238d**]

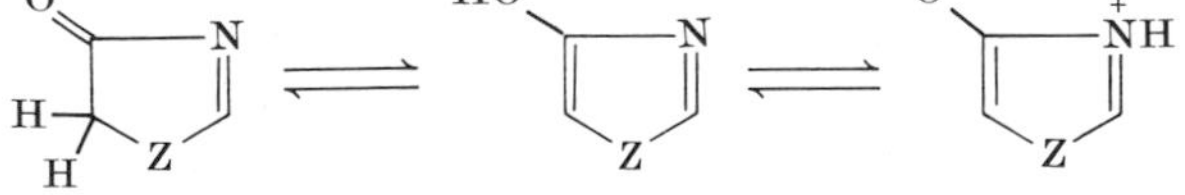

CH [**239a**] ⇌ OH [**239b**] ⇌ NH [**239c**]

therefore imidazol-4(5)-ones are considered in separate sections according to whether they carry a 1-substituent.

At the time of the previous review (II-50), a simple pattern appeared to be emerging, and it was found that the then available evidence indicated that the compounds existed in the oxo or CH forms [**238a**], [**239a**] except when an electron withdrawing group was present in the 5-(or 4-)position. Subsequent work has *not* borne out this simple picture. In particular, zwitterionic NH form [**238c**] can be important. A large amount of work is now available, but there are still significant gaps.

b. Oxazol-5-ones (II-50)

Of the four tautomeric forms [**240a–d**] for oxazol-5-ones [**240**], the previous survey (II-50) gave concrete evidence for the C4H form [**240a**] for those simple oxazol-5-ones which can be obtained optically active, and concluded that form [**240a**] was preferred unless R^4 was a strong electron-withdrawing group which then favoured the OH tautomer [**240b**]. However at the time there was no concrete evidence for this OH form [**240b**], and it now appears not to be a stable form for oxazol-5-ones and has never been conclusively identified. The tautomer [**240d**], known as pseudooxazol-5-one, is not in mobile equilibrium with the other forms. Pseudooxazolones are discussed separately in Section (Section 4-3Cc). A general survey of the chemistry of oxazol-5-ones is available (69MI77).

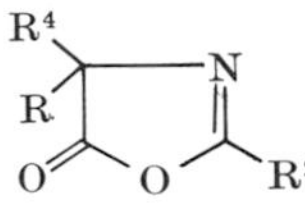

C4H

[**240a**] R = H
[**241**] R = Me

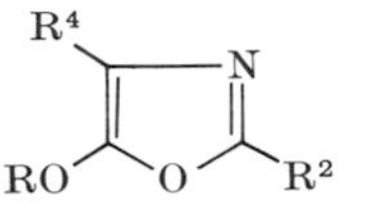

OH

[**240b**] R = H
[**242**] R = Me

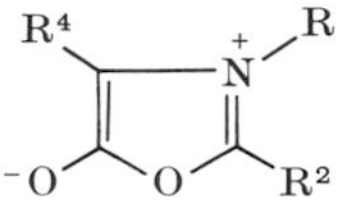

NH

[**240c**] R = H
[**243**] R = Me

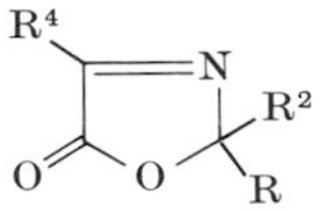

C2H

[**240d**] R = H
[**244**] R = Me

At present it appears that either the CH [**240a**] or the NH form [**240c**] can be favoured depending on the substituents R^2 and R^4 and on the medium. This situation first became apparent for 2,4-diaryloxazol-5-ones. These well known compounds ("azlactones") exist in the CH form [**240a**] in the solid state and in solution in nonpolar solvents. However, with increasing solvent polarity, increasing amounts of the zwitterionic NH form [**240c**] coexist. Kille and Fleury (68BF4636) first demonstrated the zwitterionic NH structure [**240c**] for the compound [**245a**] with Ar^2 = *p*-nitrophenyl and Ar^4 = phenyl by UV (in Ac_2O) and IR (probably solid state) comparisons with fixed derivatives of types [**242**], [**243**], [**245/1**]. The NH structure [**245a**] is suggested for Ar^2 = *p*-chlorophenyl and Ar^4 = phenyl (69TL1557) on IR evidence (KBr).

In 1970 work (70JA4340) appeared which emphasized the importance of the solvent on the equilibrium [**245a**] ⇌ [**245b**]. The characteristic features of the zwitterionic NH structure [**240c**] include low solubility in organic solvents, νC–O$^-$ at 1670–1710 cm^{-1} depending on the solvent (νC=O appears at 1820 cm^{-1} for [**245b**]), and a UV band at 403–450 nm, the precise λ_{max} depending on the solvent polarity. Huisgen *et al.* (70JA4340) evaluated the percentage of zwitterionic form [**245a**] of 2,4-diphenyloxazol-5-one in different solvents: 49% in DMF, 32% in DMSO, < 0.5% in acetone and $CHCl_3$; in dioxan no long wavelength absorption was observed (70CB2356, 70JA4340). 4-*p*-Methoxyphenyl-2-phenyloxazol-5-one [**245b**, Ar^2 = Ph, Ar^4 = *p*-MeO.C_6H_4] exists in the CH form from UV measurements in dioxan and IR in the solid state, but the zwitterionic form [**245a**] is invoked to explain the cycloadditions of oxazol-5-ones with acetylenes (70CB2356).

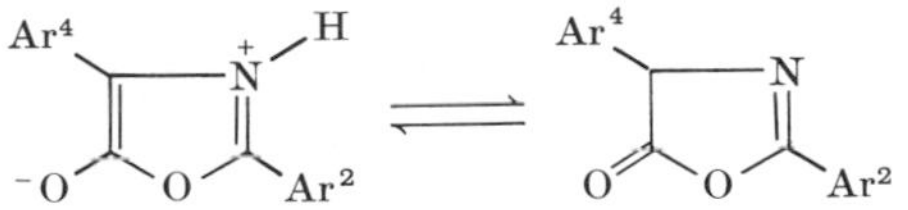

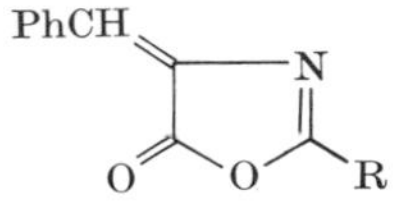

[**245a**] [**245b**] [**245/1**]

The hydroxy structure was under discussion in the previous survey (II-50) for 4-ethoxycarbonyloxazol-5-ones, but as Petersen (69TL1557) has indicated, the broad IR absorption and solubility characteristics for the 2-aryl-4-ethoxycarbonyl derivatives are best explained by the zwitterionic NH form [**240c**].

The CH structures shown for [**246**] and [**247**] were demonstrated for the solid state and solution by UV, IR and NMR on the tautomeric

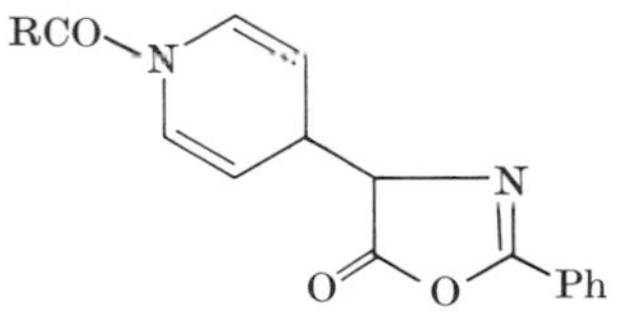

[**246**] 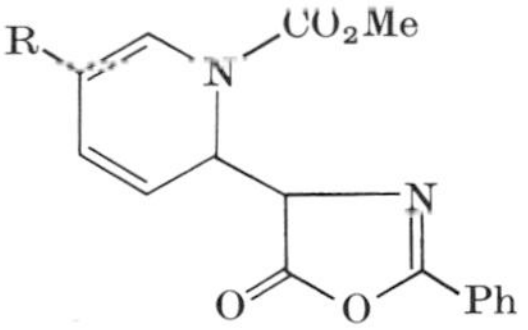 [**247**]

compounds and fixed derivatives (69CB1129). Although the OH-structure [**248a**] was assigned to [**248**], this is unlikely in view of strong absorption in the region 1800–3600 cm^{-1} in the solid state. Taking into account the other results we believe that the zwitterionic structure [**248b**] is the most probable, as it usually occurs if R^2 and R^4 are aryl

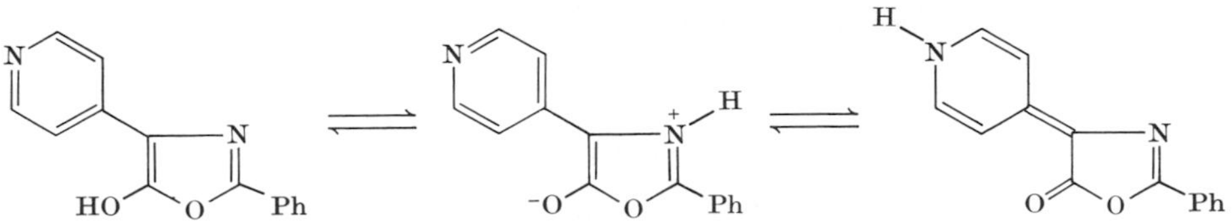

[**248a**] [**248b**] [**248c**]

groups [**245a**]. However, the OH structure [**248a**] cannot be finally eliminated without further evidence.

*c. Pseudooxazol-5-ones** (II-51) *and Oxazol-5-one Cations*

Since the previous survey (II-51), much has been accomplished in this area. Most pseudooxazol-5-ones possess additional unsaturation at the 2- or 4-position (64JO2205, 66TL383, 66TL4427, 67BJ149, 67CB-1824, 67T3363, 69MI77). The typical unsaturated pseudooxazolone structure is [**249a**]. Depending on the method of preparation and on the nature of R^1, R^2 and R^3, it is possible to get the pseudooxazolone [**249a**] alone or mixed with the oxazol-5-ones [**249b**] and [**249c**]. The

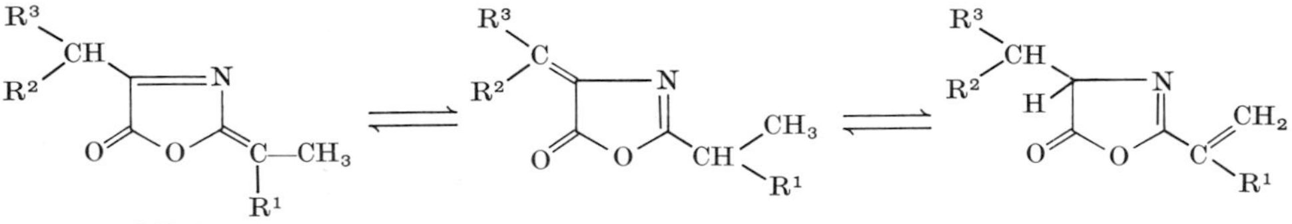

R^3 CH R^2 N O O C—CH_3 R^1 ⇌ R^3 C R^2 N O O CH CH_3 R^1 ⇌ R^3 CH R^2 H N O O C CH_2 R^1

[**249a**] [**249b**] [**249c**]

three isomers are easy to detect by IR, UV and NMR spectroscopy. A rapid dynamic equilibrium does not exist between the forms, and it is possible to separate them. Related compounds of type [**250**] (63JO98) exist as such in solution and in the solid state from UV, IR and NMR evidence. Compound [**251**] exists as such (69MI77) from UV evidence in CH_3CN. Some pseudo-oxazolones without unsaturation have been obtained, such as [**252**] (64CB2023), which was carefully studied by NMR, and [**253**] (69CB1129).

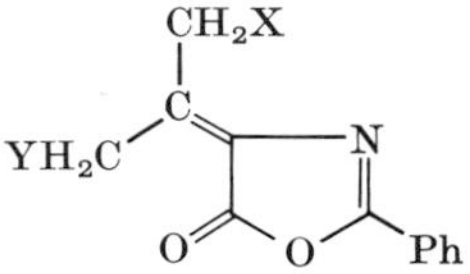

CH_2X C YH_2C N O O Ph

[**250**]

Ph N H N O O   R N H O O CF_3 Me N O O N COR

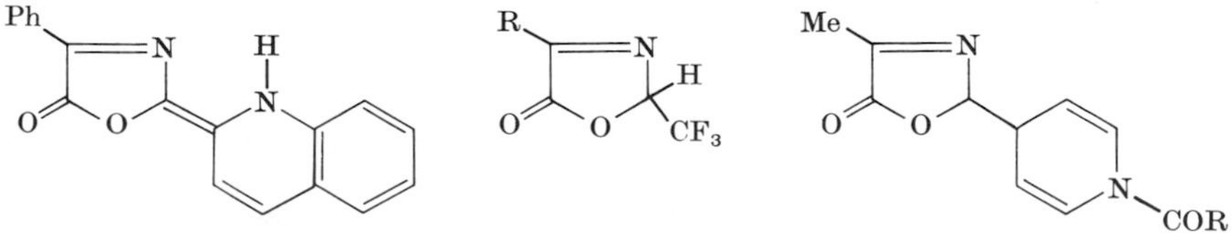

[**251**] [**252**] [**253**]

* According to Steglich (69MI77) it is preferable to name these compounds 2*H*-oxazol-5-ones to distinguish from 4*H*-oxazol-5-ones for the other tautomers.

The oxazol-5-onium cation [**254**] can be written in three tautomeric forms [**254a–c**]. Boyd and Wright (72J(PI)909) showed that whatever the substituents R^2 or R^4 (H, alkyl or aryl) these cations exist in the form [**254a**]: the IR spectra in Nujol exhibit a νC=O absorption at about 1880 cm^{-1}, and NMR measurements in CF_3COOH give the multiplicity expected for the groups in position 2 and 4. HMO calculations predict the greater stability of [**254a**] compared to [**254b**]. The salts [**255**] also exist as such.

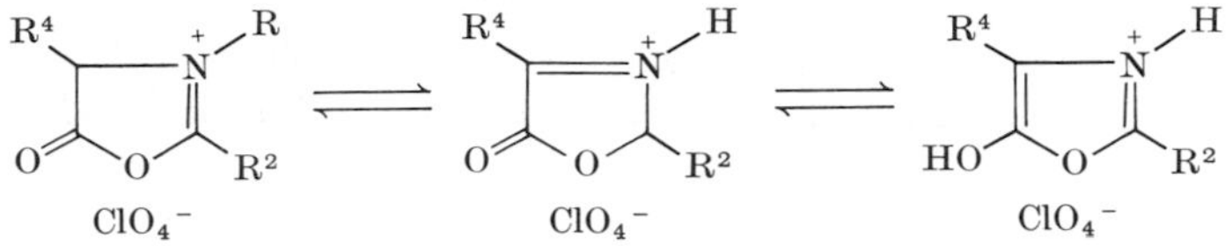

ClO_4^- ClO_4^- ClO_4^-

[**254a**] R = H [**254b**] [**254c**]
[**255**] R ≠ H

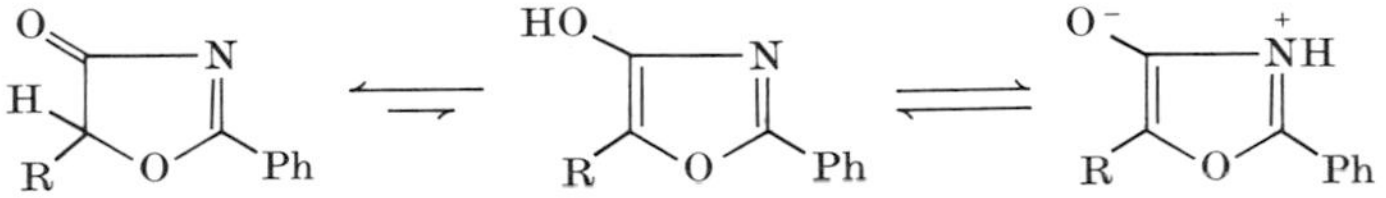

[**255/1a**] R = H [**255/1b**] [**255/1c**]
[**255/2a**] R = Ph [**255/2b**] [**255/2c**]

d. Oxazol-4-ones

Very little work is available: [**255/1**] and [**255/2**] appear on IR evidence (νC=O at 1705 cm^{-1}) to exist in the carbonyl form [**255/1a**; **255/2a**] (70CH1622); however, chemical evidence indicates that either the enol [**255/1b**] or zwitterionic [**255/1c**] form are involved in reactions (72CH1000).

e. Thiazol-5-ones (II-51)

Nothing was known at the time of the previous review about the tautomerism of simple thiazol-5-ones. The compounds studied since are mostly derivatives of 2-phenylthiazol-5-ones [**256**] which can exist in three forms* [**256a–c**].

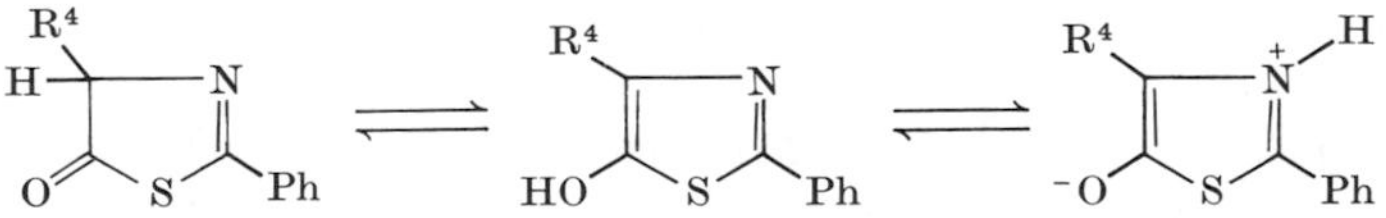

[**256a**] [**256b**] [**256c**]

The compound with $R^4 = CHPh_2$ is assigned (62JO3730) the CH form [**256a**] based on νC=O and νC=N absorption in the IR spectrum

* The pseudothiazolone form has never been implicated in the literature.

in $CHCl_3$. The νC=O frequency is decreased by about 100 cm^{-1} compared with the oxazol-5-ones, which reflects the greater polarisability of sulfur than of oxygen.

More complete work on a series of 2-phenylthiazol-5-ones (70TL169) using UV, IR and NMR is summarized in Table 4-24. The equilibrium

TABLE 4-24

TAUTOMERISM CH FORM [**256a**] TO OH FORM [**256b**] OF 2-PHENYLTHIAZOL-5-ONES (70TL169)

	Compounds (4-substituent given)					
	H	Me	CH_2Ph	CH_2CHMe_2	$CHMe_2$	CHMeEt
Solid state	CH	OH	OH	OH	—	—
Liquid	—	—	—	—	CH	CH
DMSO	CH: 25% OH: 75%	OH	OH	OH	CH: 27% OH: 73%	CH: 18% OH: 82%

depends on the nature of the 4-substituent; hydrogen and secondary alkyl tend to favour the CH form relative to primary alkyl substituents. The equilibrium also depends strongly on the solvent as observed for isoxazol-5-ones (Section 4-2Ba); in an aprotic medium solvent basicity favours the OH form [**256b**] whereas in a weakly basic solvent the CH form [**256a**] is predominant. In a protic medium the equilibrium is displaced towards the hydroxy form [**256b**]. The greater tendency for thiazol-5-ones to take up the OH tautomeric form, compared to the oxazol-5-ones (Section 4-3Cb), is attributed to a greater aromaticity of thiazoles compared with oxazoles. In polar solvents (DMSO, EtOH), the UV spectra exhibit a weak band around 420 nm (ε concentration dependent) assigned to the NH form [**256c**] the proportion of which increases with dilution. Later study (71AJ2729) of the same compounds supports these results except that the zwitterionic NH form [**256c**] is not mentioned.

Tautomeric forms [**257a**] or [**257b**] can be distinguished in the NMR by nonequivalence of the methylene group hydrogen atoms in [**257a**] but not in [**257b**] (70IS633); for instance, for [**257**, R = H, R′ = OAc] a mixture of two diastereomers is observed in $CDCl_3$; but in DMSO or pyridine, only an enantiomeric mixture appears which corresponds to the form [**257b**]. The protonation of [**257a**] with CF_3COOH gives the

[**257a**] ⇌ [**257b**] $\xrightarrow{H^+}$ [**258**]

cation [**258**]. The results are in agreement with those described in (70TL169) for [**256**, $R^4 = CH_2CHMe_2$ or CHMeEt].

Compounds of the type [**259a**] exist as such in $CHCl_3$ from IR and NMR data (72J(PI)310). In pyridine form [**259a**] may be in equilibrium with [**259b**], but this suggestion relies only on chemical observations, i.e., the rapid equilibration of the two geometric isomers of [**259a**] and easy *O*-acylation.

[**259a**] ⇌ [**259b**]

2-Alkyl-substituted isothiazol-5-ones show (72J(PI)1983) IR, UV and NMR properties analogous to those of 2-phenyl derivatives indicating predominance of the CH form in $CDCl_3$ and increase in the percentage of the OH form in DMSO.

f. 4-Thiazolones and 4-Hydroxythiazoles (II-51)

Simple thiazol-4-ones were long unknown because of their instability. Jensen and Crossland (63AS144) synthesized [**260**] and [**261**] demonstrating that in previous work compounds claimed as thiazol-4-ones were dimers. An NMR and IR study of [**260**] and [**261**] (65AS1215) shows that in DMSO and DMF 2-phenyl-4-hydroxythiazole exists to the extent of 80% in the OH form [**260b**] but in acetone the CH [**260a**] and OH forms are equally abundant. 4-Hydroxy-5-methyl-2-phenylthiazole seems to exist only in the enol form [**261b**] whatever the solvent, although nonpolar solvents have not been used because of the insolubility of these compounds. The 2-phenyl derivative [**260**] exists in the CH form [**260a**] in the solid state on IR evidence (63AS144,

[**260a**] R = H ⇌ [**260b**] R = H ⇌ [**260c**] R = H
[**261a**] R = Me [**261b**] R = Me [**261c**] R = Me

65AS1215), as do other 2-arylthiazol-4-ones (66CO(262C)1017). 2-Phenyl-4-hydroxy-5-methylthiazole showed IR bands in the solid which indicated a mixture of [**261a**] and [**261b**] (63AS144); however, later the same compound crystallized entirely in the OH form (65AS-1215). The difference in behaviour between 5-unsubstituted and 5-methyl compounds is similar to that observed in the case of thiazol-5-ones (Section 4-3Ce).

Oxyluciferin [**262**] exists predominantly in the OH form shown (72T4065) in DMSO and acetone as well as in the solid state on IR and NMR evidence.

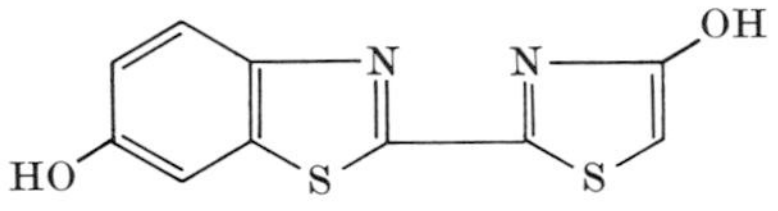

[**262**]

Structure [**263**], [**264**] and [**265**], respectively, are suggested (63AS144) for the thiazolones of Homber, Chabrier and Beyer, which were previously thought to be simple thiazol-4-ones. The hydroxy forms [**263–265**] are consistent with NMR solution data (65AS1215). Compound [**266**] has been described as such (65LA(682)201), but without any spectroscopic proof.

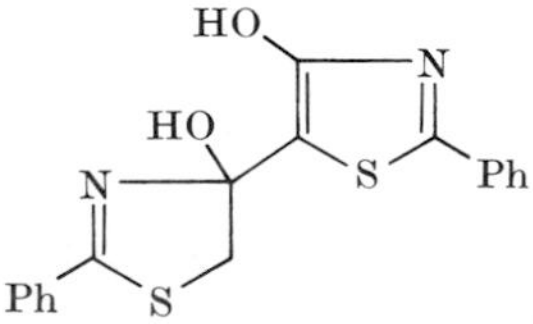

[**263**] 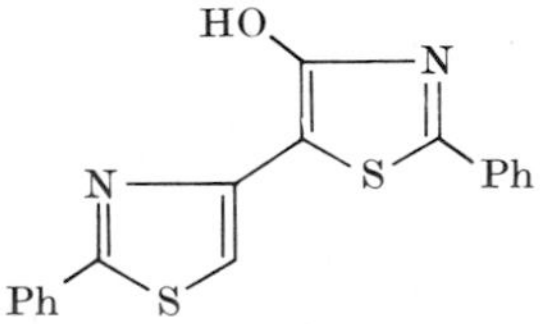 [**264**] 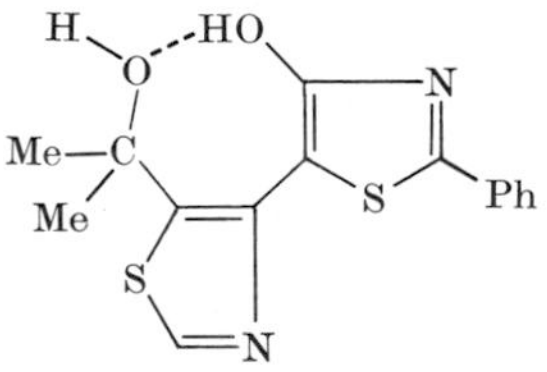 [**265**]

The benzylidene derivative [**267**] was assigned tautomeric structure [**267a**] (63AS144) because of νNH, νC=O and νC=C absorptions (at 1634 cm^{-1}) in the solid state. Taylor, (70SA(A)153, 70SA(A)165) apparently unaware of the previous paper, assigns to [**267**] structure [**267b**] in the solid state; he observes no νNH absorption, gives the same νC=O band and attributes the band at 1638 cm^{-1} to νC=N absorption (low compared with the νC=N 1685 cm^{-1} in 2-phenylthiazol-4-one). In $CDCl_3$ from NMR measurements [**267a**] is clearly predominant: this

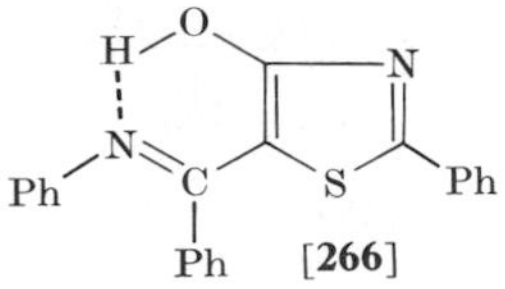

[**266**] 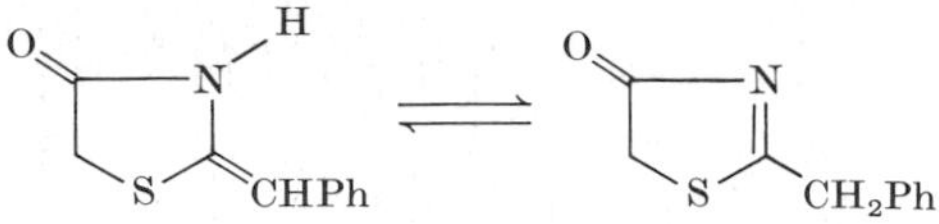 [**267a**] [**267b**]

study of many compounds of type [**268**] shows that all exist in solution in the form shown (70SA(A)153). Further work on [**268**, R = CO_2Et and R = CN] has confirmed the methylene formulation and that *cis-trans* isomers can be isolated (73AS1914, 73AS1923).

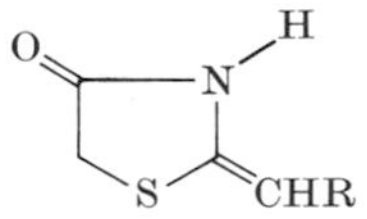

[**268**] R = CO_2R', $CONH_2$, $CONMe_2$, COPh, CN

g. 1-Substituted Imidazol-5-ones

To simplify the problem for imidazol-4(5)-ones we consider N-substituted before N-unsubstituted derivatives.

1-Methylimidazol-5-ones [**269**] can exist in the C4H [**269a**], OH [**269b**], NH [**269c**], and C2H forms [**269d**]. French authors (71BF1040) found that 1,2,4-trimethyl- [**269**, $R^2 = R^4$ = Me] and 1,4-dimethyl-2-phenyl-imidazol-5-one [**269**, R^2 = Ph, R^4 = Me] both exist in the C4H form [**269a**] under the large variety of conditions studied: UV

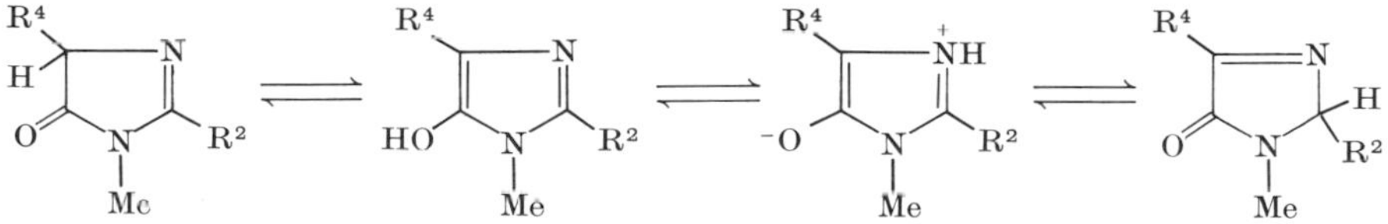

C4H [**269a**] ⇌ OH [**269b**] ⇌ NH [**269c**] ⇌ C2H [**269d**]

(in MeOH), IR (in the solid state, $CHCl_3$, DMSO and pyridine) and NMR (in $CDCl_3$, DMSO and pyridine) and comparison with 4,4-dimethyl model compounds. However, 4-phenyl analogues [**269**, R^4 = Ph] do *not* exist in the C4H form [**269a**]. The interpretation is more difficult here; model compounds are lacking except for 4,4-diphenyl derivatives (69T4265) with νC=O in KBr at 1730–1735 cm^{-1}. 1,2-Dimethyl-4-phenylimidazol-5-one [**269**, R^2 = Me, R^4 = Ph] is insoluble in IR or NMR solvents, hindering investigation in solution; the IR (KBr) shows that in the solid state it does not exist in the C4H form [**269a**] but does not distinguish between the OH [**269b**] and NH forms [**269c**]; the UV spectrum in MeOH shows considerable conjugation which again points to [**269b**] or [**269c**].

The authors conclude that 1-methyl-2,4-diphenylimidazol-5-one [**269**, $R^2 = R^4$ = Ph] exists in the OH form [**269b**] in all solvents, basing this on their NMR data; but the IR spectra in $CHCl_3$, DMSO or pyridine exhibits an absorption at 1715–1720 cm^{-1} which appears to indicate the C4H [**269a**] or perhaps the C2H [**269d**] form. The solid state IR does not show this band and in the solid it probably exists as the OH [**269b**] or NH form [**269c**]. The long wavelength (365 nm) absorption in MeOH suggests that the zwitterionic NH form [**269c**] is present in this solvent. However, CNDO calculations indicate that the zwitterionic form is the least stable one and another argument against the zwitterionic structure is the difficulty of preparing betaines in this series (70JH139).

In conclusion, it appears that whereas the simplest compounds adopt the CH form [**269a**] a 4-phenyl substituent has a large and as yet incompletely understood influence on the tautomeric equilibrium.

h. 1-Substituted 4-Hydroxyimidazoles

1-Methylimidazol-4-ones [**270**] can exist in CH [**270a**], OH [**270b**] and NH [**270c**] forms. French authors (71BF1040) studied two such compounds using UV, IR and NMR spectroscopy. 1,2-Dimethyl-5-phenylimidazol-4-one [**270**, R^2 = Me, R^5 = Ph] exists as the OH form [**270b**] in the solid state (no νC=O absorption in KBr), but a structural determination in solution was not possible because of insolubility.

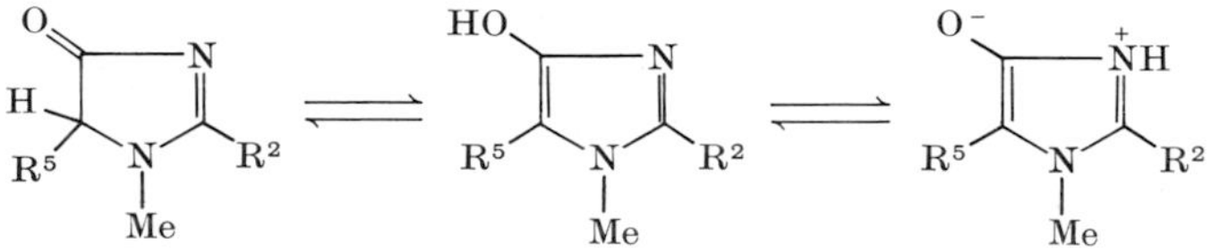

CH [**270a**] ⇌ OH [**270b**] ⇌ NH [**270c**]

1,2,5-Trimethylimidazol-4-one [**270**, R^2 = R^5 = Me] shows IR spectra in solution and in the solid state with absorption at 1700–1720 cm^{-1}. This cannot easily be assigned to νC=O of the CH form [**270a**] as the NMR spectra in the same solvents lack any signal for the 4-proton. The authors suggest the zwitterionic NH form [**270c**] but two observations are difficult to reconcile with this: first the UV spectrum in MeOH does not show any absorption above 300 nm, and second the IR band seems to be at too high frequency for a C–O^- vibration: for oxazol-5-ones the νC–O^- appears about 100 cm^{-1} lower than the νC=O of the oxo form (Section 4-3Cb). In the absence of fixed derivatives this tautomeric equilibrium cannot yet be considered solved. Compounds stated to be 1-acyl imidazol-4-ones have been formulated as 4-oxo derivatives on IR evidence (71KG801).

i. N-Unsubstituted Imidazol-4(5)-ones

We consider first the simpler 5,5-disubstituted imidazol-4-ones (58BF543, 61JO4480, 69T4265, 71BF1040) which can exist in three forms [**271a**], [**271b**] and [**271c**]. 2-Methyl-5,5-diphenyl- [**271**, R^2 = Me, R^5 = Ph] and 2,5,5-triphenyl-imidazol-4-one [**271**, R^2 = R^5 = Ph] (58BF543, 69T4265) both exhibit two series of absorption in their IR spectra in the solid state which have been shown to belong to the

[**271a**] [**271b**] [**271c**]

individual tautomers [**271a**] and [**271b**]. It is even possible to separate both "desmotropic" forms as crystalline solids, but on dissolving each gives the same equilibrium mixture and an identical spectrum. 2-Phenyl-5,5-dimethylimidazol-4-one [**271**, R^2 = Ph, R^5 = Me] (71BF-1040) in the solid state shows the same phenomenon, and the presence of about equal amounts of [**271a**] and [**271b**]. By contrast, solution IR and NMR data (71BF1040) in comparison with model *N*-methyl derivatives not only eliminate the OH form [**271c**] as a significant contributor to the 5,5-disubstituted imidazol-4-one, but also suggest that form [**271b**] is greatly predominant.

Recently Edward and Lantos (72JH363) have thoroughly studied the 5,5-diphenyl [**271**; R^5 = Ph] and the 5-spirocyclohexyl compounds [**271**; R^5, R^5 = $(CH_2)_5$] and their results are tabulated:

Solvent method	$CHCl_3$ IR	DMSO IR	H_2O UV	H_2O p*K*a	EtOH UV
[271; R^5 = Ph, R^2 = H]	**b**	**b**	80 **b**:20 **a**	80 **b**:20 **a**	
[271; R^5, R^5 = $(CH_2)_5$]	**b**	75 **b**:25 **a**	65 **b**:35 **a**	75 **b**:25 **a**	90 *b*:10 **a**

Greater stability in nonpolar solvents of the less conjugated and less polar tautomers [**271b**] is indicated, but in polar solvents the more polar tautomers [**271a**] increase in importance, especially when electron donor substituents are present in the 2-position.

2,4-Disubstituted imidazol-5-ones [**272**] can exist in eight alternative tautomeric forms [**272a–h**]. Until recently work on such compounds was scattered and incomplete but did appear to favour the CH forms

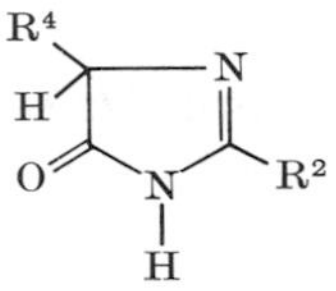

[**272a**] 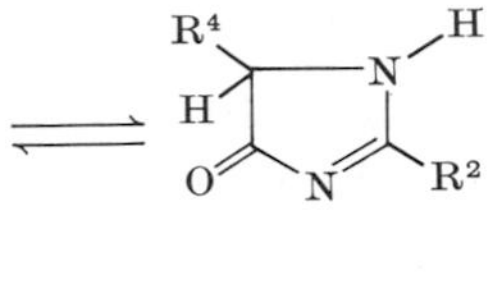 [**272b**] 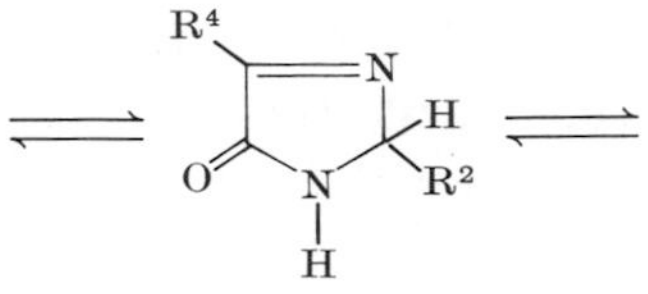 [**272c**]

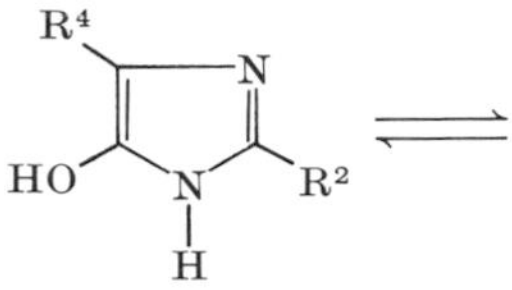

[**272d**] 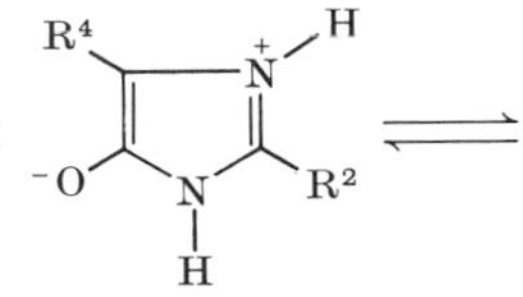 [**272e**] 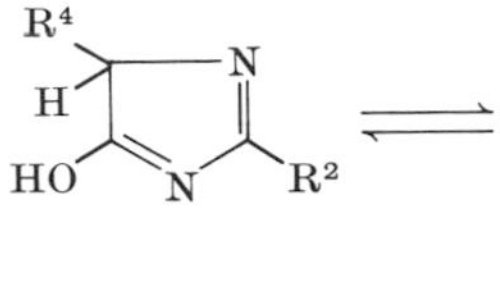 [**272f**]

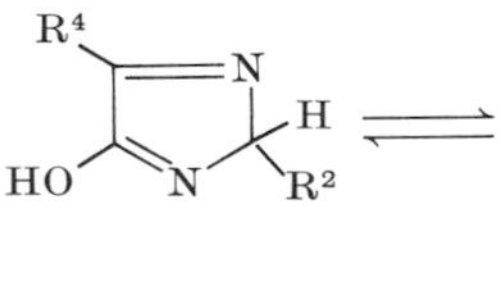

[**272g**] 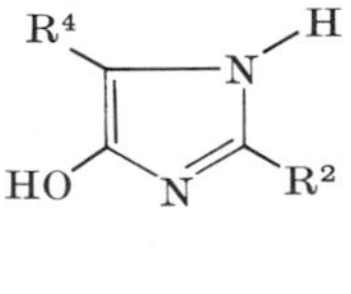 [**272h**] 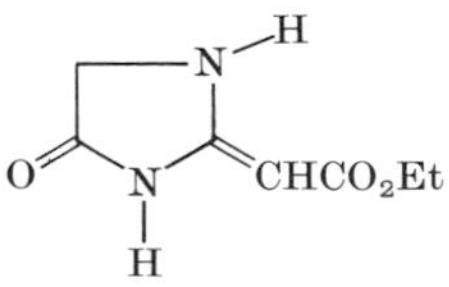 [**272/1**]

[**272a**] and/or the CH form [**272b**]; see previous survey (II-52) and for IR evidence (60JO1242). Recently, however, French authors (71BF-1040) have studied a series of 2,4-disubstituted imidazol-5-ones [**272**] by NMR, IR and UV methods, and compared their experimental conclusions with CNDO/2 type calculations. The methods used to choose among the eight possible tautomers [**272a–h**] were as follows:

i. IR spectroscopy allows a differentiation of the conjugated and unconjugated oxo forms such as [**272a**] and [**272b**] and a differentiation between CH (oxo) and OH forms.
ii. UV spectroscopy gives limited results because of the lack of fixed compounds to make comparisons, but the presence of the zwitterionic form [**272e**] is shown by long wavelength absorption.
iii. NMR spectroscopy cannot differentiate all the forms because of the fast interconversion occurring between some tautomers, e.g., [**272a**] ⇌ [**272b**] and [**272d**] ⇌ [**272e**] ⇌ [**272h**]. In spite of this, the French authors (71BF1040) were able to indicate whether [**272a**] or [**272b**] is predominant because of the existence in [**272a**] of homoallylic coupling constant between R^2 = Me and a 4-position proton: such coupling constants had already been observed in similar systems (64CB2023): oxazol-5-ones (Section 4-3Cb) and pseudooxazol-5-ones (Section 4-3Cc).

The main results can be summarized as follows:

i. If there is no substituent in position 4 or 5, the CH form [**272a**] is greatly predominant whatever the solvent and whatever the substituent at the 2-position.
ii. If a 4-phenyl group is present, one or both of the hydroxy forms [**272d**] $\rightleftharpoons$ [**272h**] predominates in the solid state and in solution in $CHCl_3$, DMSO and pyridine. In CF_3COOH the CH form [**272a**] is present in addition to the hydroxy forms.
iii. CH form [**272a**] is always more favoured than CH form [**272b**], consistent with CNDO/2 calculations.
iv. A 2-position substituent does not play an important part, whereas the influence of the 4-substituent is predominant upon the position of the equilibrium. The results again find support in CNDO calculations.
v. Solvents do not greatly affect the equilibrium. These results have recently been confirmed, and it was further shown that an alkoxycarbonylmethyl substituent in the 2-position causes methylene forms of type [**272/1**] to become favoured, as in the corresponding thiazoles (Section 4-3Cf) (73AS2221).

D. Heteroatoms-1,2,4 and a Potential Hydroxy Group

a. *Introduction*

In the previous review, all the potential hydroxyl compounds of rings containing three (and four) heteroatoms were contained in a single short section (II-54) that dealt mainly with results scattered in the literature. While the situation has not improved greatly with regard to rings containing four heteroatoms a great deal is now known about hydroxy derivatives of oxadiazoles, thiadiazoles and triazoles. We have divided the subject into ring systems where the heteroatoms occupy the 1,2,4-positions and systems in which they occupy the 1,2,3-positions: the latter are the subject of Section 4-3E, but for convenience they are also considered in this introductory section.

We first deal with compounds of type [**273**], in which the oxygen, sulphur, or pyrrole-like nitrogen is β- to the potential hydroxyl group followed by the two possible α-arrangements [**274**] and [**275**]. Finally N-unsubstituted 1,2,4-triazole derivatives are considered.

Section 4-3E deals first with oxygen, sulphur and nitrogen analogues of [**276**] and then [**277**] (only triazole derivatives known) and [**278**]

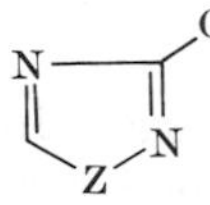

[273] 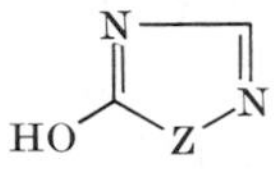 [274] 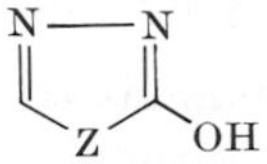 [275]

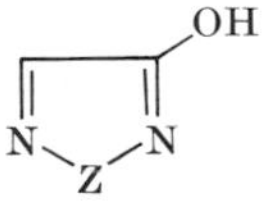

[276] 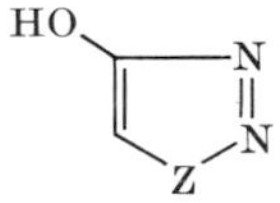 [277] 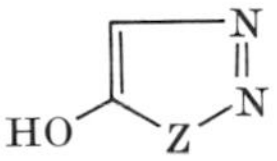 [278]

(triazole and thiadiazoles known). Section 4-3E also summarizes the small amount of work available on tetrazolones.

b. 3-Hydroxy-1,2,4-oxadiazoles

These compounds which can exist in forms denoted OH [**279a**], lactam NH [**279b**] and zwitterion NH [**279c**]. Their tautomerism was first investigated in 1965 (65T1681), IR spectra of dilute solutions in CCl_4 show that in this solvent 5-phenyl-1,2,4-oxadiazol-3-one [**279**] exists in the OH form [**279a**]. The νOH absorption changes with the concentration indicating that a monomer–dimer equilibrium, probably [**279a**] $\rightleftharpoons$ [**280**] occurs. In $CHCl_3$ the OH form also predominates;

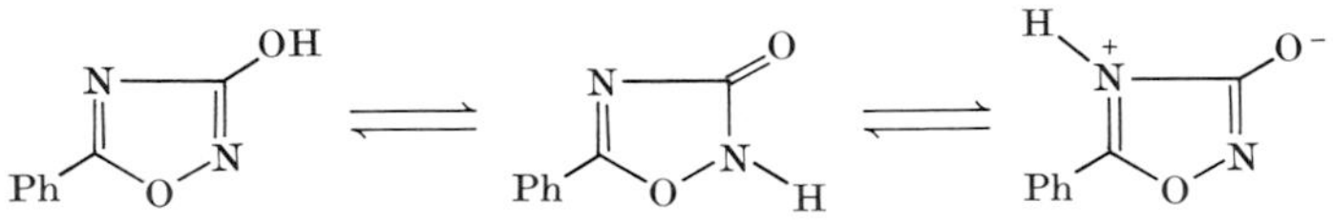

OH [**279a**] lactam NH [**279b**] zwitterion NH [**279c**]

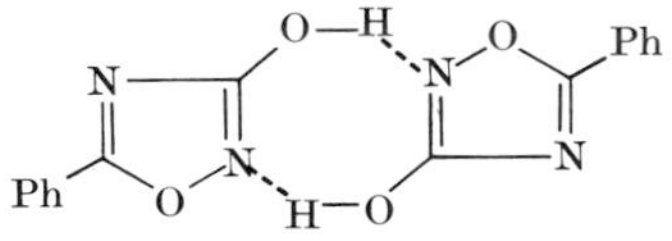

[**280**]

however, some weak bands suggest the presence of 5 to 10% of the lactam NH form [**279b**]. As the solvent polarity increases the proportion of the NH form [**279b**] evidently increases sharply, for basicity measurements show that in aqueous solution the forms [**279a**] and [**279b**] are about equally stable. UV spectroscopy does not allow any conclusion about the tautomeric equilibrium because the spectra of the tautomeric species and of its fixed derivatives are too similar.

The NMR data in $CDCl_3$ (69J(C)2794), published later, support the presence of the hydroxy form [**279a**] because of the very low field OH signal.

c. 3-Hydroxy-1,2,4-thiadiazoles (II-55)

These compounds can exist in three forms, e.g., [**281a–c**], but there is no conclusive evidence available as to which predominates. The previous review (II-55) reported tentative chemical evidence that

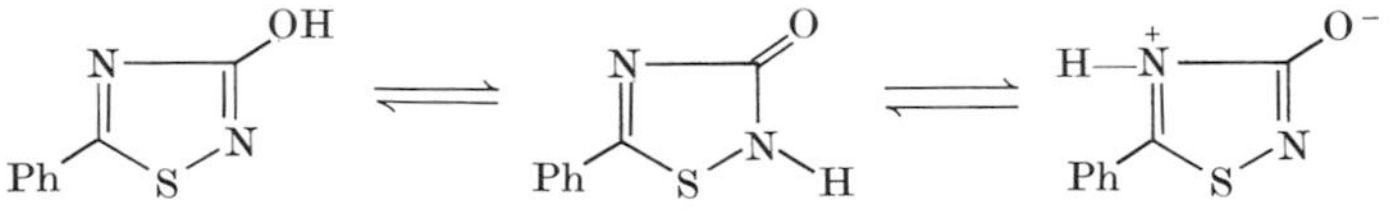

[**281a**] [**281b**] [**281c**]

1,2,4-thiadiazol-3-ones [**281**] existed in the OH form [**281a**]. However, an ultraviolet study (62J4191) was interpreted to suggest that the lactam NH form [**281b**] contributed substantially to the tautomeric equilibrium in ethanol by a comparison with the spectra of 3-amino-5-aryl- (Section 4-5Fb) and 5-anilino-3-hydroxy-1,2,4-thiadiazoles (Section 4-6Fh); see also (65HC(5)119).

d. 3-Hydroxy-1,2,4-triazoles

These compounds can exist in three forms [**282a–c**]. Though the original authors (66MI217, 68T5205) do not themselves discuss the tautomerism, the quoted IR spectra in KBr and particularly the absence of νC=O for two 1,2,4-triazol-3-ones [**282,** R^5 = *n*-Bu, or Ph] favour the OH structure [**282b**] for the solid state; νOH at 2600 cm^{-1} shows that the hydroxy group is associated [**283**] as is the case

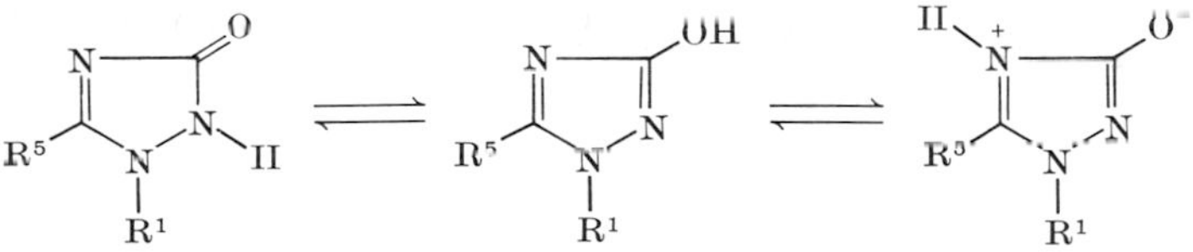

[**282a**] [**282b**] [**282c**]

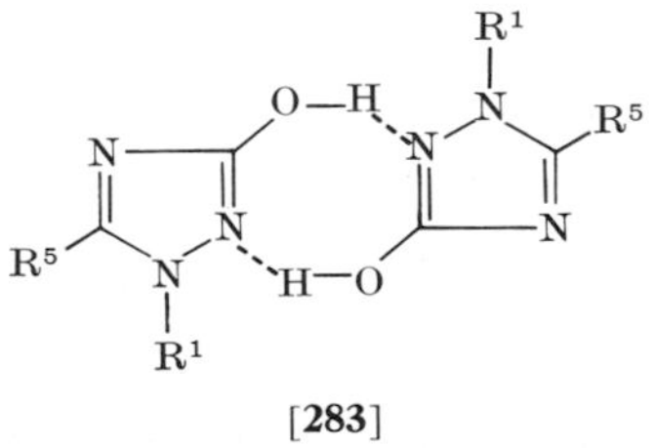

[**283**]

for 3-hydroxy-1,2,4-oxadiazoles (Section 4-3Db). In agreement, compound [**282**, R′ = Me, R^5 = Ph] shows νOH (solid and THF) and no νC=O; UV spectra also support the OH structure [**282b**] by comparison with model compounds (73CT1342).

e. 1,2,4-Oxadiazol-5-ones

No study of this tautomerism was available at the time of the previous survey. One of the present authors (65T1681) studied 3-phenyl-1,2,4-oxadiazol-5-one [**284**] using IR and UV spectroscopy and comparison with fixed methylated compounds [**285, 287**]. The IR spectra in the solid state or in $CHCl_3$ of the tautomeric compounds are similar to those of the *N*-methyl compound [**287**]*; increase in the

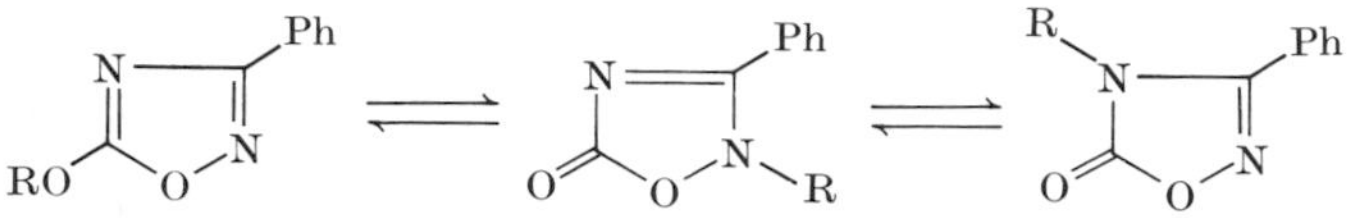

[**284a**] R = H [**285**] R = Me ⇌ [**284b**] R = H [**286**] R = Me ⇌ [**284c**] R = H [**287**] R = Me

intensity in $CHCl_3$ of the NH free absorption at 3480 cm^{-1} with dilution indicates a monomer–dimer [**288**] equilibrium. Although the UV spectrum of [**284**] in aqueous solution is closer to that of the *O*-methyl

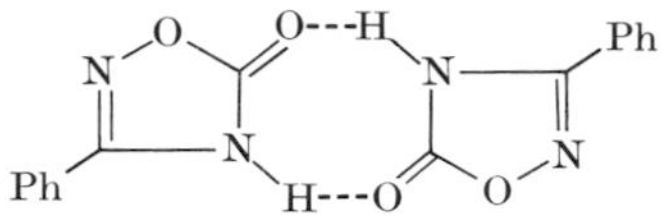

[**288**]

derivative [**285**] than to the *N*-methyl [**287**]; this probably does not indicate the OH form (65T1681); the difference between the absorption of [**284**] and the *N*–Me fixed derivative [**287**], may well occur because the predominant NH tautomer is [**284b**] which could have a chromophore different from that of the N–Me derivative [**287**].

Later, French authors (69BF823) again studied 3-phenyl-1,2,4-oxadiazol-5-one [**284**] and, using the same fixed derivatives, drew an opposite conclusion: the predominance of the OH form [**284a**] in solution. Their UV (EtOH) study is identical to that of the previous authors (65T1681), but they believe the analogy already mentioned between the UV spectra of the tautomeric species and of the O–Me derivative [**285**] is significant. However, their main argument relies on the appearance of the NMR signal arising from the phenyl group (EtOH, DMSO, acetone): a singlet for the N–Me derivative [**287**] and a multiplet for the O–Me [**285**] and the tautomeric species [**284**]. Such arguments have been criticized in Section 1-3B and are no proof of the

* Unambiguous synthesis (63AN1405) had confirmed the *N*-methyl structure [**287**].

OH structure [**284a**] for the compound [**284**] in those solvents.* In conclusion, we consider that tautomers [**284b**] and [**284c**] are still the most probable for 1,2,4-oxadiazol-5-ones, in the solid state and in solution.

f. 1-Substituted 1,2,4-Triazol-5-ones

1-Aryl-1,2,4-triazol-5-ones [**289**] could exist in the OH [**289a**] or two alternative NH forms [**289b, c**]. Compounds prepared (64LA(675)180) by ring closure of acylsemicarbazides were represented without proof in the N4H form [**289c**; R^1 = Ph, R^3 = alkyl]. Other authors (64H-1188) use νC=O at 1700–1720 cm^{-1} and νNH at 2800–3000 cm^{-1} (solvent not mentioned) to assign the same N4H structure [**289c**].

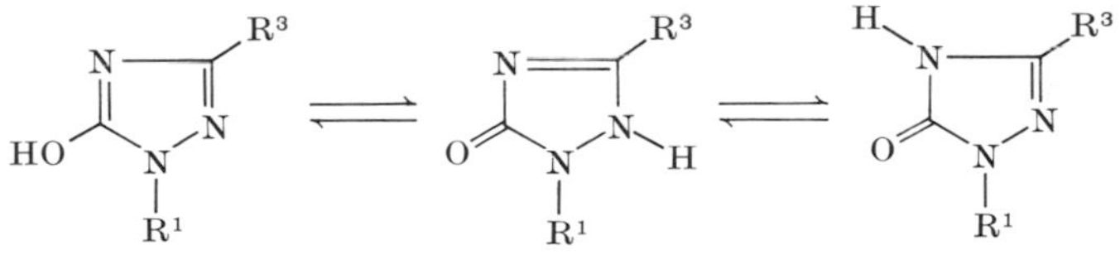

[**289a**] [**289b**] [**289c**]

These results were confirmed (66MI217) by the similarity of the IR (solid state) and UV (EtOH) spectra of the tautomeric compound [**289**; R^1 = Ph, R^3 = *n*-Bu] with the fixed compound [**290**].

In the 3-phenyl series, though UV and IR spectra were similar (68T5205) to those for the 3-alkyl compounds, the authors ascribe the structure [**289b**, R^3 = Ph]. This is unlikely: in [**289b**] the νC=O is more conjugated with R^3 = Ph and would absorb at lower frequencies than when R^3 = alkyl; further, the νNH absorption between 3000–3200 cm^{-1} is typical of associated NH; for [**289c**], dimers [**291**] are likely. Finally, structure [**289c**] is similar to that found for 1,2,4-oxadiazol-5-ones (Section 4-3De). Recently, the 4(*H*)-3-oxo form [cf. **289c**] has indeed been confirmed for the 1-methyl-3-phenyl derivative by IR and UV comparison with models (73CT1342).

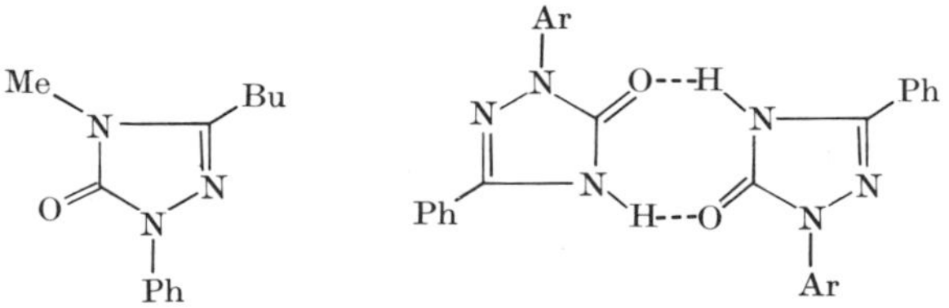

[**290**] [**291**]

* It would be useful to know whether the compound exhibits a carbonyl band in DMSO.

g. 1,3,4-Oxadiazol-2-ones (II-55)

The previous survey (II-55) cited a single example of IR evidence for the oxo form [**292**] (see also 66HC(7)183). No later work has appeared to confirm or modify this result.

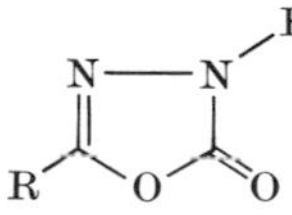

[**292**] 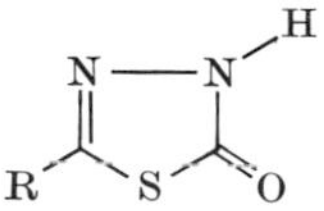 [**293**]

h. 1,3,4-Thiadiazol-2-ones (II-55)

The previous review (II-55) (see also 68HC(9)165) reported considerable IR and UV evidence that these compounds exist in the oxo form [**293**]; no further work has appeared, but this conclusion seems securely based.

i. 1-Substituted 1,3,4-Triazol-2-ones

These compounds can be written in four tautomeric forms [**294a–d**]. Nothing was known of their tautomerism at the time of the previous survey, but a series of 1,3,4-triazol-2-ones [**294**, R^1 = aryl, R^5 = alkyl] (64LA(675)180) has been assigned structure [**294a**] because they exhibit

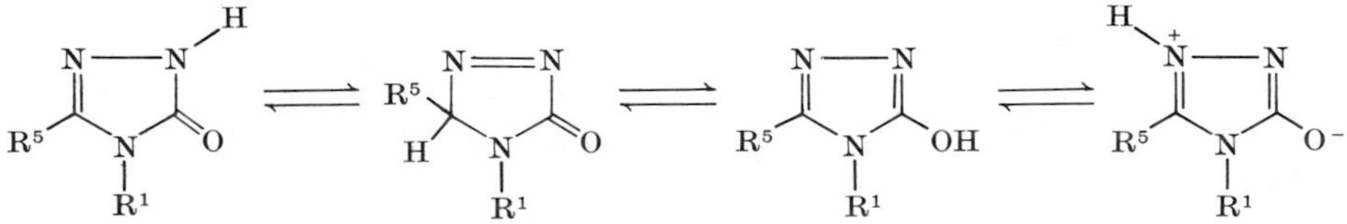

[**294a**] [**294b**] [**294c**] [**294d**]

νC=O at 1700 cm^{-1} (solid state). The IR spectrum of 1-benzyl-5-phenyl-1,3,4-triazol-2-one [**294**, R^1 = CH_2Ph, R^5 = Ph] (71T2811) in the solid state also exhibits this carbonyl band and νNH at 3190 cm^{-1} and similarly appears to exist as [**294a**]. Recently a similar structure [**294a**] has been demonstrated for the 1-methyl-5-phenyl derivative by IR and UV comparisons (73CT1342).

j. N-Unsubstituted 1,2,4- or 1,3,4-Triazolones (II-55)

Ten tautomers can be postulated for these triazolones [**295a–j**].

The previous survey (II-55) reported the considerable amount of early work, which after initial confusion, finally led to the assignment of the structure [**295a**] as the predominant form from IR results: νC=O at 1700 cm^{-1} and νNH at 2600–3100 cm^{-1} (cf. also 64H1188).

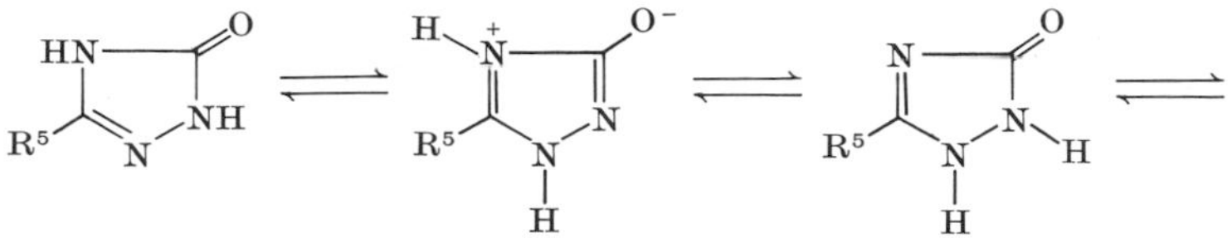

[295a] [295b] [295c]

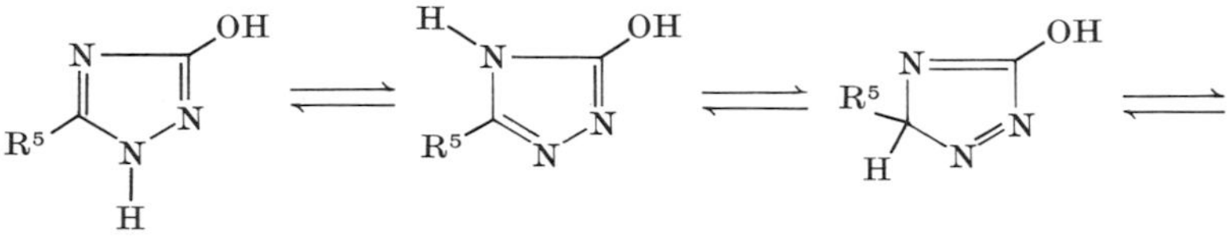

[295d] [295e] [295f]

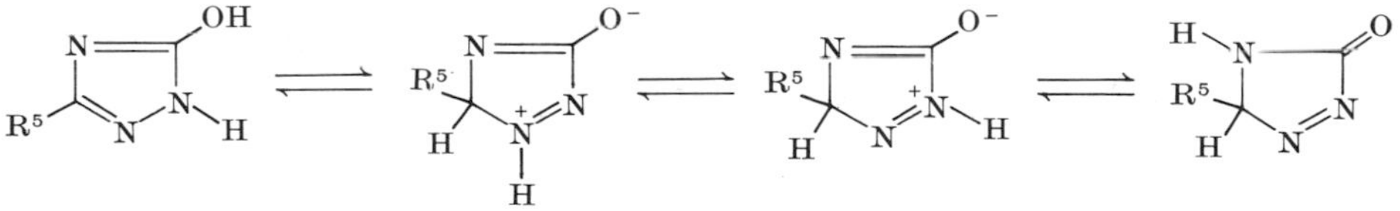

[295g] [295h] [295i] [295j]

In 1968 (68T5205), the synthesis of two individual tautomers [**295a**, R^5 = Ph] and [**295c**, R^5 = Ph] by two different methods was claimed: an unlikely result. However, the compound to which they assign structure [**295c**] had already been reported to be 2-phenyl-5-amino-1,3,4-oxadiazole (61JO1651), and the spectral data for so-called [**295c**, R^5 = Ph] agree closely with those later reported for the authentic oxadiazole (68BF4568). This conclusion is supported by the latest work on the 5-phenyl derivative (73CT1342); comparisons of both IR (solid) and UV (tetrahydrofuran) with five of the possible dimethyl derivatives supported the oxo form [**295a**] as predominant.

E. Heteroatoms-1,2,3 or -1,2,3,4 and a Potential Hydroxy Group

Introductory material to this section is included in Section 4-3Da.

a. 3-Hydroxy-1,2,5-oxadiazoles

Three tautomeric forms can be written: OH [**296a**], lactam NH [**296b**] and zwitterion NH [**296c**]. IR spectroscopy in CCl_4 shows the methoxycarbonylmethyl derivative [**296**, R = CH_2CO_2Me] to have the OH structure [**296a**], which exists in three states of association: the free monomer [**296a**], an intramolecularly *H*-bonded monomer [**297**], and a dimer [**298**] (65T1681). 3-Hydroxy-4-phenyl-1,2,5-oxadiazole

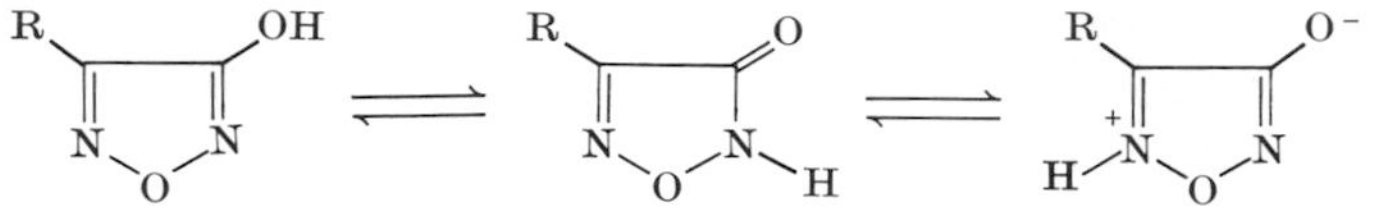

[**296a**] [**296b**] [**296c**]

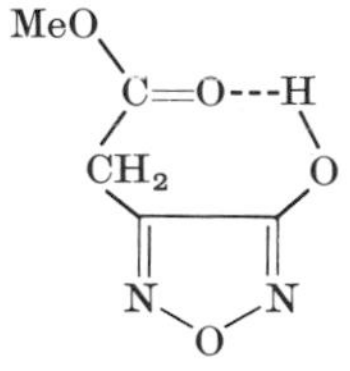

[**297**] 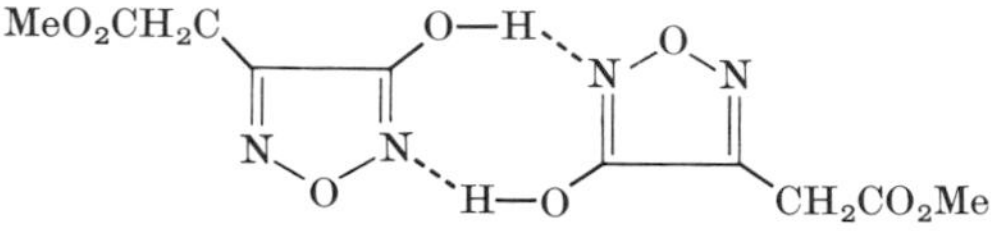 [**298**]

[**296**, R = Ph] also exists in the hydroxy form [**296a**] (69J(C)2794) from UV (EtOH), IR (Nujol) and NMR ($CDCl_3$) evidence.

b. 3-Hydroxy-1,2,5-thiadiazoles

Nothing was known of this series at the time of the previous survey. Three tautomers can be postulated for the 1,2,5-thiadiazol-3-ones [**299a–c**]. The first conclusions concerning the tautomerism of 1,2,5-thiadiazol-3-one [**299**, R^4 = H] and its 4-cyano and 4-carboxy derivatives are not very clear (64JA2861). However, the reported IR data in the solid state show that [**299**, R^4 = CN] exists predominantly in the OH form [**299a**] with strong hydrogen bondings; the authors observe "occasionally" a weak carbonyl absorption which may be due to [**299b**]. UV in EtOH shows the similarity of the tautomeric compounds with the corresponding *O*-methyl derivatives [**300**]. The NMR spectrum of [**299**, R^4 = CN] in acetonitrile shows a low field singlet which, it is alleged, "indicates the occurrence of lactam-lactim tautomerism and this is apparently favoured by solvents of high dielectric constants"; this statement is not supported by any further evidence.

However, these results are similar to those obtained for isothiazol-3-ones (Section 4-2E). The mobile proton of 3-hydroxy-1,2,5-thiadiazole [**299a**, R^4 = H: $pK_a = 5.10$] is more acidic (67JO2823) than that of 3-hydroxyisothiazole [**77a**, R^4 = H, R^5 = Me: $pK_a = 8.15$] (Section 4-2E). The influence of the 4-substituent on the pK_a of 1,2,4-thiadiazol-

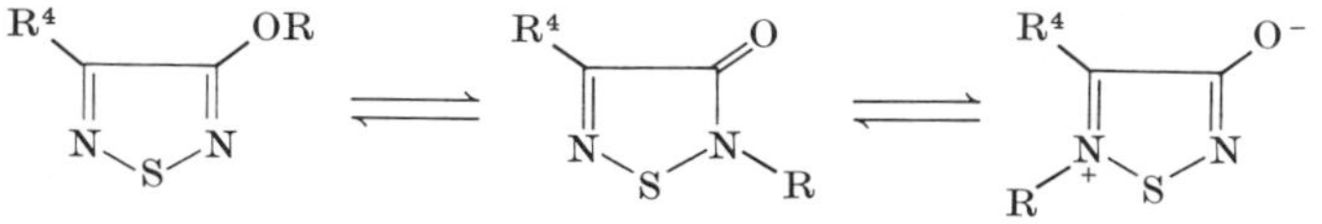

[**299a**] R = H
[**300**] R = Me

[**299b**] R = H
[**301**] R = Me

[**299c**] R = H
[**302**] R = Me

3-ones has been studied (66JO1964, 67JO2823, 68HC(9)107), but in these papers the authors do not mention the lactam NH form [**299b**] or zwitterionic tautomer [**299c**]. Comparisons including [**301**] and [**302**] are needed.

c. 3-Hydroxy-1,2,5-triazoles

Three tautomers [**303a–c**] can be postulated for these compounds. Spectral comparisons with the models [**304**] and [**305**] ([**306**] has not been synthesized) led Begtrup (69AS2025, 71AS2087, 72AS715) to the conclusion that these compounds all exist in the OH form [**303a**] (R^1 = Me, R^4 = H, or R^1 = Me, R^4 = Br or R^1 = Ph, R^4 = H) in $CHCl_3$ and in the solid state. The IR spectra ($CHCl_3$ and KBr) show no absorption above 1600 cm^{-1}, whereas the model triazolone [**305**] absorbs near 1650 cm^{-1} and zwitterionic structures analogous to [**306**] (see later) show a "pseudo carbonyl" band in the same region. Furthermore, the IR spectra of compounds [**303**] and [**304**] are rather similar in the 700–1600 region.

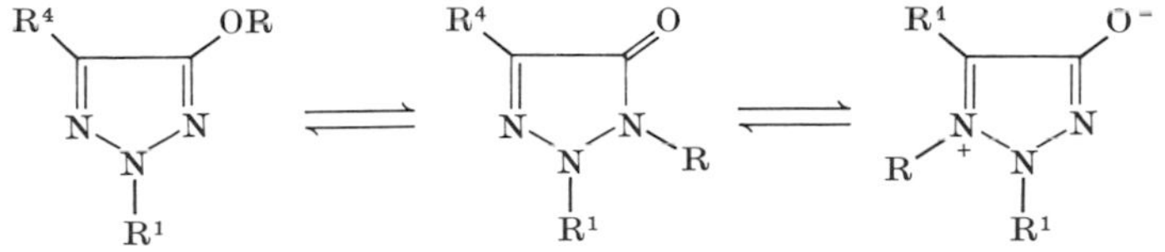

[**303a**] R = H
[**304**] R = Me

[**303b**] R = H
[**305**] R = Me

[**303c**] R = H
[**306**] R = Me

NMR spectroscopy confirms the presence of the tautomer [**303a**] in $CDCl_3$: in the 1-methyl series the 1-methyl signal appears at 4.08 ppm for [**303**], at 4.07 ppm in the case of the OMe derivative [**304**] and at 3.62 ppm for the NMe compound [**305**]. The 4-position proton has the same chemical shift, δ 7.0–7.1 ppm, in the compounds [**303, 304** and **305**; R^4 = H]; this result can be used to exclude tautomer [**303c**] which would be expected to give rise to more upfield signals (see later).

d. 1-Substituted 4-Hydroxy-1,2,3-triazoles

In 1967 Danish authors (67AS1234) prepared 1-methyl-4-hydroxy-1,2,3-triazole [**307**, R^1 = Me]. Among the four possible tautomers [**307a–d**] Begtrup (72PS9) suggests that this compound exists predominantly in the OH form [**307a**]. However, the presence of a minor amount of a zwitterionic form [**307c**] or [**307d**]* cannot be excluded. Tautomer [**307b**] can be eliminated because of the lack of a methylene

* The fixed derivative [**311**] is not known: if R = R^1 = Me and R^4 = H, then [**311**] ≡ [**306**].

signal in the PMR spectrum in $CDCl_3$. The IR spectrum (KBr and $CHCl_3$) excludes the presence of carbonyl groups as in form [**307b**] but absorption at 1625–1640 cm^{-1} could be due to traces of a zwitterionic form [**307c**] or [**307d**]: the fixed compound [**310**, R^1 = Me] shows a $C–O^-$ band at 1630 cm^{-1} (65AS2022).

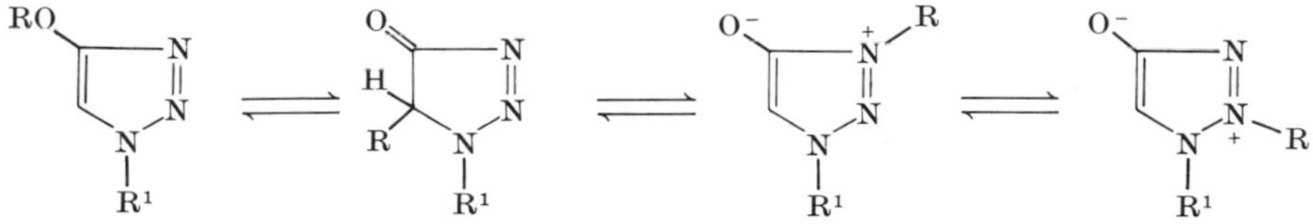

[**307a**] R = H
[**308**] R = Me

[**307b**] R = H
[**309**] R = Me

[**307c**] R = H
[**310**] R = Me

[**307d**] R = H
[**311**] R = Me

However in $CDCl_3$, [**307**] shows the CH signal at 7.00 ppm and the methyl at 4.02 ppm close to those of the OMe derivative [**308**] (7.15 and 4.03 ppm) but different from those of the NMe fixed compound [**310**] (6.65 and 3.95 ppm). Therefore the predominance of tautomer [**307a**] in solution is established. As far as the solid state is concerned, strong intermolecular *H*-bonding association occurs, and this prevents a clear distinction between OH and NH tautomers (see the analogous case of pyrazol-5-ones, Section 4-2Fc).

e. 5-Hydroxy-1,2,3-thiadiazoles

Only a single report is available on this tautomerism. German authors (66CB1618) suggest that 4-ethoxycarbonyl-1,2,3-thiadiazol-5-one [**312**] exists in the hydroxy form [**312a**]. The assignment is based on

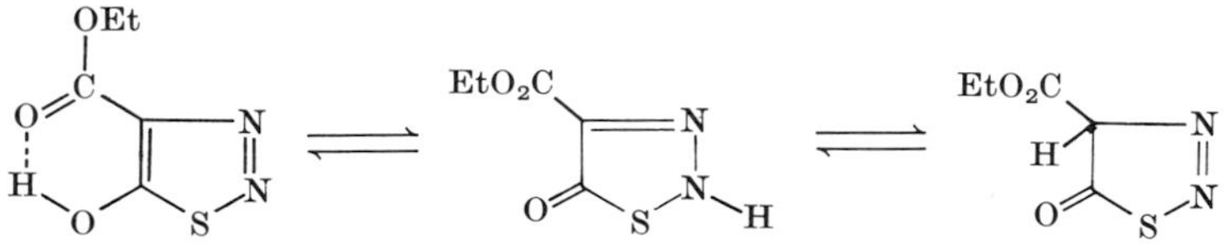

[**312a**]

[**312b**]

[**312c**]

the chelated OH absorption at 2700–3500 cm^{-1} and the associated νC=O of the carbethoxy group at 1600 cm^{-1}. A similar chelated structure is postulated for 4-carbethoxy-5-hydroxyisoxazoles [**61**] (Section 4-2Bb).

f. 5-Hydroxy-1,2,3-triazoles

Begtrup's papers on the methylation of 5-hydroxy-1,2,3-triazoles [**313**] (65AS2022, 66AS1555, 67AS633, 69AS1091, 71AS249) describe the IR and NMR properties of the fixed models OMe [**314**], NMe [**315**] and zwitterionic [**317**].*

For 1-methyl-5-hydroxy-1,2,3-triazole [**313**, R^1 = Me] (72PS9), PMR spectroscopy rules out the CH tautomer [**313c**] because a single aromatic proton and one methyl signal are observed (D_2O, MeOH, pyridine, acetone). The compound is almost insoluble in $CDCl_3$ and no spectrum was obtained in this solvent. In acetone, the PMR spectrum is similar to that of the 1-methyl-5-methoxytriazole [**314**].

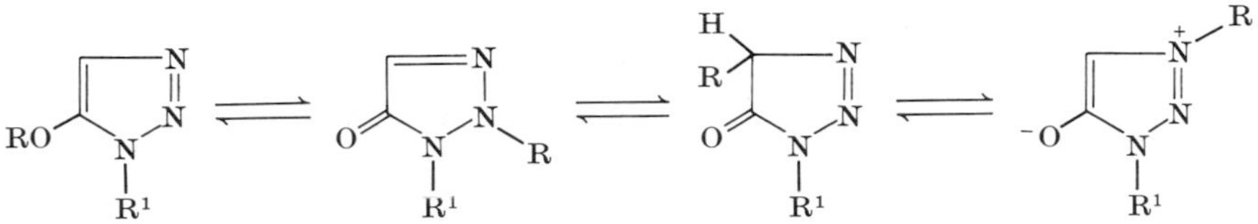

[**313a**] R = H
[**314**] R = Me

[**313b**] R = H
[**315**] R = Me

[**313c**] R = H
[**316**] R = Me

[**313d**] R = H
[**317**] R = Me

The IR spectrum of 1-methyl-5-hydroxytriazole in the solid phase, in D_2O, or in pyridine, lacks carbonyl absorption; traces of a zwitterionic form may be present (1620 cm^{-1}). Consequently the compound seems to exist mainly in the OH form [**313a**]; but the presence of a minor amount of a zwitterionic form [**313d**] cannot be excluded. To explain the exchange of the 4-position proton in D_2O, the author postulates the existence of a minor amount (< 1%) of the CH form [**313c**].

1-Benzyl- [**318**] and 1-phenyl-5-hydroxy-1,2,3-triazole [**319**] seem to behave in the same way as the 1-methyl compounds. 4-Methyl, 4-phenyl and 4-bromo substituents do not change the tautomeric equilibrium.

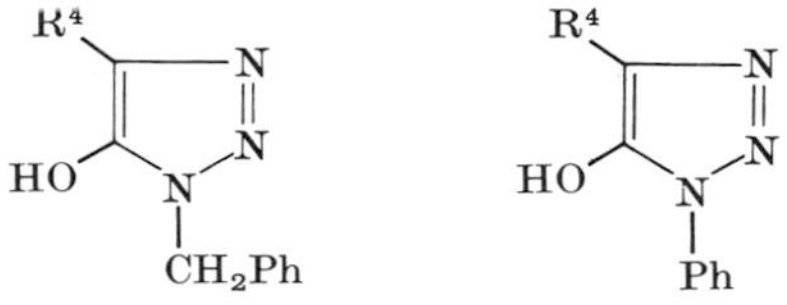

[**318**]

[**319**]

g. N-Unsubstituted Hydroxy-1,2,3-triazoles

Of the eight possible tautomers [**303a**], [**303b**], [**303c**], [**307a**], [**307b**], [**307c**], [**313a**] and [**313c**] (R^1 = H), Begtrup (72PS9) eliminates the CH forms [**307b**] and [**313c**] and zwitterionic forms [**303c**] and [**307c**] by PMR spectroscopy ($CDCl_3$, acetone, DMSO, D_2O and MeOH), because

* If R^4 = H and R = Me, then [**305**] ≡ [**315**], νC=O = 1650 cm^{-1} (71AS2087) and [**310**] ≡ [**317**], νC—O = 1630 cm^{-1} (65AS2022, 67AS1234).

he observes only one aromatic proton at ~7.3 ppm* and an OH–NH signal (in the three first solvents). In the IR spectrum (KBr or D_2O) no carbonyl band is observed, only a shoulder at 1630 cm^{-1} (zwitterionic tautomer ?); this eliminates the triazolone form [**303b**]. Finally the choice among the three OH tautomers [**303a**], [**307a**] and [**313a**] raises the problem of annular tautomerism (Section 4-1D); the author tentatively prefers the tautomer [**313a**] because of the appearance of hydrogen bonding absorption, but this last conclusion must be considered as provisional until UV and basicity measurements are undertaken.

h. Tetrazolones (II-56)

1-Aryl-1,2,3,4-tetrazol-5-ones were reported to have the NH structure [**320**] in the previous review (II-56). This was confirmed later by IR for solutions and for the solid state (67JO3580).

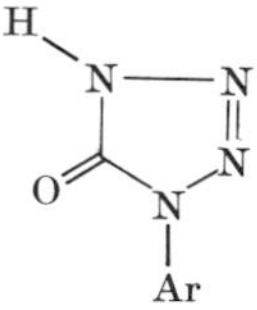

[**320**]

F. Compounds with Potential Hydroxy Groups: Conclusions

a. Introduction

Table 4-25 summarizes, in a necessarily schematic manner, all the results we just discussed together with some from Chapter 3. Compounds are divided into classes [**321**] and [**322**] where A = O, S or NR,

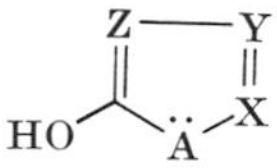

[**321**] 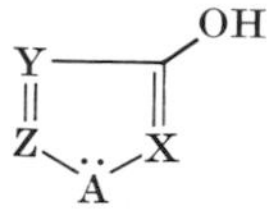 [**322**]

and where X, Y, Z can be either carbon or nitrogen atoms. A sign is assigned to each listed tautomer to indicate its probability of existence: + + +, very high; + +, medium; +, low; −, very low. A sign in parentheses indicates that the authors have not been able to choose between the two tautomers thus indicated. A question mark indicates a doubtful or debated result.

* This proton appears at 6.65 ppm ($CDCl_3$) in the fixed zwitterionic derivative [**310**] (see previously).

The compounds which remain to be studied are surrounded with a line, and the oxa- and thia-triazolones which are not mentioned in Table 4-25 should be added to this list. Among them, some have not been synthesized yet: 1,2,3-oxadiazol-4 (and 5)-ones and 1,2,3-thiadiazol-4 (and 5)-ones [the only one known is 4-carbethoxy-1,2,3-thiadiazol-5-one, an atypical case (Section 4-3Ee)]. Some others are known compounds the tautomerism of which has not been investigated: 3-ethyl-5-hydroxy-1,2,4-thiadiazole has been described (57CB182; also cited in 65HC(5)119), as well as the 1-methyl-4-hydroxy-1,2,3,5-tetrazoles (56JA411) for which fixed derivatives corresponding to the different tautomeric forms are described in the same reference.

b. *Influence of the Nature of A (O, S, NR)*

The nature of A in structures [**321**] and [**322**] has little importance, as becomes apparent from a cursory examination of Table 4-25. Some minor variations which do apparently occur may well arise only from the different conditions of study or from different substituents (this explains, for example, the different behaviour of the isothiazol-5-ones) (Section 4-2D). There is in fact no satisfactory study available on the influence of the nature of A on the tautomeric behaviour. Such a study should use the same solvents and the same concentration and the atoms X, Y and Z and the substituents on carbon should be held constant throughout.

c. *Influence of the Nature of X, Y, Z (N, CR)*

The main effect produced by the replacement of a carbon by a nitrogen atom can be represented as shown in [**323**] and [**324**], where

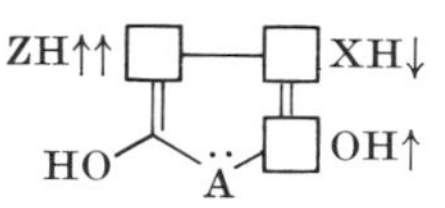

ZH↑↑ XH↓ HO A OH↑

[**323**]

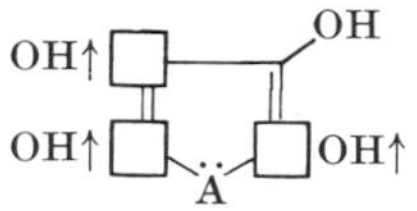

OH OH↑ OH↑ A OH↑

[**324**]

the symbol □ indicates replacement CR → N. The symbol OH ↑ indicates that the replacement increases the proportion of OH form, whereas the symbols ZH ↑ and XH ↓ indicate increase and decrease, respectively, of the corresponding NH forms.

It is logical to base the discussion of the tautomerism of hydroxy-azoles on the hydroxy-furans, -thiophenes and -pyrroles. The last compounds exist mainly in the nonaromatic carbonyl forms: the gain in aromatic resonance energy in the OH form is insufficient to outweigh

TABLE 4-25

HYDROXY-AZOLE AZOLINONE TAUTOMERISM[a]

A	X	Y	Z	Compound	HO–(A,X,Y,Z ring): H–Z, C=O	HO–C, Z=C	Z=Y, X–H, C=O	⁻O–C, Y⁺–H	Section
O	C	C	C	α-Hydroxyfurans	+	−	+ +	⊠	3-2Aa
S	C	C	C	α-Hydroxythiophenes	+	−	+ +	⊠	3-2Ca
NR	C	C	C	α-Hydroxypyrroles	+	−	+ +	⊠	3-2Ec
O	N	C	C	Isoxazol-5-ones	+ +	+	+ +	⊠	4-2B
S	N	C	C	Isothiazol-5-ones	− ?	−	+ + +	⊠	4-2D
NR	N	C	C	Pyrazol-5-ones	+ +	+	+ +	⊠	4-2F
O	C	C	N	Oxazol-2-ones	+ + +	−	−	⊠	4-3Bb
S	C	C	N	Thiazol-2-ones	+ + +	−	−	⊠	4-3Bc
NR	C	C	N	Imidazol-2-ones	+ + +	−	−	⊠	4-3Bd
O	C	N	C	Oxazol-5-ones	+ +	−	−	+ +	4-3Cb
S	C	N	C	Thiazol-5-ones	+ +	+ +	−	+	4-3Ce
NR	C	N	C	Imidazol-5-ones	+ + +	+	−	−	4-3Cg
O	N	C	N	1,2,4-Oxadiazol-5-ones	(+ +)	−	(+ +)	⊠	4-3De
S	N	C	N	1,2,4-Thiadiazol-5-ones	☐	☐	☐	⊠	
NR	N	C	N	1,2,4-Triazol-5-ones	+ + +	−	−	⊠	4-3Df
O	C	N	N	1,3,4-Oxathiazol-2-ones	+ + +	−	−	−	4-3Dg
S	C	N	N	1,3,4-Thiadiazol-2-ones	+ + +	−	−	−	4-3Dh
NR	C	N	N	1,3,4-Triazol-2-ones	+ + +	−	−	−	4-3Di
O	N	N	C	1,2,3-Oxadiazol-5-ones	☐	☐	☐	☐	
S	N	N	C	1,2,3-Thiadiazol-5-ones	☐	☐	☐	☐	4-3Ee
NR	N	N	C	5-Hydroxy-1,2,3-triazoles	−	+ + +	+ ?	+ ?	4-3Ef
NR	N	N	N	1,2,3,4-Tetrazol-5-ones	+ + +	−	−	−	4-3Eh

[a] Key to table. Probability of existence: + + +, very high; + +, high; +, low; −, very low; ( ) authors unable to choose between two tautomers indicated; ?, doubtful or debated result. ☐ Compounds still to be studied; ⊠ In this case that tautomeric form cannot exist.

TABLE 4-25 (*continued*)

Compound	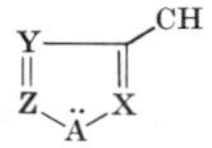 Ring (Y, Z, X, A; CH) — OH form	C=O / X–H form	$H–Y^{+}$ / O^{-} form	$H–Z^{+}$ / O^{-} form	Section
β-Hydroxyfurans	−	+ + +	×	×	3-2Ac
β-Hydroxythiophenes	+ +	+ +	×	×	3-2Cd
β-Hydroxypyrroles	−	+ + +	×	×	3-2Ee
3-Hydroxyisoxazoles	+ + +	−	×	×	4-2C
3-Hydroxyisothiazoles	+ +	+	×	×	4-2E
3-Hydroxypyrazoles	+ +	+	×	×	4-2H
4-Hydroxyisoxazoles	+ + +	−	×	−	4-3Ab
4-Hydroxyisothiazoles	+ + +	−	×	−	4-3Ac
4-Hydroxypyrazoles	+ + +	−	×	−	4-3Ad
4-Hydroxyoxazoles	+ + +	(+ ?)	(+ ?)	×	4-3Cd
4-Hydroxythiazoles	+ +	+ +	−	×	4-3Cf
4-Hydroxyimidazoles	+ +	− ?	+ ?	×	4-3Ch
3-Hydroxy-1,2,5-Oxadiazoles	+ + +	−	×	−	4-3Ea
3-Hydroxy-1,2,5-Thiadiazoles	+ +	+ ?	×	−	4-3Eb
3-Hydroxy-1,2,5-triazoles	+ + +	−	×	−	4-3Ec
4-Hydroxy-1,2,3-Oxadiazoles					
4-Hydroxy-1,2,3-thiadiazoles					
4-Hydroxy-1,2,3-triazoles	+ + +	−	+ ?	+ ?	4-3Ed
3-Hydroxy-1,2,4-oxadiazoles	+ + +	+	−	×	4-3Db
3-Hydroxy-1,2,4-thiadiazoles	+ +	+	−	×	4-3Dc
3-Hydroxy-1,2,4-triazoles	+ + +	−	−	×	4-3Dd
4-Hydroxy-1,2,3,5-tetrazoles					

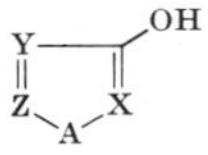

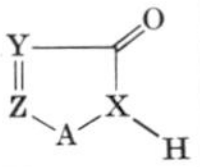

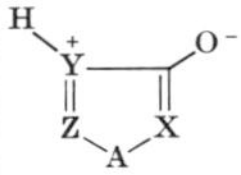

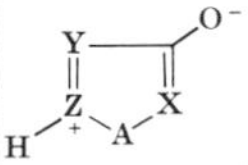

the loss in energy connected with the conversion of a carbonyl to an enol form (see Section 3-2H).

The question of the importance of zwitterionic tautomers is more difficult to generalize. It must be pointed out that authors often did not consider them; thus they possibly appear more frequently than it is indicated in Table 4-25. The probability of presence of a zwitterion will be high if the molecule has both an acidic hydrogen and a basic nitrogen: as the number of nitrogen atoms increases, the hydrogen atoms (NH as well as OH) become more acidic, but at the same time the nitrogen atoms become less basic, so that it is not possible to predict easily the importance of a zwitterionic form.

d. Influence of Benzoannelation

The effect of the replacement of a C═C double bond by a benzene ring (not shown in Table 4-25) is easy to summarize. Three cases are possible [**325**], [**326**] and [**327**].

i. The NH structure [**325a**] is more favoured than the OH form [**325b**] in which the benzene resonance is partially destroyed: the NH structure does indeed predominate for the indazolones (A = NR, Section 4-2Jc). The benzo[*c*]isoxazolones and benzo-[*c*]isothiazolones are uninvestigated but should show similar behaviour.

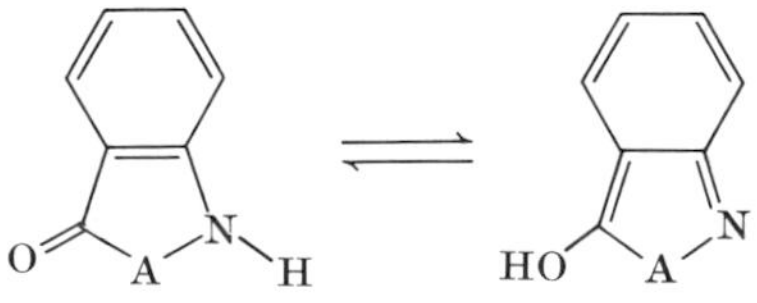

[**325a**] [**325b**]

ii. In the second case [**326**] the annelation has no significant effect; the NH tautomer [**326a**] is always predominant (Sections 4-3Bb, 4-3Bc, 4-3Bd).

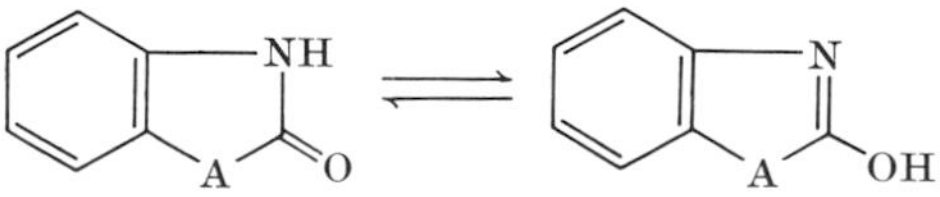

[**326a**] [**326b**]

iii. For structure of type [**327**], the OH form [**327a**] predominates for the 3-hydroxyindoxazenes (Section 4-2Cc) and the 1-substituted-3-hydroxyindazoles (Sections 4-2Hc and 4-2Jb) just as it does for the monocyclic 3-hydroxyisoxazoles and the 3-hydroxypyrazoles. However, a difference exists between the 3-hydroxyisothiazoles which prefer the OH form and the 3-hydroxybenzoisothiazoles which exist in the oxo form [**327b**] (Section 4-2E),

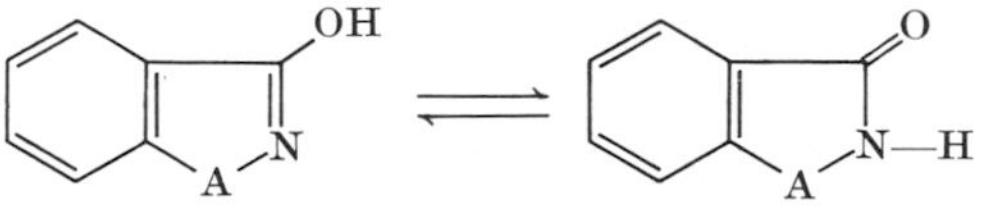

[**327a**] [**327b**]

but it must be pointed out that the structure of the benzisothiazolones [**327b**] were established in the solid state by X-ray diffraction: it would be interesting to investigate the structure of crystalline 3-hydroxyisothiazoles by X-ray analysis.

4. Compounds Containing Potential Mercapto Groups

Following the previous review (II-60), we arrange the material on mercapto azoles similarly to that on the hydroxy azoles; see the introductory section for the latter (Section 4-2A).

A. Heteroatoms-1,2 and a Potential 3-(or 5-) Mercapto Group

These compounds were completely unknown at the time of the previous review (cf. II-60), and considerable gaps still exist. Thus, simple 3-mercaptoisoxazoles, 5-mercaptoisothiazoles and 3-mercaptopyrazoles remain unknown. Nevertheless, the general pattern is already becoming clear.

a. 5-Mercaptoisoxazoles

Isoxazole-5-thione, recently prepared for the first time (67JH54), can exist in three tautomeric forms denoted SH [**328a**], NH [**328b**] and CH [**328c**]. The UV spectra in cyclohexane and comparison with the fixed derivatives [**329**] and [**330**] indicate the predominance of the SH tautomer [**328a**] for the compounds [**328**, R^3 = Ph, R^4 = H] and

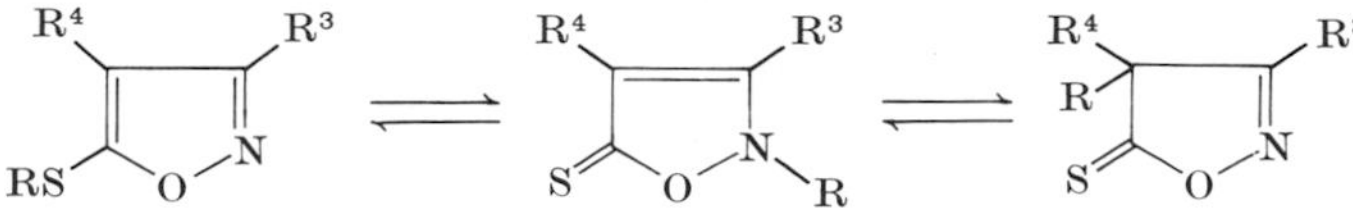

[**328a**] R = H
[**329**] R = Et

[**328b**] R = H
[**330**] R = Me

[**328c**] R = H

[**328**, R^3 = Me, R^4 = Ph] with the possible presence of some CH form [**328c**]. For 4-methyl-3-phenyl-isoxazole-5-thione [**328**, R^3 = Ph, R^4 = Me] UV spectroscopy, because of possible overlapping, is less conclusive, but the SH form [**328a**] is again probably present alone or together with the CH form [**328c**]. The UV spectra in MeOH are probably affected by dissociation of the thiols in this solvent.

In CCl_4, IR shows that the three compounds mentioned [**328**] all exist in the SH form [**328a**]. The 3-phenyl and the 3-methyl-4-phenyl derivatives also exist in the form in the solid state, with νSH at 2550 cm^{-1}. However, 4-methyl-3-phenyl-isoxazole-5-thione [**328**, R^3 = Ph, R^4 = Me] exists in the NH form [**328b**] in the crystal, with νNH at 3200–2800 cm^{-1}.

b. 4-Arylazo-5-mercaptoisoxazoles

4-(*p*-Chlorophenylazo)-3-methylisoxazole-5-thione, from UV (EtOH), IR (KBr) and NMR ($CDCl_3$) (65J3312), exists in the hydrazone form [**331a**] with intramolecular hydrogen-bonding between the C=S and N–H groups; νNH is at 3100 cm^{-1}, and in the NMR, NH at δ 15.6 ppm, independent of the concentration. This result agrees with the conclusions drawn from the more studied 4-arylazoisoxazol-5-ones (Section 4-2Bg); however, the authors do not discuss specifically the possible importance of the NH structure [**331b**] so much discussed for the 4-arylazoisoxazol-5-ones (Section 4-2Bg).

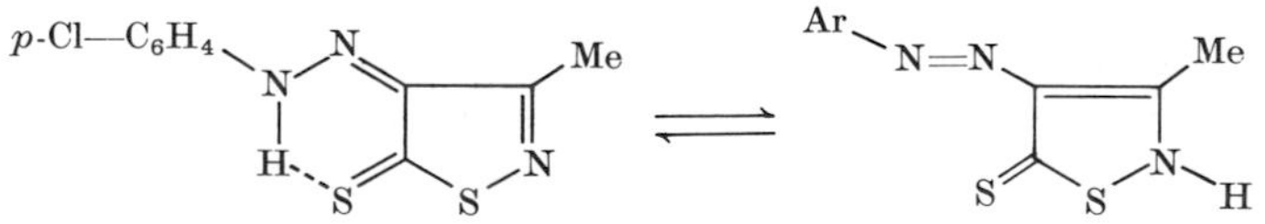

[**331a**] ⇌ [**331b**]

c. 3-Mercaptoisothiazoles

Apparently only one paper (67AN471) deals with the tautomerism of 3-mercaptoisothiazoles for which SH [**332a**] and NH forms [**332b**] can be postulated. The IR spectrum of the 5-phenyl compound [**332**] in CCl_4 exhibits no band between 3800 and 2800 cm^{-1} but shows νSH at 2575 cm^{-1}, indicating the presence of the SH form [**332a**]. Further information can be obtained from the published UV spectra. In EtOH,

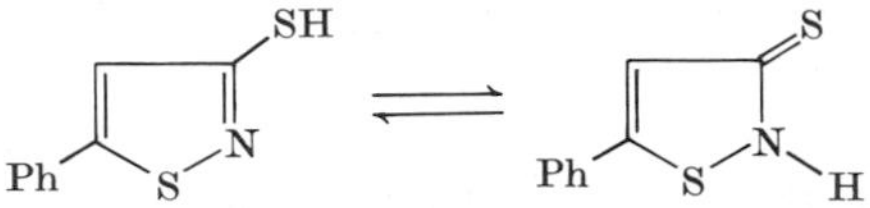

[**332a**] ⇌ [**332b**]

the tautomeric compound and its *S*-methyl derivative show similar absorption, supporting the predominance of the SH tautomer [**332a**]; however, an inflexion at longer wavelength suggests the presence of some NH form [**332b**]. The predominance of the SH form [**332a**] is consistent with the hydroxy structure found for the 3-hydroxyisothiazoles (70T2497) (Section 4-2E).

d. Pyrazole-5-thiones (II-60)

At the time of the previous survey (II-60) these compounds were unknown; since then several papers have described them and three (67MI74, 71JJ338, 74JH135) deal with the tautomerism of three such compounds [**333**], [**334**] and [**335**].

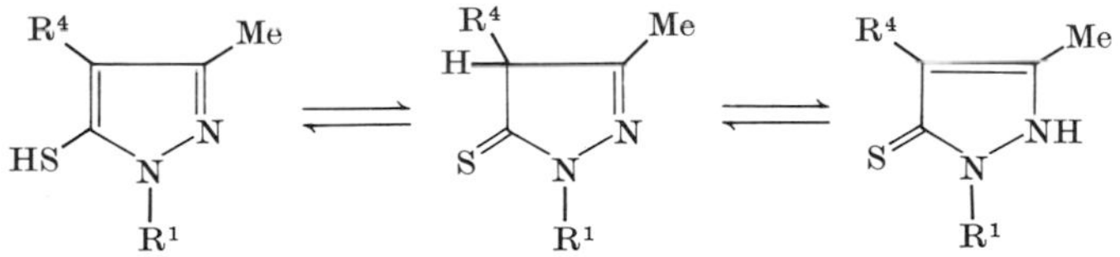

[**333a**] [**334a**] [**335a**] ⇌ [**333b**] [**334b**] [**335b**] ⇌ [**333c**] [**334c**] [**335c**]

[**333c**] R^1 = Me, R^4 = H
[**334c**] R^1 = Ph, R^4 = H
[**335c**] R^1 = Ph, R^4 = Me

We discuss initially 3-methyl-1-phenylpyrazole-5-thione [**334**]. Table 4-26 gives the characteristics* of the three fixed forms CR [**336**], SR [**337**] and NR [**338**], together with those of compound [**334**]. The conjugate acids of the SR and NR forms possess structure [**338/1**] (70KG202). By reasoning similar to that described for the pyrazol-5-ones (Section 4-2Gb), the conclusions summarized in Table 4-27 can be drawn (71JJ338).

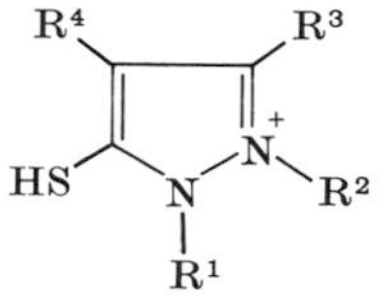

[**338/1**]

Comparing the results of Table 4-27 with those of the pyrazol-5-ones (Section 4-2Gb, Table 4-17) reveals the following points of interest:

i. Differences from pyrazolones. The CH tautomer [**333a**] is never observed even in nonpolar solvents such as CS_2 or $CHCl_3$ (see also 68CO(266C)290).

* In comparison with the pyrazol-5-ones (Section 4-2F), there is an important difference in UV spectra, a weaker basicity of the forms [**337**] and [**338**] and the higher dipole moment of the form [**338**].

TABLE 4-26

CHARACTERISTICS OF 3-METHYL-1-PHENYLPYRAZOLE-5-THIONE AND THREE FIXED DERIVATIVES

Reference: Method (solvent):	71JJ338 PMR (ppm) ($CHCl_3$)	71JJ338 IR (cm^{-1}) ($CHCl_3$)	71JJ338 UV (nm, ε) (EtOH)	71JJ338 pK_a (H_2O)	67M174 μ(D) (dioxan)
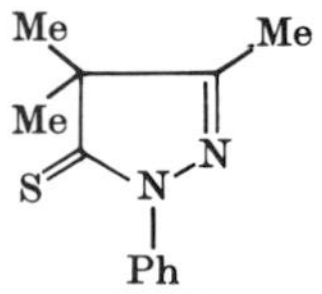 [336]	3-Me: 2.21		235 (12100) 329 (10700)	−3.50	3.16
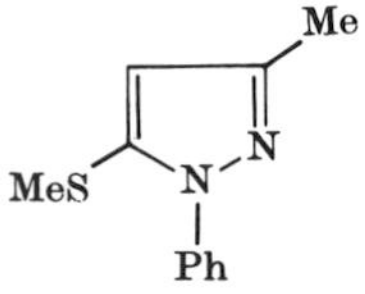 [337]	3-Me: 2.32 4-H: 6.14 J(Me3–H4) = 0.4[a] C_5 = 153.6[b]		249 (12500)	+1.30	2.80
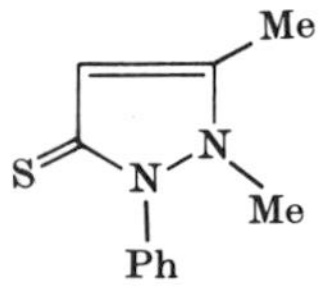 [338]	3-Me: 2.36 4-H: 6.23 J(Me3-H4) = 0.8[a] C_5 = 170.5[b]		305 (10900)	−0.35	7.60
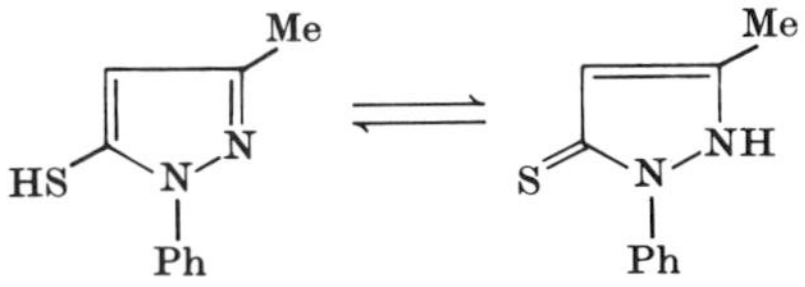 [334a] ⇌ [334c]	3-Me: 2.28 4-H: 6.20 SH: 3.28 J(Me3–H4) = 0.4[a] C_5 = 153.5[b]	CH: 2970 SH: 2540	244 (12400) 301 (3500)	−0.30	2.51

[a] Values in Hz (74JH135). [b] Values in ppm from TMS (74JH135).

TABLE 4-27

PERCENTAGES OF NH TAUTOMERS FOR THE PYRAZOLE-5-THIONES IN VARIOUS MEDIA[a]

Compound	H_2O	H_2O-EtOH (1:1)	MeOH	EtOH	Ether-EtOH (24:1)	$CHCl_3$,CS_2, C_6H_6	Solid
[**333**]	99	75	72	60	1	0[b]	NH/SH
[**334**]	98	62	36	27	0	0[c]	NH/SH
[**335**]	96	65	46	38	1	0[b]	NH/SH

[a] The remaining percentage belongs to the SH tautomer.
[b] One percent of NH form in $CHCl_3$ from UV spectroscopy.
[c] DMSO (74JH135).

ii. Analogies with pyrazolones. A methyl substituent in position 4 favours the NH tautomer [**334c**], and the N–Me series has always more NH tautomer than the N–Ph one. Basic solvents (ether) favour the SH tautomer [**334a**] whereas the acidic solvents (water) favour the NH tautomer [**334c**].

e. 4-Formylpyrazole-5-thiones

The SH structure [**339**] is suggested for 4-formyl-3-methyl-1-phenylpyrazole-5-thiol (69RO1685) by comparison of the UV (C_6H_{12}, MeOH, EtOH, *i*-PrOH) and IR (CCl_4, $CHCl_3$, THF) spectra with those of the

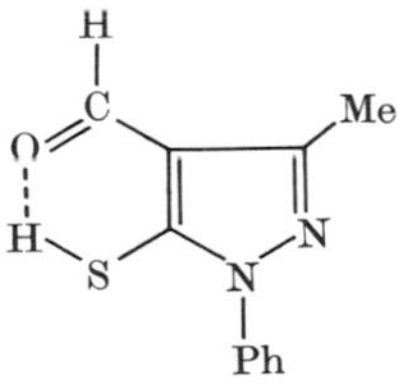

H C Me O H S N N Ph

[**339**]

corresponding fixed SMe derivative. For the corresponding oxygen compound [**153b**] (Section 4-2Gh) the same authors (64ZO3005) suggested the 4-hydroxymethylenepyrazol-5-one structure.

Recently Russian workers (73RO821) have investigated the corresponding anils, 4-phenyliminomethylpyrazole-5-thiones and -5-selenones.

f. Thioindazolones

IR and NMR study of thioindazolones in $CHCl_3$ led the authors (70AN246) to propose the SH structure [**340**, R = $PhCH_2$] for the

1-substituted derivatives, the NH structure [**341**, R = $PhCH_2$] (even in the solid state) for the 2-substituted derivatives and an equilibrium between the SH [**342a**] and NH forms [**342b**] for the parent thio-indazolone.

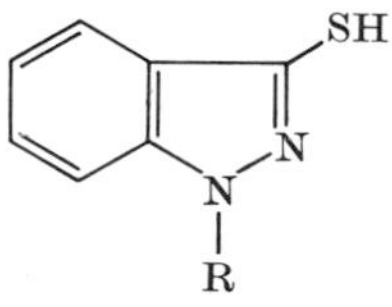

[**340**] 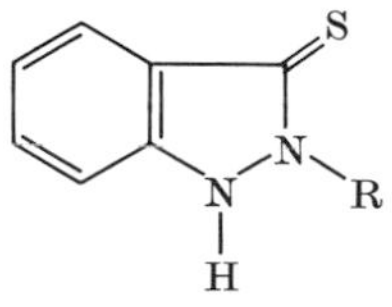 [**341**]

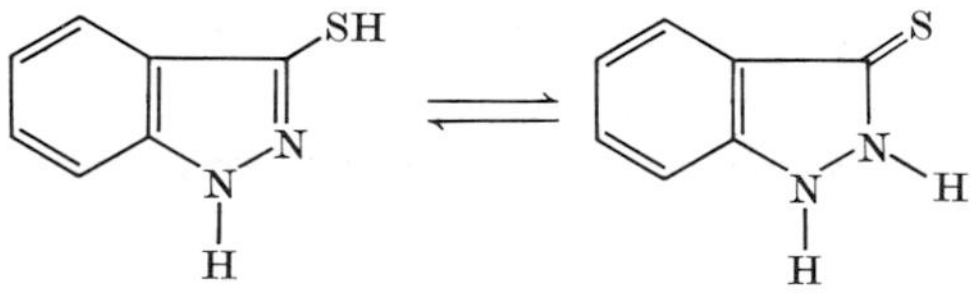

[**342a**] [**342b**]

Thioindazolones and indazolones (Sections 4-2Jb, c) thus show parallel behaviour at least as far as the *N*-substituted derivatives are concerned. As regards the *N*-unsubstituted thioindazolone the authors mention the contrast with indazolone which according to (65T3351) exists solely in the OH form; however, we have already pointed out (Section 4-2Jd) that this conclusion of (65T3351) is erroneous and that indazolone probably exists in the solid state in the NH form. It would have been surprising for a thione to be less "thiolactimized" than the oxo derivative.

B. Heteroatoms-1,2 and a Potential 4-Mercapto Group

Apparently all the 4-mercapto compounds with heteroatoms-1,2, i.e., the 4-mercapto-isoxazoles, -isothiazoles, and -pyrazoles, still remain unknown.

C. Heteroatoms-1,3 and a Potential 2-Mercapto Group

A considerable amount of work in this series had already been accomplished at the time of the earlier review, and it was clear that these compounds existed preferentially in the NH (thione) form. More recent work confirms this, but also gives some evidence for minor contributions by thiol forms.

a. Oxazole-2-thiones (II-61)

At the time of the previous survey (II-61) (see also 74HC(17)99) the thione structure [**343a**] was well established for the oxazole-2-thiones [**343**] and IR confirms this for the parent compound and its 4-methyl and 4,5-dimethyl derivatives (72CC3082).

Kjellin and Sandström (69AS2879, 69AS2888) studied quantitatively the equilibrium of oxazole-2-thiones [**343**, R^4, R^5 = Me, Me; H, Ph; Ph, H] and compared them with similar equilibria for thiazole-2-thiones (Section 4-4Cc) and imidazole-2-thiones (Section 4-4Ce). The tautomeric ratios K_T obtained from basicity measurements using Eqs. (1-14) and (1-16) (Section 1-4A) are in the range 10^5–10^6 but the quantitative significance of K_T is not quite clear because the thiones do not behave as ideal Hammett bases. The predominance of the thione [**343a**] over the thiol [**343b**] arises from the differences both of π-electron energy and of the energies of solvation between the two forms. Differences in

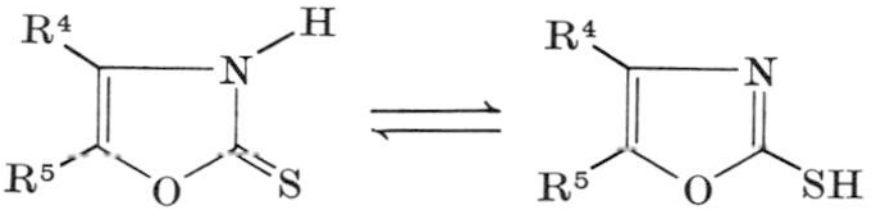

[**343a**] [**343b**]

π-electron energy between the thione and thiol forms reflect the greater stability of the thione form in five-membered rings compared with six-membered rings but do not reproduce very well the differences in behaviour between the three types of five-membered rings under study or between the differently substituted derivatives within a series. UV spectra have been compared with calculated transition energies. From these spectroscopic and basicity measurements the authors are able to estimate the position of the equilibrium in the excited state: the predominance of the thione form [**343a**] increases in this excited state.

b. Benzoxazole-2-thiones (II-61)

The previous survey (II-61) showed these compounds to exist in the thione form [**344**]. This is now confirmed for the solid state by X-ray determination (73AS945). Russian authors had also used basicity measurements (in aqueous acetone medium) (68OG665) and UV (in dioxan) (70ZO1872) to confirm the thione predominance in studies

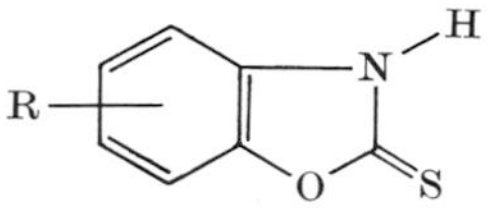

[**344**]

which parallel those on the benzoxazol-2-ones already mentioned (Section 4-3Bb). The IR work (70ZO1872) supersedes Zinner's results (60CB2035) and shows up the unreliability of the use of IR spectroscopy for tautomeric determinations in this series.

c. *Thiazole-2-thiones* (II-61)

The previous survey (II-62) summarized the evidence, clearly in favour of a thione structure [**345a**] for the thiazol-2-thiones on the basis of UV and IR spectral data. Considerable subsequent work has amply justified this conclusion, but has also demonstrated the existence of a minor proportion of thiol form [**345b**] by IR spectroscopy (68BF-2868). We discuss the tautomerism of these compounds according to the method of investigation, and note that theoretical calculations of electronic spectra (72BF1055) are also in accord with the thione formulation.

i. *Basicity Measurements*. Two papers report pK_a determinations of tautomeric thiazole-2-thiones [**345**; R^4, R^5 = H, H; Me, H; Ph, H; H, Me; Me, Me] and the corresponding fixed NMe [**346**] and SMe [**347**]

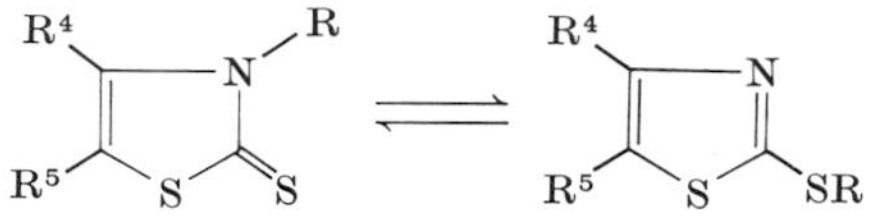

[**345a**] R = H [**345b**] R = H
[**346**] R = Me [**347**] R = Me

derivatives and derive equilibrium constants K_T (68BF2868, 69AS2888). Both groups agree on the great predominance of the thione form [**345a**]; the K_T values of ca. 10^3 (68BF2868) and ca. 10^5 (69AS2888) differ in magnitude, but, as both point out, this determination of K_T is approximate because thiazole-2-thiones, like other thioamides, are not Hammett bases. The Swedish authors (69AS2888) compare their results with the oxazole-2-thiones (Section 4-4Ca) which have a slightly higher value of K_T: this decrease in the predominance of the thione tautomer [**345a**] is reflected by the differences in the π-electron energies calculated by an HMO method taking into account the greater electron-donating capacity of a thioether sulfur atom relative to an ether oxygen in such a molecule. The site of protonation is proved to be the exocyclic sulfur atom (68BF2868).

ii. *UV Spectroscopy*. This method has been used frequently and no thiol form [**345b**] has ever been detected (66J(B)92, 66SA2005, 68BF-2868, 69AS2888). The UV spectra in H_2O and in dilute alkali or acid of thiazole-2-thione [**345**, R^4 = R^5 = H] (66J(B)92, 68BF2868) and

4-methylthiazole-2-thione [**345**; R^4 = Me, R^5 = H] (66J(B)92, 66SA-2005,* 68BF2868) agree with only slight variations in the extinction coefficients. The UV results of 4,5-dimethylthiazole-2-thione and 4-phenylthiazole-2-thione as free bases in water or as conjugated acids accord less well with the *N*-methyl models (68BF2868, 69AS2888). However, there is sufficient spectral similarity of all the tautomeric thiazole-2-thiones [**345**] investigated with the corresponding *N*-methyl derivative [**346**] to demonstrate the NH form even for the 4-phenyl series for which some doubt was left from previous studies on 5-amino-4-phenylthiazole-2-thione (49J1664). Bands were assigned to the transitions (66SA2005, 68BF2868), substituent effects were studied (68BF2868), and results were reproduced by HMO transition energy calculations (69AS2888). Ellis and Griffiths (66SA2005) explain solvent effects by formation or cleavage of hydrogen bonds, but this explanation does not hold for all cases (68BF2868) depending on the hydrogen donor or acceptor character of the tautomeric compound in a given solvent.

iii. *IR Spectroscopy*. IR spectroscopy (KBr) is the only method to have detected a small percentage of the thiol form [**345b**] (68BF2868). A band at 2500 cm^{-1} is assigned to νSH and not νNH (associated) by preparation of the deuteriated compound which shows a νSD at 1830 cm^{-1}. In the undeuteriated compounds all the strong bands show the thiazole-2-thiones to exist predominantly in the thione form as monomer or dimer.

iv. *NMR Spectroscopy*. This method is not suited for direct study of the fast equilibrium [**345a** $\rightleftharpoons$ **b**]: case 5 in Table 1-4 (Section 1-5Aa). Only averaged signals can be observed (68BF2868) which are very similar to those from the *N*-methyl derivatives [**346**]. An attempt to reveal the presence of the SH tautomer by decreasing the rate of proton exchange at low temperatures failed.

To favour the thiol tautomer [**345b**], Chanon and Metzger (72PS13) prepared a thiazole-2-thione sterically hindered at the 4-position [**345**; R^4 = *t*-Bu, R^5 = Me]. In the thermal equilibration of methyl derivatives [**346**] and [**347**] a complete shift towards [**347**] was observed if R^4 is a bulky substituent, but no corresponding shift of the NH [**345a**] towards the SH tautomer [**345b**] was found.

d. Benzothiazole-2-thiones (II-62)

The thione structure [**348**] of benzothiazolethiones in solution or in solid state was well established at the time of the previous review

* This work (66SA2005) deals not with thiazol-2-one as thought at the time, but with 4-methylthiazol-2-one (72PS10).

(II-62), (60MI98, 63MI1). Despite this, interest in them has persisted (see for example, 73BF(2)3044) because of their industrial importance.

In 1957, Russian authors (57KX38) established the thione structure [**348**] in the solid state by X-ray diffraction but they reported an unusually short intermolecular NS distance which indicated very strong hydrogen bonding. A later X-ray investigation (71CX(B)1441) confirmed the NH structure [**348**] but found a normal NS distance: the molecules form centrosymmetric hydrogen-bonded dimers.

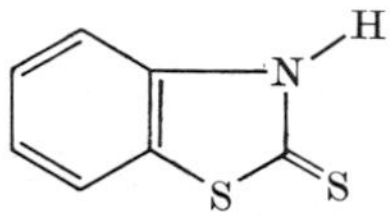

[**348**]

A complete UV study of the tautomeric benzothiazole-2-thione and its fixed methyl derivatives (66SA2005) assigns the bands and discusses solvent effects which depend mainly on hydrogen bondings. Protonation at the thione sulfur is suggested by the similar spectra of [**348**] and of its N–Me and S–Me derivatives in H_2SO_4.

The dithione structure [**349**] has been demonstrated from UV measurements on the tautomeric compound itself and its fixed derivatives in EtOH (64AN80). Compound [**350**] exists as such (64BJ1526)

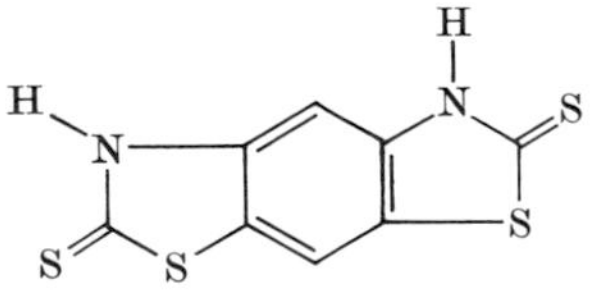

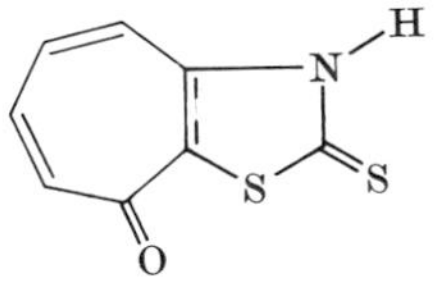

[**349**] [**350**]

from IR data and comparison of its UV spectrum with that of the *S*-methyl derivative in MeOH; the suggestion that the compound exists as a stable anion in neutral methanol is doubtful.

e. Imidazole-2-thiones (II-62)

Investigations on imidazole-2-thiones summarized in the previous review (II-62) demonstrate their thione structure [**351**] (see also 70HC(12)103).

Kjellin and Sandström (69AS2888) measured the pK_a values of a series of imidazole-2-thiones of the types [**351**], [**352**] and [**353**] with R^4, R^5: Me, Me; Ph, H; H, Ph. The pK_a values are similar for all compounds bearing no, one, or two *N*-methyl substituents,

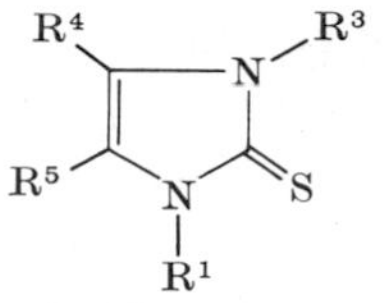

[**351**] $R^1 = R^3 = H$
[**352**] $R^1 = Me, R^3 = H$
[**353**] $R^1 = R^3 = Me$

supporting the thione structure for the *N*-unsubstituted [**351**] as well as for the *N*-monosubstituted compounds [**352**]. The authors calculated K_T values which are approximate for reasons already pointed out for the oxazole-2-thiones (Section 4-4Ca) and the thiazole-2-thiones (Section 4-4Cc), but which show that imidazole-2-thiones exist even more in the thione form than the two other heteroaromatic nuclei. The differences in π-electron energy (calculated by the HMO method) between the thiol and the thione forms do not reproduce this result: the authors assign this discrepancy to the greater energy of solvation of imidazole-2-thiones, not considered in the calculations. The UV spectra for H_2O solutions are consistent with a thione structure: a considerable hypsochromic shift observed for $R^3 = Me$, $R^4 = Ph$ is ascribed to deviation of the phenyl ring from coplanarity. There is a greater predominance of the thione form in the excited state than in the ground state.

The thione structure for [**351**, R^4 or $R^5 = o\text{-}CO_2H\text{-}C_6H_4$] has been reported from UV data (69KG719). The same authors ascribe a thione structure to 2-thioxoindeno[1,2-*d*]imidazolin-4-one [**354**] from the

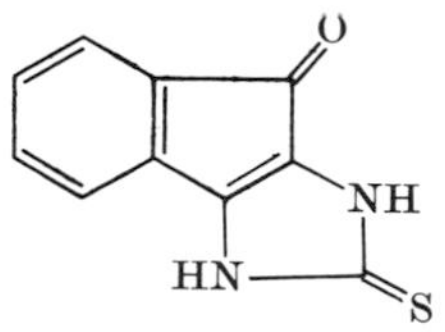

[**354**]

presence of νC=S and νNH bands. Dipole moment comparisons and X-ray studies (73MI179, 74CX(B)2348) also support the imidazole-2-thione structure (72J(PII)2045).

f. Benzimidazole-2-thiones (II-62)

In the previous review (II-62) the thione structure [**355a**] was assigned to benzimidazolethiones using UV and IR data. Recently dipole moments have also been used (72J(PII)2045).

An IR study of a series of crystalline benzimidazole-2-thiones [**355**] and [**356**] variously substituted in the benzene ring (67J(B)14) was not easy to interpret because of the difficulty in observing νC=S and νSH absorptions. Comparative study of the IR spectra with those of benzimidazoles and 2-ethylthio- or 2-phenylthio-benzimidazoles [**357**], the

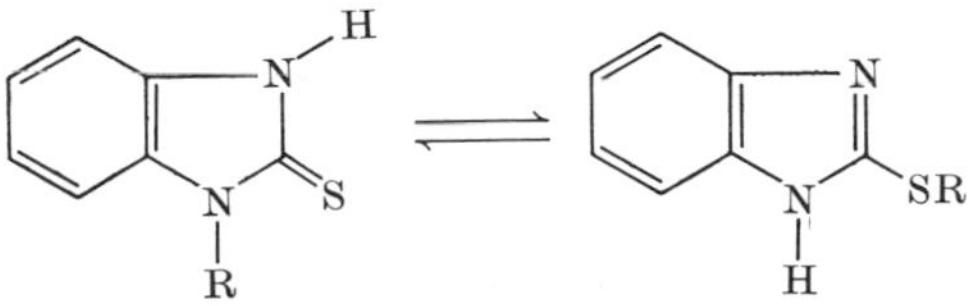

[**355a**] R = H [**355b**] R = H
[**356**] R ≠ H [**357**] R ≠ H

authors assign a strong band near 1500 cm^{-1} to the system NCSNH and a strong band near 1180 cm^{-1} for [**355a**] and 1210–1230 cm^{-1} for [**356**] to a "mixed" vibration with a major contribution from νC=S; these observations favour the thione structure [**355a**] [**356**]. An X-ray study (73CX(B)2328) confirms the thione structure of [**356**, R = β-D-ribofuranosyl].

In 1970, Russian authors (70ZO1605) assigned the thione structure [**355a**] to compounds [**355**, R = CH_2Ph, CH(Ph)Me], ascribing weak bands at 3150–3130 cm^{-1} to associated NH absorption. These same authors state very briefly that in solution they observe νSH bands from the thiol structure [**355b**].

From IR data in KBr and comparison with molecules of similar structure, compound [**358**] has been described to exist as the thione (63JO2581): strong absorptions occur at 1217 cm^{-1} and 1462 cm^{-1}.

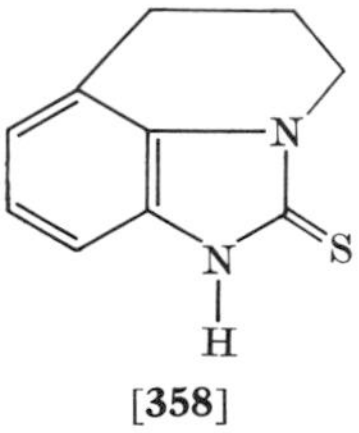

[**358**]

D. Heteroatoms-1,3 and a Potential 4- or 5-Mercapto Group

Apparently the only compounds of this type to have been described are 4,4-diphenylimidazole-5-thiones [**359**] (69T4265) together with their *N*- and *S*-methyl derivatives [**360**] and [**361**]. No tautomeric study has been reported. For the annular tautomerism of [**359a**] "desmotropy" see the work of Nyitrai and Lempert (72AH43, 73T3565).

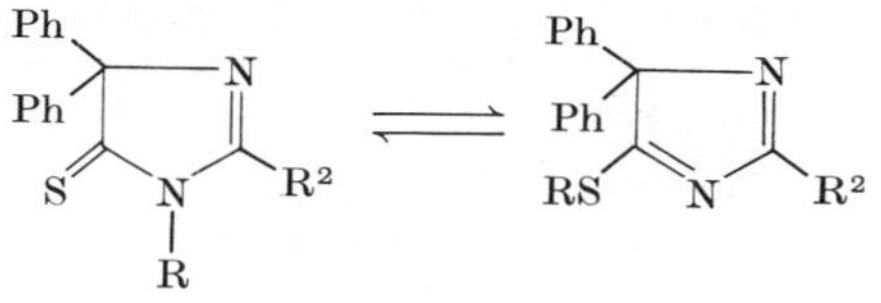

[**359a**] R = H
[**360**] R = Me

[**359b**] R = H
[**361**] R = Me

E. Heteroatoms-1,2,4 and a Potential Mercapto Group

The discussion of five-membered rings with three or four heteroatoms and a potential mercapto group has been reorganised along similar lines as used for the analogous hydroxyl compounds (see Section 4-3Da).

Since the previous review, many gaps have been filled, but a considerable number still remain. This is particularly the case for three heteroatoms-1,2,3 where most of the available evidence concerns the *v*-triazoles; little is yet known of mercapto derivatives of oxa- or thiadiazoles of this type.

a. 3-Mercapto-1,2,4-oxadiazoles

Attempts to prepare such compounds have so far failed, but *S*-alkyl derivatives [**362**] have been obtained and their UV spectra described (69J(C)2794).

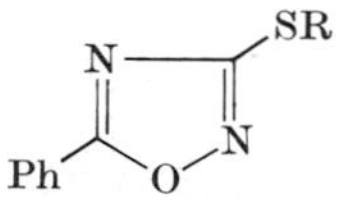

[**362**]

b. 1,2,4-Thiadiazole-3-thiones

The solid state IR of the parent [**362/1**, R = H] and 3-methyl derivatives [**362/1**, R = Me] show no clear νSH bond and the thione structure was tentatively assigned (73CC2353); further work is clearly needed.

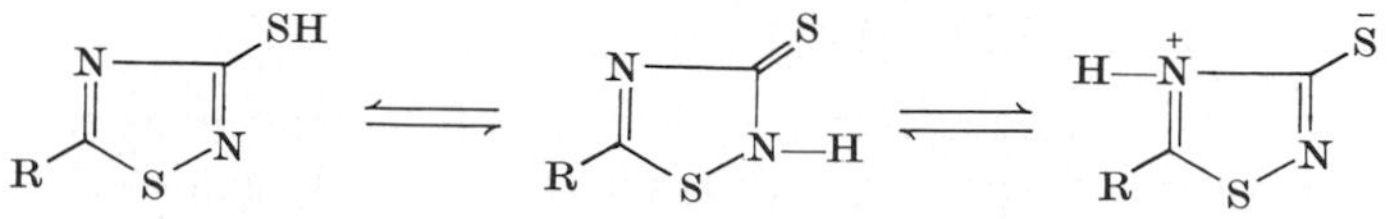

[**362/1a**] [**362/1b**] [**362/1c**]

c. 3-Mercapto-1,2,4-triazoles or 1,2,4-triazole-3-thiones

These compounds, unreported at the time of the previous survey, can exist in three forms [**363a–c**]. They have now been studied by several groups, but the situation is not yet completely clarified.

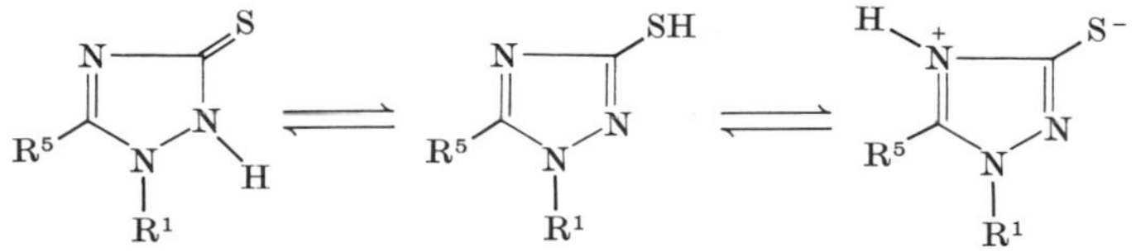

[**363a**] [**363b**] [**363c**]

Solid state IR spectra (71J(C)1016) exhibit absorption at 1150–1190 cm^{-1} assigned to νC═S and claimed to favour the thione form [**363a**]. They are at lower frequency than for other triazolethiones (see Section 4-4Ei): the authors attribute this phenomenon to a more extended conjugation of the system in [**363a**] than in [**366a**] or [**373a**]. This is also stated to explain the bathochromic effect (reported without any data) in the UV spectra; another difference from the other types of triazolethiones is the lower field chemical shift of the proton in position 5 (in DMSO with R^5 = H). All these results indicate a difference in behaviour of the 1,2,4-triazole-3-thiones from that of the 1,2,4-triazole-5-thiones or of the 1,3,4-triazole-2-thiones and are just as well explained on the thiol-structure [**363b**]; the correctness of the conclusion drawn by the authors (71J(C)1016) is critically dependent on the assignment of the band at 1150–1190 cm^{-1} to νC═S. The published data do not rule out the presence of some of the zwitterionic structure [**363c**], as a visible color is observed in solution.

However, the UV spectrum of [**363**, R^1 = Ph, R^5 = H] differs from that of the *S*-substituted fixed derivatives corresponding to [**363b**] (70KG1138) as does that of [**363**; R^1 = Me, R^5 = 2-pyridyl] (72CT2096) which again supports the thione structure [**363a**]. It seems odd, nevertheless, that the 1,2,4-triazole-3-thiones exist in the thione form whereas the 3-hydroxy-1,2,4-triazoles occur in the hydroxy form (see Section 4-3Dd), but recent work involving UV comparisons of [**363**, R' = Me, R^5 = Ph] with model compounds certainly indicates the thione (73CT1342).

d. 1,2,4-Oxadiazole-5-thiones (II-63)

No information was available at the time of the previous review (II-63) on 1,2,4-oxadiazole-5-thiones. 3-Phenyl-1,2,4-oxadiazole-5-thione [**364**], which can exist in three tautomeric forms [**364a–c**], has been examined

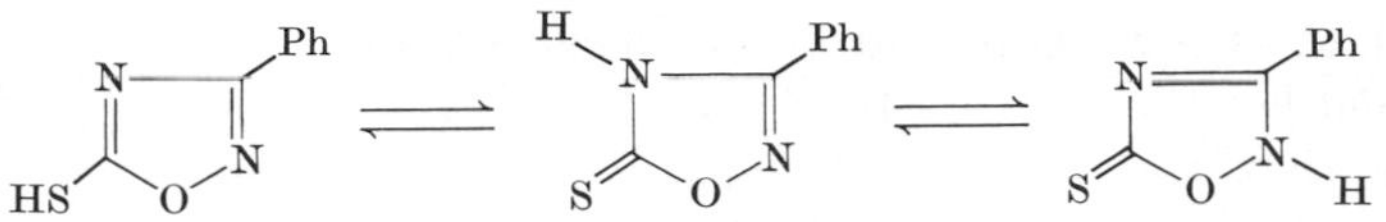

[**364a**] [**364b**] [**364c**]

by a French group (69BF823). In the solid state the authors ascribe to the compound the thione structure [**364b**] from IR measurements: no νSH band and absorptions of C═S and NH—C═S groups (they do not appear to observe any νNH band). However, they assert the predominance of the SH tautomer [**364a**] in solution (DMSO, EtOH, acetone, $CDCl_3$). This is the same paper (69BF823) that deals with 1,2,4-oxadiazol-5-ones (Section 4-3De) and the same arguments are used by the authors as were used for the oxadiazolones and which have been criticized in detail in Section 4 3De. Here again, the appearance of the phenyl signal in the NMR is not a valid criterion. Taking into account the results on 1,2,4-oxadiazol-5-ones (Section 4-3De), an NH-structure [**364b**] or [**364c**] seems more likely for the 1,2,4-oxadiazole-5-thione in solution as well as the solid state.

e. 1,2,4-Thiadiazole-5-thiones

1,2,4-Thiadiazole-5-thiones [**365**] are often described as thiols [**365a**], (see, e.g., 65HC(5)119). They are acids of considerable strength, but this is no proof of their structure (see discussion of the thiatriazole-thiones, Section 4-4Fd).

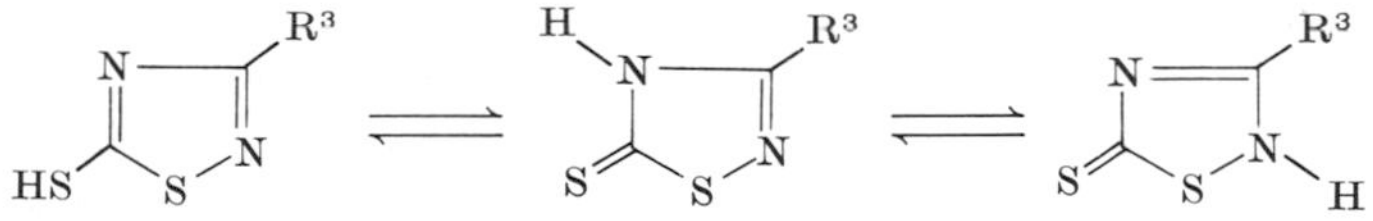

[**365a**] [**365b**] [**365c**]

A thione structure [**365b**] or [**365c**] seems more likely taking into consideration the known structures of the thiazole-2-thiones (Section 4-4Cc) and 1,2,4-triazole-5-thiones (Section 4-4Ef), and this is supported by the absence of νSH in the IR of several derivatives (73CC2353).

f. 1,2,4-Triazole-5-thiones

No compound of this series was reported in the previous survey. Three tautomers can be postulated for 1-substituted 1,2,4-triazole-5-thiones [**366a–c**].

The IR (solid state) of compounds [**366**, R^3 = H, R^1 = alkyl or aryl] (70J(C)2403, 71J(C)1016) shows bands in the 1240–1205 cm^{-1} region assigned to νC=S. This and the absence of any νSH in the region 2600–2550 led the authors to choose the thione structure [**366a**]. We consider that the evidence is equally valid for [**366b**]. The NMR spectra (DMSO) exhibit an NH signal at about 13 ppm.

UV spectral comparisons in EtOH of [**366**, R^1 = Me, R^3 = 2-pyridyl] and [**368**, R^1 = Me, R^3 = 2-pyridyl] (71CT2331, 72CT2096) or of [**366**, R^1 = Ph, R^3 = H] and [**368**, R^1 = Ph, R = CH_2Ph or CH_2CH_2X]

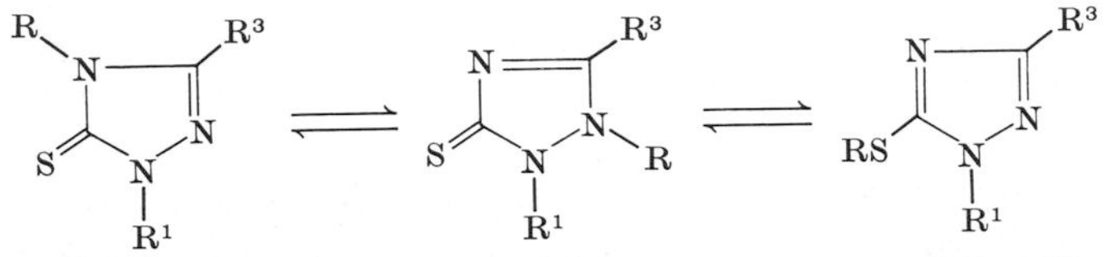

[**366a**] R = H
[**367**] R ≠ H

[**366b**] R = H

[**366c**] R = H
[**368**] R ≠ H

(70KG1138) eliminate the thiol structure [**366c**] for these compounds. In addition, the UV spectrum in EtOH of [**366**, R^1 = Ph, R^3 = H] and that of [**367**, R^1 = Ph, R = CH_2CH_2X] (70KG1138) are identical, supporting the thione structure [**366a**]. Furthermore, the νC=S absorption of the 1,2,4-triazole-5-thiones is similar to that of the 1,3,4-triazole-2-thiones, for which structure [**373a**] is securely based (Section 4-4Ei). Recently, UV comparisons with a range of model compounds has confirmed structure [**366a**, R^1 = Me, R^3 = Ph] (73CT1342).

g. 1,3,4-Oxadiazole-2-thione (II-63)

The thione structure [**369a**] of these compounds was already established at the time of the previous survey (II-63) from IR data (see also 66HC(7)183). Recently, the parent compound itself has been shown by IR and UV to be a thione (72CC3079). The tautomerism of 1,3,4-oxa- and 1,3,4-thiadiazole-2-thione (Section 4-4Eh) and 1-methyl-1,3,4-triazole-2-thione (Section 4-4Ei) have been compared (66AS57) using UV measurements in EtOH and LCAO–MO calculations. The UV spectra of the tautomeric compounds [**369**, R^5 = Me or Ph] show similarity with the *N*-methylated derivatives [**370**] supporting the thione structure [**369a**]. There is a linear correlation between calculated and experimental

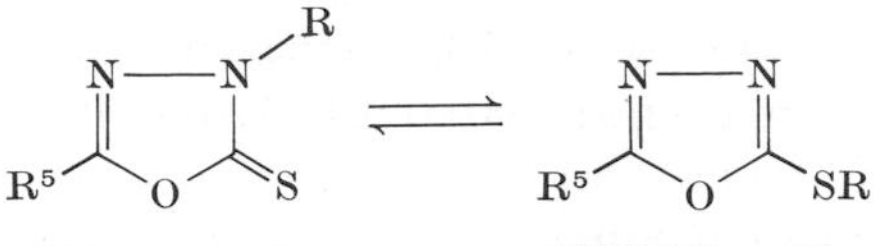

[**369a**] R = H
[**370**] R = Me

[**369b**] R = H
[**371**] R = Me

transition energies. The calculation of π-electron energies show that the thione form [**369a**] is favoured but if lone-pair energies are introduced the result is not so clear, depending on the parameter sets used. The authors measured the pK_a of their compounds and tried to correlate them with the calculated π-electron charges on the NH nitrogen atoms, but the relation was not very satisfactory.

h. 1,3,4-Thiadiazole-2-thiones (II-63)

Considerable evidence was reported for the thione structure [**372a**] of 1,3,4-thiadiazole-2-thiones in the previous review (II-63) (see also 68HC(9)165).

In subsequent work Swedish authors (66AS57) studied the equilibrium [**372a**] ⇌ [**372b**] by UV spectroscopy and LCAO–MO calculations;

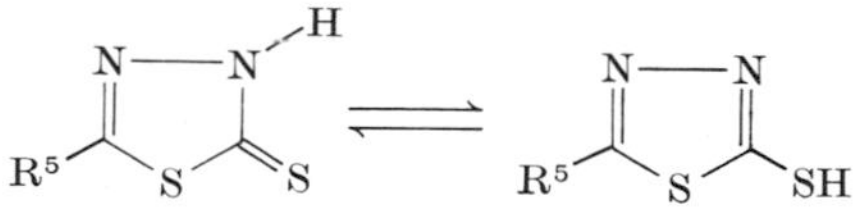

[**372a**] [**372b**]

their conclusions are similar to the ones they reported for 1,3,4-oxadiazole-2-thione (Section 4-4Eg) and support the thione structure [**372a**]. Compound [**372**, R^5 = p-Cl-C_6H_4-CO] (65AN615) exists in the thione form in the solid state from IR spectroscopy.

i. 1,3,4-Triazole-2-thiones (II-63)

In the previous survey (II-63) 1-substituted 1,3,4-triazole-2-thiones [**373**] were assigned thione structures [**373a**]. They could exist in three other forms [**373b–d**] but all the subsequent studies also support the thione structure [**373a**].

In work parallel to that carried out on 1,3,4-oxa- and 1,3,4-thiadiazole-2-thiones (Sections 4-4Eg and 4-4Eh) a similar conclusion was reached (66AS57; cf. also 73CT1342) for the thione structure [**373a**, R^5 = Me or Ph] from UV measurements in EtOH and comparison with fixed derivatives. LCAO–MO calculations of π-electron energies are in agreement.

UV spectra in EtOH of [**373**, R^5 = 2-pyridyl] and [**376**, R^5 = 2-pyridyl] (71CT2331) show that the thiol form [**373c**] can be eliminated, and further UV comparisons strongly support the thione structure [**373a**] (72CT2096). Landquist (70J(C)323) describes [**373**, R^1 = Ph, R^5 = Me] and the fixed derivatives [**374**], [**376**] and [**377**, R = CH_2OPh, R^5 = Me] but with no tautomerism investigations.

IR spectra in the solid state (70J(C)2403, 71J(C)1016) of 1,3,4-triazole-2-thiones [**373**, R^5 = H, R^1 = alkyl or aryl] show bands at

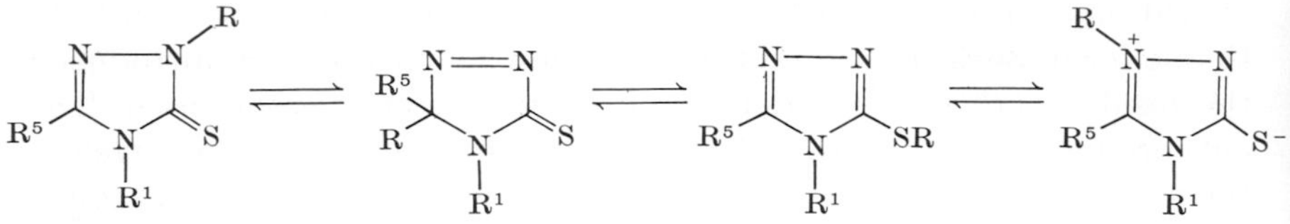

[**373a**] R = H
[**374**] R = Me

[**373b**] R = H
[**375**] R = Me

[**373c**] R = H
[**376**] R = Me

[**373d**] R = H
[**377**] R = Me

1210–1290 cm^{-1} authors assigned to νC=S of the thione structure [**373a**]. The NH signals can be observed in the NMR spectra in DMSO between δ 13.6 and 14 ppm.

The NMR spectra of a series of 1,3,4-triazole-2-thiones with various R^1 and R^5 substituents (in DMSO, $CDCl_3$, C_6H_6, CCl_4) (72AS459) show intermolecular solute–solute hydrogen-bonding in $CDCl_3$ and solvent–solute in DMSO. The anomalous behaviour of 1-(4-pyridyl)-1,3,4-triazole-2-thione has been explained by the zwitterionic structure [**378**] (71J(C)1016).

1,3,4-Triazole-2-selenones behave similarly to 1,3,4-thiazole-2-thiones and exist in the form [**379**] (72AS459).

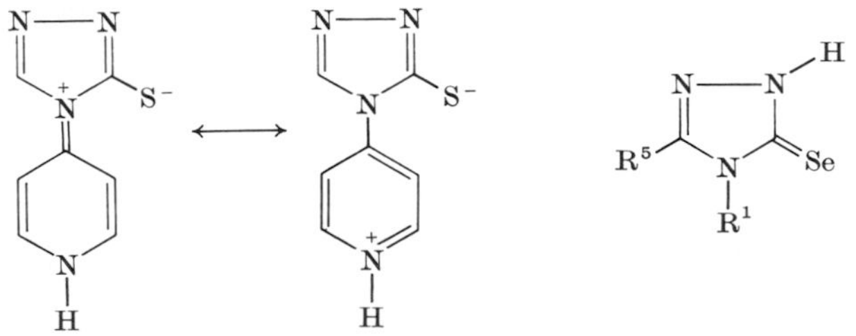

[**378**]

[**379**]

j. Unsubstituted 1,2,4- or 1,3,4-Triazolethiones

Like the corresponding triazolones (Section 4-3Dj), unsubstituted 1,2,4-triazolethiones can exist in ten forms [**380a**–**h**].

Some unsubstituted 1,2,4-triazolethiones have been studied using IR and NMR spectroscopy (71J(C)1016). The compounds [**380**, R = H, alkyl or aryl] show νC=S absorption in the solid state at 1200–1230 cm^{-1} and most of the NMR spectra in DMSO exhibit two different signals for two NH groups. The authors who do not take into consideration all the tautomeric forms conclude that the compounds exist in the thione structure [**380a**]. Japanese authors (71CT2331) studied one compound of this type [**380**, R = 2-pyridyl] and the *S*-methyl derivative corresponding to [**380g**]; later many further mono- and di-methyl derivatives were used to provide good UV evidence for form [**380a**] for both the α-pyridyl (72CT2096) and phenyl series (73CT1342): however, the authors did not consider the zwitterionic form [**380i**] nor the unlikely nonaromatic forms [**380c,e,g,j**]. A preliminary report gives $K_T = 2.6 \times 10^4$ in favour of the thione form for 1,2,4-triazole-3-thione (69CG58).

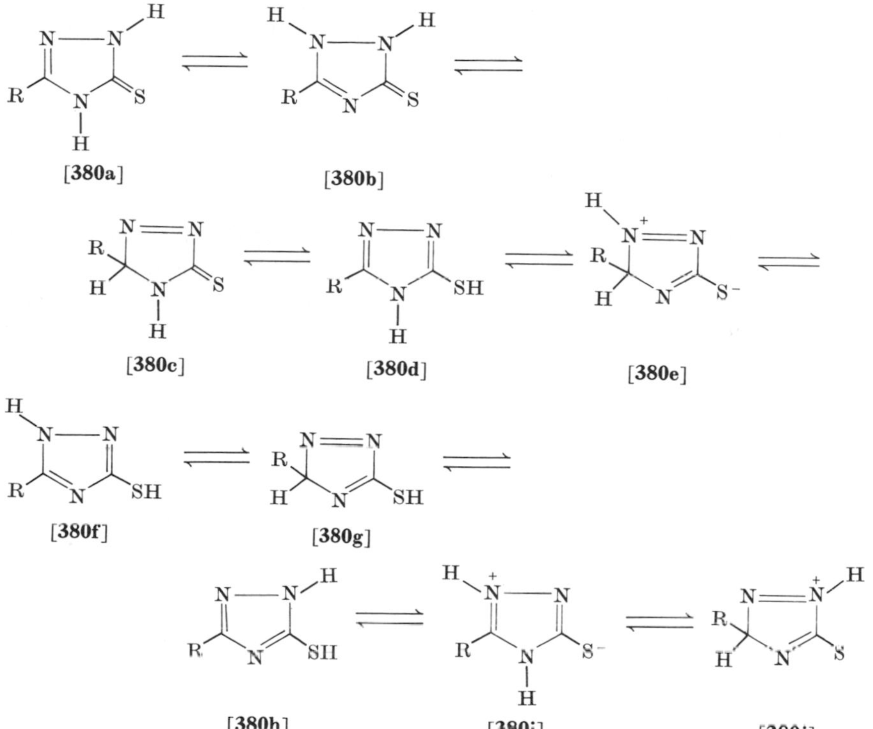

[**380a**] ⇌ [**380b**] ⇌

[**380c**] ⇌ [**380d**] ⇌ [**380e**] ⇌

[**380f**] ⇌ [**380g**] ⇌

[**380h**] ⇌ [**380i**] ⇌ [**380j**]

F. Heteroatoms-1,2,3 or -1,2,3,4 and a Potential Mercapto Group

a. 4-Mercapto-1,2,3-triazoles

None of these compounds were described in the previous survey. 4-Mercapto-1,2,3-triazoles can exist in four forms [**381a–d**], two of which are zwitterionic tautomers.

Two compounds [**381**, R^1 = Me, CH_2Ph] have been ascribed the thiol structure [**381b**]: IR (solid state) and NMR (in $CDCl_3$) measurements eliminate the thione form [**381a**] and the spectral properties of the tautomeric compounds are similar to those of the *S*-methyl compound [**383**] (72AS1243). Though the authors possess some of the fixed

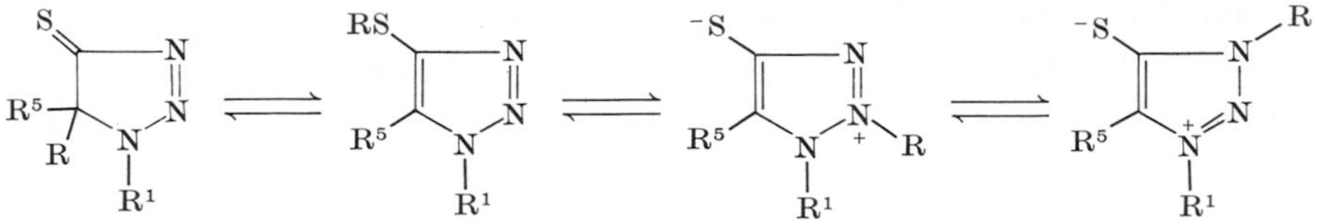

[**381a**] R = H
[**382**] R = Me

[**381b**] R = H
[**383**] R = Me

[**381c**] R = H
[**384**] R = Me

[**381d**] R = H
[**385**] R = Me

zwitterionic models [cf. **384**, **385**] (71AS3500, 72AS1243) the available data do not allow comparison with them.

b. 5-Mercapto-1,2,3-triazoles

The 1-substituted 5-mercapto-1,2,3-triazoles [**386**] can exist in four tautomeric forms [**386a–d**]. Comparison of IR and NMR data of compounds [**386**] (72AS1243) with two types of fixed derivatives [**387**] and [**390**] (but not [**388**], [**389**]) does not clearly prove the thiol-structure [**386a**] although the authors consider it the most likely.

Earlier German authors (66CB1618) had assigned the SH-structure [**386a**, R^3 = H, R^1 = Ph] from a νSH band at 2500–2590 cm^{-1} and lack of νNH absorption in $CHCl_3$.

4-Acyl- and 4-ethoxycarbonyl-5-mercapto-1,2,3-triazoles [**386**, R^3 = COX, R^1 = Me or Ph] show νSH bands the position of which is not concentration-dependent in CCl_4 (69CB417) and hence exist in the SH form [**386a**] with intramolecular hydrogen bonding between the CO and SH groups.

b/1. 1,2,5-Thiadiazole-3-thiones

These compounds, recently prepared, are unstable but the 4-carbamoyl derivative [**390/1**, R = $CONH_2$] showed no νSH in $CHCl_3$ and the thione form [**390/1b**] was assigned (73CC2349).

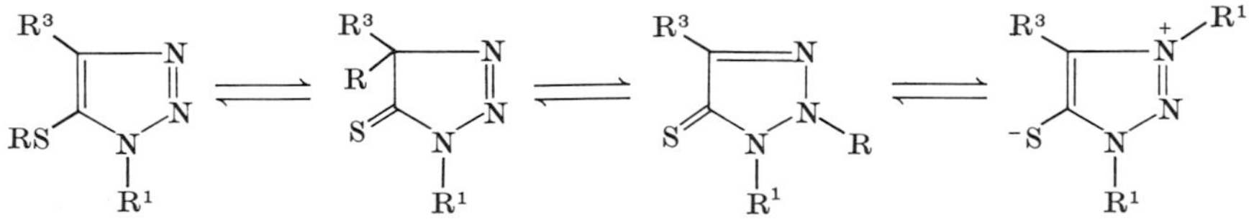

[**386a**] R = H
[**387**] R ≠ H

[**386b**] R = H
[**388**] R ≠ H

[**386c**] R = H
[**389**] R ≠ H

[**386d**] R = H
[**390**] R ≠ H

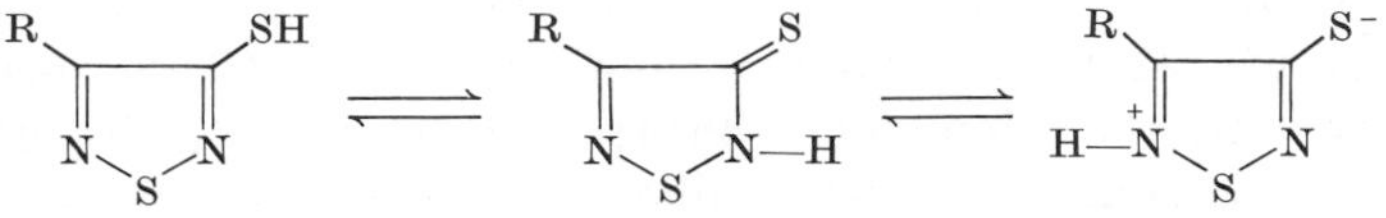

[**390/1a**]

[**390/1b**]

[**390/1c**]

c. N-Unsubstituted 5-Mercapto-1,2,3-triazoles

There is a single study (66CB1618) of the tautomerism of *N*-unsubstituted 5-mercapto-1,2,3-triazoles, synthesized earlier (62AS1800). The authors propose the thiol structure [**391a**, R^3 = H, CO_2H, CO_2Ph] because they observe a νSH band in the same region as the νSH for 5-mercapto-1-phenyl-1,2,3-triazole.

However, IR spectroscopy in $CHCl_3$ does not rule out the other possible mercapto structures [**391b**, **391c**] arising by annular tautomerism.

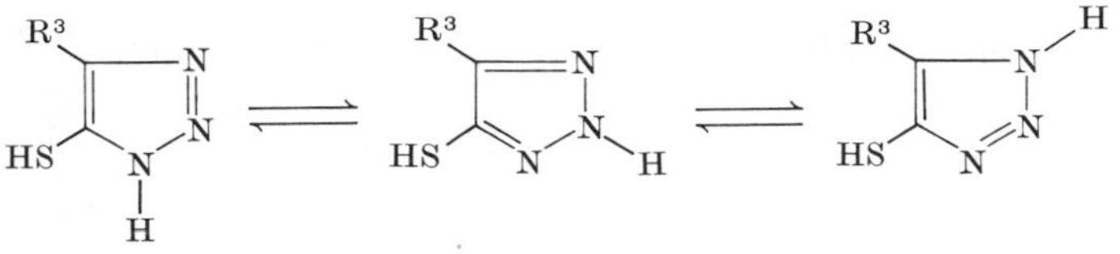

[**391a**] [**391b**] [**391c**]

d. 1,2,3,4-Thiatriazole-5-thione (II-63)

The previous survey (II-63) reported the thione structure [**392a**] for 1,2,3,4-thiatriazole-5-thione [**392**] based on work by Lieber *et al.* (57JO1750). But Jensen and Pedersen (64HC(3)263) from the same experimental data deduced that these molecules exist in the mercapto form [**392b**]. Part of the controversy concerns a band at 2533 cm^{-1} in the IR spectrum of the compound in solid state, assigned to νNH by Lieber (57JO1750) and to νSH by Jensen and Pedersen. However, a second publication by Lieber *et al.* (63JO194; see also 71PM(4)265) reports bands belonging to the group >N—C(—)=S, in the compounds [**392**] and [**393**], whereas the compounds [**394**] exhibit different bands.

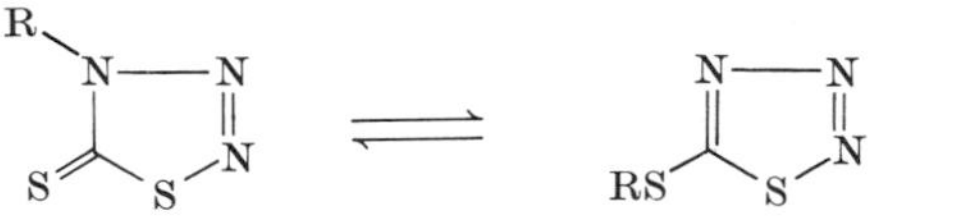

[**392a**] R = H
[**393**] R = Me, alkyl

[**392b**] R = H
[**394**] R = Me, alkyl

The main argument given by Jensen and Pedersen (64HC(3)263, p. 276) is worth quoting: "We think that the fact that this compound is a medium-strong acid is sufficient evidence that it is a thiol (a thiatriazolinethione would not be expected to be a medium-strong acid)." As tautomers [**392a**] and [**392b**] must have a common anion, it is easy to demonstrate (70C134) that the more acidic tautomer must be the less abundant* and that if the thiol is the more acidic, then the thione form [**392a**] must predominate in aqueous solution.

In conclusion, we believe that on available evidence the thione tautomeric form [**392a**] represents the structure of these compounds better than the mercapto form [**392b**].

e. 1,2,3,4-Tetrazole-5-thiones (II-63)

The previous survey (II-63) reported evidence for the thione structure [**395a**] for 1-substituted tetrazole-5-thiones [**395**] from IR, UV and pK_a measurements.†

Later results (67JO3580) on 1-aryltetrazole-5-thiones [**395**, Ar = *m*- or *p*-F–C_6H_4; *o*-, *m*- or *p*-NO_2–C_6H_4; Ph] support the thione structure [**395a**]. In solid state or in solution, UV, IR and NMR (solvent effects on the ^{19}F chemical shifts in the case of Ar = F–C_6H_4) measurements show that the tetrazolethiones lose the characteristics of the 1-aryltetrazole pattern. The authors propose the existence of a dimer [**396**] in the solid state and weakly polar solvents and the existence of a monomer [**395a**] in DMSO.

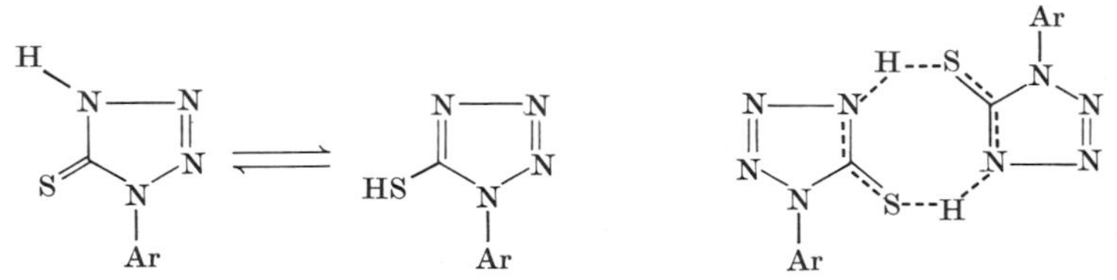

[**395a**] [**395b**] [**396**]

G. Compounds with Potential Mercapto Groups: Conclusions

The tautomerism of mercapto derivatives has been studied much less than that of the hydroxy analogues, as is obvious from a comparison of Table 4-28 with Table 4-25 (Section 4-3F). The conclusions are therefore less clear. A major difference is that no zwitterionic forms have been found in the mercapto series, not because they are less probable than for hydroxy derivatives, but simply because of our smaller knowledge of thione–mercapto tautomerism compared with that of other tautomeric functions.

* The situation was called the "Gustafsson paradox" (64AS871), and its nature was frequently emphasized (e.g., 65CI331).

† Jensen and Pedersen (64HC(3)263) for similar reasons as those described for the thiatriazolethiones (Section 4-4Fd) are in favour of a thiol structure [**395d**] but, as we pointed out (Section 4-4Fd), their arguments are not convincing.

As expected, when the SH group is in a position α relative to the O, S or NR heteroatom (denoted by Ä in the formulae in Table 4-28), the XH form is favoured in the case of X = N and Z = CR (for example, the 5-mercaptopyrazoles, Section 4-4Ad). On the other hand, the ZH form becomes predominant with Z = N and X = CR (for example, the oxazole-2-thiones, Section 4-4Ca). Finally, for X = Z = N, the tautomer ZH is generally predominant (for instance, the 1,2,4-triazole-5-thiones, Section 4-4Ef).

If the SH group is in a β position relative to Ä (see formulae in Table 4-28), tautomer SH is often predominant. However, there are apparent exceptions. The 1,2,4-triazole-3-thiones (Section 4-4Ec), and the 1,2,4-(Section 4-4Ee) and 1,2,5-thiadiazole-3-thiones (Section 4-4Fb/1) appear to favour the thione forms. As we have indicated, these results need confirmation.

More systematic work on the $>C{=}S \rightleftharpoons {=}\overset{|}{C}{-}SH$ tautomerism is necessary to allow more general conclusions to be drawn analogous to those deduced from the behaviour of hydroxy derivatives (Section 4-3F) or amino derivatives (Section 4-5G).

5. Compounds Containing Potential Amino Groups

A. Introduction

Before discussing systematically the tautomeric behaviour of amino compounds the main differences encountered in a study of the tautomerism of amino derivatives compared with that of hydroxy (Section 4-2 and 4-3) and mercapto (Section 4-4) derivatives should be pointed out.

In one way the problem is considerably more complicated because not only is the NH_2 group to be considered, but more generally the group NHR. Hence the influence of R upon the tautomeric equilibrium must be studied: it is especially significant if R is an acyl or a sulfonyl group.

In another way the problem is much simplified, because zwitterionic forms do not have to be taken into consideration. As amines are weaker acids than hydroxy and mercapto compounds* NH^- structures do not contribute significantly to the equilibrium. Although zwitterionic structures have been suggested in the literature, as mentioned in the previous review (II-74, II-75), later work has ruled them out (e.g., 67JO3580, 71J(B)2355).

* pK_a (proton loss) of thiophenol: 6.5, phenol: 10.0 and aniline: 27(62MI01).

TABLE 4-28

MERCAPTO-AZOLE-AZOLINETHIONE TAUTOMERISM[a]

A	X	Y	Z	Compound	Z—Y, HS, Ä, X (mercapto form)				Section
					H–Z—Y=X, S=C–A	Z—Y=X, HS–C–A	Z=Y–X–H, S=C–A	Z—Y⁺(H)=X, ⁻S–C–A	
O				α-Mercaptofurans	–	+++	–	×	3-3A
S	C	C	C	α-Mercaptothiophenes	–	+++	–	×	3-3B
NR				α-Mercaptopyrroles	–	+++	–	×	3-3C
O				5-Mercaptoisoxazoles	–	++	+	×	4-4Aa
S	N	C	C	5-Mercaptoisothiazoles				×	
NR				5-Mercaptopyrazoles	–	++	++	×	4-4Ad
O				Oxazole-2-thiones	+++	–	–	×	4-4Ca
S	C	C	N	Thiazole-2-thiones	+++	–	–	×	4-4Cc
NR				Imidazole-2-thiones	+++	–	–	×	4-4Ce
O				Oxazole-5-thiones					
S	C	N	C	Thiazole-5-thiones					
NR				Imidazole-5-thiones					4-4D
O				1,2,4-Oxadiazole-5-thiones	(++)	–	(++)	×	4-4Ed
S	N	C	N	1,2,4-Thiadiazole-5-thiones	(++)	–	(++)	×	4-4Ee
NR				1,2,4-Triazole-5-thiones	+++	–	–	×	4-4Ef
O				1,3,4-Oxadiazole-2-thiones	+++	–	–	–	4-4Eg
S	C	N	N	1,3,4-Thiadiazole-2-thiones	+++	–	–	–	4-4Eh
NR				1,3,4-Triazole-2-thiones	+++	–	–	–	4-4Ei
O				1,2,3-Oxadiazole-5-thiones					
S	N	N	C	1,2,3-Thiadiazole-5-thiones					
NR				5-Mercapto-1,2,3-triazoles	–	+++	–	–	4-4Fb
S	N	N	N	1,2,3,4-Thiatriazole-5-thiones	+++	–	–	–	4-4Fd
NR				1,2,3,4-Tetrazole-5-thiones	+++	–	–	–	4-4Fc

[a] See key to table in Table 4-25.

TABLE 4-28 (*continued*)

Compound	Y=C(SH)–X=… Z–Ä–X (SH)				Section
	Y–C(SH)=X, Z–A	Y–C(=S)–X–H, Z–A	H–Y⁺, C–S⁻, Z–A–X	Y–C–S⁻, H–Z⁺–A–X	
β-Mercaptofurans			×	×	3-3A
β-Mercaptothiophenes	+ + +	–	×	×	3-3B
β-Mercaptopyrroles			×	×	
3-Mercaptoisoxazoles			×	×	
3-Mercaptoisothiazoles	+ + +	–	×	×	4-4Ac
3-Mercaptopyrazoles			×	×	
4-Mercaptoisoxazoles			×		4-4B
4-Mercaptoisothiazoles			×		4-4B
4-Mercaptopyrazoles			×		4-4B
4-Mercaptooxazoles				×	
4-Mercaptothiazoles				×	
4-Mercaptoimidazoles				×	
3-Mercapto-1,2,5-oxadiazoles			×		
3-Mercapto 1,2,5 thiadiazoles	–	+ + + ?	×	–	4-4Fb/1
3-Mercapto-1,2,5-triazoles			×		
4-Mercapto-1,2,3-Oxadiazoles					
4-Mercapto-1,2,3-thiadiazoles					
4-Mercapto-1,2,3-triazoles	+ + +	–	–	–	4-4Fa
3-Mercapto-1,2,4-oxadiazoles				×	
3-Mercapto-1,2,4-thiadiazoles	–	+ + + ?	–	×	4-4Eb
1,2,4-Triazole-3-thiones	–	+ + + ?	–	×	4-4Ec
4-Mercapto-1,2,3,5-thiatriazoles					
4-Mercapto-1,2,3,5-tetrazoles					

Concerning methods of studying the tautomerism of amino derivatives, two facts are worth noting. Firstly, in the course of the synthesis of fixed derivatives with an imino structure [**397**] care is needed to avoid possible hydrolysis into a oxo structure [**398**] (see, e.g., 67BF1219). Secondly, if R in NHR contains an α-hydrogen [**399**], the presence of

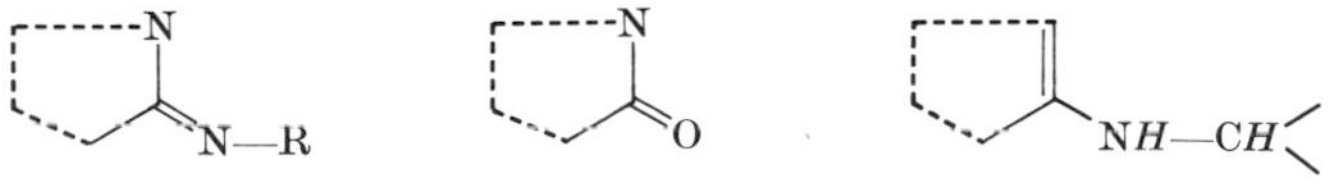

[**397**] [**398**] [**399**]

N*H*–C*H* coupling in NMR is a simple and sure proof of the existence of an amino tautomer; we find many examples of this coupling (see, e.g., 72AS459), which is usually observed in DMSO-d_6 (pay attention to the possible presence of D_2O: 71J(B)2355).

B. Heteroatoms-1,2 and a Potential 3-(or 5-)Amino Group

a. 5-Aminoisoxazoles (II-66)

Two papers published in 1961 (61SA238, 61T51), and cited in the previous review (II-66), established unambiguously the amino structure [**400a**] of 5-aminoisoxazoles from IR and NMR evidence.

Later other authors (65BS255) reached the same conclusion using other physical methods: comparative UV study (in MeOH or cyclohexane) of the tautomeric compound with the fixed derivatives corresponding to [**400a**] and [**400c**] and dipole moment measurements

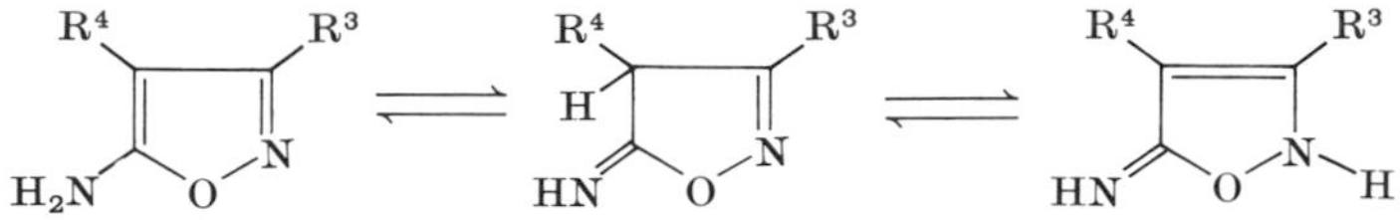

[**400a**] [**400b**] [**400c**]

(benzene). In fact, all the publications dealing with 5-aminoisoxazoles suggest form [**400a**], even if the arguments are not always convincing (66CO(263C)557).

The parent compound [**400**, $R^3 = R^4 = H$] (66CT1277) exhibits in the IR spectrum (chloroform) two νNH bands at 3520 and 3425 cm^{-1} which obey Bellamy's rule (Section 1-6Cc) (63PM(2)161, p. 326) and a $\delta\ NH_2$ band at 1634 cm^{-1}. The coupling J(3H4H) of 2.0 Hz (in CCl_4) is normal for a structure such as [**400a**], whereas in the case of [**400c**] a value of 3 Hz would be expected, see the analogous case of the isoxazol-5-ones (Section 4-2Bc).

The presence of a 4-nitro group does not modify the tautomerism: form [**401**] has been demonstrated by IR and NMR spectroscopy (68AN562).* The methylamino analogue [**402**] described in the same paper, exhibits, in $CDCl_3$, a coupling between NH and the methyl; this rules out tautomers of the type [**400b**] or [**400c**]. This same argument has been used to assign the amino structure [**403**] to a dimeric

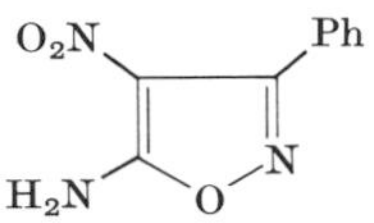

O_2N Ph H_2N O N

[**401**]

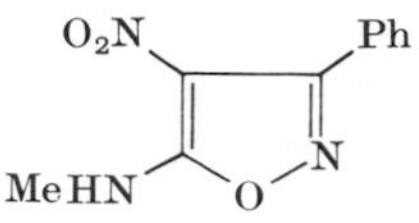

O_2N Ph MeHN O N

[**402**]

Me N O Me MeHN O N

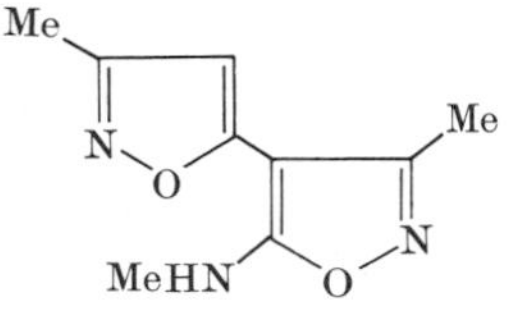

[**403**]

isoxazole for which five tautomeric forms can be written (71T379). The mass spectrometric fragmentation has been rationalized (71OM123) in terms of the amino tautomer [**400a**]. 4-Ethoxycarbonyl derivatives likewise possess the amino structure (71T5873).

The 5-hydrazinoisoxazole [**404**] has been described as such, which is probable but insufficiently demonstrated (69JH783). For 4-acetamidoisoxazoles, structure [**405**] had been previously established (II-77); this has been confirmed (66CT756, 68AN562) and extended to the sulfonamides [**406**] (66CT756).

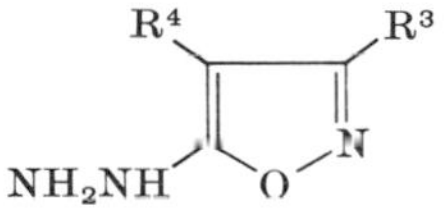

R^4 R^3 NH_2NH O N

[**404**]

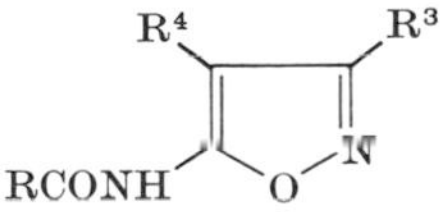

R^4 R^3 RCONH O N

[**405**]

R^4 R^3 $ArSO_2NH$ O N

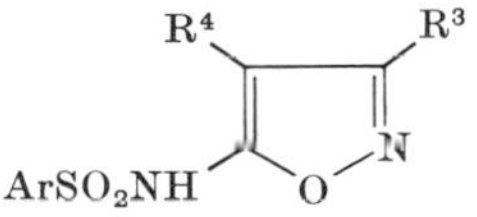

[**406**]

Comparisons of UV spectra in neutral and acid medium (61T51) shows that the 5-aminoisoxazoles are protonated on the ring nitrogen.

b. Aminobenzisoxazoles

The tautomerism of 3-aminoanthranil [**407**] has been studied in CCl_4 by IR (65CB1562) using Bellamy's rule (63PM(2)161, p. 326) and the partial deuteriation method (63PM(2)323) (Section 1-6Cc; see also 71PM(4)265, p. 428). It exists in the amino structure [**407a**]; it is interesting that the imino structure [**407b**] is not favoured, even though it retains the benzene nucleus undisturbed (compare the indazolones, Section 4-2J).

* Curiously, the same authors, describing the IR spectrum of [**401**] in Nujol, write: "3385, 3240, 3180, 3115 cm^{-1} ($NH_2 \rightleftharpoons NH$)" (68T4907). There are adequate alternative explanations for the multiplicity of bands without invoking the imino tautomer.

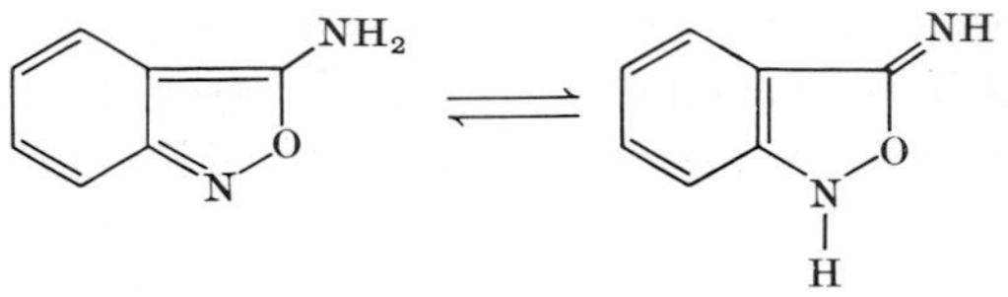

[**407a**] [**407b**]

c. 3-Aminoisoxazoles (II-67, II-77)

The amino structure [**408**] of these products and of their *N*-acyl derivatives [**409**] had already been well established at the time of the previous survey (II-67 and II-77) (61SA238, 61T51).

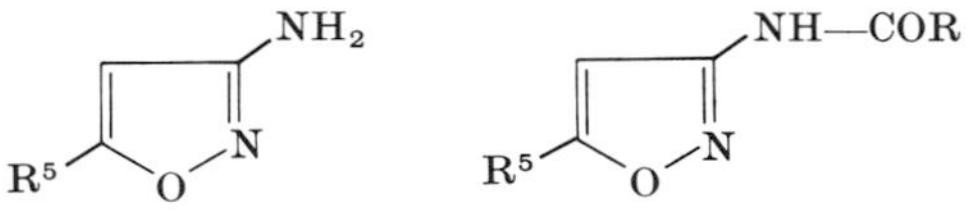

[**408**] [**409**]

Three subsequent publications confirm such structures. In UV spectroscopy (EtOH) compounds [**408**, R^5 = Ph] and [**409**, R^5 = Ph] absorb in the same way as the dialkylamino analogues [**410**] (63CB1088). In IR spectroscopy (crystalline state) the NH and ND compounds corresponding to the structures [**409**, R^5 = Me] and [**411**, R^5 = Me]

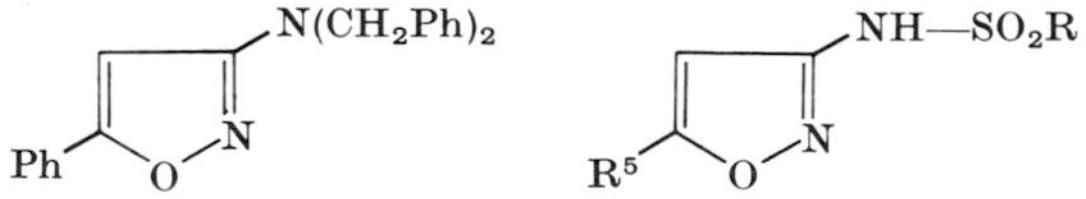

[**410**] [**411**]

have been studied (66CT756). Finally IR ($CHCl_3$) and UV (EtOH) study on a series of 3-aminoisoxazoles [**408**], including the parent molecule [**408**, R^5 = H], and of their acyl [**409**] and sulfamido [**411**] derivatives, is consistent with the amino structures shown (66CT1277).

Just as for the 5-aminoisoxazoles (Section 4-5Ba), the 3-amino isomers are protonated at the ring nitrogen (61T51).

d. 5-Aminoisothiazoles

No data were available on these compounds in the previous edition, and this tautomerism does not seem ever to have been studied specifically. Information about the electronic spectra is found in two papers cited in (65HC(4)107), dealing with 5-amino-3-phenylisothiazole [**412**] (63CB944), and with three derivatives of 5-amino-3-chloro-4-cyanoisothiazole [**413**], [**414**] and [**415**] (64JO660). Although the author (64JO660) draws no conclusion from the analogous absorptions of compounds

[**413**]–[**415**], this can be considered to be an indication of the predominance of the amino tautomers. IR and NMR demonstrate the NH_2 structure for the 4-ethoxycarbonyl-3-methyl compound (71T5873).

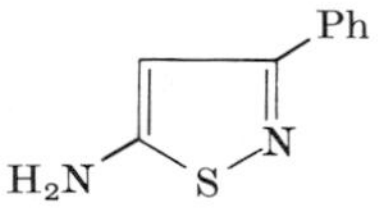

[**412**]

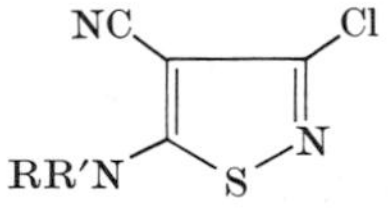

[**413**] R = R′ = H
[**414**] R = H, R′ = Me
[**415**] R = R′ = Me

e. 3-Aminoisothiazoles

3-Amino-5-phenylisothiazole [**416**] exhibits (in CCl_4) νNH_2 bands at 3490 and 3400 cm^{-1} which obey the Bellamy-Williams rule (Section 1-6Cc) (63CB944; see also 65HC(4)107). In methanol this compound absorbs at 267 and 306 nm (63CB944), close to the fixed derivative [**417**] at 265 and 308 nm (66G1009). Compound [**417**] and 3-anilino-5-phenylisothiazole [**418**] each show IR (KBr) bands at ~1492 and ~808 cm^{-1}, characteristic of the isothiazole ring according to the authors (67AN471).

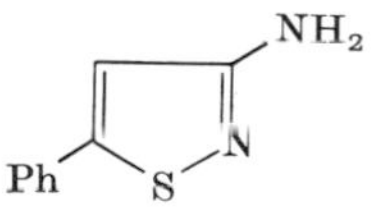

[**416**] 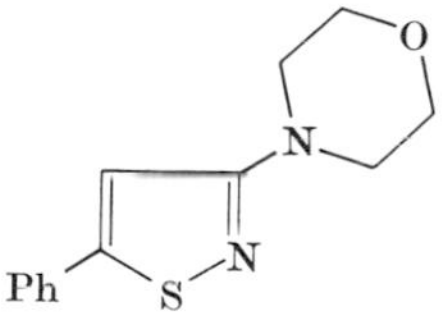 [**417**] 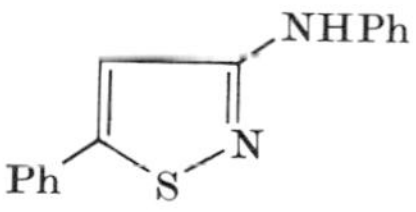 [**418**]

f. 3-Amino-1,2-benzisothiazoles

The predominance of the amino tautomers [**419**] and [**420**] is demonstrated (69CB1961) by UV absorption (MeOH) similar to the fixed compound [**421**]. In NMR spectroscopy ($CDCl_3$, DMSO, MeOD, CCl_4, C_6D_6, CD_3COCD_3) the signals arising from the aromatic protons are similar throughout the series; furthermore for [**420**, R = Me or CH_2R'], a coupling of 5 to 6 Hz is observed with NH.

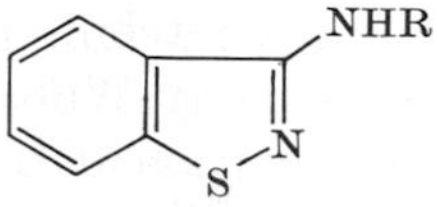

[**419**] R = H
[**420**] R ≠ H

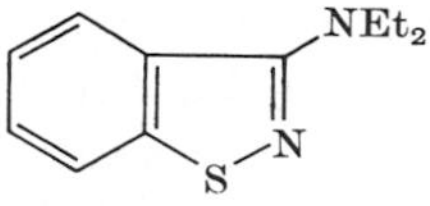

[**421**]

These compounds easily undergo Dimroth rearrangement (Section 4-9B); for compound [**419**], amino ⇌ imino tautomerism [**422**] and Dimroth rearrangement cannot be differentiated, unless the nitrogens are isotopically labelled (69CB1961).

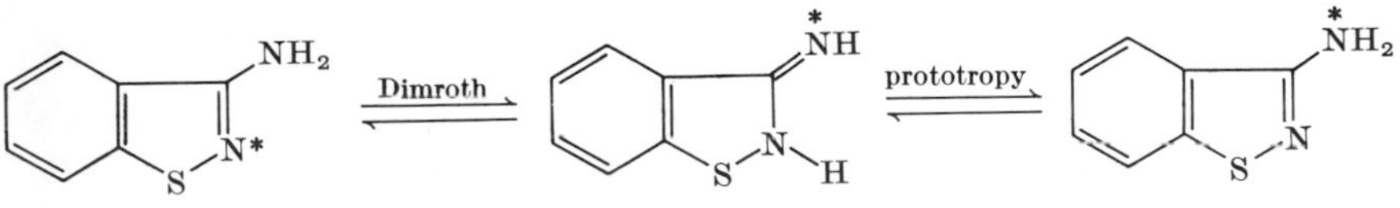

[**419**] [**422**] [**419**]

3-Amino-1,2-benzoisothiazoles are protonated at the heterocyclic nitrogen; in trifluoroacetic acid-deuteriochloroform 3:1, the cation [**423**] exhibits separate signals for the two NH and a coupling of 4.5 Hz between the methyl and one of the NH groups (69CB1961).

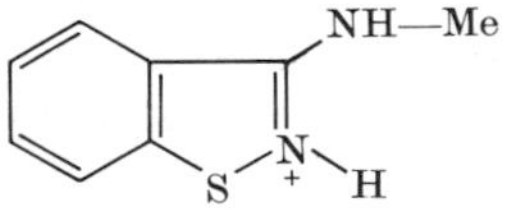

[**423**]

An IR study, based mainly on the valence vibrations of SO_2 (71PM(4)265, p. 431) but also using the νC═N and δ NH_2 (70AP(303)980) established the amino structures [**424**], [**425**] of 3-amino-1,2-benzisothiazole 1,1-dioxides. The fixed compounds [**426**] and [**427**] were used as models.

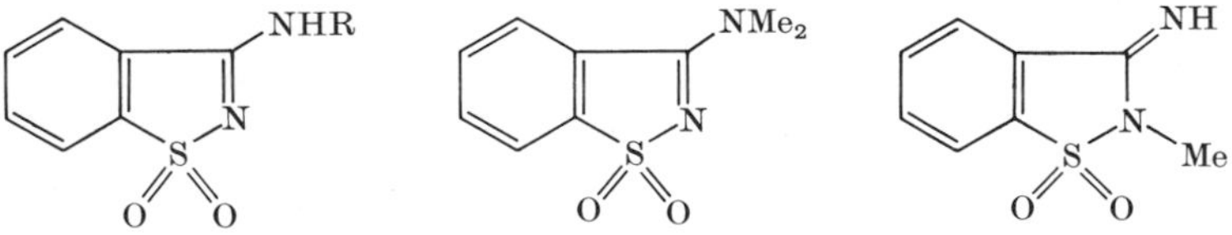

[**424**] R = H
[**425**] R ≠ H

[**426**]

[**427**]

g. 5-Aminopyrazoles (II-69)

5-Aminopyrazoles [**428**] could exist in three possible tautomeric forms [**428a–c**]. NMR studies eliminate the CH form [**428b**] (68CB3265, 70BF4436, 71CO(272C)103, 74JH423) and the amino structure [**428a**] is proved by IR spectroscopy. The two NH_2 stretching vibrations of aminopyrazoles (70CB2505) follow the Bellamy-Williams relation (Section 1-6Cc). In the solid state and in solution ($CHCl_3$, dioxan), 5-amino-1-phenylpyrazoles [**428**, R^3 = H, Me or Ph] exhibit a band at 1620 cm^{-1} which disappears on deuteriation; the authors (64ZO654)

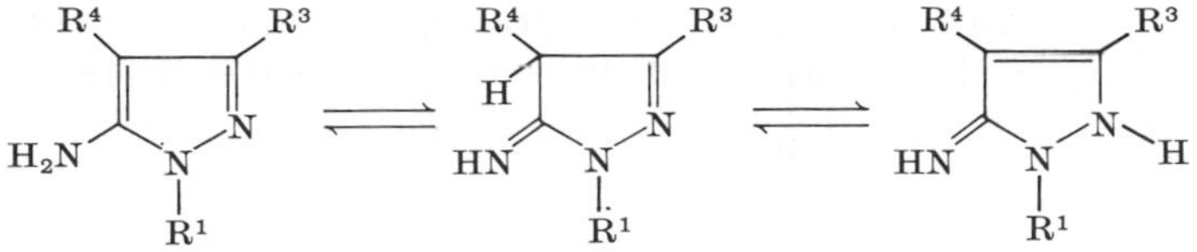

[428a] [428b] [428c]

assign it to the δNH_2 of the amino form [**428a**]. The compounds also exhibit a band at 1560 cm^{-1} attributed to the pyrazole ring (R^4 = H) (cf. Section 4-2Fc). Although NH_2 stretching bands were not mentioned, they are obvious in the published spectra. This work is confirmed by comprehensive study on the IR spectra of 1-methyl-5-amino- [**431**] and 3-amino-pyrazoles (Section 4-5Bh) in CCl_4 and in Nujol (70TH04), and by further IR and NMR work (73PC(315)382). For an X-ray study of the aminopyrazole [**428a**, R^1 = —CMe=CHCN, R^3 = Me, R^4 = H] (74JH423) see 75MIip3.

In the near IR region, 1,3-diphenyl-5-aminopyrazole [**432**] (in CCl_4) shows bands characteristic of the CH=C—NH_2 group of the amino form [**428a**] (66SA1385).

As with the OH and NH forms of pyrazolones (Section 4-2F), NMR spectroscopy cannot differentiate directly amino [**428a**] and imino [**428c**] tautomers; however, the low J(3H4H) value in [**430**] and [**429**], J = 1.9 Hz (68CB3265, 70BF4436), favours a pyrazole structure [**428a**] (71LA(750)39) (Section 4-2Fc). IR and NMR demonstrate the amino structure for the ethoxycarbonyl derivatives [**432/1**, R = H, Me, Ph] (71T5873).

[**429**] R = Me, R′ = H
[**430**] R = Ph, R′ = H
[**431**] R = Me, R′ = Ar
[**432**] R = Ph, R′ = Ph

[**432/1**]

The UV spectra and pK_a values of 5-amino-1-phenylpyrazole and the 3-methyl analogues have been interpreted on the basis of the amino form [**428a**] (66T2703). The site of protonation is not proven, but by analogy with 5-aminoisoxazoles (Section 4-5Ba) it is probably the ring nitrogen.

The suggestions of the previous review (II-69), that the amino tautomer of simple aminopyrazoles predominates, is thus amply confirmed. Despite this, these compounds are still written in the CH form [**428b**] in some publications (65CI331, 67AN471).

An IR study of sulfonamidopyrazoles (68CB3278, 68CT345) establishes the amino structure [**433**] for these compounds. However, the imino tautomer [**434b**] is preferred over [**434a**] for the dithio derivative

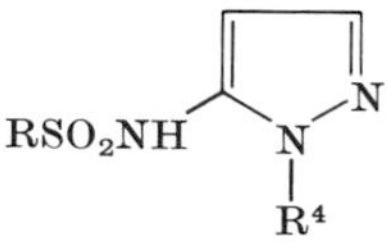

RSO_2NH N N R^4

[**433**]

SMe C S H N H N N Ph Me ⇌ SMe H S C N H N N N Ph Me

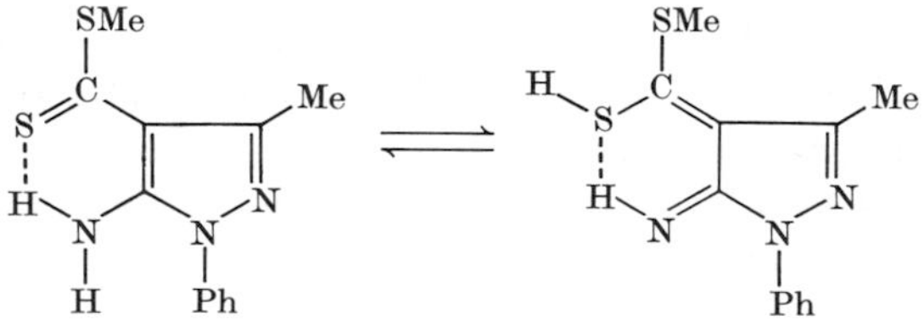

[**434a**] [**434b**]

shown (71CB1155), because of the absence of νNH_2 in the IR spectrum (KBr, $CHCl_3$) and the presence of a signal at 13.65 ppm in the NMR, assigned to the chelated NH group (the νSH cannot be observed because of its broadness, according to the authors). For comparison, compound [**435**] shows the NH_2 signal at 5.21 ppm in $CDCl_3$ (70BF4436).

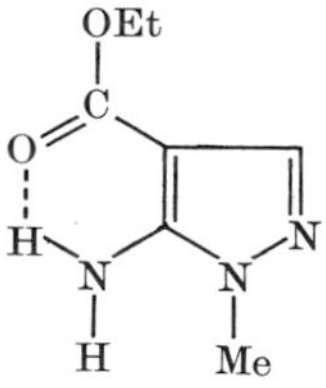

OEt C O H N H N N Me

[**435**]

h. 3-Aminopyrazoles

IR (solid state and CCl_4, $CHCl_3$) demonstrates an amino group in 3-aminopyrazoles [**436**] (64ZO654, 70TH04). This structure also explains the UV (66T2703) and has been confirmed by recent IR and NMR work (73PC(315)382).

3-Amino-4-ethoxycarbonylpyrazoles [**437**] in the IR (Nujol) exhibit three bands in the NH_2 region and a *H*-bonded $\nu C{=}O$ at 1675–1685 cm^{-1} depending on the nature of R^5 (71J(C)225). 4-Ethoxycarbonylpyrazoles show $\nu C{=}O$ at 1695–1710 cm^{-1} (71PM(4)265, p. 421). The NMR

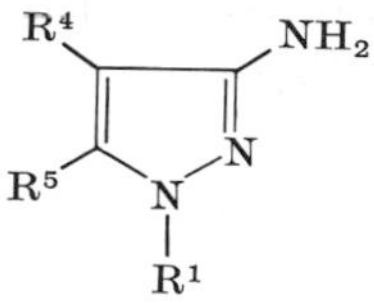

[**436**] 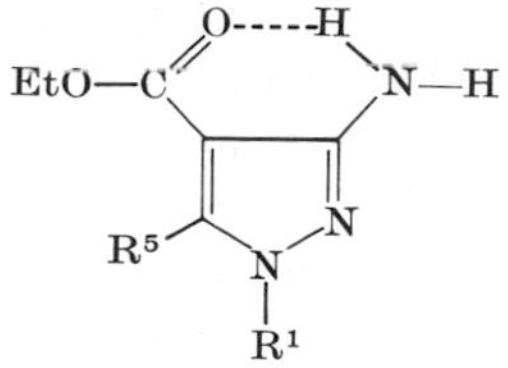 [**437**]

spectrum of [**437**, R^1 = Me, R^5 = H] (70BF4436) shows a single NH_2 signal.

i. N-Unsubstituted 3(5)-Aminopyrazoles

All the UV (63CB1088), IR (69KG312, 70TH04) and NMR (67LA(707)141, 68CB3265, 70TH04) studies conclude in favour of an amino structure for these compounds [**438**], but without specifying the ring nitrogen carrying the hydrogen [**438a**] or [**438b**] (annular tautomerism: Section 4-1B).

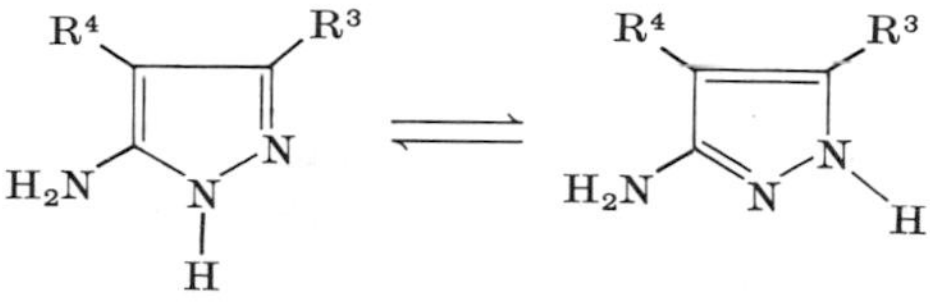

[**438a**] [**438b**]

Comparison of the coupling J(3H4H) of [**440**], 2.3 to 2.4 Hz (68CB3265), with those of 3-amino-1-methylpyrazole, 2.2 Hz (70BF4436) and of 5-amino-1-methylpyrazole, 1.8 to 1.9 Hz (68CB3265, 70BF4436), apparently indicates a predominance of the 3-amino tautomer [**440a**], but this conflicts with the theoretical prediction (Section 4-1Bbix), and the occurrence of some imino form [**440b**] would also explain the

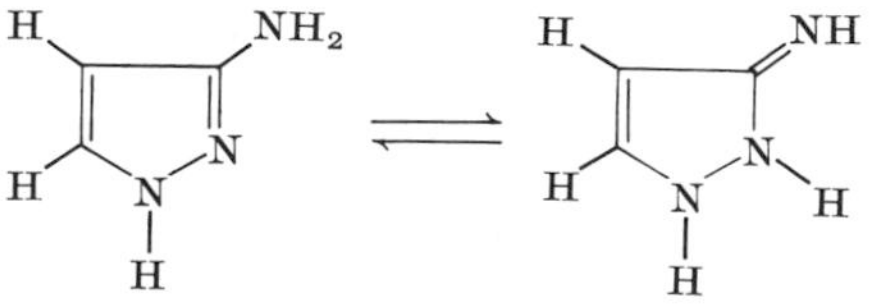

[**440a**] [**440b**]

experimental coupling. However, in a very recent detailed discussion of the NMR data, Dorn supports the 3-amino formulation (73PC(315)382, 74PC705).

The amido and sulfonamido derivatives exist in the amino structures [**441**] and [**442**] in the solid state (IR studies) (67LA(707)141). In

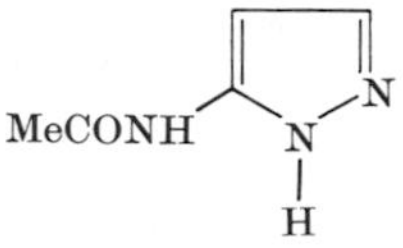

[**441**] 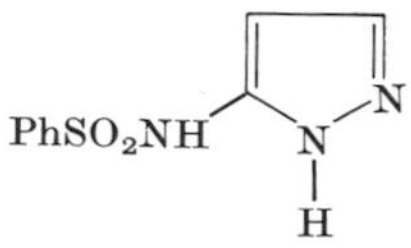 [**442**]

solution (DMSO) the coupling between the NH and CH groups establishes the structure of the formamide [**443**] and that between the two NH groups the structure of the hydrazide [**444**] (72OR733). However, the position of the "annular" proton is uncertain for all these compounds.

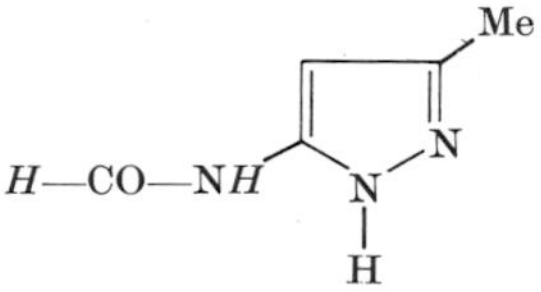

[**443**] 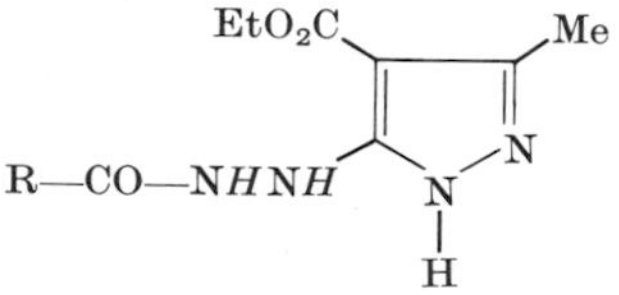 [**444**]

j. Aminoindazoles (II-70)

For aminoindazole, the amino form [**445**] is shown by IR (II-70). 3-Amino-1-benzylindazole shows bands (Nujol) at 3440, 3305, 3200, 1623, 1606 cm^{-1} (71JO1463); by analogy with the 3-aminopyrazoles (64ZO654), the two first and the fourth are assigned to the NH_2 group of the amino tautomer [**446**].

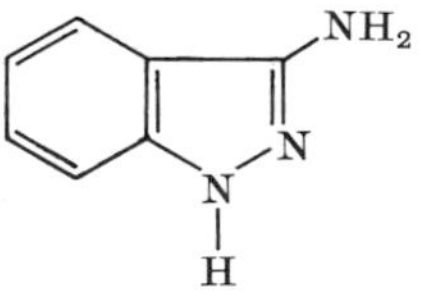

[**445**] 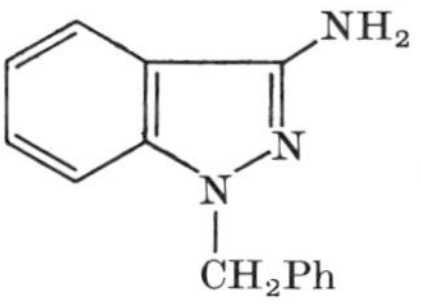 [**446**]

C. Heteroatoms-1,2 and a Potential 4-Amino Group

a. 4-Aminoisoxazoles (II-67)

The amino structure for 4-aminoisoxazoles [**447**] and their acetylated derivatives [**448**] is established by IR and NMR evidence (61SA238, 61T51, II-67). UV spectra prove that 4-aminoisoxazoles are protonated at the amino group [**449**] (61T51) unlike their 5-amino (Section 4.5Ba) or 3-amino (Section 4-5Bc) isomers.

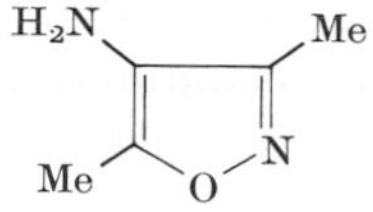

[447] 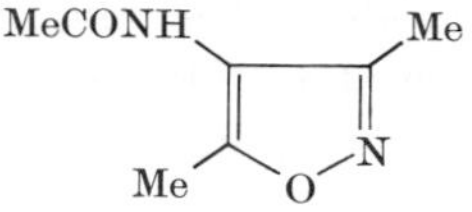 [448] 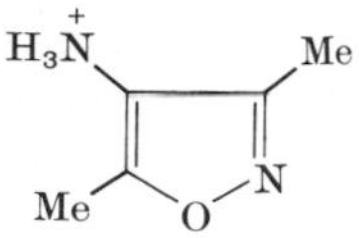 [449]

b. 4-Aminoisothiazoles

No study of the tautomerism of 4-aminoisothiazoles [**450**] has appeared (65HC(4)107). However, a 4-amino group in isothiazole causes a bathochromic effect of ca. 40 nm (63J2032), similar to those observed in the isoxazole [**447**] ($\Delta\lambda = 28$ nm) (61T51) and pyrazole series [**451**] ($\Delta\lambda = 30$ nm) (66T2703). This strongly suggests that 4-aminoisothiazoles do indeed exist in the amino form. Two recent papers (74LA1183, 75MIip4) confirmed this prediction.

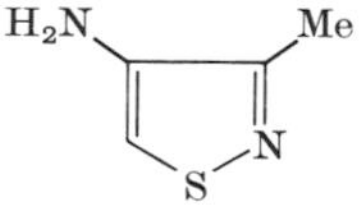

[450] 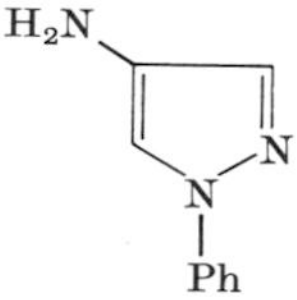 [451]

c. 4-Aminopyrazoles

The IR spectra of 1-phenyl-4-aminopyrazole in the solid state and in chloroform (64ZO654) establish the amino structure [**451**]. 4-Amino-1-phenylpyrazole [**451**] is more basic (66T2703) than its 3-amino and 5-amino isomers. The authors do not define the protonation site, but the greater basicity probably reflects protonation at the exocyclic nitrogen, by analogy with 4-aminoisoxazoles [**449**].

D. Heteroatoms-1,3 and a Potential 2-Amino Group

a. 2-Aminooxazoles (II-67)

The previous survey reported the predominance of the amino tautomer [**452**] for 2-amino-4,5-diphenyloxazole (II-67) (see also 74HC(17)99). More complete work by Najer and Menin now supports this result.

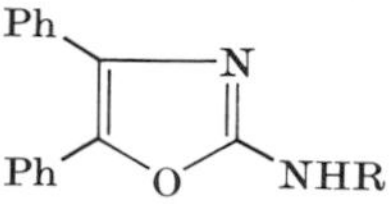

[452]

The pK_a values of the models [**454**] and [**455**] (65CO(260)4343), with Eq. (1-16) (1-4A), give K_T ca. 2.3×10^4 in favour of the amino tautomer [**453**]. The UV method fails as the absorptions of [**454**] and [**455**] are

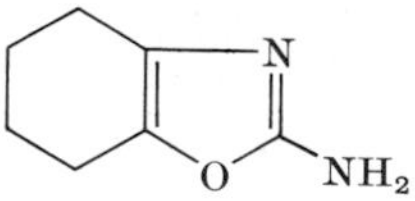

[**453**] 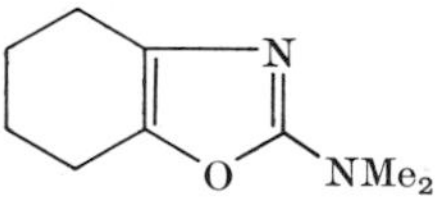 [**454**] 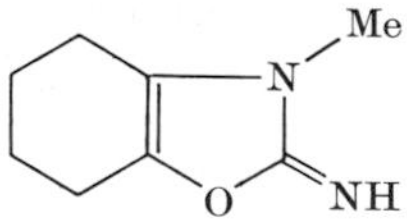 [**455**]

very close (67BF2040). However, partial deuteriation (Section 1-6Cc) also demonstrates the predominance of the amino tautomer [**453**] in CCl_4 and in CS_2 (68CO(266C)1587). The amino structure persists for derivatives of the 2-aminooxazole with substituents on the exocyclic nitrogen (69CB230).

b. 2-Aminobenzoxazoles (II-67)

The previous survey (II-67) reported for 2-aminobenzoxazole a UV study suggesting the amino structure [**456a**] but also a solid state IR study which assigned the imino structure [**456b**]: the latter conclusion was then (II-67), and is still now, considered very unlikely. However, the matter has still not been finally cleared up although 2-aminobenzoxazole and the corresponding fixed compounds are known and have been studied by mass spectrometry (70OM1341).

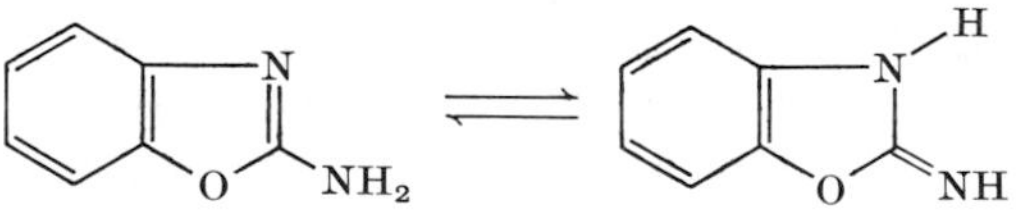

[**456a**] [**456b**]

Groth (73AS945) proposes the imino form [**456/1b**] on the basis of his X-ray measurements; however, the quoted bond lengths and angles for [**456/1**] are different from those in [**456/2**] and would fit better with the amino structure [**456/1a**, R = H]; the author explains this discrepancy by the occurrence of hydrogen bonded dimers.

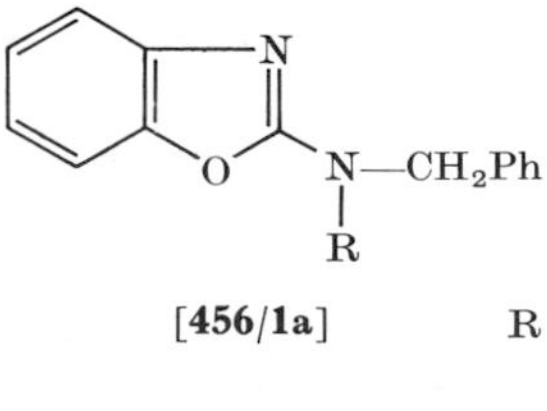

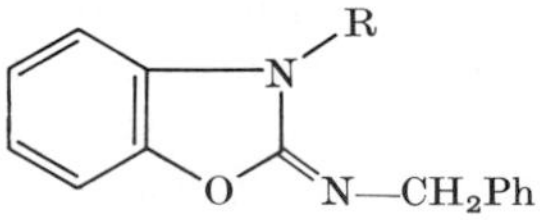

[**456/1a**] R = H [**456/1b**]
[**456/2**] R = Me

c. 2-Aminothiazoles (II-68)

Predominance of the amino tautomer for 2-aminothiazoles, reported in the previous review (II-68), has been fully confirmed by the work of Najer and Menin and of Sélim and Sélim, now to be described.

For compound [**457**] application of Eq. (1-16) (Section 1-4A) to the pK_a values of the fixed models (65CO(260)4343)* gives $K_T = 6.8 \times 10^3$. The predominance of the amino tautomer [**457**] in CCl_4 or CS_2 is shown by partial deuteriation (68CO(266C)1587); but UV spectroscopy fails because absorption in the fixed compounds is too similar (67BF2040).

Sélim and Sélim studied [**458**], [**459**] and [**460**] by UV. Only for the 4-phenyl derivative [**460**] are the spectra differentiated, but we have

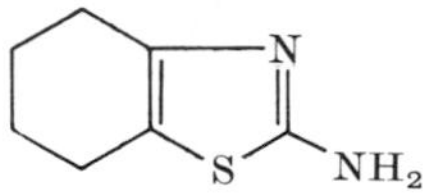

[**457**]

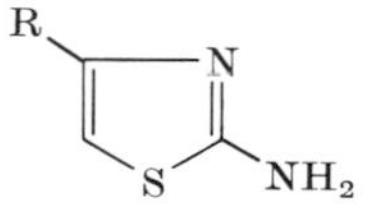

[**458**] R = H
[**459**] R = Me
[**460**] R = Ph

criticized elsewhere (Section 1-3B) the use of the appearance of the phenyl group signal in the NMR spectrum of fixed derivatives methylated on a ring nitrogen [**461**] vicinal to a phenyl group. The difficulty of applying the UV method to compound [**458**] is reported elsewhere (66BB380, 66CI1634). However, the shift of the H5 proton of [**460**] and H4 and H5 protons of [**462**] (66CI1634) are good evidence for the predominance of the amino tautomers shown. X-ray analysis (69CX(B)625) of compound [**463**] demonstrates the amino form for the solid state.

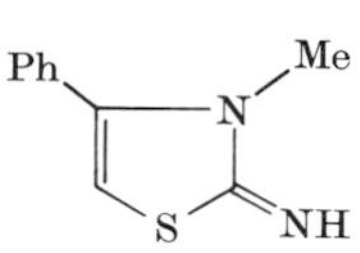

[**461**] 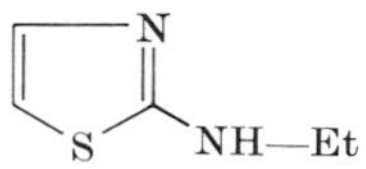 [**462**] 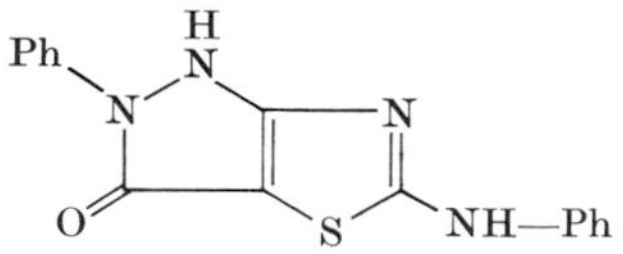 [**463**]

The amide–imide tautomerism cf. [**464a**] ⇌ [**464b**] of 2-acylaminothiazoles [**464, 465**] has been studied by Sélim and Sélim by UV and NMR spectroscopy. Unfortunately they used mainly the 4-phenyl series [**464**] which involves disadvantages (Section 1-3B). Nevertheless their general conclusion is valid; the amino tautomer [**464a**] is the more stable (see also 74RR671).

* For the structure of the common cation, see the UV study in references (65BF3527, 66BB380) and the X-ray structure determination (74CX(B)342).

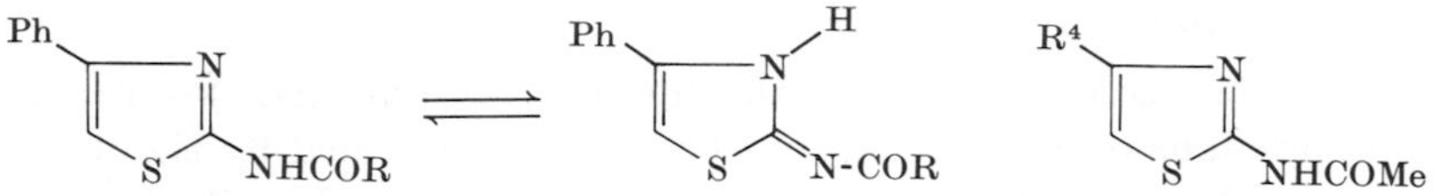

Ph N S NHCOR ⇌ Ph N H S N-COR; R^4 N S NHCOMe

[**464a**] [**464b**] [**465**]

Thus, UV of compounds [**464**, R = various] (66BF342, 66BF3403, 68BF2117) and of compounds [**465**, R^4 = H and R^4 = Me] (66BF342) show the predominance of the amide tautomer (for R^4 = H see also 66BB380). This is supported by NMR results: i.e., the appearance of the phenyl signal (but see Section 1-3B) and, more clearly, the shift of the H5 proton (68BF3270, 68BF3272).

One result of Sélim and Sélim seems most suspicious: they claim to have isolated both tautomers [**466a**] and [**466b**]. While rare authentic cases are known of similar separations (see the 2-acylaminobenzothiazoles) the reports that [**466a**] and [**466b**] possess different UV (68BF2117) and NMR (68BF3272) spectra in solution would, if true, be unprecedented in view of all available data about the activation energy of an exchange process between two nitrogen atoms (Section 1-5Aa). The so-called tautomer [**466b**] may well possess structure [**467**].

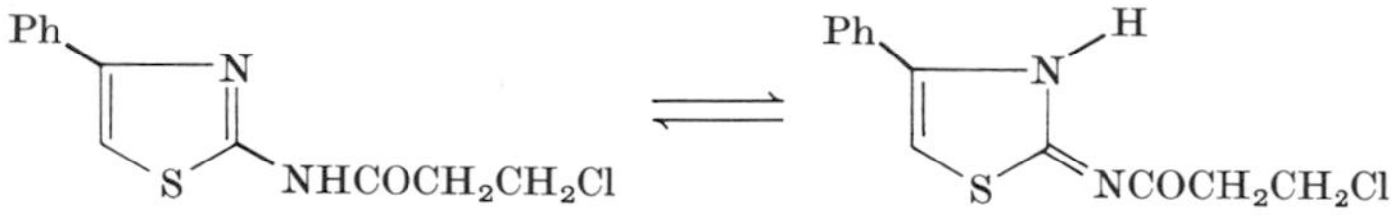

Ph N S $NHCOCH_2CH_2Cl$ ⇌ Ph N H S $NCOCH_2CH_2Cl$

[**466a**] [**466b**]

Ph N $COCH_2CH_2Cl$ S NH

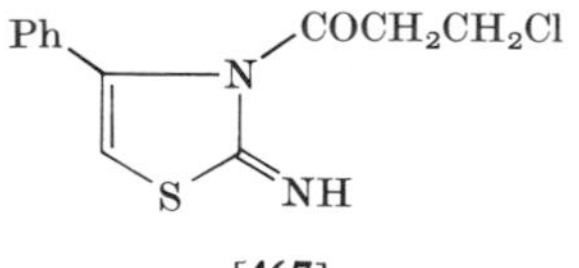

[**467**]

d. 2-Aminobenzothiazoles (II-68)

The amino structure [**468**] was already well established for 2-aminobenzothiazoles at the time of the previous review (II-68). An X-ray study (70CX(B)1736) confirms the amino structure for the NHMe derivative [**469**]. Some derivatives with two thiazole rings, such as [**470**], have been studied by UV (64AN80) without any conclusion about the tautomerism. The amino structure of compound [**471**] is demonstrated by IR spectra in the solid state (71CC4054). Compounds [**468**] and [**469**] have been studied by mass spectrometry together with the corresponding fixed derivatives (70OM1341). The 2-hydrazinobenzothiazoles exist as such and not in the hydrazone form (73BF(2)3044).

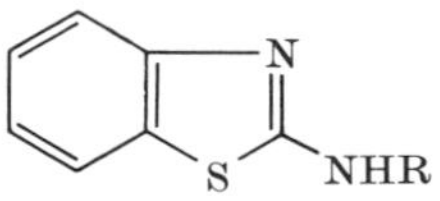

[**468**] R = H
[**469**] R = Me

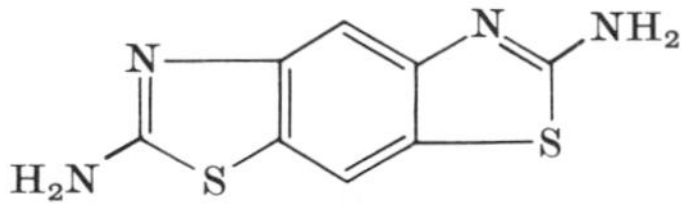

[**470**]

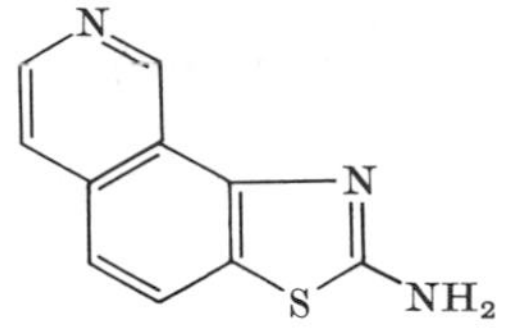

[**471**]

Tautomers [**472a**, R = H or Me] and [**472b**] have been separated (69CC4483): in solution an equilibrium is stated to occur between them (IR: $CHCl_3$ and dioxan; UV: EtOH; NMR: two NH signals, solvent not given), but in the solid state they give two different spectra.* The amide tautomer [**472a**] exhibits νC=O at 1720 cm^{-1} and the imide form [**472b**] νC=O at 1680 cm^{-1}, similar to that of the fixed compound [**473**].

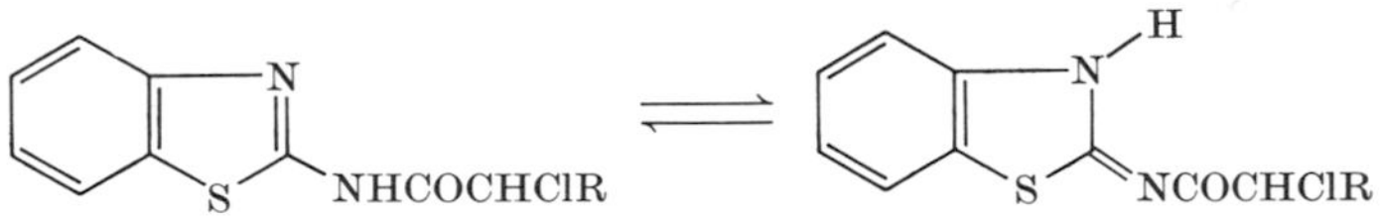

[**472a**]

[**472b**]

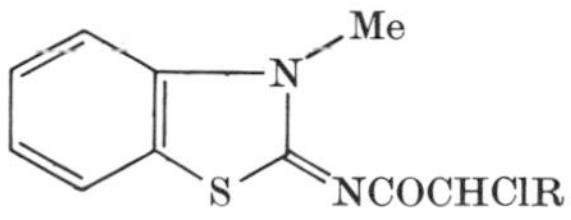

[**473**]

e. 2-Aminoimidazoles (II-71)

The previous review could report only doubtful chemical evidence on the tautomerism of 2-aminoimidazoles (II-71). Proof that the amino tautomer [**474a**] was predominant was obtained (71KG807) for 2-amino-1-methyl-4,5-diphenylimidazole [**474**]. The UV (MeOH) and IR (solid state) and pK_a values of the two corresponding fixed derivatives [**475**] and [**476**]† show that the amino tautomer [**474a**] is more stable and the last method shows K_T = [**474a**]/[**474b**] = 3×10^4.

The imino structure [**477**] was assigned on the basis of the incorrectly interpreted chemical evidence (Section 1-2), that picryl fluoride gave [**478**] (70JH1391).

* See (Section 4-7Ab) for the difficulty of differentiation by IR spectroscopy in the solid state the presence of two tautomers from two polymorphs of the same tautomer.

† It is unfortunate that this study involved a 4-phenyl derivative because of the doubtful qualities of [**476**] as a model compound (Section 1-3B).

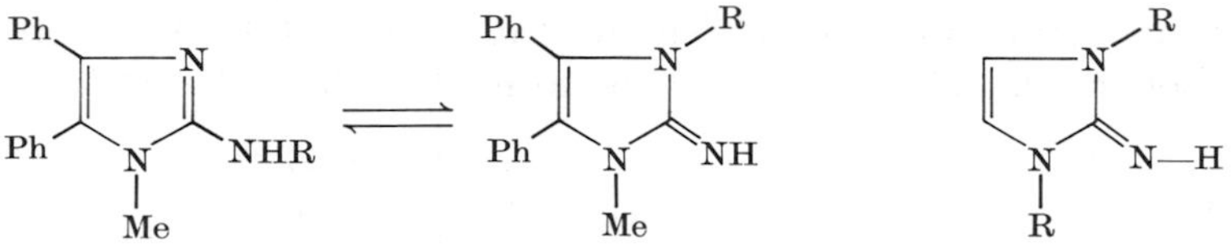

[**474a**] R = H [**474b**] R = H [**477**] R = H
[**475**] R = Me [**476**] R = Me [**478**] R = 2,4,6-trinitrophenyl

f. 2-Aminobenzimidazole

2-Aminobenzimidazole and its derivatives exist in the amino form [**479**] as demonstrated by Simonov (66RO917, 70KG1410, 71KG807): comparison with UV spectra of fixed compounds (MeOH, dioxan), IR study of the νNH_2 and δ NH_2 (solid state, $CHCl_3$), the use of Eq. (1-16) (Section 1-4A) ($K_T = 1.6 \times 10^2$ in 50% EtOH for [**479**, R = Me]) and the calculation of the π-energy of each tautomer (Section 1-7B) all favour the amino structures [**479**]–[**482**]. Another more elaborate theoretical study (74CP731) also favoured the amino tautomer and the protonation on the "pyridinic" nitrogen.

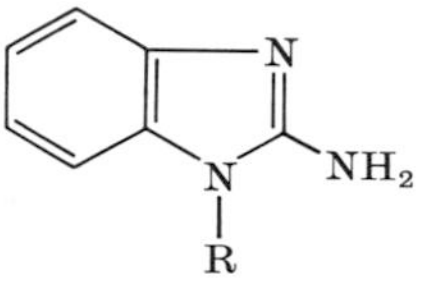

[**479**] 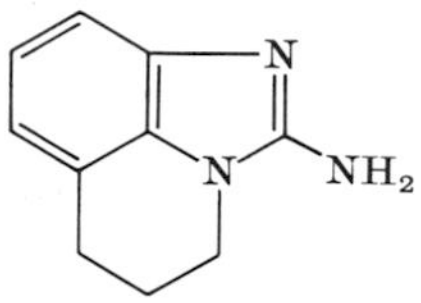 [**480**]

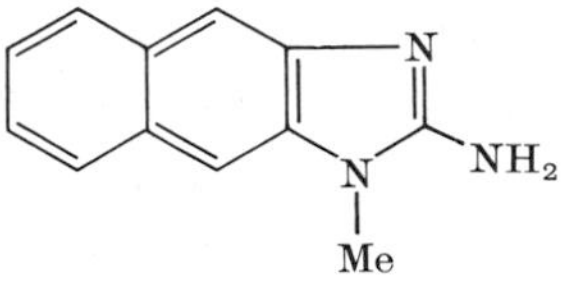

[**481**] 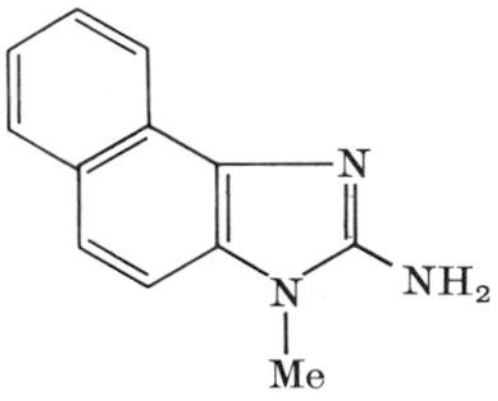 [**482**]

2-Amino-5,6-dihydro-4*H*-imidazo[4,5,1-*ij*]quinoline [**480**] was initially (63JO2581) assigned an imino structure by analogy of the UV spectra with that of the 2-oxo analogue (Section 4-3Bd), but later the amino structure [**480**] was established by IR ($CHCl_3$, partial deuteriation) (66J(B)726) and UV (EtOH) (70KG1410) spectroscopy. A UV study of the tautomerism of 2-aminobenzimidazole was recently repeated (71JO3469) and reaches the same conclusion: predominance of the amino tautomer [**479**, R = H].

The amino form [**482/1a**] persists in the system [**482/1**] where hydrogen-bonding could favour the imino form [**482/1b**] (73KG807).

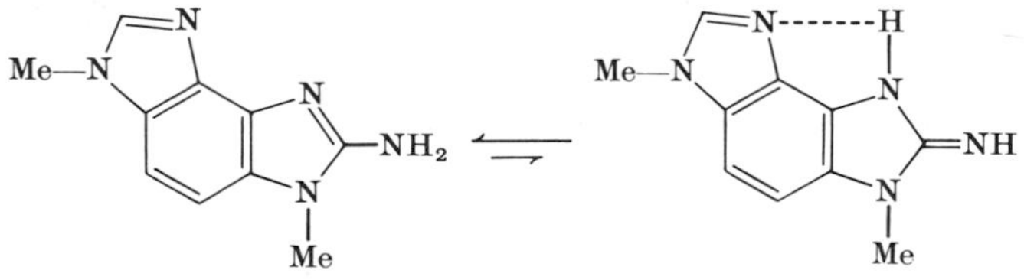

[**482/1a**] [**482/1b**]

2,3-Dihydro-1(9)*H*-imidazo[1,2-*a*]benzimidazoles can exist in two tautomeric forms [**483a**] and [**483b**]: comparison of the UV spectra (MeOH) and the determination of the pK_a values of the fixed derivatives (44% ethanol) (69JH655) show the predominance of the amino form [**483a**] ($K_T = 1.9 \times 10^2$).

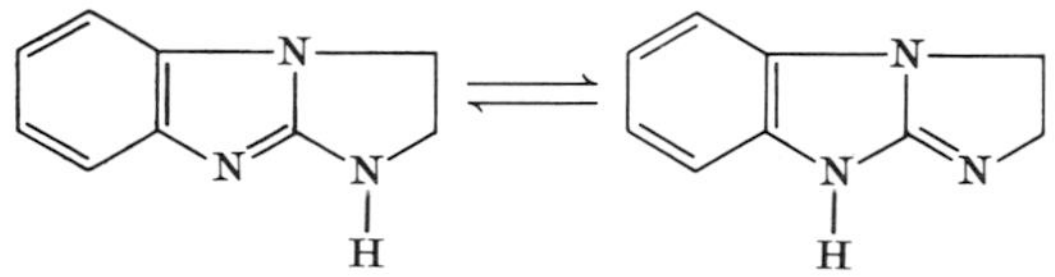

[**483a**] [**483b**]

E. Heteroatoms-1,3 and a Potential 4-(or 5-)Amino Group

a. 5-Aminooxazoles

5-Aminooxazoles, cf. [**484**], were studied by Fleury *et al.* (67BF4619, 67BF4624, 68BF4631). Of the three possible tautomers, the lack of a CH signal in NMR (DMSO) rules out forms [**484b**] and [**484c**], and thus points to the amino tautomer [**484a**], which is supported by IR spectra; νNH_2 and δNH_2 vibrations in $CDCl_3$ and KBr (67BF4619). The authors also consider a zwitterionic tautomer [**484d**], but as we have pointed out in the introduction (Section 4-5A), its presence is unlikely.

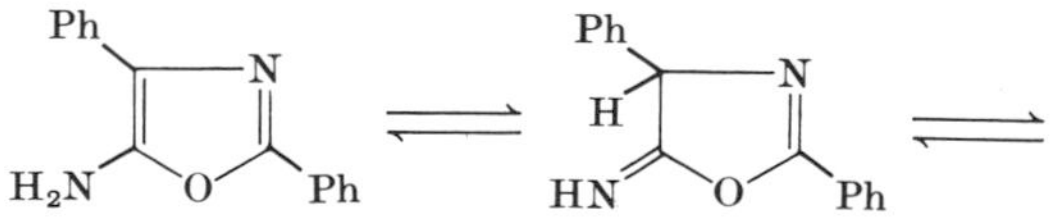

[**484a**] [**484b**]

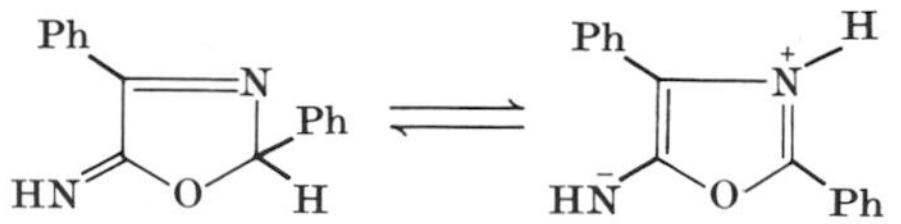

[**484c**] [**484d**]

The 5-aminooxazole conjugate acid could exist in five forms [**485a–e**] including ring–chain isomerism (Section 4-9C). NMR in trifluoroacetic acid shows the absence of CH and eliminates the structures [**485a**], [**485d**] and [**485e**] (67BF4619). UV spectra in the same solvent show

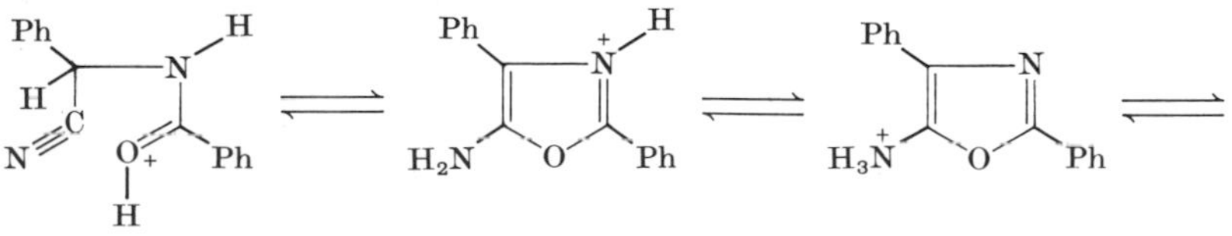

[**485a**]* [**485b**] [**485c**]

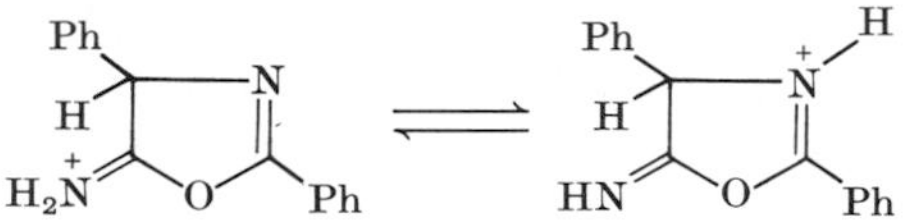

[**485d**] [**485e**]

* We have chosen to write the protonation on the amide rather than on the nitrile because of its lower basicity.

that protonation occurs on the ring nitrogen [**485b**] (as for 5-aminoisoxazoles, Section 4-5Ba), because the absorption is close to that of the base [**484**] and very different from that expected for a structure such as [**485c**] (68BF4631). However, the authors think that in chlorosulfonic acid the cation has the structure [**485c**]. They observe in the NMR a single signal at 8.5 ppm with a relative intensity of 3, attributed to the $\overset{+}{N}H_3$ of [**485c**] (67BF4619). More probably the dication [**486**] is formed in which the proton on the sp^2 ring nitrogen exchanges rapidly with the solvent. In another publication (67BF4624), the authors show that in chlorosulfonic acid the monocations [**487**] are protonated to give dications [**487/1**] ($\overset{+}{N}H_3$ at 8.5 ppm).

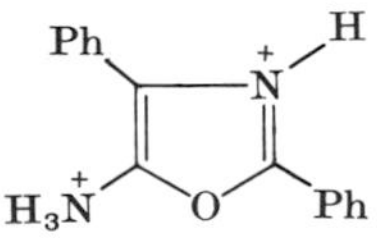

[**486**] 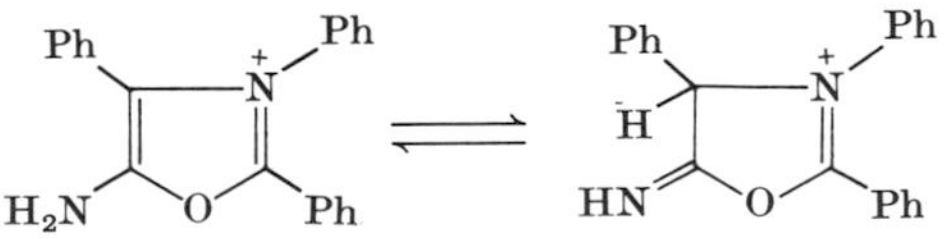 [**487a**] [**487b**]

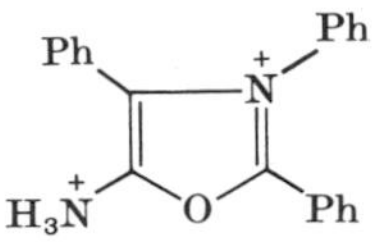

[**487/1**]

The authors write the cation [**487**] in the imino structure [**487b**] because they think that in the NMR spectrum the CH signal is hidden under the phenyl absorption. If the 4-position is unsubstituted, an amino structure [**488a**] clearly occurs (67BF4624) as no NMR signal for CH_2 in [**488b**] is found. IR study (67BF4619, 67BF4624) of the various salts does not define their structure.

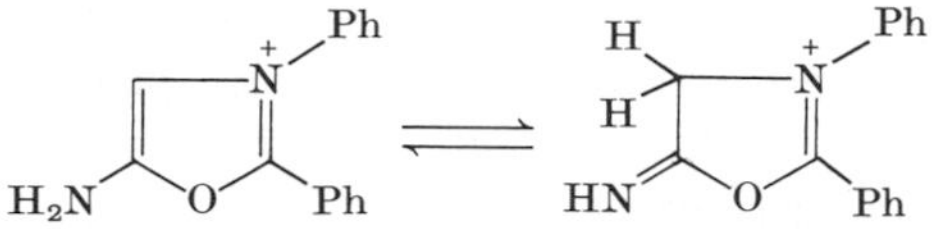

[**488a**] [**488b**]

In conclusion, it is probable that amino structures are correct for both the neutral bases [**484a**] and for conjugate acids [**485b**], [**487a**].

Very recently, a similar situation has arisen with the salts of Reissert compounds [**488/1**] and more complex structures. McEwen *et al.* (73JA2392) suggest that the equilibrium [**488/1**] contains about 50% of [**488/1a**] together with one or both of [**488/1b**] and [**488/1c**]. This most unusual situation merits further work.

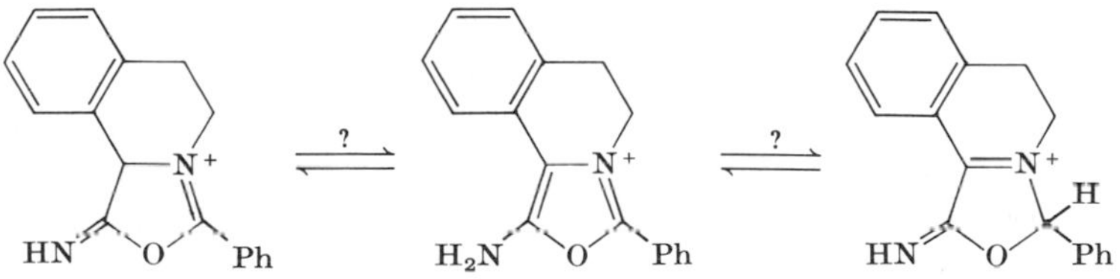

[**488/1a**] [**488/1b**] [**488/1c**]

b. 4-(or 5-)Aminoimidazoles (II-71)

No specific study is available of the tautomerism of 4- and 5-aminoimidazoles. However, published evidence (70HC(12)103) shows that they all exist in amino forms [**489a**] and [**490a**]: NMR spectra lack CH signals for tautomers [**489b**] and [**490b**] (66JA3829, 70JH1391), and IR spectra show the NH_2 bands of [**489a**] and [**490a**] (69JH53).

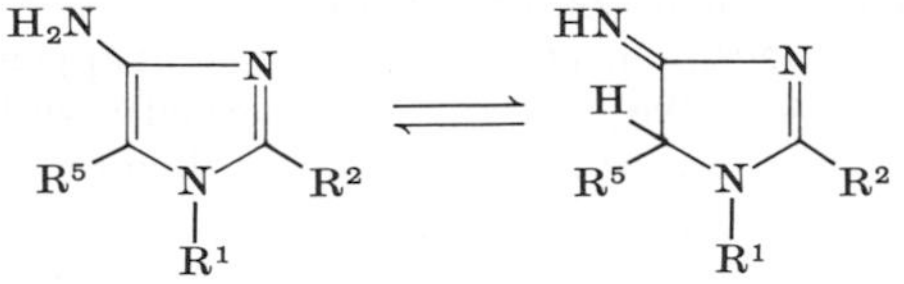

[**489a**] [**489b**]

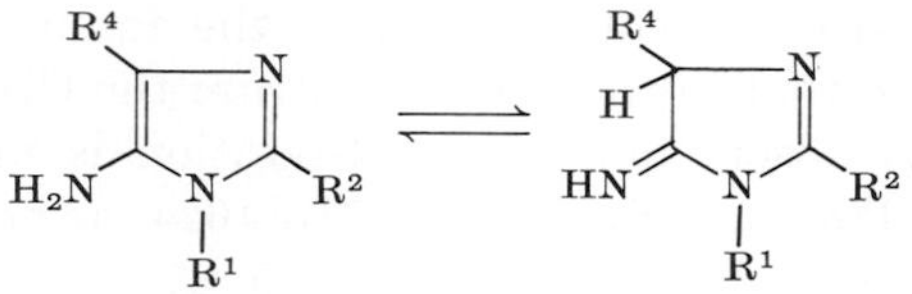

[490a] ⇌ [490b]

As for the 5-aminoisoxazoles (Section 4-5Ba) and 5-aminooxazoles (Section 4-5Ea), 4- and 5-aminoimidazoles are protonated on the ring nitrogen (70JH227). An X-ray study on compound [**491**] (71J(B)976) shows the acetamido structure.

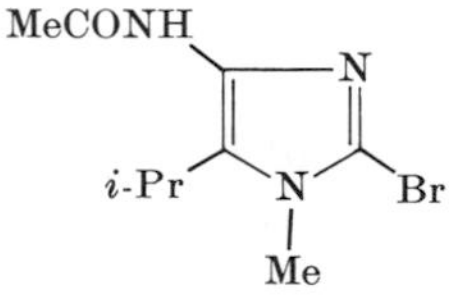

[491]

F. Compounds with Three or Four Heteroatoms

a. Aminooxadiazoles

i. *5-Amino-1,2,4-oxadiazoles.* The predominant tautomer has the amino [**492**] or methylamino [**493**] structure as shown by Najer *et al.* (68CO(266C)628, 68CO(266C)1587) and by Sélim and Sélim (67BF1219, 69BF823). The second group used the partial deuteriation IR method (solvent: $CDCl_3$) (Section 1-6C) to prove the amino form [**492**] (68CO-(266C)1587), and methyl-NH coupling in the NMR ($CDCl_3$, DMSO) to demonstrate the methylamino form [**493**] (69BF823).*

Najer *et al.* (68CO(266C)628) compared the IR (KBr), UV,† and pK_a of compounds [**492**] and [**493**] with those of the model derivatives [**494**] and [**495, 496**] to show that the amino form is predominant in the solid state, as well as in solution (EtOH or H_2O). The pK values of the fixed derivatives give K_T = [amino]/[imino] > 10^6 [Eq. (1-16), (Section 1-4Aa)]. It is necessary to assume the existence of a common cation, i.e., protonation of the amino compounds on the ring nitrogen, not proved

* These last authors consider as a proof of the existence of structures [**492**] and [**493**] the fact that in NMR spectroscopy the phenyl appears as a multiplet as does that of [**494**] (for [**495, 496**] a singlet is observed): while the conclusion is correct, this does not affect our criticisms of the method (Section 1-3B).

† The UV results are not very convincing: the hypsochromic effect observed for the N-methyl derivatives [**495**] and [**496**] in EtOH could be caused by steric interaction between the phenyl and the methyl groups (68CO(266C)628).

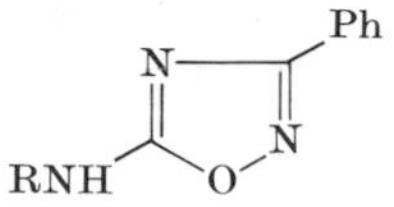

[492] R = H
[493] R = Me

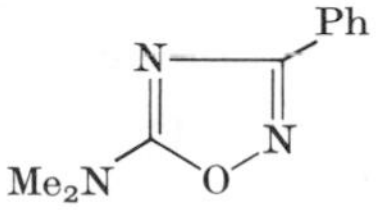

[494]

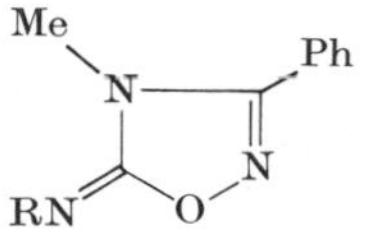

[495] R = H
[496] R = Me

by the authors. For a complicated NMR argument in favour of the NH_2 form for a complex derivative, see (72OR889).

Acylated derivatives also have an amino structure [497] as NMR spectra show an amide NH signal beyond 10 ppm (69BF823). However, for 3-phenyl-5-ureido-1,2,4-oxadiazole [498a] the authors (66J(C)1522) suggest crystallization both as the imino form [498a], with νC=O 1725 cm^{-1} (Nujol) and one of the two forms [498b] or [498c], with νC=O 1770 cm^{-1}. The evidence for these proposals is not sufficient to require this to be considered an authentic example of an imino structure.

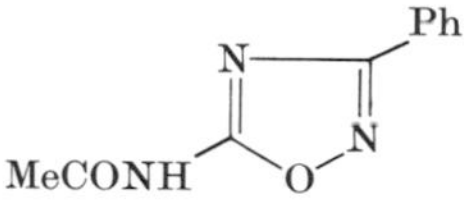

[497]

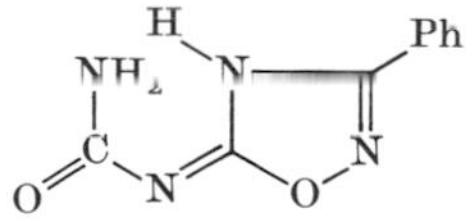

[498a] 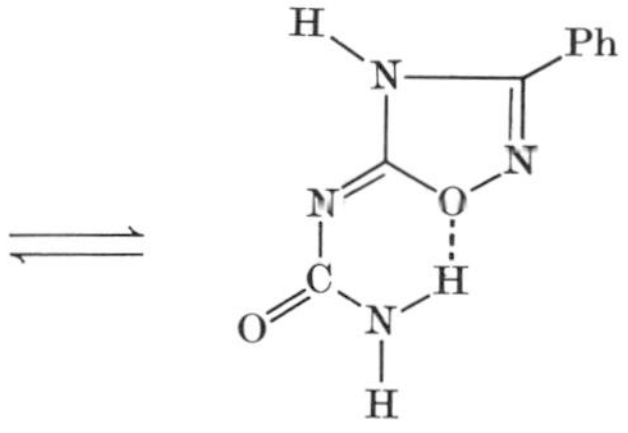 ⇌ [498b] 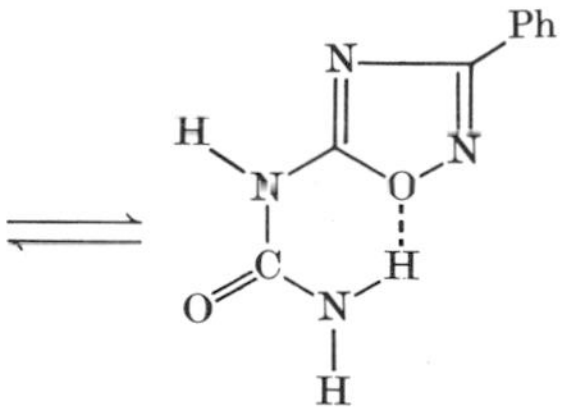 ⇌ [498c]

ii. *2-Amino-1,3,4-oxadiazoles*. The tautomerism was not considered in the previous survey, but a later review (66HC(7)183) indicates the predominance of the amino tautomer based on their UV and IR and comparison with the corresponding fixed derivatives.

Najer (68BF4568, 69BF870, 69BF874),* with the aid of fixed derivatives and using UV, fluorescence and IR spectroscopy and basicity measurements, showed that the amino forms predominate for all cases investigated: Scheme 4-14 gives the values of log K_T deduced from the basicity measurements (see also 66MC478). The amino tautomer is

* These results were the subject of preliminary communications (64CO(258)4579, 64CO(259)2868, 65CO(260)4538).

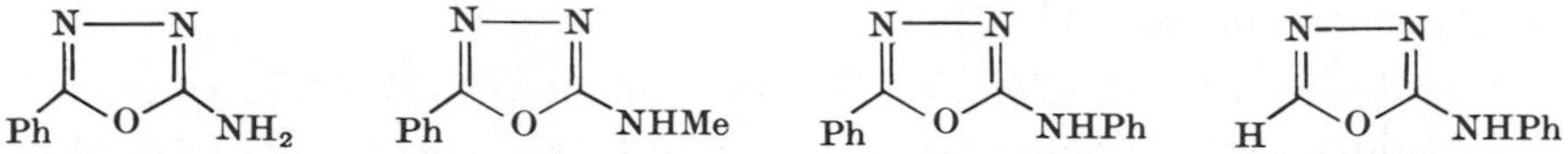

[499](*) log K_T ~5; [500] 5; [501] ~2.8; [502] ~2.9

* Partial deuteriation (I-6C) (68CO(266C)1587) supports this structure (solvent: $CDCl_3$).

SCHEME 4–14

highly favoured in the NH_2 compound, but less so for the 2-anilino-oxadiazoles [**501**] and [**502**]. A 5-phenyl substituent does not have a noticeable effect on the equilibrium. For inconclusive chemical evidence see (72JH107); we also consider that the IR and UV evidence adduced (72JH153) for a detectable proportion of the imino form of 2-amino-5-benzyl-1,3,4-oxadiazole is highly doubtful.

2-Nitrosamino compounds [**502/1**] probably exist in the amino form [**502/1c**] and/or diazohydroxide form [**502/1b**] (73J(PI)1357).

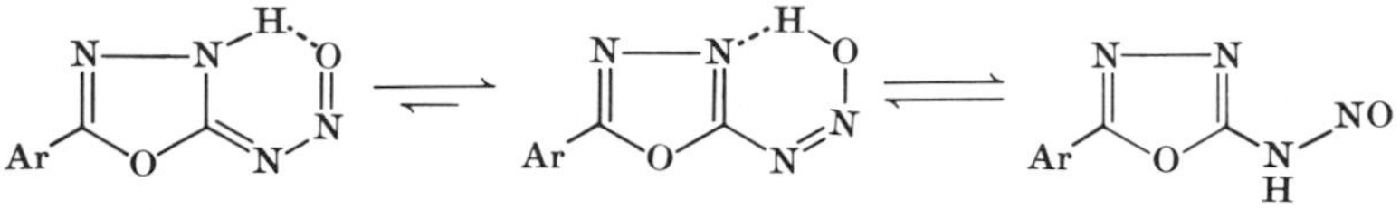

[**502/1a**] ⇌ [**502/1b**] ⇌ [**502/1c**]

iii. *3-Amino-1,2,4-oxadiazoles*. The IR spectrum of compound [**503**] (in $CHBr_3$) (66J(C)1522) exhibits two νNH_2 bands which follow Bellamy's rule (Section 1-6C).

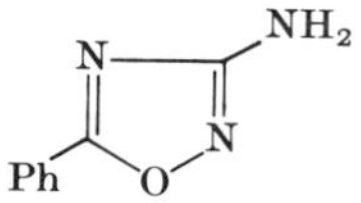

[**503**]

iv. *3-Amino-1,2,5-oxadiazoles*. IR evidence shows that a variety of 4-substituted derivatives exist in the amino form [**503/1**]; although it is implied in the publication that protonation occurs at the amino group (73PC791), this is unlikely.

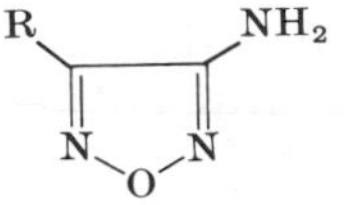

[**503/1**]

b. Aminothiadiazoles (II-72)

i. *5-Amino-1,2,4-thiadiazoles*. The previous survey (II-72) had already established the predominance of the amino tautomers [**504**] (see also 65HC(5)119). This was confirmed later by IR (solvent: CS_2) for [**504**, R = Ph] using partial deuteriation (1-6C) (68CO(266C)1587). For these compounds, two νND–H bands appear between $\nu_{as}NH_2$ and $\nu_{s}NH_2$: the authors suggest that *cis-trans* isomerism of NHD is involved.

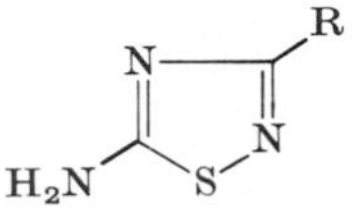

[**504**]

ii. *2-Amino-1,3,4-thiadiazoles* were previously shown (II-72) to exist in the amino form [**505**] in the solid state and in solution and basicity measurements gave K_T ca. 10^5. All the subsequent work supports amino predominance (68HC(9)165).

Najer confirmed structure [**505**] by IR ($CDCl_3$) of the partially deuteriated compound see (Section 1-6C) (68CO(266C)1587): as in the preceding case discussed (Section 4-5Fbi) the NHD group shows two νNH bands. For the NHPh compounds the proportion of amino tautomer [**506**] is less, $K_T = 10^2$ as for the 2-amino-1,3,4-oxadiazoles, (Section 4-5Fa). Replacement of a 5-phenyl by a proton does not significantly affect the equilibrium for [**507**], $K_T = 2.10^2$ (64CO(259)-

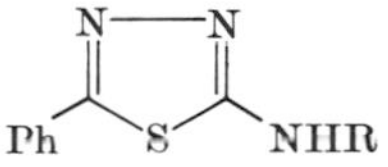

[**505**] R = H
[**506**] R = Ph

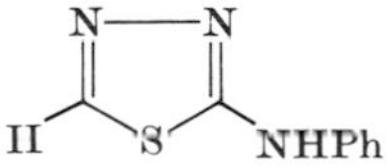

[**507**]

3563, 65CO(261)766). The K_T values are calculated from basicity measurements, with the reasonable assumption (68HC(9)165) of a common cation.

Recent work (72AS459) shows that the aminothiadiazoles [**508**] and the aminoselenadiazoles [**509**] have similar structures: NMR spectra in DMSO of both exhibit *NH*–*CH* coupling (see Introduction 4-5A).

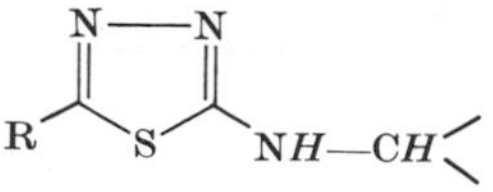

[**508**]

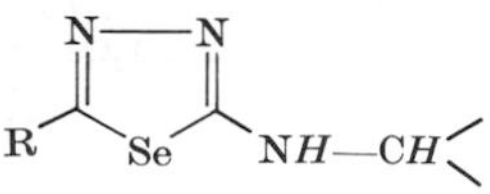

[**509**]

The amino tautomer is still predominant in the 5-alkoxy series [**510**] (Section 4-6Fh) (64AS174) or if the exocyclic nitrogen is part of a hydrazone group [**511**] as shown by UV spectroscopy and confirmed by a LCAO–MO calculation (64AS871). However, a sulfonamide group shifts the equilibrium towards the imido tautomer [**512**] (IR study in

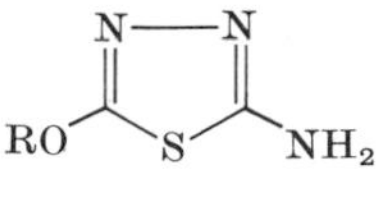

[**510**] 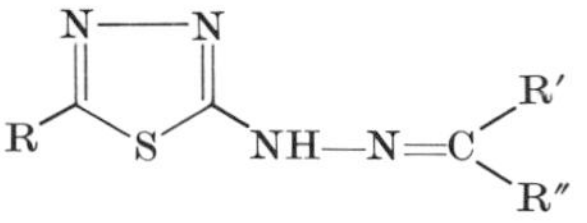 [**511**] 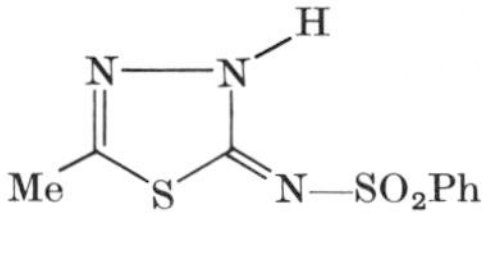 [**512**]

solid state: 66CT756), but a nitroso group apparently does not (73J-(PI)1357). Such effects had already been outlined in the previous survey (II-77) when strongly electron-withdrawing groups are bonded to the NH group.

iii. *3-Amino-1,2,4-thiadiazoles*. Goerdeler and Martens (70CB1805) demonstrated the amino form [**513**] by IR evidence (KBr, $CHCl_3$) the presence of νNH_2 and δNH_2 bands, and amino structure [**514**] by NMR (C_6H_6) detection of a Me–NH coupling (which disappears in $CDCl_3$).

iv. *3-Amino-1,2,5-thiadiazoles*. A survey by Weinstock and Pollak (68HC(9)107) gives information on these compounds from Indiana University theses that suggest the amino structure [**515**] and the ring nitrogen protonated structure for cation [**516**].

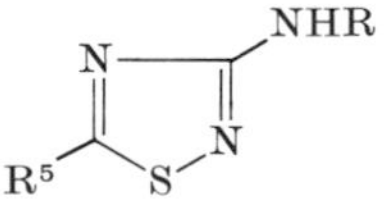

[**513**] R = H
[**514**] R = Me
[**515**] 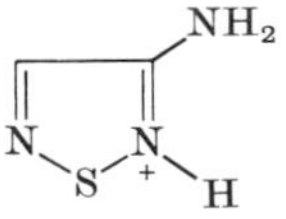 [**516**] 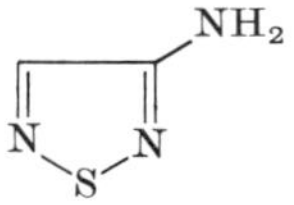

v. *5-Amino-1,2,3-thiadiazoles*. The amino form [**517**] is shown by $\nu_{as}NH$ and $\nu_s NH$ bands in the IR ($CHCl_3$), and comparison of the UV spectra of [**518**], [**519**] and [**520**] demonstrates structure [**518**] for the first (66CB1618).

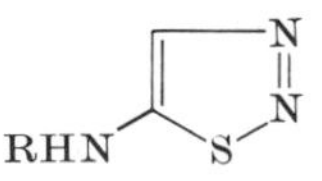

[**517**] R = H
[**518**] R = CO_2Ph
[**519**] 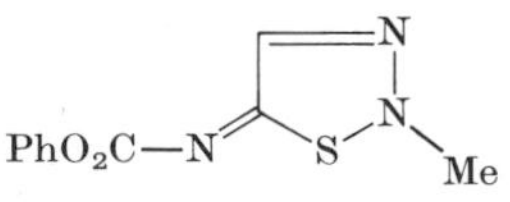 [**520**]

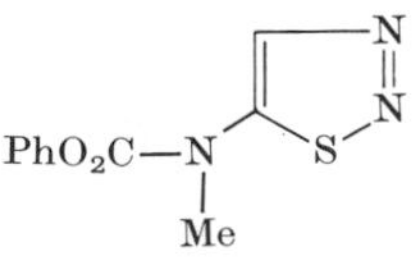

IR and NMR indicate the amino structure for 5-amino-4-ethoxy-carbonyl-1,2,3-thiadiazole (71T5873).

c. Amino-s-triazoles (II-73)

As summarized below, UV and IR studies indicate the amino forms for 5-amino-1,2,4-triazoles [**521**], 2-amino-1,3,4-triazoles [**522**] and 3-amino-1,2,4-triazoles [**523**]. Compounds [**524**] also exist in amino forms, but the annular tautomerism is not yet defined.

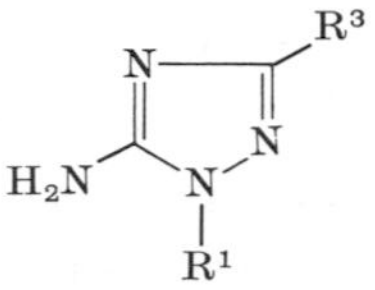

[**521**] 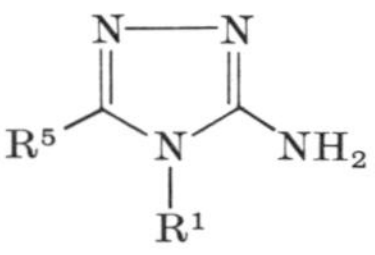 [**522**] 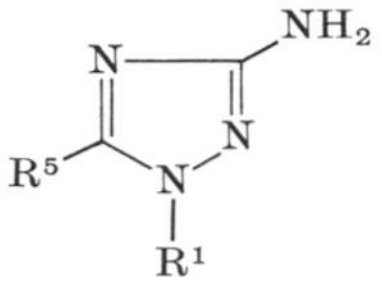 [**523**] 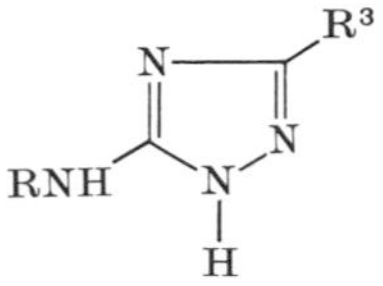 [**524**]

Similarity with the UV absorptions (H_2O, EtOH, $CHCl_3$) of NMe_2 forms is good evidence for NH_2 structures as it is known that fixed imino forms absorb completely differently (62MI393, 65AS1191). In IR spectroscopy (solid state, $CHCl_3$, DMSO) characteristic ν_{as}NH, ν_sNH and δ NH_2 bands, identified by deuteriation (62MI401, 65AS1191, 67MI186), partial deuteriation, and Bellamy's rule (Section 1-6C), (73BF(2)1849) have been used. The NMR spectrum (DMSO) of compound [**524**, R = Me] shows 5 Hz coupling between the methyl and the NH proton (73BF(2)1849). A preliminary report gives $K_T = 3 \times 10^6$ in favour of the amino form for 3-amino-1,2,4-triazole (69CG58).

Protonation on the ring nitrogen, to give cation [**525**], has been demonstrated by UV (62MI393) and IR(62MI401) spectra. The acylated derivatives [**526**] and [**527**] exist in the amino structures (62MI411); nitrosamino derivatives probably favour the amino form, cf. [**502/1c**] (73J(PI)1357).

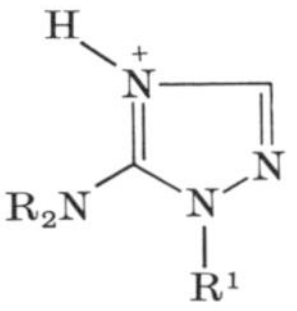

[**525**] 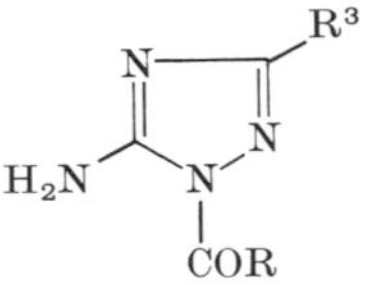 [**526**] 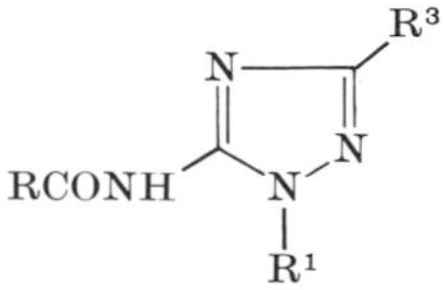 [**527**]

d. Amino-v-triazoles

Of the three possible tautomers (excluding zwitterionic forms) [**528a–c**] for 5-amino-1-substituted-1,2,3-triazoles,* NMR results rule

* Care is necessary in this series not to confuse prototropic equilibrium with Dimroth rearrangement (Section 4-9B) (II-74) [see more recently (70J(C)230)].

out the CH tautomer [**528b**] (69J(C)2379, 70J(C)230, 70JH1159, 71CB3510), and νNH_2 and δNH_2 ($CHCl_3$) indicate the amino form [**528a**] (71CB3510; see also 71T5873 and 74J(PII)1849) for the X-ray structure of [**528a**, R^1 = H, R^4 = $CONH_2$]).

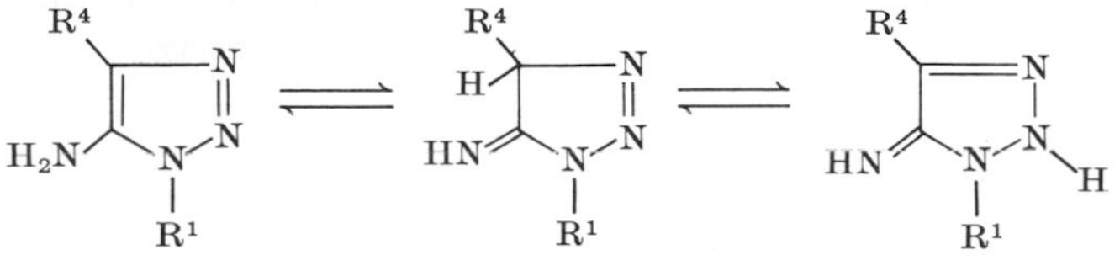

[**528a**] [**528b**] [**528c**]

The NH to CH_2 coupling in the NMR (DMSO) demonstrates an amino structure for *N*-unsubstituted alkylamino-*v*-triazoles [**529**] (70J(C)230), but the annular tautomerism remains undefined. Albert (69J(C)2379) showed by UV spectra that 5-amino-1,2,3-triazole cations were formed by protonation on a ring nitrogen.

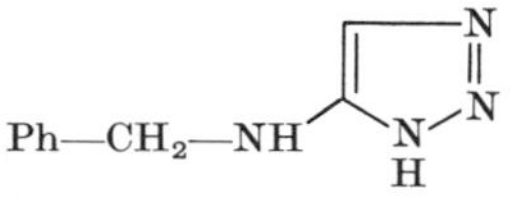

Ph—CH₂—NH

[**529**]

e. 5-Amino-1,2,3,4-thiatriazoles (II-74)

Convincing evidence in the previous survey (II-74) favoured the amino tautomers [**530**] and [**531**]. Contemporary with an NMR study (61ZA(181)447) cited in the previous survey (II-74), a paper (61ZA-(181)487; also cited in 64HC(3)263) demonstrated by NMR Me–N–H coupling the amino structure [**531**, R = Me]: and later IR work agrees (64AS566; also cited in 64HC(3)263, 71PM(4)265).

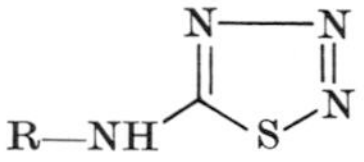

[**530**] R = H
[**531**] R = alkyl

Kauer and Sheppard (67JO3580) suggest that the imino forms [**532b**] and [**532c**] contribute significantly to the equilibrium for 5-anilinothiatriazoles. They mention similar values of σ parameters for the aminothiatriazole group and those of the dichloroazomethine (—N=CCl_2) and isocyanate (—N=C=O) groups. A strong band mentioned at 1550–1600 cm^{-1} (KBr), was later assigned to the amino form [**532a**] (71PM(4)265, p. 337). Although (see Section 4-5G) placing

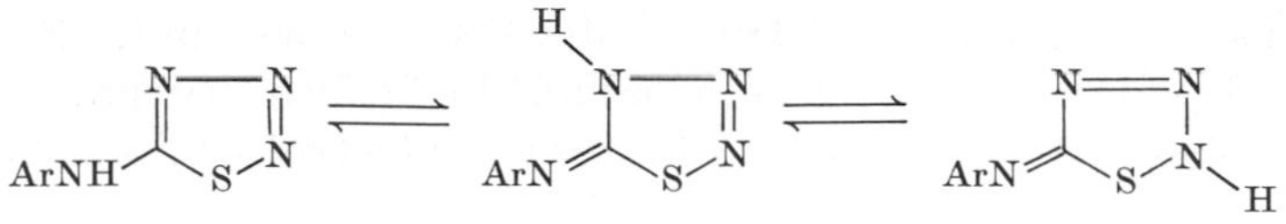

[**532a**] [**532b**] [**532c**]

an amino group between two heteroatoms, and also the conversion of NH_2 to NHPh, should lessen the preponderance of amino form [**532a**], we do not consider the arguments of Kauer and Sheppard sufficient to demonstrate the presence of an appreciable amount (> 1%) of tautomers [**532b**] and [**532c**].

f. Aminotetrazoles (II-74)

The systematic predominance of the amino form has only recently been put beyond doubt (71J(B)2355), after claims to the contrary (70CH1096).

Good evidence has long been available for the solid state; X-ray studies established the amino structures [**533**] (II-74) and [**534a**] (Section 4-1Faiii: 67CX(22)308) and an IR study (65RO2236) proved amino structures [**533**], [**534a**] and [**535a**]. The *N*-benzyl derivatives analogous to [**533**] and [**535a**] show δNH_2 bands in the IR spectra (KBr) (71J(C)703).

A dubious IR conclusion in favour of the imino forms [**534b**] and [**535b**] in Nujol (67MI89) has been criticized (71PM(4)265, p. 428).

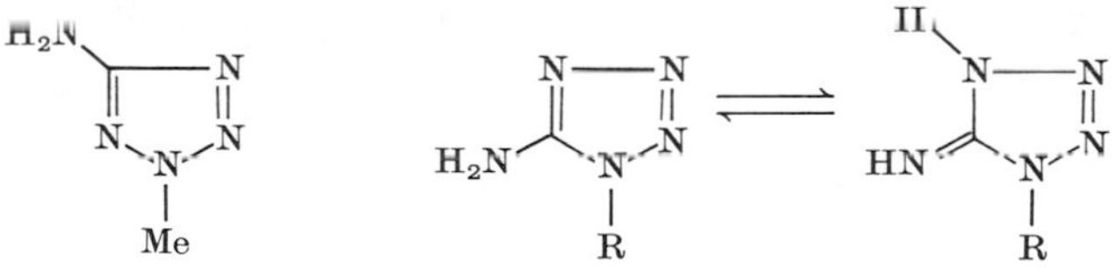

[**533**] [**534a**] R = H [**534b**] R = H
[**535a**] R = Me [**535b**] R = Me

Butler (69CH405, 70CH1096, 70J(B)138) erroneously claimed to have shown by NMR a mobile equilibrium between comparable amounts of the amino [**536a**] and the imino [**536b**] form in DMSO, addition of water was stated to shift the equilibrium towards the amino form [**536a**]. That result was surprising: [**536**] would be the only simple aminoazole to exist significantly (35%) in the imino form. It assumes unprecedented slow exchange of the proton between the two cyclic and exocyclic nitrogen atoms in the NMR time scale (Table 1-4) (Section 1-5Aa). Such a shift in tautomeric equilibrium by a small proportion of water in the DMSO-d_6 is unlikely. These contradictions were shown (70CH706, 71CH846, 71J(B)2355) to be due to D_2O in the DMSO-d_6, giving a mixture of [**536a**] and [**537**]. Basicity measurements (Section 1-4A) showed (71J(B)2355; see also 65RO2236) the surprisingly

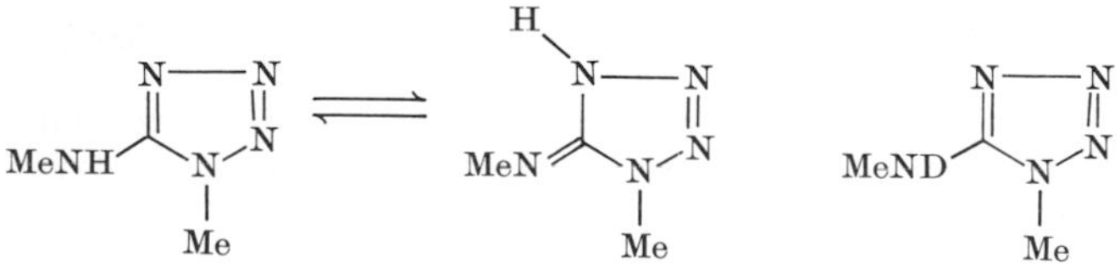

[**536a**] [**536b**] [**537**]

high value of K_T = [**536a**]/[**536b**] = 10^9. For the 1-methyl-4-amino-1,2,3,5-tetrazole [**533**] the stability of the amino form is even greater (65RO2236). For further recent spectral work, see (73BF(2)1854).

However, acyl and sulfonamido substituents at the exocyclic nitrogen do indeed favour the imino tautomer, as already shown by Sheinker at the time of the previous review (II-77). Subsequent detailed studies of Sheinker and Shchipanov *et al.* (65RO2236) and Shchipanov *et al.* (66RO350, 66RO376, 68KG215, 69KG923) show that for 1-substituted-4-acylamino-1,2,3,5-tetrazoles the 4-RCONH structure [**538**] predominates for all cases studied; 4-CH_3CONH, 4-CCl_3CONH and 4-CF_3CONH. However, for 1-substituted 5-amino-1,2,3,4-tetrazoles [**539**], either the amino [**539a**] or the imino form [**539b**] is predominant or both coexist depending on the nature of the substituent. For 5-acetamido the amino form [**539a**] predominates, the 5-CCl_3CONH compound exists as a mixture of [**539a**] and [**539b**], whereas the 5-CF_3CONH and 5-$ArSO_2NH$ derivatives occur mainly in the imino forms [**539b**]. The amido structure [**539a**, R = CH_3CO] was established

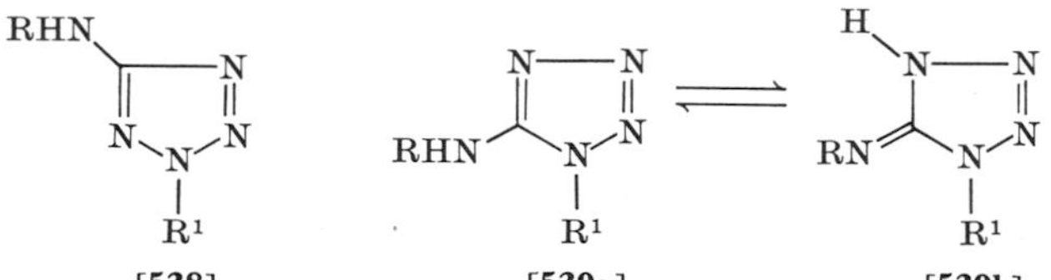

[**538**] [**539a**] [**539b**]

by IR spectroscopy for the solid state (65RO2236). In DMSO, Butler (69J(B)680) suggested a chelated imido structure [**539b**, R = CH_3CO] because of the lesser sensitivity of the NMR chemical shift of the NH proton to dilution in CCl_4 than the NH_2 protons of the aminotetrazoles [**533**] and [**535a**]; this argument is not fully convincing.

Apparently 4-nitrosamino- and 5-nitrosamino-1-alkyl-tetrazole both probably exist as the nitrosamide form, as, e.g., [**539/1c**] with contributions from the hydroxyazo form [**539/1b**], but not in the nitrosimino

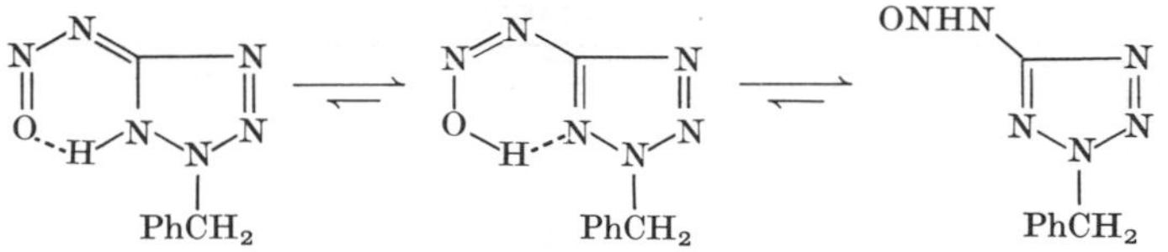

[539/1a] [539/1b] [539/1c]

form [**539/1a**] (73J(PI)1357). A cyano group stabilizes the imino form as was established (64JO650) by IR in the solid state for [**539b**, R = CN, R^1 = H, Me].

G. Compounds with Potential Amino Groups: Conclusions

The results already discussed for amino derivatives are gathered in Table 4-29. The majority of the possible cases have already been investigated (34 out of 46). The noteworthy striking result is the predominance of the amino tautomer for all the cases studied. By extrapolation, the predominance of the amino form for all aminoazole derivatives bearing NH_2 or NHR groups with R = alkyl or aryl seems justified.

Nevertheless it is possible to discern the influence of several factors upon the equilibrium constant K_T. As K_T is always $\geq 10^2$, only the basicity method (Section 1-4A) can be used for this purpose.

a. Influence of the Nature of the Heterocycle

The imino forms are relatively the most stable for those systems for which oxo and thione forms are most favoured. The 2-aminooxazoles (Section 4-5Da), 2-aminothiazoles (Section 4-5Dc) and 2-aminoimidazoles (Section 4-5De) possess K_T values around 10^4, and K_T is even smaller for the corresponding benzo derivatives: 2-aminobenzimidazole (Section 4-5Df) ca. 2×10^2. It is interesting to note the decrease in the stability of the amino tautomer from 4-amino- to 5-aminotetrazole, which can also be discerned in the corresponding acyl derivatives (Section 4-5Ff).

b. Influence of Annelation

The last section indicated the destabilization of the amino form through annelation. However, such annelation is not sufficient for an imino structure to predominate: even the amino-2,1-benzoisoxazoles [**540**] (Section 4-5Bb) exist in the amino form.

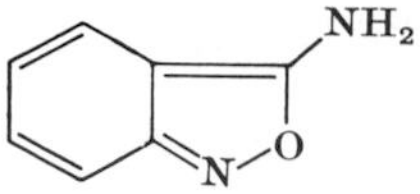

[540]

TABLE 4-29
Aminoazole Tautomerism[a]

A	X	Y	Z	Compound	H–Z—Y=X, HN=C–A (imino, H on Z)	Z—Y=X, H_2N–C–A (amino)	Z=Y–X–H, HN=C–A (imino, H on X)	Section
O				α-Aminofurans				3-4A
S	C	C	C	α-Aminothiophenes	−	+ + +	−	3-4Ba
NR				α-Aminopyrroles	−	+ + + ?	−	3-4Ca
O				5-Aminoisoxazoles	−	+ + +	−	4-5Ba
S	N	C	C	5-Aminoisothiazoles	−	+ + +	−	4-5Bd
NR				5-Aminopyrazoles	−	+ + +	−	4-5Bg
O				2-Aminooxazoles	−	+ + +	−	4-5Da
S	C	C	N	2-Aminothiazoles	−	+ + +	−	4-5Dc
NR				2-Aminoimidazoles	−	+ + +	−	4-5De
O				5-Aminooxazoles	−	+ + +	−	4-5Ea
S	C	N	C	5-Aminothiazoles				
NR				5-Aminoimidazoles	−	+ + +	−	4-5Eb
	N	C	N	5-Amino-1,2,4-oxadiazoles	−	+ + +	−	4-5Faiii
O	C	N	N	2-Amino-1,3,4-oxadiazoles	−	+ + +	−	4-5Faii
	N	N	C	5-Amino-1,2,3-oxadiazoles				
	N	C	N	5-Amino-1,2,4-thiadiazoles	−	+ + +	−	4-5Fbi
S	C	N	N	2-Amino-1,3,4-thiadiazoles	−	+ + +	−	4-5Fbii
	N	N	C	5-Amino-1,2,3-thiadiazoles	−	+ + +	−	4-5Fbv
	N	C	N	5-Amino-1,2,4-triazoles	−	+ + +	−	4-5Fc
NR	C	N	N	2-Amino-1,3,4-triazoles	−	+ + +	−	4-5Fc
	N	N	C	5-Amino-1,2,3-triazoles	−	+ + +	−	4-5Fd
S	N	N	N	5-Amino-1,2,3,4-thiatriazoles	−	+ + +	−	4-5Fe
NR	N	N	N	5-Amino-1,2,3,4-tetrazoles	−	+ + +	−	4-5Ff

[a] See key to table in Table 4-25.

TABLE 4-29 (*continued*)

Compound	Y=Z–Ä–X=C(–NH_2)–Y (ring)		Section
	Y=Z–A–X=C(–NH_2)–Y (ring)	Y=Z–A–X(–H)–C(=NH)–Y (ring)	
β-Aminofurans	+ + ?	−	3-4A
β-Aminothiophenes	+ + +	−	3-4Ba
β-Aminopyrroles			
3-Aminoisoxazoles	+ + +	−	4-5Bc
3-Aminoisothiazoles	+ + +	−	4-5Be
3-Aminopyrazoles	+ + +	−	4-5Bh
4-Aminoisoxazoles	+ + +	−	4-5Ca
4-Aminoisothiazoles	+ + +	−	4-5Cb
4-Aminopyrazoles	+ + +	−	4-5Cc
4-Aminooxazoles			
4-Aminothiazoles			
4-Aminoimidazoles	+ + +	−	4-5Eb
3-Amino-1,2,5-oxadiazoles			
4-Amino-1,2,3-oxadiazoles			
3-Amino-1,2,4-oxadiazoles	+ + +	−	4-5Faiii
3-Amino-1,2,5-thiadiazoles	+ + +	−	4-5Fbiv
4-Amino-1,2,3-thiadiazoles			
3-Amino-1,2,4-thiadiazoles	+ + +	−	4-5Fbiii
3-Amino-1,2,5-triazoles			
4-Amino-1,2,3-triazoles			
3-Amino-1,2,4-triazoles	+ + +	−	4-5Fc
3-Amino-1,2,4,5-thiatriazoles			
4-Amino-1,2,3,5-tetrazoles	+ + +	−	4-5Ff

c. Influence of the Nature of R in NHR

If the amino group is substituted by alkyl, it does not much affect the equilibrium constant. In NH-aryl compounds, an appreciable decrease in the stability of the amino form is observed, though it is still predominant [2-amino-1,3,4-oxadiazoles (Section 4-5Fa); 2-amino-1,3,4 thiadiazoles (Section 4-5Fb)]. However, if R = COR′, SO_2R', CN or NO, imino forms can become comparable and of even greater stability than the amino forms, as was already becoming clear at the time of the previous review (II-77, II-78). A more complex case of similar substituent influence has been found by Russian workers in heteroaryl formazans of type [**540/1**] (70RO619, 70RO1332, 73KG699); imino forms are implicated e.g., [**540/1b**].

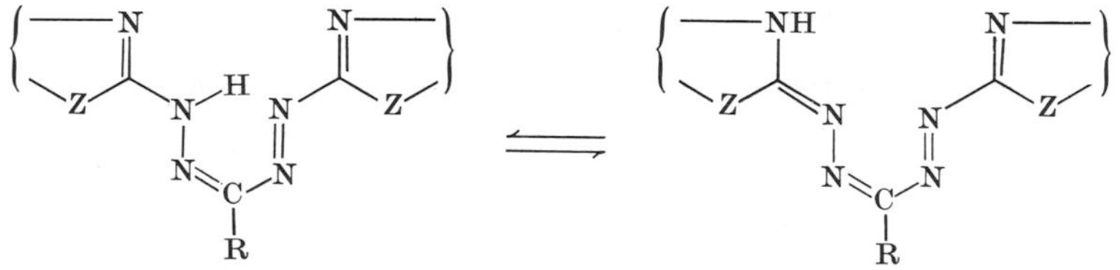

[**540/1a**] ⇌ [**540/1b**]

d. Site of Protonation of the Amino Derivatives

Protonation occurs on the ring nitrogen in all cases, unless the amino group is β to the heteroatoms (this is the case in 4-aminoisoxazoles [**541**, A = O], Section 4-5Ca).

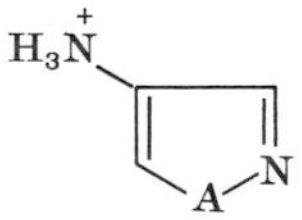

[**541**]

H. Compounds Containing Potential Hydroxy, Mercapto or Amino Groups: General Conclusions

A comparison of Tables 4-25, 4-28 and 4-29 allows some overall conclusions: the most obvious is the decrease of the relative stability of the "aromatic" tautomer when BH changes successively from CH_3 (Section 4-8), NH_2, SH to OH.

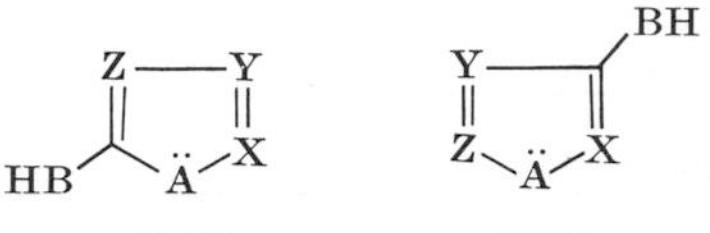

[**542**] [**543**]

The position of the mercapto group between the NH_2 and OH is not very clearly established because of the lack of results on the mercapto derivatives (Table 4-28). Comparative studies which examine the effect of the nature of BH on the tautomeric equilibrium keeping the same heterocycle and the same experimental conditions (solvent, concentration, temperature) are particularly few. Future studies of this type would be welcome and would complement those already done which have dealt with substituent effects or with changes in external factors on the same heterocycle bearing the same function.

The underlying theoretical reasons which favour one or the other tautomer had until recently received only scanty theoretical studies (Section 1-7). Recently (72UP1) a more fundamental approach has been undertaken.

As an example of this approach, Fig. 4-1 describes the case of pyrazole derivatives substituted by a group BH in the 5-position. The Y axis gives the binding energy in β units, E_f: the higher is the energy, the more stable is the tautomer; the X axis is a parameter δ_B characteristic of BH. This figure allows the following conclusions:

i. For BH = CH_3 [**544**], the methylpyrazole tautomer [**544b**] is highly predominant in agreement with experiment (Section 4-8).

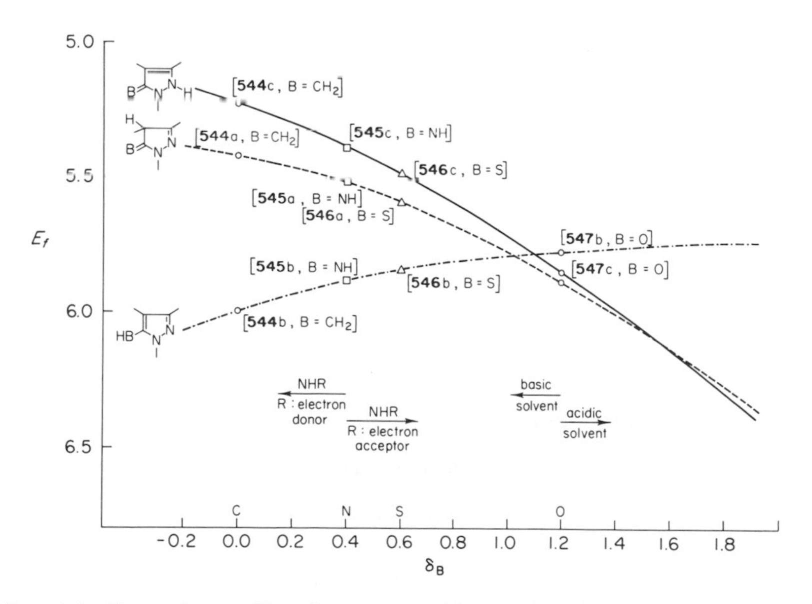

FIG. 4-1. Dependence of binding energy, E_f (β units), for pyrazole substitution with a potentially tautomeric 5-substituent, on the substituent δ_B values.

ii. For BH = NH_2 [**545**] the 5-aminopyrazole tautomer [**545b**] is still predominant (Section 4-5Bg and Table 4-29). For BH = NHR, if R is an electron donor (such as a methyl group), the stability of the amino form is further increased (δ_b decreases), whereas if R is an electron attractor (δ_b increases), the amino form, though still predominant, will be relatively less favoured (Section 4-5Bi).

iii. For BH = SH [**546**] experiment (Section 4-4Ad and Table 4-28) shows that forms [**546b**] and [**546c**] are present, whereas from Fig. 4-1 appears a sequence of decreasing stability: [**546b**] > [**546a**] > [**546c**]. It seems that the stability of the CH form [**546a**] has been overestimated by the calculation.

iv. For BH = OH [**547**] the stabilities of the three forms are comparable, the OH form [**547b**] being slightly disfavoured in agreement with experiment (Section 4-2G and Table 4-25). A solvent may be considered as a perturbation in δ_d; a basic solvent decreases δ_d, which favours the OH form [**547b**], while an acidic solvent destabilizes the OH form as δ_d increases. These effects are consistent with experiment (Section 4-2G).

Other calculations (72UP1) have shown that the influence of the nature of A on the tautomeric equilibrium is small (see Section 4-3F and Table 4-25) whereas that of X, Y and Z is important.

This global, but necessarily oversimplified, approach to azole tautomerism using semi-empirical MO methods thus affords good results. Extended to the general case of a five-membered heterocycle with a substituent BH, it should allow the calculation for each tautomer of two hypersurfaces $E = f(\text{A, B, X, Y, Z})$, according to the α or β position of the substituent BH relative to the heteroatom A. Then the stability of the various tautomers could easily be assessed, leading to specific K_T values. In this treatment, A can be NR, S or O; B to CHRR′, NHR, SH or OH, and X, Y, Z to CR or N.

6. Compounds Containing Two Potential Tautomeric Groups

A. Introduction

In the previous survey, discussion of azoles containing two potentially tautomeric groups was scattered among the discussion of compounds with a single tautomeric group. This is no longer feasible in view of the very large amount of subsequent work. Much of this later work is incomplete, and the discussion has to be considerably more tentative than for the monofunctional compounds, particularly as the number of possible forms is often quite large.

The classification adopted—by nature of the functional group—is designed to enable precise classification. Compounds which can be considered derived from difunctional tautomer derivatives by alkylation are considered under the relevant difunctional compound. Thus alkoxyisoxazolones are considered under dihydroxy derivatives, etc.

B. Two Potential Hydroxyl Groups

a. 3-Hydroxyisoxazol-5-ones

These compounds were unknown at the time of the previous survey. Recently (71IS659), 3-hydroxy-4-phenylisoxazol-5-one [**548**] and three completely fixed derivatives [**549**], [**550**] and [**551**] were described. Six tautomeric species [**548a–f**] have to be taken into consideration. The authors conclude from IR (KBr, DMSO), UV (EtOH and other solvents for the tautomeric species) and NMR (DMSO or $CDCl_3$) in favour of the zwitterionic form [**548f**] for the solid state and OH/NH forms [**548a**] or **548c**] for aqueous solution, each case in a dimeric structure. Their

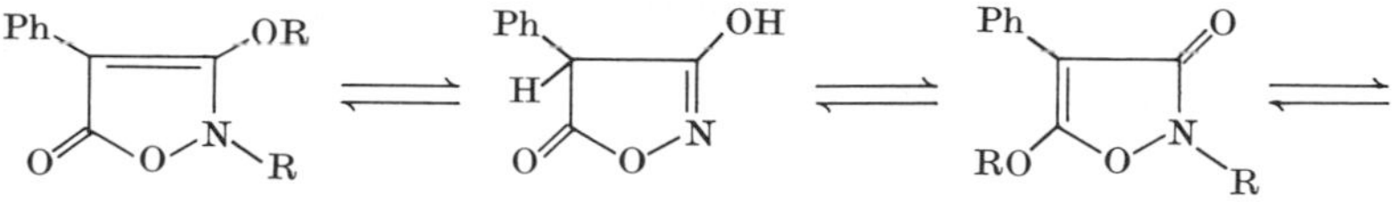

[**548a**] R = H
[**549**] R = Me

[**548b**]

[**548c**] R = H
[**550**] R = Me

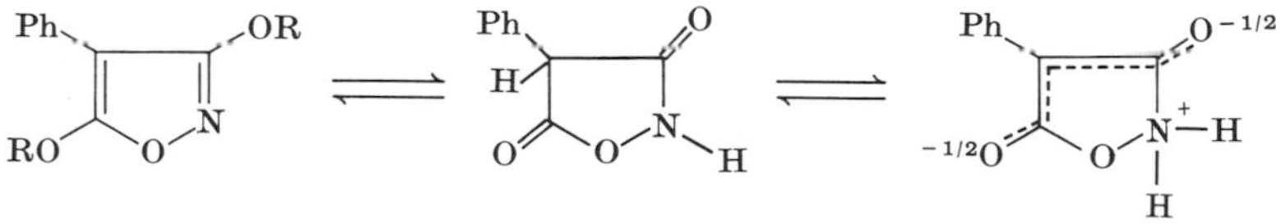

[**548d**] R = H
[**551**] R = Me

[**548e**]

[**548f**]

reasoning seems obscure, and they do not use their dimethylated derivatives for comparison; another difficulty arises because of the use of different solvents in the different methods. In the sequel, we reconsider their data in an attempt to clarify the situation.

The lack of CH proton signal eliminates structures [**548b**] and [**548e**] for DMSO. For the solid state, IR spectroscopy alone cannot give a definite answer as only associated νOH absorptions and a weak νC=O band can be clearly assigned. In DMSO the IR spectrum of [**548**] exhibits a strong νC=O band at 1715 cm^{-1} and an associated νOH absorption, eliminating the predominance of the diol structure [**548d**] in this solvent.

The fixed derivative [**550**] exhibits νC=O absorption at 1675 cm^{-1} (KBr) comparable to that of 2-methyl-4,5-disubstituted isoxazol-3-ones at 1660 cm^{-1} (Section 4-2Cb), giving a prima facie reason to eliminate [**548c**]. The fixed derivative [**549**] shows two absorptions at 1795 and 1710 cm^{-1} comparable to the absorptions at 1790 (vw) and 1726 (vs) cm^{-1} of 2,3,4-trimethylisoxazol-5-one (70BB343, Section 4-2Bf). Tautomeric 3,4-dimethylisoxazol-5-one exists almost entirely in the NH form in DMSO (Section 4-2Bf) with νC=O absorptions at 1784 (vw) and 1716 cm^{-1} (vs) (70BB343). The νC=O band at 1715 cm^{-1} of compound [**548**] favours the similar structure [**548a**]. Strong association occurs in isoxazol-5-ones (Section 4-2Bf), so that the νC=O band at 1680 cm^{-1} in the solid state could well arise from structure [**548a**]. 3,4-Dimethylisoxazol-5-one has νC=O absorption at 1690 cm^{-1} in the neat liquid. In conclusion, structure [**548a**] seems probable both for the solid state and solution. However, in view of the important solvent effects observed for isoxazol-5-ones, it would be interesting to study the 3-hydroxyisoxazol-5-ones in a wider range of solvents. The UV spectra in various solvents (71IS659) show considerable variation.

b. 3-Hydroxypyrazol-5-ones (II-46)

The complex problem of the tautomerism of 3-hydroxypyrazol-5-ones [**552a,b**] and other structures, has progressed much since the last review (II-46).*

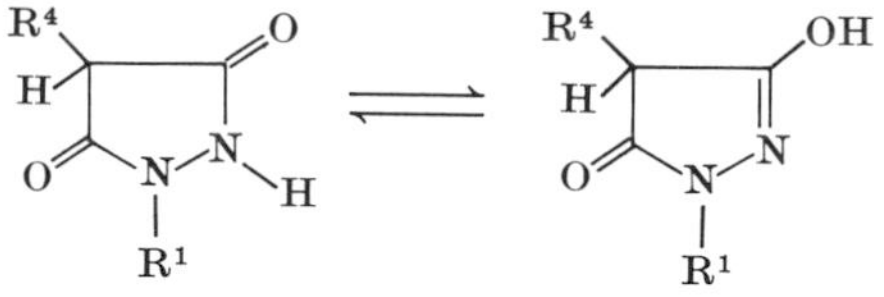

[**552a**] ⇌ [**552b**]

The most significant results concern partially fixed structures (OR and NR) for which fewer tautomers have to be considered. A study of 3-alkoxy-1-phenyl-5-pyrazolone [**553**] with the three tautomers [**553a–c**] using NMR (DMSO), IR (KBr) and UV (EtOH, C_6H_{12}) (71R601) conclude in favour of the sole presence of the CH form [**553a**] (see also 67CB1661). The same holds for the corresponding 1-(2′,4′,6′-trichloro-

* At about the same time there was also published (63RU535) a review on pyrazolinediones, which also summarized the conclusions of early studies of the tautomerism of these compounds.

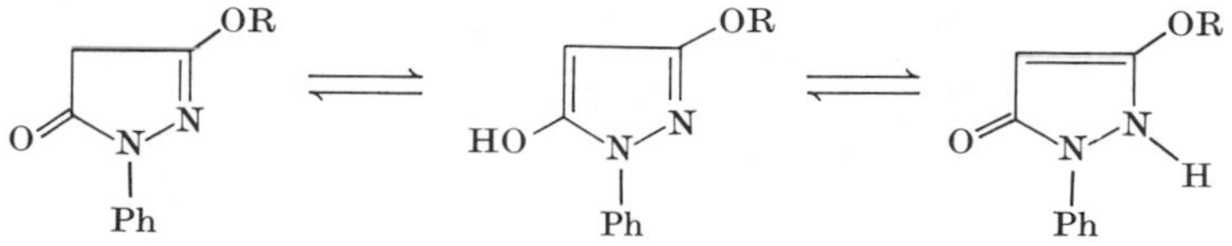

[553a] [553b] [553c]

phenyl) derivative (70T1571) from NMR ($CDCl_3$, DMSO) and IR (solid, CCl_4, $CHCl_3$, DMSO) studies, although a small amount of OH or NH form is detected in DMSO (NMR). The great stability of the CH form for pyrazol-5-ones with an electron-donor group in the 3-position has been noted previously in an earlier section (4-2Ge).

For the 5-alkoxypyrazol-3-ones, NMR ($CDCl_3$, DMSO) and IR (CCl_4, KBr) studies show the OH structure [**554b**] as predominant (71R601, 74CB1318), consistent with the other pyrazol-3-ones (Section 4-2Hc). As already mentioned (Section 4-2Hb), (and contrary to the original authors' claim) UV (EtOH: 240 nm, ε = 10,000–13,000 and 269 nm(sh), ε = 5000–6000) does not allow a choice between the NH [**554a**] and OH [**554b**] structures. In view of these results, the 5-OH

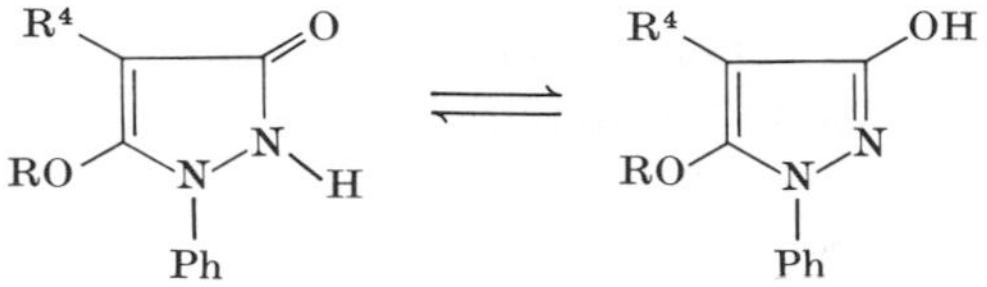

[554a] [554b]

structure [**555**] suggested (67CZ4151) (IR, absence of a carbonyl band in KBr) may have to be replaced by the isomeric hydroxy structure [**556**].

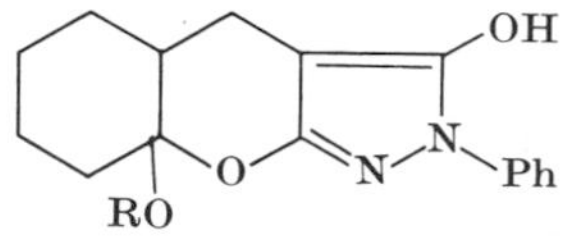

[555] 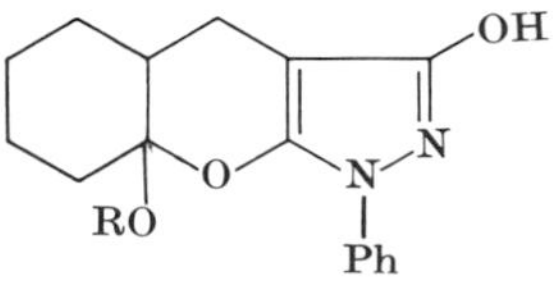 [556]

Other studies deal with semi-fixed NR structures and especially with 1,2-diphenylpyrazole-3,5-diones which can exist only in two tautomeric forms [**557a**] and [**557b**] because of symmetry. The NMR of [**557**, R^4 = H or alkyl] shows they exist exclusively in the CH form [**557a**] in $CHCl_3$ (63G570), acetone, DMSO and TFA (65G1371). It would have been interesting to use a basic solvent, such as pyridine, because (Section 4-2Gc) this solvent favours the OH form of pyrazol-5-ones. The

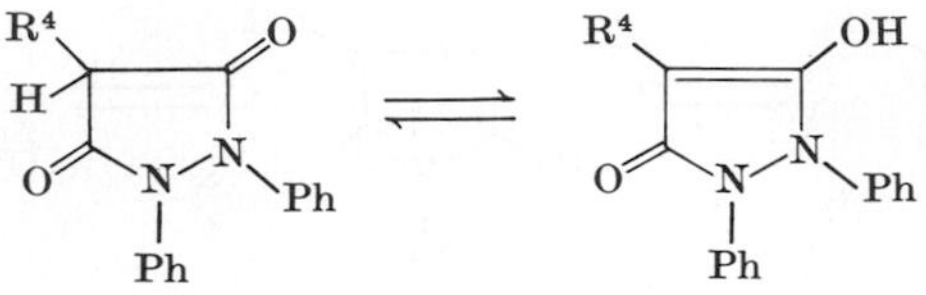

[557a] [557b]

4-nitro group induces the same effect as in the pyrazol-5-one series (Section 4-2Ge), thus [**557**, $R^4 = NO_2$] possesses the OH structure [**557b**] (65G1371).

The results obtained in the three partially fixed series of compounds, 3-OR, 5-OR and NR, indicate that the 1-substituted pyrazole-3,5-diones probably exist in tautomeric structures [**552a**] and/or [**552b**].

In $CHCl_3$ and dioxan, 1-phenyl-3-hydroxypyrazol-5-one shows two bands at 1710 and 1650 cm^{-1} which have been assigned, respectively, to the $\nu C{=}O$ and $\nu C{=}N$ absorptions of the form [**552b**, R^1 = Ph, R^4 = H] (65ZO1288).

c. Pyrazole-3,5-diones with 4-Acyl or 4-Nitroso Groups, and Bimolecular Pyrazole-3,5-diones

Available studies deal with the 1,2-diphenyl series. Mondelli and Merlini (65G320, 65G1371) found that the most stable tautomers generally retain the two 3- and 5-carbonyl groups: structures [**558**], [**559**], [**560**] and [**561**] are all the favoured forms. The phenylhydrazone structure [**561**] was also proposed later (68JO513); a signal at 13.4 ppm in $CHCl_3$ is characteristic of a chelated NH [to be compared with phenylazopyrazol-5-ones (Section 4-2Gi) and -3-ones (Section 4-2Hd)].

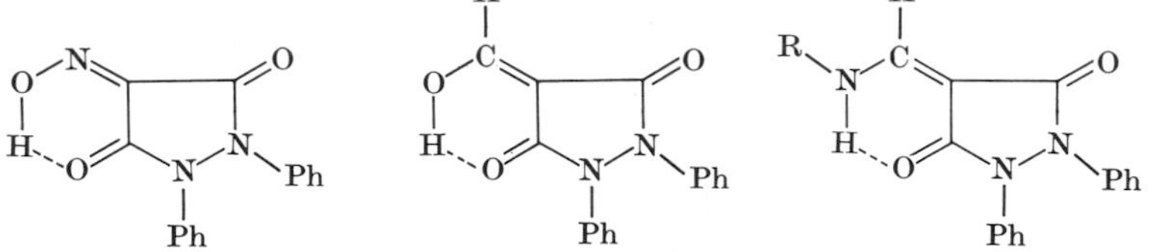

[558] [559] [560]

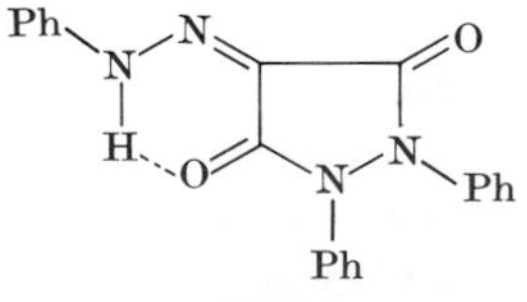

[561]

Structures such as [**562**] and [**563**], derived from the OH tautomer [**557b**], are suggested (65G1371) from IR evidence.

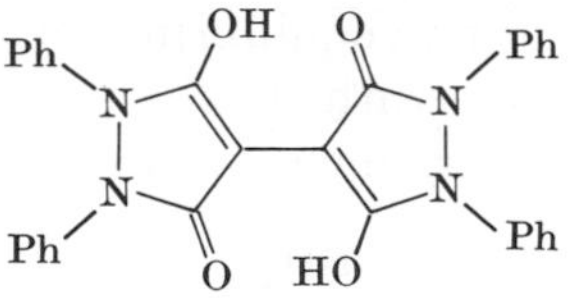

[562] 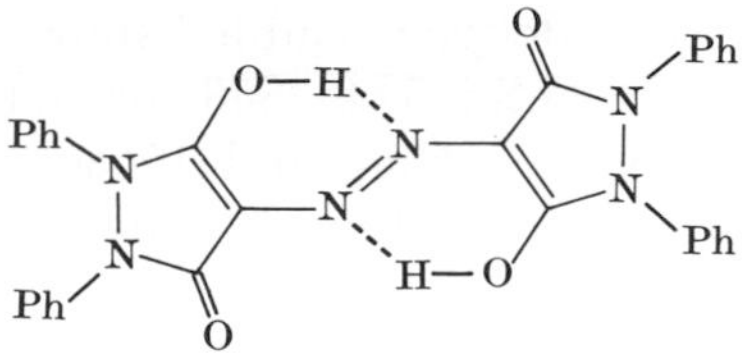 [563]

d. Thiazole-2,4-, -2,5- and -4,5-diones (II-51)

The only thiazole-2,4-dione described in the previous survey (II-51) existed in the dioxo form [**564**]. Since then, a dioxo structure has been assigned to thiazole-2,4-dione itself [**565**] based (67KG637) on νNH associated bands in the solid state at 3136 and 3048 cm^{-1} and νNH non-bonded at 3410 cm^{-1} in CCl_4. The IR spectrum of [**565**] in KBr (68BF3477) exhibits νC=O at 1680 and 1745 cm^{-1} supporting the dioxo structure; the UV maximum in EtOH is at 242 nm.

2-Isopropoxy-4-isopropylthiazol-5-one [**566**] (72J(PI)1983) according to the NMR spectrum exists only in the 5-oxo form even in DMSO

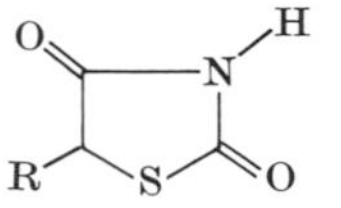

[**564**] R = CH_2CO_2H
[**565**] R = H

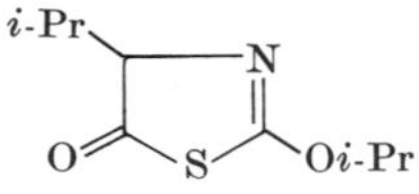

[**566**]

in which solvent 2-alkylthiazol-5-ones exist partially in the OH form (Section 4-3Ce).

No information is available for simple thiazole-4,5-diones; from IR (probably in KBr), UV (MeOH) and NMR (DMSO) dioxyluciferin probably exists as a mixture [**567a**] ⇌ [**567b**] (72T4065).

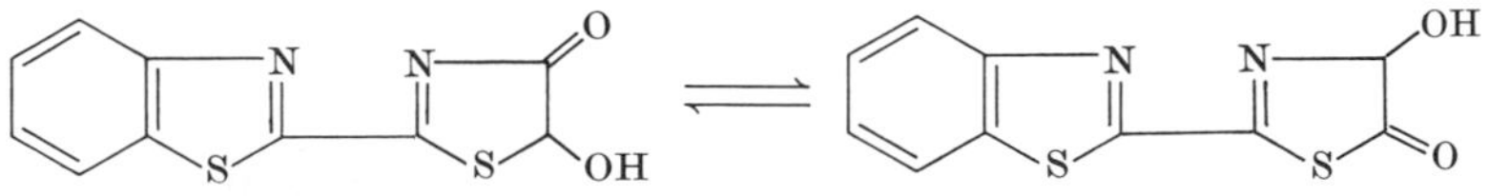

[**567a**] [**567b**]

e. Imidazole-2,4-diones (Hydantoins) (II-54)

The previous survey (II-54) reported the dioxo structure [**568a**] of hydantoins based on UV and IR measurements. No further spectroscopic data seem to have become available since then. Sörensen titrations of hydantoins (62J1462) are claimed to discriminate between imino and amino groups. According to the authors, the elevation of pH observed when formaldehyde is added shows that at least part of the hydantoin ionizes as a lactim form [**568b** or **c**]; they suggest an equilibrium between forms [**568a**], [**568b**] and [**568c**] as well as with water-associated molecules of hydantoin. We do not find these arguments convincing.

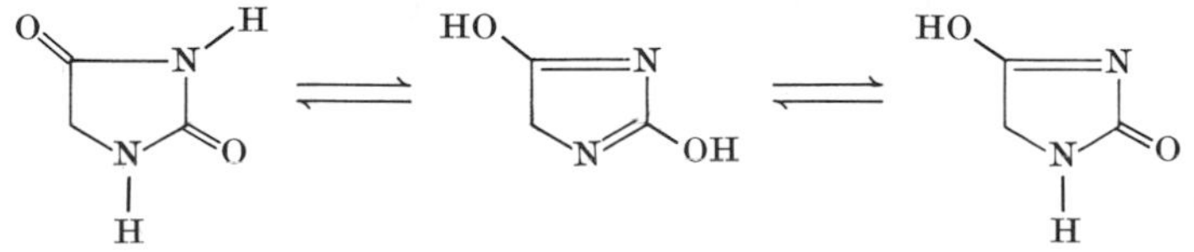

[**568a**] ⇌ [**568b**] ⇌ [**568c**]

f. Imidazole-4,5-diones

Italian authors studied 2-benzylimidazole-4,5-diones (68G458). Benzyl forms such as [**569a**] are eliminated by the NMR spectra for all the compounds.

Structures of the partially fixed compounds [**570**] and [**571**] are shown by their IR spectra in KBr and in solution, and by comparison with completely fixed compounds [**572**] and [**573**]. The *N*-monomethyl compound [**570**] exists in the dioxo form [**570a**] and the *O*-methyl derivative [**571**] as [**571a**]. Oxo–enol derivatives such as [**571**] exhibit a

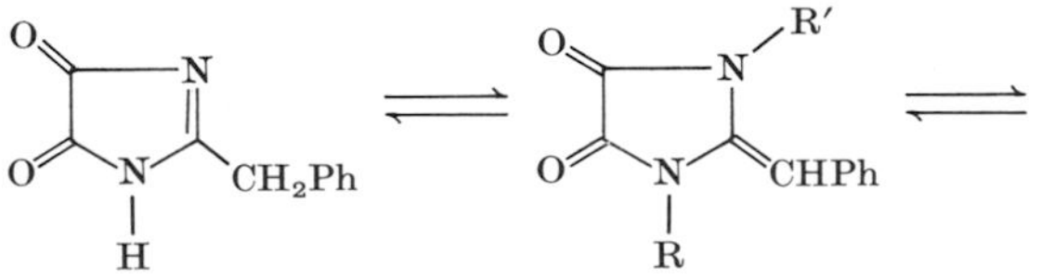

[**569a**]

[**569b**] R = R′ = H
[**570a**] R = H, R′ = Me
[**572**] R = R′ = Me

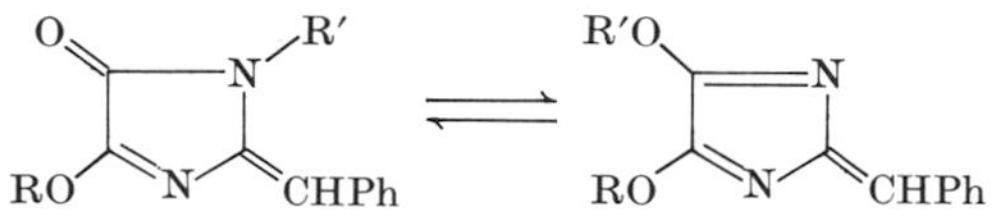

[**569c**] R = R′ = H
[**570b**] R = H, R′ = Me
[**571a**] R = Me, R′ = H
[**573**] R = R′ = Me

[**569d**] R = R′ = H
[**571b**] R = Me, R′ = H

typical band at 1595 cm^{-1}. Another feature differentiates this type of oxo–enol derivative, the phenyl group is coplanar and thus more conjugated with the imidazole ring, which shows up in the NMR spectra in DMSO (complex signal of this phenyl) and in the UV spectra in EtOH, $CHCl_3$ or DMSO (absorption at longer wavelength).

The dioxo structure [**569b**] for the *N*-unsubstituted 2-benzylidene-imidazole-4,5-dione is demonstrated by IR (absence of the band at 1595 cm^{-1} in the solid state and in solution, presence of a νNH absorption in $CDCl_3$), UV (identical absorption of [**569**] and [**570**]) and NMR (single signal for the phenyl group) evidence (68G458).

g. 1,2,4-Triazole-3,5-diones

These compounds were not described in the last review, although the tautomerism of 1-phenyl-1,2,4-triazole-3,5-diones had in fact already been investigated by chemical methods and from pK_a measurements (48MI127).

i. *1,2-Disubstituted-1,2,4-triazole-3,5-diones* exist in three tautomeric forms [**574a–c**]. Available pK_a measurements and UV spectra (in EtOH and NaOH) (64BF500) on [**574**, $R^1 = CH_2Ph$, $R^2 = Me$], [**574**, $R^1 = Ar$, $R^2 = Me$] and [**574**, $R^1 = Ar$, $R^2 = Ar$] do not allow conclusions in the absence of fixed derivatives for comparison. Later work (66T(7)213) on 1-phenyl-2-methyl-1,2,4-triazole-3,5-dione and the *N*-methyl derivative corresponding to [**574a**] produced similar results. However, IR data (Nujol or CCl_4) show that compounds [**574**] exist predominantly in the dioxo form [**574a**]: coupled νC=O in positions 3 and 5 at ~1760 cm^{-1} and 1710 cm^{-1}; νNH in position 4 at ~3475 cm^{-1} (66T(7)213).

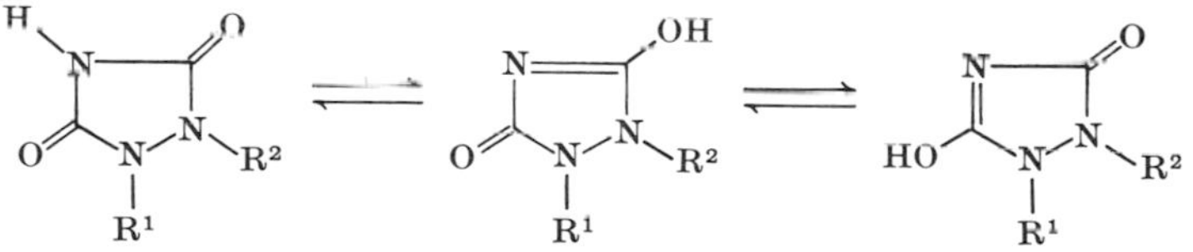

[**574a**] [**574b**] [**574c**]

ii. *1,4-Disubstituted-1,2,4-triazole-3,5-diones* can exist in three forms. Investigation (66T(7)213) of 1-phenyl-4-methyl-1,2,4-triazole-3,5-dione [**575**] and the two fixed derivatives corresponding to the two tautomers [**575a**] and [**575b**] was considered to show by comparison of the IR spectra in Nujol that [**575**] exists in the dioxo form [**575a**]: coupled νC=O in positions 3 and 5 at 1730, 1760 and 1710 cm^{-1}; νNH in position 2 at 3340 cm^{-1} in CCl_4.

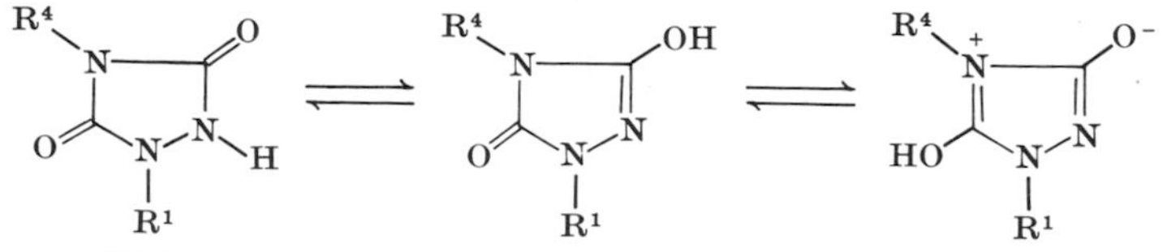

[**575a**] [**575b**] [**575c**]

iii. *1-Substituted-1,2,4-triazole-3,5-diones* can exist in six forms [**574a**, R^2 = H or **575a**, R^4 = H], [**574b**, R^2 = H], [**574c**, R^2 = H], [**575b**, R^4 = H], [**575c**, R^4 = H] and [**576**].

Early work (48MI127) concluded in favour of the oxo–hydroxy form [**575b**, R^1 = Ph, R^4 = H]. The pK_a and UV spectra of various N(1)-substituted compounds [**575**, R^1 = CH_2Ph or Ar, R^4 = H] (64BF500) are claimed (erroneously), because these compounds are more acidic than the disubstituted ones, to demonstrate the enol form [**575b**, R^4 = H]. One of the present authors (66T(7)213) from basicity measurements (in H_2SO_4) concludes that the dioxo form [**575a**, R^1 = Ph, R^4 = H] is predominant, with 10% of 3-hydroxy-5-oxo form [**575b**] in H_2O. IR spectroscopy (solid state) shows νC=O for 3- and 5-carbonyl groups. The *O*-methyl derivative [**577**] exists as such, and the predominant monoanion has the structure [**578**] from proton loss measurements in 0.01 *N* NaOH.

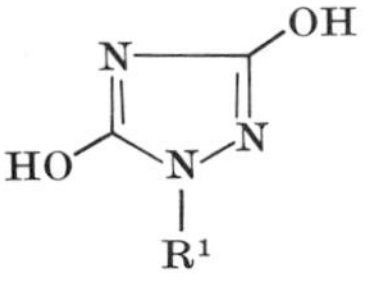

[**576**] 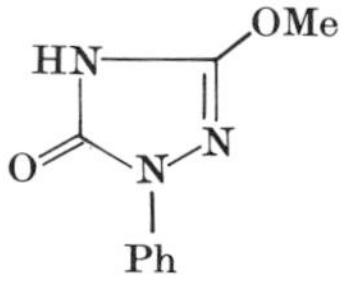 [**577**] 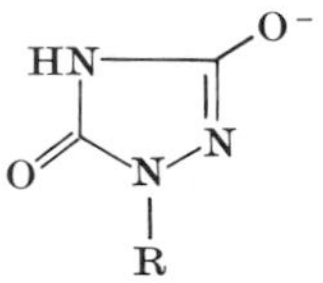 [**578**]

h. 1,3,4-Triazole-2,5-diones

Little work is available. Compound [**579**, R = NH_2] has been described as existing as such (64JO1174): double νC=O absorption between 1600 and 1650 cm^{-1} (solvent, if any, unknown) is similar to succinimide and νNH bands at 2700–3500 cm^{-1} are similar to those in dimeric lactams. No spectral data or information on the tautomerism are available (66AP(299)43) for a series of 1-substituted-1,3,4-triazole-2,5-diones [**579**, R = Ph, cyclohexyl, Et and *n*-Bu].

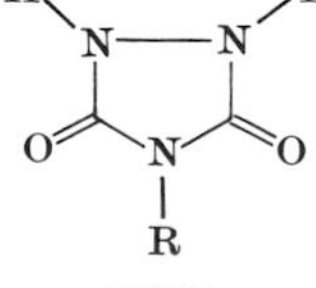

[**579**]

The pK_a = 6.25 (64BF500) of urazole [**574**, R^1 = R^2 = H] and partial IR results are available (72SA(A)855); both groups write urazole in the dioxo form [**574a**, R^1 = R^2 = H].

C. Two Potential Mercapto Groups

a. Thiazole-2,4-dithiones

Examples have been described recently (71KG192) in the dithione forms [**580**] and [**581**, R = H, Me], but no spectra are available.

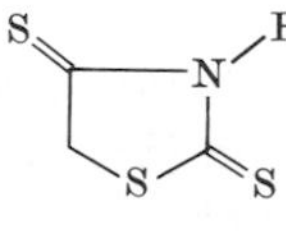

[**580**] 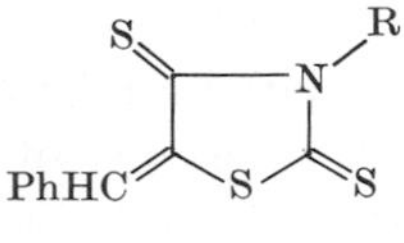 [**581**]

b. Imidazole-2,4-dithiones (II-54)

The previous survey (II-54) reported the dithione structure [**582a**] for 5-phenylimidazole-2,4-dithione described in early work (47J1598), but a later study (69CC1123) demonstrated structure [**582b**] by UV, IR and NMR spectroscopy. The UV spectrum of [**582**] in EtOH exhibits no absorption near 305 nm, where dithiones [**582a**] would be expected to absorb strongly (57J5075). The IR (Nujol) assignment of two bands at 2400 and 2355 cm^{-1} to νSH is confirmed by deuteriation. The NMR spectrum (DMSO) exhibits signals corresponding to SH, NH and aromatic protons, but none in the region where a 5*H* proton would resonate. This predominance of the thione–thiol structure [**582b**] contrasts with the dioxo and oxo–thione structures of hydantoins (Section 4-6Bd) and 5-phenyl-2-thiohydantoin (Section 4-6Ec).

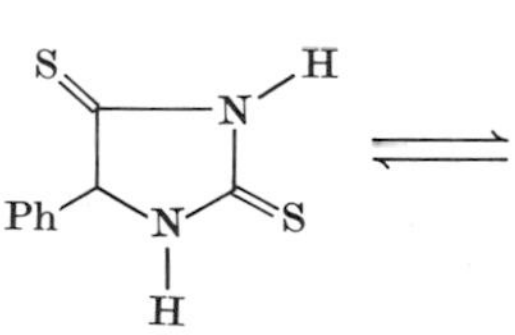

[**582a**] 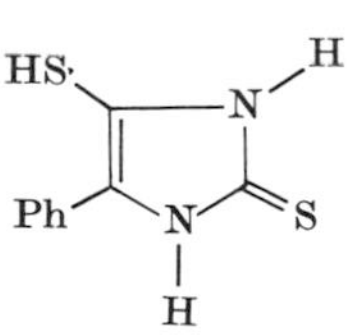 ⇌ [**582b**]

The gross structures [**583**] or [**584**], and [**585**] given for the mono- and di-benzylated derivatives (47J1598) are incorrect: these are the *S*-benzyl compounds [**586**] and [**587**], respectively (69CC1123): the UV (EtOH) absorption at 277 nm is as for [**582**]; the NMR spectra in DMSO

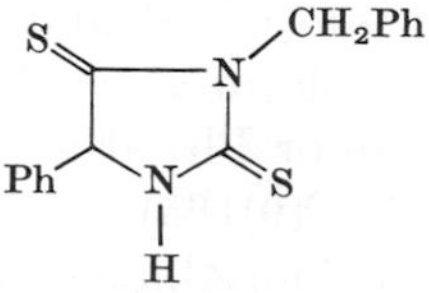

[**583**] 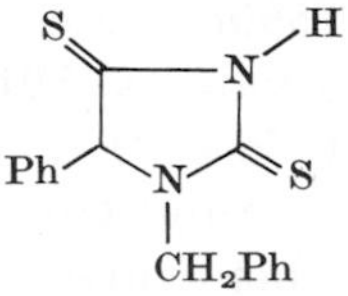 [**584**] 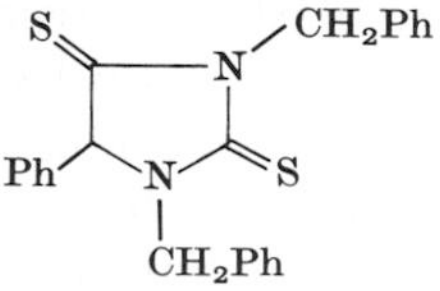 [**585**]

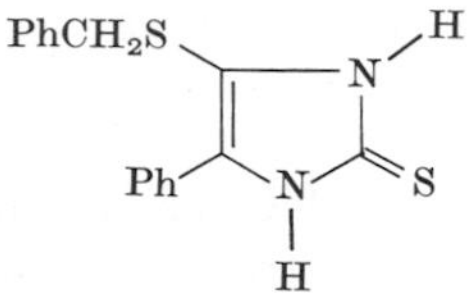

[586] 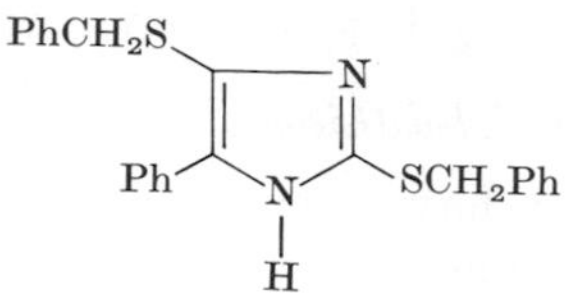 [587]

support such *S*-benzylated structures (no signal corresponding to 5*H*). In strong acid the UV spectra of [**582**] and [**586**] are similar because their cations must have the same structure (protonation on the thioureide sulfur) but in basic solutions UV spectra are different because the loss of proton takes place from SH for [**582b**] and from NH for [**586**] (69CC1123).

In the 5,5-diphenyl imidazole-2,4-dithione series, the two tautomers [**588a**, R = Ph] and [**588b**, R = Ph] have been separated (65TL1795, 69T4265): νC=N absorptions of [**588a**] at 1495 cm^{-1} and of [**588b**] at 1575 cm^{-1} in KBr. The spiro-*S*-methyl analogue [**588**; R, R = $(CH_2)_5$] exists as [**588b**] in C_6H_{12} and $CHCl_3$, 80% [**588b**] in EtOH, but only 20% [**588b**] with 80% [**588a**] in H_2O on UV evidence, supported by pK_a values, and IR (72CC2423). In the *S*-unsubstituted series, compound [**589**] is described (69T4265), but without spectroscopic data to allow a tautomerism study and comparison with the other results in this series (69CC1123).

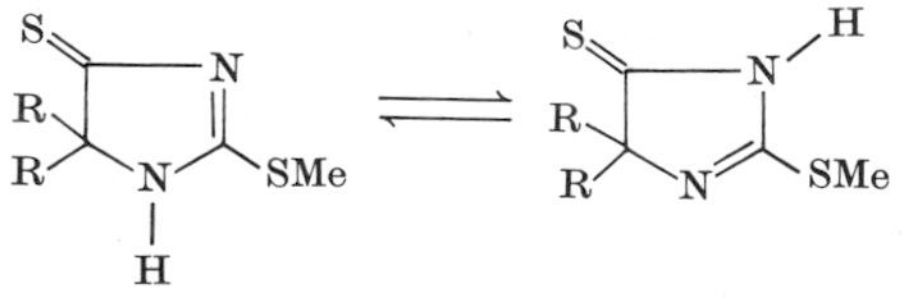

[588a] [588b] 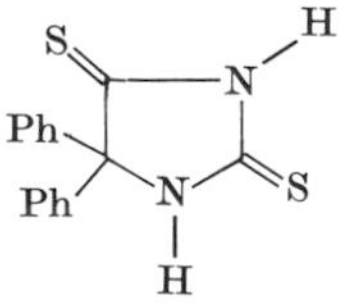 [589]

c. 5-Mercapto-1,3,4-thiadiazole-2-thiones (II-63)

In the previous survey (II-63) the structure of the unsubstituted compound was controversial: structure [**590a**] was proposed for EtOH and [**590b**] for $CHCl_3$ solution and the solid state. Later work (65MI17; see also 68HC(9)165 for a good discussion) concluded uniquely in favour of the thiol–thione structure [**590b**] by comparison of IR and pK_a of partially and totally fixed derivatives [**591**, R^3 and/or R = Me]. The partially fixed compound [**591**, R = H, R^3 = Me or Ph] also exists in the thiol–thione form shown. Sandström (68HC(9)165) rejects the earlier result in EtOH. For further evidence for a thiol–thione structure see (68CZ3880).

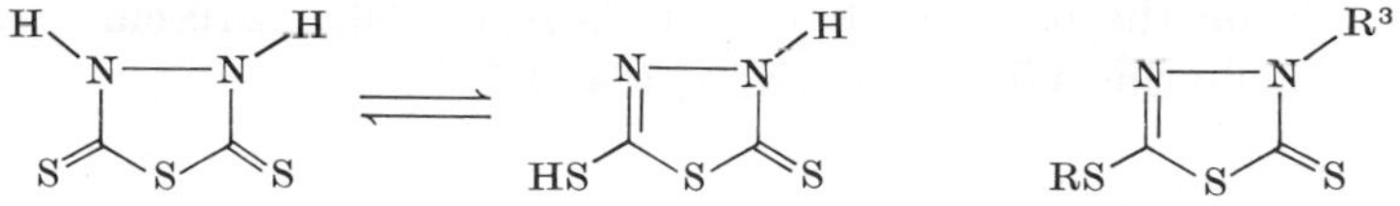

[590a] [590b] [591]

d. 1,2,4-Thiadiazole-3,5-dithiones

The gross structure of perthiocyanic acid (65HC(5)119) finally seems to be settled as the 1,2,4-thiadiazole [**592**]. IR spectra in Nujol (63J3168) are considered to indicate the dithione tautomer [**592a**] for perthiocyanic acid, but the dimercapto form [**592b**] is the only alternative considered whereas there are three possible thiol–thione structures: νNH occurs at 3010–2830 cm^{-1}; however, a medium absorption at 2560 cm^{-1} could be due to an SH group. Unfortunately no other spectroscopic data are available.

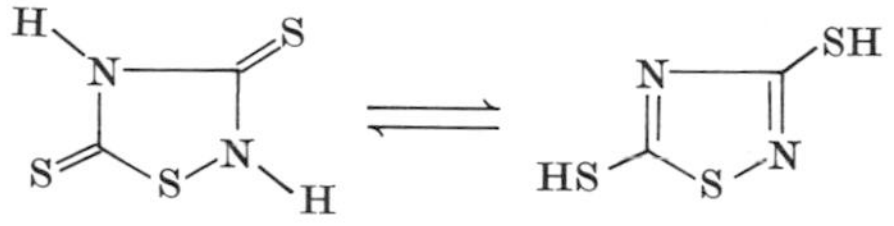

[592a] [592b]

e. 1,3,4-Triazole-2,5-dithiones

No compound of this series was described in the previous survey. A single paper (64JO1174) assigns the structures [**593**] and [**594**] to these compounds (R = H, NH_2, or N:CHPh) on the basis of IR bands in the region 1500–1550 cm^{-1}, considered to be due to CSNH. No spectroscopic data are given for the fixed derivative [**595**].

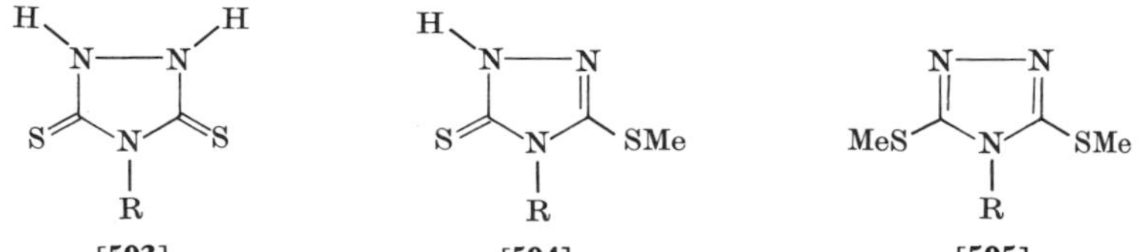

[593] [594] [595]

D. Two Potential Amino Groups

a. 3,5-Diaminoisothiazoles

3,5-Diaminoisothiazoles [**596**, R^4 = CO_2Et, R = p-NO_2–C_6H_4, COPh, PhOCO, and R = R^4 = COPh] were described in the diamino form solely on the basis of strong bands near 1500–1570 cm^{-1} due to the isothiazole ring vibrations (KBr) (64CB3106).

b. 3,5-Diaminopyrazoles

The NMR spectrum of [**597**] (DMSO) shows CH at 4.58 ppm, clearly inconsistent with any tautomer that does not possess an sp^2 carbon atom at the 4-position (68JO2606). The 4-ethoxycarbonyl derivative

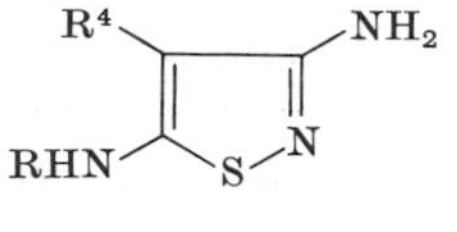

[**596**] 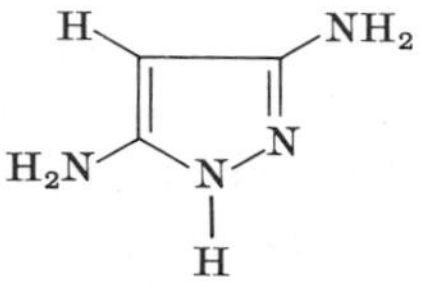 [**597**]

has the diamino structure [cf. **597**] on IR and NMR data (71T5873). Some infrared results are available on 4-arylhydrazo-3-amino-5-iminopyrazolines (73PC(315)1009).

c. 2,4-Diaminothiazoles

The instability of some 2,4-diaminothiazoles has been explained as being due to the presence of a 4-imino structure [**598b**, R = H] (50J-3491); however, the gross structure is incorrect at least for [**598**, R^5 = CO_2Et] where the isomeric 3-aryl structure [**599**] is shown by NMR spectra (69IJ884). The NMR spectrum in $CDCl_3$ of [**600**] shows the presence of two NH groups which exchange with D_2O (70J(C)1114).

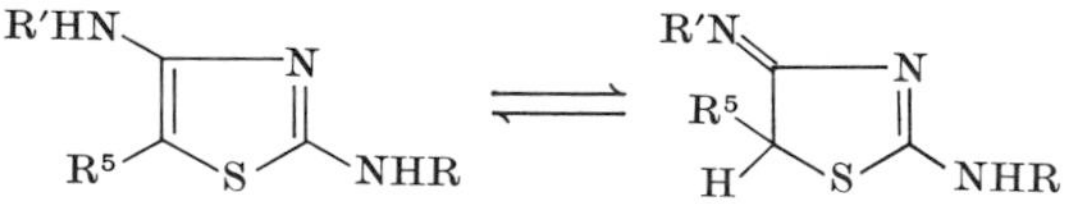

[**598a**] [**598b**] 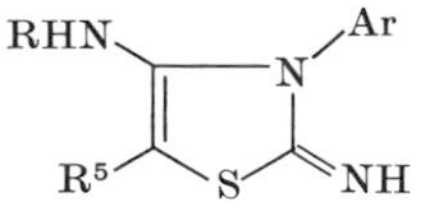 [**599**]

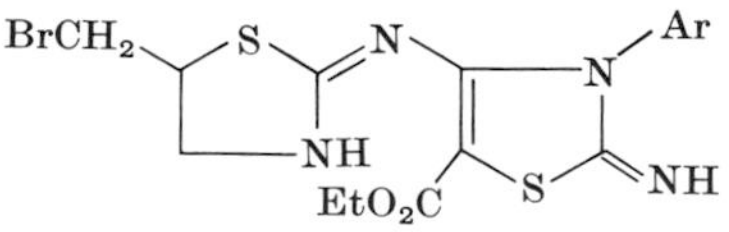

[**600**]

d. 3,5-Diamino-1,2,4-triazoles (II-73)

The previous survey (II-73) reported that for 3,5-diamino-1-phenyl-1,2,4-triazole early chemical results favoured the diimine [**601a**, R = Ph] but were corrected by UV spectroscopy which showed the predominance of a monoamino form [**601b**, R = Ph] or [**601c**, R = Ph] without excluding the presence of some diamino form [**601d**, R = Ph]. Since then considerable new work has appeared.

A double band at 1630 cm^{-1} (in KBr) for 1-methylguanazole [**601**, R = Me] was assigned to a cyclic νC=N although the compound was written (63JO2428) as [**601a**, R = Me]. However, this IR assignment was wrong.

A deeper IR study (solid state) (69KG732) including partial deuteriation for [**601**, R_1 = Me or Ph] favoured the diamino structure [**601d**]. At 3300–3000 cm^{-1}, $\nu_{as}NH_2$, ν_sNH_2 and νring–NH absorption was found and at 1640–1620 cm^{-1} δNH_2 or νC=N (exo) absorptions. After deuteriation the νNH almost disappeared and νND bands appeared near 2500 cm^{-1}. Simultaneously the δNH_2 at 1640–1620 cm^{-1} disappeared while a δND_2 appeared in the 1200–1100 cm^{-1} region.

Guanazole shows behaviour similar to the 1-substituted derivatives (63JO2428, 69KG732) and so must exist in the diamino form [**601d**, R = H] 4-Aminoguanazine hydrobromide possesses structure [**601/1**] from X-ray study (73J(PII)1).

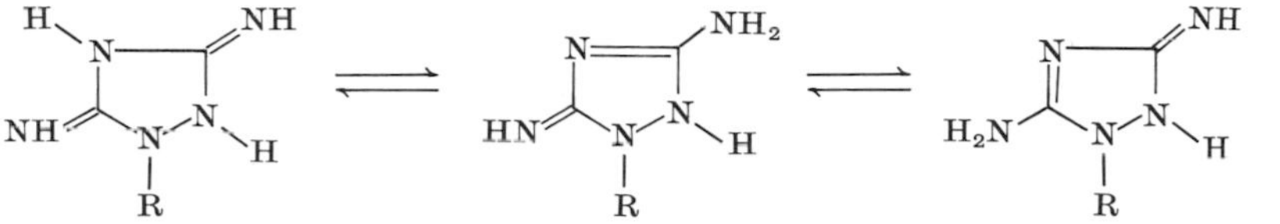

[**601a**] ⇌ [**601b**] ⇌ [**601c**]

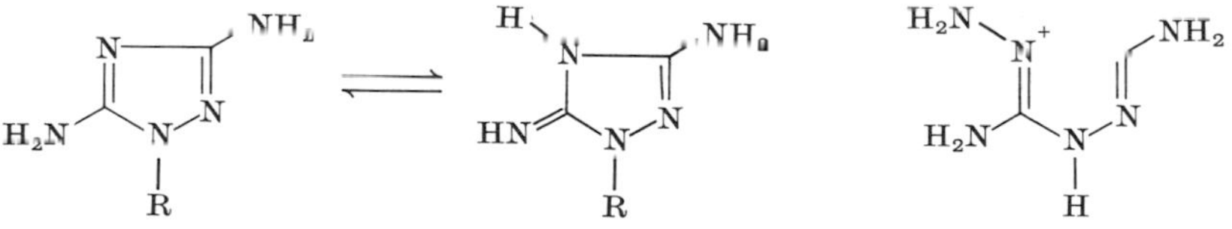

[**601d**] ⇌ [**601e**] [**601/1**]

E. One Potential Hydroxy and One Potential Mercapto Group

a. Oxazol-4-one-2-thiones

Cogrossi reported (72SA(A)855) a complete IR investigation in the (solid state) of *N*-alkyl derivatives [**602**]. The oxo–thione structure is

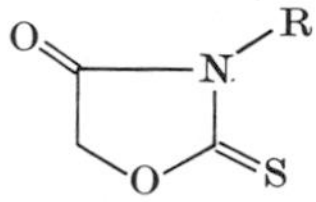

[**602**]

demonstrated by νC=O near 1750–1760 cm^{-1}, νC=S near 1080–1110 cm^{-1}, and characteristic vibrations of the thioureide group —N—C=S.

b. Thiazolonethiones and Selenazolonethiones (II-51)

i. *Thiazole-4-one-2-thiones* or rhodanines had been shown to exist in the oxo–thione structure [**603**] at the time of the previous survey [II-51]. A subsequent study (68BF3477) by UV, IR and pK_a, used fixed [**603**, **604**, R = Me or Ph; R^5, R^5 = =CHPh] and semi-fixed [**603**, **604**, R = Me or Ph; R^5 = H and R = H; R^5, R^5 = =CHPh] derivatives. Investigation of each function separately is attempted with conclusions less than clear but we consider that the results are not incompatible with the oxo–thione structure [**603**, R = H]. Further IR spectra (72SA(A)855) are interpreted on structures of type [**603**].

Russian authors have studied 5-benzylidene derivatives [**604/1**]. In 70% aqueous dioxan, the UV method shows that both the NH form [**604/1a**] and SH form [**604/1b**] occur, in proportions which depend on

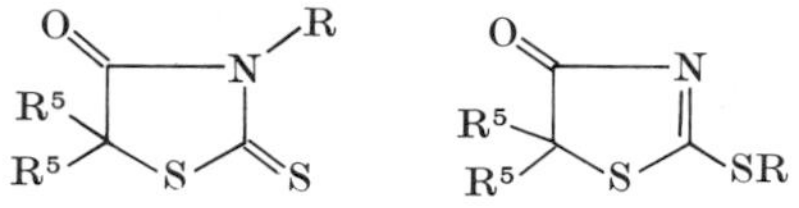

[**603**] [**604**]

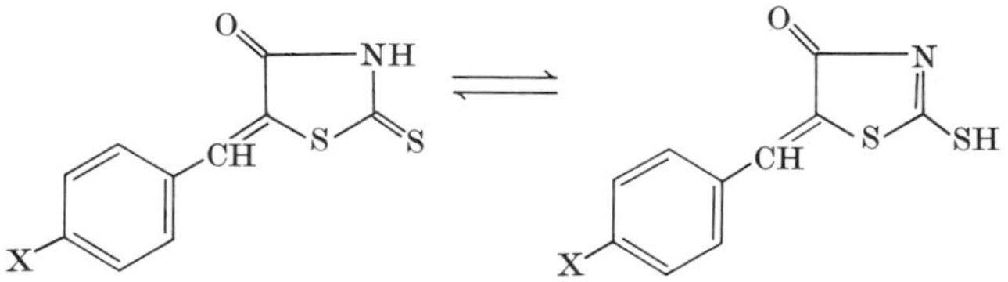

[**604/1a**] [**604/1b**]

the substituent σ value with ρ 0.38: the NH form predominates for all the compounds studied; pK_T = 0.77 for Z ≡ H (71RO1500). These results have been discussed in terms of the variation of pK_a on substitution (71ZO1815).

Partially fixed derivatives of types [**603**, R = alkyl] and [**604**, R = alkyl] have been shown to exist in the 4-oxo forms on IR and UV evidence (71KG189).

ii. *Selenazol-4-one-2-thione* [**605**] exists as such on IR evidence (72SA(A)855): νC=O at 1754 cm^{-1}, νC=S at 1162 cm^{-1}.

iii. *Thiazol-5-one-2-thiones* have not been treated subsequent to the previous survey (II-52) which gave evidence for their formulation as [**606**].

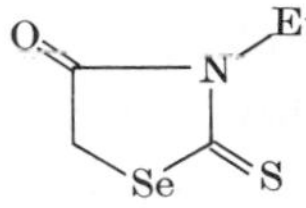

[605] 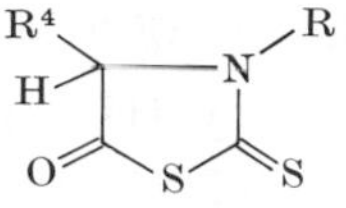 [606]

c. Imidazol-4-one-2-thione (II-54)

The previous review (II-54) reported the oxo–thione structure [**607**, $R^1 = R^3 = H$] since confirmed (68BF3477, 72SA(A)855) by IR spectra in the solid state and UV spectra in EtOH (maxima at 224 and 264 nm).

In a study of partially fixed derivatives, tautomers [**608a**] and [**608b**] were obtained individually (64TL2679, 65TL1795, 69T4265), depending on the crystallization conditions. In solution (UV, NMR) the two "desmotropes" give identical spectra; however the separate existence of the two individual tautomers in the crystalline state was established by X-ray structure determination (73T3565). In the solid state [**608a**] exhibits νC=O at 1690 cm^{-1} and νC=N near 1510–1490 cm^{-1} whereas the cross-conjugated [**608b**] shows νC=O at 1720 cm^{-1} and νC=N at 1590–1575 cm^{-1}. The spiro-cyclohexane analogue [**608/1**] is shown by UV to exist as [**608/1a**] in H_2O, as [**608/1b**]

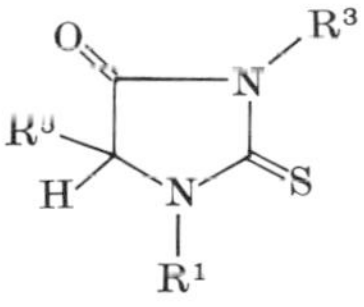

[607] 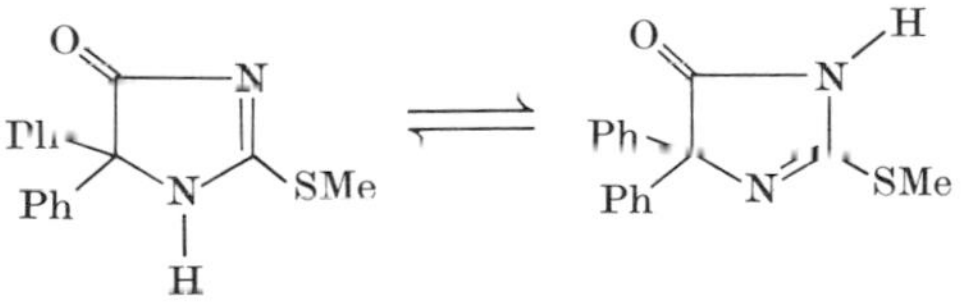 [608a] [608b]

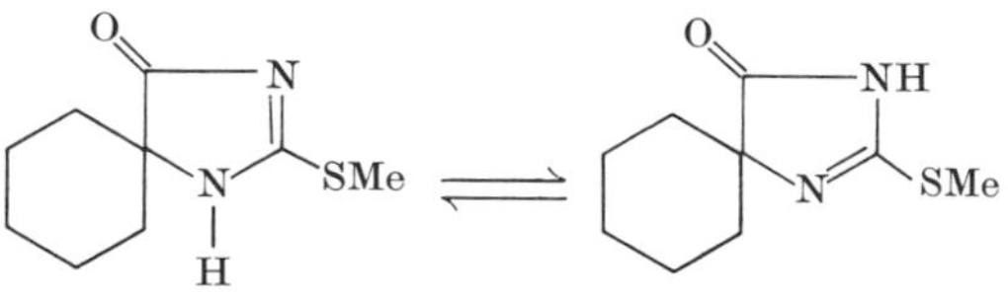

[608/1a] [608/1b]

in C_6H_{12} and as a mixture of both in EtOH, and pK_a measurements indicate K_T ca. 6 in H_2O. IR shows a similar change in tautomer composition from DMSO to $CHCl_3$ (72CC2423). Compounds [**607**, R^1 and R^3 alkyl or aryl] retain the oxo structure (72SA(A)855) with νC=O at 1756–1740 cm^{-1}.

d. 5-Mercapto-1,3,4-triazol-2-ones (II-64)

The previous survey (II-64) established the C=O group, but no subsequent work supports or rules out chemical evidence for the predominance of [**609a**] over [**609b**].

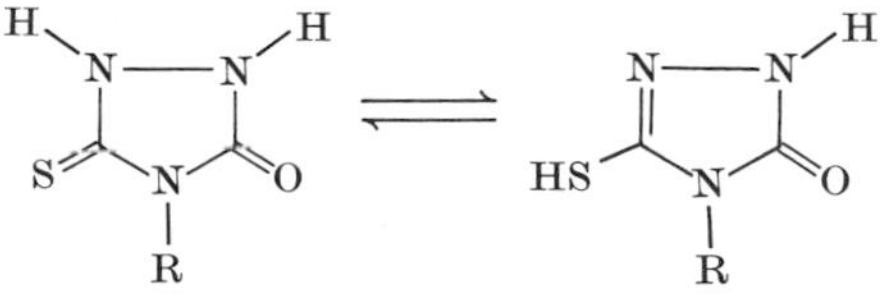

[**609a**] [**609b**]

F. One Potential Hydroxy and One Potential Amino Group

a. 3-Aminoisoxazol-5-ones

i. *2-Substituted 3-Aminoisoxazol-5-ones* [**610**, R^2 = alkyl, acyl, sulfonyl; R^4 = alkyl, aryl] have been studied as to their tautomeric structure by IR (64T165, 67T4395), UV and NMR (67T4395) spectroscopy. The νC=O absorption at 1680–1700 cm^{-1} in the solid state and at 1730–1750 cm^{-1} in CH_2Cl_2 solution and the presence of νNH_2 absorptions in all the IR spectra prove their existence as the NH_2-oxo tautomers [**610**].

4-Alkyl-3-amino-2-dialkylaminoisoxazol-5-ones show the normal νC=O but the spectra do not exhibit any free νNH_2 near 3400–3500 cm^{-1} (67T4395); the authors explain this phenomenon by hydrogen bonds between the NH_2 and the dialkylamino groups, with bands near 2850 cm^{-1} as νNH associated. UV spectra (λ_{max} 255 in EtOH) identical to the other 2,4-disubstituted 3-aminoisoxazol-5-ones support structures [**610**] for these compounds.

ii. *4,4-Disubstituted 3-Aminoisoxazol-5-ones* exist in the form [**611**; R^4 = alkyl] as shown by nonhydrogen-bonded νNH_2 (CH_2Cl_2). The

[**610**] [**611**]

νC=O appears at 1780–1790 cm^{-1}, and these derivatives do not absorb in the UV (EtOH) (67T4395).

iii. *N-Unsubstituted 3-Aminoisoxazol-5-ones.* As all the IR spectra show nonhydrogen-bonded νNH_2 and as amino tautomers are always predominant for 3-aminoisoxazoles (Section 4-5Bc) we consider further only the three 3-amino forms, namely, CH [**612a**], NH [**612b**] and OH [**612c**].

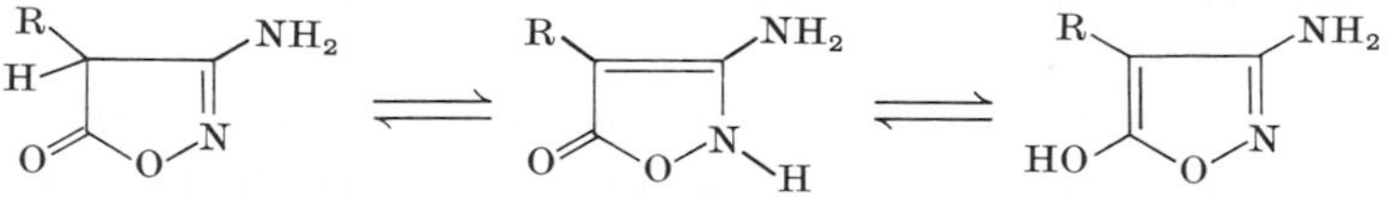

[**612a**] ⇌ [**612b**] ⇌ [**612c**]

In the solid state, the IR spectra of the 4-phenyl [**612**, R = Ph] and the 4-*n*-butyl derivatives [**612**, R = Bu] exhibit a νC=O at 1700 cm^{-1} (64T165, 67T4395) which by comparison with the 2-substituted derivatives (see above) must belong to the NH form [**612b**, R = Ph]. By contrast, the IR spectra in the solid state of [**612**, R = *i*-Pr] (64T165), and [**612**, R = Et] (67T4395) exhibit νC=O at 1765–1770 cm^{-1}, characteristic of the CH form [**612a**, R = Et, *i*-Pr]. The CH structure [**612a**] is also favoured for 3-dimethylamino compounds (69T3453) as the pure liquid or solid or in CS_2.

In solution, the CH [**612a**] and/or NH tautomers [**612b**] predominate depending on the solvent and on the nature of the 4-substituent (67T4395). The results are gathered in Table 4-30, νC=O bands clearly

TABLE 4-30

TAUTOMERIC COMPOSITION OF 3-AMINOISOXAZOL-5-ONES

<table>
<tr><th></th><th>4-Substituted</th><th>State or solvent</th><th>Method</th><th>CH [612a]</th><th>NH [612b]</th><th>Reference</th></tr>
<tr><td>3-Amino</td><td>Et</td><td>Solid state</td><td>IR</td><td>100</td><td>—</td><td>67T4395</td></tr>
<tr><td></td><td></td><td>CH_2Cl_2</td><td>IR</td><td>Major</td><td>Minor</td><td>67T4395</td></tr>
<tr><td></td><td></td><td>EtOH</td><td>UV</td><td>Minor</td><td>Major</td><td>67T4395</td></tr>
<tr><td></td><td></td><td>Acetone</td><td>NMR</td><td>60</td><td>40</td><td>67T4395</td></tr>
<tr><td></td><td>i-Pr</td><td>Solid state</td><td>IR</td><td>100</td><td>—</td><td>64T165</td></tr>
<tr><td></td><td>n-Bu</td><td>Solid state</td><td>IR</td><td>—</td><td>100</td><td>67T4395</td></tr>
<tr><td></td><td></td><td>CH_2Cl_2</td><td>IR</td><td colspan="2">100</td><td>67T4395</td></tr>
<tr><td></td><td></td><td>EtOH</td><td>UV</td><td>Minor</td><td>Major</td><td>67T4395</td></tr>
<tr><td></td><td>Ph</td><td>Solid state</td><td>IR</td><td>—</td><td>100</td><td>64T165, 67T4395</td></tr>
<tr><td></td><td></td><td>CH_2Cl_2</td><td>IR</td><td>—</td><td>100</td><td>67T4395</td></tr>
<tr><td></td><td></td><td>EtOH</td><td>UV</td><td>—</td><td>100</td><td>67T4395</td></tr>
<tr><td></td><td></td><td>Acetone</td><td>NMR</td><td>—</td><td>100</td><td>67T4395</td></tr>
<tr><td>3-Dialkylamino</td><td>Me</td><td>Solid state</td><td>IR</td><td>100</td><td>—</td><td>69T3453</td></tr>
<tr><td></td><td>Et</td><td>Solid state</td><td>IR</td><td>100</td><td>—</td><td>69T3453</td></tr>
<tr><td></td><td></td><td>CS_2</td><td>NMR</td><td>100</td><td>—</td><td>69T3453</td></tr>
</table>

distinguish between [**612a**] and [**612b**]. The existence of the 4-phenyl derivative [**612**, R = Ph] as the NH form [**612b**] is easily explained as due to the increased conjugation. The change from CH [**612a**] to NH structure [**612b**] when R changes from Et or *i*-Pr to *n*-Bu is more difficult to rationalize.

The similarity in behaviour of 3-aminoisoxazol-5-ones to the parent isoxazol-5-ones (Section 4-2Be) is apparent: 4-substitution and hydroxylic solvents favour the NH form.

b. 5-Aminoisoxazol-3-ones

IR νNH_2 and δNH_2 absorption, in the solid state and solution, of all 5-aminoisoxazol-3-ones shows that these compounds exist in an amino form [**613a**] or [**613b**], but the distinction between these two structures is difficult by IR spectroscopy. The same problem arises as for the isoxazol-3-ones (Section 4-2Cb)—differentiation of the νC═O of the NH form [**613a**] from the ν ring absorptions of [**613b**]. It is often a problem to differentiate between OH and NH forms in the solid state, cf. the discussion of pyrazolones (Section 4-2Fc).

i. *N-Substituted 5-Aminoisoxazol-3-ones* [**614**, R = COPh or *p*-Me–$C_6H_4SO_2$, R^4 = *i*-Pr] (64T165) were ascribed the amino forms [**614**]

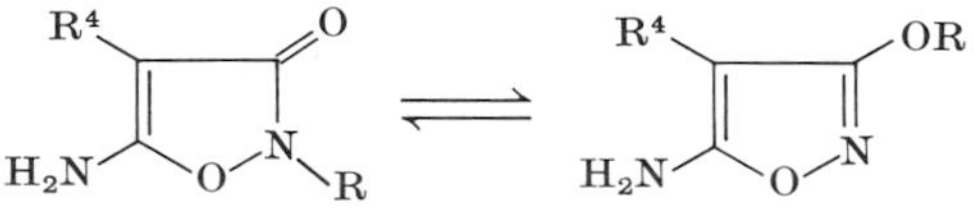

[**613a**] R = H
[**614**] R ≠ H

[**613b**] R = H

because of the IR (solid state) band near 1730 cm^{-1}; however, this band is too high for ring νC═O in [**614**] and must arise from the *N*-substituent. 2-Alkyl-5-aminoisoxazol-3-ones [**614**, R = Me, CH_2Ph] (67T4395) show, in the solid state, strong bands at ca. 1640 and 1580 cm^{-1}; the νC═O lactam bands expected near 1670 cm^{-1} (Section 4-2Cb) are probably displaced by *H*-bonding; in CH_2Cl_2 solution a broad band occurs at 1650 cm^{-1}. 5-Dialkylamino-2-methyl- and -2-phenyl-isoxazol-3-ones (69T3453) show the νC═O lactam bands as expected (in KBr) at 1675–1690 cm^{-1}.

ii. *N-Unsubstituted 5-Aminoisoxazol-3-ones*. The initial IR studies (64JO4917, 61JO4923, 64T165) did not consider the OH form [**613b**], which should not have been ruled out by these results alone. The lack of a recognizable νC═O band in the solid state was interpreted as showing their existence in a "dipolar structure" which is merely a canonical form of [**613a**]. Other authors (67T4395) rejected the OH form [**613b**] because of the difference in the IR spectra (solid state) of the sodium salt of [**613**], considered to be a good model for [**613b**], from [**613**; R = H, R^4 = Et], and also by comparison with the 3-hydroxy-isoxazoles (Section 4-2Cb). The IR spectra (solid state) of [**613**, R^4 = Et, *n*-Bu or Ph] resemble those of the *N*-substituted derivatives [**614**]. Therefore they assign the structure [**613a**] to these compounds but consider that dipolar canonical forms make a high contribution.

Compound [**613**, R^4 = $CH_2CH_2NEt_2$] (67T4395) is an exception: the IR spectrum (KBr) resembles those of 3-hydroxyisoxazoles (Section 4-2Cb) and that of the salt [**613b**, R = Na, R_4 = Et]. Broad bands at 2500–2300 cm^{-1} lead the authors to assign a zwitterionic structure [**615**] rather than the OH form [**613b**], although νOH could occur in the area found. In view of the difficulties of differentiating OH and NH structures in the solid state, this result must be treated cautiously.

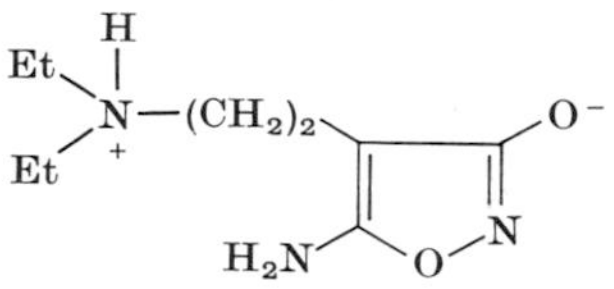

[**615**]

c. 5-Aminoisothiazol-3-ones

5-Anilino-3-hydroxyisothiazoles [**616**, R = CN, $CONH_2$, CO_2Et, COMe, COPh] are considered to exist in the OH form (64CB3106) as the IR spectra in KBr do not show any νC=O band. As we have already

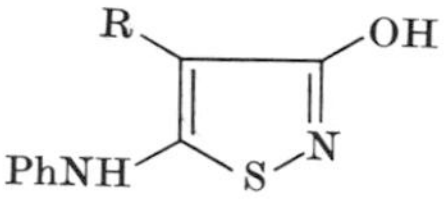

[**616**]

pointed out, differentiation of OH and NH tautomers by IR spectroscopy in the solid state is difficult (Sections 4-2Fc, 4-6Fb), and thus this result must be considered provisional.

d. Aminopyrazolones (II-70)

After a summary of Gagnon's extensive study of aminopyrazolone tautomerism, the previous review stated (II-70): "These conclusions cannot be accepted without reservation." Although no systematic study of aminopyrazolones is yet available there are well established data on their structure.

i. *1-Substituted 3-Aminopyrazol-5-ones* may exist in five tautomeric forms [**617a–e**]. Initial IR investigation in $CHCl_3$ and dioxan (65ZO1288) of 3-amino-1-phenylpyrazol-5-one [**618**] concluded in favour of the predominance of the aminopyrazolone form [**617a**, R^1 = Ph, R^4 = H], without excluding a small contribution of form [**617c**, R^1 = Ph, R^4 = H] in dioxan. Reinvestigation of [**618**] using IR and NMR spectroscopy (70T1571) confirmed that the only significant contributor is 3-amino-1-phenylpyrazol-5-one [**617a**, R^1 = Ph, R^4 = H] in solution ($CHCl_3$, DMSO) and in the solid state: the NMR signal at 3.5 ppm due to CH_2 rules out forms [**617b**], [**617c**] and [**617e**]; forms [**617b**], [**617d**] and [**617e**] are excluded by the IR bands (verified by deuteriation): 3421 ($\nu_{as}NH_2$), 3330 (ν_sNH_2), 3213 (νNH_2 bonded), 1681 (νC=O *H*-bonded) and 1635 cm^{-1} (δ NH_2) (in solid state). Later it was shown (73BB233) that the form [**617a**] persists in solvents as acidic as trifluoroethanol, shown by νC=O at: 1703 (MeCN), 1711 (THF), 1697 (DMSO), 1703 (HMPT) and 1682 cm^{-1} (CF_3CH_2OH).

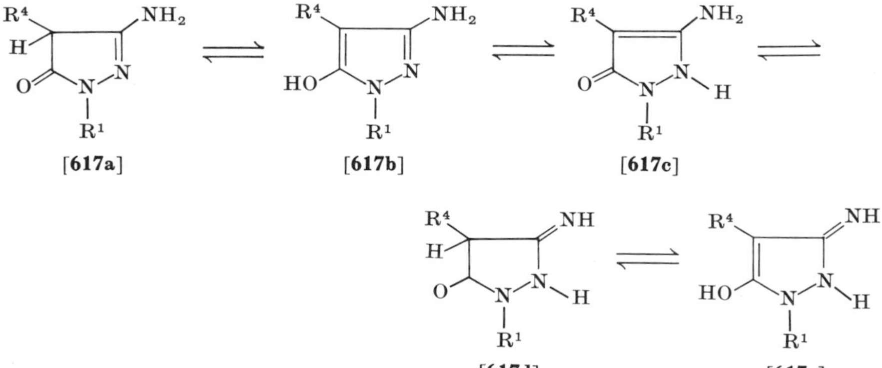

[**617a**] ⇌ [**617b**] ⇌ [**617c**] ⇌ [**617d**] ⇌ [**617e**]

The effect on the tautomerism of 1-substituted pyrazol-5-ones of a 3-NH_2 substituent was discussed in Section 4-2Ge. 3-Amino-1-phenylpyrazol-5-one [**618**] was used as a model for the CH structure in the ^{13}C NMR study of the pyrazol-5-one tautomerism (Section 4-2Fb) (70J(C)1842): a triplet is observed at 153 ppm upfield from CS_2 for the carbon in position 4.

The same structure exists for analogous compounds [**619–622**]. 3-Methylamino-1-phenylpyrazol-5-one [**619**] (74BF(2)291) has in the NMR (DMSO) a CH_2 signal at 3.63 ppm and NH at 6.80 ppm; J(NH/Me) = 6 Hz. Similarly, 3-dimethylamino-1-phenylpyrazol-5-one [**620**] (69H2641) shows ($CDCl_3$) a CH_2 signal at 3.50 ppm. 3-Anilino-1-phenylpyrazol-5-one [**621**] (68JO3336) displays ($CDCl_3$, DMSO and

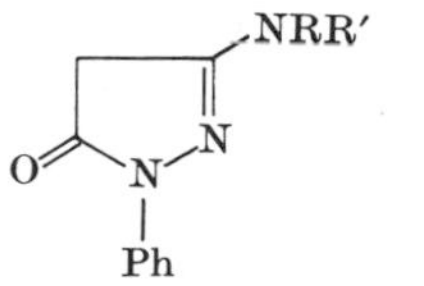

[**618**] R = R′ = H
[**619**] R = H, R′ = Me
[**620**] R = R′ = Me

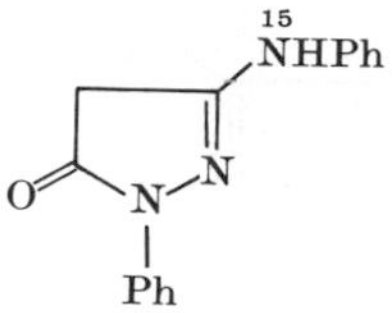

[**621**]

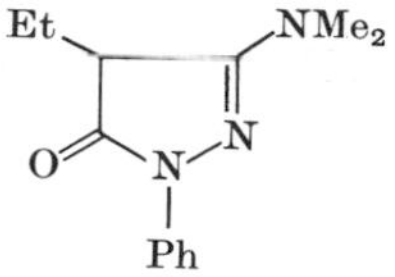

[**622**]

pyridine) a ^{15}N–H coupling 92 Hz. 3-Dimethylamino-4-ethyl-1-phenylpyrazol-5-one [**622**] (69T3453) (neat liquid) has νC=O at 1687 cm^{-1}.

In the 3-amino-1-methylpyrazol-5-one series only structure [**617a**] has been detected for [**623**] (70THO3); the NMR (DMSO) shows the signal due to CH_2 at 3.18 ppm and IR bands (assigned by deuteriation) at 3420 (NH_2as), 3340 (NH_2s), 1697 (C=O) and 1639 cm^{-1} (δ NH_2). The same applies to 4-ethyl-3-diethylamino-1-methylpyrazol-5-one [**624**] (69T3453): 3.3 ppm (H–4, $CDCl_3$) and 1681 cm^{-1} (C=O, neat liquid).

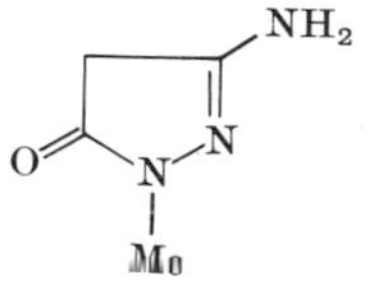

[**623**]

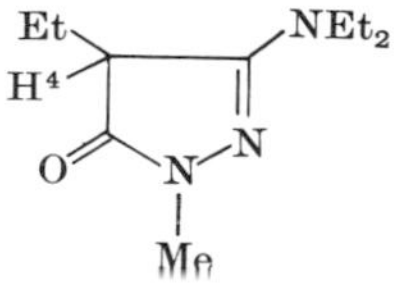

[**624**]

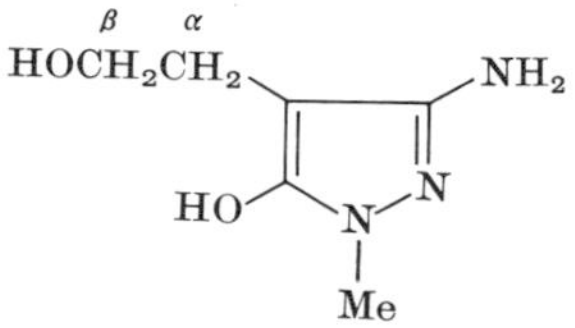

[**625**]

For a 4-β-hydroxyethyl derivative, structure [**625**] was suggested (71LA(754)113) from the following spectral data: NMR (DMSO-d_6) α-CH_2 triplet (J = 7.77 Hz) showing coupling with β-CH_2 (no signal or coupling from H(4) of form [**617a**] is observed); IR (KBr) bands at 3410, 3340, 3015 cm^{-1} (OH, NH_2); 2000–3100 (associated OH); 1655, 1640, 1600 (ring). This result is very surprising: from the method of synthesis it appears that this compound could be 5-amino-3-hydroxy-4-β-hydroxyethyl-1-methylpyrazole, which would better explain the spectral results (see following section).

Hückel calculations (71T5779) indicate the following order of stability for the tautomers of 3-amino-1-phenylpyrazol-5-one [**618**]: [**617a**] > [**617c**] > [**617d**] > [**617b**] > [**617e**].

ii. *1-Substituted 5-Amino-3-hydroxypyrazoles*, less studied than the preceding compounds, can also exist in five tautomeric forms [**626a**] to [**626e**] but seem invariably to possess the pyrazole structure [**626b**]

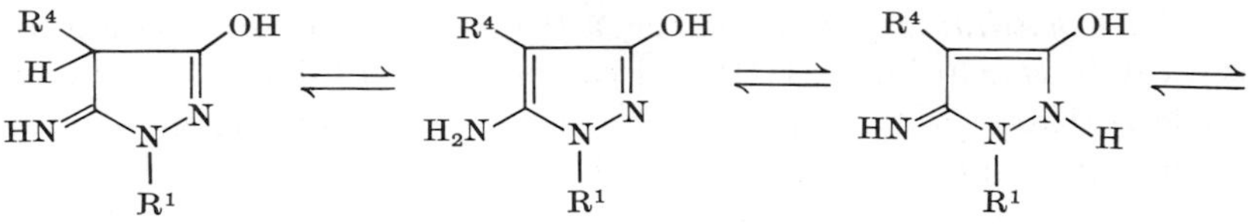

[626a] [626b] [626c]

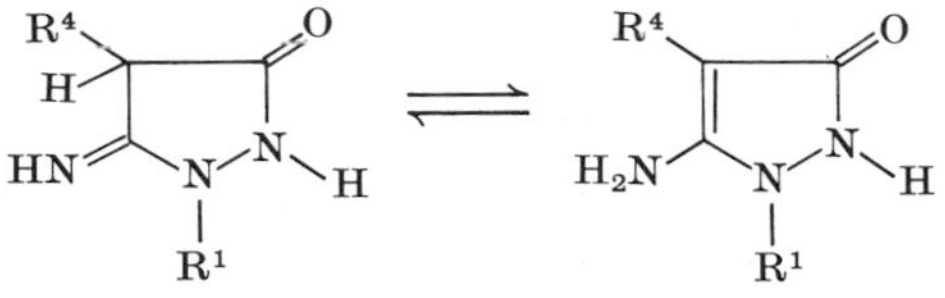

[626d] [626e]

(see, however, 70LA(734)173). Thus, 5-amino-3-hydroxy-1-methylpyrazole [**627**] (70TH03) exhibits an NMR (DMSO) signal at 4.33 ppm for one proton (comparable with the signal at 4.58 ppm of 3,5-diaminopyrazole [**597**] in Section 4-6Db) and IR (in dioxan) bands at 3400 and 3340 (νNH_2), 1635 (δ NH_2), 1560 (pyrazole) and 1530 cm^{-1}.

5-Dialkylamino derivatives [**628**], [**629**] (69T3453) can exist in only two forms, [**628a**], [**628b**]. The authors prefer the oxo forms [**628b**] and [**629b**], but we consider that their spectroscopic results are consistent

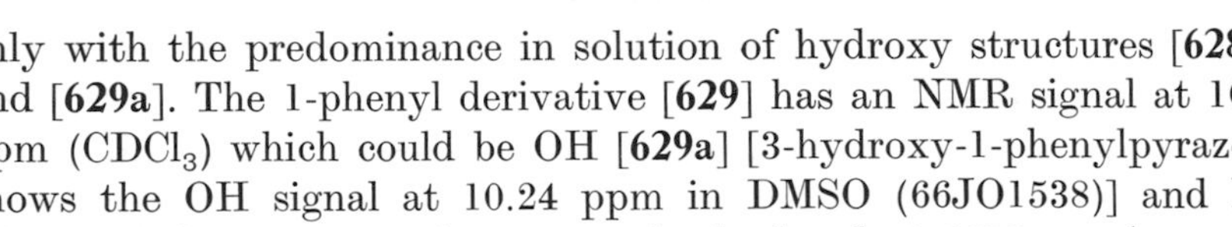

[627]

[628a] R = Me
[629a] R = Ph

[628b] R = Me
[629b] R = Ph

only with the predominance in solution of hydroxy structures [**628a**] and [**629a**]. The 1-phenyl derivative [**629**] has an NMR signal at 10.3 ppm ($CDCl_3$) which could be OH [**629a**] [3-hydroxy-1-phenylpyrazole shows the OH signal at 10.24 ppm in DMSO (66JO1538)] and IR (KBr) bands at 1610 and 1535 cm^{-1}; the band at 1610 cm^{-1} can be assigned to phenyl [3-hydroxy-1-phenylpyrazole shows it at 1600–1602 cm^{-1}(62BJ747, 63ZO2597)]. The 1-methyl derivative [**628**] shows bands at 1592 and 1535 cm^{-1} at too low a frequency for the carbonyl [the fixed N–R model for structure [**629b**], 4-methyl-1,2-diphenyl-5-piperidylpyrazolin-3-one, shows a νC═O band at 1690 cm^{-1} in KBr (69T3453)]. A remaining problem is whether in the solid state these compounds have the OH structures [**627**], [**628a**], [**629a**] or so-called NH/OH structures (Section 4-2Fc).

iii. *N-Unsubstituted Aminopyrazolones* are not enough studied to draw valid conclusions. HMO-ω calculations (71T5779) indicate greatest stability of form [**630**] and provide an explanation for the effect of the amino group on the UV spectrum (λ_{max} in *n*-heptane, 270 nm). An analogous structure was suggested (69H2641) for 3-dimethylaminopyrazol-5-one [**631**] from NMR (CH_2 at 3.21 ppm in $CDCl_3$) and UV (maxima at 247 and 279 nm in MeOH): compound [**620**] absorbs at 280 nm in MeOH; and also for the 3-hydrazinopyrazol-5-one [**632**] (67CB1661).

However, the hydroxyaminopyrazole structure [**633a**] was attributed (71LA(754)113) to a compound with the following characteristics: no NMR (DMSO) signal for a 4-position CH; IR (KBr); 3380, 3310, 3170 (OH, NH_2), 1650 and 1550 cm^{-1} (ring). This compound may possess the 5-amino-3-hydroxy structure [**633b**] as suggested earlier (Section 4-6Fdi) for the 1-methyl derivative.

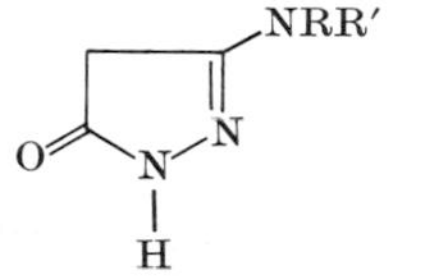

[**630**] R = R′ = H
[**631**] R = R′ = Me
[**632**] R = H, R′ = NH_2

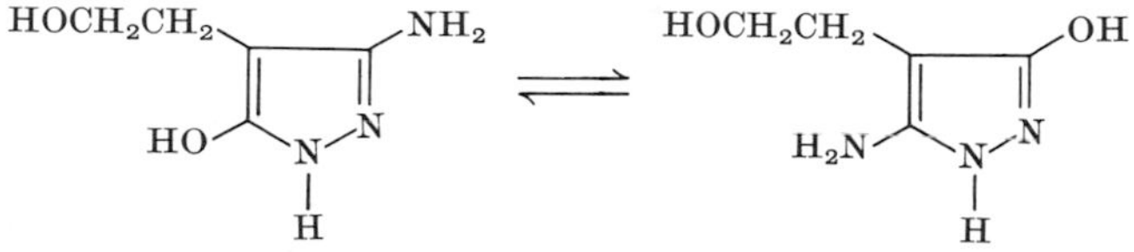

[**633a**] ⇌ [**633b**]

iv. *1-Substituted 3-Acylaminopyrazol-5-ones* Arriau (71TH01) studied 3-benzamido-1-phenylpyrazol-5-one, by the Hückel-ω method, and concluded that without chelation the CH form [**634a**] is the most stable but that chelation stabilizes the NH form [**634b**]. While the 3-amino [**618**] and 3-anilino analogues [**621**] exist exclusively in the CH form in DMSO, the percentage of the CH form [**634a**] is ca. 30% for 3-acylamino-1-phenylpyrazol-5-ones [**634**] (68JO3336). More recently (74BF(2)291) that study was repeated and a solvent- and temperature-dependent

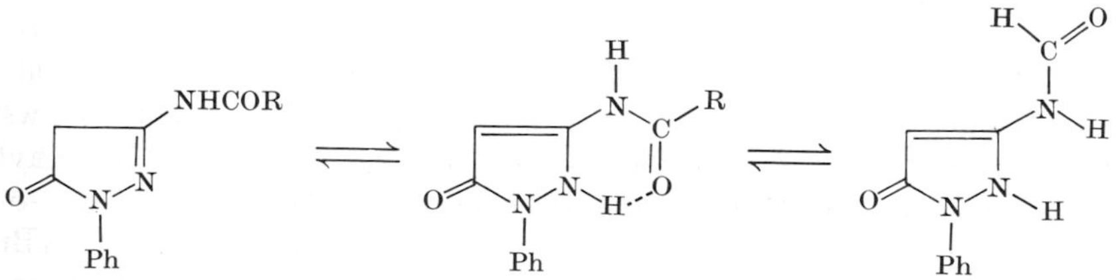

[**634a**] R = Ph
[**635a**] R = Me
[**636a**] R = H
⇌
[**634b**] R = Ph
[**635b**] R = Me
[**636b**] R = H
⇌
[**636c**]

equilibrium was found between the tautomers [**634a**] ⇌ [**634b**] and [**635a**] ⇌ [**635b**]. In the formamido compound [**636**] the three forms [**636a–c**] all occurred; e.g., in DMSO at 30°C, 15%, 60% and 25% of the three tautomers were found.

e. 2-Aminooxazol-4-ones

The tautomerism of 2-aminooxazol-4-ones had not been investigated at the time of the previous survey, but a great deal of subsequent work has been published. Five forms [**637a–e**] can be involved in the tautomerism. We consider first derivatives in which one of the two functions is fixed.

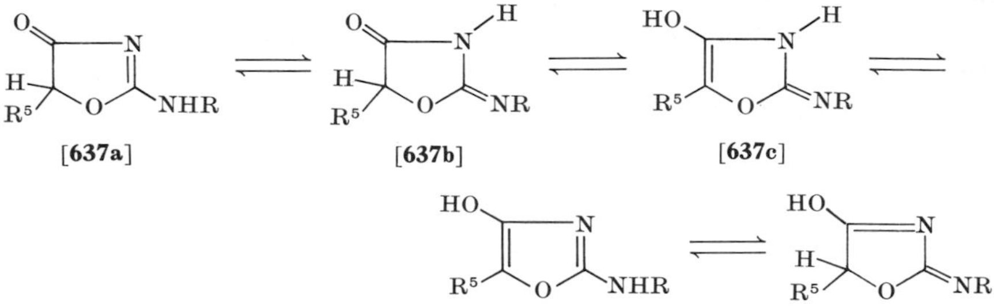

[**637a**] ⇌ [**637b**] ⇌ [**637c**] ⇌ [**637d**] ⇌ [**637e**]

i. *5,5-Disubstituted 2-Aminooxazol-4-ones* can exist in three forms corresponding to [**637a**], [**637b**] and [**637e**]. The only compound studied [**638**] exists as [**638a**] from UV (62JO1686, 70IM1126, 72JH285), IR and NMR (70IM1126, 72JH285). The UV maximum in MeOH at 215 nm compares well with that of the fixed derivative [**639**] at 225 nm (taking into account a bathochromic effect of 5 nm for each added methyl) but differs from that of [**640**] at 208 nm, considering that here

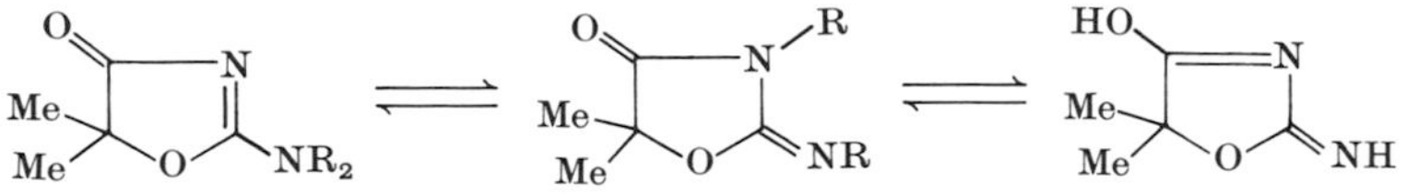

[**638a**] R = H ⇌ [**638b**] R = H ⇌ [**638c**]
[**639**] R = Me   [**640**] R = Me

also methyl groups should induce a bathochromic shift. The third possible tautomer [**638c**] would absorb in a completely different way. The IR in the solid state of [**638**] exhibits diffuse absorption at 3300–2700 cm^{-1} but in MeCN solution two bands are shown at 3340 and 3260 cm^{-1}, assigned to ν_{as} and $\nu_{s}NH_2$ absorptions; their low frequency is explained by hydrogen bonding. Two bands at 1742 cm^{-1} and

1641 cm^{-1} are assigned to νC=O and νC=N modes; they are similar to those of the fixed derivative [**639**]. The authors (72JH285) point out the unusually low intensity of the νC=O band. This is also the case for other 2-aminooxazol-4-ones, and the discussion reported is valid for them too. To confirm their assignment, the Raman spectrum in the solid state and in solution was recorded: the authors state that it is necessary to consider here the whole system rather than independent C=O and C=N group vibrations. NMR spectroscopy does not easily differentiate between structures [**638a**] or [**638b**]; the only fact supporting structure [**638a**] is the chemical shift of NH_2 in DMSO at 8.31 ppm, compared with the resonance at about 7.3 ppm for an imino proton in the same solvent.

ii. *3-Substituted 2-Iminooxazol-4-ones* can exist in the two forms [**641a,b**] analogous to [**637b**] and [**637c**]. Compounds [**641**, R = H, Me, Ph; R^5 = Ph, Ar, H] exist as [**641a**] on UV (62JO1686, 67BF207, 72JH285), IR(67BF207, 72JH285) and NMR (64JO370, 67BF207, 72JH285) evidence. The fixed model [**640**] has UV maximum in MeOH at 208 nm; all the compounds [**641**] with R = Me or H also absorb in this region, but as no model compound corresponding to [**641b**] is known it is unwise to draw a definite conclusion. The amino substituent affects the UV maximum: compounds [**641**, R = Ph] absorb at ca. 254 nm.

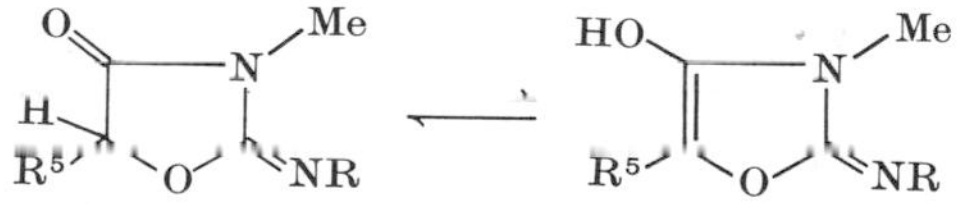

[**641a**] [**641b**]

IR spectroscopy (CCl_4, CS_2, solid state) easily differentiates [**641a**] and [**641b**]: all the spectra exhibit νC=O near 1770–1755 cm^{-1} and νC=N near 1720–1670 cm^{-1} and thus demonstrate the existence of tautomer [**641a**]. The same remarks regarding the intensity and the assignment made in the preceding section apply here. Compounds [**641**, R = H] are transparent in the 3500–3000 region ruling out form [**641b**]. For [**641**, R = H] νNH at 3340 cm^{-1} (CCl_4) and 3270 (KBr) shows hydrogen bonding.

From NMR spectroscopy one distinguishes structures [**641a**] and [**641b**] by observing the region where the 5*H* should appear: this proton signal is found in all the spectra in agreement with structure [**641a**]. For [**641**, R = H] the imino proton is observed at 7.3 ppm in DMSO (67BF207) and near 5.6–6.0 ppm in $CDCl_3$ (64JO370, 72JH285). For [**641**, R $\neq$ H] theoretically two isomers *syn* and *anti* about the C=N bond could exist; in fact NMR shows only one peak.

iii. *Fixed Amino Structures*. Although two tautomers [**642a, b**] corresponding to [**637a**] and [**637d**] can be formulated for [**642**], the oxo form [**642a**] predominates whatever the R^5, as shown by UV (62JO-1686, 67BF207, 72JH285), IR (62JO1686, 67BF207, 72JH285) and NMR (64JO370, 72JH285). In the UV all these compounds absorb at 225–230 nm in MeOH, as does the fixed derivative [**639**]; the enol structure [**642b**] should absorb at longer wavelength.

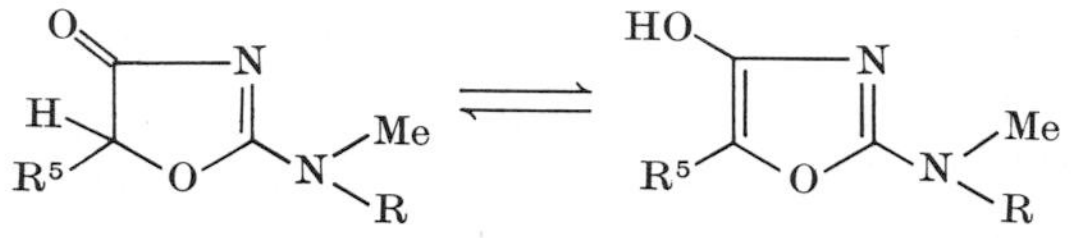

[**642a**] R = Me or Ph [**642b**]

All these compounds are transparent in the IR νOH region in CCl_4. The complex bands in the solid state near 3000 cm^{-1} are νCH of C–Me groups. In KBr νC=O at 1735–1710 cm^{-1} and νC=N 1600–1680 cm^{-1} prove the presence of tautomer [**642a**]. In CCl_4 solution [**642**; R = Me, R_5 = H, Me] exhibits νC=O at 1740 and νC=N near 1620–1630 cm^{-1}.

Signals of 5-position protons in the NMR spectra in $CDCl_3$ support the predominance of tautomer [**642a**]. Compounds [**642**, R = Me] exhibit for the two N–Me groups two different temperature-dependent signals. For R^5 = Ph the coalescence temperature is 84° and for R^5 = H about 70–75° (64JO370). This behaviour is due to the partial double-bond character of the C–NR_2 bond as in amides, i.e., the influence of the canonical form [**643**] (64JO370, 72JH285).

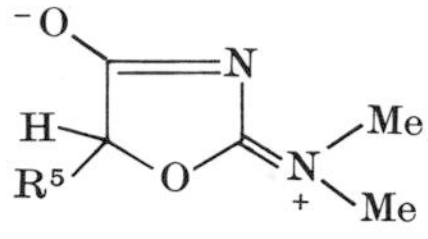

[**643**]

iv. *Unsubstituted 2-Aminooxazol-4-ones*. The initial tautomerism study of these compounds did not consider the five tautomers [**637a–e**] but only forms [**637b**] and [**637e**] for R^5 = Ar, R = H (61BF1226) and R^5 = Ar, R = alkyl, Ph (61BF1231). On this basis IR in the solid state suggested the predominance of the oxo–imino form [**637b**] whose proportion decreased as the substituent R became more electrophilic. Later this result was shown to be incorrect by the same authors (67BF207) and by others (62JO1686, 64JO370, 72JH285) using IR, UV and NMR spectroscopy and basicity measurements. In all the subsequent studies the authors used for comparison the model compounds [**639**] and [**640**] and/or the partially fixed derivatives just discussed.

It transpires that, whatever the 5-substituent, the 2-aminooxazol-4-ones exist in the amino–oxo form [**637a**] if R is an alkyl substituent; the only known exception occurs if R = Ph in which case the imino–oxo form [**637b**] is predominant.

In UV spectroscopy in MeOH all the compounds [**637**, R = alkyl] show absorption at 216–220 nm, typical of the amino–oxo structure (62JO1686, 67BF207, 72JH285); but [**637**, R = Ph] absorbs at 254 nm, similar to [**641**, R = Ph] (62JO1686). The pK_a values of these compounds [**637**, R = H, R^5 = Ar] are clearly closer to those of the derivatives [**642**, R = Me, R^5 = Ar] than to those of [**641**; R = Me, R^5 = Ar] (67BF207).

The IR spectra (solid state and MeCN) show bands in the same region as [**639**] and [**642**] (νC=O at 1715–1740 cm^{-1}, νC=N at 1620–1675 cm^{-1}) supporting structure [**637a**] and showing the occurrence of hydrogen-bonding association (νNH_2 appearing diffusely between 3300 and 2700 cm^{-1} in solid state and at 3340, 3260 cm^{-1} in MeCN) (67BF207, 72JH285).

NMR studies (64JO370, 67BF207, 72JH285) also support the presence of form [**637a**] in DMSO or $CDCl_3$, with signals attributable to 5H and amino protons. These two compounds [**637**; R = Me, R^5 = H or Ph] show all the NMR signals double in $CDCl_3$ or DMSO. Compound [**637**; R = Me, R^5 = Ph] (64JO370) exists in two crystalline forms which have different IR spectra in the solid state but identical solution IR, UV and NMR spectra. In fact the IR and NMR spectra in solution are concentration- and temperature-dependent and the authors attribute this behaviour to the existence of a monomer [**637a**] ⇌ dimer [**644**]

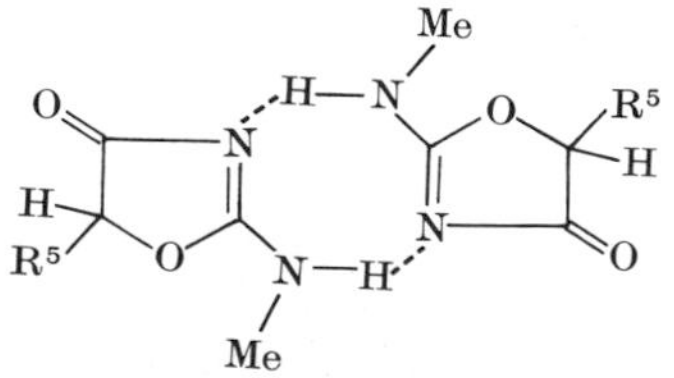

[**644**]

equilibrium. They rule out a tautomeric equilibrium because [**637a**; R = Me, R^5 = H] has similar λ and log ε values in dioxan, CH_2Cl_2, MeCN and MeOH, for example. A possible alternative explanation is the existence of *cis-trans* forms about the C–N bond (cf. the discussion of canonical form [**643**] above).

f. 2-Aminothiazol-4- and 5-ones and 2-Aminoselenazol-4-ones (II-52)

i. *2-Aminothiazol-5-ones* have not been further studied since the X-ray determination of [**645**] described in (II-52).

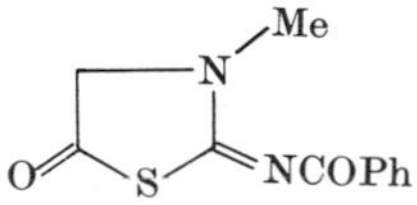

[**645**]

ii. *2-Aminothiazol-4-ones.* In the previous survey nothing was reported on 2-aminothiazol-4-ones, but since then a great deal of work has been done. As for the 2-aminooxazol-4-ones, they may exist in five tautomeric forms [**646a–e**].

The two first studies were contradictory, although both used UV comparisons with model compounds. French authors (63BF1022) concluded that all their compounds had the amino–oxo structure [**646a**, R^5 = Ph; R = H, alkyl] in EtOH except for [**646**, R = R^5 = Ph] which takes the imino–oxo structure [**646b**; R^5 = R = Ph]; these conclusions correspond well with those for the 2-aminooxazol-4-ones (Section 4-6Fe). Comrie (64J3478) claimed the presence of both tautomers [**646a**] and [**646b**] in aqueous solution for [**646**, R^5 = H, Et or

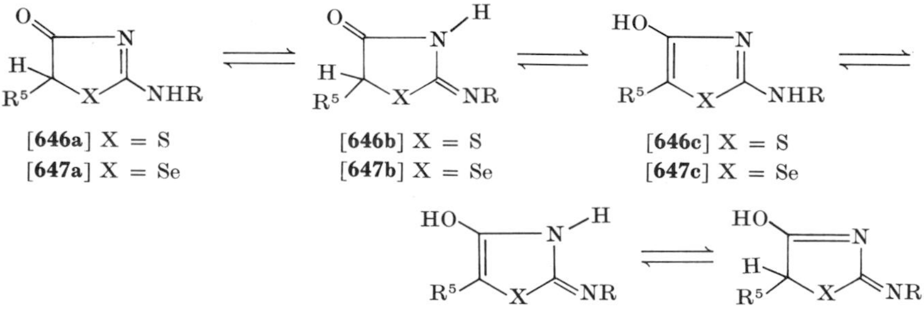

[**646a**] X = S
[**647a**] X = Se

[**646b**] X = S
[**647b**] X = Se

[**646c**] X = S
[**647c**] X = Se

[**646d**] X = S
[**647d**] X = Se

[**646e**] X = S
[**647e**] X = Se

Ph and R = H]. This confusion, as explained later (67AS1437), arose because the tautomeric species exhibit two absorptions at about 221–222 and 247–249 nm (in ether and less clearly in water). Comrie attributed the first to the imino form and the second to the amino form. The partially fixed derivatives [**648**; R = Me; R^5 = H, alkyl, Ph] absorb at 213–217 nm; however, [**649**; R = R′ = Me; R^5 = H, alkyl, Ph] absorb not only at 240–250 nm, but also at 228 nm, a fact which Comrie overlooked.* The bathochromic effect on the first maximum

* This absorption exists also in EtOH (68BF3477) but was not described by the French authors (63BF1022) either.

passing from [**646a**] to [**649**] is expected because of the influence of the methyl groups; the same effect has been observed in the 2-aminooxazol-4-ones (Section 4-6Fe).

Another difference between the work of these two groups arises in the IR data: the French (63BF1022) observe in the solid state νC=O bands at 1680–1690 cm^{-1} for [**646**] and for [**649**] and at 1705–1715 cm^{-1} for the imino derivatives [**648**]. Comrie (64J3478) claims a band at

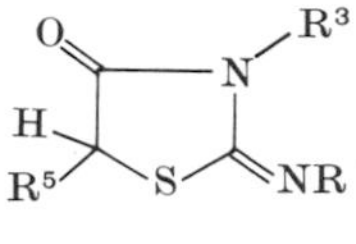

[**648**] 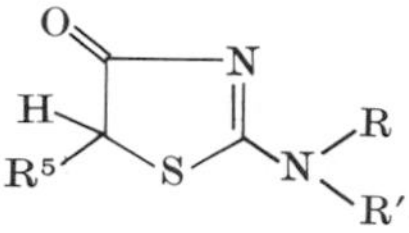 [**649**]

1740 cm^{-1} for all his compounds without mentioning the physical state. IR spectroscopy shows that 2-aminothiazol-4-ones do not exist in hydroxy forms, but conclusions about the amino–imino equilibria [**646a**] $\rightleftharpoons$ [**646b**] are not warranted because of the insufficiently different νC=O of model compounds (67AS1437). Nevertheless, some authors (67KG637) have attempted to study the tautomerism only from IR data: they claim an imino structure [**646b**; R = R^5 = H] in the solid state (which is incorrect) and an amino one [**646a**; R = R^5 = H] in $CHCl_3$ solution. In the solid state compound [**646**; R^5 = H, R = CH_2Ph] is assigned the amino form [**646a**], but the corresponding aryl derivatives are stated to exist in the imino structure [**646b**; R^5 = H; R = Ph or p-Me C_6H_4]. We see later why a solid state study of this series of compounds is inconclusive.

The 2-aminothiazol-4-ones represent a rare case of the successful use of polarography (67AS1437); the imino model [**648**; R^5 = H, R^3 = Me] gives a reduction wave but the amino model [**649**; R^5 = H, R = R′ = Me] does not. The polarographic behaviour of the tautomeric compounds provides evidence for the predominance of the amino tautomer [**646a**; R^5 = H, R = H, alkyl] and of the imino form [**646b**; R^5 = H, R = Ph].

NMR comparisons with model compounds (in D_2O) (67AS1437) give no information because of the similarity of the model compound spectra. Some authors (67JA647) used the low field positions of the two NH_2 protons as a proof of the existence of a "zwitterionic" form [**650**] in DMSO. This form is merely a canonical form of [**646a**]; further the observed shifts for NH_2 are similar to those of 2-aminooxazol-4-ones (Section 4-6Fe) except that they showed a single signal; however, hindered rotation was already pointed out for 2-dimethylaminooxazol-4-ones (Section 4-6Fe) and this sufficiently explains the two NH_2 signals.

An X-ray study of 5-phenyl-2-aminothiazol-4-one (69CH811) located the NH_2 hydrogens and found particularly short C–N distances. This was explained by the existence of the "zwitterionic form" [**650**], which

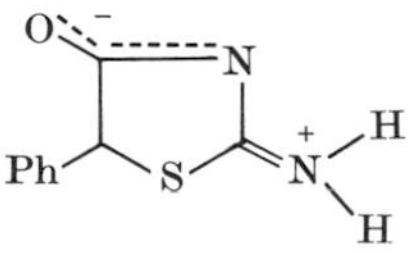

[**650**]

was erroneously claimed to have been demonstrated for the first time; as has been pointed out (70CH706), canonical forms invariably contribute to structures such as [**646a**], as they do to all normal amides. A later X-ray study (71CX(B)95), disclosed a dimer formed by two different molecules with two strong hydrogen bonds. Such dimeric forms in the solid state could explain the high melting point that some authors [67JA647) took as a proof of their "zwitterionic" form. A second crystalline modification possesses the same oxo–amino tautomeric structure (72CX(B)2074) and 2-anilino-5-phenylthiazol-4-one an analogous one [**646a**, R = R^5 = Ph] (73CX(B)1157), as does the 5-acetic acid derivative (72CX(B)2417). An X-ray determination of the parent compound [**646**, R = R^5 = H] reaches the surprising conclusion that it exists in the oxo–imino form [**646b**, R = R^5 = H])72CX(B)-2421): however, H atoms were not located, and this structure needs confirmation. The structure of the imino–oxo derivative [**648**, R^3 = Me, R = R^5 = Ph] has also been substantiated by X-ray crystallography (73CX(B)1160). Other X-ray studies by the same French group includes the compounds [**649**, R = Me, R′ = R^5 = Ph] (73CX(B)2635) and [**646a**, EtOH, R = R^5 = Ph] (73CO(276C)657).

iii. *2-Aminoselenazol-4-ones*. As for their sulfur analogues, these compounds can exist in five tautomeric forms [**647a–e**]. Initial UV comparisons (in water) with the fixed derivatives [**651**], [**652**; R = Me, R_5 = H] and [**653**; R = H, alkyl, R^3 = alkyl, R^5 = Ph, alkyl] led

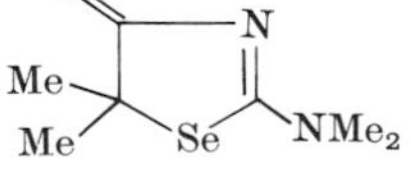

[**651**] 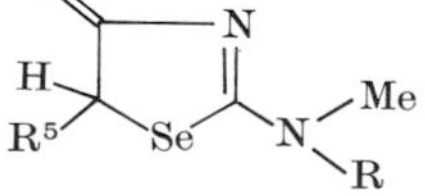 [**652**] 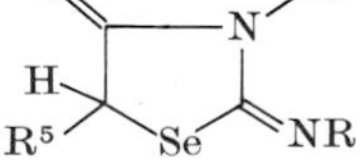 [**653**]

(erroneously) to conclusions in favour of the oxo–imino structure [**647b**] for the tautomeric species (63J5713). French authors (66CO-(262C)285, 68BF1099), surprised by the contrast with the 2-amino-oxazol- and 2-aminothiazol-4-ones, repeated this UV work. The selenazoles [**647**] do in fact exist in the oxo–amino form [**647a**] in EtOH; the confusion arose from the similarity of absorption of the fixed imino derivatives [**653**] at 224–227 nm, and the tautomeric species [**647**, R = H] at 225 nm or [**647**, R = alkyl] at 227–229 nm.* The earlier comparison did not allow for the bathochromic effect of methyl groups on the exocyclic nitrogen atoms; the assumption of an amino structure explains the results. Comparison of the pK_a values of [**647**; R = H, R^5 = Ph], [**652**; R = Me, R^5 = Ph] and [**653**; R = H, Me, R^5 = Ph] agrees with the amino structure [**647a**].

For compound [**647**, R = R^5 = Ph], where the exocyclic nitrogen bears a phenyl group (68BF1099), UV demonstrates the oxo–imino structure [**647b**], again similar to the oxazole (Section 4-6Fe) and thiazole analogues.

g. Aminoimidazolones (II-53, II-71)

i. *1-Substituted 2-Aminoimidazol-4-ones.* At the time of the last survey (II-53), studies were inconclusive. No complete investigation has yet been undertaken, but some conclusions can be drawn from fragmentary results.

NMR spectra in D_2O prove the 5-position carbon is *sp*^{3} (71JA5552, 72JH203); discussion can therefore be limited to three possible tautomers [**654a–c**]. The only available models (71JA5552) have UV maxima in EtOH at 205 nm for [**655**] and 208 nm for [**656**]. The pK_a values of 8.07 [**655**] and 9.01 [**656**] are quoted, but the difference between them

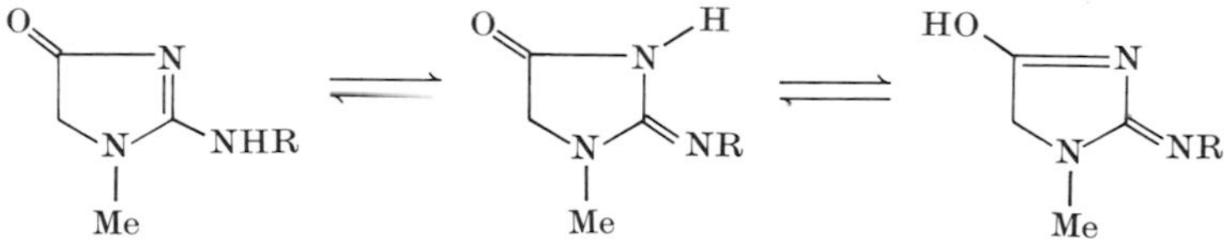

[**654a**] ⇌ [**654b**] ⇌ [**654c**]

of 0.94 pK_a unit seems rather large. Compounds [**654a**; R = H, Me] absorb at 235 nm (68JO552, 71JA5552), and [**657**] shows a second maximum at 212 nm. The pK_a values of these tautomeric compounds [**654**] are 4.55–4.80: these results rule out form [**654b**]. The authors do not consider form [**654c**]; indeed [**654c**] is not probable, and the compounds [**654**] almost certainly exist in the amino–oxo form [**654a**].

An X-ray study of 2-amino-1-ethoxy(imino)methyl-5-phenylimidazol-4-one (69CH1038) demonstrates the amino structure [**658a**]: as

* A second absorption occurs at 263 nm for [**647a**; R = H, R^5 = Ph] and 242 nm for [**652**; R = Me, R^5 = Ph].

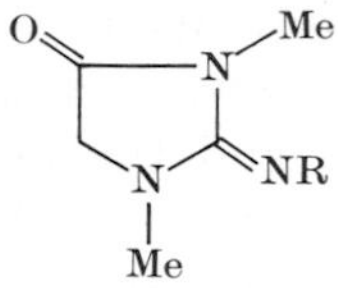

[**655**] R = H
[**656**] R = Me

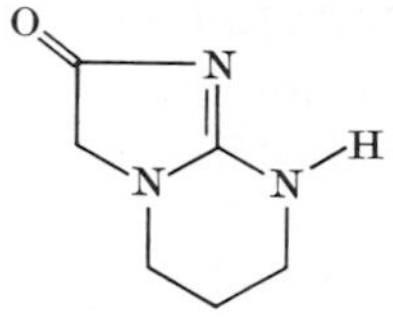

[**657**]

expected, there are important contributions from charge-separated canonical forms, but the authors claim to have "demonstrated the zwitterionic nature," particularly in view of the fact that two true zwitterionic structures exist is naïve and misleading (cf. [**658b, c**]), and the discussion on 2-amino-5-phenylthiazol-4-one (Section 4-6Ff), cf. (70CH706) is relevant here.

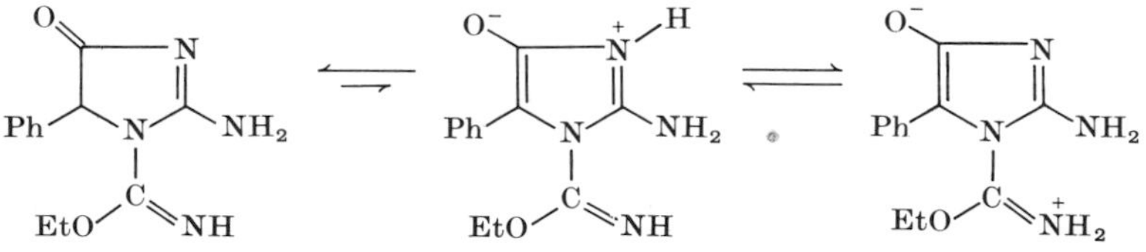

[**658a**] [**658b**] [**658c**]

ii. *2-Aminoimidazol-5-ones.* As for the previous compounds, NMR spectroscopy in D_2O shows the sp^3 nature of the 4-carbon (71JA5552), and thus only two tautomeric forms [**659a, b**] need be considered further. Compound [**655**], a model for [**659b**] was discussed in the previous section (λ_{max} 205 nm). The fixed derivative [**660**] absorbs in the UV

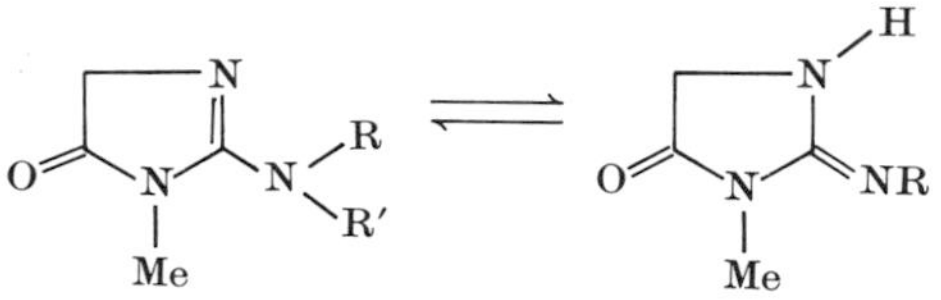

[**659a**] R′ = H
[**660**] R = R′ = Me

[**659b**]

at 208 nm (in EtOH) and has pK_a 7.57 (71JA5552). Therefore UV spectroscopy does not differentiate between [**659a**] and [**659b**]. The pK_a values of [**659**; R = H, Me] of 7.91 and 7.96 (68JO552, 71JA5552) do not allow an unambiguous choice, in the absence of exact knowledge of the cation structure, and in view of the uncertainty regarding the pK_a of compound [**655**] (see above). However, it seems probable that forms [**659a**] and [**659b**] both contribute significantly to the equilibrium. In the literature the compounds have been represented both in the imino form [**659b**; R = H or Me] (68JO552) and in the amino form [**659a**; R = H or Me] (71JA5552).

iii. *Unsubstituted 2-Aminoimidazol-4(5)-ones*. No definite conclusion was drawn in the previous survey (II-53). UV absorption at 213–225 nm (in EtOH) (68JO552, 71JA5552) and pK_a values at 4.50–4.80 led the authors to suggest an oxo–amino structure [**661a**; R, R′ = H or Me]. These compounds show, besides the functional group tautomerism, annular tautomerism. The UV results do not allow choice amongst all the possible tautomers; for [**661a**] a bathochromic effect would be expected when methyl groups are successively added on the exocyclic nitrogen (cf. Section 4-6Ff), but this is not found. NMR spectroscopy in D_2O (71JA5552) shows the 5-position carbon is sp^3-hybridized but five tautomers [**661a–e**] still have to be considered, and the problem remains unsolved.

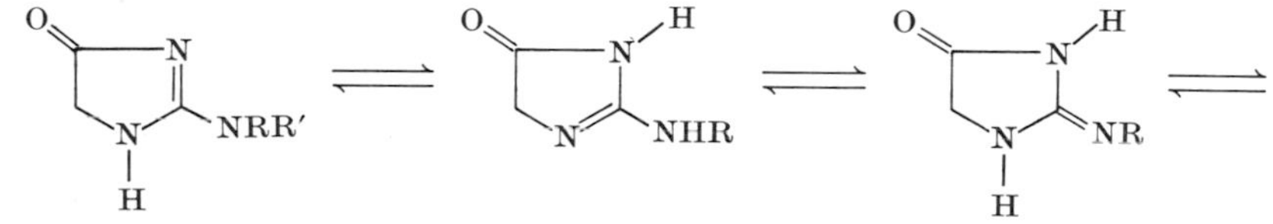

[**661a**] R′ = H
[**662**] R = R′ = Me

[**661b**] [**661c**]

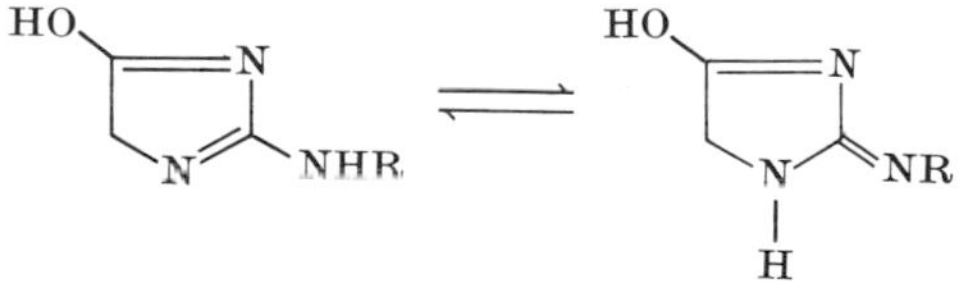

[**661d**] [**661e**]

iv. *4-Aminoimidazol-2-ones*. The previous survey (II-71) reported contradictory results on these compounds for which structures [**663a**] and [**663b**] were both postulated. No further work is available.

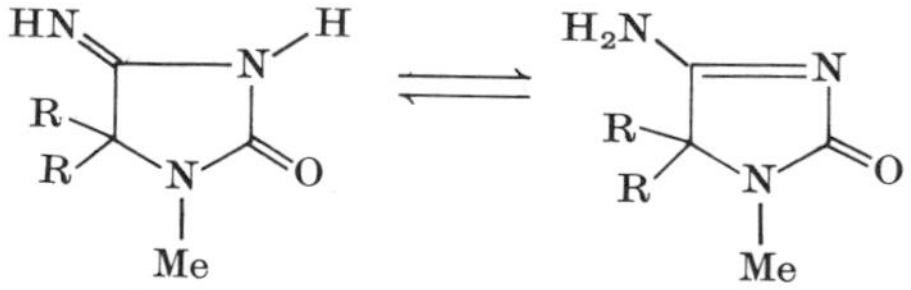

[**663a**] [**663b**]

h. Aminothiadiazolones

i. *5-Amino-1,3,4-thiadiazol-2-ones*. The 2-alkoxy derivatives [**665**, R = alkyl; R′ = H, alkyl or Ph] (64AS174) can exist in two forms [**665a,b**], and for these, IR results in the solid state (δ NH_2 at 1604–1638 cm^{-1} and νC=N at 1500–1507 cm^{-1}) suggest the amino structure [**665a**]. However, 5-amino-1,3,4-thiadiazol-2-one itself [**664**] is described as [**664c**] because of a νC=O band at 1697 cm^{-1} and a band at 1668 cm^{-1} attributed to the exo C=N bond. We do not consider these two bands as sufficient proof for the imino structure [**664c**], in view of the tendency of amino substituents to exist in the NH_2 form.

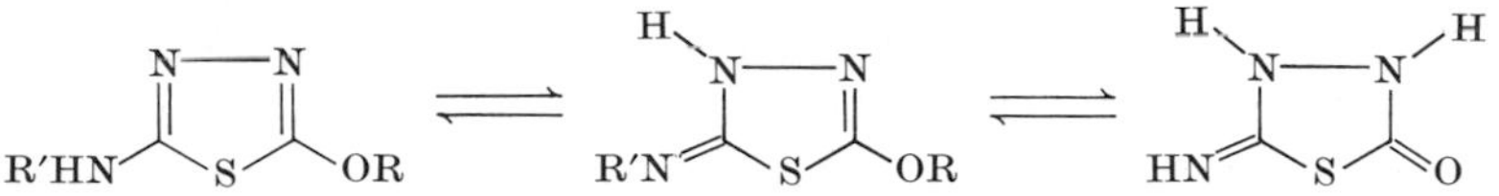

[**664a**] R = R′ = H
[**665a**] R ≠ H

[**664b**] R = R′ = H
[**665b**] R ≠ H

[**664c**]

ii. *5-Amino-1,2,4-thiadiazol-3-one.* An incomplete UV study (in EtOH) (62J4191) utilized as partially fixed models the alkoxy derivatives [**666**; R′ = alkyl, aryl] which were all attributed the amino structure [**666a**] in spite of the fact that potential NHPh derivatives often exist in the imino form [**666b**, R = Ph] as found, e.g., for 2-aminothiazol-4-ones (Section 4-6Ff). They explain the different pattern observed for the parent compounds [**667**] by the possible intervention of the structure [**667c**], but here again there is no proof that the anilino group does not exist in an imino structure.

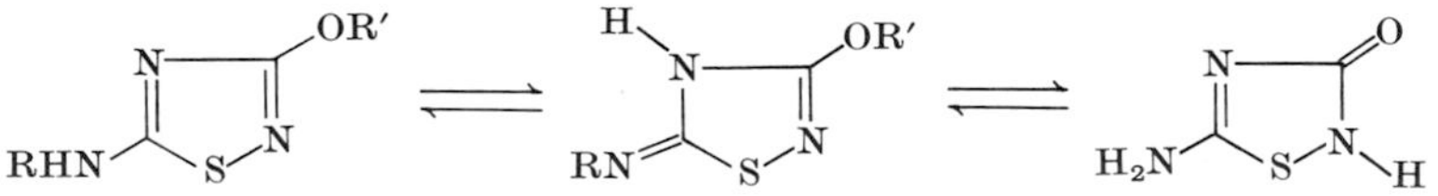

[**666a**] R′ ≠ H
[**667a**] R′ = H

[**666b**] R′ ≠ H
[**667b**] R′ = H

[**667c**]

i. 3-Amino-1,2,4-triazol-5-ones

Structures [**668**; R = COR′, H; R^4 = COMe, H] were assigned tentatively on the basis of UV, IR and NMR data (70CH866).

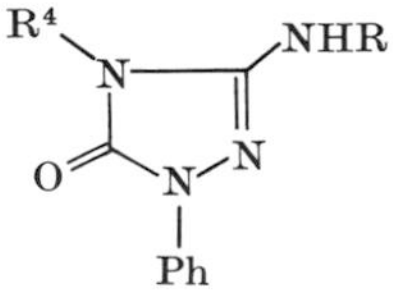

[**668**]

G. One Potential Mercapto and One Potential Amino Group

a. 5-Aminothiazole-2-thiones (II-62)

These compounds were early reported to exist as [**669a**] (47J1598, 48J2031) but the previous survey (II-62) viewed this result with caution, and indeed the earlier data have since been reinterpreted in terms of the thione form [**669b**] (66SA2005).

Ph, N, H_2N, S, SH ⇌ R, N, H, H_2N, S, S

[**669a**] [**669b**]

b. Aminothiadiazolethiones (II-64)

i. *2-Amino-1,3,4-thiadiazole-5-thiones.* Unacceptable conclusions and contradictory results were reported in the previous survey (II-64) (see also II-73). Subsequently structure [**670**] was confirmed by CSNH vibrations in the IR (64JO1174), and by X-ray analysis (72CX(B)1584).

H, N—N, S, S, NH_2

[**670**]

ii. *5-Amino-1,2,4-thiadiazole-3-thiones.* The only relevant work concerns the UV spectra (in EtOH) of S-substituted derivatives [**671**] (62J4191). As for their alkoxy analogues (Section 4-6Fh), they were all assigned the amino form [**671a**], but the anilino compounds [**671**, R = Ph] show very different UV absorptions and possibly exist in the imino structure [**671b**].

SR′, N, RHN, S, N ⇌ H, N, SR′, RN, S, N

[**671a**] [**671b**]

c. Aminomercaptotriazoles (II-65)

The previous survey reported (II-65) for 2-hydrazino-5-mercapto-1,3,4-triazole the betaine structure [**672a**] or [**672b**] as determined by X-ray diffraction; the authors (58CX808) indicated strong intermolecular hydrogen bonding and a C–S distance of 1.74 Å.

N—N, ^{-}S, N, $NH—\overset{+}{N}H_3$, H ⇌ H, N—N, ^{-}S, N, $NH—\overset{+}{N}H_3$

[**672a**] [**672b**]

Later (70CH631, 71J(B)1270) structure [**673**] was established by X-ray diffraction. Here the C–S bond length is shorter (1.68 Å) and intermolecular hydrogen bonds prevent interactions between N(1)–NH_2 and the 2-hydrazino group. X-Ray structure determinations on structures [**673/1**] (73J(PII)4), [**673/2**] (73J(PII)6) and [**673/3**] (73J(PII)9) have recently been reported.

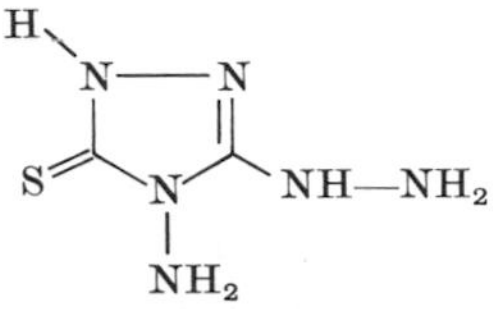

[**673**]

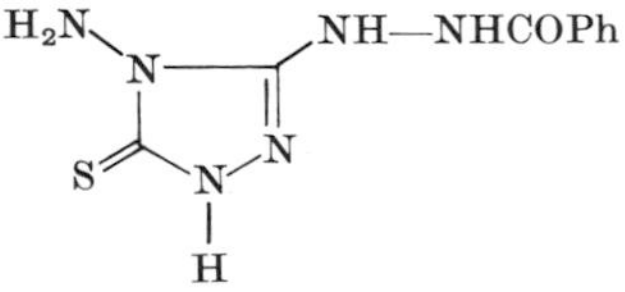

[**673/1**] 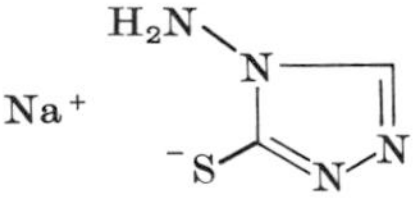 [**673/2**] 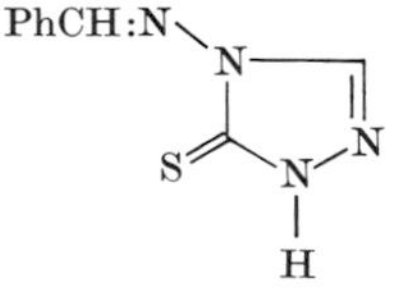 [**673/3**]

H. Compounds Containing Two Potential Tautomeric Groups: General Conclusion

The results obtained in this series are too fragmentary to justify recapitulation in a table analogous to those given for the monofunctional derivatives (Tables 4-25, 4-28, 4-29). These conclusions are also necessarily more tentative.

However two important generalizations become apparent from the data which have been discussed in the present section:

i. The tendency for compounds containing BH substituent groups to retain the endocyclic double bond decreases in the order NH_2 > SH > OH, which is consistent with the results for the monofunctional derivatives (Section 4-5H). If both BH groups are identical, the predominant tautomer most often possesses an amino–amino (Section 4-6Dd), mercaptothione (Section 4-6Cb, 4-6Cc) or dioxo (Sections 4-6Bd, e, f, g) structure.

   If the two BH groups are different, the predominant tautomer will be predictable on the assumption that the groups behave independently and follow the generalization above. Thus, for example, the amino–oxo structure is preferred to the hydroxy–imine structure (Section 4-6Fb).

ii. In comparison with the monofunctional compounds, the "fully aromatic" tautomer (with two endocyclic double bonds) is less favoured. For example, in the case of the pyrazol-5-ones the presence of a 3-amino group causes the OH tautomer to be less favoured (compare Sections 4-2Ge and 4-6Fd). Again, for the thiazol-5-ones, the presence of a 2-alkoxy group also reduces to insignificance contributions of the OH tautomer (compare Sections 4-3Ce and 4-6Bd). In the same way, whereas the anilino form was always predominant when the molecule bore only one potential tautomeric group (Section 4-5G), the phenylimino tautomer (=NPh) is present, e.g., for the 2-anilinooxazol-4-ones (Section 4-6Fe).

7. Compounds with Potential *N*-Oxide Groups (II-79)

N-Hydroxyazoles and their benzo derivatives are in tautomeric equilibrium with *N*-oxide forms. The forms can be described as OH [**674a**] and NH [**674b**] (cf. Section 1-1Ai, structures [**2a**] and [**2b**]). If two nitrogen atoms bear *N*-oxide functions, a second type of tautomerism can occur, e.g., [**675a–b**]. If there is a functional group (OH, SH, NH_2) attached to carbon in the molecule, a third type of tautomerism can occur [**676a–b**]. We examine successively these three types of tautomerism, which all involve cyclic conjugated structures; the possible nonaromatic additional tautomeric forms are of lesser importance and are not considered.

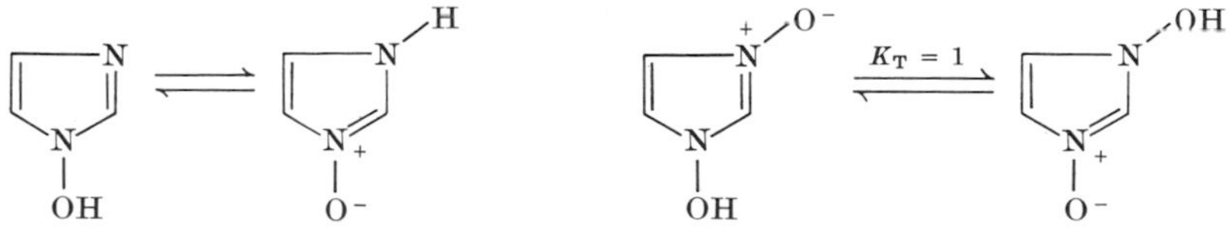

[**674a**] ⇌ [**674b**]   [**675a**] $\underset{}{\overset{K_T = 1}{\rightleftharpoons}}$ [**675b**]

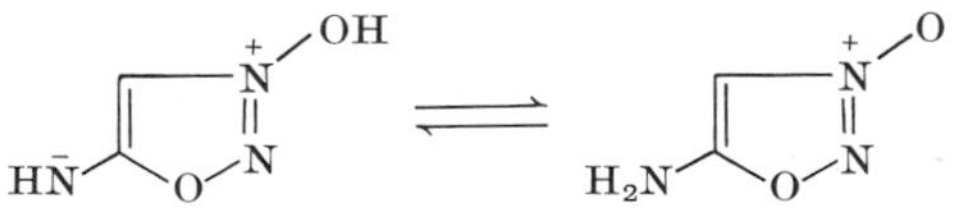

[**676a**] ⇌ [**676b**]

A. One *N*-Oxide Group

This type of tautomerism is known for the systems [**674**], [**677**]–[**680**], [**684**], [**685**, X = H] and [**688**] and has been studied except for the *N*-hydroxytetrazoles [**677**] (71MI105) and the 2-hydroxyindazoles [**678**] (67MI322).

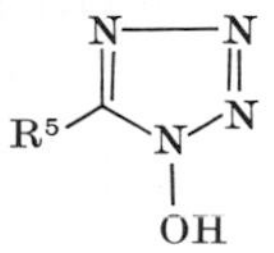

[**677**] 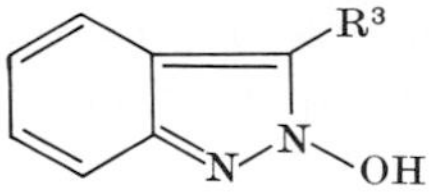 [**678**]

a. 1-Hydroxypyrazoles

1-Hydroxy-3,4,5-trimethyl pyrazole has been shown by UV spectra to exist predominantly in the OH form [**679a**]: in H_2O pK_a determination indicates ca. 3% of the *N*-oxide form [**679b**] coexisting with the OH form (73J(PII)164).

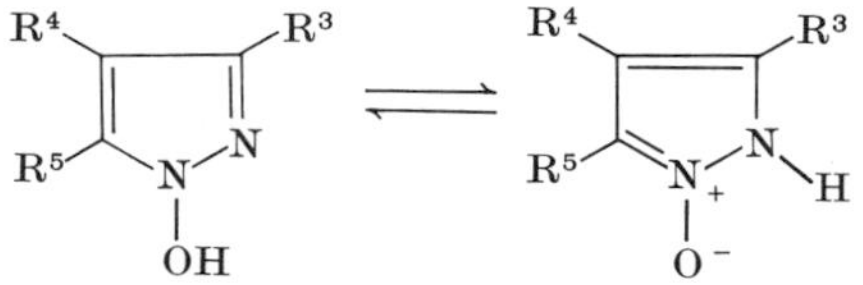

[**679a**] [**679b**]

The 1,4-dihydroxypyrazoles [**679**, R^4 = OH] were assigned the dihydroxy structure [**679a**] (69JO187) based on NMR signals at 12.5 and 8.42 ppm (solvent not given) and a strong double band at 2600 and 3000 cm^{-1} in the IR spectrum (state not given), but these data are also consistent with the NH tautomer [**679b**].

b. 1-Hydroxyimidazoles

Previous results (64CI1837, 65TL1565, 69BJ3204) have been critically summarized (70ZC338, 71J(B)2350). In aqueous solution, the equilibrium [**680a**] ⇌ [**680b**] (R^2 = R^4 = R^5 = Ph) has been studied by one of the present authors using basicity measurements (Section 1-4A) and UV spectroscopy (1-6D); PMR spectroscopy gives no useful information (71J(B)2350). The application of Eq. (1-16) (Section 1-4A) gives K_T = [**680b**]/[**680a**] as ca. 3, but the UV spectrum of compound [**680**] (R^2 = R^4 = R^5 = Ph) resembles more that of the OH model [**681**] than that of [**682**]. However, as the authors emphasize, the NR compound [**682**] is not a perfect model because of steric interactions between the *N*-methyl and the 2- and 4-phenyl groups. This leads them to write that, in spite of appearances which favour the OH tautomer [**680a**], "We believe that the best conclusion is that the two

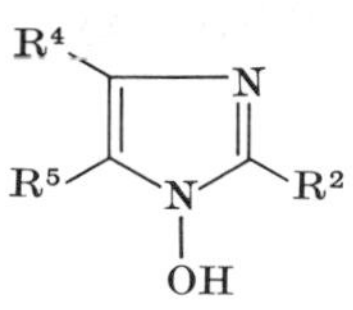

[**680a**] 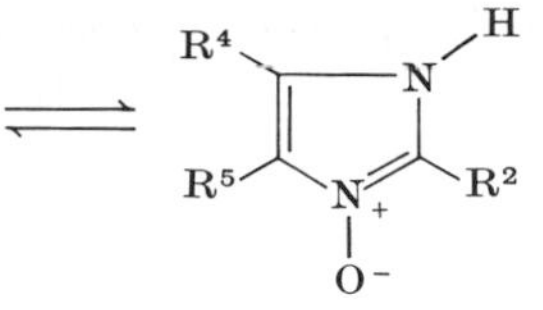 [**680b**] 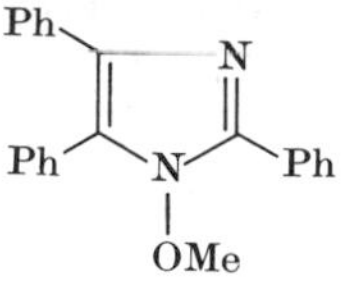 [**681**]

tautomers [**680a**] and [**680b**] exist in comparable proportions in aqueous solution; unfortunately it is not possible to be precise." In other solvents (EtOH, cyclohexane, chloroform, acetonitrile, dioxan) UV comparisons clearly favour the OH form (71J(B)2350). However, the preceding difficulty still applies; the NR derivative [**682**] gives "abnormal" spectra in all these solvents.

Another author (70ZC338) thinks that both types of tautomers exist in solution ($CHCl_3$, CH_3CN, dioxan) because the IR spectrum exhibits bands at 3410 and 3630 cm^{-1} assigned, respectively, to tautomers [**683b**] and [**683a**].

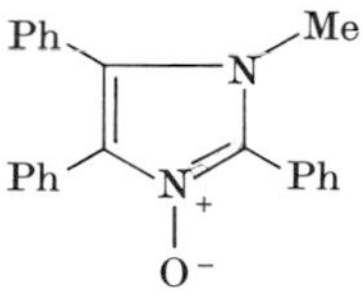

[**682**] 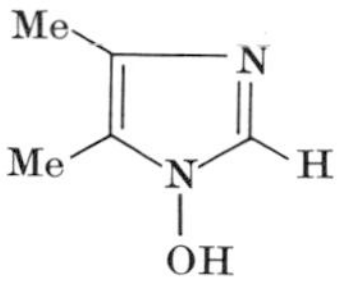 [**683a**] 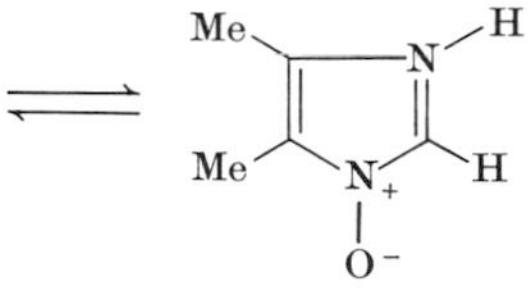 [**683b**]

These compounds are always strongly associated even in dilute solution; in the solid state those associations can lead to a state in which the distinction between the two tautomeric forms is blurred, denoted OH/NH by analogy with the pyrazol-5-ones. However, it has been claimed (69CB4177) that the two compounds [**680a**] and [**680b**] ($R^2 = R^4 = R^5$ = Ph) have been separately isolated in the crystalline state: they give different solid state IR spectra (2300–2500 cm^{-1} region) and were said to give different mass spectra (cf. Section 1-6Gb). It is not yet possible to decide finally between this point of view and the alternative explanation (71J(B)2350) of two OH/NH polymorphic forms. A later mass spectrometry study is in favour of a polymorphism of the same tautomer (probably [**680b**]) (74BB105). Crystallographic study locating the hydrogens involved in the hydrogen bondings (neutron diffraction for example), would give a definite answer.

c. 1-Hydroxy-1,2,4-triazoles

Mass spectrometry (70PC(312)869) shows that in the vapour phase these compounds [**684**] behave as *N*-oxides [**684b**], not as *N*-hydroxy derivatives [**684a**]. UV, NMR and IR spectra indicate the *N*-oxide form [**684b**] also predominates in nonpolar solvents: for the 3-phenyl derivative it is stated that UV comparison indicates 97% [**684b**]: 3% [**684a**] in dioxan (72PC(314)101).

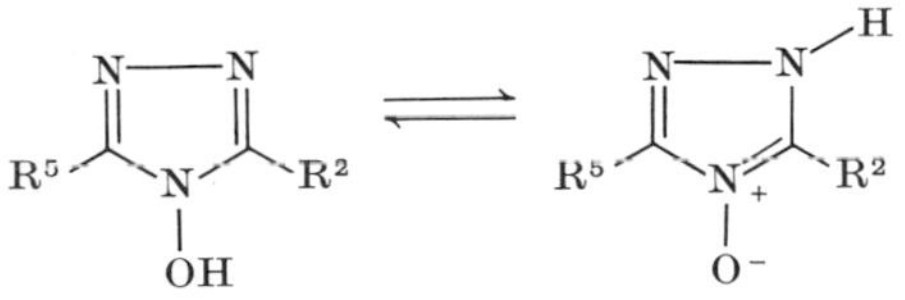

[**684a**] [**684b**]

d. 1-Hydroxybenzimidazoles

This series has also been extensively studied (for a summary of the earlier results, see 70ZC211, 71J(B)2350). UV comparison of the tautomeric species [**685**] with the fixed derivatives [**686**] and [**687**] in the same solvent must be applied carefully because very complicated UV spectra require a graphical rather than a numerical comparison of the three individual spectra to determine which form is predominant, because of the occurrence of wavelength shifts of the order of 5 nm.

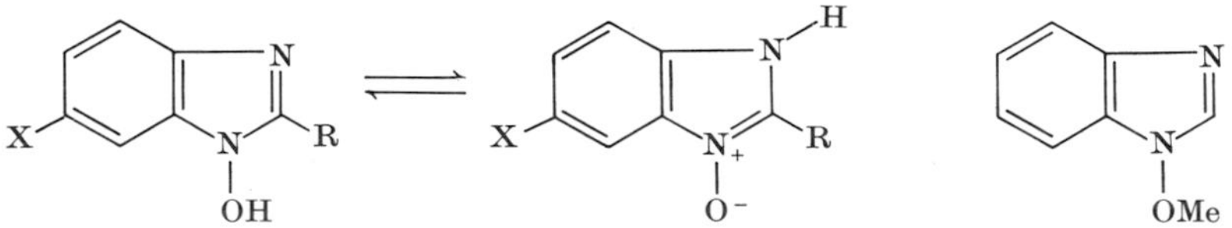

[**685a**] [**685b**] [**686**]

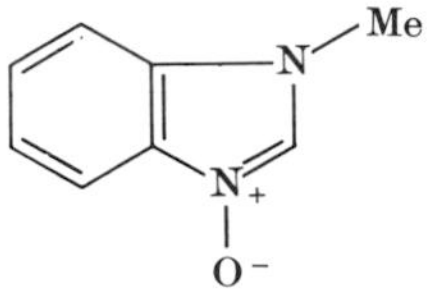

[**687**]

For 1-hydroxybenzimidazole itself [**685**, R = X = H], the OH tautomer [**685a**] predominates in EtOH but the NH form [**685b**] in water (63CT1375). Similar behaviour was found for the 6-nitro compounds [**685**, R = H or alkyl, X = NO_2]: the OH form [**685a**] in organic solvents, the NH form [**685b**] in water, and a mixture of both forms in a mixture water–EtOH (67J(C)1764). Other authors found that the OH isomer [**685a**] is always favoured (water, EtOH) for the 2-phenyl derivative [**685**, R = Ph] (66JH51).

1-Hydroxybenzimidazole has recently been reinvestigated (71J(B)-2350) using the pK_a and UV methods (PMR spectra give no useful information). Contrary to the 1-hydroxyimidazoles, the NR model [**687**] is not sterically hindered and should lead to a reliable K_T = [**685b**]/[**685a**] of ca. 12 deduced from the pK_a of [**686**] and [**687**] in H_2O. The authors show that the neutral species UV spectra, while not as clear-cut as could be wished, are not seriously at variance with this conclusion, which also agrees with the earlier work (63CT1375). In other solvents (C_6H_{12}, CH_3CN, $CHCl_3$, dioxan) UV results (71J(B)2350) indicate that the OH tautomer is predominant; in water–EtOH a mixture of both tautomers is observed: e.g., 75% [**685b**] and 25% [**685a**] in H_2O:EtOH 1:1.

e. 1-Hydroxybenzotriazoles (II-79)

This series was the earliest studied (36J111) and the only one cited (II-79) in the previous survey. In EtOH, UV spectra give the following percentages: 80–85% of OH form [**688a**] and 15–20% of NH form [**688b**]; a 6-nitro group increases the percentage of the OH form (resonance stabilization II-79). In water UV spectra indicate that the NH tautomer [**688b**] predominates (36J111).

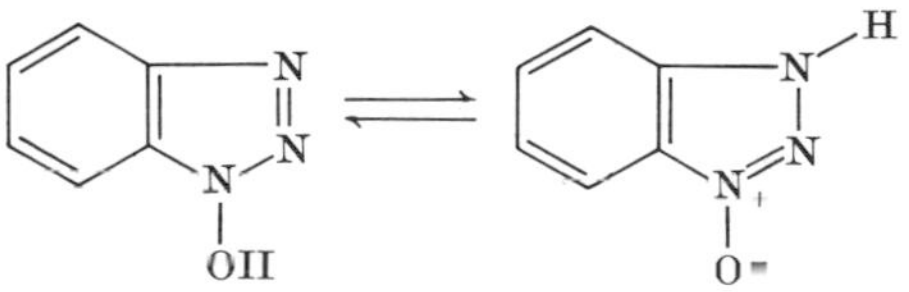

[**688a**] [**688b**]

Recently the tautomerism of benzotriazole 1-oxide and its 4- and 6-nitro derivatives has been reinvestigated by modern methods (73J(PII)160). In aqueous solution they all exist predominantly in the *N*-oxide (NH) form [**688b**] on UV evidence, although pK_a values indicate that the K_T value cannot be very great. In EtOH solution, the equilibria are displaced towards the OH form [**688a**] which predominates except for the 4-nitro compound where the two forms are present in comparable amounts.

f. General Conclusions for N-Hydroxyazoles

It seems possible now to draw more general conclusions: OH tautomers are favoured in most organic solvents and only very polar solvents (71J(B)2350) [or very acidic (73BB215)] such as water stabilize the NH form. A parallelism is shown with the solvent effects on the OH $\rightleftharpoons$ NH equilibrium of pyrazol-5-ones (Section 4-2Gc). It is less certain why the NH tautomer [**684b**] is predominant in the vapour phase.

There is no study available of the tautomerism of *N*-aminoazoles, (see 74HC(17)213) although it is possible to write two tautomeric forms, e.g., [**689a,b**] and derivatives with fixed NR forms are known, e.g., [**690**] (64MC814). However, various comparisons [e.g., with 5-aminopyrazoles (Section 4-5Bg) and pyrazol-5-ones (Section 4-2G)] indicate that form [**689a**] will be favoured.

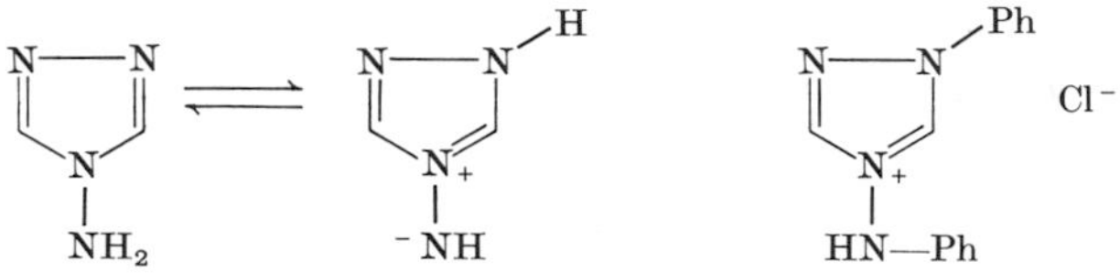

[**689a**] [**689b**] [**690**]

B. One *N*-Oxide and One *N*-Hydroxy Group

This kind of tautomerism often shows itself in NMR spectroscopy in the form of autotropic rearrangements: a single signal is observed for the 4- and 5-methyl groups of the imidazole [**691**] (64JO1620) and for the 3- and 5-methyl groups of the pyrazole [**692**] even at $-60°$ (69JO194)

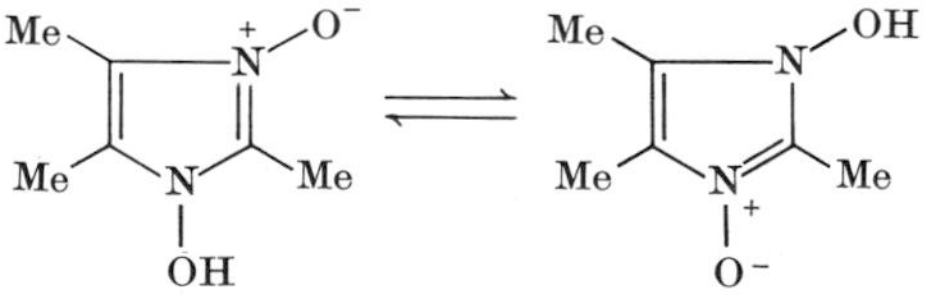

[**691a**] [**691b**]

(for the benzimidazole series, see 63TL785, 67J(B)911). That this phenomenon occurs for the imidazole derivative [**691**] rules out claims (69JO194) that the magnetic equivalence of the 3- and 5-methyls of [**692**] constitutes evidence for the chelate structure [**692c**].

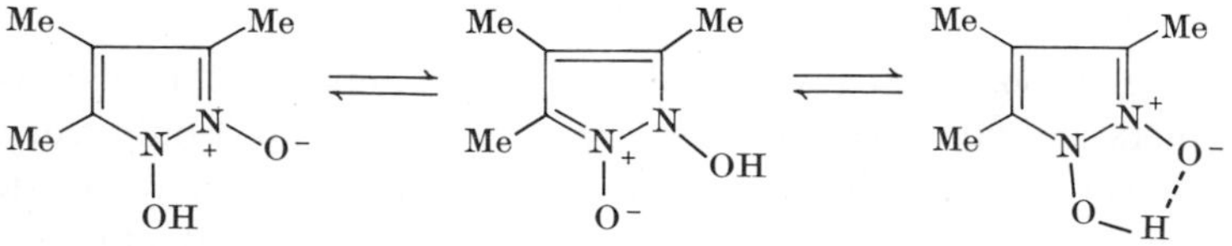

[**692a**] [**692b**] [**692c**]

C. One *N*-Oxide and One Functional BH Group

The *N*-oxide amino form [**676b**] has recently been preferred to the hydroxysydnone imine form [**676a**] by comparison with the NMe_2 model compound [**693**] using UV and IR (71T4449).

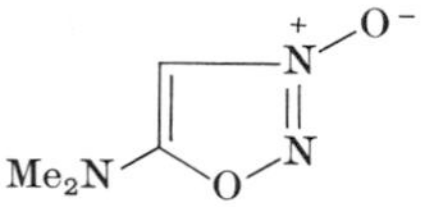

[**693**]

8. Potential Methyl or Substituted Methyl Compounds (II-80)

As already pointed out in the last review (II-80), the methyl or substituted methyl form with the endo double bond [**694a**] is almost always energetically favoured over the methylene exo form [**694b**] for aromatic heterocycles (Section 4-5H).

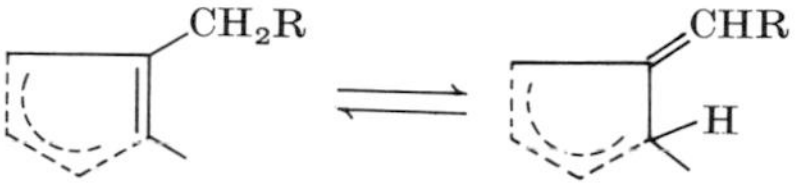

[**694a**] [**694b**]

Most known examples of the existence of forms of type [**694b**] in significant quantities can be classified into two groups: (i) the ring bears substituent C=X groups, usually oxo, which decrease the stability of the form [**694a**]; or (ii) the substituent R is an acetyl or an ethoxycarbonyl group which is conjugated with the C=C bond in [**694b**].

A. Oxo Compounds (II-51)

Compounds in which oxo groups induce potential methyl derivatives to exist in the methylene form have been discussed in the relevant chapter; oxazol-2-ones [**696**] (Section 4-3Bb), imidazol-2-ones [**697**]

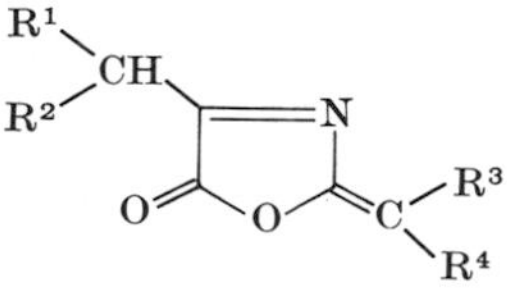

[**695**] 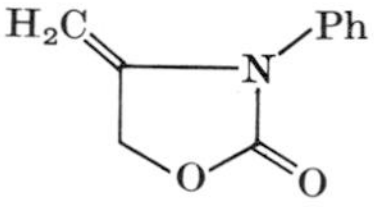 [**696**]

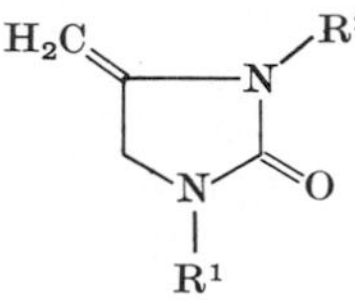

[**697**] 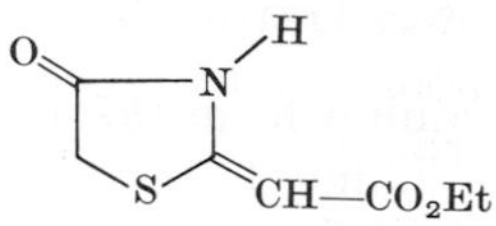 [**698**]

(Section 4-3Bd), thiazol-4-ones [**698**] (Section 4-3Cf) and, most important, pseudooxazolones [**695**] (II-51) (Section 4-3Cc) have been studied. The pseudooxazolones [**695**] must be considered as a special case since one of the two substituents must necessarily have an exo methylene structure of type [**694b**].

B. CH_2COX Derivatives

Three tautomeric forms occur [**699a–c**], but two of them [**699a,b**] are identical as regards ring tautomerism.

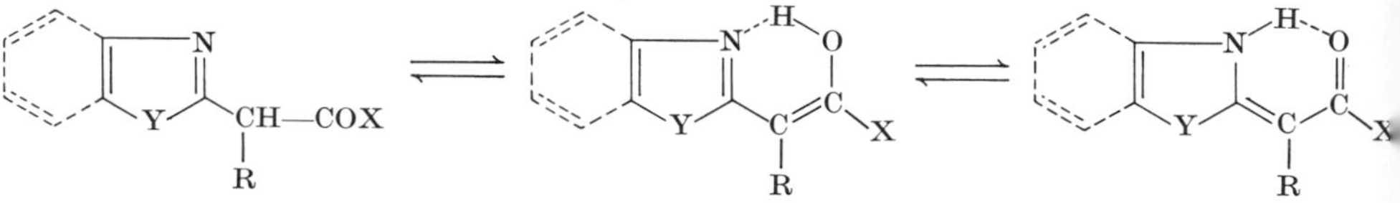

[**699a**] [**699b**] [**699c**]

In the case of the benzoxazole derivative [**700**], an NMR study in DMSO (66BA199) showed that tautomer [**700a**] is absent and the results favour structure [**700b**]. In the benzothiazole series [**701**] the authors suggest (66PC(31)262) an equilibrium between the forms [**701a**] and [**701b**] but do not provide any proof. For a complex benzisothiazole derivative which takes the methylene structure, see (72J-(PII)2125).

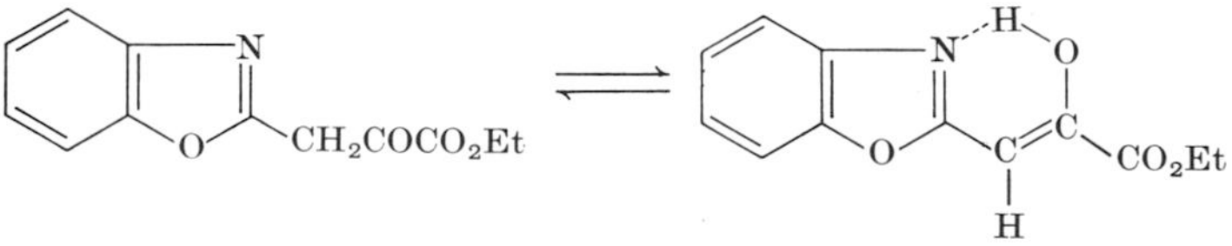

[**700a**] [**700b**]

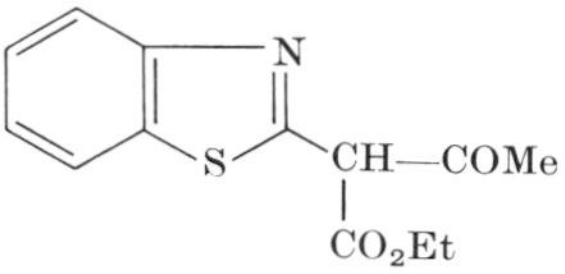

[**701a**] 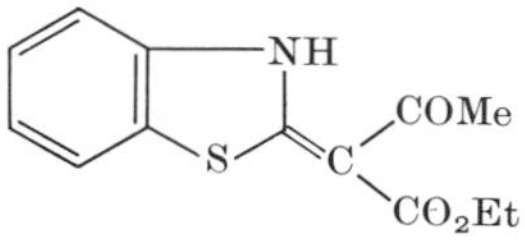 [**701b**]

Several papers deal with the 2-substituted thiazoles. Thus it has been shown that compounds [**702**] exist in solution as a mixture of enolic forms the proportions of which depend on the solvent used. The corresponding amino derivative would exist solely in the form [**703**]

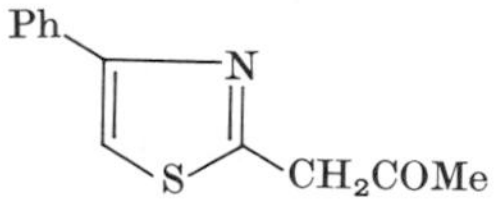

[**702**] 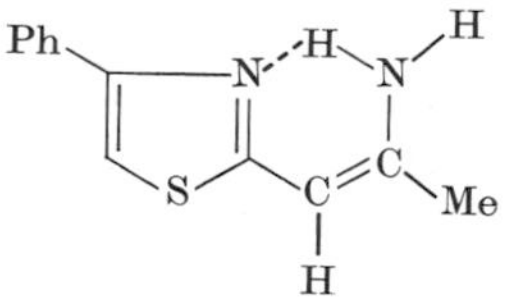 [**703**]

(68H2102). An interesting study of the acid [**704**] evaluated the ratio [**704a**]/[**704b**] = 10^5 (this ratio is even higher for the anion) (70J(B)200). Finally, the form [**705**] is stable (70SA(A)153), largely because of the presence of the oxo group in position 4 (see preceding section and Section 4-3Cf).

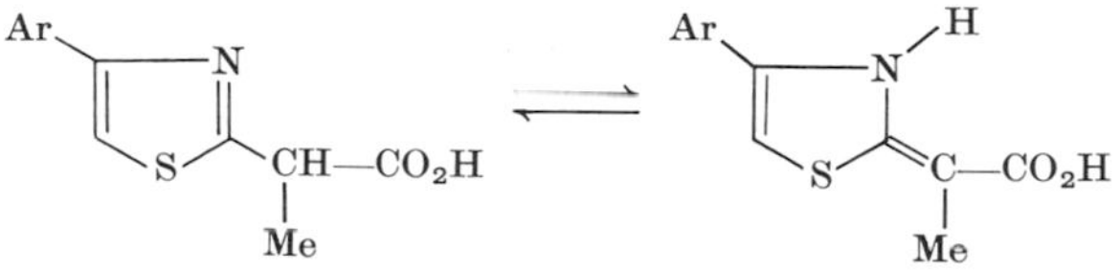

[**704a**] [**704b**] 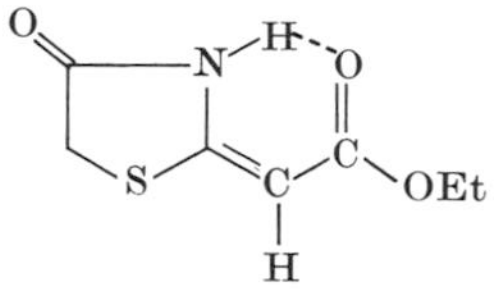 [**705**]

It appears from those results that, in spite of the conjugated chelation [**699c**], the tautomer with the exo double bond [**694b**] is rarely favoured. The phenylhydrazone [**706**] derived from 2-cyanomethylbenzimidazole (70CZ2936) can exist in two tautomeric forms [**706a**] and [**706b**]; according to Dudek and Dudek (66JA2407) (Section 1-5Ab), the measured $J(\text{H}-^{15}\text{N})$ is proportional to that coupling for [**706a**] (estimated as 92 Hz), the proportionality constant being the molar fraction of tautomer [**706a**]. The authors (70CZ2936) claimed, on this basis, that the percentage of tautomer [**706b**] increases from 30% to 45% if the temperature decreases from $-20°$ down to $-55°$ (solvent: dimethyl formamide).

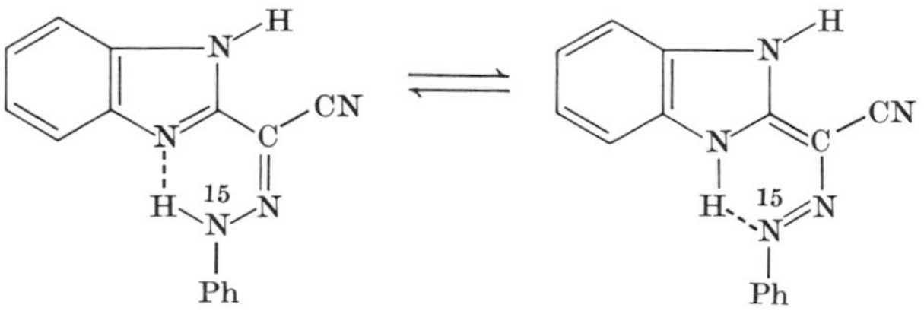

[**706a**] [**706b**]

C. Miscellaneous

The methylene form [**706/1b**] is favoured by $K_T \sim 6$ in H_2O from NMR (72JO3662).

Two isomers have been claimed for 1,5-diphenyl-3-hydroxymethyl-1,2,4-triazole (62T539); the authors assign the stable compound structure [**707a**] and to the labile compound tentatively the form [**707b**]. That work should certainly be verified by a NMR study because it could be a question of mere polymorphism.

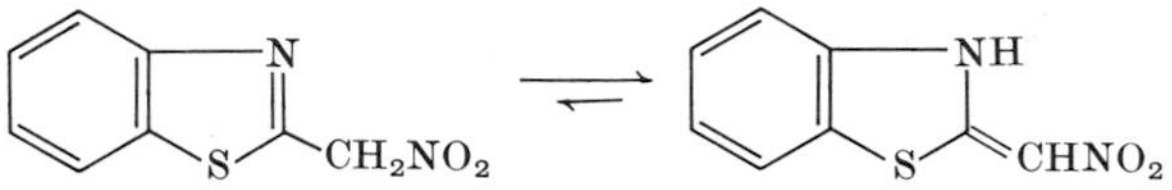

N S CH_2NO_2 ⇌ NH S $CHNO_2$

[**706/1a**] [**706/1b**]

N CH_2OH N Ph N Ph ⇌ H H N^+ C O^- Ph N N H Ph

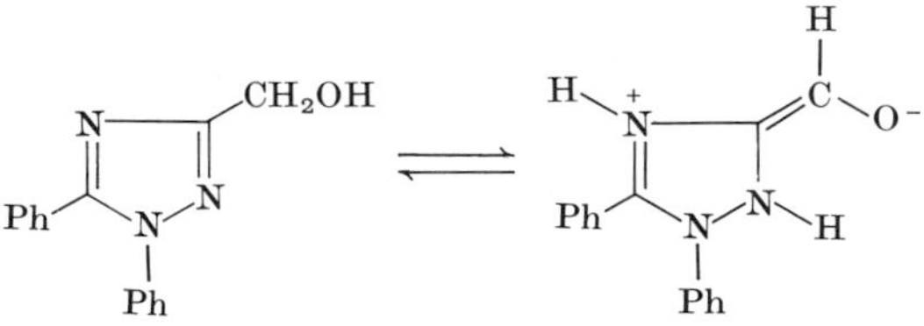

[**707a**] [**707b**]

The possibility of a tautomeric equilibrium between [**708a**] and [**708b**] has been mooted (63AG1204); in this case the CH_2R group is linked to nitrogen instead of carbon as in all previous examples.

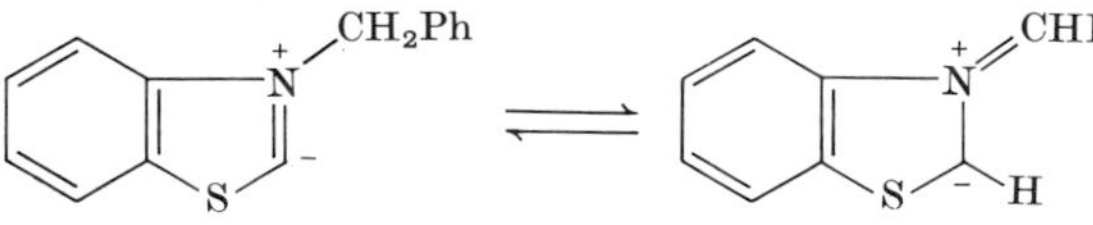

CH_2Ph N^+ S − ⇌ CHPh N^+ S − H

[**708a**] [**708b**]

9. Miscellaneous

A. Heterocycles with an S–S Linkage (II-69)

The 1,2-dithiolium ions [**709**] are heterocyclic compounds in which each of the two sulfur atoms contributes a lone electron pair to the

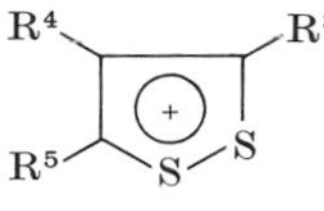

R^4 R^3 + R^5 S S

[**709**]

"aromatic" sextet (65HC(5)1, 66HC(7)39). 1,2-Dithiolium ions substituted with a BH group can exist in several tautomeric forms, one at least of which is heteroaromatic.

The cations arising from the protonation of the derivatives [**710**] were shown to possess structures [**711**], for B = O, S, NR. The same applies to [**712**] and to the thiouret salts [**713**] (66HC(7)39).

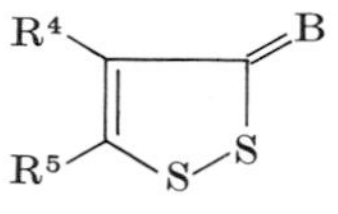

[**710**] 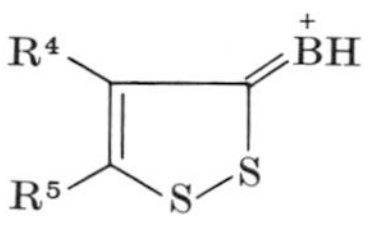 [**711**]

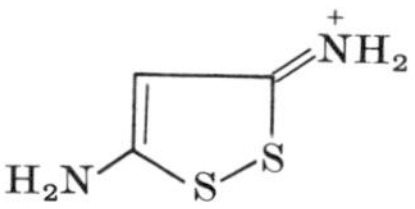

[**712**] 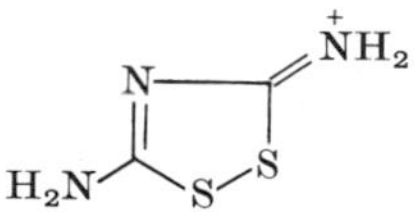 [**713**]

Perchlorate [**714**] has the OH structure [**714a**] (70CB3885) because no carbonyl band occurs in the IR spectrum (in KBr). This behaviour is analogous to that of other 4-hydroxy heterocycles with heteroatoms in the 1,2-positions (Table 4-25, Section 4-3F).

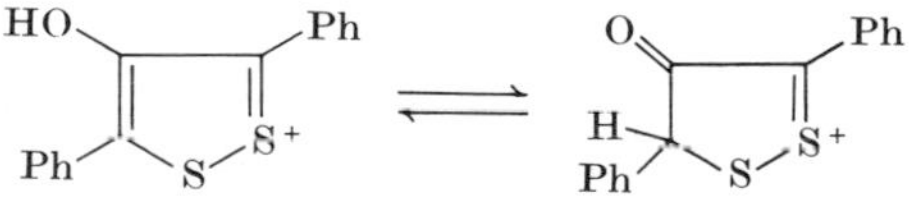

[**714a**] [**714b**]

The 5-mercapto-1,2-dithiole-3-thiones exist as such [**715a**], [**716a**], and not in the dithione form [**715b**] (68J(C)1077, 72PS11). The IR (KBr) of [**715**] shows νSH at 2380 (broad) cm^{-1}, a strong doublet νC═S at 1085 cm^{-1} and a S–S at 512 cm^{-1}. Compound [**716**] shows corresponding bands at 2420, 1180 and 508 cm^{-1} (see 63PM(2)161, p. 323, 71PM(4)265, p. 311). The NMR spectrum ($CDCl_3$) of [**716**] exhibits only the signals of the five aromatic protons.

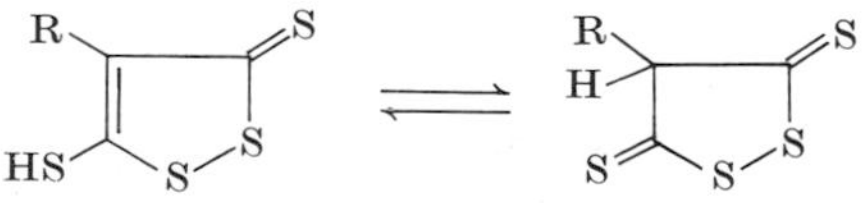

[**715a**] R = Me [**715b**] R = Me
[**716a**] R = Ph [**716b**] R = Ph

The predominance of the amino tautomers [**717**] and [**718**] (cf. X-ray in II-69) is established by careful IR study in Nujol mulls

(63J3165): $\nu_{as}NH_2$, ν_sNH_2, δNH_2, ρNH_2 and ωNH_2 were characterized, and that for [**717**] is confirmed by X-ray methods (63AS2575, 63CX1157). The corresponding *N*-acetyl derivative has been investigated (72AS-2140). These results are consistent with the behaviour of other amino–oxo (Section 4-6Ff) and aminothione (Section 4-6Ga) derivatives.

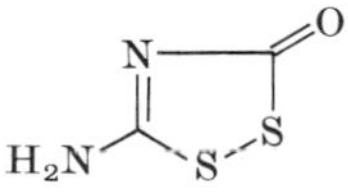

[**717**] 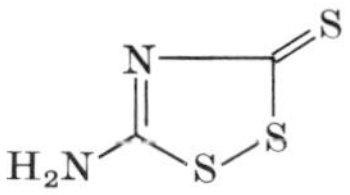 [**718**]

1,2,3-Benzodithiazole-2-oxide [**719**] has been assigned the 3*H*-structure [**719a**] (65JO2763). Although neither the chemical evidence nor the UV spectrum (in EtOH) nor the IR spectrum (KBr) is unambiguous proof, the structure [**719a**] is probably correct.

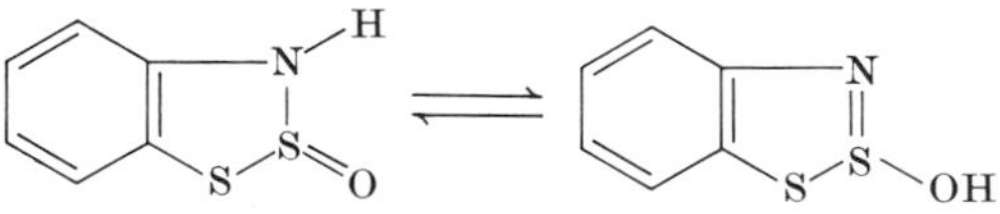

[**719a**] [**719b**]

B. Other Tautomerisms and Rearrangements (Section 1-1Avii)

We do not propose to discuss in detail cases of heteroaromatic tautomerism where no prototopy occurs or where prototropy is a secondary phenomenon, which are beyond the scope of this work on the prototropic tautomerism of heteroaromatic compounds. However, we cite some significant references.

a. No Prototropy (Valence Isomerism)

This group includes the tautomerism of benzofuroxans [**720**] ⇌ [**721**] (67J(B)914, 69HC(10)1, 70J(C)1874), the Boulton–Katritzky rearrangement type I [**722**] → [**723**] (67J(C)2005) and some cases of Dimroth

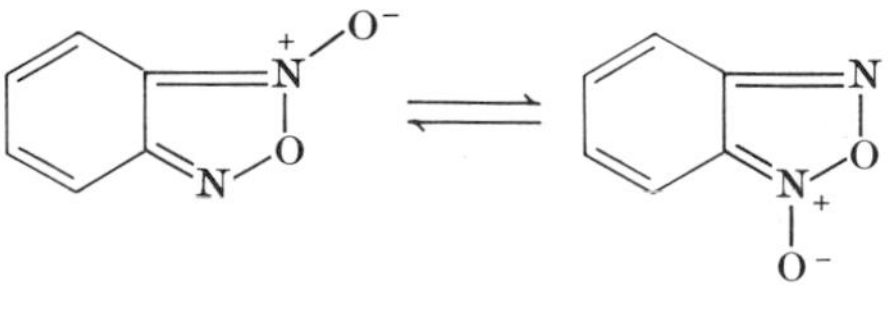

[**720**] [**721**] 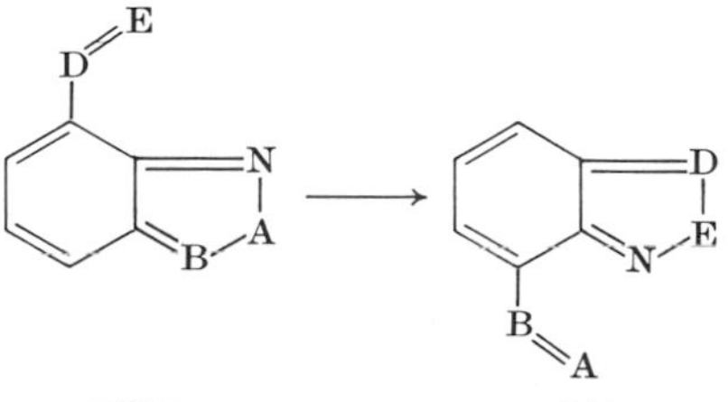 [**722**] [**723**]

rearrangements [**724**] → [**725**] (68T441, 69ZC241) and [**726**] → [**727**] (71LA(754)46).

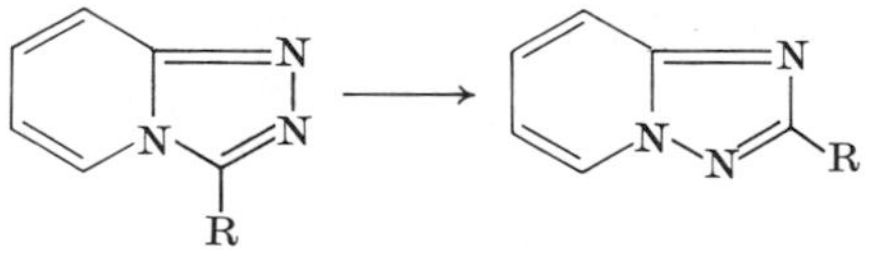

[**724**] [**725**]

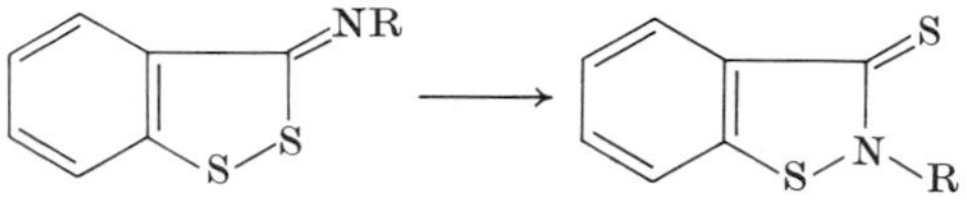

[**726**] [**727**]

b. With Prototropy (II-74)

In this group of rearrangements the migration of a proton occurs, but it is only a secondary phenomenon which arises in noncyclic species: Boulton–Katritzky rearrangement type II [**728**] → [**729**]; Dimroth rearrangement [**730**] → [**731**] (II-74) (60AG359, 68T441, 69ZC241, 70CB1900) and the rearrangement [**732**] → [**733**] (72CH52).

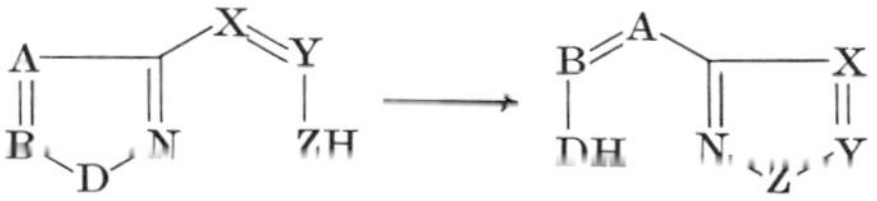

[**728**] [**729**]

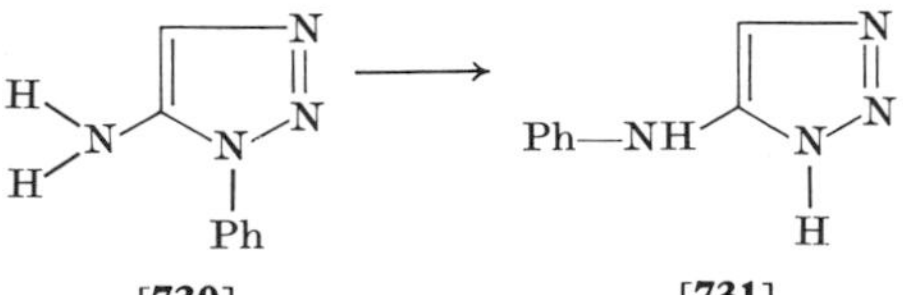

[**730**] [**731**]

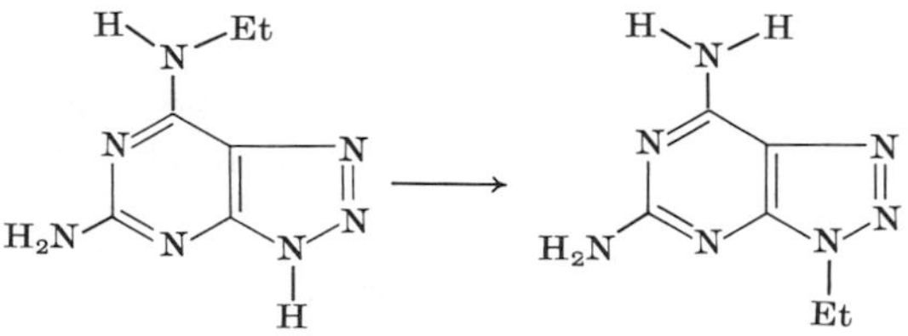

[**732**] [**733**]

C. Ring Chain Tautomerism

Ring–chain tautomerism is an important phenomenon in heterocyclic chemistry (63CR461, 63HC(1)167); however, prototropy does not play an essential part in the ring–chain tautomerism of heteroaromatic compounds, and the subject will be mentioned only briefly.

*a. No Prototropy (Valence Isomerism)**

Examples include the well-known tautomerism of tetrazole [**734**] ⇌ azidoazomethine [**735**] (70CB1900, 71BF1925, 71MI02, 71T5121, 73MI123) or the similar ring-opening of 1,2,3-triazoles [**736**] ⇌ [**737**] (67JA4760).

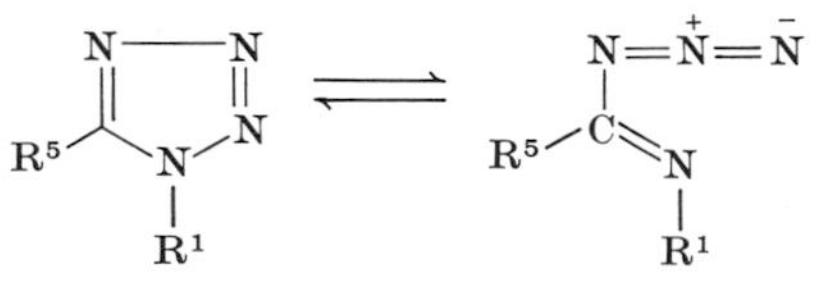

[**734**] [**735**] 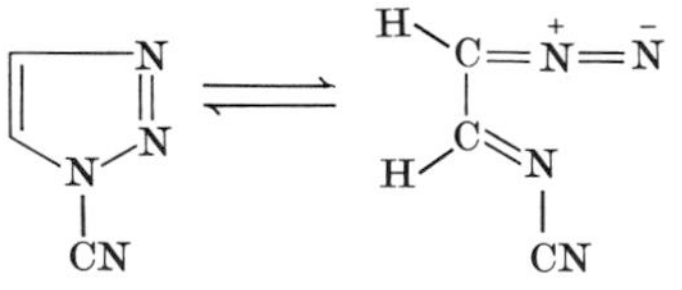 [**736**] [**737**]

If the molecule bears a group BH (OH, SH, NH_2) prototropy can occur after the ring–chain tautomerism: for example for 5-hydroxy-1,2,3-triazoles [**738**] (II-55) (60AG359). However, 4-amino-*v*-triazole does not ring-open spontaneously (73TL1137).

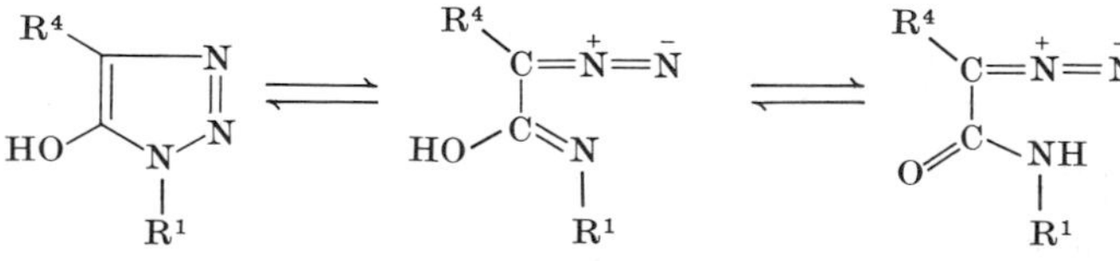

[**738**] [**739**] [**740**]

Another case of ring–chain tautomerism in which heteroaromatic compounds are concerned but without prototropy, is that of mesomeric betaines, for example [**741**]; the position of the equilibrium [**741**] ⇌ [**742**] depends on the nature of the three heteroatoms X, Y and Z.

b. Non-Heteroaromatic Compounds

We will mention only some compounds the cyclic carbinolamine form [**744**] of which affords a heteroaromatic product through dehydration, for example [**743**] → [**745**], well studied by Alper (70CH383,

* This valence isomerism is connected with some types of the Dimroth rearrangement (Section 4-9B) as the chain structure is the first step of the rearrangement; see, for example (70CB1900) for aminotetrazoles.

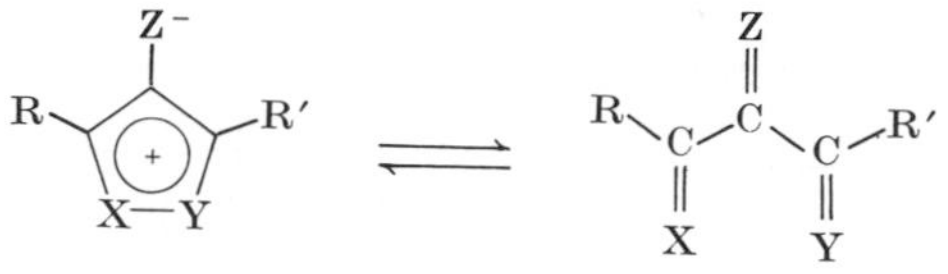

favoured for
X = Y = S, Z = 0 (70CB3885)
X = Y = NR, Z = 0 (71MI514)
X = S, Y = Z = NR (73PS1)
[**741**]

favoured for
X = Y = Z = 0
[**742**]

70JE222, 71JO1352). Further examples are provided by the 4-hydroxy-thiazolidine-2-thiones [**746**] (68BF2855) and the 4-hydroxy-2-imino thiazolidines [**747**] (71J(C)1667, 71J(C)1669).

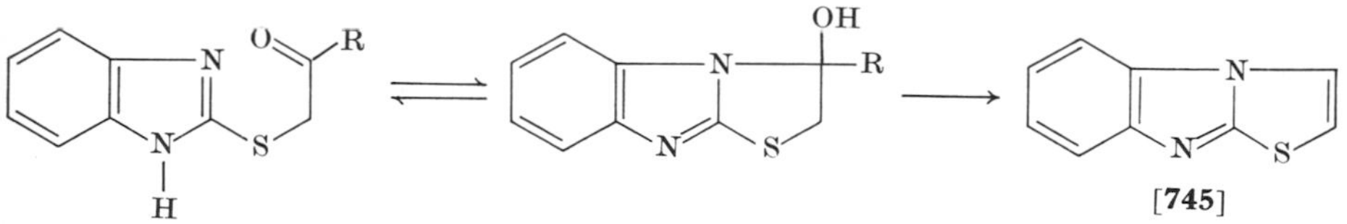

[**743**] [**744**] [**745**]

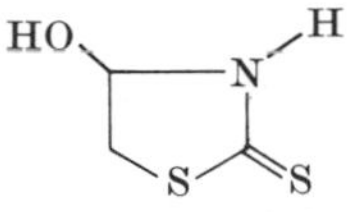

[**746**] 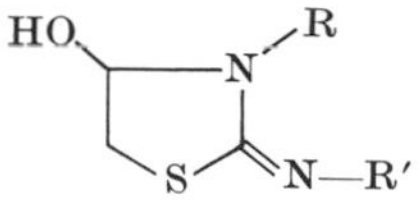 [**747**]

D. Other Types of Azole Tautomerism

A few miscellaneous examples are collected here which cannot be conveniently included in any of the preceding sections.

A UV study (65J2258) showed that the aminobenzothiazole monocations with the amino group on the homocyclic ring have the structure [**748a**], not [**748b**]; in concentrated acidic medium (20 *N* sulfuric acid), the dication [**749**] is formed.

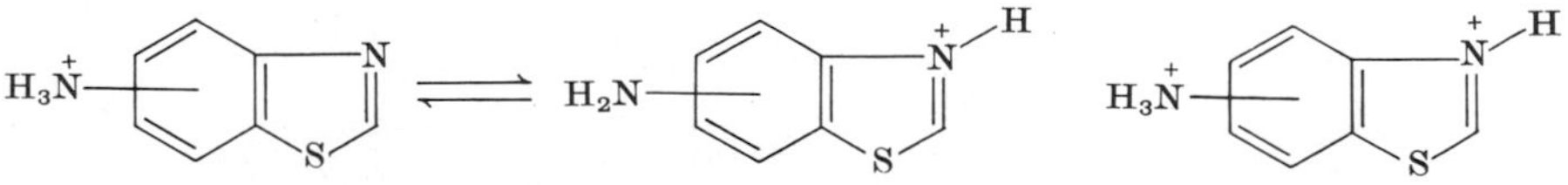

[**748a**] [**748b**] [**749**]

The addition of diazomethane to 1,4-naphthoquinone gives a 1-pyrazoline [**750**]. The structures of the isomerization product [**751a**] and the oxidation product [**752**] (arbitrary annular tautomerism for the

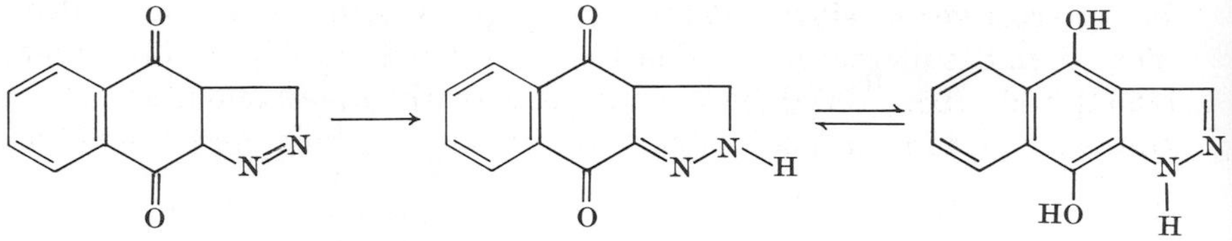

[**750**] [**751a**] [**751b**]

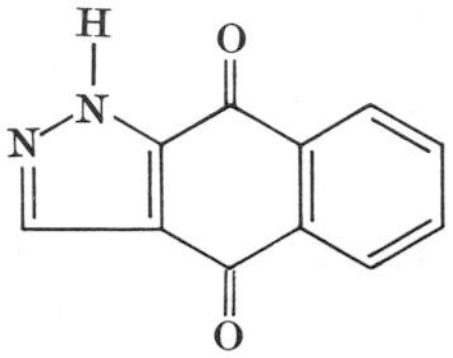

[**752**]

pyrazole) were established in UV and IR spectroscopy (63J5342): an aromatic tautomer of [**751a**], such as [**751b**], could be written.

The tautomerism of tropono[4,5-*c*]pyrazole has been studied (72J-(PI)1623, 72JO676): the NH forms [**753a**] and [**753b**] are considered to be more stable than the OH [**753c**] in the solid state (IR, KBr) and in solution (UV, methanol). By analogy with results on the annular tautomerism of indazole (Section 4-1G), tautomer [**753a**] should be more stable than [**753b**].

The tautomeric equilibrium [**754a**] ⇌ [**754b**] favours the oximino form [**754a**] in DMSO and pyridine (NMR) (72T303).

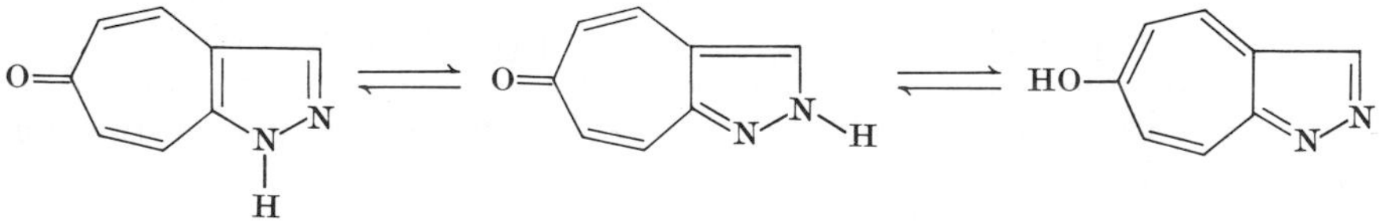

[**753a**] [**753b**] [**753c**]

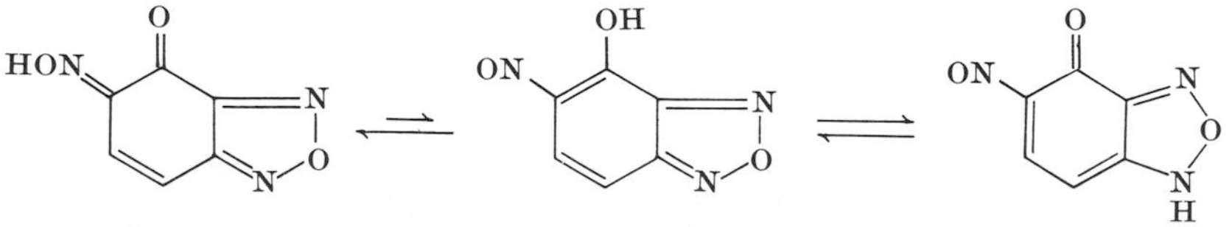

[**754a**] [**754b**] [**754c**]

An interesting case of tautomerism arises in dihydrobenz[*cd*]indazole, which has five possible tautomers: 1,2- [**755a**]; 1,3- [**755b**]; 1,5-[**755c**]; 1,8- [**755d**] and 1,6-dihydro [**755e**] (70J(C)1693, 72J(PI)68). Only tautomers [**755b**] and [**755c**] have been detected, and they were separated but interconverted slowly at room temperature and rapidly on heating alone or in the presence of acid or base. The lower stability of tautomers [**755d**] and [**755e**] could be connected with their quinonoid structure compared to the benzenoid structures of the 1,3- [**755b**] and 1,5- [**755c**]

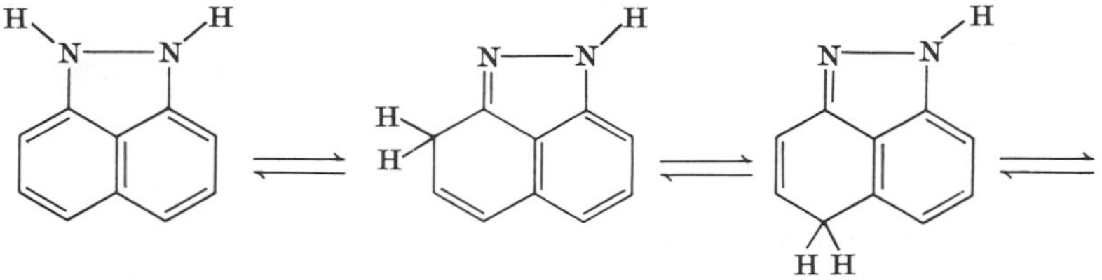

[**755a**] [**755b**] [**755c**]

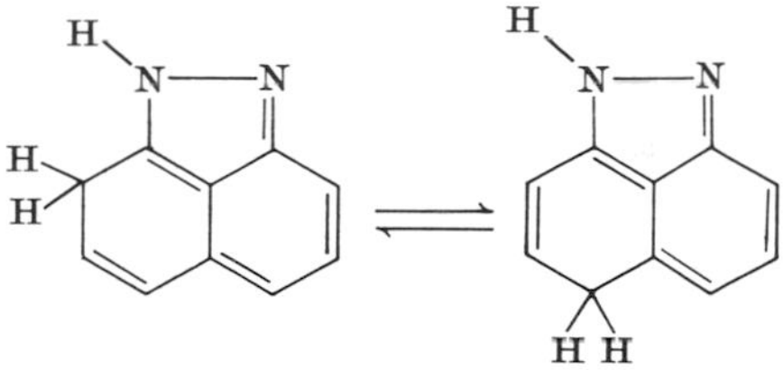

[**755d**] [**755e**]

dihydro derivatives; the same argument has explained the predominance of the 1H-tautomer for indazole (Section 4-1G). The lack of stability of tautomer [**755a**] is more surprising;* [**755a**] has 14 π-electrons and is isoelectronic with pleiadiene [**756**] (72T3587) and perimidine [**757**] (70J(C)290): apparently for four of the $4n + 2$ π-electrons to come from two adjacent heteroatoms leads to higher energy (possibly due to α-effect) (see, however, Section 6-2A). However, tautomers [**755b**] and [**755c**] are aromatic, being indazoles substituted by an alkene at the 3- or 5-position.

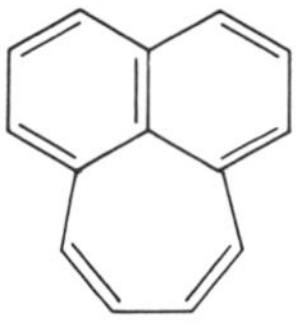

[**756**] 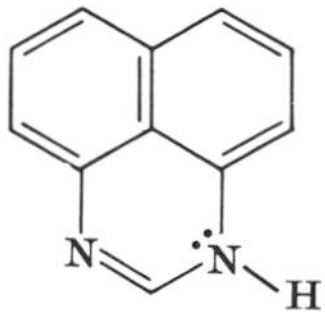 [**757**]

* The N,N′-disubstituted derivatives with a structure analogous to [**755a**] obtained when a mixture of [**755b**] and [**755c**] is treated with methylchloroformate (72J(PI)68) does not prove the presence of tautomer [**755a**] in solution. The product can be formed from [**755b**] or [**755c**], their sp^2 nitrogen being sufficiently nucleophilic.

CHAPTER 5

Purines and Other Condensed Five-Six Ring Systems with Heteroatoms in Both Rings

1. Purines

A. TAUTOMERISM INVOLVING ONLY ANNULAR NITROGEN ATOMS

a. Neutral Species (II-36)

The previous survey (II-36) merely indicated that MO calculations suggested little difference between the stability of the N(9)H [**1a**] and N(7)H [**1b**] tautomeric forms of purine; the extensive studies of the tautomerism of purine made since then are undoubtedly due in large part to the biological importance of these derivatives.

Four tautomeric forms exist for purine, which are denoted as the N(9)H [**1a**], N(7)H [**1b**], N(3)H [**1c**] and N(1)H [**1d**] forms, using the unsystematic but conventional numbering system (cf. [**1a**]). Various experimental results, which are supported by quantum-mechanical calculations, indicate that the N(3)H and N(1)H tautomers [**1c**] and [**1d**] are highly unfavoured and that the N(9)H and N(7)H tautomers [**1a**] and [**1b**] possess comparable stabilities. In solution both [**1a**] and [**1b**] are present, but purine crystallizes in the N(7)H [**1b**] tautomeric form.

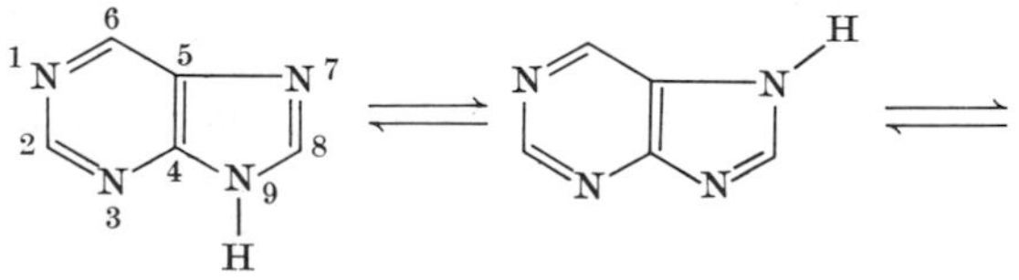

[**1a**] N(9)H

[**1b**]N(7)H

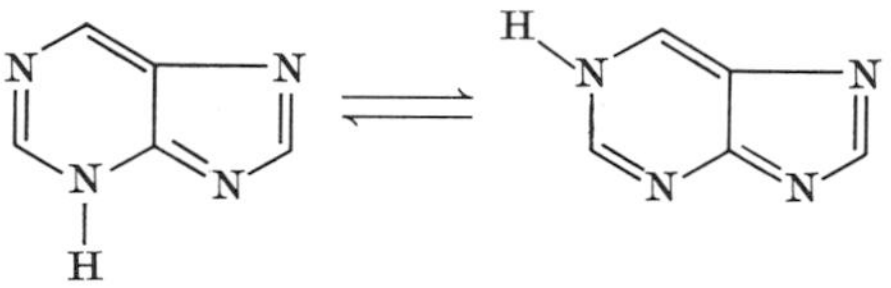

[**1c**]N(3)H

[**1d**]N(1)H

i. *Studies in Solution.* Quantitative results are given in two recent publications, using dipole moments (Section 1-4G) (70T1483) and ^{13}C NMR (Section 1-5B) (71JA1880). *N*-Methylpurines are frequently rather insoluble in the nonpolar solvents used for dipole moment determinations, and suitably substituted derivatives have had to be used. The dipole moments of 6-methylthiopurine and its 7- and 9-methyl derivatives in dioxan (Scheme 5-1) allow calculation of K_T = [N(9)H]/[N(7)H]. Provided contributions from the N(1)H and N(3)H forms are neglected, application of Eq. (1-41) with $P = \mu^2$ (Section 1-4C) gives $K_T \simeq 3$, in favour of the N(9)H tautomer. Assuming that the N-methylation increases the dipole moment by 0.3 to 0.4 D, then K_T is found to be about 1.5 to 2, again in favour of N(9)H. These results apply, of course, to 6-methylthiopurine rather than to purine itself.

N(9)H N(7)H

μ: 3.85 D 5.61 D 3.01 D

SCHEME 5-1

Recently, the tautomerism of all the possible mono- and bis-methylthiopurines and of 2,6,8-trismethylthiopurines have been studied. All appear to exist as mixtures of the 7*H* and 9*H* tautomeric forms on UV, mass spectral and dipole moment evidence. All 1- and 3-methyl derivatives are significantly stronger bases than the corresponding 7- and 9-methyl derivatives (73J(PI)793).

Comparison of the carbon-13 chemical shifts for anion formation by benzimidazole and purine in aqueous solution allows calculation of a value for K_T = [**1a**]/[**1b**] close to 1 (71JA1880). The method (Section 1-5B) utilizes the chemical shifts of the carbon atoms C(4) and C(5), and assumes that the total effect arising from the protonation of the anion is the same for benzimidazole (for which K_T = 1 as required by the symmetry) as for purine.

ii. *Studies in the Solid State*. The crystal structure of purine follows from X-ray measurements, by Watson *et al.* (65CX(19)573) and, independently, by Sweet and Marsh (65CX(19)573). A difference map enables observation of the position of the hydrogen atoms and in particular shows that a hydrogen is bonded to N(7) as shown in [**2**]. The hydrogen atom is accurately located and the hydrogen bond with N(9) is insufficient to eliminate the distinction between the two tautomers.

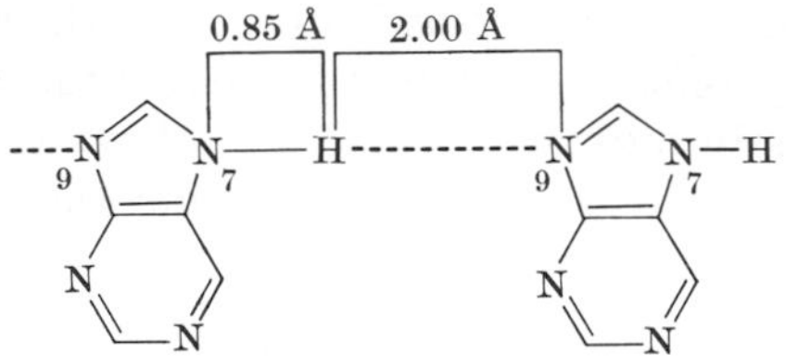

[2]

Russian workers (70IZ1735) have studied annular tautomerism in the solid state of adenine derivatives by variable temperature IR.

iii. *Results from Theoretical Calculations*. Many authors have applied quantum chemistry (Section 1-7B) to the problem of the purine tautomerism: a complete summary is found in the comprehensive survey by Pullman and Pullman (71HC(13)77). The conclusion is that tautomers N(7)H and N(9)H have comparable energies and are much more stable than tautomers N(3)H and N(1)H. For more recent work leading to the same conclusion, see (72J(PII)585, 73CT1470). The difficulty of such calculation is the choice of the correct geometry (Section 1-7A): while the geometry of the N(7)H tautomer [**1b**] is known from X-ray determination, that of the other tautomers has to be estimated. For the calculations by Pullman and Pullman (71HC(13)77), various probable geometries chosen do not affect the essential result, which classifies the tautomers into two groups. However, in the calculations by Rein (71IQ341), who uses the EHT and CNDO/2 methods, one type of geometry indicated that the N(1)H tautomer is more stable than the N(9)H, the opposite of the usual result.

Pullman and Pullman (71HC(13)77, p. 150) also discuss the reasons favouring the N(7)H tautomer in the solid state, although the intrinsic stability of the monomeric N(7)H molecule in the vapour phase is equal to, or even slightly less than, that of the N(9)H tautomer.

b. Conjugate Acids (II-36)

Previously (II-36), only predictions of protonation at N(3) or N(1) were available. Since then, the structure of the conjugated acid of purine has been investigated by NMR spectroscopy. The problem is complicated as, for the monocation alone, six tautomeric forms must be considered. Comparison of the chemical shifts (^{1}H: 64JO1988; ^{13}C: 71JA1880) of purine and of the purine monocation indicates that the preferred site of protonation is at N(1), i.e., that the cations possess structure [**3a**] and/or [**3b**]. Variations in the spin–spin coupling of the 2- and 6-protons also indicate that the monocation of purine possesses a proton at N(1) (65JO1110) and the same phenomenon is shown by

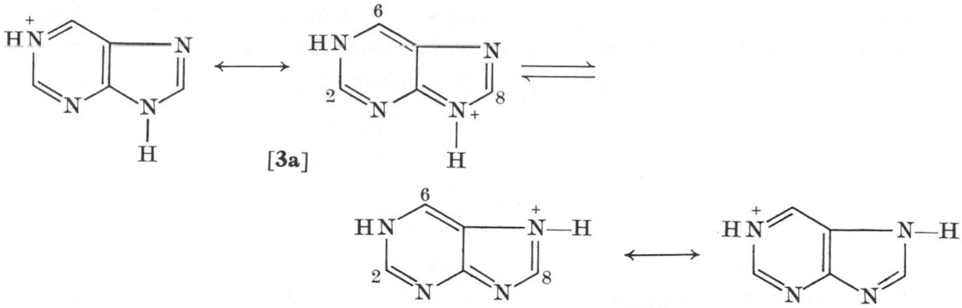

[3a]

[3b]

9-methylpurines and 8-methylthiopurines: other methylthiopurines form mixtures of cations (73J(PI)793).

The variation of ^{13}C–^{1}H coupling constants, observed on the signals arising from the 2-, 6-, and 8-hydrogen atoms, as a function of the pH (65JA3440) indicates the following percentages of protonation: N(1): 54 ± 12%; N(3): 17 ± 12%; N(7) and N(9): 29 ± 7%.

More recent NMR spectroscopy (71H1543) has established the structure of the predominant cations of purine as [**3a**] and/or [**3b**] for the monocation, as [**4**] for the dication, and as [**5**] for the trication from the position and multiplicity of the N–H signals. However, calculations indicate that both the 3*H*,7*H*- and 1*H*,7*H*-monoprotonated species should be of comparable energy (72IS819).

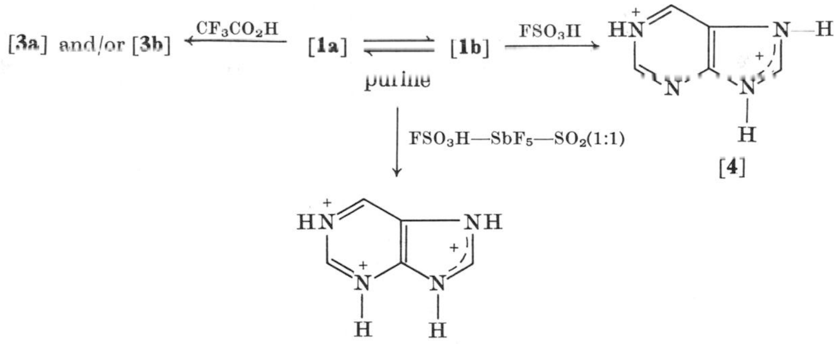

[3a] and/or [3b] ← CF_3CO_2H — [1a] ⇌ [1b] → FSO_3H [4]
purine
↓ FSO_3H—SbF_5—SO_2(1:1)
[5]

SCHEME 5–2

B. OXOPURINES (II-56)

a. Introduction

The 2-, 6- or 8-monooxopurines* exhibit two types of tautomerism, functional and annular. Already by the time of the previous survey

* For numbering see structure [**1a**], Section 5-1A. 6-Oxopurine is known as hypoxanthine.

(II-56), it was well established that these compounds exist in oxo forms (although this has not prevented most authors from naming those compounds as "hydroxypurines"). The most convincing of the earlier evidence for oxo forms was IR spectroscopy in the solid state and in solution. Misleading results can be obtained from UV spectroscopy because of the similar spectra given by the tautomeric species and the fixed derivatives in this series of compounds (66HC(6)1).

IR spectroscopy was not so conclusive regarding the positions of the protons on the nitrogen atoms; five uncharged tautomeric forms [**6a**–**e**] have to be considered for the 2-oxopurines, four [**7a**–**d**] for the 6-oxopurines and three [**8a**–**c**] for the 8-oxopurines. In addition there are charged or zwitterion (sometimes called mesoionic) tautomeric forms. Although some calculations indicate that they have a low stability (71HC(13)77), other recent HMO calculations on molecules of the type [**9**] or [**10**] show that they could be stable under normal conditions

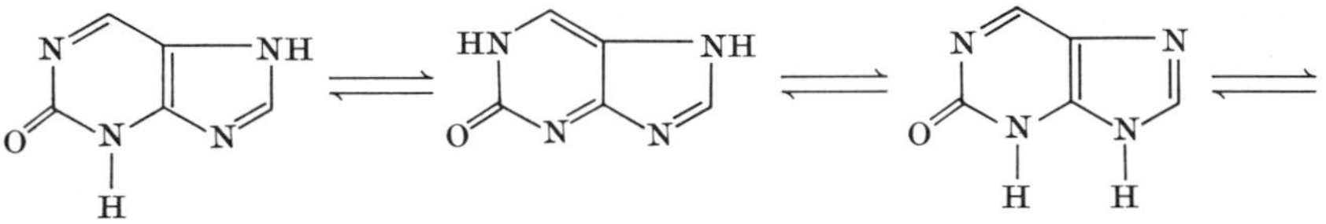

[**6a**] [**6b**] [**6c**]

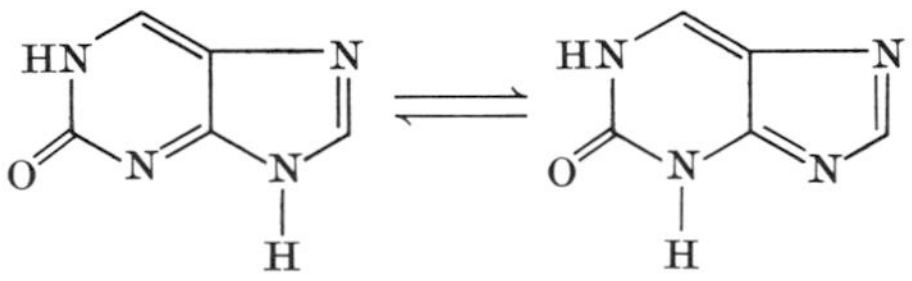

[**6d**] [**6e**]

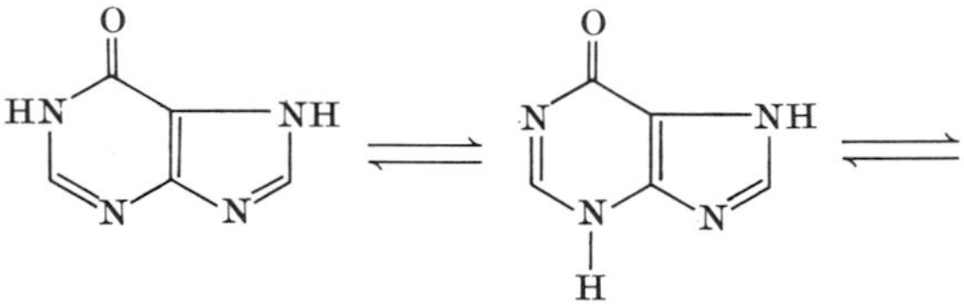

[**7a**] [**7b**]

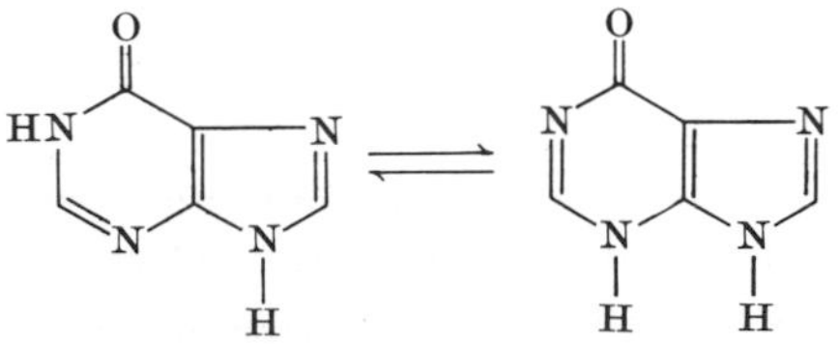

[**7c**] [**7d**]

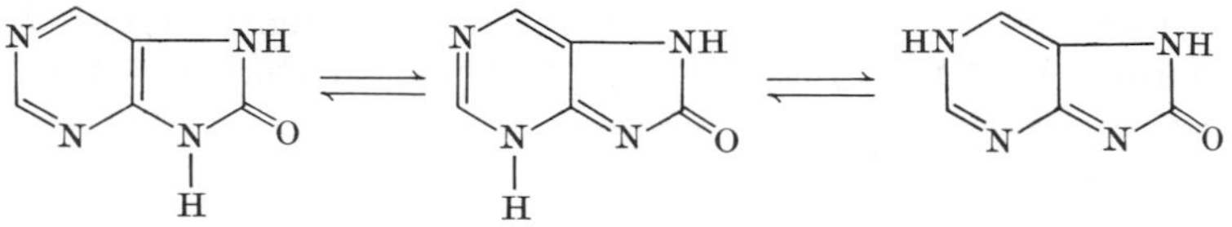

[8a] [8b] [8c]

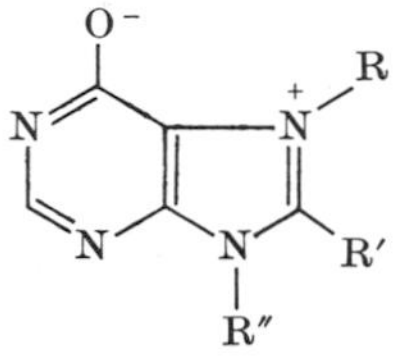

[9] 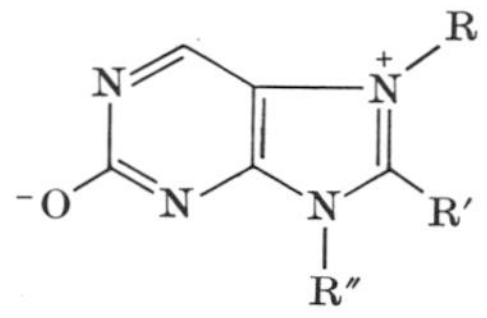 [10]

(71JH881) (see 73CT1474 for more recent work). As in all such calculations, the problem arises of choosing the correct molecular geometry.

UV and IR spectroscopy, which do distinguish between absorptions belonging to *o*-quinonoid and *p*-quinonoid forms, previously led to the conclusions (II-56) that 2- and 8-oxopurines exist, respectively, as [**6a**] and [**8a**], which are expected in view of the oxo structure for pyrimidin-2-one (I-370, Section 2-4D) and benzimidazol-2-ones (Section 4-3Bd). However, a doubt remained (II-56) between structures [**7a**] and [**7c**] for 6-oxopurine, both of which forms possess the amide rather than extended amide structures; amide structures also predominate in pyrimidin-4-one and quinazolin-4-one (Section 2-4D).

While little further experimental work has been accomplished, many theoretical studies have been undertaken. Pullman and Pullman (71HC(13)77) published in 1971 a good survey of the theoretical approach to the tautomerism of oxopurines, and the present treatment will be restricted to the main results.

b. Functional Group Tautomerism

Attempts to calculate the stabilities of two lactim–lactam tautomeric forms by the CNDO procedure often fail because of lack of information about bond lengths in the hydroxy form. Kwiatkowski (68MI365, 68TC47) attempted to interpret the UV results on 6- and 8-oxopurine by Pariser–Parr–Pople calculations, on the basis of hydroxy structures for these compounds. Their geometry for the hydroxy structure explained the spectra of 6- and 8-oxopurine but not that of 2-oxopurine. These misleading results can be explained by the similarity, mentioned above, of the UV spectra expected for the oxo and hydroxy forms. Careful and complete theoretical study on all the forms (71HC(13)77) can predict correctly the absorptions. Dipole moments should allow a differentiation of hydroxy and oxo forms in purines as theoretical calculations show, but few experimental data are available at the moment because of problems of solubility. UV and NMR spectra show that the 7*H*, 9*H* tautomer [**8a**] of purin-8-one predominates (in H_2O) (72J(PI)2950) even if there is a methylsulfonyl substituent in position 6 (74J(PI)470). However, there is no reasonable doubt but that all simple oxopurines exist as such, and not as hydroxypurines.

c. *Annular Tautomerism*

Pullman and Pullman (71HC(13)77) used MO methods for a complete study of the annular tautomerism of the 2-, 6- and 8-oxopurines and concluded that forms [**6a**], [**7a**] and [**8a**], respectively, are the most stable. These three predominant tautomeric forms each possess a proton at N(7) which seems to be general behaviour; however, the authors found that the N(9)H tautomer [**6d**] is of almost the same energy as [**6a**] and the N(9)H tautomer [**7c**] about the same as [**7a**]. This indicates the presence of comparable proportions of [**6a**] and [**6d**] in the 2-oxopurine and of [**7a**] and [**7c**] in the 6-oxopurine. The authors (71HC(13)77) point out that these theoretical results, which strictly apply to isolated molecules, are in agreement with the experimental results in solution. However, a single form may be present in the crystal; for instance, it has been shown that the 6-oxopurine contains the N(9)H tautomer [**7c**] (68CX(B)1692). By contrast, UV comparisons (in H_2O) indicate the 7*H* form for purin-6-one although the 3-methyl analogue is mainly in the 9*H* form (72IS805). 6-Methylthiopurin-8-one exists as the 7*H*,9*H* tautomer (in H_2O) by UV comparisons, and the 1-methyl derivative has the 7*H* structure (73J(PI)1225).

Calculations indicate that it would be possible to differentiate between N(7)H and N(9)H tautomers by dipole moments, but unfortunately experimental values are not yet available. Theoretical calculations of expected UV absorption of the lactam forms and comparison with the extensive experimental data (solvent: H_2O or dioxan) as summarized in (71HC(13)77) support the previous conclusions. 6-Methylthio-2-oxopurine exists in water mainly as the 3,7-diNH tautomer (73J(PI)2445) and 2-methylthiohypoxanthine probably as the 1,7-diNH tautomer (73J(PI)2647).

d. *Cation Structures* (II-57)

The previous survey (II-57) mentioned that theory predicted protonation of 2- and 6-oxopurines at N(7). Since then it has been shown that purin-8-one and its 7- and 9-methyl derivatives undergo protonation at N(1) (72J(PI)2950), but 6-methylthiopurin-8-one protonates about equally at the 1- and 3-positions (72J(PI)1225). Bergmann *et al.* have extensively researched on the structure of cations of oxopurines (73J(PI)2445, 73J(PI)2647, 74J(PI)470, 74J(PI)2229). The structure of hypoxanthine hydrochloride monohydrate has been determined (69CX-(B)1608): the nitrogen atoms carrying a proton are N(1), N(7) and N(9).

e. Anion Structures (II-56)

Monoanions derived from 2-, 6- and 8-oxopurines can exist in four tautomeric forms, e.g., [**11a–d**] for 6-oxopurine anion. The structure of such anions were not discussed in the previous treatment (cf. II-56),

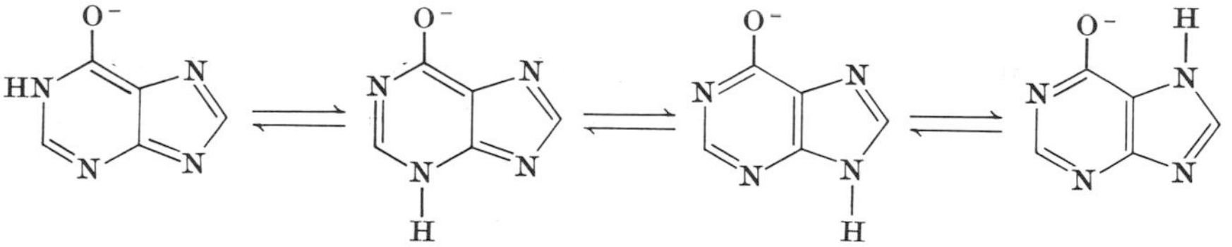

[**11a**] [**11b**] [**11c**] [**11d**]

but the subject has been reviewed by Lister (66HC(6)1), who summarized the evidence, mainly early UV work, and concluded that the 2-oxopurine anion probably exists in the 3*H* form [**12**], and the 8-oxopurine anion in the 7*H* form [**13**]. However, no clear conclusion regarding the relative importance of the forms [**11a–d**] for the 6-oxopurine anion was reached (cf. also 62JO2478).

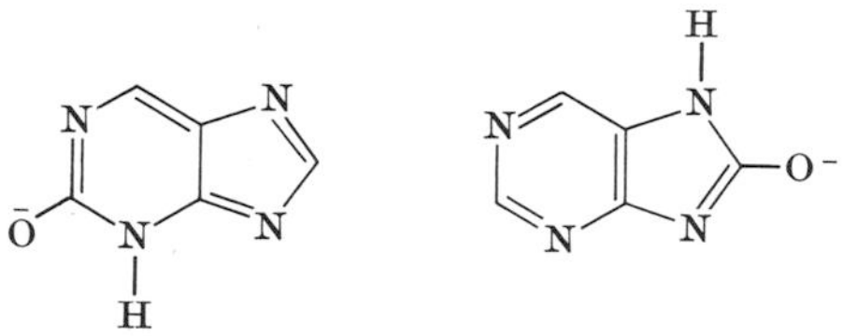

[**12**] [**13**]

More recently, the 7*H* form [**13**] for the 8-oxopurine anion (72J-(PI)2950) and for the 6-methylthiopurin-8-one anion (73J(PI)1225) have been confirmed. The purin-6-one anion appears to be a mixture of tautomers of comparable stability (72IS805).

f. Inosine

Inosine, a 9-substituted 6-oxopurine, can exist in four tautomeric forms [**14a**] to [**14d**]. Early IR studies (58BY1958) indicated the

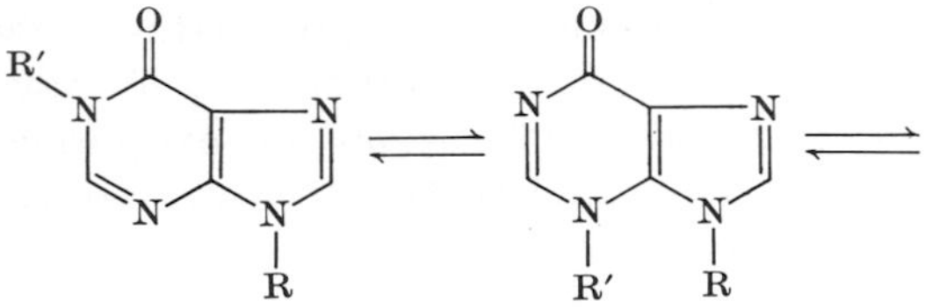

[**14a**] R′ = H
[**15**] R′ = Me

[**14b**] R′ = H
[**16**] R′ = Me

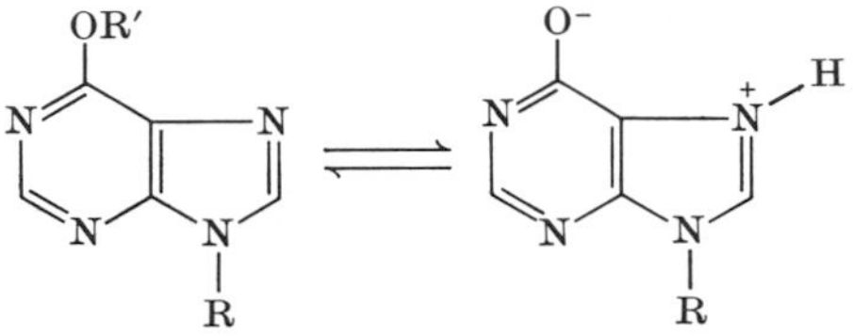

[**14c**] R′ = H
[**17**] R′ = Me

[**14d**]

[R = (ribofuranosyl: O, CH_2OH, HO, OH) in structures [**14**]–[**17**]]

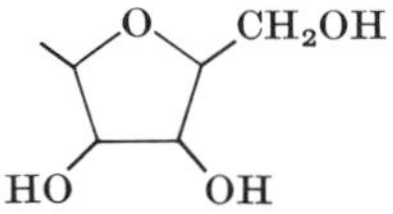

predominance of an oxo form, and this is supported by NMR work (69MI256) which suggested a lactam form in D_2O. Recently, Wolfenden (69MI307) compared the UV spectra and pK_a of inosine with those of the fixed methyl derivatives [**15**] and [**17**] and concluded that inosine exists as a mixture of forms [**14a**] and [**14c**]. We consider his reasoning to be quite unjustified: as discussed in (Section 1-4A) pK_a values can only be used directly if a common cation is involved—this is certainly not the case here and further the UV spectra of the methyl derivatives and the tautomeric compound are too similar to allow definite conclusions to be drawn.

X-ray diffraction (69NA1170) of inosine shows a proton at N(1) which indicates that [**14a**] is the structure in the crystalline state.

C. Purinethiones (II-65)

Experimentally and theoretically the purinethiones are less well studied than the oxopurine analogues (cf. 71HC(13)77). However, UV spectroscopy distinguishes more easily between the tautomeric forms of purinethiones as the spectra of the fixed derivatives are distinct. However, no complete study of all the possible forms of purinethiones is yet available for a complete discussion of their functional and annular tautomerism.

The previous survey (II-65) reported the demonstration of the thione structure [**19**] for the purine-2-thione, but the results were less conclusive for purine-6-thione [**20**], 7-methylpurine-6-thione [**21**] and 9-methylpurine-8-thione [**22**] for which the IR and UV results do not

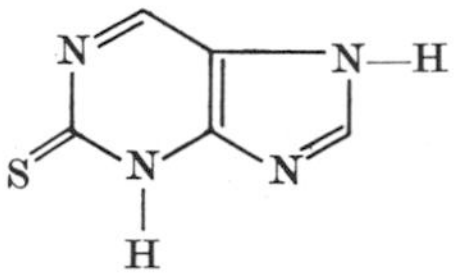

[**19**] 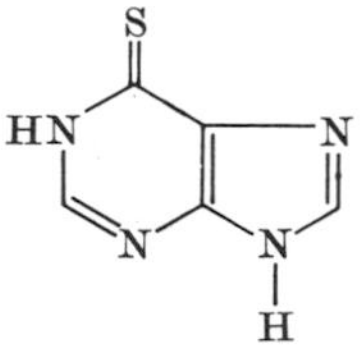 [**20**]

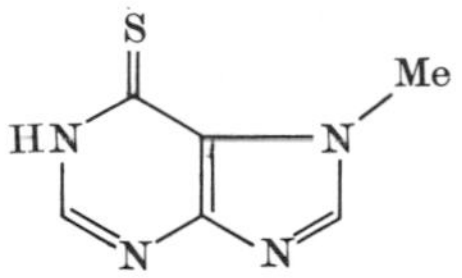

[**21**] 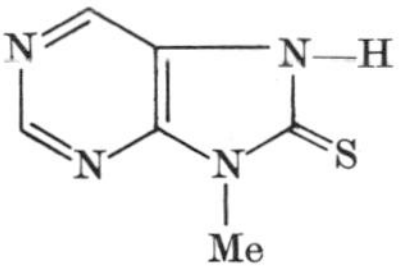 [**22**]

exclude thiol forms existing together with thione forms. Lister in his review (66HC(6)1) concluded that the thione structures predominate. Claims (60JA463) to have demonstrated the existence of thiol–thione tautomerism for 2-fluoropurine-6-thione on the basis of spectral changes with the pH probably arise from confusion of tautomerism with ionization (as discussed in 66HC(6)1).

Comparison of the UV spectra of 8-phenyl- and 2-methyl-8-phenylpurine-6-thione with the corresponding 6-methylthio derivatives demonstrates that the potentially tautomeric compounds do not exist in the 6-mercapto form (66J(C)10). Later UV and pK_a comparisons suggest the 1*H*,9*H*-tautomer as predominant for purine-6-thione and that the 9*H*-1-methyl- and 7*H*-3-methyl- are major structures for the corresponding methyl derivatives. The purine-6-thione anion is probably formed by loss of the 1-proton (72IS805).

The available (68CX(B)1698) X-ray analysis of 6-methylpurine-2-thione monohydrate is insufficiently precise to exclude the thiol form, but bond lengths do favour a thione form, such as [**23**]. Two groups of authors (69CX(B)1330, 69CX(B)1338) have reported accurate structures for purine-6-thione monohydrate [**24**] which exists in the thione form with hydrogen atoms at N(1) and N(7): this differs from the N(1,9)H form [**20**] tentatively indicated by UV experiments.

That far less complete experimental results are available for purinethiones than for oxopurines partly explains why few theoretical

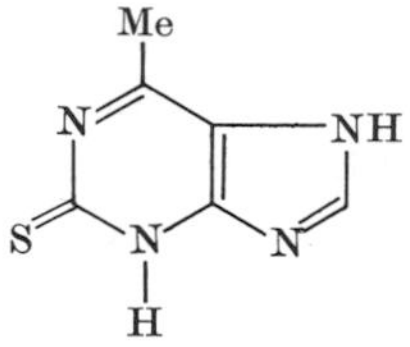

[**23**] 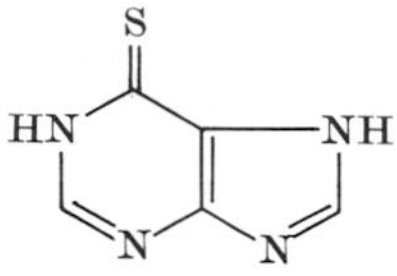 [**24**]

calculations have yet been undertaken. However, two groups of authors (71HC(13)77, 71JL(8)471) have investigated the tautomerism of purinethiones by quantum mechanics. The CNDO method (71HC(13)-77) indicates that the N(7)H tautomer [**24**] of purine-6-thione is about 3.5 kcal/mole more stable than the N(9)H structure [**20**]. The same authors used the semiempirical Pariser–Parr–Pople method to predict the electronic spectra of the individual purinethione tautomers. However, this study is incomplete as it includes only a limited number of the possible tautomeric forms: for the purine-2-thione they considered the tautomers [**25a**] and [**25b**] plus the corresponding thiol forms [**25c**] and [**25d**]; for the purine-6-thione they considered the

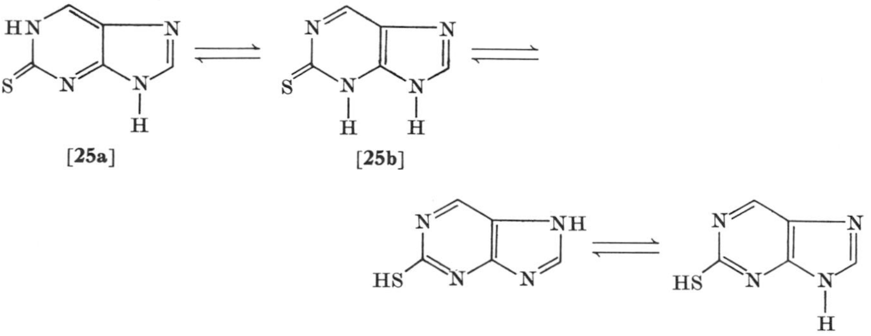

[**25a**] [**25b**]

[**25c**] [**25d**]

forms [**20**] and [**24**] and the corresponding thiol forms [**26a**] and [**26b**]; in the case of the purine-8-thione they took into consideration the tautomers [**27a**], [**27b**] and [**27c**]. While the agreement between theoretical and experimental UV absorption maxima is less satisfactory than

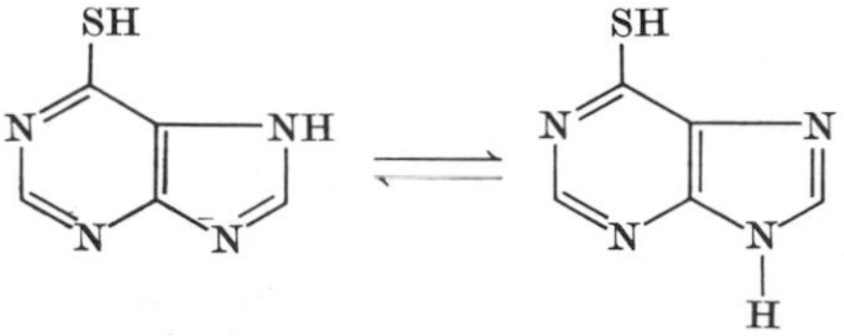

[**26a**] [**26b**]

N NH N S N H ⇌ N NH N N SH ⇌ N N N N SH H

[27a] [27b] [27c]

for the oxopurines, the theory does reproduce the correct bathochromic order purine-2-thione > purine-6-thione > purine-8-thione and the general expected bathochromic effect of thio derivatives compared with oxo ones.

From their calculations, Pullman and Pullman (71HC(13)77) predict that the dipole moments of the thione forms should be greater than those of the corresponding oxo forms with an exception for purine-6-thione [**24**], but here again no experimental data are available to allow comparisons.

Kwiatkowski (71JL(8)471) calculated the electronic absorption spectra of purinethiones using the Pariser–Parr–Pople method; he considered the same structures as Pullman and Pullman except that the purine-2-thione form [**28**] was considered instead of [**25a**]; further he did not consider any thiol forms. For purine-2-thione the experimental spectrum agrees better with that calculated for tautomer [**25b**] than that for [**28**]. For purine-6-thione the result is less clear-cut, but

N NH S N H N

[28]

its existence as tautomer [**24**] appears to be preferred to [**20**]. Unfortunately, as the author himself points out, none of these calculations consider all the possible tautomeric forms.

Consideration of the tautomerism of purinethiones is considerably assisted by results obtained in other series: (I-400), pyrimidine-4-thiones and quinazoline-4-thiones exist as mixtures of *o*- and *p*-quinonoid forms (I-400, Section 2-5E) and benzimidazole-2-thione exists as such (Section 4-4Cf).

D. Aminopurines (II-75)

The 6-aminopurines, better known as adenines, have been widely studied, with little attention being paid to the 2- and 8-amino purines.

The previous survey (II-75) reported UV and IR measurements which indicated that adenine existed in an amino structure but without locating the proton at N(7) or N(9).

Amino structures are indeed expected for aminopurines in view of the amino structure of 4-aminopyrimidines (Section 2-6C). The monocation of adenine was also previously reported (II-75) to be formed by protonation at N(1) on X-ray and NMR evidence: it is now known that for all aminopurines protonation occurs at annular nitrogen atoms which do not already bear an hydrogen atom.

a. 1-, 3-, 7- and 9-Substituted Adenines: Neutral Species and Cations

The balance between amino and imino structures for *N*-substituted adenines is more delicate. On somewhat fragmentary evidence it has been concluded that 1-methyladenine exists in the imino form [**29**] (60J539). For 3-methyladenine the evidence is more complete, and it demonstrates convincingly the amino form [**30**]. UV comparisons (in aqueous solution) with the *N*,*N*-dimethylamino analogue and IR (KBr) provide evidence for structure [**30**] (64J400) which is supported

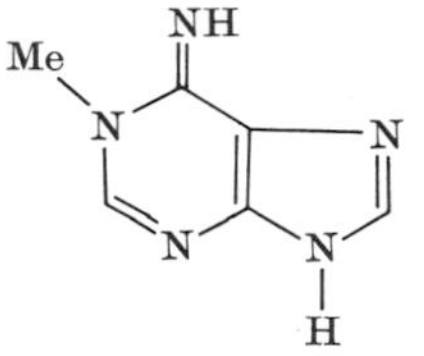

[**29**] 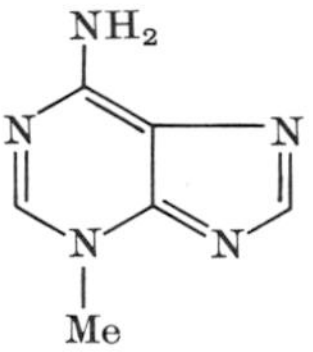 [**30**] 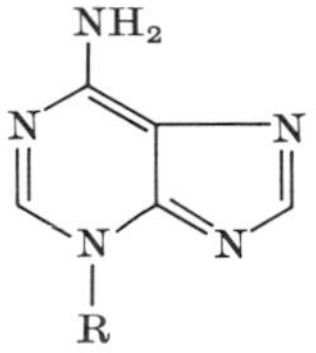 [**31**]

R = CH_2Ph
R = $CH(Ph)_2$
pK_a = 5.1
(when R = CH_2Ph)

by the pK_a values (in 95% ethanol) of various substituted adenines (64JH115) which demonstrate the corresponding amino structure for 3-benzyladenine. The last paper quoted considers the structures of the cations formed by the N-substituted adenines. Thus, the NMR spectrum of the hydrobromide of [**33**] in DMSO shows the presence of an NH_2 group; the IR spectrum (KBr) of the hydrobromide of [**33**] is different from that of the base [**33**] but similar to that of the hydrochloride of [**31**, R = CH_2Ph], which supports protonation of [**33**] at the exocyclic imino group.

The IR spectra of alkyl adenines have been extensively studied (70IZ1735).

Fluorescence spectra of the cation of 7-methyladenine [**34**] (67AS-2463) indicates the presence of more than one tautomeric form; these results are discussed in more detail for the parent adenine cation later.

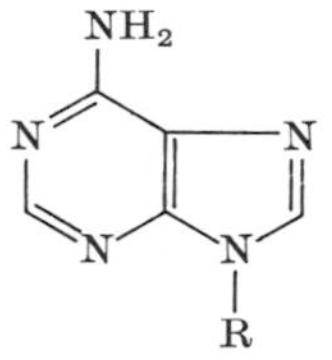

[**32**]

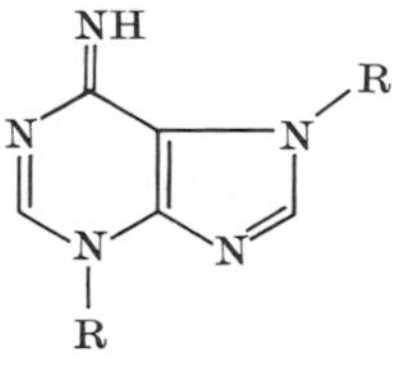

[**33**]
$pK_a = 9.6$
(when R = CH_2Ph)

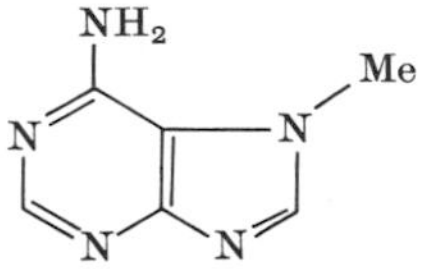

[**34**]
$pK_a = 3.6$

UV studies on adenosine [**32**, R = ribose] showed this molecule to exist in the amino form (69MI307) and pK_a values gave an approximate ratio [amino]/[imino] = 4×10^4; this result is in agreement with the presence of δNH_2 absorptions in the IR spectra in the solid state (67SA2551). Many crystal structures of 9-substituted adenines confirm the 6-amino structure, these include the following derivatives: 9-methyl- (63CX907, 64JT(40)2071, 66ZS339), 9-ethyl- (64MB(8)89, 66CX(21)754, 67MB(30)545), 8-bromo-9-ethyl- (68PN(60)402). In addition the following nucleosides have been studied by X-ray crystallography: adenosine (72CX(B)1982), deoxyadenosine (65CX(19)111), 3′-*O*-acetyladenosine (70JA4963), α-*D*-2′-amino-2′-deoxyadenosine (70-JA4056), adenosine hydrochloride (73CX(B)31, 74CX(B)2273).

UV spectroscopy at pH 1 (63JA193), fluorescence spectra (67AS2463), IR spectroscopy in acidic D_2O (67SA2551), and X-ray diffraction of suitable salts (62CX1179) show that protonation occurs on N(1) in all the N(9)-substituted adenines (see also 66HC(6)1 and 71CR439 for a discussion of the site of protonation). The second protonation of 9-methyladenine occurs at N(7) (62CX1179). Theoretical (CNDO/2) calculations have been performed on 8-substituted 9-methylguanines (neutral molecules and cations) (74JA5911).

b. Cations of N-Unsubstituted Adenines

The structure of the parent adenine cation has been the subject of considerable study and controversy. Lewin summarized earlier work, and concluded that protonation occurred at the amino group to yield [**35a**] (64J792). Early MO calculations were similarly interpreted, but later theoretical work supports the N(1) protonated species [**35b**] which is now generally accepted as predominant: for references and discussion, see (71CR439) and for an X-ray structure see (74CX(B)1528).

Børresen (67AS2463) first demonstrated the coexistence of more than one tautomeric form for the adenine cation, including at least one fluorescent and one nonfluorescent cation tautomer, from discrepancies between the absorption and fluorescence excitation spectra. By comparison with analogous methyl derivatives they propose the structures [**35a**] and [**35b**] for the fluorescent and nonfluorescent cation tautomers, respectively. However, the arguments for excluding [**35c**] as the structure of the fluorescent tautomeric cation are unconvincing, and the matter must be regarded as still open.

[**35a**] ⇌ [**35b**] ⇌ [**35c**]

Recent calculations (72TC51) of electron densities of adenine are relevant to the site of protonation or alkylation and show that the pyridine-like atoms are the most likely to undergo electrophilic attack, in agreement with the experimental data.

From low-temperature NMR spectra, Wagner and von Philipsborn (71H1543) have proposed structures for the adenine dication [**36**] and trication [**37**]. An X-ray study (74CX(B)1528, 74MI409) confirms the structure [**36**] for the dication. The adenine anion exists as expected in the amino form (73CX(B)1974).

[**36**] ← CF_3CO_2H — [**38b**] — (a) FSO_3H; (b) FSO_3H—SbF_5—SO_2 → [**37**]

c. N-Unsubstituted Adenines: Neutral Species

UV measurements (65JA11 and references therein) in trimethyl phosphate as solvent have been interpreted in terms of the 6-amino-N(9)H purine structure [**38a**] of adenines: the conclusion is probably correct, but this evidence is not conclusive. Calculations do not easily distinguish between the 7*H* and 9*H* forms of adenine (72JA7898).

Eastman (69BG407), in a study paralleling that of Børresen mentioned above, showed that the tautomerism in adenine in a neutral medium such as an alcohol could be investigated by fluorescence spectroscopy to gain evidence for the two amino forms [**38a**] and [**38b**]. The N(7)H tautomer [**38b**] is fluorescent but not the N(9)H tautomer [**38a**]; the ratio [**38b**]/[**38a**] is calculated as 0.06. The K_T is relatively insensitive to the type of solvent and temperature, but probably increases somewhat as the solvent hydrogen bonds become stronger.

The infrared and Raman spectra of polycrystalline adenine has been studied (74CP415) and the presence of an NH_2 group clearly established.

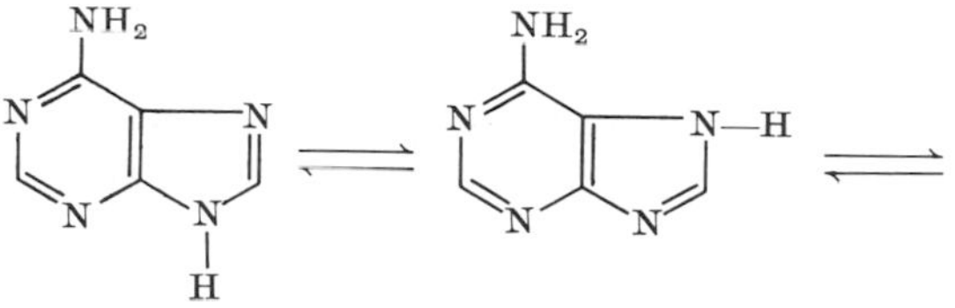

[**38a**] ⇌ [**38b**] 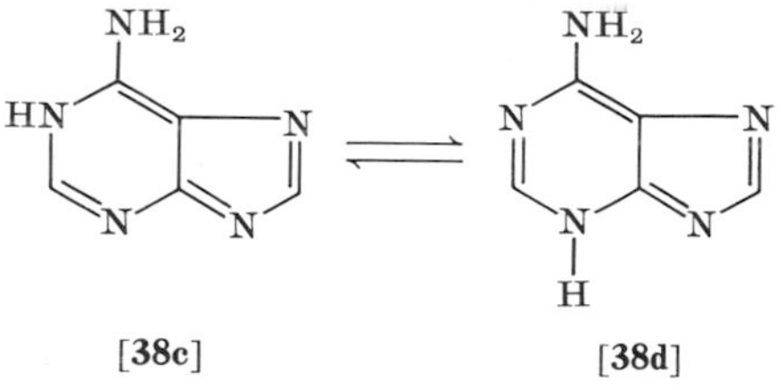 ⇌ [**38c**] ⇌ [**38d**]

d. Molecular Orbital Calculations on Adenines

Recently much theoretical work has been carried out on adenine and some of its derivatives (see, e.g., 73CT1474). Four amino forms [**38a–d**], and at least four-imino tautomers [**39a–d**] must be considered. A complete study using the CNDO/2 method (69TC(15)265) has been well summarized (71HC(13)77). The imino tautomers are usually less stable than the amino forms: this is clear for the imino tautomer [**39a**] which is less stable than the corresponding amino form [**38a**] by about 27 kcal/mole and for [**39b**] which is less stable than [**38b**] by

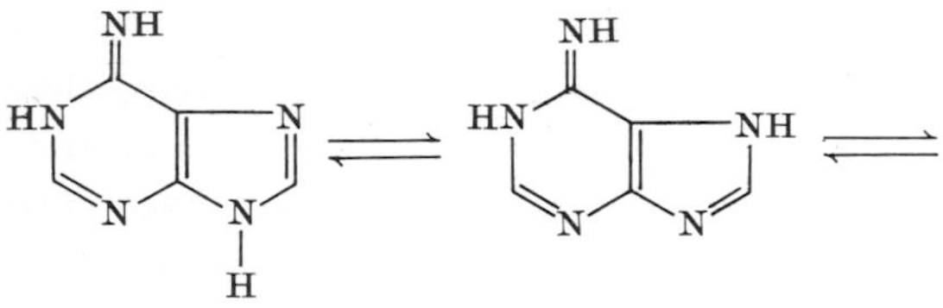

[**39a**] ⇌ [**39b**] 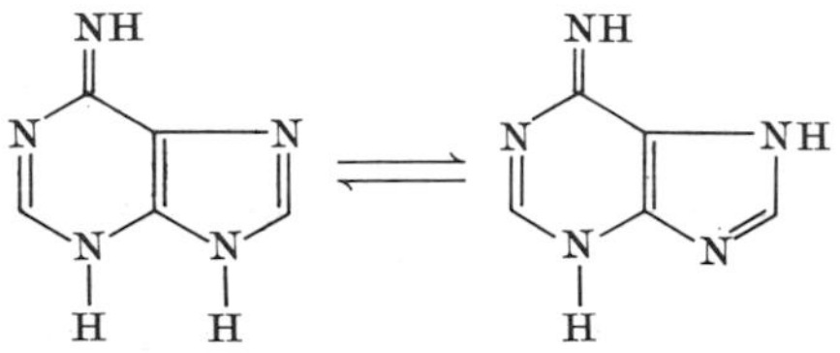 ⇌ [**39c**] ⇌ [**39d**]

24 kcal/mole: they attribute the increased stability of the amino forms to the greater π-electron delocalization energy of this form because of the Kekulé-type resonance in the pyrimidine ring. Such reasoning does not hold for [**38c**] or [**38d**] and the theoretical calculations do show that [**38c**] is less stable than the corresponding imino forms [**39a**] and [**39b**]: this result is in agreement with the previously reported result concerning the imino and amino structures of [**29**] and [**30**].

The same authors have also calculated the dipole moments and UV spectra (71HC(13)77) of all the possible adenine tautomers and compared the results with the dipole moments and UV spectra observed for all the corresponding methyl and benzyl derivatives (70J(B)1334) except those substituted at N(1) which were insufficiently soluble. Such comparisons are valid only if the substituents do not significantly affect the experimental values to be compared with unsubstituted tautomers. The fit of calculated and experimental values is good for the dipole moments but poor for the UV spectra for which the calculations give only the correct order for bathochromic shifts N(9)H < N(7)H < N(3)H. The two solvents (dioxane and benzene) used for the dipole moment measurements differed from that (EtOH) for the UV spectroscopy; the equilibrium position may vary with solvent.

The following conclusions have been drawn from this work. 3-Substituted adenines for which only the tautomeric forms corresponding to [**38d**], [**39c**] and [**39d**] are possible exist predominantly as structure [**38d**] with some contribution from form [**39c**]. 7-Substituted adenines can exist as tautomers [**38b**], [**39b**] and [**39d**], here the amino form [**38b**] predominates by a large factor. 9-Substituted adenines possess tautomeric forms [**38a**], [**39a**] and [**39c**]; again the amino form

[**38a**] predominates by a large factor. However, of the three N(1)H structures [**38c**], [**39a**] and [**39b**], one of the imino forms [**39a**] or [**39b**] almost certainly represents the lowest energy species. Heats of atomization calculated (70JA2929) for the four tautomers [**38a**], [**38b**], [**39a**] and [**39c**] show again that [**38a**] is the most stable.

Many calculations of the tautomerism of adenines and other purines and of their relative stabilities relate to their biological importance. Nucleotide bases in abnormal tautomeric forms at the time of replication can be one of the causes of spontaneous mutations (70JA2929, 71HC(13)77 and references therein). As the rate of mutation is clearly less than to that expected from the various tautomeric equilibrium constants (71JA4585 and references therein), a detailed mechanism has been proposed by Löwdin (64MI174) involving the tautomeric equilibria in the environment of the so-called DNA replication plane. A quantitative theoretical test of the Löwdin's proposal (71JA4585) shows that it can predict the proper degree of incorporation of adenine as the rare imino form [**39a**] in place of guanine.

e. 2- and 8-Aminopurines

Studies on the 2-aminopurines [**40**] are fragmentary (71HC(13)77), and nothing seems to have been done on 8-aminopurines [**41**]. By analogy with 2-aminobenzimidazole (Section 4-5De), it is expected that the predominant tautomers will be amino forms.

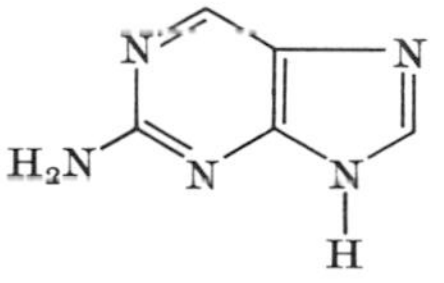

H_2N N N N N H

[**40**]

N N N N NH_2 H

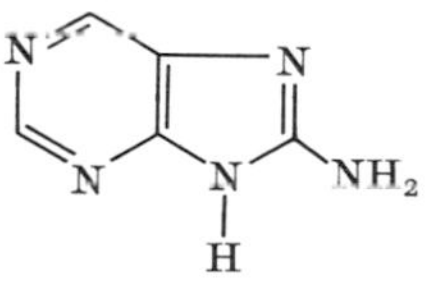

[**41**]

E. Purines Containing Two Potential Tautomeric Groups

a. 2,6-Dioxopurines (Xanthines): Neutral Species

After some early controversy, the dioxo structure [**42a**] had been established for xanthine by the time of the previous survey (II-58). The N(7)H structure of theophylline [**43**] was determined by X-ray diffraction (for discussion and references see 66HC(6)1 and 70J(B)596).

Later work (71J(C)1676) has utilized UV and NMR measurements and comparisons with methyl derivatives. UV spectra indicate that atoms N(1) and N(7) always bear a hydrogen atom but that there is

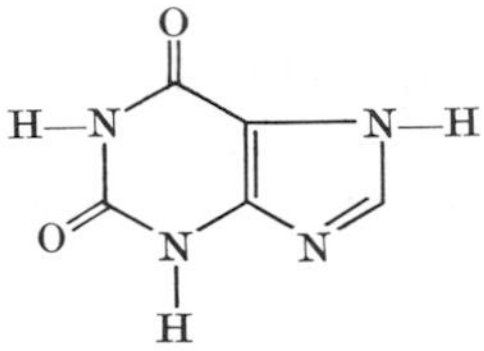

[**42a**] 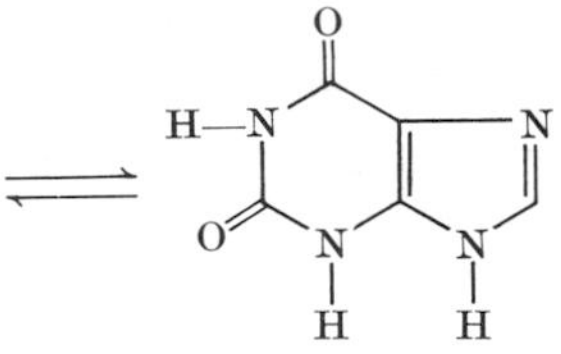 [**42b**] 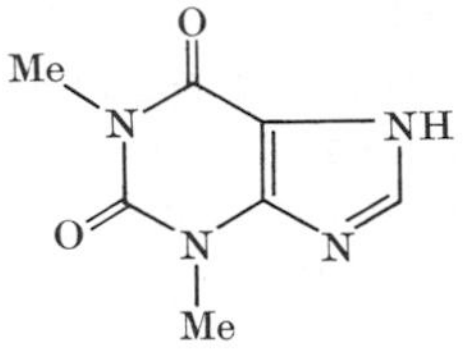 [**43**]

more likelihood of a tautomerism shift of a proton from N(3)H. However, the small UV shifts urge caution in interpreting these results, and NMR measurements in D_2O + DMSO-d_6 give a more complete picture (71J(C)1676). Comparisons of the chemical shifts of H(8) and of N–Me groups (if any) throughout a series of derivatives shows that protons are attached to the 1-, 3- and 7-nitrogen atoms, thus confirming structure [**42a**] already reported.

That xanthine and its derivatives are N(7)H tautomers is of theoretical interest because many other purines exist predominantly as N(9)H tautomers. CNDO calculations (71HC(13)77) confirm that [**42a**] is about 7 kcal/mole more stable than the N(9)H tautomer [**42b**].

Calculated values for UV absorption maxima and dipole moments have been compared with experimental data obtained for various methylxanthines in which an extra decylthio group was added at the 8-position to make the compounds soluble (69IQ103, 69JT(51)1862). The experimental results are in good agreement with the calculations based on all xanthine derivatives existing in the N(7)H form such as [**44**].

HMO calculations (71JH881) on the xanthine betaine structure [**45**] indicate that such structures should possess considerable stability.

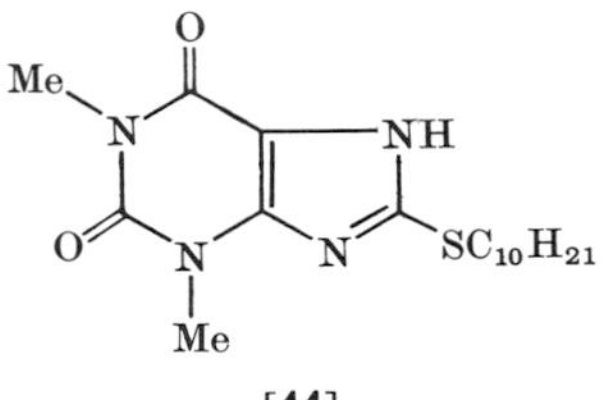

[**44**] 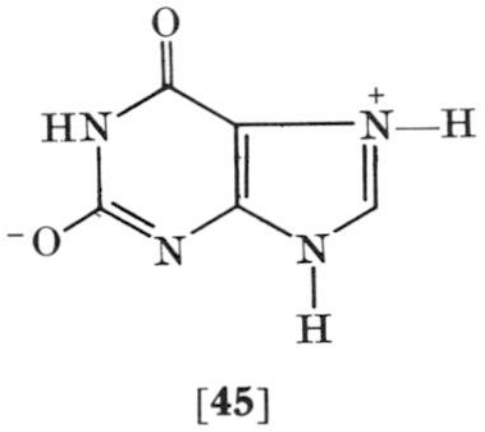 [**45**]

b. 2,6-Dioxopurines (Xanthines): Cations

NMR spectroscopy (70LA(731)174) in acidic media (from AcOH to $HClO_4$/ACOH) shows that caffeine undergoes protonation at N(9) to give [**46**] because of the downfield shift of H(8); the increase of the coupling constant between N(7)–Me and H(8) reflects the resonance contribution of the canonical form [**47**].

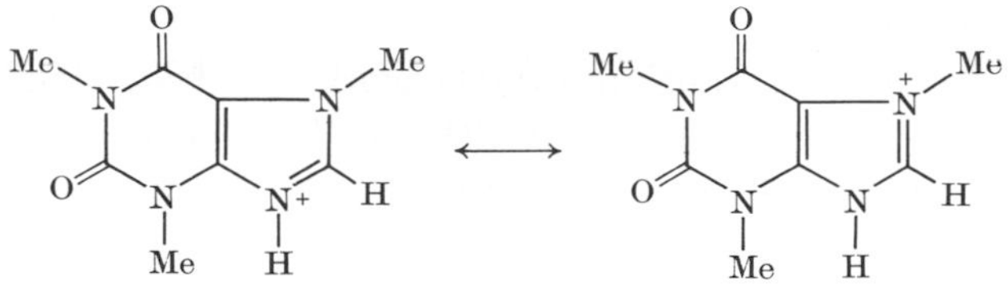

[46] [47]

NMR spectroscopy indicates that N(9)-substituted xanthines are protonated at the 7-position of the imidazole ring whereas other xanthine derivatives appear to be protonated on the 6-carbonyl group (71J(C)1676). This last conclusion is surprising, and should be confirmed.

c. 2,6-Dioxopurines (Xanthines): Anions

Early work (summarized in 66HC(6)1) indicated that the first proton loss occurs from N(3) for xanthine itself as well as for the 1- and 7-substituted derivatives, and that further dissociation occurs successively at N(7) and at N(1). If the xanthines are N(3)-substituted, the proton loss occurs first at N(7). More recent studies are substantially in agreement with these generalizations.

An X-ray study (69BJ3099) of the monosodium salt of xanthine showed that dissociation had involved the N(3)H proton to give [**48a**]. This result was criticized (71J(C)1676), because the former authors (69BJ3099) supposed xanthine to exist in the N(9)H form. The latter authors (71J(C)1676) used UV spectroscopy to show that 7- and 9-methylxanthines undergo proton loss successively at N(3)H and at N(1)H. For 3-methylxanthine the first proton loss occurs at N(7)H and the second at N(1)H; the same authors suggest that for xanthine itself and its 1-methyl derivative, the monoanion would be a mixture of forms [**48a**] and [**48b**] derived by N(3)H and N(7)H ionizations.

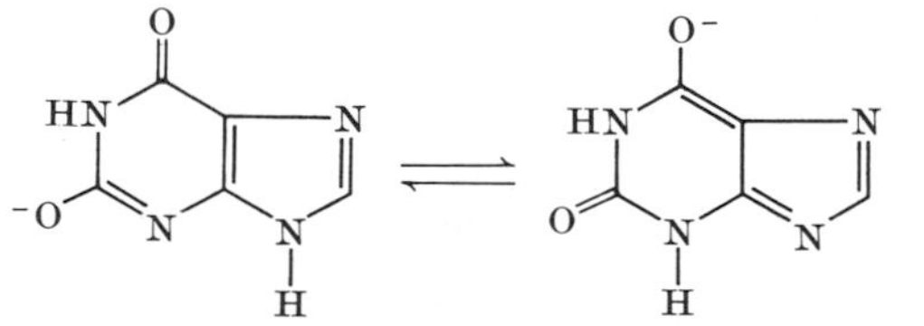

[**48a**] [**48b**]

See 74J(PI)2229 for the tautomerism of 6,8-purinediones.

d. Dithioxanthines

Compared with the xanthines, little work has been done. Dithioxanthine has been described (63AS1694) as [**49a**], but it may well exist in the N(7)H form [**49b**] as does xanthine.

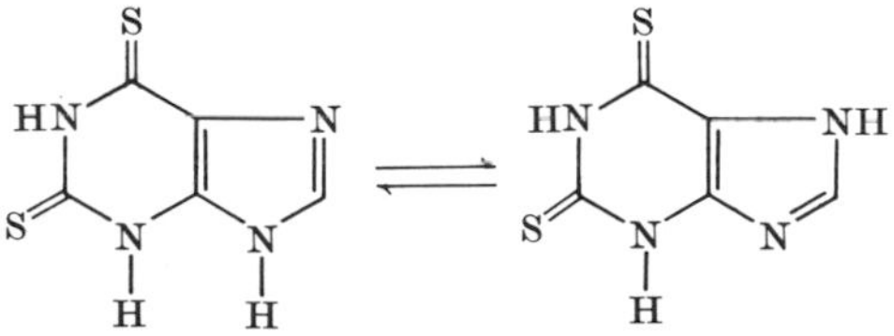

[**49a**] [**49b**]

e. Diaminopurines (II-76)

No data seem to be available on tautomerism in this series except the theoretically calculated site of protonation at N(3)H of 2,6-diaminopurines (II-76, 66HC(6)1) and the calculated electronic properties of 2,6-diaminopurine (69AS2963).

f. 6-Thioxanthines

(The tautomerism of the isomeric 2-thioxanthines has not been investigated).

A complete study of the tautomerism of 6-thioxanthine was reported in 1972 (72J(PII)1676). UV spectroscopy in aqueous solution and, more clearly, the NMR spectra (in 9:1 DMSO-d_6/D_2O) of 6-thioxanthine, of the 1-, 3-, 7- and 9-monomethyl together with several dimethyl derivatives and the 6-methylthio-2-oxo compound demonstrate that the predominant tautomers exist, where possible, in an oxothione form with the imidazole proton on N(7) [**50**; R_1, R_2 = H or Me], as found for xanthine [**42a**].

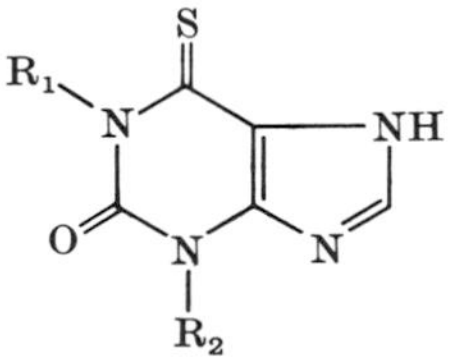

[**50**]

The structure of 6-thioxanthine cations has been investigated (72J(PII)1676) by measuring the change in chemical shift of the 8-hydrogen atom on protonation. This change is small except for 6-thioxanthines in which a 9-alkyl substituent is present. From this, and from UV spectral comparisons, the authors propose protonation at the thiocarbonyl group except for the N(9)-substituted derivatives for which some protonation is considered to occur at N(7). Additional evidence is required on this point.

Anion formation by thioxanthine derivatives (72J(PII)1676) seems to parallel that of xanthine derivatives. 7-Methyl and 9-methyl compounds undergo dissociation of the 3- and 1-position protons; in 3-methylthioxanthine the 7- and 1-protons are lost. For 1-methyl-6-thioxanthine, the dissociation occurs successively at N(3) and N(7). The monoanions of the thioxanthine itself and of its 1-methyl derivatives possess a proton at N(9).

g. Guanines (2-Amino-6-oxopurines) (II-76)

In the previous survey (II-76) it was concluded that guanine exists in an oxo–amino form [**51a**] from IR evidence, and this result was confirmed later by UV spectroscopy (65JA11). However, there was doubt left regarding the N(9)H–N(7)H tautomerism [**51a**] ⇌ [**51b**].

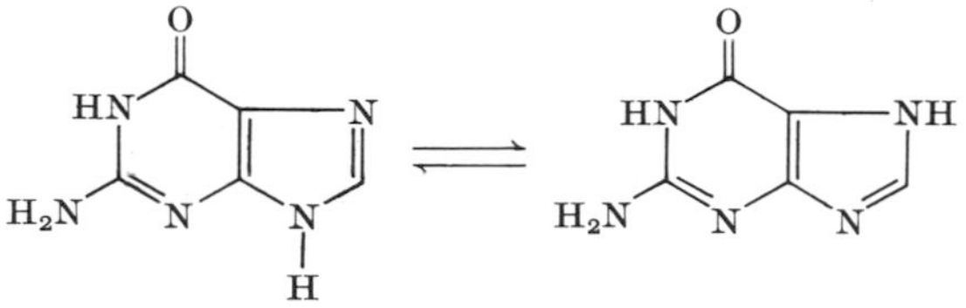

[**51a**] [**51b**]

The imidazole proton has now been located at N(9) for the solid state by X-ray diffraction (68BO436) and also for the vapour phase by UV spectroscopy (65PH3615). A survey of previous results concluded that guanine exists in solution in a mixture of the N(7)H and N(9)H tautomers (68MI73).

Calculations on heats of atomization suggest that an equilibrium should occur between the 7*H* and 9*H* tautomers of guanine but that the N(7)H tautomer should be more stable by 3.75 kcal/mole (70JA-2929). Other calculations also favour the N(7)H tautomer [**51b**] (71HC-(13)77), but only by 1 kcal/mole. However, comparisons of theoretical and experimental electronic properties (UV and dipole moments) (71HC(13)77) are not very satisfactory. The same authors made calculations to clarify the structure of N(3)-substituted derivatives; they were able to rule out significant contributions from the zwitterionic structure [**52a**] which was previously proposed, and suggested the predominance of structure [**52b**] although [**52c**] should be less stable by only 4 kcal/mole.

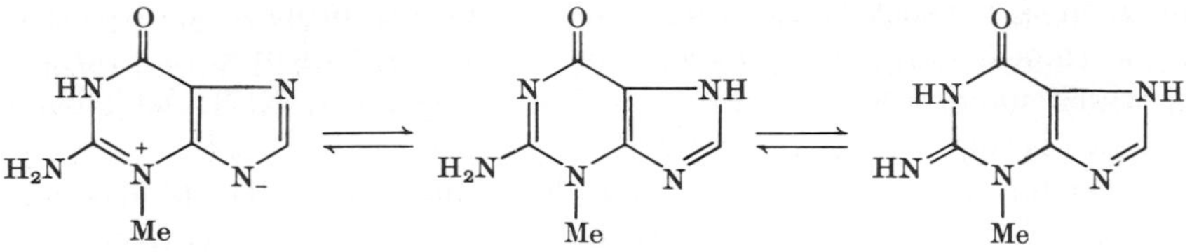

[52a] ⇌ [52b] ⇌ [52c]

The structure of the guanine cation is still not completely cleared up, although it has long been accepted that protons are attached to both the N(7) and N(9) imidazole sites (II-76, 63AS921, 66HC(6)1, 68MI73) on the basis of NMR, fluorescence and X-ray measurements. Together with the usual predominance of amino forms, this strongly suggests structure [**53a**] for the guanine cation but [**53b**] is not rigorously

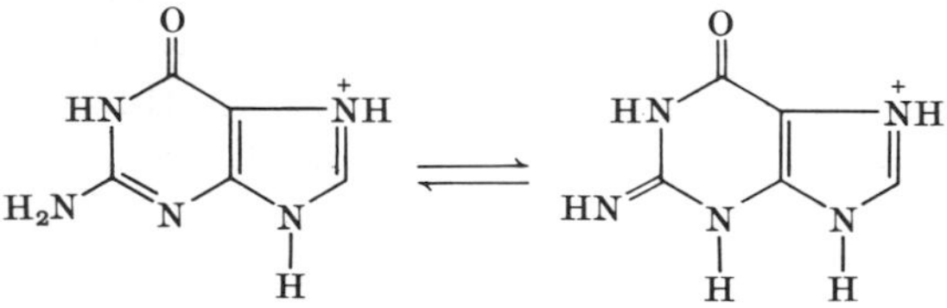

[53a] ⇌ [53b]

excluded. An accurate X-ray analysis of the 9-methylguanine hydrobromide has shown it to possess structure [**54**] (64CX126). Structure [**53a**] is also supported by X-ray measurements on the guanine cation by empirical correlations of the magnitude of the valence angles of nitrogen atoms in six-membered heterocyclic rings (65CX(19)861) depending on the presence or absence of an extra-annular hydrogen atom. For a theoretical approach to the problem of the site of protonation of 9-methylguanines see 74JA5911.

The guanine anion possesses structure [**55**], the dissociation occurring from N(1) (68M173, 71CR439).

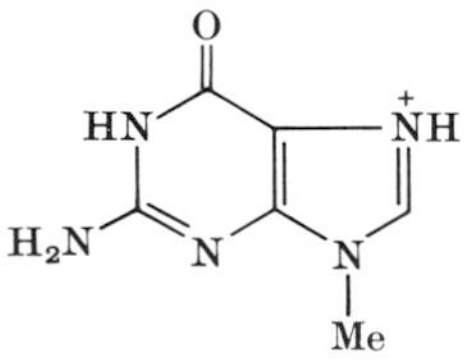

[54] 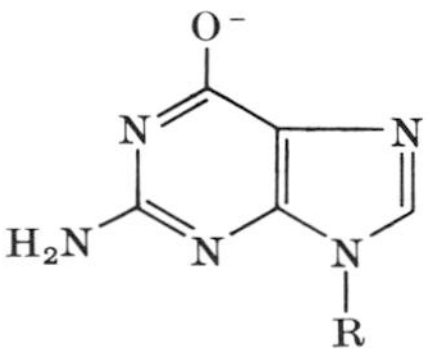 [55]

h. Guanosines

The tautomerism of guanosines has been investigated because of their biological importance. No N(7)H–N(9)H tautomerism can occur but the lactam [**56a**]–lactim [**56b**] tautomerism is of great importance as the abnormal tautomer [**56b**] is the form involved in spontaneous mutations when DNA is being replicated. This percentage was first supposed to be negligible (61PN791, 71CR439), but it has since been shown that more guanosine exists in the abnormal form than is the case for adenosine (70JA2929). In fact, the percentage of rare form is

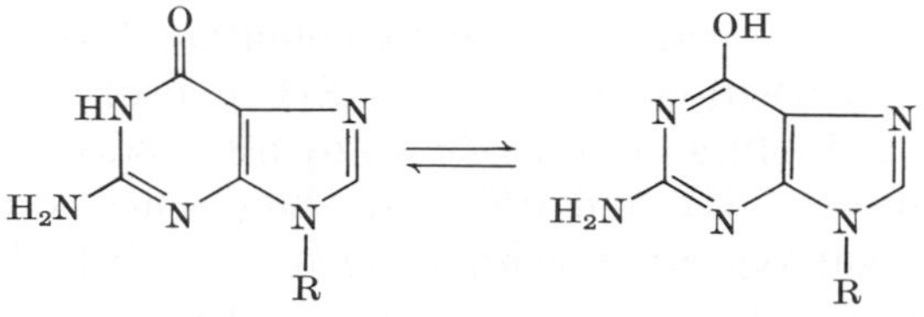

[**56a**] [**56b**]

R = ribose

too high, and would lead to a rate of mutation much greater than that observed; even Löwdin's restricted mechanism (71JA4585) does not predict the correct frequency of mutation based on pairing of rare forms.

Recently an even larger percentage of lactim form (about 16%) has been claimed (72JA3218), but the authors themselves (73JA3408) and others (73JA3511) showed later that this result was incorrect.

i. Thioguanines

The crystal structure of 6-thioguanine has been determined as [**57**] by X-ray diffraction (70JA7441). In contrast to guanine, the imidazole proton is attached to N(7) rather than to N(9).

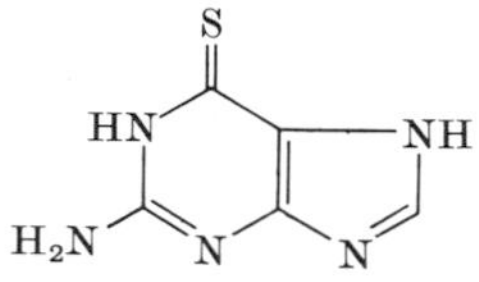

[**57**]

F. *N*-Hydroxypurines—Purine-*N*-Oxides

N-Hydroxypurines comprise a large class of compounds: the *N*-hydroxy or *N*-oxide function can be in four different positions and the purine nucleus can carry various tautomerizable groups (71MI01). Available tautomerism studies, undertaken mostly by G. B. Brown and his collaborators (66JO966, 67JO1151, 69JO978, 69JO2153, 71JO2639), concern mainly purine 1- and 3-oxides with various potential tautomeric groups. We restrict our review to the main results without giving all the reasoning. Knowledge of the *N*-hydroxy or *N*-oxide structure is important because of the possible relation between tautomeric structure and carcinogenic behaviour of some of the derivatives.

a. N-Hydroxypurines with No or One Functional Group

Potential purine 1-oxides carrying a potential C-hydroxy group in the pyrimidine 2-position exist in the N(1)-hydroxy structure, e.g., [**58**, R = H, β-D-ribofuranosyl] as shown by pK_a and IR measurements (66JO966). However, UV and pK_a data show that the corresponding N(3)-oxide derivatives often exist as such, e.g. [**59**]; the 6-methoxy derivative [**60**] has the mobile proton on the imidazole ring [**60a**] rather than, e.g., as [**60b**], the 3-position (69JO2153). Adenine 1-oxide and 3-oxide both exist in the *N*-oxide form, viz. [**61**] and [**62**] (58JA2759, 69JO2153).

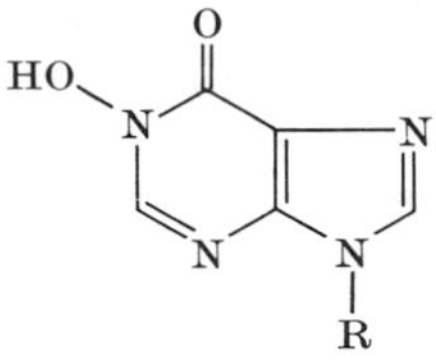

[**58**] 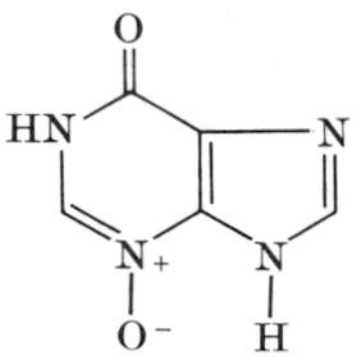 [**59**] 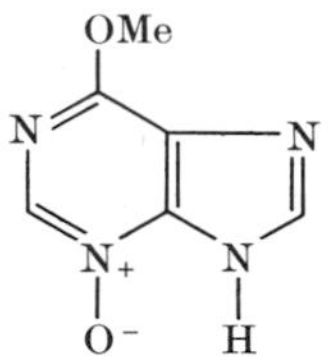 [**60a**]

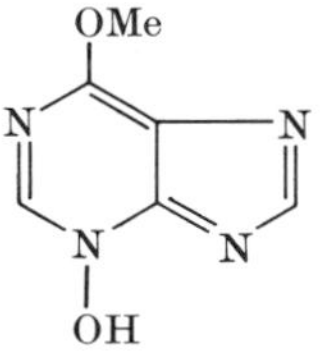

[**60b**] 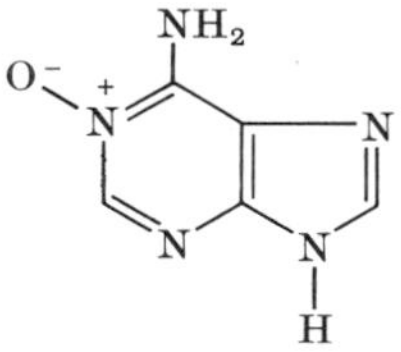 [**61**] 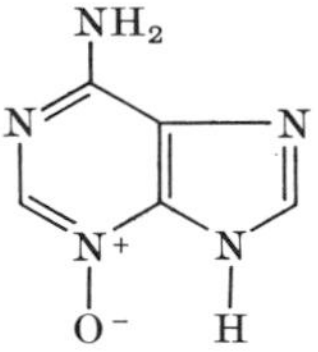 [**62**]

Potential purine 9-oxides exist mainly as 9-hydroxypurines although the 8-methyl derivative and 9-hydroxypurin-6-one do seem to contain a considerable proportion of tautomers of type [**62/1**] (72JO1867). The tautomerism of purine 7-oxides is yet unclear (66JO178).

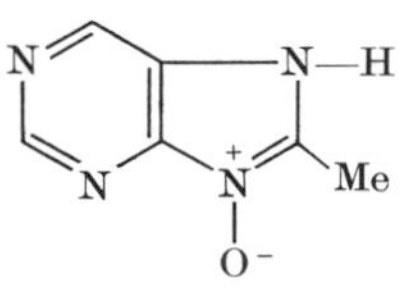

[**62/1**]

b. N-Hydroxypurines with Two Functional Groups

The structure of 1-hydroxyanthine was concluded to be [**63**] from the presence of two carbonyl bands in the IR spectrum (65JH220, 67JO-1151). Curiously this does not correspond to the usual N(7)H structure of xanthines (Section 5-1Ea). Recently (71JO2639) the structure of 3-hydroxyxanthine was reported as [**64**]: the corresponding 1-, 7- and 9-methyl derivatives and the 1,7-dimethyl derivative also all exist in 3-hydroxy forms. By contrast 2,6-diaminopurine 1-oxide possesses the *N*-oxide form [**65**] (71JO2639).

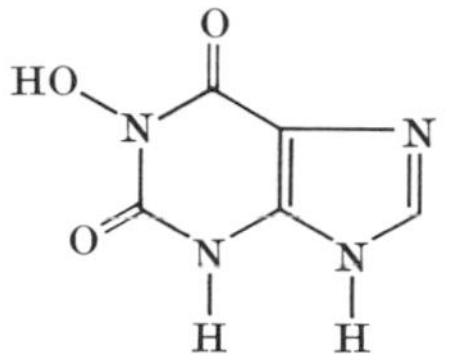

[**63**] 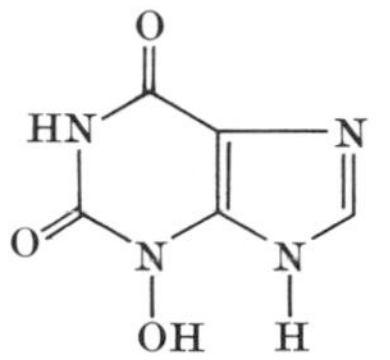 [**64**] 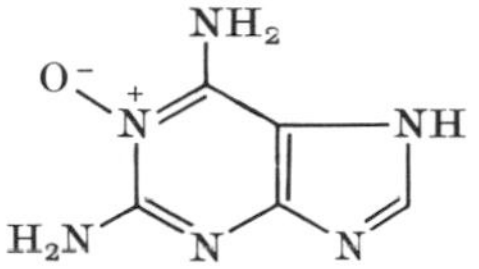 [**65**]

The tautomerism of 3-hydroxyguanines is more complicated (71JO-2639); whereas the 1-methyl and 1,7-dimethyl derivatives exist in the oxide form, cf. [**66a**], by contrast the parent compound and the 7- and 9-methyl derivatives exist as a mixture of hydroxy and oxide forms with a predominance of the hydroxy forms of type [**66b**]. These

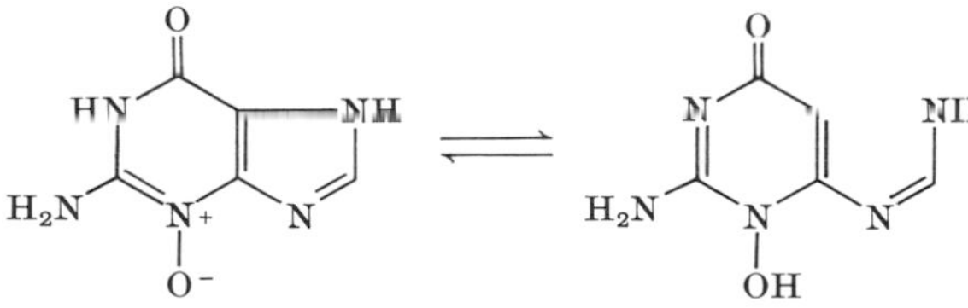

[**66a**] [**66b**]

conclusions were confirmed by solvent effects: the percentage of N–OH form increases if the water is replaced by less polar solvent such as MeOH which favours the less polar 3-hydroxy form.

From the results described, it seems that the presence of an adjacent amino group favours the *N*-oxide tautomer whereas the *N*-hydroxy tautomer is preferred when an adjacent carbonyl group is present; this pattern of behaviour is familiar for the pyridine *N*-oxides (Section 2-3F and 2-6Bc).

c. Purine N-Oxide Anions

Anion formation in 3-hydroxyxanthines occurs successively at the N(3), N(9) and N(1) positions (71JO2639) just for the xanthines themselves. The monoanion thus has structure [**67**]. The first ionization of 3-hydroxyguanines occurs from the pyrimidine ring leading to a monoanion of structure [**68**] (71JO2639).

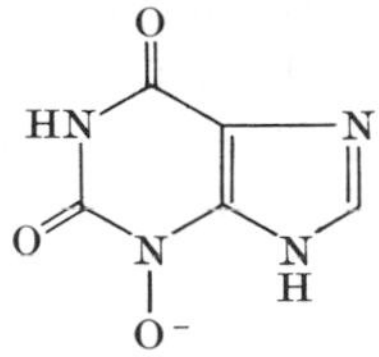

[**67**] 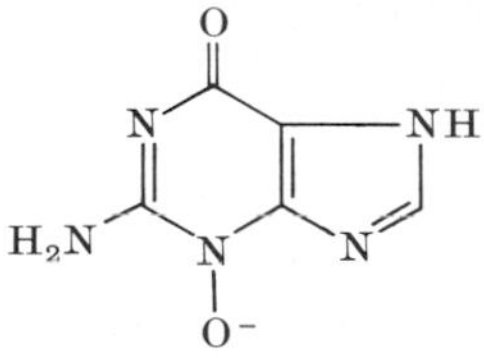 [**68**]

2. Azaindoles

A. Annular Tautomerism

The annular tautomerism of azaindoles* resembles that of benzazoles (Sections 4-1F, 4-1G, 4-1H) but is more complicated because of the possibility of the indole hydrogen shifting to a nitrogen atom of the six-membered ring. In fact such a shift never occurs to any large extent, at least in the ground state. We consider three typical cases: the monoazaindoles, imidazo [4,5-*b*]pyridines and 8-azapurines.

a. Monoazaindoles (I-426)

In a recent survey, Willette (68HC(9)27) summarized the information then available on the tautomerism of the four possible monoazaindoles [**69**]–[**72**]. All four azaindoles [**69**]–[**72**] exhibit νNH vibrations at

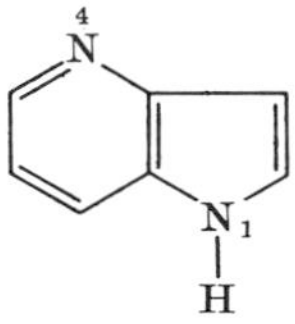

[**69**] 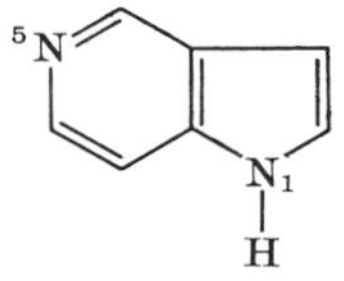 [**70**] 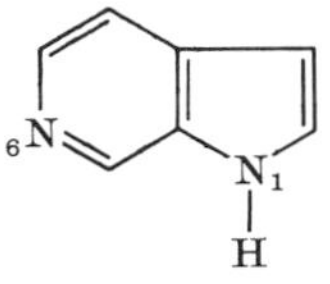 [**71**] 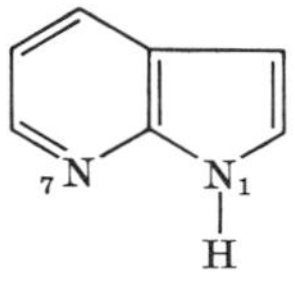 [**72**]

3472 cm^{-1} ($CHCl_3$, CCl_4) which is comparable to the νNH for indole (63PM(2)161, p. 209). The NMR spectra of the azaindoles (in $CDCl_3$), except that of 7-azaindole [**72**], show the NH/CH couplings between the NH in position 1 and the 2- and 3-hydrogens (these couplings are observed even in DCl/D_2O solution). These results confirm that [**69**]–[**72**] all possess the indole structures shown, as already concluded in the previous survey (I-426) for compound [**72**]. The lack of NH/CH coupling

* We use the term azaindole to indicate an indole derivative with one or more additional cyclic nitrogen atoms at least one of which is in the six-membered ring.

in [**72**] is ascribed to easier proton exchange because of the possibility of dimer formation [**73a**].

The dimer [**73a**] has been utilized to explain the phototautomerism of the 7-azaindole [**72**]. By comparison of the fluorescence spectra (Section 1-6E) of 7-azaindole with those of the fixed models 1-methyl-[**74**] and 7-methyl-7-azaindole [**75**], El-Bayoumi (69PN(63)253, 71JA5023, 74JA1674) demonstrated that in the lowest excited singlet electronic state, the dimer [**73b**] corresponding to the N(7)H tautomer is present (see also Section 2-5G).

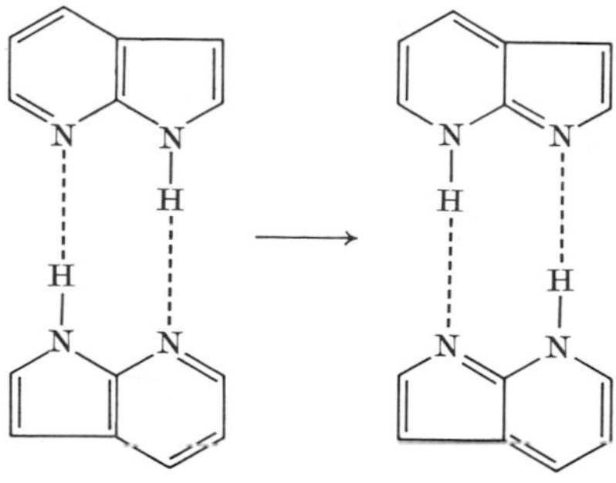

[**73a**] [**73b**] 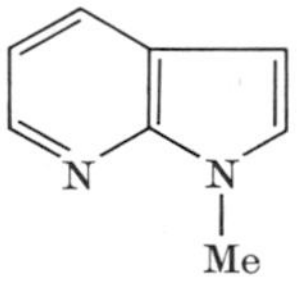 [**74**] 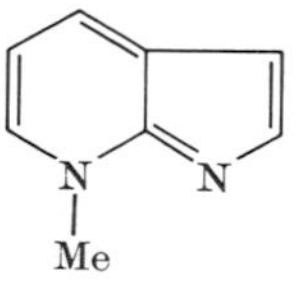 [**75**]

Monoazaindole cations are formed by protonation at the pyridine nitrogen atom (68HC(9)27). Recently, the protonation of 5-azaindole was studied in detail by NMR spectroscopy (72TL2857, 73KG767) in CF_3CO_2H, couplings (ca. 6.5 Hz) between the N(5)H and the 4- and 6-hydrogens are observed, confirming structure [**76**].

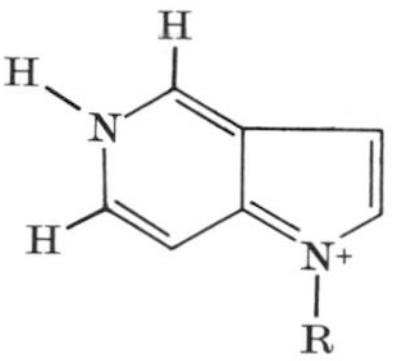

[**76**]

b. *Imidazo[4,5-b]pyridines*

Imidazo[4,5-*b*]pyridine can exist as three NH tautomers [**77a**], [**77b**] and [**77c**]. Comparison of the NH/CH coupling constants of imidazo-4,5-*b*]pyridine with coupling constants of the three fixed methyl derivatives [**78**], [**79**] and [**80**], eliminates the tautomer [**77c**] as a major contribution (72BF2916), but does not allow a choice between [**77a**] and [**77b**]. This choice was made using dipole moments (71KG1436); in benzene, the imidazo[4,5-*b*]pyridine has a dipole moment of 1.99 D, comparable to that (1.85 D) of the fixed compound [**78**], and very different from that (5.78 D) of the other fixed compound [**79**]; the authors did not measure the dipole moment of [**80**] but calculated it vectorially as ~6D.

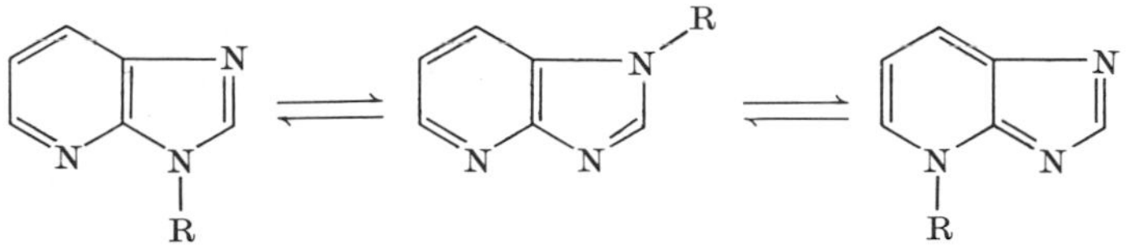

[**77a**] R = H
[**78**] R = Me

[**77b**] R = H
[**79**] R = Me

[**77c**] R = H
[**80**] R = Me

Qualitatively the result that the imidazo[4,5-*b*]pyridine exists, at least in benzene solution, mainly in the tautomeric form [**77a**] can be justified: in tautomer [**77c**], the π-system of the pyridine ring is affected and in tautomer [**77b**] there is a peri interaction between the two lone pairs, analogous to an α-effect [see discussion on 1,8-naphthyridine in [72JA2765)].

c. 8-Azapurines

The tautomerism of 8-azapurine has been investigated by semi-empirical theoretical methods by Pullman and Pullman (71HC(13)77, p. 142), who considered only the tautomers with an hydrogen atom on the five-membered ring, [**81a–c**]. The relative stabilities of the three tautomers decrease in the order N(9)H [**81a**] > N(7)H [**81b**] > N(8)H [**81c**]. The energy difference between the first two tautomers is small, whereas tautomer N(8)H [**81c**] is much less stable (about 30 kcal/mole): this agrees with the fact that quinonoid structures are always disfavoured (see the analogous case of the benzotriazoles: Section 4-1H).

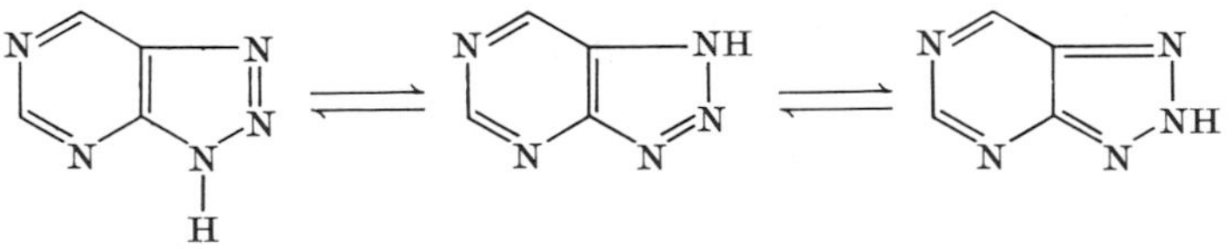

[**81a**] N(9)H

[**81b**] N(7)H

[**81c**] N(8)H

d. Pyrazolo[3,4-d]pyrimidines

Annular tautomerism in these compounds has recently been investigated by ^{13}C NMR (73JA4761, 73MI41).

B. Functional Group Tautomerism

a. Azaindoles with One Potential Tautomeric Group

The monoazaindolones have been seldom studied; see, however, 74J(PI)1531 for a paper on the influence of the position of the pyridine nitrogen on the tautomerism of azaindol-2-ones. The fragmentary results available on the tautomeric structures of azaindazolones are exclusively derived from infrared spectroscopy and are summarized in Table 5-1. It is difficult to rationalize these results completely, and there are certain doubts regarding the structures and assignments. For the pyrazolo[4,3-*c*]pyridin-3-ones [**85**] the position of the substituent R is not beyond doubt (65AJ379), and for the pyrazolo[3,4-*b*]pyridines [**86**] the assignment (70AN403, 72PS5) of the medium intensity band at 1610 cm^{-1} of the 1-benzyl isomer to an aromatic C=C vibration is simpler by analogy with the results previously obtained for the 1-benzyl-3-hydroxyindazoles (65AN116). The νC=O reported for [**87**] (Table 5-1) is low. UV spectral evidence apparently suggests that 4-azaoxindole exists as a tautomeric mixture of three forms (72JO51).

In the *N*-oxides (on the pyridine nitrogen) of pyrazolo[4,3-*c*]pyridines the carbonyl absorption disappears, but it reappears in the corresponding hydrochlorides (65AJ379). The oxo structure of the bicyclic compound [**88**] was established by X-ray crystallography (69CX(B)625); this is in agreement with the tautomeric structure of the 2-substituted indazolones (Section 4-2Jc) in the solid state.

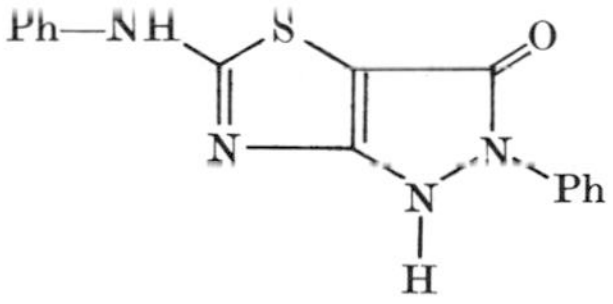

[**88**]

Examples of azaindazoles carrying a potential hydroxy group on the pyridine rings are represented by compounds of type [**89**], which is shown by IR (KBr) (68CB3265) to exist predominantly in the oxo form.

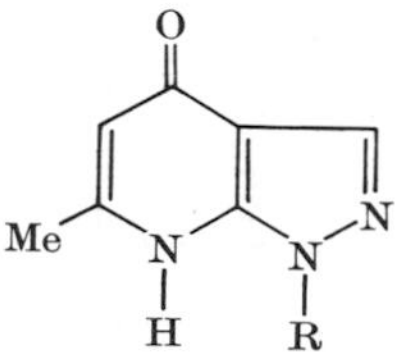

[**89**] R = H, Me, COMe

TABLE 5-1
INFRARED SPECTRA OF AZAINDAZOLONES

Compound and IR frequency			Experimental conditions	Reference
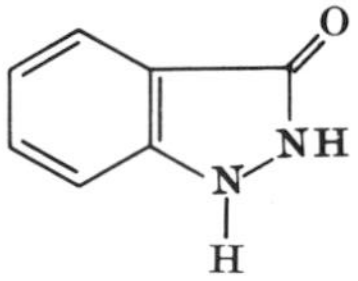 1630 (C=O)	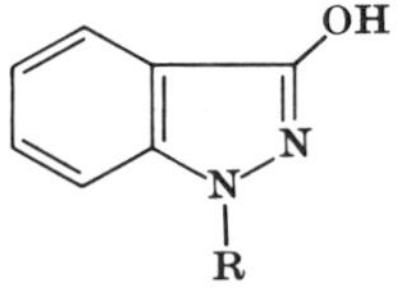 1550 (C=N)	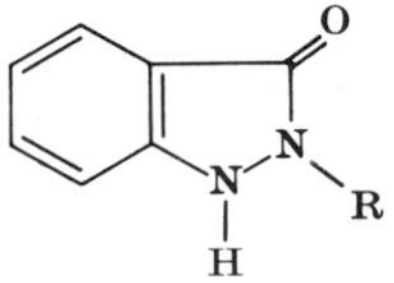 1620–1650 (C=O)	KBr Nujol	Indazolones (Section 2-Ah)
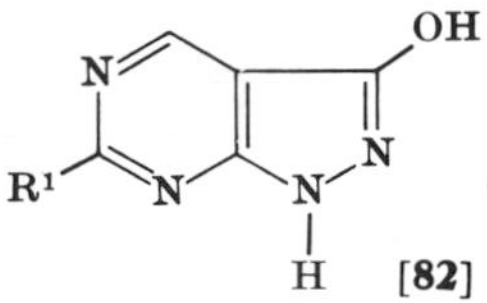 [82] $\nu_{C=O}$ absent	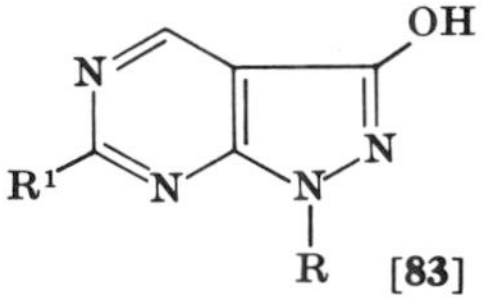 [83] $\nu_{C=O}$ absent	—	Unspecified	Pyrazolo[3,4-*d*] pyrimidines (61J0451, II-60)
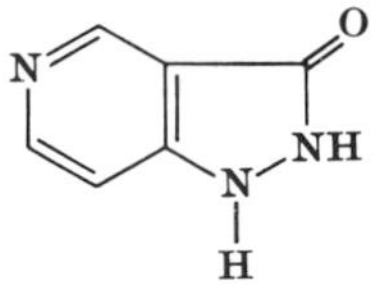 [84] 1630 (C=O)	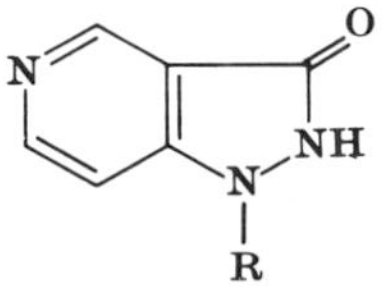 [85] 1627–1645 (C=O)	—	Nujol	Pyrazolo[4,3-*c*] pyridines (65AJ379)

Compound / IR (cm^{-1})			Medium	Remarks
[86] R = Ph: 1595(s); R = CH_2Ph {1610(m), 1585(s)}	[87] 1605(s)		KBr	Pyrazolo[3,4-*b*] pyridines (70AN403)
3400 (νOH)				Pyrazolo[3,4-*b*] pyridine (72JH235)
1673 (C=O)	1645–1657 (C=O)	R = H, R^1 = Me 1654; R = Me, R^1 = H 1632; R = R^1, = Me 1635		(67KG329)

Few azaindazoles with potential mercapto groups have been investigated. 1-Methylpyrazolo[3,4,-*d*]pyrimidine-4-thione [**90**, R = Me] was reported to exist in the thione form in the previous survey (II-65). Recently (73CO(276C)1007, 74CX(B)1598) the thione structure of the desmethyl analogue [**90**, R = H] has been established by X-ray crystallography. For the oxo-derivative (allopurinol) see (72CX(B)2148). Although analogues have been synthesized (65AJ1267) e.g., [**91**], no structural work was done; however, compounds of type [**91**] probably exist as thiones. Aminopyrazolo[3,4-*b*]pyridines exist in the amino forms (72JH235), like the aminopyrazolo[1,5-*a*]pyrimidines (74JH423, 75MIip2).

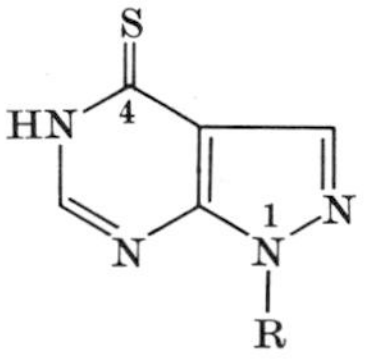

[**90**] 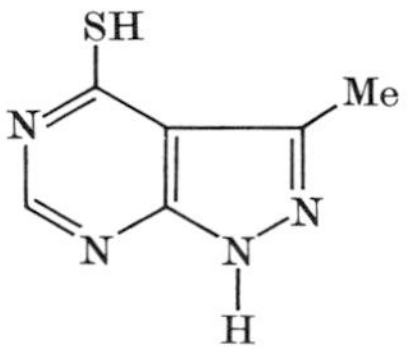 [**91**]

b. Imidazo[4,5-c]pyridines with Several Potential Tautomeric Groups

The dioxo structure of [**92**] was demonstrated from NMR measurements (63JO3041) in CF_3CO_2H and DMSO, whereas the monoanion discloses greater conjugation in the UV spectrum and is assigned structure [**93**].

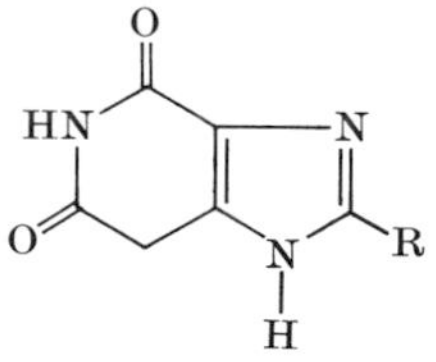

[**92**] 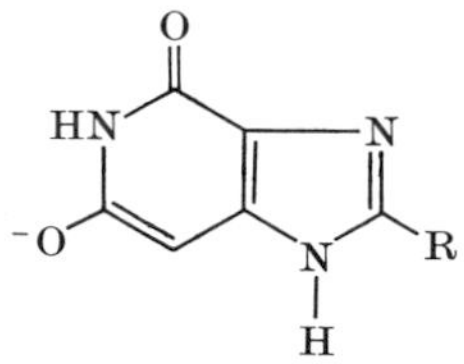 [**93**]

c. 8-Azapurines (II-59)

The previous survey (II-60) included a single example of an 8-azapurinone, the oxo structure [**94**]. Since then much work has been done because of the biological interest of these compounds. This recent work until 1970 is well reviewed by Pullman and Pullman (71HC(13)77, p. 142). For many of the known substituted and unsubstituted 8-azapurine derivatives, information is given about the functional tautomerism but not usually for the annular tautomerism. Thus the oxo structure of [**95**] and of its 7-, 8- and 9-methyl derivatives were established (68J(C)344, 68J(C)2076) by IR spectroscopy (solid state and $CHCl_3$ solution). The dilactam structures of compounds of types [**96**], [**97**] and [**98**] have been demonstrated (65CB1060) by a comparison of UV spectra (H_2O) and pK_a values of the tautomeric species with those of fixed methyl derivatives.

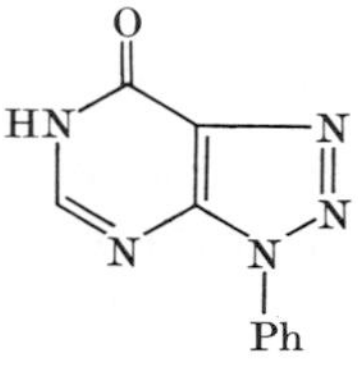

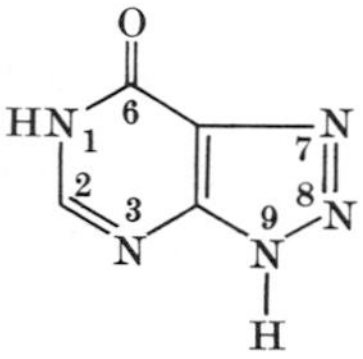

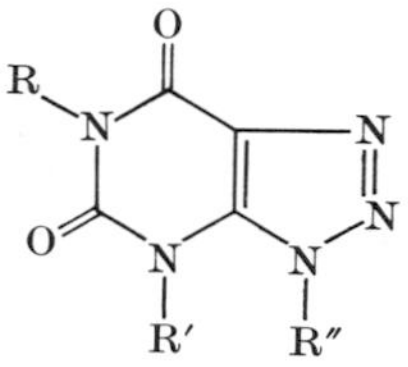

The complete tautomeric structures of 8-azaguanine [**99**] (68CX(B)-1692) and 8-azaxanthine [**100**] (69ZX(130)376) have been established from X-ray measurements of the corresponding monohydrates. These two molecules do not have the same type of triazole ring in the solid

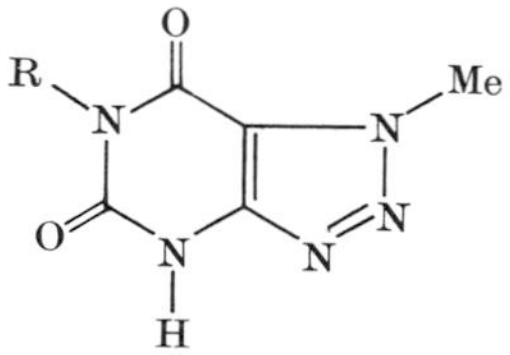

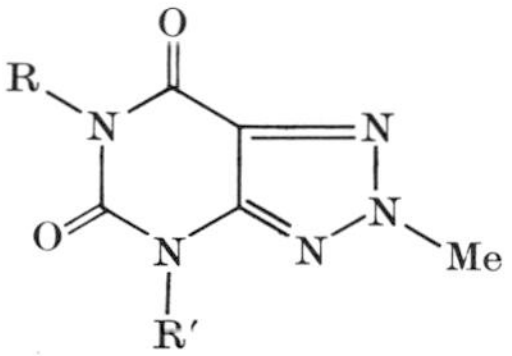

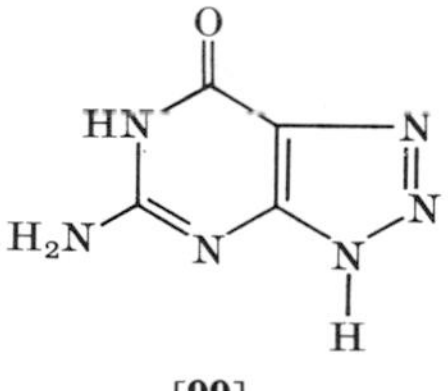

state: this different behaviour is attributed to the influence of crystal forces. The molecular structure of 8-azaguanine hydrochloride monohydrate has been determined (74CX(B)2806); the azole proton is on N(8). Pullman and Pullman (71HC(13)77, p. 142) calculated the relative stabilities of the three possible tautomers N(7)H, N(8)H and N(9)H, for 8-azaadenine, 8-azaguanine, 8-azaxanthine and 8-azahypoxanthine. All showed the same order of stability: N(9)H > N(7)H > N(8)H, with the N(8)H tautomers about 20–30 kcal/mole less stable than the others.

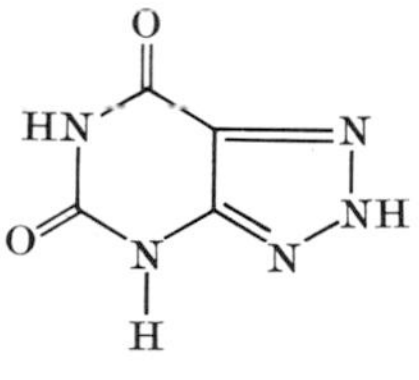

The same authors compared UV absorptions calculated for the various tautomeric species with experimental values for methyl derivatives: the agreement is quite good.

3. Azaindolizines

A. Annular Tautomerism in Conjugate Acids

For structural reasons, annular tautomerism occurs in the conjugated acids of aza derivatives of indolizine [**101**], but not in the free bases. There is a close parallelism between the behaviour towards protonation

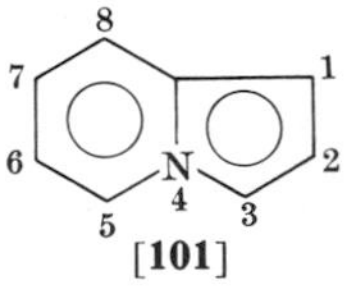

8 7 1 6 N 2 4 5 3

[**101**]

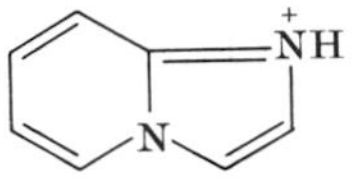

64J4226
66JO1295

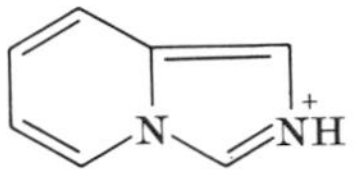

64J4226

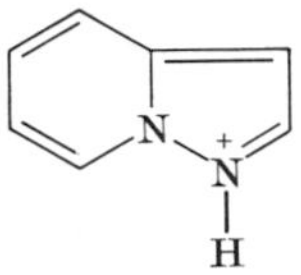

64J4226

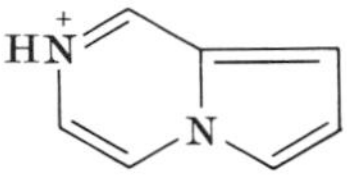

65J2778

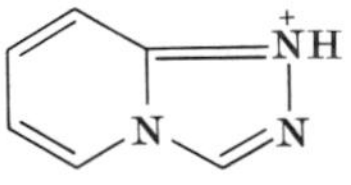

65J2778
66JO265

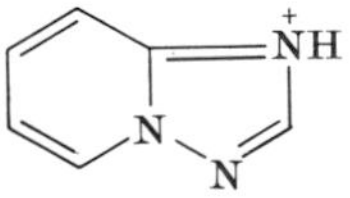

65J2778

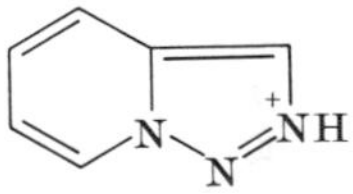

65J2778

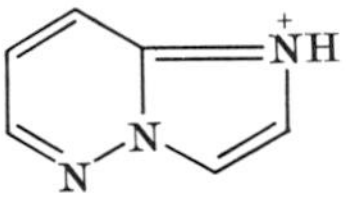

65J2778
69JH559

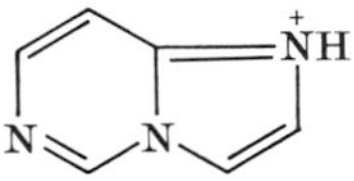

65J2778

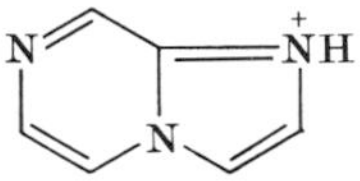

65J2778

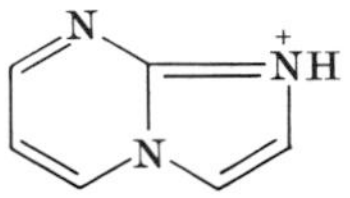

65J2778
66JO809

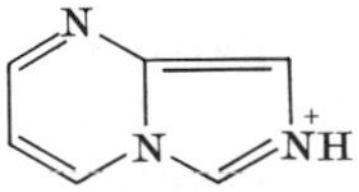

72BF2481

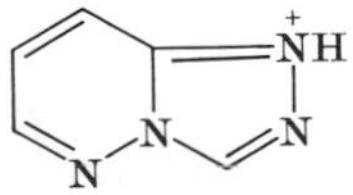

69JH559

Structure of azaindolizine monocations

Scheme 5–3

of the pairs pyrrole–azole and indolizine–azaindolizine. In no case does the predominant cation carry a proton on the pyrrole-type "bridgehead" nitrogen atom the lone pair of which is involved in the aromatic π-system. Cation formation by pyrrole and indolizine involves C-protonation (Section 3-1B), and azole (Section 4-1K) and azaindolizine cations carry the additional proton on one of the pyridine-like nitrogen atoms.

UV spectral comparisons of the bases and conjugated acids, and basicity differences, were used by Armarego (64J4226, 65J2778) to determine the structure of the azaindolizine monocations (Scheme 5-3). Other authors (66JO265, 66JO809, 66JO1295, 69JH559, 72BF2481) established the site of quaternization and used comparisons of NMR spectra with the methiodide; however, both methods give only qualitative results.

There is good agreement between the sites of protonation thus established and those predicted by frontier electron densities (66JO809, 69JH559), but not with total electron densities calculated by the CNDO/2 method (72BF2481).

B. Functional Group Tautomerism (II-59)

a. Monoazaindolizines Containing One Potential Hydroxy Group (II-59)

The oxoazaindolizines described in the previous survey (II-59) (the oxo group was always on the six-membered ring) exist in the oxo form. Since then much varied work has been carried out in this series; the results are not always easily rationalized.

From NMR evidence in DMSO (65JH53), imidazo[1,2-*a*]pyridin-5-one and imidazo[1,2-*a*]pyridin-2-one exist in the 1H-oxo [**102**] and the 3H-oxo [**103a**, R = H] forms, respectively. In the last compound, the presence of a methyl group in the 3-position modifies the equilibrium (70JJ436); in the same solvent, the NMR shows the presence of two tautomers [**103a**, R = Me] and [**103b**, R = Me]. The NMR spectra of

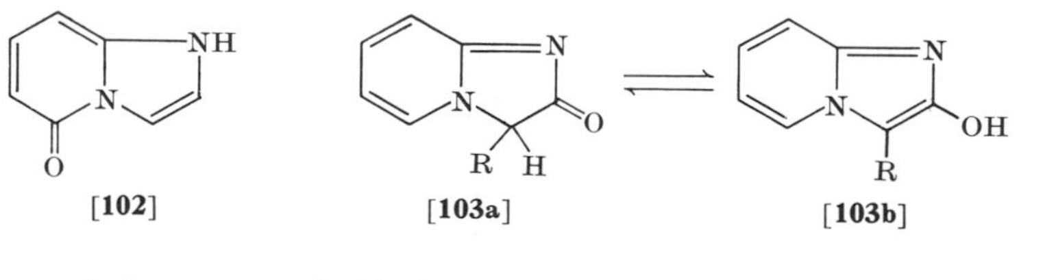

NH N O [**102**] N N O R H [**103a**] ⇌ N N OH R [**103b**]

3-oxygenated compounds [**104**, R = H, Me, Ph] provide evidence for the hydroxy form written. Derivative [**105**] exists in the oxo form in the solid state and predominantly so in DMSO solution (70LA(735)35).

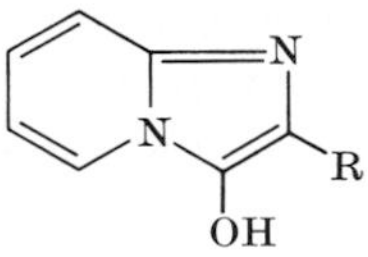

[**104**] 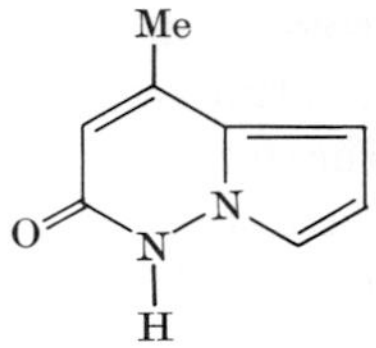 [**105**]

b. Diazaindolizines Containing One Potential Hydroxy Group

i. *Pyrazolo*[1,5-*a*]*pyrimidines.* The 5-oxopyrazolo[1,5-*a*]pyrimidines [**106**; R_2, R_7 = alkyl; R_3 = CO_2Me, H; R_6 = CN, H] exist in the solid state in the oxo form from IR evidence (70BJ849). A similar structure has been assigned to [**107**; R_2 = Me, R_3 = H, R = H, Cl] (69JH947) for the solid state as well as for DMSO solution.

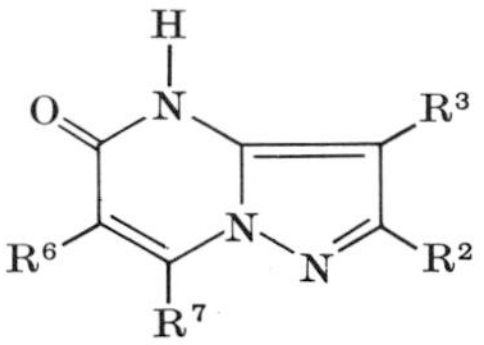

[**106**] 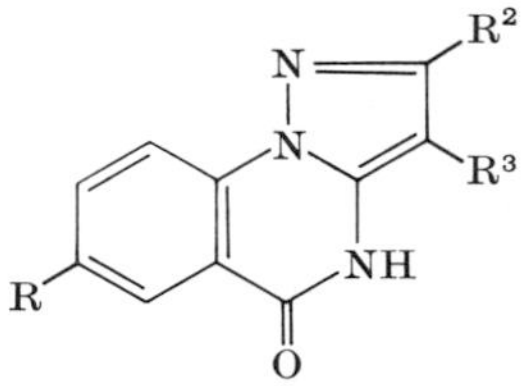 [**107**]

However, 7-oxopyrazolo[1,5-*a*]pyrimidines [**108**; R_2 = Me, R_5 = Me, Ph, R_3 = H, CO_2Et, R_6 = H, CN] appear to exist in oxo–enol mixtures, [**108a**] and [**108b**] (70BJ849), in the solid state.

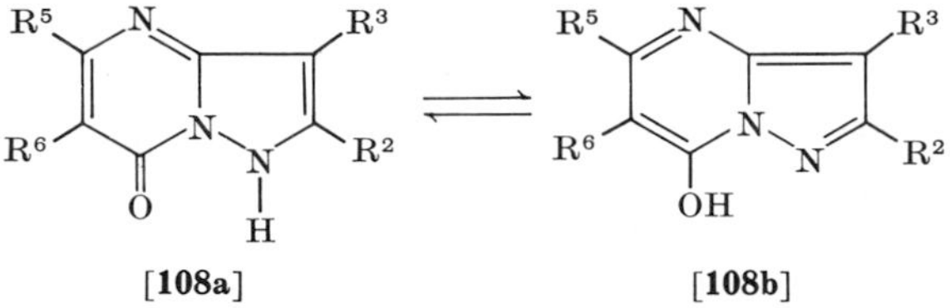

[**108a**] [**108b**]

ii. *Pyrazolo*[*1,5-c*]*pyrimidines.* IR (KBr) and NMR spectroscopy together with UV comparisons (in MeOH) of [**109**] and its *N*-methyl derivative show that the oxo form [**109**] predominates (72CB388).

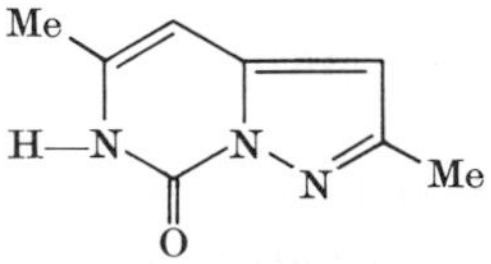

[**109**]

iii. *s-Triazolo[4,3-a]pyridines.* IR spectroscopy in the solid state shows that the 3-oxygenated derivative [**110**] exists predominantly in the oxo form [**110b**] (66JO251). However, compound [**110**] has a different fragmentation in mass spectrometry (71OM1), compared with the corresponding thione derivative [assigned a predominant thione structure (see Section 5-3Bd)], and the same authors interpret this as indicating that the product exists to some extent as a keto–enol mixture in the vapour phase.

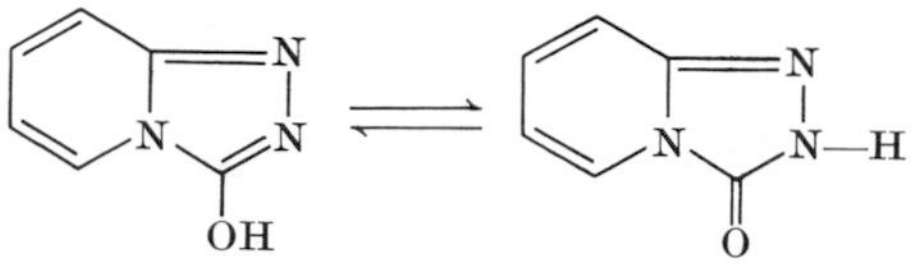

[**110a**] [**110b**] 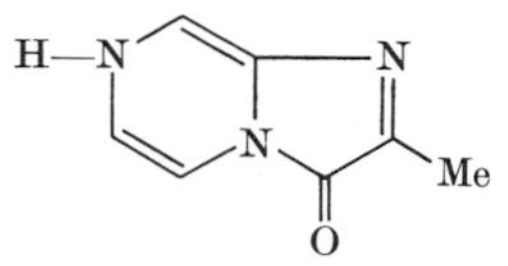 [**111**]

iv. *Imidazo-pyrazines and -pyrimidines.* The oxo structure of the species [**111**] is shown by UV (in MeOH) and NMR (in DMSO) spectroscopy and by comparison with the corresponding *N*-methyl derivative (70JJ431). Dioxo structures have been ascribed (66AH405) to compounds [**112**], [**113**] and [**114**] from the IR spectrum (KBr) which contained two bands assigned to νC=O, and by use of UV comparisons with certain *N*-alkyl derivatives. UV comparisons support structures [**114/1**] and [**114/2**] (73J(PI)1588).

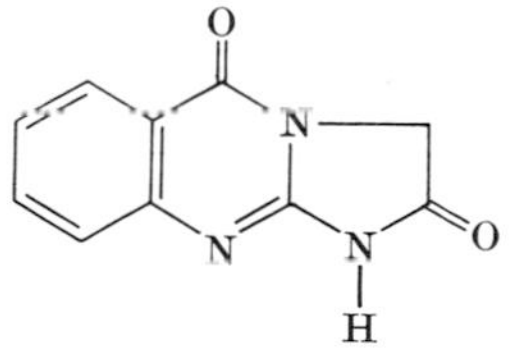

[**112**] 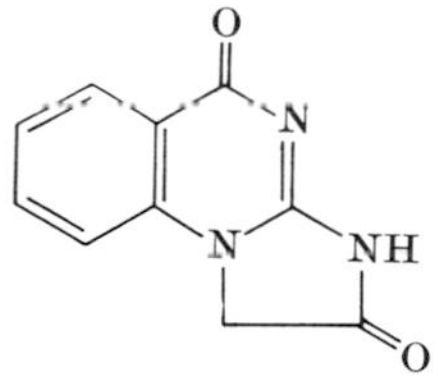 [**113**] 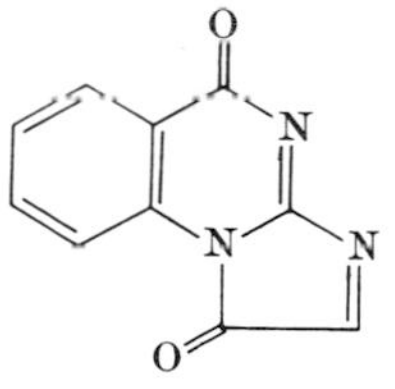 [**114**]

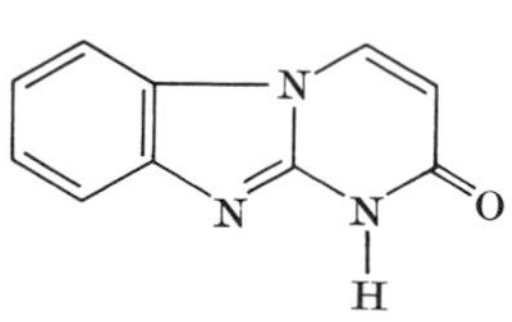

[**114/1**] 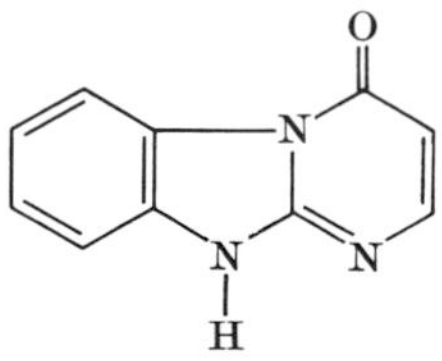 [**114/2**]

c. Tri- and Tetra-azaindolizines Containing a Potential Hydroxy Group

The *s*-triazolo[1,5-*a*]pyrimidines [**115**] and [**116**] exist in lactam forms from IR evidence in solid state (63CT67) and for similar conclusions see (59JO779, 68T2839; 63ZO2673). The possibility of annular tautomerism was not considered in this experimental work, but calculations (72T5779) do favour the pyrimidine nitrogen atom carrying the proton (as in [**115**], [**116**]). However, calculations also indicate that substituents next to the oxygen function, particularly CO_2Et but also halogen, can stabilize the hydroxy form by hydrogen bonding [**116/1**] (72T5789).

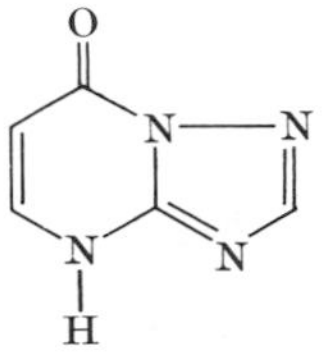

[**115**] 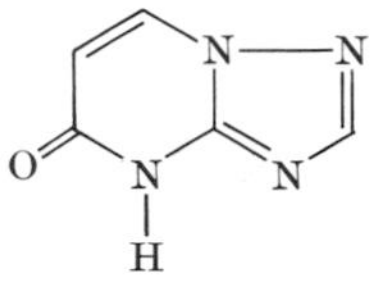 [**116**] 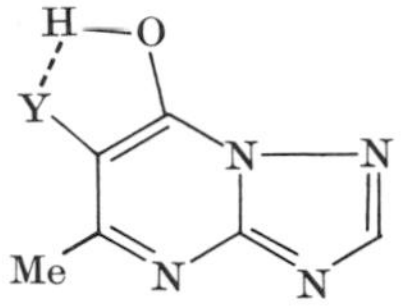 [**116/1**]

IR spectroscopy is claimed to show the occurrence of "keto–enol tautomerism" in the solid state for the *s*-triazolo[4,3-*a*]pyrazine [**117a**] ⇌ [**117b**] (71OM663). IR spectroscopy in the solid state and in DMSO shows that both the tetraazaindolizines [**118**] and [**119**] exist in the oxo forms shown (65JO2395).

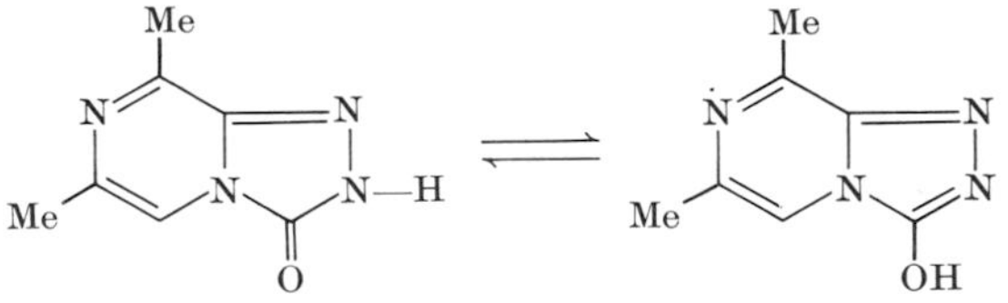

[**117a**] [**117b**]

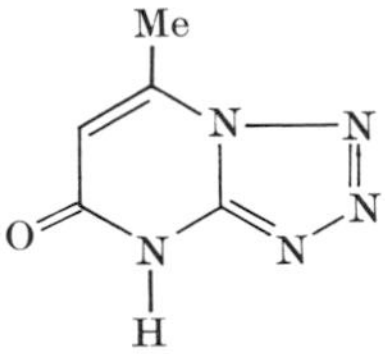

[**118**] 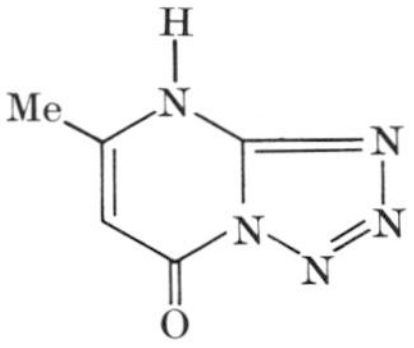 [**119**]

d. Compounds Containing One Potential Mercapto Group (II-65)

The previous survey (II-66) reported the thione structure of [**120**]. Two further examples are now available. UV comparison (MeOH) of [**121**] with the S- and N-methylated derivatives shows that the thione form is predominant (72CB388). The thione form is also predominant for [**122**] from IR and NMR (66JO251, 66JO265) and mass spectra

(71OM1). The thione form [**120**] has been further supported by calculations (72T5789), and [**122/1**] is demonstrated by X-rays (72J(PII)11).

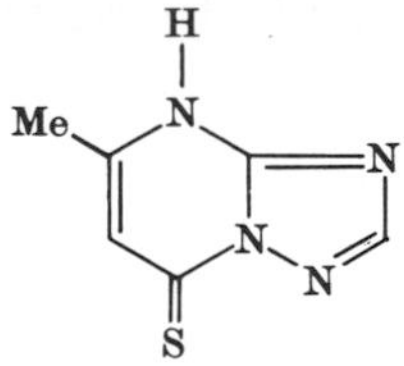

[**120**] 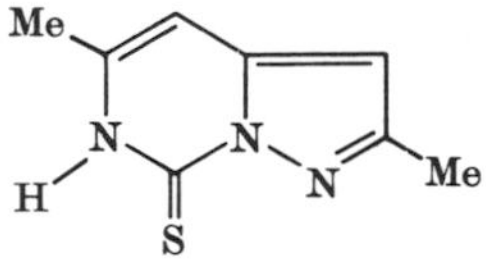 [**121**] 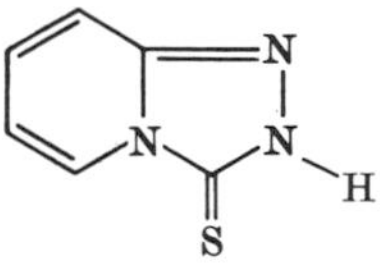 [**122**]

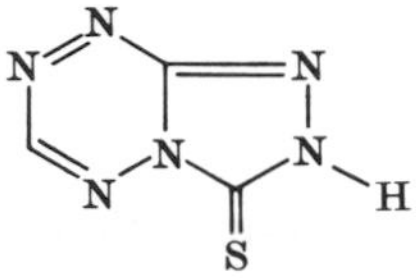

[**122/1**]

e. Compounds Containing One Potential Amino Group

All the aminoazaindolizidines so far investigated exist in the amino form; details are shown in Scheme 5-4.

The only claimed exception is in the case of [**123**] 7-aminopyrazolo-[1,5-*a*]pyrimidine (70BJ849) which is stated to exhibit νNH_2 and νNH

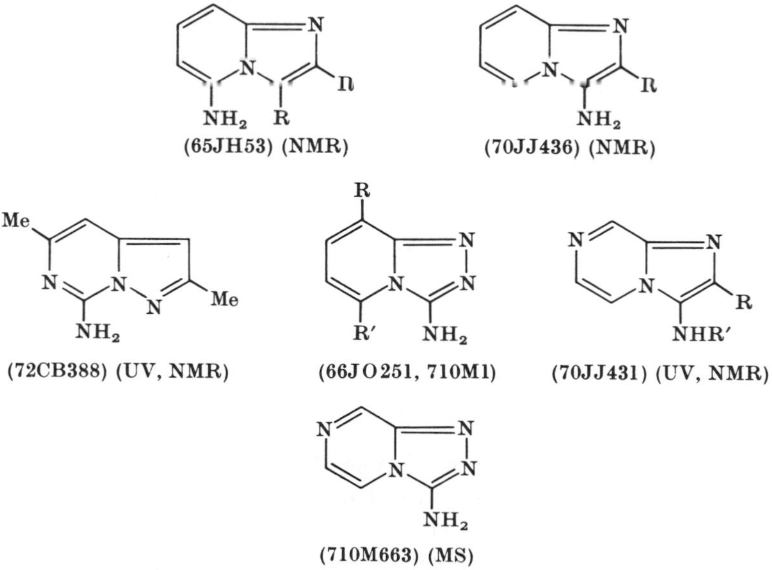

(65JH53) (NMR) (70JJ436) (NMR)

(72CB388) (UV, NMR) (66JO251, 71OM1) (70JJ431) (UV, NMR)

(71OM663) (MS)

Aminoazaindolizines

SCHEME 5–4

absorptions in the solid state and exists as an amino–imino mixture. The evidence reported for [**123b**] is very weak, and the conclusion appears highly unlikely. Earlier work in the same series had assigned the amino structure [**123a**] (62CT620).

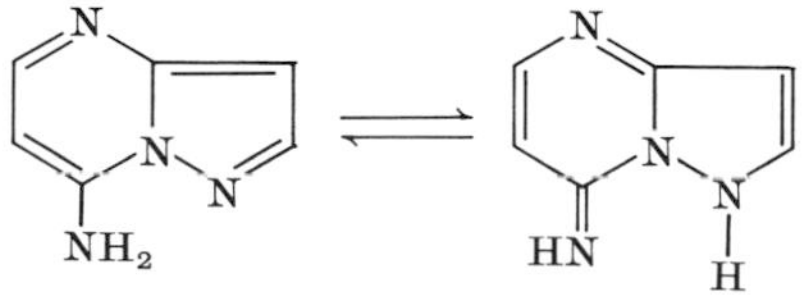

[**123a**] [**123b**]

4. [5.6]Bicyclic Compounds Containing O and S in the Five-Membered Ring

We now discuss the tautomerism of heteroaromatic [5.6]bicyclic systems with nitrogen atoms in the six-membered ring and oxygen or sulfur atoms in the five-membered ring.

A. Thiazolo-quinolines, -pyridazines, and -pyrimidines

Thiazolo[4,5-*c*]isoquinolin-2-one [**124**] and thiazolo[5,4-*c*]isoquinolin-2-one [**125**] have been studied by IR (KBr disks), UV (absolute ethanol] and NMR ($CDCl_3$, DMSO) (68CC691). Carbonyl bands in the IR spectra and comparison of the UV spectra of the tautomeric compounds with those of the corresponding fixed *N*-methyl derivatives, demonstrate the NH structures [**124**] and [**125**]. This conclusion is consistent with the tautomerism of benzothiazol-2-ones [**126**] (Section 4-3Bc).

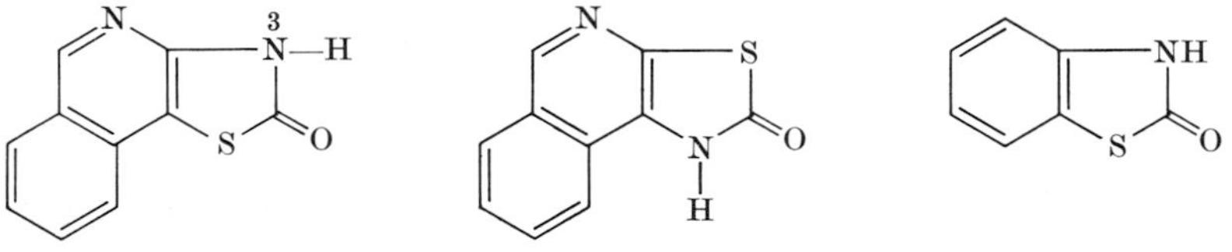

[**124**] [**125**] [**126**]

Thiazolo[4,5-*d*]pyridazinones of type [**127**] show carbonyl bands in the solid state (Nujol mulls, KBr disks) which indicates the lactam structure [**127**] (71BF3537); this is similar to phthalazinone [**128**] (I-366). The corresponding 4,7-dioxo derivative (70BF4317; Section 2-4Ce) probably exists in one of two possible oxo-hydroxy forms [**129a**] or [**129b**]; the IR evidence does not exclude the dioxo tautomer,

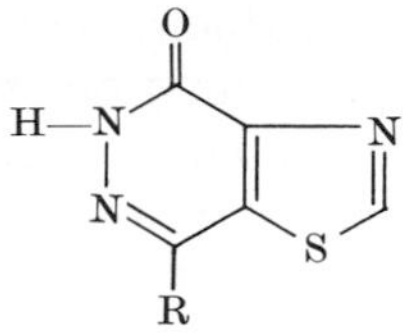

[127] 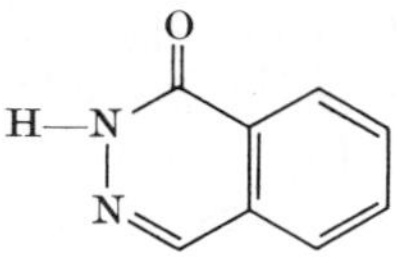 [128]

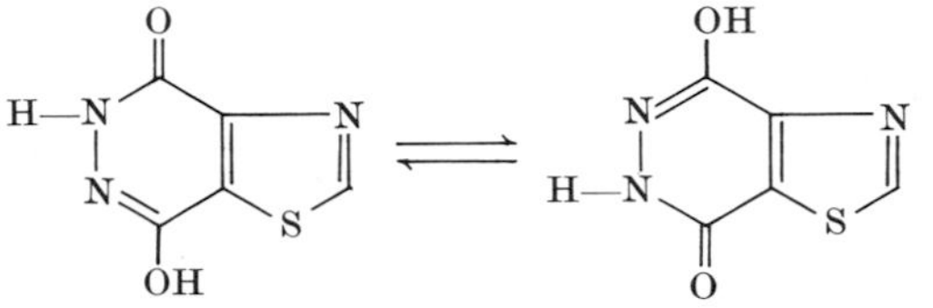

[129a] [129b]

but this is less likely, taking into account structure [130] of phthalazinedione (I-368).

7-Amino[1,2,5]thiadiazolo[3,4-*d*]pyrimidine [131] and 7-aminofurazano[3,4-*d*]pyrimidine exist in the amino form in the solid state as established by X-ray diffraction (71JA7281), a situation similar to that in adenine (Section 5-1D).

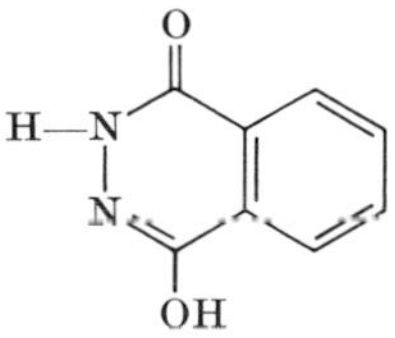

[130] 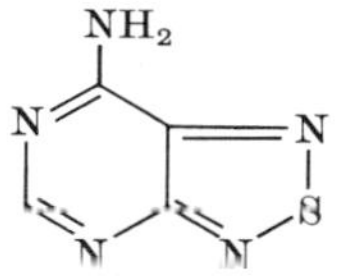 [131] 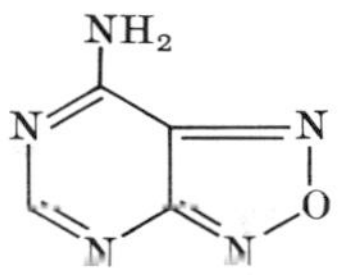 [132]

B. [1,2,3]OXADIAZOLO[3,4-*a*]QUINOXALINES

Three possible tautomeric forms [133a], [133b] and [133c] were considered for the interesting sydno[3,4-*a*]quinoxalin-4-ones studied recently (72JO1707). Tautomer [133a] is the most stable according to π-electron delocalization energies (Section 1-7B), and the UV spectrum calculated for it (Section 1-7E) agrees closely with the experimental spectrum (solvent: ethanol). Unfortunately the electronic spectrum calculated for tautomer [133b] differs considerably from that of the fixed derivative [135]. Two carbonyl bands are observed (KBr disks) at 1675 and 1800 cm^{-1} which are assigned to the quinoxalone lactam carbonyl and to the sydnone "carbonyl," respectively; in the fixed derivative [134] those bands appear at 1670 and 1790 cm^{-1}, whereas in the fixed derivative [135] only one doublet is observed at 1785–1800

cm^{-1} assigned to the "carbonyl" of the sydnone moiety. Comparison of the UV spectra of the sydno[3,4-*a*]quinoxalin-4-one with those of the fixed derivatives [**134**] and [**135**], is interpreted to show that tautomer [**133a**] also predominates in ethanol solution. However, the authors did not apparently consider tautomeric form [**133c**], which could also contribute significantly; the model compound [**136**] is not available.

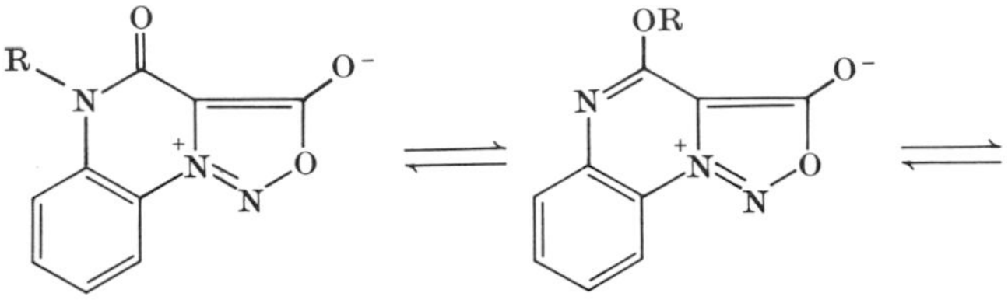

[**133a**] R = H
[**134**] R = Me

[**133b**] R = H
[**135**] R = Me

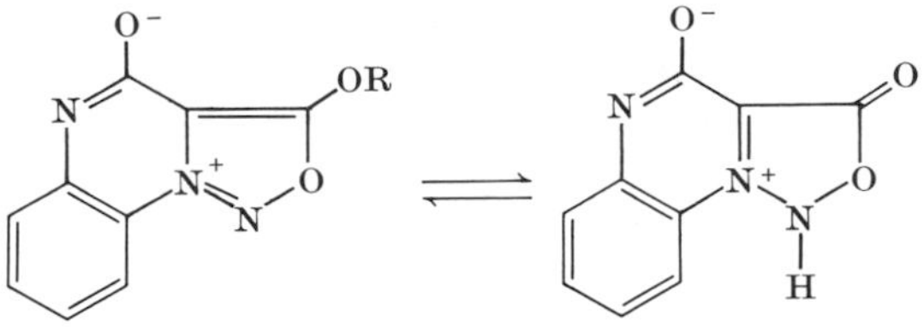

[**133c**] R = H
[**136**] R = Me

[**133d**]

5. [5.5]Bicyclic Compounds

An interesting series of heteroaromatic compounds is derived from indolizine by replacement of the 7,8- or 5,6-CH:CH groups, with a nitrogen, oxygen or sulfur atom. They may be subdivided into two groups, the pyrrolo[2,1-*b*]azole, [**137**] and the pyrrolo[1,2-*b*]azole [**138**] series. These heterocycles are isoelectronic with indolizine and other bicyclic 10 π electron structures, one pair of nonbonding electrons from each heteroatom being formally available for overlap (65J65). From the basic structures [**137**] and [**138**] two series of heterocycles can be obtained by formal replacement of a CH group by a nitrogen atom.

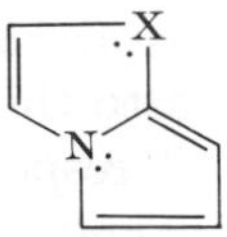

[**137**]

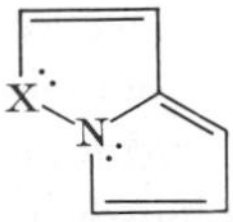

[**138**]

A. Annular Tautomerism

When X = NH, the possibility of annular tautomerism arises, if further nitrogen atoms are present then with other NH forms, and in any case with CH forms. A similar situation is encountered with azoles (Section 4-1A to K). Work on this problem is still fragmentary for the [5.5]bicyclic derivatives.

a. Pyrrolo-imidazoles and -pyrazoles (II-52)

Some relatively early work (50H273) attributed the nonaromatic structure [**139b**] to the pyrrolo[1,2-*a*]imidazole [**139**], since the IR spectrum lacks an NH band (excluding [**139a**]), but shows a C=N band at 1590 cm^{-1}. Moreover, the compound is not very basic, which was considered to eliminate the imidazole structure [**139c**]. In the previous review (II-52), it was stated that "it would be most surprising if this conclusion were correct." However, further work (see later) has now shown that CH tautomers of types [**139b,c** and **d**] are often more stable than "aromatic" structures of type [**139a**] with this class of heterocycle. Pyrrolo[1,2-*a*]imidazoles are thus more easily partially de-aromatized than simple azoles where annular tautomerism never involves nonaromatic structures (Section 4-1). The question of the tautomeric structure of [**139**] is still not satisfactorily clarified, but we would now favour [**139d**] as the most likely.

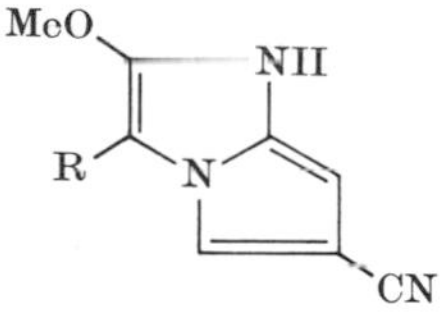

[**139a**] 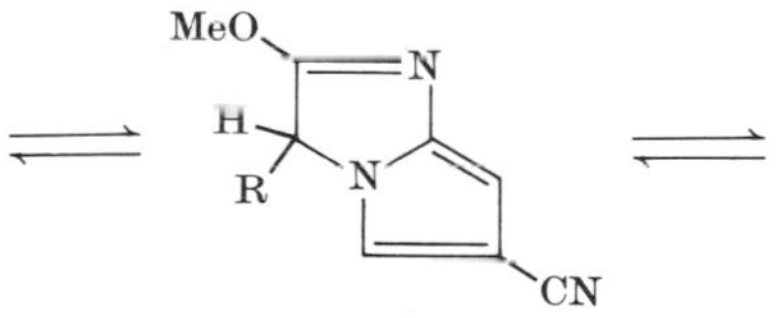 ⇌ [**139b**] 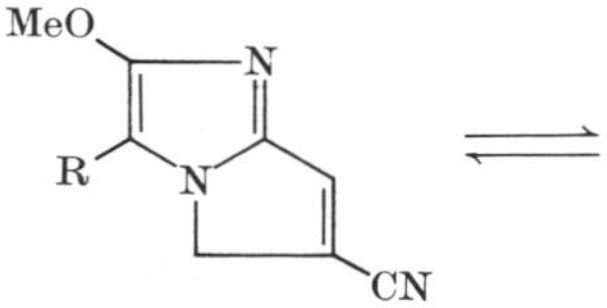 ⇌ [**139c**] 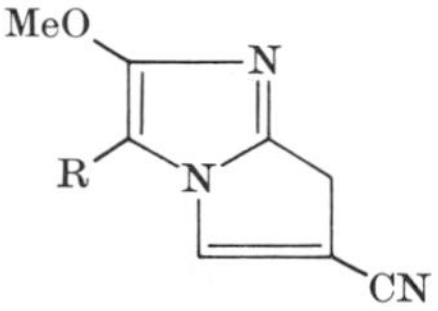 ⇌ [**139d**]

Various pyrrolo[1,2-*a*]imidazoles [**140**] and pyrrolo[1,2-*a*]benzimidazoles have been shown by PMR and IR to exist in the CH forms [**140**] and [**141**] (71KG826). Pyrrolo[1,2-*b*]pyrazoles [**142**] have also been formulated in the CH form in a patent (72CA3851). Aromatic structures corresponding to [**143**] apparently are stable only if substituted at the nitrogen atom (67KG532, 67KG536).

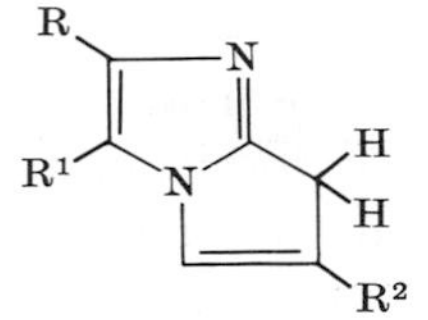

[**140**] 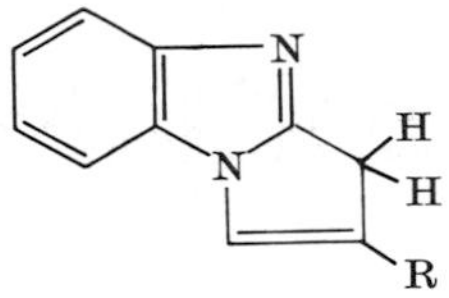 [**141**] 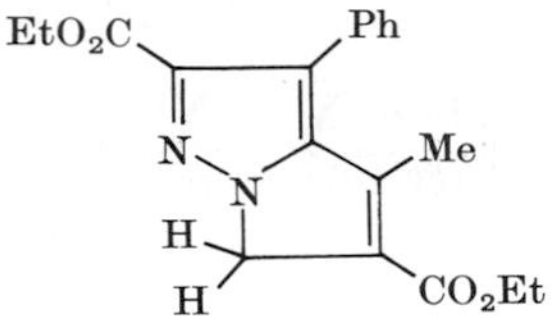 [**142**]

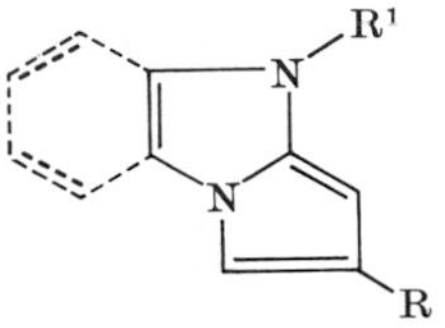

[**143**]

b. Pyrazolo- and Imidazo-triazoles and -tetrazoles

If a [5.5]bicyclic ring system possesses three (or four) nitrogen atoms, two (or three) aromatic tautomers will exist, in each of which a different nitrogen atom bears the proton. These tautomers can be identical, as in the case of autotropic rearrangement (70CB2828). Although many such compounds possessing three or more nitrogen atoms are known, information about their annular tautomerism has apparently been reported for four systems only. The NMR studies to be described do not reveal the presence of any nonaromatic tautomers; such structures are apparently favoured only for heterobicycles possessing a pyrrole or indole ring (70JO1228).

The NMR spectrum (in DMSO) of pyrazolo[5,1-*c*]-*s*-triazole [**144**] (70CB3284) shows *J*(2H3H)-2.3 Hz; by analogy with other coupling constants in pyrazoles, this has been attributed to coupling across a formal double bond and is hence evidence for tautomeric form [**144a**]. Another study based on dipole moments (75BF(2)ip1) concludes a large predominance of tautomer [**144b**] in dioxane. The *J*(2H3H) of 2.3 Hz (in DMSO or $CDCl_3$) in 6-methylimidazo[1,2-*b*]pyrazole [**145**] (73JH411) is less helpful, but the additional *J*(1H2H)-2.3 Hz seen in DMSO is consistent only with structure [**145b**]. The imidazo[1,5-*b*]-*s*-triazole [**146**] was concluded to exist primarily as the tautomer [**146a**], since

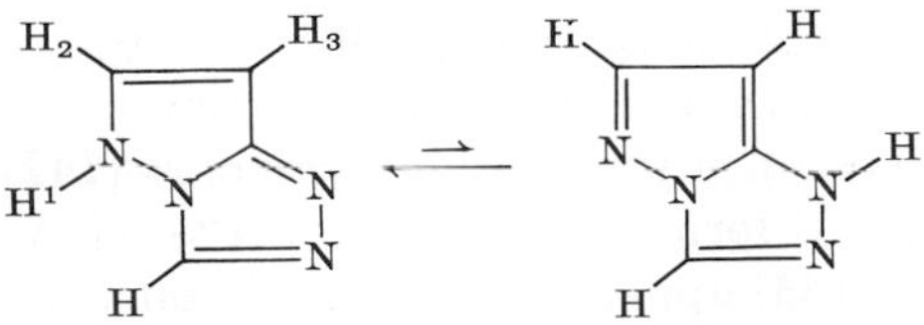

[**144a**] [**144b**]

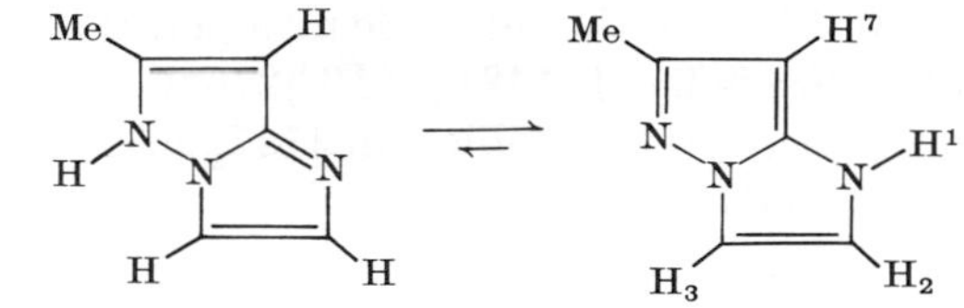

[145a] [145b]

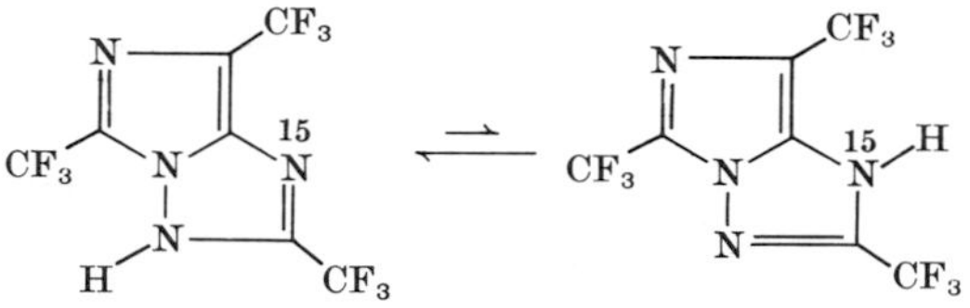

[146a] [146b]

no ^{1}H–^{15}N coupling is observable in acetone solution, even at $-120°$ (70JO3985).

The predominance of tautomer [**146/1b**] in DMSO has been established by NMR: small 3H, 2Me coupling and cross-ring 4H,8H coupling (73CO(276C)1533, 74CC2744). The imidazole nitrogen atom bears the

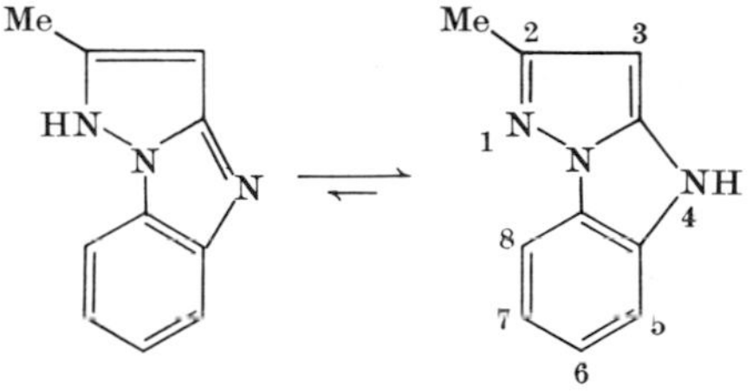

[**146/1a**] [**146/1b**]

proton in the corresponding bicyclic pyrazoloimidazoles also (73JH411, 75BF(2)ip1).

For other annular tautomerism studies on aromatic azapentalenes [**144**, **145**, **146** and aza-analogues] see references (74BF(2)1675, 75BF-(2)ip1, 75JHip1, 75MIip5).

Theoretical calculations (72PS6, 73TL2703) on the tautomeric equilibria [**144**], [**145**] and [**146**] are in good agreement with experiment. These calculations are based both on a comparison of the experimental electronic spectrum with that calculated for each tautomer (Section 1-7E), and on calculated energies of each tautomer (Section 1-7B).

When one of the two rings of a [5.5]bicyclic compound is tetrazole [**147**], the resulting system exists as the isomeric azide [**148**] (Section 4-9Ca) if the molecule contains nitrogen as the sole heteroatom (70CB-1900), and as an azide–tetrazole mixture if sulfur or oxygen is also present (68KG423). If one of the tetrazole nitrogen atoms is substituted, the tetrazole structure is fixed [**149**] (71J(C)2769).

For recent studies on the equilibrium [**147**] ⇌ [**148**] see (73MI123, 74CH411, 74JH921, 74OR485).

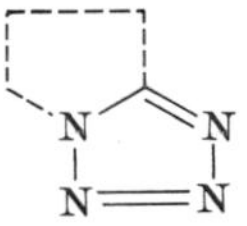

N N
N N

[**147**]

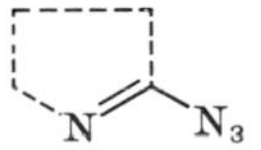

N N_3

[**148**]

N N
Ar N N Me
N N

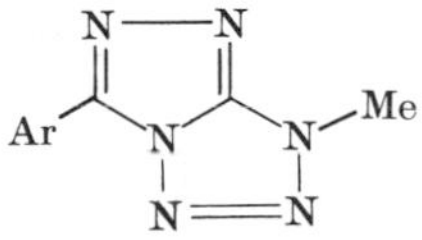

[**149**]

c. Tautomerism in Cations of [5.5]Bicycles

If the molecule possesses only two heteroatoms, protonation yields a cation which has lost the aromaticity of one ring as in pyrroles, indoles, and indolizines (Section 3-1B). NMR (TFA as solvent) has been used with success to determine the structure of the cations [**151**] of pyrrolo[2,1-*b*]thiazoles [**150**]. Usually cation [**151a**] is formed; the dimethyl derivative [**150**, R = Me] gives a mixture of [**151a**] and [**151b**] (65J4368). Protonation of pyrrolo[1,2-*a*]benzimidazoles gives cations of type [**152**](68KG905, 72KG1132); in this case the site of protonation was predicted by HMO calculations.

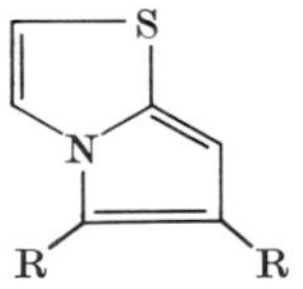

S
N
R R

[**150**]

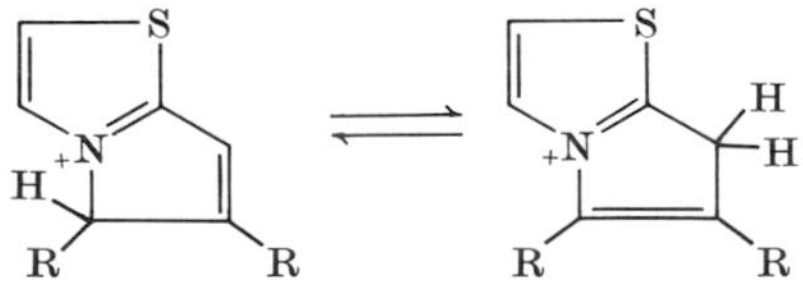

S
+N
H
R R

[**151a**]

⇌

S
H
+N H
R R

[**151b**]

Comparison of the UV spectrum of 6-phenylimidazo[2,1-*b*]thiazole hydrochloride with that of the corresponding quaternary salt demonstrates that protonation occurs on the imidazole nitrogen atom with formation of [**153**] (71AN672).

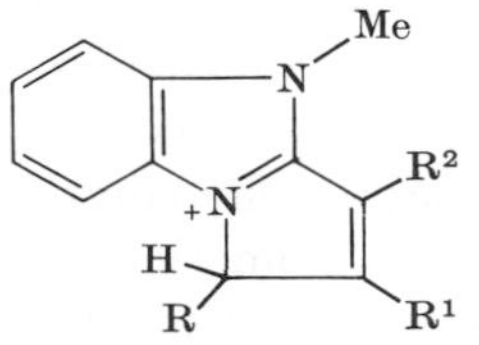

Me
N
R^2
+N
H
R R^1

[**152**]

S
+N NH
Ph

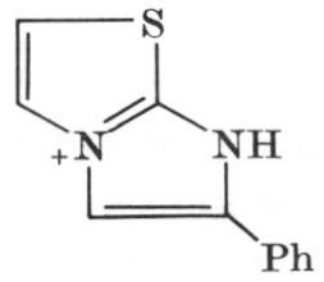

[**153**]

B. [5.5]Bicyclic Compounds with Potentially Tautomeric Groups

Although there have yet been few studies directed specifically to the tautomerism of these compounds, spectroscopic data (UV, IR, NMR) allow certain tautomeric structure assignments to be made. Despite this limited set of data, the results that have been obtained are consistent: for compounds **[154]** to **[159]** the CH tautomer is preferred to the OH structure; for compounds **[160]** and **[161]** the NH tautomer is preferred to the SH structure; and for **[162]** to **[164]**, the amino tautomer is preferred to an imino structure.

Scheme 5-5 lists the structures of the predominant tautomers for these compounds, the physical method used for structure determination

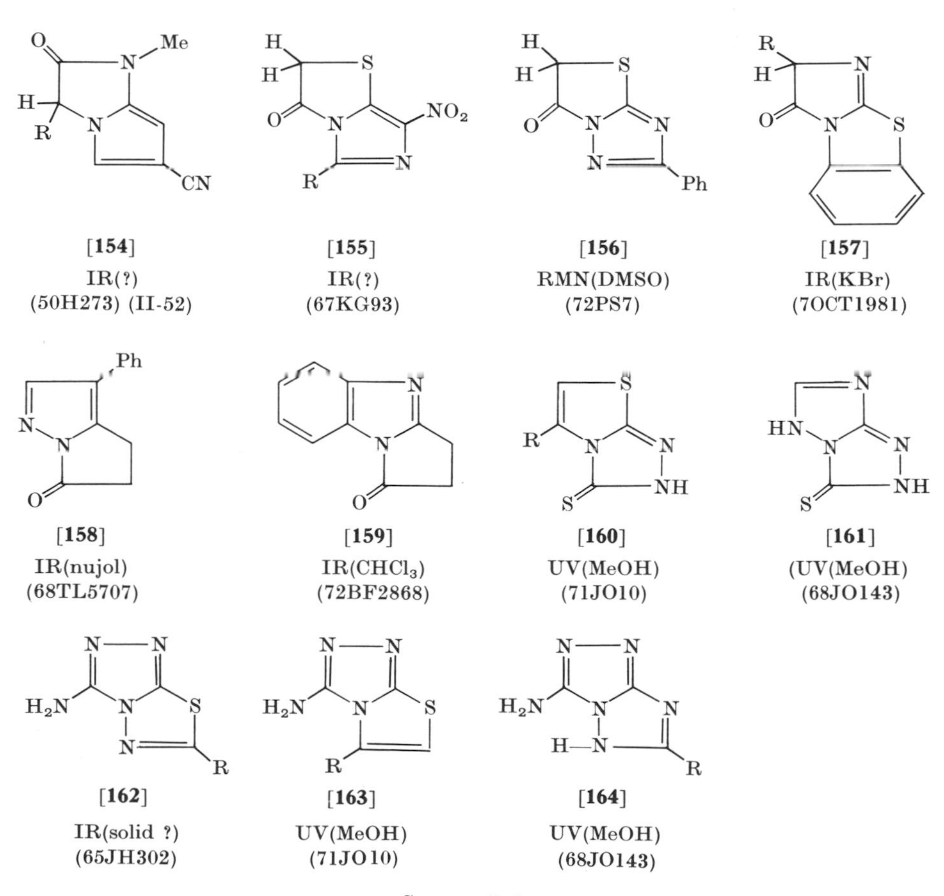

Scheme 5–5

(if there are several, the most important one), the conditions under which measurements were made (if specified) and references.

For compounds [**158**] and [**159**], fully aromatic hydroxy structures can be formed only by double prototropy. The position of the NH proton in [**161**] and [**164**], i.e., the annular tautomerism, is not known. Lastly, the preferred structures [**160**] and [**163**], are not discussed in the reference (71JO10), but deduced by us from the spectral data given.

An example of the existence of two individual tautomers [**165**] and [**166**] for 2-methylthiazolo[2,3-*b*]benzimidazole (74TL2643) has been described. That means that the sulphur ring behaves analogously to that in thiophene (Section 3-6B) and not thiazole (Section 4-8).

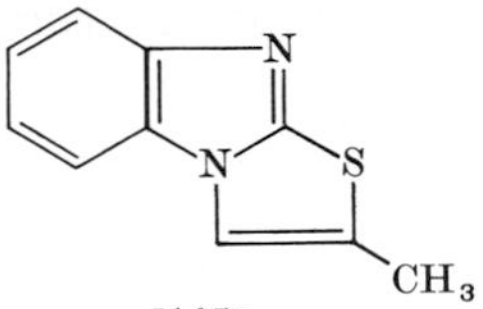

N S N CH_3

[**165**]

N S N CH_2

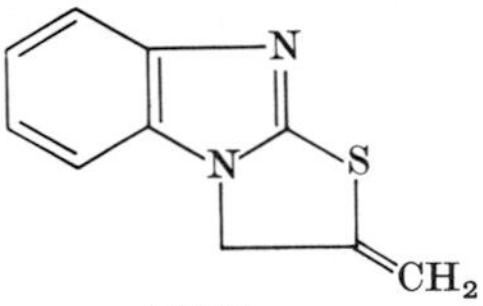

[**166**]

CHAPTER 6

Tautomerism Involving Other Than Five- and Six-Membered Rings

The heterocyclic systems considered in the preceding chapters are all potentially "aromatic"* in the Hückel sense: they possess $4n + 2$ π-electrons, 6 π-electrons for the fundamental systems, such as pyridines [**1**] (Chapter 2), thiophene [**2**] (Chapter 3), pyrazole [**3**] (Chapter 4), 10 π-electrons for bicyclic systems, such as quinoline [**4**] (Chapter 2), purine [**5**] (Chapter 5), and pyrrolo[2,1-*b*]azoles [**6**] (Chapter 5).

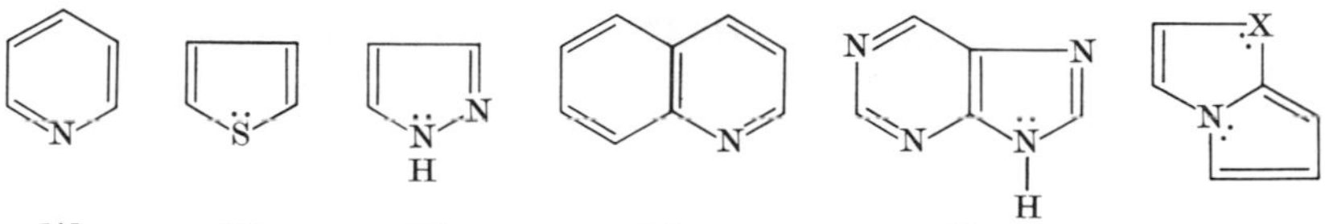

[1] [2] [3] [4] [5] [6]

For rings containing three, four, or seven, eight or more atoms, structures both with $4n + 2$ and with $4n$ electrons occur frequently; we will consider the former as "aromatic," and the latter as "antiaromatic" (65CE90). (For a discussion of the concepts of aromaticity and antiaromaticity, see the following references: 60J1274, 65BT191, 65BT250, 68AG(E)565, 68BT100, 70T5225, 70TL1311, and, in particular, 71MI01.)

1. Microcycles (Three- and Four-Membered Rings)

A. TAUTOMERISM NOT INVOLVING FUNCTIONAL GROUPS

a. Three-Membered Rings

Neither 1*H*-azirine (azirene) [**7**] nor its N-substituted derivatives are known. The system possesses 4 π-electrons and is therefore antiaromatic and unstable with respect to its nonaromatic isomer, 2*H*-azirine [**8**]. From *ab initio* MO calculations, Clark (69TC(15)225, 70MI238; see also 73CH688, 73MI1) concluded that N-substituents in 1*H*-azirines are not coplanar with the ring and that the barrier to inversion should be higher than those of aziridines (since the planar transition state will be destabilized). 1*H*-Azirine is calculated to be a considerably stronger

* Six-membered ring heterocycles with 8 π-electrons also exist, for example, 1,2-dihydropyridazines (Section 2-1B)

base than the 2*H* isomer, and the rearrangement [**7**] → [**8**] is thus thermodynamically favourable.

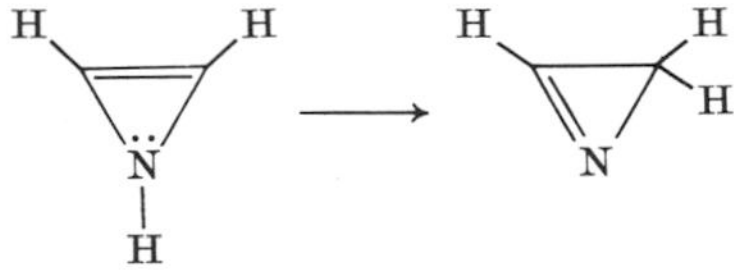

[**7**] [**8**]

The addition of nitrenes of type $R_2N \cdot N$ to acetylenes does not give the expected 1*H*-azirine [**9**], but instead the transposition product [**10**]. This result was attributed (69CH147, 73J(PI)550, 73J(PI)555, 73CH-835) to the antiaromaticity of the 1*H* tautomer [**9**]. In the same way, treatment of 2*H*-azirines [**11**] with acid chlorides does not yield *N*-acyl-1*H*-azirines (68JA2875); here again, the antiaromaticity of

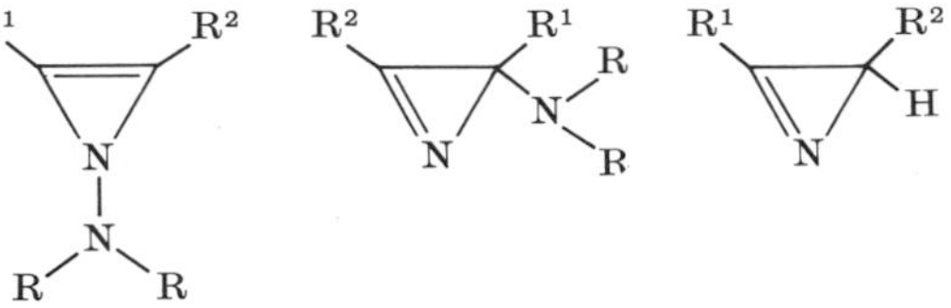

[**9**] [**10**] [**11**]

1*H*-azirines can be invoked to explain this result (for further discussion, see 71HC(13)45). A 1*H*-diazirine has been postulated as an intermediate in the photolysis of an oxadiazol-5-one (70JH71) and according to the authors this initially formed antiaromatic system first rearranges to a 3*H*-diazirine, and then gives the products actually isolated through loss of nitrogen.

b. Four-Membered Rings

Among 4-membered heterocycles, 1,2-dihydrodiazetines [**12a**] possess 6 π-electrons, and, if planar, should be aromatic (60RU153, 65RR1059), whilst the 1,4-dihydro tautomer [**12b**] would be nonaromatic. Compounds claimed to possess a stable 1,2-dihydrodiazetine structure (such as

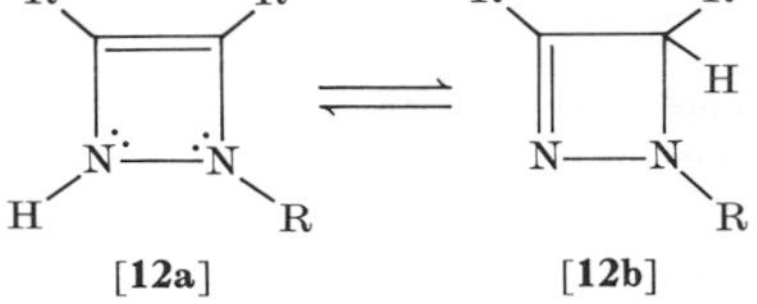

[**12a**] [**12b**]

[**13a**]) have been described (66AG389), but they may be α-diimines [**13b**]. Recently (72CH818) an authentic 1,2-dihydrodiazete has been

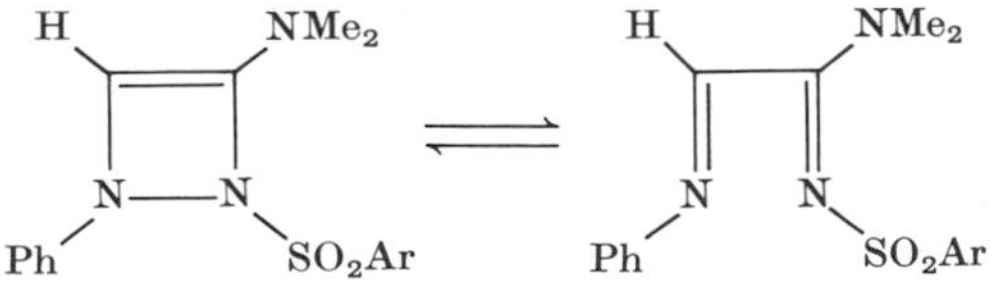

[**13a**] [**13b**]

isolated: a Δ^3-1,2-diazetine-1,2-dicarboxylate, but it is slowly transformed [$t_{1/2}(20°) = 6.9$ hours] into the corresponding α-diimine and shows no ring current effect in NMR spectroscopy. This type of ring–chain isomerism (Section 4-9Ca), formulated generally as [**14a**] $\rightleftharpoons$ [**14b**],

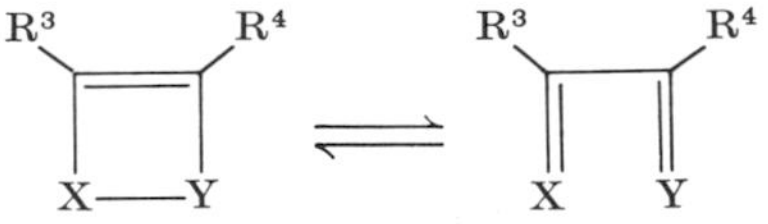

[**14a**] [**14b**]

depends on the nature of the heteroatoms X and Y. If they are both sulphur, the example ($R^3 = R^4 = CF_3$) exists as the stable cyclic dithietene (60JA1515, 61JA3434); it can be shown theoretically (65IIC(5)20) that CF_3 substituents strongly stabilize the cyclic form [**14a**, X = Y = S]. If X and Y are oxygen and sulphur, the open-chain monothiobenzil compound ($R^3 = R^4 = Ph$) is the stable form (70JO4224). These two examples again illustrate the fact that if four of the $4n + 2$ electrons come from two adjacent heteroatoms, the aromaticity of the compound is less than otherwise (Section 4-9D). However most authors consider dithiete as aromatic: there is evidence from theoretical calculations (62AK(19)75, 62AK(19)265) and the α-dithione [**14b**] $\rightleftharpoons$ dithietene [**14a**] equilibrium (73JA2383, 74JO511) for which de Mayo (73JA2383) has measured the activation energy.

B. Compounds with Potentially Tautomeric Groups

From the results of the preceding section we should expect that the CH tautomer would be preferred for aziridin-2-ones [**15a**] and diazetidin-2-ones [**16a**], rather than the OH tautomers [**15b**, **16b**], and this is in fact found. For aziridin-2-ones, several references (cited in 70CH473)

[**15a**] [**15b**] [**16a**] [**16b**]

confirm structure [**15a**], and in one case the structure has been determined by X-ray crystallography [**15a**, $R^1 = R^3 =$ 1-adamantyl] (72CH-43). For diazetidin-2-ones, NMR studies (67TL161, 68JA5273) confirm the CH structure [**16a**].

2. Macrocycles (Seven- or More-Membered Rings)

It is well known (70T5225, 70TL1311) for annulenes that the aromatic properties of the $4n + 2$ membered rings or the antiaromatic properties of the $4n$-membered rings diminish and that both series with increasing ring size tend in the limit to become nonaromatic. The same effect presumably operates for heterocycles, hence NH tautomers with 10 π-electrons are not necessarily highly stabilized, and similar tautomers with 8 π-electrons could be isolable under certain conditions.

A. Tautomerism Not Involving Functional Groups

We are concerned principally with 1*H*-azepines [**17**], 1,2-dihydrodiazocines [**18**] and 1*H*-azonines [**19**], and their aza derivatives obtained by replacing —CH= with —N=: 1,2-diazepines [**20**], 1,3-diazepines [**22**], and 1,4-diazepines [**21**]: all these systems can in theory display

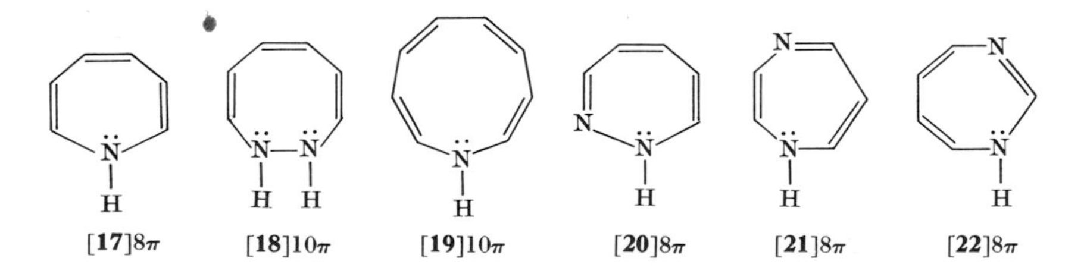

[**17**]8π [**18**]10π [**19**]10π [**20**]8π [**21**]8π [**22**]8π

annular tautomerism between one or more NH forms and various CH forms [1,2-diazepin-4-ones (Section 6-2Bb) are examples of compounds showing tautomerism between two NH forms].

Theoretical calculations (70T4269, 72T3657) predict that 1*H*-azepines should be appreciably antiaromatic, and that, of the various possible tautomers, the nonaromatic 3*H*-azepine [**23a**] should be the most stable (70T4269). Experimentally this has been confirmed, since 1*H*-azepines isomerize rapidly to 3*H*-azepines (63AG1041). The 4*H*-isomer [**23b**] can be isolated and kept, but in basic solution it isomerizes to [**23a**]; thus the stabilities of azepine tautomers decrease in the order: 3*H* > 4*H* > 1*H* (69MI249).

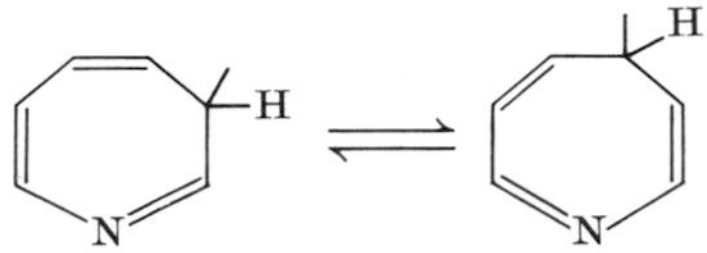

[**23a**] [**23b**]

For benzo derivatives, it has been predicted that they should be more stable than the corresponding monocycles, provided that annelation does not diminish the aromaticity of the benzene rings (70T4269). In fact, the NH tautomers are stable for the 2,3-benzazepine [**24**] (70TL4069) and for 2,3,6,7-dibenzazepine [**25**] (60JO827, 68IS507).

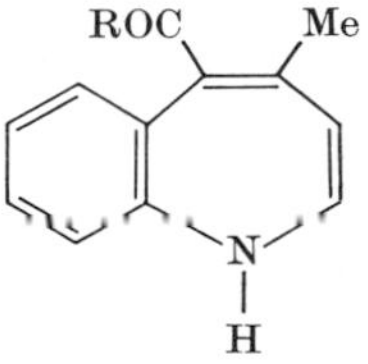

[**24**] 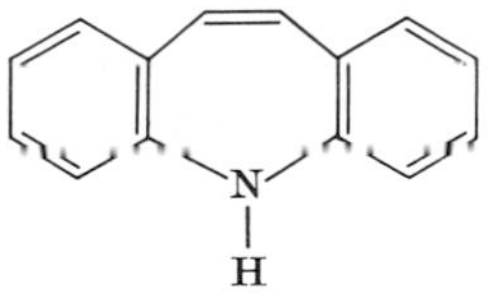 [**25**]

1*H*-1,2-Diazepines carrying a hydrogen atom at the 1-position [**20**] are unknown; compounds of this series in the 4*H*- [**26**] and 5*H*- [**27**] tautomeric form have been reported (69AS3125, 70T739).

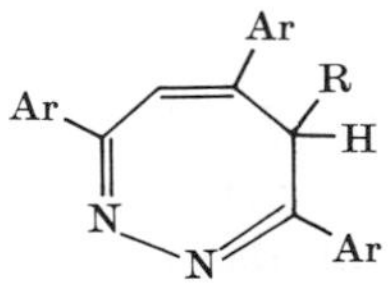

[**26**] 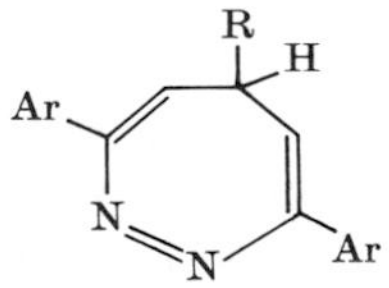 [**27**]

See also (72CH504) for a discussion on the structure of the corresponding cation and (74CH640) for the tautomerism of 1*H*-1,2-benzodiazepines. The tautomerism of several 2,3-benzodiazepines [**28**, R = H,

Ph, etc.] has received considerable investigation (70CH1197, 72CH823, 72CH827, 73J(PI)2543); thus in the phenyl series [**28**, R = Ph] all three different CH tautomers [**28a**], [**28b**] and [**28c**] have been isolated. However, antiaromatic NH tautomers [**28d**] have never been detected. Isomerizations [**28b**] into [**28a**] (70CH1197, 72CH823), [**28b**] into [**28c**] (72CH823) and [**28a**] into [**28c**] (72CH827) indicate that the stability sequence is [**28b**] $<$ [**28a**] $<$ [**28c**].

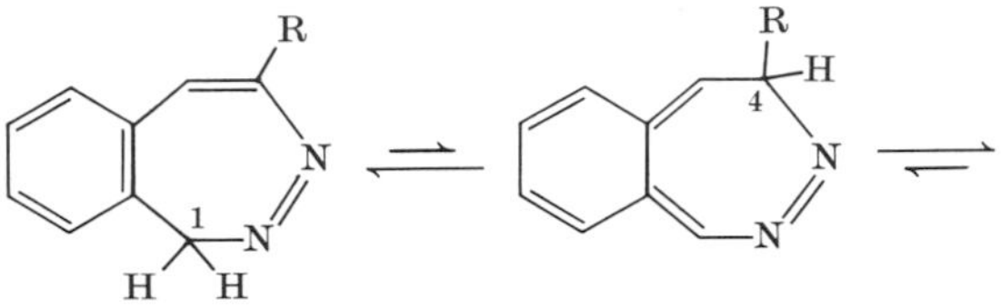

[**28a**] [**28b**]

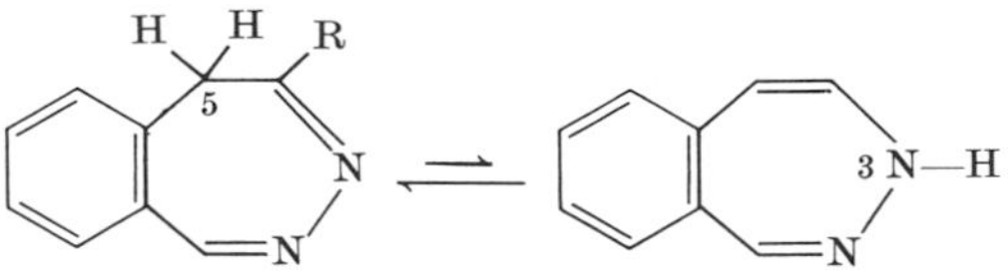

[**28c**] [**28d**]

1,4-Diazepines [**21**] are unknown both as NH and as CH tautomers. However, the corresponding 2,3-benzo-derivatives (1,5-benzodiazepines) [**29**] have been well studied; the CH tautomer [**29b**] has been shown to predominate (65J3785, 65SA1095, 67HC(8)21, 74HC(17)27) and annelation is clearly insufficient to stabilize the antiaromatic NH form [**29a**].

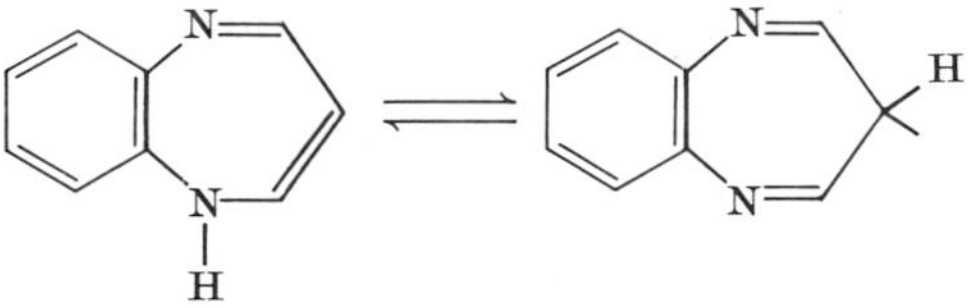

[**29a**] [**29b**]

This is in interesting contrast to the 2,3-dihydrodiazepines which exist in the NH form; here the question of antiaromaticity does not arise (65J3785). An isoxazole ring annelation also favours the CH form for the same reason, as in [**30**] (70T1393). In contrast, the 2,3-benzo-1,4-diazepine cations possess the structure [**31**], related to the NH tautomer (65J3785, 65SA1095 74HC(17)27); the higher stability of [**31**] compared to other structures is due to the vinylogous amidinium or cyanine system, as in the cations derived from 2,3-dihydrodiazepine.

1,3-Diazepine [**22**] is unknown; the corresponding dibenzo derivative exists in the NH form [**32**, X = CH]; here formation of a CH tautomer would result in loss of the aromaticity of at least one benzene ring. MO

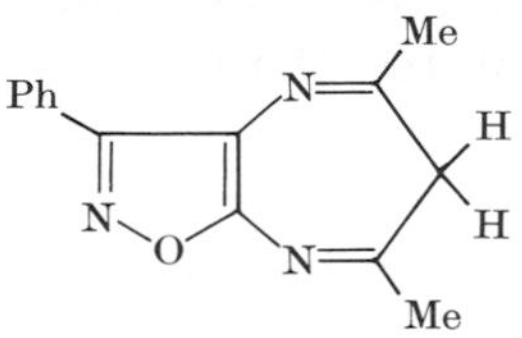

[**30**] 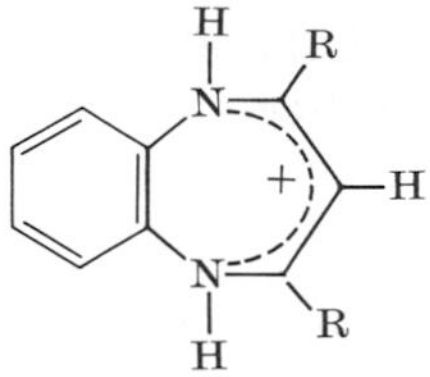 [**31**] 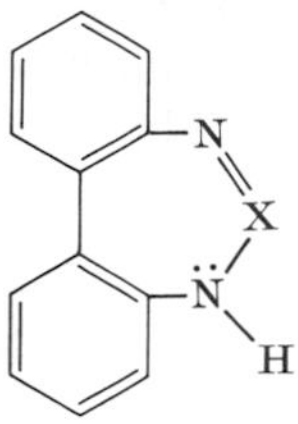 [**32**]

calculations and dipole moment measurements suggest that dibenzo-1,3-diazepines [**32**, X = CH] exist with the planes of the two benzene rings mutually twisted by 40° about the single bond joining them (70ZO1865). Similarly, the NH-proton of the dibenzo[*d*,*f*][1,2,3]-triazepine [**32**, X = N] (72CH982) should be on a terminal rather than the central nitrogen atom.

Considering now systems with eight-membered rings, the 1,2-dihydrodiazocine [**18**] ring, despite its 10 π-electrons, is apparently not very stable, because the 5,6-benzo derivative exists as the CH–CH tautomer [**33a**] rather than the NH–NH tautomer [**33b**] (60JO1509). However, for the dibenzo[*c*,*g*] derivative, the existence in any CH tautomeric form would result in destruction of the aromaticity of one or both benzeno rings, and 5,6-dihydrodibenzo[*c*,*g*][1,2]diazocine not unexpectedly exists as the stable NH–NH tautomer [**34**] (69JO3237).

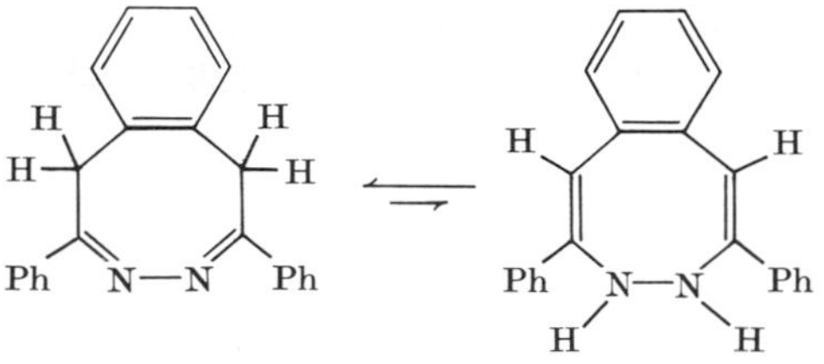

[**33a**] [**33b**] 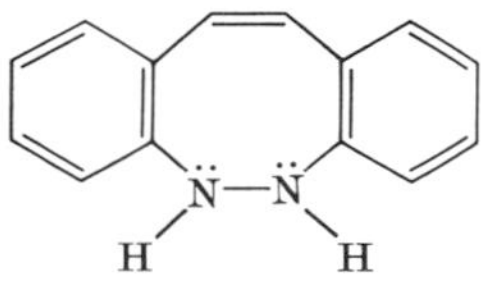 [**34**]

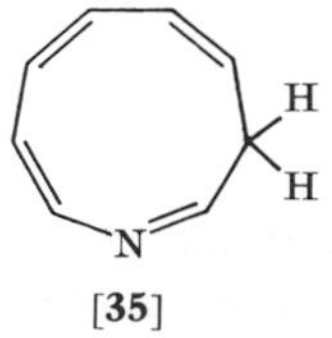

[**35**]

This is a rare example of a compound in which two adjacent heteroatoms contribute 4 π-electrons to a $(4n + 2)$ π-electron aromatic system (Sections 4-9D, 6-1A).

For azonine, the NH tautomer [**19**] is preferred to any of the CH tautomers (e.g., [**35**]) (70CH1133, 70TL825). Theoretical calculations (72T3657) predict that 1*H*-azonine is stabilized by resonance (i.e., is aromatic).

B. Compounds with Potentially Tautomeric Groups

a. Azepines and Oxepines

Paquette (69MI249) has recently reviewed the tautomerism of azepines possessing an XH substituent at the 2-position; tautomers of types [**36b**] (X = O,S) and [**36c**] (XH = NH_2) are known but "anti-aromatic" tautomers of type [**36a**] have never been observed. The tendency for the double bond of heterocyclic amidines to take up the *endo* orientation is notable for the case where XH is NH_2. Similarly, none of the OH tautomer of oxepin-3-one [**37**] has been detected by NMR (solvent unspecified) (65CI184).

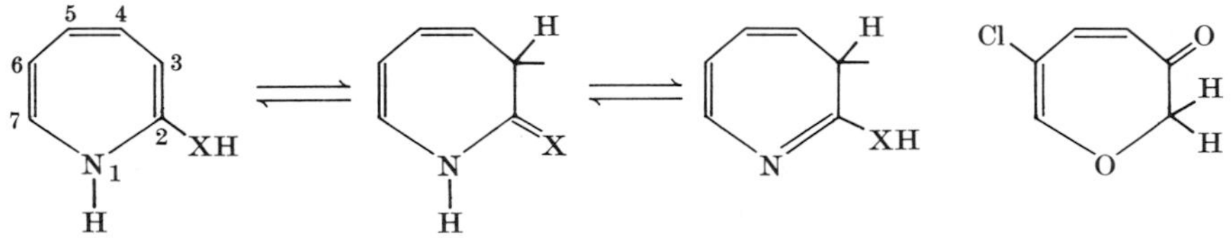

[**36a**] [**36b**] [**36c**] [**37**]

If the 4,5-double bond is replaced by a benzene ring, tautomers of the same type are observed, as in [**38**] (68BF600) and [**39**] (65JH26).

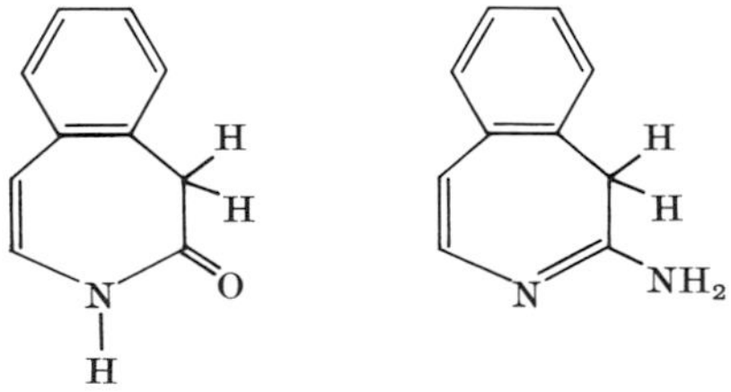

[**38**] [**39**]

b. 1,2-Diazepines

The tautomerism of 1,2-diazepin-4-ones has been thoroughly investigated by Moore (68JA1369, 73JO2939). The tautomers [**40a**] and

[**40d**] are mutually interconvertible via the OH forms [**40b**] and [**40c**], reaching the equilibrium ratio [**40d**]/[**40a**] $\simeq$ 8. The individual tautomers [**40a**] and [**40d**] are easy to isolate separately, and it is in fact hard to equilibrate them, which may be rationalized by noting that the OH or enolate forms with eight π-electrons are antiaromatic and of high energy. Moore has also demonstrated that 1,5-dihydro structures of type [**40d**] are more stable than the corresponding 1,7-dihydro tautomers by achieving the base-catalyzed conversion [**41a**] $\rightarrow$ [**41b**] (68JA-1369); [**40e**] does not appear to contribute.

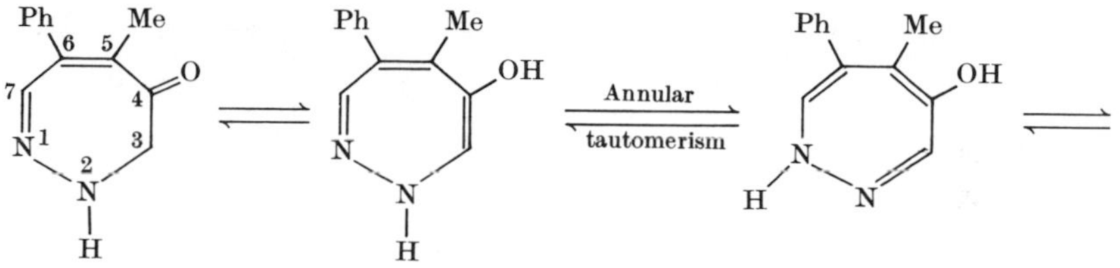

Annular tautomerism

[**40a**] [**40b**] [**40c**]

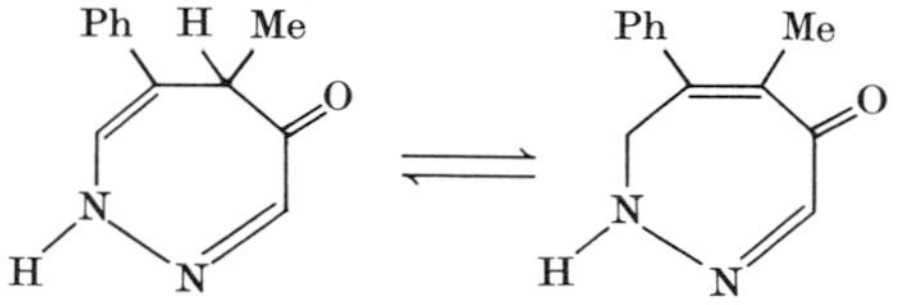

[**40d**] [**40e**]

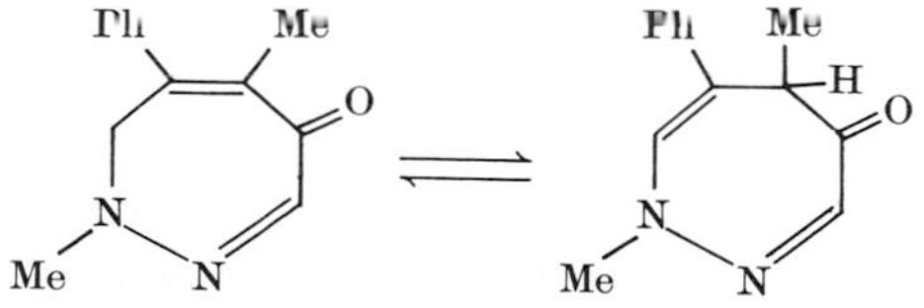

[**41a**] [**41b**]

Benzo derivatives of 1,2-diazepinones and diazepinethiones are known; their existence in the oxo and thione structures [**42**] and [**43**] was established by NMR ($CDCl_3$) and, in the case of [**42**], confirmed by IR (νC=O: 1638 cm^{-1} in KBr) (70BF2237).

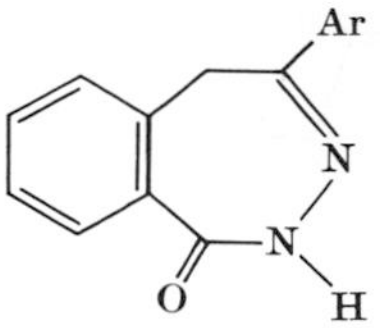

[**42**] 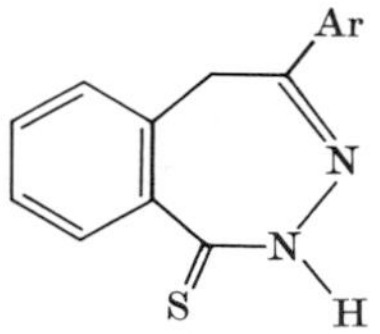 [**43**]

c. 1,3-Diazepines

For 1,3-diazepines, the oxo structures [**44**] (71CB2786) and [**45**], and the thione structure [**46**] (64MC310) are the sole tautomers detectable by IR (solid state), and NMR (in various solvents). By contrast, the amino derivative [**47**; R = R′ = H] appears to exist largely as [**47**] in EtOH since its UV spectrum is similar to those of analogues of fixed structure [**47**; R ≠ H, R′ ≠ H] and also that of [**47**; R = H, R′ ≠ H] (64MC310) (see also 67HC(8)21).

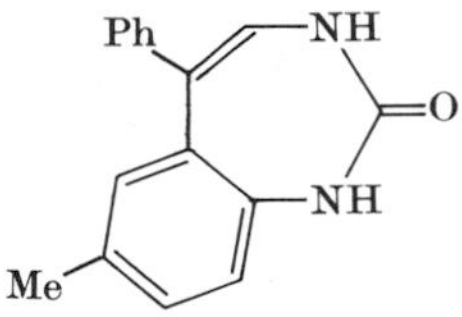

[**44**] 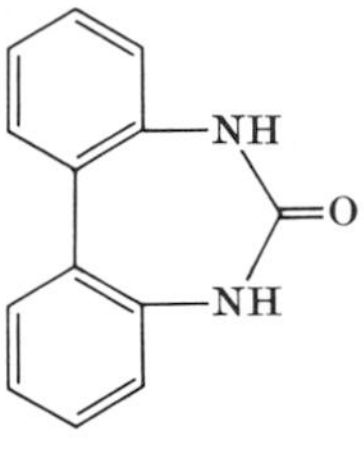 [**45**] 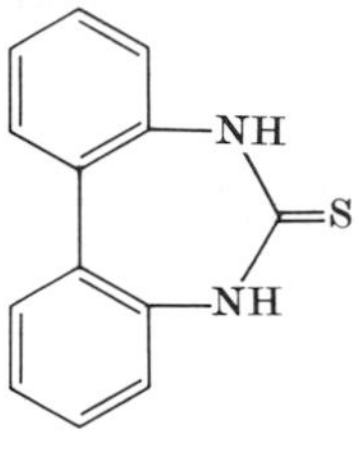 [**46**]

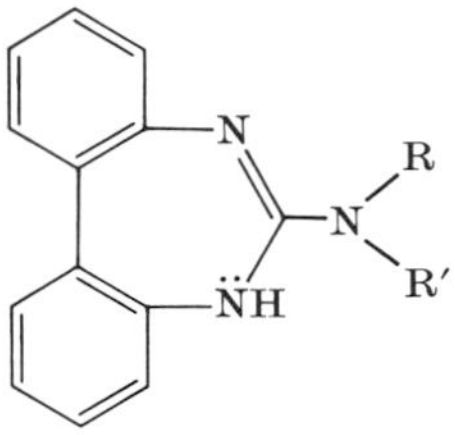

[**47**]

d. 1,4-Diazepines

The oxo structures [**48a**] and [**49**] are supported by IR carbonyl bands: [**48a**] νC=O = 1685 cm^{-1} (in $CHCl_3$) (70ZO1881); [**49**] νC=O = 1667 cm^{-1} (KBr) (71CB2786). Although it is stated in the original reference (70ZO1881) that in alkaline solution 2,3-dihydro-1*H*-1,4-benzodiazepin-2-ones exhibit lactam [**48a**] – lactim [**48b**] tautomerism, rather than keto–enol tautomerism, the authors appear to be confusing anion formation with tautomerism: the species formed in media of high pH is almost certainly the anion corresponding to [**48**].

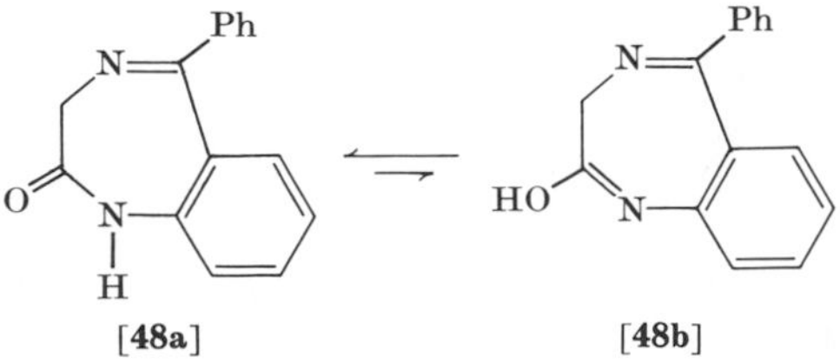

[**48a**] [**48b**] 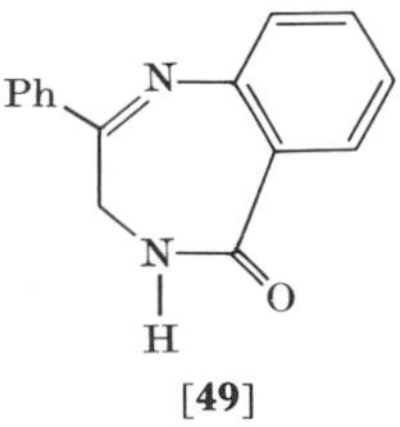 [**49**]

When the fused aromatic or azaaromatic ring is situated between the two nitrogen atoms, as in the 1,5-benzodiazepin-2-ones, five possible tautomers exists [**51a**]–[**51e**]. Of these, two ([**51d**] and [**51e**]) are anti-aromatic hydroxy derivatives of structure [**29a**] (Section 6-2A) and have never been detected. [Although structure [**52**] has been tentatively suggested on spectroscopic evidence (65J485; see also 67HC(8)21), we believe that this conclusion is probably erroneous.] Israel (67JH659) has shown that the two important contributing structures are [**51a**] and [**51b**], which we shall call the CH and NH forms, respectively. In one particular case concerning a 9-aza derivative, a lactim form of type [**51c**] was postulated to exist in D_2O-pyridine solution, since the NMR CH_2-signal is displaced from that found for the pyridine solution

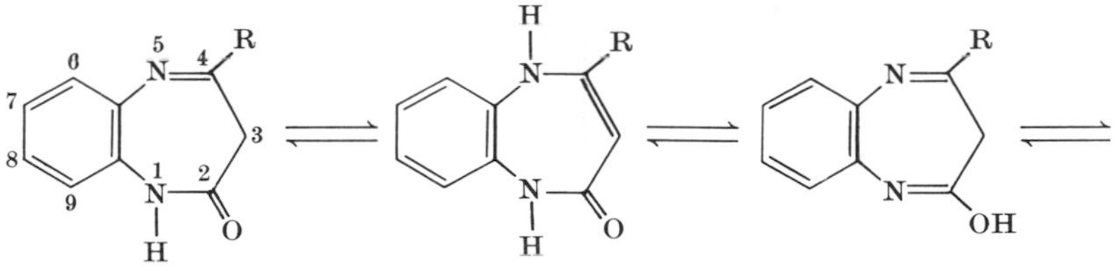

[**51a**] [**51b**] [**51c**]

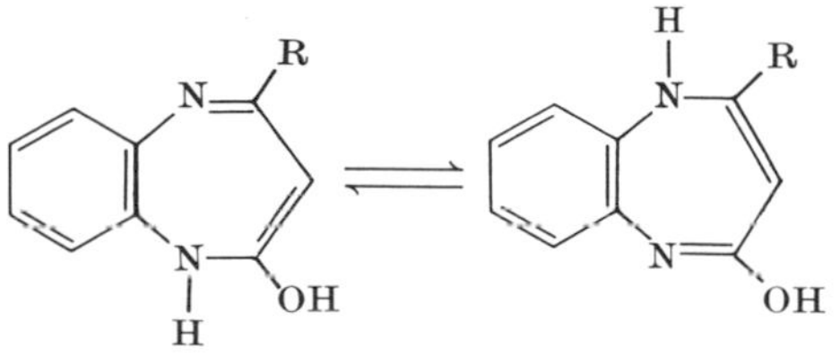

[**51d**] [**51e**]

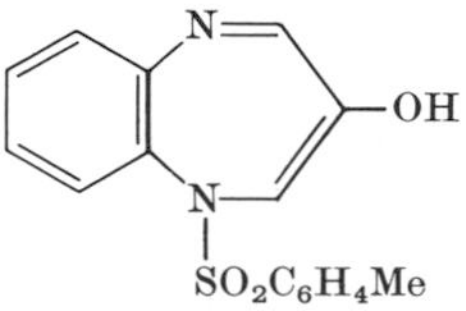

$SO_2C_6H_4Me$

[**52**]

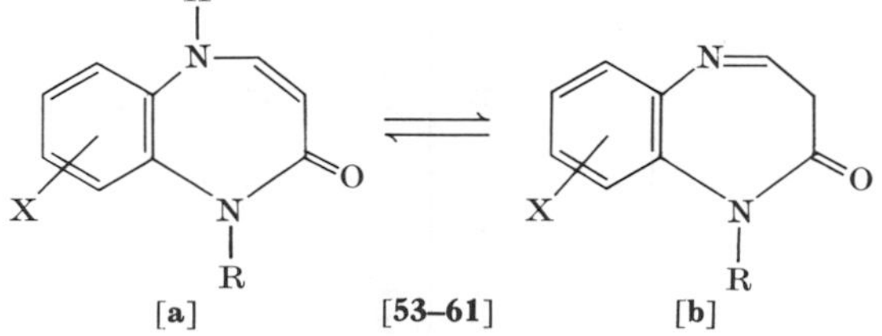

[**a**] [**53–61**] [**b**]

For substituents see Table 6-1

TABLE 6-1

TAUTOMERIC EQUILIBRIA OF 1,5-BENZODIAZEPIN-2-ONES AND OF THEIR AZA DERIVATIVES

Compound	R	Substituent on the benzene ring				CH form	NH form	Reference
		6	7	8	9			
[**53**]	Me	H	H	H	H	NMR: $CDCl_3$, DMSO; UV(MeOH): 270, 300nm; IR(?): 1680 cm^{-1}	—[a]	70J(C)1117
[**54**]	CF_3	H	Me	H	H	NMR: C_5D_5N; UV(EtOH95): 225, 303 nm; IR(KCl): 1685 cm^{-1}	—	71JH1015
		H	H	Me	H			
[**55**]	Me	N	H	H	H	NMR: C_5D_5N(90°), 25% at equilibrium	NMR: C_5D_5N (90°), 75% at equilibrium	67JH659
[**56**]	Me	H	N	H	H	NMR: C_5D_5N; UV(EtOH95): 253, 380 nm; IR(KCl): 1670 cm^{-1}	—	71JH797
[**57**]	Me	H	H	N	H	NMR: C_5D_5N, 66% at equilibrium; UV(EtOH95): 262, 293, 369(sh); IR(KCl): 1672 cm^{-1}	NMR: C_5D_5N: 33% at equilibrium; UV(EtOH95): 262, 292 nm; IR(KCl): 1680 cm^{-1}	71JH797

[58]	Me	H	H	H	N	NMR: C_5D_5N; UV(EtOH95): 252, 314 nm; IR(KCl): 1661 cm^{-1}	Very unstable in solution; UV(EtOH95): 251, 296 nm; IR(KCl): 1672 cm^{-1}	67JH659
[59]	Ph	H	H	H	N	UV(EtOH95): 251, 275, 320 nm	—	69JH735
[60]	Me	N	H	N	H	—	NMR: DMSO (65JH110); UV(EtOH95): 257, 310 nm; IR(KCl): 1695 cm^{-1}	70JH1029
[61]	Me	H	N	H	N	UV(EtOH95): 235, 320 nm; IR(KCl): 1661 cm^{-1}	Unstable in solution; UV(EtOH95): 226, 281 nm	70JH1029
[62]	Me		See text			NMR(?) IR(KBr): 1680 cm^{-1}	—	72JH153
[63]	Me		See text			—	NMR(?) IR(KBr): 1695 cm^{-1}	72JH153

[a] Contrary to Acheson and Tully (70J(C)1117), who observed only the CH form, Israel in a note from his paper (67JH659) indicates that he has observed the two forms, but the work corresponding to that note has not been published. Recently, Israel and Zoll (72JO3566), for a naphthodiazepinone of structure close to that of [**53**], observe in the NMR spectrum (solvent: perdeuteropyridine) only the signals of the CH form.

spectrum (67JH659); however, this evidence is at best tentative, and formation of some anion which would account for the observed shift is at least as likely.

The main results are summarized in Table 6-1. NMR permits easy distinction between the NH and CH structures which give distinct and easily recognizable signals, e.g., the CH_2 at the 3-position for the CH form and the CH at the 3-position in the NH form. The UV and IR spectra are less easy to interpret: rather surprisingly the CH isomer (deep yellow) absorbs at longer wavelengths than the NH form (colourless or pale yellow), whilst in the IR, the C=O of the NH form seems to be less conjugated. It is clear that more work is needed here, and the same solvent should preferably be used for the different spectroscopic methods (as in studies on pyrazolones; Section 4-2F).

Considering now the abundance of each tautomer, the results follow a rational pattern. For the unsubstituted compound [**53**] the CH form is much the most stable and introduction of nitrogen atoms at the 7- and 9-positions* does not change this situation (see data for [**56**], [**58**] and [**61**] in Table 6-1), whereas the presence of nitrogen atoms at the 6- or 8-positions does cause destabilization of the CH relative to the NH form (for compound [**56**] there is 25% CH form, and for [**57**], 66% CH). This effect is so marked that compound [**60**], substituted with nitrogen at both the 6- and 8-positions, exists entirely in the NH form. The special influence of 6- or 8-position nitrogen atoms is clearly due to the direct conjugative interaction possible between these and the NH group in the 5-position of the NH form. Such interaction is not possible for 7- or 9-nitrogen atoms. The importance of such direct conjugative interaction in amidine systems has recently been demonstrated (74J-(PII)546).

Compounds [**62**] and [**63**], studied by Kochhar (72JH153), are included in Table 6-1 as they are conveniently classed with the diazepines and azepines above. Recent results (74JH75, 75MIipl) show that the structures proposed by Kochhar for the major and minor product should be reversed. For the more complicated case of *v*-triazolo-[5,4-*b*][1,4]-diazepin-5-ones, see (72JO4124).

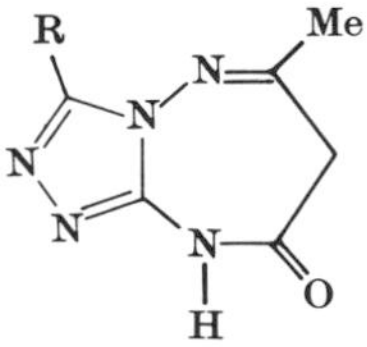

[**62**] 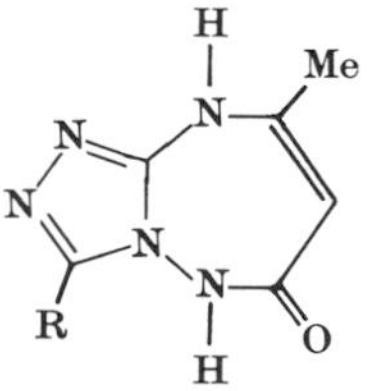 [**63**]

* For numbering system see structure [**51a**].

e. Azocines and Diazocines

Several benzazocinones and benzodiazocinediones are known [**64**–**70**]. As expected, all exist predominantly in the oxo form irrespective of whether or not the OH form possesses 12 or 14 π-electrons.

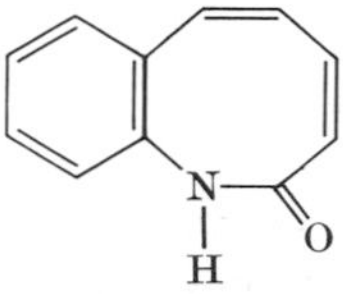

[**64**] (71JA4016) [**65**] (72JA4907) [**66**] (72JA4907)

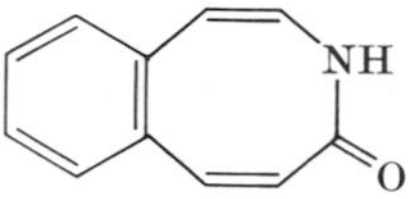

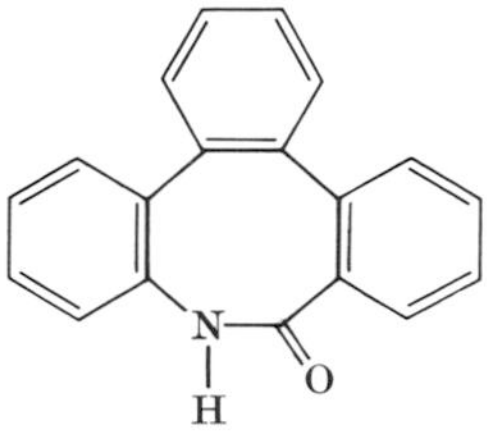

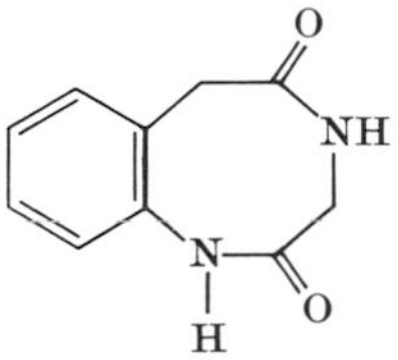

[**67**] (70CC1946) [**68**] (69JO2138, 72BF2868) [**69**] (72BF2868) [**70**] (69JO2138)

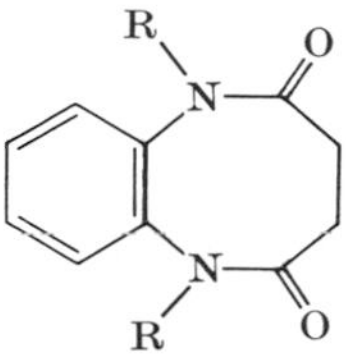

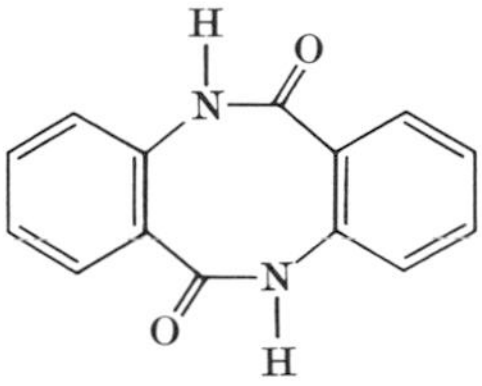

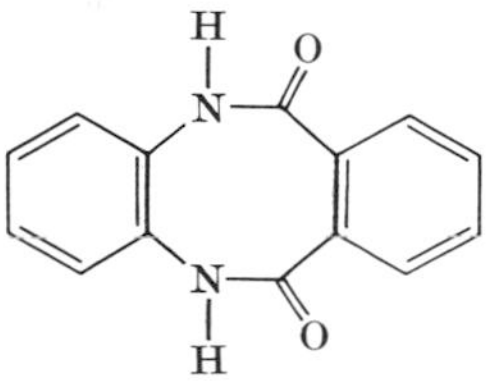

f. Summary and Conclusions

The present state of research on seven- and eight-membered heterocycles carrying an XH substituent demonstrates that there is no known stable OH or SH tautomer. In practice the only problems remaining concern the exact positions of the double bonds in certain cases. When the tautomeric group XH is amino, then the amino form of the heterocycle has so far always been found to be the most stable, although it is possible that competition could exist between the stabilizing influence of the amino group and the antiaromatic nature of the amino form. For azepines this situation never arises as there is an NH in the ring; thus we find predominance of nonaromatic tautomers

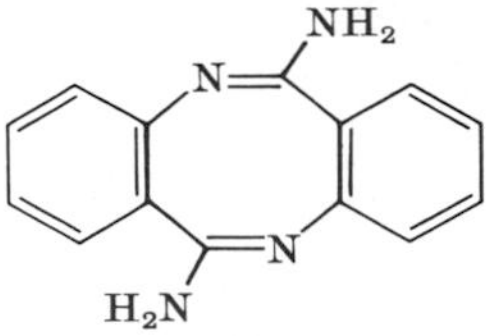

[**71**] 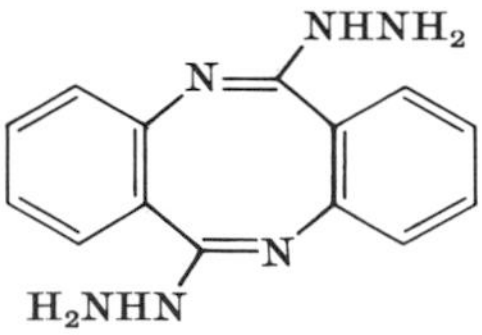 [**72**]

such as [**36c**, XH = NH_2] and [**39**]. In the case of 1,3-diazepines, the amino system with 8 π-electrons is stable probably because it is nonplanar and therefore nonconjugated [**47**, R = R′ = NH_2]. In this connection, it would be interesting to know the exact structures of [**71**] and [**72**], described in the literature as diamino (54J3429) and dihydrazino (67CH1077) derivatives, but without any proof.

3. Porphyrins

Annular tautomerism in porphyrins has recently been reviewed exhaustively (70AO105). In the solid state, X-ray crystallographic studies have eliminated structures such as [**73**] with inner hydrogens adjacent, and in solution variable temperature NMR studies (solvent: CS_2) have shown the existence of an autotropic rearrangement [**74a**] $\rightleftharpoons$ [**74b**] (72JA1745) for $\alpha,\beta,\gamma,\delta$-tetraphenylporphin with $T_c = -40°$ in CS_2 at 100 MHz, and $T_c = 0°$ in CS_2–$CDCl_3$ at 60 MHz for the dideuteriated derivative. Investigations using polarized light absorption support this conclusion for the solid state also (71PO198). The study of porphyrin

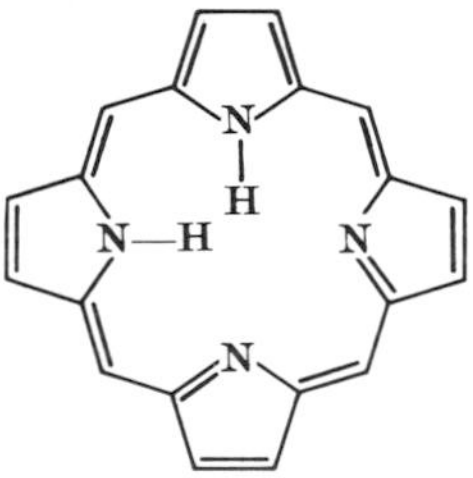

[**73**] 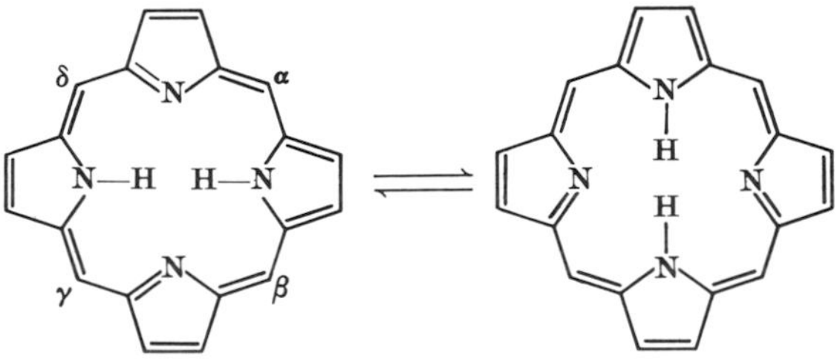 [**74a**] [**74b**]

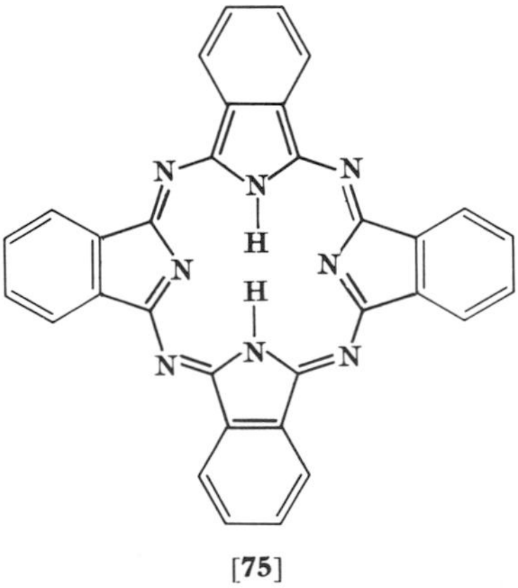

[**75**]

tautomerism by means of NMR spectroscopy (^{1}H and ^{13}C) has been completely revised by Abraham and co-workers (74TL71, 74TL1483, 75J(PII)ipl).

The situation is more complicated for the phthalocyanines [**75**]. Neutron diffraction (Section 1-4Id) (the most powerful tool for exact location of hydrogen atoms) indicates that the two central hydrogen atoms are "disordered so that on the average half a hydrogen atom is associated with each pyrrole nitrogen atom" (69CH554). This indicates that in the solid state at the temperature of the experiment either there is a rapid equilibrium of the type [**74a**] $\rightleftharpoons$ [**74b**] or the structure is random with molecules of orientations [**74a**] and [**74b**] occupying similar lattice sites.

The structure of porphyrin anions has been investigated by UV (73J(PII)414).

CHAPTER 7

Bibliography and Explanation of the Reference System

The Reference System

References are designated by a number-letter coding of which the first two digits (numbers) denote the year of publication, the next one or two digits (letters) the journal, and the final digits (numbers) the page number. Exceptionally, for journals of which more than one volume appear per year, the volume number is given in parentheses immediately after the journal letter code. A key to the journal to which letter codes refer is given in Table 7.1 and a complementary key to the letter codes for specific journals in Table 7.2.

Most journals are assigned letter codes bearing obvious similarity to their titles. Patents are listed at the end of each year and are coded similarly, e.g., 60USP2922790 means United States Patent 2,922,790 (1960). Theses are given the code TH, personal communications the code PS, and unpublished work the code UP. Work actually in press is given the year of expected publication and the journal code followed by "ip." Books, a few journals to which references are rarely made, and all other sources are coded MI (Miscellaneous) and listed under the relevant year of publication as 66MI1 etc.

Practically all sources have been checked in the original. Where a paper has been taken from a secondary source, such as *Chemical Abstracts*, the secondary source is indicated following the reference in the Bibliography.

The numbers following the references are section numbers and indicate where the reference has been cited. T, table (e.g., T4-5) S, scheme (e.g., S4-3).

TABLE 7.1

JOURNAL TITLES CORRESPONDING TO LETTER CODES

Code	Full title
AC	Analytical Chemistry
AG	Angewandte Chemie
AG(E)	Angewandte Chemie (International Edition)
AH	Acta Chimica Academiae Scientiarium Hungaricae
AJ	Australian Journal of Chemistry (Formerly Australian Journal of Scientific Research)
AK	Arkiv för Kemi
AL	Atti della Accademia Nazionale dei Lincei, Rendiconti Classe di Scienze Fisiche, Matematiche e Naturali
AN	Annali di Chimica (Rome)
AO	Accounts of Chemical Research
AP	Archiv der Pharmazie und Berichte der Deutschen Pharmazeutischen Gesellschaft
AS	Acta Chemica Scandinavica

TABLE 7.1

Code	Full title
BA	Bulletin de l'Académie Polonaise des Sciences, Série des Sciences Chimiques
BB	Bulletin des Sociétés Chimiques Belges
BF	Bulletin de la Société Chimique de France
BF(2)	Bulletin de la Société Chimique de France (Deuxième Partie)
BG	Berichte der Bunsengesellschaft für Physikalische Chemie [Zeitschrift für Elektrochemie (ZE) up to 1962]
BJ	Bulletin of the Chemical Society of Japan
BO	Biochemical and Biophysical Research Communications
BS	Bollettino Scientifico della Facoltà di Chimica Industriale di Bologna
BT	Chemistry in Britain
BY	Biochimica et Biophysica Acta
C	Chimia
CA	Chemical Abstracts
CB	Chemische Berichte
CC	Canadian Journal of Chemistry
CE	Chemical and Engineering News
CG	Mitteilungsblatt der Chemischen Gesellschaft in der DDR
CH	Chemical Communications
CI	Chemistry and Industry (London)
CO	Comptes Rendus Hebdomadaires des Séances de l'Académie des Sciences
CP	Journal de Chimie Physique et de Physicochimie Biologique
CQ	Chemica Scripta
CR	Chemical Reviews
CS	Chemical Society Reviews
CT	Chemical and Pharmaceutical Bulletin (Tokyo) [formerly Pharmaceutical Bulletin (Tokyo)]
CX	Acta Crystallographica
CX(B)	Acta Crystallographica, Section B
CZ	Collection of Czechoslovak Chemical Communications
DA	Doklady Akademii Nauk SSSR (Proceedings of the Academy of Sciences of the USSR)
E	Experientia
F	Il Farmaco (Pavia) Edizione Scientifica
G	Gazzetta Chimica Italiana
H	Helvetica Chimica Acta
HC	Advances in Heterocyclic Chemistry
IJ	Indian Journal of Chemistry

TABLE 7.1

Code	Full title
IM	Chimica e l'Industria (Milan)
IQ	International Journal of Quantum Chemistry
IS	Israel Journal of Chemistry
IZ	Izvestiya Akademii Nauk SSSR, Seriya Khimicheskaya (Bulletin of the Academy of Sciences of the USSR, Chemical Series)
J	Journal of the Chemical Society
J(A)	Journal of the Chemical Society (Section A)
J(B)	Journal of the Chemical Society (Section B)
J(C)	Journal of the Chemical Society (Section C)
J(PI)	Journal of Chemical Society (Perkin I)
J(PII)	Journal of Chemical Society (Perkin II)
JA	Journal of the American Chemical Society
JB	Journal of Biological Chemistry
JE	Journal of Chemical Education
JH	Journal of Heterocyclic Chemistry
JI	Journal of Indian Chemical Society
JJ	Journal of the Pharmaceutical Society of Japan (Yakugaku Zasshi)
JL	Journal of Molecular Structure
JM	Journal of Molecular Spectroscopy
JO	Journal of Organic Chemistry
JS	Journal of Pharmaceutical Sciences
JT	Journal of Chemical Physics
KG	Khimiya Geterotsiklicheskikh Soedinenii (Chemistry of Heterocyclic Compounds USSR)
KX	Kristallografiya
LA	Liebig's Annalen der Chemie
M	Monatshefte für Chemie
MB	Journal of Molecular Biology
MC	Journal of Medicinal and Pharmaceutical Chemistry
MI	Miscellaneous books and journals
MP	Molecular Physics
NA	Nature
NK	Nippon Kagaku Zasshi (Journal of the Chemical Society of Japan, Pure Chemistry Section)
OG	Organic Reactivity
OM	Organic Mass Spectrometry
OP	Optics and Spectroscopy (USSR) (Translation of Optika i Spektroskopiya)
OR	Organic Magnetic Resonance

TABLE 7.1

Code	Full title
P	Proceedings of the Chemical Society
PC	Journal für Praktische Chemie
PH	Journal of Physical Chemistry
PM	Physical Methods in Heterocyclic Chemistry
PN	Proceedings of the National Academy of Sciences of the USA
PP	Journal of Pharmacy and Pharmacology
PS	Personal Communication
QR	Quarterly Reviews (London)
R	Recueil des Travaux Chimiques des Pays-Bas
RC	Roczniki Chemii
RR	Revue Roumaine de Chimie
RU	Russian Chemical Reviews (Uspekhi Khimii)
RV	Pure and Applied Chemistry
RO	Russian Journal of Organic Chemistry (Zhurnal Organicheskoi Khimii)
SA	Spectrochimica Acta
SC	Science
T	Tetrahedron
TC	Theoretica Chimica Acta (Berlin)
TE	Teoreticheskaya i Eksperimental′naya Khimiya (Theoretical and Experimental Chemistry USSR)
TH	Thesis
TL	Tetrahedron Letters
UP	Unpublished Results
ZA	Zeitschrift für Analytische Chemie
ZC	Zeitschrift für Chemie
ZE	Zeitschrift für Elektrochemie [from 1962 called Berichte der Bunsengesellschaft für Physikalische Chemie (BG)]
ZF	Zhurnal Fizicheskoi Khimii (Russian Journal of Physical Chemistry)
ZN	Zeitschrift für Naturforschung
ZO	Zhurnal Obshchei Khimii (Journal of General Chemistry USSR)
ZS	Zhurnal Strukturnoi Khimii (Journal of Structural Chemistry USSR)
ZX	Zeitschrift für Kristallographie

TABLE 7.2

LETTER CODES JOURNAL TITLES

Full title	Code
Accounts of Chemical Research	AO
Acta Chemica Scandinavica	AS
Acta Chimica Academiae Scientiarum Hungaricae	AH
Acta Crystallographica	CX
Acta Crystallographica, Section B	CX(B)
Advances in Heterocyclic Chemistry	HC
Analytical Chemistry	AC
Angewandte Chemie	AG
Angewandte Chemie (International Edition)	AG(E)
Annali di Chimica (Rome)	AN
Archiv der Pharmazie und Berichte der Deutschen Pharmazeutischen Gesellschaft	AP
Arkiv för Kemi	AK
Atti della Accademia Nationale dei Lincei, Rendiconti, Classe di Science Fisiche, Matematiche e Naturali	AL
Australian Journal of Chemistry (Formerly Australian Journal of Scientific Research)	AJ
Berichte der Bunsengesellschaft für Physikalische Chemie [Zeitschrift für Elekrochemie (ZE) up to 1962]	BG
Biochimica et Biophysica Acta	BY
Biochemical and Biophysical Research Communications	BO
Bollettino Scientifico della Facoltà di Chimica Industriale di Bologna	BS
Bulletin de l'Académie Polonaise des Sciences, Série des Sciences Chimiques	BA
Bulletin de la Société Chimique de France	BF
Bulletin de la Société Chimique de France (Deuxième Partie)	BF(2)
Bulletin des Sociétés Belges	BB
Bulletin of the Chemical Society of Japan	BJ
Canadian Journal of Chemistry	CC
Chemical Abstracts	CA
Chemical Communications	CH
Chemical and Engineering News	CE
Chemical and Pharmaceutical Bulletin (Tokyo) [formerly Pharmaceutical Bulletin (Tokyo)]	CT
Chemical Reviews	CR
Chemical Society Reviews	CS
Chemica Scripta	CQ
Chemische Berichte	CB
Chemistry and Industry (London)	CI
Chemistry in Britain	BT
Chimia	C

TABLE 7.2

Full title	Code
Chimica e l'Industria (Milan)	IM
Collection of Czechoslovak Chemical Communications	CZ
Comptes Rendus Hebdomadaires des Séances de l'Académie des Sciences	CO
Doklady Akademii Nauk SSSR (Proceedings of the Academy of Sciences of the USSR)	DA
Experientia	E
Gazzetta Chimica Italiana	G
Helvetica Chimica Acta	H
Il Farmaco (Pavia) Edizione Scientifica	F
Indian Journal of Chemistry	IJ
International Journal of Quantum Chemistry	IQ
Israel Journal of Chemistry	IS
Izvestiya Akademii Nauk SSSR, Seriya Khimicheskaya (Bulletin of the Academie of Sciences of the USSR, Division of Chemical Science)	IZ
Journal of the American Chemical Society	JA
Journal of Biological Chemistry	JB
Journal of Chemical Education	JE
Journal of Chemical Physics	JT
Journal of the Chemical Society	J
Journal of the Chemical Society (Section A)	J(A)
Journal of the Chemical Society (Section B)	J(B)
Journal of the Chemical Society (Section C)	J(C)
Journal of Chemical Society (Perkin I)	J(PI)
Journal of Chemical Society (Perkin II)	J(PII)
Journal de Chimie Physique et de Physicochimie Biologique	CP
Journal of Heterocyclic Chemistry	JH
Journal of the Indian Chemical Society	JI
Journal of Medicinal and Pharmaceutical Chemistry	MC
Journal of Molecular Biology	MB
Journal of Molecular Spectroscopy	JM
Journal of Molecular Structure	JL
Journal of Organic Chemistry	JO
Journal of Pharmaceutical Sciences	JS
Journal of the Pharmaceutical Society of Japan (Yakugaku Zasshi)	JJ
Journal of Pharmacy and Pharmacology	PP
Journal of Physical Chemistry	PH
Journal für Praktische Chemie	PC

TABLE 7.2

Full title	Code
Khimiya Geterotsiklicheskikh Soedinenii (Chemistry of Heterocyclic Compounds (USSR)	KG
Kristallografiya	KX
Liebig's Annalen der Chemie	LA
Miscellaneous books and journals	MI
Mitteilungsblatt der Chemischen Gesellschaft in der DDR	CG
Molecular Physics	MP
Monatshefte für Chemie	M
Nature	NA
Nippon Kagaku Zasshi (Journal of the Chemical Society of Japan, Pure Chemistry Section)	NK
Optics and Spectroscopy (USSR) (Translation of Optika i Spektroskopiya	OP
Organic Magnetic Resonance	OR
Organic Mass Spectrometry	OM
Organic Reactivity	OG
Personal Communication	PS
Physical Methods in Heterocyclic Chemistry	PM
Proceedings of the Chemical Society	P
Proceedings of the National Academy of Sciences of the USA	PN
Pure and Applied Chemistry	RV
Quarterly Reviews (London)	QR
Recueil des Travaux Chimiques des Pays-Bas	R
Revue Roumaine de Chimie	RR
Roczniki Chemii	RC
Russian Chemical Reviews (Uspekhi Khimii)	RU
Russian Journal of Organic Chemistry (Zhurnal Organicheskoi Khimii)	RO
Science	SC
Spectrochimica Acta	SA
Tetrahedron	T
Tetrahedron Letters	TL
Theoretica Chimica Acta (Berlin)	TC
Teoreticheskaya i Eksperimental'naya Khimiya (Theoretical and Experimental Chemistry USSR)	TE
Thesis	TH

TABLE 7.2

Full title	Code
Unpublished Results	UP
Zeitschrift für Analytische Chemie	ZA
Zeitschrift für Chemie	ZC
Zeitschrift für Elektrochemie [from 1962 called Berichte der Bunsen-gesellshaft für Physikalische Chemie (BG)	ZE
Zeitschrift für Kristallographie	ZX
Zeitschrift für Naturforschung	ZN
Zhurnal Fizicheskoi Khimii (Russian Journal of Physical Chemistry)	ZF
Zhurnal Obschei Khimii (Journal of General Chemistry USSR)	ZO
Zhurnal Strukturnoi Khimii (Journal of Structural Chemistry USSR)	ZS

Bibliography

90CB2478 A. W. Day and S. Gabriel, *Chem. Ber.*, 1890, **23**, 2478. 3-4Bc

98CB2646 S. Gabriel and E. Leupold, *Chem. Ber.*, 1898, **31**, 2646. 3-4Bc

01LA(319)196 J. Thiele and N. Sultzberger, *Liebig's Ann. Chem.*, 1901, **319**, 196. 3-2Aa

13CB2107 F. Benary, *Chem. Ber.*, 1913, **46**, 2107. 3-2Db

34LA(508)51 K. von Auwers, *Liebig's Ann. Chem.*, 1934, **503**, 51. 1-4Hb(2), 1-5Ab, 4-1Aa, 4-1Bb(iv), 4-1Bb(ix)

36J111 A. K. Macbeth and J. R. Price, *J. Chem. Soc.*, 1936, 111. 4-7Ae(2)

37LA(527)291 K. von Auwers, *Liebig's Ann. Chem.*, 1937, **527**, 291. 1-4Hb(2), 4-1G

38CB604 K. von Auwers, *Chem. Ber.*, 1938, **71**, 604. 1-4Hb(2)

40CB162 K. W. F. Kohlrausch and R. Seka, *Chem. Ber.*, 1940, **73**, 162. 4-1G(iv)

41J1 H. T. Hayes and L. Hunter, *J. Chem. Soc.*, 1941, 1. 4-1Bc

41J113 I. M. Barclay, N. Campbell and G. Dodds, *J. Chem. Soc.*, 1941, 113. 4-1G

42LA(550)31 K. Fries, K. Fabel and H. Eckhardt, *Liebig's Ann. Chem.*, 1942, **550**, 31. 4-1G(v)

43H687 P. A. Plattner and L. M. Jampolsky, *Helv. Chim. Acta*, 1943, **26**, 687. 3-2Ba

43MI1 K. A. Jensen and A. Friediger, *Kgl. Dan. Vidensk. Selsk., Mat.-Fys. Medd.*, 1943, **20**, 1. 1-4G, 4-1Da(2), 4-1Fa(ii), 4-1G(ii), T4-8, S1-8

47J1598 A. H. Cook, I. Heilbron and A. L. Levy, *J. Chem. Soc.*, 1947, 1598. 4-Cb(2), 4-6Ga

48J2031 A. H. Cook, I. Heilbron and E. S. Stern, *J. Chem. Soc.*, 1948, 2031. 4-6Ga

48JA3385 E. D. Hartnell and C. E. Brilker, *J. Amer. Chem. Soc.*, 1948, **70**, 3385. 4-3Ab

48MI127 F. Arndt, L. Loewe and A. Tarlan-Akon, *Istanbul Univ. Fen. Fak. Mecm., Seri. A*, 1948, **13**, 127. 4-6Bg, 4-6Bg(iii)

49H998 F. Fleck, A. Rossi and H. Schinz, *Helv. Chim. Acta*, 1949, **32**, 998. 3-2Ba

49J118 A. W. Nineham and R. A. Raphael, *J. Chem. Soc.*, 1949, 118. 3-2Aa

49J263 L. P. Ellinger and A. A. Goldberg, *J. Chem. Soc.*, 1949, 263. 2-2D

49J1664 E. S. Stern, *J. Chem. Soc.*, 1949, 1664. 4-4Cc(ii)

50H273 C. A. Grob and P. Ankli, *Helv. Chim. Acta*, 1950, **33**, 273. S5-5

50J3491 W. Davies, J. A. MacLaren and L. R. Wilkinson, *J. Chem. Soc.*, 1950, 3491. 4-6Dc

50JA3047 V. Rousseau and H. G. Lindwall, *J. Amer. Chem. Soc.*, 1950, **72**, 3047. 4-1G(v)

51J3211 G. M. Badger and R. Pettit, *J. Chem. Soc.*, 1951, 3211. 2-6Dd

51JA3030 J. L. Rabinowitz and E. C. Wagner, *J. Amer. Chem. Soc.*, 1951, **73**, 3030. 4-1Hb(i)

51PP420 M. T. Davies, P. Mamalis, V. Pettrow and B. Sturgeon, *J. Pharm. Pharmacol.*, 1951, **3**, 420. T1-2, T4-10(2)

52J1461 S. Angyal and C. L. Angyal, *J. Chem. Soc.*, 1952, 1461. T2-5

53J1207 L. A. Duncanson, *J. Chem. Soc.*, 1953, 1207. 3-2Bb

53J3802 J. Davoll, *J. Chem. Soc.*, 1953, 3802. 3-2Ee

54CB841 H. Fiessekmann, P. Schippak and L. Zeitler, *Chem. Ber.*, 1954, **87**, 841. 3-2Cd

54CI1356 T. Nozoe, S. Seto, S. Matsumura and T. Terasawa, *Chem. Ind. (London)*, 1954, 1356. 3-2Ed

54J442 J. A. Elvidge and R. P. Linstead, *J. Chem. Soc.*, 1954, 442. 3-4Ca

54J3429 F. C. Cooper and M. W. Partridge, *J. Chem. Soc.*, 1954, 3429. 6-2Bf

54MI431 P. Maitte, *Ann. Chim. (Paris)*, 1954, **9**, 431. 2-2A

55JA1623 T. A. Geissman and A. Armen, *J. Amer. Chem. Soc.*, 1955, **72**, 1623. 3-2Ad

55MI88,99 G. W. Wheland, "Resonance in Organic Chemistry," pp. 88 and 99. Wiley, New York, 1955. 1-4D

55MI1955 A. Gomez-Sanchez and M. Yruela-Antinolo, *An. Real Soc. Espan. Fis. Quim., Ser. B*, 1955, **51**, 1955. 3-2Ac

56CB1897 H. Fiesselmann and P. Schipprak, *Chem. Ber.*, 1956, **89**, 1897. 3-2Cf

56G797 D. Dal Monte, A. Mangini and R. Passerini, *Gazz. Chim. Ital.*, 1956, **86**, 797. 4-1Aa, 4-1Bb(vii), 4-1Bb(ix)

56JA411 K. Hattori, E. Lieber and J. P. Horwitz, *J. Amer. Chem. Soc.*, 1956, **78**, 411. 4-3Fa

56JA2532 L. I. Smith, W. L. Kohlhase and R. J. Brotherton, *J. Amer. Chem. Soc.*, 1956, **78**, 2532. 4-3Ab

56JA4197 M. H. Kauffman, F. M. Ernsberger and W. S. McEwan, *J. Amer. Chem. Soc.*, 1956, **78**, 4197. 4-1Fa(ii), S1-8(2)

57CB182 J. Goerdeler, H. Groschopp and U. Sommerlad, *Chem. Ber.*, 1957, **90**, 182. 4-3Fa

57J4789 T. M. Bruice and F. K. Sutcliffe, *J. Chem. Soc.*, 1957, 4789. 3-2Ed

57J5075 J. T. Edward and S. Nielsen, *J. Chem. Soc.*, 1957, 5075. 4-6Cb

57JA1656 M. L. Bender and B. W. Turnquest, *J. Amer. Chem. Soc.*, 1957, **79**, 1656. T1-2, T4-5

57JO1750 E. Lieber, C. N. Pillai, J. Ramachandran and R. D. Hites, *J. Org. Chem.*, 1957, **22**, 1750. 4-4Fd(2)

57KX38 I. Tashpulatov, Z. V. Zvonkova and G. S. Zhdanov, *Kristallografiya*, 1957, **2**, 38. 4-4Cd

58BF543 G. Rio and A. Ranjon, *Bull. Soc. Chim. Fr.*, 1958, 543. 4-3C(i)(2)

58BY1958 H. T. Miles, *Biochim. Biophys. Acta*, 1958. **46**, 1958. 5-1Bf

58CX808 M. E. Senko and D. H. Templeton, *Acta Crystallogr.*, 1958, **11**, 808. 4-6Gc

58G977 D. Dal Monte, A. Mangini, R. Passerini and C. Zauli, *Gazz. Chim. Ital.*, 1958, **58**, 977. 1-4G, 4-1Da

58J1174 J. W. Cornforth, *J. Chem. Soc.*, 1958, 1174. 2-2D

58J1217 S. J. Holt, A. E. Kellie, D. G. O'Sullivan and P. W. Sadler, *J. Chem. Soc.*, 1958, 1217. 3-2Cf

58JA2759 M. A. Stevens and G. B. Brown, *J. Amer. Chem. Soc.*, 1958, **80**, 2759. 5-1Fa

59AS1668 B. S. Jensen, *Acta Chem. Scand.*, 1959, **13**, 1668. 4-2 Gh(2)

59BY274 H. T. Miles, *Biochim. Biophys. Acta*. 1959, **35**, 274. 2 6Cb

59JA3786 S. Y. Wang, *J. Amer. Chem. Soc.*, 1959, **81**, 3786. 2-4Dh

59JA6292 G. De Stevens, A. Halamandaris, P. Wenk and L. Dorfman, *J. Amer. Chem. Soc.*, 1959, **81**, 6292. 4-2Ha

59JO779 C. F. H. Allen, H. R. Beifuss, D. M. Burness, G. A. Reynolds J. F. Tinker and J. A. Van Allan, *J. Org. Chem.*, 1959, **24**, 779. 5-3Bc

59JO2039 F. A. Snavely and F. H. Suydam, *J. Org. Chem.*, 1959, **24**, 2039. 4-2G

59LA(623)166 S. Petersen and E. Tietze, *Liebig's Ann. Chem.*, 1959, **623**, 166. 3-1Ad

59NK402 S. Toda, *Nippon Kagaku Zasshi*, 1959, **80**, 402. 4-2Gj

59ZA(170)205 R. Gompfer and P. Altreuther, *Z. Anal. Chem.*, 1959, **170**, 205. 2-4Cd

60AG359 R. Huisgen, *Angew. Chem.*, 1960, **72**, 359. 4-9Bb, 4-9Ca

60AG967 W. Pelz, W. Püschel, H. Schellenberger and K. Löffler, *Angew. Chem.*, 1960, **72**, 967. 4-2Gj

60CB671 J. Goerdeler and H. Horstmam, *Chem. Ber.*, 1960, **93**, 671. 2-4A

60CB2035 H. Zinner, R. Reimann and A. Weber, *Chem. Ber.*, 1960, **93**, 2035. 4-4Cb

60CX946 H. W. W. Ehrlich, *Acta Crystallogr.*, 1960, **13**, 946. 4-1Ba (3), T4-1

60J539 P. Brookes and P. D. Lawley, *J. Chem. Soc.*, 1960, 539. 5-1Da

60J1274 D. Peters, *J. Chem. Soc.*, 1960, 1274.

60J1363 J. H. Ridd and B. V. Smith, *J. Chem. Soc.*, 1960, 1363. 1-4Aa, 4-1Hb(i)(3), T4-5(2), T4-10(5)

60J3278 D. G. O'Sullivan, *J. Chem. Soc.*, 1960, 3278. 4-2Jd

60JA463 J. A. Montgomery and K. Hewson, *J. Amer. Chem. Soc.*, 1960, **82**, 463. 5-1C

60JA1515 C. G. Krespan, B. C. McKusick and T. L. Cairns, *J. Amer. Chem. Soc.*, 1960, **82**, 1515. 6-1Ab

60JA5007 D. W. Moore and A. G. Whittaker, *J. Amer. Chem. Soc.*, 1960, **82**, 5007. 4-1Fa(iv)

60JO827 E. D. Bergmann and M. Rabinovitz, *J. Org. Chem.*, 1960, **25**, 827. 6-2A

60JO1242 W. I. Awad and A. E. A. G. Allah, *J. Org. Chem.*, 1960, **25**, 1242. 4-3Ci

60JO1509 N. L. Allinger and G. A. Youngdale, *J. Org. Chem.*, 1960, **25**, 1509. 6-2A

60MI25 L. C. Pauling, "The Nature of the Chemical Bond and the Structure of Molecules and Crystals: an Introduction to Modern Structural Chemistry," 3rd ed., p. 25. Cornell University Press, Ithaca, New York, 1960. 3-2Hb

60MI98 L. J. Bellamy and P. E. Rogash, *Proc. Roy. Soc., Ser. A.*, 1960, **257**, 98. 4-4Cd

60RU153 M. E. Vol'pin, *Russ. Chem. Rev.*, 1960, **20**, 153. 6-1Ab

61AK(16)459 S. Gronowitz and R. A. Hoffman, *Ark. Kemi*, 1961, **16**, 459. 2-5E

61AK(17)523 S. Forsen and M. Nillson, *Ark. Kemi*, 1961, **17**, 523. 2-3Ga

61BF1226 H. Najer, R. Guidicelli, E. Joannic-Voisinet and M. Joannic, *Bull. Soc. Chim. Fr.*, 1961, 1226. 4-6Fe(iv)

61BF1231 H. Najer and R. Giudicelli, *Bull. Soc. Chim. Fr.*, 1961, 1231. 4-6Fe(iv)

61BJ53 K. Nakanishi, N. Suzuki and F. Yamazaki, *Bull. Chem. Soc. Jap.*, 1961, **34**, 53. 2-4Dd

61CX333 R. Gerdil, *Acta Crystallogr.*, 1961, **14**, 333. 2-4Db

61CX720 R. Mason, *Acta Crystallogr.*, 1961, **14**, 720. 3-2Fb

61DA1374 O. A. Osipov, A. M. Simonov, V. I. Minkin and A. D. Garnovskii, *Dokl. Akad. Nauk SSSR*, 1961, **137**, 1374. 4-1Ca

61H1171 J. P. Wibaut and J. W. P. Boon, *Helv. Chim. Acta*, 1961, **44**, 1171. 4-1Bb(ix)

61J4827 D. Harrison and A. C. B. Smith, *J. Chem. Soc.*, 1961, 4827. 4-3Bd

61JA3434 C. G. Krespan, *J. Amer. Chem. Soc.*, 1961, **83**, 3434. 6-1Ab

61JA4034 W. E. Parham and R. Koncos, *J. Amer. Chem. Soc.*, 1961, **83**, 4034. 2-2B

61JO451 M. Hauser, E. Peters and H. Tieckelmann, *J. Org. Chem.*, 1961, **26**, 451. T5-1

61JO1651 J. C. Howard and H. A. Burch, *J. Org. Chem.*, 1961, **26**, 1651. 4-3Dj

61JO2791 B. Staskun, *J. Org. Chem.*, 1961, **26**, 2791. 2-3Bd

61JO3761 A. T. Blomquist and E. J. Moriconi, *J. Org. Chem.*, 1961, **26**, 3761. 2-3Fa

61JO4480 E. Schipper and E. Chinnery, *J. Org. Chem.*, 1961, **26**, 4480. 4-3Ci

61JO4917 L. Bauer and C. N. V. Nambury, *J. Org. Chem.*, 1961, **26**, 4917. 4-6Fb(ii)

61JO4923 C. L. Bell, C. N. V. Nambury and L. Bauer, *J. Org. Chem.*, 1961, **26**, 4923. 4-6Fb(ii)

61M1114 O. E. Polansky and G. Derflinger, *Monatsh. Chem.*, 1961, **92**, 1114. 4-1G(vi), 4-1I

61M1131 J. Derkosch, O. E. Polansky, E. Rieger and G. Derflinger, *Monatsh. Chem.*, 1961, **92**, 1131. 4-1G(v), 4-1I

61NK778 Y. Iwanami, *Nippon Kagaku Zasshi*, 1961, **82**, 778. 2-8Dd

61PN791 H. T. Miles, *Proc. Nat. Acad. Sci. U.S.*, 1961, **47**, 791. 2-6Cb(2), 5-1Eh

61SA40 S. Refn, *Spectrochim. Acta*, 1961, **17**, 40. 4-2Ha(2)

61SA238 A. R. Katritzky and A. J. Boulton, *Spectrochim. Acta*, 1961, **17**, 238. 4-5Ba, 4-5Bc, 4-5Ca

61T41 A. J. Boulton and A. R. Katritzky, *Tetrahedron*, 1961, **12**, 41. 1-6Cd

61T51 A. J. Boulton and A. R. Katritzky, *Tetrahedron*, 1961, **12**, 51. 4-5Ba(2), 4-5Bc(2), 4-5Ca(2), 4-5Cb

61T237 A. J. Owen, *Tetrahedron*, 1961, **14**, 237. 1-4G(2), 4-1Fa(ii), T4-8

61ZA(181)447 G. Englert, *Z. Anal. Chem.*, 1961, **181**, 447. 4-5Fe

61ZA(181)487 M. Kuhn and R. Mecke, *Z. Anal. Chem.*, 1961, **181**, 487. 4-5Fe

61ZE821 H. Zimmermann, *Z. Elektrochem.*, 1961, **65**, 821. 1-4G, 4-1Ca

62AG465 J. Strating, J. H. Keijer, E. Molenaar and L. Brandsma, *Angew. Chem.*, 1962, **74**, 465. 2-2A, 2-2B

62AK(19)181 G. Bergson, *Ark. Kemi*, 1962, **19**, 181. 6-1Ab

62AK(19)265 G. Bergson, *Ark. Kemi*, 1962, **19**, 265. 6-1Ab

62AJ851 D. J. Brown and J. M. Lyall, *Aust. J. Chem.*, 1962, **15**, 851. 2-6Cb

62AS789 A. B. Hörnfeldt and S. Gronowitz, *Acta Chem. Scand.*, 1962, **16**, 789. 3-2Ca(iii)

62AS1800 T. Kindt-Larsen and C. Pedersen, *Acta Chem. Scand.*, 1962, **16**, 1800. 4-4Fc

62BF1707 E. Moczar and L. Mester, *Bull. Soc. Chim. Fr.*, 1962, 1707. 4-2Ha, 4-2Hc, 4-2Hd

62BJ747 S. Kikuchi and H. Yoshida, *Bull. Chem. Soc. Jap.*, 1962, **35**, 747. 4-2Ha, 4-2Hc, 4-6Fd(ii)

62CB2195 W. Pfleiderer, *Chem. Ber.*, 1962, **95**, 2195. 2-8De(2)

62CI695 A. R. Katritzky and A. J. Waring, *Chem. Ind.* (*London*), 1962, 695. 2-4Cd

62CI1576 P. de Mayo and S. T. Reid, *Chem. Ind.* (*London*), 1962, 1576. 3-2Fb

62CT620 Y. Makisumi, *Chem. Pharm. Bull.*, 1962, **10**, 620. 5-3Be

62CX1174 R. F. Bryan and K. Tomita, *Acta Crystallogr.*, 1962, **15**, 1174. 2-6Cd

62CX1179 R. F. Bryan and K. Tomita, *Acta Crystallogr.*, 1962, **15**, 1179. 5-1Da(2)

62CZ716 J. Jonas and J. Gut, *Collect. Czech. Chem. Commun.*, 1962, **27**, 716. 2-4Fa, 2-4Fb

62CZ1886 J. Jonas and J. Gut, *Collect. Czech. Chem. Commun.*, 1962, **27**, 1886. 2-5F(2), 2-4Fb

62CZ2754 J. Jonas, M. Horak, A. Piskala and J. Gut, *Collect. Czech. Chem. Commun.*, 1962, **27**, 2754. 2-4Fa(2)

62J1462 S. Lewin, *J. Chem. Soc.*, 1962, 1462. 4-6Be

62J2606 J. A. Elvidge, *J. Chem. Soc.* 1962, 2606. 2-3Ga

62J3129 A. Albert and G. B. Barlin, *J. Chem. Soc.*, 1962, 3129. 2-5E(3)

62J3288 M. Fraser, A. Melera, B. B. Molloy and D. H. Reid, *J. Chem. Soc.*, 1962, 3288. 3-1Bc

62J3926 G. M. Badger and P. J. Nelson, *J. Chem. Soc.*, 1962, 3926. 2-6Dd

62J4191 F. Kurzer and S. A. Taylor, *J. Chem. Soc.*, 1962, 4191. 4-3Dc, 4-6Fh(ii), 4-6Gb(ii)

62JA813 W. E. Parham and L. D. Huestis, *J. Amer. Chem. Soc.*, 1962, **84**, 813. 2-2A

62JA2452 S. Masamme and N. T. Castellucci, *J. Amer. Chem. Soc.*, 1962, **84**, 2452. 2-2A

62JA2534 R. L. Hinman and E. B. Whipple, *J. Amer. Chem. Soc.*, 1962, **84**, 2534. 3-1Bb

62JA3979 C. B. Reese, *J. Amer. Chem. Soc.*, 1962, **84**, 3979. 2-8Bb

62JA4464 L. Gatlin and J. C. Davis, Jr., *J. Amer. Chem. Soc.*, 1962, **84**, 4464. 2-6Cb

62JB3573 D. W. Green, F. S. Mathews and A. Rich, *J. Biol. Chem.*, 1962, **237**, 3573. 2-4Db

62JO994 F. A. Snavely, W. S. Trahanovsky and F. H. Suydam, *J. Org. Chem.*, 1962, **27**, 994. 4-2Gj

62JO1686 C. F. Howell, N. Q. Quinones and R. A. Hardy, Jr., *J. Org. Chem.*, 1962, **27**, 1686. 4-6Fe(i), 4-6Fe(ii), 4-6Fe(iii)(2), 4-6Fe(iv)(3)

62JO2478 G. B. Elion, *J. Org. Chem.*, 1962, **27**, 2478. 5-1Be

62JO3155 J. A. Sousa and J. Weinstein, *J. Org. Chem.*, 1962, **27**, 3155. 2-8Cb

62JO3730 R. Filler and Y. S. Rao, *J. Org. Chem.*, 1962, **27**, 3730. 4-3Ce

62LA(657)131 B. Eistert, H. Fink and H. K. Werner, *Liebig's Ann. Chem.*, 1962, **657**, 131. 2-4H

62MI125 O. Serfas and G. Geppert, *Monatsber. Deut. Akad. Wiss. Berlin*, 1962, **4**, 125. 4-2Jb, 4-2Jc

62MI393 G. Cipens, V. Grinsteins and M. Tiltins, *Latv. PSR Zinat. Akad. Vestis, Kim. Ser.*, 1962, 393. 4-5Fc(2)

62MI401 G. Cipens and V. Grinsteins, *Latv. PSR Zinat. Akad. Vestis, Kim. Ser.*, 1962, 401. 4-5Fc(2)

62MI411 G. Cipens and V. Grinsteins, *Latv. PSR Zinat. Akad. Vestis, Kim. Ser.*, 1962, 411. 4-5Fc

62PH2434 J. D. Margerum, L. J. Miller, E. Saito, M. S. Brown, H. S. Mosher and R. Hardwich, *J. Phys. Chem.*, 1962, **66**, 2434. 2-8Cb(2)

62T539 E. J. Browne and J. B. Polya, *Tetrahedron*, 1962, **18**, 539. 4-8C

62T777 A. R. Katritzky, S. Øksne and A. J. Boulton, *Tetrahedron*, 1962, **17**, 777. 1-6Cd, 4-2Bb, T4-12(6)

62T853 F. Scheinmann, *Tetrahedron*, 1962, **18**, 853. 3-2Ad

62TL913 H. A. Staab and A. Mannschreck, *Tetrahedron Lett.*, 1962, 913. 4-1Ka

63AG300 H. A. Staab and A. Mannschreck, *Angew. Chem.*, 1963, **75**, 300. 4-1Cb(viii), 4-1Ka(2)

63AG1041 K. Hafner, *Angew. Chem.*, 1963, **75**, 1041. 6-2A

63AG1204 H.-W. Wanzlick and H.-J. Kleiner, *Angew. Chem.*, 1963, **75**, 1204. 4-8C

63AJ445 G. M. Badger and P. J. Neilson, *Aust. J. Chem.*, 1963, **16**, 455. 2-6Dd

63AN1405 G. D'Alo, M. Perghem and P. Grünanger, *Ann. Chim. (Rome)*, 1963, **53**, 1405. 4-3De

63AS144 K. A. Jensen and I. Crossland, *Acta Chem. Scand.*, 1963, **17**, 144. 4-3Cf(5)

63AS921 H. Ch. Börresen, *Acta Chem. Scand.*, 1963, **17**, 921. 5-1Eg

63AS1694 H. G. Mautner and G. Bergson, *Acta Chem. Scand.*, 1963, **17**, 1964. 5-1Ed

63AS2575 A. Hordvik, *Acta Chem. Scand.*, 1963, **17**, 2575. 4-9A

63BF1022 H. Najer, R. Giudicelli, C. Morel and J. Menin, *Bull. Soc. Chim. Fr.*, 1963, 1022. 4-6Ff(ii)(3)

63BF2840 E. Laviron, *Bull. Soc. Chim. Fr.*, 1963, 2840. 1-4F, 4-1Cb(ii), 4-1Cb(iv), T4-5(2)

63BG54 H-U. Schütt and H. Zimmermann, *Ber. Bunsenges. Phys. Chem.*, 1963, **67**, 54. 4-1Ka

63BG470 A. Mannschreck, W. Seitz and H. A. Staab, *Ber. Bunsenges. Phys. Chem.*, 1963, **67**, 470. 4-1Ka

63CB944 J. Goerdeler and W. Mittler, *Chem. Ber.*, 1963, **96**, 944. 4-2E, 4-5Bd, 4-5Be(2)

63CB1088 H.-D. Stachel, *Chem. Ber.*, 1963, **96**, 1088. 4-2Cb(v), 4-5Bc, 4-5Bh(i)

63CB1680 W. Logemann, G. Cavagna and G. Tosolini, *Chem. Ber.*, 1963, **96**, 1680. 3-2Ad

63CB1726 M. Ionescu, H. Mantsch and I. Gioia, *Chem. Ber.*, 1963, **96**, 1726. 2-3Fb

63CB2950 W. Pfleiderer and H. Fink, *Chem. Ber.*, 1963, **96**, 2950. 2-6De

63CB2964 W. Pfleiderer and H. Fink, *Chem. Ber.*, 1963, **96**, 2964. 2-6De

63CB2977 D. Söll and W. Pfleiderer, *Chem. Ber.*, 1963, **96**, 2977. 2-6De

63CC625 R. S. Atkinson and E. Bullock, *Can. J. Chem.*, 1963, **41**, 625. 3-2Ee

63CI1353 C. L. Bell, J. Shoffner and L. Bauer, *Chem. Ind.* (*London*), 1963, 1353. 2-3C

63CR461 P. R. Jones, *Chem. Rev.*, 1963, **63**, 461. 4-9C

63CT67 Y. Makisumi and H. Kano, *Chem. Pharm. Bull.*, 1963, **11**, 67. 5-3Bc

63CT514 T. Okamoto and H. Takayama, *Chem. Pharm. Bull.*, 1963, **11**, 514. 2-8Db(iii)

63CT669 F. Yoneda and Y. Nitta, *Chem. Pharm. Bull.*, 1963, **11**, 669. 2-4Cd

63CT744 Y. Nitta, R. Tomii and F. Yoneda, *Chem. Pharm. Bull.*, 1963, **11**, 744. 2-6Da

63CT1375 S. Takahashi and H. Kano, *Chem. Pharm. Bull.*, 1963, **11**, 1375. 4-7Ad(2)

63CX20 G. A. Jeffrey and Y. Kinoshita, *Acta Crystallogr.*, 1963, **16**, 20. 2-6Cb

63CX28 K. Hoogsteen, *Acta Crystallogr.*, 1963, **16**, 28. 2-4Db

63CX166 W. Bolton, *Acta Crystallogr.*, 1963, **16**, 166. 2-4Df

63CX318 P. Cucka, *Acta Crystallogr.*, 1963, **16**, 318. 2-4Ca

63CX520 S. G. G. MacDonald and A. B. Alleyne, *Acta Crystallogr.*, 1963, **16**, 520. 3-2Bb

63CX907 K. Hoogsteen, *Acta Crystallogr.*, 1963, **16**, 907. 5-1Da

63CX950 W. Bolton, *Acta Crystallogr.*, 1963, **16**, 950. 2-4Dg

63CX1157 R. H. Stanford, *Acta Crystallogr.*, 1963, **16**, 1157. 4-9A

63CZ1408 J. Pitha, *Collect. Czech. Chem. Commun.*, 1963, **28**, 1408. 2-3Ec(2)

63CZ1499 R. Zahradník, J. Koutecky, J. Jonáš and J. Gut, *Collect. Czech. Chem. Commun.*, 1963, **28**, 1499. 2-4Fa, 2-4Fb

63CZ1507 J. Pitha and J. Beranek, *Collect. Czech. Chem. Commun.*, 1963, **28**, 1507. 2-4Fb, 2-5F, 2-6Ec

63CZ1625 J. Pitha, *Collect. Czech. Chem. Commun.*, 1963, **28**, 1625. 2-3Ec

63CZ1651 P. Kristian, K. Autos, D. Vlochova and R. Zahradnik, *Collect. Czech. Chem. Commun.*, 1963, **28**, 1651. 2-6Ba

63CZ3392 M. Horak and J. Gut, *Collect. Czech. Chem. Commun.*, 1963, **28**, 3392. 2-4Fb, 2-5F

63G383 K. Adank, W. G. Stoll and M. Viscontini, *Gazz. Chim. Ital.*, 1963, **93**, 383. 3-2Ef

63G570 G. Pala, *Gazz. Chim. Ital.*, 1963, **93**, 570. 4-6Bb

63G964 G. Cum and G. Lo Vecchio, *Gazz. Chim. Ital.*, 1963, **93**, 964. 4-2Bb, 4-2Bd

63G1530 A. Ficalbi, *Gazz. Chim. Ital.*, 1963, **93**, 1530. 2-8Cb

63H1030 K. Adank, H. A. Pfenninger, W. G. Stoll and M. Viscontini, *Helv. Chim. Acta*, 1963, **46**, 1030. 3-2Ef

63H1259 R. E. Rosenkranz, K. Allner, R. Goos, W. von Philipsborn and C. H. Eugster, *Helv. Chim. Acta*, 1963, **46**, 1259. 3-2Ac

63H2592 W. von Philipsborn, H. Stierlin and W. Traber, *Helv. Chim. Acta*, 1963, **46**, 2592. 2-8Db(ii), 2-8Db(iv), 2-8Dd

63H2597 L. Merlini, W. von Philipsborn and M. Viscontini, *Helv. Chim. Acta*, 1963, **46**, 2597. 2-8Dd, 2-8De

63HC(1)167 D. Beke, *Advan. Heterocycl. Chem.*, 1963, **1**, 167. 4-9C

63J753 A. R. Katritzky and R. E. Reavill, *J. Chem. Soc.*, 1963, 753. 2-3C

63J2032 D. Buttimore, D. H. Jones, R. Slack and K. R. H. Wooldridge, *J. Chem. Soc.*, 1963, 2032. 4-5Cb

63J2867 L. S. Besford, G. Allen and J. M. Bruce, *J. Chem. Soc.*, 1963, 2867. 2-2D

63J3046 A. R. Katritzky and A. J. Waring, *J. Chem. Soc.*, 1963, 3046. 2-6Cd, 2-6Cb

63J3069 M. A. Butt, J. A. Elvidge and A. B. Foster, *J. Chem. Soc.*, 1963, 3069. 2-3Ec

63J3165 H. J. Eméléus, A. Haas and N. Sheppard, *J. Chem. Soc.*, 1963, 3165. 4-9A

63J3168 H. J. Emeléus, A. Haas and N. Sheppard, *J. Chem. Soc.*, 1963, 3168. 4-6Cd

63J3277 D. R. Bragg and D. G. Wibberley, *J. Chem. Soc.*, 1963, 3277. 3-2Ef

63J3855 E. Spinner, *J. Chem. Soc.*, 1963, 3855. 2-3C

63J4333 D. J. Brown and T. Teitei, *J. Chem. Soc.*, 1963, 4333. 2-5E(2)

63J4483 M. A. Butt and J. A. Elvidge, *J. Chem. Soc.*, 1963, 4483. 2-3Ga(2)

63J4778 I. Fleming and J. Harley-Mason, *J. Chem. Soc.*, 1963, 4778. 3-2Bb

63J4897 J. C. Hayloch, S. F. Mason and B. E. Smith, *J. Chem. Soc.*, 1963, 4897. 2-3D

63J4924 D. E. Ames and H. Z. Kucharska, *J. Chem. Soc.*, 1963, 4924. 2-4Cb

63J5342 F. M. Dean and P. G. Jones, *J. Chem. Soc.*, 1963, 5342. 4-9D

63J5556 N. K. Roberts, *J. Chem. Soc.*, 1963, 5556. 4-1I(2)

63J5590 J. F. W. McOmie and A. B. Turner, *J. Chem. Soc.*, 1963, 5590. 2-4Df

63J5713 A. M. Comrie, D. Dingwall and J. B. Stenlake, *J. Chem. Soc.*, 1963, 5713. 4-6Ff(iii)

63JA26 E. B. Whipple, Y. Chiang and R. L. Hinman, *J. Amer. Chem. Soc.*, 1963, **85**, 26. 3-1Ba(3)

63JA193 J. W. Jones and R. K. Robins, *J. Amer. Chem. Soc.*, 1963, **85**, 193. 5-1Da

63JA646 D. F. Veber and W. Lwowski, *J. Amer. Chem. Soc.*, 1963, **85**, 646. 3-1Ad

63JA770 E. C. Taylor and E. Smakula Hand, *J. Amer. Chem. Soc.*, 1963, **85**, 770 2-8De

63JA1007 H. T. Miles, *J. Amer. Chem. Soc.*, 1963, **85**, 1007. 2-6Cb

63JA1657 O. Jardetzky, P. Pappas and N. G. Wade, *J. Amer. Chem. Soc.*, 1963, **85**, 1657. 2-6Cb

63JA2763 Y. Chiang and E. B. Whipple, *J. Amer. Chem. Soc.*, 1963, **85**, 2763. 3-1Ba

63JA2943 P. Yates and J. A. Weisbach, *J. Amer. Chem. Soc.*, 1963, **85**, 2943. 3-2Ac

63JA4024 T. Ueda and J. J. Fox, *J. Amer. Chem. Soc.*, 1963, **85**, 4024. 2-6Cb

63JI833 S. Nanda, D. Pati, A. S. Mitra and M. K. Rout, *J. Indian Chem. Soc.*, 1963, **40**, 883. 4-2Fb

63JO98 E. Galantay, A. Szabo and J. Fried, *J. Org. Chem.*, 1963, **28**, 98. 4-3Cc

63JO194 E. Lieber, E. Oftedahl and C. N. R. Rao, *J. Org. Chem.*, 1963, **28**, 194. 4-4Fd

63JO733 W. B. Biggerstaff and K. L. Stevens, *J. Org. Chem.*, 1963, **28**, 733. 3-2Ca(iv)

63JO1886 S. Garratt, *J. Org. Chem.*, 1963, **28**, 1886. 2-3Ga, 2-3Gb

63JO1989 A. L. Bluhm, J. Weinstein and J. A. Sousa, *J. Org. Chem.*, 1963, **28**, 1989. 2-8Cb

63JO2215 E. J. Moriconi, F. J. Creegan, C. K. Donovan and F. A. Spano, *J. Org. Chem.*, 1963, **28**, 2215. 2-3Fa

63JO2313 J. G. Topliss and L. M. Konzelman, *J. Org. Chem.*, 1963, **28**, 2313. 2-4A, 2-6F

63JO2394 L. Jurd and T. A. Geissman, *J. Org. Chem.*, 1963, **28**, 2394. 2-4H

63JO2428 M. B. Frankel, E. A. Burns, J. C. Butler and E. R. Wilson, *J. Org. Chem.*, 1963, **28**, 2428. 4-6Dd(2)

63JO2581 A. Richardson, Jr., *J. Org. Chem.*, 1963, **28**, 2581. 4-3Bd, 4-4Cf, 4-5Df

63JO2883 V. R. Williams and J. G. Traynham, *J. Org. Chem.*, 1963, **28**, 2883. 2-3C

63JO3041 R. K. Robins, J. K. Horner, C. V. Greco, C. W. Noell and C. G. Beames, Jr., *J. Org. Chem.*, 1963, **28**, 3041. 5-2Bb

63LA(662)83 E. Biekert and H. Kössel, *Liebig's Ann. Chem.*, 1963, **662**, 83. 2-8Dd

63MI1 B. Stanovnik and M. Tisler, *Vestn. Slov. Kem. Drus.*, 1963, **10**, 1. 4-4Cd

63MI271 I. Murata, *Sci. Rep. Res. Inst., Tohoku Univ., Ser. A*, 1963, **12**, 271. 4-3Bd

63MI301 J. N. Murrell, "The Theory of the Electronic Spectra of Organic Molecules," p. 301. Methuen, London, 1963. 1-6E

63MI895 M. Eigen and L. de Maeyer, "Technique of Organic Chemistry" (A. Weissberger, Ed.), Vol. VIIIb, p. 895, Wiley, New York, 1963. T1-1

63NA575 G. Will, *Nature (London)*, 1963, **198**, 575. 4-1Ca

63PH721 J. B. Lounsbury, *J. Phys. Chem.*, 1963, **67**, 721. 4-1Fa(ii)

63PH874 G. Wettermark and J. Sousa, *J. Phys. Chem.*, 1963, **67**, 874. 2-8Cb

63PM(1)1 A. Albert, "Physical Methods in Heterocyclic Chemistry" (A. R. Katritzky, Ed.), Vol. 1, pp. 1–108. Academic Press, New York, 1963. T1-1, 1-4Aa

63PM(1)161 W. Cochran, "Physical Methods in Heterocyclic Chemistry" (A. R. Katritzky, Ed.), Vol. 1, pp. 161–175. Academic Press, New York, 1963. T1-1

63PM(1)189 S. Walker, "Physical Methods in Heterocyclic Chemistry" (A. R. Katritzky, Ed.), Vol. 1, pp. 189–215. Academic Press, New York, 1963. T1-1

63PM(1)217 J. Volke, "Physical Methods in Heterocyclic Chemistry" (A. R. Katritzky, Ed.), Vol. 1, pp. 217–323. Academic Press, New York, 1963. T1-1

63PM(2)1 S. F. Mason, "Physical Methods in Heterocyclic Chemistry" (A. R. Katritzky, Ed.), Vol. 2, pp. 1–88. Academic Press, New York, 1963. T1-1

63PM(2)89 E. A. C. Lucken, "Physical Methods in Heterocyclic Chemistry" (A. R. Katritzky, Ed.), Vol. 2, pp. 89–102. Academic Press, New York, 1963. T1-1

63PM(2)103 R. F. M. White, "Physical Methods in Heterocyclic Chemistry" (A. R. Katritzky, Ed.), Vol. 2, pp. 103–159. Academic Press, New York, 1963. T1-1

63PM(2)161 A. R. Katritzky and A. P. Ambler, "Physical Methods in Heterocyclic Chemistry" (A. R. Katritzky, Ed.), Vol. 2, pp. 161–360. Academic Press, New York, 1963. 3-2Ed, 1-6Cc(2), T1-1, 5-2Aa, 4-5Bb, 4-9A, 4-5Ba, 4-5Bb

63R1026 J. V. Thuijl, C. Romers and E. Havinga, *Rec. Trav. Chim. Pays-Bas*, 1963, **82**, 1026. 2-8Cc

63RU535 A. M. Khaletskii and B. L. Moldaver, *Russ. Chem. Rev.*, 1963, **32**, 535. 4-6Bb

63SC(142)1569 H. T. Miles, R. B. Bradley and E. D. Becker, *Science*, 1963, **142**, 1569. 2-6Cb

63T401 R. F. Branch, A. H. Beckett and D. B. Cowell, *Tetrahedron*, 1963, **19**, 401. T2-7

63T413 A. H. Beckett, R. F. Branch and D. B. Cowell, *Tetrahedron*, 1963, **19**, 413. T2-7

63T1011 S. C. Pakrashi, J. Bhattacharyya, L. F. Johnson, and H. Budzikiewicz, *Tetrahedron*, 1963, **19**, 1011. 2-4Da,2-4Db

63T1497 R. Jones, A. J. Ryan, S. Sternhell and S. E. Wright, *Tetrahedron*, 1963, **19**, 1497. 4-2Fd, 4-2Gj(2), 4-2Ib, T4-18(2), T4-21(2)

63T1867 H. J. Jakobsen, E. H. Larsen and S.-O. Lawesson, *Tetrahedron*, 1963, **19**, 1867. 3-2Cai(2), 3-2Caiv(3), 3-2Da, 3-2Dc, 3-3B, 3-2

63TL785 F. Minisci, R. Galli and A. Quilico, *Tetrahedron Lett.*, 1963, 785. 4-7B

63TL863 P. Beak, *Tetrahedron Lett.*, 1963, 863. 2-3Ga

63TL1027 T. L. V. Ulbricht, *Tetrahedron Lett.*, 1963, 1027. 2-6Cb

63ZO1092 V. F. Martynov and I. B. Belov, *Zh. Obshch. Khim.*, 1963, **33**, 1092. 4-2Ib, T4-21

63ZO2597 V. G. Vinokurov, V. S. Troitskaya, I. I. Grandberg and Yu. A. Pentin, *Zh. Obshch. Khim.*, 1963, **33**, 2597. 1-6Cd, 4-2Fa, 4-2Fc(2), 4-2Fd, 4-2Gc, 4-2Gd, 4-2Ge(iii), 4-2Gg, 4-2Hb(ii)(2), 4-2Hc, 4-3Ad, 4-6Fd(ii)

63ZO2673 J. A. Levin, V. A. Gulina and V. A. Kukhtin, *Zh. Obshch. Khim.*, 1963, **33**, 2673. 5-3Bc

63ZX(119)1 G. Will, *Z. Kristallogr.* 1963, **119**, 1. 4-1Ca

64AG1 H. Zimmermann, *Angew. Chem.*, 1964, **76**, 1. 4-1Aa

64AH317 P. Sohár, *Acta Chim.* (*Budapest*), 1964, **40**, 317. 2-4A, 2-4Ce

64AJ455 R. A. Jones and A. R. Katritzky, *Aust. J. Chem.*, 1964, **17**, 455. 2-8Dd, T2-8

64AJ567 D. J. Brown and T. Teitei, *Aust. J. Chem.*, 1964, **17**, 567. 2-4De

64AJ894 R. A. Jones, *Aust. J. Chem.*, 1964, **17**, 894. 3-1Ab

64AJ1438 L. K. Dalton, *Aust. J. Chem.*, 1964, **17**, 1438. 3-2Ac

64AK(22)65 S. Gronowitz, B. Normman, B. Gestblom, B. Mathiasson, and R. A. Hoffmann, *Ark. Kemi*, 1964, **22**, 65. 2-4Da-(2), 2-6Ca

64AK(21)239 A-B. Hörnfeldt and S. Gronowitz, *Ark. Kemi*, 1964, **21**, 239. 3-2Ca(iii), 3-2Da(2)

64AK(22)211 A. B. Hörnfeldt, *Ark. Kemi*, 1964, **22**, 211. 3-2Ca(ii)(5)

64AN80 A. Fravolini, G. Grandolini and G. Monzali, *Ann. Chim.* (*Rome*), 1964, **54**, 80. 4-3Bc, 4-4Cd, 4-5Dd

64AP10 K. W. Merz and H. J. Janssen, *Arch. Pharm.* (*Weinheim*), 1964, **297**, 10. 2-8Df

64AS174 E. Åkerblom and K. Skagius, *Acta Chem. Scand.*, 1964, **18**, 174. 4-5Fb(ii), 4-6Fh(i)

64AS566 K. A. Jensen, A. Holm and C. T. Pedersen, *Acta Chem. Scand.*, 1964, **18**, 566. 4-5Fe

64AS871 J. Sandström, *Acta Chem. Scand.*, 1964, **18**, 871. 4-4Fd, 4-5Fb(ii)

64BB491 M. Remson and R. Collienne, *Bull. Soc. Chim. Belg.*, 1964, **73**, 491. 3-4Bc(2)

64BF123 A. Foucauld, *Bull. Soc. Chim. Fr.*, 1964, 123. 3-4Ca

64BF500 J. Bourdais, F. Cugniet, J.-C. Prin and P. Chabrier, *Bull. Soc. Chim. Fr.*, 1964, 500. 4-6Bg(i), 4-6Bg(iii), 4-6Bh

64BF2019 N.P. Buu-Hoï, J.-P. Hoeffinger and P. Jacquignon, *Bull. Soc. Chim. Fr.*, 1964, 2019. 4-1G(iii)

64BJ1107 M. Fujisaka, Y. Ueno, H. Shinohara and E. Imoto, *Bull. Chem. Soc. Jap.*, 1964, **37**, 1107. 2-5E

64BJ1526 S. Seto and K. Ogura, *Bull. Chem. Soc. Jap.*, 1964, **37**, 1526. 4-4Cd

64BJ1740 Y. Iwanami, Y. Kenjo, K. Nishibe, M. Kajiura and S. Isoyama, *Bull. Chem. Soc. Jap.*, 1964, **37**, 1740. 2-8Dc

64BJ1745 Y. Iwanami, S. Isoyama and Y. Kenjo, *Bull. Soc. Chim. Jap.*, 1964, **37**, 1745. 2-8Dd

64CB667 H. Plieninger, W. Müller and K. Weinerth, *Chem. Ber.*, 1964, **97**, 667. 2-3Bf

64CB994 J. Gante and W. Lautsch, *Chem. Ber.*, 1964, **97**, 994. 2-4Fb, 2-5F

64CB2023 F. Weygand, W. Steglich, D. Mayer and W. von Philipsborn, *Chem. Ber.*, 1964, **97**, 2023. 4-3Cc, 4-3Ci

64CB3106 J. Goerdeler and U. Keuser, *Chem. Ber.*, 1964, **97**, 3106. 4-6Da, 4-6Fc

64CC970 O. Ohashi, M. Mashima and M. Kubo, *Can. J. Chem.*, 1964, **42**, 970. 2-4Cd

64CC2292 D. Cook, *Can. J. Chem.*, 1964, **42**, 2292. 4-1Ka

64CI1264 L. A. Summers and D. J. Shields, *Chem. Ind.* (*London*), 1964, 1264. 4-2Bg

64CI1837 F. J. Allan and G. G. Allan, *Chem. Ind.* (*London*), 1964, 1837. 4-7Ab

64CO(258)4579 H. Najer, J. Menin and J.-F. Giudicelli, *C.R. Acad. Sci.*, 1964, **258**, 4579. 1-6E, 4-5Fa(ii)

64CO(259)2868 H. Najer, J. Menin and J.-F. Giudicelli, *C.R. Acad. Sci.*, 1964, **259**, 2868. 4-5Fa(ii)

64CO(259)3385 C. Helene, A. Haug, M. Delbrück and P. Douzou, *C.R. Acad. Sci.*, 1964, **259**, 3385. 2-6Cb

64CO(259)3563 J. Menin, J.-F. Giudicelli and H. Najer, *C.R. Acad. Sci.*, 1964, **259**, 3563. 4-5Fb(ii)

64CO(259)4387 C. Helene and P. Douzou, *C.R. Acad. Sci.*, 1964, **259**, 4387. 2-6Cc

64CO(259)4853 C. Helene and P. Douzou, *C.R. Acad. Sci.*, 1964, **259**, 4853. 2-6Cb, 2-6Cc

64CR360 J. S. Rao, *Chem. Rev.*, 1964, **64**, 360. 3-2Aa

64CT100 T. Ueda and M. Furukawa, *Chem. Pharm. Bull.*, 1964, **12**, 100. 2-6Ec

64CT1329 T. Sasaki and K. Minamoto, *Chem. Pharm. Bull.*, 1964, **12**, 1329. 2-6Ec

64CX122 H. M. Sobell and K. Tomita, *Acta Crystallogr.*, 1964, **17**, 122. 2-4Dc

64CX126 H. M. Sobell and K. Tomita, *Acta Crystallogr.*, 1964, **17**, 126. 5-1Eg

64CX282 B. M. Craven, *Acta Crystallogr.*, 1964, **17**, 282. 2-4Dg

64CX891 B. M. Craven, S. Martinez-Carrera and G. A. Jeffrey, *Acta Crystallogr.*, 1964, **17**, 891. 2-4Dg

64CX1581 D. L. Barker and R. E. Marsh, *Acta Crystallogr.*, 1964, **17**, 1581. 1-4Ib, 2-6Cb

64CZ1394 J. Gut, J. Jonas and J. Pitha, *Collect. Czech. Chem. Commun.*, 1964, **29**, 1394. 2-6Ec(2)

64CZ1663 R. Lukeš, J. Němec and Jarý, *Collect. Czech. Chem. Commun.*, 1964, **29**, 1663. 3-2Aa

64CZ2060 A. Piskala and F. Sorm, *Collect. Czech. Chem. Commun.*, 1964, **29**, 2060. 2-4Fa

64DA(157)367 Yu. E. Skylar, R. P. Evstigneeva, O. D. Saralidze and N. A. Preobrazhenskii, *Dokl. Akad. Nauk SSSR*, 1964, **157**, 367. 3-1Ba

64H1188 P. Gold-Aubert, D. Melkonian and L. Toribio, *Helv. Chim. Acta*, 1964, **47**, 1188. 4-3Df, 4-3Dj

64H1748 A. Courtin, E. Class and H. Erlenmeyer, *Helv. Chim. Acta*, 1964, **47**, 1748. 3-2Cd

64H1986 J. Schmutz, F. Hunziker and W. Michaelis, *Helv. Chim. Acta*, 1964, **47**, 1986. 4-2Jb(2), 4-2Jc(3), 4-2Jd(2), S4-13

64HC(3)209 H. H. Jaffé and H. L. Jones, *Advan. Heterocycl. Chem.*, 1964, **3**, 209. 1-4C

64HC(3)263 K. A. Jensen and C. Pedersen, *Advan. Heterocycl. Chem.*, 1964, **3**, 263. 4-4Fd(2), 4-4Fe(2)

64J400 B. C. Pal and C. A. Horton, *J. Chem. Soc.*, 1964, 400. 5-1Da

64J783 G. C. Barrett, V. V. Kane and G. Lowe, *J. Chem. Soc.*, 1964, 783. 3-2Ba

64J792 S. Lewin, *J. Chem. Soc.*, 1964, 792. 5-1Db

64J915 D. J. Rabiger and M. M. Joullié, *J. Chem. Soc.*, 1964, 915. 4-1Hb(i)

64J1423 J. M. Cox, J. A. Elvidge and D. E. H. Jones, *J. Chem. Soc.*, 1964, 1423. 2-6Ba, 2-7Ed

64J1523 A. R. Katritzky and J. A. Waring, *J. Chem. Soc.*, 1964, 1523. 2-4Cd(2)

64J3005 M. J. Perkins, *J. Chem. Soc.*, 1964, 3005. 2-2F

64J3204 D. J. Brown and T. Teitei, *J. Chem. Soc.*, 1964, 3204. 2-5E

64J3478 A. M. Comrie, *J. Chem. Soc.*, 1964, 3478. 4-6Ff(ii)(2)

64J4157 J. A. Elvidge, G. T. Newbold, I. R. Senciall and T. G. Symes, *J. Chem. Soc.*, 1964, 4157. 2-6Ea

64J4226 W. L. F. Armarego, *J. Chem. Soc.*, 1964, 4226. 3-1Bc, 5-3A, S5-3(3)

64J4769 A. Stuart, D. W. West and H. C. S. Wood, *J. Chem. Soc.*, 1964, 4769. 2-6De

64J4868 R. E. Ballard and J. W. Edwards, *J. Chem. Soc.*, 1964, 4868. 1-6E, 2-3Ha

64J5200 J. D. Edwards, J. E. Page and M. Pianka, *J. Chem. Soc.*, 1964, 5200. 2-3Ga

64J(S1)5634 R. D. Chambers, J. Hutchinson and W. K. R. Musgrave, *J. Chem. Soc.*, 1964, Suppl. 1, 5634. 2-3Bc

64J5884 M. Charton, *J. Chem. Soc.*, 1964, 5884. 2-2Gb

64JA1456 J. A. Moore and C. L. Habraken, *J. Amer. Chem. Soc.*, 1964, **86**, 1456. 4-1Bb(v)

64JA2474 I. Wempen and J. J. Fox, *J. Amer. Chem. Soc.*, 1964, **86**, 2474. 2-4Db, 2-4Dd, 2-6Cb

64JA2744 E. Schweizer, *J. Amer. Chem. Soc.*, 1964, **86**, 2744. 2-2A

64JA2861 J. M. Ross and W. C. Smith, *J. Amer. Chem. Soc.*, 1964, **86**, 2861. 4-3Eb

64JA3797 R. L. Hinman and J. Lang, *J. Amer. Chem. Soc.*, 1964, **86**, 3797. 3-1Bb(2)

64JA4152 D. F. Veber and W. Lwowski, *J. Amer. Chem. Soc.*, 1964, **86**, 4152. 3-1Ad(2)

64JH13 A. I. Meyers, B. J. Betrus, N. K. Ralhan and K. B. Rao, *J. Heterocycl. Chem.*, 1964, **1**, 13. 2-2C

64JH115 J. A. Montgomery and H. J. Thomas, *J. Heterocycl. Chem.*, 1964, **1**, 115. 5-1Da

64JH221 R. R. Shoup and R. N. Castle, *J. Heterocycl. Chem.*, 1964, **1**, 221. 2-4Cb, 2-5E

64JO219 H. C. Scarborough, *J. Org. Chem.*, 1964, **29**, 219. 2-3Ea

64JO245 B. Loev, K. M. Snader and M. F. Kormendy, *J. Org. Chem.*, 1964, **29**, 245. 2-6G

64JO370 C. F. Howell, W. Fulmer, N. Q. Quinones and R. A. Hardy, Jr., *J. Org. Chem.*, 1964, **29**, 370. 4-6Fe(ii)(2), 4-6Fe(iii)(3), 4-6Fe(iv)(3)

64JO607 G. W. Stacy, A. J. Papa, F. W. Villaescusa and S. C. Ray, *J. Org. Chem.*, 1964, **29**, 607. 3-4Bc

64JO650 W. P. Norris and R. A. Henry, *J. Org. Chem.*, 1964, **29**, 650. 4-5Ff

64JO660 W. R. Hatchard, *J. Org. Chem.*, 1965, **29**, 660. 4-5Bd(2)

64JO776 B. E. Fisher and J. E. Hodge, *J. Org. Chem.*, 1964, **29**, 776. 3-2Ac

64JO862 G. G. Gallo, C. R. Pasqualucci, P. Radaelli and G. C. Lancini, *J. Org. Chem.*, 1964, **29**, 862. 4-1Cb(i), 4-1Cb(iv), T4-5(4)

64JO978 P. J. Stoffel and W. D. Dixon, *J. Org. Chem.*, 1964, **29**, 978. 4-3Bb

64JO1115 G. de Stevens, B. Smolinsky and L. Dorfman, *J. Org. Chem.*, 1964, **29**, 1115. 2-5E

64JO1174 A. W. Lutz, *J. Org. Chem.*, 1964, **29**, 1174. 4-6Bh, 4-6Ce, 4-6Gb(i)

64JO1449 R. L. Hinman and J. Lang, *J. Org. Chem.*, 1964, **29**, 1449. 3-1Bb

64JO1620 J. B. Wright, *J. Org. Chem.*, 1964, **29**, 1620. 4-7B

64JO1988 F. J. Bullock and O. Jardetzky, *J. Org. Chem.*, 1964, **29**, 1988. 5-1Ab

64JO2121 E. C. Taylor and E. E. Garcia, *J. Org. Chem.*, 1964, **29**, 2121 2-2E

64JO2205 R. Filler and E. J. Piasek, *J. Org. Chem.*, 1964, **29**, 2205. 4-3Cc

64JO2623 B. Klein, E. O'Donnel and J. M. Gordon, *J. Org. Chem.*, 1964, **29**, 2623. 2-4E

64JO2725 E. T. Holmes and H. R. Snyder, *J. Org. Chem.*, 1964, **29**, 2725. 3-2Cf

64JT(40)2071 R. F. Stewart and L. H. Jensen, *J. Chem. Phys.*, 1964, **40**, 2071. 5-1Da

64JT(41)2568 H. Hiraoka and R. Hardwick, *J. Chem. Phys.*, 1964, **41**, 2568. 2-8Cb

64LA(675)180 H. Gehlen and W. Schade, *Liebig's Ann. Chem.*, 1964, **675**, 180. 4-3Df, 4-3Di

64M147 E. Ziegler and E. Steiner, *Monatsh. Chem.*, 1964, **95**, 147. 2-4A

64M950 K. Lempert and G. Doleschall, *Monatsh. Chem.*, 1964, **95**, 950. 2-6F

64M1201 H. Junek, *Monatsh. Chem.*, 1964, **95**, 1201. 2-8Dg

64M1247 E. Ziegler and F. Hradetsky, *Monatsh. Chem.*, 1964, **95**, 1247. 2-3Ea

64M1473 H. Junek, *Monatsh. Chem.*, 1964, **95**, 1473. 2-8Dg

64MB(8)89 F. S. Mathews and A. Rich, *J. Mol. Biol.*, 1964, **8**, 89. 5-1Da

64MC310 W. E. Kreighbaum and H. C. Scarborough, *J. Med. Pharm. Chem.*, 1964, **7**, 310. 6-2Bc(2)

64MC814 C. Runti and C. Nisi, *J. Med. Pharm. Chem.*, 1964, **7**, 814. 4-7Af

64MI81 B. Stanovnik and M. Tišler, *Croat. Chem. Acta*, 1964, **36**, 81. 2-4Cd, 2-5E

64MI135 A. Pullman, "Electronic Aspects of Biochemistry" (B. Pullman, Ed.), p. 135. Academic Press, New York, 1964. 2-4Db, 2-6Cb

64MI174 P. O. Löwdin, "Electronic Aspects of Biochemistry" (B. Pullman, Ed.), p. 174. Academic Press, New York, 1964. 5-1Dd

64MI203 D. R. Harris and W. M. Macintyre, *Biophys. J.*, 1964, **4**, 203. 2-4Db

64NA(201)179 F. S. Mathews and A. Rich, *Nature* (*London*), 1964, **201**, 179. 2-6Cb

64NA(202)1206 G. C. Verschoor, *Nature* (*London*), 1964, **202**, 1206. 2-4Fa

64P368 J. Harley-Mason and T. J. Leeney, *Proc. Chem. Soc.*, 1964, 368. 3-1Ac

64PC329 H. Arold, *J. Prakt. Chem.*, 1964, **23**, 329. 4-2Jb, 4-2Jc

64PH3435 F. J. Millero, J. C. Ahluwalia and L. G. Hepler, *J. Phys. Chem.*, 1964, **68**, 3435. 2-9B

64QR295 J. Clark and D. D. Perrin, *Quart. Rev. Chem. Soc.*, 1964, **18**, 295. 2-3Bc

64R186 P. J. Van der Haak and T. J. de Boer, *Rec. Trav. Chim. Pays-Bas*, 1964, **83**, 186. 2-3C

64SA211 M. Cignitti and L. Paoloni, *Spectrochim. Acta, Sect. A*, 1964, **20**, 211. 2-4Fa

64SA1665 J. F. Corbett, *Spectrochim. Acta, Sect. A*, 1964, **20**, 1665. 2-3Hb

64T165 L. Bauer, C. N. V. Nambury and C. L. Bell, *Tetrahedron*, 1964, **20**, 165. 4-2Cb(iv), 4-6Fa(i), 4-6Fa(iii)(2), 4-6-Fb(i), 4-6Fb(ii), T4-30(2)

64T299 A. R. Katritzky and F. W. Maine, *Tetrahedron*, 1964, **20**, 299. 4-2Fa, 4-2Fb(3), 4-2Fc(4), 4-2Fd, 4-2Ga(i)(2), 4-2Ga(ii), 4-2Gc, 4-2Gf(2), 4-2Gg, 4-2Ia, T4-14, T4-16, T4-17(3), S4-4, S4-8, S4-10

64T315 A. R. Katritzky and F. W. Maine, *Tetrahedron*, 1964, **20**, 315. 4-2Hb(ii)(2), 4-2Hb(iii), 4-2Hb(iv)(2), 4-2Hc, T4-20(2)

64T531 A. Mustafa, W. Asker, A. H. Harhash, K. M. Foda, H. H. Jahine and N. A. Kassab, *Tetrahedron*, 1964, **20**, 531. 1-1C, 4-2Hc(i)

64T2835 A. J. Boulton, A. R. Katritzky, A. Majid-Hamid and S. Øksne, *Tetrahedron*, 1964, **20**, 2835. 4-2Cb(i), 4-2Cb-(iii), 4-2Cb(2)(iv), 4-2Cb(v), T4-13

64TL1477 W. D. Crow and N. J. Leonard, *Tetrahedron Lett.*, 1964, 1477. 4-2E

64TL2679 K. Lempert, J. Nyitrai, P. Sohár and K. Zauer, *Tetrahedron Lett.*, 1964, 2679. 4-6Ec

64ZN952 G. Klose, H. Müller and E. Uhlemann, *Z. Naturforsch. B*, 1964, **19**, 952. 2-8Db(ii)

64ZN962 E. Uhlemann, G. Klose and H. Müller, *Z. Naturforsch. B*, 1964, **19**, 962. 2-8Db(ii)

64ZO197 A. V. El'tsov, V. S. Kuznetsov and L. S. Efros, *Zh. Obshch. Khim.*, 1964, **34**, 197. 4-3Bd

64ZO654 V. G. Vinikurov, V. S. Troitskaya and I. I. Grandberg, *Zh. Obshch. Khim.*, 1964, **34**, 654. 4-5Bg, 4-5Bh, 4-5Bj, 4-5Cc

64ZO3005 I. Y. Kvitko and B. A. Porai-Koshits, *Zh. Obshch. Khim.*, 1964, **34**, 3005. 4-2Gh, 4-4Ae

64ZO3134 G. M. Kheifets and N. V. Khromov-Borisov, *Zh. Obshch. Khim.*, 1964, **34**, 3134. 2-4De

65AJ379 G. M. Badger and R. P. Rao, *Aust. J. Chem.*, 1965, **18**, 379. 5-2Ba(2), T5-1

65AJ559 D. J. Brown and T. Teitei, *Aust. J. Chem.*, 1965, **18**, 559. 2-6Ce

65AJ1267 G. M. Badger and R. P. Rao, *Aust. J. Chem.*, 1965, **18**, 1267. 5-2Ba

65AJ1977 P. S. Clezy and A. W. Nichol, *Aust. J. Chem.*, 1965, **18**, 1977. 5-C

65AK(23)483 B. Gestblom, S. Gronowitz, R. A. Hoffman, B. Mathiasson and S. Rodmar, *Ark. Kemi*, 1965, **23**, 483. 3-3B

65AK(23)501 B. Gestblom, S. Gronowitz, R. A. Hoffman, B. Mathiasson and S. Rodmar, *Ark. Kemi*, 1965, **23**, 501. 3-3B

65AN116 L. Baiocchi, G. Corsi and G. Palazzo, *Ann. Chim. (Rome)*, 1965, **55**, 116. 4-3Bd, 5-2Ba

65AN615 T. Bacchetti, A. Alemagna and B. Danieli, *Ann. Chim. (Rome)*, 1965, **55**, 615. 4-4Eh

65AP4 J. Schnekenburger, *Arch. Pharm. (Weinheim)*, 1965, **268**, 4. 2-3Gb

65AP411 J. Schnekenburger, *Arch. Pharm. (Weinheim)*, 1965, **298**, 411. 2-3Gb

65AP885 H. J. Roth, K. Jäger and R. Brandes, *Arch. Pharm. (Weinheim)*, 1965, **298**, 885. 2-4Dg

65AS1191 E. Åkerblom and M. Sandberg, *Acta Chem. Scand.*, 1965, **19**, 1191. 4-5Fc(2)

65AS1215 S. Gronowitz, B. Mathiasson, R. Dahlbom, B. Holmberg and K. A. Jensen, *Acta Chem. Scand.*, 1965, **19**, 1215. 4-3Cf(4)

65AS1249 A. B. Hörnfeldt, *Acta Chem. Scand.*, 1965, **19**, 1249. 3-2Cd(2)

65AS2022 M. Begtrup and C. Pedersen, *Acta Chem. Scand.*, 1965, **19**, 2022. 4-3Ed, 4-3Ef(2)

65BF52 N. P. Buu-Hoï, M. Gauthier and N. D. Xuong, *Bull. Soc. Chim. Fr.*, 1965, 52. 2-6Ba

65BF2120 H. Najer, R. Giudicelli and J. Menin, *Bull. Soc. Chim. Fr.*, 1965, 2120. 2-6F

65BF2658 N. P. Buu-Hoï, V. Bellavita, A. Ricci and G. Grandolini, *Bull. Soc. Chim. Fr.*, 1965, 2658. 3-2Cf, 3-4Bb

65BF3527 M. Sélim, M. Sélim, O. Tétu, G. Drillien and P. Rumpf, *Bull. Soc. Chim. Fr.*, 1965, 3527. 4-5Dc

65BG155 E. Lippert, D. Samuel and E. Fischer, *Ber. Bunsenges, Phys. Chem.*, 1965, **69**, 155. 1-5E

65BG190 E. Daltrozzo, G. Hohlneicher and G. Scheibe, *Ber. Bunsenges. Phys. Chem.*, 1965, **69**, 190. 2-8Cc

65BG458 H. Prigge and E. Lippert, *Ber. Bunsenges. Phys. Chem.*, 1965, **69**, 458. 1-6E, 2-4Gc(2)

65BG550 H. Deuschl, *Ber. Bunsenges. Phys. Chem.*, 1965, **6**, 550. 4-1Ea

65BS131 I. Degani, L. Lunazzi and F. Taddei, *Boll. Sci. Fac. Chim. Ind. Bologna*, 1965, **23**, 131. 2-2A, 2-2B

65BS255 G. Adembri, E. Belgodere, G. Speroni and P. Tedeschi, *Boll. Sci. Fac. Chim. Ind. Bologna*, 1965, **23**, 255. 4-5Ba

65BT191 W. Baker, *Chem. Brit.*, 1965, **1**, 191. 6

65BT250 W. Baker, *Chem. Brit.*, 1965, **1**, 250. 6

65C325 E. Daltrozzo, G. Scheibe and J. Smits, *Chimia*, 1965, **19**, 325. 2-8Cc

65CB1060 G. Nübel and W. Pfleiderer, *Chem. Ber.*, 1965, **98**, 1060. 5-2Bc

65CB1562 H. Musso and H. Schröder, *Chem. Ber.*, 1965, **98**, 1562. 4-5Bb

65CB1623 M. Eigen, G. Ilgenfritz and W. Kruse, *Chem. Ber.*, 1965, **98**, 1623. 1-4F(2), 2-4Dg

65CC749 D. Cook, *Can. J. Chem.*, 1965, **43**, 749. 2-3C

65CC3322 D. Cook, *Can. J. Chem.*, 1965, **43**, 3322. 4-2Ga(i)

65CE90 R. Breslow, *Chem. Eng. News*, 1965, **43**, 90. 6

65CI184 S. Masamune and N. T. Castellucci, *Chem. Ind. (London)*, 1965, 184. 6-2Ba

65CI331 A. R. Katritzky, *Chem. Ind. (London)*, 1965, 331. 1-1F, 4-4Fd, 4-5Bg

65CI1766 M. H. Palmer and B. Semple, *Chem. Ind. (London)*, 1965, 1766. 2-2Gb

65CO(260)4343 H. Najer, J. Armand, J. Menin and N. Voronine, *C.R. Acad. Sci.*, 1965, **260**, 4343. 4-5Da, 4-5Dc

65CO(260)4538 J.-F. Giudicelli, J. Menin and H. Najer, *C.R. Acad. Sci.*, 1965, **260**, 4538. 4-5Fa(ii)

65CO(261)766 J. Menin, J.-F. Giudicelli and H. Najer, *C.R. Acad. Sci.*, 1965, **261**, 766. 4-5Fb(ii)

65CP1334 A.-M. Bellocq, C. Perchard, A. Novak and M.-L. Josien, *J. Chim. Phys.*, 1965, **62**, 1334. 4-1Ca

65CP1344 C. Perchard, A.-M. Bellocq and A. Novak, *J. Chim. Phys.*, 1965, **62**, 1344. 4-1Ca

65CT473 H. Nakao, N. Soma, Y. Sato and G. Sunagawa, *Chem. Pharm. Bull.*, 1965, **13**, 473. 4-3Bd

65CX(18)122 T. Ashida, S. Hirokawa and Y. Okaya, *Acta Crystallogr.*, 1965, **18**, 122. 2-9A

65CX(18)203 N. Camerman and J. Trotter, *Acta Crystallogr.*, 1965, **18**, 203. 2-4Db

65CX(18)313 S. Furberg, C. S. Peterson and C. Rømming, *Acta Crystallogr.*, 1965, **18**, 313. 2-6Cb

65CX(19)111 D. G. Watson, D. J. Sutor and P. Tollin, *Acta Crystallogr.*, 1965, **19**, 111. 5-1Da

65CX(19)573 D. G. Watson, R. M. Sweet and R. E. Marsh, *Acta Crystallogr.*, 1965, **19**, 573. 1-4Ib, 5-1Aa(ii)(2)

65CX(19)698 M. Ehrenberg, *Acta Crystallogr.*, 1965, **19**, 698. 3-2Bc

65CX(19)797 B. D. Sharma and J. F. McConnell, *Acta Crystallogr.*, 1965, **19**, 797. 1-4Ib, 2-6Cc

65CX(19)861 C. Singh, *Acta Crystallogr.*, 1965, **19**, 861. 5-1Eg

65CX(19)1051 W. Bolton, *Acta Crystallogr.*, 1965, **19**, 1051. 2-4Dg

65CZ90 P. Pithova, A. Piskala, J. Pitha and F. Sorm, *Collect. Czech. Chem. Commun.*, 1965, **30**, 90. 2-4Fa

65CZ1626 P. Pithova, A. Piskala, J. Pitha and F. Sorm, *Collect. Czech. Chem. Commun.*, 1965, **30**, 1626. 2-6Ec

65DA(164)584 E. S. Levin and G. N. Rodionova, *Dokl. Akad. Nauk SSSR*, 1965, **164**, 584. 2-3Bg

65G320 G. Cardillo, L. Merlini and R. Mondelli, *Gazz. Chim. Ital.*, 1965, **95**, 320. 4-6Bc

65G583 G. Cum, G. Lo Vecchio and M.-C. Aversa, *Gazz. Chim. Ital.*, 1965, **95**, 583. 4-2Bg(2)

65G1371 R. Mondelli and L. Merlini, *Gazz. Chim. Ital.*, 1965, **95**, 1371. 4-6Bb(2), 4-6Bc(2)

65H617 A. Courtin and H. Sigel, *Helv. Chim. Acta*, 1965, **48**, 617. 3-2Cd

65H1322 A. Hofmann, W. von Philipsborn and C. H. Eugster, *Helv. Chim. Acta*, 1965, **48**, 1322. 3-2Ac(3)

65HC(4)1 A. Albert and W. L. F. Armarego, *Advan. Heterocycl. Chem.*, 1965, **4**, 1. 2-2Ga, 2-4Gb

65HC(4)43 D. D. Perrin, *Advan. Heterocycl. Chem.*, 1965, **4**, 43. 2-2Ga, 2-4Gb

65HC(4)107 R. Slack and K. R. H. Wooldridge, *Advan. Heterocycl. Chem.*, 1965, **4**, 107. 4-5Bd, 4-5Bc, 4-5Cb

65HC(5)1 R. Zahradnik, *Advan. Heterocycl. Chem.*, 1965, **5**, 1. 1-6A

65HC(5)20 R. Zahradnik, *Advan. Heterocycl. Chem.*, 1965, **5**, 20. 6-1Ab

65HC(5)119 F. Kurzer, *Advan. Heterocycl. Chem.*, 1965, **5**, 119. 4-3Dc, 4-3Fa, 4-4Ee, 4-5Fb(i), 4-6Cd

65J27 A. Albert and J. Clark, *J. Chem. Soc.*, 1965, 27. 2-5G

65J65 B. B. Molloy, D. H. Reid and F. S. Skelton, *J. Chem. Soc.*, 1965, 65. 5-5

65J485 W. Paterson and G. R. Proctor, *J. Chem. Soc.*, 1965, 485. 6-2Bd

65J575 R. E. Banks, J. E. Burgess, W. M. Cheng and R. N. Haszeldine, *J. Chem. Soc.*, 1965, 575. 1-5E, 2-3Bc

65J755 D. J. Brown and T. Teitei, *J. Chem. Soc.*, 1965, 755. 2-6Ca(2)

65J1706 D. A. Patterson and D. G. Wibberley, *J. Chem. Soc.*, 1965, 1706. 3-2Ef

65J2258 C. H. Williams, *J. Chem. Soc.*, 1965, 2258. 4-9D

65J2260 G. B. Barlin, *J. Chem. Soc.*, 1965, 2260. 2-4Cb, 2-5E

65J2543 K. Anderton and R. W. Rickards, *J. Chem. Soc.*, 1965, 2543. 3-2Ac

65J2778 W. L. F. Armarego, *J. Chem. Soc.*, 1965, 2778. 5-3A, S5-3(8)

65J2948 J. Hurst, T. Melton and D. G. Wibberley, *J. Chem. Soc.*, 1965, 2948. 3-2Ef, 3-4Ce

65J3090 S. Golding, A. R. Katritzky and H. Z. Kucharska, *J. Chem. Soc.*, 1965, 3090. 2-8Ea, T2-9

65J3093 A. R. Katritzky, H. Z. Kucharska and J. D. Rowe, *J. Chem. Soc.*, 1965, 3093. 2-8Db(ii), T2-7

65J3312 L. A. Summers, P. F. H. Freeman and D. J. Shields, *J. Chem. Soc.*, 1965, 3312. 4-2Bg, 4-4Ab

65J3785 D. Lloyd, R. H. McDougall and D. R. Marshall, *J. Chem. Soc.*, 1965, 3785. 6-2A(3)

65J3825 A. R. Katritzky and R. E. Reavill, *J. Chem. Soc.*, 1965, 3825. 2-6Bb, 2-5Ba

65J4368 B. B. Molloy, D. H. Reid and S. McKenzie, *J. Chem. Soc.*, 1965, 4368. 5-5Ac

65J4653 A. Albert, *J. Chem. Soc.*, 1965, 4653. 2-6Bb, 2-7Ea

65J5230 N. Bacon, A. J. Boulton, R. T. C. Brownlee, A. R. Katritzky and R. D. Topsom, *J. Chem. Soc.*, 1965, 5230. 2-6Be, 2-6Ec

65J5391 D. E. Ames, R. F. Chapman, H. Z. Kucharska and D. Waite, *J. Chem. Soc.*, 1965, 5391. 2-4Cb, 2-5E

65J7116 D. T. Hurst, J. F. W. McOmie and J. B. Searle, *J. Chem. Soc.*, 1965, 7116. 2-4Df

65JA11 L. B. Clark and I. Tinoco, Jr., *J. Amer. Chem. Soc.*, 1965, **87**, 11. 5-1Dc, 5-1Eg

65JA3440 J. M. Read, Jr. and J. H. Goldstein, *J. Amer. Chem. Soc.*, 1965, **87**, 3440. 5-1Ab

65JA4621 K. L. Wierzchowski, E. Litonska and D. Shugar, *J. Amer. Chem. Soc.*, 1965, **87**, 4621. 2-4Dd(2)

65JA5424 W. A. Henderson, Jr. and E. F. Ullman, *J. Amer. Chem. Soc.*, 1965, **87**, 5424. 2-3Gb

65JA5439 B. W. Roberts, J. B. Lambert and J. D. Roberts, *J. Amer. Chem. Soc.*, 1965, **87**, 5439. 1-5E(3), 2-4Db, 2-6Cd

65JA5575 E. D. Becker, H. Todd Miles and R. B. Bradley, *J. Amer. Chem. Soc.*, 1965, **87**, 5575. 1-5E(3), 2-6Cb(2), 2-6Cd

65JH26 F. Johnson and W. A. Nasutavicus, *J. Heterocycl. Chem.*, 1965, **2**, 26. 6-2Ba

65JH53 J. P. Paolini and R. K. Robins, *J. Heterocycl. Chem.*, 1965, **2**, 53. 5-3Ba, S5-4

65JH110 W. H. Nyberg, C. W. Noell and C. C. Cheng, *J. Heterocycl. Chem.*, 1965, **2**, 110. T6-1

65JH220 L. Bauer and D. Dhawan, *J. Heterocycl. Chem.*, 1965, **2**, 220. 5-1Fb

65JH302 R. G. Child and A. S. Tomcufcik, *J. Heterocycl. Chem.*, 1965, **2**, 302. S5-5

65JH447 L. Bauer, G. E. Wright, B. A. Mikrut and C. L. Bell, *J. Heterocycl. Chem.*, 1965, **2**, 447. 2-4Da

65JO243 A. L. Borror and A. F. Haeberer, *J. Org. Chem.*, 1965, **30**, 243. 2-8Cc, 2-8Dg(2)

65JO1110 W. C. Coburn, M. C. Thorpe, J. A. Montgomery and K. Hewson, *J. Org. Chem.*, 1965, **30**, 1110. 5-1Ab

65JO1255 E. E. Royals and J. C. Leffingwell, *J. Org. Chem.*, 1965, **30**, 1255. 2-3Ga

65JO1892 C. L. Habraken and J. A. Moore, *J. Org. Chem.*, 1965, **30**, 1892. 4-1Bb(v), 4-1Bb(ix), 1-5Ab(2)

65JO2395 C. Temple, Jr., W. C. Coburn, Jr., M. C. Thorpe and J. A. Montgomery, *J. Org. Chem.*, 1965, **30**, 2395. 5-3Bc

65JO2763 L. D. Huestis, M. L. Walsh and N. Hahn, *J. Org. Chem.*, 1965, **30**, 2763. 4-9A

65JO3033 P. C. Anderson and B. Staskun, *J. Org. Chem.*, 1965, **30**, 3033. 2-3Bd

65JO3341 M. Charton, *J. Org. Chem.*, 1965, **30**, 3341. 2-3D

65JO3346 M. Charton, *J. Org. Chem.*, 1965, **30**, 3346. 1-4B, 1-4Ba, 4-1Cb(ii)

65JO3377 S. Portnoy, *J. Org. Chem.*, 1965, **30**, 3377. 2-3Ec

65JO3472 J. H. Markgraf, W. T. Bachmann and D. P. Hollis, *J. Org. Chem.*, 1965, **30**, 3472. 4-1Fb(iv)

65JO3824 J. Bordner and H. Rappoport, *J. Org. Chem.*, 1965, **30**, 3824. 3-2Ec(2)

65JO4074 G. W. Stacy, F. W. Villaescusa and T. E. Wollner, *J. Org. Chem.*, 1965, **30**, 4074. 3-4Bb(2)

65KG107 S. A. Hiller, I. B. Mazheika and I. I. Grandberg, *Khim. Geterotsikl. Soedin.*, 1965, 107. 1-4G, 4-1Bb(iii)

65LA(682)201 H. Behringer and D. Weber, *Liebig's Ann. Chem.*, 1965, **682**, 201. 4-3Cf

65M2046 H. Junek, *Montash. Chem.*, 1965, **96**, 2046. 2-3Ec, 2-8Dg(2)

65MI17 B. Stanovnik and M. Tišler, *Croat. Chem. Acta*, 1965, **37**, 17. 4-6Cc

65MI117 L. Sawlewicz, *Acta Pol. Pharm.*, 1965, **22**, 117. 4-3Bd

65MI200 H. von Dobeneck and E. Brunner, *Hoppe-Seyler's Z. Physiol. Chem.*, 1965, **340**, 200. 3-2Ec

65MI237 G. Klose and E. Uhlemann, "Nuclear Magnetic Resonance in Chemistry" (B. Pesce, Ed.), p. 237, Academic Press, New York, 1965. 2-8Db(ii)(2)

65MI269 J. B. Lambert, B. W. Roberts, G. Binsch and J. D. Roberts, "Nuclear Magnetic Resonance in Chemistry" (B. Pesce, Ed.), p. 269. Academic Press, New York, 1965. 1-5E

65PC(30)163 E. Uhlemann and H. Müller, *J. Prakt. Chem.*, 1965, **30**, 163. 2-8Db(ii)

65PH3615 L. B. Clark, G. G. Perchel and I. Tinoco, *J. Phys. Chem.*, 1965, **69**, 3615. 5-1Eg

65RO2236 V. P. Shchipanov, S. L. Portnova, V. A. Krasnova, Y. N. Sheinker, and I. Y. Postovskii, *Zh. Org. Khim.*, 1965, **1**, 2236. 4-1Fa(iii), 4-5Ff(5)

65RR1059 A. T. Balaban and Z. Simon, *Rev. Roum. Chim.*, 1965, **10**, 1059. 6-1Ab

65SA1095 W. J. Barry, I. L. Finar and E. F. Mooney, *Spectrochim. Acta*, 1965, **21**, 1095. 6-2A(2)

65T1333 E. Jones and M. Moodie, *Tetrahedron*, 1965, **21**, 1333. 3-3B

65T1681 A. R. Katritzky, B. Wallis, R. T. C. Brownlee and R. D. Topsom, *Tetrahedron*, 1965, **21**, 1681. 4-3Db, 4-3De(3), 4-3Ea

65T1693 A. R. Katritzky, F. W. Maine and S. Golding, *Tetrahedron*, 1965, **21**, 1693. 4-1Bb(i), 4-1Bb(vi), 4-1Bb(vii), 4-1Bb-(ix), 4-2Ia(4), 4-2Ib(5), T4-2, T4-21(3)

65T3331 H. J. Jakobsen and S.-O. Lawesson, *Tetrahedron*, 1965, **21**, 3331. 3-2Cd(4), 3-2Cf, 3-2Da(2), 3-2Ad

65T3351 N. A. Evans, D. J. Whelan, and R. B. Johns, *Tetrahedron*, 1965, **21**, 3351. 4-2Ga(i), 4-2Gg, 4-2Ib(3), 4-2Jd(4), 4-4Af(2), T4-21(4), S4-13

65TE28 V. P. Lezina, V. F. Bystrov, L. D. Smirnov and K. M. Dyumaev, *Teor. Eksp. Khim.*, 1965, **1**, 281. 2-3D

65TH01 M. L. Roumestant, Ph.D. Thesis, University of Montpellier, France, 1965. 1-4Ab(2)

65TL1565 L. B. Volodarsky, A. N. Lisack and V. A. Koptyug, *Tetrahedron Lett.*, 1965, 1565. 4-7Ab

65TL1795 K. Lempert, J. Nyitrai and P. Sohár, *Tetrahedron Lett.*, 1965, 1795. 4-6Cb, 4-6Ec

65TL3175 R. Haller, *Tetrahedron Lett.*, 1965, 3175. 2-3Bd

65ZO1288 V. G. Vinokurov, V. S. Troitskaya and I. I. Grandberg, *Zh. Obshch. Khim.*, 1965, **35**, 1238. 4-6Bb, 4-6Fd(i)

66AC1702 P. D. Anderson and D. M. Hercules, *Anal. Chem.*, 1966, **38**, 1702. 2-5Bd

66AG389 F. Effenberger and R. Maier, *Angew. Chem.*, 1966, **78**, 390. 6-1Ab

66AH405 G. Doleschall, L. Lang and K. Lempert, *Acta Chim. (Budapest)*, 1966, **47**, 405. 5-3Bb(iv)

66AL457 S. Cabiddu and A. Ricca, *Atti Accad. Naz. Lincei, Cl. Sci. Fis., Mat. Natur., Rend.*, 1963, **35**, 530. 4-3Ab

66AP(299)43 G. Zinner and B. Böhlke, *Arch. Pharm. (Weinheim)*, 1966, **299**, 43. 4-6Bh

66AS57 J. Sandström and I. Wennerbeck, *Acta Chem. Scand.*, 1966, **20**, 57. 4-4Eg, 4-4Eh, 4-4Ei

66AS261 S. Gronowitz and A. Bugge, *Acta Chem. Scand.*, 1966, **20**, 261. 3-2Cd, 3-2Da

66AS1555 M. Begtrup and C. Pedersen, *Acta Chem. Scand.*, 1966, **20**, 1555. 4-3Ef

66AS1733 O. Meth-Cohn and S. Gronowitz, *Acta Chem. Scand.*, 1966, **20**, 1733. 3-6B

66BA199 B. Golankiewicz and K. Golankiewicz, *Bull. Acad. Pol. Sci., Ser. Sci. Chim.*, 1966, **14**, 199. 4-8B

66BB380 J. L. M. Loomans, *Bull. Soc. Chim. Belg.*, 1966, **75**, 380. 4-5Dc(3)

66BF342 O. Tétu, M. Sélim, M. Sélim and P. Rumpf, *Bull. Soc. Chim. Fr.*, 1966, 342. 4-5Dc(2)

66BF775 J. Elguero, G. Guiraud, R. Jacquier and G. Tarrago, *Bull. Soc. Chim. Fr.*, 1966, 775. 4-2Fa, T4-17

66BF2075 J. Elguero, A. Fruchier and R. Jacquier, *Bull. Soc. Chim. Fr.*, 1966, 2075. 4-1G(iii)

66BF2990 J. Elguero, R. Jacquier and G. Tarrago, *Bull. Soc. Chim. Fr.*, 1966, 2990. 4-2Gj, 4-2Ic, T4-18(4)

66BF3403 M. Sélim, O. Tétu, M. Sélim and P. Rumpf, *Bull. Soc. Chim. Fr.*, 1966, 3403. 4-5Dc

66BF3727 J. Elguero, R. Jacquier and H. C. N. Tien Duc, *Bull. Soc. Chim. Fr.*, 1966, 3727. 1-5Ab, S4-5

66BS249 A. Ricci and P. Vivarelli, *Bol. Sci. Fac. Chim. Ind. Bologna*, 1966, **24**, 249. 4-1Hb(i)

66CB445 E. Müller, R. Haller and K. W. Merz, *Chem. Ber.*, 1966, **99**, 445. 2-3Bd

66CB1002 K. Gewald, *Chem. Ber.*, 1966, **99**, 1002. 3-4A

66CB1618 J. Goerdeler and G. Gnad, *Chem. Ber.*, 1966, **99**, 1618. 4-3Ee, 4-4Fb, 4-4Fc, 4-5Fb(v)

66CB2391 E. Wittenburg, *Chem. Ber.*, 1966, **99**, 2391. 2-4Db, 2-4Dd

66CB2962 H. Wamhoff and F. Korte, *Chem. Ber.*, 1966, **99**, 2962. 4-2Bb, 4-2Gf, 4-2Ib(2), T4-21

66CB3076 A. Schönberg, K. Praefcke and J. Kohtz, *Chem. Ber.*, 1966, **99**, 3076. 3-2Ad(2)

66CB3215 E. Niwa, H. Aoki, H. Tanaka, K. Munukata and M. Namiki, *Chem. Ber.*, 1966, **99**, 3215. 3-3A

66CH198 C. O. Bender and R. Bonnett, *Chem. Commun.*, 1966, 198. 3-1Ad

66CH631 P. Beak and J. Bonham, *Chem. Commun.*, 1966, 631. 2-1C

66CI458 D. E. Ames, G. V. Boyd, A. W. Ellis and A. C. Lovesey, *Chem. Ind. (London)*, 1966, 458. 2-2Gb

66CI1634 L. M. Werbel, *Chem Ind. (London)*, 1966, 1634. 4-5Dc(2)

66CO(262C)285 J.-F. Giudicelli, J. Menin and H. Najer, *C. R. Acad. Sci., Ser. C*, 1965, **262**, 285. 4-6Ff(iii)

66CO(262C)1017 C. Broquet and A. Tchoukarine, *C. R. Acad. Sci., Ser. C*, 1966, **262**, 1017. 4-3Cf

66CO(262C)1161 J. Renault and J.-C. Cartron, *C. R. Acad. Sci., Ser. C*, 1966, **262**, 1161. 2-6Ba

66CO(262C)1204 Buu-Hoï, V. Bellavita, G. Grandolini, A. Ricci, and P. Jacquignon, *C. R. Acad. Sci., Ser. C*, 1966, **262**, 1204. 3-1Ac

66CO(263C)557 L. López and J. Barrans, *C. R. Acad. Sci., Ser. C*, 1966, **263**, 557. 4-5Ba

66CT756 T. Uno, K. Machida and K. Hanai, *Chem. Pharm. Bull.* 1966, **14**, 756. 4-5Ba(2), 4-5Bc, 4-5Fb(ii)

66CT770 Y. Maki, M. Suzuki and T. Yamada, *Chem. Pharm. Bull.* 1966, **14**, 770. 2-2E

66CT1277 I. Iwai and N. Nakamura, *Chem. Pharm. Bull.* 1966, **14**, 1277. 4-2Cb(iii), 4-2Cb(iv), 4-5Ba, 4-5Bc, T4-13

66CX(20)646 J. Gaultier and C. Hauw, *Acta Crystallogr.*, 1966, **20**, 646. 2-3Gb

66CX(20)703 G. N. Reeke and R. E. Marsh, *Acta Crystallogr.*, 1966, **20**, 703. 2-4Db

66CX(20)783 S. Martinez-Carrera, *Acta Crystallogr.*, 1966, **20**, 783. 1-4Ib, 4-1Ca

66CX(21)249 C. H. Carlisle and M. B. Hossain, *Acta Crystallogr.*, 1966, **21**, 249. 2-5Be

66CX(21)754 L. Katz, K.-I. Tomita and A. Rich, *Acta Crystallogr.*, 1966, **21**, 754. 5-1Da

66CZ1864 J. Pitha, P. Fiedler and J. Gut, *Collect. Czech. Chem. Commun.*, 1966, **31**, 1864. 2-4Fb, 2-6Ec(2)

66DA(166)635 G. M. Kheifets, N. U. Khromov-Borisov and A. I. Koltsov, *Dokl. Akad. Nauk SSSR*, 1966, **166**, 635. 2-4De

66DA(171)1101 V. M. Berezovskii and Zh. I. Akselrod, *Dokl. Akad. Nauk SSSR*, 1966, **171**, 1101. 2-4Gd

66E499 L. A. Summers, *Experientia*, 1966, **22**, 499. 4-2Bg(2)

66G1009 A. Bruno and G. Purrello, *Gazz. Chim. Ital.*, 1966, **96**, 1009. 4-5Be

66G1410 M. Colonna, P. Bruni and A. M. Guerra, *Gazz. Chim. Ital.*, 1966, **96**, 1410. 4-2Gg

66H53 A. Hofmann and C. H. Eugster, *Helv. Chim. Acta*, 1966, **49**, 53. 3-2Bd

66HC(6)1 J. H. Lister, *Advan. Heterocycl. Chem.*, 1966, **6**, 1. 5-1Ba, 5-1Be, 5-1C(2), 5-1Da, 5-1Ea, 5-1Ec, 5-1Ee, 5-1Eg

66HC(7)39 H. Prinzbach and E. Futterer, *Advan. Heterocycl. Chem.*, 1966, **7**, 39. 4-9A(2)

66HC(7)153 G. Scheibe and E. Daltrozzo, *Advan. Heterocycl. Chem.*, 1966, **7**, 153. 2-8Cc

66HC(7)183 A. Hetzheim and K. Möckel, *Advan. Heterocycl. Chem.*, 1966, **7**, 183. 4-3Dg, 4-4Eg, 4-5Fa(ii)

66HC(7)470 P. Bosshard and C. H. Eugster, *Advan. Heterocycl. Chem.*, 1966, **7**, 470. 3-4A

66J(B)44 M. Fraser, S. McKenzie and D. H. Reid, *J. Chem. Soc. B*, 1966, 44. 3-1Bc, T3-1

66J(B)92 J. L. Garraway, *J. Chem. Soc. B*, 1966, 92. 4-4Cc(ii)(3)

66J(B)191 W. L. F. Armarego, *J. Chem. Soc. B*, 1966, 191. 3-1Bc

66J(B)210 S. Lewin and D. A. Humphreys, *J. Chem. Soc. B*, 1966, 210. 2-6Cd

66J(B)436 J. W. Bunting and D. D. Perrin, *J. Chem. Soc. B*, 1966, 436. 2-2Gc

66J(B)469 W. H. Poesche, *J. Chem. Soc. B*, 1966, 469. 4-1Kc

66J(B)562 A. R. Katritzky, F. D. Popp and J. D. Rowe, *J. Chem. Soc. B*, 1966, 562. 2-3Eb(2), 2-3Ed(2), T2-1

66J(B)565 A. R. Katritzky, F. D. Popp and A. J. Waring, *J. Chem. Soc. B*, 1966, 565. 2-4De(5)

66J(B)631 A. R. Katritzky and B. Ternai, *J. Chem. Soc. B*, 1966, 631. 2-8Ec

66J(B)726 R. T. C. Brownlee, A. R. Katritzky and R. D. Topsom, *J. Chem. Soc. B*, 1966, 726. 4-5Df

66J(B)991 E. Spinner and J. C. B. White, *J. Chem. Soc. B*, 1966, 991. 2-3Bc

66J(B)996 E. Spinner and J. C. B. White, *J. Chem. Soc. B*, 1966, 996. 2-3C(2)

66J(C)10 F. Bergmann, Z. Neiman and M. Kleiner, *J. Chem. Soc. C*, 1966, 10. 5-1C

66J(C)40 J. W. Atkinson, R. S. Atkinson and A. W. Johnson, *J. Chem. Soc. C*, 1966, 40. 3-2Ec

66J(C)806 D. D. Chapman, *J. Chem. Soc. C*, 1966, 806. 2-8Dd(2)

66J(C)909 M. L. Tosato and L. Paoloni, *J. Chem. Soc. C*, 1966, 909. 2-5F

66J(C)1522 W. K. Warburton, *J. Chem. Soc. C*, 1966, 1522. 4-5Fa(i), 4-5Fa(iii)

66JA1621 G. G. Hammes and H. O. Spivey, *J. Amer. Chem. Soc.*, 1966, **88**, 1621. 2-3Ba

66JA2407 G. O. Dudek and E. P. Dudek, *J. Amer. Chem. Soc.*, 1966, **88**, 2407. 1-5Ab, 4-2Gj, 4-8B

66JA3829 J. P. Ferris and L. E. Orgel, *J. Amer. Chem. Soc.*, 1966, **88**, 3829. 4-5Eb

66JH51 G. W. Stacy, T. E. Wollner and T. R. Oakes, *J. Heterocycl. Chem.*, 1966, **3**, 51. 4-7Ad

66JH272 B. M. Baum and R. Levine, *J. Heterocycl. Chem.*, 1966, **3**, 272. 2-8Db(iii)

66JH282 H. F. Andrew and C. K. Bradsher, *J. Heterocycl. Chem.*, 1966, **3**, 282. 2-5Bb

66JO171 S. J. Rhoads, *J. Org. Chem.*, 1966, **31**, 171. 4-2Gh

66JO175 Y. Inoue, N. Furutachi and K. Nakanishi, *J. Org. Chem.*, 1966, **31**, 175. 2-4Da(2), 2-4De

66JO178 T. J. Delia and G. Bosworth Brown, *J. Org. Chem.*, 1966, **31**, 178. 5-1Fa

66JO251 K. T. Potts and H. R. Burton, *J. Org. Chem.*, 1966, **31**, 251. 5-3Bb(iii), 5-3Bd, S5-4

66JO265 K. T. Potts, H. R. Burton and S. K. Roy, *J. Org. Chem.*, 1966, **31**, 265. 5-3A, 5-3Bd, S5-3

66JO809 W. W. Paudler and J. E. Kuder, *J. Org. Chem.*, 1966, **31**, 809. 5-3A(2), S5-3

66JO966 J. C. Parham, J. Fissekis and G. Bosworth Brown, *J. Org. Chem.*, 1966, **31**, 966. 5-1F, 5-1Fa

66JO1199 S. Mizukami and E. Hirai, *J. Org. Chem.*, 1966, **31**, 1199. 1-4B, 2-6Ca(2)

66JO1295 W. W. Paudler and H. L. Blewitt, *J. Org. Chem.*, 1966, **31**, 1295. 5-3A, S5-3

66JO1538 D. F. O'Brien and J. W. Gates, Jr., *J. Org. Chem.*, 1966, **31**, 1538. 4-2Hc, 4-6Fd(ii)

66JO1722 H. Yasuda and H. Midorikawa, *J. Org. Chem.*, 1966, **31**, 1722. 4-2Gj(2), T4-18

66JO1878 L. G. Tensmeyer and C. Ainsworth, *J. Org. Chem.*, 1966, **31**, 1878. 1-5Ab, 4-1Bb(v), 4-1Bb(ix)

66JO1964 S. A. Mizsak and M. Perelman, *J. Org. Chem.*, 1966, **31**, 1964. 4-3Eb

66JO2090 E. J. Moriconi and F. J. Creegan, *J. Org. Chem.*, 1966, **31**, 2090. 2-3Bc, 2-3Fe

66JO2391 D. W. Nenry and R. M. Silverstein, *J. Org. Chem.*, 1966, **31**, 2391. 3-2Bd

66JS643 E. Shefter, M. N. G. James and H. G. Mautner, *J. Pharm. Sci.*, 1966, **55**, 643. 2-5Be

66JT759 M. J. S. Dewar and G. J. Gleicher, *J. Chem. Phys.*, 1966, **44**, 759. 1-4D, 1-7C, 4-1Eb(iv), 4-1Fa(viii)

66KG96 B. D. Chernokal'skii, A. T. Groisberg, N. F. Rakova and R. R. Shagidullin, *Khim. Geterotsikl. Soedin.*, 1966, 96. 4-2Jd

66KG101 A. S. Yelina, L. G. Tsirulnikova, E. M. Peresleni and Yu. N. Sheinker, *Khim. Geterotsikl. Soedin.*, 1966, 101. 2-7Cc

66KG149 Z. N. Nazarova and V. I. Novikov, *Khim. Geterotsikl. Soedin.*, 1966, 149. 3-3A

66KG216 Y. E. Skylar, R. P. Evstigneeva and N. A. Preobrazhenskii, *Khim. Geterotsikl. Soedin.*, 1966, 216. 3-1Ba

66KG776 I. B. Mazheika, G. I. Chipen and S. A. Hiller, *Khim. Geterotsikl. Soedin.*, 1966, 776. 4-1Eb(ii), 4-1Ec, T4-7

66LA(699)133 H. Balli and R. Gipp, *Liebig's Ann. Chem.*, 1966, **699**, 133. 4-2Gj

66M710 E. Ziegler and F. Hradetzky, *Monatsh. Chem.*, 1966, **97**, 710. 2-3Ge

66MC478 H. L. Yale and K. Losee, *J. Med. Chem.*, 1966, **9**, 478. 4-5Fa(ii)

66MI217 G. Palazzo and G. Picconi, *Boll. Chim. Farm.*, 1966, **105**, 217. 4-3Dd, 4-3Df

66MP1 G. Klose and K. Arnold, *Mol. Phys.*, 1966, **11**, 1 2-8Db(ii)

66PC(31)262 G. Barnikow, V. Kath and H. Conrad, *J. Prakt. Chem.*, 1966, **31**, 262. 4-8B

66R1072 H. Hogeveen, *Rec. Trav. Chim. Pays-Bas*, 1966, **85**, 1072. 3-1Be

66RO350 V. P. Shchipanov, Yu. N. Sheinker and I. Ya. Postovskii, *Zh. Org. Khim.*, 1966, **2**, 350. 4-5Ff

66RO376 V. P. Shchipanov, *Zh. Org. Khim.*, 1966, **2**, 376. 4-5Ff

66RO917 Y. N. Sheinker, A. M. Simonov, Y. M. Yufilov, V. N. Sheinker and E. I. Perelshtein, *Zh. Org. Khim.*, 1966, **2**, 917. 4-5Df

66SA1385 C. Cogrossi, *Spectrochim. Acta*, 1966, **22**, 1385. 4-2Fb, 4-5Bg

66SA2005 B. Ellis and P. J. F. Griffiths, *Spectrochim. Acta*, 1966, **22**, 2005. 4-4Cc(ii)(5), 4-4Cd, 4-6Ga

66T455 A. Butt, S. M. A. Hai and I. A. Akhtar, *Tetrahedron*, 1966, **22**, 455. 2-3Eg

66T1373 G. Klose and E. Uhlemann, *Tetrahedron*, 1966, **22**, 1373. 2-8Db(ii)

66T2703 S. Tabak, I. I. Grandberg and A. N. Kost, *Tetrahedron*, 1966, **22**, 2703. 4-5Bg, 4-5Bh, 4-5Cb, 4-5Cc

66T3227 M. Ionescu, A. R. Katritzky and B. Ternai, *Tetrahedron*, 1966, **22**, 3227. 2-3Fb(2), 2-5C

66T3233 L. N. Yakhontov, D. M. Krasnokutskaya, E. M. Peresleni, Y. N. Sheinker and M. V. Rubstov, *Tetrahedron*, 1966, **22**, 3233. 2-3Be, 2-3Bg, 2-6Ba, 2-7Ca

66T3253 R. Mondelli and L. Merlini, *Tetrahedron*, 1966, **22**, 3253. 2-8Db(i), 2-8Db(ii), 2-8Db(iii)(2), 2-8Db(iv), 2-8Dd(2), 2-8De(2), 2-8Dg

66T(7)213 A. A. Gordon, A. R. Katritzky and F. D. Popp, *Tetrahedron, Suppl.*, 1966, **7**, 213. 4-6Bg(i)(2), 4-6Bg(ii), 4-6Bg(iii)

66TL383 W. Steglich and R. Hurnaus, *Tetrahedron Lett.*, 1966, 383. 4-3Cc

66TL2579 A. G. Anderson, Jr. and H. L. Ammon, *Tetrahedron Lett.*, 1966, 2579. 2-8Bb

66TL3897 B. L. Kaul, P. Madharan Nair, A. V. Rama Rao and K. Venkataraman, *Tetrahedron Lett.*, 1966, 3897. 4-2Gj, T4-18

66TL3919 H. Wamhoff and F. Korte, *Tetrahedron Lett.*, 1966, 3919. 4-2Bb, 4-2Gf, 4-2Ib(2), T4-21

66TL4427 Y. Iwakura, F. Toda and Y. Torii, *Tetrahedron Lett.*, 1966, 4427. 4-3Cc

66ZS339 Y. G. Baklagina, M. D. Volkenshtein, and Y. D. Kondrashev, *Zh. Strukt. Khim.*, 1966, **7**, 399. 5-1Da

67AC877 D. N. Bailey, D. M. Hercules and T. D. Eck, *Anal. Chem.*, 1967, **39**, 877. 1-6D, 2-3Bd

67AG(E)919 A. Albert, *Angew. Chem.* (*Int. Ed. Engl.*), 1967, **6**, 919. 2-2Ga

67AJ935 R. Chong and P. S. Clezy, *Aust. J. Chem.*, 1967, **20**, 935. 3-2Ee(2)

67AL538 G. Cialdi and C. Sabelli, *Atti Accad. Naz. Lincei, Cl. Sci. Fis., Mat. Natur., Rend.*, 1967, **42**, 538. 4-2Bd

67AN471 P. Condorelli, G. Pappalardo and B. Tornetta, *Ann. Chim.* (*Rome*), 1967, **57**, 471. 4-4Ac, 4-5Be, 4-5Bg

67AS633 M. Begtrup and C. Pedersen, *Acta Chem. Scand.*, 1967, **21**, 633. 4-3Ef

67AS673 A. B. Hörnfeldt, *Acta Chem. Scand.*, 1967, **21**, 673. 3-2Cav

67AS1234 M. Begtrup, K. Hansen and C. Pedersen, *Acta Chem. Scand.*, 1967, **21**, 1234. 4-3Ed, 4-3Ef

67AS1437 E. Åkerblom, *Acta Chem. Scand.*, 1967, **21**, 1437. 1-4F, 4-6Ff(ii)(4)

67AS2463 H. C. Børresen, *Acta Chem. Scand.*, 1967, **21**, 2463. 5-1-Da(2), 5-1Db

67BF74 J. Schmitt, M. Suquet, G. Callet, J. Le Meur and P. Comoy, *Bull. Soc. Chim. Fr.*, 1967, 74. 4-2Fc

67BF207 H. Najer, R. Giudicelli, J. Menin and N. Voronine, *Bull. Soc. Chim. Fr.*, 1967, 207. 4-6Fe(ii)(4), 4-6Fe(iii)(2), 4-6-Fe(iv)(5)

67BF1219 M. Sélim and M. Sélim, *Bull. Soc. Chim. Fr.*, 1967, 1219. 4-5A, 4-5Fa(i)

67BF1296 M. Mousseron-Canet, J. P. Boca and V. Tabacik, *Bull. Soc. Chim. Fr.*, 1967, 1296. 3-2Eb(3)

67BF2040 H. Najer, R. Giudicelli and J. Menin, *Bull. Soc. Chim. Fr.*, 1967, 2040. 4-5Da, 4-5Dc

67BF2619 J. Elguero, A. Fruchier and R. Jacquier, *Bull. Soc. Chim. Fr.*, 1967, 2619. 4-1Gi, T4-9

67BF2630 R. Jacquier, M.-L. Roumestant and P. Viallefont, *Bull. Soc. Chim. Fr.*, 1967, 2630. 1-5Ac(i)

67BF2998 J. Elguero, E. Gonzalez and R. Jacquier, *Bull. Soc. Chim. Fr.*, 1967, 2998. 4-1Da, 4-1Ka, 4-1Kbi

67BF3003 R. Jacquier, C. Petrus, F. Petrus and J. Verducci, *Bull. Soc. Chim. Fr.*, 1967, 3003. 4-2Bd, 4-2Bg(i), T4-13

67BF3367 C. Fournier and J. Decombe, *Bull. Soc. Chim. Fr.*, 1967, 3367. 2-3Ec, 2-3Ea

67BF3772 J. Elguero, R. Jacquier and G. Tarrago, *Bull. Soc. Chim. Fr.*, 1967, 3772. 4-2Fa, 4-2Fc(5), 4-2Fd(4), 4-2Ga(i), 4-2Ia, T4-2, S4-8

67BF3780 J. Elguero, R. Jacquier and G. Tarrago, *Bull. Soc. Chim. Fr.* 1967, 3780. 1-4Ac, 4-2Fa, 4-2Fb(2), 4-2Fc(2), 4-2Fd(2), 4-2Ga(i), 4-2Ga(ii)(2), 4-2Gc, 4-2Gc(ii)(2), 4-2Gd, 4-2Ge(ii), 4-2Ge(iii)(2), 4-2Gg, 4-2Hb(i), 4-2Hb(ii), 4-2Hb(iii), 4-2Hb (iv)(2), 4-2Hc, 4-2Hc(iii)(2), 4-2Ib(6), T4-16, T4-17, T4-20 (7), T4-21(3), S4-12

67BF4619 G. Kille and J.-P. Fleury, *Bull. Soc. Chim. Fr.*, 1967, 4619. 4-5Ea(5)

67BF4624 P. Roesler and J.-P. Fleury, *Bull. Soc. Chim. Fr.*, 1967, 4624. 4-5Ea(4)

67BJ149 Y. Iwakura, F. Toda and Y. Torii, *Bull. Chem. Soc. Jap.*, 1967, **40**, 149. 4-3Cc

67BJ153 S. Kakimoto and S. Tonooka, *Bull. Chem. Soc. Jap.*, 1967, **40**, 153. 2-4Ga

67BJ2493 K. Nishimoto, *Bull. Chem. Soc. Jap.*, 1967, **40**, 2493. 1-7E, 2-4Gd

67BS51 I. Degani and C. Vincenzi, *Boll. Sci. Fac. Chim. Ind. Bologna*, 1967, **25**, 51. 2-2A

67C161 C. F. Kröger and W. Freiberg, *Chimia*, 1967, **21**, 161. 1-4Ab

67CB919 G. Tomaschewski and G. Geissler, *Chem. Ber.*, 1967, **100**, 919. 4-2Gg

67CB954 H. Böshagen, *Chem. Ber.*, 1967, **100**, 954. 4-2Cc(2)

67CB1661 G. Barnikow and G. Strickmann, *Chem. Ber.*, 1967, **100**, 1661. 4-6Bb, 4-6Fd(iii)

67CB1824 W. Steglich, H. Tanner and R. Hurnaus, *Chem. Ber.*, 1967, **100**, 1824. 4-3Cc

67CB2280 H. Brederech, F. Effenberger and H. G. Österlin, *Chem. Ber.*, 1967, **100**, 2280. 2-4Df

67CB3097 H. Reimlinger, *Chem. Ber.*, 1967, **100**, 3097. 4-1Aa

67CB3664 H. Bredereck, G. Simchen and H. Traut, *Chem. Ber.*, 1967, **100**, 3664. 2-4 Da,2-5E, 2-6Ca

67CB3671 E. Winterfeldt and J. M. Nelke, *Chem. Ber.*, 1967, **100**, 3671. 2-6F

67CH371 P. Hampson and A. Mathias, *Chem. Commun.*, 1967, 371. 2-3Ba, 1-5C

67CH1077 W. W. Paudler and A. G. Zeiler, *Chem. Commun.*, 1967, 1077. 6-2Bf

67CO(265C)631 M. R. Calas and J. Martinez, *C. R. Acad. Sci., Ser. C*, 1967, **265**, 631. 2-5E

67CX(22)308 K. Britts and I. L. Karle, *Acta Crystallogr.*, 1967, **22**, 308. 1-4Ib, 4-1Fa(iii), 4-5Ff

67CX(23)376 B. M. Craven, *Acta Crystallogr.*, 1967, **23**, 376. 2-4Db

67CX(23)1102 R. F. Stewart and L. H. Jensen, *Acta Crystallogr.*, 1967, **23**, 1102. 2-4Db

67CZ1637 Z. Budesinsky, J. Prikryl and E. Svatek, *Collect. Czech. Chem. Commun.*, 1967, **32**, 1637. 2-4Df

67CZ4151 L. Fisnerova, B. Kakac and O. Nemecek, *Collect. Czech. Chem. Commun.*, 1967, **32**, 4151. 4-6Bb

67DA(172)118 L. N. Yakhontov, D. M. Krasnokutskaya, E. M. Peresleni, Y. N. Sheinker and M. V. Rubtsov, *Dokl. Akad. Nauk SSSR*, 1967, **172**, 118. 2-3Be, 2-3Bg

67DA(176)613 L. N. Yakhontov, D. M. Krasnokutskaya, E. M. Peresleni, Y. N. Sheinker and M. V. Rubtsov, *Dokl. Akad. Nauk SSSR*, 1967, **176**, 613. 2-3Bg

67DA(177)592 E. M. Peresleni, L. N. Yakhontov, D. M. Krasnokutskaya and Y. N. Sheinker, *Dokl. Akad. Nauk SSSR*, 1967, **177**, 592. 2-3Be, 2-3Bg, T2-1

67G346 G. Cum, G. Lo Vecchio, M. C. Aversa and M. Crisafulli, *Gazz. Chim. Ital.*, 1967, **97**, 346. 4-2Bg(2)

67H137 H. Göth, A. R. Gagneux, C. H. Eugster and H. Schmid, *Helv. Chim. Acta*, 1967, **50**, 137. 4-2Cb(iv), 4-3Bb, T4-13

67HC(8)21 F. D. Popp and A. Catala Noble, *Advan. Heterocycl. Chem.*, 1967, **8**, 21. 6-2A, 6-2Bc, 6-2Bd

67IM1335 P. Caramella and P. Grünanger, *Chim. Ind. (Milan)*, 1967, **49**, 1335. 4-2Bg(i)

67IZ2783 V. S. Bogdanov, M. A. Kalik, Y. L. Danyushevskii and Y. L. Gol'dfarb, *Izv. Akad. Nauk SSSR, Ser. Khim.*, 1967, 2783. 3-3A, 3-3B

67J(B)14 D. Harrison and J. T. Ralph, *J. Chem. Soc. B*, 1967, 14. 4-4Cf

67J(B)84 R. A. Jones and B. D. Roney, *J. Chem. Soc. B*, 1967, 84. 2-3Bc, T2-1

67J(B)516 G. B. Barlin and T. J. Batterham, *J. Chem. Soc. B*, 1967, 516. 4-1Kbii

67J(B)590 D. A. Evans, G. F. Smith and M. A. Wahid, *J. Chem. Soc. B*, 1967, 590. 2-3Bf(3), 2-3Bg

67J(B)641 G. B. Barlin, *J. Chem. Soc. B*, 1967, 641. 4-1Cb(iv), 4-1Db, 4-1Ec, T4-5(2)

67J(B)748 D. E. Ames, G. V. Boyd, R. F. Chapman, A. W. Ellis, A. C. Lovesey and D. Waite, *J. Chem. Soc. B*, 1967, 748. 2-2Gb(2), 2-6Db

67J(B)758 A. R. Katritzky, J. D. Rowe and S. K. Roy, *J. Chem. Soc. B*, 1967, 758. 2-3Bc(3), 2-3Bg, T2-1

67J(B)885 T. Nishiwaki, *J. Chem. Soc. B*, 1967, 885 4-1Bb(viii)

67J(B)911 A. J. Boulton, A. C. Gripper Gray and A. R. Katritzky, *J. Chem. Soc. B*, 1967, 911. 4-7B

67J(B)914 A. J. Boulton, A. R. Katritzky, M. J. Sewell and B. Wallis, *J. Chem. Soc. B*, 1967, 914. 4-9Ba

67J(B)1251 J. S. G. Cox, C. Fitzmaurice, A. R. Katritzky and G. J. T. Tiddy, *J. Chem. Soc. B*, 1967, 1251. 4-1Ka

67J(B)1363 A. Zecchina, L. Cerruti, S. Coluccia and E. Borello, *J. Chem. Soc. B*, 1967, 1363. 4-1Ba

67J(C)53 R. D. Chambers, M. Hole, W. K. R. Musgrave and R. A. Storey, *J. Chem. Soc. C*, 1967, 53. 2-3Bc

67J(C)1533 A. Albert and G. Catterall, *J. Chem. Soc. C*, 1967, 1533. 2-7Ea

67J(C)1764 D. J. Neadle and R. J. Pollitt, *J. Chem. Soc. C*, 1967, 1764. 4-7Ad

67J(C)1792 G. K. J. Gibson, A. S. Lindsey and H. M. Paisley, *J. Chem. Soc. C*, 1967, 1792. 4-2Jc

67J(C)1822 R. E. Banks, D. S. Field and R. N. Haszeldine, *J. Chem. Soc. C*, 1967, 1822. 2-4Da

67J(C)2005 A. J. Boulton, A. R. Katritzky and A. Majid Hamid, *J. Chem. Soc. C*, 1967, 2005. 4-9Ba

67JA647 W. Reeve and M. Nees, *J. Amer. Chem. Soc.*, 1967, **89**, 647. 4-6Ff(ii)

67JA1183 M. Gorodetsky, Z. Luz and Y. Mazur, *J. Amer. Chem. Soc.*, 1967, **89**, 1183. 1-5E

67JA1249 E. Shefter and H. G. Mautner, *J. Amer. Chem. Soc.*, 1967, **89**, 1249. 2-5E

67JA1312 W. H. Kirchhoff, *J. Amer. Chem. Soc.*, 1967, **89**, 1312. T1-5, 4-1Ba

67JA2242 W. H. Kirchhoff, *J. Amer. Chem. Soc.*, 1967, **89**, 2242. T1-5

67JA4760 M. E. Hermes and F. D. Marsh, *J. Amer. Chem. Soc.*, 1967, **89**, 4760. 4-9Ca

67JA6835 J. E. Bloor and D. L. Breen, *J. Amer. Chem. Soc.*, 1967, **89**, 6835. 4-1Ca, 4-1Fa(vii), T4-6, T4-8

67JH54 G. Adembri and R. Nesi, *J. Heterocycl. Chem.*, 1967, **4**, 54. 4-4Aa

67JH109 S. K. Chakrabartty and R. Levine, *J. Heterocycl. Chem.*, 1967, **4**, 109. 2-8Db(iv)

67JH523 E. A. Ingalls and F. D. Popp, *J. Heterocycl. Chem.*, 1967, **4**, 523. 2-4De

67JH533 F. de Sarlo and G. Dini, *J. Heterocycl. Chem.*, 1967, **4**, 533. 2-4De, 4-2Bd, 4-2Be, 4-2Bf, T4-12(7)

67JH659 M. Israel, L. C. Jones and E. J. Modest, *J. Heterocycl. Chem.*, 1967, 659. 6-2Bd(2), T6-1(3)

67JO1106 C. W. Muth, J. R. Elkins, M. L. De Matte and S. T. Chiang, *J. Org. Chem.*, 1967, **32**, 1106. 2-3Fb(2), 2-6Bc

67JO1151 J. C. Parham, J. Fissekis and G. Bosworth Brown, *J. Org. Chem.*, 1967, **32**, 1151. 5-1F, 5-1Fb

67JO1954 H. Walba, D. L. Stiggall and S. M. Coutts, *J. Org. Chem.*, 1967, **32**, 1954. 4-1Hb(i)(2)

67JO2685 C. D. Gutsche and H. W. Voges, *J. Org. Chem.*, 1967, **32**, 2685. 2-8Dg(3)

67JO2823 L. M. Weinstock, P. Davis, B. Handelsman and R. Tull, *J. Org. Chem.*, 1967, **32**, 2823. 4-3Eb(2)

67JO3028 G. W. Stacy and T. E. Wollner, *J. Org. Chem.*, 1967, **32**, 3028. 3-2Cc

67JO3580 J. C. Kauer and W. A. Sheppard, *J. Org. Chem.*, 1967, **32**, 3580. 1-5E, 4-3Eh, 4-4Fe, 4-5A, 4-5Fe

67KG93 P. M. Kochergin, A. M. Tsyganova and L. M. Vitkova, *Khim. Geterotsikl. Soedin.*, 1967, 93. S5-5

67KG130 S. A. Hiller, I. B. Mazheika, I. I. Grandberg and L. I. Gorbacheva, *Khim. Geterosikl. Soedin.*, 1967, 130 4-1Ba, 4-1Ba, 4-1Bb(iii)

67KG329 V. S. Troitskaya, V. G. Vinokchrov, I. I. Grandberg and S. V. Tabak, *Khim. Geterotsikl. Soedin.*, 1967, 329. T5-1

67KG532 A. A. Druzina and P. M. Kochergin, *Khim. Geterotsikl. Soedin.*, 1967, 532. 5-5Aa

67KG536 R. M. Palei and P. M. Kochergin, *Khim. Geterotsikl. Soedin.*, 1967, 536. 5-5Aa

67KG637 N. N. Khovratovich and I. I. Chizhevskaya, *Khim. Geterotsikl. Soedin.*, 1967, 637. 4-6Bd

67LA(707)141 H. Dorn and H. Dilcher, *Liebig's Ann Chem.*, 1967, **707**, 141. 4-5Bh(i)(2)

67M100 H. Sterk and E. Ziegler, *Monatsh. Chem.*, 1967, **98**, 100. 2-3Ea

67M1763 H. Sterk and H. Junek, *Monatsh. Chem.*, 1967, **98**, 1763. 2-3Ec

67MB(25)67 C. E. Buff and R. E. Marsh, *J. Mol. Biol.*, 1967, **25**, 67. 2-6Cb

67MB(30)545 K.-I. Tomita, L. Katz and A. Rich, *J. Mol. Biol.*, 1967, **30**, 545. 5-1Da

67MI1 A. R. Katritzky and J. M. Lagowski, "Principles of Heterocyclic Chemistry," Methuen, London, 1967. 1-1F

67MI6 R. Fusco, "Pyrazoles, Pyrazolines, Pyrazolidines, Indazoles and Condensed Rings" (R. H. Wiley, Ed.), p. 6. Wiley (Interscience), New York, 1967. 1-4Hb

67MI74 A. D. Garnovskii, V. I. Minkin, I. I. Grandberg and T. A. Ivanova, *Khim. Geterotsikl. Soedin., Sb. I, Azot. Geterots.*, 1967, 74. 4-2Fa(2), 4-2Fc, 4-2Hb(v), 4-2Hc, 4-4Ad, T4-2b, S4-4, S4-8

67MI89 H. B. Jonassen, T. Paukert and R. A. Henry, *Appl. Spectrosc.* 1967, **21**, 89. 4-5Ff

67MI186 E. A. Lipatova, I. N. Shokhor, I. N. Lopyrev and V. A. Pevzner, *Khim. Geterotsikl. Soedin., Sb. I, Azot. Geterots.*, 1967, 186. 4-5Fc

67MI211 C. H. Jarboe, "Pyrazoles, Pyrazolines, Pyrazolidines, Indazoles and Condensed Rings" (R. H. Wiley, Ed.), p. 211. Wiley (Interscience), New York, 1967. 4-2Ia

67MI230 T. Severin and W. Seilmeyer, *Z. Lebensm.-Unters. Forsch.*, 1967, **134**, 230. 3-2Bd

67MI322 L. C. Behr, "Pyrazoles, Pyrazolines, Pyrazolidines, Indazoles and Condensed Rings" (R. H. Wiley, Ed.), p. 322. Wiley (Interscience), New York, 1967. 4-7A

67MI325 K. Schofield, "Heteroaromatic Nitrogen Compounds," p. 325. Butterworth, London, 1967. 2-8Ba, 2-8Ca

67NA1301 J. H. Griffiths, A. Wardley, V. E. Williams, N. L. Owen and J. Sheridan, *Nature (London)*, 1967, **216**, 1301. T1-5, 4-1Ca(2)

67PH1756 J. A. Caruso, P. G. Sears and A. I. Popov, *J. Phys. Chem.*, 1967, **71**, 1756. 4-1Fa(i)

67PH2375 A. D. Mighell and C. W. Reimann, *J. Phys. Chem.*, 1967, **71**, 2375. 4-1Ba(2)

67PH2668 S. Schulman and Q. Fernando, *J. Phys. Chem.*, 1967, **71**, 2668. 2-3Ha

67RC1241	E. Domagalina and I. Kurpiel, *Rocz. Chem.*, 1967, **41**, 1241. 2-4Ce
67SA717	M. Mousseron-Canet, J. P. Boca and V. Tabacik, *Spectrochim. Acta, Part A*, 1967, **23**, 717. 3-2Eb
67SA2551	R. C. Lord and G. J. Thomas, Jr., *Spectrochim. Acta, Part A*, 1967, **23**, 2551. 2-4Db, 2-6Cb, 5-1Da(2)
67T745	Y. Chiang, R. L. Hinman, S. Theodoropulos and E. B. Whipple, *Tetrahedron*, 1967, **23**, 745. 3-1Ba(2)
67T831	F. de Sarlo, *Tetrahedron*, 1967, **23**, 831. 4-2Bf, T4-12
67T871	H. J. Jakobsen and S.-O. Lawesson, *Tetrahedron*, 1967, **23**, 871. 3-2Ca(iii)
67T1197	G. M. Kheifets, N. U. Khromov-Borisov, A. I. Koltsov and M. V. Volkenstein, *Tetrahedron*, 1967, **23**, 1197. 2-4De(4)
67T2095	J. MacLeod, J. B. Thomson and C. Djerassi, *Tetrahedron*, 1967, **23**, 2095. 1-6Ga(ii)(2)
67T2657	T. Nishiwaki, *Tetrahedron*, 1967, **23**, 2657. 2-4Da
67T3363	Y. Iwakura, F. Toda and Y. Torii, *Tetrahedron*, 1967, **23**, 3363. 4-3Cc
67T3601	A. G. Anderson, Jr. and H. L. Ammon, *Tetrahedron*, 1967, **23**, 3601. 2-8Bb
67T3737	H. J. Jakobsen, *Tetrahedron*, 1967, **23**, 3737. 3-2Ca(ii)(3)
67T4395	W. Barbieri, L. Bernardi, S. Coda, V. Colo and G. Palamidessi, *Tetrahedron*, 1967, **23**, 4395. 4-6Fa(i)(3), 4-6Fa(ii), 4-6Fa(iii)(3), 4-6Fb(i), 4-6Fb(ii)(2), T4-30(11)
67T4517	L. G. Duquette and F. Johnson, *Tetrahedron*, 1967, **23**, 4517. 2-2C
67TC182	M. Gelus, P.-M. Vay and G. Berthier, *Theor. Chim. Acta*, 1967, **9**, 182. 4-1Ca
67TC259	R. D. Brown and B. A. W. Coller, *Theor. Chim. Acta*, 1967, **7**, 259. 4-1Ca
67TC342	W. Adam and A. Grimison, *Theor. Chim. Acta*, 1967, **7**, 342. 4-1Ca, 4-1Da
67TH01	L. T. Creagh, Ph.D. Thesis, North Texas State Univ., Denton, Texas, 1967. 1-5Ac(i)
67TL161	E. Fahr, W. Fischer, A. Jung, L. Sauer and A. Mannschreck, *Tetrahedron Lett.*, 1967, 161. 6-1B
67TL2109	W. Freiberg, C.-F. Kröger and R. Radeglia, *Tetrahedron Lett.*, 1967, 2109. 4-1Eb(iii)
67TL2951	U. E. Wiersum and H. Wynberg, *Tetrahedron Lett.*, 1967, 2951. 3-1Bd, 3-1Be, 5-A
67TL3669	J. Kopecky, J. E. Shields and J. Bornstein, *Tetrahedron Lett.*, 1967, 5201. 3-1Ad
67TL5201	G. W. Stacy and D. L. Eck, *Tetrahedron Lett*, 1967, 5201. 3-4Ba
67ZO2487	A. T. Prudchenko, G. S. Shchegoleva, V. A. Barkhash and N. N. Vorozhtsov, Jr., *Zh. Obshch. Khim.*, 1967, **37**, 2487. 4-2Gf
68AG(E)464	G. Simchen, *Angew. Chem.* (*Int. Ed. Engl.*), 1968, **7**, 464. 2-3Bf
68AG(E)565	R. Breslow, *Angew. Chem.* (*Int. Ed. Engl.*) **7**, 565 6

68AG(E)734 H. Bauer, *Angew. Chem. (Int. Ed. Engl.)*, 1968, **7**, 734. 3-2Ee

68AJ467 P. I. Mortimer, *Aust. J. Chem.*, 1968, **21**, 467. 2-3Bc

68AJ1113 J. R. Christie and B. Selinger, *Aust. J. Chem.*, 1968, **21**, 1113. 4-2E

68AK229 A.-B. Hörnfeldt, *Ark. Kemi*, 1968, **29**, 229. 3-2Aa, 3-2Ca(v)

68AK427 A.-B. Hörnfeldt, *Ark. Kemi*, 1968, **29**, 427. 3-2Ca(ii), 3-2Ca(v)

68AK455 A.-B. Hörnfeldt, *Ark. Kemi*, 1968, **29**, 455. 3-2Ca(v)

68AK461 A.-B. Hörnfeldt, *Ark. Kemi*, 1968, **29**, 461. 3-2Ca(v)

68AN562 G. Desimoni and G. Minoli, *Ann. Chim. (Rome)*, 1968, **58**, 562. 4-5Ba(2)

68AN664 E. Niccoli, U. Vaglini and G. Cassarelli, *Ann. Chim. (Rome)*, 1968, **58**, 664. 2-3Ga(2)

68AN1363 G. Desimoni, P. Grünanger and S. Servi, *Ann. Chim. (Rome)*, 1968, **58**, 1363. 4-3Ab

68AS1669 I. Crossland and E. Kelstruf, *Acta Chem. Scand.*, 1968, **22**, 1669. 2-2D

68AS2700 I. Crossland, *Acta Chem. Scand.*, 1968, **22**, 2700. 2-2D

68BF600 J. Gardent and G. Hazebroucq, *Bull. Soc. Chim. Fr.*, 1968, 600. 6-2Ba

68BF1099 J.-F. Giudicelli, J. Menin and H. Najer, *Bull. Soc. Chim. Fr.*, 1968, 1099. 4-6Ff(iii)(2)

68BF2117 M. Sélim, O. Tétu, M. Sélim, G. Martin and P. Rumpf, *Bull. Soc. Chim. Fr.*, 1968, 2117. 4-5Dc(2)

68BF2855 M. Chanon and J. Metzger, *Bull. Soc. Chim. Fr.*, 1968, 2855. 4-9Cb

68BF2868 M. Chanon and J. Metzger, *Bull. Soc. Chim. Fr.*, 1968, 2868. 1-3B(2), 4-4Cc, 4-4Cc(i)(3), 4-4Cc(ii)(7), 4-4Cc(iii), 4-4Cc(iv)

68BF3268 M. Sélim, G. Martin and M. Sélim, *Bull. Soc. Chim. Fr.*, 1968, 3268. 1-3B(2)

68BF3270 G. Martin, M. Sélim and M. Sélim, *Bull. Soc. Chim. Fr.*, 1968, 3270. 1-3B, 4-5Dc

68BF3272 M. Sélim, M. Sélim and G. Martin, *Bull. Soc. Chim. Fr.*, 1968, 3272. 1-3B, 4-5Dc(2)

68BF3477 M. L. Girard and C. Dreux, *Bull. Soc. Chim. Fr.*, 1968, 3477. 4-6Bd, 4-6Eb(i), 4-6Ec, 4-6Ff(ii)

68BF4203 F. Barantan, G. Fontaine and P. Maitte, *Bull. Soc. Chim. Fr.*, 1968, 4203. 2-2A

68BF4568 J.-F. Giudicelli, J. Menin and H. Najer, *Bull. Soc. Chim. Fr.*, 1968, 4568. 1-6E, 4-3Dj, 4-3Fa(ii)

68BF4631 G. Kille and J.-P. Fleury, *Bull. Soc. Chim. Fr.*, 1968, 4631. 4-5Ea(2)

68BF4636 G. Kille and J.-P. Fleury, *Bull. Soc. Chim. Fr.*, 1968, 4636. 4-3Cb

68BF5009 J. Elguero, E. Gonzalez and R. Jacquier, *Bull. Soc. Chim. Fr.*, 1968, 5009. T1-2(2), 4-1Bb(i), 4-1Bb(ii), 4-1Ka(2), T4-2(5)

68BF5019 J. Elguero, G. Guiraud, R. Jacquier and G. Tarrago, *Bull. Soc. Chim. Fr.*, 1968, 5019. 4-2Fa, 4-2Hb(iii), 4-2Hc, T4-17, S4-10(2)

68BJ959 T. Naito, S. Nakagawa, J. Okumura, K. Takahashi and K. Kasai, *Bull. Chem. Soc. Jap.*, 1968, **41**, 959. 4-3Ac

68BO436 C. E. Bugg, U. Thewalt and R. E. Marsh, *Biochem. Biophys. Res. Commun.*, 1968, **33**, 436. 5-1Eg

68BT100 R. Breslow, *Chem. Brit.*, 1968, **4**, 100. 6

68CB512 H. Bredereck, G. Simchen, R. Wahl and F. Effenberger, *Chem. Ber.*, 1968, **101**, 512. 2-6Cc, 2-7Cb

68CB1473 M. Regitz and H. J. Geelhaar, *Chem. Ber.*, 1968, **101**, 1473. 4-3Ae

68CB2679 H. J. Knackmuss, *Chem. Ber.*, 1968, **101**, 2679. 2-3Ef

68CB3265 H. Dorn and A. Zubek, *Chem. Ber.*, 1968, **101**, 3265. 4-5Bg(2), 4-5Bh(i)(3), 5-2Ba

68CB3278 H. Dorn, *Chem. Ber.*, 1968, **101**, 3278. 4-5Bg

68CC691 C. E. Hall and A. Taurins, *Can. J. Chem.*, 1968, **46**, 691. 5-4A

68CH727 P. R. Briggs, W. L. Parker and T. W. Shannon, *Chem. Commun.*, 1968, 727. 4-1Ea

68CH746 D. Rohrer and M. Sundaraligam, *Chem. Commun.*, 1968, 746. 2-4Db

68CO(266C)290 M. Saquet and A. Thuillier, *C. R. Acad. Sci., Ser. C*, 1968, **266**, 290. 4-4Ad

68CO(266C)628 H. Najer, J. Menin, D. Caillaux and G. Petry, *C. R. Acad. Sci., Ser. C*, 1968, **266**, 628. 4-5Fa(i)(3)

68CO(266C)1281 J. P. Bideau, B. Busetta and J. Housty, *C. R. Acad. Sci., Ser. C*, 1968, **266**, 1281. 2-5E

68CO(266C)1587 H. Najer, J. Menin and G. Petry, *C. R. Acad. Sci., Ser. C*, 1968, **266**, 1587. 1-6Cc, 4-5Da, 4-5Dc, 4-5Fa(i)(2), 4-5-Fb(i), 4-5Fb(ii), **S4-14**

68CO(267C)1790 G. Bravic, J. Gaultier and C. Hauw, *C. R. Acad. Sci., Ser. C*, 1968, **267**, 1790. 2-3Gb

68CT345 M. Ueda, N. Murakami and Y. Nakagawa, *Chem. Pharm. Bull.*, 1968, **16**, 345. 4-5Bg

68CT1466 A. Nakamura and S. Kamiya, *Chem. Pharm. Bull.*, 1968, **16**, 1466. 2-3Ee(2)

68CX(B)23 J. Hvoslef, *Acta Crystallogr., Sect. B*, 1968, **24**, 23. 3-2Be

68CX(B)1431 J. Hvoslef, *Acta Crystallogr., Sect. B*, 1968, **24**, 1431. 3-2Be

68CX(B)1692 J. Sletten, E. Sletten and L. H. Jensen, *Acta Crystallogr., Sect. B*, 1968, **24**, 1692. 5-1Bc, 5-2Bc

68CX(B)1698 R. Srinivasan and R. Chandrasekharan, *Acta Crystallogr., Sect. B*, 1968, **24**, 1698. 5-1C

68CZ566 P. Vetešnik, J. Kavalek, V. Beranek and O. Exner, *Collect. Czech. Chem. Commun.*, 1968, **33**, 566. 1-4Ba, 2-2Gd

68CZ2513 K. Kalfus, *Collect. Czech. Chem. Commun.*, 1968, **33**, 2513. 2-7Eb

68CZ3880 Z. Gregorowicz and Z. Klima, *Collect. Czech. Chem. Commun.*, 1968, **33**, 3880. 4-6Cc

68DA(180)473 S. V. Krivun, *Dokl. Chem.*, 1968, **180**, 473. 2-2A

68G245 P. Papini, G. Auzzi and M. Bambagiotti, *Gazz. Chim. Ital.*, 1968, **98**, 245. 4-2Gh

68G458 W. Barbieri, L. Bernardi, O. Goffredo and M. Tacchi Venturi, *Gazz. Chim. Ital.*, 1968, **98**, 458. 4-6Bf(2)

68H2102 T. R. Govindachari, S. Rajappa and K. Nagarajan, *Helv. Chim. Acta*, 1968, **51**, 2102. 4-8B

68HC(9)27 R. E. Willette, *Advan. Heterocycl. Chem.*, 1968, **9**, 27. 5-2Aa(2)

68HC(9)107 L. M. Weinstock and P. I. Pollak, *Advan. Heterocycl. Chem.*, 1968, **9**, 107. 4-3Eb, 4-5Fb(iv)

68HC(9)165 J. Sandström, *Advan. Heterocycl. Chem.*, 1968, **9**, 165. 4-3Dh, 4-4Eh, 4-5Fb(ii)(2), 4-6Cc(2)

68HC(9)211 M. Tišler and B. Stanovnik, *Advan. Heterocycl. Chem.*, 1968, **9**, 211. 2-2D

68IQ165 R. B. Hermann, *Int. J. Quantum Chem.*, 1968, **2**, 165. 3-1Be

68IS1 B. Kirson, *Isr. J. Chem.*, 1968, **6**, 1. 2-9A

68IS507 E. D. Bergmann, Z. Aizenshtat and I. Shahak, *Isr. J. Chem.*, 1968, **6**, 507. 6-2A

68IS603 W. Pfleiderer and H. Deiss, *Isr. J. Chem.*, 1968, **6**, 603. 2-4Dd, 2-4Db

68J(A)3051 S. F. Mason, J. Philp, and B. E. Smith, *J. Chem. Soc. A*, 1968, 3051. 2-3D, 2-3Ha

68J(B)556 A. Gordon, A. R. Katritzky and S. K. Roy, *J. Chem. Soc. B*, 1968, 556. 2-3Bc(2), T2-1, T2-2

68J(B)725 I. L. Finar, *J. Chem. Soc. B*, 1968, 725. 4-1Ba, 4-1Bb(ix), 4-1Bb(xi)

68J(B)1470 R. W. Baldock and A. R. Katritzky, *J. Chem. Soc. B*, 1968, 1470. 1-6Cd

68J(C)344 A. Albert and K. Tratt, *J. Chem. Soc. C*, 1968, 344. 5-2Bc

68J(C)504 R. M. Acheson, C. J. A. Brookes, D. P. Dearnaley and B. Quest, *J. Chem. Soc. C*, 1968, 504. 3-2Eb

68J(C)1045 R. M. Acheson and B. Adcock, *J. Chem. Soc. C*, 1968, 1045. 2-3Fb

68J(C)1077 J. P. Brown, *J. Chem. Soc. C*, 1968, 1077. 4-9A

68J(C)1501 D. M. O'Mant, *J. Chem. Soc. C*, 1968, 1501. 3-2Db

68J(C)1925 D. M. Brown, M. J. E. Hewlins and P. Schell, *J. Chem. Soc. C*, 1968, 1925. 2-7Eb, 2-7Ed

68J(C)2050 D. M. Brown and M. J. E. Hewlins, *J. Chem. Soc. C*, 1968, 2050. 2-6Cc

68J(C)2076 A. Albert, *J. Chem. Soc. C*, 1968, 2076. 5-2Bc

68J(C)2656 I. T. Kay and P. J. Taylor, *J. Chem. Soc. C*, 1968, 2656. 1-6Cb, 2-3Bc(2), 2-3Bd(3), T2-3

68J(C)2857 G. Adembri, F. De Sio, R. Nesi and M. Scotton, *J. Chem. Soc. C*, 1968, 2857. 2-4Ce

68J(C)2989 R. D. Chambers, J. A. H. MacBride and W. K. R. Musgrave, *J. Chem. Soc. C*, 1968, 2989. 2-4Ca

68J(C)3036 C. O. Bender and R. Bonnett, *J. Chem. Soc. C*, 1968, 3036. 3-1Ad

68JA470 S. Furberg and L. H. Jensen, *J. Amer. Chem. Soc.*, 1968, **90**, 470. 2-4Db

68JA1369 M. G. Pleiss and J. A. Moore, *J. Amer. Chem. Soc.*, 1968, **90**, 1369. 6-2Bb(2)

68JA1569 P. Beak, J. Bonham and J. T. Lee, Jr., *J. Amer. Chem. Soc.*, 1968, **90**, 1569. 2-1C

68JA2875 F. W. Fowler and A. Hassner, *J. Amer. Chem. Soc.*, 1968, **90**, 2875. 6-1Aa

68JA5273 R. B. Greenwald and E. C. Taylor, *J. Amer. Chem. Soc.*, 1968, **90**, 5273. 4-2Hc(ii), 6-1B

68JH133 G. Grandolini, A. Ricci, N. P. Buu-Hoï and F. Perin, *J. Heterocycl. Chem.*, 1968, **5**, 133. 3-2Cg

68JH533 L. Paoloni, M. L. Tosato and M. Cignitti, *J. Heterocycl. Chem.*, 1968, **5**, 533. 2-1C

68JH631 J. C. Craig, Jr. and D. E. Pearson, *J. Heterocycl. Chem.*, 1968, **5**, 631. 2-6Ba

68JO143 K. T. Potts and C. Hirsch, *J. Org. Chem.*, 1968, **33**, 143. S5-5(2)

68JO513 F. A. Snavely and C. H. Yoder, *J. Org. Chem.*, 1968, **33**, 513. 4-2Gj, 4-2Hd, 4-3Ae, 4-6Bc, T4-18

68JO552 K. Matsumoto and H. Rapoport, *J. Org. Chem.*, 1968, **33**, 552. 4-6Fg(i), 4-6Fg(ii)(2), 4-6Fg(iii)

68JO888 J. P. Paolini, *J. Org. Chem.*, 1968, **33**, 888. 2-5F

68JO2606 J. A. Settepani and J. B. Stokes, *J. Org. Chem.*, 1968, **33**, 2606. 4-6Db

68JO2956 L. T. Creagh and P. Truitt, *J. Org. Chem.*, 1968, **33**, 2956. 1-5Ac(i)(3), 4-1Eb(iii)

68JO3336 G. J. Lestina, G. P. Happ, D. P. Maier and T. H. Regan, *J. Org. Chem.*, 1968, **33**, 3336. 4-6Fd(i), 4-6Fd(iv), T4-17

68JS175 E. Shefter, *J. Pharm. Sci.*, 1968, **57**, 175. 3-2Cc

68KG215 V. P. Shchipanov, I. Y. Postovskii and E. N. Rysakova, *Khim. Geterotsikl. Soedin.*, 1968, 215. 4-5Ff

68KG423 L. F. Avramenko, T. A. Zakharova, V. Y. Pochinok and Y. S. Rozum, *Khim. Geterotsikl. Soedin.*, 1968, 423. 5-5Ab

68KG698 Y. A. Rozin, E. P. Darienko and Z. V. Pushkareva, *Khim. Geterotsikl. Soedin.*, 1968, 698. 4-3Bd(2)

68KG905 M. Yu. Kornilov, G. G. Dyadyusha and F. S. Babichev, *Khim. Geterotsikl. Soedin.*, 1968, 905. 5-5Ac

68LA(716)11 E. Müller and H. Meier, *Liebig's Ann. Chem.*, 1968, **716**, 11. 4-1I

68M2223 H. Sterk, T. Kappe and E. Ziegler, *Monatsh. Chem.*, 1968, **99**, 2223. 3-2Ba, 3-2Fa

68MC731 W. D. Podmore, *J. Med. Chem.*, 1968, **11**, 731. 2-8Dg

68MC1045 D. L. Trepanier, L. W. Rampy, K. L. Shriver, J. N. Eble and P. J. Shea, *J. Med. Chem.*, 1968, **11**, 1045. 2-4De

68MI1 J. D. Watson, "The Double Helix: a Personal Account of the Discovery of the Structure of DNA." Weidenfeld and Nicolson, London, 1968. vi

68MI73 R. Shapiro, *Progr. Nucl. Acid Res. Mol. Biol.*, 1968, **8**, 73. 5-1Eg(3)

68MI343 A.-B. Hörnfeldt, *Swed. Chem. Notes*, 1968, **80**, 343. 3-1-Ca(i), 3-2Ca(i)

68MI365 J. S. Kwiatkowski, *Acta Phys. Pol.*, 1968, **34**, 365. 5-1Bb

68MI1275 A. P. Rudko, I. N. Chernyuk, Y. S. Rozum and G. T. Pilyugin, *Ukr. Khim. Zh.*, 1968, **34**, 1275; *Chem. Abstr.*, 1969, **70**, 110146e. 2-7B

68MP73 L. Guibé and E. A. C. Lucken, *Mol. Phys.*, 1968, **14**, 73. 1-6A, 4-1Ba, 4-1Ca, 4-1Ea, 4-1Fa(v)

68OG665 N. A. Vorontsova, N. L. Poznanskaya, O. N. Vlasov and N. I. Shvetsov-Shilovskii, *Org. Reactiv.* (*USSR*), 1968, **5**, 665. 4-3Bb, 4-4Cb

68PH1642 R. H. Cox and A. A. Bothner-By, *J. Phys. Chem.*, 1968, **72**, 1642. 2-2Gd, 2-4E, 2-6Dc

68PH1646 R. H. Cox and A. A. Bothner-By, *J. Phys. Chem.*, 1968, **72**, 1646. 2-4E, 2-6Dc

68PH2092 W. F. Richey and R. S. Becker, *J. Phys. Chem.*, 1968, **49**, 2092. 2-3D

68PH3692 S. G. Schulman and H. Gershon, *J. Phys. Chem.*, 1968, **72**, 3692. 2-3Ha

68PN(60)402 S.-H. Kim and A. Rich, *Proc. Natl. Acad. Sci. USA*, 1968, **60**, 402. 5-1Da

68R1011 H. G. Peer, G. A. M. Von den Ouweland and C. N. de Groot, *Rec. Trav. Chim. Pays-Bas*, 1968, **87**, 1011. 3-2Ac 3-2Bd

68RC1867 S. Baloniak, *Rocz. Chem.*, 1968, **42**, 1867. 2-4Cd

68RR39 F. Mihai and R. Nutiu, *Rev. Roum. Chim.*, 1968, **13**, 39. 2-4Dg

68RR147 Z. Simon, F. Mihai and R. Nutiu, *Rev. Roum. Chim.*, 1968, **13**, 147. 2-4Dg

68SA237 M. M. Cordes and J. L. Walter, *Spectrochim. Acta, Part A*, 1968, **24**, 237. 4-1Ca

68SA1421 M. M. Cordes and J. L. Walter, *Spectrochim. Acta Part A*, 1968, **24**, 1421. 4-1Ha

68T151 R. Cencioni, P. F. Franchini and M. Orienti, *Tetrahedron*, 1968, **24**, 151. 1-4C, 1-7A, 1-7C, 4-2Bb, 4-2Bd, 4-2Be(3), 4-2Bf(3), T4-12(5)

68T441 M. Wahren, *Tetrahedron*, 1968, **24**, 441. 4-9Ba, 4-9Bb

68T1777 S. Schulman and Q. Fernando, *Tetrahedron*, 1968, **24**, 1777 2-3Ha(2)

68T2839 A. H. Beckett, R. G. W. Spickett and S. H. B. Wright, *Tetrahedron*, 1968, **24**, 2839. 5-3Bc

68T4477 O. A. Gansow and R. H. Holm, *Tetrahedron*, 1968, **24**, 4477. 2-3D

68T4907 G. Desimoni and G. Minoli, *Tetrahedron*, 1968, **24**, 4907. 4-5Ba

68T5205 O. Tsuge, S. Kanemasa and M. Tashiro, *Tetrahedron*, 1968, **24**, 5205. 4-3Dd, 4-3Df, 4-3Dj

68T6093 A. H. Beckett, R. W. Daisley and J. Walter, *Tetrahedron*, 1968, **34**, 6093. 3-2Ed(2)

68T6809 H. Dorn and A. Otto, *Tetrahedron*, 1968, **24**, 6809. 4-2Hc

68TC(10)47 J. S. Kwiatkowski, *Theor. Chim. Acta*, 1968, **10**, 47. 5-1Bb

68TC(11)279 H. Morita and S. Nagakura, *Theor. Chim. Acta*, 1968, **11**, 279. 2-6Cb

68TE184 M. E. Perel'son and Y. N. Sheinker, *Teor. Eksp. Khim.*, 1968, **4**, 184. 2-3Gb

68TE379 V. P. Lezina, V. F. Bystrov, L. D. Smirnov and K. M. Dyumaev, *Teor. Eksp. Khim.*, 1968, **4**, 379. 2-3D

68TH01 K. Bösl, Ph.D. Thesis, University of Karlsruhe, West Germany, 1968. 2-5Cd

68TL2747 J. Adams and R. G. Shepherd, *Tetrahedron Lett.*, 1968, 2747. 2-7De, 2-8Bd, 2-6Ea

68TL2767 A. Gordon and A. R. Katritzky, *Tetrahedron Lett.*, 1968, 2767. 2-3Bg

68TL3727 J. D. Vaughan and M. O'Donnell, *Tetrahedron Lett.*, 1968, 3727. 4-1Da, 4-1Ea(iv), 4-1Kb

68TL4593 J. L. Wong and F. N. Bruscato, *Tetrahedron Lett.*, 1968, 4593. 2-7Ea

68TL5691 E. Spinner and G. B. Yeoh, *Tetrahedron Lett.*, 1968, 5691. 2-3Bc, 2-3Be

68TL5707 A. Morimoto, K. Noda, T. Watanabe and H. Takasugi, *Tetrahedron Lett.*, 1968, 5707. S5-5

68ZC305 J. Kuthan, V. Skala and J. Palecèk, *Z. Chem.*, 1968, **8**, 305. 2-3Ac

68ZO2449 V. M. Berezovskii, Z. I. Akselrod, T. P. Fetisova, I. M. Kustanovich, and N. A. Polyakova, *Zh. Obshch. Khim.*, 1968, **38**, 2449. 2-4Gd

69AG307 G. Will, *Angew. Chem.*, 1969, **81**, 307. 1-4Ic

69AG(E)139 W. Saenger and K. H. Scheit, *Angew. Chem.* (*Int. Ed. Engl.*), 1969, **8**, 139. 2-5Be

69AG(E)986 H. U. Wagner and R. Gompper, *Angew. Chem.* (*Int. Ed. Engl.*), 1969, **8**, 986. 2-8Dg

69AH(61)181 G. Hornyak, K. Lemert and K. Zauer, *Acta Chim.* (*Budapest*), 1969, **61**, 181. 2-6Ec

69AJ563 J. H. Bowie, R. K. M. R. Kallury and R. G. Cooks, *Aust. J. Chem.*, 1969, **22**, 563. 4-2Bd

69AJ1759 D. B. Paul and H. J. Rodda, *Aust. J. Chem.*, 1969, **22**, 1759. 2-4Ga

69AJ2595 B. D. Batts and E. Spinner, *Aust. J. Chem.*, 1969, **22**, 2595, 1-6Cb, 1-6Ce, 2-3C, 2-6Bb

69AJ2611 B. D. Batts and E. Spinner, *Aust. J. Chem.*, 1969, **22**, 2611. 2-7Ea, 2-7Db(4)

69AK71 A. Kvick and I. Olovsson, *Ark. Kemi*, 1969, **30**, 71. 2-3Bc

69AP423 M. Huke and K. Gordlitzer, *Arch. Pharm.* (*Weinheim*), 1969, **302**, 423. 3-2Ad(2) 3-2Cf

69AS1091 M. Begtrup and C. Pedersen, *Acta Chem. Scand.*, 1969, **23**, 1091. 4-3Ef

69AS2025 M. Begtrup, *Acta Chem. Scand.*, 1969, **23**, 2025. 4-3Ec

69AS2031 J. Danielsen, *Acta Chem. Scand.*, 1969, **23**, 2031. 3-2Cd

69AS2879 G. Kjellin and J. Sandström, *Acta Chem. Scand.*, 1969, **23**, 2879. 4-4Ca

69AS2888 G. Kjellin and J. Sandström, *Acta Chem. Scand.*, 1969, **23**, 2888. 1-3B, 1-6E, 1-7B, 4-4Ca, 4-4Cc(i)(3), 4-4Cc(ii)(3), 4-4Ce

69AS2963 I. Fischer-Hjalmars and J. Nag-Chaudhuri, *Acta Chem. Scand.*, 1969, **23**, 2963. 5-1Ee

69AS3125 O. Buchardt, C. L. Pedersen, U. Svanholm, A. M. Duffield and A. T. Balaban, *Acta Chem. Scand.*, 1969, **23**, 3125. 6-2A

69BB407 C. Aussems, S. Jaspers, G. Leroy and F. Van Remoortere, *Bull. Soc. Chim. Belg.*, 1969, **78**, 407. 4-1Ba

69BF823 M. Sélim and M. Sélim, *Bull. Soc. Chim. Fr.*, 1969, 823. 1-3Bb, 4-3De, 4-4Ed(2), 4-5Fa(i)(3)

69BF870 J.-F. Giudicelli, J. Menin and H. Najer, *Bull. Soc. Chim. Fr.*, 1969, 870. 4-5Fa(ii)

69BF874 J.-F. Giudicelli, J. Menin and H. Najer, *Bull. Soc. Chim. Fr.*, 1969, 874. 4-5Fa(ii)

69BF1097 M. Roche and L. Pujol, *Bull. Soc. Chim. Fr.*, 1969, 1097. 1-7A, 1-7B, 4-1Ba, 4-1Ca, 4-1Da, 4-1Eb(iv), 4-1Fa(vii), T4-1, T4-6, T4-7, T4-8

69BF1715 P. Rafiteau and P. Maitte, *Bull. Soc. Chim. Fr.*, 1969, 1715. 2-2A

69BF2004 J. Schmitt, M. Langlois, C. Perrin and G. Callet, *Bull. Soc. Chim. Fr.*, 1969, 2004. 3-4Cc

69BF3133 A. Le Berre and C. Renault, *Bull. Soc. Chim. Fr.*, 1969, 3133. 2-4De

69BF4159 C. Sabaté-Alduy and J. Lematre, *Bull. Soc. Chim. Fr.*, 1969, 4159. 4-2Gf(2), 4-2Ib, T4-21(2)

69BG407 J. W. Eastman, *Ber. Bunsenges. Phys. Chem.*, 1969, **73**, 407. 5-1Dc

69BJ1467 H. Fujita, A. Imamura and C. Nagata, *Bull. Chem. Soc. Jap.*, 1969, **42**, 1467. 2-4Db

69BJ1678 H. Nakata and A. Tatematsu, *Bull. Chem. Soc. Jap.*, 1969, **42**, 1678. 1-6Ga(ii)

69BJ2690 N. Sugiyama, M. Yamamoto and C. Kashima, *Bull. Chem. Soc. Jap.*, 1969, **42**, 2690 2-3C

69BJ3099 H. Mizuno, T. Fujiwara and K. Tomita, *Bull. Chem. Soc. Jap.*, 1969, **42**, 3099. 5-1Ec(2)

69BJ3204 K. Akagane, F. J. Allan, G. G. Allan, T. Friberg, S. O. Muircheartaigh and J. B. Thomson, *Bull. Chem. Soc. Jap.*, 1969, **42**, 3204. 4-7Ab

69CB230 W. Loop, H.-J. May and H. Baganz, *Chem. Ber.*, 1969, **102** 230. 4-5Da

69CB417 M. Regitz and H. Scherer, *Chem. Ber.*, 1969, **102**, 417. 4-4Fb

69CB1129 W. Steglich and G. Höfle, *Chem. Ber.*, 1969, **102**, 1129. 4-3Cb, 4-3Cc

69CB1961 W. Geiger, H. Böshagen and H. Medenwald, *Chem. Ber.*, 1969, **102**, 1961. 4-5Bf(3)

69CB3666 G. Simchen and W. Krämer, *Chem. Ber.*, 1969, **102**, 3666. 2-3Bc

69CB3775 H. Böshagen and W. Geiger, *Chem. Ber.*, 1969, **102**, 3775. 4-2Cc

69CB4177 K. Volkamer and H. W. Zimmermann, *Chem. Ber.*, 1969, **102**, 4177. 1-6Gb, 4-7Ab

69CC743 G. A. Neville and D. Cook, *Can. J. Chem.*, 1969, **47**, 743. 2-4Dg

69CC1123 J. T. Edward and J. K. Liu, *Can. J. Chem.*, 1969, **47**, 1123. 4-6Cb(4)

69CC4483 E. Costakis, P. Canone and G. Tsatsas, *Can. J. Chem.*, 1969, **47**, 4483. 4-5Dd

69CG58 W. Freiberg and C.-F. Kröger, *Mitteilungsbl. Chem. Ges. D.D.R.*, 1969, **16**, 58. 4-4Ej, 4-5Fc

69CH147 D. J. Anderson, T. L. Gilchrist and C. W. Rees, *Chem. Commun.*, 1969, 147. 6-1Aa

69CH405 R. N. Butler, *Chem. Commun.*, 1969, 405. 4-5Ff

69CH554 B. F. Hoskins, S. A. Mason and J. C. B. White, *Chem. Commun.*, 1969, 554. 6-3

69CH811 L. A. Plastas and J. M. Stewart, *Chem. Commun.*, 1969, 811. 1-1C, 4-6Ff(ii)

69CH1038 L. A. Plastas and J. M. Stewart, *Chem. Commun.*, 1969, 1038. 1-1C, 4-6Fg(i)

69CI1077 R. E. Bowman, *Chem. Ind.*, (*London*), 1969, 1077. 3-2Ee

69CT425 A. Nakamura and S. Kamiya, *Chem. Pharm. Bull.*, 1969, **17**, 425. 2-3Ee

69CT550 T. Hino, K. Tsuneoka, M. Nakagawa and S. Akaboshi, *Chem. Pharm. Bull.*, 1969, **17**, 550. 3-3C(2)

69CT1485 M. Sano, I. Itoh, Y. Nakai and T. Naito, *Chem. Pharm. Bull.* 1969, **17**, 1485. 4-2Fc, 4-2Gg(2), 4-2Hb(i), 4-2Hb(iii), 4-2Hc(iii), T4-20(2)

69CT2266 I. Matsura and K. Okui, *Chem. Pharm. Bull.*, 1969, **17**, 2266. 2-4Ga

69CX(B)135 P. Goldstein, J. Ladell and G. Abowitz, *Acta Crystallogr., Sect. B*, 1969, **25**, 135. 1-4Ib(2), 4-1Ea

69CX(B)182 C. Sabelli and P. F. Zanazzi, *Acta Crystallogr., Sect. B*, 1969, **25**, 182. 4-2Be

69CX(B)192 C. Sabelli and P. F. Zanazzi, *Acta Crystallogr., Sect. B*, 1969, **25**, 192. 1-3B, 4-2Bd

69CX(B)362 M. Sax, R. Desiderato and T. W. Dakin, *Acta Crystallogr., Sect. B*, 1969, **25**, 362. 2-3Ha

69CX(B)625 D. L. Smith, *Acta Crystallogr., Sect. B*, 1969, **25**, 625. 4-5Dc, 5-2Ba

69CX(B)1038 K. Ozeki, N. Sakabe and J. Tanaka, *Acta Crystallogr., Sect. B*, 1969, **25**, 1038. 2-4Db

69CX(B)1050 M. Cannas, S. Biagini and G. Marongiu, *Acta Crystallogr., Sect. B*, 1969, **25**, 1050. 1-3B, 4-2Bd

69CX(B)1247 J. L. Lawrence and S. G. G. MacDonald, *Acta Crystallogr., Sect. B*, 1969, **25**, 1247. 3-2Bb

69CX(B)1330 E. Sletten, J. Sletten and L. H. Jensen, *Acta Crystallogr., Sect. B*, 1969, **25**, 1330. 5-1C

69CX(B)1608 J. Sletten and L. H. Jensen, *Acta Crystallogr. Sect. B*, 1969, **25**, 1608. 5-1Bd

69CX(B)1338 G. M. Brown, *Acta Crystallogr., Sect. B*, 1969, **25**, 1338. 5-1C

69CX(B)1423 D. W. Young, P. Tollin and H. R. Wilson, *Acta. Crystallogr., Sect. B*, 1969, **25**, 1423. 2-4Db

69CX(B)1970 B. M. Craven and T. M. Sabine, *Acta Crystallogr., Sect. B*, 1969, **25**, 1970. 2-4Dg

69CX(B)1978 B. M. Craven, E. A. Vizzini and M. M. Rodrigues, *Acta Crystallogr., Sect. B*, 1969, **25**, 1978. 2-4Dg

69CX(B)2108 S. Biagini, M. Cannas and G. Marongiu, *Acta. Crystallogr., Sect. B*, 1969, **25**, 2108. 4-2Cb(ii)

69CX(B)2144 D. J. Hunt and E. Subramanian, *Acta Crystallogr., Sect. B*, 1969, **25**, 2144. 2-4Db

69CX(B)2349 L. Cavalca, G. Fava Gasparri, A. Mangia and G. Pelizzi, *Acta Crystallogr., Sect. B*, 1969, **25**, 2349. 4-2E

69CX(B)2355 D. L. Smith and E. K. Barrett, *Acta Crystallogr., Sect. B*, 1969, **25**, 2355. 4-2Gg, 4-2Jc

69CZ3895 I. Dobas, V. Sterba and M. Vecera, *Collect. Czech. Chem. Commun.*, 1969, **34**, 3895. 4-2Ga(i)

69DA(189)326 E. S. Levin and G. N. Rodionova, *Dokl. Akad. Nauk SSSR*, 1969, **189**, 326. 2-3Bg

69H2641 H.-J. Gais, K. Hafner and M. Nevenschwander, *Helv. Chim. Acta*, 1969, **52**, 2641. 4-6Fd(i), 4-6Fd(iii)

69HC(10)1 A. J. Boulton and P. B. Ghosh, *Advan. Heterocycl. Chem.*, 1969, **10**, 1. 4-9Ba

69HC(10)132 J. D. White and M. E. Mann, *Advan. Heterocycl. Chem.*, 1969, **10**, 132. 3-1Ad(2)

69HC(10)149 W. J. Irwin and D. G. Wibberley, *Advan. Heterocycl. Chem.*, 1969, **10**, 149. 2-4Ga

69IJ884 A. Singh, H. Singh, A. S. Uppal and K. S. Narang, *Indian J. Chem.*, 1969, **7**, 884. 4-6Dc

69IM41 G. P. Marchesa and E. Panzeri, *Chim. Ind., (Milan)*, 1969, **51**, 41. 3-3C

69IQ103 B. Pullman, E. D. Bergmann, H. Weiler-Feilchenfeld and Z. Neiman, *Int. J. Quant. Chem., Symp.* 1969, 103. 1-4G, 5-1Ea

69J(B)299 J. V. Greenhill, *J. Chem. Soc. B*, 1969, **3**, 299. 2-3Ac

69J(B)307 E. Borello, A. Zecchina and E. Guglielminetti, *J. Chem. Soc. B*, 1969, 307. 4-1Da(2)

69J(B)680 R. N. Butler, *J. Chem. Soc. B*, 1969, 680. 4-5Ff

69J(B)1240 M. Charton, *J. Chem. Soc. B*, 1969, 1240. 1-2(2), 1-4B, 1-4Bb, 1-4G, 1-5B, 4-1Fa(i), 4-1Fb

69J(C)245 T. Nishiwaki, *J. Chem. Soc. C*, 1969, 245. 4-2Bg(i)

69J(C)836 F. G. Baddar, M. F. El-Newaihy and M. R. Salem, *J. Chem. Soc. C*, 1969, 836. 4-2Hc, 4-2Ib(2), T4-21

69J(C)1660 R. F. Banks, R. N. Haszeldine, D. R. Karsa, F. E. Rickett and M. Young, *J. Chem. Soc. C*, 1969, 1660. 2-5Be

69J(C)1678 D. W. Jones, *J. Chem. Soc. C*, 1969, 1678. 2-3Ec

69J(C)1729 D. W. Jones, *J. Chem. Soc. C*, 1969, 1729. 2-3Bf(2)

69J(C)2379 A. Albert, *J. Chem. Soc. C*, 1969, 2379. 4-5Fd

69J(C)2794 B. W. Nash, R. A. Newberry, R. Pickles and W. K. Warburton, *J. Chem. Soc. C*, 1969, 2794. 4-3Db, 4-3Ea, 4-4Ea

69JA706 H. Hart and J. L. Brewbaker, *J. Amer. Chem. Soc.*, 1969, **91**, 706. 4-1Aa
69JA796 M. J. S. Dewar and T. Morita, *J. Amer. Chem. Soc.*, 1969, **91**, 796. 1-7C, 4-1Ba, 4-1Eb(iv), T4-1
69JA1856 H. Feuer and J. P. Lawrence, *J. Amer. Chem. Soc.*, 1969, **91**, 1856. 2-8Eb
69JA6090 P. R. Rony, *J. Amer. Chem. Soc.*, 1969, **91**, 6090. 2-3Be
69JH53 V. Sunjic, T. Fajdiga, M. Japelj and P. Rems, *J. Heterocycl. Chem.*, 1969, **6**, 53. 4-5Eb
69JH123 T. H. Kinstle and L. J. Darlage, *J. Heterocycl. Chem.*, 1969, **6**, 123. 4-2Cc, 4-3Bb
69JH147 D. L. Eck and G. W. Stacy, *J. Heterocycl. Chem.*, 1969, **6**, 147. 3-4Ba
69JH559 M. Japelj, B. Stanovnik and M. Tišler, *J. Heterocycl. Chem.*, 1969, **6**, 559. 5-3A(2), S5-3(2)
69JH655 R. J. North and A. R. Day, *J. Heterocycl. Chem.*, 1969, **6**, 655. 4-5Df
69JH723 F. W. Short and E. J. Schoeb, *J. Heterocycl. Chem.*, 1969, **6**, 723. 4-2Gf
69JH735 M. Israel and L. C. Jones, *J. Heterocycl. Chem.*, 1969, **6**, 735. T6-1
69JH779 E. E. Gilbert and B. Veldhuis, *J. Heterocycl. Chem.*, 1969, **6**, 779. 2-4Fc, 2-5F, 2-6Ee
69JH783 G. Adembri, F. Ponticelli and P. Tedeschi, *J. Heterocycl. Chem.*, 1969, **6**, 783. 4-5Ba
69JH901 S. Biagini, M. Cannas and G. Marongiu, *J. Heterocycl. Chem.*, 1969, **6**, 901. 4-2Cb(ii)
69JH947 J. B. Wright, *J. Heterocycl. Chem.*, 1969, **6**, 947. 5-3Bb(i)
69JL(4)108 H. Hüther, F. Kajfež, L. Klasinc and V. Šunjić, *J. Mol. Struct.*, 1969, **4**, 108. 4-1Ka
69JO187 J. P. Freeman, J. J. Gannon and D. L. Surbey, *J. Org. Chem.*, 1969, **34**, 187. 4-7Aa
69JO194 J. P. Freeman and J. J. Gannon, *J. Org. Chem.*, 1969, **34**, 194. 4-7B(2)
69JO821 B. Roth and J. Z. Strelitz, *J. Org. Chem.*, 1969, **34**, 821. 2-2Gc
69JO978 V. Wolcke and G. B. Brown, *J. Org. Chem.*, 1969, **34**, 978. 5-1F
69JO1685 G. J. Lestina and T. H. Regan, *J. Org. Chem.*, 1969, **34**, 1685. 4-2Gj
69JO2125 P. Beak and J. T. Lee, Jr., *J. Org. Chem.*, 1969, **34**, 2125. 2-1C
69JO2138 W. W. Paudler and A. G. Zeiler, *J. Org. Chem.*, 1969, **34**, 2138. 6-2Be(2)
69JO2153 I. Scheinfeld, J. C. Parham, S. Murphy and G. B. Brown, *J. Org. Chem.*, 1969, **34**, 2153. 5-1F, 5-1Fa(2)
69JO3169 J. Wolinsky and H. S. Hauer, *J. Org. Chem.*, 1969, **34**, 3169. 2-2A
69JO3237 W. W. Paudler and A. G. Zeiler, *J. Org. Chem.*, 1969, **34**, 3237. 6-2A

69JO3279 P. L. Southwick, J. A. Fitzgerald, R. Madhav and D. A. Welsh, *J. Org. Chem.*, 1969, **34**, 3279. 3-2Ec, 3-2Fe

69JO3672 E. J. Moriconi and R. E. Misner, *J. Org. Chem.*, 1969, **34**, 3672. 2-2C

69JO4024 A. M. Aguiar, G. W. Prejean, J. R. Smiley Irelan and C. J. Morrow, *J. Org. Chem.*, 1969, **34**, 4024. 2-2D

69JO4164 H. E. Holmquist, *J. Org. Chem.*, 1969, **34**, 4164. 3-2Ab

69JT(51)1862 H. H. Chen and L. B. Clark, *J. Chem. Phys.*, 1969, **51**, 1862. 5-1Ea

69KG312 A. N. Borisevich and P. S. Pel'kis, *Khim. Geterotsikl. Soedin.*, 1969, 312. 4-5Bi

69KG719 A. K. Aren and D. V. Bite, *Khim. Geterotsikl. Soedin.*, 1969, 719. 4-4Ce

69KG732 V. A. Lopyrev, N. K. Beresneva and B. K. Strelets, *Khim. Geterotsikl. Soedin.*, 1969, 732. 4-6Dd(2)

69KG923 V. P. Shchipanov, *Khim. Geterotsikl. Soedin.*, 1969, 923. 4-5Ff

69KG978 B. A. Arbouzov, O. A. Erastov, A. B. Remizov and L. Z. Nikonova, *Khim. Geterotsikl. Soedin.*, 1969, 978. 3-2Ee

69LA(724)159 W. Ried and K. Wagner, *Liebig's Ann. Chem.*, 1969, **724**, 159. 4-2Gg

69MI22 H. Yasuda, *Appl. Spectrosc.*, 1969, **23**, 22. 4-2Gj, T4-18

69MI77 W. Steglich, *Fortsch. Chem. Forsch.*, 1969, **12**, 77. 1-4G, 5-1Ea, 4-3Cb, 4-3Cc(3)

69MI249 L. A. Paquette, "Nonbenzenoid Aromatics" (J. P. Snyder, Ed.), Vol. 1, p. 249. Academic Press, New York, 1969. 6-2A, 6-2Ba

69MI256 P. O. P. Ts'o, M. P. Schweizer and D. P. Hollis, *Ann. N.Y. Acad. Sci.*, 1969, **158**, 256. 5-1Bf

69MI307 R. V. Wolfenden, *J. Mol. Biol.*, 1969, **40**, 307. 5-1Bf, 5-1Da

69MI336 M. J. S. Dewar, "The Molecular Orbital Theory of Organic Chemistry," p. 336. McGraw-Hill, New York, 1969. 4-1Ab

69NA1170 P. Tollin and A. R. I. Munns, *Nature* (*London*), 1969, **222**, 1170. 5-1Bf

69NK769 K. Mamoru, *Nippon Kagaku Zasshi*, 1969, **90**, 769. 2-4Gb

69OM37 J. M. Desmarchelier and R. B. Johns, *Org. Mass Spectrom.*, 1969, **2**, 37. 4-2Jd

69OM49 L. Zamir, B. S. Jensen and E. Larsen, *Org. Mass Spectrom.*, 1969, **2**, 49. 1-6Ga

69OM433 D. M. Forkey and W. R. Carpenter, *Org. Mass Spectrom.*, 1969, **2**, 433. 4-1Fa(vi)

69OM697 J. M. Desmarchelier and R. B. Johns, *Org. Mass Spectrom.*, 1969, **2**, 697. 4-2Gb, T4-21(3)

69OR481 J. A. Glasel, *Org. Magn. Resonance*, 1969, **1**, 481. 1-5E, 2-4Dg(2)

69PH2465 R. H. Cox and A. A. Bothner-By, *J. Phys. Chem.*, 1969, **73**, 2465. 2-3Be, 2-3C

69PN(63)253 C. A. Taylor, M. A. El-Bayoumi and M. Kasha, *Proc. Nat. Acad. Sci. U.S.*, 1969, **63**, 253. 5-2Aa

69PN(63)1359 C. L. Coulter and S. W. Hawkinson, *Proc. Nat. Acad. Sci. U.S.*, 1969, **63**, 1359.

69R30 L. Brandsma and P. J. W. Schujl, *Rec. Trav. Chim. Pays-Bas*, 1969, **88**, 30. 2-2B

69R1139 J. Reedijk, *Rec. Trav. Chim. Pays-Bas*, 1969, **88**, 1139. 2-3C

69RC315 S. Baloniak, *Rocz. Chem.*, 1969, **43**, 315. 2-4Cd

69RC1187 S. Baloniak, *Rocz. Chem.*, 1969, **43**, 1187. 2-4Cd

69RO1685 I. Y. Kvitko and B. A. Porai-Koshits, *Z. Org. Khim.*, 1969, **5**, 1685. 4-4Ae

69RO1711 V. G. Kharchenko, S. K. Klimenko, V. I. Kleimenova and N. M. Kupranets, *Z. Org. Khim.*, 1969, **5**, 1711. 2-2B

69RR311 R. Nuţiu and Z. Simon, *Rev. Roum. Chim.*, 1969, **14**, 311. 2-4Dg

69RR1435 R. Nuţiu, L. Kurunczi and Z. Simon, *Rev. Roum. Chim.*, 1969, **14**, 1435. 2-4Dh

69RV187 D. M. Brown, *Pure Appl. Chem.*, 1969, **18**, 187. 2-7Ed

69T255 W. Werner, *Tetrahedron*, 1969, **25**, 255. 2-3Ba

69T517 M. Mazharuddin and G. Thyagarajan, *Tetrahedron*, 1969, **25**, 517. 2-7Ec, 2-6F, 2-2E, 2-4A, 2-5D

69T747 T. Nishiwaki, *Tetrahedron*, 1969, **25**, 747. 4-2Bd, 4-2Bg(i)

69T783 Y. Hagiwara, M. Kurihara and N. Yoda, *Tetrahedron*, 1969, **25**, 783. 2-4Da

69T1001 W. Seiffert, V. Zanker and H. Mantsch, *Tetrahedron*, 1969, **25**, 1001. 1-6E, 2-3Hb

69T2023 U. E. Matter, C. Pascual, E. Pretsch, A. Pross, W. Simon and S. Sternhell, *Tetrahedron*, 1969, **25**, 2023. 2-2A

69T2807 W. Rubaszewska and Z. R. Grabowski, *Tetrahedron*, 1969, **25**, 2807. 3-2Cf

69T3453 R. Buyle and H. G. Viehe, *Tetrahedron*, 1969, **25**, 3453. 4-6Fa(iii), 4-6Fb(i), 4-6Fd(i)(2), 4-6Fd(ii)(2), T4-30(3)

69T4265 J. Nyitrai and K. Lempert, *Tetrahedron*, 1969, **25**, 4265. 4-3Cg, 4-3C(i)(2), 4-4D, 4-6Cb(2), 4-6Ec

69T4605 G. A. Newman and P. J. S. Pauwels, *Tetrahedron*, 1969, **25**, 4605. 4-2Fa, 4-2Fc(3), 4-2Gc, 4-2Gh(2), T4-17(4)

69T4649 Y. Shvo and I. Belsky, *Tetrahedron*, 1969, **25**, 4649. 2-3Bd

69T4667 A. N. Nesmeyanov, V. N. Babin, L. A. Fedorov, M. I. Rybinskaya and E. I. Fedin, *Tetrahedron*, 1969, **25**, 4667. 1-5Ac(iii), 4-1I

69T5721 R. Fuks and H. G. Viehe, *Tetrahedron*, 1969, **25**, 5721. 3-2Ec

69TC(13)278 B. Mely and A. Pullman, *Theor. Chim. Acta*, 1969, **13**, 278. 1-6F

69TC(14)221 L. Paoloni, M. L. Tosato and M. Cignitti, *Theor. Chim. Acta*, 1969, **14**, 221. 2-3D

69TC(15)225 D. T. Clark, *Theor. Chim. Acta*, 1969, **15**, 225. 6-1Aa

69TC(15)265 B. Pullman, H. Berthod and M. Dreyfus, *Theor. Chim. Acta*, 1969, **15**, 265. 5-1Dd

69TE160 V. P. Zvolinskii, M. E. Perelson and Y. N. Sheinker, *Theor. Exp. Chem. (USSR)*, 1969, **5**, 160. 2-3Ba

69TE247 V. P. Lezina, V. F. Bystrov, B. E. Zaitsev, N. A. Andronova, L. D. Smirnov and K. M. Dyumayev, *Theor. Exp. Chem. (USSR)*, 1969, **5**, 247. 2-3D

69TH01 V. Pellegrin, Ph.D Thesis, University of Montpellier, France, 1969. 4-1Ba, 4-1Bb(vii)

69TH02 J. H. Griffiths, Ph.D Thesis University of Wales, Bangor, 1969. T1-5

69TL495 M. L. Roumestant, P. Viallefont, J. Elguero and R. Jacquier, *Tetrahedron Lett.*, 1969, 495. 1-5Ac(ii), 4-1Ba

69TL1507 E. Ager, B. Iddon and H. Suschitzky, *Tetrahedron Lett.*, 1969, 1507. 2-5Ba

69TL1557 H. J. Petersen, *Tetrahedron Lett.*, 1969, 1557. 4-3Cb(2)

69TL4173 G. Malesani, G. Rigatti and G. Rodighiero, *Tetrahedron Lett.*, 1969, 4173. 3-2G

69TL4465 R. L. N. Harris, *Tetrahedron Lett.*, 1969, 4465. 3-3C

69TL5219 S. A. Mason, J. C. B. White and A. Woodlock, *Tetrahedron Lett.*, 1969, 5219. 1-4Ic, 2-3C

69ZC241 M. Wahren, *Z. Chem.*, 1969, **9**, 241. 4-9Ba, 4-9Bb

69ZO1156 I. S. Ioffe and A. B. Tomchin, *Zh. Obshch. Khim.*, 1969, **39**, 1156. 2-7E

69ZO2339 I. S. Ioffe, A. B. Tomchin and E. N. Zhukova, *Zh. Obshch. Khim.*, 1969, **39**, 2339. 2-6Eb

69ZO2568 A. V. Bogatskii, L. Y. Glinskaya and A. I. Gren, *Zh. Obshch. Khim.*, 1969, **39**, 2568. 2-4Dg

69ZX(129)211 G. Will, *Z. Kristallogr.*, 1969, **129**, 211. 1-4Ib, 4-1Ca

69ZX(130)376 H. C. Mez and J. Donohue, *Z. Kristallogr.*, 1969, **130**, 376. 5-2Dc

70AC1178 M. Goldman and E. L. Wehry, *Anal. Chem.*, 1970, **42**, 1178. 2-3Ha

70AJ51 E. N. Cain and R. N. Warrener, *Aust. J. Chem.*, 1970, **23**, 51. 2-4A, 2-5E

70AJ619 G. E. Lewis, D. L. Lill, R. H. Prager and J. A. Reiss, *Aust. J. Chem.*, 1970, **23**, 614. 2-4H

70AN246 G. Corsi and G. Palazzo, *Ann. Chim. (Rome)*, 1970, **60**, 246. 4-4Af

70AN403 L. Baiocchi, *Ann. Chim. (Rome)*, 1970, **60**, 403. 5-2Ba, 5-T1

70AO17 A. D. Baker, *Accounts Chem. Res.*, 1970, **3**, 17. 1-6F

70AO105 E. B. Fleischer, *Accounts Chem. Res.*, 1970, **3**, 105. 6-3

70AP(303)980 G. Schwenker and K. Bösl, *Arch. Pharm. (Weinheim)*, 1970, **303**, 43. 4-5Bf

70AS1819 M. Begtrup, *Acta Chem. Scand.*, 1970, **24**, 1819. S4-5

70AS3230 S. Furberg and J. Solbakk, *Acta Chem. Scand.*, 1970, **24**, 3230. 2-4Dc

70AS3248 F. K. Larsen, M. S. Lehmann, I. Sotofte and S. E. Rasmussen, *Acta Chem. Scand.*, 1970, **24**, 3248. 1-4Ic, 4-1Ba(2), T4-1

70BB343 A. Maquestiau, Y. van Haverbeke and R. N. Muller, *Bull. Soc. Chim. Belg.*, 1970, **79**, 343. 4-2Ba(2), 4-2Bf(3), 4-6Ba(2), T4-12(2)

70BF247 R. Jacquier, C. Petrus, F. Petrus and J. Verducci, *Bull. Soc. Chim. Fr.*, 1970, 247. 4-2Ib(2), T4-21(3)

70BF273 M. Roche and L. Pujol, *Bull. Soc. Chim. Fr.*, 1970, 273. 4-1Ca, T4-6, T4-7, T4-8

70BF1590 M. Pesson and M. Antoine, *Bull. Soc. Chim. Fr.*, 1970, 1590. 2-5F

70BF1599 M. Pesson and M. Antoine, *Bull. Soc. Chim. Fr.*, 1970, 1599. 2-5F

70BF1720 H. Lumbroso, D. M. Bertin and P. Cagniant, *Bull. Soc. Chim. Fr.*, 1970, 1720. 3-3B

70BF1978 R. Jacquier, C. Petrus, F. Petrus and J. Verducci, *Bull. Soc. Chim. Fr.*, 1970, 1978. 4-2Cb(iii), 4-2Cb(iv)(5), 4-2Cb(v), T4-13

70BF1991 G. Pangon, G. Thuillier, and P. Rumpf, *Bull. Soc. Chim. Fr.*, 1970, 1991. 2-3Ec

70BF2237 L. Legrand and N. Lozac'h, *Bull. Soc. Chim. Fr.*, 1970, 2237. 6-2Bb

70BF2309 C. Katamna, *Bull. Soc. Chim. Fr.*, 1970, 2309. 3-2Ad

70BF2690 R. Jacquier, C. Petrus, F. Petrus and J. Verducci, *Bull. Soc. Chim. Fr.*, 1970, 2690. 4-2Bd(2), 4-2Bf(3), T4-12(9)

70BF3572 G. Rio and J. C. Hardy, *Bull. Soc. Chim. Fr.*, 1970, 3572. 3-2Aa, 3-2Ec

70BF4011 J. Ďaunis, R. Jacquier and M. Rigail, *Bull. Soc. Chim. Fr.*, 1970, 4011. 2-4Ca, 2-5E

70BF4317 M. Robba and Y. Le Guen, *Bull. Soc. Chim. Fr.*, 1970, 4317. 2-4Ce, 5-4A

70BF4376 A. Le Berre, B. Dumaitre and J. Petit, *Bull. Soc. Chim. Fr.*, 1970, 4376. 2-4Cd(3), 2-4Ce

70BF4436 J. Elguero, R. Jacquier and S. Mignonac-Mondon, *Bull. Soc. Chim. Fr.*, 1970, 4436. 4-5Bg(3), 4-5Bh, 4-5Bh(i)(2)

70BF4505 C. Fayat and A. Foucaud, *Bull. Soc. Chim. Fr.*, 1970, 4505. 3-2Bc

70BJ849 I. Hori, K. Saito and H. Midorikawa, *Bull. Chem. Soc. Jap.*, 1970, **43**, 849. 5-3Bb(i)(2), 5-3Be

70BJ2283 Y. Tanizaki, H. Hiratsuka and T. Hoshi, *Bull. Chem. Soc. Jap.*, 1970, **43**, 2283. 4-3Bd

70BJ3344 M. Kamiya, *Bull. Chem. Soc. Jap.*, 1970, **43**, 3344. 4-1-G(vi), 4-1I

70C134 A. R. Katritzky, *Chimia*, 1970, **24**, 134. 2-4Db, 2-6Cb, 1-1D, 1-1F, 1-2, 1-3A, 4-4Fd

70CB331 W. Ried and E. Kahr, *Chem. Ber.*, 1970, **103**, 331. 2-2E

70CB398 G. Simchen, *Chem. Ber.*, 1970, **103**, 398. 2-3Bg, T2-1

70CB722 G. Konrad and W. Pfleiderer, *Chem. Ber.*, 1970, **103**, 722. 2-6De

70CB735 G. Konrad and W. Pfleiderer, *Chem. Ber.*, 1970, **103**, 735. 2-6De

70CB1805 J. Goerdeler and P. Martens, *Chem. Ber.*, 1970, **103**, 1805. 4-5Fb(iii)

70CB1900 H. Reimlinger, *Chem. Ber.*, 1970, **108**, 1900. 4-9Bb, 4-9Ca(2), 5-5Ab

70CB2356 H. O. Bayer, H. Gotthardt and R. Huisgen, *Chem. Ber.*, 1970, **103**, 2356. 4-3Cb(2)

70CB2505 H. Dorn and A. Otto, *Chem. Ber.*, 1970, **103**, 2505. 4-5Bg

70CB2828 A. I. Hubert and H. Reimlinger, *Chem. Ber.*, 1970, **103**, 2828. 5-5Ab

70CB3205 W. Flitsch, *Chem. Ber.*, 1970, **103**, 3205. 3-2Fb

70CB3284 H. Reimlinger and R. Merenyi, *Chem. Ber.*, 1970, **103**, 3284. 5-5Ab

70CB3289 G. Häfelinger, *Chem. Ber.*, 1970, **103**, 3289. 4-1Ba

70CB3885 A. Schönberg and E. Frese, *Chem. Ber.*, 1970, **103**, 3885. 4-3Ad, 4-9A, 4-9Ca

70CC1946 J. M. Muchowski, *Can. J. Chem.*, 1970, **48**, 1946. 6-2Be

70CC2709 R. Laliberté and G. Médawar, *Can. J. Chem.*, 1970, **48**, 2709. 3-4Ba

70CC3563 N. J. Nye and W. P. Tang, *Can. J. Chem.*, 1970, **48**, 3563. 4-3Ad

70CH23 M. Barber and D. T. Clark, *Chem. Commun.*, 1970, 23. 1-6F

70CH24 M. Barber and D. T. Clark, *Chem. Commun.*, 1970, 24. 1-6F(2)

70CH25 S. H. Wilen, *Chem. Commun.*, 1970, 25. 2-2D(3)

70CH383 H. Alper, *Chem. Commun.*, 1970, 383. 4-9Cb

70CH473 E. R. Talaty and C. M. Utermoehlen, *Chem. Commun.*, 1970, 473. 6-1B

70CH631 N. W. Isaacs and C. H. L. Kennard, *Chem. Commun.*, 1970, 631. 4-6Gc

70CH706 I. J. Fletcher and A. R. Katritzky, *Chem. Commun.*, 1970, 706. 1-1C, 4-5Ff, 4-6Ff(ii), 4-6Fg(i)

70CH866 M. Ruccia and N. Vivona, *Chem. Commun.*, 1970, 866. 4-6F(1)

70CH1096 R. N. Butler, *Chem. Commun.*, 1970, 1096. 4-5Ff(2)

70CH1133 A. G. Anastassiou, S. W. Eachus, R. P. Cellura and J. H. Gebrian, *Chem. Commun.*, 1970, 1133. 6-2A

70CH1197 J. T. Sharp and P. B. Thorogood, *Chem. Commun.*, 1970, 1197. 6-2A(2)

70CH1622 Y. S. Rao and R. Filler, *Chem. Commun.*, 1970, 1622. 4-3Cd

70CO(270C)825 J. Morel, C. Paulmier, D. Semard and P. Pastour, *C. R. Acad. Sci. Ser. C*, 1970, **270**, 825. 3-2Cb, 3-2Dc

70CO(271C)1481 H. Lumbroso, D. Mazet, J. Morel and C. Paulmier, *C. R. Acad. Sci., Ser. C*, 1970, **271**, 1481. 3-2Cb

70CP951 A.-M. Bellocq and C. Garrigou-Lagrange, *J. Chim. Phys.*, 1970, **67**, 951. 4-1Ka

70CT901 M. Yamazaki, K. Noda and M. Hamana, *Chem. Pharm. Bull.*, 1970, **18**, 901. 2-8Db(iii)

70CT908 M. Yamazaki, K. Noda and M. Hamana, *Chem. Pharm. Bull.*, 1970, **18**, 908. 2-8Db(iii)

70CT1262 T. Higashino, Y. Tamura, K. Nakayama and E. Hayashi, *Chem. Pharm. Bull.*, 1970, **18**, 1262. 2-8Db(iv)

70CT1457 T. Higashino and E. Hayashi, *Chem. Pharm. Bull.*, 1970, **18**, 1457. 2-8Db(iv)

70CT1981 H. Ogura and T. Itoh, *Chem. Pharm. Bull. Jap.*, 1970, **18**, 1981. S5-5

70CX(B)1736 M. Fehlmann, *Acta Crystallogr., Sect. B*, 1970, **26**, 1736. 4-5Dd

70CX(B)1880 J. Berthou, J. Elguero and C. Rérat, *Acta Crystallogr., Sect. B*, 1970, **26**, 1880. 1-4Ib, 4-1Ba, T4-1

70CZ1406 V. Bekarek, J. Dobas, J. Socha, P. Vetesnik and M. Vecera, *Collect, Czech. Chem. Commun.*, 1970, **35**, 1406 4-2Gj

70CZ2936 V. Bekarek and J. Slouka, *Collect. Czech. Chem. Commun.*, 1970, **35**, 2936. 1-5Ab, 4-8B(2)

70DA(192)1295 Y. N. Sheinker, E. M. Peresleni, I. S. Rezchikova and N. P. Zosimova, *Dokl. Akad. Nauk SSSR*, 1970, **192**, 1295. 2-3Bg, 2-3Bh

70DA(195)868 L. E. Kholodov, E. M. Peresleni, I. F. Tishchenkova, N. P. Lostynchonko and V. G. Yashchumskii, *Dokl. Akad. Nauk SSSR*, 1970, **195**, 868. 2-8Db(iii)

70F972 G. Malesani, G. Chiarelloto, F. Marcolin and G. Rodighiero, *Farmaco, Ed. Sci.*, 1970, **25**, 972. 3-2G

70G629 L. Cavalca, A. Gaetani, A. Mangia and G. Pelizzi, *Gazz. Chim. Ital.*, 1970, **100**, 629. 4-2E

70GP1 W. J. Evers, *Ger. Patent* 2,003,525; *Chem. Abstr.*, 1970, **73**, 98783n. 3-3A

70H251 W. Skorianetz and E. Kovats, *Helv. Chim. Acta*, 1970, **53**, 251. 2-2D

70H299 R. Wagner and W. von Philipsborn, *Helv. Chim. Acta*, 1970, **53**, 299. 2-4Da, 2-4Dc(2), 2-6Cd

70H905 R. Urban, M. Grosjean and W. Arnold, *Helv. Chim. Acta*, 1970, **53**, 905. 2-4De

70H1151 M. Schellenberg, *Helv. Chim. Acta*, 1970, **53**, 1151. 2-2D

70HC(11)123 W. W. Paudler and T. J. Kress, *Advan. Heterocycl. Chem.*, 1970, **11**, 123. 2-2Ga

70HC(11)406 R. A. Jones, *Advan. Heterocycl. Chem.*, 1970, **11**, 406. 3-1Be

70HC(12)103 M. R. Grimett, *Advan. Heterocycl. Chem.*, 1970, **12**, 103. 4-4Ce, 4-5Eb

70IM1126 G. Rapi, M. Ginanneschi and E. Belgodere, *Chim. Ind. (Milan)*, 1970, **52**, 1126. 4-6Fe(i)(2)

70IS633 E. Glotter and M. D. Bachi, *Isr. J. Chem.*, 1970, **8**, 633. 4-3Ce

70IZ25 V. P. Lezina, L. D. Smirnov, K. M. Dyumaev and V. F. Bystrov, *Izv. Akad. Nauk SSSR, Ser. Khim.*, 1970, 25. 2-3D

70IZ675 V. S. Bogdanov, Y. L. Danyushevskii and Y. L. Gol'dfarb, *Izv. Akad. Nauk SSSR, Ser. Khim.*, 1970, 675. 3-3A, 3-3B

70IZ1735 Y. D. Kanaskova, B. I. Sukhorukov, Y. A. Pentin and G. V. Komarovskaya, *Izv. Akad. Nauk SSSR, Ser. Khim.*, 1970, 1735. 5-1Aa(ii), 5-1Da

70IZ2413 V. S. Bogdanov, M. A. Kalik and Y. L. Gol'dfarb, *Izv. Akad. Nauk SSSR, Ser. Khim.*, 1970, 2413. 3-2Ca(iv)

70J(B)138 R. N. Butler, *J. Chem. Soc. B*, 1970, 138. 4-5Ff

70J(B)200 J. L. Longridge, *J. Chem. Soc. B*, 1970, 200. 4-8B

70J(B)596 H. Weiler-Feilchenfeld and Z. Neiman, *J. Chem. Soc. B*, 1970, 596. 5-1Ea

70J(B)1334 E. D. Bergmann, H. Weiler-Feilchenfeld and Z. Neiman, *J. Chem. Soc. B*, 1970, 1334. 5-1Dd

70J(B)1692 R. E. Burton and I. L. Finar, *J. Chem. Soc. B*, 1970, 1692. 4-1Bb(x)

70J(C)230 A. Albert, *J. Chem. Soc. C*, 1970, 230. 4-5Fc, 4-5Fd(2)

70J(C)290 P. Flowerday, M. J. Perkins and A. R. J. Arthur, *J. Chem. Soc. C*, 1970, 290. 2-2F, 4-9D

70J(C)323 J. K. Landquist, *J. Chem. Soc. C*, 1970, 323. 4-4E(i)

70J(C)540 W. Nagata and S. Kamata, *J. Chem. Soc. C*, 1970, 540. 4-2Ia

70J(C)610 D. Lloyd and N. W. Preston, *J. Chem. Soc. C*, 1970, 610. 2-2F

70J(C)881 A. Halleux and H. G. Viehe, *J. Chem. Soc. C*, 1970, 881. 4-2Gf

70J(C)956 A. S. Bailey, W. A. Warr, G. B. Allison and C. K. Prout, *J. Chem. Soc. C*, 1970, 956. 3-4Cb

70J(C)980 K. W. Blake and P. G. Sammes, *J. Chem. Soc. C*, 1970, 980. 2-4E

70J(C)1023 C. G. Allison, R. D. Chambers, J. A. H. MacBride and W. K. R. Musgrave, *J. Chem. Soc. C*, 1970, 1023. 2-4E

70J(C)1114 A. Singh, H. Singh, C. K. Dewan and K. S. Narang, *J. Chem. Soc. C*, 1970, 1114. 4-6Dc

70J(C)1117 R. M. Acheson and W. R. Tully, *J. Chem. Soc. C*, 1970, 1117. T6-1(2)

70J(C)1251 C. O. Bender, R. Bonnett and R. G. Smith, *J. Chem. Soc. C*, 1970, 1251. 3-1Ad(3)

70J(C)1693 R. W. Alder, G. A. Niazi and M. C. Whiting, *J. Chem. Soc. C*, 1970, 1693. 4-9D

70J(C)1842 J. Feeney, G. A. Newman and P. J. S. Pauwels, *J. Chem. Soc. C*, 1970, 1842. 1-5B, 4-2Fa(2), 4-2Fb, 4-2Fc, 4-6Fd(i)

70J(C)1874 J. J. Eatough, L. S. Fuller, R. H. Good and R. K. Smalley, *J. Chem. Soc. C*, 1970, 1874. 4-9Ba

70J(C)1926 R. P. Dickinson and B. Iddon, *J. Chem. Soc. C*, 1970, 1926. 3-2Cc

70J(C)2403 A. J. Blackman and J. B. Polya, *J. Chem. Soc. C*, 1970, 2403. 4-4Ef, 4-4E(i)

70J(C)2409 I. T. Kay and N. Punja, *J. Chem. Soc. C*, 1970, 2409. 3-2Cd, 3-2Ee

70J(C)2426 I. Fleming and D. Philippides, *J. Chem. Soc. C*, 1970, **17**, 2426. 2-1C

70J(C)2431 N. B. Chapman, C. G. Hughes and R. M. Scrowston, *J. Chem. Soc. C*, 1970, 2431. 3-3B

70JA2929 N. Bodor, M. J. S. Dewar and A. J. Harget, *J. Amer. Chem. Soc.*, 1970, **92**, 2929. 1-7C, 2-3Ac, 2-3Ba, 2-3Bf, 2-3D, 2-3Ea, 2-4Da, 2-6Cb, 3-2Ac, 3-2Ec, 3-4A, 3-4Ca, 4-1Bb(ix), 4-1Cb(vi), 4-1Fa(viii), 5-1Dd(2), 5-1Eg, 5-1Eh

70JA4079 A. J. Jones, D. M. Grant, M. W. Winkley and R. K. Robins, *J. Amer. Chem. Soc.*, 1970, **92**, 4079. 2-4Db

70JA4340 H. Gotthardt, R. Huisgen and H. O. Bayer, *J. Amer. Chem. Soc.*, 1970, **92**, 4340. 4-3Cb(3)

70JA4447 J. Wallmark, H. H. Krachov, S. H. Chu and H. G. Manthner, *J. Amer. Chem. Soc.*, 1970, **92**, 4447. 3-2Ab, 3-2Cc 3-3B

70JA4956 D. C. Rohrer and M. Sundaralingam, *J. Amer. Chem. Soc.*, 1970, **92**, 4956. 5-1Da

70JA4963 S. T. Rao and M. Sundaralingam, *J. Amer. Chem. Soc.*, 1970, **92**, 4963. 5-1Da

70JA7441 C. E. Bugg and U. Thewalt, *J. Amer. Chem. Soc.*, 1970, **92**, 7441. 5-1E(i)

70JA7578 G. G. Hammes and P. J. Lillford, *J. Amer. Chem. Soc.*, 1970, **92**, 7578. 2-3Ba

70JE222 H. Alper, A. E. Alper and A. Taurins, *J. Chem. Educ.*, 1970, **47**, 222. 4-9Cb

70JH71 J. H. Boyer and P. J. A. Frints, *J. Heterocycl. Chem.*, 1970, **7**, 71. 6-1Aa

70JH139 E. B. Roche and D. W. Stansloski, *J. Heterocycl. Chem.*, 1970, **7**, 139. 4-3Cg

70JH227 N. Blazevic, F. Kajfez and V. Sunjic, *J. Heterocycl. Chem.*, 1970, **7**, 227. 4-1Cb(ii), 4-5Eb

70JH323 B. L. Currie, R. K. Robins and M. J. Robins, *J. Heterocycl. Chem.*, 1970, **7**, 323. 2-3Ee

70JH389 C. S. Wang, *J. Heterocycl. Chem.*, 9170, **7**, 389. 2-3Ea(2)

70JH479 J. P. Shoffner, L. Bauer and C. L. Bell, *J. Heterocycl. Chem.*, 1970, **7**, 479. 2-3C

70JH487 J. P. Shoffner, L. Bauer and C. L. Bell, *J. Heterocycl. Chem.*, 1970, **7**, 487. 2-4Da

70JH807 T. Kametani, K. Sota and M. Shio, *J. Heterocycl. Chem.*, 1970, **7**, 807. 4-2Jb, 4-3Bd

70JH991 L. D. Hansen, E. J. Baca and P. Scheiner, *J. Heterocycl. Chem.*, 1970, **7**, 991. 4-1Db

70JH1029 M. Israel, S. K. Tinter, D. H. Trites and E. J. Modest, *J. Heterocycl. Chem.*, 1970, **7**, 1029. T6-1(2)

70JH1113 R. Pater, *J. Heterocycl. Chem.*, 1970, **7**, 1113. 2-8Dc

70JH1159 P. N. Neuman, *J. Heterocycl. Chem.*, 1970, **7**, 1159. 4-5Fd

70JH1391 M. D. Coburn and P. N. Neuman, *J. Heterocycl. Chem.*, 1970, **7**, 1391. 4-5De, 4-5Eb

70JJ431 S. Sugiura, S. Inoue and T. Goto, *J. Pharm. Soc. Jap.*, 1970, **90**, 431. 5-3Bb(iv), S5-4

70JJ436 S. Sugiura, H. Kakoi, S. Inoue and T. Goto, *J. Pharm. Soc. Jap.*, 1970, **90**, 436. 5-3Ba, S5-4

70JM528 G. L. Blackman, R. D. Brown and F. R. Burden, *J. Mol. Spectrosc.*, 1970, **36**, 528. T1-5

70JO903 J. Pitha, *J. Org. Chem.*, 1970, **35**, 903. 2-4Db(2)

70JO1228 J. I. Sarkisian and R. W. Binkley, *J. Org. Chem.*, 1970, **35**, 1228. 5-5Ab

70JO3495 G. W. Stacy, D. L. Eck and T. E. Wollner, *J. Org. Chem.*, 1970, **35**, 3495. 3-4Bc

70JO3985 W. J. Middleton and D. Metzger, *J. Org. Chem.*, 1970, **35**, 3985. 5-5Ab

70JO4224 D. C. Dittmer and G. E. Kuhlmann, *J. Org. Chem.*, 1970, **35**, 4224. 6-1Ab

70KG191 V. P. Maksimeto and O. N. Popilin, *Khim. Geterotsikl. Soedin.*, 1970, 191. 2-5Bc

70KG202 I. I. Grandberg, V. G. Vinokurov, V. S. Troitskaya, T. A. Ivanova and V. A. Moskalenko, *Khim. Geterotsikl. Soedin.*, 1970, 202. 4-2Ga(i), 4-4Ad

70KG265 L. I. Bagal, M. S. Pevzner, N. I. Sheludyakova and A. N. Frolov, *Khim. Geterotsikl. Soedin.*, 1970, 265. 4-1Ec

70KG558 L. I. Bagal, and M. S. Pevzner, *Khim. Geterotsikl. Soedin.* 1970, 558. 4-1Ec

70KG660 A. D. Garnovskii, Y. V. Kolodyazhnyi, I. I. Grandberg, S. A. Alieva, N. F. Krokhina and N. P. Bednyagina, *Khim. Geterotsikl. Soedin.*, 1970, 660. 4-2Gg

70KG723 L. A. Boyko, Y. A. Boyko and Z. N. Nazarova, *Khim. Geterotsikl. Soedin.*, 1970, 723. 3-3A

70KG1138 Y. G. Kovalev and I. Y. Postovskii, *Khim. Geterotsikl. Soedin.*, 1970, 1138. 4-4Ec, 4-4Ef(2)

70KG1410 V. G. Poludnenko and A. M. Simonov, *Khim. Geterotsikl. Soedin.*, 1970, 1410. 4-5Df(2)

70KG1683 B. I. Khristich, *Khim. Geterotsikl. Soedin.*, 1970, 1683. 4-1Cb

70LA(731)174 H. Stamm, *Liebig's Ann. Chem.*, 1970, **731**, 174. 5-1Eb

70LA(734)173 F. Boberg and R. Schardt, *Liebig's Ann. Chem.*, 1970, **734**, 173. 4-6Fd(ii)

70LA(735)35 W. Flitsch and U. Krämer, *Liebig's Ann. Chem.*, 1970, **735**, 35. 5-3Ba

70LA(736)1 H. Bauer, *Liebig's Ann. Chem.*, 1970, **736**, 1. 3-2Ee

70MB(50)153 W. Saenger and K. H. Scheit, *J. Mol. Biol.*, 1970, **50**, 153. 2-5E

70MC773 R. P. Williams, V. J. Bauer and S. R. Safir, *J. Med. Chem.*, 1970, **13**, 773. 4-2Gf

70MI3 R. J. Sundberg, "The Chemistry of Indoles," p. 3. Academic Press, New York, 1970. 3-1Bb

70MI40 G. Berthier, L. Praud and J. Serre, "Quantum Aspects of Heterocyclic Compounds in Chemistry and Biochemistry" (E. D. Bergmann and B. Pullman, Eds.), Jerusalem Symposia on Quantum Chemistry and Biochemistry, Vol. II, p. 40. Academic Press, New York, 1970. 4-1Ca, T4-7

70MI238 D. T. Clark, "Quantum Aspects of Heterocyclic Compounds in Chemistry and Biochemistry" (E. D. Bergmann and B. Pullman, Eds.), Jerusalem Symposia on Quantum Chemistry and Biochemistry, Vol. II, p. 238. Academic Press, New York, 1970. 1-7A(2), 6-1Aa

70OM1341 H. Ogura, S. Sugimoto and T. Itoh, *Org. Mass Spectrom*, 1970, **3**, 1341. 4-5Db, 4-5Dd

70PC(312)869 H. G. O. Becker, D. Beyer and H.-J. Timpe, *J. Prakt. Chem.*, 1970, **312**, 869. 4-7Ac

70PH2133 S. T. King, *J. Phys. Chem.*, 1970, **74**, 2133. 1-6Cb, 4-1Ba, 4-1Ca

70PH2684 A. J. Jones, D. M. Grant, M. W. Winkley and R. K. Robins, *J. Phys. Chem.*, 1970, **74**, 2684. 2-4Dg, 2-6Cc

70QR433 J. Shorter, *Quart. Rev. Chem. Soc.*, 1970, **24**, 433. 1-4B

70RC1447 W. Czuba and H. Poradowska, *Rocz. Chem.*, 1970, **44**, 1447. 2-7Db

70RO619 N. P. Bednyagina, G. N. Litunova, A. P. Novikova, A. P. Zeiss and L. N. Shegoleva, *Z. Org. Khim.*, 1970, **6**, 619. 4-5G(iii)

70RO1332 A. P. Zeiss, L. N. Shegoleva, G. N. Litunova, A. P. Novikova and N. P. Bednyagina, *Z. Org. Khim.*, 1970, **6**, 1332. 4-5G(iii)

70RO1349 K. N. Zelenin and J. Dumpis, *Z. Org. Khim.*, 1970, **6**, 1349. 2-2D

70SA(A)153 P. J. Taylor, *Spectrochim. Acta, Part A*, 1970, **26**, 153. 4-3Cf(2), 4-8B

70SA(A)165 P. J. Taylor, *Spectrochim. Acta, Part A*, 1970, **26**, 165. 4-3Cf

70SA(A)825 O. L. Stiefvater, H. Jones and J. Sheridan, *Spectrochim. Acta, Part A*, 1970, **26**, 825. T1-5, 4-1Da

70T739 A. T. Balaban, *Tetrahedron*, 1970, **26**, 739. 6-2A

70T1393 G. Desimoni and G. Minoli, *Tetrahedron*, 1970, **26**, 1393. 6-2A

70T1483 B. Pullman, H. Berthod, F. Bergmann and Z. Neiman, *Tetrahedron*, 1970, **26**, 1483. 5-1Aa(i)

70T1571 G. A. Newman and P. J. S. Pauwels, *Tetrahedron*, 1970, **26**, 1571. 4-2Fa, 4-2Ge(ii)(2), 4-6Bb, 4-6Fd(i), T4-17(5)

70T2497 A. W. K. Chan, W. D. Crow and I. Gosney, *Tetrahedron*, 1970, **26**, 2497. 4-2E, 4-4Ac

70T3429 G. A. Newman and P. J. S. Pauwels, *Tetrahedron*, 1970, **26**, 3429. 4-2Fa, 4-2Fc

70T4269 M. J. S. Dewar and N. Trinajstić, *Tetrahedron*, 1970, **26**, 4269. 6-2A(3)

70T4491 M. Nakagawa and T. Hino, *Tetrahedron*, 1970, **26**, 4491. 3-1Ac

70T4777 H. Ahlbrecht and S. Fischer, *Tetrahedron*, 1970, **26**, 4777. 2-8Df

70T5225 H. P. Figeys, *Tetrahedron*, 1970, **26**, 5225. 6, 6-2

70TC(16)243 J. S. Kwiatkowski, *Theor. Chim. Acta*, 1970, **16**, 243. 2-3D, 2-4Da

70TC(16)316 G. Berthier, B. Levy and L. Paoloni, *Theor. Chim. Acta*, 1970, **16**, 316. 2-3D

70TE250 B. P. Zvolinskii, M. E. Perelson and Y. N. Sheinker, *Theor. Exp. Chem. (USSR)*, 1970, **6**, 250. 2-3Ba

70TH01 J. Verducci, Ph.D. thesis, University of Montpellier, France, 1970. 4-2Bg(i)

70TH02 M. Roche, Ph.D. thesis, University of Marseille, France, 1970. 1-7B, 4-1Da, 4-1Eb(iv), 4-1Fa(vii)

70TH03 C. Dittli, Ph.D. thesis, University of Montpellier, France, 1970. 4-6Fd(i), 4-6Fd(ii)

70TH04 L. Aspart-Pascot, Ph.D. thesis, University of Perpignan, France, 1970. 4-5Bg, 4-5Bh, 4-5Bh(i)(2)

70TL169 W. Steglich, G. Höfle, L. Wilschowitz and G. C. Barrett, *Tetrahedron Lett.*, 1970, 169. 4-3Ce(2), T4-24

70TL825 A. G. Anastassiou and J. H. Gebrian, *Tetrahedron Lett.*, 1970, 825. 6-2A

70TL1311 T. M. Krygowski, *Tetrahedron Lett.*, 1970, 1311. 6, 6-2

70TL2979 A. S. Bailey, R. Scattergood, W. A. Warr, T. S. Cameron, C. K. Prout and I. Tickle, *Tetrahedron Lett.*, 1970, 2979. 3-4Cb

70TL3155 G. H. Alt and J. P. Chupp, *Tetrahedron Lett.*, 1970, 3155. 4-2Gf, 4-2Hb(ii)(2), 4-2Hc

70TL4069 H. J. Teuber and G. Emmerich, *Tetrahedron Lett.*, 1970, 4069. 6-2A

70ZC211 H. Lettau, *Z. Chem.*, 1970, **10**, 211. 4-7Ad

70ZC338 H. Lettau, *Z. Chem.*, 1970, **10**, 338. 4-7Ab(2)

70ZC475 F. Ritschl and P. Brosche, *Z. Chem.*, 1970, **10**, 475. 4-2Fc

70ZF16 I. L. Belaits and R. N. Nurmukhametov, *Russ. J. Phys. Chem.*, 1970, **44**, 16. 3-2Ab

70ZO669 G. F. Gavrulun, V. E. Chistyakov and G. A. Kononenko, *Zh. Obshch. Khim.*, 1970, **40**, 669. 2-4Dg

70ZO1050 B. P. Lugovkin, *Zh. Obshch. Khim.*, 1970, **40**, 1050. 4-2Gf

70ZO1605 A. P. Terent'ev, I. G. Il'ina, E. G. Rukhadze and I. G. Vorontsova, *Zh. Obshch. Khim.*, 1970, **40**, 1605. 4-4Cf

70ZO1865 V. I. Minkin, I. D. Sadekov, L. L. Popova and Y. V. Kolodyazhnyi, *Zh. Obshch. Khim.*, 1970, **40**, 1865. 6-2A

70ZO1872 N. L. Aryutkina, A. F. Vasil'ev, N. A. Poznanskaya, N. I. Shvetsov-Shilovskii, S. N. Ivanova and N. N. Mel'nikov, *Zh. Obshch. Khim.*, 1970, **40**, 1872. 4-3Bb, 4-4Cb(2)

70ZO1881 S. A. Andronati, A. V. Bogatskii, Yu. I. Vikhlyaev, Z. I. Zhilina, B. M. Kats, T. A. Klygul, V. N. Khudyakova, T. K. Chumachenko and A. A. Énnan, *Zh. Obshch. Khim.*, 1970, **40**, 1881. 6-2Bd(2)

71AJ2557 E. Spinner and G. B. Yeoh, *Aust. J. Chem.*, 1971, **24**, 2557. 2-3Ec(3), 2-3Eb(5)

71AJ2729 I. D. Rae and B. N. Umbrasas, *Aust. J. Chem.*, 1971, **24**, 2729. 4-3Ce

71AN672 L. Pentimalli and G. Milani, *Ann. Chim.* (*Rome*), 1971, **61**, 672. 5-5Ac

71AN793 J. I. Degani, R. Fochi and G. Spunta, *Ann. Chim.* (*Rome*), 1971, **61**, 793. 2-2B

71AS249 M. Begtrup, *Acta Chem. Scand.*, 1971, **25**, 249. 4-3Ef

71AS2087 M. Begtrup and K. V. Poulsen, *Acta Chem. Scand.*, 1971, **25**, 2087. 4-3Ec, 4-3Ef

71AS3500 M. Begtrup, *Acta Chem. Scand.*, 1971, **25**, 3500. 4-4Fa

71BB17 A. Maquestiau, Y. van Haverbeke and R. Jacquerye, *Bull. Soc. Chim. Belg.*, 1971, **80**, 17. 4-2Fa, 4-2Fd, 4-2Gb, 4-2Gc, 4-2Ib, T4-17(3)

71BF671 T. Armand, S. Deswark, J. Pimson and H. Zamarlik, *Bull. Soc. Chim. Fr.*, 1971, 671. 3-4Ca, 3-4Cd

71BF1038 C. Dittli, J. Elguero and R. Jacquier, *Bull. Soc. Chim. Fr.*, 1971, 1038. 4-3Ad

71BF1040 R. Jacquier, J.-M. Lacombe and G. Maury, *Bull. Soc. Chim. Fr.*, 1971, 1040. 4-3Cg, 4-3Ch, 4-3C(i)(5)

71BF1858 P. Lardenois, M. Sélim and M. Sélim, *Bull. Soc. Chim. Fr.*, 1971, 1858. 2-4Da(2), 2-5E, 2-6Ca

71BF1925 J. Elguero, *Bull. Soc. Chim. Fr.*, 1971, 1925. 4-9Ca

71BF3537 M. Robba and Y. Le Guen, *Bull. Soc. Chim. Fr.*, 1971, 353. 5-4A

71BF3547 J. Morel, C. Paulmier, D. Semard and P. Pastour, *Bull. Soc, Chim. Fr.*, 1971, 3547. 3-2Cb, 3-2Ce, 3-2Dc, 3-3B

71BF3658 J. Daunis, R. Jacquier and P. Viallefont, *Bull. Soc. Chim. Fr.*, 1971, 3658. 2-5F

71BF4059 J. LeRoux, G. Letertre, P. L. Desbene and J.-J. Basselier, *Bull. Soc. Chim. Fr.*, 1971, 4059. 2-2A

71BJ1311 Y. Iwanauni, *Bull Chem. Soc. Jap.*, 1971, **44**, 1311. 2-8-Db(i), 2-8Dd(3)

71BJ1314 Y. Iwanami, *Bull. Chem. Soc. Jap.*, 1971, **44**, 1314. 2-8-Db(iv), 2-8Dd

71BJ1316 Y. Iwanami, T. Seki and T. Inagaki, *Bull. Chem. Soc. Jap.*, 1971, **44**, 1316. 2-8Db(iv)(2)

71BO435 G. C. Y. Lee, J. H. Prestegard and S. I. Chan, *Biochim. Biophys. Res. Commun.*, 1971, **43**, 435. 2-6Cb

71BY(232)1 R. Shapiro and S. Kang, *Biochim. Biophys. Acta*, 1971, **232**, 1. 2-4Dd

71CB1155 G. Matolcsy, P. Sohár and B. Bordás, *Chem. Ber.*, 1971, **104**, 1155. 4-5Bg

71CB2273 W. Pfleiderer, R. Mengel and P. Hemmerich, *Chem. Ber.* 1971, **104**, 2273. 2-6De(2)

71CB2458 U. Kraat, W. Hasenbrink, H. Warmhoff and F. Korte, *Chem. Ber.*, 1971, **104**, 2458. 3-4A

71CB2694 G. S. D. King and H. Reimlinger, *Chem. Ber.*, 1971, **104,** 2694. 4-2Fa, 4-2Gg

71CB2786 F. P. Woerner, H. Reimlinger and R. Merenyi, *Chem. Ber.* 1971, **104**, 2786. 6-2Bc, 6-2Bd

71CB2793 H. Reimlinger, W. R. F. Lingier and R. Merenyi, *Chem. Ber.*, 1971, **104**, 2793. 2-7Ea

71CB3510 H. Wamhoff and P. Sohár, *Chem. Ber.*, 1971, **104**, 3510. 4-5Fd

71CC1372 R. Laliberté and G. Médawar, *Can. J. Chem.*, 1971, **49**, 1372. 3-4Ba(2)

71CC4054 A. Taurins and R. Kang-Chuan Hsia, *Can. J. Chem.*, 1971, **49**, 4054. 4-3Bc, 4-5Dd

71CH510 M. J. Cook, A. R. Katritzky, P. Linda and R. D. Tack, *Chem. Commun.*, 1971, 510. 2-6Ba, 2-8Ba

71CH836 T. Hino, M. Nakagawa, T. Suzuki, T. Takeda, N. Kano and Y. Ishii, *Chem. Commun.*, 1971, 836. 3-3C(2)

71CH846 G. Bianchi and A. R. Katritzky, *Chem. Commun.*, 1971, 846. 4-5Ff

71CH873 K. Bolton, R. D. Brown, F. R. Burden and A. Mishra, *Chem. Commun.*, 1971, 873. T1-5, 4-1Da, T4-7(3)

71CH1099 G. B. Quistad and D. A. Lightner, *Chem. Commun.*, 1971, 1099. 2-2Ec

71CO(272C)103 L. Aspart-Pascot, J. Lematre and A. Sournia, *C. R. Acad. Sci., Ser. C*, 1971, **272**, 103. 4-5Bg

71CP465 M. Roche and L. Pujol, *J. Chim. Phys.*, 1971, **68**, 465. 1-7B, 4-1Ca, 4-1Da, 4-1Eb(iv), 4-1Fa(vii), T4-6, T4-7, T4-8

71CR295 S. D. Worley, *Chem. Rev.*, 1971, **71**, 295. 1-6F(2)

71CR439 R. M. Izatt, J. J. Christensen and J. H. Rytting, *Chem. Rev.*, 1971, **71**, 439. 5-1Da, 5-1Db, 5-1Eg, 5-1Eh

71CT564 K. I. Keda and Y. Mizuno, *Chem. Pharm. Bull.*, 1971, **19**, 564. 2-4Db

71CT2331 S. Kubota, M. Uda and M. Ohtsuka, *Chem. Pharm. Bull.*, 1971, **19**, 2331. 4-4Ef, 4-4E(i), 4-4Ej

71CX(B)95 J. P. Mornon and B. Raveau, *Acta Crystallogr., Sect. B*, 1971, **27**, 95. 4-6Ff(ii)

71CX(B)134 G. C. Verschoor and E. Keulen, *Acta Crystallogr., Sect. B*, 1971, **27**, 134. 2-4Fa

71CX(B)961 G. H. Y. Lin and M. Sundaralingam, *Acta Crystallogr.*, 1971, **27**, 961 2-5E

71CX(B)1178 W. Saenger and V. Suck, *Acta Crystallogr., Sect. B*, **27**, 1178. 2-5E

71CX(B)1201 J. Almlöf, A. Kvick and I. Olovsson, *Acta Crystallogr., Sect. B*, 1971, **27**, 1201. 2-3Bc

71CX(B)1227 W. H. DeCamp and J. M. Stewart, *Acta Crystallogr., Sect. B*, 1971, **27**, 1227. 4-2Ib, T4-21

71CX(B)1441 J. P. Chesick and J. Donohue, *Acta Crystallogr., Sect. B*. 1971, **27**, 1441. 4-4Cd

71CZ143 J. Jurina, E. Ruzicka, and L. Sobehartova, *Collect. Czech. Chem. Commun.*, 1971, **36**, 143. 2-4H

71CZ1413 J. Kuthan and M. Ichova, *Collect. Czech. Chem. Commun.*, 1971, **36**, 1413. 2-3Ac

71CZ1955 V. Uchytilova, P. Fiedler, M. Prystas and J. Gut, *Collect. Czech. Chem. Commun.*, 1971, **36**, 1955. 2-4Fb

71DA110 D. N. Kravtsov, L. A. Fedorov, A. S. Peregudov and A. N. Nesmeyanov, *Dokl. Akad. Nauk SSSR*, 1971, **196**, 110. 1-5Ac(ii)

71F1017 E. Testa and L. Fontanella, *Farmaco, Ed. Sci.*, 1971, **26**, 1017. 4-2Hc, 4-2Ib, T4-21

71H1543 R. Wagner and W. von Philipsborn, *Helv. Chim. Acta*, 1971, **54**, 1543. 5-1Ab, 5-1Db

71HC(13)45 F. W. Fowler, *Advan. Heterocycl. Chem.*, 1971, **13**, 45. 6-1Aa

71HC(13)77 B. Pullman and A. Pullman, *Advan. Heterocycl. Chem.*, 1971, **13**, 77. 1-7A, 1-7B(2), 1-7E, 5-1Aa(iii)(3), 5-1Ba(2), 5-1Bb, 5-1Bc(3), 5-1C(4), 5-1Dd(3), 5-1De, 5-1Ea, 5-1Eg(2), 5-2Ac, 5-2Bc(2)

71IJ104 M. D. Nair, *Indian J. Chem.*, 1971, **9**, 104. 4-2Gg

71IQ341 R. Rein, *Int. J. Quantum Chem.*, 1971, **4**, 341. 5-1Aa(iii)

71IS659 G. Zvilichovsky, *Isr. J. Chem.*, 1971, **9**, 659. 4-6Ba(2)

71J(B)11 U. Bressel, A. R. Katritzky and J. R. Lea, *J. Chem. Soc. B*, 1971, 11. 1-2

71J(B)279 E. Spinner and G. B. Yeoh, *J. Chem. Soc. B*, 1971, 279. 1-4Ac, 2-3Eb(2), T2-1, 2-3Be(2), 2-3Bg(3)

71J(B)289 E. Spinner and G. B. Yeoh, *J. Chem. Soc. B*, 1971, 289. 2-3Be

71J(B)296 E. Spinner and G. B. Yeoh, *J. Chem. Soc. B*, 1971, 296. 2-3Bc, T2-1

71J(B)397 P. Hampson, A. Mathias and R. Westhead, *J. Chem. Soc. B*, 1971, 397. 1-5C

71J(B)460 F. W. Fowler, A. R. Katritzky and R. J. D. Rutherford, *J. Chem. Soc. B*, 1971, 460. 2-3Bg

71J(B)976 F. P. van Remoortere and F. P. Boer, *J. Chem. Soc. B*, 1971, 976. 4-5Eb

71J(B)1261 G. B. Barlin and A. C. Young, *J. Chem. Soc. B*, 1971, 1261. 2-5E(2), 2-4Ca(2)

71J(B)1270 N. W. Isaacs and C. H. L. Kennard, *J. Chem. Soc. B*, 1971, 1270. 4-6Gc

71J(B)1425 G. B. Barlin and W. Pfleiderer, *J. Chem. Soc. B*, 1971, 1425. 2-3Bc, 2-3C, 2-4Dc, 2-6Bb, T2-1

71J(B)2339 G. P. Bean, M. J. Cook, T. M. Dand, A. R. Katritzky and J. R. Lea, *J. Chem. Soc. B*, 1971, 2339. 1-4Ac, 2-3Bd(3)

71J(B)2344 A. J. Boulton, I. J. Fletcher and A. R. Katritzky, *J. Chem. Soc. B*, 1971, 2344. 1-4Ac, 2-4Cb(2), 2-6Db(4)

71J(B)2350 S. O. Chua, M. J. Cook and A. R. Katritzky, *J. Chem. Soc. B*, 1971, 2350. 1-1D, 1-3B, 1-6Gb, 4-7Ab(3), 4-7Ad(3), 4-7Af

71J(B)2355 G. Bianchi, A. J. Boulton, I. J. Fletcher and A. R. Katritzky, *J. Chem. Soc. B*, 1971, 2355. 4-5A(2), 4-5Ff(3)

71J(C)86 F. de Sarlo, G. Dini and P. Lacrimini, *J. Chem. Soc. C*, 1971, 86. 4-2Ba, 4-2Bc, 4-2Be, 4-2Bf(2), T4-12(2)

71J(C)225 G. V. Boyd and S. R. Dando, *J. Chem. Soc. C*, 1971, 225. 4-5Bh

71J(C)703 F. L. Scott and J. C. Tobin, *J. Chem. Soc. C*, 1971, 703. 4-5Ff

71J(C)1016 A. J. Blackman and J. B. Polya, *J. Chem. Soc. C*, 1971, 1016. 4-4Ec(2), 4-4Ef, 4-4Ei(2), 4-4Ej

71J(C)1314 I. D. H. Stocks, J. A. Waite and K. R. H. Wooldridge, *J. Chem. Soc. C*, 1971, 1314. 4-2D(2)

71J(C)1667 R. S. Shadbolt, *J. Chem. Soc. C*, 1971, 1667. 4-9Cb

71J(C)1669 R. S. Shadbolt, *J. Chem. Soc. C*, 1971, 1669. 4-9Cb

71J(C)1676 D. Lichtenberg, F. Bergmann and Z. Neiman, *J. Chem. Soc. C*, 1971, 1676. 5-1Ea(2), 5-1Eb, 5-1Ec(2)

71J(C)2769 F. L. Scott, D. A. Cronin and J. K. O'Halloran, *J. Chem. Soc. C*, 1971, 2769. 5-5Ab

71J(C)3040 B. W. Bycroft, D. Cameron, and A. W. Johnson, *J. Chem. Soc. C*, 1971, 3040. 2-6Ca

71J(C)3088 D. E. Ames, H. R. Ansari, A. D. G. France, A. C. Lovesey, B. Novitt and R. Simpson, *J. Chem. Soc. C*, 1971, 3088. 2-5E

71J(C)3313 R. M. Anderson and B. R. Leverett, *J. Chem. Soc. C*, 1971, 3313. 4-2Jb

71JA1637 N. L. Allinger, M. T. Tribble, M. A. Miller and D. H. Wertz, *J. Amer. Chem. Soc.*, 1971, **93**, 1637. 1-6B

71JA1880 R. J. Pugmire and D. M. Grant, *J. Amer. Chem. Soc.*, 1971, **93**, 1880. 1-5B, 5-1Aa(i)(2), 5-1Ab

71JA4016 R. M. Coates and E. F. Johnson, *J. Amer. Chem. Soc.*, 1971, **93**, 4016. 6-2Be

71JA4297 O. A. Gansow, J. Killough and A. R. Burke, *J. Amer. Chem. Soc.*, 1971, **93**, 4297. 1-5B

71JA4585 G. E. Bass and L. J. Schaad, *J. Amer. Chem. Soc.*, 1971, **93**, 4585. 5-1Dd(2), 5-1Eh

71JA5023 K. C. Ingham, M. Abu-Elgheit and M. A. El-Bayoumi, *J. Amer. Chem. Soc.*, 1971, **93**, 5023. 1-6E, 5-2Aa

71JA5402 L. H. Vogt, Jr. and J. G. Wirth, *J. Amer. Chem. Soc.*, 1971, **93**, 5402. 2-3D

71JA5552 G. L. Kenyon and G. L. Rowley, *J. Amer. Chem. Soc.*, 1971, **93**, 5552. 4-6Fg(i)(3), 4-6Fg(ii)(4), 4-6Fg(iii)(2)

71JA6387 H. J. Gold, *J. Amer. Chem. Soc.*, 1971, **93**, 6387. 2-3Ba

71JA7281 E. Shefter, B. E. Evans and E. C. Taylor, *J. Amer. Chem. Soc.*, 1971, **93**, 7281. 5-4A

71JE712 T. L. James, *J. Chem. Educ.*, 1971, **48**, 712. 1-6F

71JH571 S. N. Lewis, G. A. Miller, M. Hausman and E. C. Szamborski, *J. Heterocycl. Chem.*, 1971, **8**, 571. 4-2E

71JH581 G. A. Miller, E. D. Weiler and M. Hausman, *J. Heterocycl. Chem.*, 1971, **8**, 581. 4-2E

71JH797 M. Israel and L. C. Jones, *J. Heterocycl. Chem.*, 1971, **8**, 797. T6-1(2)

71JH881 R. A. Coburn, *J. Heterocycl. Chem.*, 1971, **8**, 881. 5-1Ba, 5-1Ea

71JH999 M. J. Kornet, T. H. Ong and P. A. Thio, *J. Heterocycl. Chem.*, 1971, **8**, 999. 4-2Hc, 4-2Hc(iii)

71JH1015 M. Israel, L. C. Jones and M. M. Jouillié, *J. Heterocycl. Chem.*, 1971, **8**, 1015. T6-1

71JJ338 T. Tanaka, *J. Pharm. Soc. Jap.*, 1971, **91**, 338. 4-4Ad(2), T4-26(4)

71JL(8)471 J. S. Kwiatkowski, *J. Mol. Struct.*, 1971, **8**, 471. 5-1C(2)

71JL(10)245 J. S. Kwiatkowski, *J. Mol. Struct.*, 1971, **10**, 245. 1-7E, 2-3Ba, 2-4De

71JO10 K. T. Potts and S. Husain, *J. Org. Chem.*, 1971, **36**, 10. 5-5B, S5-5

71JO354 J. F. Wolfe and T. P. Murray, *J. Org. Chem.*, 1971, **36**, 354. 2-3Ea

71JO848 J. L. Wong and D. S. Fuchs, *J. Org. Chem.*, 1971, **36**, 848. 2-4Db

71JO1088 L. J. Darlage, T. H. Kinstle and C. L. McIntosh, *J. Org. Chem.*, 1971, **36**, 1088. 4-2Cc(4), 4-3Bb

71JO1165 J. E. Douglass and J. M. Wesolosky, *J. Org. Chem.*, 1971, **36**, 1165. 2-8Df

71JO1352 H. Alper, E. C. H. Keung and R. A. Partis, *J. Org. Chem.*, 1971, **36**, 1352. 4-9Cb

71JO1463 N. Finch and H. W. Gschwend, *J. Org. Chem.*, 1971, **36**, 1463. 4-5Bj

71JO2065 J. J. Eisch and F. J. Gadek, *J. Org. Chem.*, 1971, **36**, 2065. 2-8Bb(2)

71JO2639 J. C. Parham, T. G. Winn and G. Bosworth Brown, *J. Org. Chem.*, 1971, **36**, 2639. 5-1F, 5-1Fb(3), 5-1Fc(2)

71JO2986 D. L. Fields and T. H. Regan, *J. Org. Chem.*, 1971, **36**, 2986. 2-8Db(ii)

71JO3091 M. G. Reinecke, J. F. Sebastian, H. W. Johnson, Jr. and Chongsuh Pyun, *J. Org. Chem.*, 1972, **21**, 3091. 3-1Ac

71JO3372 H. Rubinstein, J. E. Skarbek and H. Feuer, *J. Org. Chem.*, 1971, **36**, 3372. 2-4Cd

71JO3469 J. Musco and D. B. Murphy, *J. Org. Chem.*, 1971, **36**, 3469. 4-5Df

71JO3921 W. W. Paudler and J. Lee, *J. Org. Chem.*, 1971, **36**, 3921. 2-4Fb, 2-8Bd

71JO3999 D. W. H. MacDowell and J. C. Wisowaty, *J. Org. Chem.*, 1971, **36**, 3999. 3-2G

71KG189 A. I. Ginak, K. A. V'yunov, V. V. Barmina and E. G. Sochilin, *Khim. Geterotsikl. Soedin.*, 1971, 189. 4-6Eb(i)

71KG192 K. A. Vyunov, A. I. Ginak and E. G. Sochilin, *Khim. Geterotsikl. Soedin.*, 1971, 192. 4-6Ca

71KG801 Y. Y. Usaevich, I. K. Feldman and Y. I. Boksinev, *Khim. Geterotsikl. Soedin.*, 1971, 801. 4-3Ch

71KG807 A. F. Pozharskii, I. S. Kasparov, Y. P. Andreichikov, A. I. Buryak, A. A. Konstantinchenko and A. M. Simonov, *Khim. Geterotsikl. Soedin.*, 1971, 807. 4-5De, 4-5Df

71KG826 P. M. Kochergin, Y. N. Sheinker, A. A. Druzhinina, R. M. Palei and L. M. Alekseeva, *Khim. Geterotsikl. Soedin.*, 1971, 826. 5-5Aa

71KG867 A. D. Garnovskii, Y. V. Kolodyazhnyi, O. A. Osipov, V. I. Minkin, S. A. Hiller, I. B. Mazheika and I. I. Grandberg, *Khim. Geterotsikl. Soedin.*, 1971, 867. 1-4G, 4-1Ca

71KG946 P. L. Ovechkin, L. A. Ignatova and B. V. Unkovskii, *Khim. Geterotsikl. Soedin.*, 1971, 946. 2-6F

71KG1313 O. A. Yerastov, S. N. Ignatyeva and T. V. Malkova, *Khim. Geterotsikl. Soedin.*, 1971, 1313. 3-2Ac

71KG1436 Y. M. Yutilov, N. R. Kal'nitskii and R. M. Bystrova, *Khim. Geterotsikl. Soedin.*, 1971, 1436. 5-2Ab

71KG1473 O. A. Yerastov and S. N. Ignatyeva, *Khim. Geterotsikl. Soedin.*, 1971, 1473. 3-2Cd

71KG1540 V. P. Lezina, N. A. Andronova, L. D. Smirnov and K. M. Dyumaev, *Khim. Geterotsikl. Soedin.*, 1971, 1540. 2-3D

71KG1640 M. M. Magdesieva, *Khim. Geterotsikl. Soedin.*, 1971, 1640. 3-2Cg

71KX115 Y. A. Omel'chenko and Y. D. Kondrashev, *Kristallografiya*, 1971, **16**, 115. 4-1Ca

71LA(750)39 H. Dorn and D. Arndt, *Liebig's Ann. Chem.*, 1971, **750**, 39. 4-2Fc, 4-5Bg, S4-5(2)

71LA(754)46 S. Hünig, G. Kiesslich, K.-H. Oette and H. Quast, *Liebig's Ann. Chem.*, 1971, **754**, 46. 4-9Ba

71LA(754)113 H. Wamhoff and C. Materne, *Liebig's Ann. Chem.*, 1971, **754**, 113. 4-6Fd(i), 4-6Fd(iii)

71M412 T. Kappe, P. F. Fritz and E. Ziegler, *Monatsh. Chem.*, 1971, **102**. 412. 2-4De

71MI01 "Aromaticity, Pseudoaromaticity, Antiaromaticity" (E. D. Bergmann and B. Pullman, Eds.), Jerusalem Symposia on Quantum Chemistry and Biochemistry, Vol. III. Academic Press, New York, 1971. 5-1F, 6

71MI02 S. Patai, "The Chemistry of the Azido Group," Wiley (Interscience), New York, 1971. 4-9Ca

71MI5 V. I. Poltev and B. I. Sukhorukov, *Biofizika*, 1971, **16**, 5; *Chem. Abstr.*, 1971, **14**, 99309. 2-6Cb

71MI39 A. R. Katritzky, *Int. Congr. Heterocycl. Chem., 3rd, Tohoku Univ., Sendai, Jap.*, 1971, 39. 2-1B

71MI105 A. R. Katritzky and J. M. Lagowski, "Chemistry of the Heterocyclic N-Oxides," p. 105. Academic Press, London, 1971. 4-7A

71MI216 H. Wamhoff, D. Schramm and F. Korte, *Synthesis*, 1971, 216. 4-2Gh, 4-2Ic

71MI234 D. T. Clark and D. M. J. Lilley, *Chem. Phys. Lett.*, 1971, **9**, 234. 1-6F(2)

71MI379 K. Harsányi, C. Gönczi and G. Horváth, *Int. Congr. Heterocycl. Chem., 3rd, Tohoku Univ., Sendai, Jap.*, 1971, 379. 2-7Ed

71MI514 M. J. Nye, M. J. O'Hare and W. P. Tang, *Int. Congr. Heterocycl. Chem., 3rd, Tohoku Univ., Sendai, Jap.* 1971, 514. 4-9Ca

71MI587 J. S. Kwiatkowski, *Acta Phys. Pol.*, 1971, **A39**, 587. 1-7E

71MI923 W. C. Johnson, P. M. Vipond and J. C. Girod, *Biopolymers*, 1971, **10**, 923; *Chem. Abstr.*, 1971, **75**, 263. 2-6Cb(3)

71OM1 K. T. Potts, E. Brugel and U. P. Singh, *Org. Mass Spectrom.*, 1971, **5**, 1. 5-3Bb(iii), 5-3Bd, S5-4

71OM123 T. Nishiwaki, *Org. Mass Spectrom.*, 1971, **5**, 123. 4-5Ba

71OM663 K. T. Potts and E. Brugel, *Org. Mass. Spectrom*, 1971, **5**, 663. 5-3Bc, S5-4

71OP198 I. A. Semenova, G. N. Sinyakov and G. P. Gurinovich, *Opt. Spectrosc. (USSR)*, 1971, **31**, 198. 6-3

71OR7 R. Mondelli, V. Bocchi, G. P. Gardini and L. Chierici, *Org. Magn. Resonance*, 1971, **3**, 7. 3-2Ec(3)

71OR689 P. Sohar, J. Reiter and L. Toldy, *Org. Mag. Resonance*, 1971, **3**, 689. 2-6Ec

71PH1129 J. C. Petersen, *J. Phys. Chem.*, 1971, **75**, 1129. 2-3Ba

71PM(3)1 A. Albert, "Physical Methods in Heterocyclic Chemistry" (A. R. Katritzky, Ed.), Vol. 3, pp. 1–26. Academic Press, New York, 1971. T1-1

71PM(3)27 P. Andersen and O. Hassel, "Physical Methods in Heterocyclic Chemistry" (A. R. Katritzky, Ed.), Vol. 3, pp. 27–51. Academic Press, New York, 1971. 1-4Ic, T1-1

71PM(3)53 G. J. Thomas, Jr., "Physical Methods in Heterocyclic Chemistry" (A. R. Katritzky, Ed.), Vol. 3, pp. 53-66. Academic Press, New York, 1971. 1-6Ce, T1-1

71PM(3)67 W. L. F. Armarego, "Physical Methods in Heterocyclic Chemistry" (A. R. Katritzky, Ed.), Vol. 3, pp. 67–222. Academic Press, New York, 1971. T1-1

71PM(3)223 G. Spiteller, "Physical Methods in Heterocyclic Chemistry" (A. R. Katritzky, Ed.), Vol. 3, pp. 223–296. Academic Press, London, 1971. T1-1

71PM(3)397 R. B. Homer, "Physical Methods in Heterocyclic Chemistry" (A. R. Katritzky, Ed.), Vol. 3, pp. 397–423. Academic Press, New York, 1971. T1-1

71PM(4)21 E. A. C. Lucken, "Physical Methods in Heterocyclic Chemistry" (A. R. Katritzky, Ed.), Vol. 4, pp. 21–53. Academic Press, New York, 1971. 1-6A, T1-1

71PM(4)121 R. F. M. White and H. Williams, "Physical Methods in Heterocyclic Chemistry" (A. R. Katritzky, Ed.), Vol. 4, pp. 121–235. Academic Press, New York, 1971. 1-5Aa, T1-1

71PM(4)237 J. Kraft and S. Walker, "Physical Methods in Heterocyclic Chemistry" (A. R. Katritzky, Ed.), Vol. 4, pp. 237–264. Academic Press, New York, 1971. 1-4G(2), 1-7D, T1-1

71PM(4)265 A. R. Katritzky and P. J. Taylor, "Physical Methods in Heterocyclic Chemistry" (A. R. Katritzky, Ed.), Vol 4, pp. 265–434. Academic Press, New York, 1971. 1-6Cc-(2), 1-6Cd, T1-1, 3-2Bc, 3-2Ed, 3-2Fb, 2-3C, 4-4Fd, 4-5Bb, 4-5Bf, 4-5Bh, 4-5Fe(2), 4-5Ff, 4-9A

71R601 G. van den Bosch, H. J. T. Bos and J. F. Arens, *Rec. Trav. Chim. Pays-Bas*, 1971, **90**, 601. 4-6Bb(2)

71RC211 M. Draminski and B. Fiszer, *Rocz. Chem.*, 1971, **45**, 211. 2-4Dd

71RO173 I. S. Ioffe, A. B. Tomchin and E. N. Zhukova, *Z. Org. Khim.*, 1971, **7**, 173. 2-7Eb

71RO179 I. S. Ioffe, A. B. Tomchin and G. A. Shurokii, *Z. Org. Khim.*, 1971, **7**, 179. 3-1Ba

71RO1500 A. I. Ginak, K. A. V'yunov and E. G. Sochilin, *Zh. Org. Khim.*, 1971, **7**, 1500. 4-6Eb(i)

71RO1953 V. S. Bogdanov, M. A. Kalik, G. M. Zhidomirov, N. D. Chuvylkin and Y. L. Goldfarb, *Z. Org. Khim.*, 1971, **7**, 1953. 3-2Ca(iv)

71RR1447 R. Bacaloglu, E. Fliegl and G. Ostrogovich, *Rev. Roum. Chim.*, 1971, **16**, 1447. 2-6Ed

71SA2119 M. Anteunis and M. Vandewalle, *Spectrochim. Acta, Part A*, 1971, **27**, 2119. 3-2Ac

71T379 P. Caramella, R. Metelli and P. Grünanger, *Tetrahedron*, 1971, **27**, 379. 4-2Bg(i), 4-5Ba

71T775 T. Hino, M. Nakagawa, T. Hashizuma, N. Yamaji, Y. Miwa, K. Tsuneoka and S. Akaboshi, *Tetrahedron*, 1971, **27**, 775. 3-1Ac(3), 3-4Cb

71T2811 M. M. Shemyakin, L. A. Neiman, S. V. Zhukova, Y. S. Nekrasov, T. J. Pehk and E. T. Lippmaa, *Tetrahedron*, 1971, **27**, 2811. 4-3D(i)

71T2921 M. H. Palmer, A. J. Gaskell, P. S. McIntyre and in part D. W. W. Anderson, *Tetrahedron*, 1971, **27**, 2921. 2-2Gb

71T3129 M. Witanowski, L. Stefaniak, H. Januszewski and G. A. Webb, *Tetrahedron*, 1971, **27**, 3129. 1-5C, 2-3Ba, 2-6Ba

71T3839 J. Z. Mortensen, B. Hedegaard and S.-O. Lawesson, *Tetrahedron*, 1971, **27**, 3839. 3-2Ba, 3-2Bb, 3-2Bd(2), 3-2Da(2), 3-2Db, 3-2Dc, 3-2Dd

71T3853 B. Hedegaard, J. Z. Mortensen and S.-O. Lawesson, *Tetrahedron*, 1971, **27**, 3853. 3-2Cd(2)

71T3861 E. B. Pedersen and S.-O. Lawesson, *Tetrahedron*, 1971, **27**, 3861. 3-2Ca(iii)

71T4407 M. E. Rennekamp, J. V. Paukstellis and R. G. Cooks, *Tetrahedron*, 1971, **27**, 4407. 1-6Ga(i)

71T4449 M. Gotz and K. Grozinger, *Tetrahedron*, 1971, **27**, 4449. 4-7C

71T4653 N. J. Corkindale and A. W. McCulloch, *Tetrahedron*, 1971, **27**, 4653. 2-3Bf, 2-3Fa

71T5121 T. Sasaki, K. Kanematsu and M. Murata, *Tetrahedron*, 1971, **27**, 5121. 4-9Ca

71T5779 J. Deschamps, J. Arriau and P. Parmentier, *Tetrahedron*, 1971, **27**, 5779. 1-7B(2), 4-1Bb(ix), 4-2Fa, 4-2Fe, 4-2Gb, 4-2Gc(iii), 4-2Ge(iii), 4-2Ia, 4-2Ib(2), 4-6Fd(i), 4-6Fd(iii)

71T5795 J. Arriau, J. Deschamps and P. Parmentier, *Tetrahedron*, 1971, **27**, 5795. 1-7B, 4-2Fa, 4-2Fd, 4-2Fe

71T5807 J. Arriau, M. Chaillet and J. Deschamps, *Tetrahedron*, 1971, **27**, 5807. 1-7B, 4-2Fa, 4-2Fc, 4-2Fe, 4-2Ga(i)(2), 4-2Gc(iii), 4-2Hb(iv)

71T5873 H. Wainhoff, H. W. Duerbeck and P. Sohar, *Tetrahedron*, 1971, **27**, 5873. 4-5Ba, 4-5Bd, 4-5Bg, 4-5Fb(v), 4-5Fd, 4-6Db

71T6011 T. Mizuno, M. Hirota, Y. Hamada and Y. Ito, *Tetrahedron*, 1971, **27**, 6011. 2-7B

71T6133 G. Bianchi, M. J. Cook and A. R. Katritzky, *Tetrahedron*, 1971, **27**, 6133. 4-3Ab, 4-3Ab(i)(2), 4-3Ab(ii), 4-3Ab(iii), 4-3Ab(iv)

71TH01 J. Arriau, Ph.D. thesis, University of Pau, France, 1971. 1-7B, 4-6Fd(iv)

71TL2467 B. Tinland and C. Decoret, *Tetrahedron Lett.*, 1971, 2467. 2-8Cb

71TL2929 J. Deschamps, H. Sauvaitre, J. Arriau, A. Maquestiau, Y. van Haverbeke and R. Jacquerye, *Tetrahedron Lett.*, 1971, 2929. 4-2Fa(2), 4-2Gb, 4-2Gc(iii), T4-17(3)

71TL3299 R. Benassi, P. Lazzeretti, L. Schenetti, F. Taddei and P. Vivarelli, *Tetrahedron Lett.*, 1971, 3299. 4-1Hb(ii)

71TL4897 H. Ahlbrecht and G. Rauchschwalbe, *Tetrahedron Lett.*, 1971, 4897. 2-8Bd

71ZF197 A. I. Finkel'shtein, T. N. Roginskaya and Y. I. Mushkin, *Zh. Fiz. Khim.*, 1971, **45**, 197. 2-6Ed

71ZO1815 A. I. Ginak, K. A. V'yunov and E. G. Sochilin, *Zh. Obshch. Khim.*, 1971, **41**, 1815. 4-6Eb(i)

71ZO2520 K. M. Dyumaev, B. A. Korolev and R. E. Lokhov, *Zh. Obshch. Khim.*, 1971, **41**, 2520. 2-3D

72AC1044 G. J. Yakatan, R. J. Juneau and S. G. Schulman, *Anal. Chem.*, 1972, **44**, 1044. 2-3Gb

72AC1240 M. P. Bratzel, J. J. Aaron, J. D. Winefordner, S. G. Schulman and H. Gershon, *Anal. Chem.*, 1972, **44**, 1240. 2-3Ha

72AC1611 P. J. Kovi, A. C. Capomacchia and S. G. Schulman, *Anal. Chem.*, 1972, **44**, 1611. 2-6Bb

72AG144 C. Nordling, *Angew Chem.*, 1972, **84**, 144. 1-6F

72AH43 J. Nyitrai and K. Lempert, *Acta Chim. Budapest*, 1972, **73**, 43. 4-4D

72AJ985 R. L. N. Harris, *Aust. J. Chem.*, 1972, **25**, 985. 3-3C

72AJ2711 D. J. Brown and R. L. Jones, *Aust. J. Chem.*, 1972, **25**, 2711. 2-4Fb

72AS31 A.-B. Hörnfeldt and P.-O. Sundberg, *Acta Chem. Scand.*, 1972, **26**, 31. 3-2Cd

72AS459 U. Svanholm, *Acta Chem. Scand.*, 1972, **26**, 459. 4-4E(i)-(2), 4-5A, 4-5Fb(ii)

72AS556 J. Skramstad, *Acta Chem. Scand.*, 1972, **26**, 556. 3-5B

72AS715 M. Begtrup, *Acta Chem. Scand.*, 1972, **26**, 715. 4-3Ec

72AS760 S. Furberg and C. S. Petersen, *Acta Chem. Scand.*, 1972, **26**, 760. 2-4Da

72AS1243 M. Begtrup, *Acta Chem. Scand.*, 1972, **26**, 1243. 4-4Fa(2), 4-4Fb

72AS1298 L. Brehm, H. Hjeds and P. Krogsgaard-Larsen, *Acta Chem. Scand.*, 1972, **26**, 1298. 4-2Ca

72AS2140 G. Eide, A. Hordvik and L. J. Sœthre, *Acta Chem. Scand.*, 1972, **26**, 2140. 4-9A

72AS2255 D. W. Aksnes and H. Kryvi, *Acta Chem. Scand.*, 1972, **26**, 2255. 2-3Ba, 2-5Ba

72BF1055 L. Bouscasse, M. Chanon, R. Phan-Tan-Luu, J. E. Vincent and J. Metzger, *Bull. Soc. Chim. Fr.*, 1972, 1055. 4-4Cc

72BF1511 J. Daunis, Y. Guindo, R. Jacquier and P. Viallefont, *Bull. Soc. Chim. Fr.*, 1972, 1511. 2-5F

72BF1903 Y. Ferré, R. Faure, E.-J. Vincent, H. Larivé and J. Metzger, *Bull. Soc. Chim. Fr.*, 1972, 1903. T2-6

72BF1975 J. Daunis, Y. Guindo, R. Jacquier and P. Viallefont, *Bull. Soc. Chim. Fr.*, 1972, 1975. 2-5F(2)

72BF2481 P. Guerret, R. Jacquier and G. Maury, *Bull. Soc. Chim. Fr.*, 1972, 2481. 5-3A(2), S5-3

72BF2868 J.-L. Aubagnac, J. Elguero and R. Jacquier, *Bull. Soc. Chim. Fr.*, 1972, 2868. S5-5, 6-2Be(2)

72BF2916 J. Elguero, A. Fruchier and S. Mignonac-Mondon, *Bull. Soc. Chim. Fr.*, 1972, 2916. 5-2Ab

72CA3851 A. Morimoto and H. Takasugi, *Chem. Abstr.*, 1972, **76**, 3851h; Jap. Patent 71 31,548(Cl. C 07*d* A 61 *K*) (1971). 5-5Aa

72CB388 E. Kranz, J. Kurz and W. Donner, *Chem. Ber.*, 1972, **105**, 388. 5-3Bb(ii), 5-3Bd, S5-4

72CC2423 J. T. Edward and J. K. Liu, *Can. J. Chem.*, 1972, **50**, 2423. 4-6Cb, 4-6Ec

72CC3079 D. E. Horning and J. M. Muchowski, *Can. J. Chem.*, 1972, **50**, 3079. 4-4Eg

72CC3082 G. Lacasse and J. M. Muchowski, *Can. J. Chem.*, 1972, **50**, 3082. 4-4Ca

72CH43 A. H.-J. Wang, J. C. Paul, E. R. Talaty and A. E. Dupuy, *Chem. Commun.*, 1972, 43. 6-1B

72CH52 C. Temple, Jr., B. H. Smith, Jr. and J. A. Montgomery, *Chem. Commun.*, 1972, 52. 4-9Bb

72CH393 R. Bonnett and R. F. C. Brown, *Chem. Commun.*, 1972, 393. 3-1Ad

72CH504 M. T. Thomas, V. Snieckus and E. Klingsberg, *Chem. Commun.*, 1972, 504. 6-2A

72CH573 F. P. Boer, J. W. Turley and F. P. van Remoortere, *Chem. Commun.*, 1972, 573. 2-3Bc, T2-1, 1-4Ib

72CH577 S. Ogawa, T. Yamaguchi and N. Gotoh, *Chem. Commun.*, 1972, 577, 2-7B

72CH771 W. J. Hehre and W. A. Lathan, *Chem. Commun.*, 1972, 771. 2-8A

72CH818 E. E. Nunn and R. N. Warrener, *Chem. Commun.*, 1972, 818. 6-1Ab

72CH823 V. I. Bendall, *Chem. Commun.*, 1972, 823. 6-2A(2)

72CH827 A. A. Reid, J. T. Sharp and S. J. Murray, *Chem. Commun.*, 1972, 827. 6-2A(2)

72CH982 S. F. Gait, M. E. Peek, C. W. Rees and R. C. Storr, *Chem. Commun.*, 1972, 982. 6-2A

72CH1000 K. T. Potts and J. Marshall, *Chem. Commun.*, 1972, 1000. 4-3Cd

72CI335 D. Lloyd and D. R. Marshall, *Chem. Ind. (London)*, 1972, 335. 3-1B

72CQ9 R. Lantz and A.-B. Hörnfeldt, *Chem. Scr.*, 1972, **2**, 9. 3-2Cd(2)

72CR1 U. Eisner and J. Kuthan, *Chem. Rev.*, 1972, **72**, 1. 2-2C

72CS355 A. D. Baker, C. R. Brundle and M. Thompson, *Chem. Soc. Rev.*, 1972, **1**, 355. 1-6F

72CT2096 S. Kubota and M. Uda, *Chem. Pharm. Bull.*, 1972, **20**, 2096. 4-1Ec, 4-4Ec, 4-4Ef, 4-4Ei, 4-4Ej

72CX(B)596 D. Suck, W. Saenger and K. Zechmeister, *Acta Crystallogr., Sect. B*, 1972, **28**, 596. 2-4Db

72CX(B)659 R. Norrestam, B. Stensland and E. Söderberg, *Acta Crystallogr., Sect. B*, 1972, **28**, 659. 2-4Gc

72CX(B)1584 T. C. Downie, W. Harrison and E. S. Raper, *Acta Crystallogr., Sect. B*, 1972, **28**, 1584. 4-6Bb(i)

72CX(B)1982 T. F. Lai and R. E. Marsh, *Acta Crystallogr., Sect. B*, 1972, **28**, 1982. 5-1Da

72CX(B)2074 J.-P. Mornon and R. Bally, *Acta Crystallogr., Sect. B*, 1972, **28**, 2074. 4-6Ff(ii)

72CX(B)2148 P. Prusiner and M. Sundaralingam, *Acta Crystallogr., Sect. B*, 1972, **28**, 2148. 5-2Ba

72CX(B)2260 A. Rahman and H. R. Wilson, *Acta Crystallogr., Sect. B*, 1972, **28**, 2260. 2-4Db

72CX(B)2417 V. Amirthalingam and K. V. Muralidharan, *Acta Crystallogr., Sect. B*, 1972, **28**, 2417. 4-6Ff(ii)

72CX(B)2421 V. Amirthalingam and K. V. Muralidharan, *Acta Crystallogr., Sect. B*, 1972, **28**, 2421. 4-6Ff(ii)

72CX(B)3405 A. Kvick and S. S. Boobes, *Acta Crystallogr., Sect. B*, 1972, **28**, 3405. 2-3Bc

72CZ656 A. Tutalkova and P. Vetesnik, *Collect. Czech. Chem. Commun.*, 1972, **37**, 656. 4-2Ga(i)(2), 4-2Ga(ii)(2)

72CZ1905 E. Ružička, V. Bekárek and P. Heinz, *Collect. Czech. Chem. Commun.*, 1972, **37**, 1905. 2-4H

72DA1366 T. S. Safonova, Y. N. Sheinker, M. P. Nemeryuk, E. M. Peresleni and T. F. Vlasova, *Dokl. Akad. Nauk SSSR*, 1972, **205**, 1366. 2-2E

72G91 G. P. Gardini and V. Bocchi, *Gazz. Chim. Ital.*, 1972, **102**, 91. 3-2Ec

72G169 G. Adembri, F. De Sio, R. Nesi and M. Scotton, *Gazz. Chim. Ital.*, 1972, **102**, 169. 2-4Cc, 2-4Ce

72G325 G. La Manna and M. Cignitti, *Gazz. Chim. Ital.*, 1972, **102**, 325. 2-4

72IS805 D. Lichtenberg, F. Bergmann and Z. Neiman, *Isr. J. Chem.*, 1972, **10**, 805. 5-1Bc, 5-1Be, 5-1C

72IS819 Z. Neiman, *Isr. J. Chem.*, 1972, **10**, 819. 5-1Ab

72J(PI)68 S. Bradbury, C. W. Rees and R. C. Storr, *J. Chem. Soc. (Perkin I)*, 1972, 68. 4-9D(2)

72J(PI)90 T. Nishiwaki and K. Kondo, *J. Chem. Soc. (Perkin I)*, 1972, 90. 4-2Bb

72J(PI)310 M. D. Bachi, *J. Chem. Soc. (Perkin I)*, 1972, 310. 4-3Cc

72J(PI)777 G. V. Boyd and S. R. Dando, *J. Chem. Soc. (Perkin I)*, 1972, 777. 4-2Hc

72J(PI)909 G. V. Boyd and P. H. Wright, *J. Chem. Soc. (Perkin I)*, 1972, 909. 4-3Cc

72J(PI)953 G. Adembri, F. De Sio, R. Nesi and M. Scotton, *J. Chem. Soc. (Perkin I)*, 1972, 953. 2-4Ce

72J(PI)1022 H. N. Al-Jallo, A. Al-Khashab and I. G. Sallomi, *J. Chem. Soc. (Perkin I)*, 1972, 1022. 4-2Ic

72J(PI)1432 M. D. Scott, *J. Chem. Soc. (Perkin I)*, 1972, 1432. 4-2E

72J(PI)1623 C. V. Greco, F. C. Pellegrini and M. A. Pesce, *J. Chem. Soc. (Perkin I)*, 1972, 1623. 4-9D

72J(PI)1924 H. Nakata, A. Tatematsu, H. Yoshizumi and S. Naga, *J. Chem. Soc. (Perkin I)*, 1972, 1924. 1-6Gb

72J(PI)1983 J. H. Davies, R. H. Davis and R. A. G. Carrington, *J. Chem. Soc. (Perkin I)*, 1972, 1983. 4-3Ce, 4-6Bd

72J(PI)2411 A. S. Bailey, A. J. Buckley, W. A. Warr and J. J. Wedgwood, *J. Chem. Soc.* (*Perkin I*), 1972, 2411. 3-4Cb, 3-4Cc

72J(PI)2820 J. A. Elvidge and A. P. Redman, *J. Chem. Soc.* (*Perkin I*), 1972, 2820. 2-6Db, 2-7Eb

72J(PI)2950 D. Lichtenberg, F. Bergmann, M. Rahat and Z. Neiman, *J. Chem. Soc.* (*Perkin I*), 1972, 2950. 5-1Bb, 5-1Bd, 5-1Be

72J(PII)11 R. C. Seccombe and C. H. L. Kennard, *J. Chem. Soc.* (*Perkin II*), 1973, 11. 5-3Bd

72J(PII)392 R. F. Cookson and G. W. H. Cheeseman, *J. Chem. Soc.* (*Perkin II*), 1972, 392. 2-2Gb, 2-4Ca, 2-4Cd

72J(PII)585 Z. Neimann, *J. Chem. Soc.* (*Perkin II*) 1972, 585. 5-1Aa(iii)

72J(PII)1295 M. J. Cook, A. R. Katritzky, P. Linda and R. D. Tack, *J. Chem. Soc.* (*Perkin II*), 1972, 1295. 2M1A, 2-1B(2), 2-1C, 2-3Bh, 2-6Ba, 2-8Ba, 2-10C(2), T2-5, T2-6

72J(PII)1459 G. B. Barlin, *J. Chem. Soc.* (*Perkin II*), 1972, 1459. 2-6Bb

72J(PII)1676 D. Lichtenberg, F. Bergmann and Z. Neiman, *J. Chem. Soc.* (*Perkin II*), 1972, 1676. 5-1Ef(3)

72J(PII)2045 C. W. N. Cumper and G. D. Pickering, *J. Chem. Soc.* (*Perkin II*), 1972, 2045. 4-3Bd, 4-4Ce, 4-4Cf

72J(PII)2125 E. Gaetani, T. Vitali, A. Mangia, M. Nardelli and G. Pelizzi, *J. Chem. Soc.* (*Perkin II*), 1972, 2125. 4-8B

72JA621 W. Saenger, *J. Amer. Chem. Soc.*, 1972, **94**, 621. 2-5E

72JA951 G. C. Y. Lee, J. H. Prestegard and S. I. Chan, *J. Amer. Chem. Soc.*, 1972, **94**, 951. 2-6Cb(2), 2-6Cc

72JA1717 N. C. Seeman, E. L. McGandy and R. D. Rosenstein, *J. Amer. Chem. Soc.*, 1972, **94**, 1717. 4-1Bb(xii)

72JA1745 C. B. Storm and Y. Teklu, *J. Amer. Chem. Soc.*, 1972, **94**, 1745. 6-3

72JA2765 J. A. Zoltewicz and L. W. Deady, *J. Amer. Chem. Soc.*, 1972, **94**, 2765. 5-2Ab

72JA3218 G. C. Y. Lee and S. I. Chan, *J. Amer. Chem. Soc.*, 1972, **94**, 3218. 5-1Eh

72JA4034 R. H. Blessing and E. L. McGandy, *J. Amer. Chem. Soc.*, 1972, **94**, 4034. 4-1Ka

72JA4907 L. A. Paquette, L. B. Anderson, J. F. Hansen, S. A. Lang and H. Berk, *J. Amer. Chem. Soc.*, 1972, **94**, 4907. 6-2Be(2)

72JA5926 F. W. Fowler, *J. Amer. Chem. Soc.*, 1972, **94**, 5926. 2-2C

72JA7898 F. Jordan and H. D. Sostman, *J. Amer. Chem. Soc.*, 1972, **94**, 7898. 5-1Dc

72JH25 K. R. Wursthorn and E. H. Sund, *J. Heterocycl. Chem.*, 1972, **9**, 25. 1-6Cd, 2-8Db(ii)(2)

72JH107 G. Werber, F. Buccheri and F. Maggio, *J. Heterocycl. Chem.*, 1972, **9**, 107. 4-5Fa(ii)

72JH153 M. M. Kochhar, *J. Heterocycl. Chem.*, 1972, **9**, 153. 4-5-Fa(ii), 6-2Bd, T6-1(2)

72JH203 G. L. Rowley and G. L. Kenyon, *J. Heterocycl. Chem.*, 1972, **9**, 203. 4-6Fg(i)

72JH225 J.-E. A. Otterstedt and R. Pater, *J. Heterocycl. Chem.*, 1972, **9**, 225. 2-8Dc

72JH235 H. Höhn, T. Denzel and W. Janssen, *J. Heterocycl. Chem.*, 1972, **9**, 235. 5-2Ba, T5-1

72JH285 G. Rapi, M. Ginanneschi, E. Belgodere and M. Chelli, *J. Heterocycl. Chem.*, 1972, **9**, 285. 1-6Ce, 4-6Fe(i)(3), 4-6Fe(ii)(4), 4-6Fe(iii)(4), 4-6Fe(iv)(4)

72JH355 J. Morel, C. Paulmier and P. Pastour, *J. Heterocycl. Chem.*, 1972, **9**, 355. 3-2Ce

72JH363 J. T. Edward and I. Lantos, *J. Heterocycl. Chem.*, 1972, **9**, 363. 4-3Ci

72JH995 J. Lee and W. W. Paudler, *J. Heterocycl. Chem.*, 1972, **9**, 995. 2-4Fb

72JH1021 R. L. Williams and M. G. ElFayoumy, *J. Heterocycl. Chem.*, 1972, **9**, 1021. 2-3Ba

72JH1039 M. D. Coburn and J. L. Singleton, *J. Heterocycl. Chem.*, 1972, **9**, 1039. 2-6Ba

72JL(12)191 J. Teysseyre, H. Sauvaitre, J. Arriau and J. Deschamps, *J. Mol. Struct.*, 1972, **12**, 191. 4-2Bc

72JO51 N. Finch, M. M. Robison and M. P. Valerio, *J. Org. Chem.*, 1972, **37**, 51. 5-2Ba

72JO221 J. Adachi and N. Sato, *J. Org. Chem.*, 1972, **37**, 221. 2-4E

72JO676 C. V. Greco and M. Pesce, *J. Org. Chem.*, 1972, **37**, 676. 4-9D

72JO1047 P. B. Ghosh and B. Ternai, *J. Org. Chem.*, 1972, **37**, 1047. 2-8Dg

72JO1707 R. A. Coburn and J. P. O'Donnell, *J. Org. Chem.*, 1972, **37**, 1707. 5-4B

72JO1867 A. A. Watson and G. Bosworth Brown, *J. Org. Chem.*, 1972, **37**, 1867. 5-1Fa

72JO2498 D. D. Chapman, *J. Org. Chem.*, 1972, **37**, 2498. 2-8Dd(3)

72JO3066 M. G. Reinecke, J. F. Sebastian, H. W. Johnson, Jr. and C. Pyun, *J. Org. Chem.*, 1972, **37**, 3066. 3-1Ac

72JO3566 M. Israel and E. Z. Zoll, *J. Org. Chem.*, 1972, **37**, 3566. T6-1

72JO3662 H. Feuer and J. P. Lawrence, *J. Org. Chem.*, 1971, **37**, 3662. 2-Eb, T2-10, 4-8C

72JO4121 C. H. Yoder, R. C. Barth, W. M. Richter, and F. A. Snavely *J. Org. Chem.*, 1972, **37**, 4121. 3-2Ab, 3-2Cf, 4-2Hd, 4-3Ae

72JO4124 C. A. Lovelette and L. Long, *J. Org. Chem.*, 1972, **37**, 4124. 6-2Bd

72JT(57)5087 R. Blinc, M. Mali, R. Osredkar, A. Prelesnik, J. Seliger, I. Zupančič and L. Ehrenberg, *J. Chem. Phys.* 1972, **57**, 5087. 2-4Db, 2-6Cb

72KG95 Y. L. Frolov, V. K. Voronov, N. M. Deriglazov, L. V. Belousova, N. M. Vitkovskaya, S. M. Tyrina and G. G. Skvortsova, *Khim. Geterotsikl. Soedin.*, 1972, 95. 2-3Ba

72KG191 B. E. Zaitsev, N. A. Andronova, V. T. Grachev, V. P. Lezina, K. M. Dyumaev and L. D. Smirnov, *Khim. Geterotsikl. Soedin.*, 1972, 191. 2-3D

72KG197 B. Y. Zaitsev, N. A. Andronova, V. T. Grachev, V. P. Lezina, K. M. Dyumaev and L. D. Smirnov, *Khim. Geterotsikl. Soedin.*, 1972, 197. 2-3D

72KG920 V. A. Bren, V. I. Usacheva and V. I. Minkin, *Khim. Geterotsikl. Soedin.*, 1972, 920. 3-2Cf

72KG962 G. N. Rodionova, R. E. Lokhov and K. M. Dyumaev, *Khim. Geterotsikl. Soedin.*, 1972, 962. 2-3Fa

72KG1011 A R. Katritzky, *Khim. Geterotsikl. Soedin.*, 1972, 1011. 2-1B

72KG1132 L. M. Alekseeva, G. G. Dvoryantseva, I. V. Persianova, Yu. N. Sheinker, R. M. Palei and P. M. Kochergin, *Khim. Geterotsikl. Soedin.*, 1972, 1132. 5-5Ac

72LA(757)100 M. Brugger, H. Wamhoff and F. Korte, *Liebig's Ann. Chem.*, 1972, **757**, 100. 2-8Dd

72LA(758)111 H. Neunhoeffer and H.-W. Frühauf, *Liebig's Ann. Chem.*, 1972, **758**, 111. 2-4Fb

72M426 T. Kappe, M. A. A. Chirazi and E. Ziegler, *Monatsh. Chem.*, 1972, **103**, 426. 2-4De

72MC727 T. Jen, B. Dienel, H. Bowman, J. Petta, A. Helt and B. Loev, *J. Med. Chem.*, 1972, **15**, 727. 2-6Cc

72MI5 B. I. Sukhorukov, A. S. Gukovskaya, L. V. Sukhoruchkina and G. I. Lavrenova, *Biofizika*, 1972, **17**, 5. 2-6Cb

72MI217 J. S. Kwiatkowski, *Acta Phys. Pol.*, 1972, **41A**, 217. 1-7E

72MI133 W. S. Powell and R. A. Heacock, *Chem. Ther.*, 1972, **7**, 133; *Chem. Abstr.*, 1972, **77**, 113250q. 3-2Ef

72MI261 C. Jamion, *Acta Biochim. Pol.*, 1972, **19**, 261. 2-7Ed

72MI265 D. T. Edmonds and P. A. Speight, *J. Magn. Resonance*, 1972, **6**, 265. 2-4Db, 2-6Cb

72MI603 T. Hino and M. Nakagawa, *Yuki Gosei Kagaku Kyokai Shi*, 1972, **30**, 603; *Chem. Abstr.*, 1972, **77**, 151747q. 3-3C

72NA(235)53 M. Nakashima, J. A. Sousa and R. C. Clapp, *Nature (London), Phys. Sci.*, 1972, **235**, 53. 2-3Gb

72OM823 T. Grønneberg and K. Undheim, *Org. Mass Spectrom.*, 1972, **6**, 823. 1-6Ga, 2-3Bg, 2-5Ba, 4-1Ea

72OM1139 A. Maquestiau, Y. van Haverbeke and R. Flammang, *Org. Mass Spectrom.*, 1972, **6**, 1139. 1-6Gb, 4-1Ea, 4-5Bh(i)

72OR733 E. Alcalde, J. de Mendoza and J. Elguero, *Org. Magn. Resonance*, 1972, **4**, 733. 4-5B(i)

72OR889 M. S. Puar and A. I. Cohen, *Org. Magn. Resonance*, 1972, **4**, 889. 4-5Fa(i)

72PC(314)101 H. G. O. Becker, G. Görmar, H. Hauffe and H.-J. Timpe, *J. Prakt. Chem.*, 1972, **314**, 101. 4-7Ac

72PM(5)1 P. J. Wheatley, "Physical Methods in Heterocyclic Chemistry" (A. R. Katritzky, Ed.), Vol. 5, pp. 1–598. Academic Press, New York, 1972. 1-4Ib(2)

72PN2488 M. Daniels, *Proc. Nat. Acad. Sci. U.S.*, 1972, **69**, 2488. 2-4Db

72PS1 M. Chanon, Personal communication to Dr. J. Elguero, 1972. 1-3B

72PS2 E. Gonzalez, Personal communication to Dr. J. Elguero, 1972. T1-2, T4-5

72PS3 L. Nygaard, Personal communication to Dr. J. Elguero, 1972. T1-5, 4-1Ba, T4-1

72PS4 A. Maquestiau, Personal communication to Dr. J. Elguero, 1972. 1-6Gb

72PS5 L. Baiocchi, Personal communication to Dr. J. Elguero, 1972. 5-2Ba

72PS6 R. Faure, Personal communication to Dr. J. Elguero, 1972. 5-5Ab

72PS7 M. I. Ali, Personal communication to Dr. J. Elguero, 1972. S5-5

72PS8 V. Tabacik, Personal communication to Dr. J. Elguero, 1972. 4-1Ea

72PS9 M. Begtrup, Personal communication to Dr. J. Elguero, 1972. 4-3Ed, 4-3Ef, 4-3Eg

72PS10 B. Ellis and P. J. F. Griffiths, Personal communication to Dr. J. Elguero, 1972. 4-4Cc(ii)

72PS11 J. P. Brouen, Personal communication to Dr. J. Elguero, 1972. 4-9A

72PS12 W. Ried, Personal communication to Dr. J. Elguero, 1972. 4-2Gg

72PS13 M. Chanon and J. Metzger, Personal communication to Dr. J. Elguero, 1972. 4-4Cc(iv)

72PS14 L. T. Creagh, Personal communication to Dr. J. Elguero, 1972. 1-5Ac(i)

72R785 D. Schuijl-Laros, P. J. W. Schuijl and L. Brandsma, *Rec. Trav. Chim. Pays-Bas*, 1972, **91**, 785. 2-2B

72RC2139 L. Skulski, *Rocz. Chem.*, 1972, **46**, 2139. 1-2

72RO2414 N. D. Agibalov, A. S. Enin, G. I. Koldobskii, B. V. Gidaspov and T. N. Timofeeva, *Z. Org. Khim.*, 1972, **8**, 2414. 4-1Kc

72RU452 A. I. Kol'tsov and G. M. Kheifets, *Russ. Chem. Rev.*, 1972, **41**, 452. 1-5Aa

72SA(A)855 C. Cogrossi, *Spectrochim. Acta, Part A*, 1972, **28**, 855. 4-6Bh, 4-6Ea, 4-6Eb(i), 4-6Eb(ii), 4-6Ec

72T303 A. S. Angeloni, V. Ceré, D. Dal Monte, E. Sandri and G. Scapini, *Tetrahedron*, 1972, **28**, 303. 4-9D

72T455 M. J. Nye and W. P. Tang, *Tetrahedron*, 1972, **28**, 455. 4-3Ab, 4-3Ab(iv), 4-3Ad

72T463 M. J. Nye and W. P. Tang, *Tetrahedron*, 1972, **28**, 463. 4-3Ab(i), 4-3Ad, 4-3Ad(iv)

72T637 W. Witanowski, L. S. Stefaniak, H. Januszewski, Z. Grabowski and G. A. Webb, *Tetrahedron*, 1972, **28**, 637. 1-5C(2), 4-1Da, 4-1Eb(iii), 4-1Fa(iv), 4-1G(iii), 4-1I

72T875 K. Hartke and F. Meissner, *Tetrahedron*, 1972, **28**, 875. 3-2Cd, 3-4Ba

72T3587 A. Dasgupta and N. K. Dasgupta, *Tetrahedron*, 1972, **28**, 3587. 4-9D(2)

72T3657 B. A. Hess, L. J. Schaad and C. W. Holyoke, *Tetrahedron*, 1972, **28**, 3657. 6-2A(2)

72T4065 N. Suzuki, M. Sato, K. Okada and T. Goto, *Tetrahedron*, 1972, **28**, 4065. 4-3Cf, 4-6Bc

72T5507 P. Beak, T. S. Woods and D. S. Mueller, *Tetrahedron*, 1972, **28**, 5507. 2-1C

72T5779 C. Glier, F. Dietz, M. Scholz and G. Fischer, *Tetrahedron*, 1972, **28**, 5779. 5-3Bc

72T5789 C. Glier, F. Dietz, M. Scholz and G. Fischer, *Tetrahedron*, 1972, **28**, 5789. 5-3Bc, 5-3Bd

72T5859 S. S. T. King, W. L. Dilling and N. B. Tefertiller, *Tetrahedron*, 1972, **28**, 5859. 2-3Bg, 2-5Ba

72TC51 R. Bonaccorsi, A. Pullman, E. Scrocco and J. Tomasi, *Theor. Chim. Acta*, 1972, **24**, 51. 5-1Db

72TE224 V. T. Grachev, V. N. Kostylev, B. E. Zaitzev, V. I. Kuz'min, K. M. Dyumaev and L. D. Smirnov, *Theor. Exp. Chem.* (*USSR*), 1972, **8**, 224. 2-3D

72TH01 R. Jacquerye, Ph.D. thesis, University of Mons, Belgium, 1972. 1-6Cb, 4-2Fc(2), 4-2Ia, 4-2Ib(7), T4-16, T4-17(3), T4-21(3), S4-10, S4-12

72TH02 B. Saint-Roch, Ph.D. thesis, University of Bordeaux (France), 1972. 4-1Ea

72TL327 J. A. Waite and K. R. H. Wooldridge, *Tetrahedron Lett.* 1972, 327. 4-2D(2)

72TL775 P. Beak and T. S. Woods, *Tetrahedron Lett.*, 1972, 775, 1-7E, 2-1C

72TL2857 G. G. Dvoryantzeva, T. N. Ulyanova, N. P. Kostyuchenko, Y. N. Sheinker, E. I. Lapan and L. N. Yakhontov, *Tetrahedron Lett.*, 1972, 2857. 5-2Aa

72TL3193 T. Grønneberg and K. Undheim, *Tetrahedron Lett.*, 1972, 3193. 1-6Ga, 2-6Ba

72TL3273 J. J. Eisch and G. Gupta, *Tetrahedron Lett.*, 1972, 3273. 2-8Bb

72TL3823 C. D. Poulter and R. B. Anderson, *Tetrahedron Lett.*, 1972, 3823. 2-4Dc

72TL3937 D. S. Iyengar, K. K. Prasad and R. V. Venkataratnam, *Tetrahedron Lett.*, 1972, 3937. 4-2Gg

72TL4295 G. M. Priestley and R. N. Warrener, *Tetrahedron Lett.*, 1972, 4295. 3-1Ad

72TL5019 M. J. Cook, A. R. Katritzky, P. Linda and R. D. Tack, *Tetrahedron Lett.*, 1972, 5019. 2-1A, 2-1B, 3-1B

72UP1 J. Arriau, J. Deschamps and J. Elguero, Unpublished results, 1972. 1-7B, 4-5H(2)

72UP2 J. Elguero and V. Pellegrin, Unpublished results, 1972. 4-2Ib

72ZC230 M. Augustin H. D. Schaedler and P. Reinemann, *Z. Chem.*, 1972, **12**, 230. 2-4Cd

73AG(E)139 H.-J. Knackmuss, *Angew. Chem.* (*Int. Ed. Engl.*,) 1973, **12**, 139. 2-3Ec, 2-3Ef

73AJ889 L. A. Summers, *Aust. J. Chem.*, 1973, **26**, 889. 4-2Bh

73AS797 T. Ottersen, *Acta Chem. Scand.*, 1973, **27**, 797. 2-4Cd

73AS835 T. Ottersen, *Acta Chem. Scand.*, 1973, **27**, 835. 2-4Cd

73AS945 P. Groth, *Acta Chem. Scand.*, 1973, **27**, 945. 4-3Bb, 4-4Cb, 4-5Db

73AS1845 T. La Cour and S. E. Rasmussen, *Acta Chem. Scand.*, 1973, **27**, 1845. 4-1Ba, T4-1

73AS1914 O. Ceder, U. Stenhede, K.-I. Dahlquist, J. M. Waisvisz and M. G. van der Hoeven, *Acta Chem. Scand.*, 1973, **27**, 1914. 4-3Cf

73AS1923 O. Ceder and U. Stenhede, *Acta Chem. Scand.*, 1973, **27**, 1923. 4-3Cf

73AS2221 O. Ceder and U. Stenhede, *Acta Chem. Scand.*, 1973, **27**, 2221. 4-3Ci

73AS2802 P. Krogsgaard-Larsen, S. B. Christensen and H. Hjeds, *Acta Chem. Scand.*, 1973, **27**, 2802. 4-2Bf

73AS3251 P. Krogsgaard-Larsen, H. Hjeds, S. B. Christensen and L. Brehm, *Acta Chem. Scand.*, 1973, **27**, 3251. 4-2Bf

73BA405 J. S. Kwiatkowski, *Bull. Acad. Pol. Sci. Seri. Sci. Chim.*, 1973, **21**, 405. 2-4E(2), 2-5E, 2-6Dc

73BB215 A. Maquestiau, Y. van Haverbeke and R. Jacquerye, *Bull. Soc. Chim. Belg.*, 1973, **82**, 215. 4-2Fa, 4-2Fb, 4-2Fc(2), 4-2Fd(3), 4-2Gb, 4-2Gc, 4-2Ge(ii), 4-7Af, T4-16, T4-17(14)

73BB233 A. Maquestiau, Y. van Haverbeke and R. Jacquerye, *Bull. Soc. Chim. Belg.*, 1973, **82**, 233. 4-2Fa, 4-2Fd, 4-2Gc, 4-2Gd, 4-2Ge(iii), 4-2Gh, 4-6Fd(i), T4-17(19)

73BB747 A. Maquestiau, Y. van Haverbeke and A. Bruyere, *Bull. Soc. Chim. Belg.*, 1973, **82**, 747. T4-15, 4-2Fa, 4-2Gb

73BB757 A. Maquestiau, Y. van Haverbeke and A. Bruyere, *Bull. Soc. Chim. Belg.*, 1973, **82**, 757. T4-15, 4-2Fa, 4-2Gb(2)

73BF(2)1849 J.-L. Barascut, R.-M. Claramunt and J. Elguero, *Bull. Soc. Chim. Fr., Part 2*, 1973, 1849. 4-5Fc(2)

73BF(2)1854 R.-M. Claramunt, R. Granados and E. Pedroso, *Bull. Soc. Chim. Fr., Part 2*, 1973, 1854. 4-5Ff

73BF(2)2039 Y. Bessière-Chrétien and H. Serne, *Bull. Soc. Chim. Fr., Part 2*, 1973, 2039. 2-8Db(iv)

73BF(2)2126 J. Daunis, *Bull. Soc. Chim. Fr., Part 2*, 1973, 2126. 2-4Fb

73BF(2)3044 M. Pays and M. Beljean, *Bull. Soc. Chim. Fr., Part 2*, 1973, 3044. 4-4Cd, 4-5Dd

73CB376 H. Böshagen and W. Geiger, *Chem. Ber.*, 1973, **106**, 376. 4-2D

73CB471 W. Merkel and W. Ried, *Chem. Ber.*, 1973, **106**, 471. 2-7Cb

73CB956 W. Merkel and W. Ried, *Chem. Ber.*, 1973, **106**, 956. 1-6D, 2-7Dc

73CB1423 H.-D. Bartfeld and W. Flitsch, *Chem. Ber.*, 1973, **106**, 1423. 3-2Fb

73CB2918 H. P. Fritz, F. H. Köhler and B. Lippert, *Chem. Ber.*, 1973, **106**, 2918. 2-4Cd(2)

73CB3533 J. Lehmann and H. Wamhoff, *Chem. Ber.*, 1973, **106**, 3533. 2-3Ea

73CB3753 H. R. Kricheldorf and W. Regel, *Chem. Ber.*, 1973, **106**, 3753. 3-2Ec

73CB3951 U. Ewers, H. Günther and L. Jaenicke, *Chem. Ber.*, 1973, **106**, 3951. 2-4Gc, 2-6De

73CC338 M. J. Nye and W. P. Tang, *Can. J. Chem.*, 1973, **51**, 338. 4-2Gf

73CC2349 D. E. Horning and J. M. Muchowski, *Can. J. Chem.*, 1973, **51**, 2349. 4-4Fb

73CC2353 G. Lacasse and J. M. Muchowski, *Can. J. Chem.*, 1973, **51**, 2353. 4-4Eb, 4-4Ee

73CH684 J. L. Wong and M. F. Zady, *Chem. Commun.*, 1973, 684. 2-7Ea

73CH688 M. J. S. Dewar and C. A. Ramsden, *Chem. Commun.*, 1973, 688. 6-1Aa

73CH835 T. L. Gilchrist, G. E. Gymer and C. W. Rees, *Chem. Commun.*, 1973, 835. 6-1Aa

73CI182 S. K. Khetan, *Chem. Ind.*, (*London*), 1973, 182. 2-3C

73CO(276C)657 G. Lepicard, J. Delettré and J.-P. Mornon, *C. R. Acad. Sci. Ser. C*, 1973, **276**, 657. 4-6Ff(ii)

73CO(276C)1007 M. Gadret, M. Goursolle and J.-M. Léger, *C. R. Acad. Sci. Ser. C.*, 1973, **276**, 1007. 5-2Ba

73CO(276C)1341 M.-T. Mussetta, M. Sélim and N. Q. Trinh, *C. R. Acad. Sci. Ser. C.*, 1973, **276**, 1341. 2-4De, 2-4Da, 2-5E

73CO(276C)1533 S. Mignonac-Mondon, J. Elguero and R. Lazaro, *C. R. Acad. Sci. Ser. C*, 1973, **276**, 1533. 5-5Ab

73CP697 M. Witanowski, L. Stefaniak, H. Januszewski and J. Elguero, *J. Chim. Phys.*, 1973, **70**, 697. 4-1Bb(v)

73CP1483 P. Mauret, J.-P. Fayet, M. Fabre, J. Elguero and M. C. Pardo, *J. Chim. Phys.*, 1973, **70**, 1483. 4-1Da(2)

73CT1342 S. Kubota and M. Uda, *Chem. Pharm. Bull.*, 1973, **21**, 1342. 4-3Dd, 4-3Df, 4-3Di, 4-3Dj, 4-4Ec, 4-4Ef, 4-4Ei, 4-4Ej

73CT1470 M. Kamiya and Y. Akahori, *Chem. Pharm. Bull.*, 1973, **21**, 1470. 5-1Aa(iii)

73CT1474 M. Kamiya, N. Haba and Y. Akahori, *Chem. Pharm. Bull.*, 1973, **21**, 1474. 2-4Da, 5-1Ba, 5-1Dd

73CT2571 T. Murata, T. Sugawara and K. Ukawa, *Chem. Pharm. Bull.*, 1973, **21**, 2571. 3-2Ee

73CX(B)31 K. Shikata, T. Ueki and T. Mitsui, *Acta Crystallogr., Sect. B*, 1973, **29**, 31. 5-1Da

73CX(B)61 C. H. Schwalbe and W. Saenger, *Acta Crystallogr., Sect. B*, 1973, **29**, 61. 2-3Ea

73CX(B)714 T. P. Singh and M. Vijayan, *Acta Crystallogr., Sect. B*, 1973, **29**, 714. 4-2Fd

73CX(B)1157 R. Bally and J.-P. Mornon, *Acta Crystallogr., Sect. B*, 1973, **29**, 1157. 4-6Ff(ii)

73CX(B)1160 R. Bally and J.-P. Mornon, *Acta Crystallogr., Sect. B*, 1973, **29**, 1160. 4-6Ff(ii)

73CX(B)1234 R. J. McClure and B. M. Craven, *Acta Crystallogr., Sect. B*, 1973, **29**, 1234. 2-6Cb

73CX(B)1259 D. W. Young and E. M. Morris, *Acta Crystallogr., Sect. B*, 1973, **29**, 1259. 2-4Db

73CX(B)1393 U. Thewalt and C. E. Bugg, *Acta Crystallogr., Sect. B*, 1973, **29**, 1393. 2-4Dh

73CX(B)1641 P. Tollin, H. R. Wilson and D. W. Young, *Acta Crystallogr., Sect. B*, 1973, **29**, 1641. 2-4Db

73CX(B)1669 J. E. Johnson and R. A. Jacobson, *Acta Crystallogr., Sect. B*, 1973, **29**, 1669. 2-7B

73CX(B)1916 J. Hjortås, *Acta Crystallogr., Sect. B*, 1973, **29**, 1916. 2-4Fc

73CX(B)1971 D. R. Carter, J. W. Turley, F. P. van Remoortere and F. P. Boer, *Acta Crystallogr., Sect. B*, 1973, **29**, 1971. 2-3Ef

73CX(B)1974 T. J. Kistenmacher, *Acta Crystallogr., Sect. B*, 1973, **29**, 1974. 5-1Db

73CX(B)2298 B. L. Trus and R. E. Marsh, *Acta Crystallogr., Sect. B*, 1973, **29**, 2298. 4-1Hb

73CX(B)2311 J. P. Declerc, G. Germain and M. H. J. Koch, *Acta Crystallogr., Sect. B*, 1973, **29**, 2311. 4-3Bd

73CX(B)2328 P. Prusiner and M. Sundaralingam, *Acta Crystallogr., Sect. B*, 1973, **29**, 2328. 4-4Cf

73CX(B)2549 L. Fallow, III, *Acta Crystallogr., Sect. B*, 1973, **29**, 2549. 2-4Db

73CX(B)2635 R. Bally, *Acta Crystallogr., Sect. B*, 1973, **29**, 2635. 4-6Ff(ii)

73G1045 M. Croci, G. Fronza and P. Vita-Finzi, *Gazz. Chim. Ital.*, 1973, **103**, 1045. 4-2Bg

73H1882 L. Re, B. Maurer and G. Ohloff, *Helv. Chim. Acta*, 1973, **56**, 1882. 3-2Bd

73H1908 C. Heizmann, P. Hemmerich, R. Mengel and W. Pfleiderer, *Helv. Chim. Acta*, 1973, **56**, 1908. 2-4Gc

73H2680 G. Müller and W. von Philipsborn, *Helv. Chim. Acta*, 1973, **56**, 2680. 2-4Gb(2), 2-4Gc(2)

73HC(15)187 F. Freeman, *Advan. Heterocycl. Chem.*, 1973, **15**, 187. 2-8Bb

73IQ27 K. L. Kapoor, *Int. J. Quantum Chem.*, 1973, **7**(1), 27. 2-6Ce

73J(PI)550 D. J. Anderson, T. L. Gilchrist, G. E. Gymer and C. W. Rees, *J. Chem. Soc.* (*Perkin I*), 1973, 550. 6-1Aa

73J(PI)555 T. L. Gilchrist, G. E. Gymer and C. W. Rees, *J. Chem. Soc.* (*Perkin I*), 1973, 555. 6-1Aa

73J(PI)793 U. Reichman, F. Bergmann, D. Lichtenberg and Z. Neiman, *J. Chem. Soc.* (*Perkin I*), 1973, 793. 5-1Aa(i), 5-1Ab

73J(PI)1225 F. Bergmann, M. Rahat and D. Lichtenberg, *J. Chem. Soc.* (*Perkin I*), 1973, 1225. 5-1Bc, 5-1Bd, 5-1Be

73J(PI)1314 R. J. Grout, B. M. Hynam and M. W. Partridge, *J. Chem. Soc.* (*Perkin I*), 1973, 1314. 2-6Bb

73J(PI)1357 R. N. Butler, T. M. Lambe, J. C. Tobin and F. L. Scott, *J. Chem. Soc.* (*Perkin I*), 1973, 1357. 4-5Fa(ii), 4-5Fb(ii), 4-5Fc, 4-5Ff

73J(PI)1432 R. Bonnett, R. F. C. Brown and R. G. Smith, *J. Chem. Soc.* (*Perkin I*), 1973, 1432. 3-1Ad

73J(PI)1588 D. W. Dunwell and D. Evans, *J. Chem. Soc.* (*Perkin I*), 1973, 1588. 5-3Bb(iv)

73J(PI)1602 A. S. Bailey and A. J. Buckley, *J. Chem. Soc.* (*Perkin I*), 1973, 1602. 3-1Ac, 3-4Cb(2)

73J(PI)2445 D. Lichtenberg, F. Bergmann and Z. Neiman, *J. Chem. Soc.* (*Perkin I*), 1973, 2445. 5-1Bc, 5-1Bd

73J(PI)2543 A. A. Reid, J. T. Sharp, H. R. Sood and P. B. Thorogood, *J. Chem. Soc.* (*Perkin I*), 1973, 2543. 6-2A

73J(PI)2647 U. Reichman, F. Bergmann and D. Lichtenberg, *J. Chem. Soc.* (*Perkin I*), 1973, 2647. 5-1Bc, 5-1Bd

73J(PI)2814 W. L. F. Armarego and B. A. Milloy, *J. Chem. Soc.* (*Perkin I*), 1973, 2814. 3-2Fa

73J(PII)1 R. C. Seccombe and C. H. L. Kennard, *J. Chem. Soc.* (*Perkin II*), 1973, 1. 4-6Dd

73J(PII)4 R. C. Seccombe and C. H. L. Kennard, *J. Chem. Soc.* (*Perkin II*), 1973, 4. 4-6Gc

73J(PII)6 R. C. Seccombe, J. V. Tillack and C. H. L. Kennard, *J. Chem. Soc.* (*Perkin II*), 1973, 6. 4-6Gc

73J(PII)9 R. C. Seccombe and C. H. L. Kennard, *J. Chem. Soc.* (*Perkin II*), 1973, 9. 4-6Gc

73J(PII)160 F. T. Boyle and R. A. Y. Jones, *J. Chem. Soc.* (*Perkin II*), 1973, 160. 4-7Ae

73J(PII)164 F. T. Boyle and R. A. Y. Jones, *J. Chem. Soc.* (*Perkin II*), 1973, 164. 4-7Aa

73J(PII)414 J. A. Clarke, P. J. Dawson, R. Grigg and C. H. Rochester, *J. Chem. Soc.* (*Perkin II*), 1973, 414. 6-3

73J(PII)557 R. G. Button and P. J. Taylor, *J. Chem. Soc.* (*Perkin II*), 1973, 557. 1-4Aa

73J(PII)1080 M. J. Cook, A. R. Katritzky, P. Linda and R. D. Tack, *J. Chem. Soc.* (*Perkin II*), 1973, 1080. 2-1B, 2-3Ac(2), 2-6Ba, 2-8Ba, T2-5, T2-6

73J(PII)2036 G. B. Ansell, *J. Chem. Soc.* (*Perkin II*), 1973, 2036. 1-1C, 1-4Ib, 4-1Fa(iii)

73J(PII)2111 S.-O. Chua, M. J. Cook and A. R. Katritzky, *J. Chem. Soc.* (*Perkin II*), 1973, 2111. 2-7B(2), 2-8Ca(2)

73JA27 P. H. Kasai and D. McLeod, Jr., *J. Amer. Chem. Soc.*, 1973, **95**, 27. 3-1Ac, 4-1Aa

73JA324 H. Saito, Y. Tanaka and S. Nagata, *J. Amer. Chem. Soc.*, 1973, **95**, 324. 4-1Eb(iii)

73JA328 W. F. Reynolds, I. R. Peat, M. H. Freedman and J. R. Lyerla, Jr., *J. Amer. Chem. Soc.*, 1973, **95**, 328. 1-5B, 4-1Cb(v)

73JA1700 P. Beak and F. S. Fry, Jr., *J. Amer. Chem. Soc.*, 1973, **95**, 1700. 2-3Ba, 1-6D

73JA2383 W. Küsters and P. de Mayo, *J. Amer. Chem. Soc.*, 1973, **95**, 2383. 6-1Ab

73JA2392 W. E. McEwen, M. A. Calabro, I. C. Mineo and I. C. Wang, *J. Amer. Chem. Soc.*, 1973, **95**, 2392. 4-5Ea

73JA3408 M. Pieber, P. A. Kroon, J. H. Prestegard and S. I. Chan, *J. Amer. Chem. Soc.*, 1973, **95**, 3408. 2-6Cb, 5-1Eh

73JA3511 Y. P. Wong, *J. Amer. Chem. Soc.*, 1973, **95**, 3511. 2-6Cb, 5-1Eh

73JA4761 T. R. Krugh, *J. Amer. Chem. Soc.*, 1973, **95**, 4761. 5-2Ad

73JA4829 J. J. Bonnet and J. A. Ibers, *J. Amer. Chem. Soc.*, 1973, **95**, 4829. 4-1Cb(vii)

73JH167 B. W. Harris, J. L. Singleton and M. D. Coburn, *J. Heterocycl. Chem.*, 1973, **10**, 167. 2-6Ba

73JH411 J. Elguero, R. Jacquier and S. Mignonac-Mondon, *J. Heterocycl. Chem.*, 1973, **10**, 411. 5-5Ab(2)

73JH1067 E. W. Brunett, D. M. Altwein and W. C. McCarthy, *J. Heterocycl. Chem.*, 1973, **10**, 1067. 3-4Ba

73JM319 V. Tabacik, H. H. Günthard and R. Jacquier, *J. Mol. Spectrosc.*, 1973, **45**, 319. 1-6Cb

73JO11 R. E. Harmon, G. Wellman and S. K. Gupta, *J. Org. Chem.*, 1973, **38**, 11. 3-4Cb(2), 3-4Cc

73JO173 R. A. Abramovitch and B. W. Cue, Jr., *J. Org. Chem.*, 1973, **38**, 173. 3-2Ea, 3-2Ec, 3-4Ca

73JO2939 J. A. Moore, W. J. Freeman, K. Kurita and M. G. Pleiss, *J. Org. Chem.*, 1973, **38**, 2939. 6-2Bb

73KG56 V. T. Grachev, B. E. Zaitsev, K. M. Dyumaev, L. D. Smirnov and M. R. Avezov, *Khim. Geterotsikl. Soedin.*, 1973, 56. 2-3Ee

73KG60 V. T. Grachev, B. E. Zaitsev, K. M. Dyumaev, L. D. Smirnov and M. R. Avezov, *Khim. Geterotsikl. Soedin.*, 1973, 60. 2-3Ee

73KG154 V. A. Bren, Zh. V. Bren and V. I. Minkin, *Khim. Geterotsikl. Soedin.*, 1973, 154. 3-2Ad

73KG398 M. M. Kaganskii, G. G. Dvoryantseva and A. S. Elina, *Khim. Geterotsikl. Soedin.*, 1973, 398. 2-2Gd

73KG535 O. N. Chupakhin, Yu. N. Sheinker, Z. V. Pushkareva, V. A. Trofimov, E. G. Kovalev, V. G. Kharchuk, *Khim. Geterotsikl. Soedin.*, 1973, 535. 2-8DC

73KG556 T. G. Koksharova, L. F. Lipatova, V. N. Konyukhov, G. N. Smotrina and Z. V. Pushkareva, *Khim. Geterotsikl. Soedin.*, 1973, 556. 2-4Ce

73KG699 N. P. Bednyagina, G. N. Lipunova, and G. M. Petrova, *Khim. Geterotsikl. Soedin.*, 1973, 699. 4-5G(iii)

73KG707 V. V. Mel'nikov, V. V. Stolpakova, M. S. Pevzner and B. V. Gidaspov, *Khim. Geterotsikl. Soedin.*, 1973, 707. 4-1-Eb(iv)

73KG767 G. G. Dvoryantseva, T. N. Ul'yanova, Yu. N. Sheinker and L. N. Yakhontov, *Khim. Geterotsikl. Soedin.*, 1973, 767. 5-2Aa

73KG807 Y. V. Koshchienko and A. M. Simonov, *Khim. Geterotsikl. Soedin.*, 1973, 807. 4-5Df

73KG810 K. M. Dyumaev and R. E. Lokhov, *Khim. Geterotsikl. Soedin.*, 1973, 810. 2-3D, 2-3Fa

73KG852 V. D. Romanenko and S. I. Burmistrov, *Khim. Geterotsikl. Soedin.*, 1973, 852. 2-8Dh

73KG1115 T. N. Ul'yanova, G. G. Dvoryantseva, Yu. N. Sheinker, A. S. Elina and I. S. Musatova, *Khim. Geterotsikl. Soedin.*, 1973, 1115. 2-6Dd

73KG1487 E. P. Nesynov, M. M. Besprozvannaya and P. S. Pel'kis, *Khim. Geterotsikl. Soedin.*, 1973, 1487. 4-6Eb(i)

73LA207 A. Treibs, L. Schulze, F.-H. Kreuzer and H.-G. Kolm, *Liebig's Ann. Chem.*, 1973, 207. 3-3C

73MI1 W. A. Lathan, L. Radom, P. C. Hariharan, W. J. Hehre and J. A. Pople, *Top. Curr. Chem.*, 1973, **40**, 1. 6-1Aa

73MI23 C. J. Dik-Edixhoven, H. Schenk and H. van der Meer, *Cryst. Struct. Commun.*, 1973, **2**, 23. 4-1Ha

73MI41 M.-T. Chenon, R. J. Pugmire, D. M. Grant, R. P. Panzica and L. B. Townsend, *Abst. Int. Conf. Heterocycl. Chem. 4th, Salt Lake City, Utah*, 1973, 41. 5-2Ad

73MI123 M. Tišler, *Synthesis*, 1973, 123. 4-9Ca, 5-5Ab

73MI130 J. P. Jacobsen, O. Snerling, E. J. Pedersen, J. T. Nielsen and K. Schaumburg, *J. Magn. Resonance*, 1973, **10**, 130. 4-1Ba

73MI179 S. Pérez-Garrido, A. López-Castro and R. Márquez, *Anal. Fís. (Spain)*, 1973, **69**, 179. 4-4Ce

73MI469 F. Bechtel, J. Gaultier and C. Hauw, *Cryst. Struct. Commun.*, 1973, **2**, 469. 1-4Ib, 4-2Fa

73MI473 F. Bechtel, J. Gaultier and C. Hauw, *Cryst. Struct. Commun.*, 1973, **2**, 473. 4-2Fa, 4-2Hb(ii)

73MI473b W. Saenger and D. Suck, *Eur. J. Biochem.*, 1973, **32**, 473. 2-5E

73OM57 A. J. Blackman and J. H. Bowie, *Org. Mass. Spectrom.*, 1973, **7**, 57. 4-1Ea

73OM271 A. Maquestiau, Y. van Haverbeke, R. Flammang and J. Elguero, *Org. Mass Spectrom.*, 1973, **7**, 271. 1-6Gb, 4-1Da

73OM1267 A. Maquestiau, Y. van Haverbeke, R. Flammang, M. C. Pardo and J. Elguero, *Org. Mass Spectrom.*, 1973, **7**, 1267. 1-6Gb, 4-1I

73OR551 U. Vögeli and W. von Philipsborn, *Org. Magn. Resonance*, 1973, **5**, 551. 2-3D

73OR573 J. Fournier, E. J. Vincent, A. M. Chauvet and A. Crevat, *Org. Magn. Resonance*, 1973, **5**, 573. 2-2Gc

73PC(315)382 H. Dorn, *J. Prakt. Chem.*, 1973, **315**, 382. 4-2Fa, 4-2Ha, 4-2Hb(i), 4-2Ia, 4-2Ib, 4-5Bg, 4-5Bh, 4-5Bi, T4-20

73PC(315)779 K. Gewald and H.-J. Jänsch, *J. Prakt. Chem.*, 1973, **315**, 779. 3-4A

73PC(315)791 G. Westphal and R. Schmidt, *J. Prakt. Chem.*, 1973, **315**, 791. 4-5Fa(iv)

73PC(315)1009 M. H. Elnagdi and S. O. Abd Allah, *J. Prakt. Chem.*, 1973, **315**, 1009. 4-6Db

73PH1595 M. P. Bratzel, J. J. Aaron, J. D. Winefordner, S. G. Schulman and H. Gershon, *J. Phys. Chem.*, 1973, **77**, 1595. 2-3Ha

73PS1 A. Ricci, Personal communication to Dr. P. Linda, 1973. 3-2Cf

73PS2 G. V. Boyd, Personal communication to Professor A. R. Katritzky, 1973. 4-9Ca

73R731 D. de Rijke and H. Boelens, *Rec. Trav. Chim. Pays-Bas*, 1973, **92**, 731. 3-2Ac(2)

73RC1735 B. Macierewicz, *Rocz. Chem.*, 1973, **47**, 1735. 3-2Db

73RO821 L. N. Kurkovskaya, N. N. Shapet'ko, I. Ya, Kvitko, Yu.

N. Koshelev and E. M. Sof'ina, *Z. Org. Khim.*, 1973, **9**, 821. 4-2G(i), 4-4Ae

73T3285 Y. Tanaka and S. I. Miller, *Tetrahedron*, 1973, **29**, 3285. 4-1Da

73T3565 K. Lempert, J. Nyitrai, K. Zauer, A. Kálmán, G. Argay, A. J. M. Duisenberg and P. Sohár, *Tetrahedron*, 1973, **29**, 3565. 4-4D, 4-6Ec

73T4291 W. Müller, U. Kraatz and F. Korte, *Tetrahedron*, 1973, **29**, 4291. 4-2Bb, 4-2Bh

73TC145 F. Török, Á. Hegedüs and P. Pulay, *Theor. Chim. Acta.*, 1973, **32**, 145. T4-1, 4-1Ca, T4-7

73TE238 Yu. L. Frolov, V. B. Mantsivoda, V. B. Modonov, S. N. Elovskii, E. S. Domnina and G. G. Skvortsova, *Theor. Exp. Chem.*, 1973, **9**, 238. 1-4G

73TH01 H. Meyer, Ph.D. thesis, University of Karlsruhe, West Germany, 1973. 2-7Ca, 2-7Da(2)

73TL1137 H. Taguchi, *Tetrahedron Lett.*, 1973, 1137. 4-9Ca

73TL2703 R. Faure, E. J. Vincent and J. Elguero, *Tetrahedron Lett.*, 1973, 2703. 5-5Ab

73TL2905 D. J. Le Count and A. T. Greer, *Tetrahedron Lett.*, 1973, 2905. 2-8Dd

73ZC298 G. Oehme, P. Neels and G. Rembarz, *Z. Chem.*, 1973, **13**, 298. 2-6Ed

73ZC375 E. Klemm, D. Klemm, J. Reichardt and H.-H. Hörhold, *Z. Chem.*, 1973, **13**, 375. 2-8Cb

73ZO1556 B. A. Korolev and M. A. Mal'tseva, *Z. Obshch. Khim.*, 1973, **43**, 1556. 2-6Ed

73ZO2730 Ya. L. Gol'dfarb, L. I. Belen'kii, I. A. Abronin, G. M. Zhidomirov and N. D. Chuvylkin, *Zhur. Obsch. Khim.*, 1973, **43**, 2730. 2-6Ba

74AS308 L. Brehm, P. Krogsgaard-Larsen and H. Hjeds, *Acta. Chem. Scand.*, 1974, **28**, 308. 4-2Ca

74AS533 P. Krogsgaard-Larsen and H. Hjeds, *Acta. Chem. Scand.*, 1974, **28**, 533. 4-2Ca

74BB105 A. Maquestiau, Y. Van Haverbeke, R. Flammang, S. O. Chua, M. J. Cook and A. R. Katritzky, *Bull. Soc. Chim. Belg.*, 1974, **83**, 105. 1-6Gb, 4-7Ab

74BB263 A. Maquestiau, Y. Van Haverbeke and R. N. Muller, *Bull. Soc. Chim. Belg.*, 1974, **83**, 263. 4-2Bb, 4-2Bf

74BF(2)291 P. Bouchet, J. Elguero, R. Jacquier and J. M. Pereillo, *Bull. Soc. Chim. Fr., Part 2*, 1974, 291. 4-6Fd(i), 4-6Fd(iv)

74BF(2)1675 J. de Mendoza and J. Elguero, *Bull. Soc. Chim. Fr., Part 2*, 1974, 1675. 5-5Ab

74CB876 U. Ewers, H. Günther and L. Jaenicke, *Chem. Ber.*, 1974, **107**, 876. 2-4Gb, 2-4Gc(2)

74CB1318 W. Sucrow, C. Mentzel and M. Slopianka, *Chem. Ber.*, 1974, **107**, 1318. 4-2Ha, 4-6Bb

74CB3275 U. Ewers, H. Günther and L. Jaenicke, *Chem. Ber.*, 1974, **107**, 3275. 2-4Gb, 2-4Gc

74CC2744	J. Elguero, A. Fruchier, L. Knutsson, R. Lazaro and J. Sandström, *Can. J. Chem.* 1974, **52**, 2744. 5-5Ab
74CH411	E. Alcalde, J. de Mendoza and J. Elguero, *Chem. Commun.*, 1974, 411. 5-5Ab
74CH605	G. O. Sørensen, L. Nygaard and M. Begtrup, *Chem. Commun.*, 1974, 605. T1-5, 4-1Da
74CH640	T. Tsuchiya, J. Kurita, H. Igeta and V. Snieckus, *Chem. Commun.*, 1974, 640. 6-2A
74CH702	M. Begtrup, *Chem. Commun.* 1974, 702. 1-5Ab, 4-1Da
74CI659	J. Clark, M. Curphey and T. Ramsden, *Chem. Ind.* (*London*), 1974, 659. 2-8Bc(2)
74CP115	P. Mauret, J. P. Fayet, M. Fabre, J. Elguero and J. de Mendoza, *J. Chim. Phys.*, 1974, **71**, 115. 4-1I
74CP415	A. Lautié and A. Novak, *J. Chim. Phys.*, 1974, **71**, 415. 5-1Dc
74CP731	Y. Ferré, M. Rajzmann, G. Pouzard and E. J. Vincent, *J. Chim. Phys.*, 1974, **71**, 731. 4-5Df
74CQ(6)184	B. Stridsberg and S. Allenmark, *Chem. Scr.*, 1974, **6**, 184. 3-2Cf
75CT207	K. Arakawa, T. Miyasaka and H. Ochi, *Chem. Pharm. Bull.*, 1974, **22**, 207. 4-2Gg, 4-2Hc
74CT1053	T. Hino, T. Suzuki and M. Nakagawa, *Chem. Pharm. Bull.*, 1974, **22**, 1053. 3-2Ed, 3-3C
74CT1239	T. Kitagawa, S. Mizukami and E. Hirai, *Chem. Pharm. Bull.*, 1974, **22**, 1239. 2-4Da(2)
74CX(B)342	G. R. Form, E. S. Raper and T. C. Downie, *Acta Crystallogr.*, *Sect. B*, 1974, **30**, 342. 4-5Dc
74CX(B)474	Å. Kvick and M. Backéus, *Acta Crystallogr.*, *Sect. B*, 1974, **30**, 474. 2-6Ba
74CX(B)1146	H. L. Ammon and G. L. Wheeler, *Acta Crystallogr.*, *Sect. B*, 1974, **30**, 1146. 2-3Bf, 2-3Eb
74CX(B)1430	P. Singh and D. J. Hodgson, *Acta Crystallogr.*, *Sect. B*, 1974, **30**, 1430. 2-4Fb
74CX(B)1490	A. Escande, J. L. Galigné and J. Lapasset, *Acta Crystallogr.*, *Sect. B*, 1974, **30**, 1490. 4-1I
74CX(B)1528	T. J. Kistenmacher and T. Shigematsu, *Acta Crystallogr.*, *Sect.*, *B*, 1974, **30**, 1528. 5-1Db
74CX(B)1598	M. Gadret, M. Goursolle and J. M. Leger, *Acta Crystallogr.*, *Sect. B*, 1974, **30**, 1598. 5-2Ba
74CX(B)1647	A. Escande and J. L. Galigné, *Acta Crystallogr.*, *Sect. B*, 1974, 30, 1647 4-1Ha
74CX(B)2009	A. Escande and J. Lapasset, *Acta Crystallogr.*, *Sect. B*, 1974, **30**, 2009. 4-1G
74CX(B)2273	A. K. Chwang, M. Sundaralingam and S. Hanessian, *Acta Crystallogr.*, *Sect. B*, 1974, **30**, 2273. 5-1Da
74CX(B)2348	S. Pérez-Garrido, A. Conde and R. Márquez, *Acta Crystallogr.*, *Sect. B*, 1974, **30**, 2348. 4-4Ce
74CX(B)2505	R. L. Harlow and S. H. Simonsen, *Acta Crystallogr.*, *Sect. B*, 1974, **30**, 2505. 4-1Bbx(ii)

74CX(B)2806 D. L. Kozlowski, P. Singh and D. J. Hodgson, *Acta Crystallogr., Sect. B*, 1974, **30**, 2806. 5-2Bc

74HC(16)33 T. L. Gilchrist and G. E. Gymer, *Advan. Heterocycl. Chem.*, 1974, **16**, 33. 4-1Da

74HC(17)27 D. Lloyd and H. P. Cleghorn, *Advan. Heterocycl. Chem.*, 1974, **17**, 27. 6-2A

74HC(17)99 R. Lakhan and B. Ternai, *Advan. Heterocycl. Chem.*, 1974, **17**, 99. 4-3Cb, 4-4Ca, 4-5Da

74HC(17)213 H.-J. Timpe, *Advan. Heterocycl. Chem.*, 1974, **17**, 213. 4-7Af

74HC(17)255 M. J. Cook, A. R. Katritzky and P. Linda, *Advan. Heterocycl. Chem.*, 1974, **17**, 255. 2-1A, 2-1B

74J(PI)470 F. Bergmann, M. Rahat and I. Tamir, *J. Chem. Soc.* (*Perkin I*), 1974, 470. 5-1Bb, 5-1Bd

74J(PI)976 S. Ogawa, T. Yamaguchi, and N. Gotoh, *J. Chem. Soc.* (*Perkin I*), 1974, 976. 2-7B

74J(PI)1531 B. A. J. Clark, M. M. S. El-Bakoush and J. Parrick, *J. Chem. Soc.* (*Perkin I*), 1974, 1531. 2-3Bi, 5-2Ba

74J(PI)2229 M. Rahat, F. Bergmann and I. Tamir, *J. Chem. Soc.* (*Perkin I*), 1974, 2229. 5-1Bd, 5-1Ec

74J(PI)2307 B. Iddon, H. Suschitzky and A. W. Thompson, *J. Chem. Soc.* (*Perkin I*), 1974, 2307. 2-3Bc, 2-5Ba

74J(PII)546 S.-O. Chua, M. J. Cook and A. R. Katritzky, *J. Chem. Soc.* (*Perkin II*), 1974, 546. 6-2Bd

74J(PII)1199 G. B. Barlin, *J. Chem. Soc.* (*Perkin II*), 1974, 1199. 2-4Ca, 2-4Cd, 2-5E

74J(PII)1849 A. Kálmán, K. Simon, J. Schawartz and G. Horváth, *J. Chem. Soc.* (*Perkin II*), 1974, 1849. 4-1Db, 4-5Fd

74JA1239 P. Singh and D. J. Hodgson, *J. Amer. Chem. Soc.*, 1974, **94**, 1239. 2-4Fb

74JA1674 K. C. Ingham and M. A. El-Bayoumi, *J. Amer. Chem. Soc.*, 1974, **96**, 1674. 5-2Aa

74JA2342 P. H. Kasai and D. McLeod, Jr., *J. Amer. Chem. Soc.*, 1974, **96**, 2342. 2-3Ba, 2-3D

74JA4319 P.-S. Song, M. Sun, A. Koziolowa and J. Koziol, *J. Amer. Chem. Soc.*, 1974, **96**, 4319. 2-4Gd

74JA4699 A. M. Trozzolo, A. Dienes and C. V. Shank, *J. Amer. Chem. Soc.*, 1974, **96**, 4699. 2-3Gb

74JA5911 F. Jordan, *J. Amer. Chem. Soc.*, 1974, **96**, 5911. 5-1Da, 5-1Eg

74JA6832 A. Psoda, Z. Kazimierczuk and D. Shugar, *J. Amer. Chem. Soc.*, 1974, **96**, 6832. 2-5E

74JH7 Z. Neiman, *J. Heterocycl. Chem.*, 1974, **11**, 7. 2-2Ga

74JH135 J. J. Bergman and B. M. Lynch, *J. Heterocycl. Chem.*, 1974, **11**, 135. 4-2Fa, 4-2Fc(2), 4-2Gc(ii), 4-4Ad, T4-26, T4-27

74JH189 G. X. Thyvelikakath, L. J. Bramlett, T. E. Snider, D. L. Morris, D. F. Haslam, W. D. White, N. Purdie, N. N. Durham and K. D. Berlin, *J. Heterocycl. Chem.*, 1974, **11**, 189. 4-1Ka

74JH423 E. Alcalde, J. de Mendoza, J. M. Garcia-Marquina, C. Almera and J. Elguero, *J. Heterocycl. Chem.*, 1974, **11**, 423. 4-5Bg, 5-2Ba

74JH751 R. M. Claramunt, J. M. Fabregà and J. Elguero, *J. Heterocycl. Chem.*, 1974, **11**, 751. 6-2Bd

74JH921 E. Alcalde, J. de Mendoza and J. Elguero, *J. Heterocycl. Chem.*, 1974, **11**, 921. 5-5Ab

74JL(22)401 L. Nygaard, D. Christen, J. T. Nielsen, E. J. Pedersen, O. Snerling, E. Vestergaard and G. O. Sørensen, *J. Mol. Struct.*, 1974, **22**, 401. T1-5, 4-1Ba, T4-1

74JM(49)423 W. D. Krugh and L. P. Gold, *J. Mol. Spectrosc.*, 1974, **49**, 423. T1-5, 4-1Fa

74JO357 J. Elguero, C. Marzin and J. D. Roberts, *J. Org. Chem.*, 1974, **39**, 357. 1-5B, 4-1Fa(iv), 4-1Fb

74JO511 S. Wawzonek and S. M. Heilmann, *J. Org. Chem.*, 1974, **39**, 511. 6-1Ab

74JO940 E. M. Burgess and J. P. Sanchez, *J. Org. Chem.*, 1974, **39**, 940. 4-1Db

74LA1183 F. Asinger and W. Leuchtenberger, *Liebig's Ann. Chem.*, 1974, 1183. 4-5Cb

74M853 H. Falk, S. Gergely and O. Hofer, *Monatsh. Chem.*, 1974, **105**, 853. 3-1Aa

74MI01 R. A. Abramovitch, "Pyridine and its Derivatives," Suppl. Part 3, Wiley (Interscience), New York, 1974. 2-T13

74MI291 E. D. Thorsett, F. R. Stermitz and C. M. O'Donnell, *Photochem. Photobiol.*, 1974, **19**, 291. 2-8Dc

74MI321 N. van den Putten, D. Heijdenrijk and H. Schenk, *Cryst. Struct. Commun.*, 1974, **3**, 321. 4-1Fa(iii)

74MI409 H. Iwasaki, *Chem. Letters* (*Tokyo*), 1974, 409. 5-1Db

74MI677 V. S. Troitskaya, Y. D. Timoshenkova, V. G. Vinokurov and V. I. Tyilin, *Vestn. Mosk. Univ. Khim.*, 1974, 677. 4-1Ba

74MI790 B. I. Sukhorukov, A. S. Gukoroskaya, G. V. Nekrasova and V. L. Antonowskii, *Biofizika*, 1974, **19**, 790. 2-4Db

74MI1269 A. M. Chauvet-Monges, Y. Martin-Borret, A. Crevat and J. Fournier, *Biochimie*, 1974, **56**, 1269. 2-2Gc

74OR224 A. Maquestiau, Y. Van Haverbeke and R. N. Muller, *Org. Magn. Resonance*, 1974, **6**, 224. 1-5Ab, 4-2Bb, 4-2Be

74OR272 J. Elguero, A. Fruchier and M. C. Pardo, *Org. Magn. Resonance*, 1974, **6**, 272. 4-1Ka

74OR469 L. Paoloni, M. Cignitti and M. L. Tosato, *Org. Magn. Resonance*, 1974, **6**, 469. 2-3D

74OR485 R. Faure, J. P. Galy, G. Giusti, E. J. Vincent and J. Elguero, *Org. Magn. Resonance*, 1974, **6**, 485. 5-5Ab

74PC469 G. W. Fischer and P. Schneider, *J. Prakt. Chem.*, 1974, **316**, 469. 2-7Ca

74PC705 F. Ritschl and H. Dorn, *J. Prakt. Chem.*, 1974, **316**, 705. 4-2Hb(i), 4-5Bh(i)

74PC970 V. S. Bogdanov, V. P. Litvinov, Y. L. Goldfarb, N. N.

Petuchova and E. G. Ostapenko, *J. Prakt. Chem.*, 1974, **316**, 970. 3-2Cc, 3-2Cf, 3-3B

74PM(6)1 E. Heilbronner, J. P. Maier and E. Haselbach, "Physical Methods in Heterocyclic Chemistry" (A. R. Katritzky, Ed.), Vol. 6, p. 1–52, Academic Press, New York, 1974. T1-1(2)

74PM(6)53 J. Sheridan, "Physical Methods in Heterocyclic Chemistry" (A. R. Katritzky, Ed.) Vol. 6, Academic Press, New York, 1974, p. 53–94. T1-1

74PM(6)147 S. G. Schulmann, "Physical Methods in Heterocyclic Chemistry" (A. R. Katritzky, Ed.), Vol. 6, pp. 147–197. Academic Press, New York, 1974. 1-6E

74PM(6)199 K. Pihlaja and E. Taskinen, "Physical Methods in Heterocyclic Chemistry" (A. R. Katritzky, Ed.), Vol. 6, pp. 199–246. Academic Press, New York, 1974. T1-1

74RO113 V. I. Slesarev and B. A. Ivin, *Z. Org. Khim.*, 1974, **10**, 113. 2-4Db

74RR671 D. Suciu and Z. Györfi, *Rev. Roum. Chim.*, 1974, **19**, 671. 2-7Ca, 4-5Dc

74RR679 R. Nuţiu, I. Sebe and M. Nuţiu, *Rev. Roum. Chim.*, 1974, **19**, 679. 2-5E

74T1225 J. Arriau, J. Deschamps, J. Teysseyre, A. Maquestiau and Y. Van Haverbeke, *Tetrahedron*, 1974, **30**, 1225. 4-2-Fa(2), 4-2Ge

74T1345 J. Arriau, J. P. Campillo, J. Elguero and J. M. Pereillo, *Tetrahedron*, 1974, **30**, 1345. 4-2Gj(2), 4-2Ic

74T2903 A. Escande, J. Lapasset, R. Faure, E.-J. Vincent and J. Elguero, *Tetrahedron*, 1974, **30**, 2903, (erratum. 1975, **31**, 2.) 1-4Ib, 4-1G, 4-1G(ii), 4-1Ha, 4-1I

74T3171 J. Daunis, R. Jacquier and C. Pigiere, *Tetrahedron*, 1974, **30**, 3171. 2-4Fb

74TL71 R. J. Abraham, G. E. Hawkes and K. M. Smith, *Tetrahedron Lett.*, 1974, 71. 6-3

74TL1483 R. J. Abraham, G. E. Hawkes and K. M. Smith, *Tetrahedron Lett.*, 1974, 1483. 6-3

74TL2643 K. K. Balasubramanian and B. Venugopalan, *Tetrahedron Lett.*, 1974, 2643. 5-5B

74ZC49 W. Liebscher, *Z. Chem.*, 1974, **14**, 49. 1-1B

75BF(2)ip1 J. P. Fayet, M. C. Vertut, R. M. Claramunt, J. M. Fabregà and L. Knutsson, *Bull. Soc. Chim. Fr. Part 2*, 1975, in press (published page 393). 5-5Ab(3)

75BF(2)ip2 P. Mauret, J. P. Fayet and M. Fabre, *Bull. Soc. Chim. Fr. Part 2*, 1975, in press. 1-4G, 4-1Eb(ii), 4-1Fa(ii) 4-1G(ii)

75CTip1 S. Kubota and M. Uda, *Chem. Pharm. Bull.*, 1975, in press. 4-1Ec

75G431 A. Cipiciani, P. Linda and G. Savelli, *Gazz. Chim. Ital.*, 1975, 431 2-3Bg

75HC(18)199 J. S. Kwiatkowski and B. Pullman, *Advan. Heterocycl. Chem.*, 1975, **18**, 199. 2-6Cb

75J(PII)ip1 R. J. Abraham, G. E. Hawkes, M. F. Hudson and K. M.

Smith, *J. Chem. Soc.* (*Perkin II*), in press (published page 204). 6-3

75JH27 A. Maquestiau, Y. Van Haverbeke and R. N. Muller, *J. Heterocycl. Chem.*, 1975, **12**, 27. 4-2Bh

75JH85 A. Maquestiau, Y. Van Haverbeke and R. N. Muller, *J. Heterocycl. Chem.*, 1975, **12**, 85. 4-2Bb, 4-2Fa, 4-2Gh(2).

75JHip1 J. P. Fayet, M. C. Vertut, P. Mauret, J. de Mendoza and J. Elguero, *J. Heterocycl. Chem.*, in press (published page 197). 5-5Ab

75MIip1 F. Leroy, J. Housty, S. Geoffre and M. Hospital, *Cryst. Struct. Commun.*, 1975, in press. 6-2Bd

75MIip2 J. Delettré, R. Bally and J.-P. Mornon, *Cryst. Struct. Commun.*, 1975, in press. 5-2Ba

75MIip3 J.-P. Mornon, J. Delettré and R. Bally, *Cryst. Struct. Commun.*, 1975, in press. 4-5Bg

75MIip4 A. Avalos, R. M. Claramunt and R. Granados, *Anal. Quím.* (*Spain*), 1975, in press. 4-5Cb

75MIip5 J. de Mendoza and M. C. Pardo, *Anal. Quím.* (*Spain*), 1975, in press (published page 434). 5-5Ab

75UP1 G. E. Hawkes, E. W. Randall, J. Elguero and C. Marzin, unpublished results. 4-2Fa(2), 4-2Fc

A 6
B 7
C 8
D 9
E 0
F 1
G 2
H 3
I 4
J 5